AF877019

Atmosphere, Earth, Ocean & Space

Editor-in-Chief

Wing-Huen Ip, National Central University, Institute of Astronomy, Jhongli District, Taoyuan City, China

Series Editors

Masataka Ando, Nagoya University, Nagoya, Japan
Chen-Tung Arthur Chen, Department of Oceanography, National Sun Yat-Sen University, Kaohsiung, Taiwan
Kaichang Di, Institute of Remote Sensing and Digital Earth, Chinese Academy of Sciences, Beijing, China
Jianping Gan, Hong Kong University of Science and Technology, Hong Kong, China
Philip L.-F. Liu, Department of Civil and Environmental Engineering, National University of Singapore, Singapore
Ching-Hua Lo, Geology, National Taiwan University, Taipei, China
James A. Slavin, Department of Atmospheric, Oceanic and Space Sciences, University of Michigan–Ann Arbor, USA
Keke Zhang, University of Exeter, UK
R. D. Deshpande, Geosciences Division, Physical Research Laboratory, Gujarat, India
A. J. Timothy Jull, Geosciences and Physics, University of Arizona AMS Laboratory, Tucson, AZ, USA

The series Atmosphere, Earth, Ocean & Space (AEONS) publishes state-of-art studies spanning all areas of Earth and Space Sciences. It aims to provide the academic communities with new theories, observations, analytical and experimental methods, and technical advances in related fields. The series includes monographs, edited volumes, lecture notes and professional books with high quality. The key topics in AEONS include but are not limited to: Aeronomy and ionospheric physics, Atmospheric sciences, Biogeosciences, Cryosphere sciences, Geochemistry, Geodesy, Geomagnetism, Environmental informatics, Hydrological sciences, Magnetospheric physics, Mineral physics, Natural hazards, Nonlinear geophysics, Ocean sciences, Seismology, Solar-terrestrial sciences, Tectonics and Volcanology.

More information about this series at http://www.springer.com/series/16015

Xueshang Feng

Magnetohydrodynamic Modeling of the Solar Corona and Heliosphere

Springer

Xueshang Feng
SIGMA Weather Group, State Key
Laboratory of Space Weather
National Space Science Center
Chinese Academy of Sciences
Beijing, China

ISSN 2524-440X ISSN 2524-4418 (electronic)
Atmosphere, Earth, Ocean & Space
ISBN 978-981-13-9080-7 ISBN 978-981-13-9081-4 (eBook)
https://doi.org/10.1007/978-981-13-9081-4

© Springer Nature Singapore Pte Ltd. 2020
This work is subject to copyright. All rights are reserved by the Publisher, whether the whole or part of the material is concerned, specifically the rights of translation, reprinting, reuse of illustrations, recitation, broadcasting, reproduction on microfilms or in any other physical way, and transmission or information storage and retrieval, electronic adaptation, computer software, or by similar or dissimilar methodology now known or hereafter developed.
The use of general descriptive names, registered names, trademarks, service marks, etc. in this publication does not imply, even in the absence of a specific statement, that such names are exempt from the relevant protective laws and regulations and therefore free for general use.
The publisher, the authors and the editors are safe to assume that the advice and information in this book are believed to be true and accurate at the date of publication. Neither the publisher nor the authors or the editors give a warranty, expressed or implied, with respect to the material contained herein or for any errors or omissions that may have been made. The publisher remains neutral with regard to jurisdictional claims in published maps and institutional affiliations.

This Springer imprint is published by the registered company Springer Nature Singapore Pte Ltd.
The registered company address is: 152 Beach Road, #21-01/04 Gateway East, Singapore 189721, Singapore

Preface

Over the past several decades, significant advances have been made in space weather modeling. As the first principle of approximation, magnetohydrodynamic (MHD) has become a basic tool in space weather model developments. This book is written as an introductory book on the numerical methods of MHD-based space weather modeling for senior undergraduate and graduate students in the area of space physics and space weather. It can also be used as a reference book for researchers and engineers who intend to use the numerical method for research in computational MHD fluid dynamics and space weather. A good portion of this book has been discussed with some senior graduate students for computational solar-interplanetary MHD modeling and used by second-year graduate students for self-study of the basics of finite volume methods (FVM). This book assumes that readers have a basic understanding of space plasma physics, and solar-terrestrial physics and some background in numerical methods. For these topics, the readers may refer to the existing excellent books: "Biskamp D. (1997): Nonlinear Magnetohydrodynamic, Cambridge University Press (ISBN 0521599180)", "Hans Goedbloed J. P., Poedts S. (2004): Principles of Magnetohydrodynamics: With Applications to Laboratory and Astrophysical Plasmas, Cambridge University Press (ISBN-10: 0521623472)", "Hans Goedbloed J. P., Keppens R., Poedts S. (2010): Advanced Magnetohydrodynamics: With Applications to Laboratory and Astrophysical Plasmas, Cambridge University Press (ISBN-10: 0521879574)", "Burlaga L. F. (1995), Interplanetary Magnetohydrodynamic, Oxford University Press, (ISBN-10: 0195084721)", "LeVeque R. J. (2004): Finite Volume Methods for Hyperbolic Problems, Cambridge University Press (ISBN 0-521-00924-3)", "Toro T. (2009): Riemann Solvers and Numerical Methods for Fluid Dynamics: A Practical Introduction, Springer (ISBN 978-3-540-25202-3)", Jardin S. (2010): Computational Methods in Plasma Physics, CRC Press (ISBN 9781439810217), "Tu C. Y. (1988): Solar Terrestrial Space Physics Volume 1, 2 (in Chinese), Science Press (ISBN 7030005813,7030000978)", "Aschwanden M. (2005): Physics of the Solar Corona: An Introduction with Problems and Solutions, Springer (ISBN 978-3-540-30766-2)", "Aschwanden M. (2019): New Millennium Solar Physics, Springer (ISBN 978-3-030-13954-4)", "Heliophysics: Space Storms and Radiation:

Causes and Effects, edited by Schrijver C. J. and Siscoe G. L. (2010), Cambridge University Press (ISBN 978-0-521-76051-5)", "Handbook of the Solar-Terrestrial Environment, edited by Kamide Y. and Chian Abraham C. L. (2007), Springer-Verlag Berlin Heidelberg (ISBN 978-3-540-46315-3)", "Howard T. (2014): Space Weather and Coronal Mass Ejections, Springer (ISBN 978-1-4614-7975-8)", and "The Scientific Foundation of Space Weather, edited by Baker D., Balogh A., Gombosi T., Koskinen H. E. J., Veronig, A. and von Steiger R. (2019), Springer Nature Switzerland (ISBN-13 978-94-024-1587-2)".

The impetus to write this book came about from three sources: The first source was the author's experience carried out during last 20 years at SIMGA Weather Group, State Key Laboratory of Space Weather, Chinese Academy of Sciences. The second source grew over the years to become more significant as it was noticed that graduates have started working on increasingly more focused areas and topics in MHD while becoming less cognizant of the general algorithmic expertise that earlier graduate students developed. It became clear that there is a need not only to cover the basis of the numerics at the core of MHD codes but also to discuss the implementation issues to ensure that all graduate students receive a robust understanding of the techniques they are working on. Finally, fast progress in advanced numerics of numerical space weather forecasting provided all the incentive to clarify and detail much of the numerical basis of the algorithms used in numerical MHD community. To this end, it was decided that the book would combine a mix of numerical and implementation details allowing the reader, if she/he desires, to fully understand and implement a robust and versatile MHD code based on the cell-centered FVM. This ambitious task is made possible only by selecting from various numerical methods in each of the topics covered in the book a handful set with which the author is intimately familiar. The result is a book that covers intimately all the topics unavoidable for the development of a robust MHD code within the framework of the cell-centered finite volume method and conservation element and solution element (CESE) method. For clarity and readability, each chapter is written as a stand-alone chapter in a self-contained way as possible as the author can, with the aim for the reader to be able to read each chapter independently.

Chapter 1 of the book presents a brief review of existing MHD models for the solar corona and heliosphere. Chapter 1 presents the state-of-the-art of the three-dimensional (3D) time-dependent MHD models of solar wind. After highlighting the influence of adverse space weather conditions via the Halloween Sun-Earth events and their associated effects in the geospace space, the author tries to summarize 3D physics-based MHD solar-interplanetary models currently used in space weather forecast. For completeness, 3D MHD models in studying the interaction of the solar wind-magnetosphere-ionosphere system are also presented in this chapter. At the end of this chapter, brief concluding remarks are given.

Chapter 2 is devoted to the concise introduction of cell-centered FVM with the techniques focused on the MHD simulations. The focus of this chapter is on the cell-centered FVM in three-dimensional computational domain, by starting with a general introduction to the first-order finite volume methods, followed by a description of high-order accurate formulations of variable reconstructions and flux functions for conservative laws and Newton-Krylov methods at the end of this very chapter. All the formulations are applicable at least to the hexahedral cell, particularly quadrilaterally-faced hexahedron (cuboid) with 6 faces, 12 edges, and 8 vertices. In this very chapter, some new numerical techniques potentially applicable to MHD simulations are provided. Most of numerical methods have been implemented in our solar wind MHD study at SIGMA weather group although no validation test results are displayed.

Chapter 3 is an application of the methods introduced in Chap. 2 to a parallel MHD code in spherical coordinates for solar wind ambient study. This code operates on a 3D spherical coordinates with both non-staggered and staggered grids on the overlapping overset grid system with hexahedral cells of quadrilateral frustum type. Some typical numerical schemes with the total variation diminishing slope limiters and approximate Riemann solvers are described in the present context of spherical grids. At the same time, alternative strategies to ensure a solenoidal magnetic field, such as projection Poisson solver, hyperbolic divergence cleaning method derived from generalized Lagrange multiplier formulation of MHD system, and constrained transport method, are employed. In this chapter, an FVM is described exemplarily on a six-component composite grid system by using a minmod limiter for oscillation control.

Starting with Chap. 4, we discuss the CESE-MHD model and its applications to background solar wind, coronal mass ejections, coronal nonlinear force-free field extrapolations, and coronal magnetic evolutions and eruptions. Chapters 4–10 are re-organized and rewritten from separate papers published at SIGMA weather group. They are presented in a self-contained manner, and each of them is devoted to a specific theme. It is hoped that the reader may understand the research trend by following each research topic and references therein. Chapter 4 describes the 3D Solar-Interplanetary space-time Conservation Element and Solution Element (SIP-CESE) MHD model and focuses on the application of the MHD model to study solar wind plasma on pentahedrons in the spherical shell domain. Chapter 5 presents the implementation of the SIP-CESE-MHD model on six-component grids. Adaptive mesh refinement implementation of SIP-CESE-MHD model is given in Chap. 6. Chapter 7 introduces data-driven solar wind modeling by CESE-MHD. Coronal nonlinear force-free magnetic field extrapolations with emphasis on the CESE-MHD-NLFFF code are discussed in Chap. 8. Chapter 9 deals with data-driven MHD Modeling of coronal magnetic evolutions and eruptions. Chapter 10 is concerned about CME propagation study with CESE-MHD. The last chapter, Chap. 11, tries to shed light on future avenue of solar-interplanetary weather forecasting. Topics discussed include physical understanding of the coronal-heliospheric system, observational data availability and application, model metrics and validation, ensemble forecasting, data assimilation, and machine learning. As space weather modeling and forecasting is an interdisciplinary field, it is difficult, if not impossible, to give a complete roadmap of future development. However, we still hope that the discussions in this chapter would be of some help to inspire new research outcome in this field.

After completing my postdoc from Beihang University, I joined Prof. Fengsi Wei's group at Center for Space Science and Applied Research (presently National Space Science Center), Chinese Academy of Sciences. He offered me a topic about which I did not have any clue at the beginning; but I liked it very much as I increased my knowledge in the field. It has been a dream opportunity for me to carry out my research with Prof. Fengsi Wei, who led me to these days not only with his scientific guidance and advice, but also by helping me keep my spirit high even in my most difficult days during these past two decades. Professor Fengsi Wei's steady encouragement, guidance, and perseverance make this book possible. His devotion and persistence in the development of space weather always provide me the energy and desire to move ahead. It is one of the greatest chances of my life to meet you, sir.

I would like to thank Profs. Shui Wang, Youqiu Hu, Xiankang Dou, Huinan Zheng, Yuming Wang, Quanming Lu, and Chenglong Shen at University of Science and Technology of China, Profs. Cheng Fang and Pengfei Chen at Nanjing University, Profs. Jingxiu Wang, Yihua Yan, Jun Zhang at Beijing Astronomical Observatory, Chinese Academy of Sciences, Profs. Xiao Zuo, Chuanyi Tu and Jianshen He at Peking University, Profs. Weixing Wan and Aimin Du at Institute of Geology and Geophysics, Chinese Academy of Sciences, Profs. Yao Chen and Bo Li at Shandong University (Weihai), Dr. Murray Dryer at Space Weather Predictions Center, National Oceanic and Atmospheric Administration, and Profs. Shi-Tsan Wu and Qiang Hu at Center for Space Plasma and Aeronomic Research (CSPAR), The University of Alabama, Huntsville, USA, and Dr. Yang Liu and Keiji Hayashi at Stanford University, have provided invaluable suggestions and discussions along the way during my past years as a professional researcher, up to the present time. I thank numerous colleagues worldwide in shaping and enriching my professional development and growth. In addition, I have been inspired by their work ethics and formidable spirit. In particular, I benefited tremendously from the collaboration with Prof. S. T. Wu, with whom the author has been working closely in the past two decades or so, till the Friday 2 days ahead of his pass on Sunday afternoon, May 21, 2017. He was a pioneer in the field of both aerospace engineering and solar-terrestrial physics. With energy and optimism, he advised and mentored many others who would go on to have successful careers in science and engineering; his impact was international. ST's wisdom, his mentorship, and his passion for science have been the objectives of my pursuit. The best way in memory of Prof. S. T. Wu is to continue to encourage and inspire future generations, as ST always did. I hope this book will contribute to that noble cause.

During the course of the last years many people have contributed in various ways to the work presented in this book. I feel privileged to have had the opportunity to further my research at State Key Laboratory of Space Weather, National Space Science Center, Chinese Academy of Sciences. The academic environment made the experience enjoyable, educational, and extremely rewarding. I would like to thank all of my colleagues at State Key Laboratory of Space Weather, particularly the members and Ph.D. students of my SIGMA weather group: Drs. Fang Shen, Changqing Xiang, Chaowei Jiang, Yufen Zhou, Yanqi Hu, Liping Yang,

Dingkun Zhong, Man Zhang, Ming Xiong, Pingbing Zuo, JianPeng Guo, Xinhua Zhao, Xi Luo, Yi Wang, Shaohua Zhang, Chaoxu Liu, Juan Wang, Huichao Li, Caixia Li, Xiaojing Liu, Lei Zhang, Jiakun Lü, and Haopeng Wang for their invaluable feedback and suggestions. I am grateful to my colleagues at SIGMA weather group for providing a homelike work environment and for the many technical and nontechnical discussions over the years. My heartfelt thanks go to Drs. Changqing Xiang, Chaowei Jiang, Yufen Zhou, Huichao Li, Caixia Li, Xiaojing Liu, Man Zhang, Juan Wang, Tianran Sun, Jiakun Lü, and Haopeng Wang for the time and effort they have spent in helping me to check the formula and re-organize the manuscript.

I would like to thank all my friends and family for their support and encouragement. I would like to thank my parents and my brothers and sisters for all their enduring support and love during my life. Special thanks to my wife, Jianlan Hu for her love, devotion, and patience over the years. Also to my sons: Qi Feng and Kang Feng for just being themselves.

Beijing, China Xueshang Feng

Acknowledgements

Research works that constitute the material of this book are jointly supported by the National Basic Research Program (973 program) under grant 2006CB806304 and 2012CB825601, the National Natural Science Foundation of China (40536029, 40621003, 40374056, 40574068, 40504020, 40523006, 41031066, 40921063, 40874091, 40890162, 40904050, 40874077, 41074121, 41074122, 41231068, 41274192, 41174150, 41374176, 41531073, 41574171, 40904049, 41004082, 40804029, 41274179, 41531073, 41731067, 41874202, and 41861164026), the Specialized Research Fund for State Key Laboratories, the International Collaboration Research Team Program of the Chinese Academy of Sciences, and the Knowledge Innovation Program of the Chinese Academy of Sciences (KZZD-EW-01-4).

Research works that constitute the material of this book utilize data provided by *BBSO, GONG, MLSO, SDO, SOHO, OMNI, STEREO, Ulysses, Wind, and WSO*. The Big Bear Solar Observatory (BBSO)/New Jersey Institute of Technology is acknowledged for the Hα image. This book utilizes the global oscillation network group (GONG) data obtained by the NSO Integrated Synoptic Program (NISP), managed by the National Solar Observatory, the Association of Universities for Research in Astronomy (AURA), Inc. under a cooperative agreement with the National Science Foundation. The data were acquired by instruments operated by the Big Bear Solar Observatory, High Altitude Observatory, Learmonth Solar Observatory, Udaipur Solar Observatory, Instituto de Astrofísica de Canarias, and Cerro Tololo Interamerican Observatory. GONG data storage is supported by the University of Colorado Boulder. The Mauna Loa Solar Observatory (MLSO) coronagraphs are operated by the High Altitude Observatory, which is sponsored by the National Science Foundation (USA). SDO is the first mission within NASA's LWS (Living With a Star) program, a space weather-focused and application-driven research program. Data from SDO observations are courtesy of NASA/SDO and the AIA, EVE, and HMI science teams. SOHO is a project of international cooperation between ESA and NASA. The SOHO/LASCO data are produced by a consortium of the Naval Research Laboratory (USA), Max-Planck-Institut fur Aeronomie

(Germany), Laboratoire de astronomie (France), and the University of Birmingham (UK). The OMNI data were obtained from the GSFC/SPDF OMNIWeb interface http://omniweb.gsfc.nasa.gov. The STEREO/SECCHI data are produced by a consortium of NRL (US), LMSAL (US), NASA/GSFC (US), RAL (UK), UBHAM (UK), MPS (Germany), CSL (Belgium), IOTA (France), and IAS (France). The Ulysses mission was a joint project between NASA and the European Space Agency. Wilcox Solar Observatory data were obtained via http://wso.stanford.edu. The Wilcox Solar Observatory is supported by NASA.

The PARAMESH software was developed at the NASA Goddard Space Flight Center and Drexel University under NASA's HPCC and ESTO/CT projects and under grant NNG04GP79G from the NASA/AISR project.

Computational resources for performing all of the calculations were provided by SIGMA Cluster at State Key Laboratory of Space Weather, Chinese Academy of Sciences, and TianHe-1 (A) at the National Supercomputer Center in Tianjin, China.

Contents

Chapter 1
Current Status of MHD Simulations for Space Weather

Abstract In this chapter, we review the status of the art of the three-dimensional (3D) time-dependent magnetohydrodynamic (MHD) models of solar wind. We first highlight the influence of adverse space weather conditions by presenting the Halloween Sun-Earth events and their associated effects in the terrestrial space and point out the importance of 3D physics-based MHD models in space weather forecast. Then we summarize three well-known frameworks, including the architectures, the characteristics of the numerical schemes, and the applications in modeling the background solar wind and solar disturbances' propagation in interplanetary space. The three architectural MHD model frameworks are CORona-HELiosphere (CORHEL), Space Weather Modeling Framework (SWMF) and Space Weather Integrated Model (SWIM). These models are developed respectively by the Center for Integrated Space Weather Modeling (CISM), the Center for Space Environment Modeling (CSEM), and the Solar-Interplanetary-GeoMAgnetic (SIGMA) Weather Group of the State Key Laboratory for Space Weather, National Space Science Center, Chinese Academy of Sciences. Next we give a brief description of hybrid empirical-MHD models for solar-interplanetary modeling, each of which is comprised of a theoretic, empirical or kinematic solar coronal model and a 3D MHD heliospheric model. Then concise summarizations are presented for other miscellaneous models for solar-interplanetary modeling. For completeness, 3D MHD models in studying the interaction between the solar wind and the magnetosphere and the ionosphere are also presented in this chapter. Finally, brief concluding remarks are given in the last section.

1.1 Introduction

The term "space weather" refers to conditions on the Sun and in the solar wind, magnetosphere, ionosphere, and thermosphere that can influence the performance and reliability of space-borne and ground-based technological systems and that can affect human life and health (e.g., [38, 40, 77, 78, 117, 154, 155, 260, 385, 434, 588]). Space weather is distinct from solar-terrestrial physics, and heliophysics more generally, because of its strong emphasis on societal impact (e.g., [38, 54, 110,

© Springer Nature Singapore Pte Ltd. 2020

X. Feng, *Magnetohydrodynamic Modeling of the Solar Corona and Heliosphere*, Atmosphere, Earth, Ocean & Space,
https://doi.org/10.1007/978-981-13-9081-4_1

155, 254, 287–289, 343, 385, 388, 426, 516, 614]). Numerous investigations have revealed that these hazard space weather events are often associated with intensive solar flares and coronal mass ejections (CMEs) (e.g., [14, 94, 101, 218, 245, 372, 433, 465, 487–489, 564, 585]). Solar flares can generate many instantaneous adverse effects on space-based technological systems. However, the most serious effects on human activity occur during major geomagnetic storms (e.g., [146]). It is now generally understood that the source of most geomagnetic storms is CMEs (e.g., [143, 176]). When CMEs are intercepted by the Earth, they will dramatically change the space weather conditions near the Earth and trigger intensive geomagnetic storms, which are responsible for spectacular aurorae and can lead to the disruption of high-frequency radio communication, the degradation of the accuracy of the global positioning system (GPS), the increased drag on satellites, and the damage of electronic instruments on satellites [15, 154, 155, 215]. Even more, the magnetic storm can induce high voltage spikes along power grids that can burn out transformers and cause black-outs in the regions of high geomagnetic latitudes [218, 389]. In addition, solar energetic particles (SEPs) associated with the initial phase of CMEs and with the shocks accelerated by CMEs can propagate along interplanetary magnetic field (IMF) and reach the Earth, which is hazardous to the astronauts in spacecraft orbiting the Earth [69], the crew and passengers of flights across polar regions [514, 515], and the instruments in satellites [175]. What's more, space weather significantly affects the reliability of the space-based systems for military use, which enable military forces to be warned of missile attacks, to communicate instantaneously, to obtain near real-time information that can be transmitted rapidly from satellite to attack platform, to navigate to a conflict area while avoiding hostile defenses along the way, and to identify and strike targets from air, land, or sea with precise and devastating effect. These adverse impacts may also challenge the decision making crucial to maintain world peace and well-being of human being. Knipp et al. [239] reported that the space weather storm of late May 1967 could have become a trip-wire at the height of the Cold War due to the disruption of radio communications without the nascent efforts of U.S. Air Force to monitor solar activities and forecast space weather.

The October–November 2003 solar eruptive events (also known as the Halloween storms) was one of the most recent extreme space weather events. During the two-week period, there were 143 well-observed flares and at least 80 CMEs originating from the three active regions with NOAA (National Oceanic and Atmospheric Administration) numbers 10484, 10486, and 10488 [139]. The SEP intensity remained above the storm level for about two weeks and the D_{st} index revealed three super-intense storms on October 30, 31, and November 20, 2003, which was the largest storm of Solar Cycle 23, associated with AR 10501 [140]. On October 29–31, 2003, geosynchronous satellites were identified to be in the magnetosheath for more than 50 intervals because the size of the dayside magnetosphere was highly decreased under the effect of very high pressure associated with high velocities and densities of the solar wind plasma, as well as high negative values of the B_z component of IMF [70]. The Halloween superstorm period also witnessed the deep ionospheric plasma density depletion in the southern hemisphere and a localized strong ionospheric plasma density enhancement over the Pacific Ocean [621]. Such kind of disproportional plasma density distribution dramatically affected the Wide Area Augmentation System (WAAS), which utilizes GPS signal to provide navigation for aircrafts, during such superstorms [612]. Owing to this for the storm of October 30, 2003 the so-called Precision Approach service of WAAS was down for more than 11 h [71]. At the same time, the Halloween superstorm had significant effects on the composition of the stratosphere and mesosphere in the polar regions [286]. Very large values of NO_x in the upper stratosphere of 180 ppbv (parts per billion by volume) were measured. The Arctic mean polar cap (>60°) NO_x enhancements of 20–70 ppbv between 40 and 60 km lasted for at least 2 weeks, while the Antarctic mean NO_x enhancement was between 10 and 35 ppbv and was halved after 2 weeks. Significant O_3 depletion was observed inside the polar vortex in a wide altitude range during this period.

Information available from NASA (National Aeronautics and Space Administration) Earth and space science missions indicates that about 59% of the spacecraft and about 18% of the instrument groups experienced some effect from the solar activity [18]. Some missions turned off their instruments or put them in a safer operating mode. Five types of space environmental effects on spacecraft were observed: electronic upsets, housekeeping noise, proton degradation to solar arrays, upper atmosphere—induced changes to orbit dynamics, and high levels of accumulated radiation. Even more, Midori, a Japanese satellite, failed to recover during this period and the Mars Radiation Environment Experiment (MARIE) instrument on Mars Odyssey was lost permanently. Other effects during the period of the Halloween storms include a power system failure in Malmo, Sweden, High-latitude GPS receiver outages, and the first alert on radiation doses received by airline passengers above 25,000 ft issued by Federal Aviation Administration [71]. More details about the Halloween event and its associated space weather effects can be referred to Tsurutani et al. [531].

All these hazard space weather effects originate from some large-scale structures of solar wind and CMEs. The solar wind is a supersonic fully ionized plasma flow emitting from the solar corona and propagating throughout the heliosphere. It carries the magnetic field of the Sun outward, because the solar wind is a highly conducting plasma in which the magnetic field is "frozen in" [36, 446]. The solar corona is highly structured and varies systematically over a period of about 11 years in relation to the solar activity cycle [36]. During solar minima, the typical pattern of the solar wind is the high-speed stream coming from the Sun's polar coronal holes and the low-speed stream emanating from low-latitudinal regions [310, 342]. Coronal holes are regions of low-density plasma on the Sun and have magnetic fields that open freely into interplanetary space [52, 427]. The regions of closed magnetic field, often interspersed with open magnetic field lines, are bright, narrow radial structures around the Sun observed by coronagraphs and during solar eclipses, which are called corona streamers [46, 246, 247, 268]. The surface where the polarity of the magnetic field changes from north to south shapes the heliospheric current sheet [58, 580], which usually circles the solar equator wavily near solar minima. With solar activity increasing, solar eruptive events such as solar flares and coronal mass ejections occur more frequently. Solar flares are explosive phenomena observed in the solar atmosphere and involve a broad spectrum of emission enhancements and energy releases of 10^{20} to 10^{25} J [152, 569]. Coronal mass ejections are enormous expulsions of plasma and magnetic field from the solar corona to the heliosphere at speeds of $\sim$50 to faster than 2000 $\mathrm{km \cdot s^{-1}}$ [39, 570]. Near solar maxima, solar winds from small coronal holes, streamers, and coronal mass ejections interleave to provide a highly structured solar wind at all heliolatitudes [311, 312]. High-speed solar streams, coronal mass ejections, and some other interplanetary structures may lead to unfavorable, and sometimes even catastrophic space weather conditions [39, 45, 52, 256, 532, 569, 570].

With modern society and military operations heavily dependent on these high-tech systems that are vulnerable to adverse space weather conditions, the economic and security consequences of the effects mentioned above are enormous (see, e.g., [56, 447, 465]). Therefore deep understanding of the CMEs' propagation in interplanetary space and predicting their impacts on the Earth are essential to deal with the threat of extreme space weather through improved operational procedures and technologies [15]. With accurate space weather forecasting and alerts, airlines can re-route flights scheduled for polar routes to preserve reliable communications; power grid operators can take defensive measures against geomagnetically induced currents (GICs); launch personnel may delay a launch; satellite operators may postpone certain operations (e.g., thruster firings). Civil aviation company can use non-GPS-based navigation systems as a backup, while other GPS users can put off operations for which precision position knowledge is required until the ionospheric disturbance is over. On October 13, 2016, President Obama of the United States of America signed an executive order to coordinate efforts to prepare American for space weather events, which recognized that extreme space weather events "could disable large portions of the electrical power grid, resulting in cascading failures that would affect key services such as water supply, healthcare, and transportation. Space weather has the potential to simultaneously affect and disrupt health and safety across entire continents." (See details at the webpage https://www.whitehouse.gov/the-press-office/2016/10/13/executive-order-coordinating-efforts-prepare-nation-space-weather-events). On March 29, 2019, the Trump Administration of the US released a new National Space Weather Strategy and Action Plan to better prepare national critical infrastructure and security for the next space weather event.

The availability of the new quality of solar corona and solar wind observational data from a huge fleet of spacecraft [e.g., ULYSSES, Solar and Heliospheric Observatory (SOHO), YOKHOH, WIND, Advanced Composition Explorer (ACE), Transition Region and Corona Explorer (TRACE), RHESSI, Hinode, Solar Terrestrial Relations Observatory(STEREO), Solar Dynamics Observatory (SDO)] has allowed us to advance our understanding of the processes involved near the Sun, in interplanetary space, and in the near-Earth environment, and thus to renew our picture of the Sun, the heliosphere, and the solar-terrestrial relationships (e.g., [53, 258, 386]). The solar and interplanetary observations from spacecraft also provide strong constrains for three-dimensional physics-based solar wind models that have become an increasingly important tool to predict the solar wind conditions in interplanetary space (e.g., [94, 101, 133, 598]). Just as weather forecasting was revolutionized by the introduction of the operational numerical weather prediction code based on the physics of the clouds and precipitation and the hydrodynamics and thermodynamics that govern the global motion of atmosphere [20], numerical space weather models incorporating basic physical understanding are expected to enable specification and forecasting of the space weather near the Earth by following the flows of mass and energy from the Sun to Earth through interplanetary space [466].

In the solar corona and heliosphere, the ideal MHD theory is largely applicable, although the Sun-to-Earth plasma system includes both microphysical and macrophysical properties. That is, realistic and accurate numerical simulation should use a theory so as to reproduce both small and large-scale plasma physics phenomena. A global simulation of the entire Sun-to-Earth system based on the kinetic theory is still beyond the computational capabilities of even the most powerful supercomputers. However, the numerical MHD modeling can provide a feasible compromise between physical content and numerical feasibility to describe the global Sun-to-Earth plasma processes. The global MHD simulations have the following merits: (1) In principle they self-consistently characterize the dynamic interaction between the magnetic field and plasma flows within the entire system, providing an effective tool of space weather modeling. (2) They can be efficiently applied to the entire Sun-to-Earth system in a practically computational manner, as powerful numerical methods are available for solving the MHD equations. (3) They can provide an instantaneous picture of the entire system, and thus can provide a global setting for a more detailed investigation, which can be complemented by observations from remote and in-situ measurements. (4) In many cases detailed and comprehensive measurements of the system can be difficult or impossible, in which case the global MHD simulations can be relied as synthetic measurements. The single-fluid MHD equations, which ignore the influence of the electrons and minority heavy ions, but are able to describe most of the large scale physical processes in the Sun-to-Earth system, are most widely used by numerical models. Besides the single-fluid MHD equations, a fully 3D MHD model of the solar corona and heliosphere can include the physical processes associated with turbulence transport and heating. In such sophisticated model, Reynolds-averaged solar wind equations are coupled with transport equations for turbulence energy, cross helicity, and correlation scale. The resulting model consists of separate energy equations for protons and electrons and accounts for the effects of electron heat conduction, radiative cooling, Coulomb collisions, Reynolds stresses, eddy viscosity, and turbulent heating of protons and electrons. Even more, the outer heliospheric model considers the effect of charge exchange between neutral and charged particles. For detailed description of such complicated MHD equations, the reader may refer to the existing literature (e.g., [84, 133, 393, 394, 544, 616]).

Based on the MHD theory, many 3D models have been developed in order to numerically study the solar wind conditions and typical space weather events and to ultimately construct operational numerical space weather models. These models often cover many different regions from the chromosphere and solar corona to interplanetary space, magnetosphere and ionosphere. In this chapter, we briefly review the recent developments of the 3D physics-based space weather modeling covering the solar corona and inner heliosphere. In order to provide a panorama of CME's initiation, propagation and impacts on Earth, 3D magnetospheric MHD models are also summarized succinctly. In the following, we first outline three architectural MHD models equipped with the associated software framework to couple different components of the modeling system, integrate different models together, and easily analyze and operate the massive observed and simulated data. Then we give a brief summary of the other 3D MHD models. For each model, we concisely describe the features of the numerical implementation and its recent developments. Next, we briefly summarize 3D magnetospheric MHD models. In the last part, short concluding remarks are given.

1.2 Architectural MHD Models for Solar-Interplanetary Modeling

This section is to briefly introduce solar-interplanetary models by focusing on three architectural framework-encapsulated MHD models.

1.2.1 CORHEL in CISM

The scientists affiliated with the National Science Foundation (NSF) Center for Integrated Space Weather Modeling (CISM) developed a loosely coupled framework for space weather prediction [137], where each model runs as a separate executable. The CISM, established in 2002, is made up of 11 member institutions and is led by Boston University. The executables in CISM are either coupled by flat files or by a general communication library intercom [265] and the grid interpolation can be handled by the Overture library [33, 169]. The use of these libraries has been demonstrated with a limited number of models and couplings so far [137]. After long-time validation, this large-scale, physics-based space weather prediction model has transitioned from research into operation and provided forecasters with a one-to-four day advance warning of high speed streams of solar plasma and Earth-directed CMEs since 2011 (see details at https://www.nsf.gov/news/news_summ.jsp?cntn_id=118395&org=nsf).

As indicated in Luhmann et al. [293], CISM uses a set of simulation codes that model both the large scale and microscale structures and dynamics of the Sun-Earth system. The fluid codes that together provide the global description of the solar terrestrial environment include the coronal and heliospheric models of magnetohydrodynamics around a sphere (MAS) developed by the Science Applications International Corporation (SAIC) (presently Predictive Science Inc.) [274, 327], the University of Colorado solar wind code (Enlil) [352, 353], the Lyon–Fedder–

Mobarry (LFM) global MHD magnetospheric code [91, 92], and the Thermosphere-Ionosphere Nested Grid (TING) [565] code. These codes are supplemented and enhanced through interaction of codes that model specific regions or local physical processes with global impact [293].

The general framework for the coronal and heliospheric solutions in CISM is described as "CORHEL" for CORona-HELiosphere. CORHEL is a coupled suite of models for quantitatively modeling the solar corona and solar wind in 3D space. It aims to supply a simple, coherent interface to access these models and includes simple coupling routines so that the output of corona model can be used to drive heliosphere model. Such a coupled suit has the ability to assist in interpreting both remote solar observations and in situ measurements within a global context. CORHEL provides three solar coronal models at present, including MAS polytropic model, MAS thermodynamic model and the potential field source surface and Wang–Sheeley–Arge (PFSS-WSA) model. The heliospheric models involved in CORHEL are the Enlil and MAS-Heliosphere models. Synoptic maps of radial magnetic field at the photosphere serve as principle observational input to CORHEL [432]. CORHEL now delivers solutions to the community via four customers at the present time: the Community Coordinated Modeling Center (CCMC), CISM, the modeling web site (http://www.predsci.com), and the Air Force Research Laboratory (AFRL) in Albuquerque. A flow chart of the key components of CORHEL is given in Fig. 1.1. In this section, we describe some details of CORHEL, including the solutions of coronal and heliospheric models and their coupling interface.

1.2.1.1 Coronal Models

The coronal code MAS solves the resistive MHD equations on a nonuniform spherical grid that allows us to concentrate grid points in regions of interest. The method of solution, including the boundary conditions, has been described in many papers [275, 279, 327]. In the implementation, the vector potential $\mathbf{A}$ is introduced to define $\mathbf{B}$, i.e. $\mathbf{B} = \nabla \times \mathbf{A}$. The dependent variables are defined on finite-difference meshes that are staggered with respect to each other, so that $\nabla \cdot \mathbf{B} = \nabla \cdot \nabla \times \mathbf{A} = 0$ is satisfied to within round-off error throughout the computation. In addition, the use of staggered meshes also leads to a natural specification of boundary conditions on the tangential electric field and the normal magnetic field. In the MAS model, a finite difference discretization is used for the radial and axial coordinates; a pseudospectral method is used for the azimuthal coordinate. The time-advance algorithm applies a semi-implicit leapfrog scheme to the wave terms, a predictor-corrector treatment to advection, and an implicit advance to the resistive and viscous diffusion terms. At the same time, the semi-implicit and implicit operators are inverted using a preconditioned conjugate gradient method. Special care is taken in maintaining the self-adjointness of the discretized operators, so that a fast inversion algorithm applicable to symmetric matrices can be used. For details of numerical scheme employed in the MAS model, readers may refer to Lionello et al. [277, 278].

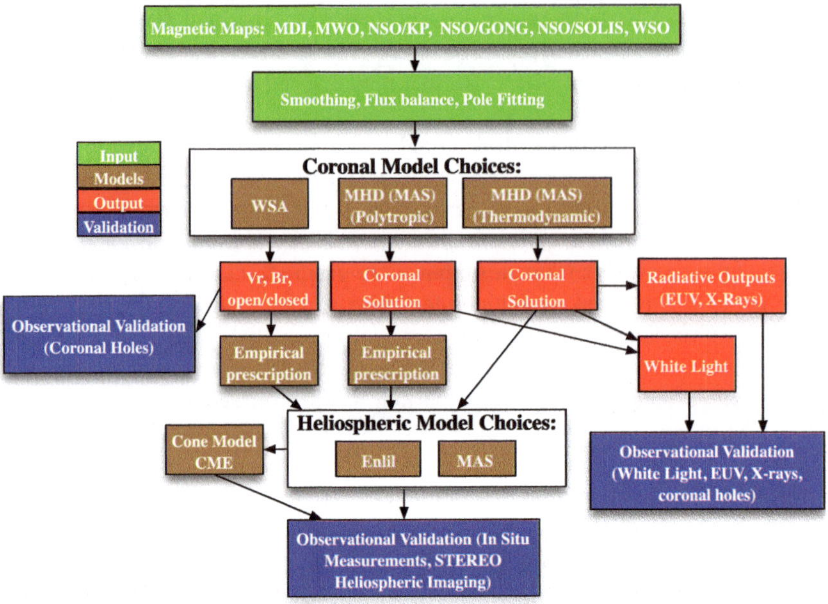

Fig. 1.1 Flow chart depicting the present state of CORHEL. Green boxes indicate the primary input: data derived from measured solar magnetic fields. The coronal model encompasses three choices: WSA runs for immediate answers from a baseline model, MAS polytropic MHD for quick turnaround runs, and MAS thermodynamic MHD for the most physically realistic description. Red indicates coronal model output, including simulated emission and white light images. Coronal models feed into the heliospheric model (two choices) using either empirical prescriptions for the velocity, density, and temperature or directly driven by the outputs from the thermodynamic coronal model. Taken from Fig. 9 of Linker [272]

To compute a coronal solution, the MAS polytropic or thermodynamic model integrates the time-dependent equations to steady state for a given boundary condition defined by magnetic maps. The raw maps must be preprocessed before used and the procedure involves smoothing the maps, balancing the flux, and extrapolating observed mid-latitude fields to the poorly or even completely obscured polar regions, a process that can significantly impact the solutions [432].

The MAS polytropic model is established based on the polytropic approximation, which neglects the complicated physics of the transition region by setting the ratio of specific heats to a reduced value. In spite of its simplicity, the polytropic approximation remains useful and has been used in many studies (e.g., [49, 166, 274]). The MAS polytropic model produces good solutions for the coronal magnetic field, including the position and shape of the streamer belt, coronal hole boundaries, and the heliospheric current sheet. Furthermore, combination of the MAS polytropic approximation and some ad hoc specifications can avoid some minor disadvantages and provide rather good input for heliospheric models (e.g., [327]).

A more advanced treatment of the energy equation takes into consideration of some important energy-transport processes (radiative losses, anisotropic thermal conduction, and coronal heating) in the transition region and solar corona. This more accurate representation of energy flow is utilized by the MAS thermodynamic model, which can obtain the simulated extreme ultraviolet (EUV) and X-ray emission as observed from the instruments aboard many spacecraft. Comparisons between the observed and simulated emission can provide strong constraints on the model. The MAS thermodynamic model accelerates solar wind by using the pressure of Alfvén wave with Wentzel–Kramers–Brillouin (WKB) approximation. In order to produce models that are consistent with observations, Lionello et al. [280] developed a parameterized approach for describing the coronal heating, which provided a pragmatic compromise of being able to include the chromosphere and transition region within the thermodynamical code and to restore the adiabatic index, γ , back to 5/3. There are three parameterized coronal heating models available: (i) an exponential decay law in the radial direction, (ii) a volumetric heating term derived from the model of Schrijver et al. [443], who calculated one-dimensional loop solutions based on the specified heat flux at the base of the corona, and (iii) an empirical heating model depending only on local variables. Lionello et al. [280] showed that heating methods (ii) and (iii) gave reasonable matches to observations. Heating model (iii) performed slightly better than heating model (ii). Other corona heating approaches under development are estimated to have fewer free parameters and eventually yield more accurate solutions than a purely empirical approach. Mikić et al.[328] used a new heating model based on the dissipation of Alfvén waves, and a new energization mechanism to twist the magnetic field in filament channels. The model uses space-borne measurements of the photospheric magnetic field from the Helioseismic and Magnetic Imager (HMI) aboard the Solar Dynamic Observatory (SDO), and a wave-turbulence-driven (WTD) model to heat the corona via low frequency Alfvén waves launched in the chromosphere [73, 282, 283]. By employing two key innovations: a physics-based MHD formulation that describes the heating of the corona over a broad range of conditions, and a novel approach for energizing the coronal magnetic field on global scales, Mikić [328] predicted what the corona would look like one week before the eclipse that occurred on 21 August 2017 across the United States. The numerical results are compared favourably with observations of the eclipse in white light and extreme ultraviolet, which improves the understanding of the relationship of observed features, including streamers, coronal holes, prominences, polar plumes and thin rays, to the magnetic field. Predictions of this kind provide opportunities to improve the models, forging the path to improve space weather prediction. Their experiment also indicates that the discrepancies between the model and observations arise from limitations in the capability of observing the Sun's magnetic field.

In the CORHEL, a semi-empirical near-Sun module (Wang–Sheeley–Arge, WSA) [4, 5] is also frequently employed. To determine solar coronal parameters and to provide inner boundary conditions for heliospheric MHD models, the WSA model is based upon a modified PFSS model [441] of the steady-state corona. WSA's coronal component can readily determine the 3D structure of the magnetic field, while its solar wind component can provide the speed and the interplanetary magnetic field polarity at any location in the inner heliosphere. Photospheric magnetic field maps are used to create Carrington maps of radial photospheric field, which serves as input to the PFSS model to determine the coronal field out to 2.5 R_s (R_s, solar radius). The output of the PFSS model serves as input to the Schatten Current Sheet (SCS) model [440], but uses the absolute value of the (radial) field there. It solves this new potential field problem between the source surface and infinity. Then it restores the proper sign to the field. The resulting solution at any radial distance beyond the original source surface can be obtained from the model, which provides a more realistic magnetic field topology of the upper corona. Empirical relationships (e.g., [4, 366, 566, 567]) are used to assign solar wind speed at a certain radial position to be a function of two coronal parameters, the flux tube expansion factor (f_s), and the minimum angular separation between an open field foot point and its nearest coronal hole boundary at the photosphere (θ_b) measured in degrees. The two parameters are determined from the coronal magnetic topology extrapolated by the PFSS and/or SCS models. McGregor et al. [314] calibrated the empirical formula of the solar wind velocity in the WSA model by using Helios perihelion observations, along with results from Riley et al. [430] and Schwadron et al. [445] as constraints, and determined a new relationship between the final solar wind velocity and the parameters of f_s and θ_b.

As the primary input to the WSA and other coronal and solar wind models, global estimates of the solar photospheric magnetic field distribution are essential for accurate modeling of the corona and solar wind. Sun et al. [496] proposed a new method to interpolate the photospheric field in polar regions of synoptic map by scaling the average flux density from the prior observations of slightly lower latitudes, which can improve the results of global models such as the WSA model and MHD simulation. They also derived a new relationship for the final solar wind speed from the observations of Michelson Doppler Imager (MDI) onboard SOHO after a trial-and-error scheme. By incorporating the various data assimilation techniques, including an ensemble Kalman filter, with a photospheric magnetic flux transport model [587], Arge et al. [7] developed the Air Force Data Assimilative Photospheric flux Transport (ADAPT) model. The ADAPT model evolves the magnetic flux on photosphere using the flux transport model when observations are not available, and then merges the modeled magnetic flux with new observations using the ensemble Kalman filter. The ensemble of model realisations with different parameter values incorporates both data and model uncertainties into the predictions of photospheric evolution. The ADAPT model provides an instantaneous snapshot of the global photospheric field distribution. Arge et al. [8] employed the global photospheric field maps generated by the ADAPT model as input to the WSA model and showed that traditional Carrington maps can not produce the same dynamical coronal structure as those obtained by using the ADAPT model. The ADAPT model has been extended to assimilate far-side active region data derived from helioseismology [9].

The three corona models in CORHEL described above can be coupled to different heliospheric models, with the results at the outer boundary of the coronal calculation supplying the inner boundary conditions for heliospheric simulations.

1.2.1.2 Heliospheric Models

For computing heliospheric solutions, two complementary approaches have been developed for the coronal models to provide the bottom boundary conditions of the heliospheric models. When the coronal state is provided by solving the MAS thermodynamic model, the heliospheric models are driven by directly using all of the magnetic and plasma variables computed in the coronal solution [281, 431].

The other approach is ad hoc techniques for deriving the inner boundary conditions for heliospheric model when the coronal state is achieved by WSA coronal model or polytropic MAS coronal model. For the MAS polytropic coronal model, the plasma radial velocity at the bottom boundary of the heliospheric models is replaced with an empirically derived velocity [366, 429]: $V_r(\theta_b) = 250 + \frac{1}{2}(650 - 250)\left[1 + \tanh\left(\frac{\theta_b - 2.86}{1.43}\right)\right]$ km s^{-1}, where θ_b is the minimum distance from an open-closed boundary, measured in unit of degrees along the surface of the photosphere. $V_r(\theta_b)$ is computed at the photosphere from a map of the open-closed boundaries generated from the MHD solution. This quantity is then mapped outward along field lines to generate V_r at the inner boundary condition of the heliospheric model (usually at 20–30 R_s). At the same time, the computed radial magnetic field from the coronal solution is used directly in the heliospheric models, and the remaining plasma quantities (density and temperature) are inferred by assuming momentum flux conservation and pressure balance over the sphere. The meridional component of the magnetic field (B_θ) is assumed to be zero, whereas the azimuthal component (B_ϕ) is derived from the rotation speed of the source surface (i.e., $B_\phi = -B_r(V_{\text{rot}}/V_r)\sin\theta$, where V_{rot} is derived from the 27.2753 day rotation period of the Sun). The nonradial components of the plasma flow are assumed to be zero in the inertial frame. For the WSA model, the distributions of radial magnetic field and radial plasma speed at the inner boundary of heliosphere model are given by WSA model output, while other MHD quantities are derived using similar empirical relationship as the polytropic coronal model.

So far, the ad hoc solutions tend to better match in situ data at Earth and Ulysses, primarily because the transition from slow to fast wind is too gradual with the present heating/acceleration model [273]. However, as the physics contained within the coronal model becomes more sophisticated, and the remaining free parameters become better constrained, the quality of the self-consistently-derived heliospheric solutions will be expected to surpass the ad hoc results.

There are two different heliospheric MHD models available in CORHEL shown in Fig. 1.1. One is a heliospheric version of the MAS (MAS-H) model formulated by Riley et al. [429] and the other is Enlil model developed by Odstrčil [350].

In the MAS-H model, the energy equation is approximated with a simple adiabatic energy equation and the polytropic index γ is assigned to be 1.5 in the hellospheric model because a polytropic relationship between pressure and density is observed to hold, with $\gamma = 1.5$ for protons [93, 529]. The MAS-H model uses the same numerical scheme as MAS coronal model. Initially, a finite difference approach was used in the radial (r) and meridional (θ) directions and the derivatives in azimuth (ϕ) were calculated pseudospectrally. It should be noted that the MAS-H model neglects the effect of pickup ions and differential rotation. The former are thought to dominate the internal energy of the solar wind beyond 6–10 astronomical unit (AU) [12] and the latter may play a role in connecting high-latitude field lines near the Sun with lower latitude interaction regions much farther away [113]. In the early development of MAS-H model, it was implemented in the inertial frame. Additionally, although the MHD model is time-dependent, the flow at the inner boundary is assumed to be time-stationary. Thus the flow at the inner boundary is "corotating", so that rotating spatial variations are responsible for the generation of dynamic phenomena in the solution. In later development [432], significant modifications have been made. First, the algorithm uses finite difference and is fully parallelized in all three dimensions, which makes it possible to run cases at significantly higher resolutions than ever. Second, the solutions can be computed in either the corotating frame of reference or in an inertial frame. Third, boundary conditions can be specified at some inner radial sphere as a function of time. A time-dependent inner boundary condition is also developed, which is able to generate evolutionary features that are not found in a steady-state approach [276].

The heliospheric model, Enlil model, solves the conservative time-dependent MHD equations in a spherical geometry using the modified high-resolution Total-Variation-Diminishing Lax-Friedrich (TVDLF) scheme [349, 350, 521]. The field-interpolated central-difference approach for solving the magnetic field is used to satisfy $\nabla \cdot \mathbf{B} = 0$. The high-resolution scheme produces second-order accuracy away from discontinuities, while simultaneously providing the stability that ensures nonoscillatory solutions. The computational domain covers the radial range from 21.5 R_s to 1 AU or beyond and the solar latitudes from $-60°$ to $+60°$. It should be noted that a monopolar IMF is imposed throughout the computational domain as the initial condition. In Enlil model, two additional continuity equations are solved simultaneously with the MHD equations for tracing the injected CME material and the magnetic field polarity [350, 353]. The magnetic field polarity is used to recover the IMF polarity when the computation is finished. In this way, the loci of the heliospheric current sheet (HCS) surface are accurately tracked and numerical reconnection is avoided in the vicinity of HCS. The lower boundary conditions for the Enlil model are provided by a coronal solution (either MAS or WSA) and has been described above. The heliospheric solution is allowed to relax to steady state. The simulation domain of Enlil is limited within 5 AU since it does not include the pick-up ion effect. The model can be used to investigate the background solar wind [357], the interaction of fast and slow solar wind streams [349], distortion of the HCS by shock waves [356], shock-induced magnetic reconnection of coronal sheets [351], propagation of transient disturbances in the heliosphere [350, 352–354], and global dynamics of magnetic clouds [552].

1.2.1.3 Simulations of Background Solar Wind

Lee et al. [264] derived the solar wind parameters at fixed radial distance of 1 AU and fixed latitude (0°) from outputs of the WSA/Enlil and MAS/Enlil models and gave a comparison of modeled and measured solar wind parameters from Carrington rotation (CR) 1999 to 2038 with a time resolution of 3.6 h. They obtained a general good agreement for large scale structures and for time scales of several days. By using the WSA/Enlil model and composite observations, McGregor et al. [313] showed that the intermediate-speed solar wind (400–500 km/s) during solar minimum represented a mix of what was originally fast and slow solar winds, which may imply a more bimodal solar wind in the corona than observed at 1.0 AU. This deepened our understanding of the radial evolution of solar wind speed, which is critical for interpreting the solar wind composition observations near Earth and throughout the inner heliosphere. McGregor et al. [314] used a recalibrated WSA model to drive the Enlil 3D MHD solar wind model and obtained solar wind parameters at Earth (1.0 AU) and Ulysses (1.4 AU). The improvements in solar wind speed, its variability, and the occurrence of high speed enhancements provided confidence that the new velocity relationship better determined the solar wind speed in the outer corona (0.1 AU). The analysis of the improved velocity field suggested the existence of two distinct mechanisms of the solar wind generation, one for fast and one for slow solar wind, and implied that a combination of present theories may be necessary to explain solar wind observations. Norquist and Meeks [347] and Norquist [346] statistically evaluated the capability of the Hakamada-Akasofu-Fry (HAF) model, the WSA one-dimensional kinematic trajectory model and the Enlil solar wind model for forecasting the solar wind speed and IMF polarity. They concluded that studies of model forecast behavior must be conducted over extensive time periods, covering as much of a solar cycle as possible. By comparing modeled results with in-situ observations at 1 AU, Gressl et al. [145] tested the performances of different combinations of coronal and heliospheric models in the CORHEL, including MAS/Enlil, WSA/Enlil and MAS/MAS-H. They found that the general structures of the ambient solar wind were reproduced well by all models. The best performance was obtained for the solar wind speed. The typical error of arrival times of high-speed solar wind was of the order of about one day. In consistent with other studies [430, 432], they found that the choice of the synoptic map had a significant effect on the model's performance, but there was no clear trend as to which synoptic map gave the best result.

Using the MAS/MAS-H model, Riley et al. [431] computed the coronal state and interplanetary solar wind during the Whole Heliosphere Interval (WHI) and presented a general analysis of the large-scale structure of the solar corona and inner heliosphere. A quantitative comparison between simulated images from EUV Imaging Telescope (EIT) and observations for CR 2068 is shown in Fig. 1.2. The modeled results tended to be slightly brighter than the observations, but, overall, the close match suggested that the parameterizations of the quiet-Sun heating was reasonably accurate. One notable mismatch was that the model failed to reproduce the easternmost of the triplet of active regions (ARs), probably because the modeled

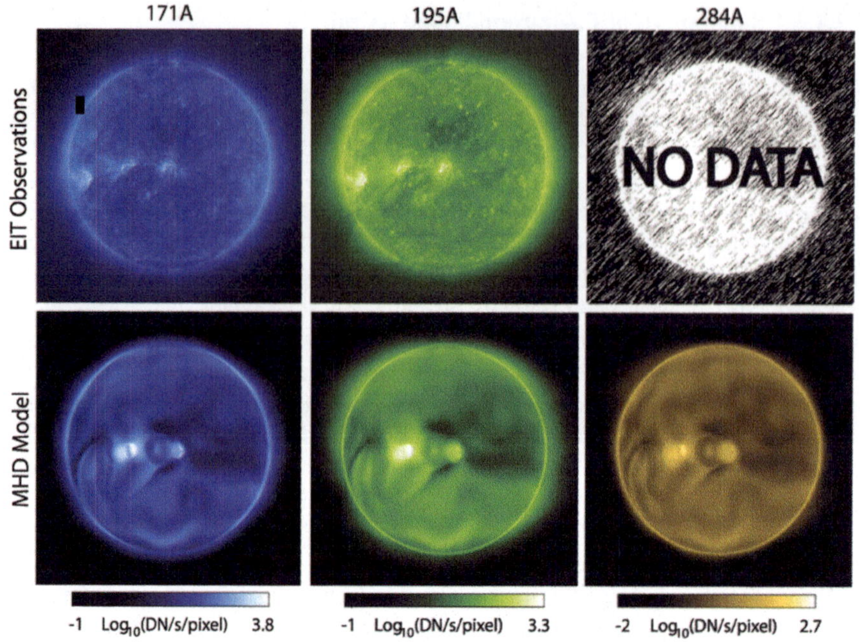

Fig. 1.2 Comparison between SOHO/EIT emission observations at 171 Å, 195 Å, and 284 Å for CR 2068 (top) with model results (bottom). There were no data from EIT at 284 Å at this time. Taken from Fig. 7 of Riley et al. [431] with permission of P. Riley

field strengths there were too low to be "lit up" by the heating function. Figure 1.3 compares modeled results computed by ad hoc solutions with observations made by the twin STEREO spacecraft. Comparisons showed that there was a basic agreement between the large-scale features in the model and observations, and there were some noteworthy disagreements. Overall, the model captured the two-stream flow during this time period, matching the minimum and maximum velocities. The phasing of the fast streams between the model and observations may be offset by a day or more from one another. The model failed to capture the sector structure of the IMF. While the sector boundary on 31 March 2008 matched well, the model erroneously predicted a return to negative polarity on April 2008.

Linker et al. [276] developed a time-dependent model of the solar wind empirically driven by magnetic maps created with the ADAPT model at a daily cadence. Potential field solutions were used to model the coronal magnetic field, and an empirical specification was used to develop boundary conditions for the MAS-H solar wind in the heliosphere. The simulation showed both classic features of stream structure in the interplanetary medium often seen in steady-state models and evolutionary features unable to be captured in a steady-state approach. The model results compared reasonably well with 1 AU OMNI observations, which suggested that the model is suitable for driving long term models of the outer heliosphere after improvements to the ingestion of magnetograms in flux transport models.

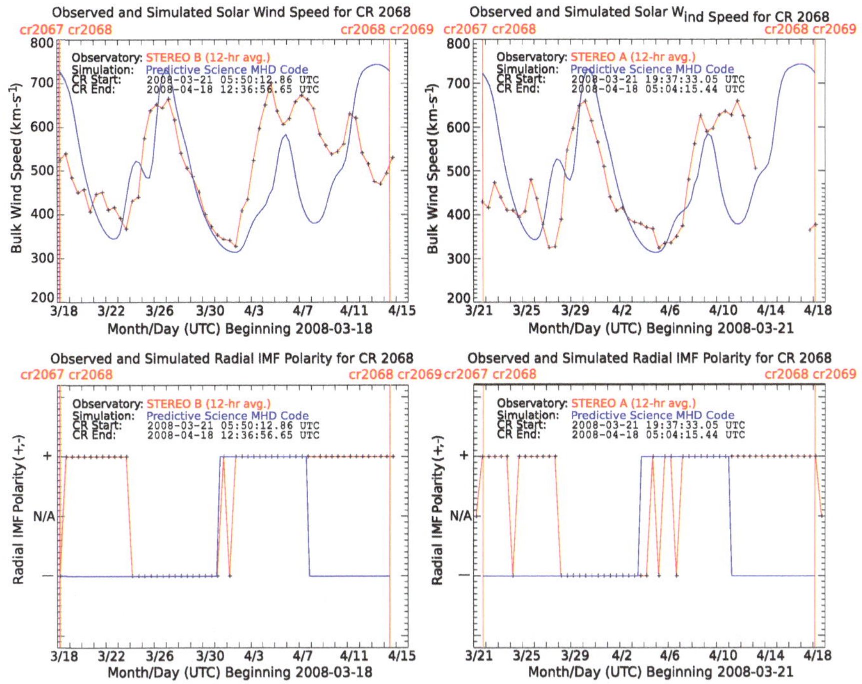

Fig. 1.3 Time series of (top) bulk solar-wind velocity and (bottom) polarity of the radial component of the IMF. Results from STEREO-B (behind) are shown on the left and from STEREO-A (ahead) are shown on the right. Model results are colored blue, while measurements are red. The boundary of the Carrington rotation is marked by the vertical red lines. Taken from Fig. 8 of Riley et al. [431] with permission of P. Riley

1.2.1.4 Simulations of CMEs

Taktakishvili et al. [497] reported the simulated results of 14 selected well-observed halo coronal mass ejection events using combined model of the WSA/Enlil and the CONE models. The CONE model [624] seeks direction, size and speed of a cone-shaped volume the best match actual CME observations. By analyzing 36 particularly strong geomagnetic storms, they tried to associate them with particular CMEs using the catalog of Large Angle Spectrometric Coronagraph (LASCO) on board SOHO, and then modeled these CMEs to demonstrate that the combination of numerical models with the observations from coronagraph as input could give reasonably good results for the CMEs' arrival times for the selected set of "geoeffective" CME events. In order to determine the heliospheric distance range where the drag becomes dominant, Temmer et al. [511] derived the evolution of the background solar wind flow using the MAS-Enlil and WSA-Enlil models and then compared the modeled background solar wind flow with the tracked observations of three CMEs (2008 June 1–6, 2009 February 13–8, and 2010 April 3–5) from the Sun to 1 AU by the Heliospheric

Imager (HI) onboard STEREO. They found that the CME speed became adjusted to the solar wind speed at very different heliospheric distances from below 30 R_s to beyond 1 AU, depending on the CME and the characteristics of ambient solar wind. The results also indicated that the Enlil model can be used to derive important information about the overall structure of the background solar wind. Besides, the Enlil model can also be used to differentiate the contribution of multiple-CME measurements in interplanetary space. Odstrčil and Pizzo [355] used the heliospheric code ENLIL to demonstrate different scenarios of CIRs from HI-2A and HI-2B onboard STEREO spacecraft, and interpreted the white-light images by HIs through simulations of the 24–25 January 2007, 31 December 2007 and 26 April 2008 CMEs. Furthermore, Webb et al. [571] tracked the propagation of multiple CMEs of late July to early August 2010 in the inner heliosphere by comparing the results from the ENLIL model, 3D reconstruction techniques based on a kinematic solar wind model, and in situ results from multiple spacecraft. Bain et al. [13] combined the WSA-ENLIL and Cone models to discuss shock connectivity in the August 2010 and July 2012 SEP events. In addition, MacNeice et al. [297] tested the accuracy of the modeled field lines connecting Earth to the Sun at the onset times of 15 impulsive solar energetic particle (SEP) bursts and found that the principal source of error resulted from the model's inability to reproduce sufficient low-latitude open flux. Therefore, time-dependent coronal models incorporating these transient events may be needed to significantly improve Earth/Sun field line forecasting.

Lionello et al. [281] improved the flexibility of the MAS-H model by involving a carefully designed boundary conditions, which enabled the model to accommodate time-dependent boundary conditions obtained from empirical coronal model, MHD coronal model, observation or a combination thereof. The boundary conditions could convert variables between the corotating and inertial frames. While the other variables necessary for specifying the boundary conditions were directly used, the tangential electric field was specially treated to avoid formation of spurious current layers near the poles. They validated the new implementation by studying the propagation of an interplanetary CME (ICME) from 18 R_S to 1.1 AU, where the CME was modeled by an out-of-equilibrium flux rope inserted in the coronal model within 7 R_S. Figure 1.4 shows the distribution of ρr^2 in the heliosphere and corona at the end of the relaxation phase in the equatorial plane and in the meridional plane perpendicular to the Sun-Earth line. Figure 1.5 displays the time history of the local plasma variables for an observer situated along its path. It should be noted that the low value of **B** within the ICME probably resulted from a considerably smaller field strength in the active regions of the model than that in the source regions of the most powerful events that produce very fast CMEs. Figures 1.4 and 1.5 show that the improved model can follow the propagation of the CME accurately. Starting from 40 h after the eruption ($t = 442$ h), the density of the plasma began to increase as the observer was swept by the ICME. Then, the observer was overtaken by the flux system and the magnetic field exhibited a typical pattern of rotation. Finally, trailing the ICME, there was a region of fast solar wind that compressed the plasma. By using the MAS/MAS-H model with a "thermodynamic" improvement of the corona and solar wind, Török et al. [518] inserted a magnetically stable flux rope along the polarity inversion line

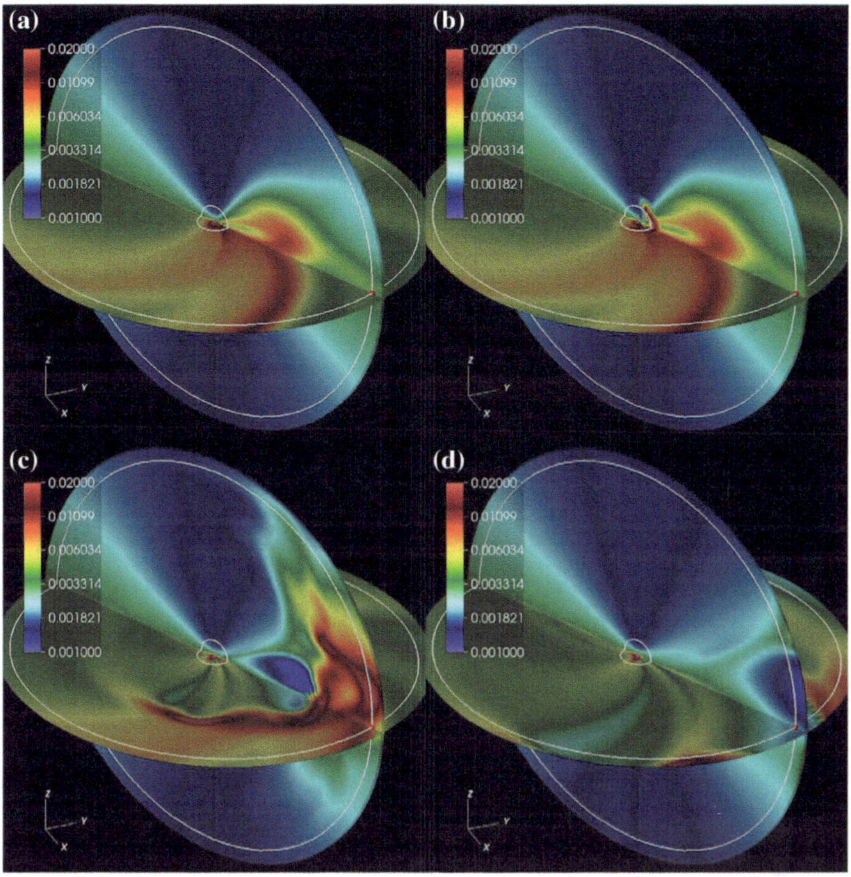

Fig. 1.4 Evolution of the scaled density (ρr^2) in two cut-planes during the propagation of an ICME from Sun to Earth (red dot). The inner white circle lies at 20 R_s. **a** Configuration in steady state just before the eruption on the solar surface (t = 402 h). **b** CME leaving the computational domain of the coronal model and entering the heliosphere (t = 408 h). **c** ICME as it reaches Earth (t = 458 h). **d** ICME leaving the computational domain of the heliospheric model. End of the calculation (t = 502 h). Taken from Fig. 7 of Lionello et al. [281] with permissions of the authors and the AAS

of the eruption's source region and initiated the solar eruption by boundary flows to generate a CME close to the observed properties of the 2000 July 14 "Bastille Day" eruption. Then, the CME was propagated to Earth, using a heliospheric MHD code. The propagating properties were discussed with regard to the development of space-weather predictions based on MHD simulations of solar eruptions.

In order to achieve more accurate solar wind parameters at Mercury's orbital location, Dewey et al. [67] integrated the CONE model into the WSA-Enlil model to include CME-related solar wind perturbations and thus enabled characterization of the effects of strong solar wind disturbances on the Mercury system. Comparisons

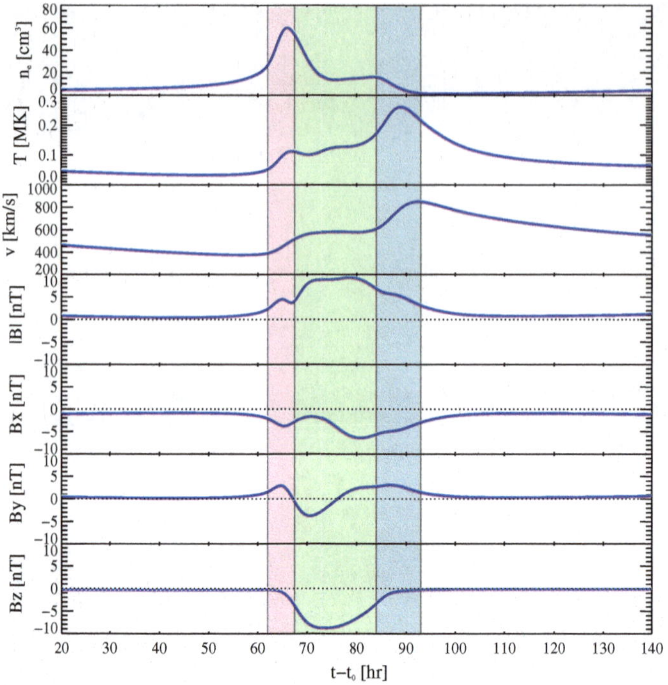

Fig. 1.5 Time history of local plasma variables at 1 AU for a point situated in the path of the ICME. The magnetic field vectors have been converted to the Geocentric solar ecliptic (GSE) coordinates. The pink highlighted region indicates the compressed solar wind upstream of the ICME. The green region indicates the flux system of the ICME. The blue region indicates the trailing plasma, which is compressed by a high speed stream behind it. Time is in hours from the insertion of the CME ($t_0 = 402$ h). Taken from Fig. 9 of Lionello et al. [281] with permissions of the authors and the AAS

of the model results with observations by the spacecraft of MErcury Surface, Space ENvironment, GEochemistry and Ranging (MESSENGER) during the period from March 2011 to December 2012 showed that the WSA-Enlil-Cone model more accurately reproduced the solar wind conditions at Mercury and demonstrated greater predictive power for magnetospheric and exospheric processes than the WSA-Enlil model alone [16]. A similar research to obtain the continuous upstream solar wind conditions at Mars has also been conducted [68]. By introducing a customized density enhancement factor in the WSA-ENLIL+Cone model based on coronagraph image observations, Werner et al. [573] modeled the multiple CME interaction event on 6–9 September 2017. The predicted arrival time of the first interplanetary shock was drastically improved, while the background solar wind preconditioned by the passage of the first interplanetary shock likely caused the last CME to experience insignificant deceleration and led to the early arrival of the second interplanetary shock.

1.2.2 BATS-R-US in SWMF

The Center for Space Environment Modeling (CSEM) is an interdisciplinary research organization of the College of Engineering at the University of Michigan, Ann Arbor. The CSEM and its collaborators developed the Space Weather Modeling Framework (SWMF), which provides a high-performance flexible framework for both physics-based space weather simulations and various space physics applications. The SWMF is now capable of simulating the Sun-Earth system from the solar corona to the upper atmosphere of the Earth faster than real time on hundreds of processors for most Sun-Earth connection events. All the source code of the SWMF and its component are freely available after registration at the web site of http://csem.engin.umich.edu/tools/swmf/index.php.

The SWMF integrates physics-based and empirical models of the Eruptive Event Generator (EE), Lower Corona (LC), Solar Corona (SC), Inner Heliosphere (IH), Outer Heliosphere (OH), Solar Energetic Particles (SEP), Global Magnetosphere (GM), Inner Magnetosphere (IM), Radiation Belt (RB), Polar Wind (PW), Ionosphere Electrodynamics (IE), and Upper Atmosphere (UA) into a high-performance coupled framework [527]. Different components can solve different equations using different spatial and temporal discretizations. The domains of two coupled models may be adjacent to each other or overlap in space. Each model corresponds to a component in the SWMF. The components may have multiple versions or can be replaced with measurements. The structure of physics component in the SWMF is shown in Fig. 1.6. A component version is a physics model plus the appropriate wrappers and couplers. The multiple component approach makes the framework more accurate, efficient, and versatile than a monolithic code. The components are compiled into libraries, and linked to the core of the framework and the shared libraries to form a single executable. The SWMF distributes the components over a parallel machine, and executes and couples them in an efficient manner using the Message Passing Interface (MPI) library for communication [520]. The SWMF is also able to provide various utilities and a shared library of general algorithms, such as linear solvers, interpolation algorithms, and timing utilities, for each model. The physics models can also be compiled into individual executables and used as stand-alone codes.

The SWMF incorporates computational physics modules with only modest modification while obtaining good parallel performance and making the SWMF as versatile as possible. In order to achieve this goal, the framework adopts a four-layered architecture: the web-based Graphical User Interface (GUI), Superstructure Layer, Physics Module Layer, and Infrastructure Layer [527]. The GUI provides a user-friendly interface via a web browser and enables the user to select components, compile the SWMF, define the layout and the input parameters, execute runs on parallel machines, monitor the progress of the runs, and post process the output data. The Superstructure Layer contains the Component Interface (wrappers and couplers) and the Framework Services. The former enables the component to accept parameters for parallel configuration, check the parameters obtained from the Control Module, initialize the session, write and read the restart files, receive and provide data to another

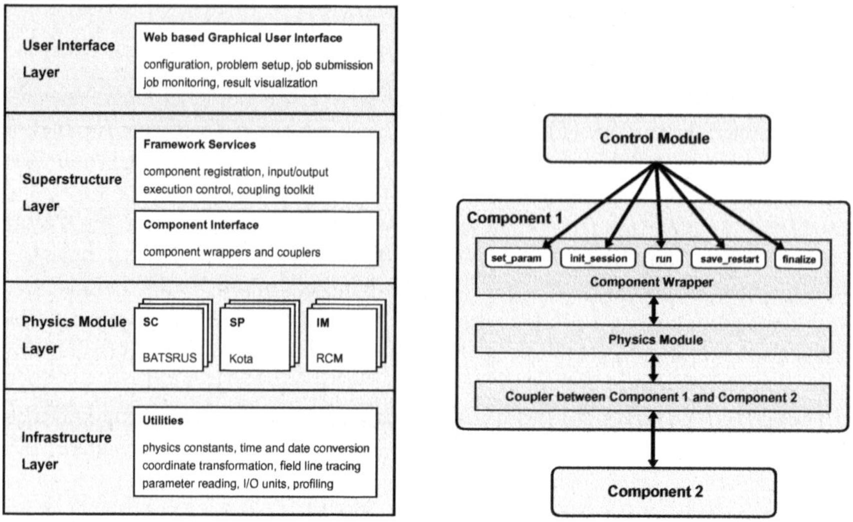

Fig. 1.6 The layered architecture of the SWMF (left) and the structure of the components (right) [527]. Reprinted from *J. Comput. Phys.*, 231, Tóth G, van der Holst B, Sokolov IV, de Zeeuw DL, Gombosi TI, Fang F, Manchester WB, Meng X, Najib D, Powell KG, Stout QF, Glocer A, Ma YJ, Opher M, Adaptive numerical algorithms in space weather modeling, 870–903. Copyright 2012, with permission from Elsevier

component via an appropriate coupler, and finalize at the end of the execution. The latter implement component registration, setting up the parallel layout, reading and distributing the input parameters, control of component execution and coupling, and the SWMF parallel coupling toolkit, which can be used by the component couplers. The Physics Module Layer contains the physics modules and provides appropriate entry points for the Component Interface. Each physics module may have multiple versions. The physics module may contain additional code for a stand-alone execution. The Infrastructure Layer consists of utilities, which define physics constants, transformation between different coordinate systems, time conversion routines, time profiling routines, and other lower level routines. The Infrastructure can be used by the physics modules as well as by the Superstructure.

In SWMF, various models are coupled together either based on simulation time or iteration number. The relevant physical quantities are passed with efficient MPI communication. Deadlocks are carefully avoided to maintain the robustness of the framework [520]. The utilities have been developed to take care of coordinate transformations and interpolation between various models. Overlapping the models is also used to couple the models when the flow is slower than the fast magnetosonic speed, which can suppress numerical artifacts due to the differences between the spatial and temporal resolutions.

The computationally most expensive domains of the SWMF are modeled by the Block-Adaptive Tree Solar wind Roe-type Upwind Scheme (BATS-R-US) code

[132, 414]. During the past decade, the BATS-R-US code has evolved from a single-purpose ideal MHD model for the study of comets [181] and the terrestrial magnetosphere into a general, highly modular space physics modeling tool. In the following subsections, we first introduce the BATS-R-US code and then briefly describe the feature of EE, SC and IH, the components associated with simulation of corona and solar wind in the SWMF.

1.2.2.1 The BATS-R-US Code

As mentioned above, many domain components are modeled by the BATS-R-US code, which plays fundamental roles in the SWMF. In these applications, the BATS-R-US code solves different sets of equations: ideal MHD, semi-relativistic MHD, Hall MHD, anisotropic MHD, MHD with Alfvén wave heating and electron thermal heat conduction, and multi-fluid MHD. The basic MHD equations used in the BATS-R-US code are the 8-wave-formulated ideal MHD equations, which is in near conservative form with the source terms proportional to $\nabla \cdot \mathbf{B}$ on the right hand sides to control the numerical error in $\nabla \cdot \mathbf{B}$ [414]. The ideal MHD equations with the source terms proportional to $\nabla \cdot \mathbf{B}$ are symmetrizable and Gallilean invariant; all waves in the system propagate at speeds $u \pm c$ (for MHD, the possible values of c are zero, and the Alfvén, magnetofast, and magentoslow speeds). Neither of these properties holds for the MHD system if the source term is ignored. In this formulation, the numerical $\nabla \cdot \mathbf{B}$ errors are propagated together with the flow, leading to improved robustness and accuracy of the numerical scheme. Moreover, this form of the MHD equations allows the derivation of an eight-wave approximate Roe-type Riemann solver that can be used in the construction of an upwind finite-volume scheme [413, 414].

The spatial discretization in the BATS-R-US code is based on total-variation-diminishing (TVD) [156, 266] type numerical schemes with the second- and third-order (in smooth monotonic regions) [243] accurate slope limiters. Sokolov et al. [474] developed a second-order accurate oscillation free interpolation scheme for the grid resolution changes. A large variety of numerical flux functions have been available, including the local Lax-Friedrichs or Rusanov [545], the Harten-Lax-van Leer-Einfeldt (HLLE) solver [85, 158], Artificial Wind [472], the Harten-Lax-van Leer-Discontinuities (HLLD) solver [329], Roe [435] and Godunov [129] fluxes. The Rusanov, HLLE and Artificial Wind schemes work for all systems of equations, the HLLD and Roe schemes are implemented for ideal MHD only, while the Godunov scheme (exact Riemann solver) can be used for hydrodynamics. The optimal combination of flux functions, limiters, and divergence control methods is highly application dependent, and a large number of choices can be regarded as an algorithmic adaptivity.

There are several alternative methods implemented in the BATS-R-US code to maintain the solenoidal constraint. The projection [30] scheme can remove the $\nabla \cdot \mathbf{B}$ error by solving an elliptic problem. The constrained transport scheme [87] generalized to adaptive mesh refinement (AMR) grids [522] maintains the $\nabla \cdot \mathbf{B} = 0$

condition to round-off error at the expense of staggering the magnetic field. For the MHD Roe solver, a new 7-wave scheme [475] has been developed to maintain the solenoidal constraint. Finally, in the hyperbolic/parabolic cleaning method of Dedner et al. [59], a new scalar field is added to the MHD variables that couples the induction equation to the $\nabla \cdot \mathbf{B}$ error.

The BATS-R-US code is designed around a 3D block-adaptive grid with an underlying tree data structure. The basic underlying unit is a block of structured grid. Each block consists of the same number of cells, typically, $4 \times 4 \times 4$ cells to $10 \times 10 \times 10$ cells. Each grid block corresponds to a node of the tree: the root of the tree is a single coarse block of structured grid covering the entire solution domain. The blocks can be split into eight children (refinement), or eight blocks can be merged into one (coarsening). Any of these blocks can in turn be refined, and so on, building up a tree of successively finer blocks. This block adaptivity has a lot of advantages: each block is a simple structured grid allowing simple numerical schemes, the fixed sized blocks are easy to load balance, the data corresponding to a block can easily fit into the cache, and the fixed length loops can be efficiently optimized (unrolled) by the compiler [486]. The block-adaptive grid algorithm has been improved and extended to handle arbitrary number of variables and generalized coordinates. By way of a continuous mapping from a logically Cartesian coordinate system into a general curved system, the BATS-R-US code can be implemented on the generalized coordinates in combination with Cartesian vector components, which means that the governing equations are kept in the Cartesian form, only the geometry of the grid is modified. This choice allows a general implementation and avoids singularities along the symmetry axis of cylindrical and spherical coordinates. In order to handle the problems caused by the converge of the cells around the symmetry axis, the BATS-R-US code also exploits a simple yet conservative "supercell" algorithm which averages the cells around the symmetry axis after every time step. To make the scheme second order accurate, a limited least-squares slope is calculated from the cells surrounding the supercell, and the cell values inside the supercell are filled accordingly. The details of the supercell algorithm can be seen in Tóth et al. [527].

The BATS-R-US code can be used to solve both steady state and time-dependent problems. Often the steady state solution is used as an initial condition for a time accurate simulation. Several measures have been exploited to accelerate convergence towards steady state, such as the methods of local time stepping and gradual grid adaptation. In some cases starting from a steady state simulation, the initial evolution only happens in a small portion of the computational domain, and thus the simulation can be significantly sped up by using a time stepping algorithm.

In time accurate simulations, the strategy of the same time step in all the grid cells with an explicit time discretization often becomes inefficient due to the adaptive grid and the physics of the simulated problem [527]. The BATS-R-US code uses the method of subcycling in the case of adaptive grids, i.e., different cells takes different number of time steps to reach the same time level. In addition, the equations can be stiff when adding various source terms to take into the consideration of different physics processes occurring in different physics domains. The numerical stability restricted by the stiffness of the equations results in much smaller time step than that required by accuracy considerations. According to the mathematical properties of the stiff terms, the BATS-R-US code can use point-implicit, semi-implicit or fully implicit time discretization schemes to allow larger time steps. When the partial differentiation equations are only stiff in a part of the computational domain, the SWMF even adopts an combination of explicit and implicit schemes, which solves some of the grid blocks explicitly, and the rest implicitly.

Since the original BATS-R-US algorithms have been hard coded for three spatial dimensions, a Block Adaptive Tree Library, BATL, is being developed to generalize the BATS-R-US algorithms to the cases of two dimensions (2D) and one dimension (1D). Tóth et al. [527] detailed on the implementation of BATL from the aspects of grid structure, grid geometry, refinement, coarsening and load balancing, and message passing. In addition, they also presented some validating examples.

1.2.2.2 Components of EE, SC, and IH

In the SWMF, the EE generator is coupled with the SC model only. The SC model drives the IH model. Both the SC and the IH models provide input to the SEP model. The IH model drives the coupled system of the GM, IM, IE, and UA models through the IH/GM interface. The models of EE, SC, IH, and several other component are modeled by the BATS-R-US code.

The EE generator component is responsible for creating a CME by employing either a physics-based model of flux emergence from the convection zone, or much simpler and less expensive empirical models that insert an unstable flux rope into the steady solar corona solution, or insert an arcade and apply shearing motion at the lower boundary of the corona model [3, 548]. The physics-based EE model only works in a stand-alone mode [89, 300], and the empirical models are used to generate CMEs in the SWMF [292, 301]. In the former, the BATS-R-US code solves the MHD equations with an optically thin radiative loss and empirical coronal heating term in a box of several megameters that should ideally include a whole active region and simulate the emergence of a flux rope from the convection zone.

The SC model describes the corona out to about 25 R_s. The two fundamental processes in the corona are coronal heating and solar wind acceleration, whose mechanisms have been challenging the theory of solar wind origin and evolution for several decades and are still open to the solar scientists. The first one is to explain how the temperature in the solar atmosphere rises by 2 orders of magnitude from the photosphere ($T \sim 10^4$ K) to the corona ($T_e \approx T_i \approx 10^6$ K) across a narrow transition region [10]. The second one needs to explain how the solar wind plasma originates from the Sun, how it is accelerated to escape the solar gravity, and how to reach the observed speed and form the bimodal structure in the inner heliosphere [10, 48]. The deposition of energy and/or momentum into the solar wind has been described by means of some empirical source terms (e.g., [147, 315, 327, 490, 535, 599]). In some more near-realistic models, the solar wind is heated and accelerated by the energy and momentum interchange between the solar plasma and large-scale Alfvén turbulence [19, 192, 539, 541]. In SWMF, several parameterized mechanisms for coronal heating and solar wind acceleration have been incorporated to obtain the near real solar wind through numerical simulations.

Originally, the physics of this domain was approximated by including additional source terms in the MHD equations, which take into account gravity, the coronal heating, and the acceleration of the solar wind [147, 533]. In some cases, the coronal heating was mimicked by solving one extra equation that describes the energy interchange between the solar wind plasma and the large-scale MHD turbulence [438]. In its later version, the SC model solved the two-temperature MHD equations with Alfvén wave heating and heat conduction on either Cartesian or spherical grid in a frame corotating with the Sun [549]. This two-temperature SC model included anisotropic thermal heat conduction for the electrons and Alfvén waves to heat and accelerate the protons. The heating of protons and the solar wind acceleration by the gradient in the wave pressure were described by the evolution of the Alfvén waves under the assumptions of the Wentzel–Kramers–Brillouin (WKB) approximation for the short wavelength Alfvén waves [192] and of Alfvén waves' propagating away from the Sun with no reflection. Following this idea, the Alfvén wave solar model has been formulated in the works of Downs et al. [72], Evans et al. [88], Meng et al. [317, 318], Sokolov et al. [476]. To move forward, van der Holst et al. [551] developed the Alfvén wave solar model named AWSoM. Compared with the two-temperature solar wind model, the AWSoM model was a three-temperature MHD model that considered the anisotropy of ion temperature in the solar corona and the inner heliosphere. The coronal heating and solar wind acceleration were addressed with low frequency Alfvén wave turbulence. The Alfvén waves are partially reflected by the Alfvén speed gradient and the vorticity along the field lines, which may be important in close proximity of the active region. At the inner boundary, they injected the Alfvén wave energy whose Poynting flux was proportional to the magnetic field strength.

The boundary conditions of the SC component are prescribed as follows. The magnetic field at the bottom boundary may be obtained from synoptic magnetograms, or a simple dipole (possibly with a few higher-order terms) may be assumed [147]. The inner boundary conditions for the temperature and mass density at the Sun may vary with longitude and latitude to achieve the most realistic solar wind near the Sun and at 1 AU. The velocity components at the inner boundary should maintain line-tying of the magnetic field. The flow at the outer boundary usually travels faster than the fast magnetosonic speed of the plasma, so no information is propagating inward. Sometimes, however, when a coronal mass ejection (CME) passes the boundary, the solar wind speed may become subfast for short periods of time. During such periods, the SC component needs to receive the outer boundary condition from the IH component. The SC component provides the plasma variables at the inner boundary of the IH domain. The SC and IH domains are allowed to overlap, which is actually numerically advantageous when the flow becomes subfast for a short time because the overlap can reduce reflections or other numerical artifacts at the inner boundary of the IH due to the use of unaligned grids and different time steps in the two domains [523]. The data transfer between SC and IH is realized using an auxiliary buffer grid.

The buffer grid is set as a spherical shell which lies in the overlap region of the SC and IH grids. Trilinear interpolation is utilized to fill the buffer grid and provide boundary condition for SC and IH, which ensures the second order accuracy in space [363].

Additionally, the split of the magnetic field into $\mathbf{B} = \mathbf{B}_0 + \mathbf{B}_1$ as done by Tanaka [498] is used in the numerical solver of the solar wind model. Therefore, only the variation of the magnetic field $\mathbf{B}_1$ is needed to numerically solve. The advantage of this splitting is that the solver is inherently more accurate for a small $\mathbf{B}_1$ than for the full magnetic field, which reduces the numerical diffusion [498]. It should be noted that in this splitting scheme, no assumption is made that $\mathbf{B}_1$ should be small compared to $\mathbf{B}_0$. The PFSS solution for $\mathbf{B}_0$ is used for the initial condition. In the SC model, the PFSS field $\mathbf{B}_0$ between the source surface radius and the outer boundary is assumed to be radial. Since the PFSS solution is not current free outside source surface, the source terms containing $\nabla \times \mathbf{B}_0$ are added to the equation. If $\mathbf{B}_0$ is time dependent, source terms containing $\frac{\partial \mathbf{B}_0}{\partial t}$ are also included [131, 527]. The inner boundary condition for $\mathbf{B}_r$ reduces to $\mathbf{B}_{1r} = 0$ at $r = 1.035\mathrm{R_s}$. Although the PFSS field for $\mathbf{B}_0$ is used as the initial value, the solar wind solution does not have a source surface, but has different-sized streamers and a Parker spiral due to the solar rotation.

The IH model typically extends from about 20 $\mathrm{R_s}$ to the orbit of the Earth. The shape of the IH domain may be rectangular and asymmetric with respect to the center of the Sun. The IH model can be well approximated with the equations of ideal MHD in an inertial (e.g., HGI) frame. The inner boundary conditions of the IH component are obtained from the SC component or measurements. The flow at the outer boundary of the IH component is always superfast. The IH component also provides the outer boundaries for the SC component when the flow at the outer boundary of SC is subfast. Finally, the IH component provides the upstream boundary conditions for the GM, whose upstream boundary is typically at about 30 $\mathrm{R_E}$ ($\mathrm{R_E}$, Earth radius) from the Earth toward the Sun. The IH model solves the single-temperature or two-temperature MHD equations on a Cartesian grid in either co-rotating or inertial frame. The grid is extensively refined near the HCS to capture the fine structure of the thin HCS. Polytropic index is taken to be 1.5 [48] or 5/3 in the IH component [549].

1.2.2.3 Simulations of Background Solar Wind

Cohen et al. [48] conducted a simulation using the SC and IH model, in which the processes of turbulent heating in the solar wind were parameterized using a phenomenological, thermodynamical model with a varied polytropic index [438] and the Bernoulli integral was used to bridge the solar wind speed derived from the WSA model [4–6] at 1 AU with the assumed distribution of the polytropic index on the solar surface. The model reproduced the mass flux from the Sun to Earth, and the large-scale structure of the magnetic field, the solar wind speed bimodal pattern in the inner heliosphere. Overall, the model produced magnitudes of the magnetohydrodynamic variables for solar minimum conditions rather well compared with 1 AU satellite data

[49]. However, the comparison demonstrated that the input magnetogram needed to be multiplied by a scaling factor in order to obtain the correct magnitude of magnetic field at 1 AU.

As mentioned in the previous part Sect. 1.2.2.2, van der Holst et al. [549] applied the two-temperature MHD to the simulation of the solar wind. The model took into account of the collisions between the electrons and protons, the anisotropic thermal heat conduction of the electrons. They assumed that the solar wind was accelerated by the Kolmogorov dissipation of the Alfvén waves in open field-line regions. The inner boundary conditions for this solar wind model were obtained from observations and the empirical WSA model, which was used to determine the Alfvén wave energy density at the inner boundary. The electron density and temperature at the inner boundary were obtained from the differential emission measure tomography (DEMT) applied to the extreme-ultraviolet images of the STEREO A and B spacecraft. They validated the solar wind model by modeling the steady state for solar minimum Carrington rotation 2077 (2008 November 20 through December 17). The results were a better agreement of the occurrence time and the density of corotating interaction regions (CIRs) with in situ ACE/WIND data than the previous semi-empirical wind model [48], in which a spatially reduced adiabatic index was used to account for the plasma heating. Figure 1.7 shows the radial velocity (left panel) and number density (right panel) in the equatorial plane during CR 2077 calculated by van der Holst et al. [549]. Three fast and slow streams are visible in alternating succession. The plasma is compressed in the regions where the fast stream catches up with the slow stream. At the rear of the fast streams, the plasma is rarified. The resulting effect gives the appearance of three dense spiral arms. These interaction regions between the fast and slow streams are known as CIRs. Figure 1.8 shows the comparison of the simulated synoptic wind with the ACE data at 1 AU. In the

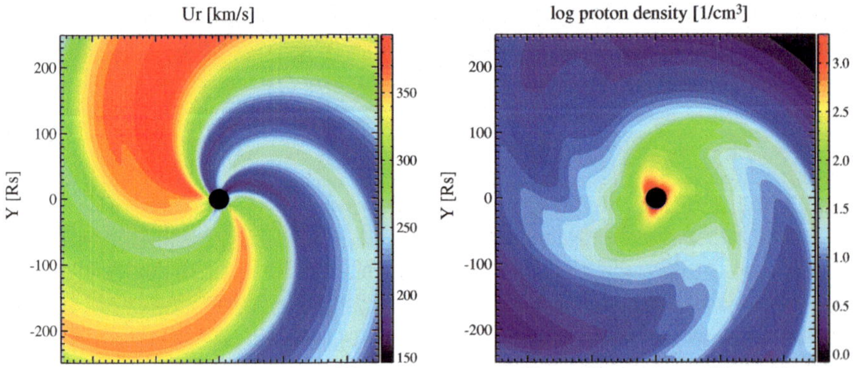

Fig. 1.7 Slow and fast speed streams in the equatorial plane (left panel) leading to CIRs with the formation of compressed region at the front of the fast stream and rarefaction at the rear as shown by the density plot (right panel). Taken from Fig. 9 of van der Holst et al. [549] with permissions of the authors and the AAS

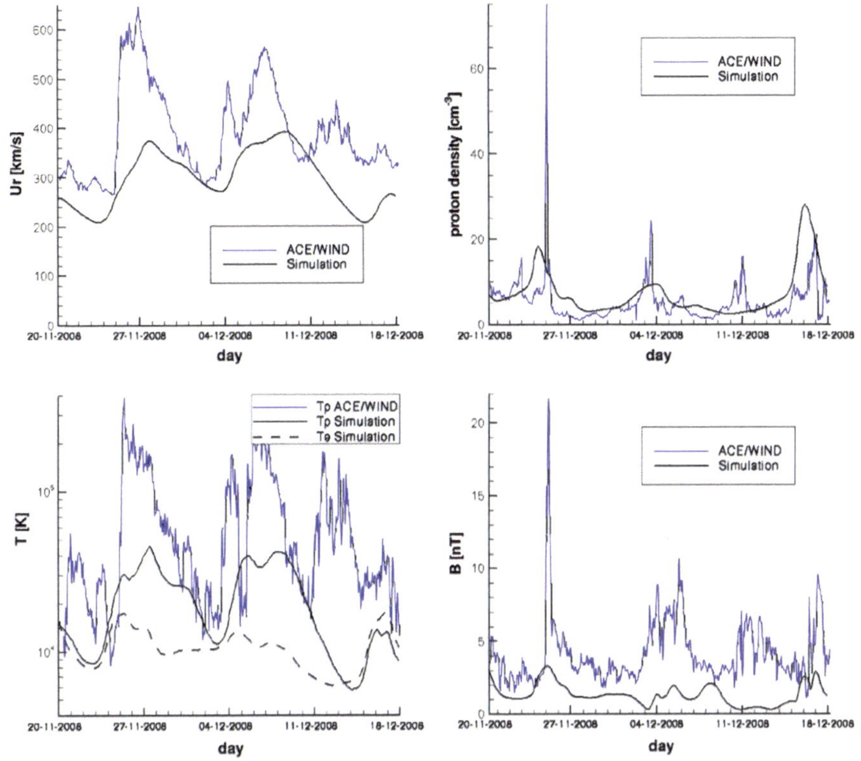

Fig. 1.8 Comparison of the simulated whole month results with ACE at 1 AU. The comparisons are for the radial solar wind velocity (top left), the proton number density (top right), the proton temperature (bottom left), and the magnetic field strength (bottom right). The simulated electron temperature is included in the temperature plot. Taken from Fig. 10 of van der Holst et al. [549] with permissions of the authors and the AAS

simulated density and radial velocity, there are three CIRs visible. The density peaks are ahead of the velocity peaks as expected and the locations of the peaks are all collocated with those observed by ACE. While the amplitudes of the density peaks match the ACE data rather well, the radial velocity peaks are relatively smaller. A similar characteristic is evident for both the proton temperature and the magnetic field strength. The relatively low magnitude in velocity and proton temperature is very likely caused by the quite extended slow wind region.

Another validation study using this model was carried out by Jin et al. [210]. In that study, they compared in situ observations made by STEREO, ACE and Venus Express with model outputs. They also compared remote-sensing observations made by SOHO/LASCO C2 and Hinode/Extreme Ultraviolet Imaging Spectrometer (EIS) with model outputs. The ionic charge states of carbon, oxygen, silicon and iron observed by the Solar Wind Ion Composition Spectrometer (SWICS) onboard ACE were compared with model predictions for the first time with the assistance of a freeze-in code. These validation results suggested that most of the model outputs for CR 2077 matched the observations very well.

Evans et al. [88] utilized a new solar corona component to self-consistently couple the Alfvén wave energy transport with the MHD equations. In this solar wind model, they not only employed the techniques similar to van der Holst et al. [549] to deal with the coronal heating and solar wind acceleration along open magnetic field lines but also introduced an additional dissipation mechanism: surface Alfvén wave (SAW) damping, which is weak in the polar regions (where Hollweg dissipation [171] is strong), and strong in subpolar latitudes and the boundaries of open and closed magnetic fields (where Hollweg dissipation is weak). The simulated results showed that SAW damping reproduced regions of enhanced temperature at the boundaries of open and closed magnetic fields seen in both tomographic reconstructions in the low corona and Ulysses data in the heliosphere. Sokolov et al. [476] presented a combined global model of the solar corona, the low corona, the transition region, and the top of the chromosphere. The realistic three-dimensional magnetic field was simulated using the data from the photospheric magnetic field measurements. The model used MHD Alfvén wave turbulence as the only momentum and energy source to heat the coronal plasma and drive the solar wind with different turbulence dissipation efficiencies in coronal holes and closed field regions. The synthetic images for multi-wavelength EUV emission were consistent with the observations from the Atmospheric Imaging Assembly (AIA) onboard SDO and the EUV imager (EUVI) onboard STEREO for CR 2107. In Meng et al. [319], the AWSoM model was applied to simulate the steady solar wind from the solar corona to 1 AU of CRs 2107 and 2123. They found that the distribution of the anisotropy ratio $p_{\perp}/p_{\parallel}$ reduces away from the sun. The corotational structures in the proton pressure anisotropy distribution is detected. The highest $p_{\perp}/p_{\parallel}$ occurred near the CIRs, while $p_{\perp}/p_{\parallel}$ in the fast solar wind beyond 40–50 R_s was less than 1 and that in the slow solar wind close to 1. They achieved a reasonable agreement with in-situ observations at 1 AU. The comparison also implied the importance of addressing the proton temperature anisotropy to capture the fast solar wind speed. Together with a coronal charge state evolution model (Michigan Ionization Code (MIC), Landi et al. [257]), Oran et al. [364] employed the AWSoM model to calculate the elemental charge-state evolution along the modeled open field lines. They initialized the charge state evolution model with the electron density, temperature, and speed simulated by the AWSoM wind model, and completed the first charge state calculation covering all latitudes in a realistic magnetic field.

1.2.2.4 Simulations of CMEs

By using the SWMF, Manchester et al. [302] numerically studied the geoeffective CME event of 2003 October 28 that erupted from AR 10486 and propagated to Earth in less than 20 hrs. They first employed synoptic magnetograms to obtain a steady state corona and solar wind through the time relaxation method. Then they initiated two CMEs from the same active region, one approximately a day earlier that preconditioned the solar wind for the much faster CME on the 28th. This second CME traveled through the corona at a rate of over 2500 km s^{-1}, driving a strong forward shock. The CMEs in this case were modeled by superposing a modified version of the Titov and Démoulin [513] flux rope to AR 10486. The model well reproduced the general morphology and the quantitative brightness recorded by SOHO. These results also showed that the CME shape was largely determined by its interaction with the ambient solar wind and was not sensitive to the initiation process.

Manchester et al. [303] conducted simulations of CMEs performed with a new two-temperature coronal model [549] to address the coupled thermodynamics of the electron and proton populations in the context of a single fluid. This model employed heat conduction for electrons, constant adiabatic index ($\gamma = 5/3$), and included Alfvén-wave pressure to accelerate the solar wind. The magnetogram data from the Global Oscillation Network Group (GONG) group of National Solar Observatory (NSO) were exploited to calculate the coronal field, and STEREO/EUVI observations to specify the density and temperature at the coronal boundary by using the DEMT method. Manchester et al. [303] simulated the propagation of fast CMEs and studied the thermodynamics of CME-driven shocks. They found that the coupling between the protons and electrons due to Coulomb collisions in the low corona was so brief that the electrons never achieved more than 10% of the maximum temperature of the protons and that heat was able to conduct on open magnetic field lines and rapidly propagated ahead of the CME to form a shock precursor of hot electrons.

During CR 2107, active region NOAA AR 11164 erupted a fast CME associated with a solar energetic particle event. Jin et al. [211] conducted simulations of this CME with one-temperature (1T) and two-temperature (2T: coupled thermodynamics of the electron and proton populations) models, both of which started from the chromosphere with heat conduction and radiative cooling. The background solar wind was driven by Alfvén-wave pressure and heated by Alfvén-wave dissipation. The magnetic field of the inner boundary was specified by using a synoptic map from HMI/SDO. The Titov-Démoulin flux-rope model was used to initiate the CME event. They compared the propagation of fast CMEs and the thermodynamics of CME-driven shocks in both the 1T and 2T CME simulations. Also, the synthesized white light images were compared with the observations from LASCO/SOHO. Due to the lack of distinction between electron and proton temperatures, heat conduction in the 1T model created an unphysical temperature precursor in front of the CME-driven shock and the incorrect shock parameters (e.g., shock Mach number, compression ratio). The results demonstrated the importance of the electron heat conduction in conjunction with proton shock heating in order to produce the physically correct CME structures and CME-driven shocks.

Figure 1.9 shows the 3D CME evolution for both 1T and 2T models at 5, 10, and 20 min after the initiation. The isosurfaces represent the radial velocity of 1000 km s^{-1}. The color scale on the isosurfaces shows the temperature (1T)/proton temperature (2T). The field lines are colored by the density so that we can roughly see the propagation of the CME material as well as the shock positions. The gray scale on the surface of the Sun shows the magnetic field strength. In the 2T model, the protons are dissipatively heated by the shock and reach 90 MK at 5 min. In addition, the morphology of the CME is quite different in the 1T and 2T models after 10 min. In the 2T model it seems that the expansion has a non-radial component caused by

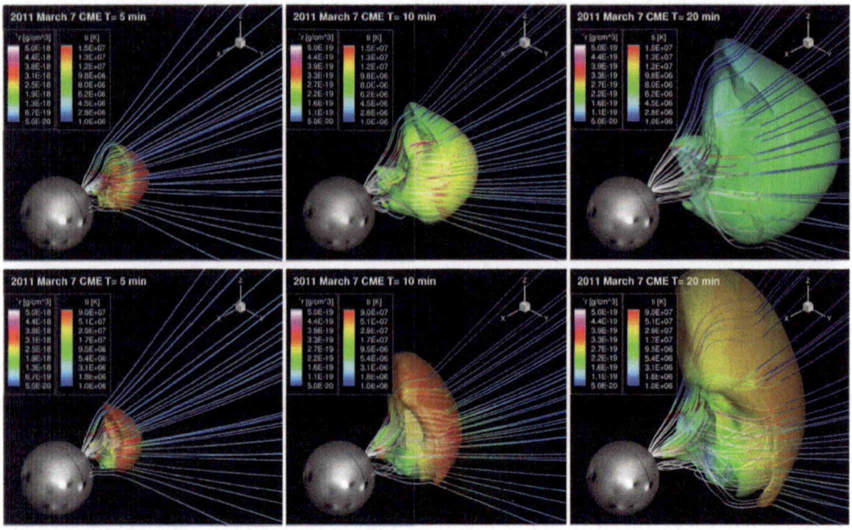

Fig. 1.9 3D CME evolution of 1T and 2T models. The isosurface represents $V_r = 1000$ km s^{-1} The color on the isosurface shows the proton temperature. The field lines are colored by plasma density. Taken from Fig. 9 of Jin et al. [211] with permissions of the authors and the AAS

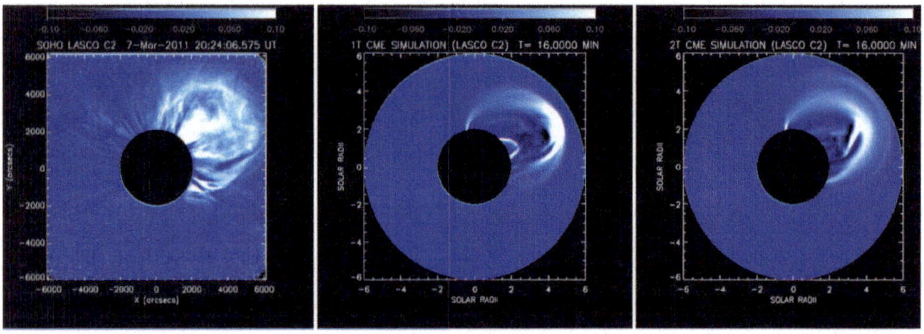

Fig. 1.10 Comparison of the SOHO/LASCO C2 CME white light image (left panel) with the 1T (middle panel) and 2T (right panel) model synthesized images for the 2011 March 7 CME event. The color scale shows the relative intensity changes. Taken from Fig. 10 of Jin et al. [211] with permissions of the authors and the AAS

the thermal pressure gradient in the CME sheath. The non-radial flows are more evident near the pole due to the simpler magnetic structure and smaller density in this region, which is consistent with the observations [138, 144, 568]. Figure 1.10 shows the comparison of the white light image of the CME recorded by SOHO/LASCO C2 (1.5-6.0 R_s) with the synthesized ones from both the 1T and 2T models. The color scale shows the relative intensity changes. Both the 1T and 2T models show the bright front related to the coronal plasma piled up at the top of the erupting flux rope. However, the faint front caused by the shock is not obvious in the 1T model but is evident in the 2T model.

Jin et al. [212] presented a numerical investigation of the impact of an erupting flux rope on surrounding solar structures through large-scale magnetic coupling. The realistic background coronal environment was constructed by running AWSoM SC model [551] to achieve a steady state MHD solution, in which the magnetic field at inner boundary was specified by the global map sampled from a time-evolving SFT model [442] on 2011 February 15 00:04:00 UT. CMEs in AR 11158 are initiated by using the analytical Gibson-Low [126] flux rope with different parameters (e.g orientation and strength). The simulation results were employed to investigate the dependence of the eruption on these parameters. They found that a CME's impact on the surrounding structures depended not only on the magnetic strength of these structures and their distance to the source region, but also on the interaction between the CME with the large-scale magnetic field.

Using a similar technique to that of Jin et al. [212, 213] presented a first-principle based CME model suitable for both scientific and operational purpose. A new data-driven tool called Eruptive Event Generator Gibson-Low (EEGGL) was developed to automatically determine the GL flux rope parameters using synoptic magnetogram data from GONG and CME speed derived from the observations of SOHO/LASCO and the inner coronagraph (COR1) and the outer coronagraph (COR2) onboard STEREO. With observation constrained EEGGL, realistic CME events were simulated self-consistently with high fidelity and forecasting capability. By combining the observation-constrained EEGGL model and the AWSoM solar wind model [363, 551], Jin et al. [214] conducted a comprehensive study of coronal mass ejection propagation on the 2011 March 7th by performing a simulation from the chromosphere to 1 AU. The simulated results reproduced many of the observed features both near the Sun and in the heliosphere. AWSoM model has been used [25] to provide a quantitative model for simulation and forecast of solar energetic particle (SEP) production during gradual events, by coupling MHD background produced from SC and IH MHD models of SWMF with the SEP model. The Multiple-Field-Line-Advection Model for Particle Acceleration (M-FLAMPA) code [473] was used to solve the kinetic transport equation for SEPs along a multitude of IMF lines originating from the Sun, using time-dependent magnetic field and plasma parameters obtained from the MHD simulation.

1.2.3 SIP–CESE MHD Model in SWIM

In a series of papers [98–100, 102–105, 107–109, 179, 196, 197, 199, 200, 202, 606, 607, 626, 628, 629], Feng and his colleagues affiliated to the Solar-Interplanetary-Geomagnetic (SIGMA) Weather Group of the state key laboratory for space weather, Chinese Academy of Sciences, have developed a 3D MHD model to investigate solar-interplanetary phenomena based on the space time conservation element and solution element (CESE) method (hereafter called SIP-CESE MHD model for brief).

In order to easily analyze and operate the massive observed and simulated data and integrate different models, the SIGMA weather group developed the system of the Space Weather Integrated Modeling (SWIM) [106]. In what follows, we give a brief introduction of the SWIM, outline the main features of the SIP-CESE MHD model and sum up its applications to investigating the background solar wind and the evolution of the solar storms from the Sun to Earth.

1.2.3.1 Space Weather Integrated Modeling

With the rapid developments of various observations and computational technologies, the higher and higher spatial resolution and temporal cadence of the observed and simulated data result in an exponential growth of the data available for space weather study and prediction. This challenges the traditional tools of data analysis and visualization due to the low efficiency of operating mass data and the difficulty in integrating different models and data, which limit further improvements on the grids' resolution of numerical models.

The SWIM system is a framework that integrates various space weather models. It provides researchers a unified environment to integrate models of different spatial regions into a workflow, to prepare initial observational data, to analyze and visualize the output of models. SWIM is programmed by using hybrid languages of C, FORTRAN and Python. SWIM is modularized into four layers: application layer (AL), execution layer (EL), calculation layer (CL) and foundation layer (FL), as shown in Fig. 1.11. The top layer is the AL, which provides command-line, graphical, and web-based user interfaces. In this layer, application programming interface is used for further SWIM-based application development and enables users to customize tasks from the FL to EL, so as to configure the problem, to visualize various data, to extend the layer with new modules, and to make prediction workflow. The second layer is the EL, which consists of MHD simulation models for steady-state solar wind, solar eruptions and their propagation in solar corona and interplanetary space, planetary magnetized environment, shock arrival prediction, and data processing. This layer also contains the couple module, which is responsible for coupling the models of different spatial regions. The CL consists of the modules that support the second layer. The layer contains different solvers of the MHD equation, such as the CESE solver, the Harten-Lax-van Leer (HLL) family solvers, and the Roe solver. This layer also contains different divergence-error controllers for the magnetic field and the module of different coronal heating/solar acceleration realizations. In addition, the layer is also devoted to various kinds of operations for data processing, assimilation and visualization, such as the surface flux transport model [66, 609], time interpolation, event tracing, data extraction of the region of interest (ROI), interactive data manipulation, slice, contour, iso-surface, streamline, volume rendering, visualization of full process of space weather events, and animation tools. The FL contains the shared library and the utilities that can be used by either the physics models or the SWIM system, as shown in Fig. 1.12. The libraries provide mesh generation and management, physical and numerical constants, and transformations among different coordinate systems

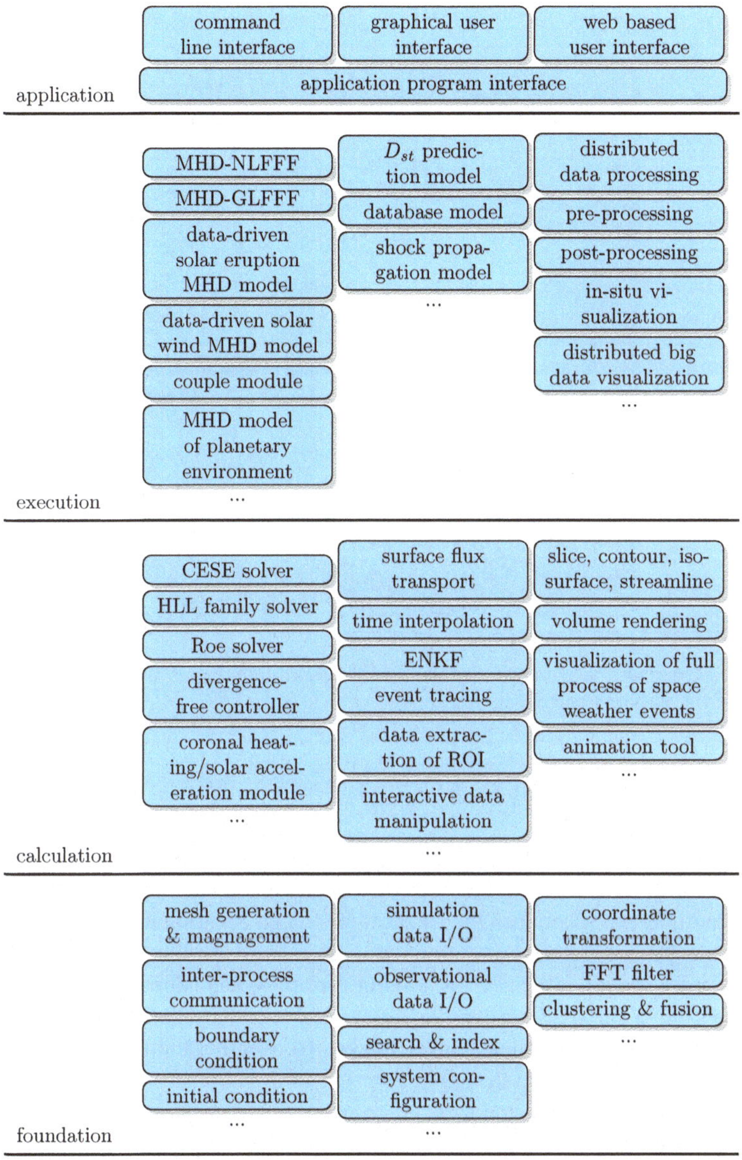

Fig. 1.11 The layered structure of SWIM

and time units. The utilities contain system configuration and various algorithms to realize inter-process communication, to deal with initial and boundary condition treatment, and to input and output data from observations and simulations.

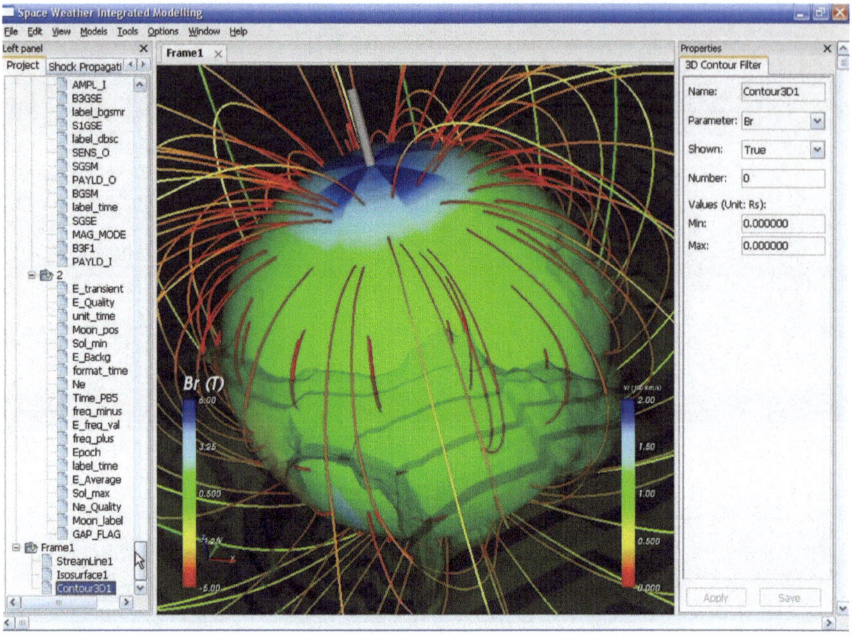

Fig. 1.12 The graphic user interface of the SWIM. Taken from Fig. 19 of Feng et al. [106] with permission of Science China

1.2.3.2 SIP–CESE MHD Model

The space time CESE method was first developed to solve linear and nonlinear convection-diffusion equations in one, two, and three dimensions [43, 44, 613, 620]. The main differences between the CESE method and traditional numerical methods lie in treating space and time as one entity in calculating the flux balance, introducing the first-order spatial derivatives as solving variables, and adopting the controllable damping effect. The CESE method is able to capture sharp discontinuities within a few grid points. The next chapters of the book will cover the details of implementation of the SIP-CESE model and the applications to study solar active regions, the global corona, the ambient solar wind and corona mass ejections. Here we presents a brief review of the model development and results obtained.

Based on the space time CESE method, the SIP-CESE MHD model was developed and used for modeling 3D solar wind background and solar transient events from the Sun to Earth on the polyhedron-splitting grid [98, 99, 179, 628]. They adopted unstructured sphere-fitting triangles on the sphere and structured grids in the radial direction and divided the sphere-shell computation domain from the Sun to Earth into non-overlapped convex pentahedrons with any two neighboring pentahedrons sharing a common surface. Part of the grids are shown in Fig. 1.13a. This kind of grid system can avoid the singularity and the convergency of the ordinary sphere

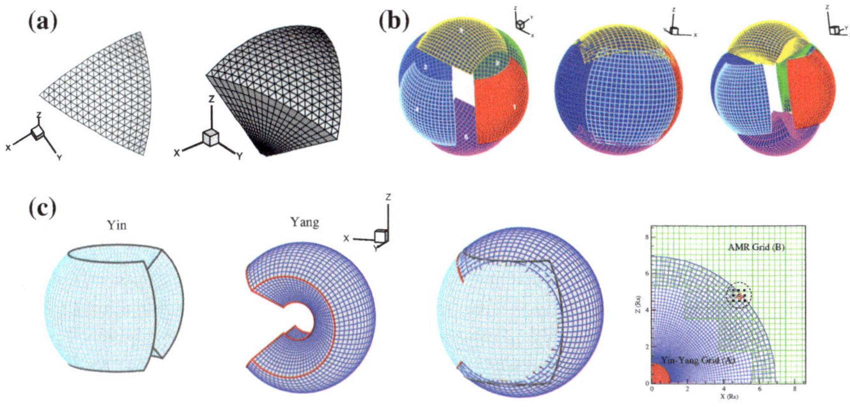

Fig. 1.13 Grid systems for the 3D CESE solvers: **a** Pentahedron grid: part of the grid structure at solar surface and in the corona [98]. Adapted from Fig. 1 of Feng et al. [98] with permissions of the authors and the AAS; **b** Six-component grid: a spherical overset grid consisting of six identical components, division of a sphere into six identical components with partial overlap, and pileup of the spherical meshes of each component in the radial direction [104]. Adapted by permission from Springer Nature Customer Service Centre GmbH: Springer *Solar Physics*, Validation of the 3D AMR SIP-CESE solar wind model for four Carrington rotations, Feng XS, Yang LP, Xiang CQ, Jiang CW, Ma X, Wu ST, Zhong DK, Zhou YF. Copyright 2012; **c** The combined grid structure of Yin-Yang grid in the near-Sun region and the AMR Cartesian grid in the off-Sun region [102]. Adapted from Figs. 1 and 2 of Feng et al. [102] with permissions of the authors and the AAS

coordinate grids near both poles and is easy to parallelize the program in the radial direction.

Feng et al. [100] employed SIP-CESE MHD model within a six-component overset grid for solar wind simulation. This composite grid system consists of six identical component meshes to envelope a spherical surface with partial overlap on their boundaries (Fig. 1.13b). Each component grid is a low-latitude spherical mesh with the same metric and the component grids can be transformed into each other by coordinate transformations such that we can make efficient and concise programs. The essential merits are as follows: (1) the removal of the singularity and mesh convergence near the poles via the use of the six-component grid system, (2) the $\nabla \cdot \mathbf{B}$ constraint error via an easy-to-use cleaning procedure by a fast multigrid Poisson solver, (3) the mitigation of the Courant-Friedrichs-Levy number disparity via the Courant-number insensitive method, (4) the use of time integration by multiple time stepping, (5) the application of time-dependent characteristic boundary condition at the subsonic region by limiting the mass flux escaping through the solar surface, (6) and the easy implementation of parallelization both in the radial direction and "(θ, ϕ)" directions. In order to produce fast and slow plasma streams of the solar wind, they include the volumetric heating source terms and momentum addition by involving the topological effct of the magnetic field expansion factor f_s and the minimum angular distance θ_b (at the photosphere) between an open field foot point and its nearest coronal hole boundary.

A hybrid 3D MHD model for solar wind study was proposed by Feng et al. [102] with combined grid systems and solvers. They decomposed the computational domain from the Sun to Earth space into the near-Sun and off-Sun domains, which were respectively partitioned with a Yin-Yang overset grid system and a Cartesian AMR grid system and coupled with a domain connection interface in the overlapping region between the near-Sun and off-Sun domains. The grids in both domains and the overlapping region are shown in Fig. 1.13c. The space-time CESE method was used in the near-Sun domain, while the HLL family methods were employed in the off-Sun domain. Just like the six-component grid, the Yin-Yang overset grid can also avoid well-known singularity and polar grid convergence problems and its body-fitting property helps achieve high-quality resolution near the solar surface. The block structured AMR Cartesian grid can automatically capture far-field plasma flow features, such as heliospheric current sheets and shock waves, and at the same time, it can save significant computational resources compared to the uniformly structured Cartesian grid. It should be mentioned that this hybridizing method allows on the off-Sun domain to use different numerical schemes (e.g., [267]), such as Roe's scheme, HLLC, HLLD, and HLLE, without any difficulty.

Feng et al. [104] carried out AMR implementation of the SIP-CESE MHD model within a six-component grid system. By transforming the governing MHD equations from the physical space (x, y, z) to the computational space (ξ, η, ζ) while retaining the conservation form, the AMR of SIP-MHD model was implemented in the reference coordinates with the aid of the parallel AMR package PARAMESH available at http://sourceforge.net/projects/paramesh/. To study the dynamic evolution of the global corona, Feng et al. [103, 105] and Yang et al. [607] developed a time-dependent MHD model driven by the magnetograms from daily-updated synoptic observations and from the surface flux transportation (SFT) model with the help of the 3D SIP-AMR-CESE MHD Model.

In order to reduce the computation hours efficiently, Feng et al. [107] explored the application of the heterogeneous clusters of multi-core central processing units (CPU) and programmable graphics processing units (GPUs) to the numerical space weather modeling for the study of solar wind background that is a crucial part in the numerical space weather modeling. GPU programming was realized for the SIP-CESE MHD model by numerically studying the solar corona/interplanetary solar wind. The preliminary tests with available hardware showed speedups of roughly $5\times$ compared to traditional software implementation. The simulated evolution of the global structures for CRs 2058 and 2062 was compared with solar remote-sensing observations and in situ solar wind measurements from spacecraft near the Earth. Comparisons revealed that the MHD results were in good overall agreement with coronal and interplanetary structures, including the sizes and distributions of coronal holes, the positions and shapes of the streamer belts, and the transitions of the solar wind speeds and magnetic field polarities.

In summary, the SIP-CESE MHD model developed by Feng and his colleagues has the following features.

(i) The adoption of the new empirical volumetric heating source term taking the topological effect of magnetic field with the expansion factor f_s and the minimum angular distance θ_b into consideration, to some extent, can effectively distinguish the high-speed solar wind from the low-speed solar wind. At the same time, the combination of the projected normal characteristic method and the mass flux limit enables the model to reproduce reasonable distributions of the plasma density, temperature and velocity on the solar surface.

(ii) The model provides a unified treatment of flow evolution in space and time and keeps the local and global space-time flux conservation in a coherent and efficient manner. Especially, the change of the solution points in SIP-CESE MHD model [104, 199] from being calculated after setting the grids into explicitly being given on the mesh nodes greatly simplifies the projection of the conservation elements (CE) onto the spatial space into a rectangular cuboid. Therefore, the fluxes at the interface of any pair of CEs can be efficiently evaluated by means of non-staggered space-time grids without using Riemann solvers or other flux models, which significantly reduces the CPU time. In addition, the treatment of time iteration by integrating two half timesteps into one full timestep leads to low-storage and makes the scheme suitable for building blocks for adaptive mesh refinement calculations.

(iii) The introduction of six-component grid for the computational domain from the Sun to Earth or beyond enables us to fit the spherical surface boundary with an easy implementation of the inner boundary conditions, and meanwhile to avoid both coordinate singularities and polar grid convergence. Particularly, it is easy to recognize the observations at the lower boundary. Thus Jiang et al. [199, 200, 202] developed a new implementation of the MHD relaxation method to reconstruct the coronal magnetic field from a photospheric vector magnetogram, which opens a new way for the study of solar active region with the help of Solar Dynamics Observatory/Helioseismic and Magnetic Imager (SDO/HMI) or SOHO/MDI observations. What's more, the incorporation of the time-dependent magnetograms into the model is preliminarily established [103, 105, 607]

(iv) Through the use of nonsingular transform from the physical space to the reference space, the quadrangular frustum pyramid cell for the spherical shell computational domain for solar wind modeling becomes the conventional rectangular box in the reference space (ξ, η, ζ), which can be seen as the usual Cartesian coordinate. Consequently, the implementation of the code follows easily from PARAMESH in Cartesian coordinate. Thus, besides the CESE scheme, many other modern numerical schemes in Cartesian coordinate such as total variation diminishing (TVD) scheme and finite volume method (FVM) can be applied directly to the transformed system. This feature provides us many flexible alternatives of solving the transformed governing equations in (ξ, η, ζ) and then we recover the solution in the physical space through the transformation to obtain the solar wind solution. What's more, the same CESE solver can apply to any coordinate system (such as Cartesian, spherical, cylindrical coordinates and any other curvilinear coordinates) with only the difference of the coordinate transformation, and consequently the solver is highly independent of the grid system.

(v) The SIP-CESE MHD model's implementation of the adaptive mesh refinement and message passing interface enables the full exploitation of the computing power in a heterogeneous CPU/GPU cluster and significantly improves the overall performance. The application of GPU to the space weather study can be a reasonable option to realize the real-time or faster-than-real-time simulations of adverse space weather events and its effects to geospace environment from the Sun to Earth.

1.2.3.3 Simulations of Background Solar Wind

Feng et al. [98, 100] used the SIP-CESE MHD model to simulate the three-dimensional coronal dynamical structure. They numerically investigated the large-scale structures of interplanetary solar wind and the evolution of the heliospheric magnetic field, and compared the simulation with the in situ measurements by the spacecraft of *Wind*. The measured solar surface magnetic fields and the empirical values of the plasma properties on the solar surface were used to prescribe the initial-boundary conditions. The results showed that the model could reproduce the bimodal structures of interplanetary solar wind at solar minima and possessed the ability to keep the divergence-free condition to a small error that has little effect on the solution accuracy without any special procedures. Both the modeled and the observed results had the consistent distribution of solar wind speed and the similar variation of the radial magnetic field polarities.

Hu et al. [179] used the SIP-CESE MHD model to numerically investigate the large-scale structures of interplanetary solar wind and the evolution of the heliospheric magnetic field throughout Solar Cycle (SC) 23. The simulation could reproduce the cyclic evolution of HCS shapes and agree with the pattern of high-density areas shown in pB measurements. In addition, the model had also been employed to carry out the exploration of the effects of different commonly-used empirical solar wind acceleration models in the MHD coronal simulation [605].

Yang et al. [606] chose Carrington rotation 2070 in 2008 to investigate the unusual properties of the background solar wind in the 2008 solar minimum by using the SIP-CESE MHD model. Figures 1.14a, b display the composite observations of Mauna Loa Solar Observatory (MLSO) Mark-IV K-coronameter and LASCO C2 board on SOHO on 14 May and on 24 May in CR 2070 in order to discern a continuation of structure from the low to high corona. Figures 1.14c, d display the derived pB images from the simulation on the meridional planes at $\phi = 270° - 90°$ and $\phi = 120° - 300°$ from 1.15 to 6 R_s, in which the pB images inside and outside 2.3 R_s are enhanced separately. Figures 1.14e, f show the magnetic field topologies projected onto the meridional planes at $\phi = 270° - 90°$ and $\phi = 120° - 300°$ from 1 to 6 R_s, where the color contours represent Br on the solar surface and the red lines denote MNL. Broadly speaking, the simulated results achieve the 3D structures of the highly inclined, broad, multiple streamers observed by LASCO/C2 and reveale that the multiple streamers result from the projection effects on the sky plane of the high-density regions originating from both the helmet streamer and the pseudostreamer located in different latitudes. Figure 1.15 compares the simulated results for CR 2070 with the observational data from OMNI and the measurements from the twin spacecraft of STEREO. Obviously, the simulation reproduces many features observed in the solar wind structures, such as two broad intermediate-speed streams

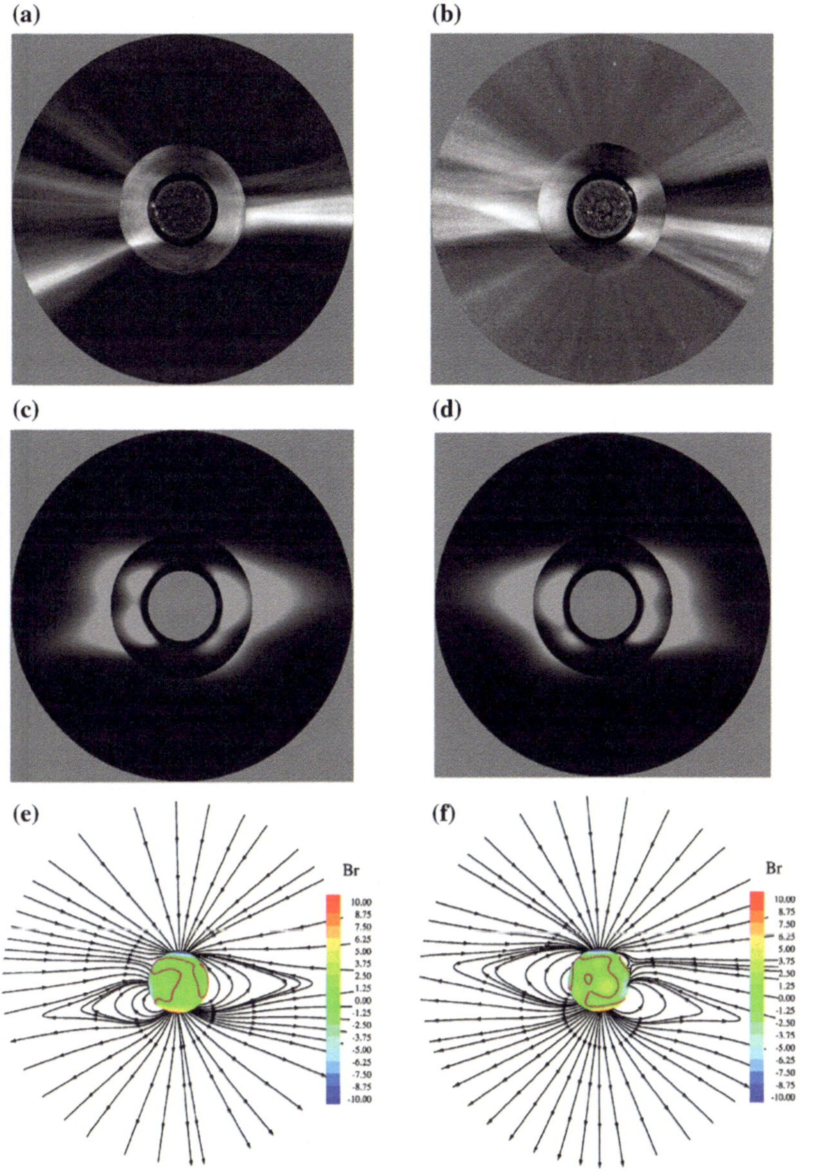

Fig. 1.14 Coronal composite images [606] on **a** 14 May and **b** 24 May and the simulated pB on the meridional planes at **c** $\phi = 270° - 90°$ and **d** $\phi = 120° - 300°$. Panels **e** and **f** show the magnetic field topology projected onto the meridional planes at **e** $\phi = 270° - 90°$ and **f** $\phi = 120° - 300°$ from 1 to 6 R_s, where the color contours represent B_r (unit: G) on the solar surface and the red lines denote magnetic neutral lines. Reprinted by permission from Springer Nature Customer Servce Centre GmbH: Springer *Solar Physics*, Simulation of the Unusual Solar Minimum with 3D SIP-CESE MHD Model by Comparison with Multi-Satellite Observations, Liping Yang, Xueshang Feng, Changqing Xiang, Shaohua Zhang, S.T. Wu. Copyright 2011

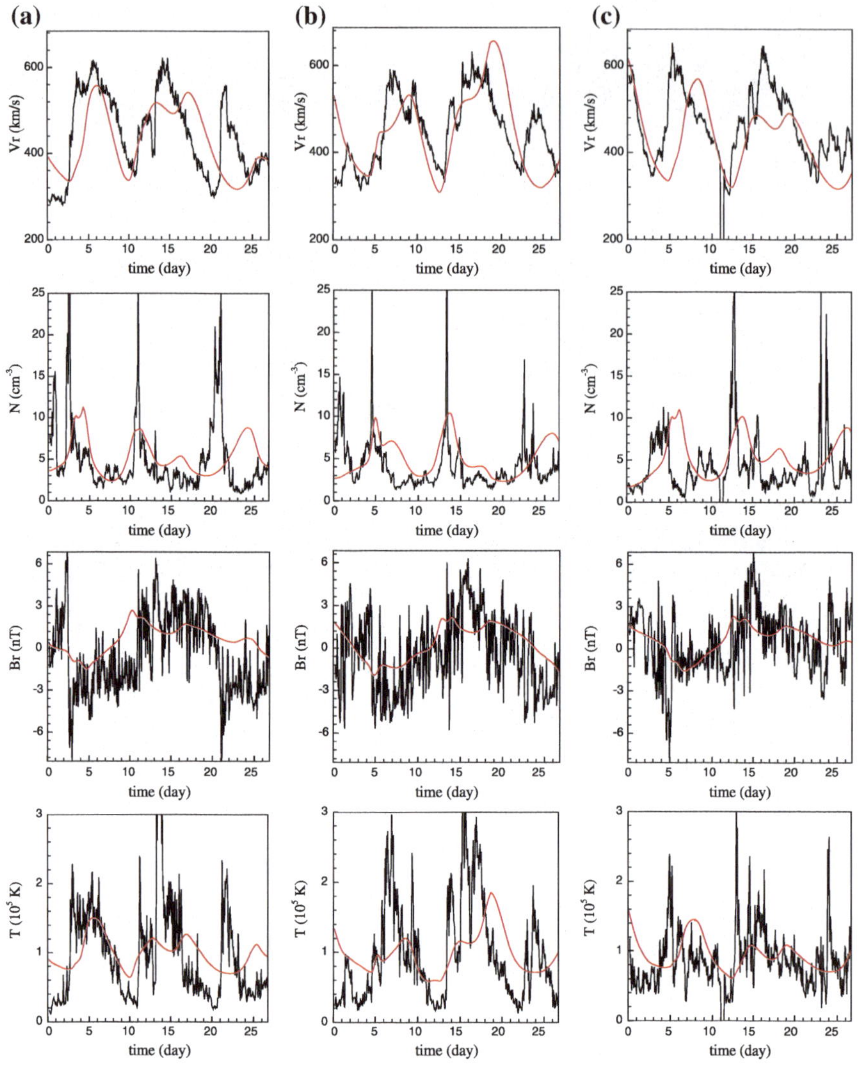

Fig. 1.15 Comparisons between the MHD results and the 1-h averaged data from the **a** STEREO-A, **b** OMNI, and **c** STEREO-B near 1 AU for the radial flow speed v_r, number density N, radial magnetic field B_r, and temperature T [606]. Reprinted by permission from Springer Nature Customer Servce Centre GmbH: Springer *Solar Physics*, Simulation of the Unusual Solar Minimum with 3D SIP-CESE MHD Model by Comparison with Multi-Satellite Observations, Liping Yang, Xueshang Feng, Changqing Xiang, Shaohua Zhang, S.T. Wu. Copyright 2011

(500–650 km s^{-1}) associated with CIRs and the latitudinal width of the high-speed streams. However, the gradients of the solar wind speed associated with CIRs are not as steep as those from the observations.

Feng et al. [104] carried out numerical studies for the solar-wind background of different solar-activity phases by using the SIP-AMR-CESE MHD model in the six-component grid system. They selected CRs 1967, 2009, 2060 and 2094 as the simulated intervals of the solar maximum, declining, minimum, and rising phases in Solar Cycles 23 and 24, respectively. The modeled results reproduced many features near the Sun and in interplanetary space, such as the distributions of polar coronal holes and mid- and low-latitude coronal holes in the corresponding solar activity phases, the changing trends of the solar-wind parameters for all four CRs, and the IMF polarities and their changes. Furthermore, Yang et al. [607] and Feng et al. [105] studied the dynamic evolution of the global corona in Year 2007 by driving the 3D SIP-AMR-CESE MHD model with the daily-updated synoptic magnetograms (MHD-DUSM). They prescribed the bottom boundary condition by exploiting the method of characteristics, the mass flux limit escaping from the bottom boundary, and the tangential component of the electric field at the lower boundary determined from the observed changes of magnetic field. Comparisons between the simulation and the observations revealed that the MHD-DUSM could well reproduce the observed coronal and interplanetary structures, including the sizes and distributions of coronal holes, the positions and shapes of the streamer belts, and the transitions of the solar wind speeds and magnetic field polarities. In addition, the MHD-DUSM also gave better results for some time-dependent quasi-steady features than the MHD simulation driven by a monthly synoptic magnetogram. Using the SIP-CESE MHD model continuously driven by the synchronic snapshots of the photospheric magnetic field derived from the model of surface flux transportation, Feng et al. [103] investigated the dynamic evolution of the global corona from 1996 September 4 to October 29 responding continuously to the changing of the photospheric magnetic field. Feng et al. [109] investigated the solar wind evolution between the solar surface to the Earth's orbit from 1 July to 11 August 2008 with the SIP-AMR-CESE MHD modell driven by the consecutive synoptic maps from GONG. Similarly, Li and Feng [267] simulated the evolution of solar wind from the solar surface to the Earth's orbit during Year 2008, and evaluated simulated results quantitatively by comparison with in-situ measurements. The data-driven simulation gave some insight into the connection between the in situ features and the global structures.

1.2.3.4 Simulations of CMEs

The first application of the SIP-CESE MHD model to study the evolution of CME is carried out by Zhou et al. [628], who studied the May 12, 1997 CME event. The main features and approximations of the numerical model included the near-real background solar wind derived from a 3D time-dependent numerical MHD model by inputting measured photospheric magnetic fields and the transient disturbances introduced by a mass flow of hot plasma. The numerical simulation enabled us to predict the arrival of the interplanetary shock and provided us with a relatively satisfactory comparison with the *Wind* spacecraft observations. Zhou and Feng [626] explored the solar-terrestrial transit process of three successive CMEs of November

4–5, 1998 originating from active region 8375 by using the SIP-CESE MHD model, in which the ejections were introduced by using the pulses of the plasma speed, the temperature, the density and the magnetic field derived from the SOHO/LASCO observations. The results showed that the interaction among the three halo CMEs produced a compound stream of the two components associated with the observation. The first halo CME formed the first component of the compound stream and the other two ones merged during the propagation in the interplanetary space and formed the second component of complicated internal structure. In general, the simulation reproduced and explained most observed features for the "complex ejecta".

By using the 3D SIP-CESE MHD model with the spherical plasmoid mimicking CME, the time-dependent propagation of the Sun-Earth connection CME events, such as 1997 November 4, 1997 May 12, and 2010 April 3 CME events are investigated [627, 629, 630]. The simulated results provided a relatively satisfactory comparison with the *Wind* spacecraft observations, such as southward interplanetary magnetic field and large-scale smooth rotation of the magnetic field associated with the CMEs. The propagation characteristics of CME in the solar-terrestrial space, including the latitudinal deflection velocity, the effects of HCS on CME's trajectory, and its disturbance amplitudes at 1 AU are systematically explored [625, 627]. Figure 1.16 exhibits the evolution of solar wind parameters at the Earth during the 12 May 1997 CME.

Wu et al. [600] presented a three-dimensional MHD simulation based on an observed eruptive twisted flux rope (sigmoid) deduced from solar vector magnetograms. This model was a combination of the two very well tested MHD models: (i) a data-driven 3D MHD active region evolution model for the reconstruction of the observed flux rope and (ii) 3D MHD global coronal-heliosphere evolution model to track the propagation of the observed flux rope. They selected the CME event from AR 11283 on 6 September 2011 to test this model. They first reproduced the formation of the flux rope (sigmoid) from AR 11283 by using the active region evolution model with input from the measured vector magnetograms given by SDO/HMI. Then these results are used as the initial boundary condition for the coronal-heliosphere evolution model for the initiation of the CME as observed. The model output suggested that the flux rope resulting from the active region evolution model produced the physical properties of a CME, and the morphology resembled the observations made by STEREO/COR-1.

1.2.3.5 Reconstructions of Coronal Magnetic Fields

In order to extrapolate the coronal magnetic field from photospheric vector magnetograms based on the nonlinear force-free method, Jiang et al. [200] and Jiang and Feng [196] exploited the CESE-MHD model to solve the zero-beta MHD equations with a fictitious frictional force. The bottom boundary condition was prescribed in a similar way as in the stress-and-relax method [437, 546], by changing the transverse field incrementally to match the magnetogram, and other boundaries of the compu-

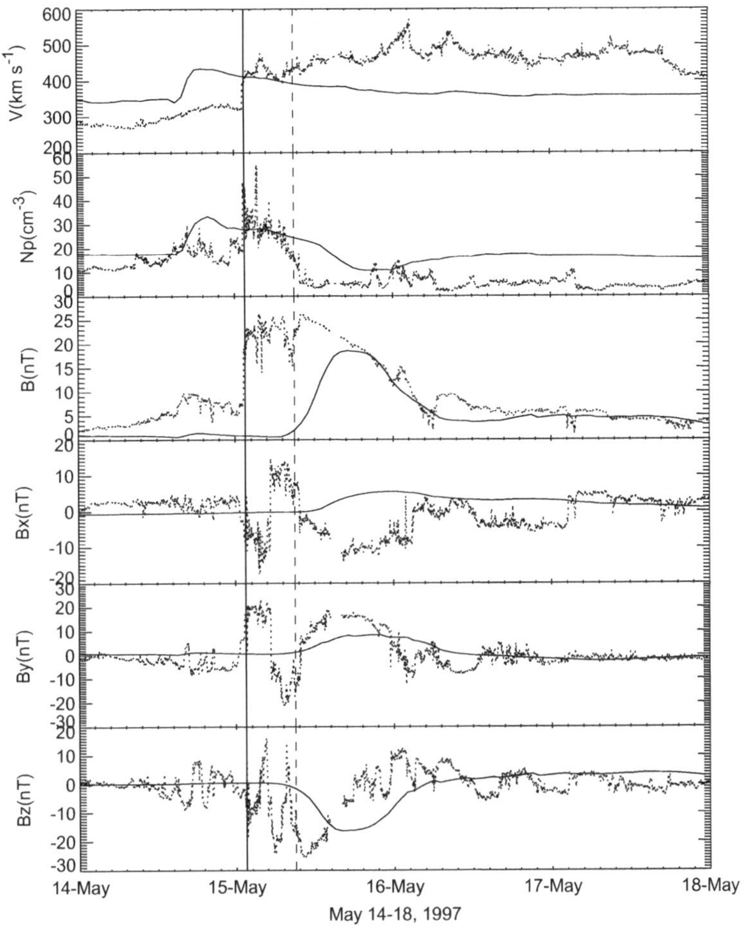

Fig. 1.16 Comparison between the in situ data obtained by WIND spacecraft and our simulation during the 12 May 1997 event. From top to bottom, shown are flow velocity, number density, magnetic field, and three components of magnetic field in geocentric solar ecliptic (GSE) coordinates. The vertical solid line indicates the time of the ICME-driven shock. The vertical dashed line indicates the starting time of the magnetic cloud. The simulated results at the Earth are shown by the solid lines. WIND observations are shown by the dotted lines. Reprinted from Fig. 13 of Zhou and Feng [627] with permission of the John Wiley & Sons, Inc

tational box were set by the nonreflecting boundary conditions. The application to the well-known benchmarks for nonlinear force-free-field reconstruction, the Low and Lou force-free equilibria, validated the method and confirmed its capability for future practical application driven by observed magnetograms. Jiang et al. [201] further used the CESE-MHD model to investigate the 3D magnetic field of NOAA AR 11117 around the time of a C-class confined flare that occurred on 2010 October 25. The model were constrained by the magnetic vector-field data from HMI/SDO around the time of the flare. The model qualitatively reproduced the basic structures of the 3D magnetic field, as supported by the visual similarity between the field lines and the coronal loops observed by AIA/SDO. The model also revealed that the small flare was correlated with a bald patch (BP), where the magnetic field is tangent to the photosphere. In addition, the computed magnetic free energy dropped during the flare by $\sim 10^{30}$ erg, which was adequate to provide the energy budget of a C-class confined flare.

Jiang et al. [204, 206] presented comprehensive studies for the formation and eruption of an active region sigmoid in AR 11283. To follow the quasi-static evolution of the coronal magnetic field, they reconstructed a time sequence of static fields using CESE-MHD nonlinear force-free field (CESE-MHD-NLFFF) model [198, 203, 205] constrained by vector magnetograms [197]. A detailed analysis of the fields compared with observations suggested a sequential scenario of a new bipole emergence, the formation of a weakly twisted flux rope through flux cancellation, a bald patch separatrix surface, and a full S-shaped structure. They investigated the whole process of the eruption of the active region sigmoid and discussed the important implications of the results for the formation and disruption of a sigmoid region with a flux rope. Jiang et al. [207] succeeded in reconstructing a large-scale pre-flare current sheet in solar active region 11967 by using an MHD-relaxation model constrained by the SDO/HMI vector magnetogram. The result demonstrated the power of a data-constrained MHD model in reproducing a current sheet in the corona as well as providing insight into the magnetic mechanism of solar flares. The MHD simulation is realized by solving the full set of time-dependent 3D ideal MHD equations with solar gravity. The initial condition for the MHD simulation consists of the magnetic field provided by the CESE-MHD-NLFFF model and a hydrostatic plasma. While the CESE-MHD-NLFFF procedure only solves for $\mathbf{B}$ and $\mathbf{v}$ in a pseudotime evolution, in the MHD model of the eruption all MHD variables $(\rho, \mathbf{v}, \mathbf{B}, p)$ are numerically calculated.

Jiang et al. [208] employed a realistic, data-constrained magnetohydrodynamic simulation to study the evolving magnetic topology for an X9.3 eruptive flare from geoeffective active region 12673 that occurred on September 6, 2017. Without any prior assumption on the magnetic configuration, Jiang et al. [208] reproduced the eruptive process of a great solar flare with a numerical MHD simulation based entirely on a pre-flare vector magnetogram on the photosphere. The numerical results reveal that the pre-flare corona contains multiple twisted flux systems with different connections, and during the eruption these twisted fluxes form a coherent flux rope through tether-cutting-like magnetic reconnection below the rope. Topological analysis shows that the rising flux rope is wrapped by a quasi-separatrix layer, which intersects itself below the rope, forming a topological structure known as a hyperbolic flux tube, where a current sheet develops, triggering the reconnection. Figure 1.17 presents the comparison of the modeled magnetic field with the observed features of the solar corona prior to the flare field lines and Fig. 1.18 illustrates the reconnection process below the rising magnetic flux rope. By mapping footpoints of the newly reconnected field lines, both the spatial location and, for the first time, the

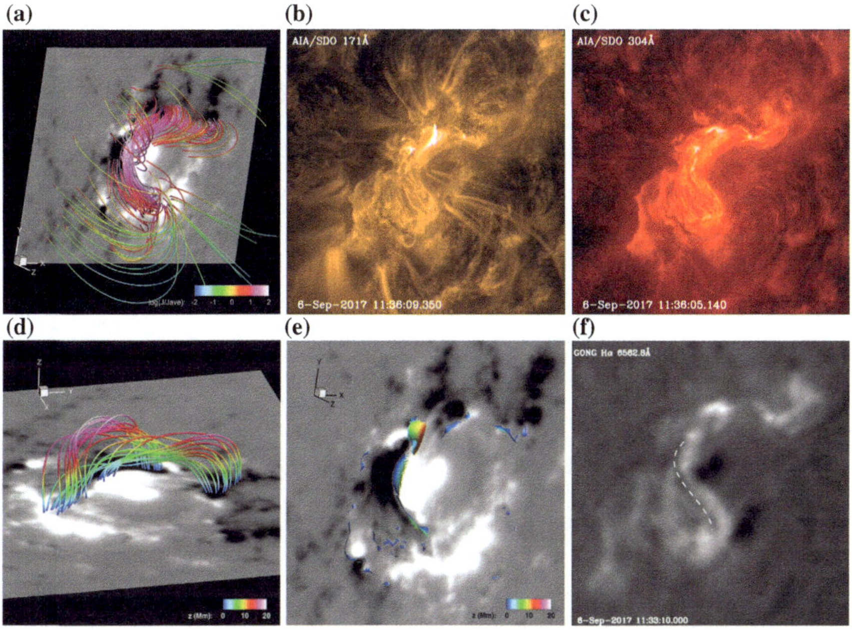

Fig. 1.17 Comparison of the reconstructed magnetic field with the observed features of the solar corona prior to the flare. **a** SDO view of sampled magnetic field lines of the CESE-MHD-NLFFF reconstruction. The color of the lines represents the value of current density $\mathbf{J}$ (normalized by its average value $\mathbf{J}_{\text{ave}}$ in the computational volume). The background is the photospheric magnetogram. **b** and **c** SDO/AIA 171 Å and 304 Å images of the pre-flare corona. **d** The low-lying magnetic field lines in the core region. The field lines are color-coded by the value of height z. **e** Locations of dips in the magnetic field lines; the color indicates the value of height z. **f** GONG H_α image of the active region. The dashed curve denotes the location of a long filament. Taken from Fig. 2 of Jiang et al. [208] with permissions of the authors and the AAS

temporal separation of the observed flare ribbons, as well as the dynamic boundary of the flux rope's feet are reproduced. For detailed analysis, the reader may refer to Jiang et al. [208].

1.3 Hybrid Empirical-MHD Models for Solar-Interplanetary Modeling

In addition to the above three models equipped with the associated software frameworks, there are many other 3D MHD models to study the corona and the solar wind. In this subsection we try to summarize a class of 3D MHD solar wind models, in which a theoretical, empirical or kinematic model is used to describe the solar corona where the plasma flow is transonic and trans-Alfvénic and a 3D MHD model

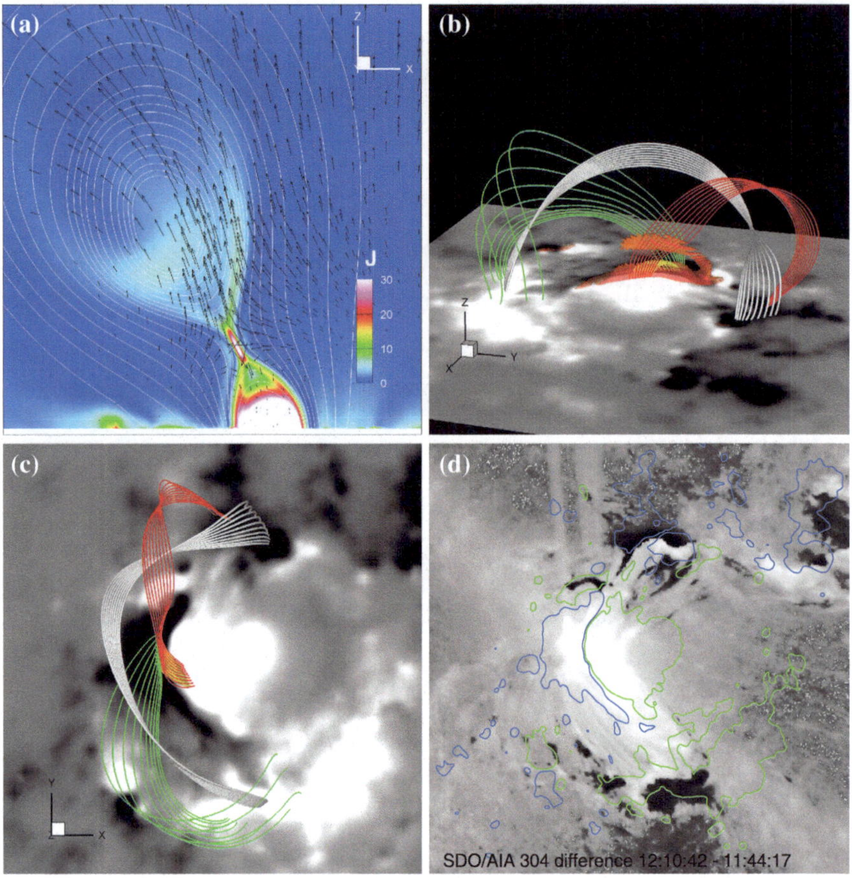

Fig. 1.18 Reconnection process below the rising magnetic flux rope. **a** Current density distribution and plasma flows (denoted by arrows) on the vertical cross section (the $y = 0$ plane) at time $t = 1.0$. The white lines are 2D field lines tracing on the plane. **b** 3D configuration of the reconnection. The white lines are within the main body of the magnetic flux rope. The red and green lines are reconnecting field lines below the rope. Their inner footpoints are sheared past each other along the polarity inversion line, and thus the field directions change abruptly across the current sheet, in which reconnection takes place, resulting in a long field line joining the magnetic flux rope and a short arcade below which the post-flare loops form (as shown by the yellow lines). The objects colored in red and orange are thin layers with the strongest current density throughout the volume, showing the current sheet in 3D. **c** Top view of the same magnetic field lines shown in **b**. **d** Difference of AIA 304 Å images before and after the flare, showing two dimming sites (or transient coronal holes) that match the locations of the magnetic flux rope's legs. The contour lines are shown for $B_z = -500$ G (colored in blue) and 500 G (colored in green). Taken from Fig. 5 of Jiang et al. [208] with permissions of the authors and the AAS

is exploited to obtain the solar wind solution in interplanetary space where the plasma flow is supersonic and super-Alfvénic. We call this class of models hybrid empirical-numerical models. Although the plasma flow and coronal magnetic field cannot be obtained in a self-consistent way in hybrid models, they are easily implemented and can achieve the solar wind solution in interplanetary space rapidly with limited computational resources. Therefore, they are of great importance in operational space weather forecasting at the present time.

1.3.1 Han and Wu's Model

In order to study interplanetary shock propagation from 18 solar radii to the Earth, Han and Wu [153] developed a 3D time-dependent MHD model. This 3D MHD code has been used for studying a number of problems in the heliosphere, for example, (1) the variations of interplanetary scintillation due to a disturbance with observational comparison of real disturbances [510], (2) the development of plasmoids' propagation outward into the interplanetary space [64], (3) interplanetary magnetic field changes at 1 AU as a consequence of simulated disturbances [590], (4) interplanetary magnetic field changes at 1 AU caused by a simulated solar disturbance and a tilted heliospheric current/plasma sheet [589], (5) evolution of the 14 April 1994 ICME and its propagation to Earth and Ulysses [75], (6) the large-scale solar wind structure observed by Ulysses based on the characteristic and representative observations of the solar magnetic field and K-coronal polarized brightness [460], (7) the propagation process of the CME on January 6, 1997 in the inner heliosphere [459], and (8) the effect of the background solar wind on the shocks' arrival at the Earth [591].

The numerical 3D MHD scheme used in this model is an extension of the two-step Lax-Wendroff finite difference method [263]. For computational stability in the MHD model, numerical Lapidus-type artificial viscosity is used by Han and Wu [153]. When using the two-step Lax-Wendroff finite difference method, the governing MHD equations are required to be written in conservation form, including the induction equation and the conservation laws of mass, momentum, and energy. The computational domain for the 3D MHD simulation is a heliocentric spherical coordinate system (r, θ, ϕ) oriented on the ecliptic plane. Earth is located at $r = 215R_s, \theta = 0°$, and $\phi = 0°$. The domain covers $-87.5° \leq \theta \leq 87.5°$; $0° \leq \phi \leq 360°$; $18R_s \leq r \leq 285R_s$. An open boundary condition at both $\theta = 87.5°$ and $\theta = -87.5°$ is used so there are no reflective disturbances. A constant grid size of $\Delta r = 3R_s$, $\Delta\theta = 5°$, and $\Delta\phi = 5°$ is used which results with $90 \times 36 \times 72$ grid points. The later application of this 3D code has been observed in a two-region setup [586, 592–594].

1.3.2 HHMS and HHMS-PI Models

In order to create a real-time, 3D MHD-based system to aid in the operational forecasting of geomagnetic activity, Detman et al. [63] introduced the Hybrid Heliospheric Modeling System (HHMS) with a source surface (potential field) current sheet model for the corona, a time-dependent 3D MHD solar wind model for interplanetary space and an empirical model for the geomagnetic A_p index. The system is driven by a sequence of photospheric magnetic maps composed of daily magnetograms. The solar wind model gives a predicted time series of MHD parameters at the location of Earth in the model grid; The predicted solar wind at Earth is used as input to the second, data-based, empirical model to predict the geomagnetic A_p index.

The HHMS model creates an empirical interface module to connect the source surface current sheet (SSCS) [440] model to the interplanetary global model, a vectorized (named IGMV) time dependent 3D MHD solar wind model [64, 153]. The empirical interface module is the keystone of the HHMS model; it connects the source surface current sheet model to the IGMV solar wind model by employing the WSA model (for details see Sect. 1.2.1.1). HHMS model adopts the solution at 0.1 AU (21.5 R_s) to prescibe the lower boundary of the time dependent 3D MHD solar wind model [64, 153]. Each coronal output gives a steady state (magnetostatic) description of the interplanetary magnetic field at 0.1 AU. However, the time series of SSCS maps can provide a pseudo steady approximation of the slowly evolving background state of the IMF at 0.1 AU.

The interface module provides the rotating time dependent lower boundary condition at 0.1 AU for the MHD solar wind model. The interface module also converts source surface map parameters to MHD parameters. From only two SSCS map parameters, the interface module must specify eight MHD parameters on the MHD model's grid boundary through eight empirical equations. For details about them and the discussion of their associated free parameters, refer to Detman et al. [63]. A linear prediction filter for A_p Index for this HHMS is the back-end of the system. It takes the MHD time series at the location of Earth in the IGMV model and predicts the geomagnetic A_p index.

By using the HHMS model, Detman et al. [63] compared test results for simulated 1 day ahead A_p forecasts for the years 1993 through 2002 with the forecast skill of the official A_p forecasts that were issued by the NOAA Space Environment Center in that time interval. Results showed that the HHMS was useful to forecasters in some years. Furthermore, Detman et al. [65] modified the HHMS model to include the physics of pickup protons (HHMS-PI) from interstellar neutral hydrogen, improve the empirical interface module, and tune its parameter to maximize model agreement with background (quiet) solar wind observations at ACE. Then they presented results of the simulation study for the famous Halloween 2003 series of solar events with shock inputs from the Fearless Forecast real-time shock arrival prediction study, and obtained good agreement between observed and simulated shock arrival times at ACE by adjusting input shock speeds.

By using the 3D time-dependent HHMS-PI model [65], Intriligator et al. [184] reported the first comparisons of pickup proton simulation results with in-situ measurements of pickup protons obtained by the SWICS instrument on Ulysses. Their simulated interval of 82 days in 2003–2004 included both the background solar wind and the October–November 2003 solar events (the "Halloween 2003" solar storms). They demonstrated that the HHMS-PI pickup proton simulations generally agreed with measurements from the instruments of Solar Wind Ion Composition Spectrometer (SWICS) and Solar Wind Observations Over the Poles of the Sun (SWOOP) onboard the Ulysses spacecraft. They spelled out the specific values of the solar input parameters for the HHMS-PI simulations that provided the best combined agreement in the times of arrival of the solar-generated shocks at both ACE and Ulysses. Intriligator et al. [185] analyzed space weather throughout the heliosphere out to Voyager 2 and beyond by using the 3D time-dependent HHMS-PI model and compared the modeled results with the available spacecraft data. They also discussed space weather throughout the heliosphere through in-depth analyses of the available simultaneous data from a number of instruments on spacecraft at various locations. By comparing HHMS-PI simulations with data from various spacecraft, and analyzing the particle and field data from Voyager 2 during the period of the Halloween 2003 solar eruptive events, they indicated that particle accelerations and other important physical processes were associated with the heliospheric propagation of these large space weather events, and concluded that the space weather event originating at the sun still had significant effects throughout the heliosphere as distant as 73 AU. By comparing in-situ observations from ACE, Ulysses, and Voyager missions with the results from the HHMS-PI model, Intriligator et al. [186] found that propagating shocks can be slowed due to pickup protons and the 3D locations of solar events and of various spacecraft are key to understanding the 3D propagation and timing of shocks, and gradients throughout the heliosphere, heliosheath, and the local interstellar medium.

1.3.3 HAF+3D MHD Codes

Wu et al. [586, 592–594] linked the 3D kinematic the Hakamada-Akasofu-Fry Version 2 (HAFv.2) model [116, 495] to the full 3D MHD model of Han and Wu [153]. In this model, the solar surface magnetograph observations are used to continuously drive the HAFv.2 model and then the Han and Wu's model throughout the heliosphere. They named this hybrid model as HAFv.2+3DMHD. Wu et al. [593] provided a detailed description of this hybrid code. The realization of the model can be split into two basic steps: (1) from the Sun to 2.5 R_s, thence to 0.08 AU (18 R_s); and (2) from 0.08 AU to the Earth and beyond.

Using a time series of photospheric magnetic maps composed from daily solar photospheric magnetograms, the solar wind structure is first simulated from the photosphere (1 solar radius, R_s) out to 2.5 R_s by the WSA model [4]. Then, HAFv.2 model uses a modified kinematic approach to simulate the solar wind conditions from 2.5 R_s out to 18 R_s with data from Carrington Rotation maps (2.5 R_s), provided by NOAA's Space Environment Center (http://www.swpc.noaa.gov/), as the input. Finally, the output of HAFv.2 at 18 R_s (0.08 AU) provides inputs for the time dependent 3D MHD solar wind model.

The HAFv.2+3DMHD model has been used to study a variety of solar eruptive events, such as the global interplanetary coronal mass ejection (ICME) from the 12 May 1997 solar event [593], multiple ICMEs' interaction and propagation from the Sun to the heliosphere following the 25–28 October 2003 solar events [592], the interplanetary evolution of the observed geoeffective coronal mass ejection during the epoch 1–4 August 2010 [594], and a coronal hole's effects on CME/shock morphology in the inner heliosphere with the help of 7 March 2011 solar events [586]. Liou et al. [284] employed the model to investigate the propagation of the extremely fast backside CME event on 23 July 2012 and the modeled results were in agreement with the in-situ measurement from STEREO-A.

Exploiting the HAFv.2+3DMHD model, a time series of synoptic photospheric magnetic maps from Wilcox Solar Observatory (WSO), and the recording of CMEs from the SECCHI/COR2A onboard STEREO, Wu et al. [595] simulated the Sun-to-Earth propagation of multiple CMEs and their associated shocks in September 2011. The simulation found that the evolution of the CME-driven shock and its interaction with the HCS and the non-uniform solar wind could explain time-intensity profile of the high-energy (>10 MeV) solar energetic particles (SEPs), and the sector boundary acted as an obstacle to the propagation of SEPs. Wu et al. [596] employed the model to study 12 CMEs and their associated shocks in September 2011. The results also demonstrated that this model is a useful tool to predict the arrival times of CME/Shock systems at Earth, the probability of observing the drivers of the interplanetary shocks, and the interaction between different solar wind structures. They also found that the background solar wind speed is an important controlling parameter in the propagation of interplanetary shocks and CMEs. In addition, a long-time simulation is needed to identify realistic associations of geomagnetic storm commencements with solar events. Specially, Wu et al. [597] investigated the CME encountered by the *Wind* spacecraft on 9 September 2011 in detail and verified the association of the short-duration ($\sim$35 min) extremely dense pulse (with a peak of $\sim$94 cm^{-3}) with the heliospheric plasma sheet compressed by the interplanetary shock.

1.3.4 LFM-Helio MHD Model

Merkin et al. [322] described a version of the Lyon–Fedder–Mobarry (LFM) model adapted to the environment of the inner heliosphere, which is referred as LFM-helio. The LFM numerical algorithms [295], which were originally developed for studying the interaction of the solar wind with the magnetosphere, are used without modification within LFM-helio. The LFM numerical algorithms use the eighth order partial donor cell method [151] for the reconstruction of differential scheme, and a water-bag beam scheme for the calculation of numerical flux for the fluid part.

Magnetic field is updated by method of constraint transport to maintain $\nabla \cdot \mathbf{B} = 0$. The model also uses the Boris correction [24] to limit the numerical propagation speed of Alfvén wave and to increase the allowable time step in lower-density regions. The inner boundary condition of LFM-helio assumes supersonic and super-Alfvénic solar wind and thus must be located beyond the critical point. Merkin et al. [322] used a regular spherical grid with the inner boundary located at 21.5 R_s (0.1 AU) and the outer boundary at 430 R_s (2 AU). The boundary condition of B_r was constrained from the solution of the WSA model [4]. The velocity, V_r, at the inner boundary is calculated from a revised version of the WSA model presented by McGregor et al. [314]. The azimuthal magnetic field is assigned as $B_\phi = (V_\phi / V_r) B_r$, where V_ϕ is the solar rotation speed at the radial distance of the boundary, $(2\pi/27\text{days}) \times 0.1$ AU. The plasma density is obtained from an empirical fit to Helios data [314]: $n[\text{cm}^{-3}] = 112.64 + 9.49 \times 10^7/(V_r[\text{km/s}])^2$, while the temperature is determined by assuming uniform thermal pressure, $nT = n_0 T_0$, where $T_0 = 8 \times 10^5$ K and $n_0 = 300\,\text{cm}^{-3}$. In Merkin et al. [322], the LFM-helio grid did not extend to the poles in order to avoid computing small time steps resulting from convergent grid lines near the axis, but Merkin et al. [324] extended the grid to include the pole and cover the whole 4π steradian computational domain and the LFM heliosphere model has also been coupled with the MAS corona model.

By using the LFM-helio model with fine resolution ($1°$, $1°$, $1 R_s$), Merkin et al. [322] simulated the solar wind and heliospheric magnetic field between 0.1 and 2 AU to study the disruption of a heliospheric current sheet fold during Carrington rotation 1892 in the decline phase of solar cycle 22. They demonstrated that the disruption constituted a process whereby the lower part of the current sheet fold separated from the main surface and, on a heliocentric spherical surface, appeared as an island of outward polarity in the sea of the field of inward polarity. They pointed out that this process is associated with non-radial motion of plasma and magnetic field induced inside a stream interaction region. The evidence of magnetic reconnection in the vicinity of the heliospheric current sheet was discussed. The simulations provided a useful global 3D context for interpreting multiple current sheet crossings commonly observed by spacecraft as well as observations of magnetic reconnection in the solar wind.

Pahud et al. [373] performed quasi steady-state simulations of Carrington rotations 2060 and 2068 by LFM-helio model and compared the numerical results with in-situ data from the ACE and MESSENGER spacecraft. The grid set and boundary condition used in this study were similar to those in Merkin et al. [322]. The model successfully reproduced the large-scale configuration of the inner heliosphere, such as timing and duration of high-speed streams and heliospheric current sheet crossings, as reflected in ACE and MESSENGER observations. The author argued that the uncertainty in the specification of the coronal conditions, rather than a poor performance of the model was responsible for the discrepancies between in-situ measurements and simulations, such as 1–2 day errors in the time of arrival of a CIR or the strength of the simulated magnetic field at the spacecraft.

Merkin et al. [324] presented a simulation study exploring heliospheric consequences of time-dependent changes at the Sun during a 2-month period in the beginning of year 2008, in which they used the Air Force Data Assimilate Photospheric Flux Transport (ADAPT) model to obtain daily updated photospheric magnetograms and drive the WSA model of the corona. The results of the WSA model was used as a time-dependent boundary condition for the LFM-helio model. They compared the simulation results with the spacecraft located near 1 AU heliocentric distance: ACE, STEREO-A and B, and MESSENGER spacecraft orbiting between 0.35 and 0.6 AU. The simulations showed that time-dependent simulations reproduced the gross-scale structure of the heliosphere with higher fidelity and moving boundaries of slow-fast wind transitions at 0.1 AU may result in the formation of inverted magnetic fields near pseudostreamers which was an intrinsically time-dependent process. The simulations demonstrated that time-dependent heliosphere modeling is a promising direction of research both for space weather applications and fundamental physics of the heliosphere.

1.3.5 CRONOS–Heliospheric Model

The CRONOS model is based on the genuinely multi-dimensional semi-discrete central scheme developed by Kurganov and Levy [251], Kurganov and Tadmor [252], and Kurganov et al. [253], and is extended to arbitrary orthogonal grids by Kissmann et al. [231]. The main advantages of this scheme are the high resolution, no Riemann solvers and characteristic decomposition involved and the upwind nature [253]. The CRONOS model was developed and is maintained by Ralf Kissmann [576]. The model is coded in C++ and fully MPI-parallel. Usage of C++ leads to a high modularity that enhances the code's extensibility and flexibility. The main features of CRONOS on numerical subjects can be seen in Kissmann et al. [230–232]. The code has already been successfully applied to simulations of the turbulent interstellar medium [230], coronal mass ejections [408], the shock propagation [409], the inner-heliospheric magnetic field [577], and the propagation of CMEs in interplanetary space [579]. The implementation of the CRONOS code with a view toward its ongoing development has been described by Kissmann et al. [232].

The CRONOS model employs a semi-discrete finite-volume scheme and integrates the governing equations forward in time by means of the Runge–Kutta method of adaptable accuracy. The core components provide many approximate Riemann solvers, such as HLL, HLLC (Hydrodynamics only) and HLLD (MHD only). Numerical problems can be solved on the Cartesian, cylindrical, and spherical grids and the cell size can be varied along each coordinated direction. Once the grid is specified by the user, it is not changed during the running time. When using spherical geometry, the code excludes the regions near the polar axes to avoid the coordinate singularities [576]. In the CRONOS model, the fluid variables are computed at the cells' centers and the magnetic field components are positioned at the corresponding orthogonal surfaces of each cell. For the magnetic field a constrained transport scheme is used to ensure the solenoidality of the magnetic field. The CRONOS model makes use of ghost cells to implement boundary conditions that can be defined for all six boundaries. Pre-implemented options are periodic, extrapolation, outflow, and reflecting boundary conditions.

Wiengarten et al. [577] employed the CRONOS model to solve the equations of ideal MHD in a one-fluid model. They followed Pomoell et al. [409] and set $\gamma = 1.05$ to accelerate the solar wind. They also used spherical coordinates (r, θ, ϕ) with the origin at the center of the Sun. For the latitudinal coordinates $\theta \in [8^\circ, 172^\circ]$, a cell size of $\Delta\theta = 2^\circ$ was used. Similarly, for the longitudinal coordinate $\phi \in [0^\circ, 360^\circ]$, $\Delta\phi = 2^\circ$ was employed. The radial length of the simulation box was determined by Earth's orbital distance ($215 R_s$) and the lower radial grid boundary (at $21.5\ R_s$), where a radial cell size of $1.5\ R_s$ was used. For the implementation of boundary conditions, CRONOS employed ghost cells, the extensions of the actual grid at its boundaries. The latitudinal boundaries were set to "reflecting", which basically ensured conservation of both mass and magnetic flux, but influenced the adjacent cells. This was compensated by a modest resolution so that there was little effect on the equatorial cells. Periodic boundary conditions were used for the ϕ-direction. The outer radial boundary condition was set to "outflow". Observationally-based inner boundary condition was used at the lower radial grid boundary $21.5\ R_s$. The input data for the magnetic field were the result of solar surface flux transport (SFT) modeling using observational data of sunspot groups [209] coupled with a current-sheet source surface (CSSS) model [622, 623]. The required initial conditions for the other MHD quantities were obtained following an empirical approach [63, 65] with improved fitting using an inverse relation between flux tube expansion and radial solar wind speed. For details refer to Wiengarten et al. [577]. They performed computations for representative solar minimum and maximum conditions, and obtained the corresponding state of the solar wind up to the Earth's orbit. After a successful comparison of the latter with observational data, the computed results could be used to drive outer-heliospheric large scale models, to simulate the evolution of CMEs in the solar wind ambient or to model solar wind plasma structures such as corotating interaction regions [577].

Wiengarten et al. [578] conducted 3D MHD simulations from 0.1 to 2 AU with the code of CRONOS to obtain realistic modeling solar wind conditions, which were used as the input to a transport code employing a stochastic differential equation (SDE) approach to investigate the 3D heliospheric transport of energetic particles in the presence of CIR-associated structures, such as compression regions and shock pairs. They derived the boundary conditions from the GONG synoptic magnetograms during CRs 2059–2061 via the WSA model, compared the simulation results of the solar wind speed, density, temperature, and the magnetic field with the spacecraft data of both the Ulysses and the two STEREO spacecraft, and demonstrated good agreements between the simulated results and the in situ measurements. By using the modeled solar wind background as input to a transport code with the SDE approach, Kopp et al. [242] further investigated the diffusion and drift behavior of Galactic cosmic rays, and the effects of CIRs on these transport processes. Additionally, Wiengarten et al. [579] incorporated turbulence transport into the Reynolds-averaged MHD equations in the framework of the CRONOS, which was used to investigate the effects on the turbulence evolution for transient events from 0.1 to 1 AU by injecting a CME from the bottom boundary. The study found that the CME-associated shock increased the turbulence levels and inhibited the cross helicity and indicated that researches on the large-scale structures associated with CMEs did not need to consider the turbulence transport effects due to the absence of strong back-reaction of the turbulence on the large scale structures.

1.3.6 SUSANOO Model

In order to provide a practical tool for real time space weather forecasting, the SUSANOO model [461, 464] is developed based on a numerical code that has been used for theoretical study of the CME [462, 463]. The ideal MHD equations in the Heliographic Inertial (HGI) coordinate is solved in a finite volume method by using the HLLD nonlinear Riemann solver with the third order Monotone Upstream-centered Schemes for Conservation Laws (MUSCL) [266, 329] and the second-order Runge–Kutta time integration. The hyperbolic divergence-cleaning method [59] is used to reduce the numerical deviation of the magnetic field divergence. Shiota et al. [463, 464] implemented this scheme on the composite Yin-Yang grid [217], a composite grid composed of two congruent spherical partial shells ($\frac{\pi}{4} < \theta < \frac{3\pi}{4}$, $-\frac{3\pi}{4} < \phi < \frac{3\pi}{4}$), each of which was orientated in a different direction. The physical quantities on the boundary of one partial shell were copied from the values at the same points on the other partial shell interpolated by a trigonometric function.

Using this code, Shiota et al. [463] studied the interaction (reconnection) between an ejecting flux rope and its ambient field in the radial range of 1.01 to 5.0 R_s, in which they mimicked the ejected flux rope formed as a result of eruption in an active region with a linear force-free spheromak-type magnetic field [219]. Their results implied that the interaction between an ejecting flux rope and its ambient field is an important process for determining the formation and orientation of the CME, and also showed that the amount and direction of the magnetic flux within the flux rope and the ambient field are key parameters for CME formation.

When considered in the heliosphere, the SUSANOO model imposes all the MHD parameters at the supersonic/superAlfvénic inner boundary as follows. Based on the synoptic map of the photospheric magnetic field, the coronal magnetic configuration is calculated using PFSS model. In order to obtain a realistic background IMF, a factor of 5 is uniformly multiplied to the magnetic field specified at the inner boundary. The solar wind speed is obtained using $V_{SW}(f_s) = 267.5 + 410/f_s^{0.4}$, i.e., the so-called WSA formula [4], where f_s is the magnetic field expansion factor. The density and temperature at the supersonic/superAlfvénic inner boundary are imposed by using the empirical relation derived from Helios observation [163]. Since the model is set in the HGI coordinate system, the relationship between the grid on the inner boundary of the model and the grid on the synoptic map is shifted with the solar rotation angular velocity in every time step.

Shiota et al. [464] exploited it to achieve the real-time space weather forecasting at the orbits of Earth and other planets from 25 to 425 R_s. There, the MHD simulation covering 3 years (2007–2009) was presented to demonstrate the ability of the model in reproducing the ambient solar wind structures. The numerical results were in reasonable agreement with in situ measurements of the solar wind at Venus and Mars by Venus Express and Mars Express, respectively.

Moreover, by conducting a simulation from 30 R_s to 430 R_s, Shiota and Kataoka [461] employed SUSANOO model to reproduce the propagation and interaction process of multiple CMEs associated with the highly complex active region NOAA 10486 in October to November 2003. The CME information is superposed on the series of maps of the MHD variables by including the internal magnetic flux rope inside the CMEs in the boundary condition at 30 R_s. The magnetic field structure takes the form of a spheromak-type magnetic flux rope [219], whose polarity is determined automatically according to the Hale-Nicholson law and the chirality law of Bothmer and Schwenn [29]. In order to mimic the propagation of the 3D structure of CME through the inner boundary, the spherical cut of the magnetic configuration is imposed on the inner boundary. The velocity of the CME is set to be the observed constant value inside each CME, which may be measured from coronagraph images. Comparing with the ambient solar wind, density within a CME is increased slightly in proportion to the magnetic field strength of the CME. Pressure inside each CME is set to be a constant value, determined by the difference in dynamic pressure from the ambient solar wind. The study by Shiota and Kataoka [461] successfully provided reasonably good results for velocity and the profile of southward magnetic field component of the Halloween Event on October 29, 2003. The simulation also indicated that for the propagation of the following CME was significantly affected by the trails of the preceding CMEs. The results demonstrated that this model is capable of predicting the magnitude of geomagnetic storms as well as predicting shock arrival times.

1.3.7 EUHFORIA

The "European heliospheric forecasting information asset" (EUHFORIA) is a space weather forecasting targeted inner heliosphere model, which is capable to provide MHD modeling of the ambient solar wind and the CME eruptions from 0.1 to 2 AU [334, 391, 406]. EUHFORIA consists of an empirical coronal model and an MHD heliosphere model. The coronal model provides plasma and magnetic field parameters at $R_b = 0.1$AU, which is then used as boundary condition to drive the heliospheric model.

The empirical coronal model extends from the photosphere up to a heliocentric distance of 0.1 AU. The coronal model follows the idea of the PFSS-WSA model (e.g., [5, 314]), where the solar wind speed is determined completely by the properties of the global three dimensional coronal magnetic field. The coronal magnetic field is constructed by PFSS+SCS model. In the lower coronal domain for $r \in [R_s, R_i]$, the PFSS model is employed to extrapolate coronal magnetic field, which is assumed to be current-free and set to be purely radial at a given source surface $R_{SS} \geq R_i$. In the upper coronal domain for $r \in [R_i, R_b]$, the SCS model is employed to calculate the coronal magnetic field by solving an additional Laplace equation similar as in the PFSS model but with different boundary condition. The main purpose of using the SCS model is to extend the magnetic field in a nearly radial fashion, and keep a thin structure for the HCS at the same time. In order to preserve the open magnetic flux and minimize the discontinuous transition at the boundary of the PFSS and HCS model, R_i and R_{SS} are set to be 2.3 R_s and 2.6 R_s respectively.

The solar wind speed at 0.1 AU, i.e., the inner boundary of the MHD heliosphere model is specified by

$$v(f, d) = v_0 + \frac{v_1}{(1+f)^\alpha}\left[1 - 0.8\exp(-(d/w)^\beta)\right]^3$$

which takes the basic form of the WSA model. Here, $v_0 = 240\,\mathrm{km/s}$, $v_1 = 675\,\mathrm{km/s}$, $\alpha = 0.222$, $\beta = 1.25$ and $w = 0.02\,\mathrm{rad}$. In this equation, f is the flux tube expansion factor (also expressed as f_s in other literatures) while d is the distance of the foot-point of the flux tube to the nearest coronal hole boundary (also expressed as θ_b in other literatures). The values of f and d are determined by the coronal magnetic field. A constant value of $50\,\mathrm{km/s}$ is subtracted from $v(f, d)$ before it is specified as the boundary condition of radial speed v_r to account for the acceleration beyond 0.1 AU in the MHD model. The whole speed map is rotated by 10° to account for solar rotation which is not included in the coronal magnetic field model. The meridional and azimuthal components of the velocity is set to be zero, i.e., the solar wind velocity is purely radial at the inner boundary.

The plasma number density at the boundary of the MHD heliosphere model is given by

$$n = n_{\mathrm{fsw}}(v_{\mathrm{fsw}}/v_r)^2$$

where $n_{\mathrm{fsw}} = 300\,\mathrm{cm}^{-3}$ and $v_{\mathrm{fsw}} = 675\,\mathrm{km/s}$. This prescription ensures a constant kinetic energy density at the inner boundary. The plasma thermal pressure is chosen to be constant and is set to be $p = 3.3\,\mathrm{nPa}$.

The radial component of the magnetic field is calculated by

$$B_r = \mathrm{sgn}(B_{\mathrm{corona}})B_{\mathrm{fsw}}(v_r/v_{\mathrm{fsw}})$$

where $\mathrm{sgn}(B_{\mathrm{corona}})$ is the sign of the magnetic field given by the PFSS+SCS model, and $B_{\mathrm{fsw}} = 300\,\mathrm{nT}$ is the magnetic field carried by the fast solar wind. By using this formula, the magnitude of the magnetic field is determined not from the PFSS+SCS model but rather from the solar wind speed, so that the underestimation of the interplanetary magnetic field may be avoided. The meridional component of the magnetic field is set to be zero, i.e., $B_\theta = 0$ and the azimuthal component is specified as $B_\phi = -(B_r/v_r)R_b\Omega \sin\theta$ in the co-rotating frame, resulting in a zero electric field.

The governing equation of the heliospheric model is the three dimensional MHD equations in the Heliocentric Earth Equatorial (HEEQ) coordinate system. The MHD equations are numerically solved in a finite volume method together with a constrained transport approach for advancing the magnetic field in an exactly (up to machine accuracy) divergence-free way. In the method, the hydrodynamical quantities are cell volume-averaged quantities, whereas the magnetic field components are cell face-averaged and the electric field components are cell edge line-averaged and are co-located in a staggered fashion on the grid. Standard piece-wise linear reconstruction and an approximate Riemann solver are employed to arrive at a robust second-order accurate scheme (e.g., [229, 407]). The computational domain extends from 0.1 to 2 AU in the radial direction, but it does not cover the whole 4π area of the spherical domain. It spans 120° in latitude and 360° in longitude. A uniform grid of 4° resolution in meridional and azimuthal directions and a radial grid spacing of $\Delta r \approx 0.0074$ AU $\approx 1.6\,R_s$ (256 cells in radial direction) in radial direction are used for the published results [406]. Since the inner boundary rotates in the HEEQ coordinate system, the electric field is determined by a method similar to Lionello et al. [281].

To obtain the steady-state ambient solar wind solution, the heliospheric MHD model is relaxed for 14 physical days. If CMEs occurs in the solar corona within several days prior to the observation time of the global magnetic synoptic map, these CMEs are inserted to the model by the cone CME model, similar to those in previous studies [352, 353]. The model treats the CME as a hydrodynamic cloud exhibiting a simple geometry as the CME evolves in the upper corona characterized by a constant angular width, propagation direction and speed. Finally, solar wind condition forecast for up to 7 days are made based on the modeled results.

In order to validate the baseline implementation of this model, Pomoell and Poedts [406] modelled the solar wind condition in the inner heliosphere during June 17–29, 2015. Five cone-model CMEs are presented in the heliosphere. The results showed that EUHFORIA successfully simulates the large-scale heliospheric dynamics including complex solar wind streams interacting with multiple eruptions. Using EUHFORIA model, Scolini et al. [448] tested the effect of different CME shapes on the simulation results. They took an artificial Earth-directed CME launched on 6 June 2008 as case study. Six runs were implemented, which corresponded to six different implementations of a spherical or spheroidal-like CME with a constant radius during insertion in the heliosphere. Their analysis showed that all the parameters specifying the CME shape in the model significantly affect simulation results at 1 AU as well as the predicted CME geoeffectiveness, confirming the pivotal role played by the shape implementation details in space weather forecasts.

1.3.8 Multi-Scale Fluid-kinetic Simulation Suite (MS-FLUKSS)

The Multi-Scale Fluid-Kinetic Simulation Suite (MS-FLUKSS) is a package of numerical codes, which has been developed at the Department of Space Science and the Center for Space Plasma and Aeronomic Research at the University of Alabama in Huntsville (UAH) [27, 28, 396, 398, 402, 403]. Originally designed to study the solar wind outflow in interplanetary space and the interaction between the solar wind (SW) and the local interstellar medium (LISM), MS-FLUKSS is an effective numerical tool for modeling the physical processes occurring throughout the heliosphere [248, 393–402, 405]. Given realistic 3D and time-dependent solar wind parameters, MS-FLUKSS is expected to deliver accurate outputs at Earth and other locations of interest in the global heliosphere.

Pogorelov et al. [396] described the structure of the MS-FLUKSS. The major constituent blocks of this suite are the following: an adaptive mesh refinement module to solve the ideal MHD equations with multiple discontinuities, multi-fluid and kinetic modules to solve the Boltzmann equation for the atom transport, and a module to calculate the cosmic ray transport. MS-FLUKSS has taken into account the momentum and energy transfer between the plasma and neutral components of the interacting SW-LISM. The collisional terms in the Boltzmann equations are first integrated in the velocity spaces of plasma and neutral particles in the assumption of the Maxwellian (or other) distribution function for all populations. Then the zeroth, first, and second moments of the source terms are taken, which provides the source terms in the mass, momentum , and energy equations of the ideal MHD or Euler system. By using a Godunov-type method of the second order of accuracy in time and space, and employing the solution of a linearized MHD Riemann problem at the interface between computational cells, MS-FLUKSS solves the system of governing equations numerically with the help of either total variation diminishing (TVD) Lax-Friedrichs or TVD Courant-Isaacson-Rees schemes [250]. Numerical magnetic field divergence $\nabla \cdot \mathbf{B}$ from computational region is eliminated by using three different techniques: (i) solution of the MHD system in the Godunov, symmetrizable Galilean invariant form, which introduces eight, nonzero characteristic velocities; (ii) a projection (artificial scalar potential) method, solution of the Poisson equation on adaptive meshes being conveniently implemented in the Chombo package; (iii) a constrained transport approach. MS-FLUKSS adds MHD components and charge-exchange/collisional source terms to the publicly available Chombo package, while the parallelization and load balancing of the AMR code are still the responsibility of Chombo. The Chombo package is a product of the community of collaborators working with the Applied Numerical Algorithms Group part of the Computational Research Division at Lawrence Berkeley National Laboratory Berkeley, CA, USA.

Exploring the global structure of the outer heliosphere requires understanding an enormous variety of physical phenomena occurring throughout the solar system, such as charge exchange processes among the LISM neutral particles and charged solar wind, the birth of pick-up ions, the origin of energetic neutral atoms, cosmic ray transport, magnetic field reconnection, particle acceleration at interplanetary shocks, and related turbulence in a collisionless plasma. MS-FLUKSS tries to solve the MHD and Boltzmann equations in a self-consistent fashion and to further use the solution obtained for the analysis of the cosmic ray propagation throughout the heliosphere, by involving multi-fluid and/or kinetic modeling of neutral particles penetrating into the heliosphere and addressing the complexity of coupling the interplanetary and interstellar magnetic fields (IMF and ISMF) at the heliospheric interface. Pogorelov et al. [394] demonstrated that the IMF line crosses the termination shock (TS) multiple times by investigating the heliospheric structure for different possible ISMF strength and orientation. Pogorelov et al. [397] used the MS-FLUKSS to investigate the effects of the 11 year solar cycle on the SW-LISM interaction, where the angle between the Sun's rotation and magnetic-dipole axes varies with solar activity and the topology of the heliospheric current sheet beyond the terminational shock has been resolved with AMR for the first time.

Further development of MS-FLUKSS permits to perform numerical analysis of the SW-LISM interaction with the realistic, based on the available observations, boundary conditions. Pogorelov et al. [403] generated a time-dependent boundary conditions from Ulysses observation to constrain the MS-FLUKSS for studying the SW-LISM interaction pattern. Simulated results demonstrate that a substantial decrease in the SW ram pressure observed by Ulysses between the TS crossings by Voyager 1 and 2 contributes significantly to the difference in the heliocentric distances at which these crossings occurred. Kim et al. [227] simulated the three-dimensional, time-dependent solar wind flow from 1 to 80 AU, using a two-fluid model in which protons and interstellar neutral hydrogen atoms are treated as separate fluids. Daily averages of the solar wind parameters from the OMNI data set are used as inner boundary conditions to reproduce large-scale fluctuations of the solar wind properties at Ulysses, Voyager, and New Horizons at various distances and latitudes between 1 and 80 AU. Kim et al. [227] showed that the model generally agrees with Ulysses data in the inner heliosphere and Voyager data in the outer heliosphere.

With further modification of MS-FLUKSS, Kim et al. [228] simulated the interaction between the SW and the partially ionized LISM (e.g., [26]) within the framework of MS-FLUKSS, by taking account of five separate fluids: one plasma fluid and four populations of neutral hydrogen atoms originating in different regions, i.e., in the undisturbed LISM, the very local interstellar medium (VLISM) around the heliopause, inner heliosheath, and the super-Alvénic SW. While the plasma flow is governed by ideal MHD equations, the flows of neutral hydrogen atoms are described hydrodynamically by means of multi-component Euler equations [394, 616], in which pickup ions resulting from the charge-exchange process between ions and neutral atoms are simply assumed to be immediately equilibrated with thermal ions to form an isotropic mixture.

For computational efficiency, Kim et al. [228] divided the computational domain from 1 to 1000 AU into two parts: In the first part, the solar wind is propagated from 1 to 12 AU using a base grid of $256 \times 128 \times 64$ cells in a spherical coordinate system (r, ϕ, θ); Subsequently, the time-dependent solutions saved at 12 AU are used as inner boundary conditions for the second part with the outer boundary at

1000 AU, where a base grid of 640 $\times$ 128 $\times$ 64 cells is employed. In both parts, the radial grid size varies with heliocentric distance such that Δr is approximately 0.03 AU at 1 AU, 0.13 AU at 12 AU, 0.4 AU at 120 AU, and 20 AU at 1000 AU, while in the non-radial directions, the base grid $\Delta\phi$ and $\Delta\theta$ are both $\sim 2.8°$. In order to resolve shocks at large distances, AMR technique is used to selectively refine the grid in the inner heliosheath and VLISM, particularly around the heliopause. With AMR, the radial grid size is reduced to $\Delta r = 0.036$ AU at 80 AU, 0.043 AU at 100 AU, 0.025 AU at 120 AU, and 0.045 AU at 140 AU, whereas the non-radial grid size becomes approximately 0.18° at 120 AU. With this modified version of MS-FLUKSS, by using OMNI daily averaged plasma and interplanetary magnetic field data as time-varying boundary conditions, Kim et al. [228] performed a global 3D time-dependent simulation to reproduce shocks propagating beyond the heliopause. The modeled shock arrival times closely match those of the late-2012 and mid-2014 shocks observed by Voyager 1, though the changes in the modeled magnetic strength $|B|$ and density across the shocks are slightly larger than observed. Furthermore, the model predicts shock arrivals at Voyager 1 around 2017.395 (Day of Year 144) and 2019.502 (Day of Year 183).

Apart from modeling the outer heliosphere, MS-FLUKSS has also been widely used in the inner heliospheric study. Kim et al. [226] carried out heliospheric modeling with boundary values derived from the Solar-Terrestrial Environment Laboratory (STEL) multistation observations of interplanetary scintillation (IPS). By employing the time-dependent tomography developed at University of California, San Diego (UCSD) (e.g., [187, 188]) to obtain the inner boundary maps on a spherical surface at 0.25 AU from the Sun, Kim et al. [226] propagated the solar wind outward and compared the simulated results with in situ measurements and also with the corresponding kinematic solution for CRs 2114–2115 and 2058–2063. They found notable differences in proton radial velocity and number density at Earth and various locations in the inner heliosphere between the MHD results and both the in situ data and the kinematic solution. The simulated results imply that fundamental differences between the MHD and kinematic models, i.e., acceleration in the MHD model and the stream interaction dynamics can cause a considerable discrepancy between the MHD solution and in situ measurements at Earth. With the boundary conditions set at 10 R_s, Pogorelov et al. [404] studied the CME events that occurred on January 23, 2012 and November 4, 1997. The boundary conditions at 10 R_s were propagated to 0.45 AU with MS-FLUKSS, and then the data generated at 0.45 AU were used to establish the solution up to 12 AU. Within MS-FLUKSS, Sarp Yalim et al. [439] presented the results of the global solar corona for CR 2145 by relaxing a PFSS magnetic field topology to a quasi-steady state that is driven by HMI line-of-sight magnetogram and horizontal velocity data.

By the way, the application of the computational MHD simulation is not limited to the area of solar-terrestrial physics. Also, the field of computational astrophysics is constantly extending existing techniques and developing innovative, new approaches to better understand how galaxies grow and how the large scale structure of the Universe evolves and to numerically capture ever more complex physics, such as cosmological structure formation, galaxy formation, reionization, core-collapse supernova explosions, accretion onto compact objects, gravitational fragmentation of molecular clouds and star formation, accretion disks, and turbulent driving and dissipation in the interstellar medium, just to name a few. Many numerical codes emerge for numerically solving hydro- and magnetohydrodynamics problems (e.g., [23]) in computational astrophysics. To provide the detailed introduction to each of them is beyond the scope of this book. The interested readers may refer to: AMRVAC [221, 410, 547, 550, 602] (http://amrvac.org), AREPO [332, 478], AstroBEAR [55] (https://astrobear.pas.rochester.edu/trac), ATHENA [471, 485] (https://trac.princeton.edu/Athena), CAFE-Q [134, 135], ECHO [35, 61, 62], ENZO [34] (http://enzo-project.org/), FLASH [118, 338] (http://flash.uchicago.edu/site/flashcode/), GIZMO [173, 174] (http://www.tapir.caltech.edu/~phopkins/Site/GIZMO.html), GOEMHD3 [470], Magnus [341], MURaM [553], NIRVANA [632–636] (http://www.aip.de/Members/uziegler/nirvana-code/), PENCIL [316] (http://pencil-code.nordita.org), PLUTO [325, 326] (http://plutocode.ph.unito.it/), RACOON [74], RAMSES [115, 512] (http://www.ics.uzh.ch/~teyssier/ramses/RAMSES.html), Surya [119] and ZEUS [481–484] (https://www.astro.princeton.edu/~jstone/zeus.html).

1.4 Other Miscellaneous Models for Solar-Interplanetary Modeling

There is quite a range of codes available for the solution of MHD problems in space physics that require the use of numerical methods-especially for cases where a direct comparison to observations is desired. Each code, however, has been developed with some focus in mind, thus leading to sometimes considerable differences in implementation and available features. Besides the architectural MHD models and hybrid empirical-MHD models, there exist other miscellaneous models for solar-interplanetary modeling that include-but are not limited to the following.

1.4.1 Usmanov's MHD Model

Usmanov [534, 535], Usmanov and Dryer [538] and Usmanov [536] proposed a fully 3D MHD for the study of solar wind background and solar disturbances from the Sun to Earth. The solar wind plasma is assumed to be a single-fluid, inviscid, polytropic gas with infinite conductivity. The computational domain is subdivided into two regions. Region I of transonic/trans-Alfvénic flow was assumed to lie between the solar surface and the sphere $r = 9.8\,R_s$, and region II of supersonic/super-Alfvénic flow extended beyond this boundary. While neglecting the solar rotation in region I, Usmanov [534] took it into account for region II. In region II, three-dimensional supersonic, super-Alfvénic, steady-state flow of an ideal fluid with a frozen-in magnetic field was considered. The method employed to construct a steady-state solution in region II was developed by Pizzo [390]: the steady-state flow governing equations were solved numerically by forward integration in the r-direction. Initial values of dependent variables were prescribed at the base of region II and then advanced along the hyperbolic radial coordinate out to 1 AU according to a two-dimensional modification of MacCormack's scheme [390].

For incorporation of the transient dynamics, Usmanov and Dryer [538] upgraded the simulation of the steady solar wind by Usmanov [534, 535] to a global fully three dimensional, time-dependent MHD model. The study was the first 3D time-dependent transonic and trans-Alfvénic simulation of solar wind that did not employ any assumptions of space symmetry and restrictions on the heliocentric extension of computational domain. Thus it enabled the modeler to initialize the disturbances at the coronal base (1 R_s) and to track their spatial and temporal evolution up to the Earth's orbit. The other unique property of the study comparing with previous researches was that the steady-state flow, which could be used as a background for simulation of transient disturbances, was inferred based on the large-scale solar magnetic field observations.

Usmanov [534] solved the 3D ideal MHD equations in Region I by using a 3D version of MacCormack's difference scheme, combined with Lapidus' artificial diffusion scheme maintaining the numerical stability of the algorithm. In Usmanov and Dryer [538], the radial distribution of mesh points in the inner region ($1 < r < 9.85\,\mathrm{R_s}$) was taken to be $r_i = r_l(1 + \Delta r/r_l)^{i-1}$, where $r_l = 1\,\mathrm{R_s}$ and $\Delta r = 0.1\,\mathrm{R_s}$, so that the radial spacing is a linear function of r ($\Delta r_i = r_i \Delta r/r_l$) and varies from 0.1 $\mathrm{R_s}$ at $r = 1\,\mathrm{R_s}$ to about 1 $\mathrm{R_s}$ near the outflow boundary (9.85 R_s). In the outer region ($9.85\mathrm{R_s} < r < 329\mathrm{R_s}$), the radial spacing was chosen to increase as $r^{1/2}$ and the mesh points were located at $r_i = r_{i-1} + \Delta r\sqrt{r_{i-1}/r_l}$, where $r_l = 9.85\,\mathrm{R_s}$ and $\Delta r = 1\mathrm{R_s}$. The mesh points were uniformly distributed from $\theta = 6°$ to $\theta = 174°$ and from $\phi = -12°$ to $\phi = 360°$ with the spacing of 12° (the two auxiliary points at $\phi = -12°$ and 360° were added in order to conveniently prescribe periodic boundary conditions). Thus, the computations were carried out on a mesh of 15 positions in θ-direction, 32 positions in ϕ-direction, and 25 (96) positions in radial direction in the inner (outer) region. Thus, a total of $15 \times 32 \times 120$ (57 600) grid points were considered in 3D, 4π steradian computational domain. As in the case of steady-state calculations, Usmanov used different values for the polytropic index γ, 1.05 and 1.17, for the inner and outer simulation region, respectively. However, unlike the steady state case [534], the numerical procedure by Usmanov and Dryer [538] of solving the MHD equations was essentially the same for both regions. The values of dependent variables obtained at the outflow boundary from the simulation in the inner region were used to prescribe the inflow boundary condition for the outer region. The interface between the regions was located in the supersonic/super-Alfvénic solar wind flow, thus an extrapolation procedure was employed to specify boundary values of dependent variables at these boundaries. In order to avoid the singularity and convergence near polar regions of the usual spherical coordinate, Usmanov [537] introduced a composite mesh with two polar caps replaced by a rotated spherical grid. This mesh consists of a usual spherical mesh with a limited extent in latitude and longitude ($42^\circ \leq \theta_1 \leq 138^\circ$, $0^\circ \leq \phi_1 \leq 360^\circ$), whose polar axis of the mesh is directed along the solar-rotation axis, and two other meshes covering the polar regions in both hemispheres. The two meshes covering both polar regions are fragments of spherical coordinates ($36^\circ \leq \theta_{2,3} \leq 144^\circ, 26^\circ \leq \phi_2 \leq 144^\circ$, $216^\circ \leq \phi_3 \leq 324^\circ$) with the polar axes laying in the equatorial plane of the coordinate system of the usual spherical mesh ($\theta_1 = 90^\circ$, $\phi_1 = 90^\circ$).

Based on this model, Usmanov and Dryer [538] performed the first simulation study of solar events in 3D case. They numerically studied eight transient disturbances associated with the major solar flares of June 1991: (i) the disturbances were originated at the coronal base (1 R_s) and their propagation through inhomogeneous ambient solar wind was simulated out to 1.5 AU; (ii) as a background for the transients, the global steady-state solar wind structure inferred from the 3D steady-state model [535] was used. The parameters of the initial pulses were prescribed in terms of the near-Sun shock velocities (as inferred from the metric Type II radio burst observations) relative to the preshock steady-state flow parameters at the flare sites. The computed parameters at the Earth's location for the period 1–18 June, 1991 were compared with the available observations of the interplanetary magnetic field, solar wind velocity, density, and with variation of the geomagnetic activity K_p index.

Following Usmanov [535], Usmanov et al. [533, 541] considered a system composed of a background coronal plasma and the Alfvén waves in the shortwave length WKB approximation. By using this steady state global axisymmetric MHD model, Usmanov et al. [533, 541] modeled quantitatively the physical properties of solar corona and solar wind on a global scale at the time of the Ulysses fast pole-to-pole scan in 1994–1995. Similarly, Usmanov and Goldstein [539] studied the three-dimensional structure of the heliosphere during solar activity minimum by comparing the output of the model with Ulysses observations obtained during the Ulysses fast pole-to-pole scan in 1994–1995. The polytropic MHD equations were solved for a steady coronal outflow by including the addition of Alfvén wave momentum and energy in the WKB approximation. In order to solve the relaxation problem in region I, Tóth [539] applied a TVD Lax-Friedrichs algorithm with the Woodward limiter [521]. The field-interpolated central difference approach suggested by Toth [519] was used to maintain the $\nabla \cdot \mathbf{B} = 0$ constraint. The code can be run on massively parallel computers using MPI. The integration along the radius in region II was performed with the MacCormack scheme.

Usmanov and Goldstein [540] expanded the two-region model of Usmanov and Goldstein [539] to include region III that extends from 1 to 100 AU and incorporated a population of interstellar neutral hydrogen and its interaction with solar wind protons. The model considered the solar wind outside 1 AU as a combination of three comoving species, solar wind protons, electrons, and pickup protons, and solved the 3D steady-state MHD equations with source terms due to photoionization and charge exchange. Separate energy equations were included for solar wind and pickup protons. The numerical simulation provided the global structure of the solar wind from the coronal base to 100 AU by comparison with Voyager 1 and 2 observations.

In order to explore the effects of interstellar pickup protons and turbulence on the structure and dynamics of the solar wind protons, Usmanov et al. [542] solved the fully three-dimensional time-dependent MHD equations in the region from 0.3 to 100 AU by using the Central Weighted Essentially Non-Oscillatory (CWENO) scheme in the composite grid [537]. Following Usmanov and Goldstein [540], the governing equations combine the mean-field equations for the solar wind plasma, magnetic field, and pickup protons and the turbulence transport equations for the turbulent energy, normalized cross-helicity, and correlation length. They then compared the modeled results with Voyager 2 observations. The model reproduced many effects caused by pickup protons and the dissipation of turbulent energy and the thermal pressure of pickup protons. Based on this work, Usmanov et al. [543] improved the formulation of the turbulence by employing an eddy viscosity approximation for the Reynolds stress tensor and the mean turbulent electric field. The improved model can account for driving of turbulence by large-scale velocity gradients. The governing equations was solved by the WENO scheme to study the global structure of the solar wind and its three-dimensional properties, including embedded turbulence, heating, and acceleration throughout the heliosphere. The model results gave a relatively satisfactory agreement with plasma and magnetic field observations on *Wind*, Ulysses, and Voyager 2 spacecraft. Subsequently, Usmanov et al. [543] modified the governing equations by employing the built-in turbulence transport equations based on Reynolds decomposition and coupled with the mean-flow Reynolds-averaged equations and introduced the fourth fluid, the interstellar hydrogen with separate continuity, momentum, and energy equations that were coupled with the charged components through photoionization and charge exchange. The simulation was in qualitative agreement with the in-situ measurements from Voyager 1. Further, Usmanov et al. [544] presented a fully three-dimensional magnetohydrodynamic model of the solar corona and solar wind by coupling Reynolds-averaged solar wind equations with transport equations for turbulence energy, cross helicity, and correlation scale. The model includes separate energy equations for protons and electrons and accounts for the effects of electron heat conduction, radiative cooling, Coulomb collisions, Reynolds stresses, eddy viscosity, and turbulent heating of protons and electrons. The computational domain extends from the coronal base to 5 AU and is divided into two regions: the inner (coronal) region, 1-30 R_s, and the outer (solar wind) region, 30 R_s-5 AU. Except for electron temperature, Usmanov et al. [544] showed that the model produced reasonable agreement with Ulysses data during its first and third fast latitude transits, and that turbulence can notably affect the Sun's loss of angular momentum.

1.4.2 Hayashi's Model

Hayashi [159] developed a 2D MHD simulation code, and Hayashi et al. [166] and Hayashi [161] extended it to a 3D MHD model for the conservative one-fluid time-dependent MHD equations governing the simulated solar wind in the spherical coordinate system rotating with a solar sidereal angular velocity. The model was based on the concept of the finite-volume method (e.g., [498, 499]) with the feature of third-order TVD numerical flux (e.g., [32, 156]) by using the approaches of the MUSCL and the minmod slope limiter. Roe's average [435] was employed to evaluate the matrices needed in flux splitting. As far as the time integration was concerned, a second order Runge–Kutta approach was used. In the code, two methods were employed to reduce the nonphysical effects of $\nabla \cdot \mathbf{B}$ [159, 161]. The artificial terms containing $\nabla \cdot \mathbf{B}$ were added to the right sides of the induction, momentum, and energy equations [414]. In addition, he exploited the pesudo-time method to solve Poisson equation for scalar potential of the numerical magnetic monopole and corrected the magnetic field according to the solved magnetic scalar potential. The grid system was constructed in the spherical coordinate system covering the heliocentric distance from 1.01 to 250 R_s. The cell sizes along the radius Δr were fixed at 0.01 R_s near the solar surface in order to treat the steep gradient of density. Beyond 1.1 $\mathrm{R_S}$, Δr was set proportional to $\ln r$ within 7 R_s and proportional to $\sqrt{r}$ in more distant regions, so that a total of 128 grid points covered from 1.01 to 250 R_s. The latitude from the north pole ($\theta = 0°$) to the south pole ($\theta = 180°$) was also covered by the grid points with the constant angular interval equal to $\pi/64$ and the grid size in the longitudinal direction was the same as that in the latitudinal direction.

Hayashi [159] proposed new boundary treatments based on the projected normal characteristic method for the MHD simulation of the trans-Alfvénic solar wind and solar corona. This new boundary treatment limited the mass flux escaping through the solar surface in order to match the solar wind mass flux Ulysses measured during its first fast-latitude scan of the heliosphere. Utilizing mass flux limit as one of the physical constrains of the projected normal characteristic method, the model produced reasonable contrasts of the plasma temperature and density between the coronal hole and streamer, and the modeled solar wind computed matched the Ulysses measurement at $r > 1$ AU quite well. Hayashi et al. [164] used this model to perform a MHD simulation for the solar corona by utilizing the temperature map derived from the multi-wavelength observation by the EUV Imaging Telescope (EIT) on SOHO and the magnetic field map from MDI/SOHO. Hayashi et al. [166] performed two 3D time-relaxation MHD simulations. One simulation used the boundary map representing the solar surface magnetic field distribution before the Halloween event in 2003, and the other adopted the map representing the post event distribution. They used the synchronic frame of the photospheric magnetic field that considered the longitudinal shift due to solar differential rotation and surface variations during the interval under investigation. The two simulations showed that the introduction of the synchronic frames of the photospheric magnetic field well reproduced the changes of the global solar corona structures sensitive to solar surface magnetic field variations.

The model has also been used in the numerical studies in super-Alfvén region, which are summarized here for convenience. Hayashi et al. [165] presented the measurement-based MHD model of the interplanetary disturbances; the quiet solar wind background was obtained with the MHD-based analysis method [162, 163] for the interplanetary radio scintillation (IPS) measurement data, and the interplanetary disturbance parameters, such as speed, direction and angular size of the expanding coronal mass ejection, were determined from the SOHO LASCO/C3 data with the cone model [603, 624]. The two successive interplanetary disturbances resulting from two flare-associated halo CMEs of April 2002 were simulated from 30 R_s to Earth and beyond. The simulated interplanetary shocks were compared with the near-Earth measurement. The results revealed that the second interplanetary disturbance propagated faster in the rarefaction region of the first event, implying that the multi-event simulation is important to enhance the simulation model.

Hayashi [160] presented a treatment of observation-based time-dependent boundary conditions for the inner boundary sphere in the time-dependent 3D MHD simulations of the global co-rotating solar wind structures. The boundary variables on the inner boundary sphere, at 50 R_s in his study, were assumed to change linearly from one instant to the next. A new feature was that, in order to maintain the divergence-free condition of the magnetic field, the changes of the time-dependent boundary magnetic field were expressed as the potential field in a thin shell volume. The solar magnetic field data from the Wilcox Solar Observatory and the solar wind speed data by the MHD-based analysis method [163] from the interplanetary scintillation (IPS) observations at Nagoya University, Japan, were used as the input boundary data. The solar wind simulated with the time-dependent boundary condition was compared with the near-Earth and Ulysses in-situ measurements and the solar wind simulated with the fixed boundary condition over a 7-month period in 1991. As shown by Hayashi [160], reasonable agreements with the in-situ measurements were obtained. The differences between the two simulations in the interplanetary field line paths were significant.

In order to adjust the model to the time-varying magnetic field on the bottom boundary, Hayashi [161] developed the model of the confined differential potential field (CDPF) to prescribe the bottom boundary condition, which assumed that the CDPF, the vector difference between the magnetic fields $\delta\mathbf{B}$ at two consecutive CRs, can be represented as a negative gradient of the scalar potential Ψ in the very thin layer above the solar photosphere. Therefore, the horizontal components of the magnetic field in the very thin layer near the bottom boundary can be obtained by solving a Laplace equation with respect to Ψ. He validated the model by presenting MHD simulations for two time intervals, from CR 2009 to 2010 and from CR 2074 to 2075. The simulations achieved some observational coronal features different from those yielded from a fixed boundary condition and slight improvements of the IMF at 1 AU, including the latitudinal component. In addition, a modified version of this model was adopted as the module of the time-dependent three-dimensional MHD simulation at the Joint Science Operation Center (JSOC) of SDO [167]. The module can routinely generate three-dimensional data of the time-relaxed minimum-energy state of the solar corona using the full-disk magnetogram data from HMI/SDO.

1.4.3 Nakamizo and Tanaka's Model

Nakamizo et al. [339] presented a 3D MHD model for the solar corona and interplanetary solar wind system, which is now operating at the Japanese space weather center and is maintained by National Institute of Information and Communications

Technology (NICT) in Japan. The governing equations are the conservative MHD system including the momentum and energy source terms. They split the magnetic field into the potential magnetic field and the perturbed component. The MHD system was solved by a finite volume method originally developed by Tanaka [498, 499] through Roe flux solver and an MUSCL interpolation with Van-Leer's differentiating limiter. In this model, Nakamizo et al. [339] developed a computational grid system based on the dodecahedron-splitting method. In the first-order splitting stage, the pentagonal face of the dodecahedron was divided into five triangles. In the second order splitting, the midpoints of each side of the triangle were connected, resulting in four smaller triangles. In each step, newly defined vertices were projected on the circumscribing sphere of the original dodecahedron, generating right triangles. This procedure was repeated until a sufficiently fine grid was obtained. The sphere were covered homogeneously by triangles of the same size without directional dependence and without apparent singularities that are present for spherical coordinates. A 3D grid system was obtained by stacking the triangle mesh in the radial direction. The resultant 3D grid system was unstructured on the sphere (2D) and structured in radial direction (1-D). With the use of the fifth-order splitting grid, there were 7682 grid points and 15360 triangles on a sphere. As a result, the angular resolution was about 2.25° both in the latitudinal and longitudinal directions. They set the innermost region at the coronal base, and the outermost region at the sphere with its radius 200 R_s. A spatial resolution in the radial direction gradually varied from 0.02 R_s at the inner boundary to 2.75 R_s at the outer boundary.

The magnetic field at the inner boundary was specified by the observational data. In order to obtain the supersonic solar wind speed, parameterized source terms were introduced into the momentum and energy equations. These source functions decay exponentially in altitude as widely used in previous studies. The absolute values of the source functions were controlled so as to reflect the topology of the coronal magnetic field. They were increased inside the magnetic flux tube with sub-radial expansion and were reduced inside the magnetic flux tube with over-radial expansion. This adjustment aimed to reproduce the variation of the solar wind speed according to the coronal magnetic structure. The simulation simultaneously reproduced the plasma structure, the high- and low-temperature regions, the open and closed magnetic field regions in the corona, the fast and slow solar wind, and the sector structures of interplanetary magnetic field. The results showed that the MHD model successfully captured many features of both the fine solar coronal features and the global solar wind structure. The resolution and speed of the model were improved in Korea Meteorological Administration [1].

1.4.4 COIN Model

Feng et al. [97] and Shen et al. [449] established the coronal and interplanetary (COIN) model, a 3D MHD model extending from the Sun to Earth beyond. The model divided the computational domain into two regions as follows: one from 1 to 22 R_s and the other from 18 Rs to 1 AU. In the transonic and trans-Alfvénic inner region, they used the three-dimensional ideal MHD equations with volumetric heating terms similar to those in Groth et al. [147] in order to describe the dynamic behavior of the coronal and solar wind plasma and produce the bimodal solar wind structure seen by Ulysses. In the supersonic and super-Alfvénic outer region, the 3D steady state nonconservative MHD equations in the inertial spherical coordinates were used in the COIN model, just as done in Han and Wu [153]. In the inner region, the time-dependent three-dimensional MHD equations were solved by a modified total variation diminishing/Lax-Friedrichs (TVDLF) scheme [95–97] with the electric field modification method [519]. This numerical model can satisfy the $\nabla \cdot \mathbf{B}$ to round-off error by using the field-interpolated central difference approach for solving the magnetic field [519] in the coronal model. For the outer region, the radial coordinate can be treated as a time-like coordinate, and a three-dimensional steady MHD flow can be solved by a two-dimensional (θ and ϕ), "time" (r)-dependent formulation. Thus they treated the radial coordinate as a time-like coordinate and applied the MacCormack II scheme.

Feng et al. [97] employed this model to simulate the solar wind background by inputting the photospheric magnetic field and adding the volumetric heating terms in to the MHD equations as done by Groth et al. [147] and Manchester [299]. The results demonstrated that the HCS ranged between the latitudes of $\pm 20°$ and the solar wind at mid-high latitudes flowed at $700 \sim 800$ km s^{-1}, which were confirmed by the Ulysses' observation in the solar minimum [342, 387]. In addition, the model also captured many details observed by Ulysses.

The model was used to investigate the propagation of coronal mass ejections in the non-homogenous background solar wind flow [449]. To establish the self-consistent structures on the source surface of 2.5 Rs for background solar wind, they employed the observational data of the solar magnetic field and K-coronal brightness and the first principle of MHD. Then they investigated the dynamical interaction of the CME on January 4, 1997 with the background solar wind flow between 2.5 R_s and 1 AU by using the self-consistent source surface structures as initial boundary conditions and introducing the time-dependent pulses of the pressure and velocity at the inner boundary to generate transient structures. The model provided a relatively satisfactory comparison with the observations.

To improve the precision in the heliosphere and save computational cost, Shen et al. [450] made a modification to the 3D COIN-TVD model mentioned above to realize the whole time-dependent simulation from the solar surface to 1 AU by applying an asynchronous and parallel time-marching method during the simulation. The modified model was used to explore the evolution of the CME on November 13, 2003 in solar-terrestrial space [450], the propagation of the CME on April 4, 2000 in the non-homogenous solar wind background [452], the kinematic features of the CME on July 12–16, 2012 in the heliosphere [456]. These simulations could obtain profiles of solar wind parameters roughly consistent with the in-situ measurements at 1AU, but more improvements were needed for the model to be used in numerical prediction. In addition, Shen et al. [451, 453, 454] employed the model to study the interaction of two CMEs' interaction in interplanetary space, analyzed variations of different forces during the interaction, and found that the momentum exchange during the collision of two CMEs was very important for the deceleration and acceleration of the CMEs. The 3D COIN-TVD model was also used to study the super-elastic collisions of CMEs in the heliosphere [455]. Results showed that the collision leaded to extra kinetic energy gain by 3–4% of the initial kinetic energy of the two CMEs, which suggested that the collision of CMEs could be superelastic. Shen et al. [457] furthered the dependence of CMEs' collision type on the ratio of the CME's kinetic energy to the CME's total energy.

For interplanetary modeling in the region from 21.5 R_s to 258 R_s, Shen et al. [458] improved 3D COIN-TVD MHD model by trying to make the boundary treatment suitable for all phases of solar cycles. Synoptic magnetogram maps from GONG and empirical WSA relation are used to assign solar wind speed at the lower boundary in the supersonic and super Alfvénic region, while temperature is specified accordingly based on its empirical relation with the solar wind speed. Magnetic field intensity and solar wind density at the inner boundary are obtained from observational data in the immediate past Carrington rotations, allowing the persistency of these two parameters in a short time period. Five free parameters for the boundary conditions are tunable to simulate the solar wind for different phases of solar cycle. This modified model was used to simulate the background solar wind from year 2007 to 2017, by capturing the time patterns of solar wind parameters well at most times.

1.4.5 VAC MHD Model

The Versatile Advection Code (VAC) is a general tool for solving MHD and hydrodynamical problems typical of astrophysical applications. A variety of numerical schemes are available for users to solve hyperbolic differential equations on one-, two- and three-dimensional spatial domain, including TVD-Roe, TVDLF, and flux correction of transport (FCT), which are all explicit, second order in space and time.

Jacobs and Poedts [189] solved the MHD equations on a 3D spherical grid of $160 \times 91 \times 180$ cells, covering a domain of $(r, \theta, \phi) \in [1.03, 20]\ R_s \times [0, \pi] \times [0, 2\pi]$. The inner boundary of the domain corresponded to the low solar corona. The mesh showed accumulation of grid cells towards the solar surface and towards the solar equator, where the cell size varied from $\Delta r = 0.01\ R_s$ to $\Delta r = 0.4\ R_s$ and from $\Delta\theta = 3.3°$ to $\Delta\theta = 1°$. In the longitudinal direction the grid kept uniform at $\Delta\phi = 2°$.

Chané et al. [42] demonstrated that two-and-a-half-dimensional (2.5D) simulations with a simple CME model, consisting of a high-density plasma blob including a magnetic flux rope, could predict the flow variables at 1 AU for a specific CME event reasonably well. Jacobs et al. [190, 191] made a detailed comparison of the results between a 2.5D and a 3D simulations and showed that the 2.5D simulation could yield acceptable results provided the momentum of the initial CME was chosen in a proper way. Poedts et al. [392] presented numerical simulations of the initial evolution of a fast CME from the lower solar corona up to 30 R_s, by superposing a magnetic flux rope on top of a stationary background solar wind with extra density added to the flux rope in order to get a fast CME forming a 3D shock wave.

Jacobs and Poedts [189] investigated the effect of new flux emergence on a magnetic system that possesses a 3D topology favourable for the breakout scenario. In a previous research by Zuccarello et al. [637], this was investigated in a 2.5D assumption. Using 2.5D version of this code, Zuccarello et al. [638] and Bemporad et al. [21] numerically studied the role of streamers in the deflection of coronal mass ejections or multiple CMEs. Their results showed that the CME deflected toward the current sheet of the larger northern helmet streamer due to an imbalance in the magnetic pressure and tension forces and finally got into the streamer. As pointed out by Zuccarello et al. [638], during solar minima, even CMEs originating from high latitude could be easily deflected toward the HCS, eventually resulting in geo-effective events, and that this latitudinal migration depended on both the strength of the large-scale coronal magnetic field and the magnetic flux of the erupting filament. These numerical results confirmed the statistical observational study made by Wei and Dryer [572].

By exploiting the loop annotation syntax (LASY), Keppens et al. [220, 221] implemented a dimension-independent block-based AMR on the parallel VAC model using the Message Passing Interface library (MPI-AMRVAC), in which the distribution of the blocks over the CPUs was fulfilled through a Morton-ordered space filling curve through the dynamically changing grid hierarchy. The MPI-AMRVAC code offers flexibility in selecting/augmenting the refinement (unrefinement) criteria, and can handle Cartesian, cylindrical and spherical geometries. Besides second order reconstruction methods available in VAC, the MPI-AMRVAC code also provides the compact stencil third-order reconstruction and the MP5 (fifth-order, monotonicity preserving) limiter. Strong stability preserving high-order Runge–Kutta time steppers are also implemented in the MPI-AMRVAC code. Those numerical schemes is able to work with different physics modules.

Zuccarello et al. [638] solved the full set of ideal MHD equations in a non-zero plasma-β environment with a potential magnetic field extrapolation of NOAA AR 9415 as the initial state by using the 3D MPI-AMRVAC code. They explored the solar coronal response to the photospheric motions observed for AR 9415, which was introduced at the inner boundary. They revealed that the photospheric converging shearing motions formed a quickly outward flux rope that carried away the plasma confined inside the flux rope against the gravitational attraction by the Sun. Moreover, the motions also drove a compressed leading edge propagating at a speed of about 550 km s^{-1} and preceding the CME. Their simulation demonstrated that both the initial magnetic field configuration and the plasma-magnetic-field interaction are key to understand the CME initiation and early evolution phenomenon. By combing the MPI-AMRVAC code and other fluid, particle and kinetic models, Lapenta et al. [262] performed detailed simulations in the settings of various plasma configurations relevant for coronal mass ejections in the solar corona, heliospheric current sheet conditions and magnetospheric reconnection sites. To derive the information on the 3D structure of CMEs from white-light (total and polarized brightness) images from the Multi Element Telescope for Imaging and Spectroscopy (METIS) coronagraph [112] on board the Solar Orbiter, Pagano et al. [371] performed a 3D MHD simulation of a flux rope ejection where a CME is produced by using the MPI-AMRVAC software to solve the MHD equations with the source term of external gravity. They found that the polarization ratio technique reproduced with high accuracy the position of the centre of mass along the line of sight and allowed information to be derived on the real 3D direction of propagation of the CME. Porth et al. [410] presented an update to the open source MPI-AMRVAC simulation toolkit for solar and non-relativistic astrophysical magnetofluid dynamics that are related to hydrodynamics with dust coupling and the conservative implementation of Hall magnetohydrodynamics. They discussed the parallel scaling of the code and demonstrated excellent weak scaling up to 30,000 processors, which enables the model to exploit modern peta-scale infrastructure. It should be noted that the model has also been employed to investigate the problem of two-dimensional resistive magnetohydrodynamic reconnection [222], the evolution of solar prominences [223, 601], and extended to multidimensional general-relativistic magnetohydrodynamics for studying the issue of black hole accretion [411]. Xia et al. [602] reported on the development of MPI-AMRVAC version 2.0 in hydrodynamic and MHD astrophysical applications.

MPI-AMRVAC code has been used to provide interplanetary space weather forecasting models with accurate time dependent boundary conditions of erupting magnetic flux ropes in the upper solar corona. In order to model the ejection of magnetic flux ropes in the global corona, Pagano et al. [368–370, 372] employed a dual modeling technique of MHD simulations coupled with a quasi-static non-potential global model [610]. Both are applied on a spatial domain that covers the entire solar surface, although they extend over a different radial distance. The non-potential model uses a time series of observed synoptic magnetograms to drive the non-potential quasi-static evolution of the coronal magnetic field. This technique is able to follow the formation and loss of equilibrium of magnetic flux ropes. Following this, an MHD simulation captures the dynamic evolution of the erupting flux rope, when it is ejected into interplanetary space. Pagano et al. [372] focused on the MHD simulations that follow the ejection of magnetic flux ropes to 4 R_s, and showed that the dual use of quasi-static non-potential magnetic field simulations and full time dependent MHD simulations can produce realistic inhomogeneous boundary conditions for space weather forecasting tools. By using MPI-AMRVAC MHD code to the reduced MHD equations with only the density, velocity, and the magnetic field, and without the gradient of gas pressure and gravity, and the energy equation, Guo et al. [150] developed a data-driven MHD model using the zero-β approximation. The initial condition is provided by a nonlinear force-free field derived from the magnetofrictional method based on vector magnetic field observed by the Helioseismic and Magnetic Imager on board the Solar Dynamics Observatory. The bottom boundary uses observed time series of the vector magnetic field and the vector velocity derived by the Differential Affine Velocity Estimator for Vector Magnetograms (e.g., [444]). The data-driven MHD simulation was carried out for active region 11123 observed on 2010 November 11 and reproduced the eruption process of the magnetic flux rope.

1.4.6 WENO MHD Model

Kleimann et al. [238] used the third-order CWENO scheme proposed by Kleimann et al. [237] to advance the equations of single-fluid time-dependent MHD for a problem related to space physics. They run the code on a 3D Cartesian, non-staggered grid, and showed that the code can handle shock like gradients without producing spurious oscillations.

The CWENO scheme [237] is a 3D variant of the scheme by Kurganov and Levy [251] with third-order Runge–Kutta time stepping. This third-order numerical scheme combines the techniques of WENO and AMR,which enables it to efficiently capture strong gradients on spatial scales being small compared to the overall scale of the plasma system considered. The advantages are its third-order accuracy in smooth regions (which automatically becomes second order near strong gradients to minimize spurious oscillations) and an easy generalization to multi-dimensional systems of equations due to the fact that no (exact or approximate) Riemann solver is needed. The CWENO scheme thus allows simultaneously to achieve high shock resolution comparable with the best shock capturing schemes and high-order convergence in smooth regions dominated by plasma waves. Both the Generalized Lagrange Multiplier (GLM) approach by Dedner et al. [59] and a classical projection scheme [30] can be used to maintain the solenoidality condition for magnetic field.

To demonstrate the suitability of the code, Kleimann et al. [238] conducted the simulation of CMEs and similar heliospheric transients and performed studies of the solar wind expansion during phases of solar activity minimum. In the study, the computational volume consisted of a brick-shaped region of space covering $100 \times 70 \times 50$ cells in the x, y, and z directions, respectively. Each cell was a cube with a side length of typically $\delta x = \delta y = \delta z = 0.1$, implying a coverage of $[10, 7, 5]\ R_s$ of real space. The Sun's center was located at the origin, with the dipolar axis pointing into the positive z direction. The computational domain was surrounded by two layers of ghost cells whose values were updated after each time step either from symmetry considerations (for 'mirror' boundaries intersecting the origin), or use of outflowing boundary conditions (at the actual 'outer' boundaries). The solar surface did not coincide with any of the Cartesian coordinate surfaces, and therefore requires special treatment. They adopted a carefully-designed weighted interpolation method to prescribe physical quantities at the inner boundary [238]. The results demonstrated convergence of the system into a stable Parker-like steady state for both hydrodynamic and MHD winds and confirmed the CME's nearly self-similar evolution close to the Sun, thus highlighting the importance of detailed modeling especially at small heliospheric radii. Besides, Kleimann et al. [111] gave a brief overview of the motivation for MHD simulations of coronal mass ejections and discussed some of the present problems with the modeling. With the help of CWENO MHD model, they presented CWENO-based CME simulations within solar corona and related them to ACE observations near 1 AU.

1.5 Global MHD Models of Magnetosphere

The magnetosphere is the lens through which solar space weather phenomena are focused and directed towards the Earth. In particular, the non-linear interaction of the solar wind with the Earth's magnetic field leads to the formation of highly inhomogenous electrical currents in the ionosphere which can ultimately result in damage to and problems with the operation of space-borne technological system and ground-based distribution networks. As for the physics of the magnetosphere in the frame of space weather forecasting, it is very crucial to understand the energy transfer process in geospace or the interaction of the solar wind-magnetosphere-ionosphere (S-M-I) system not only because it is the last ring in the Sun-Earth connection chain, but also because the magnetosphere and ionosphere are the places where hazard space weather events exert the adverse effects to the technology systems and human activities.

In general, the solar wind and the magnetosphere are described with ideal MHD equations, particularly, the solar wind, the magnetosheath, the outer tail including the lobes and the plasma sheet are well described by the ideal MHD equations. In the outer magnetosphere, the dynamics are often reproduced quite reliably, as verified by comparisons with in situ spacecraft measurements, while the inner magnetosphere is problematic, because the overlapping different ring current and plasmasphere plasmas populations with vastly different temperatures cannot be correctly represented by the single-temperature MHD equations (e.g., [416]). Thin boundaries (such as the magnetopause) where particle kinetic properties become important, can also pose a problem. In the solar wind-magnetosphere-ionosphere system, the ionosphere is in most cases electrostatic solving the potential distribution in the ionosphere, which couples to the magnetosphere via field-aligned currents.

As an initial and boundary value problem, global MHD simulation of solar wind-magnetosphere-ionosphere system employs solar wind input variables (ρ, $\mathbf{v}$, T, $\mathbf{B}$) as boundary conditions in the sunward wall of the computational domain, and apply outflow conditions on other walls of the computational domain. A dipole is placed in the origin empty space. Initially, a mirror dipole sunward of the Earth's dipole is implemented to obtain a "seed" magnetopause. However, the ideal MHD equations are not applicable in the resistive ionosphere, and therefore the ionospheric domain uses electrostatic equations. Usually, the ionosphere is assumed to be a spherical shell at $\sim 1.017\ R_E$, which is connected with the magnetospheric inner boundary (at $2 \sim 3\ R_E$) along the dipole magnetic field lines. In fact, the region between the ionospheric shell and the magnetospheric inner boundary is a passive medium, which only transmits electric effects, and where the perpendicular currents along the magnetic field lines are neglected in order to get the field aligned currents.

The M-I coupling process used may differ slightly in different MHD magnetospheric models. Generally, the global magnetosphere is coupled with the ionosphere during the MHD simulations in the following way. First, the ionospheric conductivity is calculated on the basis of the solar zenith angle, the field aligned current (FAC), the plasma pressure and temperature at the inner boundary of the simulation domain (e.g., [506]). The detailed equation regarding the ionospheric conductivity is provided by, say [82, 92, 581]. The FAC on the ionosphere (at $\sim$1.017 R_E) is calculated by projecting the FAC from the inner boundary along the dipole magnetic field lines. Second, the ionospheric electric potential is solved from the resulting 2D Poisson equation for given FAC and ionospheric conductivity. The associated Poisson equation is given below by Eq. (1.1). Third, the electric potential is mapped from the ionosphere to the magnetospheric inner boundary. And then, the components of the velocity (perpendicular to the magnetic field) at the inner boundary is determined by the projection of the electric potential ($-\nabla\Phi \times \mathbf{B}/\mathbf{B}^2$), where Φ and $\mathbf{B}$ are the electric potential and magnetic field, respectively. The above procedure is the M-I coupling process, and the feed-back loop is calculated once in every fixed time interval, say every time step, during the MHD simulations.

The inner boundary of the magnetosphere domain is connected with the ionosphere surface by assuming a dipole magnetic field. The ionospheric electric potential is calculated for given field aligned current with the ionospheric conductivity. On the assumed 2D ionospheric shell, the electric potential can be connected with the field aligned current through Ohm's law with given conductances

$$\nabla \cdot (\Sigma \cdot \nabla\Phi) = -J_{||}\,|\sin I| \tag{1.1}$$

where Φ is the ionospheric potential, $J_{||}$ is the field aligned current (positive while entering the ionosphere from the inner boundary of magnetosphere), and I is the inclination angle of the magnetic field. Σ denotes the tensor of the ionospheric conductances written as

$$\Sigma = \begin{pmatrix} \Sigma_{\theta\theta} & \Sigma_{\theta\phi} \\ -\Sigma_{\theta\phi} & \Sigma_{\phi\phi} \end{pmatrix}$$

which contains components from EUV ionization and particle precipitation. According to Tsunomura [530], the conductivity terms are written as the following functions:

$$\Sigma_{\theta\theta} = k_1(\lambda)\frac{\sigma_0\sigma_1}{\sigma_1\cos^2 I + \sigma_0\sin^2 I},$$

$$\Sigma_{\theta\phi} = k_2(\lambda)\frac{\sigma_0\sigma_2\sin I}{\sigma_1 cos^2 I + \sigma_0\sin^2 I},$$

$$\Sigma_{\phi\phi} = k_3(\lambda)\frac{\sigma_1\sigma_0\sin^2 I + (\sigma_1^2 + \sigma_2^2)\cos^2 I}{\sigma_1\cos^2 I + \sigma_0\sin^2 I}$$

Here, σ_0 is the parallel conductivity, σ_1 is the Pedersen conductivity, σ_2 is the Hall conductivity. The amplification factors, k_1, k_2 and k_3, depend on the latitude λ, so as to represent large height-integrated conductivities near the equator. And I is the inclination of the magnetic field. Therefore, a Poisson equation for the ionospheric electric potential Φ can be deduced from Eq. (1.1). This equation is solved on an ionospheric grid built by mapping the inner boundary of the MHD calculation along dipole field lines to the ionosphere. The field aligned currents are calculated at the inner boundary by evaluating, $-\nabla \times \mathbf{B} \cdot \mathbf{B}/\mathbf{B}^2$, which can be mapped directly to the ionosphere. Once the electric potential is obtained, the potential itself is mapped out to the inner boundary of the MHD domain where it prescribes a boundary condition for the drift velocity. For details, the reader may refer to the existing literature (e.g., [82, 92, 178, 194, 340, 417, 418, 498, 581]).

While the current global MHD codes use different computational schemes for solving the MHD equations, the results in the magnetosphere (MHD domain) are often quite similar. Typically the boundary conditions as well as the grid implementation vary; for example, most of the global MHD codes limit their inner boundary of the MHD domain to $\sim 3\ R_E$ distance in order to save computation time. In the global MHD simulations of the solar wind-magnetosphere-ionosphere system, most of the current MHD simulations use fixed grids in the MHD simulation domain, although the grids are typically refined in the near Earth region (e.g., [91, 504]). AMR technique has been used (e.g., [414, 562]); however, GUMICS-4 [193, 374] uses subcycling (variable time step) to save computation time and the adaptation in blocks rather than cell-by-cell is implemented in the MHD domain.

As a powerful modeling tool for understanding space weather, MHD approach makes up the limitations of single-point measurements provided by spacecraft. Numerous investigations have been carried out for the simulations of the dynamics and configurations of magnetosphere under due northward and southward interplanetary magnetic fields (IMFs) since the seminal paper of Dungey [76]. After decades' development, several global magnetospheric MHD models have been established.

The main global magnetospheric MHD models summarized here are the Integrated Space weather prediction Model (ISM) [477, 575], the Geospace Environment Data Analysis System (GEDAS) model [358, 359], the Tanaka model [498], the BATS-R-US model [132, 344, 414, 523, 527], the Grand Unified Magnetosphere-Ionosphere Coupling Simulation, version 4 (GUMICS-4) [193, 194], the Lyon–Fedder–Mobarry (LFM) model [295], the Open General Geospace Circulation Model (Open GGCM) [417], the Piecewise Parabolic Method with a Lagrangian Remap MHD (PPMLR-MHD) model [178], HLLD model [330], AMR-CESE-MHD model [561, 562], and HLLC model [148]. Table 1.1 shows these models' main properties, in which the resolution order means the higher order scheme in those regions where the solution is smooth. Columns 1 to 7 represent the authors, code names, the numerical schemes, the orders of accuracy of numerical methods, the coordinated systems of the spacing grids, the $\nabla \cdot \mathbf{B}$ controllers, and the states signing whether the coupling between the magnetosphere and ionosphere is considered, respectively. Besides the models listed in Table 1.1, there are other global magnetospheric MHD models [224, 259, 583, 604]. All these models have been set up by solving certain forms of the MHD equations with somewhat different numerical techniques.

Table 1.1 Global magnetospheric MHD models

Authors	Code name	Numerical scheme	Order	Grid	$\nabla \cdot \mathbf{B}$ controller	M-I coupling
Lyon et al.[a]	LFM	TVD	8	Distorted spherical geometry	Constrained transport	Yes
Raeder et al.[b]	Open GGCM	TVD	4	Stretched Cartesian	Constrained transport	Yes
Powell et al.[c]	BATS-R-US	TVD+Roe	2	Adaptive Cartesian	8-wave	Yes
Ogino et al.[d]	GEDAS	Modified leap-frog method	2	Cartesian	No	Yes
Tanaka et al.[e]	Tanaka	TVD+MUSCL	3	Spherical geometry	projection method	Yes
Janhunen et al.[f]	GUMICS-4	Godunov-type scheme+HLL/Roe solver	1	adaptive Cartesian	Projection method	Yes
Sonnerup et al.[g]	ISM	A combination of semi-implicit finite-difference and time-step control with partial donor cell terms	2	Mixed	No	Yes
Hu et al.[h]	PPMLR-MHD	PPMLR	3	Cartesian	8-wave	Yes
Miyoshi et al.[i]	HLLD	Upwind FVM+ Harten-Lax-van Leer-Discontinuities (HLLD)	2	Spherical geometry	No	No
Wang et al.[j]	AMR-CESE-MHD	Space-time conservation element and solution element	2	Spherical geometry	8-wave+ diffusion	Yes
Guo[k]	HLLC	Upwind FVM+ Harten-Lax-van Leer-contact (HLLC)	2	Cartesian	The hyperbolic cleaning method	No

[a][295]
[b][417]
[c][414, 523]
[d][359, 361]
[e][499, 501]
[f][194]
[g][477, 574, 575]
[h][178]
[i][330]
[j][561, 562]
[k][148]

1.5.1 ISM Model

The ISM was developed [477, 575] at Mission Research Corporation (MRC), Nashua, New Hampshire. Besides its standard MHD ingredient, ISM code in its most general form incorporates a number of phenomena essential for realistic space weather prediction. Included among these are particle drift physics in the inner magnetosphere; magnetosphere-ionosphere-thermosphere coupling and associated chemistry, described by equations that grade continuously from their form in the upper atmosphere (height >100 km) to their MHD form in the magnetosheath and solar wind; and basic Vlasov physics to describe current driven field-aligned potential drops in the auroral zones.

ISM is based on the MHD equations for plasma augmented with hydrodynamic equations for a collisionally coupled neutral thermosphere [574, 575]. The combined equations transition continuously from pure MHD for plasma in the solar wind and magnetosphere to ionospheric/thermospheric equations at low altitudes. Finite-difference grid resolution varies from a few hundred kilometers in the ionosphere to several Earth radii at the outer boundary of the computational domain. At the magnetopause, resolution ranges from 0.2 to 0.8 R_E. The Earth does not rotate in these simulations.

The simulation domain is cylindrical with a radius of 60 R_E and its axis along the GSE x-axis, that is, along the line pointing from the Earth to the Sun. Its upstream and downstream ends are at $x = 40\ R_E$ and $x = -300\ R_E$, respectively. The computational grid is orthogonal and axisymmetric around the x-axis. For $R > 3.5\ R_E$ it is generated analytically by superposition of a uniform field and a dipole field with a spherical boundary formed at 3.5 R_E, where it connects to an inner, purely spherical grid [574, 575].

While keeping transported quantities monotone and minimizing diffusion, the numerical code consists of a second-order finite-difference scheme with partial donor cell method (PDM) terms [151] to maintain numerical stability in regions containing large gradients, and an artificially enhanced dielectric constant in the displacement current to restrict the speed of light c. With the later feature, the effective Alfvén wave speed v_A would be limited to a maximum of 600 km/s so as to avoid unacceptably small time step in the integration above the auroral zones, where very large values of v_A occur in the real system.

The ISM model treats the magnetosphere and the ionosphere together as a single domain governed by a single set of equations. This single set of equations moves continuously from the ionosphere to the solar wind in the form of continuum mechanics equations appropriate to each domain. As pointed out by Siscoe et al. [467], the advantage is that whereas other codes can project cusp plasma parameters only down to the inner boundary of the MHD grid (usually an Earth-centered spherical surface with a radius between 3 and 4 R_E), the ISM model can project them continuously down to the ionosphere. This allows the cusp's image to be mapped at the ionospheric level.

By including a modified displacement current along with a simplified model of the ionosphere, Sonnerup et al. [477] employed the MHD spine of ISM to examine the steady state properties of the magnetosphere for zero interplanetary magnetic field. Siscoe et al. [468] used the ISM code to compute the distribution of $\mathbf{E}_{\parallel}$ on the face of the magnetopause, and found that a substantial parallel potential (between 10 and 15 kV) exists between the magnetopause and the ionosphere in northern dawn and southern dusk sectors. By discussing flow through reconnection for strictly northward interplanetary magnetic field, Siscoe et al. [469] reported properties of a little-studied but possibly common mode of magnetic reconnection flow characterized by plasma flowing at a high speed right through a stationary reconnection site.

Maynard et al. [307] employed the ISM model to show merging in the magnetosheath between IMF field lines on opposite sides of a directional discontinuity. By correlating measurements from diverse locations and using ISM's MHD models to put the measurements in context, Maynard [306] carried out a try to constrain the interpretations of the global and meso-scale dynamics of magnetic merging, which is a primary means for coupling energy transfer from the solar wind into the magnetosphere-ionosphere system. By using the steady state magnetosphere configuration obtained from ISM, Maynard et al. [305] interpreted Cluster observations of fast shocks in the magnetosheath launched as a tangential discontinuity with a pressure increase crossed the bow shock. Maynard et al. [308] probed the interactions of the heliospheric current and plasma sheets with the bow shock by using the ISM model and the observations from the spacecraft of Cluster and Polar. To understand the dynamical process on the flanks of the magnetopause, Maynard et al. [309] employed both spacecraft data and the ISM code to provide contexts for interpreting Cluster measurements during a 90 min period on 20 November 2003 when Cluster crossed the magnetopause four times in the vicinity of the sash.

With ISM simulations, Siscoe et al. [467] showed that the latitude of the high-altitude cusp decreases as the IMF swings from north to south, that there is a pronounced dawn-dusk asymmetry at high-altitude associated with a dawn-dusk component of the IMF, and that at the same time there is also a pronounced dawn-dusk asymmetry at low-altitude. Ober et al. [348] reported results of the ISM simulation of the magnetospheric response to a step-function increase in solar-wind density pulse similar to that observed by WIND on 10 January 1997.

1.5.2 GEDAS Model

The GEDAS model [358–360, 554, 555] solves the normalized primitive variable non-conservative MHD equations as an initial-boundary value problem by using a modified version of the leapfrog scheme, which is a combination of the leapfrog scheme and the two-step Lax-Wendroff scheme. The ionospheric boundary condition imposed near the Earth is determined by requiring a static equilibrium [358]. The inner boundary of the model is located at around 3.0 R_E and the simulation domain is a rectangular box with an uniform Cartesian grid whose resolution is 0.3 R_E [225].

The GEDAS model was then used to investigate the subjects such as the substorms [340], the dayside magnetopause reconnection [378–381] and the shock wave event [361]. Kim et al. [225] performed numerical studies to understand the magnetospheric responses to a sudden change in the solar wind dynamic pressure D_p by using the solar wind conditions observed by the Wind satellite as the model input. Meanwhile, by using the GEDAS model, Fukazawa et al. [122] investigated the responses of the reconnection electric field and polar cap potential to the IMF and solar wind velocity.

1.5.3 Tanaka's Model (Also REPPU Model)

By using finite volume total-variation diminishing scheme, Tanaka [498, 499, 502] developed a 3D MHD model for the coupling of the solar wind-magnetosphere-ionosphere (S-M-I) system. They adopted an unstructured grid with the simulation domain given by $3.0\ R_\mathrm{E} \leq r \leq 200.0\ R_\mathrm{E}$. The basic equations for the magnetosphere are the modified MHD equations written by dividing the magnetic field $\mathbf{B}$ as $\mathbf{B} = \mathbf{B}_0 + \mathbf{B}_1$, where $\mathbf{B}_0$ is a dipole field with $\mathbf{B}_1$ time-dependent deviation from $\mathbf{B}_0$ [498]. The numerical scheme is based on the finite volume method with an upwinding numerical flux calculated by the linearized Riemann solver. To obtain a higher order of accuracy, the upwinding flux is extended to the second-order and third-order TVD numerical flux in the calculation of the finite volume method, through the MUSCL approach and Van Leer's differentiable limiter. Not only numerical roundoff errors but also the use of upwind fluxes and a non-Cartesian grid system makes it a serious issue to fulfill the $\nabla \cdot \mathbf{B} = 0$ condition automatically. Tanaka's model uses the so-called projection method [30] to clean the $\nabla \cdot \mathbf{B}$ error, that is, the variable components of magnetic field $\mathbf{B}$ are replaced every several time steps, with a new field by solving the Poisson equation with the conjugate residual method.

Tanaka's model was used to study the field-aligned currents system (FAC) [499], the convection in the magnetosphere-ionosphere (M-I) system [500, 501, 503], and the formation of the theta aurora [505]. Moreover, real-time global simulations were carried out by comparing the modeled results with the observations to investigate the substorm mechanism [82, 233, 506], the magnetospheric response to a geomagnetic sudden commencement [120, 121, 249, 337] and the sudden enhancement of the plasma pressure in the near-Earth plasma sheet [608].

After various improvements through applications, Tanaka's global magnetospheric MHD model has been drastically improved to high-resolution and high-speed computation, and named REPPU (REProduce Plasma Universe) since Tanaka [504]. The REPPU code is developed by paying the most attention to how to resolve in detail the FAC [82], which is believed to the most important point in realistically reproducing the substorm observations by the M-I coupling simulation.

The REPPU code uses an unstructured grid system without apparent singularity. The generation of this grid is initiated from a dodecahedron (e.g., [81, 83, 333, 339, 504, 507, 508]). The grid system in the magnetospheric domain is based on a triangle. First, a sphere at the inner boundary (2.6 R_E) is divided into 12 pentagons.

Each pentagon is further divided into five triangles, and then each triangle is divided into four smaller triangles, to cover the spherical surface with 60 triangles. Continuing this dividing process repeatedly, 30,722 grid points (61,440 triangles) after six rounds of division and 122,886 grid points (245,760 triangles) after seven rounds of division are obtained respectively on a sphere, which correspond to the minimum grid size 0.052 R_E and 0.0226 R_E near the inner boundary. The 3D grid system is generated by radially stacking 360 spheres. The outer boundary of the simulation domain is, for instance, located at 200 R_E at midnight and 600 R_E at noon. The grid structure is unchanged throughout the calculation.

Currently, REPPU code is capable of reproducing many features recognized in substorms, including sudden intensification of the auroral electrojet [504], a positive bay at midlatitude [504], a westward traveling surge (e.g., [80, 504]), and counter electrojet at magnetic equator [82]. To give a recent exemplary studies, Ebihara and Tanaka [81] performed the global MHD simulation and found that the amount of the energy released from the lobe is comparable to that continuously supplied from the solar wind. This could partially explain the reason why the ionospheric energy dissipation rate is correlated with the energy input rate from the solar wind. In order to further investigate what determines the amount of the magnetic energy stored in the lobe, Ebihara et al. [83] performed global magnetohydrodynamics simulation with REPPU model for different solar wind conditions. They found that about 33–88% of the magnetic energy entering the magnetosphere comes from the kinetic energy of the solar wind. The kinetic energy is efficiently converted to the magnetic energy just outside the magnetosphere. The amount of energy consumed in the ionosphere is found to depend on the solar wind parameters. The intensity of the auroral electrojet increases with the southward component of interplanetary magnetic field (IMF) and the solar wind velocity, which is consistent with observations. This may imply that the magnitude of the substorm is determined by the solar wind parameters and is predictable.

1.5.4 BATS-R-US Code

As a critical component of SWMF mentioned in Sect. 1.2.2, the BATS-R-US [414, 523] magnetosphere code solves various forms of the MHD equations, which include Hall [296, 525], semi-relativistic [131], multi-species [285, 296] and multi-fluid MHD [127], and anisotropic pressure [526]. For magnetospheric simulations, the inner boundary is usually taken at 2.5 R_{E} centered around the Earth. The detail of the BATS-R-US code is introduced in Sect. 1.2.2. Gombosi et al. [130] reported on the first comprehensive numerical simulation of a space weather event, starting with the generation of a CME and subsequently following this transient solar wind disturbance as it evolves into a magnetic cloud and travels through interplanetary space towards Earth where it interacts with the terrestrial magnetosphere-ionosphere system. With the BATS-R-US code, Ridley et al. [428] tried to determine whether the quasi-steady state of the magnetosphere from their MHD simulation could be taken as a prediction for the true state of the magnetosphere during an interval of relatively steady IMF from observation. Tóth et al. [524] studied the evolution of the global magnetosphere due to the sudden changes in the solar wind density and magnetic field. Wang et al. [560] compared ionospheric outputs from the BATS-R-US code with magnetic and plasma observations from the satellites of Challenging Minisatellite Payload (CHAMP) and Defense Meteorological Satellite Program (DMSP) under both quiet and storm conditions, with emphasis focused on the dependence of the model's performance for various magnetic local times, solar wind conditions, and seasons. Ilie et al. [182] investigated the role of IMF B_z fluctuations periodicity in the transfer of solar wind mass and energy to the magnetosphere during the co-rotating interaction region/high-speed stream event of 10 November 2003 through global modelling simulations using the BATS-R-US. Lu et al. [290, 291] examined the relationship between the shape and size of the magnetopause and the solar wind conditions, and studied the energy flow from the solar wind to the magnetosphere for northward and southward interplanetary magnetic fields. The BATS-R-US code has been used to understand reconnection at the magnetopause (e.g., [128, 240, 241] and references therein). Ngwira et al. [344] introduced a global 3D specially adapted BATS-R-US MHD model setup for simulating extreme space weather events of the Carrington superstorm of September 1859 and Halloween storm October 2003. They presented the numerical results for "very extreme" constructed/idealized solar wind boundary conditions driving the magnetosphere, and characterized the reaction of the magnetosphere-ionosphere system and the associated induced geoelectric field on the ground to such extreme driving conditions. In order to understand how the various internally and externally driven processes affect the Saturnian magnetosphere, Jia et al. [195] utilized the code to self-consistently couple the sparse solar wind and the Saturnian magnetosphere and ionosphere and incorporate mass-loading processes associated with Enceladus and its extended neutral cloud.

Tóth et al. [528] developed a new modeling capability to embed the implicit particle-in-cell (PIC) model [183, 304] into the BATS-R-US code. As one of the very first applications of the algorithm of the MHD with embedded PIC domains (MHD-EPIC) [57], they simulated the interaction between Jupiter's magnetospheric plasma and Ganymede's magnetosphere by comparing the MHD-EPIC simulations with pure Hall MHD simulations and juxtaposed both modeled results with Galileo observations to assess the importance of kinetic effects in controlling the configuration and dynamics of Ganymede's magnetosphere. By employing the MHD-EPIC code, Chen et al. [47] performed a 3D global simulation to study the dayside magnetic reconnection in flux transfer events (FTEs) occurring in the Earth's magnetosphere.

1.5.5 GUMICS-4 Model

The GUMICS-4 model [194] is a global magnetosphere-ionosphere coupling model that integrates a 3D MHD model of the magnetosphere with a high-resolution ionospheric model. GUMICS-4 solves the ideal, conservative,non-relativistic MHD equations by using a first order Godunov-type scheme. GUMICS-4 primarily uses the 7-wave Riemann solver to preserve $\nabla \cdot \mathbf{B} = 0$ to truncation error, and periodically removes the divergence of magnetic field with the projection method to avoid error accumulation. GUMICS-4 employs an cell-by-cell adapted Cartesian grid for the magnetosphere and the simulation box extends from -224.0 to $+32.0$ R_{E} in x and from -64.0 to $+64.0$ R_{E} in y and z. The magnetosphere feeds the inputs to the ionospheric model through the field-aligned current, electron precipitation and solar EUV ionization, while the ionospheric field-aligned current and potential from the ionospheric model are mapped along the dipole field lines to the inner magnetospheric boundary at 3.7 R_{E} [194].

By using GUMICS, a series of investigations have been carried out, such as the stormtime magnetospheric energy input [375], the magnetopause energy and mass transfer [376], the shock propagation in the magnetosphere [2], the spatial variation of the magnetospheric energy transfer [377], and the magnetopause reconnection and energy conversion [170, 255]. Gordeev et al. [141] presented an approach for systematic and quantitative testing of GUMICS-4 code performance based on statistical empirical dependencies of the key magnetospheric parameters obtained from observations. By means of a large set of nearly stationary solutions (162 runs altogether) with different stationary IMF and solar wind inputs, they found that the GUMICS-4 stationary solutions generally fitted the statistical models' relations, however, some discrepancies existed such that the size of the tail magnetopause and the lobe magnetic field magnitude were systematically lower and the ionospheric potential was smaller in magnitude compared to their empirical counterparts. Juusola et al. [216] compared statistically ionospheric seasonal variations in GUMICS-4 model with in-situ solar wind measurements and ionospheric convection measurements by the Super Dual Auroral Radar Network (SuperDARN) radars, and electric currents derived from the magnetic field measured by the CHAMP satellite.

1.5.6 LFM Model

The LFM model [90, 295] is developed by solving the semi-conservative MHD equations on a static and deformed spherical mesh. The high Alfvén speed problem is solved in two ways: firstly by restricting the computational region to some distance greater than $2 \sim 3$ $\mathrm{R_E}$ from the Earth, thus cutting out the highest speed and smallest cell regions, and secondly by the implementation of the Boris (Alfvén) correction [24]. The low β issue is also addressed by using an equation for the plasma total energy (thermal plus kinetic) instead of one for the total energy. The semi-conservative formulation chosen by the LFM model allows for finite difference schemes that numerically conserve mass (ρ), momentum ($\rho\mathbf{v}$), and plasma energy ($e_p = \frac{p}{\gamma-1} + \frac{1}{2}\rho\mathbf{v}^2$), but with no strict conservation of total energy ($e = \frac{p}{\gamma-1} + \frac{1}{2}\rho\mathbf{v}^2 + \frac{1}{2\mu}\mathbf{B}^2$). The semi-conservative form avoids the difficulty that fully conservative schemes often suffer from instability in low β regions where the pressure must be computed as the difference of two large quantities (e and $\frac{1}{2\mu}\mathbf{B}^2$). The LFM is coded in the basic cell structure within a distorted cube, a general hexahedron updated by an upwind TVD scheme with high order accuracy of flux interpolation. The numerical flux of fluid part is evaluated by using a water-bag beam scheme model. The magnetic field uses the constrained transport algorithms proposed by Yee [611] and later by Evans and Hawley [87] to maintained $\nabla \cdot \mathbf{B} = 0$ to roundoff errors. Besides being used in the magnetosphere, the LFM code has also been applied to simulate the ambient solar wind and interplanetary magnetic field structure, which has been illustrated in Sect. 1.3.4.

The LFM model includes a self-consistent ionosphere model and its inner boundary is at about 3.5 R_E. The LFM has been used to simulate the magnetospheric structure [90], the equatorial plasma convection [331] for idealized cases, and geomagnetic storms [180, 294, 416, 617]. At the same time, a series of investigations have been given by simulating real events and comparing the modeled result with the observations, such as magnetic storm event [180], Sawtooth events and steady convection intervals [136], magnetic cloud events [244], seasonal variations of field-aligned currents, cross polar cap potential, and precipitating electron energy flux [581], magnetic reconnection at the dayside magnetopause under solar wind conditions with interplanetary magnetic field (IMF) clock angles [365] and evolution of Birkeland currents [321]. Additionally, the LFM model has been employed for a subject of statistical calibration and emulation based on approximated features [168, 234, 235]. Bhattarai and Lopez[22] analysed the results obtained from the LFM model during periods of purely northward IMF with different solar wind velocity and ionospheric conductivity to show a reduction of the viscous potential with increasing magnitude of northward IMF, and also found that the magnetosphere during northward IMF is mostly closed, with the exception of a localized region where the magnetic reconnection occurs, creating an open magnetosphere. By using the data available from the Active Magnetosphere and Planetary Electrodynamics Response Experiment (AMPERE) and from the Iridium satellite constellation, Merkin et al. [323] presented the first results of assimilation of low-altitude ionospheric magnetic perturbations into the LFM global magnetospheric model coupled with the Rice Convection Model (RCM). The synthetic observations were produced from the LFM-RCM model with one set of parameters, while a model version with different parameters assimilated the ionospheric observations and calculated the magnetospheric pressure corrections for reproducing the synthetic observations. The results showed that the model with assimilated synthetic data responded correctly and expectedly by modifying ionospheric currents and magnetic perturbations. Merkin et al. [323] employed the LFM-RCM model with different parameters to assimilate the ionospheric observations and calculate the magnetospheric pressure corrections to reproduce the synthetic observations. By analysing the results from the LFM model, Lyons [345] suggested that localized flow channels can propagate tailward across the lobe and drive localized magnetotail reconnection, that the cross-tail width of reconnection and resulting plasma sheet flow channels and dipolarization fronts are related to the width of inflow from the lobe, and that IMF discontinuities drive lobe flow channels. Zhang et al. [618] investigated the effects of magnetospheric mass loading on the control of dayside magnetic reconnection, and found from global MHD simulations that if the magnetospheric mass density is large enough, there is a transition from "global control"(in which the integrated reconnection rate is insensitive to ionospheric mass) to "local control" (in which the integrated reconnection rate decreases with magnetospheric mass density). Under idealized solar wind conditions, simulations were performed [582] to reproduce the observed qualitative features of bursty bulk flows. The simulation was done with a high resolution grid of $(212, 192, 256)$ in radial, polar and azimuthal directions respectively. The LFM-Magnetosphere-Ionosphere Coupler/Solver (MIX) version of the code that treats the inner boundary at 2 R_E as a spherical surface with known conductivity is employed [320]. Wiltberger et al. [582] reproduced the observed qualitative features of the flow and the magnetic field pattern of bursty bulk flows. Their numerical results show earthward propagating bursty bulk flows emerging from the magnetotail after about 5:00 h of simulation time, which proliferate later. Further, by tracing electron trajectories using MHD field output from the LFM code, Eshetu et al. [86] investigated electron injection and energization by bursty bulk flows to find that bursty bulk flows can inject energetic electrons of few to 100 keV from the magnetotatail beyond -24 R_E to inward of geosynchronous, while accelerating them in the process.

1.5.7 Open GGCM Model

The Open GGCM is a global coupled model of Earth's magnetosphere, ionosphere, and thermosphere [417, 422]. The magnetosphere part solves the MHD equations as an initial boundary value problem. The MHD calculation only extends to 3 R_E from Earth. At that inner boundary the MHD part of the model is coupled with the ionosphere, mainly by the closure of field-aligned currents (FACs) in the ionosphere. The region within 3 R_E is treated as a magnetosphere-ionosphere coupling region where physical processes that couple the magnetosphere to the ionosphere-thermosphere system are parameterized using simple models and relations. The ionosphere-thermosphere system is modeled using the NOAA Coupled Thermosphere Ionosphere Model (CTIM) [420, 421]. CTIM computes self-consistently the ionospheric Pedersen and Hall conductances, which are then used to solve the ionospheric potential equation (e.g., [419]). The typical computational domain is partitioned in the geocentric solar ecliptic coordinate system. The grid extends to $\sim$20 R_E in the sunward direction, several hundreds of R_E in the anti-sunward direction, and $\sim$20 R_E in the transverse directions. The grid resolution is 0.1–0.2 R_E at the sub-solar magnetopause and 0.2–0.3 R_E in the near-Earth tail [422].

In the outer magnetosphere, the MHD equations are formulated as semi-conservative form just as done by Lyon et al. [295], and the solution is accomplished using an explicit second-order predictor-corrector finite difference time stepping scheme. The spatial derivatives are also computed using finite differences. However, for simulating super-magnetosonic flows and shocks, a hybrid scheme originally proposed by Harten and Zwas [157] is used, where a fourth-order scheme with a minimal diffusion error [615] with the diffusive first-order Rusanov scheme is employed. The numerical switch ensures that a high-order solution is obtained in regions of smooth variation of the flow, i.e., where there are no discontinuities, while the solution degrades to low order at discontinuities, such as shocks and contacts, where the high-order scheme would fail due to numerical dispersion.

The Open GGCM uses the Constrained Transport (CT) method introduced by Yee [611] and later by Evans and Hawley [87], which employs a staggered grid that allows near perfect (to roundoff error) preservation of $\nabla \cdot \mathbf{B}$. The OpenGGCM employs a stretched Cartesian grid, ranging from equidistant Cartesian grids to structured adaptive mesh refinement grids (see Raeder [417], for an overview and discussion of grids). As usual, the solar wind and IMF conditions are the outer boundary conditions on the dayside, while those on the other five outer boundaries are free. The inner boundary conditions are derived from the ionospheric model. The coupling between the magnetosphere and the ionosphere is an essential part of the model because the ionosphere controls, to a large extent, magnetospheric convection by providing the resistive closure of the FACs that are generated from the interaction of the solar wind with the magnetosphere [418].

The Open GGCM MHD model has been used to simulate the space weather events such as substorms and storms [417, 420, 422], the cold dense plasma sheet (CDPS) event [269], cusp-crossing events [51], and auroral signatures near substorm onset [424]. Using solar wind data from WIND satellite observation for the substorm event as an input at dayside, Hu et al. [177], Raeder et al. [423], and Zhu et al. [631] reconstructed a sequence of global magnetospheric configurations around the expansion onset by means of OpenGGCM simulation, and then analyzed the plasma instability prior to the initial expansion onset of the 23 March 2007 THEMIS substorm event. Ge et al. [124, 125] studied the interaction of dipolarization fronts within multiple bursty bulk flows of a substorm on 27 February 2009, and dipolarization fronts and associated auroral activities by conjugate observations and perspectives from global MHD simulations of OpenGGCM. In obtaining the Kp and Dst predictions of the high-speed CME on 23 July 2012, Bala et al. [17] used the Open GGCM to understand the global magnetospheric process in time sequence.

1.5.8 PPMLR-MHD Model

Casting MHD equations in the solar wind-magnetosphere region into a conservation form in Lagrangian coordinates, Hu et al. [178] established a global MHD simulation model named the PPMLR-MHD model by employing an extension of the Lagrangian version of the piecewise parabolic method (PPMLR) developed by Colella andWoodward [50]. The PPMLR-MHD algorithm has a formal accuracy of the third order in space and the second order in time. The simulation domain is taken to be $-300.0\ R_{\mathrm{E}} \leq x \leq 30.0\ R_{\mathrm{E}}$, $-150.0\ R_{\mathrm{E}} \leq y, z \leq 150.0\ R_{\mathrm{E}}$ in GSE coordinate. As most of other models, the inner boundary is set to be at 3.0 $\mathrm{R_E}$ for the magnetosphere. A uniform fine mesh is laid out in the near Earth domain of $-10.0\ R_{\mathrm{E}} \leq x, y, z \leq 10.0\ R_{\mathrm{E}}$, and the grid spacing outside this domain increases according to a geometrical series of common ratio 1.05 along each axis.

The PPMLR-MHD model has been used to study the magnetospheric response to the solar wind dynamic pressure pulses [556], the ionospheric response to the interplanetary shocks [149], the global features of Kelvin-Helmholtz waves at the magnetopause [270], the magnetosphere under the radial IMF [509], and the effects of the northward turning of the IMF to the magnetosphere [383]. Peng et al. [384] investigated three auroral brightening events caused by dynamic pressure enhancement under different IMF conditions. The simulation results suggested that the enhancement of dynamic pressure can increase the ionospheric transpolar potential and the corresponding field-aligned currents, leading to the observed auroral brightening. Particularly, by using the global PPMLR-MHD model, Wang et al. [568] carried out a comparison study of equivalent current systems for the substorm event of 8 March 2008, and Wang et al. [557] investigated the impact of a super solar storm on the magnetosphere and the ionosphere based on the geomagnetic field observation during Carrington event in 1859. Based on 240 numerical test runs, Wang et al. [558] suggested that the energy coupling function between the solar wind and the magnetosphere is proportional to $N_{sw}^{0.24} V_{sw}^{1.47} B^{0.86}$ of density number N_{sw}, solar wind speed V_{sw} and magnetic field strength B. This complicated form of the coupling function can be reasonably explained by the contribution from the two types of energy flux (magnetic energy flux being proportional to $V_{sw} B^2$ and kinetic energy flux being proportional to $N_{sw} V_{sw}^3$). Wang et al. [558] also estimated that about 13% of the solar wind kinetic energy is transferred into the magnetosphere and about 35% of the input energy is dissipated in the ionosphere. Using the solar wind observation as input, Zhang et al. [619] employed the PPMLR-MHD model to simulate the geomagnetically induced currents and their associated sudden commencement event monitored at two low-latitude 500 kV substations of China during the large storm of 17 March 2015. Sun et al. [492] investigated the linkage between the main impulse phase variation of the magnetic field on the ground and the magnetospheric negative response to an interplanetary shock. Their study revealed the intrinsic physically related chain response of the magnetospheric and ground magnetic field to interplanetary shocks, and thus linkage between the magnetospheric sudden impulse (SI) and ground SI together. Sun et al. [493] studied the flux transfer events (FTEs) on the dayside magnetopause and found that the azimuthal lengths of FTEs can have large variability under constant solar wind conditions, as a result of the time-dependent magnetic reconnection process. The PPMLR-MHD global simulation code was also used to study the X-ray emissions in the magnetosheath, which helps with the proposal of the Solar wind-Magnetosphere-Ionosphere Link Explorer (SMILE) mission [491, 494]. SMILE mission is jointly proposed by the European Space Agency and the Chinese Academy of Sciences to achieve the objective of remotely detecting the large scale magnetopause boundary [31, 559].

1.5.9 HLLD Model

Miyoshi et al. [330] developed an MHD algorithm for global simulations of planetary magnetospheres based on an approximate nonlinear Riemann solver, the so-called HLLD approximate Riemann solver. Under the assumptions that the normal velocity and the background potential magnetic field in the Riemann fan are constant, an approximate nonlinear solution of the MHD Riemann problem, in which the contributions of the background potential magnetic field are subtracted and multi-species plasmas as well as general equation of state are included, can be algebraically obtained. Besides focusing on some theoretical aspects of the HLLD Riemann solver, they displayed an example of the solar wind-magnetosphere interaction using the first-order HLLD Riemann solver on a cubed-sphere grid system [436].

1.5.10 CESE-AMR Model

Wang et al. [561, 562] developed a physics-based global MHD model by using the space-time conservation element and solution element (CESE) method in general curvilinear coordinates on a six-component grid system with adaptive mesh refinement [100, 104]. The magnetosphere part solves the MHD equations as an initial-boundary-value problem in the region from 3 R_E to 156 R_E, while the region within 3 R_E is treated as a magnetosphere-ionosphere coupling region. The coupling between the magnetosphere and the ionosphere was established by mapping FACs from the magnetosphere to the ionosphere [418]. The ionosphere is treated as a two-dimensional spherical shell at 1.017 R_E. As usual, a potential equation for the ionosphere is then solved according to Raeder [417].

Wang et al. [561] employed the CESE AMR model to simulate the interaction between the solar wind with due southward interplanetary magnetic field conditions and Saturn's magnetosphere. The numerical results showed the effect of the rotation of Saturn and solar wind on the global Saturn's magnetosphere and the formation of the plasmoids in the magnetotail. Wang et al. [562] used the model to explore the Earth's magnetosphere response to the solar wind with due northward interplanetary magnetic field. The modeled results were consistent with those provided by the CCMC at Goddard Space Flight Center through their public runs on request system. Wang et al. [563] carried out a parametric study with different values of the IMF B_z and the upstream solar wind dynamic pressure D_p under northward and southward IMFs to investigate their influences on the Earth's magnetopause and bow shock. Their results showed that the increase of the southward IMF B_z could result in an earthward movement of the magnetopause, and the displacement could increase with the intensity of the IMF B_z. The increase of the northward IMF B_z could also bring the magnetopause to move earthward but with a small distance. The subsolar bow shock during southward IMF is much closer to the Earth than during northward IMF.

1.5.11 HLLC Model

By revisiting the previous derivation of the HLLC Riemann solver, Guo [148] presented an extended HLLC Riemann solver specifically designed for the MHD system with the magnetic field decomposed into a strong internal magnetic field and an external component. The magnetic field **B** is decomposed into the internal magnetic field $\mathbf{B}_0$ and the external field $\mathbf{B}_1$. Here $\mathbf{B}_0$ is modeled as a magnetic dipole field, with a strength of 3.12×10^{-5} T at the equator of the Earth. At the same time, the hyperbolic cleaning method [60] is used to clean the $\nabla \cdot \mathbf{B}$ during the simulation. The simulation box is set to be -300 $R_E \leq x \leq 30$ R_E, and -150 $R_E \leq y, z \leq 150$ R_E with the inner boundary being a sphere with a radius of $r = 3$ R_E. Under the case of neglecting the magnetosphere-ionosphere electrostatic coupling model, the numerical tests are carried out under the typical interplanetary conditions, with a due northward interplanetary magnetic field (IMF). Without loss of high numerical resolution, the numerical scheme recovers the previously developed HLLC Riemann solver for the MHD when the internal field is set to zero, and thus suitable for the MHD simulations no matter whether a strong internal magnetic field is included or not.

In the early stage, the MHD simulations of the solar wind interaction with magnetosphere and ionosphere were carried out theoretically by using constant solar wind and IMF as input. But in recent years, researchers have tried to model the responses of the magnetosphere to real-time observations by using the global MHD simulations with realistic observed parameters as input. In general, the MHD approach is believed to correctly describe the large-scale magnetospheric equilibria in the outer magnetosphere, such as magnetospheric size, magnetic field/flux, and pressure parameters, and reasonably well predicts the absolute values of some of the key magnetospheric parameters, as well as their variations controlled by the varying solar wind input. The MHD results for some other parameters, such as global convection, total field-aligned current, or magnetotail's ability of loading the magnetospheric flux after the turning of north-south IMF, may present different answers [142].

As pointed out by Gordeev et al. [142], systematic validations of the global MHD models against observations still continue to be a challenge, as well as comparative benchmarking of different models against each other. Honkonen et al. [172] have studied the performance of four MHD models (BATS-R-US, GUMICS-4, LFM, Open GGCM) for the Earth's magnetosphere by comparing model predictions to the magnetic field measurements of Cluster 1, Geotail, and Wind spacecraft during a multiple substorm event. Gordeev et al. [141, 142] suggested a novel approach for the validation and benchmarking of the global MHD component of the space weather models in magnetospheric modeling. For the first time Gordeev et al. [142] showed a systematic evaluation of four first-principle magnetospheric models (BATS-R-US, Open GGCM, LFM, and GUMICS) by quantitatively demonstrating validation results characterizing statistically the absolute values and variations of basic magnetospheric variables, such as the scale size of the magnetosphere, the intensity of global magnetospheric convection, and the amount of magnetic flux and plasma pressure in the magnetotail. By validating the Lyon–Fedder–Mobarry (LFM) model against a set of the AMPERE-derived FACs, Kleiber et al. [236] developed a validation framework for the global magnetospheric models of the near-Earth space environment with particular focus on FACs. Their results supported the scientific assessment that the higher resolution run is a better representation of the satellite-estimated process than the lower resolution version. According to the scores obtained by the global MHD models of BATS-R-US, LFM, GUMICS, and Open GGCM, Gordeev et al. [142] suggested that there be no best one to provide satisfactory scores in all parameters. One model can lead in describing some parameters but may fail in producing other ones.

As it is known, driving magnetospheric models with the outputs from solar wind models is a necessary part of the whole Sun-to-Earth space weather modeling system. However, small-scale solar wind "noise", resulting primarily from stochastic processes, is usually below present typical solar wind model resolution. Both large-scale solar wind structures and small-scale solar wind "noise" can affect the Earth's magnetosphere. By proposing statistical "downscaling" of solar wind model results prior to their use as input to a magnetospheric model, Owens et al. [367] showed that an ensemble of results from the LFM model by the simple downscaling scheme could add value to the magnetospheric forecast, both improving the best estimate and quantifying the uncertainty.

In addition, many different efforts have been made to incorporate a more sophisticated physics by adding particle physics to the large-scale response description constrained by the global MHD [11, 382, 517]. However, merging particle physics into the model does not always provide better agreement to reality [172, 415, 425]. Eastwood et al. [79] attempted to summarise the scientific foundations of magnetospheric space weather forecasting by reviewing the physics of the magnetosphere in the frame of space weather forecasting, focusing on recent results, current understanding, and an assessment of probable future developments. Ganushkina et al. [123] discussed the basic structure and dynamics of the primary electric current systems in the Earth's magnetosphere, and the implications of understanding magnetospheric current systems, including paths forward for new investigations with the robust set of observations being produced by the numerous scientific and commercial satellites orbiting Earth. The ground-based magnetometer index of Dst is a commonly used measure of near-Earth current systems, in particular the storm time inner magnetospheric current systems. The ability of a large-scale, physics-based model to reproduce, or even predict, this index is therefore a tangible measure of the overall validity of the code for space weather research and space weather operational usage. Liemohn et al. [271] presented experimental real-time simulations of SWMF at the CCMC by assessing the Dst output from continuous operational simulations.

Overall, further quantitative assessment and validation of the global MHD component of the prediction models against the reality will be still one of the critical issues requiring for a definite answer in the future study.

1.6 Concluding Remarks

We have given a brief retrospect on the 3D MHD modeling from the solar corona to 1 AU. It is now established that 3D MHD modeling has become a very important tool to study the quasi-steady structures and eruptive phenomena in the solar corona and interplanetary space, and that data-driven physics-based coronal and interplanetary models will provide more accurate and longer lead-time predictions on global, regional and local scales. Till now, 3D MHD modeling can reproduce the large-scale structures and some important details of the solar wind background at solar minima and decline phases. For solar wind at maxima, current models need further improvements to capture the transitions with reasonable accuracy. Furthermore, most numerical models can basically demarcate the variation trends of many solar wind parameters when simulating CMEs' propagations in the solar corona and interplanetary space, thus provide an invaluable tool to gain an insight into the connection between the solar corona and interplanetary space. Although hybrid models employ the empirical and/or kinematic models to describe the solar corona, the high stability, relatively low computational cost and few computational hours enable them to be attractive choices for operational numerical space weather prediction. Even more, empirical models could have more to contribute now than ever before. Besides producing quick, and independent forecasts, simplified empirical models assist numerical model validation and verification, and append value to numerical model forecasts through parameterization, uncertainty estimation and downscaling of sub-grid processes. On the other hand, 3D solar-interplanetary MHD models can give the solar wind parameters and the entire evolution of CMEs in a self-consistent way, which enables them to reproduce both the coronal observations and the in-situ measurements in interplanetary space. However, more efforts need to be made for practical use. Physics, rather than statistics, must be at the base of the developing of space weather reliable forecasting tools. A better understanding of space weather processes will inevitably result in a greater understanding of model development. Feng et al. [101] have discussed the problems that 3D MHD simulations must face from the aspects of physics, numerical schemes, and computational requirements. To some extent, the breakthrough of the problems, such as better numerics, better determination of the boundary conditions, better representation of the underlying physical processes and speedup with lead time, will pave the way for the 3D physics-based MHD models to be put into operational use. Besides, motivated by the applications of space weather modeling, numerical algorithms based on the hybrid Particle-In-Cell (PIC) ion and fluid electron model have been developed (e.g., [37, 41, 183, 261, 262, 304, 362, 479, 584]) in simulating space weather processes in many space weather events: the conversion of magnetic energy to kinetic energy, and the interaction of the Earth magnetosphere with the solar wind.

Now space weather prediction is undergoing the transition from empirical-model-based objective analysis to physics-based numerical prediction (e.g., [114, 298, 335, 336, 412, 466, 480]). Attention should be paid to both the researches on numerical models and the operational ones so as to promote a coordinated development of the researches and applications of physics-based space weather modeling. In addition, an organic combination of observations, empirical models and MHD simulations will be an incentive to motivate rapid development of numerical modeling and the advent of physics-based numerical space weather predictions, which will help the physics-based models to track and predict the propagation of the solar storms from the Sun to Earth with acceptable accuracy, and thus benefit the systems of society, economics, and national securities.

References

1. Ahn BH (2011) South Korea's renewed focus on space weather. Space Weather 9:S10010
2. Andréeová K, Pulkkinen T, Laitinen TV, Přech L (2008) Shock propagation in the magnetosphere: observations and MHD simulations compared. J Geophys Res 113(A9):A09224
3. Antiochos SK, DeVore CR, Klimchuk JA (1999) A model for solar coronal mass ejections. Astrophys J 510:485–493
4. Arge CN, Pizzo VJ (2000) Improvement in the prediction of solar wind conditions using near-real time solar magnetic field updates. J Geophys Res 105:10,465–10,480
5. Arge CN, Odstrčil D, Pizzo VJ, Mayer LR (2003) Improved method for specifying solar wind speed near the Sun. In: Velli M, Bruno R, Malara F, Bucci B (eds) Solar Wind Ten, American institute of physics conference series, vol 679, pp 190–193 (2003)
6. Arge CN, Luhmann JG, Odstrčil D, Schrijver CJ, Li Y (2004) Stream structure and coronal sources of the solar wind during the May 12th, 1997 CME. J Atmos Solar Terr Phys 66:1295–1309
7. Arge CN, Henney CJ, Koller J, Compeau CR, Young S, MacKenzie D, Fay A, Harvey JW (2010) Air force data assimilative photospheric flux transport (ADAPT) model. Twelfth Int Solar Wind Conf 1216:343–346
8. Arge CN, Henney CJ, Koller J, Toussaint WA, Harvey JW, Young S (2011) Improving data drivers for coronal and solar wind models. In: Pogorelov NV, Audit E, Zank GP (eds) 5th international conference of numerical modeling of space plasma flows (ASTRONUM 2010), Astronomical society of the Pacific conference series, vol 444, p 99
9. Arge CN, Henney CJ, Hernandez IG, Toussaint WA, Koller J, Godinez HC (2013) Modeling the corona and solar wind using ADAPT maps that include far-side observations. Thirteen Int Solar Wind Conf 1539:11–14
10. Aschwanden MJ (2004) Physics of the solar corona. an introduction. Praxis Publishing Ltd, Chichester, UK
11. Ashour-Abdalla M, Bosqued JM, El-Alaoui M, Peroomian V, Walker R (2010) Observations and simulations of a highly structured plasma sheet during northward IMF. J Geophys Res: Space Phys 115(A10):A10227. https://doi.org/10.1029/2009JA015135
12. Axford WI (1972) The interaction of the solar wind with the interstellar medium. NASA Spec Publ 308:609
13. Bain HM, Mays ML, Luhmann JG, Li Y, Jian LK, Odstrčil D (2016) Shock connectivity in the August 2010 and July 2012 solar energetic particle events inferred from observations and ENLIL modeling. Astrophys J 825:1. https://doi.org/10.3847/0004-637X/825/1/1
14. Baker DN (2002) How to cope with space weather. Science 297(5586):1486–1487
15. Baker DN, Balstad R, Mand BEJ, Cameron FJF, Fisher GM, Forbes KF, Kintner PM, Leffler LG (2008) Severe space weather events-understanding societal and economic impacts: a workshop report. The National Academies Press, Washington
16. Baker DN, Poh G, Odstrčil D, Arge CN, Benna M, Johnson CL, Korth H, Gershman DJ, Ho GC, McClintock WE, Cassidy TA, Merkel A, Raines JM, Schriver D, Slavin JA, Solomon SC, Trávníček PM, Winslow RM, Zurbuchen TH (2013) Solar wind forcing at Mercury:

WSA-ENLIL model results. J Geophys Res (Space Phys) 118:45–57. https://doi.org/10.1029/2012JA018064
17. Bala R, Reiff P, Russell CT (2015) Testing the estimated hypothetical response of a major CME impact on Earth and its implications to space weather. J Geophys Res: Space Phys 120(5):3432–3443. https://doi.org/10.1002/2014JA020739
18. Barbieri LP, Mahmot RE (2004) October-November 2003's space weather and operations lessons learned. Space Weather 2(9):S09002
19. Barnes A (1992) Theory of magnetohydrodynamic waves - The WKB approximation revisited. J Geophys Res 97(12):105
20. Bauer P, Thorpe A, Brunet G (2015) The quiet revolution of numerical weather prediction. Narure 525:47–55. https://doi.org/10.1038/nature14956
21. Bemporad A, Zuccarello FP, Jacobs C, Mierla M, Poedts S (2012) Study of multiple coronal mass ejections at solar minimum conditions. Solar Phys 281:223–236
22. Bhattarai SK, Lopez RE (2013) Reduction of viscous potential for northward interplanetary magnetic field as seen in the LFM simulation. J Geophys Res (Space Phys) 118:3314–3322. https://doi.org/10.1002/jgra.50368
23. Bodenheimer P, Laughlin GP, Rózyczka M, Yorke HW (2007) Numerical methods in astrophysics: an introduction. Taylor and Francis, Boca Raton
24. Boris JP (1970) A physically motivated solution of the Alfvén problem. Technical Report 2167, NRL Memorandum Report, Naval Research Laboratory, Washington DC
25. Borovikov D, Sokolov IV, Roussev II, Taktakishvili A, Gombosi TI (2018) Toward a quantitative model for simulation and forecast of solar energetic particle production during gradual events. I. Magnetohydrodynamic background coupled to the SEP model. Astrophys J 864:88. https://doi.org/10.3847/1538-4357/aad68d
26. Borovikov SN, Pogorelov NV (2014) Voyager 1 near the Heliopause. Astrophys J Lett 783:L16. https://doi.org/10.1088/2041-8205/783/1/L16
27. Borovikov SN, Kryukov IA, Pogorelov NV (2009) Adaptive mesh refinement on curvilinear grids. In: Pogorelov NV, Audit E, Colella P, Zank GP (eds) Numerical modeling of space plasma flows: ASTRONUM-2008, Astronomical society of the Pacific conference series, vol 406, p 127
28. Borovikov SN, Heerikhuisen J, Pogorelov NV (2013) Hybrid parallelization of adaptive MHD-kinetic module in Multi-Scale Fluid-Kinetic Simulation Suite. In: Pogorelov NV, Audit E, Zank GP (eds) Numerical modeling of space plasma flows (ASTRONUM2012), Astronomical society of the Pacific conference series, vol 474, p 219
29. Bothmer V, Schwenn R (1998) The structure and origin of magnetic clouds in the solar wind. Ann Geophys 16:1–24. https://doi.org/10.1007/s00585-997-0001-x
30. Brackbill JU, Barnes DC (1980) The effect of nonzero $\nabla \cdot \mathbf{B}$ on the numerical solution of the magnetohydrodynamic equations. J Comput Phys 35(3):426–430
31. Branduardi-Raymont G, Wang C (2016) Smile Team: SMILE (Solar wind Magnetosphere Ionosphere Link Explorer): X-ray imaging of the Sun-Earth connection. In: XMM-Newton: the next decade, p 81
32. Brio M, Wu C (1988) An upwind differencing scheme for the equations of ideal magnetohydrodynamics. J Comput Phys 75(2):400–422
33. Brown D, Henshaw W, Quinlan D (1997) Overture: an object-oriented framework for solving partial differential equations. In: Ishikawa Y, Oldehoeft R, Reynders J, Tholburn M (eds) Scientific computing in object-oriented parallel environments. Lecture notes in computer science, vol 1343. Berlin, Heidelberg, Germany, pp 177–184
34. Bryan GL, Norman ML, O'Shea BW, Abel T, Wise JH, Turk MJ, Reynolds DR, Collins DC, Wang P, Skillman SW, Smith B, Harkness RP, Bordner J, Kim Jh, Kuhlen M, Xu H, Goldbaum N, Hummels C, Kritsuk AG, Tasker E, Skory S, Simpson CM, Hahn O, Oishi JS, So GC, Zhao F, Cen R, Li Y (2014) ENZO: an adaptive mesh refinement code for astrophysics. Astrophys J Suppl Ser 211:19. https://doi.org/10.1088/0067-0049/211/2/19
35. Bucciantini N, Del Zanna L (2011) General relativistic magnetohydrodynamics in axisymmetric dynamical spacetimes: the X-ECHO code. Astron Astrophys 528:A101. https://doi.org/10.1051/0004-6361/201015945

36. Burlaga LF (1995) Interplanetary magnetohydrodynamics. Oxford University Press, Oxford
37. Cannon P, Angling M, Barclay L, Curry C, Dyer C, Edwards R, Greene G, Hapgood M, Horne R, Jackson D et al (2002) The hybrid multiscale simulation technology: an introduction with application to astrophysical and laboratory plasmas scientific computation. Springer, Berlin
38. Cannon P, Angling M, Barclay L, Curry C, Dyer C, Edwards R, Greene G, Hapgood M, Horne R, Jackson D et al (2013) Extreme space weather: impacts on engineered systems and infrastructure. Royal Academy of Engineering
39. Cargill PJ, Lk H (2007) Coronal mass ejection. In: Kamide Y, Chian A (eds) Handbook of the solar-terrestrial environment. Springer, Berlin, pp 117–132
40. Cassak PA, Emslie AG, Halford AJ, Baker DN, Spence HE, Avery SK, Fisk LA (2017) Space physics and policy for contemporary society. J Geophys Res: Space Phys 122(4):4430–4435. https://doi.org/10.1002/2017JA024219
41. Chac6n L, Chen G (2016) A curvilinear, fully implicit, conservative electromagnetic PIC algorithm in multiple dimensions. J Comput Phys 316:578–597. https://doi.org/10.1016/j.jcp.2016.03.070
42. Chané E, van der Holst B, Jacobs C, Poedts S, Kimpe D (2006) Inverse and normal coronal mass ejections: evolution up to 1 AU. Astron Astrophys 447:727–733
43. Chang SC (1995) The method of space-time conservation element and solution element-A new approach for solving the Navier-Stokes and Euler equations. J Comput Phys 119:295
44. Chang SC, Wang XY, To WM (2000) Application of the space-time conservation element and solution element method to one-dimensional convection-diffusion problems. J Comput Phys 165(1):189–215
45. Chen PF (2011) Coronal mass ejections: models and their observational basis. Living Rev Solar Phys 8:1
46. Chen Y, Li X, Song HQ, Shi QQ, Feng SW, Xia LD (2009) Intrinsic instability of coronal streamers. Astrophys J 691:1936–1942
47. Chen Y, Tóth G, Cassak P, Jia X, Gombosi TI, Slavin JA, Markidis S, Peng IB, Jordanova VK, Henderson MG (2017) Global three-dimensional simulation of Earth's dayside reconnection using a two-way coupled magnetohydrodynamics with embedded particle-in-cell model: initial results. J Geophys Res (Space Phys) 122(A11):10. https://doi.org/10.1002/2017JA024186
48. Cohen O, Sokolov IV, Roussev II, Arge CN, Manchester WB, Gombosi TI, Frazin RA, Park H, Butala MD, Kamalabadi F, Velli M (2007) A semiempirical magnetohydrodynamical model of the solar wind. Astrophys J Lett 654:L163–L166
49. Cohen O, Sokolov IV, Roussev II, Gombosi TI (2008) Validation of a synoptic solar wind model. J Geophys Res 113(A12):A03104
50. Colella P, Woodward PR (1984) The piecewise parabolic method (PPM) for gas-dynamical simulations. J Comput Phys 54(1):174–201
51. Connor H, Raeder J, Trattner K (2012) Dynamic modeling of cusp ion structures. J Geophys Res 117(A4):A04203. https://doi.org/10.1029/2011JA017203
52. Cranmer SR (2009) Coronal holes. Living Rev Solar Phys 6:3
53. Crooker NU (2000) Solar and heliospheric geoeffective disturbances. J Atmos Solar Terr Phys 62:1071–1085
54. Crowley G, Azeem I (2018) Chapter 23 - extreme ionospheric storms and their effects on gps systems. In: Buzulukova N (ed) Extreme events in geospace. Elsevier, pp 555–586. https://doi.org/10.1016/B978-0-12-812700-1.00023-6
55. Cunningham AJ, Frank A, Varnière P, Mitran S, Jones TW (2009) Simulating magnetohydrodynamical flow with constrained transport and adaptive mesh refinement: algorithms and tests of the AstroBEAR code. Astrophys J Suppl. Ser 182:519–542. https://doi.org/10.1088/0067-0049/182/2/519
56. Daglis I, Baker D, Kappenman J, Panasyuk M, Daly E (2004) Effects of space weather on technology infrastructure. Space Weather 2(S02):S02004. https://doi.org/10.1029/2003SW000044
57. Daldorff LKS, Tóth G, Gombosi TI, Lapenta G, Amaya J, Markidis S, Brackbill JU (2014) Two-way coupling of a global Hall magnetohydrodynamics model with a local implicit

particle-in-cell model. J Comput Phys 268:236–254. https://doi.org/10.1016/j.jcp.2014.03.009
58. D'Amicis R, Bruno R, Pallocchia G, Bavassano B, Telloni D, Carbone V, Balogh A (2010) Radio evolution of solar wind turbulence during Earth and Ulysses alignment of 2007 August. Astrophys J 717:472–480
59. Dedner A, Kemm F, Kröner D, Munz C, Schnitzer T, Wesenberg M (2002) Hyperbolic divergence cleaning for the MHD equations. J Comput Phys 175:645
60. Dedner A, Kemm F, Kröner D, Munz CD, Schnitzer T, Wesenberg M (2002) Hyperbolic divergence cleaning for the MHD equations. J Comput Phys 175(2):645–673
61. Del Zanna L, Zanotti O, Bucciantini N, Londrillo P (2007) ECHO: a Eulerian conservative high-order scheme for general relativistic magnetohydrodynamics and magnetodynamics. Astron Astrophys 473:11–30. https://doi.org/10.1051/0004-6361:20077093
62. Del Zanna L, Bugli M, Bucciantini N (2014) High-order schemes for non-ideal 3+1 GRMHD: a study of the kinematic dynamo process in accretion tori. In: Pogorelov NV, Audit E, Zank GP (eds) 8th International conference of numerical modeling of space plasma flows (ASTRONUM 2013). Astronomical society of the Pacific conference series, vol 488, p 217 (2014)
63. Detman T, Smith Z, Dryer M, Fry CD, Arge CN, Pizzo V (2006) A hybrid heliospheric modeling system: background solar wind. J Geophys Res (Space Phys) 111:A07102
64. Detman TR, Dryer M, Yeh T, Han SM, Wu ST (1991) A time-dependent, three-dimensional MHD numerical study of interplanetary magnetic draping around plasmoids in the solar wind. J Geophys Res 96:9531–9540
65. Detman TR, Intriligator DS, Dryer M, Sun W, Deehr CS, Intriligator J (2011) The influence of pickup protons, from interstellar neutral hydrogen, on the propagation of interplanetary shocks from the Halloween 2003 solar events to ACE and Ulysses: A 3-D MHD modeling study. J Geophys Res (Space Phys) 116:A03105
66. DeVore CR, Boris JP, Sheeley NR Jr (1984) The concentration of the large-scale solar magnetic field by a meridional surface flow. Solar Phys 92:1–14
67. Dewey RM, Baker DN, Anderson BJ, Benna M, Johnson CL, Korth H, Gershman DJ, Ho GC, McClintock WE, Odstrčil D, Philpott LC, Raines JM, Schriver D, Slavin JA, Solomon SC, Winslow RM, Zurbuchen TH (2015) Improving solar wind modeling at Mercury: incorporating transient solar phenomena into the WSA-ENLIL model with the Cone extension. J Geophys Res (Space Phys) 120:5667–5685. https://doi.org/10.1002/2015JA021194
68. Dewey RM, Baker DN, Mays ML, Brain DA, Jakosky BM, Halekas H, Connerney JE, Odstrčil D, Luhmann JG, Lee CO (2016) Continuous solar wind forcing knowledge: providing continuous conditions at Mars with the WSA+ENLIL+Cone model. J Geophys Res: Space Phys 121(7):6207–6222. https://doi.org/10.1002/2015JA021941
69. Di Fino L, Zaconte V, Stangalini M, Sparvoli R, Picozza P, Piazzesi R, Narici L, Larosa M, Del Moro D, Casolino M, Berrilli F, Scardigli S (2014) Solar particle event detected by ALTEA on board the international space station. The March 7th, 2012 X5.4 flare. J Space Weather Space Clim 4(27):A19
70. Dmitriev AV, Suvorova AV (2004) Geosynchronous magnetopause crossings on October 29 31, 2003. Cosmic Res 42:551–560
71. Doherty P, Coster AJ, Murtagh W (2004) Space weather effects of October-November 2003. GPS Solut 8(4):267–271
72. Downs C, Roussev II, van der Holst B, Lugaz N, Sokolov IV, Gombosi TI (2010) Toward a realistic thermodynamic magnetohydrodynamic model of the global solar corona. Astrophys J 712(2):1219
73. Downs C, Lionello R, Mikić Z, Linker JA, Velli M (2016) Closed-field coronal heating driven by wave turbulence. Astrophys J 832:180. https://doi.org/10.3847/0004-637X/832/2/180
74. Dreher J, Grauer R (2005) Racoon: a parallel mesh-adaptive framework for hyperbolic conservation laws. Parallel Comput 31(8):913–932. https://doi.org/10.1016/j.parco.2005.04.011
75. Dryer M, Wu CC, Smith ZK (1997) Three-dimensional MHD simulation of the April 14, 1994, interplanetary coronal mass ejection and its propagation to Earth and Ulysses. J Geophys Res 102:14,065–14,074

76. Dungey JW (1961) Interplanetary magnetic field and the auroral zones. Phys Rev Lett 6(2):47
77. Eastwood JP (2008) The science of space weather. Philos Trans R Soc Lond A: Math Phys Eng Sci 366(1884):4489–4500. https://doi.org/10.1098/rsta.2008.0161
78. Eastwood JP, Biffis E, Hapgood MA, Green L, Bisi MM, Bentley RD, Wicks R, McKinnell LA, Gibbs M, Burnett C (2017) The economic impact of space weather: where do we stand? Risk Anal 37(2):206–218. https://doi.org/10.1111/risa.12765
79. Eastwood JP, Nakamura R, Turc L, Mejnertsen L, Hesse M (2017) The scientific foundations of forecasting magnetospheric space weather. Space Sci Rev 212(3):1221–1252. https://doi.org/10.1007/s11214-017-0399-8
80. Ebihara Y, Tanaka T (2015) Substorm simulation: formation of westward traveling surge. J Geophys Res: Space Phys 120(12):10,466–10,484. https://doi.org/10.1002/2015JA021697
81. Ebihara Y, Tanaka T (2017) Energy flow exciting field-aligned current at substorm expansion onset. J Geophys Res: Space Phys 122(12):12,288–12,309. https://doi.org/10.1002/2017JA024294
82. Ebihara Y, Tanaka T, Kikuchi T (2014) Counter equatorial electrojet and overshielding after substorm onset: global MHD simulation study. J Geophys Res: Space Phys 119(9):7281–7296
83. Ebihara Y, Tanaka T, Kamiyoshikawa N (2019) New diagnosis for energy flow from solar wind to ionosphere during substorm: global MHD simulation. J Geophys Res: Space Phys 124(1):360–378. https://doi.org/10.1029/2018JA026177
84. Echim MM, Lemaire J, Lie-Svendsen Ø (2011) A review on solar wind modeling: kinetic and fluid aspects. Surv Geophys 32(1):1–70. https://doi.org/10.1007/s10712-010-9106-y
85. Einfeldt B, Roe PL, Munz CD, Sjogreen B (1991) On Godunov-type methods near low densities. J Comput Phys 92:273–295
86. Eshetu WW, Lyon JG, Hudson MK, Wiltberger MJ (2019) Simulations of electron energization and injection by BBFs using high resolution LFM MHD fields. J Geophys Res: Space Phys 124:1222–1238. https://doi.org/10.1029/2018JA025789
87. Evans CR, Hawley JF (1988) Simulation of magnetohydrodynamic flows - A constrained transport method. Astrophys J 332:659–677
88. Evans RM, Opher M, Oran R, van der Holst B, Sokolov IV, Frazin R, Gombosi TI, Vásquez A (2012) Coronal heating by surface Alfvén wave damping: implementation in a global magnetohydrodynamics model of the solar wind. Astrophys J 756:155
89. Fang F, Manchester W, Abbett WP, van der Holst B (2010) Simulation of flux emergence from the convection zone to the corona. Astrophys J 714:1649–1657
90. Fedder JA, Lyon JG (1995) The Earth's magnetosphere is 165 R_E long: self-consistent currents, convection, magnetospheric structure, and processes for northward interplanetary magnetic field. J Geophys Res 100:3623–3635
91. Fedder JA, Lyon JG, Slinker SP, Mobarry CM (1995a) Topological structure of the magnetotail as a function of interplanetary magnetic field direction. J Geophys Res 100:3613–3621
92. Fedder JA, Slinker SP, Lyon JG, Elphinstone RD (1995b) Global numerical simulation of the growth phase and the expansion onset for a substorm observed by Viking. J Geophys Res 100:19,083–19,094
93. Feldman WC, Barraclough BL, Gosling JT, McComas DJ, Riley P, Goldstein BE, Balogh A (1998) Ion energy equation for the high-speed solar wind: Ulysses observations. J Geophys Res 103:14,547–14,558
94. Feng XS, Xiang CQ (2013) Numerical study of interplanetary solar storms (in Chinese). Sci Sin-Terrae 43(6):912–933
95. Feng XS, Wei FS, Zhong DK (2003a) A Class of two-step TVD MacCormack type numerical scheme for MHD equations. Chin J Space Sci 23(6):401–402
96. Feng XS, Wu ST, Wei FS, Fan QL (2003b) A class of TVD type combined numerical scheme for MHD equations with a survey about numerical methods in solar wind simulations. Space Sci Rev 107:43–53
97. Feng XS, Xiang CQ, Zhong DK, Fan QL (2005) A comparative study on 3-D solar wind structure observed by Ulysses and MHD simulation. Chin Sci Bull 50(7):672–676

98. Feng XS, Zhou YF, Wu ST (2007) A novel numerical implementation for solar wind modeling by the modified conservation element/solution element method. Astrophys J 655:1110
99. Feng XS, Zhang Y, Yang LP, Wu ST, Dryer M (2009) An operational method for shock arrival time prediction by one-dimensional CESE-HD solar wind model. J Geophys Res 114(A13):A10103
100. Feng XS, Yang LP, Xiang CQ, Wu ST, Zhou YF, Zhong DK (2010) Three-dimensional solar wind modeling from the Sun to Earth by a SIP-CESE MHD model with a six-component grid. Astrophys J 723:300
101. Feng XS, Xiang CQ, Zhong DK (2011) The state-of-art of three-dimensional numerical study for corona-interplanetary process of solar storms (in chinese). Sci Sin-Terrae 41(6):1–28
102. Feng XS, Zhang SH, Xiang CQ, Yang LP, Jiang CW, Wu ST (2011) A hybrid solar wind model of the CESE+HLL method with a Yin-Yang overset grid and an AMR grid. Astrophys J 734:50
103. Feng XS, Jiang CW, Xiang CQ, Zhao XP, Wu ST (2012a) A data-driven model for the global coronal evolution. Astrophys J 758(1):62
104. Feng XS, Yang LP, Xiang CQ, Jiang CW, Ma X, Wu ST, Zhong DK, Zhou YF (2012b) Validation of the 3D AMR SIP-CESE solar wind model for four Carrington rotations. Solar Phys 279:207–229
105. Feng XS, Yang LP, Xiang CQ, Liu Y, Zhao XP, Wu ST (2012c) Numerical study of the global corona for CR 2055 driven by daily updated synoptic magnetic field. In: Pogorelov NV, Font JA, Audit E, Zank GP (eds) Numerical modeling of space plasma slows (ASTRONUM 2011), Astronomical society of the Pacific conference series, vol 459, p 202
106. Feng XS, Xiang CQ, Zhong DK (2013a) Numerical Study of Interplanetary solar storms (in Chinese). Sci Sinica Terrae 43(6):912–933
107. Feng XS, Zhong DK, Xiang CQ, Zhang Y (2013b) GPU-accelerated computing of three-dimensional solar wind background. Sci China Earth Sci 56(11):1864–1880
108. Feng XS, Xiang CQ, Zhong DK, Zhou YF, Yang LP, Ma XP (2014) SIP-CESE MHD model of solar wind with adaptive mesh refinement of hexahedral meshes. Comput Phys Commun 185(7):1965–1980
109. Feng XS, Ma XP, Xiang CQ (2015) Data-driven modeling of the solar wind from 1 R_s to 1 AU. J Geophys Res 120:10,159–10,174
110. Ferguson DC (2018) Chapter 15 - Extreme space weather spacecraft surface charging and arcing effects. In: Buzulukova N (ed) Extreme events in geospace. Elsevier, pp 401–418. https://doi.org/10.1016/B978-0-12-812700-1.00015-7
111. Fichtner H, Kopp A, Kleimann J, Grauer R (2008) On MHD modeling of coronal mass ejections. In: Pogorelov NV, Audit E, Zank GP (eds) Numerical modeling of space plasma flows, Astronomical society of the Pacific conference series, vol 385, p 151
112. Fineschi S, Antonucci E, Romoli M, Bemporad A, Capobianco G, Crescenzio G, Nicolini G, Massone G, Telloni D, Focardi M, Landini F, Pancrazzi M, Poletto L, Pelizzo MG, Da Deppo V, Moses JD, Andretta V, Naletto G, Nicolosi P, Spadaro D, Berlicki A, Uslenghi M, Malvezzi M, Teriaca L, Abbo L, Magli E (2013) Novel space coronagraphs: METIS, a flexible optical design for multi-wavelength imaging and spectroscopy. In: Solar Physics and Space Weather Instrumentation V, Proceedings of the SPIE, vol 8862, p 88620G, https://doi.org/10.1117/12.2028544
113. Fisk LA (1996) Motion of the footpoints of heliospheric magnetic field lines at the Sun: implications for recurrent energetic particle events at high heliographic latitudes. J Geophys Res 101:15,547–15,554
114. Folini D (2018) Climate, weather, space weather: model development in an operational context. J Space Weather Space Clim 8:A32. https://doi.org/10.1051/swsc/2018021
115. Fromang S, Hennebelle P, Teyssier R (2006) A high order Godunov scheme with constrained transport and adaptive mesh refinement for astrophysical magnetohydrodynamics. Astron. Astrophys. 457:371–384. https://doi.org/10.1051/0004-6361:20065371
116. Fry CD, Sun W, Deehr CS, Dryer M, Smith Z, Akasofu SI, Tokumaru M, Kojima M (2001) Improvements to the HAF solar wind model for space weather predictions. J Geophys Res 106:20,985–21,002, https://doi.org/10.1029/2000JA000220

117. Fry EK (2012) The risks and impacts of space weather: policy recommendations and initiatives. Space Policy 28(3):180–184. https://doi.org/10.1016/j.spacepol.2012.06.005
118. Fryxell B, Olson K, Ricker P, Timmes FX, Zingale M, Lamb DQ, MacNeice P, Rosner R, Truran JW, Tufo H (2000) Flash: an adaptive mesh hydrodynamics code for modeling astrophysical thermonuclear flashes. Astrophys J Suppl Ser 131(1):273. http://stacks.iop.org/0067-0049/131/i=1/a=273
119. Fuchs FG, McMurry AD, Mishra S, Waagan K (2011) Simulating waves in the upper solar atmosphere with SURYA: a well-balanced high-order finite-volume code. Astrophys J 732:75. https://doi.org/10.1088/0004-637X/732/2/75
120. Fujita S, Tanaka T, Kikuchi T, Fujimoto K, Hosokawa K, Itonaga M (2003) A numerical simulation of the geomagnetic sudden commencement: 1. Generation of the field-aligned current associated with the preliminary impulse. J Geophys Res 108(A12):1417. https://doi.org/10.1029/2002JA009407
121. Fujita S, Tanaka T, Kikuchi T, Tsunomura S (2004) A numerical simulation of a negative sudden impulse. Earth Planets Space 56(4):463–472
122. Fukazawa K, Aoyama T, Ogino T, Yumoto K (2010) Response of the reconnection electric field and polar cap potential to the IMF and velocity of solar wind. J Atmos Solar-Terr Phys 72(13):1019–1023
123. Ganushkina NY, Liemohn MW, Dubyagin S (2018) Current systems in the Earth's magnetosphere. Rev Geophys 56(2):309–332. https://doi.org/10.1002/2017RG000590
124. Ge YS, Raeder J, Angelopoulos V, Gilson ML, Runov A (2011) Interaction of dipolarization fronts within multiple bursty bulk flows in global MHD simulations of a substorm on 27 February 2009. J Geophys Res (Space Phys) 116:A00I23. https://doi.org/10.1029/2010JA015758
125. Ge YS, Zhou XZ, Liang J, Raeder J, Gilson ML, Donovan E, Angelopoulos V, Runov A (2012) Dipolarization fronts and associated auroral activities: 1. Conjugate observations and perspectives from global MHD simulations. J Geophys Res (Space Phys) 117(A16):A10226
126. Gibson SE, Low BC (1998) A time-dependent three-dimensional magnetohydrodynamic model of the coronal mass ejection. Astrophys J 493:460–473
127. Glocer A, Tóth G, Ma Y, Gombosi T, Zhang JC, Kistler L (2009) Multifluid block-adaptive-tree solar wind Roe-type upwind scheme: magnetospheric composition and dynamics during geomagnetic storms-Initial results. J Geophys Res 114(A12):203
128. Glocer A, Dorelli J, Toth G, Komar CM, Cassak PA (2016) Separator reconnection at the magnetopause for predominantly northward and southward IMF: techniques and results. J Geophys Res: Space Phys 121(1):140–156. https://doi.org/10.1002/2015JA021417
129. Godunov SK (1959) A difference method for numerical calculation of discontinuous solutions of the equations of hydrodynamics. Mat Sb 47(3):271–306
130. Gombosi TI, DeZeeuw DL, Groth C, Powell KG, Stout QF (2000) Multiscale MHD simulation of a coronal mass ejection and its interaction with the magnetosphere-ionosphere system. J Atmos Solar-Terr Phys 62(16):1515–1525
131. Gombosi TI, Tóth G, De Zeeuw DL, Hansen KC, Kabin K, Powell KG (2002) Semirelativistic magnetohydrodynamics and physics-based convergence acceleration. J Comput Phys 177(1):176–205
132. Gombosi TI, Powell KG, De Zeeuw DL, Clauer CR, Hansen KC, Manchester WB, Ridley AJ, Roussev II, Sokolov IV, Stout Quentin F, QF, Tóth G, (2004) Solution-adaptive magnetohydrodynamics for space plasmas: sun-to-earth simulations. Comput Sci Eng 6(2):14–35
133. Gombosi TI, van der Holst B, Manchester WB, Sokolov IV (2018) Extended MHD modeling of the steady solar corona and the solar wind. Living Rev Solar Phys 15(1):4. https://doi.org/10.1007/s41116-018-0014-4
134. González-Avilés JJ, Guzmán FS (2018) CAFE-Q: code designed to solve the resistive MHD equations with thermal conductivity. IEEE Trans Plasma Sci 46(7):2378–2385. https://doi.org/10.1109/TPS.2018.2796312
135. González-Avilés JJ, Cruz-Osorio A, Lora-Clavijo FD, Guzmán FS (2015) Newtonian CAFE: a new ideal MHD code to study the solar atmosphere. Mon Not R Astron Soc 454:1871–1885. https://doi.org/10.1093/mnras/stv2013

136. Goodrich C, Pulkkinen T, Lyon J, Merkin V (2007) Magnetospheric convection during intermediate driving: sawtooth events and steady convection intervals as seen in Lyon-Fedder-Mobarry global MHD simulations. J Geophys Res 112(A08):201
137. Goodrich CC, Sussman AL, Lyon JG, Shay MA, Cassak PA (2004) The CISM code coupling strategy. J Atmos Solar Terr Phys 66:1469–1479
138. Gopalswamy N, Hanaoka Y, Hudson HS (2000) Structure and dynamics of the corona surrounding an eruptive prominence. Adv Space Res 25:1851–1854
139. Gopalswamy N, Yashiro S, Liu Y, Michalek G, Vourlidas A, Kaiser ML, Howard RA (2005) Coronal mass ejections and other extreme characteristics of the 2003 October-November solar eruptions. J Geophys Res 110:A09S15
140. Gopalswamy N, Yashiro S, Michalek G, Xie H, Lepping RP, Howard RA (2005) Solar source of the largest geomagnetic storm of cycle 23. Geophys Res Lett 32:L12S09
141. Gordeev E, Facskó G, Sergeev V, Honkonen I, Palmroth M, Janhunen P, Milan S (2013) Verification of the GUMICS-4 global MHD code using empirical relationships. J Geophys Res: Space Phys 118(6):3138–3146. https://doi.org/10.1002/jgra.50359
142. Gordeev E, Sergeev V, Honkonen I, Kuznetsova M, Rastätter L, Palmroth M, Janhunen P, Tóth G, Lyon J, Wiltberger M (2015) Assessing the performance of community-available global MHD models using key system parameters and empirical relationships. Space Weather 13:868–884. https://doi.org/10.1002/2015SW001307
143. Gosling JT (1993) The solar flare myth. J Geophys Res 98(A11):18,937–18,949
144. Gosling JT, Riley P, McComas DJ, Pizzo VJ (1998) Overexpanding coronal mass ejections at high heliographic latitudes - Observations and simulations. J Geophys Res 103:1941
145. Gressl C, Veronig AM, Temmer M, Odstrčil D, Linker JA, Mikić Z, Riley P (2014) Comparative study of MHD modeling of the background solar wind. Solar Phys 289:1783–1801
146. Gritzner CF (2010) Changing Climates (Global Connections). Chelsea House Pub (L), New York, USA
147. Groth CPT, De Zeeuw DL, Gombosi TI, Powell KG (2000) Global three-dimensional MHD simulation of a space weather event: CME formation, interplanetary propagation, and interaction with the magnetosphere. J Geophys Res 105:25,053–25,078
148. Guo XC (2015) An extended HLLC Riemann solver for the magneto-hydrodynamics including strong internal magnetic field. J Comput Phys 290:352–363
149. Guo XC, Hu YQ (2007) Response of Earth's ionosphere to interplanetary shocks. Chin J Geophys 50(4):817–823
150. Guo Y, Xia C, Keppens R, Ding MD, Chen PF (2019) Solar magnetic flux rope eruption simulated by a data-driven magnetohydrodynamic model. Astrophys J 870(2):L21. https://doi.org/10.3847/2041-8213/aafabf
151. Hain KH (1987) The partial donor cell method. J Comput Phys 73(1):131–147. https://doi.org/10.1016/0021-9991(87)90110-0
152. Haisch B, Strong KT, Rodono M (1991) Flares on the Sun and other Stars. Ann Rev Astron Astrophys 29:275–324
153. Han SM, Wu ST (1988) A three-dimensional, time-dependent numerical modeling of supersonic, super-Alfvén MHD flow. Comput Fluids 16(1):81–103
154. Hapgood M (2018) Chapter 1 - Linking space weather science to impacts-the view from the Earth. In: Buzulukova N (ed) Extreme events in geospace. Elsevier, pp 3–34. https://doi.org/10.1016/B978-0-12-812700-1.00001-7
155. Hapgood M (2018) Chapter 1 - Societal and economic importance of space weather. In: Camporeale E, Wing S, Johnson JR (eds) Machine learning techniques for space weather. Elsevier, pp 3–26. https://doi.org/10.1016/B978-0-12-811788-0.00001-9
156. Harten A (1983) On the symmetric form of systems of conservation laws with entropy. J Comput Phys 49(1):151–164
157. Harten A, Zwas G (1972) Self-adjusting hybrid schemes for shock computations. J Comput Phys 9(3):568–583
158. Harten A, Lax P, Leer B (1983) On upstream differencing and Godunov-type schemes for hyperbolic conservation laws. SIAM Rev 25(1):35–61

159. Hayashi K (2005) Magnetohydrodynamic simulations of the solar corona and solar wind using a boundary treatment to limit solar wind mass flux. Astron Astrophys Suppl 161(2):480
160. Hayashi K (2012) An MHD simulation model of time-dependent co-rotating solar wind. J Geophys Res (Space Phys) 117(A8):A08,105
161. Hayashi K (2013) An MHD simulation model of time-dependent global solar corona with temporally varying solar-surface maps. J Geophys Res (Space Phys) 118(A12):6889–6906
162. Hayashi K (2016) MHD-IPS analysis of relationship among solar wind density, temperature, and flow speed. J Geophys Res (Space Phys) 121(A7)121:7367–7384. https://doi.org/10.1002/2016JA022750
163. Hayashi K, Kojima M, Tokumaru M, Fujiki K (2003) MHD tomography using interplanetary scintillation measurement. J Geophys Res (Space Phys) 108(A3):1102. https://doi.org/10.1029/2002JA009567
164. Hayashi K, Benevolenskaya E, Hoeksema T, Liu Y, Zhao XP (2006a) Three-dimensional magnetohydrodynamic simulation of a global solar corona using a temperature distribution map obtained from SOHO EIT measurements. Astrophys J Lett 636(2):L165
165. Hayashi K, Zhao XP, Liu Y (2006) MHD simulation of two successive interplanetary disturbances driven by cone-model parameters in IPS-based solar wind. Geophys Res Lett 33(20):L20103. https://doi.org/10.1029/2006GL027408
166. Hayashi K, Zhao XP, Liu Y (2008) MHD simulations of the global solar corona around the Halloween event in 2003 using the synchronic frame format of the solar photospheric magnetic field. J Geophys Res 113:A07104
167. Hayashi K, Hoeksema JT, Liu Y, Bobra M, XD S, Norton A, (2015) The helioseismic and magnetic imager (HMI) vector magnetic field pipeline: magnetohydrodynamics simulation module for the global solar corona. Solar Phys. 290(5):1507–1529
168. Heaton MJ, Kleiber W, Sain SR, Wiltberger M (2015) Emulating and calibrating the multiple-fidelity Lyon-Fedder-Mobarry magnetosphere-ionosphere coupled computer model. J R Stat Soc: Ser C (Appl Stat) 64(1):93–113
169. Henshaw WD, Schwendeman DW (2003) An adaptive numerical scheme for high-speed reactive flow on overlapping grids. J Comput Phys 191(2):420–447
170. Hoilijoki S, Souza VM, Walsh BM, Janhunen P, Palmroth M (2014) Magnetopause reconnection and energy conversion as influenced by the dipole tilt and the IMF Bx. J Geophys Res: Space Phys 119(6):4484–4494
171. Hollweg JV (1986) Transition region, corona, and solar wind in coronal holes. J Geophys Re. 91:4111–4125
172. Honkonen I, Rastätter L, Grocott A, Pulkkinen A, Palmroth M, Raeder J, Ridley A, Wiltberger M (2013) On the performance of global magnetohydrodynamic models in the Earth's magnetosphere. Space Weather 11(5):313–326
173. Hopkins PF (2016) A constrained-gradient method to control divergence errors in numerical MHD. Mon Not R Astron Soc 462:576–587. https://doi.org/10.1093/mnras/stw1578
174. Hopkins PF, Raives MJ (2015) Accurate, meshless methods for magnetohydrodynamics. Mont Not R Astron Soc 455:51–88. https://doi.org/10.1093/mnras/stv2180
175. Horne RB, Glauert SA, Meredith NP, Boscher D, Maget V, Heynderickx D, Pitchford D (2013) Space weather impacts on satellites and forecasting the Earth's electron radiation belts with SPACECAST. Space Weather 11:169–186
176. Howard T (2011) Coronal mass ejections: an introduction. Springer, New York
177. Hu B, Wolf RA, Toffoletto FR, Yang J, Raeder J (2011) Consequences of violation of frozen in-flux: evidence from OpenGGCM simulations. J Geophys Res (Space Phys) 116:A06223. https://doi.org/10.1029/2011JA016667
178. Hu YQ, Guo XC, Li GQ, Wang C, Huang ZH (2005) Oscillation of quasi-steady Earth's magnetosphere. Chin Phys Lett 22(10):2723–2726. https://doi.org/10.1088/0256-307x/22/10/073
179. Hu YQ, Feng XS, Wu ST, Song WB (2008) Three-dimensional MHD modeling of the global corona throughout solar cycle 23. J Geophys Res 113(A12):A03106

180. Huang CL, Spence HE, Lyon JG, Toffoletto FR, Singer HJ, Sazykin S (2006) Storm-time configuration of the inner magnetosphere: Lyon-Fedder-Mobarry MHD code, Tsyganenko model, and GOES observations. J Geophys Res 111(A11): A11S16. https://doi.org/10.1029/2006JA011626
181. Huang ZG, Tóth G, I GT, Jia XZ, Rubin M, Combi MR, Bieler KC A Hansen, Shou YS, Altwegg K, (2016) Four-fluid MHD simulations of the plasma and neutral gas enviornment of Comet 67P/Churyumpv-Gerasimenko near perihelion. J Geophys Res (Space Phys) 121:4247–4268
182. Ilie R, Liemohn MW, Kozyra J, Borovsky J (2010) An investigation of the magnetosphere-ionosphere response to real and idealized co-rotating interaction region events through global magnetohydrodynamic simulations. Proc R Soc Lond A: Math Phys Eng Sci 466(2123):3279–3303. https://doi.org/10.1098/rspa.2010.0074
183. Innocenti ME, Johnson A, Markidis S, Amaya J, Deca J, Olshevsky V, Lapenta G (2017) Progress towards physics-based space weather forecasting with exascale computing. Adv Eng Softw 111:3–17. https://doi.org/10.1016/j.advengsoft.2016.06.011
184. Intriligator DS, Detman T, Gloecker G, Gloeckler C, Dryer M, Sun W, Intriligator J, Deehr C (2012) Pickup protons: comparisons using the three-dimensional MHD HHMS-PI model and Ulysses SWICS measurements. J Geophys Res (Space Phys) 117(A6):A06104. https://doi.org/10.1029/2011JA017424
185. Intriligator DS, Detman T, Intriligator J, Dryer M, Sun W, Deehr C, Webber WR, Decker RB, McPherron RL (2012) Space weather throughout the heliosphere. In: Hu Q, Li G, Zank GP, Ao X, Verkhoglyadova O, Adams JH (eds) American institute of physics conference series, vol 1500, pp 100–108
186. Intriligator DS, Sun W, Detman T, Miller WD, Intriligator J, Dryer M, Webber W, Deehr C, Gloeckler G (2016) The Sun's dynamic influence on the outer heliosphere, the heliosheath, and the local interstellar medium. In: Journal of physics conference series, vol 767, p 012013. https://doi.org/10.1088/1742-6596/767/1/012013
187. Jackson BV, Hick PP, Kojima M, Yokobe A (1998) Heliospheric tomography using interplanetary scintillation observations: 1. Combined Nagoya and Cambridge data. J Geophys Res: Space Phys 103(A6):12,049–12,067. https://doi.org/10.1029/97JA02528
188. Jackson BV, Hick PP, Buffington A, Kojima M, Tokumaru M, Fujiki K, Ohmi T, Yamashita M (2003) Time-dependent tomography of hemispheric features using interplanetary scintillation (IPS) remote-sensing observations. AIP conference proceedings 679(1):75–78. https://doi.org/10.1063/1.1618545
189. Jacobs C, Poedts S (2012) A numerical study of the response of the coronal magnetic field to flux emergence. Solar Phys 280:389–405
190. Jacobs C, Poedts S, van der Holst B (2006) 3D evolution of a "density-driven" CME event. In: SOHO-17. 10 years of SOHO and beyond, ESA special publication, vol 617
191. Jacobs C, van der Holst B, Poedts S (2007) Comparison between 2.5D and 3D simulations of coronal mass ejections. Astron Astrophys 470:359–365
192. Jacques SA (1977) Momentum and energy transport by waves in the solar atmosphere and solar wind. Astrophys J 215:942–951
193. Janhunen P, Koskinen KEJ, Pulkinen TI (1996) A new global ionosphere-magnetosphere coupling simulation utilizing locally varying time step. In: International conference on substorms, vol 389, p 205
194. Janhunen P, Palmroth M, Laitinen T, Honkonen I, Juusola L, Facsko G, Pulkkinen TI (2012) The GUMICS-4 global MHD magnetosphere-ionosphere coupling simulation. J Atmos Solar-Terr Phys 80:48–59
195. Jia XZ, Kivelson MG, Gombosi TI (2016) Global MHD modeling of the coupled magnetosphere-ionosphere system at Saturn, American geophysical union (AGU), chap 25, pp 319–334. https://doi.org/10.1002/9781119066880.ch25
196. Jiang CW, Feng XS (2012) A New implementation of the magnetohydrodynamics-relaxation method for nonlinear force-free field extrapolation in the solar corona. Astrophys J 749(2):135

197. Jiang CW, Feng XS (2013) Extrapolation of the solar coronal magnetic field from SDO/HMI magnetogram by a CESE-MHD-NLFFF code. Astrophys J 769:144
198. Jiang CW, Feng XS (2014) Preprocessing the Photospheric Vector Magnetograms for an NLFFF Extrapolation Using a Potential-Field Model and an Optimization Method. Solar Phys. 289:63–77. https://doi.org/10.1007/s11207-013-0346-0
199. Jiang CW, Feng XS, Zhang J, Zhong DK (2010) AMR simulations of magnetohydrodynamic problems by the CESE method in curvilinear coordinates. Solar Phys 267(2):463–491. https://doi.org/10.1007/s11207-010-9649-6
200. Jiang CW, Feng XS, Fan YL, Xiang CQ (2011) Reconstruction of the coronal magnetic field using the CESE-MHD method. Astrophys J 727:101
201. Jiang CW, Feng XS, Wu ST, Hu Q (2012a) Study of the three-dimensional coronal magnetic field of active region 11117 around the time of a confined flare using a data-driven CESE-MHD model. Astrophys J 759:85
202. Jiang CW, Feng XS, Xiang CQ (2012b) A new code for nonlinear force-free field extrapolation of the global corona. Astrophys J 755(1):62
203. Jiang CW, Feng XS, Wu ST, Hu Q (2013) Magnetohydrodynamic simulation of a sigmoid eruption of active region 11283. Astrophys J Lett 771:L30. https://doi.org/10.1088/2041-8205/771/2/L30
204. Jiang CW, Wu ST, Feng XS, Hu Q (2014) Formation and eruption of an active region sigmoid. I. A study by nonlinear force-free field modeling. Astrophys J 780:55
205. Jiang CW, Wu ST, Feng XS, Hu Q (2014) Nonlinear force-free field extrapolation of a coronal magnetic flux rope supporting a large-scale solar filament from a photospheric vector magnetogram. Astrophys J Lett 786:L16. https://doi.org/10.1088/2041-8205/786/2/L16
206. Jiang CW, Wu ST, Feng XS, Hu Q (2016) Data-driven magnetohydrodynamic modelling of a flux-emerging active region leading to solar eruption. Nat Commun 7:11522. https://doi.org/10.1038/ncomms11522
207. Jiang CW, Yan XL, Feng XS, Duan AY, Hu Q, Zuo PB, Wang Y (2017) Reconstruction of a large-scale pre-flare coronal current sheet associated with a homologous X-shaped flare. Astroph J 850:8. https://doi.org/10.3847/1538-4357/aa917a
208. Jiang CW, Zou P, Feng XS, Hu Q, Liu R, Vemareddy P, Duan AY, Zuo PB, Wang Y, Wei FS (2018) Magnetohydrodynamic simulation of the X9.3 flare on (2017) September 6: evolving magnetic topology. Astrophys J 869(1):13. https://doi.org/10.3847/1538-4357/aaeacc
209. Jiang J, Cameron R, Schmitt D, Schessler M (2010) Modeling the sun's open magnetic flux and the heliospheric current sheet. Astrophys J 709(1):301
210. Jin M, Manchester WB, van der Holst B, Gruesbeck JR, Frazin RA, Landi E, Vasquez AM, Lamy PL, Llebaria A, Fedorov A, Toth G, Gombosi TI (2012) A global two-temperature corona and inner heliosphere model: a comprehensive validation study. Astrophys J 745:6
211. Jin M, Manchester WB, van der Holst B, Oran R, Sokolov I, Toth G, Liu Y, Sun XD, Gombosi TI (2013) Numerical simulations of coronal mass ejection on 2011 March 7: one-temperature and two-temperature model comparison. Astrophys J 773:50
212. Jin M, Schrijver CJ, Cheung MCM, DeRosa ML, Nitta NV, Title AM (2016) A numerical study of long-range magnetic impacts during coronal mass ejections. Astrophys J 820:16
213. Jin M, Manchester WB, van der Holst B, Sokolov I, Tóth G, Mullinix RE, Taktakishvili A, Chulaki A, Gombosi TI (2017a) Data-constrained coronal mass ejections in a global magnetohydrodynamics model. Astrophys J 834:173. https://doi.org/10.3847/1538-4357/834/2/173
214. Jin M, Manchester WB, van der Holst B, Sokolov I, Tóth G, Vourlidas A, de Koning CA, Gombosi TI (2017) Chromosphere to 1 AU simulation of the 2011 March 7th event: a comprehensive study of coronal mass ejection propagation. Astrophys J 834:172. https://doi.org/10.3847/1538-4357/834/2/172
215. Juan JM, Sanz J, Rovira-Garcia A, Gonzíćlez-Casado G, Ibánez D, Perez RO (2018) AATR an ionospheric activity indicator specifically based on gnss measurements. J Space Weather Space Clim 8:A14. https://doi.org/10.1051/swsc/2017044

216. Juusola L, Facskø G, Honkonen I, Janhunen P, Vanhamäki H, Kauristie K, Laitinen TV, Milan SE, Palmroth M, Tanskanen EI, Viljanen A (2014) Statistical comparison of seasonal variations in the GUMICS-4 global MHD model ionosphere and measurements. Space Weather 12(10):582–600
217. Kageyama A, Sato T (2004) Yin-yang grid: an overset grid in spherical geometry. Geochem Geophys Geosyst 5:9005
218. Kappenman JG (2001) An introduction to power grid impacts and vulnerabilities from space weather. In: Daglis I (ed) Space storms and space weather hazards, NATO science series, vol 38. Springer, Netherlands, pp 335–361
219. Kataoka R, Ebisuzaki T, Kusano K, Shiota D, Inoue S, Yamamoto TT, Tokumaru M (2009) Three-dimensional MHD modeling of the solar wind structures associated with 13 December 2006 coronal mass ejection. J Geophys Res (Space Phys) 114(A10):A10102. https://doi.org/10.1029/2009JA014167
220. Keppens R, Nool M, Tóth G, Goedbloed JP (2003) Adaptive mesh refinement for conservative systems: multi-dimensional efficiency evaluation. Comput Phys Commun 153:317–339
221. Keppens R, Meliani Z, van Marle AJ, Delmont P, Vlasis A, van der Holst B (2012) Parallel, grid-adaptive approaches for relativistic hydro and magnetohydrodynamics. J Comput Phys 231:718–744. https://doi.org/10.1016/j.jcp.2011.01.020
222. Keppens R, Porth O, Galsgaard K, Frederiksen JT, Restante AL, Lapenta G, Parnell C (2013) Resistive magnetohydrodynamic reconnection: resolving long-term, chaotic dynamics. Phys Plasmas 20(9):092109. https://doi.org/10.1063/1.4820946
223. Keppens R, Xia C, Porth O (2015) Solar prominences: "double, double... boil and bubble". Astrophys J Lett 806:L13. https://doi.org/10.1088/2041-8205/806/1/L13
224. Kholodov AS, Vasilév MO, Molokov EA (2011) Magnetohydrodynamic models of the upper atmosphere of Earth and some applications thereof. Izvestiya, Atmos Ocean Phys 47(1):87–107. https://doi.org/10.1134/S0001433810061015
225. Kim KH, Park KS, Ogino T, Lee DH, Sung SK, Kwak YS (2009) Global MHD simulation of the geomagnetic sudden commencement on 21 October 1999. J Geophys Res 114(A8):A08212. https://doi.org/10.1029/2009JA014109
226. Kim TK, Pogorelov NV, Borovikov SN, Jackson BV, Yu HS, Tokumaru M (2014) MHD heliosphere with boundary conditions from a tomographic reconstruction using interplanetary scintillation data. J Geophys Res: Space Phys 119(10):7981–7997. https://doi.org/10.1002/2013JA019755
227. Kim TK, Pogorelov NV, Zank GP, Elliott HA, McComas DJ (2016) Modeling the solar wind at the Ulysses, Voyager, and new horizons spacecraft. Astrophys J 832(1):72. https://doi.org/10.3847/0004-637x/832/1/72
228. Kim TK, Pogorelov NV, Burlaga LF (2017) Modeling shocks detected by Voyager 1 in the local interstellar medium. Astrophys J 843(2):L32. https://doi.org/10.3847/2041-8213/aa7b2b
229. Kissmann R, Pomoell J (2012) A semidiscrete finite volume constrained transport method on orthogonal curvilinear grids. SIAM J Sci Comput 34(2):A763–A791. https://doi.org/10.1137/110834329
230. Kissmann R, Kleimann J, Fichtner H, Grauer R (2008) Local turbulence simulations for the multiphase ISM. Mon Not R Astron Soc 391(4):1577–1588
231. Kissmann R, Pomoell J, Kley W (2009) A central conservative scheme for general rectangular grids. J Comput Phys 228(6):2119–2131
232. Kissmann R, Kleimann J, Krebl B, Wiengarten T (2018) The CRONOS code for astrophysical magnetohydrodynamics. Astrophys J Suppl Ser 236(2):53. http://stacks.iop.org/0067-0049/236/i=2/a=53
233. Kitamura K, Shimazu H, Fujita S, Watari S, Kunitake M, Shinagawa H, Tanaka T (2008) Properties of AE indices derived from real-time global simulation and their implications for solar wind-magnetosphere coupling. J Geophys Res 113(A3):A03S10. https://doi.org/10.1029/2007JA012514

234. Kleiber W, Sain SR, Heaton MJ, Wiltberger M, Reese CS, Bingham D (2013) Parameter tuning for a multi-fidelity dynamical model of the magnetosphere. Ann Appl Stat 7(3):1286–1310
235. Kleiber W, Sain SR, Wiltberger MJ (2014) Model calibration via deformation. SIAM/ASA J Uncertain Quantif 2(1):545–563
236. Kleiber W, Hendershott B, Sain SR, Wiltberger M (2016) Feature-based validation of the Lyon-Fedder-Mobarry magnetohydrodynamical model. J Geophys Res: Space Phys 121:1192–1200. https://doi.org/10.1002/2015JA021825
237. Kleimann J, Kopp A, Fichtner H, Grauer R, Germaschewski K (2004) Three-dimensional MHD high-resolution computations with CWENO employing adaptive mesh refinement. Comput Phys Commun 158:47–56
238. Kleimann J, Kopp A, Fichtner H, Grauer R (2009) A novel code for numerical 3-D MHD studies of CME expansion. Ann Geophys 27:989–1004
239. Knipp DJ, Ramsay AC, Beard ED, Boright AL, Cade WB, McFadden Hewins R I M, Denig WF, Kilcommons LM, Shea MA, Smart DF (2016) The May 1967 great storm and radio disruption event: extreme space weather and extraordinary responses. Space Weather 14(9):614–633. https://doi.org/10.1002/2016SW001423
240. Komar CM, Cassak PA, Dorelli JC, Glocer A, Kuznetsova MM (2013) Tracing magnetic separators and their dependence on IMF clock angle in global magnetospheric simulations. J Geophys Res: Space Phys 118(8):4998–5007. https://doi.org/10.1002/jgra.50479
241. Komar CM, Fermo RL, Cassak PA (2015) Comparative analysis of dayside magnetic reconnection models in global magnetosphere simulations. J Geophys Res (Space Phys) 120:276–294. https://doi.org/10.1002/2014JA020587
242. Kopp A, Wiengarten T, Fichtner H, Effenberger F, Kühl P, Heber B, Raath JL, Potgieter MS (2017) Cosmic-ray transport in heliospheric magnetic structures. II. Modeling particle transport through corotating interaction regions. Astrophys J 837:37. https://doi.org/10.3847/1538-4357/aa603b
243. Koren B (1993) A robust upwind discretisation method for advection, diffusion and source terms. In: Vreugdenhil CB, Koren B (eds) Numerical methods for advection-diffusion problems. Vieweg, Braunschweig, p 117
244. Korth H, Anderson BJ, Lyon JG, Wiltberger M (2008) Comparison of Birkeland current observations during two magnetic cloud events with MHD simulations. Ann Geophys 26(3):499–516
245. Koskinen HEJ, Baker DN, Balogh A, Gombosi T, Veronig A, von Steiger R (2017) Achievements and challenges in the science of space weather. Space Sci Rev 212(3):1137–1157. https://doi.org/10.1007/s11214-017-0390-4
246. Koutchmy S, Livshits M (1992) Coronal streamers. Space Sci Rev 61(3):3923–417
247. Koutchmy S, Zirker JB, Steinolfson RS, Zhugzda JD (1991) Coronal activity. In: Cox AN, Livingston WC, Matthews RS (eds) Solar interior and atmosphere. The University of Arizona Press, Tuscon, AZ, pp 1044–1086
248. Kryukov IA, Pogorelov NV, Zank GP, Borovikov SN (2012) Numerical modeling of the solar wind turbulence. AIP Conference proceedings 1436(1):48–54. https://doi.org/10.1063/1.4723589
249. Kubota Y, Kataoka R, Den M, Tanaka T, Nagatsuma T, Fujita S (2013) Global MHD simulation of the magnetospheric response to large and sudden enhancement of the solar wind dynamic pressure. In: AGU fall meeting abstracts, vol 1, p 2230
250. Kulikovskii AG, Pogorelov NV, Semenov AY (2001) Mathematical aspects of numerical solution of hyperbolic systems. Chapman & Hall, London
251. Kurganov A, Levy D (2000) A third-order semidiscrete central scheme for conservation laws and convection-diffusion equations. SIAM J Sci Comput 22(4):1461–1488
252. Kurganov A, Tadmor E (2000) New high-resolution central schemes for nonlinear conservation laws and convection-diffusion equations. J Comput Phys 160(1):241–282
253. Kurganov A, Noelle S, Petrova G (2001) Semidiscrete central-upwind schemes for hyperbolic conservation laws and Hamilton-Jacobi equations. SIAM J Sci Comput 23(3):707–740

254. Lai ST, Cahoy K, Lohmeyer W, Carlton A, Aniceto R, Minow J (2018) Chapter 16 - Deep dielectric charging and spacecraft anomalies. In: Buzulukova N (ed) Extreme events in geospace. Elsevier, pp 419–432. https://doi.org/10.1016/B978-0-12-812700-1.00016-9
255. Laitinen T, Palmroth M, Pulkkinen T, Janhunen P, Koskinen H (2007) Continuous reconnection line and pressure-dependent energy conversion on the magnetopause in a global MHD model. J Geophys Res: Space Phys 112(A11):A11201. https://doi.org/10.1029/2007JA012352
256. Lakhina GS, Tsurutani BT (2016) Geomagnetic storm: historical perspective to modern view. Geosci Lett 3:5
257. Landi E, Gruesbeck JR, Lepri ST, Zurbuchen TH, Fisk LA, Frazin R, Manchester W, Sokolov I, Gombosi TI (2012) Charge state evolutionin the solar wind II. Plasma charge state composition in the inner corona and accelerating fast solar wind. Astrophys J 761(1):48
258. Lang KR (2009) The Sun for space. Springer, Berlin
259. Lani A, Yalim MS, Poedts S (2014) A GPU-enabled finite volume solver for global magnetospheric simulations on unstructured grids. Comput Phys Commun 185(10):2538–2557
260. Lanzerotti LJ (2017) Space weather: historical and contemporary perspectives. Space Sci Rev 212(3):1253–1270. https://doi.org/10.1007/s11214-017-0408-y
261. Lapenta G (2012) Particle simulations of space weather. J Comput Phys 231(3):795–821. https://doi.org/10.1016/j.jcp.2011.03.035
262. Lapenta G, Pierrard V, Keppens R, Markidis S, Poedts S, Šebek O, Trávníček PM, Henri P, Califano F, Pegoraro F, Faganello M, Olshevsky V, Restante AL, Nordlund Å, Trier Frederiksen J, Mackay DH, Parnell CE, Bemporad A, Susino R, Borremans K (2013) SWIFF: space weather integrated forecasting framework. J Space Weather Space Clim 3(27):A05
263. Lax P, Wendroff B (1960) Systems of conservation laws. Commun Pure Appl Math 13(2):217–237
264. Lee CO, Luhmann JG, Odstrčil D, MacNeice PJ, de Pater I, Riley P, Arge CN (2009) The solar wind at 1 AU during the declining phase of solar cycle 23: comparison of 3D numerical model results with observations. Solar Phys 254:155–183
265. Lee JY, Sussman A (2005) High performance communication between parallel programs. In: Proceedings of the 19th IEEE international parallel and distributed processing symposium (IPDPS'05) - workshop 4 - vol 05. IEEE Computer Society, Washington, DC, USA, Paper No. 1530-2075/05. https://doi.org/10.1109/IPDPS.2005.243
266. van Leer B (1979) Towards the ultimate conservative difference scheme. V. A second-order sequel to godunov's method. J Comput Phys 32(1):101–136
267. Li HC, Feng XS (2018) CESE-HLL magnetic field-driven modeling of the background solar wind during Year 2008. Journal of Geophysical Research: Space Phys 123(6):4488–4509. https://doi.org/10.1029/2017JA025125
268. Li J, Raymond JC, Acton LW, Kohl JL, Romoli M, Noci G, Naletto G (1998) Physical structure of a coronal streamer in the closed-field region as observed from UVCS/SOHO and SXT/YOHKOH. Astrophys J 506:431–438
269. Li W, Raeder J, Dorelli J, Øieroset M, Phan TD (2005) Plasma sheet formation during long period of northward IMF. Geophys Res Lett 32(12):L12S08
270. Li WY, Wang C, Tang BB, Guo XC, Lin D (2013) Global features of Kelvin-Helmholtz waves at the magnetopause for northward interplanetary magnetic field. J Geophys Res: Space Phys 118(8):5118–5126
271. Liemohn M, Ganushkina NY, De Zeeuw DL, Rastaetter L, Kuznetsova M, Welling DT, Toth G, Ilie R, Gombosi TI, Holst B (2018) Real-time SWMF at CCMC: assessing the Dst output from continuous operational simulations. Space Weather 16(10):1583–1603. https://doi.org/10.1029/2018SW001953
272. Linker JA (2011) A next-generation model of the corona and solar wind. Technical Report AFRL-OSR-VA-TR-2012-0199, Air Force Force of Scientific Research, Arlington VA
273. Linker JA (2011) A next-generation model of the corona and solar wind. Technical Report FCM-P30-1995, Air Force Office of Scientific Research, 875 N Randolf St Arlinton, VA 22203

274. Linker JA, Mikić Z, Biesecker DA, Forsyth RJ, Gibson SE, Lazarus AJ, Lecinski A, Riley P, Szabo A, Thompson BJ (1999) Magnetohydrodynamic modeling of the solar corona during Whole Sun Month. J Geophys Res 104:9809–9830
275. Linker JA, Lionello R, Mikić Z, Amari T (2001) Magnetohydrodynamic modeling of prominence formation within a helmet streamer. J Geophys Res 106:25,165–25,176
276. Linker JA, Caplan RM, Downs C, Lionello R, Riley P, Mikić Z, Henny CJ, Arge CN, Kim T, Pogorelov N (2016) An empirically driven time-dependent model of the solar wind. J Phys: Conf Ser 719:012012
277. Lionello R, Mikić Z, Schnack DD (1998) Magnetohydrodynamics of solar coronal plasmas in cylindrical geometry. J Comput Phys 140:172–201
278. Lionello R, Mikić Z, Linker JA (1999) Stability of algorithms for waves with large flows. J Comput Phys 152:346–358
279. Lionello R, Linker JA, Mikić Z (2001) Including the transition region in models of the large-scale solar corona. Astrophys J 546:542–551
280. Lionello R, Linker JA, Mikić Z (2009) Multispectral emission of the Sun during the first whole Sun month: magnetohydrodynamic simulations. Astrophys J 690:902–912
281. Lionello R, Downs C, Aand LTJ, Török RP, Mikić Z (2013) Magnetohydrodynamic simulations of interplanetary coronal mass ejections. Astrophys J 777(1):76
282. Lionello R, Velli M, Downs C, Linker JA, Mikić Z (2014) Application of a Solar Wind Model Driven by Turbulence Dissipation to a 2D Magnetic Field Configuration. Astrophys. J. 796:111. https://doi.org/10.1088/0004-637X/796/2/111
283. Lionello R, Velli M, Downs C, Linker JA, Mikić Z, Verdini A (2014) Validating a time-dependent turbulence-driven model of the solar wind. Astrophysical J 784(2):120. http://stacks.iop.org/0004-637X/784/i=2/a=120
284. Liou K, Wu CC, Dryer M, Wu ST, Rich N, Plunkett S, Simpson L, Fry CG, Schenk K (2014) Global simulation of extremely fast coronal mass ejection on July 2012. J Atmos Solar Terr Phys 121:32–41
285. Liu Y, Nagy AF, Kabin K, Combi MR, DeZeeuw DL, Gombosi TI, Powell KG (2000) Two-species, 3D, MHD simulation of Europa's interaction with Jupiter's magnetosphere. Geophys Res Lett 27(12):1791–1794
286. López-Puertas M, Funke B, Gil-López S, von Clarmann T, Stiller GP, HöPfner M, Kellmann S, Fischer H, Jackman CH (2005) Observation of NO_x enhancement and ozone depletion in the Northern and Southern Hemispheres after the October–November 2003 solar proton events. J Geophys Res 110:A09S43
287. Love JJ, Bedrosian PA (2018) Chapter 9 - Extreme-event geoelectric hazard maps. In: Buzulukova N (ed) Extreme events in geospace. Elsevier, pp 209–230, https://doi.org/10.1016/B978-0-12-812700-1.00009-1
288. Love JJ, Lucas GM, Kelbert A, Bedrosian PA (2017) Geoelectric hazard maps for the mid-atlantic united states: 100 year extreme values and the 1989 magnetic storm. Geophys Res Lett 45(1):5–14. https://doi.org/10.1002/2017GL076042
289. Love JJ, Lucas GM, Kelbert A, Bedrosian PA (2018) Geoelectric hazard maps for the pacific northwest. Space Weather 16(8):1114–1127. https://doi.org/10.1029/2018SW001844
290. Lu JY, Liu ZQ, Kabin K, Zhao MX, Liu DD, Zhou Q, Xiao Y (2011) Three dimensional shape of the magnetopause: global MHD results. J Geophys Res: Space Phys 116(A9):A09237. https://doi.org/10.1029/2010JA016418
291. Lu JY, Jing H, Liu ZQ, Kabin K, Jiang Y (2013) Energy transfer across the magnetopause for northward and southward interplanetary magnetic fields. J Geophys Res: Space Phys 118(5):2021–2033
292. Lugaz N, Manchester WB IV, Roussev II, Tóth G, Gombosi TI (2007) Numerical investigation of the homologous coronal mass ejection events from active region 9236. Astrophys J 659:788–800
293. Luhmann JG, Solomon SC, Linker JA, Lyon JG, Mikic Z, Odstrčil D, Wang W, Wiltberger M (2004) Coupled model simulation of a Sun-to-Earth space weather event. J Atmos Solar Terr Phys 66:1243–1256

294. Lyon J, Lopez R, Goodrich C, Wiltberger M, Papadopoulos K (1998) Simulation of the March 9, 1995, substorm: auroral brightening and the onset of lobe reconnection. Geophys Res Lett 25(15):3039–3042
295. Lyon JG, Fedder JA, Mobarry CM (2004) The Lyon-Fedder-Mobarry (LFM) global MHD magnetospheric simulation code. J Atmos Solar Terr Phys 66:1333–1350
296. Ma YJ, Nagy AF, Toth G, Cravens TE, Russell CT, Gombosi TI, Wahlund JE, Crary FJ, Coates AJ, Bertucci CL et al (2007) 3D global multi-species Hall-MHD simulation of the Cassini T9 flyby. Geophys Res Lett 34(24):L24S10. https://doi.org/10.1029/2007GL031627
297. MacNeice P, Elliott B, Acebal A (2011) Validation of community models: 3. Tracing field lines in heliospheric models. Space Weather 9:S10003
298. MacNeice P, Jian LK, Antiochos SK, Arge CN, Bussy-Virat CD, DeRosa ML, Jackson BV, Linker JA, Mikic Z, Owens MJ, Ridley AJ, Riley P, Savani N, Sokolov I (2018) Assessing the quality of models of the ambient solar wind. Space Weather 16:1644–1667. https://doi.org/10.1029/2018SW002040
299. Manchester W (2003) Buoyant disruption of magnetic arcades with self-induced shearing. J Geophys Res (Space Phys) 108:1162
300. Manchester W IV, Gombosi T, DeZeeuw D, Fan Y (2004a) Eruption of a buoyantly emerging magnetic flux rope. Astrophys J 610:588–596
301. Manchester WB, Gombosi TI, Roussev I, Ridley A, de Zeeuw DL, Sokolov IV, Powell KG, Tóth G (2004b) Modeling a space weather event from the Sun to the Earth: CME generation and interplanetary propagation. J Geophys Res (Space Phys) 109:A02107
302. Manchester WB IV, Vourlidas A, Tóth G, Lugaz N, Roussev II, Sokolov IV, Gombosi TI, De Zeeuw DL, Opher M (2008) Three-dimensional MHD simulation of the 2003 October 28 coronal mass ejection: comparison with LASCO coronagraph observations. Astrophys J 684:1448–1460
303. Manchester WB IV, van der Holst B, Tóth G, Gombosi TI (2012) The coupled evolution of electrons and ions in coronal mass ejection-driven shocks. Astrophys J 756:81
304. Markidis S, Lapenta G, Rizwan-uddin, (2010) Multi-scale simulations of plasma with iPIC3D. Math Comput Simul 80(7):1509–1519. https://doi.org/10.1016/j.matcom.2009.08.038
305. Maynard N, Farrugia C, Ober D, Burke W, Dunlop M, Mozer F, Rème H, Décréau P, Siebert K (2008) Cluster observations of fast shocks in the magnetosheath launched as a tangential discontinuity with a pressure increase crossed the bow shock. J Geophys Res: Space Phys 113:A10212. https://doi.org/10.1029/2008JA013121
306. Maynard NC (2005) Coupling the solar-wind/IMF to the ionosphere through the high latitude CUSPS. Surv Geophys 26(1):255–280. https://doi.org/10.1007/s10712-005-1882-4
307. Maynard NC, Sonnerup BUÖ, Siscoe GL, Weimer DR, Siebert KD, Erickson GM, White WW, Schoendorf JA, Ober DM, Wilson GR, Heinemann MA (2002) Predictions of magnetosheath merging between IMF field lines of opposite polarity. J Geophys Res (Space Phys) 107:1456. https://doi.org/10.1029/2002JA009289
308. Maynard NC, Farrugia CJ, Burke WJ, Ober DM, Scudder JD, Mozer FS, Russell CT, Réme H, Mouikis C, Siebert KD (2011) Interactions of the heliospheric current and plasma sheets with the bow shock: Cluster and Polar observations in the magnetosheath. J Geophys Res (Space Phys) 116:A01212. https://doi.org/10.1029/2010JA015872
309. Maynard NC, Farrugia CJ, Burke WJ, Ober DM, Mozer FS, Réme H, Dunlop M, Siebert KD (2012) Cluster observations of the dusk flank magnetopause near the sash: ion dynamics and flow-through reconnection. J Geophys Res (Space Phys) 117(A16):A10201. https://doi.org/10.1029/2012JA017703
310. McComas DJ, Barraclough BL, Fusten HO, Gosling JT, Santiago-Muñoz E, Skoug RM, Goldstein BE, Neugebaur M, Riley P, Balogh A (2000) Solar wind observations over Ulysses' first full polar orbit. J Geophys Res 105(A5):10,419–10,433
311. McComas DJ, Goldstein BE, Gosling JT, Skoug RM (2001) Ulysses' second fast-latitude scan: remarkably different solar wind. Space Sci Rev 97:99–103
312. McComas DJ, Elliott HA, Gosling JT, Reisenfeld DB, Skoug RM, Goldstein BE, Neugebaur M, Balogh A (2002) Ulysses' second fast-latitude scan: complexity near solar maximum and the reformation of polar coronal holes. Geophys Res Lett 29(9):4

313. McGregor SL, Hughes WJ, Arge CN, Odstrčil D, Schwadron NA (2011a) The radial evolution of solar wind speeds. J Geophys Res 116(A15):A03106
314. McGregor SL, Hughes WJ, Arge CN, Owens MJ, Odstrčil D (2011b) The distribution of solar wind speeds during solar minimum: calibration for numerical solar wind modeling constraints on the source of the slow solar wind. J Geophys Res 116:A03101
315. McKenzie JF, Axford WI, Banaszkiewicz M (1997) The fast solar wind. Geophys Res Lett 24(22):2877–2880
316. McNally CP, Lyra W, Passy JC (2012) A well-posed kelvin-helmholtz instability test and comparison. Astrophys J Suppl Ser 201(2):18. https://doi.org/10.1088/0067-0049/201/2/18
317. Meng X, Tóth G, Liemohn MW, Gombosi TI, Runov A (2012) Pressure anisotropy in global magnetospheric simulations: a magnetohydrodynamics model. J Geophys Res 117(A8):A08216. https://doi.org/10.1029/2012JA017791
318. Meng X, Tóth G, Sokolov IV, Gombosi TI (2012) Classical and semirelativistic magnetohydrodynamics with anisotropic ion pressure. J Comput Phys 231:3610–3622
319. Meng X, van der Holst B, Tóth G, Gombosi TI (2015) Alfvén wave solar model (AWSoM): proton temperature anisotropy and solar wind acceleration. Mon Not R Astron Soc 454:3697–3709
320. Merkin V, Lyon J (2010) Effects of the low-latitude ionospheric boundary condition on the global magnetosphere. J Geophys Res 115(A10):A10202. https://doi.org/10.1029/2010JA015461
321. Merkin V, Anderson B, Lyon J, Korth H, Wiltberger M, Motoba T (2013) Global evolution of Birkeland currents on 10 min timescales: MHD simulations and observations. J Geophys Res 118(8):4977–4997
322. Merkin VG, Lyon JG, McGregor SL, Pahud DM (2011) Disruption of a heliospheric current sheet fold. Geophys Res Lett 38:L14107
323. Merkin VG, Kondrashov D, Ghil M, Anderson BJ (2016) Data assimilation of low-altitude magnetic perturbations into a global magnetosphere model. Space Weather 14(2):165–184. https://doi.org/10.1002/2015SW001330
324. Merkin VG, Lyon JG, Lario D, Arge CN, Henney CJ (2016b) Time-dependent magnetohydrodynamic simulations of the inner heliosphere. J Geophys Res: Space Phys 121:2866–2890
325. Mignone A, Bodo G, Massaglia S, Matsakos T, Tesileanu O, Zanni C, Ferrari A (2007) PLUTO: A numerical code for computational astrophysics. Astrophys J Suppl Ser 170:228
326. Mignone A, Zanni C, Tzeferacos P, van Straalen B, Colella P, Bodo G (2012) The PLUTO code for adaptive mesh computations in astrophysical fluid dynamics. Astrophys J Suppl Ser 198:7. https://doi.org/10.1088/0067-0049/198/1/7
327. Mikić Z, Linker JA, Schnack DD, Lionello R, Tarditi A (1999) Magnetohydrodynamic modeling of the global solar corona. Phys Plasmas 6:2217–2224
328. Mikić Z, Downs C, Linker JA, Caplan RM, Mackay DH, Upton LA, Riley P, Lionello R, Török T, Titov VS, Wijaya J, Druckmüller M, Pasachoff JM, Carlos W (2018) Predicting the corona for the 21 August 2017 total solar eclipse. Nat Astron 2:913–921. https://doi.org/10.1038/s41550-018-0562-5
329. Miyoshi T, Kusano K (2005) A multi-state HLL approximate Riemann solver for ideal magnetohydrodynamics. J Comput Phys 208:315–344
330. Miyoshi T, Terada N, Matsumoto Y, Fukazawa K, Umeda T, Kusano K (2010) The hlld approximate Riemann solver for magnetospheric simulation. IEEE Trans Plasma Sci 38(9):2236–2242. https://doi.org/10.1109/TPS.2010.2057451
331. Mobarry C, Fedder J, Lyon J (1996) Equatorial plasma convection from global simulations of the Earth's magnetosphere. J Geophys Res 101(A4):7859–7874
332. Mocz P, Pakmor R, Springel V, Vogelsberger M, Marinacci F, Hernquist L (2016) A moving mesh unstaggered constrained transport scheme for magnetohydrodynamics. Mon Not R Astron Soc 463(1):477–488. https://doi.org/10.1093/mnras/stw2004
333. Moriguchi T, Nakamizo A, Tanaka T, Obara T, Shimazu H (2008) Current systems in the jovian magnetosphere. J Geophys Res: Space Phys 113(A5):A05204. https://doi.org/10.1029/2007JA012751

334. Moschou SP, Pierrard V, Keppens R, Pomoell J (2017) Interfacing MHD single fluid and kinetic exospheric solar wind models and comparing their energetics. Solar Phys 292(9):139. https://doi.org/10.1007/s11207-017-1164-6
335. Möstl C, Isavnin A, Boakes PD, Kilpua EKJ, Davies JA, Harrison RA, Barnes D, Krupar V, Eastwood JP, Good SW, Forsyth RJ, Bothmer V, Reiss MA, Amerstorfer T, Winslow RM, Anderson BJ, Philpott LC, Rodriguez L, Rouillard AP, Gallagher P, Nieves-Chinchilla T, Zhang TL (2017) Modeling observations of solar coronal mass ejections with heliospheric imagers verified with the heliophysics system observatory. Space Weather 15(7):955–970. https://doi.org/10.1002/2017SW001614
336. Möstl C, Amerstorfer T, Palmerio E, Isavnin A, Farrugia CJ, Lowder C, Winslow RM, Donnerer JM, Kilpua EKJ, Boakes PD (2018) Forward modeling of coronal mass ejection flux ropes in the inner heliosphere with 3DCORE. Space Weather 16(3):216–229. https://doi.org/10.1002/2017SW001735
337. Motoba T, Fujita S, Kikuchi T, Tanaka T (2007) Solar wind dynamic pressure forced oscillation of the magnetosphere-ionosphere coupling system: a numerical simulation of directly pressure-forced geomagnetic pulsations. J Geophys Res 112(A11):204
338. Murawski K, Ballai I, Srivastava AK, Lee D (2013) Three-dimensional numerical simulation of magnetohydrodynamic-gravity waves and vortices in the solar atmosphere. Mon Not R Astron Soc 436:1268–1277. https://doi.org/10.1093/mnras/stt1653
339. Nakamizo A, Tanaka T, Kubo Y, Kamei S, Shimazu H, Shinagawa H (2009) Development of the 3-D MHD model of the solar corona-solar wind combining system. J Geophys Res (Space Phys) 114(A7):A07109. https://doi.org/10.1029/2008JA013844
340. Nakata H, Shinohara M, Kamide Y, Ogino T (2004) Ionospheric electric potentials for substorms calculated from a solar wind-magnetosphere MHD simulation and a magnetogram inversion technique. J Geophys Res 109(A1):A01211. https://doi.org/10.1029/2002JA009743
341. Navarro A, Lora-Clavijo FD, González GA (2017) Magnus: a new resistive MHD code with heat flow terms. Astrophys J 844:57. https://doi.org/10.3847/1538-4357/aa7a13
342. Neugebauer M (1999) The three-dimensional solar wind at solar activity minimum. Rev Geophys 37:107–126
343. Ngwira CM, Pulkkinen AA (2018) Chapter 8 - An overview of science challenges pertaining to our understanding of extreme geomagnetically induced currents. In: Buzulukova N (ed) Extreme events in geospace. Elsevier, pp 187–208, https://doi.org/10.1016/B978-0-12-812700-1.00008-X
344. Ngwira CM, Pulkkinen A, Kuznetsova MM, Glocer A (2014) Modeling extreme "Carrington-type" space weather events using three-dimensional global MHD simulations. J Geophys Res: Space Phys 119(6):4456–4474
345. Nishimura Y, Lyons LR (2016) Localized reconnection in the magnetotail driven by lobe flow channels: global MHD simulation. J Geophys Res: Space Phys 121:1327–1338. https://doi.org/10.1002/2015JA022128
346. Norquist DC (2013) Forecast performance assessment of a kinematic and a magnetohydrodynamic solar wind model. Space Weather 11:17–33
347. Norquist DC, Meeks WC (2010) A comparative verification of forecasts from two operational solar wind models. Space Weather 8:S12005
348. Ober DM, Wilson GR, Burke WJ, Maynard NC, Siebert KD (2007) Magnetohydrodynamic simulations of transient transpolar potential responses to solar wind density changes. J Geophys Res (Space Phys) 112(A11):A10212
349. Odstrčil D (1994) Interactions of solar wind streams and related small structures. J Geophys Res 99(17):653
350. Odstrčil D (2003) Modeling 3-D solar wind structure. Adv Space Res 32:497–506
351. Odstrčil D, Karlicky M (1997) Triggering of magnetic reconnection in the current sheet by shock waves. Astron Astrophys 326:1252–1258
352. Odstrčil D, Pizzo VJ (1999) Distortion of the interplanetary magnetic field by three-dimensional propagation of coronal mass ejections in a structured solar wind. J Geophys Res 104:28,225–28,240

353. Odstrčil D, Pizzo VJ (1999) Three-dimensional propagation of CMEs in a structured solar wind flow: 1. CME launched within the streamer belt. J Geophys Res 104:483–492
354. Odstrčil D, Pizzo VJ (1999c) Three-dimensional propagation of coronal mass ejections in a structured solar wind flow 2. CME launched adjacent to the streamer belt. J Geophys Res 104:493–504
355. Odstrčil D, Pizzo VJ (2009) Numerical heliospheric simulations as assisting tool for interpretation of observations by STEREO heliospheric imagers. Solar Phys 259:297–309. https://doi.org/10.1007/s11207-009-9449-z
356. Odstrčil D, Dryer M, Smith Z (1996) Propagation of an interplanetary shock along the heliospheric plasma sheet. J Geophys Res 101:19,973–19,986
357. Odstrčil D, Linker JA, Lionello R, Mikic Z, Riley P, Pizzo VJ, Luhmann JG (2002) Merging of coronal and heliospheric numerical two-dimensional MHD models. J Geophys Res 107:1493
358. Ogino T (1986) A three-dimensional MHD simulation of the interaction of the solar wind with the Earth's magnetosphere: the generation of field-aligned currents. J Geophys Res 91(A6):6791–6806
359. Ogino T, Walker RJ, Ashour-Abdalla M (1992) A global magnetohydrodynamic simulation of the magnetosheath and magnetosphere when the interplanetary magnetic field is northward. Plasma Sci 20(6):817–828
360. Ogino T, Walker RJ, Ashour-Abdalla M (1994) A global magnetohydrodynamic simulation of the response of the magnetosphere to a northward turning of the interplanetary magnetic field. J Geophys Res 99(A6):11,027–11,042
361. Ogino T, Kajiwara Y, Nakao M, Park KS, Fukazawa K (2003) Yi Y (2007) MHD simulation of the shock wave event on October 24. Acta Astronautica 61(10):923–931
362. Olshevsky V, Bacchini F, Poedts S, Lapenta G (2019) Slurm: fluid particle-in-cell code for plasma modeling. Comput Phys Commun 235:16–24. 10.1016/j.cpc.2018.06.014, http://www.sciencedirect.com/science/article/pii/S0010465518302224
363. Oran R, van der Holst B, landi E, Jin M, Sokolov IV, Gombosi TI, (2013) A global wave-driven magnetohydrodynamic solar model with a unified treatment of open and closed magnetic field topologies. Astrophys J 778(2):176
364. Oran R, Landi E, van der Holst B, Lepri ST, Vásquez AM, Neuvo FA, Frazin R, Manchester W, Sokolov I, Gombosi TI (2015) A steady-state picture of solar wind acceleration and charge state composition derived from a global wave-driven MHD model. Astrophys J 806(1):55
365. Ouellette JE, Rogers BN, Wiltberger M, Lyon JG (2010) Magnetic reconnection at the dayside magnetopause in global Lyon-Fedder-Mobarry simulations. J Geophys Res (Space Phys) 115:A08222. https://doi.org/10.1029/2009JA014886
366. Owens MJ, Spence HE, McGregor S, Hughes WJ, Quinn JM, Arge CN, Riley P, Linker J, Odstrčil D (2008) Metrics for solar wind prediction models: comparison of empirical, hybrid, and physics-based schemes with 8 years of L1 observations. Space Weather 6:S08001
367. Owens MJ, Horbury TS, Wicks RT, McGregor SL, Savani NP, Xiong M (2014) Ensemble downscaling in coupled solar wind-magnetosphere modeling for space weather forecasting. Space Weather 12(6):395–405
368. Pagano P, Mackay DH, Poedts S (2013a) Effect of gravitational stratification on the propagation of a CME. A&A 560:A38. https://doi.org/10.1051/0004-6361/201322036
369. Pagano P, Mackay DH, Poedts S (2013) Magnetohydrodynamic simulations of the ejection of a magnetic flux rope. A&A 554:A77. https://doi.org/10.1051/0004-6361/201220947
370. Pagano P, Mackay DH, Poedts S (2014) Simulating AIA observations of a flux rope ejection. A&A 568:A120. https://doi.org/10.1051/0004-6361/201424019
371. Pagano P, Bemporad A, Mackay DH (2015) Future capabilities of CME polarimetric 3D reconstructions with the METIS instrument: a numerical test. Astron Astrophys 582:A72. https://doi.org/10.1051/0004-6361/201425462
372. Pagano P, Mackay DH, Yeates AR (2018) A new technique for observationally derived boundary conditions for space weather. J Space Weather Space Clim 8:A26. https://doi.org/10.1051/swsc/2018012

373. Pahud DM, Merkin VG, Arge CN, Hughes WJ, McGregor SM (2012) An MHD simulation of the inner heliosphere during Carrington rotations 2060 and 2068: comparison with MESSENGER and ACE spacecraft observations. J Atmos Solar Terr Phys 83:32–38
374. Palmroth M, Janhunen P, Pulkkinen TI, Peterson WK (2001) Cusp and magnetopause locations in global MHD simulation. J Geophys Res 106(A12):29,435–29,450
375. Palmroth M, Pulkkinen TI, Janhunen P, Wu CC (2003) Stormtime energy transfer in global MHD simulation. J Geophys Res 108(A1):1048. https://doi.org/10.1029/2002JA009446
376. Palmroth M, Laitinen TV, Pulkkinen TI (2006) Magnetopause energy and mass transfer: results from a global MHD simulation. Ann Geophys 24:3467–3480
377. Palmroth M, Fear RC, Honkonen I (2012) Magnetopause energy transfer dependence on the interplanetary magnetic field and the Earth's magnetic dipole axis orientation. Ann Geophys 30(3):515–526
378. Park KS, Ogino T (2005) MHD simulation of the effects of the IMF and dipole tilt on the dayside magnetic reconnection. Proc ISSS 7:26–31
379. Park KS, Ogino T (2006) A study of the storm event on October 21–22, 1999 by the MHD simulation. Earth Planets Space 58(5):633–643
380. Park KS, Ogino T, Kim YH (2010) Effects of the dipole tilt and northward and duskward IMF on dayside magnetic reconnection in a global MHD simulation. J Geophys Res 115(A2):A02208. https://doi.org/10.1029/2009JA014212
381. Park KS, Lee DY, Ogino T, Lee DH (2015) MHD simulations using average solar wind conditions for substorms observed under northward IMF conditions. J Geophys Res (Space Phys) 120:7672–7686
382. Pembroke A, Toffoletto F, Sazykin S, Wiltberger M, Lyon J, Merkin V, Schmitt P (2012) Initial results from a dynamic coupled magnetosphere-ionosphere-ring current model. J Geophys Res: Space Phys 117(A2):A02211. https://doi.org/10.1029/2011JA016979
383. Peng Z, Wang C (2013) Effect of the northward turning of the interplanetary magnetic field on the dipolarization. J Geophys Res 118(12):7605–7610
384. Peng Z, Wang C, Hu YQ, Kan JR, Yang YF (2011) Simulations of observed auroral brightening caused by solar wind dynamic pressure enhancements under different interplanetary magnetic field conditions. J Geophys Res: Space Phys 116(A6):A06217. https://doi.org/10.1029/2010JA016318
385. Pescaroli G, Wicks R, Giacomello G, Alexander D (2018) Increasing resilience to cascading events: the M.OR.D.OR. scenario. Saf Sci 110:131–140. https://doi.org/10.1016/j.ssci.2017.12.012
386. Pesnell WD, Thompson BJ, Chamberlin PC (2012) The Solar Dynamics Observatory (SDO). Solar Phys 275(1):3–15. https://doi.org/10.1007/s11207-011-9841-3
387. Phillips JL, Bame SJ, Barnes A, Barraclough BL, Feldman WC, Goldstein BE, Gosling JT, Hoogeveen GW, McComas DJ, Neugebauer M, Suess ST (1995) Ulysses solar wind plasma observations from pole to pole. Geophys Res Lett 22:3301–3304
388. Piccinelli R, Krausmann E (2018) North europe power transmission system vulnerability during extreme space weather. J Space Weather Space Clim 8:A03. https://doi.org/10.1051/swsc/2017033
389. Pirijola R (2007) Space weather effects on power grids. Praxis Publishing Ltd., Chichester, UK, p 269
390. Pizzo VJ (1982) A three-dimensional model of corotating streams in the solar wind: 3. Magnetohydrodynamic streams. J Geophys Res (Space Phys) 87(A6):4374–4394
391. Poedts S, Pomoell J (2017) EUHFORIA: a solar wind and CME evolution model. In: EGU general assembly conference abstracts, vol 19, p 7396
392. Poedts S, Jacobs C, van der Holst B, Chané E, Keppens R (2009) Numerical simulations of the solar corona and coronal mass ejections. Earth, Planets, and Space 61:599–602
393. Pogorelov NV, Zank GP, Ogino T (2004) Three-dimensional features of the outer heliosphere due to coupling between the interstellar and interplanetary magnetic fields. I. Magnetohydrodynamic model: interstellar perspective. Astrophys J 614:1007–1021. https://doi.org/10.1086/423798

394. Pogorelov NV, Zank GP, Ogino T (2006) Three-dimensional features of the outer heliosphere due to coupling between the interstellar and interplanetary magnetic fields. II. The presence of neutral hydrogen atoms. Astrophys J 644:1299–1316. https://doi.org/10.1086/503703
395. Pogorelov NV, Zank GP, Borovikov SN, Florinski V, Heerikhuisen J, Kryukov IA (2008) MHD-neutral Model of the outer heliosphere. In: Pogorelov NV, Audit E, Zank GP (eds) Numerical modeling of space plasma flows, Astronomical society of the Pacific conference series, vol 385, p 180
396. Pogorelov NV, Borovikov SN, Florinski V, Heerikhuisen J, Kryukov IA, Zank GP (2009) Multi-scale fluid-kinetic simulation suite: a tool for efficient modeling of space plasma flows. In: Pogorelov NV, Audit E, Colella P, Zank GP (eds) Numerical modeling of space plasma flows: ASTRONUM-2008, Astronomical society of the Pacific conference series, vol 406, p 149
397. Pogorelov NV, Borovikov SN, Zank GP, Ogino T (2009) Three-dimensional features of the outer heliosphere due to coupling between the interstellar and interplanetary magnetic fields. III. The effects of solar rotation and activity cycle. Astrophys J 696(2):1478–1490. https://doi.org/10.1088/0004-637x/696/2/1478
398. Pogorelov NV, Borovikov SN, Heerikhuisen J, Kryukov IA, Zank GP (2010) Modeling heliospheric phenomena with the multi-scale fluid-kinetic simulation suite. In: Pogorelov NV, Audit E, Zank GP (eds) Numerical modeling of space plasma flows, Astronum-2009, Astronomical society of the Pacific conference series, vol 429, p 266
399. Pogorelov NV, Borovikov SN, Heerikhuisen J, Kim T, Kryukov IA, Zank GP (2011) Heliospheric modeling with MS-FLUKSS: interpretation of the IBEX measurements and time-dependent simulations. In: Pogorelov NV, Audit E, Zank GP (eds) 5th international conference of numerical modeling of space plasma flows (ASTRONUM 2010), Astronomical society of the Pacific conference series, vol 444, p 130
400. Pogorelov NV, Borovikov SN, Burlaga LF, Ebert RW, Heerikhuisen J, Hu Q, Kryukov IA, Suess ST, Zank GP (2012) Numerical modeling of transient phenomena in the distant solar wind and in the heliosheath. AIP conference proceedings 1436(1):321–330. https://doi.org/10.1063/1.4723626
401. Pogorelov NV, Borovikov SN, Burlaga LF, Ebert RW, Heerikhuisen J, Kim TK, Kryukov IA, Suess ST, Wu ST, Zank GP (2012) Numerical modeling of the solar wind flow with observational boundary conditions. AIP conference proceedings 1500(1):134–139. https://doi.org/10.1063/1.4768756
402. Pogorelov NV, Borovikov SN, Bedford MC, Heerikhuisen J, Kim TK, Kryukov IA, Zank GP (2013) Modeling solar wind flow with the multi-scale fluid-kinetic simulation suite. In: Pogorelov NV, Audit E, Zank GP (eds) Numerical modeling of space plasma flows (ASTRONUM2012), Astronomical society of the Pacific conference series, vol 474, p 165
403. Pogorelov NV, Suess ST, Borovikov SN, Ebert RW, McComas DJ, Zank GP (2013) Three-dimensional features of the outer heliosphere due to coupling between the interstellar and interplanetary magnetic fields. IV. Solar cycle model based on ulysses observations. Astrophys J 772(1):2. https://doi.org/10.1088/0004-637x/772/1/2
404. Pogorelov NV, Borovikov SN, Kryukov IA, Wu ST, Yalim MS, Colella PC, Van Straalen B (2017) Modeling coronal mass ejections with the multi-scale fluid-kinetic simulation suite. In: Journal of physics conference series, vol 837, p 012014. https://doi.org/10.1088/1742-6596/837/1/012014
405. Pogorelov NV, Heerikhuisen J, Roytershteyn V, Burlaga LF, Gurnett DA, Kurth WS (2017) Three-dimensional features of the outer heliosphere due to coupling between the interstellar and heliospheric magnetic field. V. The bow wave, heliospheric boundary layer, instabilities, and magnetic reconnection. Astrophys J 845:9. https://doi.org/10.3847/1538-4357/aa7d4f
406. Pomoell J, Poedts S (2018) EUHFORIA: European heliospheric forecasting information asset. J Space Weather Space Clim 8(27):A35. https://doi.org/10.1051/swsc/2018020
407. Pomoell J, Vainio R (2012) Influence of solar wind heating formulations on the properties of shocks in the corona. Astrophys J 745(2):151. http://stacks.iop.org/0004-637X/745/i=2/a=151

408. Pomoell J, Vainio R, Kissmann R (2008) MHD modeling of coronal large-amplitude waves related to CME lift-off. Solar Phys 253:249–261
409. Pomoell J, Vainio R, Kissmann R (2011) MHD simulation of the evolution of shock structures in the solar corona: implications for coronal shock acceleration. Astrophys Space Sci Trans 7(3):387–394
410. Porth O, Xia C, Hendrix T, Moschou SP, Keppens R (2014) MPI-AMRVAC for solar and astrophysics. Astrophys J Suppl Ser 214:4. https://doi.org/10.1088/0067-0049/214/1/4
411. Porth O, Olivares H, Mizuno Y, Younsi Z, Rezzolla L, Moscibrodzka M, Falcke H, Kramer M (2017) The black hole accretion code. Comput Astrophys Cosmol 4:1. https://doi.org/10.1186/s40668-017-0020-2
412. Posner A, Hesse M, St Cyr OC (2014) The main pillar: assessment of space weather observational asset performance supporting nowcasting, forecasting, and research to operations. Space Weather 12(4):257–276. https://doi.org/10.1002/2013SW001007
413. Powell KG (1997) An approximate Riemann solver for magnetohydrodynamics. Springer, Berlin, pp 570–583. https://doi.org/10.1007/978-3-642-60543-7_23
414. Powell KG, Roe PL, Linde TJ, Gombosi TI, de Zeeuw DL (1999) A solution-adaptive upwind scheme for ideal magnetohydrodynamics. J Comput Phys 154:284–309
415. Pulkkinen A, Kuznetsova M, Ridley A, Raeder J, Vapirev A, Weimer D, Weigel RS, Wiltberger M, Millward G, Rastätter L et al (2011) Geospace environment modeling 2008–2009 challenge: ground magnetic field perturbations. Space Weather 9(2):S02004
416. Pulkkinen TI, Baker DN, Wiltberger M, Goodrich C, Lopez RE, Lyon JG (1998) Pseudobreakup and substorm onset: observations and MHD simulations compared. J Geophys Res: Space Phys 103(A7):14,847–14,854. https://doi.org/10.1029/97JA03244
417. Raeder J (2003) Global magnetohydrodynamics - a tutorial. Springer, Berlin, pp 212–246
418. Raeder J, Berchem J, Ashour-Abdalla M (1998) The geospace environment modeling grand challenge: results from a global geospace circulation model. J Geophys Res 103(A7):14,787–14,797
419. Raeder J, McPherron RL, Frank LA, Kokubun S, Lu G, Mukai T, Paterson WR, Sigwarth JB, Singer HJ, Slavin JA (2001) Global simulation of the geospace environment modeling substorm challenge event. J Geophys Res 106:381–396. https://doi.org/10.1029/2000JA000605
420. Raeder J, Wang Y, Fuller-Rowell TJ (2001) Geomagnetic storm simulation with a coupled magnetosphere-ionosphere-thermosphere model. Geophys Monogr Ser 125:377–384
421. Raeder J, Wang YL, Fuller-Rowell TJ, Singer HJ (2001) Global simulation of magnetospheric space weather effects of the bastille day storm. Solar Phys 204:323–337. https://doi.org/10.1023/A:1014228230714
422. Raeder J, Larson D, Li W, Kepko EL, Fuller-Rowell T (2008) OpenGGCM simulations for the THEMIS mission. Space Sci Rev 141(1):535–555
423. Raeder J, Zhu P, Ge Y, Siscoe G (2010) Open Geospace General Circulation Model simulation of a substorm: axial tail instability and ballooning mode preceding substorm onset. J Geophys Res (Space Phys) 115:A00I16. https://doi.org/10.1029/2010JA015876
424. Raeder J, Zhu P, Ge YS, Siscoe G (2013) Auroral signatures of ballooning mode near substorm onset: OpenGGCM simulations. Auroral Phenomenol Magnetos Process: Earth Other Planets 197:389–396
425. Rastätter L, Kuznetsova M, Vapirev A, Ridley A, Wiltberger M, Pulkkinen A, Hesse M, Singer H (2011) Geospace environment modeling 2008–2009 challenge: geosynchronous magnetic field. Space Weather 9(4):S04005
426. Redmon RJ, Denig WF, Loto'aniu PT, Fuller-Rowell D (2018) Chapter 24 - Recent geoeffective space weather events and technological system impacts. In: Buzulukova N (ed) Extreme events in geospace. Elsevier, pp 587–609. https://doi.org/10.1016/B978-0-12-812700-1.00024-8
427. Reiss MA, Hofmeister SJ, De Vischer R, Temmer M, Veronig AM, Delouille V, Mampaey B, Ahammer H (2015) Improvements on coronal hole detection in SDO/AIA images using supervised classification. J Space Weather Space Clim 5:A23

428. Ridley A, De Zeeuw D, Gombosi T, Powell K (2001) Using steady state MHD results to predict the global state of the magnetosphere-ionosphere system. J Geophys Res 106(A12):30,067–30,076
429. Riley P, Linker JA, Mikić Z (2001) An empirically-driven global MHD model of the solar corona and inner heliosphere. J Geophys Res 106:15,889–15,902
430. Riley P, Linker JA, Mikić Z, Odstrčil D, Zurbuchen TH, Lario D, Lepping RP (2003) Using an MHD simulation to interpret the global context of a coronal mass ejection observed by two spacecraft. J Geophys Res 108:1272
431. Riley P, Lionello R, Linker JA, Mikić Z, Luhmann J, Wijaya J (2011) Global MHD modeling of the solar corona and inner heliosphere for the whole heliosphere interval. Solar Phys 274(1–2):361–377
432. Riley P, Linker JA, Lionello R, Mikić Z (2012) Corotating interaction regions during the recent solar minimum: the power and limitations of global MHD modeling. J Atmos Solar Terr Phys 83:1–10
433. Riley P, Baker D, Liu YD, Verronen P, Singer H, Güdel M (2018) Extreme space weather events: from cradle to grave. Space Sci Rev 214:21. https://doi.org/10.1007/s11214-017-0456-3
434. de la Roche AB (2013) The merger of two global commons: the need for new governance. Space Policy 29(2):159–163. https://doi.org/10.1016/j.spacepol.2013.03.013
435. Roe P (1981) Approximate Riemann solvers, parameter vectors, and difference schemes. J Comput Phys 43(2):357–372
436. Ronchi C, Iacono R, Paolucci P (1996) The cubed sphere: a new method for the solution of partial differential equations in spherical geometry. J Comput Phys 124(1):93–114
437. Roumeliotis G (1996) The "stress-and-relax" method for reconstructing the coronal magnetic field from vector magnetograph data. Astrophys J 473:1095–1104
438. Roussev II, Gombosi TI, Sokolov IV, Velli M, Manchester W IV, DeZeeuw DL, Liewer P, Tóth G, Luhmann J (2003) A three-dimensional model of the solar wind incorporating solar magnetogram observations. Astrophys J Lett 595:L57–L61
439. Sarp Yalim M, Pogorelov N, Liu Y (2017) A data-driven MHD model of the global solar corona within multi-scale fluid-kinetic simulation suite (MS-FLUKSS). J Phys Conf Ser 837:012015. https://doi.org/10.1088/1742-6596/837/1/012015
440. Schatten KH (1971) Current sheet magnetic model for the solar corona. Cosmic Electrodyn 2:232–245
441. Schatten KH, Wilcox JM, Ness NF (1969) A model of interplanetary and coronal magnetic fields. Solar Phys 6:442–455
442. Schrijver CJ, De Rosa ML (2003) Photospheric and heliospheric magnetic fields. Solar Phys 212:165–200
443. Schrijver CJ, Sandman AW, Aschwanden MJ, De Rosa ML (2004) The coronal heating mechanism as identified by full-sun visualizations. Astrophys J 615:512–525
444. Schuck PW (2008) Tracking vector magnetograms with the magnetic induction equation. Astrophys J 683:1134–1152. https://doi.org/10.1086/589434
445. Schwadron NA, McComas DJ, Elliott HA, Gloeckler G, Geiss J, von Steiger R (2005) Solar wind from the coronal hole boundaries. J Geophys Res 110:A04104
446. Schwenn R (1990) Large-scale structure of the interplanetary medium. In: Scwenn R, Marsch E (eds) Physics of the inner heliosphere I. Large-scale phenomena. Physics and chemistry in space and solar physics, vol 20. Springer, Berlin, pp 99–171
447. Schwenn R (2006) Space weather: the solar perspective. Living Rev Solar Phys 3:1–72
448. Scolini C, Verbeke C, Poedts S, Chané E, Pomoell J, Zuccarello FP (2018) Effect of the initial shape of coronal mass ejections on 3-D MHD simulations and geoeffectiveness predictions. Space Weather 16:754–771. https://doi.org/10.1029/2018SW001806
449. Shen F, Feng XS, Wu ST, Xiang CQ (2007) Three-dimensional MHD simulation of CMEs in three-dimensional background solar wind with the self-consistent structure on the source surface as input: numerical simulation of the January 1997 Sun-Earth connection event. J Geophys Res (Space Phys) 112(A11):A06109

450. Shen F, Feng XS, Song WB (2009) An asynchronous and parallel time-marching method: application to three-dimensional MHD simulation of solar wind. Sci China Ser E: Tech Sci 52(10):2895–2902
451. Shen F, Feng XS, Wang Y, Wu ST, Song WB, Guo JP, Zhou YF (2011a) Three-dimensional MHD simulation of two coronal mass ejections' propagation and interaction using a successive magnetized plasma blobs model. J Geophys Res (Space Phys) 116(A15):A09103
452. Shen F, Feng XS, Wu ST, Xiang CQ, Song WB (2011b) Three-dimensional MHD simulation of the evolution of the April 2000 CME event and its induced shocks using a magnetized plasma blob model. J Geophys Res (Space Phys) 116(A15):A04102
453. Shen F, Feng XS, Wu ST (2012) 3D MHD numerical study of two CMEs' evolution and their interaction. In: Pogorelov NV, Font JA, Audit E, Zank GP (eds) Numerical modeling of space plasma slows (ASTRONUM 2011), Astronomical society of the Pacific conference series, vol 459, p 247
454. Shen F, Wu ST, Feng XS, Wu CC (2012) Acceleration and deceleration of coronal mass ejections during propagation and interaction. J Geophys Res (Space Phys) 117(A16):A11101
455. Shen F, Shen CL, Wang YM, Feng XS, Xiang CQ (2013) Could the collision of CMEs in the heliosphere be superelastic? Validation through three-dimensional simulations. Geophys Res Lett 40(8):1457–1461
456. Shen F, Shen CL, Zhang J, Hess P, Wang YM, Feng XS, Cheng HZ, Yang Y (2014) Evolution of the 12 July 2012 CME from the Sun to the Earth: data-constrained three-dimensional MHD simulations. J Geophys Res 119:7128–7141
457. Shen F, Wang YM, Shen CL, Feng XS (2016) Turn on the super-elastic collision nature of coronal mass ejections through low approaching speed. Sci Rep 6:19576
458. Shen F, Yang Z, Zhang J, Wei W, Feng XS (2018) Three-dimensional MHD simulation of solar wind using a new boundary treatment: comparison with in situ data at Earth. Astrophys J 866(1):18. http://stacks.iop.org/0004-637X/866/i=1/a=18
459. Shi Y, Wei FS, Feng XS, Ye Z (2001a) Numerical modeling for the propagation of interplanetary disturbance in the three -dimensional solar wind background during the January 1997 event. Chin J Geophys 44(3):294–301
460. Shi Y, Wei FS, Feng XS, Ye Z (2001) Three-dimensional MHD simulation for the solar wind structure observed by Ulysses. Chin Sci Bull 46(16):1352–1355. https://doi.org/10.1007/BF03183386
461. Shiota D, Kataoka R (2016) Magnetohydrodynamic simulation of interplanetary propagation of multiple coronal mass ejections with internal magnetic flux rope (SUSANOO-CME). Space Weather 14(2):56–75
462. Shiota D, Kusano K, Miyoshi T, Nishikawa N, Shibata K (2008) A quantitative MHD study of the relation among arcade shearing, flux rope formation, and eruption due to the tearing instability. J Geophys Res (Space Phys) 113:A03S05. https://doi.org/10.1029/2007JA012516
463. Shiota D, Kusano K, Miyoshi T, Shibata K (2010) Magnetohydrodynamic modeling for a formation process of coronal mass ejections: interaction between an ejecting flux rope and an ambient field. Astrophys J 718:1305–1314
464. Shiota D, Kataoka R, Miyoshi Y, Hara T, Tao C, Masunaga K, Futaana Y, Terada N (2014) Inner heliosphere MHD modeling system applicable to space weather forecasting for the other planets. Space Weather 12:187–204
465. Siscoe G (2000) The space-weather enterprise: past, present, and future. J Atmos Solar Terr Phys 62:1223–1232
466. Siscoe G (2007) Space weather forecasting historically viewed through the lens of meteorology, Praxis Publishing Ltd., Chichester, UK, p 5
467. Siscoe G, Crooker N, Siebert K, Maynard N, Weimer D, White W (2005) Cusp geometry in MHD simulations. Surv Geophys 26:387–407. https://doi.org/10.1007/s10712-005-1902-4
468. Siscoe GL, Erickson GM, Sonnerup BUO, Maynard NC, Siebert KD, Weimer DR, White WW (2001) Global role of $E_{\parallel}$ in magnetopause reconnection: an explicit demonstration. J Geophys Res: Space Phys 106(A7):13,015–13,022. https://doi.org/10.1029/2000JA000062

469. Siscoe GL, Erickson GM, Sonnerup BUO, Maynard NC, Schoendorf JA, Siebert KD, Weimer DR, White WW, Wilson GR (2002) Flow-through magnetic reconnection. Geophys Res Lett 29(13):1626. https://doi.org/10.1029/2001GL013536
470. Skála J, Baruffa F, Büchner J, Rampp M (2015) The 3D MHD code GOEMHD3 for astrophysical plasmas with large Reynolds numbers. Code description, verification, and computational performance. Astron Astrophys 580:A48. https://doi.org/10.1051/0004-6361/201425274
471. Skinner MA, Ostriker EC (2010) The Athena astrophysical magnetohydrodynamics code in cylindrical geometry. Astrophys J Suppl Ser 188:290–311. https://doi.org/10.1088/0067-0049/188/1/290
472. Sokolov IV, Timofeev EV, Sakai J, Takayama K (2002) Artificial wind - A new framework to construct simple and efficient upwind shock-capturing schemes. J Comput Phys 181(1):354–393
473. Sokolov IV, Roussev II, Gombosi TI, Lee MA, Kóta J, Forbes TG, Manchester WB, Sakai JI (2004) A new field line advection model for solar particle acceleration. Astrophys J Lett 616:L171–L174
474. Sokolov IV, Powell KG, Gombosi TI, Roussev II (2006) A TVD principle and conservative TVD schemes for adaptive Cartesian grids. J Comput Phys 220:1–5
475. Sokolov IV, Powell KG, Cohen O, Gombosi TI (2008) Computational MagnetoHydrodynamics, Based on solution of the well-posed Riemann problem. In: Pogorelov NV, Audit E, Zank GP (eds) Numerical modeling of space plasma flows, Astronomical society of the pacific conference series, vol 385, p 291
476. Sokolov IV, van der Holst B, Oran R, Downs C, Roussev II, Jin M, Manchester WB IV, Evans RM, Gombosi TI (2013) Magnetohydrodynamic waves and coronal heating: unifying empirical and MHD turbulence models. Astrophys J 764:23
477. Sonnerup BU, Siebert KD, White WW (2001) Simulations of the magnetosphere for zero interplanetary magnetic field: the ground state. J Geophys Res: Space Phys 106(A12):29,419–29,434
478. Springel V (2010) E pur si muove: Galilean-invariant cosmological hydrodynamical simulations on a moving mesh. Mon Not R Astron Soc 401:791–851. https://doi.org/10.1111/j.1365-2966.2009.15715.x
479. Stanier A, Chacón L, Chen G (2019) A fully implicit, conservative, non-linear, electromagnetic hybrid particle-ion/fluid-electron algorithm. J Comput Phys 376:597–616. https://doi.org/10.1016/j.jcp.2018.09.038
480. Steenburgh RA, Biesecker DA, Millward GH (2014) From predicting solar activity to forecasting space weather: practical examples of research-to-operations and operations-to-research. Solar Phys 289(2):675–690. https://doi.org/10.1007/s11207-013-0308-6
481. Stone JM, Norman ML (1992a) ZEUS-2D: a radiation magnetohydrodynamics code for astrophysical flows in two space dimensions. I - The hydrodynamic algorithms and tests. Astrophys J Suppl Ser 80:753–790. https://doi.org/10.1086/191680
482. Stone JM, Norman ML (1992) ZEUS-2D: a radiation magnetohydrodynamics code for astrophysical flows in two space dimensions. II. The magnetohydrodynamic algorithms and tests. Astrophys J Suppl Ser 80:791. https://doi.org/10.1086/191681
483. Stone JM, Norman ML (2013) ZEUS-2D: simulation of fluid dynamical flows. Astrophysics source code library. http://ascl.net/1306.014
484. Stone JM, Mihalas D, Norman ML (1992) ZEUS-2D: a radiation magnetohydrodynamics code for astrophysical flows in two space dimensions. III - The radiation hydrodynamic algorithms and tests. Astrophys J Suppl Ser 80:819–845. https://doi.org/10.1086/191682
485. Stone JM, Gardiner TA, Teuben P, Hawley JF, Simon JB (2008) Athena: a new code for astrophysical MHD. Astrophys J Suppl Ser 178:137
486. Stout QF, De Zeeuw DL, Gombosi TI, Groth CPT, Marshall HG, Powell KG (1997) Adaptive blocks: a high performance data structure. In: Proceedings of supercomputing'97, pp 1–10
487. Strong K, Saba J, Kucera T (2012) Understanding space weather: the Sun as a variable Star. Bull Am Meteorol Soc 93(9):1327–1335. https://doi.org/10.1175/BAMS-D-11-00179.1

488. Strong K, Viall N, Schmelz J, Saba J (2017) Understanding space weather: Part III: the Sun domain. Bull Am Meteorol Soc 98(12):2593–2602. https://doi.org/10.1175/BAMS-D-16-0204.1
489. Strong KT, Schmelz JT, Saba JLR, Kucera TA (2017) Understanding space weather: Part II: the violent Sun. Bull Am Meteorol Soc 98(11):2387–2396. https://doi.org/10.1175/BAMS-D-16-0191.1
490. Suess ST, Wang AH, Wu ST, Poletto G, McComas DJ (1999) A two-fluid, MHD coronal model. J Geophys Res 104(A3):4697–4708
491. Sun TR, Wang C, Wei F, Sembay S (2015a) X-ray imaging of Kelvin-Helmholtz waves at the magnetopause. J Geophys Res: Space Phys 120(1):266–275. https://doi.org/10.1002/2014JA020497
492. Sun TR, Wang C, Zhang JJ, Pilipenko VA, Wang Y, Wang JY (2015) The chain response of the magnetospheric and ground magnetic field to interplanetary shocks. J Geophys Res: Space Phys 120(1):157–165. https://doi.org/10.1002/2014JA020754
493. Sun TR, Tang BB, Wang C, Guo XC, Wang Y (2019) Large scale characteristics of flux transfer events on the dayside magnetopause. J Geophys Res: Space Phys 124:2425–2434. https://doi.org/10.1029/2018JA026395
494. Sun TR, Wang C, Sembay S, Lopez RE, Escoubet P, Branduardi-Raymont G, Zheng JH, Yu XZ, Guo XC, Dai L, Liu ZQ, Wei F, Guo YH (2019) Soft X-ray imaging of the magnetosheath and cusps under different solar wind conditions: MHD simulations. J Geophys Res: Space Phys 124:2435–2450. https://doi.org/10.1029/2018JA026093
495. Sun W, Dryer M, Fry CD, Deehr CS, Smith Z, Akasofu SI, Kartalev MD, Grigorov KG (2002) Evaluation of solar Type II radio burst estimates of initial solar wind shock speed using a kinematic model of the solar wind on the April 2001 solar event swarm. Geophys Res Lett 29:1171. https://doi.org/10.1029/2001GL013659
496. Sun X, Liu Y, Hoeksema JT, Hayashi K, Zhao X (2011) A new method for polar field interpolation. Solar Phys 270:9–22
497. Taktakishvili A, Pulkkinen A, MacNeice P, Kuznetsova M, Hesse M, Odstrčil D (2011) Modeling of coronal mass ejections that caused particularly large geomagnetic storms using ENLIL heliosphere cone model. Space Weather 9:S06002
498. Tanaka T (1994) Finite volume TVD scheme on an unstructured grid system for three-dimensional MHD simulation of inhomogeneous systems including strong background potential fields. J Comput Phys 111:381–390
499. Tanaka T (1995) Generation mechanisms for magnetosphere-ionosphere current systems deduced from a three-dimensional MHD simulation of the solar wind-magnetosphere-ionosphere coupling processes. J Geophys Res (Space Phys) 100(A7):12,057–12,074
500. Tanaka T (1999) Configuration of the magnetosphere-ionosphere convection system under northward IMF conditions with nonzero IMF B_y. J Geophys Res 104(A7):14,683–14,690
501. Tanaka T (2003) Formation of magnetospheric plasma population regimes coupled with the dynamo process in the convection system. J Geophys Res 108(A8):1315. https://doi.org/10.1029/2002JA009668
502. Tanaka T (2003) Finite volume TVD Schemes for magnetohydrodynamics on unstructered grids. Space plasma simulation. Springer, Berlin, pp 275–295
503. Tanaka T (2007) Magnetosphere-ionosphere convection as a compound system. Space Sci Rev 133(1):1–72. https://doi.org/10.1007/s11214-007-9168-4
504. Tanaka T (2015) Substorm auroral dynamics reproduced by advanced global magnetosphere-ionosphere (M-I) coupling simulation, American Geophysical Union (AGU) and Wiley. Washington, DC, chap 13:177–190. https://doi.org/10.1002/9781118978719.ch13
505. Tanaka T, Obara T, Kunitake M (2004) Formation of the theta aurora by a transient convection during northward interplanetary magnetic field. J Geophys Res 109(A9):201
506. Tanaka T, Nakamizo A, Yoshikawa A, Fujita S, Shinagawa H, Shimazu H, Kikuchi T, Hashimoto K (2010) Substorm convection and current system deduced from the global simulation. J Geophys Res 115(A5):220

507. Tanaka T, Ebihara Y, Watanabe M, Den M, Fujita S, Kikuchi T, Hashimoto KK, Kataoka R (2017) Global simulation study for the time sequence of events leading to the substorm onset. J Geophys Res: Space Phys 122(6):6210–6239. https://doi.org/10.1002/2017JA024102
508. Tanaka T, Obara T, Watanabe M, Fujita S, Ebihara Y, Kataoka R (2017) Formation of the sun-aligned arc region and the void (polar slot) under the null-separator structure. J Geophys Res: Space Phys 122(4):4102–4116. https://doi.org/10.1002/2016JA023584
509. Tang BB, Wang C, Li WY (2013) The magnetosphere under the radial interplanetary magnetic field: a numerical study. J Geophys Res: Space Phys 118(12):7674–7682
510. Tappin S, Dryer M, Han S, Wu S (1988) Expected IPS variations due to a disturbance described by a 3-D MHD model. Planet Space Sci 36(11):1155–1163
511. Temmer M, Rollett T, Möstl C, Veronig AM, Vršnak B, Odstrčil D (2011) Influence of the ambient solar wind flow on the propagation behavior of interplanetary coronal mass ejections. Astrophys J 743:101. https://doi.org/10.1088/0004-637X/743/2/101
512. Teyssier R (2002) Cosmological hydrodynamics with adaptive mesh refinement. A new high resolution code called RAMSES. Astron Astrophys 385:337–364. https://doi.org/10.1051/0004-6361:20011817
513. Titov VS, Démoulin P (1999) Basic topology of twisted magnetic configurations in solar flares. Astron Astrophys 351:707–720
514. Tobiska WK, Atwell W, Beck P, Benton E, Copeland K, Dyer C, Gersey B, Getley I, Hands A, Holland M, Hong S, Hwang J, Jones B, Malone K, Meier MM, Mertens C, Phillips T, Ryden K, Schwadron N, Wender SA, Wilkins R, Xapsos MA (2015) Advances in atmospheric radiation measurements and modeling needed to improve air safety. Space Weather 13:202–210
515. Tobiska WK, Bouwer D, Smart D, Shea M, Bailey J, Didkovsky L, Judge K, Garrett H, Atwell W, Gersey B, Wilkins R, Rice D, Schunk R, Bell D, Mertens C, Xu X, Wiltberger M, Wiley S, Teets E, Jones B, Hong S, Yoon K (2016) Global real-time dose measurements using the automated radiation measurements for aerospace safety (ARMAS) system. Space Weather 14(11):1053–1080. https://doi.org/10.1002/2016SW001419
516. Tobiska WK, Meier MM, Matthiae D, Copeland K (2018) Chapter 18 - Characterizing the variation in atmospheric radiation at aviation altitudes. In: Buzulukova N (ed) Extreme events in geospace. Elsevier, pp 453–471. https://doi.org/10.1016/B978-0-12-812700-1.00018-2
517. Toffoletto F, Sazykin S, Spiro R, Wolf R (2003) Inner magnetospheric modeling with the rice convection model. Space Sci Rev 107(1):175–196. https://doi.org/10.1023/A:1025532008047
518. Török T, Downs C, Linker JA, Lionello R, Titov VS, Mikić Z, Riley P, Caplan RM, Wijaya J (2018) Sun-to-Earth MHD simulation of the 2000 July 14 "Bastille Day" eruption. Astrophys J 856:75. https://doi.org/10.3847/1538-4357/aab36d
519. Tóth G (2000) The $\nabla \cdot \mathbf{B} = 0$ constraint in shock-capturing magnetohydrodynamics codes. J Comput Phys 161:605–652
520. Tóth G (2006) Flexible, efficient and robust algorithm for parallel execution and coupling of components in a framework. Comput Phys Commun 174:793–802
521. Tóth G, Odstrčil D (1996) Comparison of some flux corrected transport and total variation diminishing numerical schemes for hydrodynamic and magnetohydrodynamic problems. J Comput Phys 128(1):82–100
522. Tóth G, Roe PL (2002) Divergence- and curl-preserving prolongation and restriction formulas. J Comput Phys 180:736–750
523. Tóth G, Sokolov IV, Gombosi TI, Chesney DR, Clauer CR, De Zeeuw DL, Hansen KC, Kane KJ, Manchester WB, Oehmke RC, Powell KG, Ridley AJ, Roussev II, Stout QF, Volberg O, Wolf RA, Sazykin S, Chan A, Yu B, Kóta J (2005) Space weather modeling framework: a new tool for the space science community. J Geophys Res (Space Phys) 110(A9):A12226
524. Tóth G, Volberg O, Ridley AJ, Gombosi TI, De Zeeuw DL, Hansen KC, Chesney DR, Stout QF, Powell KG, Kane KJ, et al (2005) A physics-based software framework for Sun-Earth connection modeling. In: Multiscale coupling of Sun-Earth processes, vol 1. Elsevier, p 383
525. Tóth G, Ma Y, Gombosi TI (2008) Hall magnetohydrodynamics on block-adaptive grids. J Comput Phys 227(14):6967–6984

526. Tóth G, Meng X, van der Holst B, Gombosi T (2010) Magnetosphere modeling using MHD with anisotropic pressure. In: 38th COSPAR scientific assembly, vol 38, p 2066
527. Tóth G, van der Holst B, Sokolov IV, de Zeeuw DL, Gombosi TI, Fang F, Manchester WB, Meng X, Najib D, Powell KG, Stout QF, Glocer A, Ma YJ, Opher M (2012) Adaptive numerical algorithms in space weather modeling. J Comput Phys 231:870–903
528. Tóth G, Jia X, Markidis S, Peng IB, Chen Y, Daldorff LKS, Tenishev VM, Borovikov D, Haiducek JD, Gombosi TI, Glocer A, Dorelli JC (2016) Extended magnetohydrodynamics with embedded particle-in-cell simulation of Ganymede's magnetosphere. J Geophys Res: Space Phys 121(2):1273–1293. https://doi.org/10.1002/2015JA021997
529. Totten TL, Freeman JW, Arya S (1996) Application of the empirically derived polytropic index for the solar wind to models of solar wind propagation. J Geophys Res 101:15,629–15,636
530. Tsunomura S (1999) Numerical analysis of global ionospheric current system including the effect of equatorial enhancement. Ann Geophys 17:692–706. https://doi.org/10.1007/s00585-999-0692-2
531. Tsurutani BT, Mannucci AJ, Iijima B, Guarnieri FL, Gonzalez WD, Judge DL, Gangopadhyay P, Pap J (2006) The extreme Halloween 2003 solar flares (and Bastille Day, 2000 Flare), ICMEs, and resultant extreme ionospheric effects: a review. Adv Space Res 37:1583–1588
532. Tsurutani BT, Verkhoglyadova OP, Jand Mannucci GS A, Lakhina Li G, Zank GP (2009) A brief review of "solar flares" on the ionosphere. Radio Sci 44:RS0A17. https://doi.org/10.1029/2008RS004029
533. Usmanov A, Besser B, Fritzer J, Goldstein M (2000a) Simulation of a coronal streamer: Alfvén wave acceleration. Adv Space Res 25(9):1897–1900
534. Usmanov AV (1993a) A global numerical 3-D MHD model of the solar wind. Solar Phys 146:377–396
535. Usmanov AV (1993b) The global structure of the solar wind in June 1991. Solar Phys 148:371–382
536. Usmanov AV (1995) Numerical 3-D time-dependent MHD model of the solar wind. Space Sci Rev 72:121–124
537. Usmanov AV (1996) A global 3-D MHD solar wind model with Alfvén waves. In: Winterhalter D, Gosling JT, Habbal SR, Kurth WS, Neugebauer M (eds) American institute of physics conference series, vol 382, pp 141–144. https://doi.org/10.1063/1.51468
538. Usmanov AV, Dryer M (1995) A global 3-D simulation of interplanetary dynamics in June 1991. Solar Phys 159:347–370
539. Usmanov AV, Goldstein ML (2003) A tilted-dipole MHD model of the solar corona and solar wind. J Geophys Res (Space Phys) 108:1354
540. Usmanov AV, Goldstein ML (2006) A three-dimensional MHD solar wind model with pickup protons. J Geophys Res (Space Phys) 111:A07101
541. Usmanov AV, Goldstein ML, Besser BP, Fritzer JM (2000) A global MHD solar wind model with WKB Alfvén waves: comparison with Ulysses data. J Geophys Res 105:12,675–12,696
542. Usmanov AV, Goldstein ML, Matthaeus WH (2012) Three-dimensional magnetohydrodynamic modeling of the solar wind including pickup protons and turbulence transport. Astrophy J 754:40. https://doi.org/10.1088/0004-637X/754/1/40
543. Usmanov AV, Goldstein ML, Matthaeus WH (2014) Three-fluid, three-dimensional magnetohydrodynamic solar wind model with eddy viscosity and turbulent resistivity. Astrophys J 788:43. https://doi.org/10.1088/0004-637X/788/1/43
544. Usmanov AV, Matthaeus WH, Goldstein ML, Chhiber R (2018) The steady global corona and solar wind: a three-dimensional mhd simulation with turbulence transport and heating. Astrophys J 865(1):25. http://stacks.iop.org/0004-637X/865/i=1/a=25
545. Rusanov V (1961) Calculation of interaction of non-steady shock waves with obstacles. J Comput Math Phys 1:267–279
546. Valori G, Kleim B, Keppens R (2005) Extrapolation of a nonlinear force-free field containg a highly twisted magnetic loop. Astrophys J 433:335–347
547. van der Holst B, Keppens R, Meliani Z (2008) A multidimensional grid-adaptive relativistic magnetofluid code. Comput Phys Commun 179:617–627. https://doi.org/10.1016/j.cpc.2008.05.005

548. van der Holst B, Manchester W IV, Sokolov IV, Tóth G, Gombosi TI, DeZeeuw D, Cohen O (2009) Breakout Coronal mass ejection or streamer blowout: the bugle effect. Astrophys J 693:1178
549. van der Holst B, Manchester WB IV, Frazin RA, Vásquez AM, Tóth G, Gombosi TI (2010) A data-driven, two-temperature solar wind model with Alfvén waves. Astrophys J 725:1373
550. van der Holst B, Keppens R, Meliani Z, Porth O, van Marle AJ, Delmont P, Xia C (2012) MPI-AMRVAC: MPI-adaptive mesh refinement-versatile advection code. Astrophysics source code library. http://ascl.net/1208.014
551. van der Holst B, Sokolov IV, Meng X, Jin M, Manchester WB IV, Tóth G, Gombosi TI (2014) Alfvén wave solar model (AWSoM): coronal heating. Astrophys J 782:81
552. Vandas M, Odstrčil D (2000) Magnetic cloud evolution: a comparison of analytical and numerical solutions. J Geophys Res 105:12,605–12,616
553. Vögler A, Shelyag S, Schüssler M, Cattaneo F, Emonet T, Linde T (2005) Simulations of magneto-convection in the solar photosphere. Equations, methods, and results of the MURaM code. Astron Astrophys 429:335–351. https://doi.org/10.1051/0004-6361:20041507
554. Walker RJ, Ogino T (1989) Global magnetohydrodynamic simulations of the magnetosphere. IEEE Trans Plasma Sci 17:135–149. https://doi.org/10.1109/27.24618
555. Walker RJ, Ogino T, Raeder J, Ashour-Abdalla M (1993) A global magnetohydrodynamic simulation of the magnetosphere when the interplanetary magnetic field is southward: the onset of magnetotail reconnection. J Geophys Res 98:17,235–17,250. https://doi.org/10.1029/93JA01321
556. Wang C, Liu JB, Huang ZH, Richardson JD (2007) Response of the magnetic field in the geosynchronous orbit to solar wind dynamic pressure pulses. J Geophys Res 112(A12):A12,210
557. Wang C, Li H, Guo X, Ding K, Huang Z (2012) Numerical study on the response of the Earth's magnetosphere-ionosphere system to a super solar storm. Sci China Ser D-Earth Sci 55:1037–1042
558. Wang C, Han JP, Li H, Peng Z, Richardson JD (2014) Solar wind-magnetosphere energy coupling function fitting: results from a global MHD simulation. J Geophys Res: Space Phys 119(8):6199–6212. https://doi.org/10.1002/2014JA019834
559. Wang C, Li ZJ, Sun TR, Liu ZQ, Liu J, Wu Q, Zheng JH, Li J (2017) SMILE satellite mission survey. Space Int (in Chinese) 464(8):13–16. http://caj.d.cnki.net//KDoc/docdown/pubdownload.aspx?dk=kdoc%3apdfdown%3a0241a47e9da3945da7587644d3f6b25e
560. Wang H, Ridley AJ, Lühr H (2008) Validation of the space weather modeling framework using observations from champ and dmsp. Space Weather 6(3):S03001. https://doi.org/10.1029/2007SW000355
561. Wang J, Feng XS, Du AM, Ge YS (2014) Modeling the interaction between the solar wind and Saturn's magnetosphere by the AMR-CESE-MHD method. J Geophys Res: Space Phys 119(12):9919–9930
562. Wang J, Du AM, Zhang Y, Zhang TL, Ge YS (2015) Modeling the Earth's magnetosphere under the influence of solar wind with due northward IMF by the AMR-CESE-MHD model. Sci China Earth Sci 58(7):1235–1242
563. Wang J, Guo Z, Ge YS, Du A, Huang C, Qin P (2018) The responses of the Earth's magnetopause and bow shock to the IMF B_z and the solar wind dynamic pressure: a parametric study using the AMR-CESE-MHD model. J Space Weather Space Clim 8:A41. https://doi.org/10.1051/swsc/2018030
564. Wang S, Wei FS (2007) Progress of space weather research in China (in Chinese). Prog Geophys 22(4):1025–1029
565. Wang W, Killeen TL, Burns AG, Roble RG (1999) A high-resolution, three-dimensional, time dependent, nested grid model of the coupled thermosphere-ionosphere. J Atmos Solar Terr Phys 61:385–397
566. Wang YM, Sheeley NR Jr (1990) Solar wind speed and coronal flux-tube expansion. Astrophys J 355:726–732

567. Wang YM, Sheeley NR Jr, Phillips JL, Goldstein BE (1997) Solar wind stream interactions and the wind speed-expansion factor relationship. Astrophys J Lett 488:L51
568. Wang YM, Chen C, Gui B, Shen C, Ye P, Wang S (2011) Statistical study of coronal mass ejection source locations: understanding CMEs viewed in coronagraphs. J Geophys Res 116:A04104
569. Webb DF, Howard TA (2011) Solar flares: magnetohydrodynamic process. Living Rev Solar Phys 8:6
570. Webb DF, Howard TA (2012) Coronal mass ejections: observations. Living Rev Solar Phys 9:3
571. Webb DF, Möstl C, Jackson BV, Bisi MM, Howard TA, Mulligan T, Jensen EA, Jian LK, Davies JA, de Koning CA, Liu Y, Temmer M, Clover JM, Farrugia CJ, Harrison RA, Nitta N, Odstrčil D, Tappin SJ, Yu HS (2013) Heliospheric imaging of 3D density structures during the multiple coronal mass ejections of late July to early August 2010. Solar Phys 285:317–348. https://doi.org/10.1007/s11207-013-0260-5
572. Wei F, Dryer M (1991) Propagation of solar flare-associated interplanetary shock waves in the heliospheric meridional plane. Solar Phys 132:373–394
573. Werner ALE, Yordanova E, Dimmock AP, Temmer M (2019) Modeling the multiple CME interaction event on 6–9 September 2017 with WSA-ENLIL+Cone. Space Weather 17:357–369. https://doi.org/10.1029/2018SW001993
574. White WW, Siscoe GL, Erickson GM, Kaymaz Z, Maynard NC, Siebert KD, Sonnerup BUO, Weimer DR (1998) The magnetospheric sash and the cross-tail S. Geophys Res Lett 25(10):1605–1608. https://doi.org/10.1029/98GL50865
575. White WW, Schoendorf JA, Siebert KD, Maynard NC, Weimer DR, Wilson GL, Sonnerup BUO, Siscoe GL, Erickson GM (2001) MHD simulation of magnetospheric transport at the mesoscale. Am Geophys Union 229–240: https://doi.org/10.1029/GM125p0229
576. Wiengarten T (2011) MHD simulation of the inner-heliospheric magnetic field. Master's thesis, the Fakultät fur Physik und Astronomie of the Ruhr-Universität, Germany
577. Wiengarten T, Kleimann J, Fichtner H, Cameron R, Jiang J, Kissmann R, Scherer K (2013) MHD simulation of the inner-heliospheric magnetic field. J Geophys Res 29–44
578. Wiengarten T, Kleimann J, Fichtner H, Kühl P, Kopp A, Heber B, Kissmann R (2014) Cosmic ray transport in heliospheric magnetic structures. I. Modeling background solar wind using the CRONOS magnetohydrodynamic code. Astrophys J 788:80. https://doi.org/10.1088/0004-637X/788/1/80
579. Wiengarten T, Fichtner H, Kleimann J, Kissmann R (2015) Implementing turbulence transport in the CRONOS framework and application to the propagation of CMEs. Astrophys J 805:155. https://doi.org/10.1088/0004-637X/805/2/155
580. Wilcox JM, Hokesema JT, Scherrer PH (1980) Origin of the warped heliospheric current sheet. Science 209:603–605
581. Wiltberger M, Weigel RS, Lotko W, Fedder JA (2009) Modeling seasonal variations of auroral particle precipitation in a global-scale magnetosphere-ionosphere simulation. J Geophys Res (Space Phys) 114:A01204. https://doi.org/10.1029/2008JA013108
582. Wiltberger M, Merkin V, Lyon JG, Ohtani S (2015) High-resolution global magnetohydrodynamic simulation of bursty bulk flows. J Geophys Res: Space Phys 120(6):4555–4566. https://doi.org/10.1002/2015JA021080
583. Winglee R, Menietti J (1998) Auroral activity associated with pressure pulses and substorms: a comparison between global fluid modeling and Viking UV imaging. J Geophys Res: Space Phys 103(A5):9189–9205
584. Winske D, Yin L, Omidi N, Karimabadi H, Quest K (2003) Hybrid simulation codes: past, present and future-a tutorial. Springer, Berlin, pp 136–165
585. Wold AM, Mays ML, Taktakishvili A, Jian LK, Odstrčil D, MacNeice P (2018) Verification of real-time WSA-ENLIL+Cone simulations of CME arrival-time at the CCMC from 2010 to 2016. J Space Weather Space Clim 8:A17. https://doi.org/10.1051/swsc/2018005
586. Wood BE, Wu CC, Rouillard AP, Howard RA, Socker DG (2012) A coronal hole's effects on coronal mass ejection shock morphology in the inner heliosphere. Astrophys J 755(1):43

587. Worden J, Harvey J (2000) An evolving synoptic magnetic flux map and implications for the distribution of photospheric magnetic flux. Solar Phys 195:247–268
588. Wright Jr JM, Lennon BGJ, Corell RW, Ostenso NA, Huntress Jr WT, Harrison JB (1995) The national space weather program strategic plan. Technical Report FCM-P30-1995, the Working Group for the National Space Weather Program, Washington DC
589. Wu CC, Dryer M (1997) Three-dimensional MHD simulation of interplanetary magnetic field changes at 1 AU caused by a simulated solar disturbance and a tilted heliospheric current/plasma sheet. Solar Phys 173:391–408
590. Wu CC, Dryer M, Wu ST (1996) Three-dimensional MHD simulation of interplanetary magnetic field changes at 1 AU as a consequence of simulated solar flares. Ann Geophys 14:383–399
591. Wu CC, Fry CD, Berdichevsky D, Dryer M, Smith Z, Detman T (2005) Predicting the arrival time of shock passages at Earth. Solar Phys 227:371–386
592. Wu CC, Fry CD, Dryer M, Wu ST, Thompson B, Liou K, Feng XS (2007a) Three-dimensional global simulation of multiple ICMEs' interaction and propagation from the Sun to the heliosphere following the 25–28 October 2003 solar events. Adv Space Res 40:1827–1834
593. Wu CC, Fry CD, Wu ST, Dryer M, Liou K (2007b) Three-dimensional global simulation of interplanetary coronal mass ejection propagation from the Sun to the heliosphere: solar event of 12 May 1997. J Geophys Res (Space Phys) 112:A09104
594. Wu CC, Dryer M, Wu ST, Wood BE, Fry CD, Liou K, Plunkett S (2011) Global three-dimensional simulation of the interplanetary evolution of the observed geoeffective coronal mass ejection during the epoch 1–4 August 2010. J Geophys Res (Space Phys) 116(A15):A12103
595. Wu CC, Dryer M, Wu ST, Wood BE, Fry CD, Liou K, Plunkett S (2016a) Global magnetohydrodynamic simulation of the March 15, 2013 coronal mass ejection event -Interpretation of the 30–80 MEV proton flux. J Geophys Res (Space Phys) 121(A01):56–76
596. Wu CC, Liou K, Vourlidas A, Plunkett S, Dryer M, Wu ST, Socker D, Wood BE, Hutting L, Howard RA (2016b) Numerical simulation of multiple CME-driven shocks in the month of 2011 September. J Geophys Res (Space Phys) 121(A03):1839–1856
597. Wu CC, Liou K, Lepping RP, Vourlidas A, Plunkett S, Socker D, Wu ST (2017) Observation of an extremely large-density heliospheric plasma sheet compressed by an interplanetary shock at 1 AU. Solar Phys 292:109. https://doi.org/10.1007/s11207-017-1114-3
598. Wu ST, Dryer M (2015) Comparative analyses of current three-dimensional numerical solar wind models. Sci China: Earth Sci 58:839–858
599. Wu ST, Guo WP, Michels DJ, Burlaga LF (1999) MHD description of the dynamical relationships between a flux rope, streamer, coronal mass ejection, and magnetic cloud: an analysis of the January 1997 Sun-Earth connection event. J Geophys Res 104(A7):14,789–14,801
600. Wu ST, Zhou YF, Jiang CW, Feng XS, Hu Q (2016) A data-constrained three-dimensional magnetohydrodynamic simulation model for a coronal mass ejection initiation. J Geophys Res: Space Phys 121:1009–1023. https://doi.org/10.1002/2015JA021615
601. Xia C, Keppens R, Antolin P, Porth O (2014) Simulating the in situ condensation process of solar prominences. Astrophys J Lett 792:L38. https://doi.org/10.1088/2041-8205/792/2/L38
602. Xia C, Teunissen J, Mellah IE, Chané E, Keppens R (2018) MPI-AMRVAC 2.0 for solar and astrophysical applications. Astrophys J Suppl Ser 234(2):30. http://stacks.iop.org/0067-0049/234/i=2/a=30
603. Xie H, Ofman L, Lawrence G (2004) Cone model for halo CMEs: application to space weather forecasting. J Geophys Res (Space Phys) 109(A3):A03109. https://doi.org/10.1029/2003JA010226
604. Yalim MS (2008) An artificial compressibility analogy approach for compressible ideal MHD: application to space weather simulation. (PhD thesis), Université libre de Bruxelles
605. Yang LP, Feng XS, Xiang CQ, Jiang CW (2011) Numerical validation and comparison of three solar wind heating methods by the SIP-CESE MHD model. Chin Phys Lett 28(3):039,601
606. Yang LP, Feng XS, Xiang CQ, Zhang SH, Wu ST (2011) Simulation of the unusual solar minimum with 3D SIP-CESE MHD model by comparison with multi-satellite observations. Solar Phys 271:91–110

607. Yang LP, Feng XS, Xiang CQ, Liu Y, Zhao XP, Wu ST (2012) Time-dependent MHD modeling of the global solar corona for year 2007: driven by daily-updated magnetic field synoptic data. J Geophys Res 117(A16):A08110
608. Yao Y, Ebihara Y, Tanaka T (2015) Sudden pressure enhancement and tailward retreat in the near-Earth plasma sheet: THEMIS observation and MHD simulation. J Geophys Res: Space Phys 120(1):201–211
609. Yeates AR, Mackay DH, van Ballegooijen AA (2007) Modelling the global solar corona: filament chirality observations and surface simulations. Solar Phys 245:87–107
610. Yeates AR, Attrill GDR, Nandy D, Mackay DH, Martens PCH, van Ballegooijen AA (2010) Comparison of a global magnetic evolution model with observations of coronal mass ejections. Astrophys J 709(2):1238. http://stacks.iop.org/0004-637X/709/i=2/a=1238
611. Yee K (1966) Numerical solution of initial boundary value problems involving Maxwell's equations in isotropic media. IEEE Trans Antennas Propag 14(3):302–307. https://doi.org/10.1109/TAP.1966.1138693
612. Yizengaw E, Moldwin MB, Dyson PL, Immel TJ (2005) Southern Hemisphere ionosphere and plasmasphere response to the interplanetary shock event of 29–31 October 2003. J Geophys Res 110:A09S30
613. Yu ST, Chang SC (1997) Treatments of stiff source terms in conservation laws by the method of space-time conservation element and solution element. AIAA paper 97-0435
614. Yue X, Wan W, Yan L, Sun W, Hu L, Schreiner WS (2018) Chapter 22 - The effect of solar radio bursts on gnss signals. In: Buzulukova N (ed) Extreme events in geospace. Elsevier, pp 541–554, https://doi.org/10.1016/B978-0-12-812700-1.00022-4
615. Zalesak ST (1979) Fully multidimensional flux-corrected transport algorithms for fluids. J Comput Phys 31(3):335–362
616. Zank GP, Pauls HL, Williams LL, Hall DT (1996) Interaction of the solar wind with the local interstellar medium: a multifluid approach. J Geophys Res 101:21,639–21,656 https://doi.org/10.1029/96JA02127
617. Zhang B, Brambles O, Lotko W, Dunlap-Shohl W, Smith R, Wiltberger M, Lyon J (2013) Predicting the location of polar cusp in the Lyon-Fedder-Mobarry global magnetosphere simulation. J Geophys Res 118(10):6327–6337
618. Zhang B, Brambles OJ, Wiltberger M, Lotko W, Ouellette JE, Lyon JG (2016) How does mass loading impact local versus global control on dayside reconnection? Geophys Res Lett 43(5):1837–1844. https://doi.org/10.1002/2016GL068005
619. Zhang JJ, Wang C, Sun TR, Liu CM, Wang KR (2015) GIC due to storm sudden commencement in low-latitude high-voltage power network in China: observation and simulation. Space Weather 13(10):643–655. https://doi.org/10.1002/2015SW001263
620. Zhang ZC, Yu ST (1999) Shock capturing without Riemann solver-A modified space-time CE/SE method for conservation laws. AIAA paper 99-0904
621. Zhao B, Wan W, Tschu K, Igarashi K, Kikuchi T, Nozaki K, Watari S, Li G, Paxton LJ, Liu L, Ning B, Liu JY, Su SY, Bulanon HP (2008) Ionosphere disturbances observed throughout southeast asia of the superstorm of 20–22 November 2003. J Geophys Res: Space Phys 113(A3):A00A04
622. Zhao X, Hoeksema J (1995) Predicting the heliospheric magnetic field using the current sheet-source surface model. Adv Space Res 16(9):181–184
623. Zhao X, Hoeksema J, Rich N (2002a) Modeling the radial variation of coronal streamer belts during sunspot ascending phase. Adv Space Res 29(3):411–416
624. Zhao XP, Plunkett SP, Liu W (2002) Determination of geometrical and kinematical properties of halo coronal mass ejections using the cone model. J Geophys Res (Space Phys) 107(A8):1223. https://doi.org/10.1029/2001JA009143
625. Zhou Y, Feng X (2017) Numerical study of the propagation characteristics of coronal mass ejections in a structured ambient solar wind. J Geophys Res (Space Phys) 122:1451–1462. https://doi.org/10.1002/2016JA023053
626. Zhou YF, Feng XS (2008) Numerical study of successive CMEs during November 4–5, 1998. Sci China Ser E: Technol Sci 51:1600–1610

627. Zhou YF, Feng XS (2013) MHD numerical study of the latitudinal deflection of coronal mass ejection. J Geophys Res: Space Phys 118(10):6007–6018
628. Zhou YF, Feng XS, Wu ST (2008) Numerical simulation of the 12 May 1997 CME event. Chin Phys Lett 25:790–793
629. Zhou YF, Feng XS, Wu ST, Du D, Shen F, Xiang CQ (2012) Using a 3-D spherical plasmoid to interpret the Sun-to-Earth propagation of the 4 November 1997 coronal mass ejection event. J Geophys Res 117(A16):A01102
630. Zhou YF, Feng XS, Zhao XH (2014) Using a 3-D MHD simulation to interpret propagation and evolution of a coronal mass ejection observed by multiple spacecraft: the 3 April 2010 event. J Geophys Res: Space Phys 119:9321–9333
631. Zhu P, Raeder J, Germaschewski K, Hegna CC (2009) Initiation of ballooning instability in the near-earth plasma sheet prior to the 23 March 2007 THEMIS substorm expansion onset. Ann. Geophys. 27(3):1129–1138. https://doi.org/10.5194/angeo-27-1129-2009
632. Ziegler U (1998) NIRVANA: an adaptive mesh refinement code for gas dynamics and MHD. Comput Phys Commun 109:111
633. Ziegler U (2004) An ADI-based adaptive mesh Poisson solver for the MHD code NIRVANA. Comput Phys Commun 157:207
634. Ziegler U (2008) The NIRVANA code: parallel computational MHD with adaptive mesh refinement. Comput Phys Commun 179(4):227–244. https://doi.org/10.1016/j.cpc.2008.02.017
635. Ziegler U (2011) NIRVANA: a numerical tool for astrophysical gas dynamics. Astrophysics source code library. arXiv:1101.006
636. Ziegler U (2011) A semi-discrete central scheme for magnetohydrodynamics on orthogonal-curvilinear grids. J Comput Phys 230(4):1035–1063. https://doi.org/10.1016/j.jcp.2010.10.022
637. Zuccarello FP, Jacobs C, Soenen A, Poedts S, van der Holst B, Zuccarello F (2009) Modelling the initiation of coronal mass ejections: magnetic flux emergence versus shearing motions. A&A 507(1):441–452
638. Zuccarello FP, Bemporad A, Jacobs C, Mierla M, Poedts S, Zuccarello F (2012) The role of streamers in the deflection of coronal mass ejections: comparison between STEREO three-dimensional reconstructions and numerical simulations. Astrophys J 744(1):66

Chapter 2
Cell-Centered Finite Volume Methods

Abstract This chapter is devoted to the description of finite volume method (FVM). FVM is a discretization technique for partial differential equations, especially those that arise from physical conservation laws. FVM uses a volume integral formulation of the problem with a finite partitioning set of volumes to discretize the equations. Currently, FVM is a common practice for discretizing computational fluid/MHD dynamics equations. Here we devote our attention to the cell-centered FVM in three-dimensional computational domain. We begin with a general introduction to the first order finite volume methods, followed by a description of high-resolution formulations for conservative laws. It is mentioned that all the formulations are aimed to be applicable at least to the hexahedral cell, particularly quadrilaterally-faced hexahedron (cuboid) with 6 faces, 12 edges, and 8 vertices.

2.1 Brief Overview of Finite Volume Methods

In a finite-volume approach, the governing partial differential equations are solved on a domain discretized into series of contiguous control volumes, or cells making up the computational grid for the geometry of interest. The equations are solved in integral form enforcing both locally and globally conservation of key flow properties, such as mass, momentum, and energy.

Finite volume schemes can be categorized, in the main, as of the cell-centered (e.g., [139, 145, 208, 210, 211, 214, 219, 220, 367, 368, 389, 454, 475]) or the node-centered (also referred as vertex-centered) type (e.g., [5, 110, 112, 126, 129, 131, 345, 348–350, 355, 420]). A third approach also exists, the so called FVM of the edge-type (e.g., [40, 66, 67]). A fourth approach may be the barycentric finite volume (e.g., [9, 114]). For the cell-centered approach, the (finite) volumes used to satisfy the integral form of the equation are the mesh elements themselves. For the node-centered approach, the finite volumes are elements of the mesh dual to the computational mesh. For the edge-type control volumes the nodes are placed on the edges of an original triangulation. For the barycentric finite volume, the average of solution is computed on control volumes that are defined by joining the barycenter

© Springer Nature Singapore Pte Ltd. 2020

X. Feng, *Magnetohydrodynamic Modeling of the Solar Corona and Heliosphere*, Atmosphere, Earth, Ocean & Space,
https://doi.org/10.1007/978-981-13-9081-4_2

of each volume of the primal mesh with its vertices: each control volume is then associated with a hypersurface of the initial mesh.

The cell-centered finite volume method utilizes any given cell itself as the corresponding control volume and the numerical fluxes are evaluated at each cell-to-cell boundary faces by using corresponding quadrature formula. The possible candidates of the cells are tetrahedrons, pyramids, prisms, hexahedrons (e.g., [4, 235]), or any types of polyhedrons (e.g., [5]) in three dimensions can be utilized. The cells can be even locally concave. The order of accuracy is also obtainable for arbitrarily high orders based on high-order solution reconstruction. For the simplicity of the control volume definition and generality of the high-order solution strategy, the current chapter briefly describes the standard cell-centered finite volume method; for more detailed explanations of the principles of finite-volume methods the reader may refer to the textbooks by Toro [469], and LeVeque [271].

Essential equations in hydrodynamics and magnetohydrodynamics (MHD) can be written in the form of hyperbolic conservation laws in Cartesian coordinates as follows:

$$\frac{\partial \mathbf{U}}{\partial t} + \nabla \cdot \mathbf{F} = \mathbf{S}(\mathbf{U}) \tag{2.1}$$

where $\mathbf{U}$ is the vector of conserved quantities, $\mathbf{F}$ is a nonlinear flux depending on the state vector $\mathbf{U}$, and $\mathbf{S}$ is the source vector. In integrating Eq. (2.1), one derives the conventional finite-volume methods based on Reynolds transport theorem [486], in which space and time are treated separately. By Reynolds transport theorem, using Gauss's law in three-dimensional cell V_i, the integration form of Eq. (2.1) can be written as

$$\frac{\partial \mathbf{U}_i}{\partial t} V_i = -\oint\limits_{\partial V_i} \mathbf{F} \cdot \mathbf{n} dA + \mathbf{S}_i V_i \tag{2.2}$$

where V_i is the cell volume and ∂V_i is the cell boundary. $\mathbf{U}_i = \frac{1}{V_i}\int_{V_i} \mathbf{U} dV$ and $\mathbf{S}_i = \frac{1}{V_i}\int_{V_i} \mathbf{S} dV$ are the cell-averaged conserved variables and source terms. Here, dV and dA are the volume and surface element of a fixed spatial control volume V_i with the centroid indexed by i and $\mathbf{n}$ is the unit outward vector normal to the surface of the control volume. The flux, $\mathbf{F} \cdot \mathbf{n}$, represents the flux of solution quantities through the boundaries of the control volume. These fluxes can be generally categorized as either arising from wave propagation phenomena (hyperbolic fluxes) or from diffusion processes (elliptic fluxes). For instance, a 3D hexahedral computational cell is depicted in Fig. 2.1 for a fixed three-dimensional control volume that does not vary in time. Alternatively, integrate Eq. (2.2) with respect to time from t_1 to t_2 to arrive at

$$\int\limits_{V_i} \mathbf{U} dV \Bigg|_{t_1}^{t_2} = \int_{t_1}^{t_2} dt \left(-\oint\limits_{\partial V_i} (\mathbf{F} \cdot \mathbf{n}) dA + \int\limits_{V_i} \mathbf{S} dV \right) \tag{2.3}$$

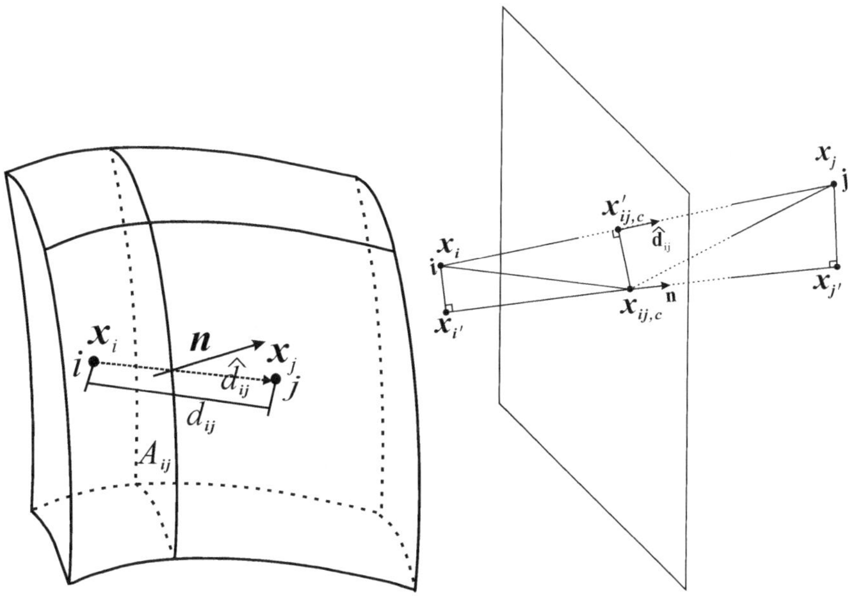

Fig. 2.1 Three-dimensional hexahedral cell and cell face gradient reconstruction

As such, the conservation laws state that the rate of change of the total amount of a substance contained in a fixed spatial control volume V_i is equal to the combination of the following two factors: (i) the flux of that substance across the boundary ∂V_i of the control volume V_i, and (ii) the integration of the source term over the fixed spatial domain. The conventional finite volume methods concentrate on the evaluation of the right side of Eq. (2.2).

Numerical methods for conservation laws can in general be divided into two approaches: a one-step approach, where time and space discretization is coupled via the Cauchy–Kowalewski procedure and semi-discrete schemes (see e.g., [271, 469]), where time and space integrations are separated. Both methods have been successful towards high accuracy and non-oscillatory transitions, by using strong stability preserving time-marching schemes, combined with polynomial reconstruction value of $\mathbf{U}$ and the flux $\mathbf{F}$ at the interface from a given cell average.

Rewrite the integral form of the conservation equations given by Eq. (2.2) to arrive at the semi-discrete form of the conservation equations as follows:

$$\frac{\partial \mathbf{U}_i}{\partial t} = -\mathbf{R}(\mathbf{U}_i) \tag{2.4}$$

where $\mathbf{R}(\mathbf{U})$ is the residual operator for the control volume. Equation (2.2) or equivalently Eq. (2.4), is a non-linear system of first-order ordinary differential equations describing the time evolution of the average value of the solution vector in each computational cell. In order to solve the resulting system of equations, the various terms appearing in the right-hand side of Eq. (2.2) must be evaluated. Then, the temporal derivative in Eq. (2.4) is usually discretized by a finite difference time-marching method, such as the Runge–Kutta methods. Due to the above two factors of spatial and temporal treatments, the fractional step (or splitting) methods resort to strong measurements of numerically segregating the two factors and treating them in a sequential manner. The flux, $\mathbf{F}$, at the cell boundary is evaluated as a function of the discontinuous states on either side of the cell interfaces, where discontinuities arise due to the piecewise approximations for $\mathbf{U}$ within each control volume. That is, the numerical fluxes, $(\mathbf{F} \cdot \mathbf{n})$, at the quadrature point of each face of cell V_i are determined from the solution of a Riemann problem. Given the left and right interface solution states, $\mathbf{U}_L$ and $\mathbf{U}_R$, an upwind numerical flux is evaluated by solving a Riemann problem or approximate Riemann problem in the direction defined by the normal to the face. In a word, the finite volume method evaluates the change of the conservative values of all control volumes by integrating their interface or edge fluxes. From the point view of programming it consists of two steps, calculating fluxes at every interface and gathering interface fluxes for every control volume. Therefore, the data structure is designed as having two primary arrays, one for control volumes, another for faces, with a bi-directional reference between control volumes and faces.

In general, the governing equations are solved to achieve steady state as well as time accurate solutions. In time accurate simulations the simplest time stepping scheme uses the same time step in all the grid cells with an explicit time discretization like Runge–Kutta method. In some cases the equations are stiff, i.e., the numerical stability restricts the time step to be much smaller than what would be dictated by accuracy considerations. Depending on the mathematical properties of the stiff terms, point-implicit, semi-implicit or fully implicit time discretization schemes can sometimes allow larger time steps (e.g., [3, 68, 121, 209, 213, 214, 227, 234, 244, 297, 398, 400, 440, 450, 473, 475]).

This chapter is organized as follows. Section 2.2 describes the implementation of first order FVM. Some basics of MHD equations are introduced in Sect. 2.3. Magnetic field divergence cleaning methods given in Sect. 2.4 are crucial in MHD simulations. Section 2.5 is devoted to high resolution practice of MHD equations. As key steps of high-order FVM, cell-centered state variable reconstructions, transformation of cell-centered reconstructions from conservative to primitive variables and numerical flux functions are provided in Sects. 2.6, 2.7 and 2.8. Section 2.9 presents Newton–Krylov methods of solving non-linear equations resulted from the FVM.

2.2 FVM: First Order Practice

In practice, if the flux dyad, $\mathbf{F} = \mathbf{F}_{\text{inv}} - \mathbf{F}_{\text{diff}}$, consists of convective (hyperbolic or inviscid) flux, $\mathbf{F}_{\text{inv}}$, diffusion (elliptic) flux, $\mathbf{F}_{\text{diff}}$, the governing equations (2.1) can be written out in essentially convection-diffusion form

$$\frac{\partial \mathbf{U}}{\partial t} + \nabla \cdot \mathbf{F}_{\text{inv}} = \nabla \cdot \mathbf{F}_{\text{diff}} + \mathbf{S} \tag{2.5}$$

As done for Eq. (2.2), using a cell centered finite volume formulation, spatial discretization of Eq. (2.5) results in semi-discrete form or ordinary differential equation (ODE) as follows:

$$\frac{\partial \mathbf{U}_i}{\partial t} V_i + \sum_{j=1}^{\mathrm{N_{faces}}} (\mathbf{F}_{\text{inv}} \cdot \mathbf{n})_{ij} A_{ij} = \sum_{j=1}^{\mathrm{N_{faces}}} (\mathbf{F}_{\text{diff}} \cdot \mathbf{n})_{ij} A_{ij} + \mathbf{S}_i V_i \tag{2.6}$$

Here, $\mathbf{U}_i$ denotes the vector of cell-averaged conserved variables associated with the finite volume cell indexed using the subscript i, which has a volume V_i. A_{ij} is the area of the face f that separates cells i and j, $\mathbf{n}$ is the unit normal to the face directed from cell i to cell j, and $\mathrm{N_{faces}}$ is the number of faces surrounding cell i. The computational domain over which the function is discretised by a grid into a number of non-overlapping finite volumes, or cells. A cell can be arbitrarily shaped, but its boundary must consist of a number of faces, each separating it from a single other cell, see Fig. 2.1 for reference to a 3D hexahedral computational cell. These faces are assumed to be straight (planar, in three dimensions) so that the normal unit vector $\mathbf{n}$ has a constant value $\mathbf{n}_f$ along each face f. We focus on cell-centred finite volume methods, meaning that the values of $\mathbf{U}$ are sought only at the centroids of the cells. Within the finite-volume framework, to achieve a first order spatial accuracy, the vector of conservative variables $\mathbf{U}_i$ and source terms $\mathbf{S}_i$ are assumed to be constant in the cell i. The convective and resistive diffusion fluxes $(\mathbf{F}_{\text{inv}} \cdot \mathbf{n})_{ij}$ and $(\mathbf{F}_{\text{diff}} \cdot \mathbf{n})_{ij}$ are reconstructed at a face f using the data from the two cells that lie adjacent to the face. Since face f belongs to the cell i, one of these adjacent cells will be cell i. The other adjacent cell, indexed using the subscript j, is a cell that shares face f with cell i. The numerical fluxes, $(\mathbf{F}_{\text{inv}} \cdot \mathbf{n})_{ij}$, at the each face of cell i are determined from the solution of a Riemann problem. Given the left and right interface solution states, $\mathbf{U}_L = \mathbf{U}_i$ and $\mathbf{U}_R = \mathbf{U}_j$, an upwind numerical flux is evaluated by solving a Riemann problem or approximate Riemann problem (Fig. 2.2) in the direction defined by the normal to the face. If $\mathbf{U}_i$ is constant in each cell i, the numerical procedure outlined above results in a first-order finite-volume scheme. If the initial solution states, $\mathbf{U}_L$ and $\mathbf{U}_R$, are determined by performing the solution variable reconstruction as discussed in Sect. 2.6, the numerical procedure results in a second-order or high order upwind finite-volume scheme.

The evaluation of the convective components of the numerical flux in Eq. (2.6) depends on the solution state at the cell face with the form $(\mathbf{F}_{\text{inv}} \cdot \mathbf{n})(\mathbf{U}_{ij})$ or equivalently $(\mathbf{F}_{\text{inv}} \cdot \mathbf{n})(\mathbf{W}_{ij})$, where $\mathbf{U}_{ij}$ and $\mathbf{W}_{ij}$ are respectively the conservative solution variables and the primitive solution variables, both evaluated at the cell interface by some interpolation methods of the left and the right solution states $\mathbf{U}_L = \mathbf{U}_i$ and $\mathbf{U}_R = \mathbf{U}_j$, or equivalently $\mathbf{W}_L = \mathbf{W}_i$ and $\mathbf{W}_R = \mathbf{W}_j$.

Generally, the diffusion flux $\mathbf{F}_{\text{diff}}$ in Eq. (2.5) takes the form of $\mathbf{F}_{\text{diff}} = \mathbf{F}_{\text{diff}}(\mathbf{U}, \nabla \mathbf{U})$ consisting of viscous terms, magnetic diffusion as well thermal conduction terms. The gradients of the conservative variables at both the cell centers and the faces are

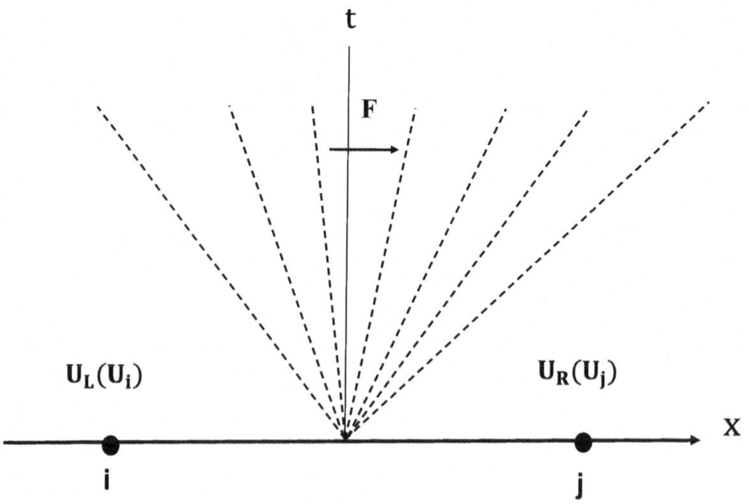

Fig. 2.2 The Riemann problem

required to be computed for the evaluation of diffusion flux $\mathbf{F}_{\text{diff}}$. The evaluation of the gradients for the primitive variables $(\nabla\mathbf{W})_{ij}$ or the conservative variables $(\nabla\mathbf{U})_{ij}$ at the cell interface, requires some additional work. Typically, particularly on unstructured grids, reconstruction of the diffusive flux involves a form of the unlimited cell-centered gradient reconstruction procedure such as the least-squares or Green-Gauss gradient methods, followed by averaging the mutually adjacent cell-centered gradients to obtain the face gradient.

Some numerical flux functions of the MHD system will be introduced in Sect. 2.8 in detail. In this section, we will assume that $\mathbf{F}_{\text{inv}}$ is already known and introduce the rest of the FVM procedure. The following part presents some cell-face interpolation criteria of dependent variables and their gradients at the cell centers, the internal cell faces and at the boundaries required for the discretization of diffusive and convective terms of Eq. (2.6).

2.2.1 *Calculation of Inviscid and Diffusion Fluxes at the Cell Interface*

Here, X represents a nominal variable, $X|_i$ and $(\nabla X)|_i$ denote the cell-centered value and cell-centered gradient (usually situated at the centroid $\mathbf{x}_i$ of cell i). The cell centroid $\mathbf{x}_i$ of cell i is

$$\mathbf{x}_i = (x_i, y_i, z_i) = \frac{1}{V_i}\int_{V_i}(x, y, z)dV = \frac{1}{V_i}\left(\int_{V_i} xdV, \int_{V_i} ydV, \int_{V_i} zdV\right),$$

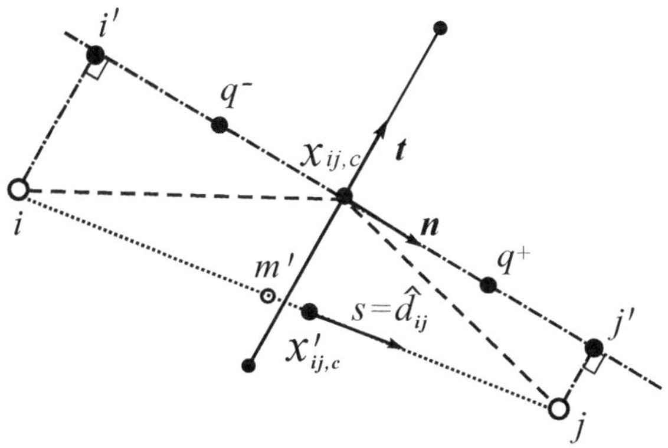

Fig. 2.3 Geometric elements of two cells i and j sharing the same interface A_{ij}

and the face centroid $\mathbf{x}_{ij,c}$ is

$$\mathbf{x}_{ij,c} = (x_{ij,c}, y_{ij,c}, z_{ij,c}) = \frac{1}{A_{ij}} \int_{A_{ij}} (x, y, z) dS = \frac{1}{A_{ij}} \left(\int_{A_{ij}} x dS, \int_{A_{ij}} y dS, \int_{A_{ij}} z dS \right).$$

Also, we denote $X_i = \frac{1}{V_i} \int_{V_i} X(\mathbf{x}) dV$ the cell-averaged of variable X in a finite volume cell V_i. Up to second order accuracy, $X_i = X|_i \equiv X(\mathbf{x}_i)$.

The simplest criterion of assigning at the cell-face point $\mathbf{x}_{ij,c}$ (usually situated at the face centroid) the value is to linearly interpolate cell-centered values somewhere on the line $\overline{\mathbf{x}_i \mathbf{x}_j}$ (point $\mathbf{x}'_{ij,c}$ in Figs. 2.1 and 2.3). That is,

$$X(\mathbf{x}_{ij,c}) \approx X(\mathbf{x}'_{ij,c}) = (1 - \alpha_f) X_i + \alpha_f X_j \tag{2.7}$$

where $\alpha_f = \frac{|\mathbf{x}'_{ij,c} - \mathbf{x}_i|}{|\mathbf{x}_j - \mathbf{x}_i|}$ is the interpolation factor. The evaluation of this factor depends on how to define $\mathbf{x}'_{ij,c}$ point position. For example, $\mathbf{x}'_{ij,c}$ can be defined in such a way that the distance from $\mathbf{x}'_{ij,c}$ to $\mathbf{x}_{ij,c}$ is minimum. Choose the point $\mathbf{x}'_{ij,c}$ on the line $\overline{\mathbf{x}_i \mathbf{x}_j}$ such that $\overline{\mathbf{x}_{ij,c} \mathbf{x}'_{ij,c}}$ is perpendicular to the line $\overline{\mathbf{x}_i \mathbf{x}_j}$, see Figs. 2.1 and 2.3. In that case, $\alpha_f = \frac{(\mathbf{x}_{ij,c} - \mathbf{x}_i) \cdot \mathbf{s}}{|\mathbf{x}_j - \mathbf{x}_i|}$ with $\mathbf{s} = \hat{d}_{ij} = \frac{\mathbf{x}_j - \mathbf{x}_i}{|\mathbf{x}_j - \mathbf{x}_i|}$. However, it is very common to define this point using $\alpha_f = \frac{|\mathbf{x}_{ij,c} - \mathbf{x}_i|}{|\mathbf{x}_{ij,c} - \mathbf{x}_i| + |\mathbf{x}_j - \mathbf{x}_{ij,c}|}$. The differences between the two choices of α_f decrease when it approaches 0.5.

Another approach to obtain the value at the cell face centroid $\mathbf{x}_{ij,c}$ relies on the interpolated values of the dependent variable at the corners or vertices. Thus, the value of the dependent variable (or its gradient) at the face point $\mathbf{x}_{ij,c}$ is obtained by arithmetic averaging process: $X(\mathbf{x}_{ij,c}) = \sum_{v\ell=1}^{N_v} X_{v\ell}/N_v$, where the summation is extended to all the N_v vertexes of the interface A_{ij} and $X_{v\ell} = X|_{\mathbf{x}_{v\ell}}$ is the dependent variable at the cell vertex ($N_v = 4$ for 3D hexahedral computational cells as shown in Fig. 2.1). The dependent variable at the cell vertex $\mathbf{x}_{v\ell}$ can be approximated by assuming

$$a_I X_{v\ell} = \sum a_k X_k \tag{2.8}$$

where $X_{v\ell}$ is the value to be interpolated at a given vertex $\mathbf{x}_{v\ell}$, X_k are the values of the dependent variable at the surrounding cell centroids ($k = 1, 2, \ldots, 8$ for 3D hexahedral computational cells as shown in Fig. 2.1) sharing the same vertex $\mathbf{x}_{v\ell}$ in common. Here, $a_k = \omega_k / |\mathbf{x}_k - \mathbf{x}_{v\ell}|$ with ω_k being the solid angle and $a_I = \sum a_k$. Even though the solid angle is not very difficult to estimate, the coefficients in Eq. (2.8) have been approximated in a simpler manner as $a_k = V_k/|\mathbf{x}_k - \mathbf{x}_{v\ell}|$ or $a_k = V_k/|\mathbf{x}_k - \mathbf{x}_{v\ell}|^2$, where V_k is the cell volume associated to the grid cell k.

With known cell-centered values of dependent variables or cell-face interpolated values of dependent variables available such as Eqs. (2.7) and (2.8), we can employ least squares method or Green-Gauss theorem (See Sect. 2.6 for more details) to obtain the gradient values $\nabla X|_i = \nabla X|_{\mathbf{x}_i}$ at the cell centroids. After the gradient values at the cell centroids are known, we can use the interpolated gradient at $\mathbf{x}'_{ij,c}$ to obtain the cell face value

$$X(\mathbf{x}_{ij,c}) \approx X(\mathbf{x}'_{ij,c}) + \nabla X|_{\mathbf{x}'_{ij,c}} \cdot \left(\mathbf{x}_{ij,c} - \mathbf{x}'_{ij,c}\right)$$

where $\nabla X|_{\mathbf{x}'_{ij,c}} = (1 - \alpha_f)\, \nabla X|_i + \alpha_f\, \nabla X|_j$ and $X(\mathbf{x}'_{ij,c}) = (1 - \alpha_f)X_i + \alpha_f X_j$.

Interface values can be obtained directly using the projected cell-centered values on the surface normal direction. The linear interpolation gives

$$X(\mathbf{x}_{ij,c}) \approx (1 - \alpha'_f)X_{i'} + \alpha'_f X_{j'}$$

where $\alpha'_f = \left|\mathbf{x}_{ij,c} - \mathbf{x}_{i'}\right| / \left|\mathbf{x}_{j'} - \mathbf{x}_{i'}\right|$, $X_{i'} = X(\mathbf{x}_{i'}) = X_i + (\nabla X)|_i \cdot (\mathbf{x}_{i'} - \mathbf{x}_i)$, and $X_{j'} = X(\mathbf{x}_{j'}) = X_j + (\nabla X)|_j \cdot \left(\mathbf{x}_{j'} - \mathbf{x}_j\right)$. Here, as shown in Figs. 2.1 and 2.3, the points $\mathbf{x}_{i'}$ and $\mathbf{x}_{j'}$ are situated on the line passing through the interface centroid and along the face normal $\mathbf{n}$. The line $\overline{\mathbf{x}_i\mathbf{x}_{i'}}$ is perpendicular to the line $\overline{\mathbf{x}_{i'}\mathbf{x}_{j'}}$ with the perpendicular foot point $\mathbf{x}_{i'}$. Similarly, the line $\overline{\mathbf{x}_j\mathbf{x}_{j'}}$ is perpendicular to the line $\overline{\mathbf{x}_{i'}\mathbf{x}_{j'}}$ with the perpendicular foot point $\mathbf{x}_{j'}$.

Interface value can be obtained from cell-centered values of both the dependent variable and its gradient. Obviously, the estimated approximation from i or from j may be different:

$$\begin{aligned} X(\mathbf{x}_{ij,c}) &\approx X(\mathbf{x}^-_{ij,c}) = X_i + (\nabla X)|_i \cdot \left(\mathbf{x}_{ij,c} - \mathbf{x}_i\right), \\ X(\mathbf{x}_{ij,c}) &\approx X(\mathbf{x}^+_{ij,c}) = X_j + (\nabla X)|_j \cdot \left(\mathbf{x}_{ij,c} - \mathbf{x}_j\right) \end{aligned}$$

This approach is more accurate when the distance ($\left|\mathbf{x}_{ij,c} - \mathbf{x}_i\right|$ or $\left|\mathbf{x}_{ij,c} - \mathbf{x}_j\right|$) is shorter. For this reason, the interface value can be estimated using some kind of interpolation depending on the distance to the cell centers. For example

$$X(\mathbf{x}_{ij,c}) \approx (1 - \alpha_f)X\left(\mathbf{x}^-_{ij,c}\right) + \alpha_f X\left(\mathbf{x}^+_{ij,c}\right)$$

Evidently, two choices can be used to define α_f, $\alpha_f = \frac{|\mathbf{x}'_{ij,c}-\mathbf{x}_i|}{|\mathbf{x}_j-\mathbf{x}_i|}$ and $\alpha_f = \frac{|\mathbf{x}_{ij,c}-\mathbf{x}_i|}{|\mathbf{x}_{ij,c}-\mathbf{x}_j|+|\mathbf{x}_{ij,c}-\mathbf{x}_i|}$. Commonly, a simple arithmetic average: $X(\mathbf{x}_{ij,c}) \approx \frac{1}{2}\left(X\left(\mathbf{x}^-_{ij,c}\right) + X\left(\mathbf{x}^+_{ij,c}\right)\right)$ is used [306].

A more accurate procedure for the evaluation of the diffusion term can be formulated as follows (e.g., [381]). Introduce two auxiliary points, q^- and q^+, which are situated somewhere in the surface normal direction (see Figs. 2.1 and 2.3). Then, the following can be written:

$$
\begin{aligned}
X_{q^-} &\approx X_i + (\nabla X)|_i \cdot \left(\mathbf{x}_{q^-} - \mathbf{x}_i\right)\\
X_{q^-} &\approx X\left(\mathbf{x}_{ij,c}\right) + (\nabla X)_{ij} \cdot \left(\mathbf{x}_{q^-} - \mathbf{x}_{ij,c}\right) = X\left(\mathbf{x}_{ij,c}\right) - (\nabla X)_{ij} \cdot \mathbf{n}\left|\mathbf{x}_{q^-} - \mathbf{x}_{ij,c}\right|\\
X_{q^+} &\approx X(\mathbf{x}_{ij,c}) + (\nabla X)_{ij} \cdot \left(\mathbf{x}_{q^+} - \mathbf{x}_{ij,c}\right) = X(\mathbf{x}_{ij,c}) + (\nabla X)_{ij} \cdot \mathbf{n}\left|\mathbf{x}_{q^+} - \mathbf{x}_{ij,c}\right|\\
X_{q^+} &\approx X_j + (\nabla X)|_j \cdot \left(\mathbf{x}_{q^+} - \mathbf{x}_j\right)
\end{aligned}
$$

From these equations,

$$
\begin{aligned}
X(\mathbf{x}_{ij,c}) = {} & \frac{\left|\mathbf{x}_{q^-} - \mathbf{x}_{ij,c}\right| X_j + \left|\mathbf{x}_{q^+} - \mathbf{x}_{ij,c}\right| X_i}{\left|\mathbf{x}_{q^-} - \mathbf{x}_{ij,c}\right| + \left|\mathbf{x}_{q^+} - \mathbf{x}_{ij,c}\right|}\\
& + \frac{\left|\mathbf{x}_{q^-} - \mathbf{x}_{ij,c}\right| (\nabla X)|_j \cdot \left(\mathbf{x}_{q^+} - \mathbf{x}_j\right) + \left|\mathbf{x}_{q^+} - \mathbf{x}_{ij,c}\right| (\nabla X)|_i \cdot \left(\mathbf{x}_{q^-} - \mathbf{x}_i\right)}{\left|\mathbf{x}_{q^-} - \mathbf{x}_{ij,c}\right| + \left|\mathbf{x}_{q^+} - \mathbf{x}_{ij,c}\right|}
\end{aligned}
\tag{2.9}
$$

$$
(\nabla X)_{ij} \cdot \mathbf{n} = \frac{X_j - X_i}{\left|\mathbf{x}_{q^-} - \mathbf{x}_{ij,c}\right| + \left|\mathbf{x}_{q^+} - \mathbf{x}_{ij,c}\right|} + \frac{(\nabla X)|_j \cdot \left(\mathbf{x}_{q^+} - \mathbf{x}_j\right) - (\nabla X)|_i \cdot \left(\mathbf{x}_{q^-} - \mathbf{x}_i\right)}{\left|\mathbf{x}_{q^-} - \mathbf{x}_{ij,c}\right| + \left|\mathbf{x}_{q^+} - \mathbf{x}_{ij,c}\right|}
\tag{2.10}
$$

The two extra points, q^- and q^+, are defined in such a way that $\left|\mathbf{x}_{q^-} - \mathbf{x}_i\right| = \left|\mathbf{x}_{q^-} - \mathbf{x}_{ij,c}\right|$ and $\left|\mathbf{x}_{q^+} - \mathbf{x}_j\right| = \left|\mathbf{x}_{q^+} - \mathbf{x}_{ij,c}\right|$. Equation (2.10) also gives an expression for the variable gradient at the cell interface along the normal direction.

The interface gradient can benefit from the decomposition of the face normal vector $\mathbf{n}$ into the (in general) nonorthogonal $\mathbf{s}$, $\mathbf{t}$, and $\mathbf{r}$ directions: $\mathbf{n} = \beta_s\mathbf{s} + \beta_t\mathbf{t} + \beta_r\mathbf{r}$. The unit vector $\mathbf{s}$ is defined in the $\overline{\mathbf{x}_i\mathbf{x}_j}$ direction (i.e., $\mathbf{s} = \hat{d}_{ij} = \frac{\mathbf{x}_j-\mathbf{x}_i}{|\mathbf{x}_j-\mathbf{x}_i|}$, see Fig. 2.1), while the other two unit vectors are chosen arbitrarily, with the conditions of being orthogonal ($\mathbf{t}\cdot\mathbf{r} = 0$) and situated on the face surface (then $\mathbf{t}\cdot\mathbf{n} = \mathbf{r}\cdot\mathbf{n} = 0$). The coefficients β_s, β_t, β_r can easily be obtained: $\beta_s = 1/(\mathbf{s}\cdot\mathbf{n})$ (from a dot product of $\mathbf{n}$ and $\mathbf{n}$), $\beta_t = -(\mathbf{s}\cdot\mathbf{t})/(\mathbf{s}\cdot\mathbf{n})$ (from $\mathbf{n}\cdot\mathbf{t} = 0$), and $\beta_r = -(\mathbf{s}\cdot\mathbf{r})/(\mathbf{s}\cdot\mathbf{n})$ (from $\mathbf{n}\cdot\mathbf{r} = 0$). Then

$$
\begin{aligned}
(\nabla X)_{ij} \cdot \mathbf{n} &= (\nabla X)_{ij} \cdot (\beta_s\mathbf{s} + \beta_t\mathbf{t} + \beta_r\mathbf{r})\\
&\approx \frac{X_j - X_i}{\left|\mathbf{x}_j - \mathbf{x}_i\right| (\mathbf{s}\cdot\mathbf{n})} - \overline{(\nabla X)}_{ij} \cdot \left(\frac{(\mathbf{s}\cdot\mathbf{t})\mathbf{t}}{\mathbf{s}\cdot\mathbf{n}} + \frac{(\mathbf{s}\cdot\mathbf{r})\mathbf{r}}{\mathbf{s}\cdot\mathbf{n}}\right)
\end{aligned}
\tag{2.11}
$$

The interface gradient $\overline{(\nabla X)_{ij}}$ on the right hand of the Eq. (2.11) can be estimated using the cell center values, with the interpolation methods similar to those of the interface values. For example, $\overline{(\nabla X)_{ij}} \approx (1-\alpha_f)\,(\nabla X)|_i + \alpha_f\,(\nabla X)|_j$ with $\alpha_f = \frac{\left|\mathbf{x}'_{ij,c}-\mathbf{x}_i\right|}{\left|\mathbf{x}_j-\mathbf{x}_i\right|}$. Equation (2.11) can be simplified to

$$(\nabla X)_{ij} \cdot \mathbf{n} \approx \frac{X_j - X_i}{\left|\mathbf{x}_j - \mathbf{x}_i\right|(\mathbf{s}\cdot\mathbf{n})} - \overline{(\nabla X)_{ij}} \cdot \left(\frac{\mathbf{s}}{\mathbf{s}\cdot\mathbf{n}} - \mathbf{n}\right)$$

since

$$\frac{(\mathbf{s}\cdot\mathbf{t})\mathbf{t} + (\mathbf{s}\cdot\mathbf{r})\mathbf{r}}{\mathbf{s}\cdot\mathbf{n}} = \frac{\mathbf{s} - (\mathbf{s}\cdot\mathbf{n})\mathbf{n}}{\mathbf{s}\cdot\mathbf{n}} = \frac{\mathbf{s}}{\mathbf{s}\cdot\mathbf{n}} - \mathbf{n}$$

holds obviously (decomposition of vector $\mathbf{s}$ on $(\mathbf{n}, \mathbf{t}, \mathbf{r})$ directions, $\mathbf{s} = (\mathbf{s}\cdot\mathbf{t})\mathbf{t} + (\mathbf{s}\cdot\mathbf{r})\mathbf{r} + (\mathbf{s}\cdot\mathbf{n})\mathbf{n}$).

Interface gradients are directly evaluated by interpolating cell-centered values with a deferred-correction approach:

$$(\nabla X)_{ij} \cdot \mathbf{n} \approx \frac{X_j - X_i}{\left|\mathbf{x}_j - \mathbf{x}_i\right|}\alpha + \overline{(\nabla X)_{ij}} \cdot (\mathbf{n} - \mathbf{s}\alpha) \tag{2.12}$$

As suggested by Perez-Segarra et al. [381], the coefficient α has several choices: (1) $\alpha = 1$; (2) $\alpha = 1/(\mathbf{n}\cdot\mathbf{s})$; (3) $\alpha = \mathbf{n}\cdot\mathbf{s}$. It is evident that the choice $\alpha = 1/(\mathbf{n}\cdot\mathbf{s})$ gives the same results as Eq. (2.11) if the interface gradient $\overline{(\nabla X)_{ij}}$ is evaluated in the same manner. In that case, both expressions, Eq. (2.11) and Eq. (2.12), are algebraically equivalent.

In practice, taking the average of the two adjacent cell gradients obtained by least squares or Green-Gauss theorem is an easy way to construct the interface gradients (e.g., [62, 220]). The most obvious candidate is to choose equal weights of averaging, i.e., arithmetic averaging, for the two cell gradients:

$$(\nabla X)_{\text{ave}} = \frac{(\nabla X)|_i + (\nabla X)|_j}{2} \tag{2.13}$$

Here, $(\nabla X)_{\text{ave}}$ denotes the gradient obtained by averaging the cell-centered reconstructed gradients on the cells i and j that lie adjacent to the face. Except for the arithmetic averaging, we can set the weight of averaging by considering some geometrical elements within the cells. The volume (area in 2D) weighted averaging or linear interpolation based on the distance of the opposite cell centroid to the interface center can be used as alternatives:

$$(\nabla X)_{\text{ave}} = \frac{V_i\,(\nabla X)|_i + V_j\,(\nabla X)|_j}{V_i + V_j}, \tag{2.14}$$

$$(\nabla X)_{\text{ave}} = \frac{\left|\mathbf{x}'_{ij,c} - \mathbf{x}_j\right| (\nabla X)|_i}{\left|\mathbf{x}_j - \mathbf{x}_i\right|} + \frac{\left|\mathbf{x}'_{ij,c} - \mathbf{x}_i\right| (\nabla X)|_j}{\left|\mathbf{x}_j - \mathbf{x}_i\right|} \tag{2.15}$$

where $\mathbf{x}'_{ij,c}$ is the closest point to the face center $\mathbf{x}_{ij,c}$ on the line segment connecting centroids $\mathbf{x}_i$ and $\mathbf{x}_j$ of cells i and j. The line $\overline{\mathbf{x}_{ij,c}\mathbf{x}'_{ij,c}}$ is perpendicular to the line $\overline{\mathbf{x}_i\mathbf{x}_j}$ with $\mathbf{x}'_{ij,c}$ the perpendicular foot point, see Fig. 2.1.

However, it is known that the above averaging procedure could lead to checkerboarding effects that pollute the final solution. To combat any possible checkerboarding, the addition of an extra term is taken into account [194, 493] in the performance of the interface gradient reconstruction to notionally provide the following scheme

$$(\nabla X)_{\text{ave}} = \frac{(\nabla X)|_i + (\nabla X)|_j}{2},\ (\nabla X)_{ij} = (\nabla X)_{\text{ave}} - \left((\nabla X)_{\text{ave}} \cdot \hat{d}_{ij}\right)\hat{d}_{ij} + \frac{X_j - X_i}{d_{ij}}\hat{d}_{ij} \tag{2.16}$$

Here, $\hat{d}_{ij} = \frac{\mathbf{x}_j - \mathbf{x}_i}{|\mathbf{x}_j - \mathbf{x}_i|}$ represents the unit vector from the centroid of cell i to cell j and $d_{ij} = |\mathbf{x}_j - \mathbf{x}_i|$ is the distance between the respective cell centroids. For reference to a 3D hexahedral computational cell, see Fig. 2.1. This modification results in a stable discretization of the viscous terms, and does not affect the accuracy as long as the interface integration point is located halfway between the two cell-centers, such as structured grids. When combining the finite difference term $\frac{X_j - X_i}{d_{ij}}$ with the averaged components for unstructured grid, special care should be taken not to destroy the consistency of discretization. Cary et al. [62] showed that dropping the averaged component may result in inconsistent schemes for unstructured triangular grids. Thus, the averaged component obtained by one of the three methods mentioned in Eqs. (2.13)–(2.15) should be combined with finite difference term. There are two well-known methods commonly used for interface gradient constructions: edge-normal and face-tangent. Conveniently, the edge-normal/face-tangent interface gradients can be implemented without the edge-normal/face-tangent vector both in two and three dimensions [344].

In the edge-normal gradient construction, the interface gradient $(\nabla X)_{ij}$ is defined with the edge normal derivative (along $\hat{d}_{ij}^{\perp}$, i.e., the unit vector perpendicular to the edge vector $\hat{d}_{ij}$) specified by the averaged cell gradient [112, 220, 344] as:

$$(\nabla X)_{ij} = \frac{X_j - X_i}{d_{ij}}\hat{d}_{ij} + \left((\nabla X)_{\text{ave}} \cdot \hat{d}_{ij}^{\perp}\right)\hat{d}_{ij}^{\perp} \tag{2.17}$$

This kind of edge-normal gradient construction requires that the gradient projection along the edge normal direction equals $\left((\nabla X)_{\text{ave}} \cdot \hat{d}_{ij}^{\perp}\right)\hat{d}_{ij}^{\perp}$ [112, 462]. In three dimensions, the definition of $\hat{d}_{ij}^{\perp}$ is ambiguous. To eliminate the term $\hat{d}_{ij}^{\perp}$, we can use the following relation:

$$\left((\nabla X)_{\text{ave}} \cdot \hat{d}_{ij}^{\perp}\right) \hat{d}_{ij}^{\perp} = (\nabla X)_{\text{ave}} - \left((\nabla X)_{\text{ave}} \cdot \hat{d}_{ij}\right) \hat{d}_{ij}$$

Then Eq. (2.17) becomes:

$$(\nabla X)_{ij} = \frac{X_j - X_i}{d_{ij}} \hat{d}_{ij} + (\nabla X)_{\text{ave}} - \left((\nabla X)_{\text{ave}} \cdot \hat{d}_{ij}\right) \hat{d}_{ij} \tag{2.18}$$

which is consistent with the interface gradient Eq. (2.16) as used in the former references (e.g., [112, 194, 440, 493]). The deduction from Eqs. (2.17) to (2.18) shows that the edge-normal interface gradient construction can be implemented without the edge-normal vector. The interface gradient, constructed in Eqs. (2.17) and (2.18), recovers the edge-term along $\hat{d}_{ij}$ and the average least-squares or Green-Gauss gradient, $(\nabla X)_{\text{ave}}$, along the direction normal to $\hat{d}_{ij}$. The edge-term $\frac{X_j - X_i}{d_{ij}}$ is important for high-frequency damping, which is therefore deliberately incorporated to improve the h-ellipticity [346].

Alternatively, the so-called face-tangent interface gradient construction requires that the projection of the constructed gradient along the face-tangent direction is $((\nabla X)_{\text{ave}} \cdot \mathbf{t})\mathbf{t}$ [112, 462], which reads as follows:

$$(\nabla X)_{ij} = a_n \mathbf{n} + ((\nabla X)_{\text{ave}} \cdot \mathbf{t})\mathbf{t} \tag{2.19}$$

where a_n is a component to be determined, and $\mathbf{t}$ denotes a unit vector along the face tangent direction. Following Nishikawa [344], a_n is obtained by assuming that the edge-derivative $\frac{X_j - X_i}{d_{ij}}$ is equal to the interface gradient when projected along $\hat{d}_{ij}$, i.e.,

$$(\nabla X)_{ij} \cdot \hat{d}_{ij} = a_n(\mathbf{n} \cdot \hat{d}_{ij}) + ((\nabla X)_{\text{ave}} \cdot \mathbf{t})(\mathbf{t} \cdot \hat{d}_{ij}) = \frac{X_j - X_i}{d_{ij}}$$

Then we have

$$a_n = \frac{1}{\mathbf{n} \cdot \hat{d}_{ij}} \left(\frac{X_j - X_i}{d_{ij}} - (\mathbf{t} \cdot \hat{d}_{ij})((\nabla X)_{\text{ave}} \cdot \mathbf{t}) \right)$$

Obviously, $a_n \mathbf{n}$ in Eq. (2.19) incorporates the finite difference term in the face normal direction. The term $(\mathbf{t} \cdot \hat{d}_{ij})((\nabla X)_{\text{ave}} \cdot \mathbf{t})$ in a_n can be rewritten in the form independent of the face-tangent direction [344] according to the following geometric identity:

$$\begin{aligned}(\mathbf{t} \cdot \hat{d}_{ij})((\nabla X)_{\text{ave}} \cdot \mathbf{t}) &= \left((\mathbf{t} \cdot \hat{d}_{ij})\mathbf{t}\right) \cdot (\nabla X)_{\text{ave}} = \left(\hat{d}_{ij} - (\mathbf{n} \cdot \hat{d}_{ij})\mathbf{n}\right) \cdot (\nabla X)_{\text{ave}} \\ &= (\nabla X)_{\text{ave}} \cdot \hat{d}_{ij} - (\mathbf{n} \cdot \hat{d}_{ij})(\mathbf{n} \cdot (\nabla X)_{\text{ave}})\end{aligned} \tag{2.20}$$

Considering Eq. (2.20), we substitute a_n into Eq. (2.19) to obtain the final interface gradient:

$$
\begin{aligned}
(\nabla X)_{ij} &= \frac{\mathbf{n}}{\mathbf{n}\cdot\hat{d}_{ij}}\left(\frac{X_j - X_i}{d_{ij}} - (\nabla X)_{\text{ave}}\cdot\hat{d}_{ij}\right) + ((\nabla X)_{\text{ave}}\cdot\mathbf{n})\mathbf{n} + ((\nabla X)_{\text{ave}}\cdot\mathbf{t})\mathbf{t} \\
&= \frac{\mathbf{n}}{\mathbf{n}\cdot\hat{d}_{ij}}\left(\frac{X_j - X_i}{d_{ij}} - (\nabla X)_{\text{ave}}\cdot\hat{d}_{ij}\right) + (\nabla X)_{\text{ave}}
\end{aligned}
\tag{2.21}
$$

Hence, the face-tangent interface gradient construction can be implemented without any face-tangent vector for both two- and three-dimensional grids. In this way, the face-tangent interface gradient is constructed such that it recovers the edge-directional derivative $\frac{X_j - X_i}{d_{ij}}$ and the average least-squares or Green-Gauss gradient along the direction tangential to the face. This interface gradient form can also be seen in the literature [306, 344, 361] and references therein. From the interface gradient in Eq. (2.21), we can see that the damping coefficient generally depends on the inner-product $\mathbf{n}\cdot\hat{d}_{ij}$, which may be seen as a skewness measure [346].

It is noted that the contribution of damping for face tangent construction is much larger than the edge-normal formulation in a finite difference term on skewed grids [344, 462]. Thus the face tangent construction is more robust. For a grid with no skewness ($\mathbf{n}\cdot\hat{d}_{ij} = 1$), such as a structured grid, edge-normal and face tangent constructions Eqs. (2.18) and (2.21) become identical. Furthermore, the damping role vanishes on highly stretched triangles in the edge-normal construction and simplifies the construction of finite difference term, as pointed out by Nishikawa [344].

Besides the edge-normal and face tangent constructions, there are many suggestions in the context of finite volume methods for the discretization of diffusive fluxes appearing in viscous flow simulations. Cary et al. [62] and Jalali et al. [220] introduced an interface gradient algorithm with a finite difference term in the so-called corrected normal direction $\mathbf{n}$. Based on the reconstructed solutions along the face normal direction, the interface gradient along the normal direction is directly obtained by a finite difference approximation independent of the averaged component. The normal direction finite difference correction reads:

$$
(\nabla X)_{ij}\cdot\mathbf{n} = \frac{1}{\left(\mathbf{n}\cdot\hat{d}_{ij}\right)d_{ij}}\left(X_j^c - X_i^c\right) \tag{2.22}
$$

where $X_i^c = X_i + ((\nabla X)|_i\cdot\mathbf{t})(\mathbf{x}_{ic}\cdot\mathbf{t})$ and $X_j^c = X_j + \left((\nabla X)|_j\cdot\mathbf{t}\right)\left(\mathbf{x}_{jc}\cdot\mathbf{t}\right)$ with $\mathbf{x}_{ic} = \mathbf{x}_{ij,c} - \mathbf{x}_i$ and $\mathbf{x}_{jc} = \mathbf{x}_{ij,c} - \mathbf{x}_j$. Evidently, the terms $((\nabla X)|_i\cdot\mathbf{t})(\mathbf{x}_{ic}\cdot\mathbf{t})$ and $\left((\nabla X)|_j\cdot\mathbf{t}\right)\left(\mathbf{x}_{jc}\cdot\mathbf{t}\right)$ can be rewritten as follows

$$
((\nabla X)|_i\cdot\mathbf{t})(\mathbf{x}_{ic}\cdot\mathbf{t}) = ((\mathbf{t}\cdot\mathbf{x}_{ic})\mathbf{t})\cdot(\nabla X)|_i = (\mathbf{x}_{ic} - (\mathbf{n}\cdot\mathbf{x}_{ic})\mathbf{n})\cdot(\nabla X)|_i
$$

$$
\left((\nabla X)|_j\cdot\mathbf{t}\right)\left(\mathbf{x}_{jc}\cdot\mathbf{t}\right) = \left((\mathbf{t}\cdot\mathbf{x}_{jc})\mathbf{t}\right)\cdot(\nabla X)|_j = \left(\mathbf{x}_{jc} - (\mathbf{n}\cdot\mathbf{x}_{jc})\mathbf{n}\right)\cdot(\nabla X)|_j
$$

which do not depend on the face-tangent direction. Then Eq. (2.22) becomes

$$(\nabla X)_{ij} \cdot \mathbf{n} = \frac{1}{\left(\mathbf{n} \cdot \hat{d}_{ij}\right) d_{ij}} \left(X_j + \left(\mathbf{x}_{jc} - (\mathbf{n} \cdot \mathbf{x}_{jc})\mathbf{n}\right) \cdot (\nabla X)|_j \right.$$
$$\left. -X_i - (\mathbf{x}_{ic} - (\mathbf{n} \cdot \mathbf{x}_{ic})\mathbf{n}) \cdot (\nabla X)|_i \right)$$

In addition to the finite difference term, a jump term coming from the discontinuous solution at the face center can be added to the interface gradient. The jump term (e.g., [220, 418]) can be defined as follows:

$$(\nabla X)_{ij,\mathrm{jump}} = \frac{\alpha \mathbf{n}}{|\hat{d}_{ij} \cdot \mathbf{n}| d_{ij}} (X_R - X_L) \tag{2.23}$$

where formally we have $X_R = X_j + \left(\mathbf{x}_{ij,c} - \mathbf{x}_j\right) \cdot (\nabla X)|_j$, $X_L = X_i + \left(\mathbf{x}_{ij,c} - \mathbf{x}_i\right) \cdot (\nabla X)|_i$ and the jump term coefficient α is an arbitrary value. The combination of the jump term Eq. (2.23) and the averaged gradient $(\nabla X)_{\mathrm{ave}}$ yields the following evaluation of the interface gradient

$$(\nabla X)_{ij} = (\nabla X)_{\mathrm{ave}} + (\nabla X)_{ij,\mathrm{jump}} \tag{2.24}$$

which is employed [344, 346, 347] to calculate the gradients in the viscous flux in cell-centered diffusion schemes. Through numerical tests, Jalali et al. [220] suggested that $\alpha = \frac{4}{3}$ gives the most accurate solution for the combination of jump term with the averaged gradient Eq. (2.13), which was also taken by Nishikawa [346] in his fourth-order accurate flux integral for one-dimensional and multi-dimensional Cartesian grids. While, the numerical experiment suggests $\alpha = 1$ [220] when combining the jump term with the averaged gradient Eq. (2.15). As claimed by Nishikawa [347], this particular formulation allows a simple extension to the Navier–Stokes schemes. According to Cary et al. [62], Nishikawa [344] and Jalali et al. [220], the jump term $(\nabla X)_{ij,\mathrm{jump}}$ can be simply added to other components of the interface gradient like Eq. (2.22) without violating the consistency of discretization and has the role of stabilizing by damping high-frequency errors. For more combinations of Eq. (2.23) with other interface gradient estimates, see Jalali et al. [220]. Even more, Sejekan and Ollivier-Gooch [418] suggested the calculation of second-order gradients at flux quadrature points using a linear interpolation scheme along with a high-accuracy jump term to enhance stability of the system.

Moreover, Nishikawa [344] found that $\alpha = (\mathbf{n} \cdot \hat{d}_{ij})|\mathbf{n} \cdot \hat{d}_{ij}|$ and $\alpha = \frac{|\mathbf{n} \cdot \hat{d}_{ij}|}{\mathbf{n} \cdot \hat{d}_{ij}}$ for their diffusive flux calculation correspond to their so-called edge normal average-least-squares (Avg-LSQ-EN) and face-tangent average-least-squares (Avg-LSQ-FT) schemes. As commented by Jalali et al. [220], this choice of α can be used in Eq. (2.23). In this case, Eq. (2.23) can be written as:

$$(\nabla X)_{ij,\text{jump}} = \mathbf{n}(\mathbf{n}\cdot\hat{d}_{ij})\frac{X_R - X_L}{d_{ij}},\ (\nabla X)_{ij,\text{jump}} = \frac{\mathbf{n}}{\mathbf{n}\cdot\hat{d}_{ij}}\frac{X_R - X_L}{d_{ij}}$$

which become identical to Eq. (2.23) with $\alpha = 1$ for a grid of no skewness ($\mathbf{n}\cdot\hat{d}_{ij} = 1$).

Also, the interface gradient can be approximated by the finite-difference to the normal derivative [215, 333, 351]:

$$(\nabla X)_{ij}\cdot\mathbf{n} = \frac{X_{j'} - X_{i'}}{d_{i'j'}} \tag{2.25}$$

where $d_{i'j'}$ is the distance between two points $\mathbf{x}_{i'}$ and $\mathbf{x}_{j'}$ forming a line passing through the interface centroid and perpendicular to the interface. The line $\overline{\mathbf{x}_i\mathbf{x}_{i'}}$ is perpendicular to the line $\overline{\mathbf{x}_{i'}\mathbf{x}_{j'}}$ with the perpendicular foot point $\mathbf{x}_{i'}$. Similarly, the line $\overline{\mathbf{x}_j\mathbf{x}_{j'}}$ is perpendicular to the line $\overline{\mathbf{x}_{i'}\mathbf{x}_{j'}}$ with the perpendicular foot point $\mathbf{x}_{j'}$, see Figs. 2.1 and 2.3. The values $X_{i'}$ and $X_{j'}$ can be formally approximated as

$$\begin{aligned}X_{i'} &= X_i + (\nabla X)|_i\cdot(\mathbf{x}_{i'} - \mathbf{x}_i) = X_i + (\nabla X)|_i\cdot(\mathbf{x}_{i'} - \mathbf{x}_{ij,c}) + (\nabla X)|_i\cdot(\mathbf{x}_{ij,c} - \mathbf{x}_i)\\ &= X_L + (\nabla X)|_i\cdot(\mathbf{x}_{i'} - \mathbf{x}_{ij,c}) = X_L + (\nabla X)|_i\cdot(-\mathbf{n})|\mathbf{x}_{i'} - \mathbf{x}_{ij,c}|\end{aligned}$$

$$\begin{aligned}X_{j'} &= X_j + (\nabla X)|_j\cdot(\mathbf{x}_{j'} - \mathbf{x}_j) = X_j + (\nabla X)|_j\cdot(\mathbf{x}_{j'} - \mathbf{x}_{ij,c}) + (\nabla X)|_j\cdot(\mathbf{x}_{ij,c} - \mathbf{x}_j)\\ &= X_R + (\nabla X)|_j\cdot(\mathbf{x}_{j'} - \mathbf{x}_{ij,c}) = X_R + (\nabla X)|_j\cdot\mathbf{n}|\mathbf{x}_{j'} - \mathbf{x}_{ij,c}|\end{aligned}$$

where $|\mathbf{x}_{i'} - \mathbf{x}_{ij,c}|$ and $|\mathbf{x}_{j'} - \mathbf{x}_{ij,c}|$ stand for the distances between the interface centroid and the points $\mathbf{x}_{i'}$ and $\mathbf{x}_{j'}$. Thus we have $d_{i'j'} = |\mathbf{x}_{i'} - \mathbf{x}_{ij,c}| + |\mathbf{x}_{j'} - \mathbf{x}_{ij,c}|$. After a few simple calculations, we obtain the final expression of this interface gradient:

$$(\nabla X)_{ij}\cdot\mathbf{n} = \frac{(\nabla X)|_j\,|\mathbf{x}_{j'} - \mathbf{x}_{ij,c}| + (\nabla X)|_i\,|\mathbf{x}_{i'} - \mathbf{x}_{ij,c}|}{|\mathbf{x}_{i'} - \mathbf{x}_{ij,c}| + |\mathbf{x}_{j'} - \mathbf{x}_{ij,c}|}\cdot\mathbf{n} + \frac{X_R - X_L}{d_{i'j'}} \tag{2.26}$$

From the above formula, it is obvious that the interface gradient has two parts: inverse distance weighted least-squares gradient projected along the face normal vector and a high-frequency damping term $\frac{X_R - X_L}{d_{i'j'}}$ with the damping coefficient given by $\frac{1}{d_{i'j'}}$.

As we know, terms like $\nu\rho\nabla\mathbf{v}$ (ν is the kinematic plasma viscosity coefficient), $\nabla(\rho\mathbf{v})$, $\nabla\rho$, $\nabla\mathbf{B}$, $k\nabla T$ (k is the thermal conductivity), τ (τ is the viscous stress tensor whose elements $\tau_{ij} = \mu\left(\frac{\partial v_i}{\partial x_j} + \frac{\partial v_j}{\partial x_i}\right) - \frac{2\mu}{3}\frac{\partial v_k}{\partial x_k}\delta_{ij}$ with μ being the coefficient of viscosity and v_i, x_i (i = 1 to 3) representing the three components of velocity and coordinate.), $\tau\cdot\mathbf{v}$, are usually typical terms appearing in resistive diffusion flux $\mathbf{F}_{\text{diff}}\cdot\mathbf{n} = \mathbf{F}_{\text{diff}}(\mathbf{U}, \nabla\mathbf{U})\cdot\mathbf{n}$ (e.g., [6, 51, 60, 172, 210, 234, 258, 311, 314, 339, 340, 361, 398, 440, 443, 450]). General terms involved in the diffusion fluxes are of the gradient type $\Gamma\nabla X$, where Γ represents the effective diffusion coefficient, possibly depending on the dependent variables, and ∇X (X denotes one of the conservative or primitive variables) is an element of the diffusion flux $\mathbf{F}_{\text{diff}}$. Hence, across the cell face we have

$$\int_{V_i} \nabla \cdot (\Gamma \nabla X) dV = \sum_j \Gamma_{ij} (\nabla X)_{ij} \cdot \mathbf{n} A_{ij}$$

where Γ_{ij} is the diffusion coefficient taking its value at the $\mathbf{x}_{ij,c}$ point of the cell face, and j runs i's neighborhoods sharing face A_{ij} with cell i. The main task to evaluate the diffusion terms at the internal cell faces is to calculate $(\nabla X)_{ij} \cdot \mathbf{n}$. For instance, from Eq. (2.25), interface gradients can be evaluated directly at the desired cell face centroid $\mathbf{x}_{ij,c}$ (Fig. 2.3).

$$(\nabla X)_{ij} \cdot \mathbf{n} A_{ij} = \left(\frac{\partial X}{\partial \mathbf{n}}\right)_{\mathbf{x}_{ij,c}} A_{ij} \approx \frac{(X_{j'} - X_{i'})}{(\mathbf{x}_j - \mathbf{x}_i) \cdot \mathbf{n}} A_{ij}$$

Here, $X_{i'}$ and $X_{j'}$ are the values at the projected positions i' and j' of the centroids i and j on the surface normal vector direction $\mathbf{n}$. These values can be estimated using the gradient at the centroid position: $X_{i'} \approx X_i + (\nabla X)|_i \cdot (\mathbf{x}_{i'} - \mathbf{x}_i)$, $X_{j'} \approx X_j + (\nabla X)|_j \cdot (\mathbf{x}_{j'} - \mathbf{x}_j)$. Then,

$$\begin{aligned}(\nabla X)_{ij} \cdot \mathbf{n} A_{ij} &= \left(\frac{\partial X}{\partial \mathbf{n}}\right)_{\mathbf{x}_{ij,c}} A_{ij} \approx \frac{(X_{j'} - X_{i'})}{(\mathbf{x}_j - \mathbf{x}_i) \cdot \mathbf{n}} A_{ij} \\ &= \frac{X_j - X_i}{(\mathbf{x}_j - \mathbf{x}_i) \cdot \mathbf{n}} A_{ij} + \frac{(\nabla X)|_j \cdot (\mathbf{x}_{j'} - \mathbf{x}_j) - (\nabla X)|_i \cdot (\mathbf{x}_{i'} - \mathbf{x}_i)}{(\mathbf{x}_j - \mathbf{x}_i) \cdot \mathbf{n}} A_{ij}\end{aligned}$$

When the above equation is deployed in the implicit discretized equation, the first term is evaluated implicitly, while the second one (containing derivative term) is evaluated explicitly in the source term. In principle, the evaluation of such diffusion flux at the interface ij can be approximated as $(\nabla X)_{ij} \cdot \mathbf{n}$ with $(\nabla X)_{ij}$ being one of the appropriate interface gradient listed above and $\mathbf{n}$ the unit normal direction. That is, the interface gradient approximation formula $(\nabla X)_{ij}$ (such as Eqs. (2.10), (2.11), (2.12), and (2.26)) and $(\nabla X)_{ij} \cdot \mathbf{n}$ (such as Eqs. (2.18), (2.21), (2.22), and (2.24)) can be used for the evaluation of diffusion flux at the cell face.

A distinguished feature of the above proposed interface gradient evaluation is that it automatically introduces a damping term into the resulting viscous scheme, which is essential for effective high-frequency error damping and, in some cases, for consistency also. Meanwhile, for diffusive flux approximation in unstructured finite volume solvers (e.g., [130, 217, 218, 220, 257, 265, 335, 343, 344, 346, 349, 351, 357, 381, 418, 456]), many interface gradient construction variants are compared in terms of discretization and truncation errors, the accuracy of the derived diffusion schemes are numerically verified and the importance of the damping term for highly-skewed typical grids is highlighted. For highly non-uniform grids, viscous discretization is highly mesh dependent. Nishikawa [346] introduced a general principle for constructing robust and accurate viscous discretization, and Syrakos et al. [456] performed a critical analysis of some popular methods for the discretization

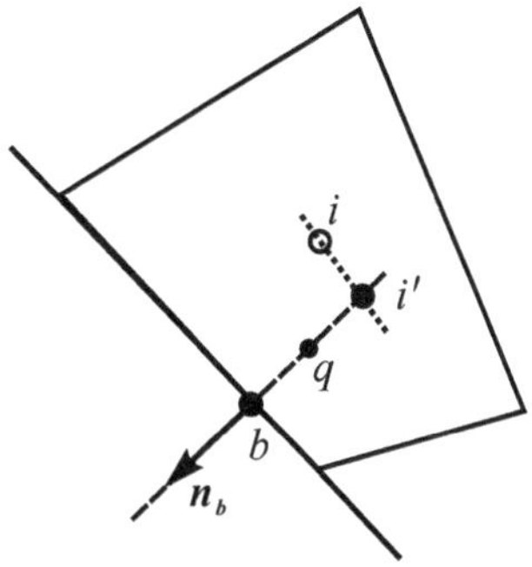

Fig. 2.4 Boundary face, where a boundary condition is imposed through a Gaussian point b on the boundary face, and $\mathbf{n}_b$ denotes the outer normal to the boundary face

of the gradient operator in finite volume methods. For detailed discussions about viscous discretization schemes, we can refer to the literature mentioned just above.

The evaluation of the diffusion fluxes at the boundaries is also of importance. Figure 2.4 describes a case of boundary cell, where i denotes the centroid of the boundary element with one boundary face of centroid b and of surface vector $\mathbf{n}_b$ pointing outward. The specification of boundary conditions involves either specifying the unknown boundary value X_b, or alternatively, the boundary flux $(\Gamma \nabla X)_b \cdot \mathbf{n}_b$ with Γ the diffusion coefficient or simply $(\nabla X)_b \cdot \mathbf{n}_b$. Usually, Neumann and Dirichlet boundary conditions are involved. In case of the Neumann boundary condition, normal fluxes $(\nabla X)_b \cdot \mathbf{n}_b$ are known at the boundary, which can be used directly in the discretized equation. However, the value of the variable at the boundary is usually needed to be evaluated. In case of the Dirichlet boundary, the dependent variable itself, X_b, is known at the centroid b of the boundary face. For the boundary face shown in Fig. 2.4, the diffusion fluxes at the boundary are evaluated using the known value of X_b at point b.

As shown in Fig. 2.4, the dashed line passes through the point $\mathbf{x}_b$ and is orthogonal to the face. $\mathbf{x}_{i'}$ is the projected point of $\mathbf{x}_i$ onto the dashed line $\overline{\mathbf{x}_{i'}\mathbf{x}_b}$, i.e., the dotted line $\overline{\mathbf{x}_i\mathbf{x}_{i'}}$ is perpendicular to the dashed line $\overline{\mathbf{x}_{i'}\mathbf{x}_b}$ with the perpendicular foot point $\mathbf{x}_{i'}$. Since $\mathbf{n}_b$ points from $\mathbf{x}_{i'}$ to $\mathbf{x}_b$, we have

$$(\nabla X)_b \cdot \mathbf{n}_b A_{ib} \approx \frac{X_b - X_{i'}}{|\mathbf{x}_{i'} - \mathbf{x}_b|} A_{ib} \approx \left(\frac{X_b - X_i}{|\mathbf{x}_{i'} - \mathbf{x}_b|} - \frac{(\nabla X)_i \cdot (\mathbf{x}_{i'} - \mathbf{x}_i)}{|\mathbf{x}_{i'} - \mathbf{x}_b|} \right) A_{ib}$$

with A_{ib} being the surface area of the boundary face element. Consider an extra point q somewhere along the line $\overline{\mathbf{x}_b\mathbf{x}_{i'}}$ (Fig. 2.4) such that we can write

$$\begin{aligned} X_q &\approx X_i + (\nabla X)_i \cdot (\mathbf{x}_q - \mathbf{x}_i),\ X_q \approx X_b + (\nabla X)_b \cdot (\mathbf{x}_q - \mathbf{x}_b) \\ &= X_b - \left|\mathbf{x}_q - \mathbf{x}_b\right| (\nabla X)_b \cdot \mathbf{n}_b, \end{aligned}$$

which gives

$$(\nabla X)_b \cdot \mathbf{n}_b A_{ib} \approx \frac{X_b - X_q}{\left|\mathbf{x}_q - \mathbf{x}_b\right|} A_{ib} \approx \left(\frac{X_b - X_i}{\left|\mathbf{x}_q - \mathbf{x}_b\right|} - \frac{(\nabla X)_i \cdot \left(\mathbf{x}_q - \mathbf{x}_i\right)}{\left|\mathbf{x}_q - \mathbf{x}_b\right|} \right) A_{ib}$$

The distance between points b and q can be scaled by $\left|\mathbf{x}_q - \mathbf{x}_b\right| = C\left|\mathbf{x}_{i'} - \mathbf{x}_b\right|$, where C is a parameter that determines the position of point q. To define this point position the following condition is prescribed: $\left|\mathbf{x}_q - \mathbf{x}_i\right| = \left|\mathbf{x}_q - \mathbf{x}_b\right|$. Based on this criterion, it is required [381] that

$$C = \min\left(1, \frac{\left|\mathbf{x}_{i'} - \mathbf{x}_i\right|^2 + \left|\mathbf{x}_{i'} - \mathbf{x}_b\right|^2}{2\left|\mathbf{x}_{i'} - \mathbf{x}_b\right|^2}\right)$$

by limiting the value of C to 1 (i.e., $\left|\mathbf{x}_q - \mathbf{x}_b\right| \leq \left|\mathbf{x}_{i'} - \mathbf{x}_b\right|$).

In general, diffusion fluxes at the boundaries are evaluated using all the cell-centered gradient criteria as for the internal faces, as introduced above. In these equations, the gradient at the boundary is assumed to be the same as the gradient at the nearest grid node (see Fig. 2.4): $(\nabla X)_j \equiv (\nabla X)_b = (\nabla X)_i$.

For more detailed description refer to the associated discussions about boundary conditions, such as Sects. 8.3, 15.6, 16.8, and Chap. 18, of Moukalled et al. [330], for instance.

The performance of finite-volume solvers depends heavily on gradient algorithms, which are required in the inviscid flux reconstruction, the evaluations of viscous fluxes, and source terms, especially on unstructured grids typical of applications involving complex geometries. We will return to such topic in Sect. 2.6.

2.2.2 FVM: Temporal Discretization

In order to set up the spatially discretized system subjected to temporal discretization, the system ODE is first formulated. This is the ODE which governs the time evolution of conservative variables in all the cells associated with the finite-volume discretization. In order to obtain such an equation, the spatially discretized equations (2.6) for each cell are written out in a residual form like Eq. (2.4). This form is obtained by summing the face fluxes over all the faces associated with a particular cell. The face-fluxes associated with the convective and resistive diffusion fluxes are inserted into respective residual operators. Since the fluxes are obtained using data from the two cells adjacent to a face, the residual operators for cell i are a function of the cell data and the data of neighboring cells that share a common face. Symbolically, we can rewrite Eq. (2.6) as follows:

$$\frac{\partial \mathbf{U}_i}{\partial t} + \mathbf{R}^i_{\text{inv}}(\mathbf{U}_i, \mathbf{U}_{nbs}) = \mathbf{R}^i_{\text{diff}}(\mathbf{U}_i, \mathbf{U}_{nbs}) + \mathbf{R}^i_s(\mathbf{U}_i) \tag{2.27}$$

Here, $\mathbf{R}^i_{\text{inv}}$ is the convective residual, $\mathbf{R}^i_{\text{diff}}$ is the diffusion residual, and $\mathbf{R}^i_s$ is the source residual. $\mathbf{U}_{\text{nbs}}$ are the conservative variables that belong to the cell neighbors of cell i. Besides the evaluation of the residuals, the numerical time integration of the system ODE (2.27) is a crucial step in carrying out the simulations. Evidently, a simple explicit temporal discretization reads as follows

$$\frac{\mathbf{U}_i^{n+1} - \mathbf{U}_i^n}{\Delta t} + \mathbf{R}^i_{\text{inv}}(\mathbf{U}_i^n, \mathbf{U}_{nbs}^n) = \mathbf{R}^i_{\text{diff}}(\mathbf{U}_i^n, \mathbf{U}_{nbs}^n) + \mathbf{R}^i_s(\mathbf{U}_i^n)$$

which is the standard forward Euler method.

Ignoring the subscripts in Eq. (2.27), the system ODE is obtained by collecting the residual form ODEs that are written out for individual cells and creating a single ODE system given by

$$\frac{\partial \mathbf{U}}{\partial t} + \mathbf{R}_{\text{inv}}(\mathbf{U}) = \mathbf{R}_{\text{diff}}(\mathbf{U}) + \mathbf{R}_s(\mathbf{U}) \tag{2.28}$$

In order to integrate the residual operators at a time step, a fully implicit time integration scheme for Eq. (2.28)

$$\frac{\mathbf{U}^{n+1} - \mathbf{U}^n}{\Delta t} + \mathbf{R}_{\text{inv}}(\mathbf{U}^{n+1}) = \mathbf{R}_{\text{diff}}(\mathbf{U}^{n+1}) + \mathbf{R}_s(\mathbf{U}^{n+1})$$

or equivalently for Eq. (2.6)

$$\frac{\partial \mathbf{U}_i}{\partial t} V_i + \sum_{j=1}^{N_{\text{faces}}} (\mathbf{F}_{\text{inv}} \cdot \mathbf{n})_{ij}^{n+1} A_{ij} = \sum_{j=1}^{N_{\text{faces}}} (\mathbf{F}_{\text{diff}} \cdot \mathbf{n})_{ij}^{n+1} A_{ij} + \mathbf{S}_i^{n+1} V_i \tag{2.29}$$

can be employed. The scheme produces an nonlinear equation at every time step with the following symbolical expression

$$\mathbf{G}(\mathbf{U}) = \mathbf{0},\, \mathbf{G}(\mathbf{U}) = \frac{\mathbf{U} - \mathbf{U}^n}{\Delta t} + \mathbf{R}(\mathbf{U}),\, \mathbf{R}(\mathbf{U}) = \mathbf{R}_{\text{inv}}(\mathbf{U}) - \mathbf{R}_{\text{diff}}(\mathbf{U}) - \mathbf{R}_s(\mathbf{U}) \tag{2.30}$$

which can be solved by the Newton method. This iterative technique obtains a new approximate solution $\mathbf{U}^{k+1}$ to the system of equations $\mathbf{G} = \mathbf{0}$ by setting the first order Taylor expansion of the function at the current iterate to zero.

$$\mathbf{G}(\mathbf{U}^k) + \frac{\partial \mathbf{G}(\mathbf{U}^k)}{\partial \mathbf{U}^k}(\mathbf{U}^{k+1} - \mathbf{U}^k) = \mathbf{0}$$

The conservative variables at the new iterate is obtained by solving a large system of sparse linear equations. Expanding out the function $\mathbf{G}$ in Eq. (2.30), the functional form of the linear system at every iterate leads to the new approximate solution

$$\mathbf{U}^{k+1} = \mathbf{U}^k - \left[\frac{I}{\Delta t} + \frac{\partial \mathbf{R}_{\text{inv}}}{\partial \mathbf{U}} - \frac{\partial \mathbf{R}_{\text{diff}}}{\partial \mathbf{U}} - \frac{\partial \mathbf{R}_s}{\partial \mathbf{U}}\right]^{-1} \times \left(\frac{\mathbf{U}^k - \mathbf{U}^n}{\Delta t} + \mathbf{R}(\mathbf{U}^k)\right) \quad (2.31)$$

In Eq. (2.31), the partial derivatives of the residual operators with respect to the vector of conservative variables are referred to as the residual Jacobians. The inversion of these residual Jacobians represents the coupling between the conservative variables. This coupling allows for the numerical time integration to be performed at a physical time step Δt that is larger than the limiting explicit time step restrictions imposed by the various residual operators. The residual Jacobians are analytically derived by differentiating the underlying fluxes [Eq. (2.6)] that constitute each residual operator. It is noted that an exact differentiation of the resistive diffusion residual is also involved. The inclusion of these terms induces a stronger coupling between the conservative variables, thereby giving faster convergence of the Newton iterations and improving its stability properties. The convergence of the Newton iteration is defined by a relative tolerance criterion, the system being deemed as converged if the L_2-norm of the function $\mathbf{G}$ reduces by certain orders of magnitude. The sparse linear system inversion at each Newton step can be carried out by using, for instance, the Generalized Minimum Residual (GMRES) algorithm, implemented in the Portable and Extensible Toolkit for Scientific Computing (PETSc) library [16, 17, 444].

The basic technique to find a steady state solution to Eq. (2.6) is through time stepping. An implicit time stepping scheme can be formulated for each cell i in an unstructured grid framework as shown in Eq. (2.29). As the finite volume method is conservative, we need only determine the Jacobian of the residual $\mathbf{R}$ of Eq. (2.31) with respect to the variables of cell i and then for each face, add this Jacobian to the Jacobian of the neighbouring cell with respect to the variables of cell i. Symbologically, the linearization of the inviscid flux, diffusion and source terms from Eq. (2.29) for a particular face f that separates cells i and j (Fig. 2.1) can be written as

$$\begin{aligned}
(\mathbf{F}_{\text{inv}} \cdot \mathbf{n})_{ij}^{n+1} &= (\mathbf{F}_{\text{inv}} \cdot \mathbf{n})_{ij}^{n} + \left(\frac{\partial\left(\mathbf{F}_{\text{inv}} \cdot \mathbf{n}\right)_{ij}}{\partial \mathbf{U}_i}\right)^n \Delta \mathbf{U}_i^n + \left(\frac{\partial\left(\mathbf{F}_{\text{inv}} \cdot \mathbf{n}\right)_{ij}}{\partial \mathbf{U}_j}\right)^n \Delta \mathbf{U}_j^n \\
(\mathbf{F}_{\text{diff}} \cdot \mathbf{n})_{ij}^{n+1} &= (\mathbf{F}_{\text{diff}} \cdot \mathbf{n})_{ij}^{n} + \left(\frac{\partial\left(\mathbf{F}_{\text{diff}} \cdot \mathbf{n}\right)_{ij}}{\partial \mathbf{U}_i}\right)^n \Delta \mathbf{U}_i^n + \left(\frac{\partial\left(\mathbf{F}_{\text{diff}} \cdot \mathbf{n}\right)_{ij}}{\partial \mathbf{U}_j}\right)^n \Delta \mathbf{U}_j^n \\
\mathbf{S}_i^{n+1} &= \mathbf{S}_i^n + \left(\frac{\partial \mathbf{S}_i}{\partial \mathbf{U}_i}\right)^n \Delta \mathbf{U}_i^n + \left(\frac{\partial \mathbf{S}_i}{\partial \mathbf{U}_j}\right)^n \Delta \mathbf{U}_j^n
\end{aligned} \quad (2.32)$$

with $\Delta \mathbf{U}_i = \mathbf{U}_i^{n+1} - \mathbf{U}_i^n$. Note that the evaluation of these terms in Eq. (2.32) at the face that separates cells i and j depends only on the values of conservative variables on its left and right cells.

The terms on the right hand side of Eq. (2.32) involve the convective flux Jacobians associated with each cell interface. One approach to calculating the Jacobian is through a finite difference method which involves perturbing each of the conservative variables and calculating the change in the inviscid flux. This is computationally expensive for large scale three dimensional MHD calculations. An alternative technique is to derive the flux Jacobian analytically.

Maybe, the simple reconstruction of convective flux is due to a Lax–Friedrichs scheme [271, 440, 469, 472]. This reconstruction can be physically interpreted as averaging the exact solution of a shock tube (Riemann) problem with the discontinuous cell data forming the initial states. This method is known as the local Lax–Friedrichs method where the numerical viscosity is chosen as the maximum wavespeed associated with the face centered Riemann problem [271]. The convective flux $(\mathbf{F}_{\text{inv}} \cdot \mathbf{n})_{ij}$ at the face can be written as

$$(\mathbf{F}_{\text{inv}} \cdot \mathbf{n})_{ij} = \frac{1}{2}\left((\mathbf{F}_{\text{inv}} \cdot \mathbf{n})_i + (\mathbf{F}_{\text{inv}} \cdot \mathbf{n})_j\right) - \frac{1}{2}\lambda_{\max}(\mathbf{U}_j - \mathbf{U}_i) \tag{2.33}$$

Here, $(\mathbf{F}_{\text{inv}} \cdot \mathbf{n})_i$ and $(\mathbf{F}_{\text{inv}} \cdot \mathbf{n})_j$ denote the cell-centered convective fluxes that are calculated using the data from the cells i and j, respectively. $\lambda_{\max}$ denotes the maximum wavespeed associated with the discontinuous face-centered shock tube problem. Typically, this is taken to be the absolute maximum value of the fastest characteristic wave speeds at the interface. Basic characteristics and flux function formulations of MHD equations are provided in Sects. 2.3, 2.5 and 2.8.

When using a Lax-Friedrich's method [271, 440, 469, 472] to MHD equations, the inviscid flux at face f can be written as shown in Eq. (2.33), for simplicity of calculation $(\mathbf{F}_{\text{inv}} \cdot \mathbf{n})_i$ and $(\mathbf{F}_{\text{inv}} \cdot \mathbf{n})_j$ represent the component of inviscid flux along the face normal computed from cell center values of conservative variables, and $\lambda_{\max}$ is the fastest wave speed which is the sum of the contravariant velocity and the fast plasma wave speed at face f

$$\lambda_{\max} = |\mathbf{v} \cdot \mathbf{n}| + \sqrt{\frac{1}{2}\left(a^2 + \frac{|\mathbf{B}|^2}{\rho} + \sqrt{\left(a^2 + \frac{|\mathbf{B}|^2}{\rho}\right)^2 - 4\frac{B_n^2}{\rho}a^2}\right)}$$

$\lambda_{\max}$ is also a function of the conservative variables at cells i and j. The contravariant velocity $\mathbf{v} \cdot \mathbf{n}$, acoustic wave speed a, magnitude of magnetic field $|\mathbf{B}|$ and its normal component B_n are obtained as arithmetic averages of values from either cell.

$$\begin{aligned}\left(\frac{\partial\,(\mathbf{F}_{\text{inv}} \cdot \mathbf{n})_{ij}}{\partial \mathbf{U}_i}\right) &= \frac{1}{2}\left(\frac{\partial\,(\mathbf{F}_{\text{inv}} \cdot \mathbf{n})_i}{\partial \mathbf{U}_i}\right) + \frac{|\lambda_{\max}|}{2}, \quad \left(\frac{\partial\,(\mathbf{F}_{\text{inv}} \cdot \mathbf{n})_{ij}}{\partial \mathbf{U}_j}\right) \\ &= \frac{1}{2}\left(\frac{\partial\,(\mathbf{F}_{\text{inv}} \cdot \mathbf{n})_j}{\partial \mathbf{U}_j}\right) - \frac{|\lambda_{\max}|}{2}\end{aligned} \tag{2.34}$$

Equation (2.34) represents the Jacobian obtained from differentiation of numerical flux shown in Eq. (2.33). Note that $\lambda_{\max}$ in Eq. (2.33) is a function of the conservative variables at cells i and j and hence should also be differentiated further to obtain the true Jacobian associated with the numerical flux. For simplicity, it is assumed to be locally constant in the derivation of Eq. (2.34). This approximation can potentially restrict the maximum Courant number for stable calculations as well as steady state convergence rates [30, 31]. The above equations are written for an interior face which has a left cell and a right cell adjacent to it. At boundary cells, only the diagonal contribution from the first formula in Eq. (2.34) is used. The addition on the diagonal tends to stabilize the matrix solve calculations. The Jacobian matrices in Eq. (2.34) can be computed by writing the cell-centered inviscid flux in terms of the conservative variables $\mathbf{U}$, if the variables in the original governing equations are defined in terms of the conservative variables as shown in a concrete MHD equation (2.37).

In the equation for the residual, the viscous flux $\mathbf{F}_{\text{diff}} \cdot \mathbf{n}$ can be evaluated as follows. First, the cell-centered conserved variables, $\mathbf{U}_i$, are converted to cell-centered primitive variables, $\mathbf{W}_i$. Then approximate the primitive variables and their gradients on a given cell interface respectively by interpolation such as arithmetic average and by Eq. (2.16), and store them in an extended solution vector as $\mathbf{E}_f$. The viscous flux is finally evaluated. Thus $\frac{\partial(\mathbf{F}_{\text{diff}} \cdot \mathbf{n})}{\partial \mathbf{U}}$ is performed through a three-step application of the chain rule:

$$\frac{\partial(\mathbf{F}_{\text{diff}} \cdot \mathbf{n})}{\partial \mathbf{U}_i} = \frac{\partial(\mathbf{F}_{\text{diff}} \cdot \mathbf{n})}{\partial \mathbf{E}_f} \frac{\partial \mathbf{E}_f}{\partial \mathbf{W}_i} \frac{\partial \mathbf{W}_i}{\partial \mathbf{U}_i}$$

$\frac{\partial \mathbf{W}_i}{\partial \mathbf{U}_i}$ is the standard Jacobian of the primitive variables with respect to the conserved variables and is readily available. The Jacobian, $\frac{\partial \mathbf{E}_f}{\partial \mathbf{W}_i}$, is due to geometry only, for it only depends on the interpolation used in the viscous flux evaluation described in Eqs. (2.10), (2.11), (2.12), (2.18), (2.21), (2.22), (2.24) and (2.26). In the 3D case, the face variables at the cell interface are given by the extended solution vector $\mathbf{E}_f = (\mathbf{W}, \partial_x \mathbf{W}, \partial_y \mathbf{W}, \partial_z \mathbf{W})$. Evidently, $\frac{\partial(\mathbf{F}_{\text{diff}} \cdot \mathbf{n})}{\partial \mathbf{E}_f}$ and $\frac{\partial \mathbf{E}_f}{\partial \mathbf{W}_i}$ can be easily determined from the known $\mathbf{F}_{\text{diff}}$. The Jacobian components related to the known source terms is straight forward since they usually rely only on the solution values for local cell and its neighbors. The other terms in the equation for the residual are the cell volume and area of each face which are constant with respect to the solution variables.

The implicit method can become extremely dissipative for large time steps resulting in large errors for a time accurate calculation. The implicit method developed is extremely good for steady state problems where accuracy in time is immaterial. To solve the equations accurately in time, a dual-time algorithm [45, 440, 523] can be used to modify the existing implicit method. The algorithm essentially consists of performing sub-iterations for every physical time step. The original finite volume approximation given in Eq. (2.29) can be written as below

$$\frac{\partial \mathbf{U}_i}{\partial t} + \mathbf{R}(\mathbf{U}_i) = 0 \tag{2.35}$$

Here $\mathbf{R}$ represents the residual term which consists of the convective and diffusive fluxes as well as the source term contributions. This equation is inherently non-linear that requires either Newton's method or a time-stepping method to solve it accurately in time. The time stepping method can be used to let the second term in Eq. (2.36) go to zero at steady state. Here τ is referred to as the pseudo-time variable.

$$\frac{\partial \mathbf{U}_i}{\partial \tau} + \left(\frac{\partial \mathbf{U}_i}{\partial t} + \mathbf{R}(\mathbf{U}_i) \right) = 0 \tag{2.36}$$

Therefore the time accurate calculation is reduced to solving a steady state problem in τ at each physical time step. Assume that m denotes time level in pseudo-time while n is the time level in actual physical time. The implicit method for Eq. (2.36) can be formulated as follows:

$$\frac{\partial \mathbf{U}_i}{\partial \tau} + \left(\frac{\partial \mathbf{U}_i}{\partial t} + \mathbf{R}(\mathbf{U}_i) \right)^{(m+1)} = 0$$

where, for simplicity,

$$\left(\frac{\partial \mathbf{U}}{\partial t} + \mathbf{R}(\mathbf{U}) \right)^{(m+1)} = \frac{\mathbf{U}^{(m+1)} - \mathbf{U}^n}{\Delta t} + \mathbf{R}^{(m)} + \left(\frac{\partial \mathbf{R}}{\partial \mathbf{U}} \right)^{(m)} \Delta \mathbf{U}$$

$$\left(\left(\frac{1}{\Delta \tau} + \frac{1}{\Delta t} \right) \mathbf{I} + \left(\frac{\partial \mathbf{R}}{\partial \mathbf{U}} \right)^{(m)} \right) \Delta \mathbf{U} = \frac{\mathbf{U}^n - \mathbf{U}^{(m)}}{\Delta t} - \mathbf{R}^{(m)}$$

Note that $\Delta \mathbf{U} = \mathbf{U}^{(m+1)} - \mathbf{U}^{(m)}$ denotes the difference in the conservative variables between successive pseudo-time steps within a given physical time step. Here, $\mathbf{U}^{(0)} = \mathbf{U}^n$. For more discussions, refer to Kifonidis and Müller [248] for instance.

2.3 FVM: Some Basics of MHD Equations

This section is mainly devoted to some basics of ideal MHD equations. The ideal MHD equations are a classical model from plasma physics used to study the dynamics of electrically conducting fluids. There are numerous phenomena for which these are the first-principal appropriate equations, such as laboratory plasmas under certain conditions, the Earth's magnetosphere, the Sun, and other astrophysical objects. MHD has much in common with ordinary hydrodynamics, to which it reduces in the limit of small magnetic field, but it also contains some additional features. The MHD equations are a single set of fluid equations for the total mass density, center-of-mass momentum density, total energy density, and the magnetic field of the system (e.g., [55, 87, 127, 187–189, 388, 422, 460]). The ideal MHD equations can be written as a system of conservation laws

$$\frac{\partial \mathbf{U}}{\partial t} + \nabla \cdot \mathbf{F}(\mathbf{U}) = \mathbf{0}, \quad \nabla \cdot \mathbf{B} = 0 \tag{2.37}$$

with conservative variables $\mathbf{U} = (\rho, \rho\mathbf{v}, \mathbf{B}, e)^T$, and the flux vectors $\mathbf{F}(\mathbf{U})$ given by

$$\mathbf{F}(\mathbf{U}) = \left(\rho\mathbf{v}, \rho\mathbf{v}\mathbf{v} + \left(p + \frac{1}{2}\mathbf{B}\cdot\mathbf{B} \right)\mathbf{I} - \mathbf{B}\mathbf{B}, \mathbf{v}\mathbf{B} - \mathbf{B}\mathbf{v}, \right.$$
$$\left. \left(e + p + \frac{1}{2}\mathbf{B}\cdot\mathbf{B} \right)\mathbf{v} - \mathbf{B}\left(\mathbf{B}\cdot\mathbf{v}\right) \right)^T$$

where $\mathbf{I}$ is a 3×3 identity matrix, ρ is the mass density, $\mathbf{v} = (u, v, w)$ is the velocity vector. Here, p is the thermal pressure, $\mathbf{B}$ is the magnetic field vector scaled by $\sqrt{\mu_0}$ with μ_0 being the magnetic permeability, and e is the total energy, defined as $e = \frac{p}{\gamma - 1} + \frac{1}{2}\rho \mathbf{v}^2 + \frac{\mathbf{B}^2}{2}$. Furthermore, γ stands for the ratio of specific heats of the plasma. The plasma is assumed to obey the ideal gas law and to be calorically perfect, which is a reasonable approximation for most space and solar plasmas. The ideal gas law together with the total energy $e = \frac{p}{\gamma - 1} + \frac{1}{2}\rho \mathbf{v}^2 + \frac{\mathbf{B}^2}{2}$ are necessary constitutive relations to close the set of ideal MHD equations. Meanwhile, the solenoidal constraint ($\nabla \cdot \mathbf{B} = 0$) should be satisfied, which can be recovered analytically from Eq. (2.37) by taking the divergence of the magnetic induction equation, supposing divergence free initial conditions.

The governing equations of ideal magnetohydrodynamics (2.37) describe the physics of a conducting fluid with the following assumptions [187, 188, 226, 417]:

$$\frac{\lambda}{L} << 1, \frac{\epsilon}{\tau\sigma} << 1, \left(\frac{V_c}{c}\right)^2 << 1, \frac{\nu}{V_c L} << 1, \frac{\eta}{\mu_0 V_c L} << 1$$

where V_c, τ, L and ν are, respectively, speed, time, length scales for the problem, and the kinetic viscosity. c is the speed of light, σ represents the electric conductivity of the fluid, and ϵ is the dielectric constant. μ_0 and η are the magnetic permeability and the resistivity, respectively. λ is the Debye length characterizing the typical size of a region over which charge imbalance due to thermal fluctuations may occur.

For convenience, we sometimes write the tensor of fluxes $\mathbf{F}(\mathbf{U})$ as $(\mathbf{f}(\mathbf{U}), \mathbf{g}(\mathbf{U}), \mathbf{h}(\mathbf{U}))$ with the flux functions $\mathbf{f} = \left(f_\rho, \mathbf{f}_{\rho\mathbf{v}}, \mathbf{f}_\mathbf{B}, f_e\right)^T$, $\mathbf{g} = \left(g_\rho, \mathbf{g}_{\rho\mathbf{v}}, \mathbf{g}_\mathbf{B}, g_e\right)^T$, and $\mathbf{h} = \left(h_\rho, \mathbf{h}_{\rho\mathbf{v}}, \mathbf{h}_\mathbf{B}, h_e\right)^T$ in respective (x, y, z) directions, whose components with the subscripts, ρ, $\rho\mathbf{v}$, $\mathbf{B}$, e, correspond to the flux functions of continuity equation for density, momentum equations, magnetic induction equations, and energy equation.

There are a rich set of references available for theoretical background of MHD. A good MHD monograph, both for an introduction to MHD and for further study of the basic phenomena is provided by Biskamp [43], which is a very excellent book containing reconnection and MHD turbulence. A more introductory text is the book by Freidberg [143], see also concise introductory ones (e.g., [359, 417]). The standard text for astrophysical MHD applications is cosmical magnetic fields by Parker [379] and Hu [204]. A tokamak-oriented text, including the basic MHD problems as well as the more complicated phenomena, is the one by White [495]. The most representative source on two-fluid dynamics is perhaps the review article by Braginskii [49], which contains a systematic derivation of the equations from kinetic theory as well as a clear introduction to the effects represented by the individual terms (and why they appear). Through the review article by Braginskii [49], we can better understand the dependence of both the two-fluid and MHD models on the kinetic theory of plasmas, and their range of validity. Besides, the interested readers will no doubt find it stimulating to explore various MHD phenomena further from other literature (e.g., [11, 12, 44, 55, 86, 187–189, 363, 391, 478, 479]).

By the way, except FVM methods, there is a vast literature on application of finite difference schemes, particle-in-cell methods, finite element methods to plasma simulations (e.g., [38, 42, 70, 75, 89, 156, 207, 226, 303, 332, 441, 442]). The introductions to these simulation techniques are beyond the scope of this chapter and the interested reader may refer to the existing literature.

2.3.1 Linearized Riemann Solvers for Conservative MHD Equations

Finite volume methods commonly in use rely on the solution of Riemann problems at cell faces to determine the fluxes of conserved quantities. A Riemann solver naturally embeds the concept of upwinding: the discretization stencil should be biased towards the direction from which characteristic waves propagate. Upwind methods provide a built-in mechanism for numerical dissipation which is ultimately responsible for the stability of the method. It should be emphasized that almost exclusively these Riemann problems are one-dimensional. A Riemann problem is a special form of a one-dimensional initial value problem having discontinuous initial states. This is the situation at the interface between two computational cells, if $\mathbf{U}_i$ is considered the left initial state, $\mathbf{U}_L$, and $\mathbf{U}_j$ considered the right, $\mathbf{U}_R$, as depicted on Fig. 2.2 for a simple one-dimensional computational domain in the direction normal to the cell face. In this case, the inviscid equations describing the solution transport through the face can be reduced to the one-dimensional form

$$\frac{\partial \mathbf{U}}{\partial t} + \frac{\partial \mathbf{f}(\mathbf{U})}{\partial x} = \frac{\partial \mathbf{U}}{\partial t} + \hat{\mathbf{A}}(\mathbf{U}) \frac{\partial \mathbf{U}}{\partial x} = 0 \tag{2.38}$$

where x represents the coordinate direction normal to the face and $\hat{\mathbf{A}}(\mathbf{U}) = \frac{\partial \mathbf{f}}{\partial \mathbf{U}}$ is the Jacobian of the inviscid flux vector with respect to the solution conservative vector $\mathbf{U}$. The one-dimensional Riemann problem is to solve Eq. (2.38) with the following initial data:

$$\mathbf{U}(x, 0) = \mathbf{U}_L = \text{constant} \quad \text{for } x < 0; \quad \mathbf{U}(x, 0) = \mathbf{U}_R = \text{constant} \quad \text{for } x > 0 \tag{2.39}$$

The Roe approximate Riemann or linearized Riemann solver does not set out to solve the Riemann problem exactly, but instead it seeks a good estimate of the numerical flux at the interface based on a local linearization of the governing equations. It does this by locally (i.e, at each cell interface) replacing the Jacobian matrix $\hat{\mathbf{A}}(\mathbf{U}) = \frac{\partial \mathbf{f}}{\partial \mathbf{U}}$ in Eq. (2.38) above with a constant matrix $\hat{\mathbf{A}} = \mathbf{A}(\mathbf{U}_L, \mathbf{U}_R)$ as a function of the local left and right states. Usually, $\mathbf{A}$ is held a constant matrix for the solution and is constructed using some averaging procedure for $\mathbf{U}_L$ and $\mathbf{U}_R$ (or equivalently some averaging procedure for primitive variables $\mathbf{W}_L$ and $\mathbf{W}_R$ corresponding to $\mathbf{U}_L$ and $\mathbf{U}_R$, respectively). For gas dynamics $\hat{\mathbf{A}}$ should satisfy the following property [402]:

$$\mathbf{f}(\mathbf{U}_R) - \mathbf{f}(\mathbf{U}_L) = \hat{\mathbf{A}}(\mathbf{U}_L, \mathbf{U}_R)(\mathbf{U}_R - \mathbf{U}_L) \tag{2.40}$$

which ensures that the solution to the Riemann problem is exact for a single shock.

Consider the ideal MHD equations (2.37) in one dimension

$$\frac{\partial}{\partial t}\begin{pmatrix} \rho \\ \rho u \\ \rho v \\ \rho w \\ B_x \\ B_y \\ B_z \\ e \end{pmatrix} + \frac{\partial}{\partial x}\begin{pmatrix} \rho u \\ \rho u^2 + p + \frac{1}{2}\mathbf{B}^2 - B_x^2 \\ \rho uv - B_x B_y \\ \rho uw - B_x B_z \\ 0 \\ uB_y - vB_x \\ uB_z - wB_x \\ u(e + p + \frac{1}{2}\mathbf{B}^2) - B_x(\mathbf{v}\cdot\mathbf{B}) \end{pmatrix} = 0 \tag{2.41}$$

where the fifth equation $(B_x)_t = 0$, combined with a consequence of $\nabla\cdot\mathbf{B} = 0$ in one-dimensional case, show that $B_x = \text{const}$.

These equations can be transformed to the quasilinear form as

$$\frac{\partial \mathbf{W}}{\partial t} + \bar{\mathbf{A}}\frac{\partial \mathbf{W}}{\partial x} = 0 \tag{2.42}$$

where $\mathbf{W} = (\rho, u, v, w, B_x, B_y, B_z, p)^T$ are known as the primitive variables. The matrix $\bar{\mathbf{A}}$ is given by

$$\bar{\mathbf{A}} \equiv \frac{\partial \mathbf{W}}{\partial \mathbf{U}}\frac{\partial \mathbf{f}}{\partial \mathbf{U}}\frac{\partial \mathbf{U}}{\partial \mathbf{W}} \equiv \frac{\partial \mathbf{W}}{\partial \mathbf{U}}\hat{\mathbf{A}}\frac{\partial \mathbf{U}}{\partial \mathbf{W}} \tag{2.43}$$

$$= \begin{bmatrix} u & \rho & 0 & 0 & 0 & 0 & 0 & 0 \\ 0 & u & 0 & 0 & -B_x/\rho & B_y/\rho & B_z/\rho & 1/\rho \\ 0 & 0 & u & 0 & -B_y/\rho & -B_x/\rho & 0 & 0 \\ 0 & 0 & 0 & u & -B_z/\rho & 0 & -B_x/\rho & 0 \\ 0 & 0 & 0 & 0 & 0 & 0 & 0 & 0 \\ 0 & B_y & -B_x & 0 & -v & u & 0 & 0 \\ 0 & B_z & 0 & -B_x & -w & 0 & u & 0 \\ 0 & \gamma p & 0 & 0 & (\gamma-1)\mathbf{B}\cdot\mathbf{v} & 0 & 0 & u \end{bmatrix}$$

with $\hat{\mathbf{A}} = \frac{\partial \mathbf{f}}{\partial \mathbf{U}}$.

The eigenvalues of $\bar{\mathbf{A}}$ are: $(u, u \pm c_f, u \pm c_a, u \pm c_s, 0)^T$ where c_s, c_a, c_f are the slow magnetosonic, Alfvén, and fast magnetosonic sound speeds, respectively. These speeds are defined below:

$$c_a = \frac{|B_x|}{\sqrt{\rho}} \tag{2.44}$$

$$c_{f,s}^2 = \frac{1}{2}\left[a_t^2 \pm \left(a_t^4 - 4a^2c_a^2\right)^{\frac{1}{2}}\right] \tag{2.45}$$

where a is the usual sound speed defined as $a^2 = \gamma p/\rho$, and $a_t^2 = (\gamma p + (B_x^2 + B_y^2 + B_z^2))/\rho$. In the definitions above, the plus (minus) sign is for the fast (slow) magnetosonic sound speeds. Thus, in MHD, disturbances travel at three different characteristic speeds as opposed to gas dynamics in which there is only one sound speed. Furthermore, in MHD, disturbances propagate anisotropically depending on the orientation with respect to the magnetic field.

The left $\boldsymbol{\ell}_k$ and right $\boldsymbol{r}_k$ eigenvectors of $\bar{\mathbf{A}}$ related to the eigenvalue $\lambda_k \in \{u, u \pm c_f, u \pm c_a, u \pm c_s\}$ can be found in the former works (e.g., [52, 128, 515, 516]). Note that the 5th row from the top of $\bar{\mathbf{A}}$ contains zeros and thus we have one zero eigenvalue $\lambda = 0$. This zero eigenvalue is non-physical as the eigenvalues should appear either singly as the x-component of the flow, u, or in pairs symmetric about u. Corresponding to $\lambda = 0$, we set its left and right eigenvectors as $\boldsymbol{\ell} = (0, 0, 0, 0, 1, 0, 0, 0)$, $\boldsymbol{r} = (0, 0, 0, 0, 1, 0, 0, 0)^T$ in order to form bi-orthogonal eigenvector system with other eigenvectors associated to nonzero eigenvalues. Once the eigensystem of the matrix $\bar{\mathbf{A}}$ is known, the left and right conservative eigenvectors associated with $\hat{\mathbf{A}}$ can be easily obtained via $\mathbf{R}_k = \frac{\partial \mathbf{U}}{\partial \mathbf{W}} \boldsymbol{r}_k$, and $\mathbf{L}_k = \boldsymbol{\ell}_k \frac{\partial \mathbf{W}}{\partial \mathbf{U}}$. That is, $\mathbf{L}_k \hat{\mathbf{A}} = \lambda_k \mathbf{L}_k, \;\; \hat{\mathbf{A}} \mathbf{R}_k = \lambda_k \mathbf{R}_k$.

The solution to the one-dimensional Riemann problem satisfies Eq. (2.41) with the initial data Eq. (2.39). Due to the non-hyperbolic constraint on the divergence of the magnetic field, $\nabla \cdot \mathbf{B} = 0$, the eight MHD conservation laws (2.41) are not fully independent. As a result, for each direction there are only seven physical characteristic waves. At the same time, since $\nabla \cdot \mathbf{B} = 0$ requires the continuity of the magnetic field normal component, B_x, through any surface, the Riemann problem for the MHD equations is ill-posed if a jump across the face of the control volume occurs in B_x in the above x-dimensional case. Otherwise, if B_x is continuous, then the initial jumps in seven other MHD variables determine seven amplitudes of the characteristic waves originating from the initial jumps, so that the Riemann problem is well-posed. The solution to the Riemann problem is self-similar in x/t. The solution contains one contact discontinuity, Alfvén shocks, and magnetosonic shocks or rarefactions.

Taking all these considerations above, we solve one-dimensional Riemann problem (2.41) and (2.39) in a form of eight-like characteristic waves by formally assuming $\lambda = 0$ with its corresponding left and right eigenvectors $\boldsymbol{\ell} = (0, 0, 0, 0, 1, 0, 0, 0)$, $\boldsymbol{r} = (0, 0, 0, 0, 1, 0, 0, 0)^T$ as a complementary to the seven physical characteristic waves. Let's look for a solution of the form

$$\mathbf{U}(x, t) = \sum_{i=1}^{8} v_i(x, t) \mathbf{R}_i$$

with $v_i = \mathbf{L}_i \cdot \mathbf{U}$. Substituting this into Eq. (2.38) follows

$$\sum_{i=1}^{8} \left(\frac{\partial v_i}{\partial t} + \lambda_i \frac{\partial v_i}{\partial x} \right) \mathbf{R}_i = 0$$

Multiplying by $\mathbf{L}_j$ and using the bi-orthogonality property $\mathbf{L}_j \cdot \mathbf{R}_i = \delta_{ji}$, we have

$$\frac{\partial v_j}{\partial t} + \lambda_j \frac{\partial v_j}{\partial x} = 0$$

which is a linear advection equation with explicit solution

$$v_j(x,t) = \mathbf{L}_j \cdot \mathbf{U}_0(x - \lambda_j t)$$

if the initial value is $\mathbf{U}(x,0) = \mathbf{U}_0(x)$. Hence, the full solution is

$$\mathbf{U}(x,t) = \sum_{i=1}^{8} \mathbf{L}_i \cdot \mathbf{U}_0(x - \lambda_i t)\mathbf{R}_i$$

Evidently, the solution consists of seven waves moving with the constant wave-speeds, $\lambda_i \in \{u, u \pm c_f, u \pm c_a, u \pm c_s\}$. For the discontinuous initial data given by Eq. (2.39), each of these waves is a discontinuity at which the solution jumps by $\mathbf{L}_i \cdot (\mathbf{U}_R - \mathbf{U}_L)$ as they travel from left to right. With the zero eigenvalue $\lambda = 0$, we always have $B_x = \text{const}$. The solution of $\mathbf{U}$ at $x = 0$ is represented by $\mathbf{U}^*$, which takes the following form,

$$\mathbf{U}^* = \mathbf{U}_L + \sum_{\lambda_i < 0} \mathbf{L}_i \cdot (\mathbf{U}_R - \mathbf{U}_L)\mathbf{R}_i \quad \text{left to right}$$

$$\mathbf{U}^* = \mathbf{U}_R + \sum_{\lambda_i \geq 0} \mathbf{L}_i \cdot (\mathbf{U}_L - \mathbf{U}_R)\mathbf{R}_i \quad \text{right to left}$$

where $\lambda = 0$ is included for consistently conforming to the later description. That is,

$$\mathbf{U}^* = \frac{1}{2}\left[\mathbf{U}_L + \mathbf{U}_R - \sum_{i=1}^{8} \text{sign}(\lambda_i)\mathbf{L}_i \cdot (\mathbf{U}_R - \mathbf{U}_L)\mathbf{R}_i\right]$$

For a linear system with $\mathbf{f}(\mathbf{U}) = \hat{\mathbf{A}}\mathbf{U}$, we see that

$$\begin{aligned}
\mathbf{f}(\mathbf{U}^*) = \hat{\mathbf{A}}\mathbf{U}^* &= \hat{\mathbf{A}}\left[\mathbf{U}_L + \sum_{\lambda_i < 0} \mathbf{L}_i \cdot (\mathbf{U}_R - \mathbf{U}_L)\mathbf{R}_i\right] \\
&= \mathbf{f}(\mathbf{U}_L) + \sum_{\lambda_i < 0} \lambda_i \mathbf{L}_i \cdot (\mathbf{U}_R - \mathbf{U}_L)\mathbf{R}_i \\
&= \hat{\mathbf{A}}\left[\mathbf{U}_R + \sum_{\lambda_i \geq 0} \mathbf{L}_i \cdot (\mathbf{U}_L - \mathbf{U}_R)\mathbf{R}_i\right]
\end{aligned}$$

$$= \mathbf{f}(\mathbf{U}_R) + \sum_{\lambda_i \geq 0} \lambda_i \mathbf{L}_i \cdot (\mathbf{U}_L - \mathbf{U}_R)\mathbf{R}_i$$

Hence,

$$\mathbf{f}(\mathbf{U}^*) = \frac{1}{2}\left[\mathbf{f}(\mathbf{U}_L) + \mathbf{f}(\mathbf{U}_R) + \sum_{i=1}^{8} |\lambda_i| \mathbf{L}_i \cdot (\mathbf{U}_L - \mathbf{U}_R)\mathbf{R}_i\right] \tag{2.46}$$

Or equivalently,

$$\mathbf{f}(\mathbf{U}^*) = \frac{1}{2}\left[\mathbf{f}(\mathbf{U}_L) + \mathbf{f}(\mathbf{U}_R) - |\hat{\mathbf{A}}|(\mathbf{U}_R - \mathbf{U}_L)\right] \tag{2.47}$$

with the dissipation matrix $|\hat{\mathbf{A}}| = \mathbf{R}|\Lambda|\mathbf{R}^{-1}$.

Note that in a numerical scheme it is important to compute the flux at $x = 0$ from Eq. (2.46) rather than first calculating $\mathbf{U}^*$ and then using $\mathbf{f} = \mathbf{f}(\mathbf{U}^*)$. This is because (2.46) is a continuous function of $\mathbf{U}_L$ and $\mathbf{U}_R$, whereas $\mathbf{U}^*$ is not.

2.3.2 *Linearized Riemann Solvers for Non-conservative MHD Equations*

Godunov [166] observed that the conservative form of the ideal MHD equations leads to an eigenvalue degeneracy. This implies that while the ideal MHD equations are hyperbolic they are not strictly hyperbolic. Adding a source vector $\mathbf{S} = \left(S_\rho, \mathbf{S}_{\rho\mathbf{v}}, \mathbf{S}_\mathbf{B}, S_e\right)^T = -\nabla \cdot \mathbf{B}\,(0, \mathbf{B}, \mathbf{v}, \mathbf{B} \cdot \mathbf{v})^T$ to the conservative equations (2.37),

$$\frac{\partial \mathbf{U}}{\partial t} + \nabla \cdot \mathbf{F}(\mathbf{U}) = \mathbf{S}, \tag{2.48}$$

one can then remove the eigenvalue degeneracy. Adding this source term also makes the equations Galilean invariant. This is also the rationale adopted by Powell et al. [389] for including the non-conservative source term in the equations. If the divergence of the induction equation, which includes the source term, is taken, then we get the following equation

$$\frac{\partial \nabla \cdot \mathbf{B}}{\partial t} + \nabla \cdot (\mathbf{v} \nabla \cdot \mathbf{B}) = 0 \tag{2.49}$$

This implies that $\nabla \cdot \mathbf{B}/\rho$, if non-zero (presumably due to numerical error), will get convected along streamlines.

Written out in the one-dimensional case, Eq. (2.48) becomes

$$\frac{\partial \mathbf{U}}{\partial t}+\frac{\partial \mathbf{f}}{\partial x}=-\frac{\partial B_x}{\partial x}\begin{pmatrix}0\\B_x\\B_y\\B_z\\u\\v\\w\\\mathbf{v}\cdot\mathbf{B}\end{pmatrix}=-\begin{pmatrix}0\,0\,0\,0 & 0 & 0\,0\,0\\0\,0\,0\,0 & B_x & 0\,0\,0\\0\,0\,0\,0 & B_y & 0\,0\,0\\0\,0\,0\,0 & B_z & 0\,0\,0\\0\,0\,0\,0 & u & 0\,0\,0\\0\,0\,0\,0 & v & 0\,0\,0\\0\,0\,0\,0 & w & 0\,0\,0\\0\,0\,0\,0 & \mathbf{v}\cdot\mathbf{B} & 0\,0\,0\end{pmatrix}\frac{\partial \mathbf{U}}{\partial x}$$

which can be rewritten as follows:

$$\frac{\partial \mathbf{U}}{\partial t}+\hat{\mathbf{A}}\frac{\partial \mathbf{U}}{\partial x}=0 \tag{2.50}$$

where $\hat{\mathbf{A}}$ is the sum of the flux Jacobian $\frac{\partial \mathbf{f}}{\partial \mathbf{U}}$ for the homogeneous ideal MHD equations and $\mathbf{A}_P$ due to the Powell source term [389] written in the above matrix form, i.e., $\hat{\mathbf{A}}=\frac{\partial \mathbf{f}}{\partial \mathbf{U}}+\mathbf{A}_P$.

When using the non-conservative source term (2.48), the one-dimensional quasi-linear form in the primitive variables $\mathbf{W}$ is

$$\frac{\partial \mathbf{W}}{\partial t}+\bar{\mathbf{A}}\frac{\partial \mathbf{W}}{\partial x}=-\begin{pmatrix}0\,0\,0\,0 & 0 & 0\,0\,0\\0\,0\,0\,0 & \frac{B_x}{\rho} & 0\,0\,0\\0\,0\,0\,0 & \frac{B_y}{\rho} & 0\,0\,0\\0\,0\,0\,0 & \frac{B_z}{\rho} & 0\,0\,0\\0\,0\,0\,0 & u & 0\,0\,0\\0\,0\,0\,0 & v & 0\,0\,0\\0\,0\,0\,0 & w & 0\,0\,0\\0\,0\,0\,0 & -(\gamma-1)\mathbf{v}\cdot\mathbf{B} & 0\,0\,0\end{pmatrix}\frac{\partial \mathbf{W}}{\partial x}$$

i.e.,

$$\frac{\partial \mathbf{W}}{\partial t}+\mathbf{A}\frac{\partial \mathbf{W}}{\partial x}=0 \tag{2.51}$$

where

$$\mathbf{A}=\begin{bmatrix}u & \rho & 0 & 0 & 0 & 0 & 0 & 0\\0 & u & 0 & 0 & 0 & B_y/\rho & B_z/\rho & 1/\rho\\0 & 0 & u & 0 & 0 & -B_x/\rho & 0 & 0\\0 & 0 & 0 & u & 0 & 0 & -B_x/\rho & 0\\0 & 0 & 0 & 0 & u & 0 & 0 & 0\\0 & B_y & -B_x & 0 & 0 & u & 0 & 0\\0 & B_z & 0 & -B_x & 0 & 0 & u & 0\\0 & \gamma p & 0 & 0 & 0 & 0 & 0 & u\end{bmatrix} \tag{2.52}$$

While $\bar{\mathbf{A}}$ in Eq. (2.43) has an eigenvalue degeneracy, $\mathbf{A}$ in Eq. (2.52) does not. It is clear that the zero row has disappeared and we have the advection equation $(B_x)_t + u(B_x)_x = 0$. As this wave carries non-zero magnetic field divergence it is nicknamed the divergence wave. Corresponding to this divergence wave, we have $\lambda = u$, $\boldsymbol{\ell} = (0, 0, 0, 0, 1, 0, 0, 0)$, $\boldsymbol{r} = (0, 0, 0, 0, 1, 0, 0, 0)^T$. The eigenvalues λ_k, eigenvectors $\boldsymbol{r}_k$ and $\boldsymbol{\ell}_k$ for the eight-wave matrix $\mathbf{A}$ can be found in the work of Powell et al. [389] or Murawski [334] for details. In this case, $\hat{\mathbf{A}}$ in Eq. (2.52) gives us physical eigenvalues and eigenvectors.

Proceeding as before, we can solve the linearized Reimann problem (2.50) with the initial data (2.39). We can easily obtain $\mathbf{U}^*$

$$\mathbf{U}^* = \frac{1}{2}\left[\mathbf{U}_L + \mathbf{U}_R - \sum_{i=1}^{8} \text{sign}(\lambda_i)\mathbf{L}_i \cdot (\mathbf{U}_R - \mathbf{U}_L)\mathbf{R}_i\right]$$

and the flux $\mathbf{f}(\mathbf{U}^*)$. This, of course, is not strictly valid for linearized Riemann solvers, but it is, nonetheless, still used. Alternatively, following Roe's method [389, 402], the flux can be written as:

$$\mathbf{f}(\mathbf{U}^*) = \frac{1}{2}\left[\mathbf{f}(\mathbf{U}_L) + \mathbf{f}(\mathbf{U}_R) - |\hat{\mathbf{A}}|(\mathbf{U}_R - \mathbf{U}_L)\right] \tag{2.53}$$

with the dissipation matrix $|\hat{\mathbf{A}}| = \mathbf{R}|\Lambda|\mathbf{R}^{-1}$, and $\Lambda = \text{diag}(\lambda_k)$ consists of the eigenvalues λ_k of matrix $\hat{\mathbf{A}}$, and $\mathbf{R}$ and $\mathbf{R}^{-1}$ are, right and left conservative eigenvectors, respectively. The conservative eigenvectors $\mathbf{R} = \frac{\partial \mathbf{U}}{\partial \mathbf{W}}\boldsymbol{r}_k$ and $\mathbf{L} = \boldsymbol{\ell}_k\frac{\partial \mathbf{W}}{\partial \mathbf{U}}$ as mentioned above.

Furthermore, it is convenient to solve the Riemann problems in terms of the primitive variables. Riemann solvers can be developed based on either Eq. (2.42) or Eq. (2.52). That is, we can directly consider the quasilinear form $\frac{\partial \mathbf{W}}{\partial t} + \mathbf{A}_x\frac{\partial \mathbf{W}}{\partial x} = 0$ in terms of the primitive variables, with $\mathbf{A}_x = \bar{\mathbf{A}}$ in Eq. (2.43) or $\mathbf{A}_x = \mathbf{A}$ in Eq. (2.52), and the initial data $\mathbf{W}_L$ and $\mathbf{W}_R$ derived from $\mathbf{U}_L$ and $\mathbf{U}_R$. Depending upon whether we use $\mathbf{A}_x = \mathbf{A}$ or $\bar{\mathbf{A}}$ we get an eight-wave or a seven-wave method. The solution to this Riemann problem is then given by:

$$\mathbf{W}^* = \mathbf{W}_L + \sum_{\lambda_k<0}(\boldsymbol{\ell}_k \cdot (\mathbf{W}_R - \mathbf{W}_L))\boldsymbol{r}_k = \mathbf{W}_R - \sum_{\lambda_k\geq 0}(\boldsymbol{\ell}_k \cdot (\mathbf{W}_R - \mathbf{W}_L))\boldsymbol{r}_k \tag{2.54}$$

where λ_k, $\boldsymbol{\ell}_k$, and $\boldsymbol{r}_k$ are the eigenvalues, left and right eigenvectors, respectively, of matrix $\mathbf{A}_x$. Thus,

$$\mathbf{W}^* = \frac{1}{2}\left[\mathbf{W}_L + \mathbf{W}_R - \sum_{i=1}^{8} \text{sign}(\lambda_i)\boldsymbol{\ell}_i \cdot (\mathbf{W}_R - \mathbf{W}_L)\boldsymbol{r}_i\right]$$

where i is an index for the loop over the entropy, divergence, Alfvén, magneto-acoustic waves. From $\mathbf{W}^*$ in Eq. (2.54) we can easily obtain $\mathbf{U}^*$ and then the flux $\mathbf{f}(\mathbf{U}^*)$ as in Eqs. (2.47) and (2.53).

In Eqs. (2.47) and (2.53), the terms denoted with subscripts L and R are evaluated from the face-midpoint states just to the left and right of the interface. The eigenvalues and eigenvectors are evaluated at an "interface" state that is some combination of the L and R states. In order to obtain accurate numerical solutions for the ideal MHD equations, Brio and Wu [52] first developed the Roe-type Riemann solver [402] for MHD with $\gamma = 2$. For general γ, construction of the Roe-type matrix and corresponding Roe-type average has been pursued (e.g., [18, 61]). Practically, an arithmetic average is often utilized instead of the Roe average, and it seems to work well [407]. It is preferable to average these by the primitive variables (for example, $\hat{\mathbf{A}} = \hat{\mathbf{A}}(\frac{1}{2}(\mathbf{W}_L + \mathbf{W}_R))$) rather than by the conservative variables because this ensures that the mean state is physically realizable. This does not have the Roe property, but it is nevertheless able to handle strong shocks. In practice, a simple arithmetic averaging of the primitive variables is done to compute the interface state. Vector variables (velocity, magnetic field) are averaged component by component.

Although it is not practicable to use an exact MHD Riemann solver in a numerical scheme, a simple linear Riemann solver is accurate and robust enough and also efficient in comparison with the nonlinear Riemann solver [88, 407]. Presently, the linearized Riemann solver becomes a standard solver for the practical MHD simulations (e.g., [254, 323, 389, 460]).

As shown above, the Powell source term is necessary to restore a symmetric MHD system. This symmetrizable form of Eq. (2.48) allows the derivation of an eight-wave approximate Riemann solver for constructing an upwind solution scheme for multi-dimensional flows. It is suggested that dissipative matrix $D = |\hat{\mathbf{A}}|$ with the Powell source term is used in any consideration of numerical flux for dissipative term. This altered flux Jacobian is used only to derive the dissipation term. Just as Lax–Friedrichs scheme differs from Roe scheme in the construction of a dissipation term, the Powell source term is only used to build the dissipation term. Thus, no inconsistency with the previous conserving flux derivations is introduced.

2.3.3 Complexity of MHD Equations

Due to the complex wave structure appearing in the dynamics, the MHD equations are neither strictly hyperbolic nor strictly convex (e.g., [52, 94, 95, 358, 404, 421, 499, 516]). The MHD equations form a non-strictly hyperbolic system as some eigenvalues may coincide at some points and that compound wave structures, involving both shocks and rarefactions, may sometimes develop. In the context of Lax [260], a characteristic field k is said to be linearly degenerate if $\frac{\partial \lambda_k(\mathbf{U})}{\partial \mathbf{U}} \cdot \mathbf{r}_k(\mathbf{U}) = 0$, and genuinely nonlinear (convex) if $\frac{\partial \lambda_k(\mathbf{U})}{\partial \mathbf{U}} \cdot \mathbf{r}_k(\mathbf{U}) \neq 0$ for all $\mathbf{U}$. Serna [421] performed an analysis of the wave structure of the ideal magnetohydrodynamic (MHD) equations by presenting an analytical expression of the nonlinearity term associated to each characteristic field [52, 389, 404]. According to the analysis of Serna [421], the characteristic fields associated with Alfvén waves and entropy waves or contacts are linearly degenerate. However, a nonlinear characteristic field (fast or slow) may become nonconvex in terms of the change of sign of the components of the magnetic field. That is, the rotation of the magnetic field induces non-strict hyperbolicity of the MHD system and non-genuinely nonlinearity in some of the local wavefields [52, 499]. It turns out that the remedy of this problem is to consider a form of the equations that is not strictly in a conservation form Eq. (2.50).

Compared to the Euler equations of gas dynamics, the complexity of the ideal MHD wave dynamics also lies on the presence of the magnetic field that generates new magnetic and magnetoacoustic waves [29, 94, 95, 358, 366, 382, 457–459, 470]. It is clear that the MHD Riemann problem is considerably more complicated than that of the ordinary gas dynamics, not just due to its more waves [229, 384, 385, 445].

The eigenvalues are degenerate in the following two cases:

$$B_x = 0, c_f = \sqrt{a^2 + \frac{B_t^2}{\rho}}, c_s = c_a = 0$$

$$B_x \neq 0, B_t = 0, c_f = \max(a, c_a), c_s = \min(a, c_a)$$

For the transverse magnetic field component $B_t = \sqrt{B_y^2 + B_z^2} = 0$, either $c_s = c_a$ if $c_a < a$, or $c_f = c_a$ if $c_a > a$. Thus, if $c_a < a$ (resp. $c_a > a$) we have slow wave (resp. fast wave) degeneracy. The slow-wave and fast-wave eigenvectors contain terms proportional to $B_y = B_t$ and $B_z = B_t$ as $B_t \to 0$, under which the eigenvectors are still linearly independent irrespective of the direction of the transverse magnetic field component. For the case of $c_a = a$, both the fast and slow waves are degenerate. When $B_x = 0$ the fast waves cannot change the transverse velocity components, and both the Alfvén and the slow wave merge with the contact discontinuity. The equations then reduce to those of gas dynamics with pressure replaced by the sum of the gas and the magnetic pressures. It occurs that when the magnetic field components disappear the eigenvectors become singular. By renormalizing the eigenvectors, these singularities can be removed (e.g., [52, 234, 334, 388, 389, 404]), such that the eigenvectors can be defined in such a way that they form an independent set.

Since the MHD equations possess non-convexity as well as non-strict hyperbolicity, a solution of the Riemann problem may be composed not only of ordinary shock and rarefaction waves but also other waves as compound waves and overcompressible shocks [52, 229, 324, 445]. To solve the Riemann problem for such MHD system in general is complicated and computationally expensive, if not impossible, which is the reason why the linearized Riemann solver is commonly used. However, the discontinuous solutions formed by the regular waves can be discussed easily using the Rankine–Hugoniot relations, $S[\mathbf{U}] = [\mathbf{f}]$, where S denotes the speed of the discontinuities and $[\cdot]$ indicates the jump of the states ahead of and behind the discontinuities,

$\mathbf{U}_1$ and $\mathbf{U}_2$. When solutions for $u_1, u_2 \neq S$ are considered, i.e., there is net particle transport across the discontinuities, and $B_x \neq 0$, three types of shock solutions exist. In the cases of the compressible solutions, that is, $\rho_1 < \rho_2$ and $u_1 > u_2$, the fast and slow shocks are achieved, in which both shocks conserve the signs of the tangential components of the magnetic field. The fast and slow shock are associated with the fast wave c_f and slow wave c_s respectively. While the magnetic field is strengthened behind the fast shock, it is weakened behind the slow shock. In the cases of the incompressible condition, that is, $\rho_1 = \rho_2$ and $u_1 = u_2$, the following relations

$$[\rho] = [p] = [B_y^2 + B_z^2] = 0, \pm\sqrt{\rho}[v] = [B_y], \pm\sqrt{\rho}[w] = [B_z]$$

hold across the discontinuity which moves with the Alfvén wave speed $u \mp c_a$. This solution is called the Alfvén shock. It is also called the rotational discontinuity, since thermodynamical quantities, ρ and p, keep unchanged across the discontinuity. Meanwhile, the discontinuity moving with the entropy wave speed u must satisfy

$$[B_y] = [B_z] = [v] = [w] = [p] = 0$$

for the case of $B_x \neq 0$, or

$$\left[p + \frac{B_y^2 + B_z^2}{2}\right] = 0,$$

for $B_x = 0$. The former discontinuity for the case of $B_x \neq 0$ is called the contact discontinuity. At the contact discontinuity, the tangential components of the velocity and magnetic field must be continuous. On the other hand, the latter relation for $B_x = 0$, indicates that the tangential velocity and the tangential magnetic field admit a jump across the discontinuity. Hence, it is called the tangential discontinuity. For a systematic discussion of evolutionary properties of MHD discontinuities, refer to Somov [445] for instance.

Takahashi and Yamada [458, 459] built a code to obtain the exact solutions of Riemann problems in ideal MHD for an arbitrary initial condition to investigate a variety of solutions more thoroughly. The code can handle not only regular waves but also switch-on/off rarefactions and all types of non-regular shocks: intermediate shocks and switch-on/off shocks. Furthermore, the initial conditions with vanishing normal or transverse magnetic fields can be handled. Promisingly, the solver may be applied to numerical MHD schemes like the Godunov scheme in the future.

Ledentsov and Somov [261, 262] considered the boundary conditions for the equations of ideal MHD at the surfaces of discontinuities of various types, and established the specific form of transition solutions for possible pairs of discontinuous flows, and set up the correspondence between the standard classification of discontinuous flows in ideal MHD and the characteristic parameter of plasma flow—the value of the mass flow through the discontinuity. Based on them, they constructed a generalized scheme of permitted or continuous transitions between MHD discontinuities that contains not only evolutionary types of discontinuities but also non-evolutionary ones:

the contact discontinuity, the switch-on and switch-off, Alfvén, and trans-Alfvén shocks. By the way, they investigated the possibility of plasma heating by different types of MHD discontinuities. The larger the jumps in plasma density and magnetic energy density at the discontinuity, the stronger the plasma heating. The formation of non-evolutionary trans-Alfvén shocks also contributes to the plasma heating. Such conditions occur near a region of the reverse currents in the reconnection process, for example, in solar flares.

Another important issue is ensuring $\nabla \cdot \mathbf{B} = 0$ in multidimensions. Clearly, the discretization errors lead to non-zero divergence over time. Physically, this means that nothing maintains conservation of a magnetic flux in the Gauss' law. The divergence constraint is intrinsic since it follows from the evolution equation and must not be viewed as an additional equation. Because usually the discrete divergence of the discrete curl is not exactly zero, $\nabla \cdot \mathbf{B}$ errors arise in numerical simulations and may increase with time. If this happens, the behavior of the system can become unphysical: Magnetic field lines may have wrong topologies, leading to plasma transport orthogonal to the magnetic field. Several remedies are provided in Sect. 2.4.

2.3.4 *Entropy Fix*

In ideal MHD wave dynamics the points of acoustic causality are not only sonic points but the points where nonconvexity occurs too. Then, an additional entropy fix in the neighborhood of points where nonlinear characteristic fields are nonconvex is necessary to ensure correct formation of the peculiar complex wave structure of MHD equations as the so-called compound waves. As pointed out by Harten [190], an addition of so-called "entropy fix" is necessary to the Roe's scheme, in order to avoid expansion shocks. In the Roe's scheme of the MHD equations, Powell et al. [389] suggested that the entropy fix is applied to the magnetosonic waves to bound those eigenvalues away from zero during the flow expansion. This is achieved by replacing $|\lambda_k|$ in Eqs. (2.47) and (2.53) for the values of k corresponding to the magnetoacoustic waves c_f and c_s only with $|\lambda_k^*|$

$$|\lambda_k^*| = \begin{cases} |\lambda_k| & |\lambda_k| \geq \frac{\delta\lambda_k}{2} \\ \frac{|\lambda_k|^2}{\delta\lambda_k} + \frac{\delta\lambda_k}{4} & |\lambda_k| < \frac{\delta\lambda_k}{2} \end{cases}$$

with $\delta\lambda_k = \max(4(\lambda_{kR} - \lambda_{kL}), 0)$. Serna [421] suggested a characteristic-based nonconvex entropy-fix upwind scheme, which is able to detect local regions containing nonconvex singularities and to handle an entropy correction through a prescription of a local viscosity ensuring convergence to the entropy solution. It should be mentioned that the entropy conserving flux construction [107–109, 496, 497] for the MHD equations has also been introduced.

2.3.5 *Artificial Dissipation*

It is well known that upwind schemes can produce unphysical results when the numerical dissipation becomes small as it does for stationary or slowly moving shocks that are aligned with the grid. The usual remedy is to employ a more dissipative Riemann solver, such as HLL (e.g., [192, 469]). However, it is naturally simple to just add some dissipation to the linear Riemann solver described in Eqs. (2.47) and (2.53). As formerly suggested [127, 128], by setting $c_m = \max(c_f(\mathbf{U}_L), c_f(\mathbf{U}_R))$, $\rho_m = \max(\rho(\mathbf{U}_L), \rho(\mathbf{U}_R))$ we can add the dissipative fluxes

$$\mathbf{f}_{\rho\mathbf{v}} = \alpha_p c_m \rho_m (\mathbf{v}_L - \mathbf{v}_R), f_e = \alpha_e c_m \rho_m (T_L - T_R), \mathbf{f}_\mathbf{B} = \alpha_m c_m (\mathbf{B}_L - \mathbf{B}_R)$$

to the Roe flux given by Eqs. (2.47) and (2.53). Here $\mathbf{f}_{\rho\mathbf{v}}$ is the viscous momentum flux, f_e is the energy flux due to thermal conduction (T is the temperature), and $\mathbf{f}_\mathbf{B}$ is magnetic diffusion. α_p, α_e are user defined parameters ($\alpha_p = \alpha_e = \alpha_m = 0.2$ is fine according to Falle [127]). This avoids the problems associated with stationary or slowly moving shocks, and it does not change the order of scheme if the spatial accuracy is of second order, since dissipative terms are $O(\Delta x^2)$ in smooth regions [127, 128].

2.4 Magnetic Field Divergence Cleaning Methods

A consequence of Faraday's induction law is that an initially divergence free magnetic field remains divergence free for all time. However, applying the divergence operator ($\nabla\cdot$) to the right hand side of the induction equation ($\nabla \times (\mathbf{v} \times \mathbf{B})$) does not ensure that the solenoidal property of the magnetic field is satisfied. In order to single out physically relevant solutions, the MHD system has to be augmented with additional admissibility criteria of the divergence-free condition $\nabla \cdot \mathbf{B} = 0$, which means that magnetic field lines neither begin nor end, but only present in loops, possibly extending to infinity. The implementation of $\nabla \cdot \mathbf{B} = 0$ into a numerical approximation is a major difficulty. A non-zero divergence leads to an anomalous Lorentz force parallel to the magnetic field [48]. The violation of the divergence constraint is due to the nonphysical intermediate state within a numerical discontinuity profile, and this violation may frequently lead to severe stability problems. Thus, the development of divergence cleaning numerical techniques and their comparative studies never stop.

Researchers have tried to enforce the divergence-free constraint in their MHD formulations such as the projection method [48, 134, 421, 460, 471, 516, 524], constraint transport method [77, 125, 137, 510, 511, 524, 531], eight-wave solution method [157, 213, 234, 389, 471], and hyperbolic divergence cleaning method [96–99, 185, 209, 212, 214, 232, 318–320, 362, 431–433, 454, 502, 504, 524].

The following methods, to be described briefly below, are some techniques to ensure that some discrete form of the $\nabla \cdot \mathbf{B} = 0$ is satisfied within machine round-off error or truncation-level error.

2.4.1 Projection Method

This method forces the divergence-free constraint by solving a Poisson equation to subtract off the portion of the magnetic field that leads to non-zero divergence [48, 460, 461]. Suppose that after a timestep the field is $\mathbf{B}^*$ with $\nabla \cdot \mathbf{B}^* \neq 0$. The field at the next timestep is then given by

$$\mathbf{B} = \mathbf{B}^* - \nabla \phi$$

where ϕ is the solution of the Poisson equation

$$\nabla^2 \phi = \nabla \cdot \mathbf{B}^*$$

This clearly ensures $\nabla \cdot \mathbf{B} = 0$. This works well, but it does come at the cost of solving a Poisson equation at each timestep.

2.4.2 Hyperbolic/Parabolic Cleaning Method

Dedner et al. [96] describe a rather neat procedure in which the monopoles associated with non-zero $\nabla \cdot \mathbf{B}$ are destroyed by a combination of advection and decay. Often, this procedure is called the generalized Lagrange multiplier (GLM) formulation of MHD equations. In practice, an additional scalar evolution equation is solved and coupled with the induction equation as follows:

$$\frac{\partial \mathbf{B}}{\partial t} + \nabla \cdot (\mathbf{vB} - \mathbf{Bv} + \psi \mathbf{I}) = 0$$

$$\frac{\partial \psi}{\partial t} + c_h^2 \nabla \cdot \mathbf{B} = -\frac{c_h^2}{c_p^2} \psi \tag{2.55}$$

where c_h and c_p indicate uniformly constant free parameters. From these equations, a telegraph equation

$$\frac{\partial^2 \psi}{\partial t^2} + \frac{c_h^2}{c_p^2} \frac{\partial \psi}{\partial t} = c_h^2 \Delta \psi$$

is obtained, which implies that divergence errors are propagated to the domain boundaries at finite speed c_h and decay with time and distance. The constant ratio c_h^2/c_p^2, which has the dimension of inverse time, sets the damping rate. In the limiting case of

$c_p \to \infty$, one retrieves the simple hyperbolic correction by reducing to an ordinary wave equation. The main advantage of this method is that Eq. (2.55) already possesses the correct conservative form, allowing for direct treatment with any existing schemes or codes.

In GLM formulation, this procedure is very easy to implement: in one dimension one simply adds the equations

$$\frac{\partial B_x}{\partial t} + \frac{\partial \psi}{\partial x} = 0, \quad \frac{\partial \psi}{\partial t} + c_h^2 \frac{\partial B_x}{\partial x} = -\frac{c_h^2}{c_p^2}\psi \tag{2.56}$$

to the system (2.41). Note that one no longer assumes that B_x is constant, but B_x and ψ are decoupled from the other variables. The Riemann problem for these variables is described by Eq. (2.56) without the source term and is extremely simple.

There remains the choice of c_h, which is the wavespeed associated with ψ, and the decay constant c_p. Since one wants these waves to be advected away as fast as possible, c_h should simply be the fastest wavespeed in the domain. c_p should be such that waves with wavelength of the order of the grid spacing decay rapidly. Dedner et al. [96] chose $c_p = \sqrt{0.18 c_h}$, depending on their numerical experiment (See also Dedner et al. [97, 98] for reference). A dimensionless parameter $\alpha = \frac{\Delta h c_h}{c_p^2} \in [0, 1]$ is suggested [137, 318, 319, 454] with Δh being the mesh spacing. The third choice $c_p = \sqrt{-\frac{\Delta t}{\ln c_d} c_h^2}$ with $c_d \in (0, 1)$ has been used [96, 185, 232, 502, 524]. Since $c_h = \frac{\text{CFL}}{\Delta t}\Delta h$ with CFL the Courant number, the later two choices become the same if we set $c_d = \exp(-\alpha \text{CFL})$. Susanto et al. [454] verified that, for the choice of $c_p = \sqrt{0.18 c_h}$, the hyperbolic time step restriction was dominant for all their problems and grids tested.

It seems clear that strong divergence errors should be treated with high correction speeds while only small errors do not require going beyond the largest system speed of the MHD system. Altmann [6] applied two different speed steps for the correction by setting

$$c_h = \begin{cases} c_{\max}, & |\nabla \cdot \mathbf{B}| < 5 \times 10^{-3} \\ 0.5 c_{\text{ref}}, & 5 \times 10^{-3} \le |\nabla \cdot \mathbf{B}| < 5 \times 10^{-2} \\ c_{\text{ref}}, & |\nabla \cdot \mathbf{B}| > 5 \times 10^{-2} \end{cases}$$

where $c_{\max}$ denotes the maximum physical speed within the cell, determined by the CFL condition and c_{ref} a user defined maximum correction speed, $c_{\text{ref}} > c_{\max}$. The introduction of speed steps is balancing areas of slightly varying divergence errors. Altmann [6] showed that this strategy is superior to a direct mapping of the local divergence to the correction speed.

2.4.3 The Unstaggered Constrained Transport Method

The unstaggered constraint transport method solves all cell-centered quantities, including all components of the magnetic field and magnetic flux function that expresses the magnetic field. In practice, the magnetic field is expressed as $\mathbf{B} = \nabla \times \mathbf{A}$ (e.g., [77, 125, 199, 327, 405, 503]) so that the condition $\nabla \cdot \mathbf{B} = 0$ is automatically satisfied. However, $\mathbf{A} \to \mathbf{A} - \nabla\psi$ which is a gauge transformation. In other words, the vector potential is undetermined to within the gradient of a scalar function. Different choices of ψ represent different gauge condition choices. For the vector potential, the most commonly chosen gauge condition is to make use of the Coulomb gauge, and then we get

$$\mathbf{J} = \nabla \times \mathbf{B} = \nabla \times (\nabla \times \mathbf{A}) = -\nabla^2 \mathbf{A} \tag{2.57}$$

Usually, satisfying the gauge conditions requires the solution of an elliptic equation. Obviously, the difficulty with this approach is that a representation of the Lorentz force requires taking a second derivative of the flux function, $\mathbf{A}$, although the divergence-free condition is satisfied automatically. That forces an application of the higher order numerical schemes for evolving the magnetic flux function. Even then, one can encounter serious problems due to anomalous Lorentz force which apparently reveals itself in the neighborhood of large spatial gradients. By the way, Tóth [471] and Chacón [68] introduced a finite-volume interpretation of the CT schemes that place all of the variables at the cell center.

2.4.4 *The Staggered Constraint Transport*

In the staggered-grid approach, the divergence free constraint is satisfied by placing the magnetic field components at the centroids of appropriate cell faces and remaining plasma variables such as mass, momentum and energy are stored at the centroids of computational cells. On such a grid the MHD equations can be approximated in a way that preserves solenoidality of discrete magnetic field [23, 25, 263, 291, 530, 531]. Sometimes, these methods are called flux-CT [25, 160], upwinding constrained transport (UCT) [291], field-interpolated CT [88], transport-flux-interpolated CT [408], corner transport upwind (CTU) [159, 160, 199, 318], a stable upwind scheme [158], HLL-flux CT [325, 531], unsplit staggered mesh scheme (USM) [263].

The main advantages of the staggered-grid CT approach are that there is no elliptic solve and there are no free parameters to choose such as in the hyperbolic divergence-cleaning technique. This approach comes from incompressible fluid mechanics where the velocity field must be kept divergence free. However, staggered grids are expensive for storage and handling on meshes with hanging nodes that are common to unstructured grid methods.

2.4.5 *Source Term Approaches*

Besides Powell's eight-wave formulation, there are other local source term treatments advocated to mitigate the $\nabla \cdot \mathbf{B}$ error. Janhunen [225] and Dellar [102] formulated various arguments to only add the $\nabla \cdot \mathbf{B}$ related source term to the induction equation, which restores momentum and energy conservation. These two local source term approaches can be described with the following respective sources $(S_\rho, \mathbf{S}_{\rho\mathbf{v}}, \mathbf{S}_\mathbf{B}, S_e)^T$

$$\begin{pmatrix} S_\rho \\ \mathbf{S}_{\rho\mathbf{v}} \\ S_e \\ \mathbf{S}_\mathbf{B} \end{pmatrix} = \begin{pmatrix} 0 \\ -(\nabla \cdot \mathbf{B})\mathbf{B} \\ -(\nabla \cdot \mathbf{B})\mathbf{B} \cdot \mathbf{v} \\ -(\nabla \cdot \mathbf{B})\mathbf{v} \end{pmatrix}_{\text{pow}}, \quad \begin{pmatrix} S_\rho \\ \mathbf{S}_{\rho\mathbf{v}} \\ S_e \\ \mathbf{S}_\mathbf{B} \end{pmatrix} = \begin{pmatrix} 0 \\ 0 \\ 0 \\ -(\nabla \cdot \mathbf{B})\mathbf{v} \end{pmatrix}_{\text{jan}} \tag{2.58}$$

Both Powell's and Janhunen's source terms simply modify the MHD equations so that they remain Galilean invariant even when $\nabla \cdot \mathbf{B} \neq 0$ [102, 225, 389]. It is found by Janhunen [225] and Dellar [102] that for the existence of nonzero $\nabla \cdot \mathbf{B}$ the only source term required is the $\mathbf{v}(\nabla \cdot \mathbf{B})$ term in the induction equation, while the energy and momentum conservation equations are unchanged. Powell's interest is to keep $\nabla \cdot \mathbf{B}$ small, but this is not very satisfactory in enforcing the constraint $\nabla \cdot \mathbf{B} = 0$ in multidimensions [390, 471]. Instead, Janhunen's focus [102, 225] was on deriving a positive method, as well as local energy and momentum conservation, which could be regained by discarding the source terms in the energy and momentum equations of Powell's eight-wave equations. The linearized eigenstructure of Janhunen's equations is quite similar to Powell's [389]. In these two formulations, the eigenvalues are the same, and the "divergence wave" exists. The divergence wave eigenvectors differ; the other seven eigenvectors are the same.

In numerical simulations magnetic monopoles tend to appear in fairly compact regions with discontinuities or sharp gradients of magnetic fields. This means that the generation of magnetic monopoles is a local process on grid-scale. In general, built-in diffusion is insufficient to act as such a $\nabla \cdot \mathbf{B}$ suppression mechanism. Addition of numerical diffusion to the MHD solvers seems an appealing way of reducing $\nabla \cdot \mathbf{B}$. Marder [302] proposed a parabolic approach to diffuse the numerically generated divergence by adding a diffusive term to the induction equation

$$\frac{\partial \mathbf{B}}{\partial t} = \nu \nabla \nabla \cdot \mathbf{B}$$

or equivalently, by updating the following source term

$$\mathbf{B} = \mathbf{B} + \nu \Delta t \nabla \nabla \cdot \mathbf{B}$$

with the maximal allowed diffusive time step equal to the CFL time step. Here ν is an artificial diffusion parameter. This technique has been used widely (e.g., [98, 134, 135, 246, 482]). See also Langdon [259] for suggested improvements of Marder's parabolic method.

Keppens et al. [245] also proposed a diffusive source term approach by setting the respective sources $(S_\rho, \mathbf{S}_{\rho\mathbf{v}}, S_e, \mathbf{S}_\mathbf{B})$ as follows:

$$\begin{pmatrix} S_\rho \\ \mathbf{S}_{\rho\mathbf{v}} \\ S_e \\ \mathbf{S}_\mathbf{B} \end{pmatrix} = \begin{pmatrix} 0 \\ 0 \\ \nu\mathbf{B}\cdot\nabla(\nabla\cdot\mathbf{B}) \\ \nu\nabla(\nabla\cdot\mathbf{B}) \end{pmatrix}_{\text{diff}} \tag{2.59}$$

Motivated by (2.58) and (2.59), the following local source term approaches to handle the $\nabla\cdot\mathbf{B}$ constraint can be proposed

$$\begin{pmatrix} S_\rho \\ \mathbf{S}_{\rho\mathbf{v}} \\ S_e \\ \mathbf{S}_\mathbf{B} \end{pmatrix} = \begin{pmatrix} 0 \\ -(\nabla\cdot\mathbf{B})\mathbf{B} \\ -(\nabla\cdot\mathbf{B})\mathbf{B}\cdot\mathbf{v} \\ -(\nabla\cdot\mathbf{B})\mathbf{v} \end{pmatrix}_{\text{pow}} + \begin{pmatrix} 0 \\ 0 \\ \nu\mathbf{B}\cdot\nabla(\nabla\cdot\mathbf{B}) \\ \nu\nabla(\nabla\cdot\mathbf{B}) \end{pmatrix}_{\text{diff}} \tag{2.60}$$

$$\begin{pmatrix} S_\rho \\ \mathbf{S}_{\rho\mathbf{v}} \\ S_e \\ \mathbf{S}_\mathbf{B} \end{pmatrix} = \begin{pmatrix} 0 \\ 0 \\ 0 \\ -(\nabla\cdot\mathbf{B})\mathbf{v} \end{pmatrix}_{\text{jan}} + \begin{pmatrix} 0 \\ 0 \\ \nu\mathbf{B}\cdot\nabla(\nabla\cdot\mathbf{B}) \\ \nu\nabla(\nabla\cdot\mathbf{B}) \end{pmatrix}_{\text{diff}} \tag{2.61}$$

After adding a source term proportional to $\nabla(\nabla\cdot\mathbf{B})$, the divergence satisfies an advection-diffusion equation instead of a pure advection equation. Note that the source term $\mathbf{B}\cdot\nabla(\nabla\cdot\mathbf{B})$ in the energy equation provides only a correction to the non-conservative system of Powell's and Janhuen's MHD formulations, and conservation is lost whenever the solenoidality of the magnetic field is destroyed. Rewrite $\mathbf{B}\cdot\nabla(\nabla\cdot\mathbf{B}) = \nabla\cdot(\mathbf{B}(\nabla\cdot\mathbf{B})) - (\nabla\cdot\mathbf{B})^2$, where the first term $\nabla\cdot(\mathbf{B}(\nabla\cdot\mathbf{B}))$ is in divergence form, and only the second one $-(\nabla\cdot\mathbf{B})^2$ is responsible for energy dissipation. Therefore, the total contribution from the second term $-(\nabla\cdot\mathbf{B})^2$ rapidly decreases as magnetic monopoles become weaker. The two local source terms $\nu\mathbf{B}\cdot\nabla(\nabla\cdot\mathbf{B})$ and $\nu\nabla(\nabla\cdot\mathbf{B})$ have the form of artificial-like diffusion, rather than being used in a conventional sense. Practically, the purpose of these terms is not to regularize otherwise singular solutions in the vicinity of discontinuities, but to enforce physical consistency of numerical computations. The contribution of these terms will be pronounced only when the solenoidality of magnetic field is lost. Thus, no overly excessive diffusion of the magnetic field itself should be introduced.

In the above equations, ν plays the role of artificial magnetic viscosity. From numerical considerations, it is most reasonable to use scaling $\nu \approx \lambda\Delta h$, where λ is the fastest characteristic speed in the plasma and Δh is the cell size. The actual value of the scaling constant ν should be determined upon the tradeoff between the desired $\nabla\cdot\mathbf{B}$ damping rate and the CFL time step restrictions. A natural choice of the value ν would correspond to a situation in which convective and diffusive time scales would be of the same order. In practical simulations, it may be sufficient to use $\nu = \lambda\Delta h$. In this case, the diffusive term acts to locally damp errors at a rate that does not introduce a stricter CFL limit.

The simple diffusive source term treatments offer a computationally cheap alternative allowing for direct implementation with any existing schemes. This diffusive control of $\nabla \cdot \mathbf{B}$, has the same advantages and disadvantages as its former counterparts. This technique is not strictly conservative, but appears to keep the level of $\nabla \cdot \mathbf{B}$ lower than the former approaches do, as pointed out by Powell et al. [390].

2.5 FVM: High Resolution Practice of MHD Equations

The compressible MHD equations are written in non-dimensional conservative differential form as follows:

$$\frac{\partial \mathbf{U}}{\partial t} + \nabla \cdot \mathbf{F}(\mathbf{U}) = \frac{\partial \mathbf{U}}{\partial t} + \frac{\partial \mathbf{f}(\mathbf{U})}{\partial x} + \frac{\partial \mathbf{g}(\mathbf{U})}{\partial y} + \frac{\partial \mathbf{h}(\mathbf{U})}{\partial z} = \mathbf{S}(\mathbf{U}), \tag{2.62}$$

where $\mathbf{U} = (\rho, \rho\mathbf{v}, \mathbf{B}, e)^T$ is the vector of conservative variables. $\mathbf{S} = \big(S_\rho, \mathbf{S}_{\rho\mathbf{v}}, \mathbf{S}_\mathbf{B}, S_e)^T$ denotes the source term vector. $\mathbf{F}(\mathbf{U}) = (\mathbf{f}, \mathbf{g}, \mathbf{h})$ are the flux vectors in the x-, y-, and z-directions, respectively. In practical application, $\mathbf{F}(\mathbf{U}) = \mathbf{F}_{\text{inv}}(\mathbf{U}) - \mathbf{F}_{\text{diff}}(\mathbf{U}, \nabla\mathbf{U})$, may consist of the non-linear hyperbolic and elliptic fluxes. Equation (2.62) becomes ideal MHD case if $\mathbf{F}(\mathbf{U})$ is given as in Eq. (2.37).

In what follows we present the implementation of FVM for Eq. (2.62) in the 3D computational domain decomposed by a finite set of cells V_i's, such as hexahedral cells shown in Fig. 2.1. By integrating Eq. (2.62) over a cell V_i, we get the following semi-discrete scheme

$$\frac{d\mathbf{U}_i}{dt} = -\frac{1}{V_i}\oint_{\partial V_i} (\mathbf{f}n_x + \mathbf{g}n_y + \mathbf{h}n_z)dA + \mathbf{S}_i. \tag{2.63}$$

In Eq. (2.63), as mentioned above, V_i is the cell volume and ∂V_i is the cell boundary. $\mathbf{U}_i = \frac{1}{V_i}\int_{V_i} \mathbf{U} dV$ is the cell-averaged conserved solution vector. $\mathbf{S}_i = \frac{1}{V_i}\int_{V_i} \mathbf{S} dV$ is the cell-averaged source term vector. $\mathbf{n} = (n_x, n_y, n_z)^T$ is a unit normal vector, which points outward from the boundary ∂V_i, and dA is the surface element. Approximating the surface integral numerically we obtain

$$\frac{d\mathbf{U}_i}{dt} = -\frac{1}{V_i}\sum_{j=1}^{N_{\text{faces}}}\sum_{m=1}^{N_g} (\mathbf{f}n_x + \mathbf{g}n_y + \mathbf{h}n_z)_{ij,m}\omega_{ij,m}A_{ij} + \mathbf{S}_i, \tag{2.64}$$

where m indicates the Gaussian quadrature point over each interface and $\omega_{ij,m}$ represents the associated Gaussian quadrature weight to each Gauss point. N_g is the number of Gauss quadrature points on each cell face. If the flux is based on the solution of a local Riemann problem with discontinuous initial data defined by left and right solution quantities, $\mathbf{U}_L$ and $\mathbf{U}_R$, the hyperbolic (inviscid) numerical flux at each quadrature point is then given by $\mathbf{F}_{\text{inv}} \cdot \mathbf{n} = (\mathbf{f}, \mathbf{g}, \mathbf{h}) \cdot \mathbf{n} = \mathcal{F}_{\text{inv}}(\mathbf{U}_L, \mathbf{U}_R, \mathbf{n})$, where the flux function $\mathcal{F}_{\text{inv}}$ conveys the solution to the Riemann problem in a direction defined by the cell face normal, $\mathbf{n}$, with left and right solution values, $\mathbf{U}_L$ and $\mathbf{U}_R$. Approximate Riemann solvers are often used to solve the Riemann problem and evaluate the numerical flux. Some flux functions applicable to the context described here are provided in Sect. 2.8. The left and right solution states, $\mathbf{U}_L$ and $\mathbf{U}_R$, are determined by performing cell-centred, piecewise, K-exact, polynomial reconstruction within each computational cell in Sect. 2.6. The order of the polynomial reconstruction determines the spatial accuracy of the solution by providing more accurate approximations of the solution values at the Gauss quadrature points for flux calculation. As the truncation error for K-degree exact reconstruction is $O(\Delta x^{K+1})$, a $(K+1)$-order accurate spatial discretization can be achieved for smooth hyperbolic problems. Meanwhile, the number of Gaussian quadrature points at which the numerical flux is integrated on each cell face should match the order of the reconstruction. It is well known that an N_g-point Gaussian quadrature rule on straight line interval integrates polynomials of degree $2N_g - 1$ exactly, yielding a $2N_g$-order accurate formula. As pointed out by Barth and Frederickson [36], Gaussian quadrature points $N_g \geq \frac{K+1}{2}$ are adequate for evaluating the inviscid flux integral to achieve the full accuracy of the K-exact reconstruction. In the practice of K-exact polynomial reconstruction (e.g., [144, 146, 210, 211, 214]), $\mathtt{int}((K+2)/2)$ quadrature points per dimension are used to ensure that the order of accuracy of the schemes is preserved. For example, one Gauss quadrature point $N_g = 1$ per face for second-order schemes (piecewise linear, $K = 1$, reconstruction) or lower, but four Gauss quadrature points $N_g = 4$ per face (two quadrature points per dimension) for third- and fourth-order schemes (piecewise quadratic, $K = 2$, and cubic, $K = 3$ reconstruction).

Again, Gaussian quadrature is used to evaluate the cell-average value of the source vector, $\mathbf{S}_i = \frac{1}{V_i}\int_{V_i}\mathbf{S}(\mathbf{U})dV$, and its contribution to the solution residual for each cell. As with the cell face flux integration, K-exact reconstruction requires a $(K+1)$-order accurate quadrature rule for the source vector. Thus, for piecewise linear, $K = 1$, reconstruction, a single quadrature point is employed in evaluating $\mathbf{S}_i$ to second order accuracy, whereas, for piecewise quadratic, $K = 2$, and piecewise cubic, $K = 3$, reconstruction, eight Gaussian points are used to obtain third and fourth order accuracy for the source vector, $\mathbf{S}_i$. For high-order accurate volumetric integrals and surface integrals on 3D hexahedral cell at Gaussian quadrature points, refer to Sect. 2.5.3.

In the evaluation of inviscid fluxes, in order to control spurious oscillations at discontinuities, the reconstructed variables will be bounded or limited (e.g., [322]) such that K-order multi-dimensional polynomials $\mathbf{U}_L^K(\mathbf{x})$ and $\mathbf{U}_R^K(\mathbf{x})$ must not introduce new extrema, that is, $\mathbf{U}_L^K(\mathbf{x})$, $\mathbf{U}_R^K(\mathbf{x})$ should lie between $\mathbf{U}_i$ and $\mathbf{U}_j$. Here, $\mathbf{U}_L^K(\mathbf{x})$ and $\mathbf{U}_R^K(\mathbf{x})$, depending on the control volume averages $\mathbf{U}_i$ and $\mathbf{U}_j$, are used as a representation of the spatial variation within cells, and $\mathbf{U}_i$ and $\mathbf{U}_j$ are the cell averages used in the left and right reconstructions. Here, $\mathbf{U}_L^K(\mathbf{x})$ is the state in the interior of the finite volume V_i and $\mathbf{U}_R^K(\mathbf{x})$ is the state outside V_i separated from $\mathbf{U}_L^K(\mathbf{x})$ by the intercell surface A_{ij}. Some common polynomial reconstructions are provided in Sect. 2.6. At any point along the control volume boundary or cell interface, the reconstructed solution and gradient are double valued, because the values from the two adjacent control volumes may be different. For advective fluxes, we use an upwind flux approximation. For diffusive fluxes, we average the gradients computed in each control volume at the control volume boundary or cell interface.

If the governing equations contain some viscous (elliptic) terms, then $\mathbf{F} \cdot \mathbf{n} = \mathbf{F}_{\text{inv}}(\mathbf{U}) \cdot \mathbf{n} - \mathbf{F}_{\text{diff}}(\mathbf{U}, \nabla\mathbf{U}) \cdot \mathbf{n}$, which is the sum of the non-linear hyperbolic and elliptic fluxes. In this situation, both the hyperbolic flux $\mathbf{F}_{\text{inv}}(\mathbf{U}) \cdot \mathbf{n}$ and the elliptic flux $\mathbf{F}_{\text{diff}}(\mathbf{U}, \nabla\mathbf{U}) \cdot \mathbf{n}$ must be approximated at each quadrature point, m, of a cell face, f, with a flux function that approaches the true flux at the rate imposed by the expected order of accuracy in the asymptotic limit of sufficiently small mesh size. This requirement implies a similar set of conditions on the accuracy of the solution vector, $\mathbf{U}$, and its gradient, $\nabla\mathbf{U}$, used in calculating the hyperbolic and elliptic fluxes. In general, the elliptic fluxes, $\mathbf{F}_{\text{diff}} \cdot \mathbf{n}$, will have the form $\mathbf{F}_{\text{diff}} \cdot \mathbf{n} = \mathcal{F}_{\text{diff}}(\mathbf{U}, \nabla\mathbf{U}, \mathbf{n})$, where $\mathcal{F}_{\text{diff}}$ is the corresponding elliptic flux function that provides the flux in direction $\mathbf{n}$ given the cell interface values of the solution, $\mathbf{U}$, and its gradient $\nabla\mathbf{U}$. As suggested (e.g., [210, 211, 214, 322]), the unlimited, cell-centred, K-exact, polynomial reconstruction used in the hyperbolic flux evaluation is used again (to get rid of considerable additional computational effort) to obtain a $(K + 1)$-order accurate value of the interface solution, $\mathbf{U}$. Direct differentiation of this K-exact reconstructed solution is also utilized to obtain a K-order accurate value of the interface gradient, $\nabla\mathbf{U}$ (i.e., one order less accurate due to differentiation). For instance, given left and right $(K + 1)$-order accurate unlimited reconstructions, $\mathbf{U}_L^K(\mathbf{x})$ and $\mathbf{U}_R^K(\mathbf{x})$, for the variables, $\mathbf{U}$, a K-order solution gradient at the cell face is calculated as the arithmetic mean of the left and right reconstructed gradients, and then elliptic fluxes are evaluated as $\mathbf{F}_{\text{diff}}(\mathbf{U}, \nabla\mathbf{U}) \cdot \mathbf{n} = \mathcal{F}_{\text{diff}}(\mathbf{U}, \nabla\mathbf{U}, \mathbf{n}) = \mathbf{F}_{\text{diff}}\left(\frac{1}{2}\left(\mathbf{U}_L^K(\mathbf{x}) + \mathbf{U}_R^K(\mathbf{x})\right), \frac{1}{2}\left(\nabla\mathbf{U}_L^K(\mathbf{x}) + \nabla\mathbf{U}_R^K(\mathbf{x})\right)\right) \cdot \mathbf{n}$, where the reconstructed gradient, $\nabla\mathbf{U}^K(\mathbf{x})$, at location, $\mathbf{x}$, is obtained by $\nabla\mathbf{U}^K(\mathbf{x}) = \left((\partial_x, \partial_y, \partial_z)\mathbf{U}^K\right)|_{\mathbf{x}}$. The evaluation of the elliptic fluxes in terms of simple arithmetic means of the unlimited cell-centred K-exact reconstructed solutions and their gradients in the cells to the right and left of the interface is straightforward and offers some computational merits. In particular, a K-order accurate flux is obtained by using just cell centered reconstructions. However, an important issue left for concern is to assess other properties of the discretized elliptic operator, such as positivity of the operator about local satisfaction of a discrete maximum principle as well as the susceptibility of the operator to odd-even solution decoupling, as similar formulations for standard second-order discretizations can lead to numerical difficulties related to these issues [81, 210, 216, 307, 309, 456]. The reader may refer to these cited papers for the related discussions. In addition, other appropriate choices for the numerical evaluation of the elliptic flux function, $\mathcal{F}_{\text{diff}}$, at the cell face would also be possible, such as the method based on the solution of the generalized Riemann problem for diffusion equations [161, 292] or the direct discontinuous Galerkin (DDG) elliptic flux of Liu and Yan [286].

If the source vector, $\mathbf{S}_i = \frac{1}{V_i}\int_{V_i} \mathbf{S}(\mathbf{U}, \nabla\mathbf{U})dV$, contains the gradient terms, such as Powell's terms, the evaluation of the gradient terms can also be calculated with $\nabla\mathbf{U}^K(\mathbf{x})$ in the same way by directly differentiating the unlimited reconstructed solutions as done in evaluating those in the elliptic terms.

In a word, to obtain a consistent K-order accurate scheme for both hyperbolic and elliptic operators, with $(K + 1)$-order solution reconstruction and K-order gradient representation following from K-exact reconstruction, requires a K-order accurate quadrature rule. The use of K-exact reconstruction is particularly helpful in ensuring uniform accuracy on irregular and solution adapted mesh (e.g., [211]).

Note that piecewise linear, $K = 1$, reconstruction with one quadrature point is appropriate for a second-order treatment of the hyperbolic fluxes, but $(K = 1)$-exact reconstruction is problematic for the evaluation of elliptic fluxes. For the latter case, second-order accuracy is only formally achieved by relying on error cancellation on uniform regular grids (e.g., [210]). It should be mentioned that the discretization of the gradient operator has received less attention despite its fundamental importance with regards to the accuracy of the FVM. The most popular gradient schemes are the divergence theorem (DT) (or Green-Gauss) scheme, and the least-squares (LS) scheme. Both are widely believed to be second order accurate, but Syrakos et al. [456] showed that in fact the common variant of the DT gradient is second-order accurate only on structured meshes whereas it is zeroth-order accurate on general unstructured meshes, and the least-squares gradient is second-order and first-order accurate, respectively. Through a theoretical analysis and numerical tests, their results indicated that the zeroth-order accuracy of the DT gradient is inherited by the FVM as a whole, and the discretization error does not reduce with grid refinement. As validated by Syrakos et al. [456], use of the least-squares gradient leads to second-order accurate results for the popular public domain partial differential equation solver OpenFOAM.

2.5.1 Flux Evaluation and Rotational Invariance

Now we return to the flux reformulation for Eq. (2.62). The flux $\mathbf{F}(\mathbf{U}) \cdot \mathbf{n}$ can be evaluated by $\mathbf{F}(\mathbf{U}) \cdot \mathbf{n} = \mathbf{f}(\mathbf{U})n_x + \mathbf{g}(\mathbf{U})n_y + \mathbf{h}(\mathbf{U})n_z$, which, however, would require three different flux evaluations. According to the rotational invariance property [96, 389, 397, 426, 460, 504], we have

$$\mathbf{f}(\mathbf{U})n_x + \mathbf{g}(\mathbf{U})n_y + \mathbf{h}(\mathbf{U})n_z = \mathbf{T}^{-1}\mathbf{f}(\mathbf{T}\mathbf{U}) \tag{2.65}$$

for $\mathbf{U} = (\rho, \rho\mathbf{v}, \mathbf{B}, e)^T$ with

$$\mathbf{T} = \begin{pmatrix} 1 & 0 & 0 & 0 & 0 & 0 & 0 & 0 \\ 0 & n_x & n_y & n_z & 0 & 0 & 0 & 0 \\ 0 & t_{1x} & t_{1y} & t_{1z} & 0 & 0 & 0 & 0 \\ 0 & t_{2x} & t_{2y} & t_{2z} & 0 & 0 & 0 & 0 \\ 0 & 0 & 0 & 0 & n_x & n_y & n_z & 0 \\ 0 & 0 & 0 & 0 & t_{1x} & t_{1y} & t_{1z} & 0 \\ 0 & 0 & 0 & 0 & t_{2x} & t_{2y} & t_{2z} & 0 \\ 0 & 0 & 0 & 0 & 0 & 0 & 0 & 1 \end{pmatrix}, \tag{2.66}$$

$$\mathbf{T}^{-1} = \begin{pmatrix} 1 & 0 & 0 & 0 & 0 & 0 & 0 & 0 \\ 0 & n_x & t_{1x} & t_{2x} & 0 & 0 & 0 & 0 \\ 0 & n_y & t_{1y} & t_{2y} & 0 & 0 & 0 & 0 \\ 0 & n_z & t_{1z} & t_{2z} & 0 & 0 & 0 & 0 \\ 0 & 0 & 0 & 0 & n_x & t_{1x} & t_{2x} & 0 \\ 0 & 0 & 0 & 0 & n_y & t_{1y} & t_{2y} & 0 \\ 0 & 0 & 0 & 0 & n_z & t_{1z} & t_{2z} & 0 \\ 0 & 0 & 0 & 0 & 0 & 0 & 0 & 1 \end{pmatrix}. \tag{2.67}$$

When $\sqrt{n_x^2 + n_y^2} \rightarrow 0$,

$$\begin{cases} t_{1x} = \frac{n_z}{\sqrt{n_x^2+n_z^2}} \\ t_{1y} = 0 \\ t_{1z} = -\frac{n_x}{\sqrt{n_x^2+n_z^2}} \end{cases}, \quad \begin{cases} t_{2x} = -\frac{n_x n_y}{\sqrt{n_x^2+n_z^2}} \\ t_{2y} = \sqrt{n_x^2 + n_z^2} \\ t_{2z} = -\frac{n_y n_z}{\sqrt{n_x^2+n_z^2}} \end{cases}, \tag{2.68}$$

otherwise,

$$\begin{cases} t_{1x} = \frac{n_y}{\sqrt{n_x^2+n_y^2}} \\ t_{1y} = -\frac{n_x}{\sqrt{n_x^2+n_y^2}} \\ t_{1z} = 0 \end{cases}, \quad \begin{cases} t_{2x} = \frac{n_x n_z}{\sqrt{n_x^2+n_y^2}} \\ t_{2y} = \frac{n_y n_z}{\sqrt{n_x^2+n_y^2}} \\ t_{2z} = -\sqrt{n_x^2 + n_y^2} \end{cases}. \tag{2.69}$$

Here, $\mathbf{T}$, which rotates the (x, y, z) coordinate system to a new $(\hat{x}, \hat{y}, \hat{z}) = (n, t_1, t_2)$ coordinate system, is the rotational matrix with $\mathbf{T}^{-1}$ being its inverse. $\mathbf{n}$ is the direction normal to the cell face, $\mathbf{t}_1$ and $\mathbf{t}_2$ are tangential directions to the cell face, respectively. The unit vectors $\mathbf{n}$, $\mathbf{t}_1$ and $\mathbf{t}_2$ form an orthogonal system, i.e $\mathbf{n} \cdot \mathbf{t}_1 = 0$, $\mathbf{n} \cdot \mathbf{t}_2 = 0$ and $\mathbf{t}_1 \cdot \mathbf{t}_2 = 0$. Note that, $\mathbf{n} = \mathbf{t}_1 \times \mathbf{t}_2$.

Thus, the flux can be evaluated by exploiting the rotational invariance property of the MHD equations (e.g., [389, 460, 504]), which states that the flux $\mathbf{F}(\mathbf{U}) \cdot \mathbf{n}$ can be evaluated by first applying the rotation matrix $\mathbf{T}$ to the original column conserved variables $\mathbf{U}$ yielding the rotated conserved variables $\hat{\mathbf{U}} = \mathbf{TU}$. Note that $\widehat{\mathbf{U}}$ is aligned with a new Cartesian frame $(\hat{x}, \hat{y}, \hat{z})$, where the coordinate $\hat{x}$ is in the direction $\mathbf{n}$ normal to the surface A_{ij} and $\hat{y}, \hat{z}$ are in the directions tangential to the surface A_{ij}. Then, the interface flux $\hat{\mathbf{f}} = \mathbf{f}(\hat{\mathbf{U}})$ is evaluated and rotated back to the (x, y, z) coordinate system by applying the inverse of the rotation matrix. Equation (2.64) can now be written as:

$$\frac{d\mathbf{U}_i}{dt} = -\frac{1}{V_i} \sum_{j=1}^{N_{\text{faces}}} \sum_{m=1}^{N_g} \left(\mathbf{T}^{-1}\mathbf{f}(\hat{\mathbf{U}})\right)_{ij,m} \omega_{ij,m} A_{ij} + \mathbf{S}_i \tag{2.70}$$

in which the face flux $\left(\mathbf{T}^{-1}\mathbf{f}(\hat{\mathbf{U}})\right)_{ij,m}$ is conventionally obtained by handling the so-called split Riemann problem initialized by the reconstructed left state $\mathbf{U}_L$ and right state $\mathbf{U}_R$. Because of this rotational invariance property, the numerical flux at each of the N_g Gaussian quadrature points on each face is determined by solving a Riemann problem of a specific flux $\hat{\mathbf{f}}$ of the $\hat{x}$-direction in the rotated MHD equations. In what follows, for simplicity, we only consider the evaluation of numerical flux in $\hat{x}$-direction. In order to define numerical fluxes $\hat{\mathbf{f}} = \mathbf{f}(\hat{\mathbf{U}})$ across the intercell surfaces we use an augmented one-dimensional system in the rotated frame $(\hat{x}, \hat{y}, \hat{z})$

$$\frac{\partial \hat{\mathbf{U}}}{\partial t} + \frac{\partial \hat{\mathbf{f}}}{\partial \hat{x}} = 0 \tag{2.71}$$

This augmented one-dimensional system is identical to the split multidimensional systems in Cartesian coordinates. We use approximate Riemann solvers to define the numerical flux $\hat{\mathbf{f}} = \mathbf{f}(\hat{\mathbf{U}})$ at each Gaussian point at the interface A_{ij}. Denote the location of the Gaussian point m at the interface A_{ij} by $\mathbf{x}_{ij,m}$. Now, the interface A_{ij}, at which $\mathbf{x}_{ij,m}$ is located, corresponds to $\hat{x} = 0$ in terms of the rotated coordinates $(\hat{x}, \hat{y}, \hat{z})$. We solve the approximate Riemann problem for Eq. (2.71) with given initial data (e.g., [41, 375, 466, 469]). The Riemann problem of Eq. (2.71) at time level t_n at the interface A_{ij} is denoted by $\mathrm{RP}^n_{ij,m}(\hat{\mathbf{U}}^n_L, \hat{\mathbf{U}}^n_R)$ with the initial conditions

$$\hat{\mathbf{U}}(\hat{x}, t^n) = \begin{cases} \hat{\mathbf{U}}^n_L = \mathbf{T}\mathbf{U}^n_L & \text{if } \hat{x} < 0 \\ \hat{\mathbf{U}}^n_R = \mathbf{T}\mathbf{U}^n_R & \text{if } \hat{x} > 0 \end{cases}$$

where $\mathbf{U}^n_L$ and $\mathbf{U}^n_R$ depend on the control volume averages $\mathbf{U}^n_i$ and $\mathbf{U}^n_j$. Here $\mathbf{U}^n_L$ is the state in the interior of the finite volume V_i and $\mathbf{U}^n_R$ is the state outside V_i separated from $\mathbf{U}^n_L$ by the intercell surface A_{ij}.

For adequate large $t - t^n$, the approximate Riemann problem $\mathrm{RP}^n_{ij,m}(\hat{\mathbf{U}}^n_L, \hat{\mathbf{U}}^n_R)$ has a similarity solution that is constant along plane manifolds $s = \hat{x}/(t - t^n) = \text{const.}$ in the $(\hat{x}, t)$ plane. Now denote this approximate Riemann solution by $\hat{\mathbf{U}}^n_{\mathrm{RP},m}(s)$. The value along $s = 0$ gives the solution at the interface A_{ij} and thus the numerical flux $\hat{\mathbf{f}} = \mathbf{f}(\hat{\mathbf{U}}^n_{\mathrm{RP},m}(0))$ at Gaussian point m. Then, the corresponding face flux $\left(\mathbf{T}^{-1}\mathbf{f}(\hat{\mathbf{U}})\right)_{ij,m}$ in Eq. (2.70) is approximately set as $\mathbf{T}^{-1}\hat{\mathbf{f}}$ at $\hat{x} = 0$. Because this flux is obtained by a one-dimensional system in local grid coordinate frame, the resulting flux is also called the grid-aligned flux.

Note that, when $\mathbf{U}_L = \mathbf{U}_i$ and $\mathbf{U}_R = \mathbf{U}_j$ the finite volume scheme is first-order accurate when an approximate Riemann solver is used. Some higher order linear reconstructions are presented in Sect. 2.6. Also, when $\mathbf{U}_L$ and $\mathbf{U}_R$ depend on the control volume averages at the previous time level(s) only, the scheme is explicit. When $\mathbf{U}_L$ and $\mathbf{U}_R$ also depend on the control volume averages at the current time level, the scheme is implicit. Some explicit and implicit time integration schemes are provided in Sect. 2.5.5.

2.5.2 Numerical Flux Computation of Dedner et al.'s GLM-MHD System

Dedner's formulation of MHD system reads as follows:

$$\frac{\partial \mathbf{U}}{\partial t} + \nabla \cdot \mathbf{F}(\mathbf{U}) = \mathbf{S}, \quad \nabla \cdot \mathbf{B} = 0 \tag{2.72}$$

with conservative variables $\mathbf{U} = (\rho, \rho\mathbf{v}, \mathbf{B}, e, \psi)^T$, and the flux vectors $\mathbf{F}(\mathbf{U})$ given by

$$\mathbf{F}(\mathbf{U}) = \Bigg(\rho\mathbf{v}, \rho\mathbf{v}\mathbf{v} + \left(p + \frac{1}{2}\mathbf{B}\cdot\mathbf{B}\right)\mathbf{I} - \mathbf{B}\mathbf{B}, \mathbf{v}\mathbf{B} \\ -\mathbf{B}\mathbf{v} + \mathbf{I}\psi, \left(e + p + \frac{1}{2}\mathbf{B}\cdot\mathbf{B}\right)\mathbf{v} - \mathbf{B}\left(\mathbf{B}\cdot\mathbf{v}\right), c_h^2\mathbf{B}\Bigg)^T$$

with

$$\begin{pmatrix} S_\rho \\ \mathbf{S}_{\rho\mathbf{v}} \\ \mathbf{S}_{\mathbf{B}} \\ S_e \\ \mathbf{S}_\psi \end{pmatrix} = -\begin{pmatrix} 0 \\ \mathbf{0} \\ \mathbf{0} \\ 0 \\ \frac{c_h^2}{c_p^2}\psi \end{pmatrix}_{\mathrm{GLM}} ; -\begin{pmatrix} 0 \\ (\nabla\cdot\mathbf{B})\mathbf{B} \\ \mathbf{0} \\ \mathbf{B}\cdot(\nabla\psi) \\ \frac{c_h^2}{c_p^2}\psi \end{pmatrix}_{\mathrm{EGLM}} ; -\begin{pmatrix} 0 \\ (\nabla\cdot\mathbf{B})\mathbf{B} \\ (\nabla\cdot\mathbf{B})\mathbf{v} \\ (\nabla\cdot\mathbf{B})\mathbf{B}\cdot\mathbf{v} + \mathbf{B}\cdot(\nabla\psi) \\ \mathbf{v}\cdot(\nabla\psi) + \frac{c_h^2}{c_p^2}\psi \end{pmatrix}_{\mathrm{GEGLM}}$$

which are introduced by Dedner et al. [96] to allow the existence of the monopoles associated with non-zero $\nabla\cdot\mathbf{B}$ that are destroyed by a combination of advection and decay. Here, a new unknown function ψ which couples the divergence constraint to the hyperbolic system is introduced with the generalized Lagrange multiplier (GLM) formulation of the MHD equations. The extended GLM (EGLM) formulation of the MHD equations has additional terms on the right-hand side different from the GLM formulation. And further Galilean invariance form of EGLM formulation (GEGLM) is achieved by adding some hyperbolic terms to the EGLM system. The coefficients c_p and c_h control the amount of diffusion in ψ and the advection speed, respectively. The 'purely hyperbolic' correction can be obtained by taking c_p to infinity (no diffusion). Their choice is given in Sect. 2.4.2.

Obviously, Eq. (2.72) can be solved as a whole for the full system with nine variables by using the method in Sect. 2.5.1. Another option is following Dedner et al. [96], by decoupling the equations for B_n and ψ from the rest of the system, so that the numerical fluxes are applied only to the other seven variables. Remind that the modified induction and divergence equations [96–98] from the hyperbolic part are

$$\frac{\partial \mathbf{B}}{\partial t} + \nabla\cdot(\mathbf{v}\mathbf{B} - \mathbf{B}\mathbf{v}) + \nabla\psi = 0, \; \frac{\partial \psi}{\partial t} + c_h^2\nabla\cdot\mathbf{B} = 0 \tag{2.73}$$

with some $c_h > 0$. With a slight modification of adding an unit vector to the rotation matrix $\mathbf{T}$ in (2.66), the rotation matrix $\mathcal{T}$ for Eq. (2.72) reads as follows:

$$\mathcal{T} = \begin{pmatrix} 1 & 0 & 0 & 0 & 0 & 0 & 0 & 0 & 0 \\ 0 & n_x & n_y & n_z & 0 & 0 & 0 & 0 & 0 \\ 0 & t_{1x} & t_{1y} & t_{1z} & 0 & 0 & 0 & 0 & 0 \\ 0 & t_{2x} & t_{2y} & t_{2z} & 0 & 0 & 0 & 0 & 0 \\ 0 & 0 & 0 & 0 & n_x & n_y & n_z & 0 & 0 \\ 0 & 0 & 0 & 0 & t_{1x} & t_{1y} & t_{1z} & 0 & 0 \\ 0 & 0 & 0 & 0 & t_{2x} & t_{2y} & t_{2z} & 0 & 0 \\ 0 & 0 & 0 & 0 & 0 & 0 & 0 & 1 & 0 \\ 0 & 0 & 0 & 0 & 0 & 0 & 0 & 0 & 1 \end{pmatrix},$$

which transforms the x-direction onto the unit vector in $\hat{x}$-direction (i.e., the normal direction $\mathbf{n}$). Here, t_{1x}, t_{1y}, t_{1z} and t_{2x}, t_{2y}, t_{2z} are the same as given by (2.66). As done previously for system (2.62), the numerical flux function $\mathbf{F}(\mathbf{U}) \cdot \mathbf{n}$ at the interface A_{ij} is estimated with the idea of rotated one-dimensional Riemann solvers. Then, according to Dedner et al. [98], we arrive at a 2×2 linear system from subsystem (2.73) for $\mathbf{B} \cdot \mathbf{n}$ and ψ,

$$\frac{\partial(\mathbf{B} \cdot \mathbf{n})}{\partial t} + \frac{\partial \psi}{\partial \hat{x}} = 0, \; \frac{\partial \psi}{\partial t} + c_h^2 \frac{\partial(\mathbf{B} \cdot \mathbf{n})}{\partial \hat{x}} = 0, \tag{2.74}$$

which does not depend on the other MHD quantities. Without loss of generality we proceed as in Sect. 2.5.1 to solve the Riemann problem to Eq. (2.74) at each Gaussian point at the interface A_{ij}. The Riemann problem to Eq. (2.74) at $\hat{x} = 0$ can be solved with left-hand state $(\hat{B}^n_{\hat{x},L}, \psi_L) = (\mathbf{B}^n_L \cdot \mathbf{n}, \psi^n_L)$ and right-hand state $(\hat{B}^n_{\hat{x},R}, \psi_R) = (\mathbf{B}^n_R \cdot \mathbf{n}, \psi^n_R)$. It is easy to find that the solution to the local Riemann problem with left-hand state $(\mathbf{B}^n_L \cdot \mathbf{n}, \psi^n_L)$ and right-hand state $(\mathbf{B}^n_R \cdot \mathbf{n}, \psi^n_R)$ has the value

$$B^n_{ij} = \frac{1}{2}(\mathbf{B}^n_L + \mathbf{B}^n_R) \cdot \mathbf{n} - \frac{1}{2c_h}(\psi^n_R - \psi^n_L), \; \psi^n_{ij} = \frac{1}{2}(\psi^n_L + \psi^n_R) - \frac{c_h}{2}(\mathbf{B}^n_R - \mathbf{B}^n_L) \cdot \mathbf{n} \tag{2.75}$$

on the cell interface $\hat{x} = 0$. Therefore, we can employ ψ_{ij} and $c_h^2 B^n_{ij}$ as numerical flux at the cell interface.

In one-dimension the divergence constraint and the evolution equation for the first component of the magnetic field only allow for a constant $B_x (\hat{B}_{\hat{x}} = \mathbf{B} \cdot \mathbf{n})$. Thus a consistent application of the one-dimensional solver would require $(\mathbf{B} \cdot \mathbf{n})_{ij}$ to be constant over the cell interface A_{ij}. In the case of the GLM correction B_x can vary in space and time. System (2.73) and analytical expression (2.75) offer a consistent choice for $(\mathbf{B} \cdot \mathbf{n})_{ij}$, which can be used in the computation of the fluxes associated with the other MHD quantities. That is, in the GLM formulation, B^n_{ij} in Eq. (2.75) is a natural choice for such computation. But, in the absence of a better choice, i.e., in the case of the uncorrected MHD system or of the source term approach like Powell's source terms in Eq. (2.48), the average $B^n_{ij} = \frac{1}{2}(\mathbf{B}^n_L + \mathbf{B}^n_R) \cdot \mathbf{n}$ can be taken as a reasonable choice for the calculation of the MHD flux, since in this case no natural choice is available for the normal component of the magnetic field on A_{ij}, and the flux for $\mathbf{B}$ in the normal direction is zero due to the divergence condition.

2.5.3 High-Order Integration on Hexahedral Cells

In obtaining high-order FVM, volumetric integrals and surface integrals at Gaussian quadrature points must be involved. Achieving high-order accuracy on 3D hexahedral grids has to properly handle the cell faces of the grid cells. An isoparametric transformation representation for the cell faces can be exploited to obtain sufficiently accurate volume and surface integrals (e.g., [92, 132, 273, 525]), especially for the nonplanar cell faces. The isoparametric transformation can facilitate high-order accurate numerical integration in a transformed domain (e.g., [132, 212, 214]) so as to establish a general framework for developing standard finite elements in multi-dimensions. The isoparametric transformation can effectively transform a control domain V in the physics space into a control domain V' in the reference space. The possible candidates of the cells are tetrahedrons, pyramids, prisms, hexahedrons (e.g., [4, 235]), or any types of polyhedrons (e.g., [5]) in three dimensions can be utilized. The control volume can be convex polyhedron or even locally concave.

Basically, the isoparametric transformation can be expressed as [273]

$$\mathbf{x} = \sum_{k=1}^{P} \mathbf{x}_{cp}^{(k)} v_k(x', y', z') \tag{2.76}$$

where (x', y', z') is the coordinates in reference domain, (x, y, z) the coordinates in physics domain, $\mathbf{x}_{cp}^{(k)}$ the position vector of control points, and P the number of control points. Shape function $v_k(x', y', z')$ satisfies:

$$\text{(I)}\quad v_i(x'_j, y'_j, z'_j) = \begin{cases} 1 & i = j \\ 0 & i \neq j \end{cases}$$

$$\text{(II)}\quad v_k(x'_j, y'_j, z'_j) = \begin{cases} \delta_{kj} & (x'_j, y'_j, z'_j) \in V' \\ 0 & (x'_j, y'_j, z'_j) \notin V' \end{cases}$$

Shape functions are local functions restricted to an element, and are of vital importance to discontinuous finite element approximations. The shape functions of any high orders can be constructed readily, for instance, by Lagrangian interpolation polynomials. The shape functions $v_i(x'_j, y'_j, z'_j)$ display two important properties: (I) $v_i(x'_j, y'_j, z'_j)$ is unit at the control point j or (x'_j, y'_j, z'_j), and (II) it is non-zero on the control domain V' and is zero everywhere else. These conditions mean that the isoparametric transformation on the face/edge only relies on the local control points on the face/edge so as to make the isoparametric transformation of element locality coincide on the cell interface for the adjacent cells. For detailed descriptions and discussions about such transformations, see [273, Chap. 3: Shape Functions and Elemental Calculations].

In what follows we provide several implementations of isoparametric transformation from the physics domain to the reference cube.

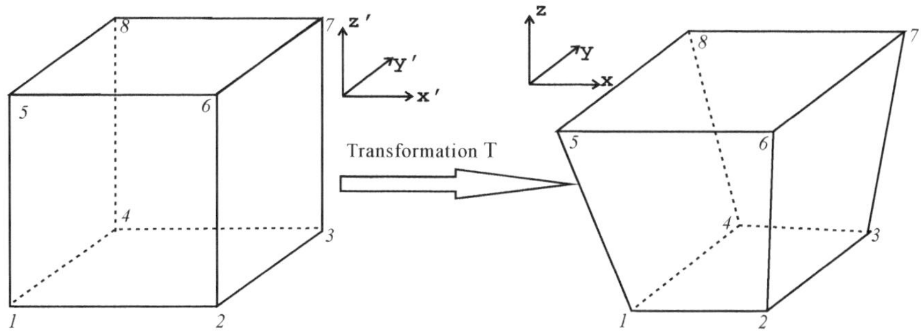

Fig. 2.5 Transformation from unit cube to 3D hexahedral cell

Transformation From Unit Cube to 3D Hexahedral Cell with Plane Faces

Consider a hexahedral cell V in physical space with control points defined by position vectors of vertices $\mathbf{x}_k (k = 1, 2, \ldots, 8)$, and $V' = [0, 1] \times [0, 1] \times [0, 1]$ is the reference unit cube. In this case, cell faces are planes as shown in Fig. 2.5, and then a trilinear one-to-one mapping, from a reference unit cube $V' = \{(x', y', z') : 0 \leq x', y', z' \leq 1\}$ to a hexahedral cell V, can be defined as

$$\begin{aligned}
\mathbf{T}(x', y', z') = \frac{1}{2} \big[&(1-y')(1-z')\mathbf{f}(x', 0, 0) + (1-y')z'\mathbf{f}(x', 0, 1) + y'(1-z')\mathbf{f}(x', 1, 0) \\
&+ y'z'\mathbf{f}(x', 1, 1) + (1-x')(1-z')\mathbf{f}(0, y', 0) + (1-x')z'\mathbf{f}(0, y', 1) \\
&+ x'(1-z')\mathbf{f}(1, y', 0) + x'z'\mathbf{f}(1, y', 1) + (1-x')(1-y')\mathbf{f}(0, 0, z') \\
&+ (1-x')y'\mathbf{f}(0, 1, z') + x'(1-y')\mathbf{f}(1, 0, z') + x'y'\mathbf{f}(1, 1, z') \\
&- (1-x')(1-y')(1-z')\mathbf{f}(0, 0, 0) - x'y'(1-z')\mathbf{f}(1, 1, 0) - x'(1-y')z'\mathbf{f}(1, 0, 1) \\
&- (1-x')y'z'\mathbf{f}(0, 1, 1) - x'(1-y')(1-z')\mathbf{f}(1, 0, 0) - x'y'z'\mathbf{f}(1, 1, 1) \\
&- (1-x')y'(1-z')\mathbf{f}(0, 1, 0) - (1-x')(1-y')z'\mathbf{f}(0, 0, 1) \big]
\end{aligned} \tag{2.77}$$

where $\mathbf{f}$ is a continuous mapping from the edge of V' to that of V. On the edge, we have $\mathbf{T} = \mathbf{f}$, i.e.,

$$\begin{cases}
\mathbf{T}(x', 0, 0) = \mathbf{f}(x', 0, 0), \\
\mathbf{T}(x', 1, 0) = \mathbf{f}(x', 1, 0), \ 0 \leq x' \leq 1. \\
\mathbf{T}(x', 0, 1) = \mathbf{f}(x', 0, 1), \\
\mathbf{T}(x', 1, 1) = \mathbf{f}(x', 1, 1),
\end{cases}$$

$$\begin{cases}
\mathbf{T}(0, y', 0) = \mathbf{f}(0, y', 0), \\
\mathbf{T}(1, y', 0) = \mathbf{f}(1, y', 0), \ 0 \leq y' \leq 1. \\
\mathbf{T}(1, y', 1) = \mathbf{f}(1, y', 1), \\
\mathbf{T}(0, y', 1) = \mathbf{f}(0, y', 1),
\end{cases}$$

$$\begin{cases}
\mathbf{T}(0, 0, z') = \mathbf{f}(0, 0, z'), \\
\mathbf{T}(1, 0, z') = \mathbf{f}(1, 0, z'), \ 0 \leq z' \leq 1. \\
\mathbf{T}(1, 1, z') = \mathbf{f}(1, 1, z'), \\
\mathbf{T}(0, 1, z') = \mathbf{f}(0, 1, z'),
\end{cases}$$

Given transformations $\mathbf{f}(x', 0, 0), \mathbf{f}(x', 1, 0), \mathbf{f}(x', 0, 1), \mathbf{f}(x', 1, 1), 0 \le x' \le 1$; $\mathbf{f}(0, y', 0)$, $\mathbf{f}(1, y', 0)$, $\mathbf{f}(1, y', 1)$, $\mathbf{f}(0, y', 1), 0 \le y' \le 1$; $\mathbf{f}(0, 0, z'), \mathbf{f}(1, 0, z')$, $\mathbf{f}(1, 1, z'), \mathbf{f}(0, 1, z'), 0 \le z' \le 1$, transformation $\mathbf{T}$ transforms a unit cube into a control volume V bordered by the edges transformed by $\mathbf{T}$ from those of the cube [514, 525].

For convex polyhedron with straight line edge, we can use every two vertices forming an edge to determine these transforms:

$$\begin{aligned}
\mathbf{f}(x', 0, 0) &= (1 - x')\mathbf{f}(0, 0, 0) + x'\mathbf{f}(1, 0, 0) \\
\mathbf{f}(x', 1, 0) &= (1 - x')\mathbf{f}(0, 1, 0) + x'\mathbf{f}(1, 1, 0) \\
\mathbf{f}(x', 0, 1) &= (1 - x')\mathbf{f}(0, 0, 1) + x'\mathbf{f}(1, 0, 1) \\
\mathbf{f}(x', 1, 1) &= (1 - x')\mathbf{f}(0, 1, 1) + x'\mathbf{f}(1, 1, 1)
\end{aligned} \tag{2.78}$$

$$\begin{aligned}
\mathbf{f}(0, y', 0) &= (1 - y')\mathbf{f}(0, 0, 0) + y'\mathbf{f}(0, 1, 0) \\
\mathbf{f}(1, y', 0) &= (1 - y')\mathbf{f}(1, 0, 0) + y'\mathbf{f}(1, 1, 0) \\
\mathbf{f}(0, y', 1) &= (1 - y')\mathbf{f}(0, 0, 1) + y'\mathbf{f}(0, 1, 1) \\
\mathbf{f}(1, y', 1) &= (1 - y')\mathbf{f}(1, 0, 1) + y'\mathbf{f}(1, 1, 1)
\end{aligned} \tag{2.79}$$

$$\begin{aligned}
\mathbf{f}(0, 0, z') &= (1 - z')\mathbf{f}(0, 0, 0) + z'\mathbf{f}(0, 0, 1) \\
\mathbf{f}(0, 1, z') &= (1 - z')\mathbf{f}(0, 1, 0) + z'\mathbf{f}(0, 1, 1) \\
\mathbf{f}(1, 0, z') &= (1 - z')\mathbf{f}(1, 0, 0) + z'\mathbf{f}(1, 0, 1) \\
\mathbf{f}(1, 1, z') &= (1 - z')\mathbf{f}(1, 1, 0) + z'\mathbf{f}(1, 1, 1)
\end{aligned} \tag{2.80}$$

Substitute (2.78), (2.79), and (2.80) into (2.77) to obtain

$$\begin{aligned}
\mathbf{T}(x', y', z') = &(1 - x')(1 - y')(1 - z')\mathbf{f}(0, 0, 0) + x'(1 - y')(1 - z')\mathbf{f}(1, 0, 0) \\
&+ (1 - x')(1 - y')z'\mathbf{f}(0, 0, 1) + x'(1 - y')z'\mathbf{f}(1, 0, 1) \\
&+ (1 - x')y'(1 - z')\mathbf{f}(0, 1, 0) + x'y'(1 - z')\mathbf{f}(1, 1, 0) \\
&+ (1 - x')y'z'\mathbf{f}(0, 1, 1) + x'y'z'\mathbf{f}(1, 1, 1)
\end{aligned} \tag{2.81}$$

Equation (2.81) can be reduced to

$$\begin{aligned}
\mathbf{T}(x', y', z') = &[\mathbf{f}(1, 0, 0) + \mathbf{f}(0, 0, 1) + \mathbf{f}(0, 1, 0) + \mathbf{f}(1, 1, 1) \\
&- \mathbf{f}(0, 0, 0) - \mathbf{f}(1, 0, 1) - \mathbf{f}(1, 1, 0) - \mathbf{f}(0, 1, 1)]\, x'y'z' \\
&+ [\mathbf{f}(0, 0, 0) + \mathbf{f}(1, 1, 0) - \mathbf{f}(1, 0, 0) - \mathbf{f}(0, 1, 0)]\, x'y' \\
&+ [\mathbf{f}(0, 0, 0) + \mathbf{f}(1, 0, 1) - \mathbf{f}(1, 0, 0) - \mathbf{f}(0, 0, 1)]x'z' \\
&+ [\mathbf{f}(0, 0, 0) + \mathbf{f}(0, 1, 1) - \mathbf{f}(0, 0, 1) - \mathbf{f}(0, 1, 0)]y'z' \\
&+ [\mathbf{f}(1, 0, 0) - \mathbf{f}(0, 0, 0)]x' + [\mathbf{f}(0, 1, 0) - \mathbf{f}(0, 0, 0)]y' \\
&+ [\mathbf{f}(0, 0, 1) - \mathbf{f}(0, 0, 0)]z' + \mathbf{f}(0, 0, 0)
\end{aligned} \tag{2.82}$$

Denote eight vertices of a polyhedron by $\mathbf{x}_1 = \mathbf{f}(0,0,0)$, $\mathbf{x}_2 = \mathbf{f}(1,0,0) - \mathbf{x}_1$, $\mathbf{x}_3 = \mathbf{f}(0,1,0) - \mathbf{x}_1$, $\mathbf{x}_4 = \mathbf{f}(0,0,1) - \mathbf{x}_1$, $\mathbf{x}_5 = \mathbf{f}(1,1,0) - \mathbf{x}_1 - \mathbf{x}_2 - \mathbf{x}_3$, $\mathbf{x}_6 = \mathbf{f}(1,0,1) - \mathbf{x}_1 - \mathbf{x}_2 - \mathbf{x}_4$, $\mathbf{x}_7 = \mathbf{f}(0,1,1) - \mathbf{x}_1 - \mathbf{x}_3 - \mathbf{x}_4$, $\mathbf{x}_8 = \mathbf{f}(1,1,1) - \mathbf{x}_1 - \mathbf{x}_2 - \mathbf{x}_3 - \mathbf{x}_4 - \mathbf{x}_5 - \mathbf{x}_6 - \mathbf{x}_7$. Then, the trilinear mapping from a reference unit cube can be defined by

$$\mathbf{T}(x',y',z') = \mathbf{x}_1 + \mathbf{x}_2 x' + \mathbf{x}_3 y' + \mathbf{x}_4 z' + \mathbf{x}_5 x'y' + \mathbf{x}_6 x'z' + \mathbf{x}_7 y'z' + \mathbf{x}_8 x'y'z'$$

and the six trilinear surfaces are described by fixing one of x', y' and z' to 0 or 1. The tangent vectors to the coordinate lines are defined by

$$\begin{aligned}
\frac{\partial \mathbf{T}}{\partial x'} &= \mathbf{T}_{x'}(y',z') = \mathbf{x}_2 + \mathbf{x}_5 y' + \mathbf{x}_6 z' + \mathbf{x}_8 y'z' \\
\frac{\partial \mathbf{T}}{\partial y'} &= \mathbf{T}_{y'}(x',z') = \mathbf{x}_3 + \mathbf{x}_5 x' + \mathbf{x}_7 z' + \mathbf{x}_8 x'z' \\
\frac{\partial \mathbf{T}}{\partial z'} &= \mathbf{T}_{z'}(x',y') = \mathbf{x}_4 + \mathbf{x}_6 x' + \mathbf{x}_7 y' + \mathbf{x}_8 x'y'
\end{aligned} \tag{2.83}$$

The determinants of the Jacobians for volume and surface integration are given by

$$\begin{aligned}
\det \mathbf{J}(x',y',z') &= \left| \frac{\partial(x,y,z)}{\partial(x',y',z')} \right| = \mathbf{T}_{x'}(y',z') \cdot \left(\mathbf{T}_{y'}(x',z') \times \mathbf{T}_{z'}(x',y') \right) \\
\det \mathbf{J}_{x'}(y',z') &= \| \mathbf{T}_{y'}(x',z') \times \mathbf{T}_{z'}(x',y') \| \big|_{x'=\text{const}} \\
\det \mathbf{J}_{y'}(x',z') &= \| \mathbf{T}_{z'}(x',y') \times \mathbf{T}_{x'}(y',z') \| \big|_{y'=\text{const}} \\
\det \mathbf{J}_{z'}(x',y') &= \| \mathbf{T}_{x'}(y',z') \times \mathbf{T}_{y'}(x',z') \| \big|_{z'=\text{const}}
\end{aligned} \tag{2.84}$$

and the normal vectors to the coordinate planes by

$$\begin{aligned}
\mathbf{n}_{x'}(y',z') &= - \left. \frac{\mathbf{T}_{y'} \times \mathbf{T}_{z'}}{\det \mathbf{J}_{x'}} \right|_{x'=\text{const}} \\
\mathbf{n}_{y'}(x',z') &= - \left. \frac{\mathbf{T}_{z'} \times \mathbf{T}_{x'}}{\det \mathbf{J}_{y'}} \right|_{y'=\text{const}} \\
\mathbf{n}_{z'}(x',y') &= - \left. \frac{\mathbf{T}_{x'} \times \mathbf{T}_{y'}}{\det \mathbf{J}_{z'}} \right|_{z'=\text{const}}
\end{aligned} \tag{2.85}$$

Alternatively, define shape functions of control points $\mathbf{x}_{cp}^{(k)}$ $(k = 1, 2, \ldots, 8)$ as follows

$$\begin{aligned}
&v_1 = (1-x')(1-y')(1-z'),\ v_2 = x'(1-y')(1-z'), \\
&v_3 = x'y'(1-z'),\ v_4 = y'(1-x')(1-z'), \\
&v_5 = z'(1-x')(1-y'),\ v_6 = x'z'(1-y'), \\
&v_7 = x'y'z',\ v_8 = y'z'(1-x')
\end{aligned}$$

Then we have the same expression

$$\mathbf{T}(x', y', z') = \sum_{k=1}^{8} v_k(x', y', z')\mathbf{x}_{cp}^{(k)}$$

Consider a volumetric integral of a continuous smooth function, $g(\mathbf{x})$, over a control volume V. In order to evaluate the volumetric integral $\int_V g(\mathbf{x})dV$, we change the variables and integration domain to those of the reference unit cube by utilizing the trilinear transformation, $(x, y, z) = \mathbf{T}(x', y', z')$, and its transformation Jacobian determinant, $\det \mathbf{J}$ (e.g., [132, 273]). Thus, the volumetric integral is calculated in the reference space (x', y', z') as

$$\int_V g(x, y, z)\, dV = \int_0^1 \int_0^1 \int_0^1 g(\mathbf{T}(x', y', z')) \det \mathbf{J}\, dV'$$
$$\simeq \sum_{m=1}^{N_g} g\left(\mathbf{T}(x'_m, y'_m, z'_m)\right) (\det \mathbf{J})_m \omega_m = \sum_{m=1}^{N_g} g(\mathbf{T}_m)\hat{\omega}_m$$

where $dV = dxdydz$ and $dV' = dx'dy'dz'$ are the volume elements in the physics space and reference space. Here, N_g is the number of volumetric Gaussian points (x'_m, y'_m, z'_m) in the cube with weights ω_m, while $\mathbf{T}_m = \mathbf{T}(x'_m, y'_m, z'_m)$ are their corresponding Gaussian points in the physics volume V with weights $\hat{\omega}_m = (\det\mathbf{J})_m\omega_m$.

Surface integrals are evaluated in a similar fashion. For example, to evaluate a surface integral along a surface with constant x'-coordinate, $\mathcal{A}_{x'}$, the following expression is used:

$$\int_{\mathbf{T}(\mathcal{A}_{x'})} g(\mathbf{T})\, dA = \int_0^1 \int_0^1 g(\mathbf{T}(x', y', z')) \det \mathbf{J}_{x'}\, dy'dz'$$
$$\simeq \sum_{m=1}^{N_g} g(\mathbf{T}(x', y'_m, z'_m))(\det \mathbf{J}_{x'})_m \omega_m = \sum_{m=1}^{N_g} g(\mathbf{T}_m)\hat{\omega}_m$$

and N_g is the number of Gaussian points on the surfaces. Here, (y'_m, z'_m) are Gaussian points in the quadrilateral plane $[0, 1]^2$ with weights ω_m, while $\mathbf{T}_m = \mathbf{T}(x', y'_m, z'_m)$ are their corresponding Gaussian points in the physics surface of volume V with weights $\hat{\omega}_m = (\det\mathbf{J}_{x'})_m\omega_m$.

One-dimensional numerical integration primarily consists of finding the area under the curve defined by a function $g(s)$. An integral over $[a, b]$ can be changed into the following equality

$$\int_a^b g(s)\, ds = \frac{b-a}{2} \int_{-1}^1 g\left(\frac{b-a}{2}s + \frac{a+b}{2}\right) ds$$

which leads to an integral over the standard interval $[-1, 1]$. We only need to consider the integral of a function $g(s)$ over the standard interval $[-1, 1]$. Applying the Gaussian quadrature rule then results in the following approximation:

$$\int_{-1}^{1} g(s)\,ds \approx \sum_{i=1}^{N_g} w_i g(s_i)$$

where s_i denotes the integration points or Gaussian quadrature points, w_i are the associated weighting parameters, and N_g is the total number of integration points. The error associated with the above quadrature is $O(d^{2N_g} g(s)/ds^{2N_g})$. This means that an exact integration can be obtained for a polynomial $g(s)$ of up to $(2N_g - 1)$ degree, if N_g integration points are used. The general N_g-point Gauss-Legendre rule is exact for polynomial functions of degree $\leq 2N_g - 1$ or less. Some low-order Gauss-Legendre quadrature rules [2, 92] for solving the integration problem (over interval $[-1, 1]$) are listed in Table 2.1.

The above 1-D integration quadrature formulae can be easily extended to multidimensions. For Gaussian quadrature on the standard quadrilateral element $[-1, 1]^2$, we can integrate numerically with respect to s by first fixing t,

$$\int_{-1}^{1}\int_{-1}^{1} g(s, t) ds dt \approx \int_{-1}^{1}\left(\sum_{i=1}^{M_g} w_i g(s_i, t)\right) dt$$

where s_i and w_i are Gaussian quadrature points and weights of order M_g in the s-direction. Next integrating numerically with respect to t we have

$$\int_{-1}^{1}\int_{-1}^{1} g(s, t)\, ds dt \approx \sum_{i=1}^{M_g}\sum_{j=1}^{N_g} w_i \hat{w}_j g(s_i, t_j)$$

where t_j and $\hat{w}_j$ are Gaussian quadrature points and weights of order N_g in the t direction. This integration rule is exact if the integrand $g(s, t)$ contains only the monomials $s^i t^j (i = 0, 1, \ldots, 2M_g - 1, j = 0, 1, \ldots, 2N_g - 1)$. Usually $M_g = N_g$ (so $t_i = s_i$, $\hat{w}_i = w_i$), and we have

Table 2.1 Gaussian quadrature

Number of points, N_g	Points, s_i	Weights, ω_i
2	$\pm\frac{1}{\sqrt{3}}$	1
3	0	$\frac{8}{9}$
	$\pm\sqrt{\frac{3}{5}}$	$\frac{5}{9}$
4	$\pm\sqrt{\frac{3}{7} - \frac{2}{7}\sqrt{\frac{6}{5}}}$	$\frac{18+\sqrt{30}}{36}$
	$\pm\sqrt{\frac{3}{7} + \frac{2}{7}\sqrt{\frac{6}{5}}}$	$\frac{18-\sqrt{30}}{36}$

$$\int_{-1}^{1}\int_{-1}^{1} g(s,t)\,dsdt \approx \sum_{i=1}^{N_g}\sum_{j=1}^{N_g} w_i w_j g(s_i,t_j)$$

for Gaussian quadrature of order N_g for the standard quadrilateral element.

In the same manner, for Gaussian quadrature of order N_g for the standard cube $[-1,1]^3$ element, we have

$$\int_{-1}^{1}\int_{-1}^{1}\int_{-1}^{1} g(r,s,t)\,drdsdt \approx \sum_{i=1}^{N_g}\sum_{j=1}^{N_g}\sum_{k=1}^{N_g} w_i w_j w_k g(r_i,s_j,t_k)$$

The same integration points listed in Table 2.1 can be used to carry out the numerical integration in each of the dimensions.

In one-dimensional case, at least $\text{int}(\frac{N+1}{2})$ integration points are needed for N-order accuracy of integral approximation. In three-dimensional case, Gaussian quadrature points in a control volume consist of Gaussian quadrature points r_i, s_j, t_k from individual direction. Assume that $N_{gx'}, N_{gy'}, N_{gz'}$ are Gaussian quadrature points along x', y', z' directions, at least $n = \text{int}(\frac{N+1}{2})$ Gaussian quadrature points are entailed along every direction in order to achieve N-order approximation, i.e., $N_{gx'} \geq n, N_{gy'} \geq n, N_{gz'} \geq n$. There are at least $N_{gx'} \times N_{gy'} \times N_{gz'} \geq n^3$ Gaussian quadrature points in a control volume, while Gaussian quadrature points $N_{gx'} \times N_{gy'}, N_{gx'} \times N_{gz'}, N_{gy'} \times N_{gz'}$ larger than or at least equal to n^2 are required to attain N-order accuracy for each canonical face integral. Some relations between accuracy order and number of Gauss quadrature points for solving the integration problem are listed in Table 2.2.

Along with the above rule, the basic abscissae and weights for three Gaussian points generated for a [0, 1] domain (e.g., [2, 92]) are $[0.5, 0.5 \pm \sqrt{3/5}]$ and $[8/18, 5/18, 5/18]$, respectively. For a fifth-order accurate scheme, 27 Gaussian points in the volume of the hexahedral cell, and 9 Gaussian points on each cell face can be used. All volumetric integrals are computed with a fifth-order Gaussian integration rule derived with three points in each canonical direction (i.e., a $3 \times 3 \times 3$ product rule (e.g., [132, 273])) and $N_g = 27$ volumetric Gaussian points.

Under the isoparametric transformation, the volume of hexahedral cell V can be easily calculated by

Table 2.2 Accuracy order and number of Gaussian quadrature points

Accuracy order of integral approximation	2	3	4	5
Number of Gaussian points over 1D interval	1	2	2	3
Number of face Gaussian points	1	4	4	9
Number of volumetric Gaussian points	1	8	8	27

$$\int_V dV = \int_0^1 \int_0^1 \int_0^1 \det \mathbf{J}\, dx' dy' dz'$$

with Gaussian quadrature.

In K-exact polynomial reconstruction, the integrated expression usually includes control volume moments of the form $\int_V x^n y^m z^k\, dV$. These moments are computed by applying Gauss's theorem

$$\begin{aligned}\int_V x^n y^m z^k\, dV &= \int_V \nabla \cdot \left(\frac{1}{n+1} x^{n+1} y^m z^k \hat{x}\right) dV \\ &= \int_{\partial V} \left(\frac{1}{n+1} x^{n+1} y^m z^k \hat{x}\right) \cdot \mathbf{n}\, dS = \int_{\partial V} \frac{1}{n+1} x^{n+1} y^m z^k n_x\, dS\end{aligned} \tag{2.86}$$

$$\int_V x^n y^m z^k\, dV = \frac{1}{3} \int_{\partial V} \left(\frac{1}{n+1} x^{n+1} y^m z^k, \frac{1}{m+1} x^n y^{m+1} z^k, \frac{1}{k+1} x^n y^m z^{k+1}\right) \cdot \mathbf{n}\, dS \tag{2.87}$$

where $\mathbf{n} = (n_x, n_y, n_z)$ is the face normal to the boundary surface ∂V of control volume V, and the last integral exactly around each control volume is evaluated by using Gauss quadrature with an appropriate number of quadrature points; as many as $\left(\text{int}(\frac{n+m+k+3}{2})\right)^2$ points are required for $n + m + k$ moments. While the flux form on the right of Eqs. (2.86) and (2.87) can be extended to curved boundaries.

Obviously, the volume of hexahedral cell V is defined by

$$\int_V dV = \frac{1}{3} \int_V \nabla \cdot (x, y, z)\, dV = \frac{1}{3} \int_{\partial V} (x, y, z) \cdot \mathbf{n}\, dS$$

and the centroid $\mathbf{x}_{\text{centroid}}$ of a control volume V can be obtained by the following formula

$$\mathbf{x}_{\text{centroid}} = \frac{1}{V} \int_V (x, y, z) dV = \frac{1}{V} \int_0^1 \int_0^1 \int_0^1 \mathbf{T}(x', y', z') \det \mathbf{J}\, dx' dy' dz'$$

Besides the above transformation from unit cube to 3D hexahedral cell with planar faces, 3D hexahedral cell with non-planar faces are frequently used in solar wind simulation and in the following the approximate transformations from the cube $[-1, 1] \times [-1, 1] \times [-1, 1]$ to two kinds of specified cells with non-planar faces are provided. After the establishment of the approximate transformation, the volume and surface integrals given above apply to such specified cells.

Transformation From Unit Cube to 3D Hexahedral Cell with One Non-planar Face

When one of cell face is non-planar, we can map the cube $[-1, 1] \times [-1, 1] \times [-1, 1]$ in the reference space to a 3D hexahedral cell with one non-planar face,

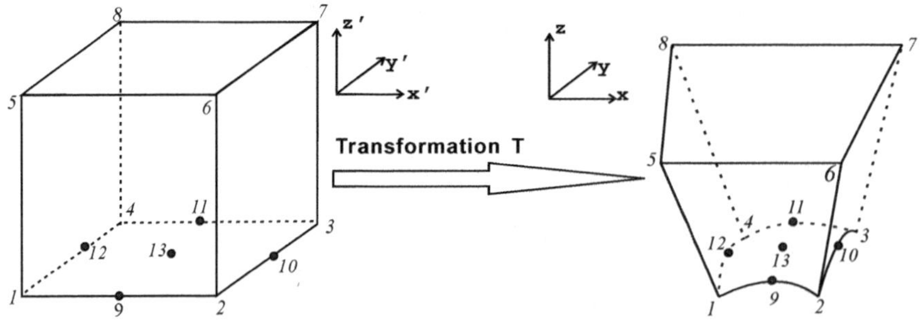

Fig. 2.6 Transformation from cube to 3D hexahedral cell with one non-planar face

say, a part of spherical surface with $\sqrt{x^2+y^2+z^2}=R_1$. This 3D hexahedral cell with one non-planar face is naturally used in the solar coronal study (e.g., [134, 136, 137, 214]) to fit the inner boundary at circled solar surface. For such hexahedral cell, besides shape functions of eight vertices we further need shape functions at the centers of curved edges and non-planar surface. In total, we require thirteen control points as shown in Fig. 2.6 to construct the transform mapping. Local shape functions at thirteen control points $\mathbf{x}_{cp}^{(k)}(k=1,2,\ldots,13)$ read

$$
\begin{aligned}
&v_1 = x'(x'-1)y'(y'-1)(1-z')/8,\ v_2 = x'(x'+1)y'(y'-1)(1-z')/8,\\
&v_3 = x'(x'+1)y'(y'+1)(1-z')/8,\ v_4 = x'(x'-1)y'(y'+1)(1-z')/8,\\
&v_5 = (1-x')(1-y')(1+z')/8,\ v_6 = (1+x')(1-y')(1+z')/8,\\
&v_7 = (1+x')(1+y')(1+z')/8,\ v_8 = (1-x')(1+y')(1+z')/8,\\
&v_9 = (1-x'^2)y'(y'-1)(1-z')/4,\ v_{10} = x'(x'+1)(1-y'^2)(1-z')/4,\\
&v_{11} = (1-x'^2)y'(y'+1)(1-z')/4,\ v_{12} = x'(x'-1)(1-y'^2)(1-z')/4,\\
&v_{13} = (1-x'^2)(1-y'^2)(1-z')/2
\end{aligned}
$$

Then, it follows that

$$\mathbf{T}(x',y',z') = \sum_{k=1}^{13} v_k(x',y',z')\mathbf{x}_{cp}^{(k)}$$

It should be noted that more control points (thus more shape functions) on the canonical plane $z'=-1$ can fit the corresponding non-planar plane more accurately.

Transformation From Cube to 3D Hexahedral Cell with Two Non-planar Faces

When two cell faces are non-planar, we can similarly map the cube $[-1,1]\times[-1,1]\times[-1,1]$ in the reference space to a 3D hexahedral cell with two non-planar faces. This 3D hexahedral cell with two non-planar faces is naturally used in the spherical shell computational domain from the Sun to Earth space [134, 136, 137, 214]. For such hexahedral cell, besides shape functions of eight vertices we further

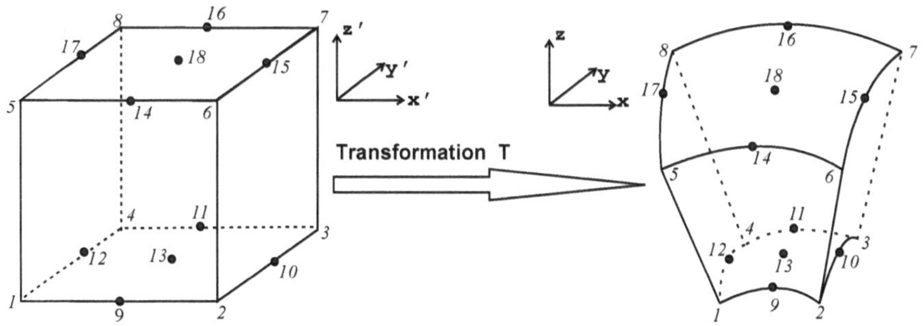

Fig. 2.7 Transformation from cube to 3D hexahedral cell with two non-planar faces

need shape functions at the centers of curved edges and these two non-planar surfaces. In total, we require eighteen control points as shown in Fig. 2.7 to construct the transform mapping. Here, we assume that two non-planar surfaces are parts of spherical surfaces $\sqrt{x^2+y^2+z^2}=R_1$ and $\sqrt{x^2+y^2+z^2}=R_2$ with $R_1<R_2$. Then, local shape functions at eighteen control points $\mathbf{x}_{cp}^{(k)}$ $(k=1,2,\ldots,18)$ read

$$
\begin{aligned}
&v_1=x'(x'-1)y'(y'-1)(1-z')/8,\ v_2=x'(x'+1)y'(y'-1)(1-z')/8,\\
&v_3=x'(x'+1)y'(y'+1)(1-z')/8,\ v_4=x'(x'-1)y'(y'+1)(1-z')/8,\\
&v_5=x'(x'-1)y'(y'-1)(1+z')/8,\ v_6=x'(x'+1)y'(y'-1)(1+z')/8,\\
&v_7=x'(x'+1)y'(y'+1)(1+z')/8,\ v_8=x'(x'-1)y'(y'+1)(1+z')/8,\\
&v_9=(1-x'^2)y'(y'-1)(1-z')/4,\ v_{10}=x'(x'+1)(1-y'^2)(1-z')/4,\\
&v_{11}=(1-x'^2)y'(y'+1)(1-z')/4,\ v_{12}=x'(x'-1)(1-y'^2)(1-z')/4,\\
&v_{13}=(1-x'^2)(1-y'^2)(1-z')/2,\ v_{14}=(1-x'^2)y'(y'-1)(1+z')/4,\\
&v_{15}=x'(x'+1)(1-y'^2)(1+z')/4,\ v_{16}=(1-x'^2)y'(y'+1)(1+z')/4,\\
&v_{17}=x'(x'-1)(1-y'^2)(1+z')/4,\ v_{18}=(1-x'^2)(1-y'^2)(1+z')/2
\end{aligned}
$$

Then, it follows that

$$\mathbf{T}(x',y',z')=\sum_{k=1}^{18}v_k(x',y',z')\mathbf{x}_{cp}^{(k)}$$

Similarly, more control points on the canonical planes $z'=\pm1$ can approximate the corresponding non-planar planes more accurately.

Integration of the Source Terms

In the high-order MHD scheme, high-order accurate integration of the source terms $\mathbf{Q}$ can be realized by using Gaussian quadrature with the help of the isoparametric transformation as discussed above. The cell-averaged source terms $\mathbf{Q}_i$ in cell V_i are calculated as

$$\mathbf{Q}_i = \frac{1}{V_i}\int_{V_i} \mathbf{Q} dV \approx \frac{1}{V_i}\sum_{m=1}^{N_g} \mathbf{Q}|_{\mathbf{x}_m}\,\widetilde{\omega}_m, \tag{2.88}$$

where N_g is the number of the volumetric Gauss points $\mathbf{x}_m$ and $\widetilde{\omega}_m$ is the associated weight.

When the MHD equations take the Powell's form [389] or the generalized Lagrange multiplier (GLM) formulation [96] or even contain other divergence cleaning terms as described in Sect. 2.4, the source terms may involve $-\mathbf{B}(\nabla\cdot\mathbf{B})$, $-(\mathbf{u}\cdot\mathbf{B})(\nabla\cdot\mathbf{B})$, $-\mathbf{u}(\nabla\cdot\mathbf{B})$, $-\mathbf{u}\cdot(\nabla\psi)$, $\mathbf{B}\cdot\nabla(\nabla\cdot\mathbf{B})$, and $\nabla(\nabla\cdot\mathbf{B})$, and then their cell averages can be discretized as follows

$$\frac{1}{V_i}\int_{V_i}(-\mathbf{B})(\nabla\cdot\mathbf{B})dV \approx \frac{1}{V_i}(-\mathbf{B}_i)\sum_{f=1}^{N_{\text{faces}}}\sum_{m=1}^{N_g}(\mathbf{B}\cdot\mathbf{n})_{f,m}\widetilde{\omega}_{f,m},$$

$$\frac{1}{V_i}\int_{V_i}(-\mathbf{u})(\nabla\cdot\mathbf{B})dV \approx \frac{1}{V_i}(-\mathbf{u}_i)\sum_{f=1}^{N_{\text{faces}}}\sum_{m=1}^{N_g}(\mathbf{B}\cdot\mathbf{n})_{f,m}\widetilde{\omega}_{f,m},$$

$$\frac{1}{V_i}\int_{V_i}-(\mathbf{u}\cdot\mathbf{B})(\nabla\cdot\mathbf{B})dV \approx \frac{1}{V_i}\left(-\left(\mathbf{u}_i\cdot\mathbf{B}_i\right)\right)\sum_{f=1}^{N_{\text{faces}}}\sum_{m=1}^{N_g}(\mathbf{B}\cdot\mathbf{n})_{f,m}\widetilde{\omega}_{f,m},$$

$$\frac{1}{V_i}\int_{V_i}(-\mathbf{u})\cdot(\nabla\psi)dV \approx \frac{1}{V_i}(-\mathbf{u}_i)\cdot\left(\sum_{f=1}^{N_{\text{faces}}}\sum_{m=1}^{N_g}(\psi\mathbf{n})_{f,m}\widetilde{\omega}_{f,m}\right)$$

$$\frac{1}{V_i}\int_{V_i}\nabla(\nabla\cdot\mathbf{B})\,dV = \frac{1}{V_i}\sum_{f=1}^{N_{\text{faces}}}\sum_{m=1}^{N_g}(\nabla\cdot\mathbf{B})_{f,m}\mathbf{n}_f\,\widetilde{\omega}_{f,m}$$

$$\frac{1}{V_i}\int_{V_i}(-\mathbf{B})\cdot\nabla(\nabla\cdot\mathbf{B})dV \approx \frac{1}{V_i}(-\mathbf{B}_i)\cdot\left(\sum_{f=1}^{N_{\text{faces}}}\sum_{m=1}^{N_g}(\nabla\cdot\mathbf{B})_{f,m}\mathbf{n}_f\widetilde{\omega}_{f,m}\right)$$

where the subscripts f, m stand for the values evaluated at Gaussian point m on the cell surface f, $\mathbf{n}_f$ is the face normal, $\widetilde{\omega}_{f,m}$ is the associated weight over the cell surface, and N_g is the number of the surface Gauss points.

Alternatively, in K-exact polynomial reconstruction, we can directly differentiate the unlimited reconstructed polynomials of state variables to obtain their derivatives and substitute these derivatives into the integrand, then evaluate the integrand by using Gauss quadrature integral as defined by Eq. (2.88).

2.5.4 Boundary Conditions

At boundary cells, domain boundaries align with cell faces. Consistent accuracy treatment of boundary conditions is a crucial element for numerical methods. In order to maintain consistent accuracy at boundaries, all geometric data (i.e., cell area, centroid, geometric moments, normals, edge lengths, locations of the Gauss quadrature integration points) should be evaluated to the same order of accuracy as that of the interior scheme and an appropriate accurate representation of the boundary geometry is required. In addition, a definite-order prescription for solution values at the boundary based on the boundary equations is needed such that the order of accuracy of the interior scheme is not affected. One approach to imposing boundary data is to make use of extra rows of ghost cells that are added beyond the geometric boundary of the computational domain. Solution values are then imposed in the ghost cells in such a way that the reconstructed solution values and/or fluxes at the boundary approximates the required boundary condition.

In the ghost-cell approach, ghost cells live outside of the domain, and are placed so that the domain boundary is equidistant between the ghost-cell and the boundary-cell centers. Boundary conditions are prescribed by extrapolation across the relevant boundary to the ghost cells surrounding the computational domain. Under most circumstances of interest, this provides a second-order approximation of the boundary condition at the boundary face. Before each step in the multi-step time integration schemes, the ghost cells are updated based on **U** in the computational domain and on the type of the boundary condition represented by the ghost cell. Typically, the variables in the ghost cell are either independent of **U** (e.g., inflow), or functions of the neighbouring cells inside the computational domain (e.g., solid wall or outflow). A special case arises for periodic boundary conditions that relate the ghost cell to the opposite side of the computational domain. The implicit schemes are designed to be compatible with any of these boundary conditions.

In general, boundary conditions can be of two categories: (I) physical boundary conditions, prescribed by the nature of the problem under study, and (II) numerical boundary conditions, vital to make the problem numerically well-posed. The physical conditions are put forward by the problem itself, but they may not be sufficient to numerically solve the problem. Meanwhile, the number of necessary and sufficient boundary conditions can be less than the number of primitive variables. In such cases, even if the physical boundaries make the mathematical problem well-posed, then the numerical problem may be ill-posed due to the discretization of the computational domain [364]. If this is the case, it is required to specify numerical boundary conditions in addition to the physical ones. The numerical boundaries can be characterized by compatibility relations that prevent the uncertainty in the variables that are not specified by the physical conditions and they avoid numerical effects like wave reflections from a nonreflecting boundary. Usually, in the incompressible fluid dynamic case, the physical boundary conditions are sufficient to provide enough conditions for all the physical quantities, except the pressure, which has to be calculated to insure the solenoidality of the velocity field [173]. But, in the compressible case, the pressure does not rely on the velocity field and additional boundary conditions should be provided for the density and the pressure (or temperature). The numerical boundary conditions can be calculated according to a local one-dimensional inviscid (LODI) set of the governing equations expressed at the boundary in projected normal characteristic form (e.g., [138, 371, 383, 392, 451, 463, 509]). The projected normal characteristic equations are obtained by the decomposition of solutions of the governing equations into characteristic modes that cross the domain boundary propagating from inside to outside, or the reverse. The method of characteristics used to establish boundary conditions are consistent with the hyperbolic structure of fluid or MHD equations.

In the ghost-cell approach, we have to typically use one-sided differences near the boundaries, because grid points are available only on the interior side of the boundary. Theoretical approaches and tests [84, 179, 383, 463] indicate that the global order is the same as that of the scheme away from the boundaries when the order of the scheme away from the boundaries is one higher than the order at and near the boundaries, and that sometimes using the same order of boundary approximations as the interior scheme (second or higher) will lead to an unstable solution. Therefore, If the order of approximation near the boundary is equal to the interior scheme order minus one, the overall accuracy of the scheme is not affected. Lu et al. [293] introduced some new high-order scheme for boundary points. For the Neumann (and mixed) boundary conditions, they derived high-order equations to determine the values of the function at the boundary points, before the primitive function reconstruction method is applied. With Dirichlet, Neumann, or mixed boundary conditions, they demonstrated that the derived high-order scheme for boundary points could show its advantages over the standard lower-order boundary scheme.

There exist huge literatures dealing with boundary conditions. For instance, a method for imposing boundary conditions in the context of hyperbolic systems of conservation laws in the finite volume framework has been reported, and is particularly well suited for approximations in the framework of finite volume methods in the sense that it computes directly the normal flux at the boundary with using just the hyperbolic nature of the system and nothing else (e.g., [142, 163, 164, 196–198, 266]). Some usual boundary conditions for MHD have been specified in the existing literature (e.g., [68, 79, 138, 371, 392, 504, 509]).

If additional layers of ghost cells are used for specifying boundary conditions, the reconstruction algorithm for the cells near boundaries is the same as that for an interior cell. During the implementation of boundary conditions with additional ghost cells, the reconstruction stencil remains unmodified near boundaries and the order of the polynomial interpolation is unchanged. Thus, the only requirements for implementing high-order boundary conditions with ghost cells are to ensure that sufficient layers of ghost cells are used in order to carry out the reconstruction and to specify accurate solution data in each ghost cell. For instance, in a standard second-order scheme with a least-squares reconstruction, a minimum of two ghost cells is required: the first one is used for computing the flux through the domain boundary and the second one for the reconstruction of the former.

An alternative boundary condition enforcement by constraining the least-squares reconstruction in control volumes adjacent to the boundary is suggested for a high-order-accurate mesh finite-volume scheme (e.g., [208, 210, 221, 367]). Constrained reconstruction approach prescribes boundary condition in a more accurate and straight forward manner, particularly for curved boundary geometries. In this method, the boundary data is determined using only information from the interior of the computational domain and by the imposition of boundary conditions at the Gaussian quadrature integration points of cell faces lying along the boundary of the domain. This yields additional constraints on the cell-centred reconstruction procedure for interior cells adjacent to the boundary. By constraining the least-squares reconstruction in control volumes near boundaries, complex boundary data can be enforced to give the cell gradients that conserve the mean and satisfy the boundary conditions.

It should be noted that both procedures (i.e., ghost cells and constrained reconstruction) or their combination can be employed for boundary-condition prescription. In Sects. 2.6.2 and 2.6.3 below, three typical boundary conditions of Dirichlet boundary conditions, Neumann boundary conditions and Robin boundary conditions are provided via the help of K-exact and variational reconstruction techniques.

2.5.5 Time Integration

To advance the solution from time-level t^n to time-level t^{n+1}, the semi-discretization Eq. (2.64) is again reduced to solving the following ordinary differential equations (ODEs)

$$\frac{\partial \mathbf{U}_i}{\partial t} = -\mathbf{R}(\mathbf{U}_i) \tag{2.89}$$

This coupled non-linear system of ODEs describes the time evolution of cell-average $\mathbf{U}_i$. Usually, the finite-volume scheme involves three main steps when integrating the ODEs forward in time: (i) reconstruction in which an approximation for $\mathbf{U}(\mathbf{x})$ is obtained within each computational cell; (ii) residual evaluation in which the residual vector, $\mathbf{R}$, is evaluated for each cell based on the reconstructed solution; and (iii) time integration, in which a time-marching scheme is used to advance the solution to the next time level based on the residual.

Conventionally, many standard time integration methods can be employed. For discussion and implementation of diversified time integration schemes, the interested readers may refer to a fairly sizable literature of articles and books (e.g., [45–47, 56, 57, 141, 170, 171, 184, 222, 233, 243, 244, 275, 305, 337, 469, 473, 475, 532]). In the following, we just list some typical time integration methods.

In order to solve Eq. (2.89) numerically, one usually employ a predictor-corrector explicit time stepping scheme. To say a few,

1. The second order two-stage predictor-corrector Runge–Kutta scheme:

$$\mathbf{U}_i^{n+\frac{1}{2}} = \mathbf{U}_i^n - \frac{\Delta t}{2}\mathbf{R}(\mathbf{U}_i^n), \mathbf{U}_i^{n+1} = \mathbf{U}_i^n - \Delta t\mathbf{R}(\mathbf{U}_i^{n+\frac{1}{2}})$$

2. The second order accurate low-storage four-stage Runge–Kutta method [45]:

$$\mathbf{U}^{(0)} = \mathbf{U}_i^n, \mathbf{U}^{(k)} = \mathbf{U}^{(0)} - \alpha_k \Delta t \mathbf{R}_i(\mathbf{U}^{(k-1)}), \mathbf{U}_i^{n+1} = \mathbf{U}_i^{(4)}$$

with the coefficients $\alpha_k (k = 1, \ldots, 4)$ being equal to $[0.1084, 0.2602, 0.5052, 1.0]$.

3. The third order accurate strong-stability-preserving (SSP) low-storage Runge–Kutta scheme [170, 171]:

$$\mathbf{U}' = \mathbf{U}_i^n - \Delta t \mathbf{R}(\mathbf{U}_i^n), \mathbf{U}'' = \frac{3}{4}\mathbf{U}_i^n + \frac{1}{4}\left(\mathbf{U}' - \Delta t \mathbf{R}(\mathbf{U}')\right),$$
$$\mathbf{U}_i^{n+1} = \frac{1}{3}\mathbf{U}_i^n + \frac{2}{3}\left(\mathbf{U}'' - \Delta t \mathbf{R}(\mathbf{U}'')\right)$$

For unsteady flow calculations, the time step, Δt, is a global time step defined as the minimum of the local time steps of all control cells $\Delta t = \min_i \Delta t_i$, where the local time step is according to the Courant–Friedrichs–Levy (CFL) stability criterion defined by $\Delta t_i = \mathrm{CFL}\frac{\Delta h_i}{|\mathbf{v}_i| + c_{fi}}$ with Δh_i a characteristic length of control cell V_i defined as the diameter of the smallest inscribed sphere of control volume V_i. Meanwhile, $\mathbf{v}_i$ and c_{fi} are the local velocity vector and speed of fast magnetosonic wave in control cell V_i, respectively. CFL is the user-definable CFL coefficient, and is usually set to be a positive number less than 1.

However, in such explicit schemes, the numerical stability conditions prescribe an upper limit $\Delta t_{\max}$ on the allowed time step. When only a steady state solution is expected as the final solution by time marching, the explicit approach can be very inefficient. If the convergence towards the steady state stagnates, the physical time related to the final solution can become significantly huge relative to $\Delta t_{\max}$, therefore, an extra number of explicit time steps are required. Even for time-accurate calculations, the variables $\mathbf{U}(t)$ may evolve on a time-scale much longer than $\Delta t_{\max}$. For instance, even if the fastest wave of the hyperbolic system is not induced during the simulation, the time step is still severely constrained. Another case is that a strongly elliptic or diffusive source term dictates an additional limiting diffusion time $\Delta t_{\max}$, while in practice it is balanced by another term and $\mathbf{U}(t)$ evolves slowly.

Fortunately, an implicit treatment of some or all of the variables in system (2.89) can relieve the most severe stability restrictions on the time step, or can even permit utilizing arbitrarily large time steps to reach a steady state (e.g., [244, 473]). The right choice of temporal discretization for the most efficient numerical solution mainly relies on the specific problem at hand. Some widely used implicit time stepping schemes may be as follows:

1. The second-order Crank–Nicolson method:

$$\mathbf{U}_i^{n+1} = \mathbf{U}_i^n - \frac{\Delta t}{2}\left[\mathbf{R}(\mathbf{U}_i^{n+1}) + \mathbf{R}(\mathbf{U}_i^n)\right]$$

2. The second-order backward differentiation formula (BDF2):

$$\mathbf{U}_i^{n+1} = \mathbf{U}_i^n + \Delta t \left[-\frac{2}{3}\mathbf{R}(\mathbf{U}_i^{n+1}) + \frac{1}{3}\frac{\mathbf{U}_i^n - \mathbf{U}_i^{n-1}}{\Delta t} \right]$$

In the first time step of BDF2, when $\mathbf{U}_i^{n-1}$ is not available, one could use the second-order Crank–Nicolson method, or the more stable backward Euler scheme $\mathbf{U}_i^{n+1} = \mathbf{U}_i^n - \Delta t\ \mathbf{R}(\mathbf{U}_i^{n+1})$, which introduces a first-order error in the first time step.

3. The second order semi-implicit time discretization:

$$\mathbf{U}_{\text{expl}}^{n+1} = \mathbf{U}_{\text{expl}}^n - \Delta t \mathbf{R}_{\text{expl}} \left(\mathbf{U}^n - \frac{\Delta t}{2}\mathbf{R}(\mathbf{U}^n) \right)$$

$$\mathbf{U}_{\text{impl}}^{n+1} = \mathbf{U}_{\text{impl}}^n - \frac{\Delta t}{2} \left(\mathbf{R}_{\text{impl}}(\mathbf{U}_{\text{expl}}^n, \mathbf{U}_{\text{impl}}^n) + \mathbf{R}_{\text{impl}}(\mathbf{U}_{\text{expl}}^{n+1}, \mathbf{U}_{\text{impl}}^{n+1}) \right)$$

where $\mathbf{U}_{\text{expl}}, \mathbf{U}_{\text{impl}}$ contain all explicitly and implicitly treated variables of $\mathbf{U}$, hence symbolically, $\mathbf{U} = (\mathbf{U}_{\text{expl}}, \mathbf{U}_{\text{impl}})$. That is, when only some $\mathbf{U}_{\text{impl}}$ of the variables $\mathbf{U}$ are treated implicitly, we first advance the explicitly treated $\mathbf{U}_{\text{expl}}$ using a second-order explicit time integration scheme, e.g., a two-stage Runge–Kutta scheme and subsequently update the implicitly treated variables $\mathbf{U}_{\text{impl}}$ (e.g., [244, 473]). In the semi-implicit approach, the time step must be chosen in accordance with the limitations imposed by the explicitly treated terms.

For time-accurate problems, second-order temporal accuracy can be maintained under various semi-implicit discretizations where only some of the variables or only some source terms are treated implicitly (e.g., [244, 473]). Alternatively, as mentioned in Sect. 2.2.2, dual time techniques work well for time-accurate problems (e.g., [195, 222, 248, 524]).

Generally speaking, solving the MHD equations involves the following steps (not necessarily in order).

1. Set the time step based on accuracy, efficiency and robustness requirements, usually by the Courant-Friedrichs-Lewy stability condition.
2. Reconstruct a high-order approximation to the solution, based on the current average values in local control volumes, and interpolate the conserved (or primitive or characteristic) variables at the left and right sides of each cell face using one limited reconstruction method.
3. Using the left and right states, determine the fluxes through each cell face. This may involve solving an approximate Riemann problem at the cell interface. Meanwhile, volume average of the source vector is calculated on each cell.
4. Update the cell averages using a stable ODE time-integration method.
5. Perform a $\nabla \cdot \mathbf{B} = 0$ preserving step or a divergence cleaning step depending upon the divergence-free method chosen to deal with $\nabla \cdot \mathbf{B}$.
6. Update ghost cells or boundary values.
7. Repeat all steps above until desired time or steady-state solution is reached.

2.6 State Variable Reconstructions

In the FVM for the numerical solution of hyperbolic conservation laws, the true flux function at the cell interfaces is approximated by a numerical flux function. This numerical flux function is typically evaluated by solving a local Riemann problem at each Gaussian quadrature point on each interface, taking as input the left- and right-sided limit of reconstructed variables. If this reconstruction returns simply the adjacent cell average values, this leads to a first order scheme.

Before we proceed, we would like to clarify order of accuracy and high order. Mathematically, a numerical approximation is said to be kth order (or order k) if the approximation error is proportional to Δh^k with Δh the local mesh size. It is routinely accepted [106, 492] that high order usually means third order or higher, while low order schemes stand for the first or second-order schemes. However, in what follows we use higher or high order indistinguishably to mean order $k > 1$, that is, second or even higher than second order.

To obtain higher order accuracy (second or even higher than second order), the left and right states $\mathbf{U}_L$ and $\mathbf{U}_R$ can be taken from higher order polynomials $\mathbf{U}_i(\mathbf{x})$ representing the reconstructed solution in control volume V_i. To this end, one needs to define higher order reconstruction functions based on the average values of the cell of interest and its associated neighbors. Evaluating these polynomials at the cell interfaces prescribes input values for the numerical flux function, which results in a higher order accurate scheme. On fixed grids, as pointed out by Godunov [165], the higher order approximation of hyperbolic conservation laws with linear schemes is impossible without creating spurious oscillations. To remedy this situation, the order of reconstruction has to be limited at solution discontinuities. There are a large variety of such non-linear limiters developed to achieve higher order accuracy without creating oscillations.

The first such kind of limiter functions may be due to van Leer limiter [267, 269] and Roe superbee limiter [403]. These limiter functions establish second order accurate reconstructions in smooth parts of the solution and lie in the second order total variation diminishing (TVD) region [455] by giving design rules for limiters. Afterwards come a vast number of limiters, such as the minmod limiter [271], the superbee limiter [240, 469], the van Leer limiter [240, 269], the van Albada limiter [228, 240, 481], Barth-Jespersen limiter [37], Venkatakrishnan limiter [485], the biased averaging procedure (BAP) [76, 281], and the multi-dimensional limiting process (MLP) [236, 249]. In most cases, these classical approaches result in the second order accuracy. However, they become first order near smooth extrema due to the TVD property [255, 271]. Higher orders of accuracy can be achieved only via wider stencils. Utilizing a four-point centered stencil, the piecewise parabolic method (PPM) of a third order reconstruction was proposed [83, 498] with its later improvements [82, 386, 419]. Suresh and Huynh [453] used a five-point stencil for even higher order reconstruction.

The second category via the use of non-polynomial reconstructions comes into being since Harten et al. [193] presented essentially non-oscillatory (ENO) schemes.

The idea inherited in the ENO schemes (e.g., [438, 439]) is to divide the stencil of interest into smaller sub-stencils with piecewise polynomial reconstructions on each sub-stencil. The scheme then employs an optimal stencil selection procedure by choosing the locally smoothest stencil for the reconstruction of state variables; while other stencils containing discontinuities are all discarded. In such a way, the final approximation to the scheme does not contain all available data since it avoids interpolation across discontinuities. To get over this shortcoming, weighted essentially non-oscillatory (WENO) schemes were developed [148, 230, 231, 288, 435] by combining all of the stencils and assigning a nonlinear weight coefficient to each reconstruction stencil based on the local smoothness of flowfield. Unlike ENO, WENO schemes use a convex combination of all candidate stencils to obtain the reconstruction of the interface values, with higher weights for cells with smoother solutions and lower weights for cells near discontinuities. As a result, WENO methods can achieve high-order accuracy at smooth regions while are non-oscillatory near discontinuities. Thus, many high-order WENO-type schemes have also been established (e.g., [27, 115, 205, 256, 264, 279, 296, 304, 319, 393, 428, 436, 437, 465, 477]), and a brief survey of several developments has been given by Shu [437] for high order WENO and DG methods for time-dependent convection-dominated partial differential equations. Both ENO and WENO schemes show superior advantages in capturing shocks and preserving numerical accuracy, however, it is a enormous task most of the time to select admissible and proper stencils from a large number of cells, especially for high-order and multi-dimensional cases on unstructured grids. Nevertheless, ENO/WENO schemes have always been the object of attention [152–155, 203, 527, 529]. There are also other highly non-linear limiters for state variable reconstructions. Artebrant and Schroll [10] developed a local double-logarithmic reconstruction that only entails three input values. Based on this local double-logarithmic reconstruction, Čada and Torrilhon [58] designed a limiter function to yield third order accuracy in smooth parts of the solution and at extrema. They also suggest a criterion to distinguish between smooth extrema and discontinuities in order to avoid extrema clipping. Furthermore, a parameter-free smoothness criterion has been investigated [416].

Various higher order K-exact polynomial reconstructions can be easily created by employing data reconstructions and spurious oscillations are avoided by constraining the reconstruction with the monotone enforcement. Often, piecewise K-exact polynomial reconstructions are established for multi-dimensional problems based on the least squares technique and monotonicity enhancement, especially for unstructured mesh. Barth et al. [33, 34, 36] proposed a high-order reconstruction of state variables for unstructured mesh based on K-exact least-squares reconstruction. Venkatakrishnan [484] introduced a smooth alternative to the Barth-Jespersen limiter. Ollivier-Gooch [370] proposed a data-dependent weighted least-squares reconstruction in the K-exact framework (called DD-L2-ENO) with a fixed stencil. The weights are chosen to strongly emphasize smooth data in the reconstruction, satisfying the weighted ENO conditions of Liu et al. [288]. Because DD-L2-ENO is designed in the framework of K-exact reconstruction, such reconstructions on arbitrary meshes can be easily implemented. Liu and Jiao [285] presented a family of weighted-least-squares based essentially non-oscillatory (WLS-ENO) reconstruction methods over unstructured meshes. Unlike the WENO scheme, WLS-ENO approach uses a generalized finite difference formulation based on weighted least squares, rather than a weighted averaging of traditional finite differences. The generalized finite difference method is derived rigorously from Taylor series, and hence can deliver the same order of accuracy as traditional finite differences. In WLS-ENO, the convexity requirement can be satisfied automatically so as to construct non-oscillatory schemes in a more systematic way, since the weights are specified a priori. These methods (e.g., [285, 370]) suggest alternative approaches to achieving ENO behavior by modifying the weighted least squares (WLSQ) method itself, and assign a small weight to cells that are likely to produce oscillatory solutions. Thus, the least squares reconstruction itself possesses ENO behavior. These ENO-weighted least squares schemes require only a single stencil for reconstruction and maintain a smooth transition between high order accurate reconstructions and shock capturing.

Haider et al. [182, 183] presented a algorithm for high order piecewise K-exact polynomial reconstruction (e.g., [36]), where the $(m+1)$th pointwise derivative is recursively calculated from the mth derivative on a neighborhood of that point $(m \leq K-1)$. This algorithm computes a high order approximate in a given cell using data from the direct neighborhood cells, eliminating the need to handle directly large reconstruction stencils. Since the algorithm accesses data from grid cells beyond the adjacent cells in an implicit and natural way, the resulting high order finite volume method is well suited for general unstructured polyhedral grids and for an efficient parallel implementation. Also, in order to make the stencil as compact as possible, Yang et al. [506] introduced a technique for achieving high order accuracy without explicitly constructing the large stencil through successive differentiation [54]. This method constructs progressively higher derivatives using the same face-neighbor stencil and previously reconstructed derivatives to build a higher order accurate stencils through implicit connectivity to further neighbors. Such a successive differentiation method is particularly attractive to implementation on solution adaptive meshes as distant neighbor information does not have to be recomputed after each solution adaption cycle. Yang et al.'s method is also particularly well suited for implementation in a distributed memory parallel environment, as a given cell does not have to explicitly know on which processor distant-neighboring cells reside. According to the conception of successive differentiation, unstructured MUSCL techniques have been proposed [104, 508].

The BAP limiter was extended to a high-order finite volume scheme on unstructured grids based on a secondary reconstruction and successive limiting process [278, 280, 489], in which the higher order derivatives' coefficients of the reconstruction polynomials are firstly limited by the WBAP, then the recomputed lower order derivatives' coefficients in the secondary reconstruction based the limited higher order coefficients are prescribed by the WBAP. The successive limiting procedure, together with the WBAP, can effectively control the numerical oscillations and results in an essentially monotone shock transition. A distance weighted BAP limiter was also developed [289, 290] for the second- and high order finite volume method on unstructured grids in the two- and three-dimensional cases, where the standard deviation coefficients of the flow variables are used as the smooth indicators, and the derivatives of the flow variables are limited by weighting all derivatives of each cell according to smoothness.

Michalak and Ollivier-Gooch [313] established an unstructured grid limiter broadly based on the framework introduced by the Barth-Jespersen limiter specifically for high order schemes. Like the Venkatakrishnan limiter, it is designed to be differentiable to enable good convergence to steady state by implicit schemes, which is realized by the construction of a new function defining the relation between the limiter value and the unlimited reconstruction at Gauss points. Ivan and Groth [211] proposed a high-order central ENO (CENO) reconstruction on fixed central stencils, which involves an unlimited K-exact reconstruction in smooth regions, and switches a limited piecewise linear reconstruction when discontinuities are identified. The CENO scheme is also extended to three-dimensional cases, in which the robustness and the high-order accuracy are validated by numerical results (e.g., [74, 214, 454]).

In orthogonal curvilinear coordinates, Mignone [316] employed a piecewise polynomial approximation to the cell average values to reconstruct left and right interface states from within a computational cell to arbitrary order of accuracy by inverting a Vandermonde-like linear system of equations with spatially varying coefficients, where limiting techniques and monotonicity constraints are revised for conventional reconstruction schemes, namely, the piecewise linear method (PLM), third-order WENO scheme and the piecewise parabolic method (PPM). As required in practical implementation, explicit analytical expressions for second up to fifth-order accuracy have been derived by Mignone [316] for uniform radial grids in cylindrical and spherical coordinates, with the reconstruction weights depending only on the grid index rather than on the physical location of the cell.

TVD-schemes for multi-dimensional structured meshes can be constructed much easier by dimensional splitting. However, most TVD limiters give rise to large spurious oscillations near discontinuities for multi-dimensional problems. Within the framework of the finite volume methods, the MUSCL approach [267] adopts a slope limiter to control the numerical oscillations near the discontinuities. The limiters of the MUSCL schemes are usually designed to satisfy the TVD conditions (e.g., [190, 455]). As proved by Suresh [452], the MUSCL scheme preserves positivity if the reconstruction in each cell is limited by the cell averages of first neighbors. Therefore, limiters are used to suppress these oscillations while keeping second-order reconstruction in the smooth region of flow field. On structured grids, the finite difference method (FDM) and finite volume method (FVM) have been applied along with mature limiting method based on solid theories. However, these structured schemes can not be extended directly to multi-dimensional unstructured grids due to various reasons. Firstly, the schemes for structured grids are usually developed based on one-dimensional analysis and extended to multi-dimensional structured grids by dimensional-splitting, which is infeasible for unstructured grids. Secondly, one-dimensional principles, for instance, the TVD condition, are not necessary feasible on multi-dimensional unstructured grids. Darwish and Moukalled [91] gave an excellent overview on TVD schemes for unstructured grids. In unstructured grid cases, it is difficult to implement and enforce a monotonicity criterion that depends on direct neighbor information. Unlike structured grid procedures where there is an implied structured connectivity between neighboring grid points, it is more difficult to compute higher derivatives on unstructured grids due to a lack of explicit connectivity beyond the first neighboring cells. Furthermore, a scheme applying TVD condition will cause accuracy deterioration at extrema even in smooth regions, and thus the total variation bounding (TVB) [434], ENO/WENO [193, 230, 526], and multi-dimensional optimal order detection (MOOD) [80] schemes were developed.

By extending Spekreijse's monotone condition [447], Barth and Jespersen designed a limiter on unstructured grids [37], which modifies the piecewise linear distribution at each control volume. Barth-Jespersen limiter removes local extrema and insures stability. However, this limiter shows similar effects as that of TVD condition, which reduces accuracy at smooth extrema. Furthermore, Barth-Jespersen limiting function is non-differentiable, and thus the convergence is less satisfactory. As an improvement, Venkatakrishnan limiter [484] is a differentiable limiting function similar to that of van Albada limiter [481] and it achieves better convergence compared with Barth-Jespersen limiter. Whereas, Venkatakrishnan limiter is not strictly monotone, and thus it might produce oscillations across a shock wave. Anyway, van Albada limiter, Barth-Jespersen limiter and Venkatakrishnan limiter are the forerunners of limiting functions that have been successfully applied on unstructured grids. The designing conceptions of these limiters have served as a guide to subsequent development of high-order accurate limiting procedures.

Generally speaking, one of major tasks in high-order FVM is to design an efficient and accurate limiting strategy for the multidimensional hyperbolic conservation laws applicable to both structured and unstructured grids, so as to preserve numerical accuracy in smooth regions, and be robust and effective in coping with discontinuities, such as shock waves. In what follows we are mainly interested in briefly introducing some piecewise polynomial reconstructions.

2.6.1 Second-Order Reconstruction

Suppose that the scalar variable X stands for either the conserved variables ρ, ρu_x, ρu_y, ρu_z, e, B_x, B_y and B_z or the primitive variables ρ, u_x, u_y, u_z, p, B_x, B_y and B_z. Assume that $X_i(\mathbf{x})$ denotes the reconstructed solution of $X(\mathbf{x})$ in control volume V_i at a position $\mathbf{x}$ and $X_i = \frac{1}{V_i}\int_{V_i} X(\mathbf{x})dV$ is the control volume averaged value of $X(\mathbf{x})$. Then $X_i(\mathbf{x})$ can be written as

$$X_i(\mathbf{x}) = X_i + (\nabla X)|_i \cdot (\mathbf{x} - \mathbf{x}_i), \tag{2.90}$$

where $(\nabla X)|_i = (\nabla X)|_{\mathbf{x}_i}$ is the gradient of the scalar variable X at the centroid $\mathbf{x}_i$ of control volume V_i, and $\mathbf{x}$ is the location at which the solution is sought. The transformation of cell-centered reconstructions from conservative to primitive variables is discussed in Sect. 2.7.

To approximate the gradient $(\nabla X)|_i$ at the centroid $\mathbf{x}_i$ (of the quantity X in question), either a least-squares technique (e.g., [33]), or the Green-Gauss method (e.g., [37]) can be employed. At local extrema and discontinuities the reconstruction polynomial may generate new extrema and therefore cause oscillations in the numerical solution. In order to circumvent this phenomenon, various slope limiters are designed to obtain the required reconstruction polynomials.

Gradients are required for reconstructing values of a scalar at the cell faces and for computing viscous fluxes. There are vast amount of literatures about gradient calculations (e.g., [149, 150, 200, 238, 251, 395, 456, 519]). Basically, cell-based Green-Gauss, vertex/node-based Green-Gauss, cell-based least-squares, vertex/node-based least-squares gradient evaluation methods are available (e.g., [37, 110, 112, 446, 456]).

2.6.1.1 Green-Gauss Gradient Evaluation

The gradient $(\nabla X)|_i$ is computed using the divergence theorem or Green-Gauss method from approximate forms of the exact relation

$$\int_{V_i} \nabla X \, dV = \int_{\partial V_i} X \mathbf{n} \, dS. \tag{2.91}$$

where V_i is the volume of control volume V_i, and ∂V_i is the boundary of the control volume V_i. According to the midpoint integration rule [53], the mean value of a quantity over cell V_i (or over face $f = A_{ij}$) is equal to its value at the centroid $\mathbf{x}_i$ of the cell (or center $\mathbf{x}_{ij,c}$ of the face $f = A_{ij}$), up to second-order accuracy. That is, $(\nabla X)|_i = \frac{1}{V_i} \int_{V_i} \nabla X dV$, and $\frac{1}{A_{ij}} \int_{A_{ij}} X dA = X(\mathbf{x}_{ij,c})$. Then we have

$$\frac{1}{V_i} \int_{\partial V_i} X \mathbf{n} \, dS = \frac{1}{V_i} \sum_{j=1}^{N_{\text{faces}}} \left(\mathbf{n}_{ij} \int_{A_{ij}} X dA \right) = \frac{1}{V_i} \sum_{j=1}^{N_{\text{faces}}} X(\mathbf{x}_{ij,c}) A_{ij} \mathbf{n}_{ij},$$

so that

$$(\nabla X)|_i = \frac{1}{V_i} \int_{V_i} \nabla X \, dV = \frac{1}{V_i} \int_{\partial V_i} X \mathbf{n} \, dS = \frac{1}{V_i} \sum_{j=1}^{N_{\text{faces}}} X(\mathbf{x}_{ij,c}) A_{ij} \mathbf{n}_{ij}, \tag{2.92}$$

where A_{ij} is the area of interface A_{ij} of the control volume V_i, and $\mathbf{x}_{ij,c}$ is the face center of interface A_{ij}. $\mathbf{n}_{ij}$ is the unit normal vector pointing outward from the boundary ∂V_i to its neighbor V_j, and the summation is taken over all the faces enclosing the cell. These faces are assumed to be planar in three dimensions, as in Fig. 2.1, so that the normal unit vector $\mathbf{n}$ has a constant value $\mathbf{n}_{ij}$ along each face $f = A_{ij}$.

However, this approximation cannot be the final formula for the gradient because Eq. (2.92) contains $X(\mathbf{x}_{ij,c})$, the X values at the face centers, whereas we need a formula that uses only the known values at the cell centroids. There are many face interpolation methods (FIMs) to obtain $X(\mathbf{x}_{ij,c})$. In what follows we list several such common approaches available.

In a common practice, the value $X(\mathbf{x}_{ij,c})$ in the rightmost hand side of Eq. (2.92) can be taken simply as a volume weighted mean of the control-volume averaged values X_i and X_j at both sides of the face:

$$\text{FIM-I:}\ X(\mathbf{x}_{ij,c}) = \frac{V_i X_i + V_j X_j}{V_i + V_j}.$$

As shown in Fig. 2.1, $\mathbf{x}_{ij,c}$ is the centroid of face $f = A_{ij}$ and $\mathbf{x}'_{ij,c}$ is its closest point on the line segment connecting $\mathbf{x}_i$ and $\mathbf{x}_j$, the line $\overline{\mathbf{x}_{ij,c}\mathbf{x}'_{ij,c}}$ is perpendicular to the line $\overline{\mathbf{x}_i\mathbf{x}_j}$ with $\mathbf{x}'_{ij,c}$ the perpendicular foot point. Usually, face $f = A_{ij}$ of cell i exhibits: non-orthogonality if $\mathbf{x}_j - \mathbf{x}_i$ is not parallel to $\mathbf{n}_{ij}$; unevenness if the midpoint $(\mathbf{x}_i + \mathbf{x}_j)/2$ of the line segment joining $\mathbf{x}_i$ and $\mathbf{x}_j$ does not coincide with $\mathbf{x}'_{ij,c}$ (i.e., the cell centroids are unequally spaced on either side of the face); skewness if $\mathbf{x}'_{ij,c}$ does not coincide with $\mathbf{x}_{ij,c}$ (i.e., the line joining the cell centroids does not pass through the face center $\mathbf{x}_{ij,c}$). Another practice is to approximate $X(\mathbf{x}_{ij,c})$ by $X(\mathbf{x}'_{ij,c})$, the exact values of X at points $\mathbf{x}'_{ij,c}$ rather than $\mathbf{x}_{ij,c}$; these values also do not belong to the set of cell centered values, but since $\mathbf{x}'_{ij,c}$ lies on the line segment joining cell centroids $\mathbf{x}_i$ and $\mathbf{x}_j$, the value $X(\mathbf{x}'_{ij,c})$ can in turn be approximated by linear interpolation between X_i and X_j, say $\overline{X}(\mathbf{x}'_{ij,c})$ (the overbar denotes linear interpolation):

$$\text{FIM-II:}\ \overline{X}(\mathbf{x}'_{ij,c}) = \frac{|\mathbf{x}'_{ij,c} - \mathbf{x}_j|}{|\mathbf{x}_j - \mathbf{x}_i|} X_i + \frac{|\mathbf{x}'_{ij,c} - \mathbf{x}_i|}{|\mathbf{x}_j - \mathbf{x}_i|} X_j$$

By using $\overline{X}(\mathbf{x}'_{ij,c})$ instead of $X(\mathbf{x}_{ij,c})$ in the right hand side of Eq. (2.92), one obtains an approximation to the gradient:

$$(\nabla X)|_i = \frac{1}{V_i} \sum_{j=1}^{N_{\text{faces}}} \overline{X}(\mathbf{x}'_{ij,c}) A_{ij} \mathbf{n}_{ij}, \tag{2.93}$$

which now depends only on cell centered values of X. A crucial question is whether and how the replacement of $X(\mathbf{x}_{ij,c})$ by $\overline{X}(\mathbf{x}'_{ij,c})$, changing Eqs. (2.92) to (2.93), has affected the accuracy, which is discussed by Syrakos et al. [456] by considering structured grids and unstructured grids separately.

Sozer et al. [446] conducted a survey of some gradient reconstruction methods for cell-centered data on unstructured meshes within the scope of accuracy assessment. Through consecutive local scaling of an analytical test function, formal order of accuracy, as well as error magnitudes for each of several methods, is evaluated on a complex mesh of various cell types. Selectively taken from the survey of Sozer et al. [446], inverse distance weighted (IDW) face interpolation and least-squares method are presented below.

Inverse distance weighted (IDW) face interpolation is a popular approach for face averaging. IDW method makes use of the entire neighbor stencil $k(k = 1, 2, \ldots, N_k)$ around face A_{ij}:

$$\text{FIM-III:}\ X(\mathbf{x}_{ij,c}) = \sum_{k=1}^{N_k} \frac{X_k}{|\mathbf{x}_k - \mathbf{x}_{ij,c}|^2} \Bigg/ \sum_{k=1}^{N_k} \frac{1}{|\mathbf{x}_k - \mathbf{x}_{ij,c}|^2}$$

where $|\mathbf{x}_k - \mathbf{x}_{ij,c}|$ is the distance between the stencil centroid $\mathbf{x}_k$ and the current face center $\mathbf{x}_{ij,c}$, and the summation is taken over the entire stencil. For 3D hexahedral computational cells as shown in Fig. 2.1, $N_k = 18$.

Moreover, the average of the scalar X over the face A_{ij} can be obtained by using the weighted least-squares (WLSQ) method. Considering the Taylor series expansion of X_k at the stencil centroid $\mathbf{x}_k$ around the face center $\mathbf{x}_{ij,c}$:

$$X_k = X(\mathbf{x}_{ij,c}) + (\nabla X)_{ij} \cdot (\mathbf{x}_k - \mathbf{x}_{ij,c})$$

where $(\nabla X)_{ij} = (\nabla X)|_{\mathbf{x}_{ij,c}}$ is the gradient of the interpolated scalar X at face center $\mathbf{x}_{ij,c}$. For simplicity, we mark $X(\mathbf{x}_{ij,c})$ as X_{ij} in the following description. Define a functional as follows

$$f(\omega_k, X_k, X_{x,ij}, X_{y,ij}, X_{z,ij}) = \sum_{k=1}^{N_k} \omega_k \left[X_k - X_{ij} - X_{x,ij}\Delta x_k - X_{y,ij}\Delta y_k - X_{z,ij}\Delta z_k\right]^2$$

where $X_{x,ij}$, $X_{y,ij}$, and $X_{z,ij}$ are the three components of $(\nabla X)_{ij}$; $\omega_k = \frac{1}{|\mathbf{x}_k - \mathbf{x}_{ij,c}|}$ [446] are the weights for each stencil point k, $\Delta x_k = x_k - x_{ij,c}$, $\Delta y_k = y_k - y_{ij,c}$, $\Delta z_k = z_k - z_{ij,c}$, with $\mathbf{x}_k = (x_k, y_k, z_k)$ and $\mathbf{x}_{ij,c} = (x_{ij,c}, y_{ij,c}, z_{ij,c})$. In order to derive the stencil coefficients, the functional $f(\omega_k, X_k, X_{x,ij}, X_{y,ij}, X_{z,ij})$ is minimized by setting the gradient to zero

$$\frac{\partial f}{\partial X_{ij}} = -\sum_{k=1}^{N_k} 2\omega_k \left[X_k - X_{ij} - X_{x,ij}\Delta x_k - X_{y,ij}\Delta y_k - X_{z,ij}\Delta z_k\right] = 0$$

$$\frac{\partial f}{\partial X_{x,ij}} = -\sum_{k=1}^{N_k} 2\omega_k \Delta x_k \left[X_k - X_{ij} - X_{x,ij}\Delta x_k - X_{y,ij}\Delta y_k - X_{z,ij}\Delta z_k\right] = 0$$

$$\frac{\partial f}{\partial X_{y,ij}} = -\sum_{k=1}^{N_k} 2\omega_k \Delta y_k \left[X_k - X_{ij} - X_{x,ij}\Delta x_k - X_{y,ij}\Delta y_k - X_{z,ij}\Delta z_k\right] = 0$$

$$\frac{\partial f}{\partial X_{z,ij}} = -\sum_{k=1}^{N_k} 2\omega_k \Delta z_k \left[X_k - X_{ij} - X_{x,ij}\Delta x_k - X_{y,ij}\Delta y_k - X_{z,ij}\Delta z_k\right] = 0$$

Utilizing equations above, a linear system of equations in terms of the stencil values X_k and X_{ij}, and the derivatives $X_{x,ij}$, $X_{y,ij}$, and $X_{z,ij}$ can be written as $\mathcal{M}\left[X_{ij}, X_{x,ij}, X_{y,ij}, X_{z,ij}\right]^T = \mathcal{B}\left[X_1, X_2, \ldots, X_{N_k}\right]^T$ with

$$\mathcal{M} = \begin{pmatrix} \sum\omega_k & \sum\omega_k\Delta x_k & \sum\omega_k\Delta y_k & \sum\omega_k\Delta z_k \\ \sum\omega_k\Delta x_k & \sum\omega_k\Delta x_k^2 & \sum\omega_k\Delta x_k\Delta y_k & \sum\omega_k\Delta x_k\Delta z_k \\ \sum\omega_k\Delta y_k & \sum\omega_k\Delta x_k\Delta y_k & \sum\omega_k\Delta y_k^2 & \sum\omega_k\Delta y_k\Delta z_k \\ \sum\omega_k\Delta z_k & \sum\omega_k\Delta x_k\Delta z_k & \sum\omega_k\Delta y_k\Delta z_k & \sum\omega_k\Delta z_k^2 \end{pmatrix}$$

$$\mathcal{B} = \begin{pmatrix} \omega_1 & \omega_2 & \cdots & \omega_{N_k} \\ \omega_1\Delta x_1 & \omega_2\Delta x_2 & \cdots & \omega_{N_k}\Delta x_{N_k} \\ \omega_1\Delta y_1 & \omega_2\Delta y_2 & \cdots & \omega_{N_k}\Delta y_{N_k} \\ \omega_1\Delta z_1 & \omega_2\Delta z_2 & \cdots & \omega_{N_k}\Delta z_{N_k} \end{pmatrix}$$

The stencil coefficients C_k for each stencil point are contained in the first row of $\mathcal{M}^{-1}\mathcal{B}$. The coefficients can be expressed shortly as: $C_k = \mathcal{M}^{-1}_{1,\ell}\mathcal{B}_{\ell,k}$, where $\mathcal{M}^{-1}_{1,\ell}$ is the value in the first row and ℓth column of $\mathcal{M}^{-1}$. The interpolated value at the face is then expressed as

$$\text{FIM-IV:}\ X_{ij} = \sum_{k=1}^{N_k} C_k X_k$$

For detailed accuracy study of gradient calculation methods about FIM-III and FIM-IV, refer to Sozer et al. [446].

The Green-Gauss gradient estimation methods introduced above are usually called cell-based Green-Gauss gradient evaluation. There is another kind of Green-Gauss gradient calculation named vertex/node-based Green-Gauss gradient method. The nodal Green-Gauss procedure involves estimating the gradient in a cell as

$$(\nabla X)|_i = \frac{1}{V_i}\sum_m \widetilde{X}_m S_m \mathbf{n}_m \tag{2.94}$$

where the summation is performed over the faces of each control volume surface with area S_m and face normal $\mathbf{n}_m$. If the face is a quadrilateral, the face is first split into two triangles and treated as two faces. This is just the case for hexahedral cell as shown in Fig. 2.1. Here, $\widetilde{X}_m$ is the average value of X on the face m, and may be computed as

$$\widetilde{X}_m = \frac{1}{3}(X_{vm1} + X_{vm2} + X_{vm3})$$

where $X_{v\ell}(\ell = m1, m2, m3)$ represent the values of X at the nodes comprising face m. Thus, the nodal Green-Gauss procedure involves first computing nodal values of X from surrounding cell center data, followed by the trapezoidal integration procedure of Eq. (2.94).

To obtain $(\nabla X)_i$ in Eq. (2.94), the main task is to compute the nodal value $X_{v\ell}$ in $\widetilde{X}_m$. In what follows we present several possible ways to obtain the nodal values from surrounding cell-center data. The first one is the "nodal volume" scheme [223] that involves a volume weighted average of the cell values X_j's with cells j's surrounding each node $v\ell$:

$$X_{v\ell} = \sum_{j=1}^{N_\ell} V_j X_j \Bigg/ \sum_{j=1}^{N_\ell} V_j$$

where N_ℓ is the connected cell number of vertex $v\ell$.

The nodal values $X_{v\ell}$ can also be determined in the manner described by Frink et al. [150] and Frink [151]. A weighted average of the cell-centered solution quantities is taken to obtain $X_{v\ell}$ at each node $v\ell$ of the vertices that span the cell of interest. It is assumed that the contribution from the cell-centered solution value (already known in each surrounding cell) to a vertex belonging to the cell is inversely proportional to the distance between the vertex and the centroid of that surrounding cell according to Shepard's formula [427],

$$X_{v\ell} = \sum_{j=1}^{N_\ell} \frac{X_j}{r_j} \Bigg/ \sum_{j=1}^{N_\ell} \frac{1}{r_j}$$

where $r_j = |\mathbf{x}_j - \mathbf{x}_{v\ell}| = \sqrt{(x_j - x_{v\ell})^2 + (y_j - y_{v\ell})^2 + (z_j - z_{v\ell})^2}$. Similarly, the subscripts $v\ell$ and j refer to the node and its surrounding cell, respectively. Note that this process utilizes information from all of the cells surrounding the candidate cell, which thus can be used to produce a truly multidimensional higher-order reconstruction [150].

The third one is the "nodal least squares" scheme [238], in which the vertex value $X_{v\ell}(\ell = m1, m2, m3)$ at the nodes comprising face m can be calculated by solving a vertex-weighted-least-squares (VWLSQ) Eq. (2.129). In the "nodal least squares" scheme, we are concerned with interpolating data to the nodes, instead of computing derivatives directly as for the vertex least squares procedure [238, 519].

Similar to VWLSQ method, pseudo-Laplacian (PL) procedure is also an effective alternative to calculate vertex values [149, 200, 251, 395]. The vertex value $X_{v\ell}$ is computed from a weighted sum of the values at the surrounding cell centers, i.e., at the cells j's sharing the vertex $v\ell$.

$$X_{v\ell} = \sum_{j=1}^{N_\ell} \omega_{c,j} X_j \Bigg/ \sum_{j=1}^{N_\ell} \omega_{c,j}$$

The weighting factors $\omega_{c,j}$ are chosen to satisfy the condition of zero pseudo-Laplacian:

$$L(x_{v\ell}) = \sum_{j=1}^{N_\ell} \omega_{c,j}(x_j - x_{v\ell}) = 0$$
$$L(y_{v\ell}) = \sum_{j=1}^{N_\ell} \omega_{c,j}(y_j - y_{v\ell}) = 0 \qquad (2.95)$$
$$L(z_{v\ell}) = \sum_{j=1}^{N_\ell} \omega_{c,j}(z_j - z_{v\ell}) = 0$$

with the notations $\mathbf{x}_{v\ell} = (x_{v\ell}, y_{v\ell}, z_{v\ell})$ and $\mathbf{x}_j = (x_j, y_j, z_j)$. This is a desirable property since the Laplacian of a linear function is exactly zero. The weights are determined by defining

$$\omega_{c,j} = 1 + \Delta\omega_{c,j}$$

The cost-function $C = \sum_{j=1}^{N_\ell} \left(r_j \Delta\omega_{c,j}\right)^2$ is minimized by solving an optimization problem subject to the constraints Eq. (2.95). It should be noted [251] that the factor r_j adopted in the cost-function for the optimization problem effectively allows larger values of weight $\Delta\omega_{c,j}$ for those cells closer to the node in question. Frink et al. [149] used the same implementation without weight r_j.

Employing the method of Lagrangian multipliers, the original optimization problem of minimizing the cost-function C with the constraints in Eq. (2.95), is equivalent to minimizing the function $\mathcal{F}$ defined by

$$\mathcal{F} = C - 2\lambda_x \sum_{j=1}^{N_\ell} \omega_{c,j}(x_j - x_{v\ell}) - 2\lambda_y \sum_{j=1}^{N_\ell} \omega_{c,j}(y_j - y_{v\ell}) - 2\lambda_z \sum_{j=1}^{N_\ell} \omega_{c,j}(z_j - z_{v\ell})$$

which leads to

$$\Delta\omega_{c,j} = \frac{1}{r_j^2}\left(\lambda_x(x_j - x_{v\ell}) + \lambda_y(y_j - y_{v\ell}) + \lambda_z(z_j - z_{v\ell})\right)$$

The solution to the constrained optimization problem yields the Lagrange multipliers

$$\lambda_x = \left[-R_x(I_{yy}I_{zz} - I_{yz}^2) + R_y(I_{xy}I_{zz} - I_{xz}I_{yz}) - R_z(I_{xy}I_{yz} - I_{yy}I_{xz})\right]/D$$

$$\lambda_y = \left[R_x(I_{xy}I_{zz} - I_{xz}I_{yz}) - R_y(I_{xx}I_{zz} - I_{xz}^2) + R_z(I_{xx}I_{yz} - I_{xy}I_{xz})\right]/D$$

$$\lambda_z = \left[-R_x(I_{xy}I_{yz} - I_{yy}I_{xz}) + R_y(I_{xx}I_{yz} - I_{xy}I_{xz}) - R_z(I_{xx}I_{yy} - I_{xy}^2)\right]/D$$

where

$$D = I_{xx}(I_{yy}I_{zz} - I_{yz}^2) - I_{xy}(I_{xy}I_{zz} - I_{xz}I_{yz}) + I_{xz}(I_{xy}I_{yz} - I_{yy}I_{xz})$$

and

$$R_A = \sum_{j=1}^{N_\ell}(A_j - A_{v\ell}),\, I_{AB} = \sum_{j=1}^{N_\ell}\frac{1}{r_j^2}(A_j - A_{v\ell})(B_j - B_{v\ell}),$$

with $A, B \in \{x, y, z\}$. These weights are constructed entirely from geometric information.

For the above gradient calculation procedures, boundary points are treated in the same manner as interior points, possibly through the use of ghost cells. However, the concrete implementations are diversified and depend heavily on the geometrical shape of computational domain and grid type. The interested reader may refer to the references cited in the context.

2.6.1.2 Implicit Green-Gauss Gradient Algorithms

The basic Green-Gauss method is formulated by the midpoint quadrature approximation to the integral definition of the gradient operator with the face-midpoint solution value evaluated by an average of the solutions in the two cells sharing the face, or a distance-weighted interpolation (e.g., [329]). For cell-centered schemes, the method may lose consistency on irregular grids, where the interpolation misses the face-midpoint as the line between the centroids of the two cells does not pass through the face-midpoint (e.g., [130, 381, 446]). Consistency can be recovered by considering the values somewhere on the line linking cell's centroids or evaluating the face value by the average of two values linearly interpolated from the cells using the gradients (which are unknowns). It is consistent and exact for linear functions on arbitrary grids, but it requires iterations to obtain the gradients. Such implicit Green-Gauss gradient algorithms have been created (e.g., [130, 329, 381]). The main difference of such gradient formulation with the standard least-squares method is its implicitness in that the gradient at surrounding point j is needed in order to obtain the gradient at the centroid of current cell i of interest, thus an iterative process is a must (estimated gradients are obtained from previous iterations).

Perez-Segarra et al. [381] proposed a modified least-squares method. The gradient is evaluated minimizing the sum of the squares of the differences of the extrapolated variable values at the midpoint m' between the centroid where the gradient is to be evaluated i and a stencil of surrounding points j, that is, $\mathbf{x}_{m'}$ is the midpoint lying along the line $\overline{\mathbf{x}_i\mathbf{x}_j}$, as shown in Fig. 2.3. First, the value of the variable at the midpoint m' along the line $\overline{\mathbf{x}_i\mathbf{x}_j}$, is extrapolated from i and j:

$$X_{m'^-} = X_i + (\nabla X)|_i \cdot (\mathbf{x}_{m'} - \mathbf{x}_i),\;\; X_{m'^+} = X_j + (\nabla X)|_j \cdot (\mathbf{x}_{m'} - \mathbf{x}_j)$$

Then, the difference of the variable X at point $\mathbf{x}_{m'}$ is

$$D_j = X_{m'^-} - X_{m'^+} = \left(X_i - X_j\right) + (\nabla X)|_i \cdot (\mathbf{x}_{m'} - \mathbf{x}_i) - (\nabla X)|_j \cdot \left(\mathbf{x}_{m'} - \mathbf{x}_j\right)$$

In order to find the gradient at $\mathbf{x}_i$, we minimize the sum of the squares of the weighted differences, i.e.,

$$\frac{\partial \sum_{j\in\{V_j\}_i} \left(\omega_j D_j\right)^2}{\partial \left(\frac{\partial X}{\partial \kappa}\Big|_{\mathbf{x}_i} \right)} = 0, \quad \kappa \in \{x, y, z\}$$

which provides us the following set of equations with three unknowns

$$\mathcal{M} \begin{pmatrix} \frac{\partial X}{\partial x}\Big|_{\mathbf{x}_i} \\ \frac{\partial X}{\partial y}\Big|_{\mathbf{x}_i} \\ \frac{\partial X}{\partial z}\Big|_{\mathbf{x}_i} \end{pmatrix} = \begin{pmatrix} \sum_{j\in\{V_j\}_i} \omega_j^2 \left(x_j - x_i\right) \left(2\left(X_j - X_i\right) - (\nabla X)|_j \cdot \left(\mathbf{x}_j - \mathbf{x}_i\right)\right) \\ \sum_{j\in\{V_j\}_i} \omega_j^2 \left(y_j - y_i\right) \left(2\left(X_j - X_i\right) - (\nabla X)|_j \cdot \left(\mathbf{x}_j - \mathbf{x}_i\right)\right) \\ \sum_{j\in\{V_j\}_i} \omega_j^2 \left(z_j - z_i\right) \left(2\left(X_j - X_i\right) - (\nabla X)|_j \cdot \left(\mathbf{x}_j - \mathbf{x}_i\right)\right) \end{pmatrix} \tag{2.96}$$

with

$$\mathcal{M} = \begin{pmatrix} I_{xx} & I_{xy} & I_{xz} \\ I_{xy} & I_{yy} & I_{yz} \\ I_{xz} & I_{yz} & I_{zz} \end{pmatrix}, \quad I_{AB} = \sum_{j\in\{V_j\}_i} \omega_j^2 \Delta A_j \Delta B_j$$

for instance, $I_{xx} = \sum_{j\in\{V_j\}_i} \omega_j^2 (x_j - x_i)^2$, $I_{xy} = \sum_{j\in\{V_j\}_i} \omega_j^2 (x_j - x_i)(y_j - y_i)$. The weighting factor is taken as the inverse of the distance, $\omega_j = 1/\left|\mathbf{x}_j - \mathbf{x}_i\right|$ (e.g., [308, 381]). Here, $\{V_j\}_i$ is the set of the surrounding neighbors of current cell i. As in the implemented least-squares method, the stencil of points chosen as a standard approximation is the neighbors of the centroid where the gradient is evaluated (see Figs. 2.1 and 2.3).

As can be observed in Eq. (2.96), the array of coefficients can be precomputed, stored and inverted such that the evaluation of the gradient during calculation can be fast. The right-hand side of Eq. (2.96) should be computed at each iteration. Evidently, since the gradient at point j is needed, an iterative process is necessary (estimated gradients are obtained from previous iterations).

Wang et al. [490] proposed the variational reconstruction (VR) method, in which a globally coupled system of linear equations for the gradients is derived by minimizing the solution jumps at faces. This leads to a system of linear equations, which can be iteratively solved along with a finite-volume flow solver by using the LU-SGS method. If one iteration is performed for the gradients per solver iteration, then the cost per iteration is almost the same as the explicit methods such as the GG and LSQ methods, or more efficient than LSQ methods with extended stencils beyond face neighbors. The VR method outwits the stability issue effectively by the gradient stencil spanning the entire grid [276, 287]. By minimizing the jumps of high-order polynomials [276], the VR method can be easily extended to high-order cases. Hexahedral mesh, prismatic mesh and tetrahedral mesh are tested by Li et al. [276] to illustrate that this method can be applied to 3D arbitrary grids. A two-dimensional description was provided in Appendix B of Nishikawa [352]. Accordingly, the VR method is briefly stated below.

Consider a linear polynomial defined over a cell i

$$X_i(\mathbf{x}) = X_i + \nabla X|_i \cdot (\mathbf{x} - \mathbf{x}_i)$$

where X_i is a given cell-averaged solution value, $\mathbf{x}_i$ is the centroid coordinates of the cell, and $\nabla X|_i$ is the gradient to be determined. The polynomials are defined independently at cells, and therefore they do not match over the face that divides two adjacent cells, creating a jump. In the VR method [276, 352, 490], $\nabla X|_i$ is determined by minimizing the following global functional:

$$\mathcal{F} = \sum_{j\in\{V_j\}_i} \frac{1}{2L_{ij}} \int_{A_{ij}} \Delta\mathbf{V}^T \Delta\mathbf{V} dS$$

where $\{V_j\}_i$ is the set of its surrounding neighbors of current cell i, and the integrand denotes the interface jump integration function for a given internal interface A_{ij} separating the left element i and the right element j. L_{ij} is a length scale to be defined at the face. $\Delta\mathbf{V}$ is a weight jump vector defined by

$$\Delta\mathbf{V} = \boldsymbol{\Omega}\Delta\mathbf{X}, \quad \Delta\mathbf{X} = \left(X_R - X_L, \nabla X|_j - \nabla X|_i\right)$$

with X_L and X_R linearly interpolated to the face from cells i and j, respectively, and the diagonal matrix $\boldsymbol{\Omega} = \text{diag}\left(1, \omega_x L_x, \omega_y L_y, \omega_z L_z\right)$ serves to weight the components of $\Delta\mathbf{X}$, where L_x, L_y and L_z are length scales, and ω_x, ω_y and ω_z are weights. If we evaluate the integral over the face by the midpoint rule, it follows that

$$\mathcal{F} = \frac{1}{2} \sum_{j\in\{V_j\}_i} \Delta\mathbf{X}^T \mathbf{Q} \Delta\mathbf{X}, \quad \mathbf{Q} = \frac{\boldsymbol{\Omega}^T \boldsymbol{\Omega} A_{ij}}{L_{ij}}$$

which means that the jumps are evaluated at the face midpoint (face centroid) $\mathbf{x}_{ij,c} = (x_{ij,c}, y_{ij,c}, z_{ij,c})$, as shown in Fig. 2.3. The polynomial coefficients $\mathbf{g}_i = \nabla X|_i = \left.\left(\frac{\partial X}{\partial x}, \frac{\partial X}{\partial y}, \frac{\partial X}{\partial z}\right)\right|_i$ are constrained by the stationary condition:

$$\frac{\partial \mathcal{F}}{\partial \mathbf{g}_i} = \sum_{j\in\{V_j\}_i} \left(\mathbf{Q}\frac{\partial \Delta\mathbf{X}}{\partial \mathbf{g}_i}\right)^T \Delta\mathbf{X} = 0 \tag{2.97}$$

To write out the vector $\Delta\mathbf{X}$, we get

$$\Delta\mathbf{X} = \mathcal{M}_{jm}\mathbf{g}_j - \mathcal{M}_{im}\mathbf{g}_i + (X_j - X_i)\mathbf{e}_1$$

with $\mathbf{e}_1 = (1, 0, 0, 0)$, and

$$\mathcal{M}_{km} = \begin{pmatrix} \Delta x_{km} & \Delta y_{km} & \Delta z_{km} \\ 1 & 0 & 0 \\ 0 & 1 & 0 \\ 0 & 0 & 1 \end{pmatrix} = \begin{pmatrix} x_m - x_k & y_m - y_k & z_m - z_k \\ 1 & 0 & 0 \\ 0 & 1 & 0 \\ 0 & 0 & 1 \end{pmatrix},$$

where $k \in \{i, j\}$, and $\mathbf{x}_m = (x_m, y_m, z_m) = \mathbf{x}_{ij,c}$. After some algebraic manipulation, it follows from Eq. (2.97) that

$$\mathcal{A}_{ii}\mathbf{g}_i - \sum_{j\in\{V_j\}_i} \mathcal{A}_{ij}\mathbf{g}_j = \mathbf{b}_i \tag{2.98}$$

where

$$\mathcal{A}_{ii} = \sum_{j\in\{V_j\}_i} \mathcal{M}_{ij}^T \mathbf{Q} \mathcal{M}_{ij} = \sum_{j\in\{V_j\}_i} \frac{A_{ij}}{L_{ij}} \begin{pmatrix} \Delta x_{im}^2 + \omega_x^2 L_x^2 & \Delta x_{im}\Delta y_{im} & \Delta x_{im}\Delta z_{im} \\ \Delta x_{im}\Delta y_{im} & \Delta y_{im}^2 + \omega_y^2 L_y^2 & \Delta y_{im}\Delta z_{im} \\ \Delta x_{im}\Delta z_{im} & \Delta y_{im}\Delta z_{im} & \Delta z_{im}^2 + \omega_z^2 L_z^2 \end{pmatrix}$$

$$\mathcal{A}_{ij} = \mathcal{M}_{ij}^T \mathbf{Q} \mathcal{M}_{ij} = \frac{A_{ij}}{L_{ij}} \begin{pmatrix} \Delta x_{im}\Delta x_{jm} + \omega_x^2 L_x^2 & \Delta x_{im}\Delta y_{jm} & \Delta x_{im}\Delta z_{jm} \\ \Delta y_{im}\Delta x_{jm} & \Delta y_{im}\Delta y_{jm} + \omega_y^2 L_y^2 & \Delta y_{im}\Delta z_{jm} \\ \Delta x_{jm}\Delta z_{im} & \Delta y_{jm}\Delta z_{im} & \Delta z_{jm}\Delta z_{im} + \omega_z^2 L_z^2 \end{pmatrix}$$

$$\mathbf{b}_i = \sum_{j\in\{V_j\}_i} \frac{A_{ij}}{L_{ij}} \begin{pmatrix} (X_j - X_i)\Delta x_{im} \\ (X_j - X_i)\Delta y_{im} \\ (X_j - X_i)\Delta z_{im} \end{pmatrix}$$

This shows that Eq. (2.98) is in a similar form to the right hand side of a LSQ method. For the 3D hexahedral cell, $(L_x, L_y, L_z) = (x_j - x_i, y_j - y_i, z_j - z_i)$, and $\omega_x = \omega_y = \omega_z = 1$, and $L_{ij} = \max(\text{Rad}(A_{ij}), \sqrt{A_{ij}})$, where $\text{Rad}(A_{ij})$ is the radius of the circumcircle of the face A_{ij}. Alternatively, $L_{ij} = d_{ij} = |\mathbf{x}_i - \mathbf{x}_j|$ (e.g., [276, 373]).

In both Eqs. (2.96) and (2.98), the iterative algorithm for computing gradients can start with setting iteration counter, $k = 0$ and cell-center gradients, $\nabla X|_i^{(0)} = \left(\frac{\partial X}{\partial x}, \frac{\partial X}{\partial y}, \frac{\partial X}{\partial z}\right)\Big|_i^{(0)} = \mathbf{0}$ at all cells or alternatively employing the gradients available at the previous outer iteration/time step in time-dependent finite volume simulations. The iterations are considered to be converged when the L_2 norm of the gradient correction, $||\ \nabla X|^{(k+1)} - \nabla X|^{(k)}\ ||_2$ becomes lesser than a pre-defined tolerance ϵ, say, 10^{-8}.

In the variational reconstruction (e.g., [276, 373, 490]), an interfacial jump integration is defined to measure the jumps of the reconstruction polynomial and its spatial derivatives on each cell interface by minimizing the total interfacial jump integration in the computational domain. On each control volume, the derived system of linear equations to determine the reconstruction polynomials is implicit relations between the coefficients of the reconstruction polynomials defined on a compact stencil involving only the current cell and its direct face neighbors. The remarkable features of the variational reconstruction are that there exists a unique solution to the reconstruction problem and the block Jacobi, Gauss-Seidel, and the successive over-relaxation iterations can be proved to be convergent.

Inspired by the variational reconstruction method, Nishikawa [352] established a similar implicit gradient methodology, but with a totally different approach to deriving an implicit gradient system. By utilizing a linear system of equations for gradients from a hyperbolic diffusion scheme, Nishikawa [352] proposed an implicit gradient algorithm for unstructured grids, with the gradients coupled with face-neighbors on the left hand side, and the classical Green-Gauss gradient formula on the right hand side. The resulting gradient system can be solved efficiently by the block Jacobi iteration or Gauss-Seidel scheme.

Instead of using face values of quantities as required in the standard Green-Gauss (GG) gradient reconstruction, Deka et al. [100] proposed the modified Green-Gauss (MGG) reconstruction that achieves the gradients using the normal derivatives at the cell interfaces. Due to its simplicity and capability of simultaneously achieving cell-centered gradients and face normal derivatives, the MGG procedure is formulated below according to Deka et al. [100].

Apply the Gauss divergence theorem to $\mathbf{u} = (\mathbf{a} \cdot \mathbf{x})\mathbf{q}$, where $\mathbf{a}$ is an arbitrary constant vector, $\mathbf{x}$ the position vector in a stationary reference frame and $\mathbf{q}$ a vector field. By using the standard Green-Gauss method, we can obtain

$$\int_V \nabla \cdot \left[(\mathbf{a} \cdot \mathbf{x})\mathbf{q}\right] dV = \int_{\partial V} \left[(\mathbf{a} \cdot \mathbf{x})\mathbf{q}\right] \cdot \mathbf{n} dA = \int_{\partial V} (\mathbf{a} \cdot \mathbf{x})\mathbf{q} \cdot \mathbf{n} dA \tag{2.99}$$

where ∂V is the boundary surface of control volume V, and $\mathbf{n}$ is the outer unit normal to ∂V. Expanding the left hand-side of Eq. (2.99) leads to,

$$\int_V (\mathbf{a} \cdot \mathbf{x})\nabla \cdot \mathbf{q} dV + \int_V \mathbf{q} \cdot \nabla(\mathbf{a} \cdot \mathbf{x}) dV = \int_{\partial V} (\mathbf{a} \cdot \mathbf{x})\mathbf{q} \cdot \mathbf{n} dA$$

Keep in mind that $\mathbf{a}$ represents an arbitrary constant vector so as to know that the above expression further reduces to,

$$\int_V \mathbf{q} dV + \int_V (\nabla \cdot \mathbf{q})\mathbf{x} dV = \int_{\partial V} (\mathbf{q} \cdot \mathbf{n})\mathbf{x} dA$$

and then rearranging gives,

$$\frac{1}{V}\int_V \mathbf{q} dV = \frac{1}{V}\int_{\partial V} (\mathbf{q} \cdot \mathbf{n})\mathbf{x} dA - \frac{1}{V}\int_V (\nabla \cdot \mathbf{q})\mathbf{x} dV \tag{2.100}$$

Set $\mathbf{x} = \mathbf{x}_i + \mathbf{x}_r$, where $\mathbf{x}_i$ is the position vector of the cell centroid of control volume V_i, so that Eq. (2.100) is rewritten as,

$$\frac{1}{V_i}\int_{V_i}\mathbf{q}dV = \frac{1}{V_i}\int_{\partial V_i}(\mathbf{q}\cdot\mathbf{n})(\mathbf{x}-\mathbf{x}_i)dA + \frac{1}{V_i}\int_{\partial V_i}(\mathbf{q}\cdot\mathbf{n})\mathbf{x}_i dA$$
$$-\frac{1}{V_i}\int_{V_i}(\nabla\cdot\mathbf{q})\mathbf{x}_i dV - \frac{1}{V_i}\int_{V_i}(\nabla\cdot\mathbf{q})\mathbf{x}_r dV \tag{2.101}$$

The integral in the left hand-side of Eq. (2.101) is a volume integral that precisely defines the cell-averaged value of the vector $\mathbf{q}$. The first integral in the right hand-side of Eq. (2.101) is a surface integral that can be approximated by

$$\frac{1}{V_i}\int_{\partial V_i}(\mathbf{q}\cdot\mathbf{n})(\mathbf{x}-\mathbf{x}_i)\,dA = \frac{1}{V_i}\sum_{j\in\{V_j\}_i}\left(\mathbf{x}_{ij,c}-\mathbf{x}_i\right)q_{ij}A_{ij}$$

over the faces forming the cell, where $q_{ij} = \mathbf{q}_{ij}\cdot\mathbf{n}_{ij}$ is a scalar quantity of the inner product between vectors $\mathbf{q}_{ij}$ and $\mathbf{n}_{ij}$, $\mathbf{q}_{ij}$ is the interpolated value at the face centroid $\mathbf{x}_{ij,c} = (x_{ij,c}, y_{ij,c}, z_{ij,c})$, as shown in Fig. 2.3. Here, A_{ij} is the internal interface area of ∂V_i separating the left element V_i and the right element V_j, $\mathbf{n}_{ij}$ is the outer unit normal to ∂V_i pointing from V_i to V_j, and $\{V_j\}_i$ are the direct face-sharing neighbor cells of cell i.

In the right hand-side of Eq. (2.101), the second and third integrals cancel one another, while the fourth integral is again a volume integral for further evaluation. Evidently, this fourth integral is zero if and only if $\nabla\cdot\mathbf{q}$ is a constant, such as the case of a constant vector or a solenoidal field. However, for arbitrary $\mathbf{q}$, this integral has a non-zero contribution, and at the cell-center it follows approximately,

$$\nabla\cdot\mathbf{q} = (\nabla\cdot\mathbf{q})|_i + O(\Delta h) \tag{2.102}$$

from Taylor's series, where Δh is the characteristic length of the cell i defined as the diameter of the smallest inscribed sphere of the cell volume. Substituting Eq. (2.102) into the fourth integral and simplifying by invoking the definition of the centroid lead to, $\frac{1}{V_i}\int_{V_i}(\nabla\cdot\mathbf{q})\mathbf{x}_r dV \sim O(\Delta h^2)$, which can be ignored in a second order accuracy scheme. From Eq. (2.101), the MGG reconstruction is ready to be expressed as

$$\mathbf{q}_i = \frac{1}{V_i}\sum_{j\in\{V_j\}_i} q_{ij}\left(\mathbf{x}_{ij,c}-\mathbf{x}_i\right)A_{ij} \tag{2.103}$$

where all the geometric quantities have the same meanings as defined previously. By utilizing the vector reconstruction expressed by Eq. (2.103), a formula for the centroidal gradients of any scalar ϕ can be derived if taking $\mathbf{q} = \nabla\phi$. The MGG approximate gradient reconstruction then reads,

$$(\nabla\phi)|_i = \frac{1}{V_i}\sum_{j\in\{V_j\}_i}\left.\frac{\partial\phi}{\partial n}\right|_{ij}\left(\mathbf{x}_{ij,c}-\mathbf{x}_i\right)A_{ij} \tag{2.104}$$

while the normal derivative is defined as,

$$\left.\frac{\partial \phi}{\partial n}\right|_{ij} = (\nabla\phi)|_{ij} \cdot \left(\alpha \hat{d}_{ij}\right) + (\nabla\phi)|_{ij} \cdot \left(\mathbf{n}_{ij} - \alpha \hat{d}_{ij}\right) \tag{2.105}$$

where, $\hat{d}_{ij} = \frac{\mathbf{x}_j - \mathbf{x}_i}{|\mathbf{x}_j - \mathbf{x}_i|}$ is an unit vector from i to the cell center of the face-sharing neighbor cell denoted as j and $\alpha = (\mathbf{n}_{ij} \cdot \hat{d}_{ij})$ is a scaling factor. The orthogonal and non-orthogonal contributions in Eq. (2.105) may be numerically evaluated as,

$$\begin{aligned}(\nabla\phi)|_{ij} \cdot \left(\alpha \hat{d}_{ij}\right) &= \alpha \left(\frac{\phi_j - \phi_i}{|\mathbf{x}_j - \mathbf{x}_i|}\right) \\ (\nabla\phi)|_{ij} \cdot \left(\mathbf{n}_{ij} - \alpha \hat{d}_{ij}\right) &= \frac{1}{2}\left((\nabla\phi)|_i + (\nabla\phi)|_j\right) \cdot \left(\mathbf{n}_{ij} - \alpha \hat{d}_{ij}\right)\end{aligned} \tag{2.106}$$

Because the non-orthogonal correction requires the use of gradients at the cell-centers that are in turn the sought quantities themselves, Eq. (2.106) shows that the gradient computation becomes implicit on non-orthogonal meshes. As before, this situation can be resolved with an iterative approach where the gradients appearing in the correction term are lagged. That is, the normal derivative calculations then become,

$$\left.\frac{\partial \phi}{\partial n}\right|_{ij}^{(k+1)} = \alpha \left(\frac{\phi_j - \phi_i}{|\mathbf{x}_j - \mathbf{x}_i|}\right) + \frac{1}{2}\left((\nabla\phi)|_i^{(k)} + (\nabla\phi)|_j^{(k)}\right) \cdot \left(\mathbf{n}_{ij} - \alpha \hat{d}_{ij}\right) \tag{2.107}$$

The iterative algorithm for computing gradients is summarized as below: (1) Set iteration counter, $k = 0$ and cell-center gradients, $\nabla\phi^{(0)} = 0$ at all cells; (2) Compute the terms in Eq. (2.106); (3) Obtain the face normal derivatives by employing the quantities from step 2 in Eq. (2.107); (4) Reconstruct the cell center gradients using Eq. (2.104); (5) Check if convergence is achieved. If yes, stop the iterations else set $k = k + 1$ and go to step 2.

The iterations are assumed to be converged when the L_2 norm of the gradient correction, $||\nabla\phi^{(k+1)} - \nabla\phi^{(k)}||_2$ become lesser than a pre-defined tolerance ϵ, say, 10^{-8}. For time-dependent finite volume simulations, where gradients are to be repetitively computed, the cost of MGG reconstruction due to the iterative approach can be significantly diminished by using a "non-iterative piggybacking" in which the non-orthogonal correction employs the gradients available at the previous outer iteration/time step. This does not, however, compromise on the advantages that MGG reconstruction can offer for practical simulations as seen from the investigated problems [100]. Thus, the computational cost associated with this MGG approach could become reduced in three-dimensional problems if "non-iterative piggybacking" is employed in the time-marching process.

Similarly, both Green-Gauss and MGG reconstructions are derived from the Gauss divergence theorem, however, Green-Gauss reconstruction uses interpolated values of the scalar at the faces, while MGG reconstruction employs the normal derivatives at the faces. Meanwhile, MGG gradient reconstruction requires an iterative approach on non-orthogonal grids, while Green-Gauss reconstruction is a one-step process.

In general context of state variable reconstruction, the gradients are calculated for each state variable individually. The gradients of each element of a solution vector can be achieved by evaluating those of each element of the vector separately. However, for a divergence-free vector solution variable such as magnetic field vector **B**, it is helpful to formulate its gradients simultaneously so that the elements can constrain each other and meet the divergence constraint. A constrained-gradient method was proposed [201, 202] to control magnetic field divergence errors in numerical MHD.

As it is known, the value of $\nabla \cdot \mathbf{B}$ at the centroid of cell i can be expressed by

$$(\nabla \cdot \mathbf{B})_i = \frac{1}{V_i}\int_{V_i} \nabla \cdot \mathbf{B} dV = \frac{1}{V_i}\oint_{V_i} \mathbf{B} \cdot d\mathbf{A} = \frac{1}{V_i}\sum_j \mathbf{B}_{ij} \cdot \mathbf{n}_{ij} A_{ij}, \tag{2.108}$$

where j denotes the neighbor cell sharing the same face with i, $\mathbf{n}_{ij}$ is the face normal pointing from cell i to its neighbor cell j, and A_{ij} is an face area vector (see Figs. 2.1 and 2.3). While, $\mathbf{B}_{ij}$ stands for the face value of the magnetic field at the face centroid $\mathbf{x}_{ij,c}$, which is calculated by $\mathbf{B}_{ij} = \frac{\mathbf{B}_L + \mathbf{B}_R}{2}$, with $\mathbf{B}_L = \mathbf{B}\left(\mathbf{x}_{ij,c}\right) = \mathbf{B}_i + \nabla \otimes \mathbf{B}|_i \cdot \left(\mathbf{x}_{ij,c} - \mathbf{x}_i\right)$ and $\mathbf{B}_R = \mathbf{B}\left(\mathbf{x}_{ij,c}\right) = \mathbf{B}_j + \nabla \otimes \mathbf{B}|_j \cdot \left(\mathbf{x}_{ij,c} - \mathbf{x}_j\right)$. Then, Eq. (2.108) can be rewritten as

$$\begin{aligned}(\nabla \cdot \mathbf{B})_i = \frac{1}{2V_i} \sum_{j \in \{V_j\}_i} &\left(\mathbf{B}_i + (\nabla \otimes \mathbf{B})|_i \cdot \left(\mathbf{x}_{ij,c} - \mathbf{x}_i\right)\right.\\ &\left. + \mathbf{B}|_j + (\nabla \otimes \mathbf{B})_j \cdot \left(\mathbf{x}_{ij,c} - \mathbf{x}_j\right)\right) \cdot \mathbf{n}_{ij} A_{ij},\end{aligned} \tag{2.109}$$

where '$\otimes$' denotes the outer product, and $(\nabla \otimes \mathbf{B})|_i$ and $(\nabla \otimes \mathbf{B})|_j$ are the gradient tensors calculated at cell i and j, respectively

$$(\nabla \otimes \mathbf{B})|_i \cdot \left(\mathbf{x}_{ij,c} - \mathbf{x}_i\right) \equiv (\nabla \otimes \mathbf{B})|_i^T \left(\mathbf{x}_{ij,c} - \mathbf{x}_i\right) = \left.\begin{pmatrix} \frac{\partial B_x}{\partial x} & \frac{\partial B_x}{\partial y} & \frac{\partial B_x}{\partial z} \\ \frac{\partial B_y}{\partial x} & \frac{\partial B_y}{\partial y} & \frac{\partial B_y}{\partial z} \\ \frac{\partial B_z}{\partial x} & \frac{\partial B_z}{\partial y} & \frac{\partial B_z}{\partial z} \end{pmatrix}\right|_i \begin{pmatrix} x_{ij,c} - x_i \\ y_{ij,c} - y_i \\ z_{ij,c} - z_i \end{pmatrix}$$

From Eq. (2.109), we see that $(\nabla \cdot \mathbf{B})_i = 0$ holds, provided that $(\nabla \otimes \mathbf{B})|_i$ satisfies:

$$\begin{aligned}&\sum_{j \in \{V_j\}_i} \left[(\nabla \otimes \mathbf{B})|_i \cdot (\mathbf{x}_{ij,c} - \mathbf{x}_i)\right] \cdot \mathbf{n}_{ij} A_{ij} \\ &\quad = - \sum_{j \in \{V_j\}_i} \left[\mathbf{B}_i + \mathbf{B}_j + (\nabla \otimes \mathbf{B})|_j \cdot \left(\mathbf{x}_{ij,c} - \mathbf{x}_j\right)\right] \cdot \mathbf{n}_{ij} A_{ij}.\end{aligned} \tag{2.110}$$

In order to simplify Eq. (2.110), we introduce the double-dot product, ':' (i.e., $\mathbf{A} : \mathbf{B} = \sum_i \sum_j A_{ij} B_{ij}$), so that Eq. (2.110) becomes

$$(\nabla \otimes \mathbf{B})|_i : \sum_{j\in\{V_j\}_i} \left(\mathbf{x}_{ij,c} - \mathbf{x}_i\right) \otimes \mathbf{n}_{ij} A_{ij}$$
$$= - \sum_{j\in\{V_j\}_i} \left[\mathbf{B}_i + \mathbf{B}_j + (\nabla \otimes \mathbf{B})\,|_j \cdot \left(\mathbf{x}_{ij,c} - \mathbf{x}_j\right)\right] \cdot \mathbf{n}_{ij} A_{ij}. \tag{2.111}$$

By setting

$$S_0 \equiv \sum_{j\in\{V_j\}_i} \left[\mathbf{B}_i + \mathbf{B}_j + (\nabla \otimes \mathbf{B})|_j \cdot \left(\mathbf{x}_{ij,c} - \mathbf{x}_j\right)\right] \cdot \mathbf{n}_{ij} A_{ij}, \tag{2.112}$$

Equation (2.111) is then rewritten as

$$(\nabla \otimes \mathbf{B})|_i : \mathbf{Q} = -S_0, \tag{2.113}$$

where $\mathbf{Q} \equiv \sum_{j\in\{V_j\}_i} \left(\mathbf{x}_{ij,c} - \mathbf{x}_i\right) \otimes \mathbf{n}_{ij} A_{ij}$ only involves the geometrical factors of the cell. After the centroid gradient values of magnetic field vector $\mathbf{B}$ are known, the aim of constrained-gradient method is to further constrain the gradients of $\mathbf{B}$ so as to satisfy $\nabla \cdot \mathbf{B} = 0$ constraint. The specific process is as follows.

After the 'initial' gradient is $(\nabla \otimes \mathbf{B})|_{i,0}$, define

$$(\nabla \otimes \mathbf{B})|_i = (\nabla \otimes \mathbf{B})|_{i,0} + \mathbf{G} \tag{2.114}$$

where $\mathbf{G}$ is a correction term that ensures $(\nabla \otimes \mathbf{B})|_i$ satisfies Eq. (2.113). Substituting Eq. (2.114) into Eq. (2.113), we get an equation for $\mathbf{G}$

$$\mathbf{G} : \mathbf{Q} = -\left[S_0 + (\nabla \otimes \mathbf{B})_{i,0} : \mathbf{Q}\right]. \tag{2.115}$$

In order to make $(\nabla \otimes \mathbf{B})|_i$ as close to $(\nabla \otimes \mathbf{B})|_{i,0}$ as possible, we can choose $\mathbf{G}$ based on minimizing some "penalty function", for example,

$$f_{\text{penalty}} \equiv \mathbf{G} : \mathbf{G}. \tag{2.116}$$

Divide both sides of Eq. (2.115) by $|\mathbf{Q}| = \sqrt{\mathbf{Q} : \mathbf{Q}}$ to obtain

$$\mathbf{G} : \widehat{\mathbf{Q}} = g_0, \tag{2.117}$$

where $\widehat{\mathbf{Q}} = \frac{\mathbf{Q}}{|\mathbf{Q}|}$, and $g_0 = -|\mathbf{Q}| \frac{S_0 + (\nabla\otimes\mathbf{B})|_{i,0} : \mathbf{Q}}{\mathbf{Q}:\mathbf{Q}}$. Further suppose $\mathbf{G} = g_0 \mathcal{X}$, where $\mathcal{X}$ is a quantity to be solved, and then Eq. (2.117) reduces to $\mathcal{X} : \widehat{\mathbf{Q}} = 1$. After these operations, minimizing the "penalty function" in Eq. (2.116) is converted to minimizing the function $f_{\mathcal{X}} = \mathcal{X} : \mathcal{X}$. Assuming that

$$\mathcal{X} = \begin{pmatrix} X_{11} & X_{12} & X_{13} \\ X_{21} & X_{22} & X_{23} \\ X_{31} & X_{32} & X_{33} \end{pmatrix}, \quad \widehat{\mathbf{Q}} = \begin{pmatrix} \widehat{Q}_{11} & \widehat{Q}_{12} & \widehat{Q}_{13} \\ \widehat{Q}_{21} & \widehat{Q}_{22} & \widehat{Q}_{23} \\ \widehat{Q}_{31} & \widehat{Q}_{32} & \widehat{Q}_{33} \end{pmatrix},$$

then we transform the problem into the optimization problem with conditional constraints:

$$\begin{cases} \min(\mathcal{X} : \mathcal{X}) \\ \mathcal{X} : \widehat{\mathbf{Q}} = 1 \end{cases} \tag{2.118}$$

The optimization problem (2.118) can be solved by the Lagrange multiplier method, and the optimum solution is $\mathcal{X} = \widehat{\mathbf{Q}}$. Therefore, $\mathbf{G}$ is expressed as

$$\mathbf{G} = -\mathbf{Q}\left(\frac{S_0 + (\nabla \otimes \mathbf{B})|_{i,0} : \mathbf{Q}}{\mathbf{Q} : \mathbf{Q}}\right). \tag{2.119}$$

By considering Eqs. (2.114) and (2.119) we are now left to solve Eq. (2.114) to obtain $(\nabla \otimes \mathbf{B})|_i$. However, note that the unknown solution $(\nabla \otimes \mathbf{B})|_i$ depends on $\mathbf{G}$, and hence on the neighbouring gradients $(\nabla \otimes \mathbf{B})|_j$ through Eq. (2.112). This leads to a system of linear equations $(\nabla \otimes \mathbf{B})|_i$, which can be calculated through an iterative approach:

$$(\nabla \otimes \mathbf{B})|_i^{(k)} = (\nabla \otimes \mathbf{B})|_{i,0} - \mathbf{Q}\left(\frac{S_0^{(k)} + (\nabla \otimes \mathbf{B})|_{i,0} : \mathbf{Q}}{\mathbf{Q} : \mathbf{Q}}\right), \tag{2.120}$$

$$S_0^{(k)} \equiv \sum_{j \in \{V_j\}_i} \left[\mathbf{B}_i + \mathbf{B}_j + (\nabla \otimes \mathbf{B})|_j^{(k-1)} \cdot \left(\mathbf{x}_{ij,c} - \mathbf{x}_j\right)\right] \cdot \mathbf{n}_{ij} A_{ij}.$$

During the correction procedure, the following steps are taken. First calculate the 'initial' gradients $(\nabla \otimes \mathbf{B})|_{i,0}$ by Green-Guass or least squares method or simply using the estimated gradients obtained from previous time steps along with a finite-volume flow solver. Second, iterate Eq. (2.120) to achieve the targeted $(\nabla \otimes \mathbf{B})|_i$. This converges to the gradient $(\nabla \otimes \mathbf{B})|_i$ satisfying the divergence-free reconstruction of the variables.

In the above description, we just state it for the purpose of evaluating the gradient of magnetic field $\mathbf{B}$. In the implementation of constrained-gradient method, Hopkins [201] also considered the slope limiter involved in the magnetic field reconstruction.

2.6.1.3 Least Squares Gradient Evaluation

Alternatively, the gradient of a solution variable, ∇X, can be determined by applying a least-squares method (e.g., [33, 299, 300, 322, 367, 370]). This technique is suitable for both structured and unstructured mesh and relies on a stencil formed by the nearest and possibly next to nearest neighbouring cells. The basic thought of the least-square reconstruction is to use a series of stencils by Taylor series expansion to get an over-determined equations. Then using the least-squares method to solve this over-determined equations, the solution is just the gradients that we need to carry out the state variable reconstruction. Usually, two styles of stencil are mainly used: edge-based stencil and vertex-based stencil. For a cell, the edge-based stencil data points include cell center itself, its face neighbors' and edge neighbors' cell centers, while the vertex-based stencils contain the vertex neighbors' cell centers. For standard second-order schemes, a edge-based stencil is a natural choice. For the second-order or third-order current schemes, a vertex-based stencil is suggested. In a word, stencils are constructed both in the interior and near boundaries by adding layers of neighbors until the required number of total neighbors is reached.

For a cell-centered discretization of hexahedral grid in three dimensions (3D), the edge-based stencil includes six neighbouring cell centroids sharing a face with cell V_i of interest, which implies to employ total of 7 hexahedral cells including cell V_i of interest and its direct face-sharing neighbors V_j's for the reconstruction of gradients at cell i. While, the vertex-based stencil is formed by joining the nearest twenty-six neighbouring cell centroids, that is, in 3D computational domain there are a total of 27 hexahedral cells including cell V_i of interest and its direct neighbors V_j's for the reconstruction of gradients at cell i.

The unknown gradient $(\nabla X)_i$ in Eq. (2.90) can be computed using the least-squares method. It is expected that Eq. (2.90) conserves the cell average value on every cell of the stencils, which yields

$$\frac{1}{V_i}\int_{V_i} X_i(\mathbf{x})dV = X_i \tag{2.121}$$

$$\frac{1}{V_j}\int_{V_j} X_i(\mathbf{x})dV = X_j \tag{2.122}$$

It can be seen that Eq. (2.90) satisfies Eq. (2.121) automatically and Eq. (2.122) constitutes a set of over-determined linear equation systems which can be solved by the least-squares method. In practice, the geometric weights $\omega_j = \frac{1}{|\mathbf{x}_j - \mathbf{x}_i|}$ are used to specify the relative importance of the control volumes in the stencil, and Eq. (2.122) is modified as

$$\frac{\omega_j}{V_j}\int_{V_j} X_i(\mathbf{x})dV = \omega_j X_j, j \in \{V_j\}_i$$

where $\{V_j\}_i$ is the set of its surrounding neighbors of current cell i utilized for computing gradient $(\nabla X)_i$. The gradient, which is constant at the cell V_i of interest, is found from the following formula:

$$\omega_j(\mathbf{x}_j - \mathbf{x}_i) \cdot (\nabla X)|_i = \omega_j(X_j - X_i), \tag{2.123}$$

From Eq. (2.123), with the notations $\mathbf{x}_i = (x_i, y_i, z_i)$ and $\mathbf{x}_j = (x_j, y_j, z_j)$ we obtain the following over-determined linear system

$$\begin{pmatrix} \omega_1(x_1 - x_i) & \omega_1(y_1 - y_i) & \omega_1(z_1 - z_i) \\ \omega_2(x_2 - x_i) & \omega_2(y_2 - y_i) & \omega_2(z_2 - z_i) \\ \vdots & \vdots & \vdots \\ \omega_N(x_N - x_i) & \omega_N(y_N - y_i) & \omega_N(z_N - z_i) \end{pmatrix} \begin{pmatrix} \left.\frac{\partial X}{\partial x}\right|_{\mathbf{x}_i} \\ \left.\frac{\partial X}{\partial y}\right|_{\mathbf{x}_i} \\ \left.\frac{\partial X}{\partial z}\right|_{\mathbf{x}_i} \end{pmatrix} = \begin{pmatrix} \omega_1(X_1 - X_i) \\ \omega_2(X_2 - X_i) \\ \vdots \\ \omega_N(X_N - X_i) \end{pmatrix}, \tag{2.124}$$

where N is the supporting number of neighbours $\{V_j\}_i$ used in the reconstruction stencil. The cell stencil used for reconstruction may consist of the cell and its face and vertex neighbours, which depends on the grids and flows investigated. Evidently, Eq. (2.124) is an over-determined equations. Namely, the number N of equations is greater than the number of unknowns: N equations but three unknowns. To solve this equation system, we need to use a least-squares method. This equation system (2.124) can be rewritten in matrix form as: $\boldsymbol{\omega}\mathcal{A}\boldsymbol{\chi} = \boldsymbol{\omega}\mathbf{b}$ with $\boldsymbol{\omega} = \text{diag}(\omega_1, \omega_2, \ldots, \omega_N)$ the diagonal weight matrix. The gradient is computed by minimizing the sum of the squared errors shown below

$$||\boldsymbol{\omega}\mathcal{A}\boldsymbol{\chi} - \omega\mathbf{b}||_2 = \sum_j \left(\omega_j \, (\nabla X)|_i \cdot \Delta\mathbf{x}_j - \omega_j \Delta X_j\right)^2,$$

with $\Delta\mathbf{x}_j = \mathbf{x}_j - \mathbf{x}_i$ and $\Delta X_j = X_j - X_i$. According to the mathematical theory, $(\boldsymbol{\omega}\mathcal{A})^T\boldsymbol{\omega}\mathcal{A}\boldsymbol{\chi} = (\boldsymbol{\omega}\mathcal{A})^T\boldsymbol{\omega}\mathbf{b}$ is the normal equations of over-determined equations $\boldsymbol{\omega}\mathcal{A}\boldsymbol{\chi} = \boldsymbol{\omega}\mathbf{b}$. The solutions of the normal equations is also the least square solver of the over-determined equations. Then, just as done by Yalim [505], from the minimization problem one obtains a 3×3 system of linear algebraic equations

$$\mathcal{M} \begin{pmatrix} \left.\frac{\partial X}{\partial x}\right|_{\mathbf{x}_i} \\ \left.\frac{\partial X}{\partial y}\right|_{\mathbf{x}_i} \\ \left.\frac{\partial X}{\partial z}\right|_{\mathbf{x}_i} \end{pmatrix} = \begin{pmatrix} I_{xX} \\ I_{yX} \\ I_{zX} \end{pmatrix}, \tag{2.125}$$

with

$$\mathcal{M} = \begin{pmatrix} I_{xx} & I_{xy} & I_{xz} \\ I_{xy} & I_{yy} & I_{yz} \\ I_{xz} & I_{yz} & I_{zz} \end{pmatrix}, \quad I_{AB} = \sum_{j=1}^{N} \omega_j^2 \Delta A_j \Delta B_j$$

for example,

$$I_{xx} = \sum_{j=1}^{N} \omega_j^2 (x_j - x_i)^2, \qquad I_{xy} = \sum_{j=1}^{N} \omega_j^2 (x_j - x_i)(y_j - y_i),$$

and

$$I_{xX} = \sum_{j=1}^{N} \omega_j^2 (x_j - x_i)(X_j - X_i).$$

The other terms have a similar form. The above terms only depend on grid geometry and hence can be pre-computed and stored.

Solutions to the 3×3 linear system represented by Eq. (2.125) can be readily obtained using Cramer's rule. That is, taking the inverse of the resulting matrix $\mathcal{M}$, and then multiplying both sides of the resulting equation by the inverse matrix, the gradients at $\mathbf{x}_i$ of the solution variables are found as follows:

$$\begin{aligned}
\frac{\partial X}{\partial x} &= \frac{1}{\Delta}\left[(I_{yy}I_{zz} - I_{yz}^2)I_{xX} + (I_{xz}I_{yz} - I_{xy}I_{zz})I_{yX} + (I_{xy}I_{yz} - I_{xz}I_{yy})I_{zX}\right] \\
\frac{\partial X}{\partial y} &= \frac{1}{\Delta}\left[(I_{xz}I_{yz} - I_{xy}I_{zz})I_{xX} + (I_{xx}I_{zz} - I_{xz}^2)I_{yX} + (I_{xy}I_{xz} - I_{xx}I_{yz})I_{zX}\right] \\
\frac{\partial X}{\partial z} &= \frac{1}{\Delta}\left[(I_{xy}I_{yz} - I_{xz}I_{yy})I_{xX} + (I_{xy}I_{xz} - I_{xx}I_{yz})I_{yX} + (I_{xx}I_{yy} - I_{xy}^2)I_{zX}\right]
\end{aligned} \tag{2.126}$$

where $\Delta = |\mathcal{M}| = I_{xx}I_{yy}I_{zz} + 2(I_{xy}I_{xz}I_{yz}) - I_{xx}I_{yz}^2 - I_{zz}I_{xy}^2 - I_{yy}I_{xz}^2$ is the determinant of coefficient matrix of Eq. (2.125). Note that the determinant of the coefficient matrix in Eq. (2.125) can be zero for some cells although the largest stencil possibility is chosen for the reconstruction. For those cells, the gradient is assigned to zero.

Alternatively, we can solve the equation system (2.124) directly through a Gram–Schmidt process, just as Barth did [32, 34, 35]. Rewrite Eq. (2.124) as

$$\begin{pmatrix}\mathbf{L}_1 & \mathbf{L}_2 & \mathbf{L}_3\end{pmatrix}\begin{pmatrix}\left.\frac{\partial X}{\partial x}\right|_{\mathbf{x}_i} \\ \left.\frac{\partial X}{\partial y}\right|_{\mathbf{x}_i} \\ \left.\frac{\partial X}{\partial z}\right|_{\mathbf{x}_i}\end{pmatrix} = \boldsymbol{\omega}\mathbf{b} \tag{2.127}$$

According to Gram–Schmidt process, we can find $\begin{pmatrix}\mathbf{V}_1 & \mathbf{V}_2 & \mathbf{V}_3\end{pmatrix}^T$ such that

$$\begin{pmatrix}\mathbf{V}_1 \\ \mathbf{V}_2 \\ \mathbf{V}_3\end{pmatrix}\begin{pmatrix}\mathbf{L}_1 & \mathbf{L}_2 & \mathbf{L}_3\end{pmatrix} = \begin{pmatrix}1 & 0 & 0 \\ 0 & 1 & 0 \\ 0 & 0 & 1\end{pmatrix}$$

The row vectors $\mathbf{V}_i$ (i=1,2,3) are given by $\mathbf{V}_i = \frac{\mathbf{U}_i}{\mathbf{L}_i \cdot \mathbf{U}_i}$, where

$$\begin{aligned}
\mathbf{U}_1 &= (l_{33}l_{22} - l_{23}l_{23})\mathbf{L}_1 - (l_{33}l_{12} - l_{23}l_{13})\mathbf{L}_2 - (l_{22}l_{13} - l_{23}l_{12})\mathbf{L}_3 \\
\mathbf{U}_2 &= (l_{33}l_{11} - l_{13}l_{13})\mathbf{L}_2 - (l_{33}l_{12} - l_{13}l_{23})\mathbf{L}_1 - (l_{11}l_{23} - l_{13}l_{12})\mathbf{L}_3 \\
\mathbf{U}_3 &= (l_{11}l_{22} - l_{12}l_{12})\mathbf{L}_3 - (l_{22}l_{13} - l_{12}l_{23})\mathbf{L}_1 - (l_{11}l_{23} - l_{12}l_{13})\mathbf{L}_2
\end{aligned}$$

and $l_{ij} = \mathbf{L}_i \cdot \mathbf{L}_j$. Then, the gradients of the solution variables are $\begin{pmatrix}\mathbf{V}_1 & \mathbf{V}_2 & \mathbf{V}_3\end{pmatrix}^T \boldsymbol{\omega}\mathbf{b}$, and the gradient is assigned to zero when $\mathbf{L}_i \cdot \mathbf{U}_i = 0$.

Certainly, Eq. (2.124) can also be solved by adopting other approaches such as singular value orthogonal decomposition (SVD) and Householder QR factorization. These methods are available in LAPACK routines [8] to solve the full rank least-squares problem.

Besides cell-based weighted-least-squares gradient method outlined above, Zhang [519] introduced a vertex-based weighted-least-squares (VWLSQ) gradient reconstruction method for cell-centered FVM, by first calculating vertex gradients and then computing the cell-centered gradients with an arithmetic averaging procedure. In this vertex-based gradient reconstruction method, a weighted-least-squares interpolation is performed at each vertex. It is known that the cell-centered variable X_j could be extrapolated from the variable $X_{v\ell}$ ($\equiv X|_{\mathbf{x}_{v\ell}}$) at a vertex of cell by using the vertex gradient $(\nabla X)_{v\ell}$ ($\equiv (\nabla X)|_{\mathbf{x}_{v\ell}}$),

$$X_j = X_{v\ell} + (\nabla X)_{v\ell} \cdot (\mathbf{x}_j - \mathbf{x}_{v\ell}) \tag{2.128}$$

where $\mathbf{x}_j$ is the cell center connected to the vertex $v\ell$ of cell i of interest, and $\mathbf{x}_{v\ell}$ is the vertex position of cell i. For each cell j connected to vertex $v\ell$, one can establish a linear equation as Eq. (2.128), and then obtain an over-determined weighted system of equations as follows

$$\begin{pmatrix} \omega_1 & \omega_1(x_1 - x_{v\ell}) & \omega_1(y_1 - y_{v\ell}) & \omega_1(z_1 - z_{v\ell}) \\ \vdots & \vdots & \vdots & \vdots \\ \omega_{N_\ell} & \omega_{N_\ell}(x_{N_\ell} - x_{v\ell}) & \omega_{N_\ell}(y_{N_\ell} - y_{v\ell}) & \omega_{N_\ell}(z_{N_\ell} - z_{v\ell}) \end{pmatrix} \begin{pmatrix} X|_{\mathbf{x}_{v\ell}} \\ \frac{\partial X}{\partial x}\Big|_{\mathbf{x}_{v\ell}} \\ \frac{\partial X}{\partial y}\Big|_{\mathbf{x}_{v\ell}} \\ \frac{\partial X}{\partial z}\Big|_{\mathbf{x}_{v\ell}} \end{pmatrix} = \begin{pmatrix} \omega_1 X_1 \\ \vdots \\ \omega_{N_\ell} X_{N_\ell} \end{pmatrix} \tag{2.129}$$

where N_ℓ is the number of cells connected to vertex $v\ell$. The unknown vertex value $X_{v\ell}$ and gradient $(\nabla X)_{v\ell}$ can be achieved by solving the over-determined system (2.129) via a least squares method. In Eq. (2.129), $\omega_j = \frac{1}{|\mathbf{x}_{v\ell} - \mathbf{x}_j|^q}$ $(q \geq 0)$ is inversely proportional to the distance between vertex $v\ell$ and its connected cell j.

The VWLSQ method focuses on the first order spatial derivative $(\nabla X)_{v\ell}$ instead of the vertex value $X_{v\ell}$. Once the vertex gradients are available, the gradients in cells and interfaces can both be obtained by the following arithmetic averaging procedure

$$(\nabla X)|_i = \sum_{v\ell=1}^{N_v} (\nabla X)_{v\ell}/N_v, \; (\nabla X)_{ij} = \sum_{v\ell=1}^{N_v} (\nabla X)_{v\ell}/N_v$$

where N_v is the adjoining vertex number of cell i or interface A_{ij}.

For viscous flow simulations, the facial gradient reconstruction is a necessary step for the computation of viscous fluxes (e.g., [112]). The VWLSQ scheme becomes more efficient for viscous flow simulations because the calculated gradients can be used for computations of both the inviscid and viscous fluxes. For the second-order spatial discretization, this kind of gradient estimation is expected to be first-order accuracy, or constant. Therefore, a directly arithmetic averaging procedure of second-order accuracy is sufficient for the computation of the gradients. In addition, Zhang [519] discussed the efficiency advantage of the vertex-based reconstruction processes by showing how many least-squares problems should be solved on different grid types such as triangle, tetrahedron, quadrangle, and hexahedron.

2.6.1.4 Green-Gauss/Weighted-Least-Squares Hybrid Gradient Reconstruction

Gradient calculation strategies on unstructured meshes may be broadly classified as least-squares based approaches and Green-Gauss based approaches [45]. The least-squares based approaches achieve the derivatives by minimizing the sum of squares of errors between the true and estimated values in a suitably defined neighborhood. These approaches involve the solution to a system of equations in each control volume depending on the order of derivatives involved in the calculations. The gradients on three-dimensional meshes require that the neighborhood be defined with care, particularly on non-uniform and unstructured meshes in order to ensure non-singular matrices in the least-squares reconstruction. There have been extensive investigations (e.g., [100, 110, 111, 308, 456]) of the least-squares approach on unstructured meshes for different grid topologies, with and without weights to assess the accuracy of estimated gradients and their impact on convective flux discretisation, and it has been also reported [308] that both the unweighted and weighted least-squares constructions may produce inaccurate gradients on highly stretched curved meshes.

Green-Gauss based approaches, derived from a variant of the Gauss divergence theorem, typically involve a smaller compact stencil and give closed-form expressions for the solution gradients. These methods require estimates of the solution at the face, which has to use cell-to-face interpolation strategies. However, the accuracy of the interpolation determines the accuracy of the estimated gradients. As shown by Sozer et al. [446], Green-Gauss based approaches with volume (and inverse volume)-weighted interpolation may generate inconsistent gradients on arbitrary polygonal meshes. The so-called consistent gradient approaches may include those using a cell-to-node interpolation based on the linearity-preserving pseudo-Laplacian strategy [62, 200], the nodal projection scheme proposed by Katz and Sankaran [237], the auxiliary control volume approach [37] and the iterative Green-Gauss reconstruction [328]. However, the practical implementation of these consistent reconstruction approaches can not guarantee no problems, as discussed formerly (e.g., [110, 111, 456]).

In current situations, it is difficult to answer which class of approaches is the "best" for gradient estimation and it is only possible to compare the advantages and limitations of various approaches based on the performance for canonical problems. Compared to Green-Gauss based methods, the least-squares approaches have been shown to be inaccurate [308], while Correa et al. [85] found the opposite. Syrakos et al. [456] presented a comprehensive overview of existing gradient reconstruction methods with a critical analysis of some popular reconstruction schemes with an equal emphasis on least-squares and Green-Gauss based approaches. It is beneficial to devise alternate approaches for gradient estimation that harness the merits of least-squares and Green-Gauss based approaches.

By combining the strengths of the cell-based Green-Gauss (GG) with cell-based weighted-least-squares (WLSQ) formula, two gradient reconstruction methods for unstructured meshes are presented by Shima et al. [429, 430]. First, the weight function of WLSQ is obtained from the analogy of Green-Gauss to yield WLSQ(G) method; second, WLSQ(G) and Green-Gauss are smoothly switched using a blending function based on the geometry of each cell to achieve Green-Gauss/Weighted-Least-Squares (GLSQ) method.

Recall that the gradient in the WLSQ is computed to minimize the sum of the squared errors below

$$\sum_j \varpi_j \left((\nabla X)|_i \cdot \Delta \mathbf{x}_j - \Delta X_j \right)^2, \varpi_j > 0$$

Thereafter, denote WLSQ(q) by the method of weighted least squares with the weight function, $\varpi_j = d_{ij}^{-q}, q \geq 0$, using cell center distance $d_{ij} = |\mathbf{x}_j - \mathbf{x}_i|$ between the respective cell centroids.

Under the stationary condition, the following linear equation for the gradient is obtained

$$\mathcal{M} \, (\nabla X)|_i = \sum_j \varpi_j \Delta X_j \Delta \mathbf{x}_j = \sum_j \varpi_j d_{ij} \Delta X_j \hat{d}_{ij} \tag{2.130}$$

or equivalently

$$\mathcal{M} \begin{pmatrix} \left.\frac{\partial X}{\partial x}\right|_{\mathbf{x}_i} \\ \left.\frac{\partial X}{\partial y}\right|_{\mathbf{x}_i} \\ \left.\frac{\partial X}{\partial z}\right|_{\mathbf{x}_i} \end{pmatrix} = \begin{pmatrix} I_{xX} \\ I_{yX} \\ I_{zX} \end{pmatrix} = \begin{pmatrix} \sum_j \varpi_j \Delta x_j \Delta X_j \\ \sum_j \varpi_j \Delta y_j \Delta X_j \\ \sum_j \varpi_j \Delta z_j \Delta X_j \end{pmatrix}, \tag{2.131}$$

with

$$\mathcal{M} = \begin{pmatrix} I_{xx} & I_{xy} & I_{xz} \\ I_{xy} & I_{yy} & I_{yz} \\ I_{xz} & I_{yz} & I_{zz} \end{pmatrix}, \quad I_{AB} = \sum_j \varpi_j \Delta A_j \Delta B_j$$

Here, $\hat{d}_{ij} = \frac{\mathbf{x}_j - \mathbf{x}_i}{|\mathbf{x}_j - \mathbf{x}_i|}$ again represents the unit vector from the centroid of cell i to cell j. For reference to a 3D hexahedral computational cell, see Fig. 2.1.

According to Green-Gauss theory, the cell center gradient can be computed simply as

$$(\nabla X)|_i = \frac{1}{V_i} \sum_{j=1}^{N_{\text{faces}}} \overline{X}_{ij} A_{ij} \mathbf{n}_{ij}$$

where A_{ij} is the surface area for interface A_{ij} of the control volume V_i, and $\mathbf{n}_{ij}$ is the unit normal vector pointing outward from the boundary ∂V_i to its neighbor V_j, and the summation is taken over all the faces enclosing the cell. In numerical computations, cell face values $\overline{X}_{ij}$ can be obtained by the following interpolated value of neighboring cells

$$\overline{X}_{ij} = (1 - \alpha_j) X_i + \alpha_j X_j = X_i + \alpha_j \Delta X_j$$

with $\Delta X_j = X_j - X_i$. Usually α_j is taken to be a half. With these values, the gradient using Green-Gauss is obtained as

$$(\nabla X)|_i = \frac{1}{V_i}\sum_{j=1}^{N_{\text{faces}}} \alpha_j A_{ij} \Delta X_j \mathbf{n}_{ij} \tag{2.132}$$

Comparing the gradient formula (2.130) of WLSQ(q) and (2.132) of Green-Gauss, the terms concerning a surrounding (jth) cell, ΔX_j, in each scheme has the following form:

$$\begin{aligned} &\text{WLSQ}: \mathcal{M}^{-1}\varpi_j d_{ij}\Delta X_j \hat{d}_{ij} \\ &\text{Green-Gauss}: \frac{1}{V_i}\alpha_j A_{ij}\Delta X_j \mathbf{n}_{ij} \end{aligned} \tag{2.133}$$

If the mesh is nearly orthogonal, there holds the approximate relation: $\mathbf{n}_{ij} \approx \hat{d}_{ij}$. With focus on the weight of the jth cell, the constant coefficients $\mathcal{M}^{-1}$ and $\frac{1}{V_i}$ can be neglected. Therefore, one can use the weight function $\varpi_j = \alpha_j A_{ij}/d_{ij}$ for WLSQ, so that the similar surface area weight from the jth cell in Green-Gauss is realized.

When α_j in $\varpi_j = \alpha_j A_{ij}/d_{ij}$ is defined as $\alpha_{wj} = \left(\frac{2l_j}{d_{ij}}\right)^2$, where subscript w stands for WLSQ, and $l_j = |\mathbf{x}_i - \mathbf{x}_{ij,c}|$ is the distance between the cell center and the face center ($\mathbf{x}_{ij,c}$ is the centroid of cell face A_{ij}), the weight function using this factor becomes $\varpi_j = \left(\frac{2l_j}{d_{ij}}\right)^2 A_{ij}/d_{ij} = 4A_{ij}l_j^2 d_{ij}^{-3}$. Obviously, it is equivalent with the weight function of WLSQ(3). Shima et al. [429] proved that WLSQ (3) achieves second order spatial accurate gradient computation in linear coordinates systems including non-orthogonal and non-uniform ones (See Appendix C of Shima et al. [429]).

For a nonorthogonal cell, α_{wj} is extended using the projections of d_{ij} and l_j to the cell face normal $\mathbf{n}_{ij}$, $d'_{ij} = |(\mathbf{x}_j - \mathbf{x}_i)\cdot\mathbf{n}_{ij}|$ and $l'_j = |(\mathbf{x}_i - \mathbf{x}_{ij,c})\cdot\mathbf{n}_{ij}|$ as $\alpha_{wj} = \left(\frac{2l'_j}{d'_{ij}}\right)^2$. By using the weight function $\varpi_j = \left(\frac{2l'_j}{d'_{ij}}\right)^2 A_{ij}/d_{ij}$ in (2.131), a gradient reconstruction named WLSQ(G) is obtained [429, 430]. This WLSQ(G) preserves linear distribution and gives second-order spatial accuracy for a parallel linear mesh. WLSQ(G) exploits the favorable natures of Green-Gauss that come from usage of cell face areas as appropriate weights for the center values of the surrounding cells.

As commented by Shima et al. [429, 430], all WLSQs including WLSQ(G) can exhibit huge reconstruction error in a thin and distorted mesh even if cell center gradient is calculated properly. However, Green-Gauss is more stable in such meshes. Furthermore, they suggested a mixed formula switching between Green-Gauss and WLSQ

$$\left[\beta_i\mathcal{M} + 2(1-\beta_i)V_i\mathbf{I}\right](\nabla X)|_i = \beta_i\sum_j \varpi_j d_{ij}\hat{d}_{ij}\Delta X_j + 2(1-\beta_i)\sum_j \alpha_{Gj}A_{ij}\mathbf{n}_{ij}\Delta X_j \tag{2.134}$$

with $\beta_i \in [0, 1]$ and $\alpha_{Gj} = \frac{1}{2}$. This formula gives Green-Gauss when $\beta_i = 0$ and WLSQ when $\beta_i = 1$. Equation (2.134) is the stationary solution to the following relation

$$\Pi_i = \beta_i \sum_j \varpi_j((\nabla X)|_i \,\Delta\mathbf{x}_j - \Delta X_j)^2 + 2(1-\beta_i)\left\{\Pi_{G-Gx} + \Pi_{G-Gy} + \Pi_{G-Gz}\right\}$$

where $\Pi_{G-Gx} = \sum_j A_{ij} n_{x,ij}(x_{ij,c} - x_i)\left(\alpha_j \frac{\Delta X_j}{x_{ij,c}-x_i} - \frac{\partial X}{\partial x}\right)^2$ with $\mathbf{n}_{ij} = (n_{x,ij}, n_{y,ij}, n_{z,ij})$. The stationary value of Π_{G-Gx} just formally gives the gradient by Green-Gauss in x-direction.

Then, the so-called GLSQ(Green-Gauss/Weighted Least-Squares) is obtained with the following form:

$$\left[\beta_i \mathcal{M} + 2(1-\beta_i)V_i\mathbf{I}\right](\nabla X)|_i = \sum_j \left(\beta_i \alpha_{wj}\hat{d}_{ij} + 2(1-\beta_i)\alpha_{Gj}\mathbf{n}_{ij}\right) A_{ij}\Delta X_j$$

GLSQ can be easily calculated if the matrix is once set and inverted before flow simulations for a time-independent mesh. The small interpolation parameter β_i is expected to be used on thin-and-distorted cells, leading to Green-Gauss.

The choice of β_i should satisfy a kind of monotonicity condition, which depends only on mesh geometry. In Green-Gauss, WLSQ and GLSQ methods, the difference between the cell interface value X_{ij} and the cell center value X_i can be expressed by a linear combination of the finite difference $\Delta X_j = X_j - X_i$ of the cell center values as follows

$$X_{ij} - X_i = \sum_j C_{ij}\Delta X_j$$

For example, the coefficient matrix C_{ij} of GLSQ is

$$C_{ij} = (\mathbf{x}_{ij,c} - \mathbf{x}_i)\left[\beta_i \mathcal{M} + 2(1-\beta_i)V_i\mathbf{I}\right]^{-1}\sum_j \left(\beta_i \alpha_{wj}\hat{d}_{ij} + 2(1-\beta_i)\alpha_{Gj}\mathbf{n}_{ij}\right) A_{ij}$$

Here, the coefficient matrix is required to satisfy the monotonicity condition as

$$|C_{ij}| = \max(|C_{ij}|) < 1$$

In order to choose the interpolation value β_i so as to satisfy the monotonicity condition, certain approximate relation is used in WLSQ(G) to get

$$\beta_i < \min\left(1, \frac{2}{\mathrm{AR}}\right), \quad \mathrm{AR} = \frac{2\max_j|\mathbf{x}_{ij,c} - \mathbf{x}_i| \times \max_j(A_{ij})}{V_i}$$

where AR has been defined as an "effective" aspect ratio for arbitrary polyhedra and polygons, meaning (maximum side length)/(minimum side length). For details, refer to Shima et al. [429, 430].

In summary, the gradient formulation by GLSQ ($\alpha_{Gj} = \frac{1}{2}$) is given by solving the following linear equation (Appendix A of Shima et al. [429])

$$\left[\beta_i \mathcal{M} + 2(1-\beta_i)V_i\mathbf{I}\right] (\nabla X)|_i = \beta_i \sum_j \varpi_j d_{ij}\hat{d}_{ij} \Delta X_j + (1-\beta_i) \sum_j A_{ij}\mathbf{n}_{ij}\Delta X_j$$

$$\beta_i = \min\left(1, \frac{V_i}{\max_j |\mathbf{x}_{ij,c} - \mathbf{x}_i| \times \max_j (A_{ij})}\right),$$

with

$$\mathcal{M} = \begin{pmatrix} I_{xx} & I_{xy} & I_{xz} \\ I_{xy} & I_{yy} & I_{yz} \\ I_{xz} & I_{yz} & I_{zz} \end{pmatrix}, \quad I_{AB} = \sum_j \varpi_j \Delta A_j \Delta B_j, \quad \varpi_j = \left(\frac{2l'_j}{d'_{ij}}\right)^2 A_{ij}/d_{ij}$$

As discussed by Shima et al. [429, 430], only Green-Gauss and GLSQ satisfy the geometrical monotonicity condition, which should be required by a gradient calculation method for robustness. They suggest that the gradient computed by the other methods should be used better with some slope limiter for robustness even if flows are smooth. GLSQ owns the capability by achieving second-order spatial accuracy for cell center gradient in linear grid systems and nearly second-order otherwise; showing exact value for linear distribution except for a very thin mesh; providing reasonable gradient in a very thin mesh.

In the above gradient evaluation methods associated with LSW in Sects. 2.6.1.3 and 2.6.1.4, for the boundary stencil, boundary condition information is used to generalize the procedure without reducing the reconstruction stencil. That is, for the cells sharing one side with the boundary, the use of only the immediate neighbors is not sufficient to overcome the stiffness in the reconstruction. Therefore, for such cells, all cells that share at least one vertex with the current cell can be chosen as the reconstruction stencil. For some concrete implementations, refer to Sect. 2.6.2.

2.6.1.5 Slope Limiting Procedures

To avoid spurious oscillations, the extrapolated boundary values $X_{ij,gp}$ obtained from Eq. (2.90) must be constrained with the application of a limiter function (e.g., [322]). The limited extrapolated boundary values at Gaussian point gp are then used for the left and right states $X_{ij,gp,L}$ and $X_{ij,gp,R}$

$$X_{ij,gp,L} = X_i + \Phi_i\ (\nabla X)|_i \cdot (\mathbf{x}_{ij,gp} - \mathbf{x}_i), X_{ij,gp,R} = X_j + \Phi_j\ (\nabla X)|_j \cdot (\mathbf{x}_{ji,gp} - \mathbf{x}_j), \tag{2.135}$$

where $\Phi \in [0, 1]$ is the limiter function yet to be defined. $\mathbf{x}_{ij,gp}$ is the position vector of the quadrature point gp on the faces A_{ij} of cell V_i, which shares cell i and cell j and should be the same as $\mathbf{x}_{ji,gp}$.

The limited reconstruction of the solution variables to the Gauss quadrature points on the cell faces in a cell, V_i, should be constrained, such that the reconstructed quantities will be bounded. According to Barth and Jespersen [37], the monotonicity criterion is enforced by requiring that the values of the reconstructed polynomials within a control volume should not exceed the maximum and minimum control-volume-averaged solutions of the neighbors of the control volumes including the control volume V_i itself. To this end, researchers have proposed many monotone limiters, among which several such limiters are presented as follows.

(1) Barth-Jespersen limiter (e.g., [37, 313, 504])

Barth and Jespersen suggested that the formation of local extrema at the flux integration Gauss points $\mathbf{x}_{ij,gp}$ should be prevented, i.e., $X_i(\mathbf{x}_{ij,gp}) \in [X_i^{\min}, X_i^{\max}]$, where $X_i^{\min} = \min_j(X_i, X_j)$, $X_i^{\max} = \max_j(X_i, X_j)$. Such limiter function Φ_i, satisfying the condition $X_i(\mathbf{x}_{ij,gp}) \in [X_i^{\min}, X_i^{\max}]$, is equal to $\Phi_i = \min_j(\Phi_{ij})$, with Φ_{ij} the limiter function at the face A_{ij}. The limiter Φ_{ij} corresponding to a solution variable X in cell, V_i, is computed as the minimum of the limiter values corresponding to that variable in all the quadrature points located on the faces of that cell. Barth-Jespersen limiter function reads

$$\Phi_{ij,gp} = \begin{cases} \min\left(1, \frac{X_i^{\max} - X_i}{(\nabla X)|_i \cdot (\mathbf{x}_{ij,gp} - \mathbf{x}_i)}\right), & \text{if } (\nabla X)|_i \cdot (\mathbf{x}_{ij,gp} - \mathbf{x}_i) > 0, \\ \min\left(1, \frac{X_i^{\min} - X_i}{(\nabla X)|_i \cdot (\mathbf{x}_{ij,gp} - \mathbf{x}_i)}\right), & \text{if } (\nabla X)|_i \cdot (\mathbf{x}_{ij,gp} - \mathbf{x}_i) < 0, \\ 1, & \text{if } (\nabla X)|_i \cdot (\mathbf{x}_{ij,gp} - \mathbf{x}_i) = 0. \end{cases} \tag{2.136}$$

Then, the final limiter Φ_i is given as follows:

$$\Phi_i = \min_j(\Phi_{ij}), \; \Phi_{ij} = \min_{gp}(\Phi_{ij,gp}), \tag{2.137}$$

It is noted that Eq. (2.136) has another version with the vertex $\mathbf{x}_{v\ell}$ of cell i instead of Gaussian point $\mathbf{x}_{ij,gp}$. That is, one can obtain $\Phi_i = \min_{v\ell}(\Phi_{v\ell})$, where $\Phi_{v\ell}$ comes from Eq. (2.136) with Gaussian point $\mathbf{x}_{ij,gp}$ replaced by the vertex $\mathbf{x}_{v\ell}$. In this way, the limiter acts to ensure that the values of X at vertices $\mathbf{x}_{v\ell}$ of the cell for which the gradient is being calculated are bounded by the values of X that are used in calculating the gradient.

Slightly different from Barth and Jespersen [37], an alternative expression can be presented by following Kitamura and Shima [252]

$$\Phi_i = \min\left[1, \left(\frac{\max_j |X_j - X_i|}{\max_{ij,gp} \left|(\nabla X)|_i \cdot (\mathbf{x}_{ij,gp} - \mathbf{x}_i)\right|}\right), \left(\frac{\min_j |X_j - X_i|}{\min_{ij,gp} \left|(\nabla X)|_i \cdot (\mathbf{x}_{ij,gp} - \mathbf{x}_i)\right|}\right)\right],$$

and also Wang [491] suggested

$$\Phi_i = \min\left[1, \left(\frac{\max_j |X_j - X_i|}{\max_{v\ell} |(\nabla X)|_i \cdot (\mathbf{x}_{v\ell} - \mathbf{x}_i)|}\right), \left(\frac{\min_j |X_j - X_i|}{\min_{v\ell} |(\nabla X)|_i \cdot (\mathbf{x}_{v\ell} - \mathbf{x}_i)|}\right)\right],$$

where the minimum and maximum over the neighbors j's are found by examining the values of X used in the reconstruction such that a single limiter is then taken to be the minimum of the limiters determined for each variable to satisfy the non-oscillatory criterion for all variables. Further explanations can be found in Refs. [252, 491].

(2) Venkatakrishnan limiter (e.g., [313, 484, 485, 504])

Clearly, the Barth-Jespersen limiter introduces non-differentiability in the computation of the reconstruction polynomials. This non-differentiability will cause the greatest degradation in convergence performance. For this reason, Venkatakrishnan [484] proposed a differentiable alternative by replacing $\min(1, y)$ in Eq. (2.136) with

$$\psi(y) = \frac{y^2 + 2y}{y^2 + y + 2}. \tag{2.138}$$

Now the formula for the Venkatakrishnan limiter is written as

$$\Phi_{ij,gp} = \begin{cases} \psi\left(\frac{X_i^{\max} - X_i}{(\nabla X)|_i \cdot (\mathbf{x}_{ij,gp} - \mathbf{x}_i)}\right), & \text{if } (\nabla X)|_i \cdot (\mathbf{x}_{ij,gp} - \mathbf{x}_i) > 0, \\ \psi\left(\frac{X_i^{\min} - X_i}{(\nabla X)|_i \cdot (\mathbf{x}_{ij,gp} - \mathbf{x}_i)}\right), & \text{if } (\nabla X)|_i \cdot (\mathbf{x}_{ij,gp} - \mathbf{x}_i) < 0, \\ 1, & \text{if } (\nabla X)|_i \cdot (\mathbf{x}_{ij,gp} - \mathbf{x}_i) = 0. \end{cases} \tag{2.139}$$

Just as done for the Barth-Jespersen limiter, the minimum of the limiter values corresponding to a solution variable in all the quadrature points of cell, V_i, is taken to be the limiter corresponding to that variable in that cell. However, one shortcoming of this limiter is that it will lead to accuracy loss relative to the unlimited scheme in regions of nearly uniform flow and smooth extrema. In order to improve accuracy in these regions, Venkatakrishnan uses a method inspired by van Albada et al. [481] to make the switch between limited and unlimited regions of flow using a differentiable function so that convergence properties are not adversely affected (e.g., [313]). For the case $(\nabla X)|_i \cdot (\mathbf{x}_{ij,gp} - \mathbf{x}_i) > 0$, the limiter becomes

$$\Phi_{ij,gp} = \frac{1}{\Delta_-}\left[\frac{(\Delta_+^2 + \epsilon^2)\Delta_- + 2\Delta_-^2\Delta_+}{\Delta_+^2 + 2\Delta_-^2 + \Delta_-\Delta_+ + \epsilon^2}\right], \tag{2.140}$$

with $\Delta_- = (\nabla X)|_i \cdot (\mathbf{x}_{ij,gp} - \mathbf{x}_i)$, $\Delta_+ = X_i^{\max} - X_i$ and $\epsilon^2 = (\mathcal{K}\Delta h)^3$. Δh is the local grid size or characteristic length of the cell V_i, and $\mathcal{K}$ is a tunable positive constant. If the control volume can be approximated as a circle, Δh can be taken to be the diameter of this circle.

Venkatakrishnan takes $\mathcal{K} = 0.3$ and Δh a relevant local grid size. Venkatakrishnan mentions that $\epsilon^2 = 10^{-12}$ is a valid option to revert to the unlimited scheme for smooth solutions. Taking a value of $\mathcal{K} = 0.0$ is known to stall convergence to steady-state, whereas a higher value, such as $\mathcal{K} = 5.0$, leads to oscillations near discontinuities, see Blazek [45] for more details. To prevent division by a very small value, Δ_- in the term $\frac{1}{\Delta_-}$ should be replaced by $\text{sign}(\Delta_-)(|\Delta_-| + \omega)$ with $\omega = 10^{-12}$.

The non-differentiability of a slope limiter usually presents a more unfavorable convergence in uniform regions (than in non-uniform regions). By effectively disabling the limiter in uniform regions, parameter ϵ in Eq. (2.140) indeed improves convergence. Large values of $\mathcal{K}$ are advantageous to accuracy in smooth regions and good convergence, but can lead to significant overshoots near solution discontinuities since the limiter no longer strictly enforces monotonicity for any $\mathcal{K} > 0$. Michalak and Ollivier-Gooch [313] discussed the non-differentiability of the Barth-Jespersen limiter [37] and how Venkatakrishnan limiter [484] only partially addresses the problem. Venkatakrishnan limiter can not provide sufficient accuracy even in smooth regions without any local extrema. While the Barth-Jespersen limiter does satisfy these conditions, its lack of differentiability will make it difficult to achieve a steady-state solution.

To deal with the stagnation region, Kitamura and Shima [252] proposed a simple method that limiters be activated only at shock discontinuities, that is, limiters are turned off at subsonic speeds. By using Barth-Jespersen and Venkatakrishnan limiters in Eqs. (2.136) and (2.139) as the first limiters (denoted by Φ_1), Kitamura and Shima [252] suggested a secondary limiting procedure (denoted by Φ_2) such that limiters be activated only at shock discontinuities. In other words, Kitamura and Shima [252] proposed to turn off the limiter (i.e., $\Phi_2 = 1$) at subsonic speeds. Thus, the modification to the limiter function Φ_1, can be obtained in the following way

$$\Phi_2 = \max\left(\Phi_1, f\left(M_{\max}\right)\right), M_{\max} = \max_j(M_i, M_j)$$

by specifying a function $f(M)$, where j stands for any surrounding cells of the current cell i. Here, M_i, M_j are the Mach numbers in the current cell i and neighbor cell j. The function $f(M)$ should be (1) $0 \leq f(M) \leq 1$ (so that $\Phi_2 \in [0, 1]$), (2) $f(M) = 1$ at subsonic (then, Φ_1 is replaced by $f(M)$) and near zero for the others (then, the original Φ_1 is recovered), (3) continuous and differentiable even at $M = 1$. To satisfy those requirements, the following function

$$f(M) = 0.5\{1 - \tanh[5\pi(M - 1)]\}$$

is adopted among infinite number of choices. Then, the secondary limiting for the existing limiter is given as

$$\Phi_2 = \max\left(\Phi_1, 0.5\left\{1 - \tanh[5\pi(M_{\max} - 1)]\right\}\right), M_{\max} = \max_j(M_i, M_j)$$

Such an implementation is called the second limiter Φ_2 because it is applied after the first limiter Φ_1. As pointed out by Kitamura and Shima [252], this modification is in principle applicable to any given slope limiters and is used right after the usual calculation of the limiter function Φ_1 without further requirement of any prescribed parameters in specific limiters.

(3) Multi-dimensional limiting process (MLP) (e.g., [377, 378])

Along with the maximum principle of multi-dimensional monotonic condition [37, 447, 484], multi-dimensional limiting process (MLP) limiter was introduced on structured grids [249, 513]. MLP limiter properly absorbs multidimensional information by using the MLP condition that satisfies maximum/minimum principles (refer to Sect. 3.2 of Yoon et al. [513]). MLP has been extended to unstructured grids (e.g., [113, 162, 236, 378, 522]). The MLP condition can guarantee high-order spatial accuracy and improve convergence without yielding spurious oscillations. Starting from the MUSCL type reconstruction on unstructured grids followed by the efficient implementation of the MLP condition, Park et al. [378] designed an unstructured MLP limiter. Thereby, Park and Kim [377] constructed three-dimensional unstructured MLP limiter and proved that the limiter obeys local extremum diminishing condition. Especially, well-controlled vertex values at interpolation stage make it possible to produce multi-dimensional monotonic distribution of cell-centered values.

The basic idea of the MLP condition is to control the distribution of both cell-centered and cell-vertex physical properties to mimic the multi-dimensional nature of flow physics. The vertex values $X_{v\ell}$, the reconstructed (or interpolated) values at vertex, are required to satisfy the following MLP condition

$$X_{v\ell} \in [X_{v\ell}^{\min}, X_{v\ell}^{\max}]$$

where $X_{v\ell}^{\min} = \min_{j=1}^{N_\ell} X_j$ and $X_{v\ell}^{\max} = \max_{j=1}^{N_\ell} X_j$ are the minimum and maximum of the cell-averaged values among all neighboring cells sharing the same vertex $v\ell$, and N_ℓ is the number of cells connected to vertex $v\ell$. In principle, the MLP condition can be implemented regardless of grid topology. Since local extrema may appear at cell-vertex, each vertex value should be monitored by the MLP condition. Thus, the MLP range for the multi-dimensional slope limiter should be confined by

$$0 \le \phi \le \max\left(\frac{X_{v\ell}^{\min} - X_i}{(\nabla X)|_i \cdot (\mathbf{x}_{v\ell} - \mathbf{x}_i)}, \frac{X_{v\ell}^{\max} - X_i}{(\nabla X)|_i \cdot (\mathbf{x}_{v\ell} - \mathbf{x}_i)}\right)$$

around the vertex $v\ell$ itself. The gradient $(\nabla X)|_i$ can be evaluated by least-squares approach and Green-Gauss approach.

With this MLP limiting range, the MLP slope limiter can be generally formulated as follows.

$$\Phi_i = \min_{v\ell=1}^{N_v} \begin{cases} \Phi(r_{i,v\ell}), & \text{if } (\nabla X)|_i \cdot (\mathbf{x}_{v\ell} - \mathbf{x}_i) \neq 0 \\ 1, & \text{otherwise} \end{cases}$$

where N_v is the adjoining vertex number of cell i. Here, $r_{i,v\ell}$ is the ratio of the minimum or maximum allowable variation to the estimated variation at the vertex $v\ell$,

$$r_{i,v\ell} = \max\left(\frac{X_{v\ell}^{\min} - X_i}{(\nabla X)|_i \cdot (\mathbf{x}_{v\ell} - \mathbf{x}_i)}, \frac{X_{v\ell}^{\max} - X_i}{(\nabla X)|_i \cdot (\mathbf{x}_{v\ell} - \mathbf{x}_i)}\right)$$

For monotonicity enforcement, Φ should be in the range of $0 \le \Phi(r) \le \min(r_{i,v\ell}, 1)$. It is natural for Φ to take the upper bound of the limiting region, i.e., $\Phi(r) = \min(r, 1)$. Evidently, this limiter is non-differentiable, since it inherits the basic characteristics from Barth-Jespersen limiter. Adapting Venkatakrishnan limiter, the following limiter

$$\Phi\left(\frac{\Delta_+}{\Delta_-}\right)=\frac{1}{\Delta_-}\left[\frac{(\Delta_+^2+\epsilon^2)\Delta_-+2\Delta_-^2\Delta_+}{\Delta_+^2+2\Delta_-^2+\Delta_-\Delta_++\epsilon^2}\right],$$

can also be proposed. Here, $\Delta_-=(\nabla X)|_i\cdot(\mathbf{x}_{v\ell}-\mathbf{x}_i)$, $\Delta_+=X_{v\ell}^{\max}-X_i$ if $\Delta_->0$, and $\Delta_+=X_{v\ell}^{\min}-X_i$ if $\Delta_-<0$. Meanwhile, $\epsilon^2=(\mathcal{K}\Delta h)^3$, which is the same as that of Eq. (2.140), plays the role to distinguish a nearly smooth region from a fluctuating one.

In order to reduce dissipation, Kang et al. [236] suggested that the MLP limiter be only invoked in the vicinity of shock waves/nonlinear discontinuities. For improving shock stability and reducing numerical dissipation in smooth flows, Zhang et al. [522] proposed a MLP-pw (pressure-weighted) limiter by combining weak-MLP (for reducing dissipation) and strict-MLP (for enhancing the numerical stability) conditions with a differentiable pressure weight function.

(4) Weighted biased averaging procedure (WBAP) limiter (e.g., [278, 281])

Choi and Liu [76] created a biased averaging procedure (BAP) to limit the gradient in Eq. (2.90). To improve the BAP limiter, a weighted biased averaging procedure (WBAP) is proposed by Li et al. [281] and Li [278]. WBAP introduces a new biased function to control dissipation property of the biased functions. To enhance its robustness in capturing strong shock waves, WBAP adds nonlinear weight functions in the biased averaging procedure. Also, the WBAP, acting on the scaled differences or gradients of the flow variables, makes the limiting procedure preserve self-similarity and behave exactly as the slope limiter in the MUSCL approach.

To state the WBAP limiter, we write the known gradient, $(\nabla X)|_i$ simply as

$$(\nabla X)|_i=\left(\left.\frac{\partial X}{\partial x}\right|_{\mathbf{x}_i},\left.\frac{\partial X}{\partial y}\right|_{\mathbf{x}_i},\left.\frac{\partial X}{\partial z}\right|_{\mathbf{x}_i}\right)^T\equiv\left((\partial_x X)_i\,,(\partial_y X)_i,(\partial_z X)_i\right)^T$$

Then the WBAP adjusts each derivative component $(\partial_\kappa X)_i$, $\kappa\in\{x,y,z\}$, separately in the following way

$$\widetilde{(\partial_\kappa X)_i}=L\left((\partial_\kappa X)_i\,,(\partial_\kappa X)_1\,,(\partial_\kappa X)_2\,,\ldots,(\partial_\kappa X)_M\right)\tag{2.141}$$

where $(\partial_\kappa X)_j$ ($j=1,2,\ldots,M$) are the first order derivatives of X at the centroids of the direct neighbor cells j's of current cell i. It is noted that the current cell i and its M immediate neighbors sharing face are sufficient for the limiting procedure. In case of 3D hexahedral grid as shown in Fig. 2.1, $M=6$. According to Li et al. [281] and Li [278], the WBAP limiting function L takes the following form

$$L(a_0,a_1,a_2,\ldots,a_M)=a_0\mathcal{W}\left(1,\frac{a_1}{a_0},\frac{a_2}{a_0},\ldots,\frac{a_M}{a_0}\right)\tag{2.142}$$

where a_0 is the first argument of L, and the function $\mathcal{W}$ is called the WBAP limiter which is functionally identical to the limiters of the TVD-MUSCL schemes. The function $\mathcal{W}$ takes the following form:

$$\mathcal{W}^{L1}(1, \theta_1, \ldots, \theta_M) = \begin{cases} \sqrt{\frac{n+M}{n+\sum_{m=1}^{M} \frac{1}{\theta_m^2}}}, & \text{if } \theta_1, \ldots, \theta_M > 0 \\ 0, & \text{otherwise,} \end{cases} \tag{2.143}$$

or

$$\mathcal{W}^{L2}(1, \theta_1, \ldots, \theta_M) = \begin{cases} \frac{n+\sum_{m=1}^{M} \frac{1}{\theta_m}}{n+\sum_{m=1}^{M} \frac{1}{\theta_m^2}}, & \text{if } \theta_1, \ldots, \theta_M > 0 \\ 0, & \text{otherwise.} \end{cases} \tag{2.144}$$

which are called the WBAP-L1 limiter (Eq. (16) of Li et al. [281]), or the WBAP-L2 limiter (Eq. (24) of Li et al. [281]). With the above statements, by using (2.90) we can write the WBAP reconstruction as

$$X_i(\mathbf{x}) = X_i + (\widetilde{\nabla X})_i \cdot (\mathbf{x} - \mathbf{x}_i),$$

with $(\widetilde{\nabla X})_i = (\widetilde{(\partial_x X)_i}, \widetilde{(\partial_y X)_i}, \widetilde{(\partial_z X)_i})^T$ defined by (2.141).

Li et al. [281] pointed out that there has not been theoretical basis for choosing the value of n in Eqs. (2.143) and (2.144) for the multidimensional cases. But a larger n will emphasize more contribution of the current cell (when $n \to \infty$ the limited gradients will be reduced to the unlimited ones). In practice, $n \leq 5$ is sufficient to capture the discontinuities.

The WBAP in Eqs. (2.143) and (2.144) is continuously differentiable and parameter-free. According to Eq. (2.142), only the gradients of the current cell and its immediate neighbors are used in the limiting procedure. Thus, the WBAP limiter is very compact. Of course, if the least squares reconstruction of the gradients is used, more cells may be involved, which makes the actual stencil of the limiting procedure much larger. However, since the gradients are computed and stored at a previous stage of the simulation, the compactness of this limiting procedure can not be affected.

Since the WBAP limiter is applied to the components of the gradients, it is not rotationally invariant. While the Barth-Jespersen limiter [37] and the MLP limiter [378] satisfy the rotational invariance property for scalar quantities but do not for the vector valued quantities. In fact, it is quite difficult to design a finite volume scheme with fully rotational invariance if the governing equations are in terms of the Cartesian velocity components. Therefore, being not rotational invariant is not a particular drawback unique to the WBAP limiting procedure itself.

2.6.2 K-Exact Polynomial Reconstruction

Robust and efficient high resolution schemes are one of the key ingredients for computing large scale hyperbolic conservation laws. High-order accurate reconstruction schemes make it quite feasible to accurately capture complex flow structure with acceptable computational cost. At the same time, non-linearity and hyperbolic feature of these equations require sophisticated numerical treatments across physical discontinuities, particularly in multidimensional flows. The high-order solution accuracy of cell-centered finite volume is usually based on K-exact solution reconstruction, which is the solution exactly represented by Kth degree polynomials (e.g., [34, 36]). The objective of K-exact reconstruction is to find a representation of the solution based on data from neighboring control volumes where the induced truncation error of the solution inside each control volume is on the order of $O(h^{K+1})$, where h is some measure of local grid spacing. Barth and Frederickson [36] presented three criteria for the reconstruction operator: conservation in the mean, K-exact reconstruction, and compact support.

As usual, we denote the cell centroid $\mathbf{x}_i = (x_i, y_i, z_i)$ for each control volume V_i in cell-centered FVM. According to three criteria for the reconstruction [36], for any smooth function $X(\mathbf{x})$ and its control volume averaged values X_i, the K-exact least-squares reconstruction uses a compact stencil in the neighborhood of control volume i to compute an expansion $X_i(\mathbf{x})$ that conserves the cell mean in control volume i and reconstructs exactly polynomials of degree $\leq K$ (equivalently, $X_i(\mathbf{x}) - X(\mathbf{x}) = O(h^{K+1})$). The criterion of compact support enforces locality into the stencil selection. The stencil of control volumes surrounding V_i consists of neighbors that are both spatially near V_i and express topological locality to V_i. The reconstruction procedure must have compact support that involves the number and locality of the neighbouring solution states used in the cell reconstruction.

Conservation of the mean requires that the average of the reconstructed function $X_i(\mathbf{x})$ and the original function $X(\mathbf{x})$ over control volume i be the same:

$$X_i = \frac{1}{V_i}\int_{V_i} X_i(\mathbf{x})d\mathbf{x} = \frac{1}{V_i}\int_{V_i} X(\mathbf{x})d\mathbf{x}$$

The expansion $X_i(\mathbf{x})$ at the reference point $\mathbf{x}_i$ (here the centroid of control volume V_i is used) can be written as

$$\begin{aligned}
X_i(\mathbf{x}) = X|_{\mathbf{x}_i} &+ \left.\frac{\partial X}{\partial x}\right|_{\mathbf{x}_i}(x-x_i) + \left.\frac{\partial X}{\partial y}\right|_{\mathbf{x}_i}(y-y_i) + \left.\frac{\partial X}{\partial z}\right|_{\mathbf{x}_i}(z-z_i) \\
&+ \frac{1}{2}\left.\frac{\partial^2 X}{\partial x^2}\right|_{\mathbf{x}_i}(x-x_i)^2 + \frac{1}{2}\left.\frac{\partial^2 X}{\partial y^2}\right|_{\mathbf{x}_i}(y-y_i)^2 + \frac{1}{2}\left.\frac{\partial^2 X}{\partial z^2}\right|_{\mathbf{x}_i}(z-z_i)^2 \\
&+ \left.\frac{\partial^2 X}{\partial x\partial y}\right|_{\mathbf{x}_i}(x-x_i)(y-y_i) + \left.\frac{\partial^2 X}{\partial x\partial z}\right|_{\mathbf{x}_i}(x-x_i)(z-z_i) \\
&+ \left.\frac{\partial^2 X}{\partial y\partial z}\right|_{\mathbf{x}_i}(y-y_i)(z-z_i) + \cdots \\
&+ \sum_{\substack{p_1=0\\ p_1+p_2+p_3=K}}^{K}\sum_{p_2=0}^{K}\sum_{p_3=0}^{K}\frac{1}{p_1!p_2!p_3!}\left.\frac{\partial^K X}{\partial x^{p_1}\partial y^{p_2}\partial z^{p_3}}\right|_{\mathbf{x}_i}(x-x_i)^{p_1}(y-y_i)^{p_2}(z-z_i)^{p_3},
\end{aligned} \tag{2.145}$$

Assume that the total number of terms $(x - x_i)^{p_1}(y - y_i)^{p_2}(z - z_i)^{p_3}$ is $N_D = \frac{1}{d!}\prod_{n=1}^{d}(K + n)$ with d the number of space dimensions, and N_D takes discrete values of 4, 10, 20, 35, …, for first, second, third, fourth, …, -order accurate spatial derivatives. The order of accuracy of the scheme depends upon the number of terms kept on the right hand side of Eq. (2.145). For example, first term alone leads to first order scheme. Whereas, $K = 1$ leads to second order scheme, which requires the evaluation of first order spatial derivatives of X at the cell centers. The third order scheme is obtained by setting $K = 2$, which requires evaluation of both first and second order derivatives of X with respect to space variable at the cell centers. Thus, quadratic reconstruction in three dimensions requires the calculation of nine derivatives

$$\frac{\partial X}{\partial x}, \frac{\partial X}{\partial y}, \frac{\partial X}{\partial z}, \frac{\partial^2 X}{\partial x^2}, \frac{\partial^2 X}{\partial y^2}, \frac{\partial^2 X}{\partial z^2}, \frac{\partial^2 X}{\partial x \partial y}, \frac{\partial^2 X}{\partial x \partial z}, \frac{\partial^2 X}{\partial y \partial z}$$

at the cell centers.

The basic idea of a high-order solution reconstruction is to find the K-exact polynomial for cell i that fits best with its neighbor cells in the sense of volume-averaged values and also satisfies the constraint that the volume average of the reconstructed polynomial has to be exactly the same as the finite volume solution. Taking the control volume average of the above expansion over control volume i and equating it to the volume average value X_i gives

$$\begin{aligned} X_i = X|_{\mathbf{x}_i} &+ \left.\frac{\partial X}{\partial x}\right|_{\mathbf{x}_i}\overline{x}_i + \left.\frac{\partial X}{\partial y}\right|_{\mathbf{x}_i}\overline{y}_i + \left.\frac{\partial X}{\partial z}\right|_{\mathbf{x}_i}\overline{z}_i \\ &+ \frac{1}{2}\left.\frac{\partial^2 X}{\partial x^2}\right|_{\mathbf{x}_i}\overline{x^2}_i + \frac{1}{2}\left.\frac{\partial^2 X}{\partial y^2}\right|_{\mathbf{x}_i}\overline{y^2}_i + \frac{1}{2}\left.\frac{\partial^2 X}{\partial z^2}\right|_{\mathbf{x}_i}\overline{z^2}_i \\ &+ \left.\frac{\partial^2 X}{\partial x \partial y}\right|_{\mathbf{x}_i}\overline{xy}_i + \left.\frac{\partial^2 X}{\partial x \partial z}\right|_{\mathbf{x}_i}\overline{xz}_i + \left.\frac{\partial^2 X}{\partial y \partial z}\right|_{\mathbf{x}_i}\overline{yz}_i + \cdots \\ &+ \sum_{\substack{p_1=0 \\ p_1+p_2+p_3=K}}^{K}\sum_{p_2=0}^{K}\sum_{p_3=0}^{K}\frac{1}{p_1!p_2!p_3!}\left.\frac{\partial^K X}{\partial x^{p_1}\partial y^{p_2}\partial z^{p_3}}\right|_{\mathbf{x}_i}\overline{x^{p_1}y^{p_2}z^{p_3}}_i, \end{aligned} \tag{2.146}$$

where

$$\overline{x^{p_1}y^{p_2}z^{p_3}}_i = \frac{1}{V_i}\int_{V_i}(x - x_i)^{p_1}(y - y_i)^{p_2}(z - z_i)^{p_3}dV$$

is referred to geometry moment of K-exact reconstruction. Unlike the geometrical cell center, the centroid of volume is of importance in that, up to second-order accuracy, the volume average of any function $X(\mathbf{x})$ can be interchanged with the function point-value evaluated at the centroid $\mathbf{x}_i$. That is, the cell centered value $X(\mathbf{x}_i)$ approximates to the control volume average X_i according to $X_i - X(\mathbf{x}_i) = O(h^2)$. This is just the second-order reconstruction case by the least-squares technique previously introduced in Sect. 2.6.1.3.

To seek these derivatives, we can minimize the error in predicting the mean value of the function for neighboring control volumes in the reconstruction stencil $\{V_j\}_i$. From the reconstructed function $X_i(\mathbf{x})$, we can obtain the volume average value in a control volume V_j,

$$
\begin{aligned}
&\frac{1}{V_j}\int_{V_j} X_i(\mathbf{x})dV \\
&= X|_{\mathbf{x}_i} + \left.\frac{\partial X}{\partial x}\right|_{\mathbf{x}_i}\frac{1}{V_j}\int_{V_j}(x-x_i)dV \\
&\quad + \left.\frac{\partial X}{\partial y}\right|_{\mathbf{x}_i}\frac{1}{V_j}\int_{V_j}(y-y_i)dV + \left.\frac{\partial X}{\partial z}\right|_{\mathbf{x}_i}\frac{1}{V_j}\int_{V_j}(z-z_i)dV \\
&\quad + \left.\frac{\partial^2 X}{\partial x^2}\right|_{\mathbf{x}_i}\frac{1}{2V_j}\int_{V_j}(x-x_i)^2dV + \left.\frac{\partial^2 X}{\partial y^2}\right|_{\mathbf{x}_i}\frac{1}{2V_j}\int_{V_j}(y-y_i)^2dV \\
&\quad + \left.\frac{\partial^2 X}{\partial z^2}\right|_{\mathbf{x}_i}\frac{1}{2V_j}\int_{V_j}(z-z_i)^2dV \\
&\quad + \left.\frac{\partial^2 X}{\partial x\partial y}\right|_{\mathbf{x}_i}\frac{1}{V_j}\int_{V_j}(x-x_i)(y-y_i)dV + \left.\frac{\partial^2 X}{\partial x\partial z}\right|_{\mathbf{x}_i}\frac{1}{V_j}\int_{V_j}(x-x_i)(z-z_i)dV \\
&\quad + \left.\frac{\partial^2 X}{\partial y\partial z}\right|_{\mathbf{x}_i}\frac{1}{V_j}\int_{V_j}(y-y_i)(z-z_i)dV + \cdots \\
&\quad + \sum_{\substack{p_1=0\\p_1+p_2+p_3=K}}^{K}\sum_{p_2=0}^{K}\sum_{p_3=0}^{K}\frac{1}{p_1!p_2!p_3!}\left.\frac{\partial^K X}{\partial x^{p_1}\partial y^{p_2}\partial z^{p_3}}\right|_{\mathbf{x}_i} \\
&\quad \times \frac{1}{V_j}\int_{V_j}(x-x_i)^{p_1}(y-y_i)^{p_2}(z-z_i)^{p_3}dV,
\end{aligned}
$$

To avoid computing moments of each control volume in $\{V_j\}_i$ about V_i, we replace $x-x_i, y-y_i$ and $z-z_i$ with $(x-x_j)+(x_j-x_i)$, $(y-y_j)+(y_j-y_i)$ and $(z-z_j)+(z_j-z_i)$, respectively, and then we obtain the following compact form

$$
\begin{aligned}
\frac{1}{V_j}\int_{V_j} X_i(\mathbf{x})dV &= X|_{\mathbf{x}_i} + \left.\frac{\partial X}{\partial x}\right|_{\mathbf{x}_i}\widehat{x}_{ij} + \left.\frac{\partial X}{\partial y}\right|_{\mathbf{x}_i}\widehat{y}_{ij} + \left.\frac{\partial X}{\partial z}\right|_{\mathbf{x}_i}\widehat{z}_{ij} \\
&\quad + \left.\frac{\partial^2 X}{\partial x^2}\right|_{\mathbf{x}_i}\frac{\widehat{x^2}_{ij}}{2} + \left.\frac{\partial^2 X}{\partial y^2}\right|_{\mathbf{x}_i}\frac{\widehat{y^2}_{ij}}{2} + \left.\frac{\partial^2 X}{\partial z^2}\right|_{\mathbf{x}_i}\frac{\widehat{z^2}_{ij}}{2} \\
&\quad + \left.\frac{\partial^2 X}{\partial x\partial y}\right|_{\mathbf{x}_i}\widehat{xy}_{ij} + \left.\frac{\partial^2 X}{\partial x\partial z}\right|_{\mathbf{x}_i}\widehat{xz}_{ij} + \left.\frac{\partial^2 X}{\partial y\partial z}\right|_{\mathbf{x}_i}\widehat{yz}_{ij} + \cdots \\
&\quad + \sum_{\substack{p_1=0\\p_1+p_2+p_3=K}}^{K}\sum_{p_2=0}^{K}\sum_{p_3=0}^{K}\frac{1}{p_1!p_2!p_3!}\left.\frac{\partial^K X}{\partial x^{p_1}\partial y^{p_2}\partial z^{p_3}}\right|_{\mathbf{x}_i}\widehat{x^{p_1}y^{p_2}z^{p_3}}_{ij},
\end{aligned}
\tag{2.147}
$$

where the geometric terms are of the general form

$$\widehat{x^{p_1}y^{p_2}z^{p_3}}_{ij} = \sum_{l=0}^{p_1}\sum_{m=0}^{p_2}\sum_{n=0}^{p_3}\left(C_{p_1}^{l}C_{p_2}^{m}C_{p_3}^{n}(x_j-x_i)^l(y_j-y_i)^m(z_j-z_i)^n\overline{x^{p_1-l}y^{p_2-m}z^{p_3-n}}_j\right)$$

with $C_p^q = \frac{p!}{q!(p-q)!}$. Equation (2.147) evaluates the average value of the reconstruction $X_i(\mathbf{x})$ for a control volume j. The difference between this evaluation and the actual control volume average X_j is easy to assess. The derivatives at $\mathbf{x}_i$ are chosen to minimize this difference over the stencil $\{V_j\}_i$ in a least-squares sense. The least-squares problem reads

$$\begin{pmatrix} \Gamma_{\text{mean}} \\ \Gamma_1 \\ \Gamma_2 \\ \vdots \\ \Gamma_N \end{pmatrix} \chi = \begin{pmatrix} X_i \\ \omega_1 X_1 \\ \omega_2 X_2 \\ \omega_3 X_3 \\ \vdots \\ \omega_N X_N \end{pmatrix} \tag{2.148}$$

where, $\chi = \left(X\big|_{\mathbf{x}_i}, \frac{\partial X}{\partial x}\big|_{\mathbf{x}_i}, \frac{\partial X}{\partial y}\big|_{\mathbf{x}_i}, \frac{\partial X}{\partial z}\big|_{\mathbf{x}_i}, \frac{1}{2}\frac{\partial^2 X}{\partial x^2}\big|_{\mathbf{x}_i}, \frac{1}{2}\frac{\partial^2 X}{\partial y^2}\big|_{\mathbf{x}_i}, \frac{1}{2}\frac{\partial^2 X}{\partial z^2}\big|_{\mathbf{x}_i}, \frac{\partial^2 X}{\partial x \partial y}\big|_{\mathbf{x}_i},\right.$ $\left.\frac{\partial^2 X}{\partial x \partial z}\big|_{\mathbf{x}_i}, \frac{\partial^2 X}{\partial y \partial z}\big|_{\mathbf{x}_i}, \cdots\right)^T$, $\Gamma_{\text{mean}} = \left(1, \overline{x}_i, \overline{y}_i, \overline{z}_i, \overline{x^2}_i, \overline{y^2}_i, \overline{z^2}_i, \overline{xy}_i, \overline{xz}_i, \overline{yz}_i, \ldots\right)$, and $\Gamma_j = (\omega_j,\ \omega_j\widehat{x}_{ij},\ \omega_j\widehat{y}_{ij},\ \omega_j\widehat{z}_{ij},\ \omega_j\widehat{x^2}_{ij},\ \omega_j\widehat{y^2}_{ij},\ \omega_j\widehat{z^2}_{ij},\ \omega_j\widehat{xy}_{ij},\ \omega_j\widehat{xz}_{ij},\ \omega_j\widehat{yz}_{ij},$ $\cdots)$. Here, N is the number of nearby control volumes in the stencil. The stencil must be large enough to compute the unknown values of polynomial. For the general case of a K-exact polynomial reconstruction, the size of the stencil must be at least $N_D = \frac{1}{6}(K+1)(K+2)(K+3)(\leq N+1)$. That is, it is necessary to have more neighbors in the stencil than unknown derivatives so as to increase the robustness of the reconstruction scheme.

To emphasize geometrically nearby data, the geometric weights ω_j are usually chosen as follows

$$\omega_j = \frac{1}{|\mathbf{x}_j - \mathbf{x}_i|^q}$$

where typically $q \in [0, 2]$. Geometric weights ω_j are used to specify the relative importance of good prediction for various control volumes in the stencil, with the weights based on distance between control volume reference points. Evidently, the weights ω_j, emphasizing neighbors that are geometrically close to the control volume i, have the advantage of reducing the influence of data farther from the reconstruction control volume. It is suggested that q has values of zero, one, or two. Usually, the choice of q does not impact the accuracy of the scheme for isotropic or lightly anisotropic grids but sometimes it does affect its conditioning and stability (e.g., [144, 208, 210, 219, 312]). By the way, Michalak and Ollivier-Gooch [312] proposed that $q = 1$ for the inviscid cases and $q = 0$ for viscous cases.

The first equation in the linear system is the mean constraint, which can be removed with Gauss elimination by replacing the first column of the left hand side with zeroes. Then we obtain an unconstrained least-squares problem of reduced size:

$$\boldsymbol{\omega}\mathcal{A}\boldsymbol{\chi} = \boldsymbol{\omega}\mathbf{b} \tag{2.149}$$

where $\boldsymbol{\omega} = \text{diag}(\omega_1, \omega_2, \ldots, \omega_N)$ is the diagonal weight matrix, and

$$\mathcal{A} = \begin{pmatrix} \mathcal{L}_1 \\ \mathcal{L}_2 \\ \vdots \\ \mathcal{L}_N \end{pmatrix}, \mathbf{b} = \begin{pmatrix} (X_1 - X_i) \\ (X_2 - X_i) \\ \vdots \\ (X_N - X_i) \end{pmatrix}$$

Here, $\boldsymbol{\chi} = \left(\frac{\partial X}{\partial x}\Big|_{\mathbf{x}_i}, \frac{\partial X}{\partial y}\Big|_{\mathbf{x}_i}, \frac{\partial X}{\partial z}\Big|_{\mathbf{x}_i}, \frac{1}{2}\frac{\partial^2 X}{\partial x^2}\Big|_{\mathbf{x}_i}, \frac{1}{2}\frac{\partial^2 X}{\partial y^2}\Big|_{\mathbf{x}_i}, \frac{1}{2}\frac{\partial^2 X}{\partial z^2}\Big|_{\mathbf{x}_i}, \frac{\partial^2 X}{\partial x \partial y}\Big|_{\mathbf{x}_i}, \frac{\partial^2 X}{\partial x \partial z}\Big|_{\mathbf{x}_i}, \frac{\partial^2 X}{\partial y \partial z}\Big|_{\mathbf{x}_i}, \cdots \right)^T$, $\mathcal{L}_j = \left(\left(\widehat{x}_{ij} - \overline{x}_i\right), \left(\widehat{y}_{ij} - \overline{y}_i\right), \left(\widehat{z}_{ij} - \overline{z}_i\right), \left(\widehat{x^2}_{ij} - \overline{x^2}_i\right), \left(\widehat{y^2}_{ij} - \overline{y^2}_i\right), \left(\widehat{z^2}_{ij} - \overline{z^2}_i\right), \left(\widehat{xy}_{ij} - \overline{xy}_i\right), \left(\widehat{xz}_{ij} - \overline{xz}_i\right), \left(\widehat{yz}_{ij} - \overline{yz}_i\right), \cdots \right)$ with its general element $\widehat{x^{p_1}y^{p_2}z^{p_3}}_{ij} - \overline{x^{p_1}y^{p_2}z^{p_3}}_i$, and $j = 1, 2, \ldots, N$. This means that the derivatives $\boldsymbol{\chi}$ at $\mathbf{x}_i$ are chosen to minimize this error over the stencil $\{V_j\}_i$ in a least-squares sense. The least-squares problem (2.149) can be solved for the derivatives of X by using any least-squares technique. If needed, the constant term in the expansion, $X|_{\mathbf{x}_i}$, is evaluated during the back-substitution phase of the least-squares solution, using the constraint equation and the derivatives.

The data matrix $\mathcal{A}$ is the $N \times (N_D - 1)$ coefficient matrix, and the solution $\boldsymbol{\chi}$ of equation $\boldsymbol{\omega}\mathcal{A}\boldsymbol{\chi} = \boldsymbol{\omega}\mathbf{b}$ must be sought in the least-squares sense, i.e., finding an estimate of $\boldsymbol{\chi}$ that renders

$$\min ||\boldsymbol{\omega}\mathcal{A}\boldsymbol{\chi} - \boldsymbol{\omega}\mathbf{b}||_2$$

By rewriting the square of this norm as $(\boldsymbol{\omega}\mathcal{A}\boldsymbol{\chi} - \boldsymbol{\omega}\mathbf{b})^T(\boldsymbol{\omega}\mathcal{A}\boldsymbol{\chi} - \boldsymbol{\omega}\mathbf{b})$, differentiating, and setting it equal to 0, we see that the minimum (of both the norm and its square) occurs at the $\boldsymbol{\chi}$ that satisfies the square system $(\boldsymbol{\omega}\mathcal{A})^T\boldsymbol{\omega}\mathcal{A}\boldsymbol{\chi} = (\boldsymbol{\omega}\mathcal{A})^T\boldsymbol{\omega}\mathbf{b}$, which is called the normal equations. Because the condition number of $(\boldsymbol{\omega}\mathcal{A})^T\boldsymbol{\omega}\mathcal{A}$ is the square of the condition number of $\boldsymbol{\omega}\mathcal{A}$ [167], it may be better to work directly on $\boldsymbol{\omega}\mathcal{A}$ rather than to use the normal equations. There are a variety of ways to efficiently solve the solution of the least-squares problem (2.149), such as, QR factorization and singular value decomposition (SVD) (e.g., [167, 210, 313, 367, 413]).

QR factorization decomposes the left-hand side matrix $\boldsymbol{\omega}\mathcal{A}$ into a product of orthogonal matrix $\mathbf{Q}(\mathbf{Q}^T\mathbf{Q} = \mathbf{I})$ and an upper triangular matrix $\mathbf{R}$, i.e., $\boldsymbol{\omega}\mathcal{A} = \mathbf{QR}$. There are three methods for obtaining the QR factorization: Householder transformations or reflections; Givens transformations or rotations; and the (modified)

Gram–Schmidt procedure (e.g., [105, 315, 449]). After the upper-triangularization is complete, back-substitution yields the required derivatives.

The SVD approach [167] computes a pseudo-inverse matrix, $(\omega\mathcal{A})^\dagger$, with which the solution to the least-squares problem is then given simply by the matrix-vector product $\boldsymbol{\chi} = (\omega\mathcal{A})^\dagger \boldsymbol{\omega}\mathbf{b}$. To be more specific, the singular value decomposition is performed on the coefficient matrix $\boldsymbol{\omega}\mathcal{A}$ and this results in $\boldsymbol{\omega}\mathcal{A} = \mathcal{U}\Sigma\mathbf{V}^T$, where $\mathcal{U}$ is orthogonal $N \times N$ unitary matrix whose columns are left-singular vectors of $\boldsymbol{\omega}\mathcal{A}$, Σ is $N \times (N_D - 1)$ diagonal matrix, i.e., $\text{diag}(\sigma_1, \sigma_2, \ldots, \sigma_{N_D-1})$, whose entries are the singular values of $\boldsymbol{\omega}\mathcal{A}$, and $\mathbf{V}$ is orthogonal $(N_D - 1) \times (N_D - 1)$ unitary matrix whose columns are the right-singular vectors of $\boldsymbol{\omega}\mathcal{A}$. The original over-determined system can be transformed to the normal equation by the pre-multiplication of $(\boldsymbol{\omega}\mathcal{A})^T$, and this results in $(\boldsymbol{\omega}\mathcal{A})^T(\boldsymbol{\omega}\mathcal{A})\boldsymbol{\chi} = (\boldsymbol{\omega}\mathcal{A})^T\boldsymbol{\omega}\mathbf{b}$. Note that the coefficient of the normal equation can be reduced as $(\boldsymbol{\omega}\mathcal{A})^T\boldsymbol{\omega}\mathcal{A} = \mathbf{V}\Sigma\mathbf{V}^T$ by orthogonality of $\mathcal{U}$, and

$$\left((\boldsymbol{\omega}\mathcal{A})^T(\boldsymbol{\omega}\mathcal{A})\right)^{-1} = \mathbf{V}\left(\Sigma^T\Sigma\right)^{-1}\mathbf{V}^T = \mathbf{V}\left(\text{diag}(\frac{1}{\sigma_1^2}, \frac{1}{\sigma_2^2}, \ldots, \frac{1}{\sigma_{N_D-1}^2})\right)\mathbf{V}^T,$$

is also easily computed. This makes possible the pre-computation of $(N_D - 1) \times N$ pseudoinverse matrix $(\boldsymbol{\omega}\mathcal{A})^\dagger = \left((\boldsymbol{\omega}\mathcal{A})^T(\boldsymbol{\omega}\mathcal{A})\right)^{-1}(\boldsymbol{\omega}\mathcal{A})^T$, and finally the solution of the least squares problem can be obtained simply by a matrix-vector multiplication, i.e., $\boldsymbol{\chi} = (\boldsymbol{\omega}\mathcal{A})^\dagger\boldsymbol{\omega}\mathbf{b}$. In the SVD approach, if a fixed stencil is used for all dependent variables, the pseudoinverse matrix $(\boldsymbol{\omega}\mathcal{A})^\dagger$ can be stored and re-used in the reconstruction of all variables, thereby avoiding being repeatedly evaluated.

Additionally, in K-exact reconstruction we are faced with conventional issues related to conditioning and/or invertibility that generally increase with the order of the scheme as well as can rely on mesh features, such as cell size, aspect ratio, and topology. In particular, for highly anisotropic meshes, ill-conditioning issue exacerbates for higher-order computations as the presence of the higher moments results in a considerable difference in the order of magnitude of matrix columns. Jalali and Ollivier Gooch [219] showed that a simple column-scaling procedure can be applied to the least-squares problem to improve the conditioning and make it independent of the mesh size and aspect ratio. This column-scaling conditioning of the least-squares problem is introduced by defining a diagonal column-scaling matrix, $\mathcal{D}_C$, with diagonal entries composed by the inverse of the maximum absolute value of the entries of each column of matrix $\boldsymbol{\omega}\mathcal{A}$. Post multiplication of $\boldsymbol{\omega}\mathcal{A}$ by $\mathcal{D}_C$ yields the product, $\boldsymbol{\omega}\mathcal{A}\mathcal{D}_C$, whose entries are all situated in the range $[-1, 1]$. With the column scaled, the least-squares problem (2.149) can be modified as

$$\boldsymbol{\omega}\mathcal{A}\mathcal{D}_C\Upsilon(\equiv \boldsymbol{\omega}\mathcal{A}\mathcal{D}_C\mathcal{D}_C^{-1}\boldsymbol{\chi}) = \boldsymbol{\omega}\mathbf{b},$$

and the solution for the reconstruction coefficients, $\boldsymbol{\chi}$, results then from the scaled solution of the least-squares problem, Υ, by using $\boldsymbol{\chi} = \mathcal{D}_C\Upsilon$. For instance, when the least-squares problem is solved by the SVD approach, the solution is then given simply by $\boldsymbol{\chi} = \mathcal{D}_C(\boldsymbol{\omega}\mathcal{A}\mathcal{D}_C)^\dagger\boldsymbol{\omega}\mathbf{b}$ where the matrix $\mathcal{D}_C(\boldsymbol{\omega}\mathcal{A}\mathcal{D}_C)^\dagger\boldsymbol{\omega}$ can be pre-computed and stored for re-use. This column-scaling conditioning procedure can be applied to the least-squares problem so as to significantly improve the conditioning, make it virtually independent of the mesh size and aspect ratio, and afford robust and reliable solutions to the least-squares problem (e.g., [144, 210, 219]).

All the methods mentioned above perform well when solving the least squares problem, while the SVD is believed to be quite robust and stable and hence is preferred (e.g., [210, 312, 413]).

Evidently, fixed compact stencil favors the computation of derivatives when using the least squares. The size of the compact stencil is determined by the number of required derivatives. In K-exact polynomial reconstruction, compact support implies that the reconstruction $X_i(\mathbf{x})$ will not require data that is too far physically from $\mathbf{x}_i$. This in turn implies, for reasonable meshes, that data used in constructing $X_i(\mathbf{x})$ can not be topologically far from control volume V_i; that is, there will be a small number of levels of neighbors used to compute $X_i(\mathbf{x})$. Conversely, using only a small number of levels of topologically-connected neighbors in the reconstruction ensures compact physical support as well. If the number of edge-based neighbours is found to be insufficient for proper least squares computations, extended (vertex-based) stencil consisting of vertex neighbours, is used for the WLSQ formulation. Stencils are determined iteratively. The initial stencil consists of first neighbors. For control volumes that need a larger stencil, second neighbors are added. Additional layers of neighbors are added until a large enough stencil has been found for each control volume. Nearly all interior points use first neighbors (control volumes adjacent to V_i) for second order to form the set $\{V_j\}_i$ of control volumes in the reconstruction stencil, and add second neighbors in $\{V_j\}_i$ for third and fourth order. Boundary points often add another layer of neighbors. Because stencil determination is done as a preprocessing step and stored for later use, it is possible to add members to the stencils of control volumes which are too near the boundary to have sufficient support for the desired order of accuracy of reconstruction.

For high-order numerical methods, consistent high-order treatment of boundary conditions is fundamental. In order to achieve high accuracy at boundaries, a high-order prescription for solution values at the boundary is required. By constraining the least-squares reconstruction in control volumes adjacent to the boundary, the boundary data is determined by the interior information of the computational domain and the imposition of boundary conditions at the Gaussian quadrature integration points of cell faces lying along the boundary of the computational domain.

In the following we present boundary condition prescription via constrained reconstruction by including (i) Dirichlet boundary conditions; (ii) Neumann boundary conditions; (iii) Robin boundary conditions; (iv) linear boundary relations between variables that can act as coupled constraints to a set of reconstructed variables. The boundary formulation follows that as described in the literature [208, 210, 221, 265, 340, 367].

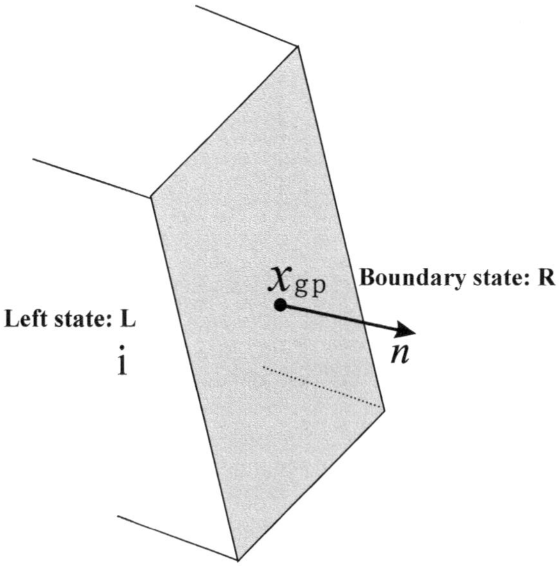

Fig. 2.8 Boundary face, where a boundary condition is imposed through the right state R. $\mathbf{x}_{gp}$ stands for a Gaussian point on the boundary face, and $\mathbf{n}$ denotes the outer normal to the boundary face

(i) Dirichlet boundary conditions

If the solution satisfies a Dirichlet boundary condition $X(\mathbf{x}) = f_{\mathrm{DB}}(\mathbf{x})$ along part of the boundary, one has the following enforcement

$$
\begin{aligned}
f_{\mathrm{DB}}(\mathbf{x}_{gp}) &= X_i(\mathbf{x}_{gp}) \\
&= X|_{\mathbf{x}_i} + \left.\frac{\partial X}{\partial x}\right|_{\mathbf{x}_i}(x_{gp} - x_i) + \left.\frac{\partial X}{\partial y}\right|_{\mathbf{x}_i}(y_{gp} - y_i) + \left.\frac{\partial X}{\partial z}\right|_{\mathbf{x}_i}(z_{gp} - z_i) \\
&+ \frac{1}{2}\left.\frac{\partial^2 X}{\partial x^2}\right|_{\mathbf{x}_i}(x_{gp} - x_i)^2 + \frac{1}{2}\left.\frac{\partial^2 X}{\partial y^2}\right|_{\mathbf{x}_i}(y_{gp} - y_i)^2 + \frac{1}{2}\left.\frac{\partial^2 X}{\partial z^2}\right|_{\mathbf{x}_i}(z_{gp} - z_i)^2 \\
&+ \left.\frac{\partial^2 X}{\partial x \partial y}\right|_{\mathbf{x}_i}(x_{gp} - x_i)(y_{gp} - y_i) + \left.\frac{\partial^2 X}{\partial x \partial z}\right|_{\mathbf{x}_i}(x_{gp} - x_i)(z_{gp} - z_i) \\
&+ \left.\frac{\partial^2 X}{\partial y \partial z}\right|_{\mathbf{x}_i}(y_{gp} - y_i)(z_{gp} - z_i) + \cdots
\end{aligned}
\tag{2.150}
$$

at each Gauss integration point $\mathbf{x}_{gp} = (x_{gp}, y_{gp}, z_{gp})$ on the boundary, as shown in Fig. 2.8. For each control volume adjacent to the boundary, one constraint is added to the least-squares reconstruction problem for each boundary Gauss integration point. As an example, if two Gaussian points $\mathbf{x}_{gp,a} = (x_{gp,a}, y_{gp,a}, z_{gp,a})$ and $\mathbf{x}_{gp,b} = (x_{gp,b}, y_{gp,b}, z_{gp,b})$ reside on the boundary face, the enforcement of the boundary constraint (2.150) at these two Gaussian points yields the resulting equalities

$$\begin{pmatrix} \Gamma_{\text{mean}} \\ \Gamma_{\text{DB},gp,a} \\ \Gamma_{\text{DB},gp,b} \\ \Gamma_1 \\ \Gamma_2 \\ \vdots \\ \Gamma_{N_{bd}} \end{pmatrix} \chi = \begin{pmatrix} X_i \\ f_{\text{DB}}(\mathbf{x}_{gp,a}) \\ f_{\text{DB}}(\mathbf{x}_{gp,b}) \\ \omega_1 X_1 \\ \omega_2 X_2 \\ \vdots \\ \omega_{N_{bd}} X_{N_{bd}} \end{pmatrix}$$

which is an enlarged system of the reconstruction problem (2.148). Here, $\Gamma_{\text{DB},gp,a/b} = (1,\ x_{gp,a/b} - x_i,\ y_{gp,a/b} - y_i,\ z_{gp,a/b} - z_i,\ (x_{gp,a/b} - x_i)^2,\ (y_{gp,a/b} - y_i)^2,\ (z_{gp,a/b} - z_i)^2,\ (x_{gp,a/b} - x_i)(y_{gp,a/b} - y_i),\ (x_{gp,a/b} - x_i)(z_{gp,a/b} - z_i),\ (y_{gp,a/b} - y_i)(z_{gp,a/b} - z_i),\ \cdots)$. Meanwhile, χ, Γ_{mean} and Γ_j are given in Eq. (2.148) with $j = 1, 2, \cdots, N_{bd}$. N_{bd} is the number of the neighbours adjacent to the boundary cell i. Obviously, there are three equality constraints on the reconstruction for this control volume: conservation of the mean and two boundary constraints.

(ii) Neumann boundary conditions

Neumann boundary conditions can be implemented in an analogous way. If the solution satisfies a Neumann boundary condition $\frac{\partial X}{\partial \mathbf{n}}(\mathbf{x}) = f_{\text{NB}}(\mathbf{x})$ along part of the boundary, then the following condition

$$\begin{aligned} f_{\text{NB}}(\mathbf{x}_{gp}) &= \frac{\partial X_i(\mathbf{x}_{gp})}{\partial \mathbf{n}} = \nabla X_i(\mathbf{x}_{gp}) \cdot \mathbf{n} \\ &= n_x \left(\left.\frac{\partial X}{\partial x}\right|_{\mathbf{x}_i} + \left.\frac{\partial^2 X}{\partial x^2}\right|_{\mathbf{x}_i} (x_{gp} - x_i) \right. \\ &\qquad \left. + \left.\frac{\partial^2 X}{\partial x \partial y}\right|_{\mathbf{x}_i} (y_{gp} - y_i) + \left.\frac{\partial^2 X}{\partial x \partial z}\right|_{\mathbf{x}_i} (z_{gp} - z_i) + \cdots \right) \\ &+ n_y \left(\left.\frac{\partial X}{\partial y}\right|_{\mathbf{x}_i} + \left.\frac{\partial^2 X}{\partial y^2}\right|_{\mathbf{x}_i} (y_{gp} - y_i) \right. \\ &\qquad \left. + \left.\frac{\partial^2 X}{\partial x \partial y}\right|_{\mathbf{x}_i} (x_{gp} - x_i) + \left.\frac{\partial^2 X}{\partial y \partial z}\right|_{\mathbf{x}_i} (z_{gp} - z_i) + \cdots \right) \\ &+ n_z \left(\left.\frac{\partial X}{\partial z}\right|_{\mathbf{x}_i} + \left.\frac{\partial^2 X}{\partial z^2}\right|_{\mathbf{x}_i} (z_{gp} - z_i) \right. \\ &\qquad \left. + \left.\frac{\partial^2 X}{\partial x \partial z}\right|_{\mathbf{x}_i} (x_{gp} - x_i) + \left.\frac{\partial^2 X}{\partial y \partial z}\right|_{\mathbf{x}_i} (y_{gp} - y_i) + \cdots \right) \end{aligned}$$

can be enforced at each Gauss integration point $\mathbf{x}_{gp}$ on the boundary face. As usual, $\mathbf{n} = (n_x, n_y, n_z)$ is the unit boundary normal. For reference, see Fig. 2.8. Completely analogous to the Dirichlet case, the constrained least-squares system is now given by

$$\begin{pmatrix} \Gamma_{\text{mean}} \\ \Gamma_{\text{NB},gp,a} \\ \Gamma_{\text{NB},gp,b} \\ \Gamma_1 \\ \Gamma_2 \\ \vdots \\ \Gamma_{N_{bd}} \end{pmatrix} \boldsymbol{\chi} = \begin{pmatrix} X_i \\ f_{\text{NB}}(\mathbf{x}_{gp,a}) \\ f_{\text{NB}}(\mathbf{x}_{gp,b}) \\ \omega_1 X_1 \\ \omega_2 X_2 \\ \vdots \\ \omega_{N_{bd}} X_{N_{bd}} \end{pmatrix}$$

which is an enlarged system of the reconstruction problem (2.148). Here, $\Gamma_{\text{NB},gp,a/b} = (0, n_x, n_y, n_z, \; 2n_x(x_{gp,a/b}-x_i), 2n_y(y_{gp,a/b}-y_i), 2n_z(z_{gp,a/b}-z_i), n_x(y_{gp,a/b}-y_i) + n_y(x_{gp,a/b}-x_i), n_x(z_{gp,a/b}-z_i) + n_z(x_{gp,a/b}-x_i), n_y(z_{gp,a/b}-z_i) + n_z(y_{gp,a/b}-y_i), \cdots)$. At the same time, $\boldsymbol{\chi}$, Γ_{mean} and Γ_j are given in Eq. (2.148) with $j = 1, 2, \ldots, N_{bd}$. N_{bd} is the number of the neighbours adjacent to the boundary cell i.

(iii) Robin boundary conditions

Robin boundary conditions are a weighted combination of Dirichlet boundary conditions and Neumann boundary conditions. Robin boundary data for a dependent variable, X, in terms of a K-exact reconstruction polynomial, $X(\mathbf{x})$, at a given quadrature point, $\mathbf{x}_{gp}$, is given by

$$f_{\text{RB}}(\mathbf{x}_{gp}) = g_1(\mathbf{x}_{gp})X_i(\mathbf{x}_{gp}) + g_2(\mathbf{x}_{gp})\frac{\partial X_i(\mathbf{x}_{gp})}{\partial \mathbf{n}}$$

where $g_1(\mathbf{x}_{gp})$ and $g_2(\mathbf{x}_{gp})$ are the coefficients defining the contributions of the Dirichlet and Neumann components to the boundary value, $f_{\text{RB}}(\mathbf{x}_{gp})$, respectively. Robin boundary conditions are imposed by

$$\begin{aligned} f_{\text{RB}}(\mathbf{x}_{gp}) &= g_1(\mathbf{x}_{gp})X_i(\mathbf{x}_{gp}) + g_2(\mathbf{x}_{gp})\frac{\partial X_i(\mathbf{x}_{gp})}{\partial \mathbf{n}} \\ &= g_1(\mathbf{x}_{gp})\left(X|_{\mathbf{x}_i} + \left.\frac{\partial X}{\partial x}\right|_{\mathbf{x}_i}(x_{gp}-x_i) + \left.\frac{\partial X}{\partial y}\right|_{\mathbf{x}_i}(y_{gp}-y_i) + \left.\frac{\partial X}{\partial z}\right|_{\mathbf{x}_i}(z_{gp}-z_i) \right. \\ &+ \frac{1}{2}\left.\frac{\partial^2 X}{\partial x^2}\right|_{\mathbf{x}_i}(x_{gp}-x_i)^2 + \frac{1}{2}\left.\frac{\partial^2 X}{\partial y^2}\right|_{\mathbf{x}_i}(y_{gp}-y_i)^2 + \frac{1}{2}\left.\frac{\partial^2 X}{\partial z^2}\right|_{\mathbf{x}_i}(z_{gp}-z_i)^2 \\ &+ \left.\frac{\partial^2 X}{\partial x \partial y}\right|_{\mathbf{x}_i}(x_{gp}-x_i)(y_{gp}-y_i) + \left.\frac{\partial^2 X}{\partial x \partial z}\right|_{\mathbf{x}_i}(x_{gp}-x_i)(z_{gp}-z_i) \\ &+ \left.\left.\frac{\partial^2 X}{\partial y \partial z}\right|_{\mathbf{x}_i}(y_{gp}-y_i)(z_{gp}-z_i) + \cdots \right) \\ &+ g_2(\mathbf{x}_{gp})\left(n_x \left(\left.\frac{\partial X}{\partial x}\right|_{\mathbf{x}_i} + \left.\frac{\partial^2 X}{\partial x^2}\right|_{\mathbf{x}_i}(x_{gp}-x_i) \right.\right. \end{aligned} \tag{2.151}$$

$$
\begin{aligned}
&+ \left.\frac{\partial^2 X}{\partial x \partial y}\right|_{\mathbf{x}_i} (y_{gp} - y_i) + \left.\frac{\partial^2 X}{\partial x \partial z}\right|_{\mathbf{x}_i} (z_{gp} - z_i) + \cdots \Bigg) \\
&+ n_y \left(\left.\frac{\partial X}{\partial y}\right|_{\mathbf{x}_i} + \left.\frac{\partial^2 X}{\partial y^2}\right|_{\mathbf{x}_i} (y_{gp} - y_i) \right. \\
&+ \left. \left.\frac{\partial^2 X}{\partial x \partial y}\right|_{\mathbf{x}_i} (x_{gp} - x_i) + \left.\frac{\partial^2 X}{\partial y \partial z}\right|_{\mathbf{x}_i} (z_{gp} - z_i) + \cdots \right) \\
&+ n_z \left(\left.\frac{\partial X}{\partial z}\right|_{\mathbf{x}_i} + \left.\frac{\partial^2 X}{\partial z^2}\right|_{\mathbf{x}_i} (z_{gp} - z_i) + \left.\frac{\partial^2 X}{\partial x \partial z}\right|_{\mathbf{x}_i} (x_{gp} - x_i) \right. \\
&+ \left. \left.\frac{\partial^2 X}{\partial y \partial z}\right|_{\mathbf{x}_i} (y_{gp} - y_i) + \cdots \right) \Bigg)
\end{aligned}
$$

By adding the constraint of Eq. (2.151) to the system of linear equations for the reconstruction problem (2.148), one can obtain the following least-squares problem for a boundary cell

$$
\begin{pmatrix} \Gamma_{\text{mean}} \\ \Gamma_{\text{RB},gp,a} \\ \Gamma_{\text{RB},gp,b} \\ \Gamma_1 \\ \Gamma_2 \\ \vdots \\ \Gamma_{N_{bd}} \end{pmatrix} \boldsymbol{\chi} = \begin{pmatrix} X_i \\ f_{\text{RB}}(\mathbf{x}_{gp,a}) \\ f_{\text{RB}}(\mathbf{x}_{gp,b}) \\ \omega_1 X_1 \\ \omega_2 X_2 \\ \vdots \\ \omega_{N_{bd}} X_{N_{bd}} \end{pmatrix}
$$

where $\Gamma_{\text{RB},gp,a/b} = g_1(\mathbf{x}_{gp,a/b})\Gamma_{\text{DB},gp,a/b} + g_2(\mathbf{x}_{gp,a/b})\Gamma_{\text{NB},gp,a/b}$. Again, $\boldsymbol{\chi}$, Γ_{mean} and Γ_j are the same as those in Eq. (2.148) with $j = 1, 2, \ldots, N_{bd}$.

(iv) Imposition of boundary conditions via coupled linear constraints

Boundary condition prescription via constrained reconstruction can be used to impose boundary conditions via coupled linear constraints. To illustrate this, consider the specification of boundary data for $\nabla \cdot (\rho \mathbf{v}) = 0$, which means that solar plasma is permitted to freely leave the reservoir, but no "backflow" is allowed [175]. This condition also contributes one additional coupled linear constraint at every Gauss integration point to the associated reconstruction problem (2.148) of each component of $\rho \mathbf{v}$, which leads to simultaneous over-determined system of equations for the three moment components $\rho \mathbf{v}$. Least-squares system about the imposition of boundary conditions via coupled linear constraints for several types of physical boundary can be found in the literature [208, 210, 221, 339, 340].

By constraining the least-squares reconstruction in control volumes near boundaries, boundary condition prescription via constrained reconstruction yields additional constraints on the cell-centred reconstruction procedure for interior cells adjacent to the boundary. The resulting over-determined system can then be solved in a least-squares sense while ensuring that the constraint is exactly satisfied. Gauss elimination is used for all constraint rows, including elimination from the least-squares rows of the problem; with multiple constraints, pivoting is an essential adjunct to the Gauss elimination. After the Gauss elimination, the remaining least-squares problem can again be solved by QR factorization, with back substitution including the constraint rows. This gives a least-squares reconstruction to achieve the cell gradients that conserve the mean and satisfy the boundary conditions. Note that on the Gauss point of boundary faces, there is only one solution and gradient reconstructed from one side for each variable and thus the normal fluxes can be directly found.

However, a major obstacle in developing a high-order solver is that there are many more details that must be exactly right for a nominally-high-order scheme to be genuinely-high-order than for a second-order scheme to be second-order. In many cases, these details are either omitted or glossed over in the literature. For unstructured-mesh finite volume solver, a high-order-accurate reconstruction, flux integration, and boundary treatment must be correctly implemented. For such discussion, refer to Ollivier-Gooch et al. [368].

When the reconstruction stencil contains control volumes on opposite sides of the discontinuity-data that are not smoothly connected, K-exact least-squares polynomial reconstruction of piecewise constant control volume averaged data may produce spurious oscillations in the reconstruction in the vicinity of sharp gradients and discontinuities, which can in turn produce stability problems in the numerical schemes. In order to suppress spurious oscillations resulting from reconstruction over a compact stencil with discontinuity-data, monotonicity enforcement must be taken by reducing the computed gradients (slope limiting). For such purposes, many monotonicity procedures have been developed, and several of them are presented below.

2.6.2.1 Michalak–Ollivier-Gooch Limiter

Following the framework introduced by the Barth-Jespersen limiter and Venkatakrishnan limiter, there are some improvements. Nejat and Ollivier-Gooch [340] introduced hyperbolic tangent function in their application of Venkatakrishnan limiter, by which the limiter only activates in limited region. Michalak and Ollivier-Gooch [313] further improved this method. Here we follow the high-order accurate reconstruction procedure documented by Nejat and Ollivier-Gooch [340] and Michalak and Ollivier-Gooch [313].

Keep in mind that the control volume average values X_i and reference-point values $X|_{\mathbf{x}_i}$ are different when devising a limiter for high-order accuracy reconstructions. For convenience, the reconstruction polynomial in Eq. (2.145) is rewritten in terms of the control volume average with the help of Eq. (2.146) to yield:

$$
\begin{aligned}
X_i(\mathbf{x}) =& X_i + \left(\left.\frac{\partial X}{\partial x}\right|_{\mathbf{x}_i} ((x - x_i) - \overline{x}_i) \right. \\
&\left. + \left.\frac{\partial X}{\partial y}\right|_{\mathbf{x}_i} ((y - y_i) - \overline{y}_i) + \left.\frac{\partial X}{\partial z}\right|_{\mathbf{x}_i} ((z - z_i) - \overline{z}_i) \right)
\end{aligned}
$$

$$+\left(\frac{1}{2}\left.\frac{\partial^2 X}{\partial x^2}\right|_{\mathbf{x}_i}\left((x-x_i)^2-\overline{x^2}_i\right)\right.$$
$$+\frac{1}{2}\left.\frac{\partial^2 X}{\partial y^2}\right|_{\mathbf{x}_i}\left((y-y_i)^2-\overline{y^2}_i\right)+\frac{1}{2}\left.\frac{\partial^2 X}{\partial z^2}\right|_{\mathbf{x}_i}\left((z-z_i)^2-\overline{z^2}_i\right)$$
$$+\left.\frac{\partial^2 X}{\partial x\partial y}\right|_{\mathbf{x}_i}\left((x-x_i)(y-y_i)-\overline{xy}_i\right)+\left.\frac{\partial^2 X}{\partial x\partial z}\right|_{\mathbf{x}_i}\left((x-x_i)(z-z_i)-\overline{xz}_i\right)$$
$$+\left.\frac{\partial^2 X}{\partial y\partial z}\right|_{\mathbf{x}_i}\left((y-y_i)(z-z_i)-\overline{yz}_i\right)+\cdots$$
$$+\sum_{p_1=0}^{K}\sum_{p_2=0}^{K}\sum_{\substack{p_3=0\\ p_1+p_2+p_3=K}}^{K}\frac{1}{p_1!p_2!p_3!}\left.\frac{\partial^K X}{\partial x^{p_1}\partial y^{p_2}\partial z^{p_3}}\right|_{\mathbf{x}_i}$$
$$\left.\times\left((x-x_i)^{p_1}(y-y_i)^{p_2}(z-z_i)^{p_3}-\overline{x^{p_1}y^{p_2}z^{p_3}}_i\right)\right), \tag{2.152}$$

This implies that the reconstructed solution at any point is the control volume average plus second and high-order contributions from the reconstruction:

$$X_i(\mathbf{x})=X_i+S(\mathbf{x}-\mathbf{x}_i)+H(\mathbf{x}-\mathbf{x}_i)$$

where $S(\mathbf{x}-\mathbf{x}_i)$ includes only linear terms for second-order reconstruction, $H(\mathbf{x}-\mathbf{x}_i)$ contains only quadratic terms for third-order reconstruction, and both quadratic and cubic terms for fourth-order reconstruction, even quadratic up to K-degree terms for $(K+1)$-order reconstruction. As in the work of Barth [33, 34], Nejat and Ollivier-Gooch [340] and Michalak and Ollivier-Gooch [313], the high-order accurate reconstruction can be limited by

$$X_i(\mathbf{x})=X_i+\Phi_i\left(S(\mathbf{x}-\mathbf{x}_i)+H(\mathbf{x}-\mathbf{x}_i)\right) \tag{2.153}$$

with Φ_i certain limiter function to ensure no new extrema. The limited extrapolated boundary values $X_{ij,gp}$ obtained from Eq. (2.153) at Gaussian point gp are then used for the left and right states $X_{ij,gp,L}$ and $X_{ij,gp,R}$

$$X_{ij,gp,L}=X_i+\Phi_i\left(S(\mathbf{x}_{ij,gp}-\mathbf{x}_i)+H(\mathbf{x}_{ij,gp}-\mathbf{x}_i)\right),$$
$$X_{ij,gp,R}=X_j+\Phi_j\left(S(\mathbf{x}_{ji,gp}-\mathbf{x}_j)+H(\mathbf{x}_{ji,gp}-\mathbf{x}_j)\right).$$

with $\mathbf{x}_{ij,gp}=\mathbf{x}_{ji,gp}$. Evidently, Barth-Jespersen limiter and Venkatakrishnan limiter introduced above can be applied.

Because $\Phi_i=1+O(\Delta x^2)$ in smooth regions for Venkatakrishnan limiter, this application of the limiter degrades solution accuracy from third- or fourth-order to second-order. To overcome this default, (2.153) can be modified [101, 340] as follows

$$X_i(\mathbf{x})=X_i+[(1-\sigma_i)\Phi_i+\sigma_i]S(\mathbf{x}-\mathbf{x}_i)+\sigma_i H(\mathbf{x}-\mathbf{x}_i)$$

by a discontinuity detector σ_i, which is zero near discontinuities and one in smooth regions of the flow. Nejat and Ollivier-Gooch [340] took $\sigma_i = \frac{1-\tanh(S(\Phi_0-\Phi_i))}{2}$, where S determines the sharpness of the step function, adjusting the rapidity of the transition between zero and one, while Φ_0 defines the limiter value that activates the switch. As discussed by Nejat and Ollivier-Gooch [340], too small a value for S or too large a value for Φ_0 can reduce accuracy in smooth regions by invoking limiting unnecessarily. Meanwhile, too large a value for S or too small a value for Φ_0 can produce spurious oscillations because the unlimited high-order reconstruction is sometimes used for cases where limiting is required for monotonicity. Thus, this approach may violate the monotonicity requirement as the high-order terms may have contributed to reducing the overshoot in the unlimited reconstruction used in determining the value of Φ_i.

For the improvement of the limiting function Φ_i in Eq. (2.153), Michalak and Ollivier-Gooch [313] replace $\min(1, y)$, given in Eq. (2.136), by a new approximation $\widetilde{\min}(1, y)$, with the following form of function

$$\widetilde{\min}(1, y) = \begin{cases} P(y), & y < y_t \\ 1, & y \geq y_t \end{cases}$$

with function $P(y)$ satisfying

$$P|_0 = 0,\ P|_{y_t} = 1,\ \left.\frac{\partial P}{\partial y}\right|_0 = 1,\ \left.\frac{\partial P}{\partial y}\right|_{y_t} = 0$$

Like Venkatakrishnan's function, $\widetilde{\min}(1, y)$ is differentiable at all points and $P(y) \leq \widetilde{\min}(1, y), y \in [0, y_t]$. However, unlike Eq. (2.138), this new limiting function $\widetilde{\min}(1, y)$ has a value of exactly 1 for a range of values $y \geq y_t$ with a threshold value of $y_t \in (1, 2)$.

This function allows to preserve high-order accuracy on uniform grids by satisfying $|\widetilde{\min}(1, y) - 1| \leq O(\Delta x^3)$. Additionally, this function is effective in maintaining high order accuracy in regions of mild mesh non-uniformity by setting the threshold value y_t to accommodate the degree of non-uniformity. Smaller values of y_t are likely to activate the limiter on non-uniform meshes but result in a limiter that permits non-differentiability. Like the parameter $\mathcal{K}$ given in Eq. (2.140), the choice of y_t is a compromise between maintaining good accuracy on non-uniform grids and maintaining good convergence properties. Michalak and Ollivier-Gooch [313] use $P(y) = -\frac{4}{27}y^3 + y$ with $y_t = 1.5$ in their numerical tests.

Now we can use $\widetilde{\min}(1, y)$ to compute a maximum allowable value of $\Phi_{ij,gp}$ for each Gauss point $\mathbf{x}_{ij,gp}$

$$\Phi_{ij,gp} = \begin{cases} \widetilde{\min}\left(1, \frac{X_i^{\max}-X_i}{(\nabla X)|_i\cdot(\mathbf{x}_{ij,gp}-\mathbf{x}_i)}\right), & \text{if } (\nabla X)|_i \cdot (\mathbf{x}_{ij,gp} - \mathbf{x}_i) > 0, \\ \widetilde{\min}\left(1, \frac{X_i^{\min}-X_i}{(\nabla X)|_i\cdot(\mathbf{x}_{ij,gp}-\mathbf{x}_i)}\right), & \text{if } (\nabla X)|_i \cdot (\mathbf{x}_{ij,gp} - \mathbf{x}_i) < 0, \\ 1, & \text{if } (\nabla X)|_i \cdot (\mathbf{x}_{ij,gp} - \mathbf{x}_i) = 0. \end{cases} \tag{2.154}$$

In the same way, set the limiter $\Phi_i = \min_j(\Phi_{ij})$, $\Phi_{ij} = \min_{gp}(\Phi_{ij,gp})$ as the minimum of the limiter values corresponding to that variable in all the quadrature points located on all the faces of that cell.

To maintain high-order accuracy permits the probability of violating solution monotonicity near smooth extrema. Furthermore, rapid changes of the limiter value and its non-differentiability in uniform regions prevent good convergence of the solver. Therefore, analogous to the second-order limiter of Venkatakrishnan, to eliminate the effect of the limiter when the local solution variation is $O(\Delta x^2)$ or smaller, Michalak and Ollivier-Gooch [313] proposed to disable the limiter when $\delta X = \delta X_i^{\max} - \delta X_i^{\min} < (\mathcal{K}\Delta x)^{\frac{3}{2}}$, where $\mathcal{K}$ is a tunable parameter. Instead of a simple switch, the following limiting procedure (hereafter called Michalak–Ollivier-Gooch's limiter) is proposed by Michalak and Ollivier-Gooch [313] to maintain differentiability

$$\tilde{\Phi}_i = \tilde{\sigma}_i + (1 - \tilde{\sigma}_i)\Phi_i \tag{2.155}$$

where Φ_i is defined by Eq. (2.154) and

$$\tilde{\sigma}_i = \begin{cases} 1, & \delta X^2 < (\mathcal{K}\Delta x)^3 \\ s\left(\frac{(\delta X)^2 - (\mathcal{K}\Delta x)^3}{(\mathcal{K}\Delta x)^3}\right), & (\mathcal{K}\Delta x)^3 < \delta X^2 < 2(\mathcal{K}\Delta x)^3 \\ 0, & \delta X^2 > 2(\mathcal{K}\Delta x)^3 \end{cases}$$

with the transition function $s(y)$ defined by $s(y) = 2y^3 - 3y^2 + 1$. The limited reconstruction is then computed for each Gauss point $\mathbf{x}_{ij,gp}$, and can be used for the left and right states $X_{ij,gp,L}$ and $X_{ij,gp,R}$

$$\begin{aligned} X_{ij,gp,L} &= X_i + \tilde{\Phi}_i \left(S(\mathbf{x}_{ij,gp} - \mathbf{x}_i) + H(\mathbf{x}_{ij,gp} - \mathbf{x}_i) \right), \\ X_{ij,gp,R} &= X_j + \tilde{\Phi}_j \left(S(\mathbf{x}_{ji,gp} - \mathbf{x}_j) + H(\mathbf{x}_{ji,gp} - \mathbf{x}_j) \right). \end{aligned} \tag{2.156}$$

Venkatakrishnan limiter and Michalak–Ollivier-Gooch limiter both require a tuning parameter $\mathcal{K}$ to detect regions of uniform flow. Ideally the solutions produced by the schemes should be relatively insensitive to the choice of this parameter. For sensitivity to this tuning parameter $\mathcal{K}$, one can refer to Michalak and Ollivier-Gooch [313]. Venkatakrishnan limiter and Michalak–Ollivier-Gooch limiter differ significantly in their action. Venkatakrishnan's ϵ^2 term modifies the limiter value for all cases, and increasing the value of $\mathcal{K}$ allows a progressively larger overshoot in the solution at shocks. In Michalak–Ollivier-Gooch limiter, near shocks the transition function is exactly one, and the basic limiter enforces monotonicity regardless of the value of $\mathcal{K}$, which makes the scheme less sensitive than Venkatakrishnan's to the choice of $\mathcal{K}$.

Like the Venkatakrishnan limiter, Michalak–Ollivier-Gooch limiter $\tilde{\Phi}_i$ is designed to be differentiable, which improves convergence to steady state by implicit schemes. To preserve third and fourth order accuracy at smooth extrema, a smooth switch between the limiter value and the unlimited reconstruction at Gauss points has been constructed to disable the limiter in regions of nearly uniform flow. In this way, as claimed by Michalak and Ollivier-Gooch [313], the greatest gains in accuracy are seen when using a third or fourth order scheme, and Michalak–Ollivier-Gooch limiter $\tilde{\Phi}_i$ is also effective in reducing the dissipation and increasing the accuracy of second-order schemes relative to Venkatakrishnan limiter.

Furthermore, an additional limiter modification was made using two prescribed Mach numbers in the supersonic case to accurately treat the stagnation region. Detailed implementation procedure can be found in the original paper of Michalak and Ollivier-Gooch [313].

Following in the footsteps of the high-order Michalak and Ollivier-Gooch limiter, Tsoutsanis [476] explored the impact of the definition of the bounds of the limiter, for higher-order MUSCL numerical schemes on 2D unstructured meshes in the FVM framework. The key characteristics of the modified limiters focus on the stencil elements that are used to establish the bounds of the reconstructed solution. Numerical tests there show that using the direct side neighbours for establishing the bounds results in sub-optimal order of accuracy, since the limiter can be mistakenly activated due to the grid-topology and grid-quality.

2.6.2.2 Data-Dependent Weighted Least-Squares Reconstruction

By exploiting the best properties of both limited K-exact reconstruction schemes and ENO schemes, Ollivier-Gooch [369, 370] presented a data-dependent weighted least-squares reconstruction (called DD-L2-ENO) with a fixed stencil, satisfying the weighted ENO conditions of Liu et al. [288].

When using Householder transformations to reduce the left-hand side of Eq. (2.149), after the upper-triangularization of the least-squares problem, the problem looks like:

$$\begin{pmatrix} * & * & * & \cdots & * \\ & * & * & \cdots & * \\ & & * & \cdots & * \\ & & & \ddots & * \\ & & & & * \\ & 0 & & & \end{pmatrix}_{N\times M} \chi = \begin{pmatrix} r_1 \\ r_2 \\ \vdots \\ r_M \\ r_{M+1} \\ \vdots \\ r_{N-1} \\ r_N \end{pmatrix}$$

before back-substitution. Here, $\chi = \left(\left.\frac{\partial X}{\partial x}\right|_{\mathbf{x}_i}, \left.\frac{\partial X}{\partial y}\right|_{\mathbf{x}_i}, \left.\frac{\partial X}{\partial z}\right|_{\mathbf{x}_i}, \left.\frac{1}{2}\frac{\partial^2 X}{\partial x^2}\right|_{\mathbf{x}_i}, \left.\frac{1}{2}\frac{\partial^2 X}{\partial y^2}\right|_{\mathbf{x}_i}, \left.\frac{1}{2}\frac{\partial^2 X}{\partial z^2}\right|_{\mathbf{x}_i}, \left.\frac{\partial^2 X}{\partial x \partial y}\right|_{\mathbf{x}_i}, \left.\frac{\partial^2 X}{\partial x \partial z}\right|_{\mathbf{x}_i}, \left.\frac{\partial^2 X}{\partial y \partial z}\right|_{\mathbf{x}_i}, \cdots \right)^T$, $M = \frac{1}{d!}\prod_{n=1}^{d}(K+n) - 1$ is the total number of the derivatives needed to calculate with d the number of space dimensions, N is the number of nearby control volumes in the stencil. Clearly, the first M equations are satisfied exactly, while the remaining $N-M$ equations are not. The residual $\hat{\mathcal{R}}$ for the solution is defined by $\hat{\mathcal{R}} = \sqrt{\sum_{\ell=M+1}^{N} r_\ell^2}$. In order to remove the effects of the geometric weighting term, Ollivier-Gooch [369, 370] used $\mathcal{R} = \Delta h^2 \hat{\mathcal{R}}$ to compute data-dependent weights to enforce the ENO property, where Δh is a length scale for the control volume i. $\mathcal{R}$ also is a good measure of how well the solution is approximated locally, and thus can be used as a possible candidate for a grid refinement measure.

As we know, the reconstruction scheme by K-exact least-squares technique is designed for smooth functions. For non-smooth functions—those with $O(1)$ discontinuities, such a scheme allows unphysical overshoots of $O(1)$. To overcome this, Ollivier-Gooch [369, 370] proposed to reconstruct by using only data that is smoothly connected to data in V_i. This idea is taken directly by ENO schemes, which by design search for a smooth stencil and completely exclude non-smooth data from the reconstruction. WENO schemes work less directly, using all possible stencils and weighting those containing non-smooth data with a weight that is of the order of the truncation error.

By assigning weights to control volumes rather than to stencils, DD-L2-ENO's goal is to weight non-smooth data with $O(\Delta h^{K+1})$ to satisfy the ENO condition of Liu et al. [288]. This is realized by designing a data-dependent weight that multiplies the previously calculated geometric weight. The data-dependent weight construction is based on the following two facts [369, 370]:

1. If the function is non-smooth and the neighborhood of V_i crosses a discontinuity, then a modified Taylor expansion does not adequately describe the function locally and there will exist at least one control volume j for which

$$\frac{1}{V_j}\int_{V_j} X_i(\mathbf{x})dV - X_j = O(1)$$

This means that the residual of the least-squares problem will be $O(1)$. On the other hand, for stencils with only smooth regions of the function, one has

$$X_i(\mathbf{x}) - X(\mathbf{x}) = O(\Delta h^{K+1})$$

and $\mathcal{R}$ is also $O(\Delta h^{K+1})$. Therefore, $\mathcal{R}$ can be used as a measure of the smoothness of the data for the entire stencil $\{V_j\}_i$.

2. Whether the data in V_j is smoothly connected to data in V_i can be determined by the following asymptotic evaluation

$$\frac{X|_j - X|_i}{|\mathbf{x}_j - \mathbf{x}_i|} = \begin{cases} O(1), & \text{smoothly connected} \\ O(\Delta h^{-1}), & \text{not smoothly connected} \end{cases}$$

In DD-L2-ENO scheme, the data-dependent weighting uses $\mathcal{R}$ to detect stencils with non-smooth data and $\frac{X|_j - X|_i}{|\mathbf{x}_j - \mathbf{x}_i|}$ to select which data within that stencil should be excluded and which included. One appropriate data-dependent weighting from Ollivier-Gooch [370] is:

$$\omega'_j = W_j^{\mathrm{DD}} \omega_j,\; W_j^{\mathrm{DD}} = \frac{1}{\epsilon + \mathcal{R}\left|\frac{X|_j - X|_i}{|\mathbf{x}_j - \mathbf{x}_i|}\right|^{K+1}},\; \omega_j = \frac{1}{|\mathbf{x}_j - \mathbf{x}_i|^2}$$

where $\epsilon = 10^{-10}$ prevents division by zero; $\mathcal{R}\left|\frac{X|_j - X|_i}{|\mathbf{x}_j - \mathbf{x}_i|}\right|^{K+1}$ is the smoothness indicator; $\mathcal{R}$ is computed from a data-independent least-squares reconstruction, as described above. The asymptotic behavior of this weighting for the three important cases is

$$W_j^{DD} = \begin{cases} \max\left(\epsilon^{-1}, O(\Delta h^{-(k+1)})\right), & \text{smooth function} \\ O(1), & \text{smooth data in stencil with non-smooth data} \\ O(\Delta h^{k+1}) & \text{non-smooth data} \end{cases}$$

Another data-dependent weighting from Ollivier-Gooch [369] reads:

$$\omega'_j = W_j^{\mathrm{DD}} \omega_j,\; W_j^{\mathrm{DD}} = \frac{1}{1 + \delta\mathcal{R}_1\left|\frac{X|_j - X|_i}{|\mathbf{x}_j - \mathbf{x}_i|}\right|^{K+1}},\; \omega_j = \frac{1}{|\mathbf{x}_j - \mathbf{x}_i|^2}$$

where $\mathcal{R}_1$ can be $\mathcal{R}$ or alternatively be chosen as

$$\mathcal{R}_1 = \frac{\hat{\mathcal{R}}}{\sqrt{\frac{1}{N}\sum_{j=1}^{N}\omega_j^2}}, \quad \omega_j = \frac{1}{|\mathbf{x}_j - \mathbf{x}_i|^2} \tag{2.157}$$

Here, the value of δ is not critical since only the asymptotic behavior of W_j^{DD} matters, and the value $\delta = 0.1$ is suggested [369]. Similarly, the asymptotic behavior of this weighting for the three important cases is

$$W_j^{DD} = \begin{cases} 1 + O(\Delta h^{k+1}), & \text{smooth stencil} \\ O(1), & \text{smooth data in stencil with non-smooth data} \\ O(\Delta h^{-k-1}), & \text{non-smooth data} \end{cases}$$

From the asymptotic behavior of weighting W_j^{DD}, the data-dependent weights are all of same order for stencils containing only smoothly connected data, resulting in the good qualities of the data-independent reconstruction for smooth functions. For stencils that are not entirely smooth, the data-dependent weight for nonsmooth data becomes smaller than that for smooth data by a factor of the order of truncation error. These weightings satisfy the ENO conditions of Liu et al. [288].

Different from standard K-exact schemes with geometric weights $\omega_j = \frac{1}{|\mathbf{x}_j - \mathbf{x}_i|^2}$, DD-L2-ENO schemes [369, 370] solves the least-squares reconstruction Eq. (2.149) with a fixed stencil, with the geometric weights replaced by the data-dependent weights $\omega'_j = W_j^{\mathrm{DD}} \omega_j$. In this way, each point in a reconstruction stencil is assigned a data-dependent weight that strongly reflects smooth data in the reconstruction. Specifically, the weight assigned to nonsmooth data is on the order of the truncation error. In this respect, DD-L2-ENO is similar to WENO schemes, except that the point-wise nature of the weighting with DD-L2-ENO imparts some additional flexibility in selecting the smooth data to include in the reconstruction. In both DD-L2-ENO and WENO schemes, the weights change smoothly with data variations, which favors steady-state convergence in comparison to stencil-searching ENO schemes. DD-L2-ENO scheme automatically allows graceful degradation of accuracy locally in regions where insufficient smooth data exists for reconstruction of the required high order; this helps in enhancing robustness in application. The K-exact least-squares polynomial reconstruction heritage of these DD-L2-ENO schemes allows them to be applied easily to function reconstruction on unstructured meshes in multiple dimensions.

2.6.2.3 Solution Dependent Weighted Least-Squares (SDWLS) Gradients

Although the possibility of solution-dependent weights are indicated by Barth [32], there is no mention about the way to choose (the expression of) the weights for TVD (non-oscillatory) solutions. DD-L2-ENO scheme can be one successful practice of solution-dependent weights.

Motivated by considering the natural ability of least squares formulation to handle arbitrary sets of data points arising in unstructured grid based flow computations, Mandal and Rao [299] achieved high resolution in inviscid computations by using linear and quadratic variable reconstructions based on solution dependent weighted least squares (SDWLS) gradients (e.g., [299–301]). As function of the solutions, the weights of the weighted least squares formulation are generated in such a way that the SDWLS scheme satisfies TVD criterion in one-dimension [300]. Thus, the nonlinearity in the discretization is introduced through solution dependent weights to achieve non-oscillatory behavior. In what follows we abide by the description of SDWLS in Mandal and Rao [299].

Considering that a cell i contains N vertex neighbouring cells ($j = 1, \ldots, N$), values at these N vertex neighbors can be expressed in terms of solution variable at the cell center i using Taylor's series expansion. Using quadratic representation, the weighted least squares formulation corresponding to Eq. (2.149) becomes $\omega\mathcal{A}\boldsymbol{\chi} = \omega\mathbf{b}$ with $\omega = \mathrm{diag}(\omega_1, \ldots, \omega_N)$ and solution dependent weights ω_j. According to Mandal and Subramanian [300] and Mandal and Rao [299], an influence coefficient based concept are used to derive the weights in the weighted least squares formulation to calculate the derivatives. The possible simple expressions for weights [300] can be

$$\omega_j = \frac{1}{\left|X|_j - X|_i\right|^2}, \quad \text{and} \quad \omega_j = \frac{1}{\left|X|_j - X|_i\right|},$$

which correspond to van Albada limiter and Van Leer limiter, respectively. Other weights given in Mandal and Subramanian [300] can also be used.

In the SDWLS formulation for gradient calculations (e.g., [299–301]), both edge-based stencil and vertex-based stencil can be used to yield the weighted least squares system, which can be solved by the QR method. In case of quadratic reconstruction in three-dimensions, the number of edge-based neighbours may be insufficient for proper least squares computations, and vertex-based stencil is suggested for the SDWLS formulation. The interpretation of the weights as influence coefficient makes the SDWLS approach in multi-dimensions simple and easily applicable to any kind of unstructured grids.

Mandal et al. [301] discussed a link between the solution dependent weights for weighted least-squares formulation and various independently proposed limiter functions, such as van Albada, Barth, van Leer, minmod, and Venkatakrishnan. This link helps in implementing weighted least-squares reconstruction directly into a higher-order scheme without needing limiter function. The implementation of the SDWLS approach thus becomes simple and computationally efficient. It is also shown in Table 2 of Mandal et al. [301] for the first time, how several independently proposed limiter functions available in the literature belongs to a class of solution dependent weighted least-squares (SDWLS) formulations.

2.6.2.4 U-MUSCL

In higher order finite volume methods, the reconstruction of a high order polynomial within each computational cell is required by using large stencils consisting of a neighborhood of nearby cells. Commonly, this method may involve a large stencil consisting not only of direct neighboring cells but also of neighbors-of-neighbors to construct the gradients using a weighted least squares (WLSQ) method, which are then limited to avoid spurious oscillations near discontinuities. Without enlarging stencil, the unstructured MUSCL (U-MUSCL) technique has been proposed to achieve high-order accurate reconstruction schemes [54, 104, 508]. Following these papers, the U-MUSCL is documented as below.

Burg [54] developed an U-MUSCL scheme for variable extrapolation formulation for unstructured finite volume codes, which closely resembles the MUSCL scheme used within structured finite volume solvers. The formulation is based on information of the solution and its gradients already known. The interpolation function in the U-MUSCL scheme reads

$$X_{ij}(\theta) = X|_i + \Phi_i \left[\frac{\theta}{2} \left(X|_j - X|_i \right) + (1-\theta)\, (\nabla X)|_i \cdot (\mathbf{x}_{ij} - \mathbf{x}_i) \right] \tag{2.158}$$

where $(\nabla X)|_i = (\nabla X)|_{\mathbf{x}_i}$ is the gradient at the center or centroid $\mathbf{x}_i$ of cell i, and X_{ij} is the state value at position $\mathbf{x}_{ij}$ located at the cell face between cell i and its neighbor j. While, θ is the U-MUSCL parameter, and Φ_i is a limiter function in order to avoid numerical oscillations in the regions of steep gradients, such as the Barth-Jesperson limiter [37] and the Venkatakrishnan limiter [485]. At Gauss points $\mathbf{x}_{ij,gp}$ or equivalently $\mathbf{x}_{ji,gp}$ on the cell face A_{ij}, the second-order scheme (2.158) can be expressed as:

$$X_{ij,gp,L}(\theta) = X|_i + \Phi_i \left[\frac{\theta}{2} \left(X|_j - X|_i\right) + (1-\theta)\,(\nabla X)|_i \cdot (\mathbf{x}_{ij,gp} - \mathbf{x}_i) \right],$$
$$X_{ij,gp,R}(\theta) = X|_j + \Phi_j \left[\frac{\theta}{2} \left(X|_i - X|_j\right) + (1-\theta)\,(\nabla X)|_j \cdot (\mathbf{x}_{ji,gp} - \mathbf{x}_j) \right]$$

This U-MUSCL extrapolation formulation represents a one-parameter family of schemes. Under certain circumstances, it is entirely equivalent to the MUSCL-scheme. For instance, in one-dimensional case, if the central difference approximation is used for the gradient (i.e., $(\nabla X)|_{x_i} = (X|_{i+1} - X|_{i-1})/(2\Delta x)$), then this scheme reduces to the standard MUSCL scheme.

As commented by Dement and Ruffin [104] and Yang et al. [508] (particularly, Table 2 of Yang et al. [508]), one can vary parameter θ to achieve multiple different schemes: an upwind scheme as in the classic unstructured formulation for second-order variable extrapolation, with $\theta = 0$; the second-order fully upwind MUSCL-type variable extrapolation, with $\theta = -1$; a third-order variable extrapolation to the cell face with $\theta = 1/2$ for regular hexahedral grids; a third-order approximation to the derivative at the node, with $\theta = 1/3$; and a central difference scheme, with $\theta = 1$. This formula is an upwind formula as long as $\theta < 1$. Note that only face neighbor information is needed, and the only state variables required are the cell averages associated to those face neighbor cells, implying that no interpolation or averaging is necessary as required for the structured grid-based method.

The U-MUSCL scheme was extended [104, 506–508] to third order accuracy by introducing an additional parameter, θ_3:

$$\begin{aligned} X_{ij}(\theta) = & X|_i + \Phi_i \left[\frac{\theta}{2}(X|_j - X|_i) + (1-\theta)\,(\nabla X)|_i \cdot (\mathbf{x}_{ij} - \mathbf{x}_i) \right. \\ & + \frac{1}{2}\left(\frac{\theta_3}{2}\left((\nabla X)|_j \cdot (\mathbf{x}_{ij} - \mathbf{x}_i) - (\nabla X)|_i \cdot (\mathbf{x}_{ij} - \mathbf{x}_i)\right) \right. \\ & \left.\left. + (1-\theta_3)(\mathbf{x}_{ij} - \mathbf{x}_i)^T \cdot \left(\nabla^2 X\right)|_i \cdot (\mathbf{x}_{ij} - \mathbf{x}_i)\right)\right] \end{aligned} \tag{2.159}$$

where $\left(\nabla^2 X\right)|_i = \left(\nabla^2 X\right)|_{\mathbf{x}_i} \equiv \left((\nabla \frac{\partial X}{\partial x})|_{\mathbf{x}_i},\, \left(\nabla \frac{\partial X}{\partial y}\right)\Big|_{\mathbf{x}_i},\, (\nabla \frac{\partial X}{\partial z})|_{\mathbf{x}_i} \right)$ is the Hessian matrix of the state variable X taking its value at the center or centroid $\mathbf{x}_i$ of cell i. Variations in θ and θ_3 can produce a scheme from second to fourth order accurate, which yields a fourth order extrapolation to the cell face on a regular grid if $\theta = -\frac{1}{6}$ and $\theta_3 = -\frac{4}{3}$ are chosen [507]. Correspondingly, at Gauss points $\mathbf{x}_{ij,gp}$ or equivalently $\mathbf{x}_{ji,gp}$ on the cell face A_{ij}, the scheme (2.159) can be expressed as:

$$\begin{aligned} X_{ij,gp,L}(\theta) = & X|_i + \Phi_i \left[\frac{\theta}{2}(X|_j - X|_i) + (1-\theta)\,(\nabla X)|_i \cdot (\mathbf{x}_{ij,gp} - \mathbf{x}_i) \right. \\ & + \frac{1}{2}\left(\frac{\theta_3}{2}\left((\nabla X)|_j \cdot (\mathbf{x}_{ij,gp} - \mathbf{x}_i) - (\nabla X)|_i \cdot (\mathbf{x}_{ij,gp} - \mathbf{x}_i)\right) \right. \end{aligned}$$

$$+ (1-\theta_3)(\mathbf{x}_{ij,gp}-\mathbf{x}_i)^T \cdot \left(\nabla^2 X\right)\big|_i \cdot (\mathbf{x}_{ij,gp}-\mathbf{x}_i)\Big)\Big],$$

$$X_{ij,gp,R}(\theta) = X|_j + \Phi_j \left[\frac{\theta}{2}(X|_i - X|_j) + (1-\theta)\,(\nabla X)|_j \cdot (\mathbf{x}_{ji,gp}-\mathbf{x}_j)\right.$$

$$+ \frac{1}{2}\left(\frac{\theta_3}{2}\left((\nabla X)|_i \cdot (\mathbf{x}_{ji,gp}-\mathbf{x}_j) - (\nabla X)|_j \cdot (\mathbf{x}_{ji,gp}-\mathbf{x}_j)\right)\right.$$

$$\left.\left. + (1-\theta_3)(\mathbf{x}_{ji,gp}-\mathbf{x}_j)^T \cdot \left(\nabla^2 X\right)\big|_j \cdot (\mathbf{x}_{ji,gp}-\mathbf{x}_j)\right)\right],$$

To compute the higher derivatives, the Green-Gauss theorem or least squares method can be applied [506, 508]. For a function X, using Green-Gauss theorem Eq. (2.92), there follows that

$$(\nabla X)|_i = \frac{1}{V_i}\int_{V_i} \nabla X\, dV = \frac{1}{V_i}\int_{\partial V_i} X\,\mathbf{n}\, dS = \frac{1}{V_i}\sum_{j=1}^{N_{\text{faces}}} X(\mathbf{x}_{ij,c}) A_{ij}\mathbf{n}_{ij},$$

Since the first derivative is already known from the existing computation in the second-order scheme, it is easy to determine the second derivatives directly by using Green-Gauss theorem as follows:

$$\left(\nabla \frac{\partial X}{\partial \kappa}\right)\bigg|_{\mathbf{x}_i} = \frac{1}{V_i}\int_{V_i} \nabla \frac{\partial X}{\partial \kappa}\, dV = \frac{1}{V_i}\int_{\partial V_i} \frac{\partial X}{\partial \kappa}\mathbf{n}\, dS, \quad \kappa \in \{x, y, z\}$$

Another approach for evaluating the gradient, ∇X, is to employ a least squares procedure [508] like Eq. (2.124). Once the first derivatives are computed with the least squares approach, the same least squares procedure can be used to calculate the second derivatives, i.e., $\nabla \frac{\partial X}{\partial \kappa}$, $\kappa \in \{x, y, z\}$, based on the values of the first derivatives $\frac{\partial X}{\partial \kappa}$. In a word, the general procedure can be described as follows: first, the components of the gradient of X are computed by the least square method such that the gradient of X is known in each cell. Then, repeating the procedure with the gradient of X components as arguments yields second order derivatives of X with the same least square method. That is, by replacing the existing variable of X in the right-hand side of Eq. (2.124) with the first derivative in x ($\frac{\partial X}{\partial x}$, for example), the second derivatives $\left(\frac{\partial^2 X}{\partial x^2}, \frac{\partial^2 X}{\partial x \partial y}, \frac{\partial^2 X}{\partial x \partial z}\right)$ can be computed. Similarly, by calling the same routine using $\frac{\partial X}{\partial y}$ and $\frac{\partial X}{\partial z}$, one can find $\left(\frac{\partial^2 X}{\partial y \partial x}, \frac{\partial^2 X}{\partial y^2}, \frac{\partial^2 X}{\partial y \partial z}\right)$ and $\left(\frac{\partial^2 X}{\partial z \partial x}, \frac{\partial^2 X}{\partial z \partial y}, \frac{\partial^2 X}{\partial z^2}\right)$. Using the symmetry property, only six additional variables $\left(\frac{\partial^2 X}{\partial x^2}, \frac{\partial^2 X}{\partial x \partial y}, \frac{\partial^2 X}{\partial x \partial z}, \frac{\partial^2 X}{\partial y^2}, \frac{\partial^2 X}{\partial y \partial z}, \frac{\partial^2 X}{\partial z^2}\right)$ need to be found.

As suggested by Dement and Ruffin [104], these derivatives can be computed by the so called WLSQ(G) weights [429, 430]. In order to modify the weights of the WLSQ method in a data-dependent way, Dement and Ruffin [104] followed the idea of ENO-weighted least squares method [369] to obtain a non-oscillatory weight. In Dement and Ruffin's implementation, the data-dependent ENO weights

$$W_j^{\mathrm{DD}} = \frac{1}{1 + \delta \mathcal{R}_1 \left| \frac{X|_j - X|_i}{|\mathbf{x}_j - \mathbf{x}_i|} \right|^{K+1}}$$

from Eq. (25) in Ollivier-Gooch [369] are used, where δ is a parameter (chosen to be 0.1 by Dement and Ruffin [104]), and $\mathcal{R}_1$ is defined in Eq. (2.157), which is obtained by the residual of the QR decomposition of the non-data dependent WLSQ procedure when the WLSQ matrix system is solved by the Householder QR decomposition algorithm. The redefined ENO weights for the scheme read as follows:

$$\omega_j' = W_j^{\mathrm{DD}} \omega_j$$

Here, $\omega_j = \frac{1}{|\mathbf{x}_j - \mathbf{x}_i|^2}$ [369] based on the local topology of the mesh can be used, while Dement and Ruffin [104] employed $\omega_j = \left(\frac{2l_j'}{d_{ij}'}\right)^2 A_{ij}/d_{ij}$ as in WLSQ(G) [429, 430] introduced in Sect. 2.6.1.4. Then, the least squares procedure is recomputed with these modified weights. Similarly, the successive differentiation procedure can be repeated as well to obtain higher order moments as required. As pointed out by Dement and Ruffin [104], the calculated gradients and second derivatives are strongly dependent on the choice of these data dependent weights, and thus they must be chosen technically to enhance robust shock capturing and smooth flow resolving capabilities.

It is important to note that these new second derivatives will likely be at most second order accurate, having inherited error from the second order accurate computation of the first derivatives. As claimed by Dement and Ruffin [104], in practice, this effect does not emerge significantly in the solutions, and the simplicity and compactness of the method make it an attractive one for extending unstructured mesh-based codes to higher order accuracy. The U-MUSCL approach has salient features that Green-Gauss theorem and the existing linear least-squares implementation are utilized to compute the second derivative through successive differentiation, and that it is innovative in that it can achieve high-order accuracy without using excessive storage and reduce costs of solving WLSQ problem required for usual higher-order K-exact least-squares approaches.

2.6.2.5 High-Order Version of WBAP Approach

The BAP limiter firstly proposed by Choi and Liu [76] transforms gradients of state variables with biased averaging function, and then converts the real gradients of state variables by inverse transforming with the self-similar function. This limiter is simple to implement and parameter-free, and it is also differentiable and easy to be extended to unstructured grids. The robustness and fast convergence of BAP limiter are verified by Choi and Liu, while they did not provide detailed numerical accuracy for high-order scheme. By utilizing the key characteristics of both the BAP limiter and WENO scheme, the BAP approach has been implemented straightforwardly in higher order reconstructions [278, 280, 289, 290, 489] for unstructured grids. By employing several candidate K-exact reconstructions on a given cell via a secondary reconstruction (SR), the higher order WBAP limiter has been proposed [278, 280, 489] to firstly limit the coefficients corresponding to the higher order derivatives of the reconstruction polynomials, and then apply the WBAP limiter to the coefficients of the lower order derivatives in the SR that are recomputed using the limited higher order coefficients. Meanwhile, the distance weighted BAP (DWBAP) and the distance derivative weighted ENO (DDWENO) limiters for the second- and high-order finite volume method have been suggested for unstructured grids [289, 290]. They introduce the standard deviation coefficients of the state variables as the smooth indicators, and obtain the limited derivatives of the state variables with the inverse distance weighting for all derivatives of each cell according to smoothness. In comparison to the WENO schemes, without identifying several candidate stencils, these BAP approaches can achieve the similar effect of WENO schemes in the fixed stencil with high computational efficiency.

In what follows these WBAP limiters are presented briefly according to the existing literature [280, 289, 290].

(I) WBAP limiter with secondary reconstruction and successive limiting procedure

Li and Ren [280] proposed a high order version of the WBAP limiter firstly given by Li et al. [281]. The high-order WBAP limiter depends on two key techniques: the secondary reconstruction and the successive limiting procedure. Here, these two key techniques are described according to Li and Ren [280].

Recall that the K-exact reconstruction polynomial Eq. (2.152) of degree K, $X_i(\mathbf{x})$, can be reformulated as follows:

$$X_i(\mathbf{x}) = X_i + \boldsymbol{\chi}_i^T \boldsymbol{\varphi}_i, \tag{2.160}$$

where $\boldsymbol{\chi}_i = (\chi_{i,1}, \chi_{i,2}, \dots, \chi_{i,N_D-1})^T$, and $\chi_{i,\ell}$ is the coefficient (also called the scaled derivative) with the general form of $\chi_{i,\ell} = \frac{1}{p_1!p_2!p_3!} \frac{\partial^{p_1+p_2+p_3} X}{\partial x^{p_1} \partial y^{p_2} \partial z^{p_3}}\Big|_{\mathbf{x}_i}$ and $p_1 + p_2 + p_3 > 0$, corresponding to derivatives of the reconstruction polynomial; $\boldsymbol{\varphi}_i = (\varphi_{i,1}, \varphi_{i,2}, \dots, \varphi_{i,N_D-1})^T$ is the vector of the basis function on V_i, and $\varphi_{i,\ell}$ is the zero-mean basis with the general form of $(x - x_i)^{p_1}(y - y_i)^{p_2}(z - z_i)^{p_3} - \overline{x^{p_1} y^{p_2} z^{p_3}}_i$. $\ell = 1, 2, \dots, N_D - 1$, where $N_D - 1$ is the number of coefficients of Kth degree reconstruction polynomial except the constant term, with $N_D = \frac{(K+1)(K+2)(K+3)}{6}$ in three dimensional case. To be more precise, $\boldsymbol{\chi}_i$ and $\boldsymbol{\varphi}_i$ are

$$\begin{aligned}
\boldsymbol{\chi}_i = &\left(\frac{\partial X}{\partial x}\Big|_{\mathbf{x}_i}, \frac{\partial X}{\partial y}\Big|_{\mathbf{x}_i}, \frac{\partial X}{\partial z}\Big|_{\mathbf{x}_i}, \frac{1}{2}\frac{\partial^2 X}{\partial x^2}\Big|_{\mathbf{x}_i}, \frac{1}{2}\frac{\partial^2 X}{\partial y^2}\Big|_{\mathbf{x}_i}, \frac{1}{2}\frac{\partial^2 X}{\partial z^2}\Big|_{\mathbf{x}_i}, \right. \\
&\left. \frac{\partial^2 X}{\partial x \partial y}\Big|_{\mathbf{x}_i}, \frac{\partial^2 X}{\partial x \partial z}\Big|_{\mathbf{x}_i}, \frac{\partial^2 X}{\partial y \partial z}\Big|_{\mathbf{x}_i}, \cdots \right)^T, \\
\boldsymbol{\varphi}_i = &(\delta x_i, \delta y_i, \delta z_i, \delta x_i^2 - \overline{x^2}_i, \delta y_i^2 - \overline{y^2}_i, \delta z_i^2 - \overline{z^2}_i, \delta x_i \delta y_i - \overline{xy}_i, \\
&\delta x_i \delta z_i - \overline{xz}_i, \delta y_i \delta z_i - \overline{yz}_i, \cdots)^T,
\end{aligned}$$

where $\delta x_i = x - x_i, \delta y_i = y - y_i, \delta z_i = z - z_i$, and $\overline{x^{p_1} y^{p_2} z^{p_3}}_i = \frac{1}{V_i} \int_{V_i} (x - x_i)^{p_1} (y - y_i)^{p_2} (z - z_i)^{p_3} dV$. In this way, one can obtain $X_i(\mathbf{x})$ in all the cells i's of computational domain, and hereafter $X_i(\mathbf{x})$ is called the primary reconstruction (PR) in V_i.

To proceed with the secondary reconstruction (SR), we consider a specified cell i with its stencil of direct neighboring cells j's ($j = 1, 2, \ldots N$). For 3D hexahedral computational cells as shown in Fig. 2.1, $N = 6$ is a natural choice. Besides the PR, $X_i(\mathbf{x})$, of V_i, the PR, $X_j(\mathbf{x})$, on a neighboring cell j of V_i can be considered as one candidate reconstruction of V_i, i.e.,

$$X_{j \to i}(\mathbf{x}) = X_j(\mathbf{x}), \quad \mathbf{x} \in V_i. \tag{2.161}$$

But, the candidate reconstruction $X_{j \to i}(\mathbf{x})$ in Eq. (2.161) does not conserve the mean of V_i. In order to restore the conservation of the mean, $X_{j \to i}(\mathbf{x})$ needs to be modified. For a third order polynomial reconstruction, Eq. (2.161) is written as

$$X_{j \to i}(\mathbf{x}) = X_j + \boldsymbol{\chi}_j^T \boldsymbol{\varphi}_j, \quad \mathbf{x} \in V_i, \tag{2.162}$$

where $\boldsymbol{\chi}_j = (\chi_{j,1}, \chi_{j,2}, \ldots, \chi_{j,9})^T$, and $\boldsymbol{\varphi}_j = (\varphi_{j,1}, \varphi_{j,2}, \ldots, \varphi_{j,9})^T$. If $\boldsymbol{\varphi}_j$ is expressed by the vector of the basis function $\boldsymbol{\varphi}_i$ on V_i, then one easily arrives at

$$\boldsymbol{\varphi}_j = \mathbf{A} \boldsymbol{\varphi}_i + \mathbf{b}, \tag{2.163}$$

where

$$\mathbf{A} = \begin{pmatrix} 1 & 0 & 0 & 0\,0\,0\,0\,0\,0 \\ 0 & 1 & 0 & 0\,0\,0\,0\,0\,0 \\ 0 & 0 & 1 & 0\,0\,0\,0\,0\,0 \\ 2\Delta x & 0 & 0 & 1\,0\,0\,0\,0\,0 \\ 0 & 2\Delta y & 0 & 0\,1\,0\,0\,0\,0 \\ 0 & 0 & 2\Delta z & 0\,0\,1\,0\,0\,0 \\ \Delta y & \Delta x & 0 & 0\,0\,0\,1\,0\,0 \\ \Delta z & 0 & \Delta x & 0\,0\,0\,0\,1\,0 \\ 0 & \Delta z & \Delta y & 0\,0\,0\,0\,0\,1 \end{pmatrix}_{9\times 9}, \quad \mathbf{b} = \begin{pmatrix} \Delta x \\ \Delta y \\ \Delta z \\ \Delta x^2 + \overline{\delta x_i^2} - \overline{\delta x_j^2} \\ \Delta y^2 + \overline{\delta y_i^2} - \overline{\delta y_j^2} \\ \Delta z^2 + \overline{\delta z_i^2} - \overline{\delta z_j^2} \\ \Delta x \Delta y + \overline{\delta x_i \delta y_i} - \overline{\delta x_j \delta y_j} \\ \Delta x \Delta z + \overline{\delta x_i \delta z_i} - \overline{\delta x_j \delta z_j} \\ \Delta y \Delta z + \overline{\delta y_i \delta z_i} - \overline{\delta y_j \delta z_j} \end{pmatrix}_{9\times 1}$$

and $\Delta x = x_i - x_j$, $\Delta y = y_i - y_j$, $\Delta z = z_i - z_j$. According to Eq. (2.162) and (2.163), one gets

$$X_{j \to i}(\mathbf{x}) = X_j + \boldsymbol{\chi}_j^T (\mathbf{A} \boldsymbol{\varphi}_i + \mathbf{b}) = X_j + \boldsymbol{\chi}_j^T \mathbf{b} + \left(\boldsymbol{\chi}_j^T \mathbf{A}\right) \boldsymbol{\varphi}_i, \quad \mathbf{x} \in V_i. \tag{2.164}$$

A straightforward computation shows

$$\frac{1}{V_i} \int_{V_i} X_{j \to i}(\mathbf{x}) dV = X_j + \boldsymbol{\chi}_j^T \mathbf{b},$$

which implies that $X_{j\to i}(\mathbf{x})$ is not conservative on cell i as long as $X_j + \boldsymbol{\chi}_j^T\mathbf{b} \neq X_i$. On the other hand, we have

$$X_j(\mathbf{x}) = X_i(\mathbf{x}) + O(\Delta h^{K+1}), \quad \mathbf{x} \in V_i, \tag{2.165}$$

due to the K-exactness property, and the cell average of Eq. (2.165) yields

$$\frac{1}{V_i}\int_{V_i} X_{j\to i}(\mathbf{x})dV = \frac{1}{V_i}\int_{V_i} X_j(\mathbf{x})dV = \frac{1}{V_i}\int_{V_i} X_i(\mathbf{x})dV + O(\Delta h^{K+1}),$$

which is equivalent to

$$X_j + \boldsymbol{\chi}_j^T\mathbf{b} = X_i + O(\Delta h^{K+1}). \tag{2.166}$$

Equation (2.166) indicates that the conservation error is very small although $X_{j\to i}(\mathbf{x})$ is not conservative on cell i. Hence, in order to recover the conservation of the mean, a simple modification to Eq. (2.164) is made to replace $X_j + \boldsymbol{\chi}_j^T\mathbf{b}$ with X_i, and thus the practical candidate reconstruction on cell V_i can be determined by

$$X_{j\to i}(\mathbf{x}) = X_i + (\boldsymbol{\chi}_j^T\mathbf{A})\boldsymbol{\varphi}_i, \quad \mathbf{x} \in V_i.$$

which holds exactly in the sense of $O(\Delta h^{K+1})$. Now we are in a position to state the secondary reconstruction

$$X_{j\to i}(\mathbf{x}) = X_i + \boldsymbol{\chi}_{j\to i}^T\boldsymbol{\varphi}_i, \quad \mathbf{x} \in V_i, \tag{2.167}$$

where $\boldsymbol{\chi}_{j\to i} = \boldsymbol{\chi}_j^T\mathbf{A} = (\chi_{j\to i,\ell})^T$ $(\ell = 1, 2, \ldots, 9)$ with its elements explicitly given by

$$\begin{aligned}
\chi_{j\to i,1} &= \chi_{j,1} + 2\Delta x\chi_{j,4} + \Delta y\chi_{j,7} + \Delta z\chi_{j,8},\\
\chi_{j\to i,2} &= \chi_{j,2} + 2\Delta y\chi_{j,5} + \Delta x\chi_{j,7} + \Delta z\chi_{j,9},\\
\chi_{j\to i,3} &= \chi_{j,3} + 2\Delta z\chi_{j,6} + \Delta x\chi_{j,8} + \Delta y\chi_{j,9},\\
\chi_{j\to i,4} &= \chi_{j,4},\ \chi_{j\to i,5} = \chi_{j,5},\ \chi_{j\to i,6} = \chi_{j,6},\\
\chi_{j\to i,7} &= \chi_{j,7},\ \chi_{j\to i,8} = \chi_{j,8},\ \chi_{j\to i,9} = \chi_{j,9},
\end{aligned} \tag{2.168}$$

with $\Delta x = x_i - x_j$, $\Delta y = y_i - y_j$, $\Delta z = z_i - z_j$. This finishes the formulation of the secondary reconstruction (SR) $X_{j\to i}(\mathbf{x})$ of V_i.

With the PR of the current cell in Eq. (2.160) and the SRs in Eq. (2.167) available, we turn to the successive limiting procedure. In order to make the limiting procedure compact, only the PRs for the direct face neighbors of cell V_i are used to construct the SRs for cell V_i. The successive limiting procedure takes the advantage of the analytical formulations of the SR: first, the coefficients corresponding to the higher order derivatives of the reconstruction polynomials are limited by WBAP, and then the coefficients corresponding to the lower order derivatives in the SR are computed using the limited higher order coefficients before applying the WBAP. To explain this idea, based on the SRs in Eq. (2.167), the successive limiting procedure for the third-order scheme is exemplarily given as follows:

(1) On cell V_i, the second-order scaled derivatives of the SRs are computed in terms of Eq. (2.168), namely

$$\hat{\chi}_{j\to i,\ell} = \chi_{j,\ell}, \quad \ell = 4, 5, 6, 7, 8, 9, \quad j = 1, 2, \ldots, N.$$

The second-order scaled derivatives on cell V_i are limited by

$$\tilde{\chi}_{i,\ell} = L(\chi_{i,\ell}, \hat{\chi}_{1\to i,\ell}, \hat{\chi}_{2\to i,\ell}, \hat{\chi}_{3\to i,\ell}, \hat{\chi}_{4\to i,\ell}, \hat{\chi}_{5\to i,\ell}, \hat{\chi}_{6\to i,\ell}), \quad \ell = 4, 5, 6, 7, 8, 9; \tag{2.169}$$

which is carried out on all the computational cells.

(2) By inserting the limited second-order scaled derivatives $\tilde{\chi}_{i,\ell}$ ($\ell = 4, 5, 6, 7, 8, 9$) given in Eq. (2.169) into Eq. (2.168), the first-order scaled derivatives of the SRs are recomputed through

$$\begin{aligned} \hat{\chi}_{j\to i,1} &= \chi_{j,1} + 2\Delta x\tilde{\chi}_{j,4} + \Delta y\tilde{\chi}_{j,7} + \Delta z\tilde{\chi}_{j,8}, \\ \hat{\chi}_{j\to i,2} &= \chi_{j,2} + 2\Delta y\tilde{\chi}_{j,5} + \Delta x\tilde{\chi}_{j,7} + \Delta z\tilde{\chi}_{j,9}, \quad j = 1, 2, \ldots, N. \\ \hat{\chi}_{j\to i,3} &= \chi_{j,3} + 2\Delta z\tilde{\chi}_{j,6} + \Delta x\tilde{\chi}_{j,8} + \Delta y\tilde{\chi}_{j,9}, \end{aligned}$$

Then, these recomputed first-order scaled derivatives are limited by

$$\tilde{\chi}_{i,\ell} = L(\chi_{i,\ell}, \hat{\chi}_{1\to i,\ell}, \hat{\chi}_{2\to i,\ell}, \hat{\chi}_{3\to i,\ell}, \hat{\chi}_{4\to i,\ell}, \hat{\chi}_{5\to i,\ell}, \hat{\chi}_{6\to i,\ell}), \quad \ell = 1, 2, 3$$

After all these scaled derivatives are limited, one finally obtains a newly reconstructed polynomial

$$X_i(\mathbf{x}) = X_i + \tilde{\boldsymbol{\chi}}_i\boldsymbol{\varphi}_i = X_i + \sum_{\ell=1}^{9} \tilde{\chi}_{i,\ell}\varphi_{i,\ell}, \quad \mathbf{x} \in V_i.$$

which is the targeted reconstruction polynomial of cell V_i.

The specific form of WBAP limiter, i.e., $L(\cdot)$ in Eq. (2.169), is given by

$$L(a_0, a_1, a_2, \ldots, a_M) = a_0\mathcal{W}(1, \frac{a_1}{a_0}, \frac{a_2}{a_0}, \ldots, \frac{a_M}{a_0}),$$

where $a_0, a_1, a_2, \ldots a_M$ are the variables to be limited. For the high-order finite-volume schemes, the first argument a_0 is specifically related to the PR of the current cell, and a_1 through a_M are the scaled derivatives of the SRs on the current cell. Li and Ren [280] employed the WBAP-L2 limiter

$$\mathcal{W}^{L2}(1, \sigma_1, \ldots, \sigma_M) = \begin{cases} \frac{n+\sum_{k=1}^{M}\frac{1}{\sigma_k^{q-1}}}{n+\sum_{k=1}^{M}\frac{1}{\sigma_k^{q}}}, & \text{if } \sigma_1, \ldots, \sigma_M > 0; \\ 0, & \text{else.} \end{cases}$$

where $\sigma_M = a_M/a_0$ ($\sigma_0 = 1$), and $n = 5, q = 4$ are chosen by Li and Ren [280] for controlling the numerical oscillations.

It is noted that the key element of the successive limiting procedure [278, 280, 489] is to apply the limiter to the high-order scaled derivatives of the SRs first over all the cells of the computational domain and the lower order scaled derivatives are evaluated again according to the already limited high-order scaled derivatives.

By the way, the conception of the primary K-exact polynomial reconstruction and the secondary reconstructions derived from the primary K-exact polynomial reconstructions was also used by Li and Ren [279] to construct the K-exact WENO scheme. By utilizing the advantages of both the K-exact and WENO reconstructions, this procedure performs the primary K-exact reconstruction for every control volume on a systematically chosen central stencil. Based on the primary reconstruction in a cell of interest, additional candidate reconstruction polynomials are derived by means of very simple and efficient SRs. The weighted average procedure of the WENO scheme is then applied to the primary and secondary K-exact polynomial reconstructions to ensure the shock-capturing capability of the scheme. This procedure ingeniously combines the simplicity of the K-exact reconstruction with the robustness of the WENO schemes and paves a systematic and unified way to construct high-order accurate shock capturing schemes. After an efficient problem-independent shock detector is introduced for the purpose of further improving the efficiency, numerical experiments show that in most regions of the smooth flow field, only the primary K-exact reconstruction is actually used without using any limiting procedure.

(II) Distance derivative weighted ENO (DDWENO)

The process of the DDWENO limiter proposed by Liu et al. [290] can be divided into two steps: first, K-exact reconstruction with a fixed stencil is implemented, and meanwhile the smoothness indicator of each stencil is computed; then, all the derivatives in the same fixed stencil are limited and weighted according to the smoothness of each cell.

The smoothness indicator based on the standard deviation coefficients is defined to measure smoothness of state variables for a stencil. To introduce the standard deviation coefficients, consider a specified cell i with its stencil of direct neighboring cells j's. The standard deviation coefficient ζ_i is defined as the ratio of the standard deviation σ_i to the average value $\overline{X}$ of the stencil,

$$\zeta_i = \frac{\sigma_i}{\left|\overline{X}\right|}, \quad \sigma_i = \sqrt{\frac{1}{N+1}\left(\left(X_i - \overline{X}\right)^2 + \sum_{j=1}^{N}\left(X_j - \overline{X}\right)^2\right)},$$

where the average value $\overline{X}$ of the stencil is defined by $\overline{X} = \frac{1}{N+1}\left(X_i + \sum_{j=1}^{N} X_j\right)$, and N is the number of neighbor cells in a reconstruction stencil. The value of the standard deviation coefficients stand for the smoothness of a stencil. In smooth regions, the coefficients are very small, approaching to zero. But in discontinuous regions, it is relatively large. Since only cell averaged values in the fixed stencil are needed, the value of the standard deviation coefficients ζ_i is simple and easy to be calculated. In the limiting step, the standard deviation coefficients ζ_i is used as the smoothness indicator. As pointed out by Liu et al. [290], the standard deviation coefficients can accurately detect shock discontinuities and have high efficiency, especially on unstructured grids.

The DDWENO limiter weights all the derivatives of state variables of each cell in the fixed reconstruction stencil according to the smoothness indicator or the standard deviation coefficients. That is, the derivatives $\chi_{i,\ell}$ in Eq. (2.160) of the central cell or current cell i in the stencil are limited

$$\tilde{\chi}_{i,\ell} = L\left(\chi_{i,\ell}, \chi_{1,\ell}, \chi_{2,\ell}, \ldots, \chi_{N,\ell}\right),$$

where $L(\cdot)$ is the limiting function, and $\chi_{j,\ell}$ ($\ell = 1, 2, \ldots, N_D - 1$) denote the derivatives of state variables of neighbor cells in the fixed stencil.

Considering that the larger the standard deviation coefficient is, the stronger discontinuity the cell approaches, hence, one weight the derivatives by the inverse of the standard deviation coefficients with the nonlinear weighted coefficients of each cell expressed as

$$\begin{aligned}
\omega_i^{(1)} &= \frac{\alpha_i}{\sum_{m=1}^{N+1}\alpha_m}, \quad \alpha_i = \frac{1}{\zeta_i + \delta}, \\
\omega_j^{(1)} &= \frac{\alpha_j}{\sum_{m=1}^{N+1}\alpha_m}, \quad \alpha_j = \frac{1}{\zeta_j + \delta}, \quad j = 1, 2, \ldots, N,
\end{aligned}$$

where δ is a tolerance to avoid being divided by zero, and ζ_i denotes the standard deviation coefficient of cell i.

The nonlinear weighted coefficients are very effective in capturing shock waves and suppressing numerical oscillations near discontinuities. However, it cannot preserve numerical accuracy in smooth regions for high-order schemes. Therefore, in order to effectively preserve numerical accuracy, Liu et al. [290] further introduce another weighted coefficients related to geometric distance, $\omega_i^{(2)}$, which can be expressed as follows

$$\begin{aligned}
\omega_i^{(2)} &= \left(\frac{1}{D_i/D_{avg}}\right)^{S(\zeta_i)}, \quad D_i = \frac{1}{N_{\text{faces}}}\sum_{m=1}^{N_{\text{faces}}}\left|\mathbf{x}_i - \mathbf{x}_m\right|, \\
\omega_j^{(2)} &= \left(\frac{1}{D_j/D_{avg}}\right)^{S(\zeta_i)}, \quad D_j = \frac{1}{N_{\text{faces}}}\sum_{m=1}^{N_{\text{faces}}}\left|\mathbf{x}_j - \mathbf{x}_m\right|, \quad j = 1, 2, \ldots, N, \\
D_{avg} &= \frac{1}{N}\sum_{j=1}^{N}\left|\mathbf{x}_j - \mathbf{x}_i\right|, \quad S(\zeta_i) = \min\left(\frac{\varepsilon}{\zeta_i + \delta}, 10\right),
\end{aligned}$$

where D_i, D_j denote the average distance of the central cell center $\mathbf{x}_i$, and other neighbor cell center $\mathbf{x}_j$ in the reconstruction stencil to the face centers $\mathbf{x}_m$ of the central cell. D_{avg}, the local distance of the stencil, is defined by the average distance of the central cell center $\mathbf{x}_i$ and other neighbor cell centers $\mathbf{x}_j$ in the stencil. N_{faces} denotes the number of faces for cell i, and N is the number of cells in a stencil except the central cell. As suggested by Liu et al. [290], $\delta = 10^{-8}$ is set to avoid division by zero, and ε is a free control parameter in $[0, 0.1]$. Due to the maximum digit of the computer, the upper bound of $S(\zeta_i)$ is limited to be 10 in order to avoid data overflow. In smooth regions of the flowfield, the standard deviation coefficients are very small, and the values of $S(\zeta_i)$ are close to 10. Therefore, the proportion of the central cell in the reconstruction stencil is relatively large so as to effectively preserve numerical accuracy in smooth regions. In non-smooth regions, the values of $S(\zeta_i)$ are very small, and the weighted coefficients $\omega_i^{(2)}$ have small effect, which can suppress the oscillations near discontinuities.

Referring to the conception of the BAP limiter, Liu et al. [290] carried out an extra nonlinear transformation on all the derivatives to reach the final expression of the so-called DDWENO limiter as follows

$$\tilde{\chi}_{i,\ell} = L\left(\chi_{i,\ell}, \chi_{1,\ell}, \chi_{2,\ell}, \ldots, \chi_{N,\ell}\right) = B^{-1}\left(\omega_i^{(1)}\omega_i^{(2)}B(\chi_{i,\ell}) + \sum_{j=1}^{N}\omega_j^{(1)}\omega_j^{(2)}B(\chi_{j,\ell})\right), \tag{2.170}$$

where the nonlinear transformation function and its self-similar function adopted are

$$B(x) = \frac{x}{\sqrt{1+x^2}}, \quad B^{-1}(x) = \frac{x}{\sqrt{1-x^2}}.$$

That is, one first transforms the derivatives with a kind of biased function, and then reduce the real derivatives by the inverse transformation with the self-similar function. The nonlinear transformation does not degenerate numerical accuracy, on the contrary, it can improve the stability of the flow solver.

By using the inverse distance weighting and the standard deviation coefficient, a distance weighted BAP (DWBAP) limiter for the high order finite volume method was proposed by Liu and Zhang [289] for unstructured grids. The DWBAP limiter is defined as

$$\tilde{\chi}_{i,\ell} = L\left(\chi_{i,\ell}, \chi_{1,\ell}, \chi_{2,\ell}, \ldots, \chi_{N,\ell}\right) = B^{-1}\left(\omega_i B(\chi_{i,\ell}) + \sum_{j=1}^{N}\omega_j B(\chi_{j,\ell})\right) \tag{2.171}$$

where $B(x) = \arctan(x)$ is the biased function, and the weighting coefficient ω for the biased function is expressed as follows

$$\omega_i = \frac{\beta_i}{\beta_i + \sum_{j=1}^{N}\beta_j}, \quad \beta_i = \frac{1}{\left(\frac{1}{N_{\text{faces}}}\sum_{m=1}^{N_{\text{faces}}}|\mathbf{x}_i - \mathbf{x}_m|\right)^{S(\zeta_i)}}$$

$$\omega_j = \frac{\beta_j}{\beta_i + \sum_{j=1}^{N} \beta_j}, \quad \beta_j = \frac{1}{\left(\frac{1}{N_{\text{faces}}} \sum_{m=1}^{N_{\text{faces}}} \left|\mathbf{x}_j - \mathbf{x}_m\right|\right)^{S(\zeta_i)}}$$

Similarly, $S(\zeta_i) = \frac{1}{\sqrt{\zeta_i + \delta} + 0.05}$ is a smooth function of the coefficient of standard deviation ζ_i. Again, $\delta = 10^{-8}$, and the value 0.05 is the maximum control parameter for limiting the upper bound of $S(\zeta_i)$ to avoid data overflow [289].

Different from the direct slope limiter, inverse distance weights adopted in DDWENO limiter in Eq. (2.170) and DWBAP limiter in Eq. (2.171) distribute the weighting coefficient among each cell in the reconstruction stencil that depends on the non-linear function of distance, and ultimately limit the derivatives of central cell by combining derivatives with different weights.

2.6.2.6 CENO (Central ENO)

In a series of papers (e.g., [144–147, 208–214, 454]), Ivan and his co-authors proposed the high order central essentially non-oscillatory (CENO) scheme. CENO scheme is a hybrid scheme of switching between the high order K-exact reconstruction and a limited piecewise linear ($K = 1$) reconstruction. In the abbreviation "CENO", "C" means "central" since both the high-order and the low-order stencils are central with respect to the cell, and "ENO" implies that it satisfies the ENO property of Harten [190].

CENO scheme is constructed on a fixed central stencil that avoids the complexity of considering multiple non-central stencil configurations that characterizes traditional ENO/WENO schemes. The fixed stencil used during the CENO reconstruction procedure makes it easily applicable to meshes having irregular and unstructured topologies. CENO maintains solution monotonicity in regions of large gradients or discontinuities by reverting the high order K-exact reconstruction to a limited piecewise linear ($K = 1$) reconstruction. During its implementation, CENO detects regions where the order of the reconstruction should be reduced and the limiters applied, by computing a smoothness indicator for every variable individually within each cell as part of a post-analysis step after the unlimited K-exact reconstruction has been built. Then, the smoothness indicator is used to ensure that the limited linear reconstruction procedure is applied to computational cells with under-resolved and/or non-smooth solution content and the unlimited K-exact reconstruction scheme where the solution is fully resolved. It should be emphasized that local switching to lower order is only performed for those reconstructed variables that are deemed non-smooth or under-resolved. The limiting procedure used in the limited piecewise linear ($K = 1$) reconstruction of the CENO scheme can be the multi-dimensional limiting process (MLP) [378], Barth-Jesperson limiter and Venkatakrishnan limiter.

Here we briefly describe the CENO scheme according to Ivan et al. [214]. This hybrid reconstruction method combines the unlimited K-exact least-square reconstruction $X^K_{\text{unlimit}}(\mathbf{x})$ in smooth regions with a limited second order accuracy least-square reconstruction $X^1_{\text{limit}}(\mathbf{x})$ in the discontinuities regions. To switch between the two reconstructions, a smoothness indicator

$$\mathbb{S} = \frac{\alpha}{\max(1-\alpha, \varepsilon)} \frac{N_{sos} - N_D}{N_D - 1}, \alpha = 1 - \frac{\sum_j \left(X_j^K(\mathbf{x}_j) - X_i^K(\mathbf{x}_j)\right)^2}{\sum_j \left(X_j^K(\mathbf{x}_j) - X_i\right)^2}$$

is used as a criteria, where the ranges of the indices j are taken to include either the whole or a subset of the supporting reconstruction stencil for current cell i of interest, and $\mathbf{x}_j$ is the centroid of cell j in the stencil of cell i. $N_D = \frac{1}{6}(K+1)(K+2)(K+3)$ (number of degrees of freedom in 3D) denotes the number of unknown polynomial coefficients (e.g., 20 for $K = 3$) of the Taylor expansion for $X_i(\mathbf{x})$. ε is introduced to avoid division by zero, such as $\epsilon = 10^{-8}$. Here, N_{SOS} denotes the size of the stencil used for computing the smoothness indicator. The stencil used for computing the smoothness indicator can be chosen smaller than the reconstruction stencil. For instance, Ivan et al. [214] computed the smoothness indicator using a 27-cell stencil (with 26 first-degree neighbors for hexahedral cell in 3D case).

The parameter, $\alpha \in (-\infty, 1]$, measures how accurately centroidal solution values of neighboring cells can be reproduced using the reconstruction for cell i: in smooth flow α is close to one, while at cells with a discontinuity or an under-resolved feature, α tends away from one. The smoothness indicator $\mathbb{S}$ has range $-\frac{N_{sos}-N_D}{N_D-1} < \mathbb{S} < \frac{N_{sos}-N_D}{N_D-1}/\epsilon$: for smooth variation $\mathbb{S}$ is large, and for non-smooth or under-resolved features $\mathbb{S}$ is small. Typically, a cut-off value, $\mathbb{S}_C$, is selected in the range of 1000–5000 [209, 214, 454] to measure the smoothness. Then, the high-order reconstruction is used if $\mathbb{S} > \mathbb{S}_C$, while the limited low-order reconstruction is invoked if $\mathbb{S} < \mathbb{S}_C$. The form of the smoothness indicator, detailed by Ivan [208], is inspired by the definition of multiple-correlation coefficients and least-squares goodness-of-fit testing.

The above procedure can be carried out in an alternative manner [214, 454]. Using standard Taylor approximation theory, K-exactness, a continuous and differentiable reconstructed function satisfies

$$X_j^K(\mathbf{x}_j) - X_i^K(\mathbf{x}_j) = O(\Delta h^{K+1}), X_j^K(\mathbf{x}_j) - X_i = O(\Delta h)$$

with Δh the local mesh spacing. This implies that $\alpha = 1 - O(\Delta h^{2K})$ for smooth resolved solution content. In this case, it is evident that $\alpha \to 1$ and $\mathbb{S} \to \infty$ as $\Delta h \to 0$ and for non-smooth solutions one expects that

$$X_j^K(\mathbf{x}_j) - X_i^K(\mathbf{x}_j) = O(1), X_j^K(\mathbf{x}_j) - X_i = O(1)$$

and therefore α will generally not be close to one. By taking into account derivatives at the centroid of cell i, a variability measurement

$$\zeta_i = \sqrt{\frac{1}{N_D' - 1} \sum_{\substack{p_1=0 \\ 0<p_1+p_2+p_3\le K'}}^{K'} \sum_{p_2=0}^{K'} \sum_{p_3=0}^{K'} \left(\frac{1}{p_1!p_2!p_3!} \frac{\partial^{p_1+p_2+p_3} X}{\partial x^{p_1} \partial y^{p_2} \partial z^{p_3}} \bigg|_{\mathbf{x}_i} \right)^2 \xi_V(p_1, p_2, p_3)}, \tag{2.172}$$

is introduced [214, 454] to measure the variability of solution variable $X(\mathbf{x})$ in cell i, where $\xi_V(p_1, p_2, p_3) = \left(V_i^{\frac{2}{3}}\right)^{p_1+p_2+p_3}$, $N_D' = \frac{1}{6}(K'+1)(K'+2)(K'+3)$ and $1 \leq K' \leq K$. Note that the term ζ_V provides an appropriate length scaling for each derivative based on the cell volume, and it has only K' distinct values due to the common variation range for p_1, p_2 and p_3.

When ζ_i is smaller than a threshold value (low variability), high-order reconstruction is always used, and only when ζ_i is greater than the threshold the smoothness indicator is computed and the CENO switching mechanism is activated. In particular, the smoothness indicator for the solution variable X is evaluated in cell i when

$$\zeta_i > \epsilon_A + \epsilon_R X_i$$

where ϵ_A and ϵ_R represent absolute and relative variability thresholds, chosen to be 10^{-5} [214]. It is mentioned that K' can be equal to K or less. As suggested by Ivan et al. [214], using only the first order derivatives is sufficient to calculate ζ_i, this simplification is obtained in Eq. (2.172) by computing the summation terms with $K' = 1$ and taking $N_D' = 4$.

The implementation procedure runs as follows:

1. If $\zeta_i < \epsilon_A + \epsilon_R X_i$, use the unlimited K-exact least-square reconstruction $X_i^K(\mathbf{x})$;
2. If $\zeta_i > \epsilon_A + \epsilon_R X_i$, calculate $\mathbb{S}$ to make a check, if $\mathbb{S} > \mathbb{S}_c$, use the unlimited K-exact least-square reconstruction $X_i^K(\mathbf{x})$; otherwise use the limited low-order reconstruction $X_i^1(\mathbf{x})$.

The CENO scheme [208, 211] overcomes some of the disadvantages associated with ENO and WENO schemes. The major difference between CENO and other ENO or WENO is that CENO uses just a single fixed central stencil rather than multiple stencils, which can reduce the computational amount. The central stencil (also the most accurate stencil) provides the most accurate reconstruction due to cancellation of truncation errors, and thus the CENO scheme is relatively-easily extended to multi-dimensional unstructured meshes. By a solution smoothness indicator, the switching in the hybrid technique makes the CENO reconstruction an effective method for eliminating the appearance of spurious numerical oscillations in under-resolved regions and in regions that contain strong discontinuities and/or shocks.

2.6.3 High-Order Variational Reconstruction Approach

Keep in mind that, for a degree K polynomial in high-order K-exact reconstruction, the number of unknown coefficients is $N_D - 1 = \frac{(K+3)(K+2)(K+1)}{6} - 1$ for three-dimensional problems. With an increase of K, the number of unknown coefficients increases very fast. For example, for the cubic reconstruction ($K = 3$), the number of unknowns is 19 and thus at least 19 cells or control volumes have to be included in the reconstruction stencil to determine the unknowns in the 3-exact reconstruction. In addition, the matrix of the reconstruction may be singular. To improve the stability of the schemes, the reconstruction stencil should have a number of cells at least 1.5 times of the number of unknowns, as recommended by Li [278], say, at least 28 cells for the cubic reconstruction. What is more, the large stencil will reduce the computational efficiency especially in parallel computing and bring difficulties in designing high-order treatment of boundary conditions.

To overcome this deficiency brought by the large stencil in the high-order K-exact reconstruction, the variational reconstruction (VR) has been suggested [276, 373, 490] to reconstruct arbitrarily high-order polynomials by using a compact stencil involving only face-neighboring cells. In Eq. (2.97) of Sect. 2.6.1.2, second order accurate VR method has been introduced. In this subsection, high-order VR approach is formulated.

To begin with, following the K-exact reconstruction polynomial in Eq. (2.152) of degree K, we consider a piecewise degree K polynomial defined on V_i in the form

$$X_i(\mathbf{x}) = X_i + \boldsymbol{\chi}_i^T \boldsymbol{\varphi}_i, \tag{2.173}$$

where $\boldsymbol{\chi}_i = (\chi_{i,1}, \chi_{i,2}, \ldots, \chi_{i,D_N-1})^T$, and $\chi_{i,\ell}$ ($\ell = 1, 2, \ldots, D_N - 1$) is the coefficient corresponding to the derivatives of the reconstruction polynomial; $\boldsymbol{\varphi}_i = (\varphi_{i,1}, \varphi_{i,2}, \ldots, \varphi_{i,D_N-1})^T$ is the vector of the basis function on V_i, and $\varphi_{i,\ell}$ ($\ell = 1, 2, \ldots, D_N - 1$) is the zero-mean basis sharing the form of $\frac{1}{\Delta h_i^{p_1+p_2+p_3}} \left((x - x_i)^{p_1}(y - y_i)^{p_2}(z - z_i)^{p_3} - \overline{x^{p_1}y^{p_2}z^{p_3}}_i\right)$ with $\overline{x^{p_1}y^{p_2}z^{p_3}}_i = \frac{1}{V_i}\int_{V_i}(x - x_i)^{p_1}(y - y_i)^{p_2}(z - z_i)^{p_3}dV$ defined previously. Δh_i is the length scale for the non-dimensionalization of the basis functions, and for the hexahedral cell, $\Delta h_i = \max(\mathrm{Rad}(V_i), \sqrt[3]{V_i})$, where $\mathrm{Rad}(V_i)$ is the radius of the inscribed sphere of control volume V_i. This length scale is used to avoid growth of the condition number of the reconstruction matrix with grid refinement [148, 489]. It should be noted that the difference between the reconstruction polynomial in Eq. (2.173) and that in Eq. (2.152) is the scaling factor $\frac{1}{\Delta h_i^{p_1+p_2+p_3}}$ used in the zero-mean basis function.

The principal idea of VR is to compute the unknown coefficients $\boldsymbol{\chi}_i = (\chi_{i,1}, \chi_{i,2}, \ldots, \chi_{i,D_N-1})^T$ by constructing a specific functional form based on Eq. (2.173) and minimizing the resulting functional. This functional may be designed as follows:

$$\mathcal{F} = \sum_{j\in\{V_j\}_i} \mathcal{F}_{ij} \tag{2.174}$$

where $\{V_j\}_i$ is the set of surrounding neighbors of the current cell i, and $\mathcal{F}_{ij}$ is a measure of the interface jump of the reconstruction polynomial and its various order of derivatives across the cell interface. According to Pan et al. [373], $\mathcal{F}_{ij}$ is defined as the following the interface jump integration function

$$\mathcal{F}_{ij} = \frac{\omega_{ij}}{A_{ij}} \sum_{\substack{p_1=0 \\ p_1+p_2+p_3 \le K}}^{K} \sum_{p_2=0}^{K} \sum_{p_3=0}^{K} \int_{A_{ij}} \left[\frac{L_x^{p_1} L_y^{p_2} L_z^{p_3}}{p_1! p_2! p_3!} \left(\frac{\partial^{p_1+p_2+p_3} X_i(\mathbf{x})}{\partial x^{p_1} \partial y^{p_2} \partial z^{p_3}} - \frac{\partial^{p_1+p_2+p_3} X_j(\mathbf{x})}{\partial x^{p_1} \partial y^{p_2} \partial z^{p_3}} \right) \right]^2 dA \quad (2.175)$$

where A_{ij} stands for the interface shared by control volumes V_i and V_j, and also A_{ij} is the area of A_{ij}. Here, ω_{ij} is the weight of surface A_{ij}, and for the hexahedral cell $\omega_{ij} = \max(\mathrm{Rad}(A_{ij}), \sqrt{A_{ij}})$, where $\mathrm{Rad}(A_{ij})$ is the radius of the circumcircle of the face A_{ij}. L_x, L_y, and L_z are the scaling factors to make all terms in the summation to have the same dimension. In practice, choices of the scaling factors can be made as follows

$$L_x = (x - x_i), L_y = (y - y_i), L_z = (z - z_i),$$

or

$$L_x = \frac{1}{2}(\Delta h_{i,x} + \Delta h_{j,x}), L_y = \frac{1}{2}(\Delta h_{i,y} + \Delta h_{j,y}), L_z = \frac{1}{2}(\Delta h_{i,z} + \Delta h_{j,z}),$$

in which $\Delta h_{i,x} = \max(x_{i,v\ell}) - \min(x_{i,v\ell})$, $\Delta h_{i,y} = \max(y_{i,v\ell}) - \min(y_{i,v\ell})$, $\Delta h_{i,z} = \max(z_{i,v\ell}) - \min(z_{i,v\ell})$, where subscript $v\ell$ stands for each of the vertex $\mathbf{x}_{i,v\ell} = (x_{i,v\ell}, y_{i,v\ell}, z_{i,v\ell})$ of the control volume V_i. The latter choice of the scaling factors has been used by Pan et al. [373].

To compute the unknown coefficients $\boldsymbol{\chi}_i = (\chi_{i,1}, \chi_{i,2}, \ldots, \chi_{i,D_N-1})^T$ in Eq. (2.173), first substitute $X_i(\mathbf{x})$, $X_j(\mathbf{x})$ into Eq. (2.175), and then minimize $\mathcal{F}$ by requiring

$$\frac{\partial \mathcal{F}}{\partial \chi_{i,m}} = 0, \quad m = 1, 2, \ldots, D_N - 1 \quad (2.176)$$

In Eq. (2.175), only $\mathcal{F}_{ij}$ with its one neighboring cell being V_i will have none-zero terms, which means that Eq. (2.175) is only related to the information on V_i and its face-neighboring cells. Therefore, the reconstruction relation Eq. (2.175) uses a compact stencil, in other words, $\{V_j\}_i$ only includes the direct face-sharing neighbor cells of cell i.

After defining the reconstruction stencil, Eq. (2.176) can be written as

$$\sum_{j\in\{V_j\}_i} \left(\sum_{r=1}^{D_N-1} \frac{\omega_{ij}}{A_{ij}} \chi_{i,r} \sum_{\substack{p_1=0 \\ p_1+p_2+p_3 \le K}}^{K} \sum_{p_2=0}^{K} \sum_{p_3=0}^{K} \int_{A_{ij}} \left[\frac{L_x^{p_1} L_y^{p_2} L_z^{p_3}}{p_1! p_2! p_3!} \right]^2 \left(\frac{\partial^{p_1+p_2+p_3} \varphi_{i,m}(\mathbf{x})}{\partial x^{p_1} \partial y^{p_2} \partial z^{p_3}} \frac{\partial^{p_1+p_2+p_3} \varphi_{i,r}(\mathbf{x})}{\partial x^{p_1} \partial y^{p_2} \partial z^{p_3}} \right) dA \right)$$

$$= \sum_{j\in\{V_j\}_i} \left(\sum_{r=1}^{D_N-1} \frac{\omega_{ij}}{A_{ij}} \chi_{j,r} \sum_{\substack{p_1=0 \\ p_1+p_2+p_3 \le K}}^{K} \sum_{p_2=0}^{K} \sum_{p_3=0}^{K} \int_{A_{ij}} \left[\frac{L_x^{p_1} L_y^{p_2} L_z^{p_3}}{p_1! p_2! p_3!} \right]^2 \right.$$

$$\left(\frac{\partial^{p_1+p_2+p_3}\varphi_{i,m}(\mathbf{x})}{\partial x^{p_1}\partial y^{p_2}\partial z^{p_3}} \frac{\partial^{p_1+p_2+p_3}\varphi_{j,r}(\mathbf{x})}{\partial x^{p_1}\partial y^{p_2}\partial z^{p_3}} \right) dA \Bigg)$$

$$+\sum_{j\in\{V_j\}_i}\left(\frac{\omega_{ij}}{A_{ij}}(X_j - X_i)\int_{A_{ij}}\varphi_{i,m}(\mathbf{x})dA\right), \quad m = 1, 2, \ldots, D_N - 1$$

Corresponding to cell V_i, the above equations can be written locally in a matrix form as

$$\mathcal{A}_{ii}\boldsymbol{\chi}_i = \sum_{j\in\{V_j\}_i}\mathcal{A}_{ij}\boldsymbol{\chi}_j + \mathbf{b}_i \tag{2.177}$$

where $\boldsymbol{\chi}_i = (\chi_{i,1}, \chi_{i,2}, \ldots, \chi_{i,D_N-1})^T$, and

$$\mathcal{A}_{ii} = \left[\sum_{j\in\{V_j\}_i}\left(\frac{\omega_{ij}}{A_{ij}}\sum_{\substack{p_1=0\\p_1+p_2+p_3\le K}}^{K}\sum_{p_2=0}^{K}\sum_{p_3=0}^{K}\int_{A_{ij}}\left[\frac{L_x^{p_1}L_y^{p_2}L_z^{p_3}}{p_1!p_2!p_3!}\right]^2 \left(\frac{\partial^{p_1+p_2+p_3}\varphi_{i,m}(\mathbf{x})}{\partial x^{p_1}\partial y^{p_2}\partial z^{p_3}} \frac{\partial^{p_1+p_2+p_3}\varphi_{i,r}(\mathbf{x})}{\partial x^{p_1}\partial y^{p_2}\partial z^{p_3}} \right) dA\right)\right]_{(D_N-1)\times(D_N-1)}$$

$$\mathcal{A}_{ij} = \left[\frac{\omega_{ij}}{A_{ij}}\sum_{\substack{p_1=0\\p_1+p_2+p_3\le K}}^{K}\sum_{p_2=0}^{K}\sum_{p_3=0}^{K}\int_{A_{ij}}\left[\frac{L_x^{p_1}L_y^{p_2}L_z^{p_3}}{p_1!p_2!p_3!}\right]^2 \frac{\partial^{p_1+p_2+p_3}\varphi_{i,m}(\mathbf{x})}{\partial x^{p_1}\partial y^{p_2}\partial z^{p_3}} \frac{\partial^{p_1+p_2+p_3}\varphi_{j,r}(\mathbf{x})}{\partial x^{p_1}\partial y^{p_2}\partial z^{p_3}} dA\right]_{(D_N-1)\times(D_N-1)}$$

$$\mathbf{b}_i = \left[\sum_{j\in\{V_j\}_i}\left(\frac{\omega_{ij}}{A_{ij}}(X_j - X_i)\int_{A_{ij}}\varphi_{i,m}(\mathbf{x})dA\right)\right]_{(D_N-1)\times 1}$$

These face integral terms in the reconstruction matrices can be computed exactly using Gaussian quadrature formulas with sufficient precision. Over all control volumes, Eq. (2.177) can be assembled into a large system of linear equations. It can be proven [373, 490] that there exists a unique solution for this system of equations, since the corresponding matrix $\mathbf{A}$ is symmetric and positive definite if Eq. (2.177) is assembled in a compact form $\mathbf{A}\boldsymbol{\chi} = \mathbf{b}$, $\boldsymbol{\chi} = (\boldsymbol{\chi}_1, \boldsymbol{\chi}_2, \ldots, \boldsymbol{\chi}_{D_N-1})$, $\mathbf{b} = (\mathbf{b}_1, \mathbf{b}_2, \ldots, \mathbf{b}_{D_N-1})$. For detailed proof, refer to Wang et al. [490]. After the unknown coefficients in Eq. (2.173) are available by solving Eq. (2.177), reconstructed polynomials on all control volumes can be obtained. The simplest block-Jacobian iterative solver

$$\boldsymbol{\chi}_i^{(k+1)} = \mathcal{A}_{ii}^{-1}\left(\sum_{j\in\{V_j\}_i}\mathcal{A}_{ij}\boldsymbol{\chi}_j^{(k)} + \mathbf{b}_i\right), i = 1, 2, \ldots, \text{number\_cells} \tag{2.178}$$

where k is the index of iteration. It is indicated [490] that the iterative methods, such as the block Jacobi, Gauss-Seidel and successive over-relaxation (SOR) (with the over-relaxation factor $0 < \omega < 2$) iterations, are convergent in solving the linear system Eqs. (2.177) and (2.178) of the variational reconstruction. Also, a use of the Gauss-Seidel technique with over relaxation can significantly improve the convergence rate [373, 490].

Alternatively, Wang et al. [490] suggested the following interface jump integration

$$\mathcal{F}_{ij} = \frac{1}{d_{ij}} \sum_{p=0}^{K} \int_{A_{ij}} \left[\frac{d_{ij}^{p}}{p!} \left(\frac{\partial^p X_i(\mathbf{x})}{\partial \mathbf{n}^p} - \frac{\partial^p X_j(\mathbf{x})}{\partial \mathbf{n}^p} \right) \right]^2 dA$$

where $d_{ij} = |\mathbf{x}_i - \mathbf{x}_j|$ is the distance between the centroids of the two cells, and $\mathbf{n}$ denotes the coordinate along the normal direction. Correspondingly, associated to Eq. (2.176) there has the following matrix form

$$\mathcal{A}_{ii} = \left[\sum_{j \in \{V_j\}_i} \sum_{p=0}^{K} \int_{A_{ij}} \frac{d_{ij}^{2p-1}}{(p!)^2} \frac{\partial^p \varphi_{i,m}(\mathbf{x})}{\partial \mathbf{n}^p} \frac{\partial^p \varphi_{i,r}(\mathbf{x})}{\partial \mathbf{n}^p} dA \right]_{(D_N-1)\times(D_N-1)}$$

$$\mathcal{A}_{ij} = \left[\sum_{p=0}^{K} \int_{A_{ij}} \frac{d_{ij}^{2p-1}}{(p!)^2} \frac{\partial^p \varphi_{i,m}(\mathbf{x})}{\partial \mathbf{n}^p} \frac{\partial^p \varphi_{j,r}(\mathbf{x})}{\partial \mathbf{n}^p} dA \right]_{(D_N-1)\times(D_N-1)}$$

$$\mathbf{b}_i = \left[\sum_{j \in \{V_j\}_i} \left(\frac{1}{d_{ij}} (X_j - X_i) \int_{A_{ij}} \varphi_{i,m}(\mathbf{x}) dA \right) \right]_{(D_N-1)\times 1}$$

The variational reconstruction provides a general framework of reconstructing piecewise polynomials on unstructured grids. Various definitions of the jump integration can be created for the variational reconstruction. The jump integrations are in principle selected such that the resulting reconstruction polynomials should be K-exact and relying on compact stencils, which are satisfied by all the designed jump integrations. For example, the jump integration $\mathcal{F}_{ij}$ can be defined [490] as

$$\mathcal{F}_{ij} = \frac{1}{d_{ij}} \sum_{p=0}^{K} \int_{\partial V_i} \left[\frac{d_{ij}^{p}}{p!} \left(\frac{\partial^p X_i(\mathbf{x})}{\partial \mathbf{n}^p} - \frac{\partial^p X_j(\mathbf{x})}{\partial \mathbf{n}^p} \right) \right]^2 dA + \frac{1}{d_{ij}} \sum_{p=0}^{K} \int_{\partial V_j} \left[\frac{d_{ij}^{p}}{p!} \left(\frac{\partial^p X_i(\mathbf{x})}{\partial \mathbf{n}^p} - \frac{\partial^p X_j(\mathbf{x})}{\partial \mathbf{n}^p} \right) \right]^2 dA$$

or

$$\mathcal{F}_{ij} = \frac{1}{d_{ij}} \sum_{\substack{p_1=0\\p_1+p_2+p_3\le K}}^{K} \sum_{p_2=0}^{K} \sum_{p_3=0}^{K} \frac{d_{ij}^{p_1+p_2+p_3}}{p_1!p_2!p_3!} \times \left(\frac{1}{V_i} \int_{V_i} \frac{\partial^{p_1+p_2+p_3} X_i(\mathbf{x})}{\partial x^{p_1} \partial y^{p_2} \partial z^{p_3}} dV - \frac{1}{V_i} \int_{V_i} \frac{\partial^{p_1+p_2+p_3} X_j(\mathbf{x})}{\partial x^{p_1} \partial y^{p_2} \partial z^{p_3}} dV \right)^2 + \frac{1}{d_{ij}} \sum_{\substack{p_1=0\\p_1+p_2+p_3\le K}}^{K} \sum_{p_2=0}^{K} \sum_{p_3=0}^{K} \frac{d_{ij}^{p_1+p_2+p_3}}{p_1!p_2!p_3!} \times \left(\frac{1}{V_j} \int_{V_j} \frac{\partial^{p_1+p_2+p_3} X_i(\mathbf{x})}{\partial x^{p_1} \partial y^{p_2} \partial z^{p_3}} dV - \frac{1}{V_j} \int_{V_j} \frac{\partial^{p_1+p_2+p_3} X_j(\mathbf{x})}{\partial x^{p_1} \partial y^{p_2} \partial z^{p_3}} dV \right)^2$$

Besides, Li et al. [276] employed the following interface jump integration

$$\mathcal{F}_{ij} = \frac{1}{d_{ij}} \sum_{\substack{p_1=0\\p_1+p_2+p_3\le K}}^{K} \sum_{p_2=0}^{K} \sum_{p_3=0}^{K} \int_{A_{ij}} \left[\frac{d_{ij}^{p_1+p_2+p_3}}{p_1!p_2!p_3!} \left(\frac{\partial^{p_1+p_2+p_3} X_i(\mathbf{x})}{\partial x^{p_1} \partial y^{p_2} \partial z^{p_3}} - \frac{\partial^{p_1+p_2+p_3} X_j(\mathbf{x})}{\partial x^{p_1} \partial y^{p_2} \partial z^{p_3}} \right) \right]^2 dA$$

In the variational reconstruction, the WBAP limiter [280, 281] is used to suppress non-physical oscillations on the trouble cells. On each control volume, the construction of candidate polynomials for the WBAP limiter are obtained through the so-called secondary reconstruction technique. The piecewise reconstruction polynomials obtained from the variational reconstruction procedure is utilized as the primary reconstruction polynomials. Following the derivation of Eq. (2.167) in Sect. 2.6.2, the secondary reconstruction uses the continuous extensions of the primary reconstruction polynomials on face-neighbors of the current cell as the additional candidate polynomials. As done in Sect. 2.6.2.5, the WBAP limiter with secondary reconstruction and successive limiting procedure can be easily adopted for the variational reconstruction.

Additionally, to improve the efficiency of the limiting procedure, Li and Ren [279] introduced a problem-independent shock detector. The shock detector relies on the smoothness indicator IS_i defined by

$$\mathrm{IS}_i = \frac{\sum\limits_{j\in\{V_j\}_i} |X_i(\mathbf{x}_i) - X_j(\mathbf{x}_i)|}{N_j (\Delta h_i)^{\frac{K+1}{2}} \max(|X_j|, |X_i|)},$$

where N_j stands for the total number of direct face-neighboring cells $\{V_j\}_i$ and $N_j = 6$ is a natural choice for the hexahedral cell. Here, Δh_i is the local mesh spacing, say, taking the scale length $\Delta h_i = \sqrt[3]{V_i}$ of cell V_i. Obviously, $|X_i(\mathbf{x}_i) - X_j(\mathbf{x}_i)| = O(\Delta h_i^{K+1})$

in smooth region, and $|X_i(\mathbf{x}_i) - X_j(\mathbf{x}_i)| = O(1)$ near discontinuity. Therefore, as the cell size $\Delta h_i \rightarrow 0$, $\text{IS}_i \rightarrow 0$ in smooth regions and $\text{IS}_i \rightarrow \infty$ near the discontinuities. A threshold $\tilde{S}_{\text{dis}}$ is defined to distinguish smooth regions and shock regions. If the shock detector $\text{IS}_i < \tilde{S}_{\text{dis}}$ the unlimited variational reconstruction is used, otherwise the limited reconstruction is invoked. In solving two-dimensional gas dynamic Euler equations on unstructured grids [279, 280], $\tilde{S}_{\text{dis}} = 1$ is chosen for the second-order and third-order schemes. The shock detector is used solely for improving the computational efficiency as the WBAP limiter is accuracy-preserving for smooth numerical solutions [280].

To maintain the accuracy of reconstructions up to the boundary, the boundary conditions need to be considered in the reconstruction procedure. In the K-exact reconstructions, implementation of the boundary conditions solves the complicated constrained least-squares problem, as done in Sect. 2.6.2 for Dirichlet boundary conditions, Neumann boundary conditions and Robin boundary conditions. In the variational construction, the boundary conditions can be directly considered [490] by the introduction of proper boundary jump integration so that the cost function is given by

$$\mathcal{F} = \sum_{j \in \{V_j\}_i} \mathcal{F}_{ij} + \sum_{\text{bdf}=1}^{N_{\text{bdf}}} \mathcal{F}_{i,\text{bdf}} \tag{2.179}$$

where $\{V_j\}_i$ is the set of surrounding internal neighbors of the current boundary cell i, N_{bdf} is the number of boundary surfaces associated to the current boundary cell i, and $\mathcal{F}_{ij}$ can be any internal interface jump integration defined previously.

For Dirichlet boundary conditions, if $X(\mathbf{x}) = f_{\text{DB}}(\mathbf{x})$ holds along the boundary $A_{i,\text{bdf}}$, we can define the boundary surface jump integration

$$\mathcal{F}_{i,\text{bdf}} = \frac{1}{d_{i,\text{bdf}}} \sum_{p=0}^{K} \int_{A_{i,\text{bdf}}} \left[\frac{d_{ij}^p}{p!} \left(\frac{\partial^p X_i(\mathbf{x})}{\partial \mathbf{n}^p} - \frac{\partial^p f_{\text{DB}}(\mathbf{x})}{\partial \mathbf{n}^p} \right) \right]^2 dA$$

where $d_{i,\text{bdf}}$ is the distance between the centroid of boundary cell i and the face centroid of the boundary face $A_{i,\text{bdf}}$.

For a Neumann boundary condition, if $\frac{\partial X}{\partial \mathbf{n}}(\mathbf{x}) = f_{\text{NB}}(\mathbf{x})$ holds along the boundary $A_{i,\text{bdf}}$, we can define the boundary surface jump integration

$$\mathcal{F}_{i,\text{bdf}} = \frac{1}{d_{i,\text{bdf}}} \sum_{p=0}^{K-1} \int_{A_{i,\text{bdf}}} \left[\frac{d_{ij}^p}{p!} \left(\frac{\partial^{p+1} X_i(\mathbf{x})}{\partial \mathbf{n}^{p+1}} - \frac{\partial^p f_{\text{NB}}(\mathbf{x})}{\partial \mathbf{n}^p} \right) \right]^2 dA$$

where $\mathbf{n}$ is the unit normal vector to the boundary face $A_{i,\text{bdf}}$.

Robin boundary conditions for a dependent variable, $X(\mathbf{x})$, are imposed by

$$g_1(\mathbf{x})X(\mathbf{x}) + g_2(\mathbf{x})\frac{\partial X(\mathbf{x})}{\partial \mathbf{n}} = f_{\text{RB}}(\mathbf{x})$$

along the boundary $A_{i,\mathrm{bdf}}$, where $g_1(\mathbf{x})$ and $g_2(\mathbf{x})$ are the coefficients defining the contributions of Dirichlet and Neumann components to the boundary value, $f_{\mathrm{RB}}(\mathbf{x})$, respectively. Then, we can define the boundary surface jump integration

$$\mathcal{F}_{i,\mathrm{bdf}} = \frac{1}{d_{i,\mathrm{bdf}}} \int_{A_{i,\mathrm{bdf}}} \left(g_1(\mathbf{x}) X_i(\mathbf{x}) + g_2(\mathbf{x}) \frac{\partial X_i(\mathbf{x})}{\partial \mathbf{n}} - f_{\mathrm{RB}}(\mathbf{x}) \right)^2 dA +$$

$$\frac{1}{d_{i,\mathrm{bdf}}} \sum_{p=1}^{K-1} \int_{A_{i,\mathrm{bdf}}} \left[\frac{d_{ij}^p}{p!} \left(\frac{\partial^p \left(g_1(\mathbf{x}) X_i(\mathbf{x}) + g_2(\mathbf{x}) \frac{\partial X_i(\mathbf{x})}{\partial \mathbf{n}} \right)}{\partial \mathbf{n}^p} - \frac{\partial^p f_{\mathrm{RB}}(\mathbf{x})}{\partial \mathbf{n}^p} \right) \right]^2 dA$$

for sufficiently smooth $g_1(\mathbf{x})$, $g_2(\mathbf{x})$, and $f_{\mathrm{RB}}(\mathbf{x})$.

As we can see, various boundary conditions can be implemented in a straightforward manner by properly specifying the interface jump integration at the boundary. The variational reconstruction treatment of the boundary conditions is more accurate and flexible than the constrained least-squares reconstruction treatment introduced in Sect. 2.6.2. In multi-dimensional least-squares reconstruction, the reconstruction polynomial at boundary cells must be one-order lower in accuracy than that at inner cells because there are no enough constitutive equations to achieve a stable reconstruction scheme without decreasing spatial accuracy at the boundary cells. However, the variational reconstruction does not deteriorate the degrees of reconstruction polynomials at boundary cells. Therefore, the variational reconstruction can achieve higher order of accuracy with physical boundaries.

2.7 Transformation Between Conservative and Primitive Variables for Cell-Centered Reconstruction

Since only the cell-averages of the conservative variables are available from cell-centered FVM schemes, the cell-averages of the primitive and characteristic variables need to be derived from the cell-averages of the conservative variables. As it is known, the transformation of cell-averages from conservative to primitive variables are only second-order accurate. The reconstruction in characteristic variables is not suitable for FV schemes on unstructured grids and is ruled out, because there is not a certain characteristic direction for the entire domain, and characteristic variables are based upon a one-dimensional diagonalization of the system. Conservative variables $\mathbf{U} = (\rho, \rho\mathbf{v}, \mathbf{B}, e)^T$ (density, momentum, magnetic field and total energy) have the property that the discrete values $\mathbf{U}_i$ represent cell averages of the conserved quantities over cell V_i. This makes them natural candidates for higher than second order schemes, where the reconstruction is not piecewise linear anymore, and one has to distinguish carefully between cell averages and point-values at the cell centroid. Beyond second-order accuracy, the midpoint approximation traditionally used in finite volume methods must be abandoned. The algorithm must distinguish all cell-

/face-/edge-averaged quantities from pointwise cell-/face-/edge-centered approximations, however, products of averages are only equal to averages of products to within a second order approximation, that is, for hyperbolic systems in which the conversion from conservative to primitive variables is not analytic but only available numerically. Moreover, there are systems of magnetohydrodynamics equations with diffusion terms such as anisotropic thermal conduction for which the physical fluxes can only be written in terms of the primitive variables. As a result, a large number of such expensive conversions from the conservative to the primitive variables must be performed, namely one for each quadrature point for the integration of the numerical fluxes by solving (approximate) Riemann problems at the cell interfaces. Furthermore, cell-centered piecewise reconstructions in terms of the conservative variables provides the lower-quality shock-capturing fidelity when compared to the reconstruction performed in terms of the primitive variables (e.g., [517]). Hence for higher than first-order accuracy, the primitive variable reconstructions at the interface by means of an appropriate spatial interpolation are preferred, since it is convenient to compute the averaged fluxes and the averaged sources directly from the primitive variables $\mathbf{W}$. As well, the primitive variables $\mathbf{W} = (\rho, u, v, w, B_x, B_y, B_z, p)^T$ have desirable properties: the limited spatial reconstruction in these variables guarantees a nonnegative internal energy in the spatial reconstruction. Moreover, velocities and pressure remain constant across a contact discontinuity, while all components of the conservative variables have jumps across such waves. However, the underlying finite volume scheme is still advanced in time to update the cell averages of the conserved variables $\mathbf{U}_i$, which are the only known input quantities at the reference time level. Hence, combining the advantages of conservative and primitive variables the whole reconstruction strategy may run through the following procedure (e.g., [387, 517]):

(1) Obtain a reconstructed polynomial $\mathbf{U}_i(\mathbf{x})$ in conserved variables valid within each cell V_i by performing a first standard spatial reconstruction $\mathbf{U}_i(\mathbf{x})$ of the conserved variables starting from the cell averages $\mathbf{U}_i$.

(2) Since $\mathbf{U}_i(\mathbf{x})$ is defined at any point inside the cell V_i, simply evaluate $\mathbf{U}_i(\mathbf{x})$ at the cell centroid $\mathbf{x}_i$ so as to obtain the point-value $\mathbf{U}|_i = \mathbf{U}_i(\mathbf{x}_i) = \mathbf{U}_i(x_i, y_i, z_i)$ by converting from cell averages $\mathbf{U}_i$ to point-values $\mathbf{U}|_i$, since the simple identity $\mathbf{U}|_i = \mathbf{U}_i$ is valid only up to second order of accuracy! Afterwards, perform a conversion from the point-values of the conserved variables to the point-values in primitive variables, i.e.,

$$\begin{aligned}\mathbf{W}(\mathbf{U})|_i &= (\rho, u, v, w, B_x, B_y, B_z, p)\big|_i^T \\ &= \left(\rho|_i, \frac{(\rho u)|_i}{\rho|_i}, \frac{(\rho v)|_i}{\rho|_i}, \frac{(\rho w)|_i}{\rho|_i}, \mathbf{B}|_i, (\gamma - 1)\left(e|_i - \frac{1}{2}\frac{((\rho \mathbf{v})|_i)^2}{\rho|_i} - \frac{\mathbf{B}|_i^2}{2}\right)\right)^T,\end{aligned} \tag{2.180}$$

thus obtaining the corresponding primitive variables $\mathbf{W}(\mathbf{U})|_i$ from $\mathbf{U}|_i$ at each cell centroid. That is, by Eq. (2.180) the point-values of primitive variables $\mathbf{W}$ are derived from those of the conserved variables $\mathbf{U}$ located at the cell centroid. This is the only step in the entire algorithm that entails a conversion from the conservative to the primitive variables.

(3) From the point-values $\mathbf{W}(\mathbf{U})|_i$ of the primitive variables at the cell centroid, a second spatial reconstruction is invoked to obtain a reconstruction $\mathbf{W}_i(\mathbf{x})$ in primitive variables. Afterwards, use the reconstruction formulae $\mathbf{W}_i(\mathbf{x})$ to obtain two representations for the state at the interface, one with a left-biased reconstruction stencil $\mathbf{W}_L$ and the other with a right-biased stencil $\mathbf{W}_R$, and subsequently convert the now pointwise values back to their conserved states $\mathbf{U}_L$ and $\mathbf{U}_R$, such that the latter two states then serve as input for the approximate Riemann solver.

Due to the nature of the finite-volume scheme and use of Riemann-solver-based flux functions, the application of the primitive variable reconstruction assures the conservation properties of the scheme, and is prone to maintain tighter control on the positivity of key flow variables such as density and pressure.

Following the idea of Noelle [360], the piecewise polynomial reconstructions in primitive variables can be derived from those of the conservative variables over cell V_i based on given cell averages. To proceed, recall that in this chapter $X_i, X|_i$, and $X_i(\mathbf{x})$ denote the cell-averaged value in cell V_i, the point-value at the cell centroid $\mathbf{x}_i$ of cell V_i, and the reconstructed polynomial of state variable X in cell V_i. From this, a reconstruction of the primitive variables can be achieved by matching the expansion coefficients of these variables with those of the conservative variables. If, for example, K-exact polynomial reconstructions of the conservative variables $\rho_i(\mathbf{x})$, $(\rho u)_i(\mathbf{x})$, $(\rho v)_i(\mathbf{x})$, $(\rho w)_i(\mathbf{x})$, $(B_x)_i(\mathbf{x})$, $(B_y)_i(\mathbf{x})$, $(B_z)_i(\mathbf{x})$ and $e_i(\mathbf{x})$ of density, momentum, magnetic field and total energy are given, the derived polynomial reconstructions $\widetilde{u}_i(\mathbf{x}), \widetilde{v}_i(\mathbf{x}), \widetilde{w}_i(\mathbf{x}), \widetilde{p}_i(\mathbf{x})$ of the primitive variables are defined by prescribing that for all $\mathbf{x} \in V_i$,

$$\begin{aligned}
&\rho_i(\mathbf{x})\widetilde{u}_i(\mathbf{x}) = (\rho u)_i(\mathbf{x}) + O(\Delta h_i^{K+1})\\
&\rho_i(\mathbf{x})\widetilde{v}_i(\mathbf{x}) = (\rho v)_i(\mathbf{x}) + O(\Delta h_i^{K+1})\\
&\rho_i(\mathbf{x})\widetilde{w}_i(\mathbf{x}) = (\rho w)_i(\mathbf{x}) + O(\Delta h_i^{K+1})\\
&\frac{\widetilde{p}_i(\mathbf{x})}{\gamma-1} + \frac{1}{2}\rho_i(\mathbf{x})\left(\widetilde{u}_i(\mathbf{x})^2 + \widetilde{v}_i(\mathbf{x})^2 + \widetilde{w}_i(\mathbf{x})^2\right) +\\
&\frac{1}{2}\left((B_x)_i(\mathbf{x})^2 + (B_y)_i(\mathbf{x})^2 + (B_z)_i(\mathbf{x})^2\right) = e_i(\mathbf{x}) + O(\Delta h_i^{K+1})
\end{aligned} \tag{2.181}$$

where Δh_i denotes the grid size such as $\sqrt[3]{V_i}$. Since $\rho_i(\mathbf{x})$, $(B_x)_i(\mathbf{x})$, $(B_y)_i(\mathbf{x})$, $(B_z)_i(\mathbf{x})$ are already known from the conserved variable reconstructions, the derived polynomial reconstructions $\widetilde{u}_i(\mathbf{x}), \widetilde{v}_i(\mathbf{x}), \widetilde{w}_i(\mathbf{x}), \widetilde{p}_i(\mathbf{x})$ can be easily obtained by balancing the coefficients in Eq. (2.181). Hence, the set of the derived polynomials $\widetilde{\rho}_i(\mathbf{x}) = \rho_i(\mathbf{x}), \widetilde{u}_i(\mathbf{x}), \widetilde{v}_i(\mathbf{x}), \widetilde{w}_i(\mathbf{x}), \widetilde{(B_x)}_i(\mathbf{x}) = (B_x)_i(\mathbf{x}), \widetilde{(B_y)}_i(\mathbf{x}) = (B_y)_i(\mathbf{x}), \widetilde{(B_z)}_i(\mathbf{x}) = (B_z)_i(\mathbf{x}), \widetilde{p}_i(\mathbf{x})$ leads to a reconstruction $\widetilde{X}_i(\mathbf{x}) \in \left\{\widetilde{\rho}_i(\mathbf{x}), \widetilde{u}_i(\mathbf{x}), \widetilde{v}_i(\mathbf{x}), \widetilde{w}_i(\mathbf{x}), \widetilde{(B_x)}_i(\mathbf{x}), \widetilde{(B_y)}_i(\mathbf{x}), \widetilde{(B_z)}_i(\mathbf{x}), \widetilde{p}_i(\mathbf{x})\right\}$ for each primitive variable $X \in \{\rho, u, v, w, B_x, B_y, B_z, p\}$ over cell V_i.

Next, this derived reconstruction $X_i(\mathbf{x})$ from Eq. (2.181) is limited as follows. Let $X(\mathbf{U})_i$ be the value of primitive variable X obtained from the cell-average value $\mathbf{U}_i$ of the conservative variables, that is, $X(\mathbf{U})_i$ is derived from $\mathbf{U}_i$ analogously according to Eq. (2.180). Then the limited reconstruction is given by

$$X_i(\mathbf{x}) = \alpha_i \widetilde{X}_i(\mathbf{x}) + (1 - \alpha_i) X(\mathbf{U})_i \tag{2.182}$$

where $\alpha_i = \min_j \alpha_{ij}$ and $\alpha_{ij} = \min_{gp}(\alpha_{ij,gp})$ with $\alpha_{ij,gp}$ defined by

$$\alpha_{ij,gp} = \begin{cases} 1, \text{ if } \widetilde{X}_i(\mathbf{x}_{ij,gp}) = X(\mathbf{U})_i \\ \min\left(1, \frac{X(\mathbf{U})_i^{\max} - X(\mathbf{U})_i}{\widetilde{X}_i(\mathbf{x}_{ij,gp}) - X(\mathbf{U})_i}\right), \text{ if } \widetilde{X}_i(\mathbf{x}_{ij,gp}) > X(\mathbf{U})_i \\ \min\left(1, \frac{X(\mathbf{U})_i^{\min} - X(\mathbf{U})_i}{\widetilde{X}_i(\mathbf{x}_{ij,gp}) - X(\mathbf{U})_i}\right), \text{ if } \widetilde{X}_i(\mathbf{x}_{ij,gp}) < X(\mathbf{U})_i \end{cases}$$

at all quadrature points $\mathbf{x}_{ij,gp} \in \partial V_i$, and

$$X(\mathbf{U})_i^{\max} = \max\left\{\max_{j \in \{V_j\}_i} X(\mathbf{U})_j, X(\mathbf{U})_i\right\}, X(\mathbf{U})_i^{\min} = \min\left\{\min_{j \in \{V_j\}_i} X(\mathbf{U})_j, X(\mathbf{U})_i\right\}$$

Obviously, $\alpha_i \in [0, 1]$. Here, $\{V_j\}_i$ is the set of the surrounding neighbors of current cell i used for the reconstruction. It is noted that

$$\min_{V_j \bigcap V_i \neq 0} X(\mathbf{U})_j \leq X_i(\mathbf{x}_{ij,gp})\big|_{\mathbf{x}_{ij,gp} \in \partial V_i} \leq \max_{V_j \bigcap V_i \neq 0} X(\mathbf{U})_j$$

holds componentwise. This limiting procedure is inherently multidimensional. Up to now, Eq. (2.182) achieves the limited polynomial reconstructions of primitive variables derived from conservative ones. Of course, other standard limiting procedures can be employed in obtaining such polynomial reconstructions of primitive variables. As remarked on the piecewise quadratic reconstruction by Noelle [360], it does not enforce any monotonicity constraints on the reconstruction, such a weak form of a limiter leads to a stable scheme even in the presence of strong shocks and the piecewise quadratic reconstruction is allowed to possess extrema in the interior of the cell V_i.

Note that nonlinear, invertible transformation (2.180) from the pointwise cell-centered values of the conserved variables $\mathbf{U}$ to those of primitive variables $\mathbf{W}$ has been utilized [133, 180, 310] to provide a fourth-order accurate conversion between the primitive and conservative solution variables. Zanotti and Dumbser [517] presented a version of conservative ADER-WENO finite volume schemes, in which both the high order spatial reconstruction as well as the time evolution of the reconstruction polynomials in the local space-time predictor stage are performed in primitive variables, rather than in conservative ones. The resulting space-time polynomials of the primitive variables can then be directly employed as input for the numerical fluxes at the cell interfaces in the underlying conservative finite volume scheme, and hence, the primitive variables are converted in just one single conversion at each cell center from the conservative ones. In the CENO scheme implementation, Freret et al. [147] offered a scheme of combined reconstruction variables: the unlimited high-order reconstruction is performed in terms of the conservative variables, while for the treatment of non-smooth solution content, the scheme reverts to a limited linear reconstruction, and the reconstruction of primitive variables can be adopted.

2.8 Numerical Flux Functions

There are many commonly used flux functions to obtain the inviscid flux. Due to the rotational invariance property, the numerical flux at each of the N_g Gaussian quadrature points on each face is determined by solving a Riemann problem of a specific flux of the normal direction. We use the rotational invariance of the equations to formulate the flux functions along arbitrary normal directions for the boundaries of a finite volume but without labeling the face normal **n** and Gaussian point anymore for simplicity. For brevity of description, the derivation is done here for **n** along x-direction without any ambiguity; results for an arbitrarily aligned face can be obtained by use of a simple rotation matrix.

Since the derivation of the flux functions depends on the solution of Riemann problem, they are also called "Riemann" solvers in FVM nomenclature. In MHD, the solution to the Riemann problem requires the solution of nonlinear systems of equations, and the decay of the initial discontinuity gives rise to a self-similar wave pattern in the $x - t$ plane where seven wave modes can develop, as shown in Fig. 2.9. For the mathematical description of the Riemann problem, see Sect. 2.3 of this Chapter.

In MHD system, fast waves are the outermost ones and enclose a pair of Alfvén waves, a pair of slow waves bounding a tangential discontinuity in the middle. Fast and slow waves can be either shock or rarefaction waves where primary flow quantities (density, pressure, velocity and magnetic field) change discontinuously (for the shock) or smoothly (for the rarefaction waves), depending on the pressure jump and magnetic field orientation. Across the Alfvén wave, thermodynamic quantities remain continuous while the tangential components of magnetic field trace circles for MHD system [317]. Finally, through the contact mode, only density exhibits a jump while thermal pressure, velocity and magnetic field remain continuous.

Across each wave, the Rankine–Hugoniot jump conditions must be satisfied

$$\lambda_\ell \left(\mathbf{U}_L^{(\ell)} - \mathbf{U}_R^{(\ell)}\right) = \mathbf{F}(\mathbf{U}_L^{(\ell)}) - \mathbf{F}(\mathbf{U}_R^{(\ell)}) \tag{2.183}$$

where $\ell = 1, \ldots, 7$ stands for the waves. An exact Riemann solver demands the simultaneous solution of the Rankine–Hugoniot jump conditions (2.183) across each wave through a self-consistent procedure that resolves fast, slow and Alfvén waves to the left and to the right of the contact (tangential in the degenerate case) discontinuity, always located at the center of the structure. This task is computationally expensive in MHD although the exact solutions of Riemann problems in ideal magnetohydrodynamics (MHD) for an arbitrary initial condition can be obtained and may be implemented to evaluate numerical fluxes for numerical simulations [458, 459]. For

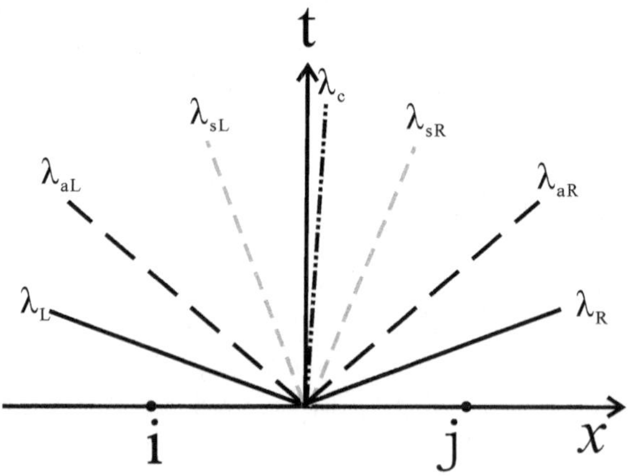

Fig. 2.9 MHD Riemann fan structure with two initial constant states. The configuration includes seven waves corresponding to a pair of fast magnetosonic waves with speed $\lambda_{L,R} = u \mp c_f$, a pair of Alfvén waves with speed $\lambda_{aL,aR} = u \mp c_a$, a pair of slow magnetosonic waves with speed $\lambda_{sL,sR} = u \mp c_s$ and a contact (or tangential) discontinuity in the middle propagating at the fluid speed $\lambda_c = u$

this reason, approximate solutions of the Riemann problems are usually preferred to acquire the flux functions. Such flux functions are called approximate Riemann solvers, which can be applied much more cheaply than the exact Riemann solver and yet give results that in many cases are satisfactory. In Sect. 2.3, we have described the Roe approximate Riemann solver, which replaces the nonlinear Riemann problem by a linearized Riemann problem. In this section, we first introduce the classical Osher–Solomon Riemann solver. Second, we present HLL type of approximate Riemann solvers, which are based on the simplification of the wave structures in the Riemann fan and the integral average of the solution over the Riemann fan. Third, some remarks are given about the properties of Riemann solvers. Finally, the hybrid Riemann solvers are introduced, which combine the classical approximate Riemann solvers to take the advantage of each solver's own strength.

2.8.1 Osher–Solomon Solver

As a generalization of the Enquist–Osher scheme [124] to systems of conservation laws, the Osher–Solomon scheme [372] was designed with two major goals: one is to prevent the sonic glitch and another is to represent physically steady states as discrete steady states. For such purpose, the Rankine–Hugoniot condition at shocks has to be abandoned as a design principle. While this would lead to a slightly increased numerical viscosity at shocks, it also allows for stable steady state representations of steady shocks. The Osher–Solomon numerical flux function reads as

$$\mathbf{f}_{\text{Osher}} = \frac{1}{2}(\mathbf{f}_R + \mathbf{f}_L) - \frac{1}{2}\int_{\mathbf{U}_L}^{\mathbf{U}_R} |\mathbf{A}| d\mathbf{U}$$

where the integral needs to be evaluated in phase space. However, the integral depends on the integration path chosen in order to make the actual integration tractable [372]. If $\Psi(s)$ is the path that connects the left state $\mathbf{U}_L$ and the right state $\mathbf{U}_R$ in phase space, the Osher–Solomon numerical flux explicitly reads

$$\mathbf{f}_{\text{Osher}} = \frac{1}{2}(\mathbf{f}_R + \mathbf{f}_L) - \frac{1}{2}\int_0^1 |\mathbf{A}(\Psi(s))| \frac{\partial \Psi}{\partial s} ds, \tag{2.184}$$

where $\Psi(s)$ is a Lipschitz continuous function linking the left and right states, i.e., $\Psi(0) = \mathbf{U}_L$ and $\Psi(1) = \mathbf{U}_R$. Usually, a simple family of straight line segments is usually used [1, 63–65, 116–118, 139, 270, 341, 412]. With a simple straight line segment path chosen as follows

$$\Psi(s) \equiv \Psi(\mathbf{U}_L, \mathbf{U}_R, s) = \mathbf{U}_L + s(\mathbf{U}_R - \mathbf{U}_L), \text{ with } 0 \le s \le 1,$$

the Osher–Solomon interface flux according to Eq. (2.184) is defined by

$$\mathbf{f}_{\text{Osher}} = \frac{1}{2}(\mathbf{f}_R + \mathbf{f}_L) - \frac{1}{2}\int_0^1 |\mathbf{A}\left(\mathbf{U}_L + s(\mathbf{U}_R - \mathbf{U}_L)\right)| ds(\mathbf{U}_R - \mathbf{U}_L). \tag{2.185}$$

The key problem with the general definition (2.185) is how to compute the integral for a given hyperbolic system, and the absolute value operator of the Jacobian is very difficult to integrate analytically; see Toro [469] for discussions. To avoid the analytical integration of the viscosity matrix $|\mathbf{A}(\mathbf{U}_L + s(\mathbf{U}_R - \mathbf{U}_L))|$, it is suggested [65, 117, 118, 139] that the integral (2.185) be numerically evaluated along the segment path with a Gauss-Legendre quadrature rule of sufficient accuracy defined by

$$\mathbf{f}_{\text{Osher}} = \frac{1}{2}(\mathbf{f}_R + \mathbf{f}_L) - \frac{1}{2}\left(\sum_{\ell=1}^{q} \omega_\ell \left|\mathbf{A}\left(\mathbf{U}_L + s_\ell\left(\mathbf{U}_R - \mathbf{U}_L\right)\right)\right|\right)(\mathbf{U}_R - \mathbf{U}_L),$$

where such a Gaussian rule is given with q points $s_\ell \in [0, 1]$ and associated weights ω_ℓ of the quadrature formula in the unit interval $[0, 1]$.

2.8.2 HLL Family Solver

In this subsection, we introduce HLL family solvers by outlining HLL, HLLC, HLLD, and HLLEM. The Harten-Lax-Leer (HLL) approximate Riemann solver approximates the solution of the Riemann problem with a two-wave model assumption. The HLLC approximate Riemann solver assumes a three-wave model and yields better resolution of the isolated contact discontinuities. The HLLD approximate Riemann solver possesses five three-wave model and has the ability to resolve all isolated discontinuities as well as isolated fast shocks in the MHD system. The HLLEM scheme attempts to accommodate as many intermediate characteristic fields as one wishes.

2.8.2.1 HLL Approximate Riemann Solver

The so-called HLL approximate Riemann solver, proposed by Harten et al. [192], may be one of the simplest Godunov-type scheme. The HLL Riemann solver is constructed by assuming an average intermediate state between the fastest and slowest waves. Consider a subsonic solution of the single-state approximate Riemann problem at the interface between the left and right states, $\mathbf{U}_L$ and $\mathbf{U}_R$, where the minimum signal speed S_L and the maximum signal speed S_R are negative and positive, respectively (Fig. 2.10a). To calculate the intermediate flux $\mathbf{f}^*$, integrating conservation laws (2.41) over the Riemann fan, $(\Delta t S_L, \Delta t S_R) \times (0, \Delta t)$ yields

$$\begin{aligned}\int_{\Delta t S_L}^{\Delta t S_R} \mathbf{U}(x, \Delta t) dx - \int_{\Delta t S_L}^{0} \mathbf{U}(x, 0) dx - \int_{0}^{\Delta t S_R} \mathbf{U}(x, 0) dx \\ + \int_0^{\Delta t} \mathbf{f}\left(\mathbf{U}(\Delta t S_R, t)\right) dt - \int_0^{\Delta t} \mathbf{f}\left(\mathbf{U}(\Delta t S_L, t)\right) dt = 0\end{aligned}$$

which results in

$$\Delta t (S_R - S_L)\mathbf{U}^* - \Delta t (S_R \mathbf{U}_R - S_L \mathbf{U}_L) + \Delta t (\mathbf{f}_R - \mathbf{f}_L) = 0$$

with $\mathbf{f}_L = \mathbf{f}(\mathbf{U}_L)$ and $\mathbf{f}_R = \mathbf{f}(\mathbf{U}_R)$. The intermediate state is obtained as

$$\mathbf{U}^* = \frac{S_R \mathbf{U}_R - S_L \mathbf{U}_L - (\mathbf{f}_R - \mathbf{f}_L)}{S_R - S_L} \tag{2.186}$$

Then, perform the integral conservation laws (2.41) over $(\Delta t S_L, 0) \times (0, \Delta t)$

$$\int_{\Delta t S_L}^{0} \mathbf{U}(x, \Delta t) dx - \int_{\Delta t S_L}^{0} \mathbf{U}(x, 0) dx + \int_0^{\Delta t} \mathbf{f}\left(\mathbf{U}(0, t)\right) dt - \int_0^{\Delta t} \mathbf{f}\left(\mathbf{U}(\Delta t S_L, t)\right) dt = 0$$

to know that

$$-\Delta t S_L (\mathbf{U}^* - \mathbf{U}_L) + \Delta t (\mathbf{f}^* - \mathbf{f}_L) = 0$$

Together with Eq. (2.186) we may obtain

$$\mathbf{f}^* = \frac{S_R \mathbf{f}_L - S_L \mathbf{f}_R + S_R S_L (\mathbf{U}_R - \mathbf{U}_L)}{S_R - S_L} \tag{2.187}$$

Alternatively, across each wave, the Rankine–Hugoniot jump conditions (2.183) must be satisfied, that is, $S_\kappa \mathbf{U}^* - \mathbf{f}^* = S_\kappa \mathbf{U}_\kappa - \mathbf{f}_\kappa$ with $\kappa = L, R$, from which the expression of $\mathbf{U}^*$ in (2.186) and $\mathbf{f}^*$ in (2.187) can also be derived from the jump conditions across S_L and S_R.

If both signal speeds are of the same sign, the fluxes must be evaluated only from the upstream side. Therefore, in general, the HLL fluxes become

$$\mathbf{f}_{\mathrm{HLL}}(\mathbf{U}_L, \mathbf{U}_R) = \begin{cases} \mathbf{f}_L & S_L > 0, \\ \mathbf{f}^* & S_L \leq 0 \leq S_R, \\ \mathbf{f}_R & S_R < 0, \end{cases} \tag{2.188}$$

It should be pointed out that to have a consistent system of equations, $\mathbf{f}^*$ and $\mathbf{U}^*$ are independent variables, that is, $\mathbf{f}^* \neq \mathbf{f}(\mathbf{U}^*)$.

The minimum signal speed S_L and the maximum signal speed S_R can be estimated from the initial left and right input states. A very common recipe is

$$S_L = \min\left(u_L - c_f(\mathbf{U}_L), u_R - c_f(\mathbf{U}_R)\right), S_R = \max\left(u_L + c_f(\mathbf{U}_L), u_R + c_f(\mathbf{U}_R)\right) \tag{2.189}$$

which are the smallest and largest eigenvalues of the leftmost and rightmost eigenmodes, and c_f is the fast magnetosonic speed (e.g., [93, 137, 531]).

Another choice for S_L and S_R is following Einfeldt et al. [120]:

$$S_L = \min\left(u_L - c_f(\mathbf{U}_L), u_{\mathrm{Roe}} - c_f(\mathbf{U}_{\mathrm{Roe}})\right), S_R = \max\left(u_{\mathrm{Roe}} + c_f(\mathbf{U}_{\mathrm{Roe}}), u_R + c_f(\mathbf{U}_R)\right) \tag{2.190}$$

$\mathbf{U}_{\mathrm{Roe}}$ in Eq. (2.190) represents the Roe average of $\mathbf{U}_L$ and $\mathbf{U}_R$. The HLL solver with the wave speed estimate of Eq. (2.190) is called HLLE solver [448, 469]. The calculation method of Roe average for ideal MHD may be found in Cargo and Gallice [61]. Multidimensional HLLE solvers based on the multidimensional Riemann problem can be found in the literature (e.g., [19, 22, 169, 336]).

For MHD, Janhunen [225] proposed a novel way to estimate S_L and S_R by the following formulas:

$$\begin{aligned} \hat{a}^2 &= \gamma \frac{\max(p_L, p_R)}{\min(\rho_L, \rho_R)} \\ \hat{v}_A^2 &= \frac{\max\left(\mathbf{B}_L^2, \mathbf{B}_R^2\right)}{\min(\rho_L, \rho_R)} \\ \hat{v}_{Ax}^2 &= \frac{\max\left(B_{xL}^2, B_{xR}^2\right)}{\min(\rho_L, \rho_R)} \\ \hat{c}_f^2 &= \frac{1}{2}\left[\hat{a}^2 + \hat{v}_A^2 + \sqrt{(\hat{a}^2 - \hat{v}_A^2)^2 + 4\hat{a}^2(\hat{v}_A^2 - \hat{v}_{Ax}^2)}\right] \\ S_L &= \min(u_L, u_R) - \hat{c}_f, S_R = \max(u_L, u_R) + \hat{c}_f \end{aligned} \tag{2.191}$$

According to the numerical verification of Janhunen [225], this set of wave speeds guarantee positivity of the solver.

If S_L and S_R are bounded by zero, i.e., we first calculate S_L and S_R by Eqs. (2.189), (2.190) or (2.191), and then constrain their values by $S_L = \min(S_L, 0)$ and $S_R = \max(S_R, 0)$ respectively, Eq. (2.188) can be unified with Eq. (2.187)

$$\mathbf{f}_{\text{HLL}}(\mathbf{U}_L, \mathbf{U}_R) = \frac{S_R \mathbf{f}_L - S_L \mathbf{f}_R + S_R S_L (\mathbf{U}_R - \mathbf{U}_L)}{S_R - S_L}$$

There are also some one-wave models that can be deduced from HLL Riemann solver. Consider a one-wave model with single speed $S_+ > 0$, where S_+ is determined by $S_+ = \max\left(|u_L| + c_f(\mathbf{U}_L), |u_R| + c_f(\mathbf{U}_R)\right)$. By choosing $S_L = -S_+$ and $S_R = S_+$ in $\mathbf{f}_{\text{HLL}}$ one obtains the Rusanov solver

$$\mathbf{f}_{\text{Rusanov}} = \frac{1}{2}(\mathbf{f}_L + \mathbf{f}_R) - \frac{1}{2} S_+ (\mathbf{U}_R - \mathbf{U}_L)$$

which is derived by Rusanov [406]. Customarily, the Rusanov solver $\mathbf{f}_{\text{Rusanov}}$ is also called local Lax–Friedrichs (LLF) solver $\mathbf{f}_{\text{LLF}}$, i.e., $\mathbf{f}_{\text{LLF}} \equiv \mathbf{f}_{\text{Rusanov}}$. The LLF solver is considered to be more diffusive than the HLL solver [474]. The (true) Lax–Friedrichs (LF) solver can be obtained from Rusanov solver by setting $S_+ = \frac{\Delta x}{\Delta t}$, that is

$$\mathbf{f}_{\text{LF}} = \frac{1}{2}(\mathbf{f}_L + \mathbf{f}_R) - \frac{1}{2}\frac{\Delta x}{\Delta t}(\mathbf{U}_R - \mathbf{U}_L)$$

Despite its simplicity, LF solver is even more diffusive than the LLF solver [467].

2.8.2.2 HLLC Riemann Solver

The HLLC solver, originally proposed by Toro et al. [468] for Euler equations, restores the tangential discontinuity in the Riemann fan. For HLLC, the solution of the Riemann problem is approximated by three waves separating four constant states, as illustrated in Fig. 2.10b. The single intermediate state $\mathbf{U}^*$ of the HLL solver is split into two intermediate states $\mathbf{U}_L^*$ and $\mathbf{U}_R^*$, which are separated by the tangential discontinuity S_M. For MHD, the HLLC solver has been discussed by a variety of authors (e.g., [169, 178, 277, 283, 321, 336]). The differences among these HLLC

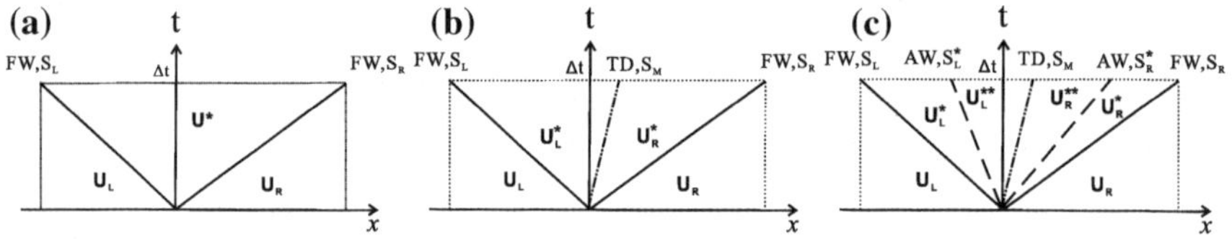

Fig. 2.10 Schematic structure of the Riemann fan in the **a** HLL solver with one intermediate state, **b** HLLC solver with two intermediate states, and **c** HLLD solver with four intermediate states

solvers lie in the specific choices of the intermediate states. Here, we only introduce the HLLC Riemann solver derived by Li [277].

Assume that face normal **n** is along x-direction, that is, the velocity component along **n** is $u = \mathbf{v} \cdot \mathbf{n}$. In the following, the total pressure

$$p_T = p + \frac{1}{2}\mathbf{B}^2, \tag{2.192}$$

which consists of the thermal pressure and the magnetic pressure, is used. The whole set of expressions for the intermediate states $\mathbf{U}_L^*$ and $\mathbf{U}_R^*$ are described as follows. According to the assumption made by Toro et al. [468] and Batten et al. [39], it is specified that

$$S_M = u_L^* = u_R^* = u^*,$$

where u^* is the average velocity in the intermediate states and can be extracted from the HLL approximation of $\mathbf{U}^*$ (Eq. (2.186)), namely

$$S_M = \frac{\rho_R u_R(S_R - u_R) - \rho_L u_L(S_L - u_L) + p_{TL} - p_{TR} - B_{xL}^2 + B_{xR}^2}{\rho_R(S_R - u_R) - \rho_L(S_L - u_L)}.$$

The remaining variables are calculated by applying the Rankine–Hugoniot conditions across the S_κ waves

$$\mathbf{f}_\kappa - \mathbf{f}_\kappa^* = S_\kappa(\mathbf{U}_\kappa - \mathbf{U}_\kappa^*) \tag{2.193}$$

where $\kappa = L, R$. Note that the expansion form of Eq. (2.193) may be found in Eq. (13) of Li [277]. Here, we only present the expressions of $\mathbf{U}_L^*$. The right intermediate state $\mathbf{U}_R^*$ can be derived similarly. After some manipulations, we obtain

$$p_{TL}^* = \rho_L(S_L - u_L)(S_M - u_L) + p_{TL} - B_{xL}^2 + (B_{xL}^*)^2.$$

To satisfy the pressure balance condition $p_{TL}^* = p_{TR}^*$, we get

$$B_{xL}^* = B_{xR}^*. \tag{2.194}$$

This is automatically satisfied in the one-dimensional case, i.e., $B_{xL}^* = B_{xR}^* = B_{xL} = B_{xR}$. For the multi-dimensional cases, the HLL average from $\mathbf{U}^*$ (Eq. (2.186)) is used

$$B_{xL}^* = B_{xR}^* = B_x^{\mathrm{HLL}} \tag{2.195}$$

Other components derived from Eq. (2.193) are

$$\rho_L^* = \rho_L \frac{S_L - u_L}{S_L - u^*}$$

$$(\rho u)_L^* = \rho_L^* u^*$$

$$(\rho v)_L^* = (\rho v)_L \frac{S_L - u_L}{S_L - u^*} - \frac{B_{xL}^* B_{yL}^* - B_{xL} B_{yL}}{S_L - u^*}$$

$$(\rho w)_L^* = (\rho w)_L \frac{S_L - u_L}{S_L - u^*} - \frac{B_{xL}^* B_{zL}^* - B_{xL} B_{zL}}{S_L - u^*}$$

Furthermore, the integral conservation laws (2.41) over the Riemann fan $(\Delta t S_L, \Delta t S_R) \times (0, \Delta t)$

$$\int_{\Delta t S_L}^{\Delta t S_M} \mathbf{U}(x, \Delta t) dx + \int_{\Delta t S_M}^{\Delta t S_R} \mathbf{U}(x, \Delta t) dx - \int_{\Delta t S_L}^{0} \mathbf{U}(x, 0) dx - \int_{0}^{\Delta t S_R} \mathbf{U}(x, 0) dx$$
$$+ \int_0^{\Delta t} \mathbf{f}\left(\mathbf{U}(\Delta t S_R, t)\right) dt - \int_0^{\Delta t} \mathbf{f}\left(\mathbf{U}(\Delta t S_L, t)\right) dt = 0$$

should be complied so as to obtain

$$\frac{u^* - S_L}{S_R - S_L} \mathbf{U}_L^* + \frac{S_R - u^*}{S_R - S_L} \mathbf{U}_R^* = \frac{S_R \mathbf{U}_R - S_L \mathbf{U}_L - (\mathbf{f}_R - \mathbf{f}_L)}{S_R - S_L}, \tag{2.196}$$

From this formula, we get

$$B_{xL}^* B_{yL}^* = B_{xR}^* B_{yR}^*, \quad B_{xL}^* B_{zL}^* = B_{xR}^* B_{zR}^*$$

Together with Eq. (2.194) or Eq. (2.195), we obtain

$$B_{yL}^* = B_{yR}^*, \quad B_{zL}^* = B_{zR}^*$$

which can be assigned the single HLL average state values from $\mathbf{U}^*$ (Eq. (2.186)),

$$B_{yL}^* = B_{yR}^* = B_y^{\mathrm{HLL}}, \quad B_{zL}^* = B_{zR}^* = B_z^{\mathrm{HLL}}$$

The energy density in the middle state should also satisfy Eq. (2.196), which results in

$$e_L^* = e_L \frac{S_L - u_L}{S_L - u^*} + \frac{(p_{TL}^* u^* - p_{TL} u_L) - [B_{xL}^* (\mathbf{B}^{\mathrm{HLL}} \cdot \mathbf{v}^{\mathrm{HLL}}) - B_{xL} (\mathbf{B}_L \cdot \mathbf{v}_L)]}{S_L - u^*}$$

With $\mathbf{U}_L^*$ and $\mathbf{U}_R^*$, the HLLC flux is finally obtained with the following form:

$$\mathbf{f}_{\mathrm{HLLC}}(\mathbf{U}_L, \mathbf{U}_R) = \begin{cases} \mathbf{f}_L, & \text{if } S_L > 0, \\ \mathbf{f}_L^* = \mathbf{f}_L + S_L(\mathbf{U}_L^* - \mathbf{U}_L), & \text{if } S_L \le 0 < S_M, \\ \mathbf{f}_R^* = \mathbf{f}_R + S_R(\mathbf{U}_R^* - \mathbf{U}_R), & \text{if } S_M \le 0 \le S_R, \\ \mathbf{f}_R, & \text{if } S_R < 0, \end{cases} \tag{2.197}$$

The multidimensional HLLC solvers are developed for MHD based on the multidimensional MHD Riemann problem [20, 26]. For problem containing strong background magnetic field, the magnetic field can be decomposed into a strong internal magnetic field and an external component to improve robustness and accuracy (e.g., [460]). The HLLC solver specifically designed for such applications is derived by Guo [176]. Efforts have also been taken to make the HLLC solver more robust in the presence of strong shocks (e.g., [250, 426]).

2.8.2.3 HLLD Riemann Solver

The Harten-Lax-van Leer-Discontinuities (HLLD) solver for MHD was firstly developed by Miyoshi and Kusano [324] and it can be considered as an extension of the Harten-Lax-van Leer (HLL) solver presented by Harten and Hyman [191]. The HLLD solver approximates the solution of the MHD Riemann problem by four intermediate states, i.e., $\mathbf{U}_L^*$, $\mathbf{U}_L^{**}$, $\mathbf{U}_R^{**}$, and $\mathbf{U}_R^*$, which are separated by two Alfvén waves with speeds S_L^* and S_R^*, and the contact discontinuity with speed S_M, as shown in Fig. 2.10c. Considering the property of the Alfvén wave and the tangential discontinuity, the HLLD is derived under the assumption that the normal velocity is constant over the Riemann fan

$$u_L^* = u_L^{**} = u_R^{**} = u_R^* = u^* = S_M,$$

which is acquired by the HLL average

$$S_M = \frac{\rho_R u_R (S_R - u_R) - \rho_L u_L (S_L - u_L) + p_{TL} - p_{TR}}{\rho_R (S_R - u_R) - \rho_L (S_L - u_L)}.$$

The total pressure defined by Eq. (2.192) is also assumed to be constant

$$p_T^* = p_{TL}^* = p_{TL}^{**} = p_{TR}^{**} = p_{TR}^* \tag{2.198}$$

which gives

$$p_T^* = \frac{\rho_R p_{TL} (S_R - u_R) - \rho_L p_{TR} (S_L - u_L) + \rho_L \rho_R (S_R - u_R)(S_L - u_L)(u_R - u_L)}{\rho_R (S_R - u_R) - \rho_L (S_L - u_L)}$$

when combined with the jump conditions of the normal momentum across S_R and S_L.

After obtaining S_M and p_T^*, the states $\mathbf{U}_\kappa^*$ adjacent to $\mathbf{U}_\kappa$ are solved from the jump conditions across S_κ

$$\mathbf{f}_\kappa - \mathbf{f}_\kappa^* = S_\kappa (\mathbf{U}_\kappa - \mathbf{U}_\kappa^*) \tag{2.199}$$

with $\kappa = L, R$. The expansion of Eq. (2.199) may be found in Eq. (42) of Miyoshi and Kusano [324]. Solving Eq. (2.199), we get

$$\rho_\kappa^* = \rho_\kappa \frac{S_\kappa - u_\kappa}{S_\kappa - S_M} \tag{2.200}$$

$$v_\kappa^* = v_\kappa - B_x B_{y\kappa} \frac{S_M - u_\kappa}{\rho_\kappa (S_\kappa - u_\kappa)(S_\kappa - S_M) - B_x^2} \tag{2.201}$$

$$w_\kappa^* = w_\kappa - B_x B_{z\kappa} \frac{S_M - u_\kappa}{\rho_\kappa (S_\kappa - u_\kappa)(S_\kappa - S_M) - B_x^2} \tag{2.202}$$

$$B_{y\kappa}^* = B_{y\kappa} \frac{\rho_\kappa (S_M - u_\kappa)^2 - B_x^2}{\rho_\kappa (S_\kappa - u_\kappa)(S_\kappa - S_M) - B_x^2} \tag{2.203}$$

$$B_{z\kappa}^* = B_{z\kappa} \frac{\rho_\kappa (S_M - u_\kappa)^2 - B_x^2}{\rho_\kappa (S_\kappa - u_\kappa)(S_\kappa - S_M) - B_x^2} \tag{2.204}$$

$$e_\kappa^* = \frac{(S_\kappa - u_\kappa) e_\kappa - p_{T\kappa} u_\kappa + p_T^* S_M + B_x (\mathbf{v}_\kappa \cdot \mathbf{B}_\kappa - \mathbf{v}_\kappa^* \cdot \mathbf{B}_\kappa^*)}{S_\kappa - S_M} \tag{2.205}$$

In cases when the operations of 0/0 appear in Eqs. (2.200)–(2.204), these formulas are replaced by $v_\kappa^* = v_\kappa$, $w_\kappa^* = w_\kappa$ and $B_{y\kappa}^* = B_{z\kappa}^* = 0$. The propagation speeds of the Alfén waves in the intermediate states are prescribed by

$$S_\kappa^* = S_M \mp \frac{|B_x|}{\sqrt{\rho_\kappa^*}} \tag{2.206}$$

where '−' and '+' correspond to $\kappa = L$ and R, respectively. The jump conditions across S_κ^*

$$\mathbf{f}_\kappa^{**} - \mathbf{f}_\kappa^* = S_\kappa^* (\mathbf{U}_\kappa^{**} - \mathbf{U}_\kappa^*) \tag{2.207}$$

lead to

$$\rho_\kappa^{**} = \rho_\kappa^* \tag{2.208}$$

and

$$p_{T\kappa}^{**} = p_{T\kappa}^*$$

For the rest components of $\mathbf{U}_\kappa^{**}$, the derivation is a bit complicated. Firstly, the jump condition across S_M is utilized

$$\mathbf{f}_R^{**} - \mathbf{f}_L^{**} = S_M^* (\mathbf{U}_R^{**} - \mathbf{U}_L^{**}) \tag{2.209}$$

From the jump conditions of Eq. (2.209) for tangential components of magnetic field and velocity, we are able to derive the following equalities

$$v_L^{**} = v_R^{**} \equiv v^{**} \tag{2.210}$$

$$w_L^{**} = w_R^{**} \equiv w^{**} \tag{2.211}$$

$$B_{yL}^{**} = B_{yR}^{**} \equiv B_y^{**} \tag{2.212}$$

$$B_{zL}^{**} = B_{zR}^{**} \equiv B_z^{**} \tag{2.213}$$

However, as illustrated in Miyoshi and Kusano [324], the values of v^{**}, w^{**}, B_y^{**} and B_z^{**} can not be solved from Eqs. (2.207) and (2.209). Fortunately, the integral conservation laws (2.41) over the Riemann fan $(\Delta t S_L, \Delta t S_R) \times (0, \Delta t)$ would provide enough information. The integral

$$\begin{aligned}
&\int_{\Delta t S_L}^{\Delta t S_L^*} \mathbf{U}(x, \Delta t)dx + \int_{\Delta t S_L^*}^{\Delta t S_M} \mathbf{U}(x, \Delta t)dx + \int_{\Delta t S_M}^{\Delta t S_R^*} \mathbf{U}(x, \Delta t)dx \\
&+ \int_{\Delta t S_R^*}^{\Delta t S_R} \mathbf{U}(x, \Delta t)dx - \int_{\Delta t S_L}^{0} \mathbf{U}(x, 0)dx - \int_{0}^{\Delta t S_R} \mathbf{U}(x, 0)dx \\
&+ \int_0^{\Delta t} \mathbf{f}\left(\mathbf{U}(\Delta t S_R, t)\right) dt - \int_0^{\Delta t} \mathbf{f}\left(\mathbf{U}(\Delta t S_L, t)\right) dt = 0
\end{aligned}$$

results in

$$\begin{aligned}
&\Delta t(S_L^* - S_L)\mathbf{U}_L^* + \Delta t(S_M - S_L^*)\mathbf{U}_L^{**} + \Delta t(S_R^* - S_M)\mathbf{U}_R^{**} + \Delta t(S_R - S_R^*)\mathbf{U}_R^* \\
&+\Delta t S_L \mathbf{U}_L - \Delta t S_R \mathbf{U}_R + \Delta t \mathbf{f}_R - \Delta t \mathbf{f}_L = 0
\end{aligned} \tag{2.214}$$

Substituting Eqs. (2.206), (2.208) and (2.210)–(2.213) into Eq. (2.214), it can be derived that

$$v^{**} = \frac{\sqrt{\rho_L^*}v_L^* + \sqrt{\rho_R^*}v_R^* + (B_{yR}^* - B_{yL}^*)\text{sign}(B_x)}{\sqrt{\rho_L^*} + \sqrt{\rho_R^*}}$$

$$w^{**} = \frac{\sqrt{\rho_L^*}w_L^* + \sqrt{\rho_R^*}w_R^* + (B_{zR}^* - B_{zL}^*)\text{sign}(B_x)}{\sqrt{\rho_L^*} + \sqrt{\rho_R^*}}$$

$$B_y^{**} = \frac{\sqrt{\rho_L^*}B_{yR}^* + \sqrt{\rho_R^*}B_{yL}^* + \sqrt{\rho_L^*\rho_R^*}(v_R^* - v_L^*)\text{sign}(B_x)}{\sqrt{\rho_L^*} + \sqrt{\rho_R^*}},$$

$$B_z^{**} = \frac{\sqrt{\rho_L^*}B_{zR}^* + \sqrt{\rho_R^*}B_{zL}^* + \sqrt{\rho_L^*\rho_R^*}(w_R^* - w_L^*)\text{sign}(B_x)}{\sqrt{\rho_L^*} + \sqrt{\rho_R^*}},$$

where $\text{sign}(B_x)$ is the sign of B_x. Finally, the energy of $\mathbf{U}_\kappa^{**}$ is solved from Eq. (2.207)

$$e_\kappa^{**} = e_\kappa^* \mp \sqrt{\rho_\kappa^*}(\mathbf{v}_\kappa^* \cdot \mathbf{B}_\kappa^* - \mathbf{v}^{**} \cdot \mathbf{B}^{**})\text{sign}(B_x)$$

where '−' and '+' correspond to $\kappa = L$ and R, respectively. Up to now, we have finished deriving the complete set of the intermediate states $\mathbf{U}_L^*$, $\mathbf{U}_L^{**}$, $\mathbf{U}_R^{**}$ and $\mathbf{U}_R^*$ as well as the wave speed of S_L^*, S_M and S_R^*. The lower and upper bounds of the signal speeds for the Riemann problem, i.e., S_L and S_R, may be calculated using Eqs. (2.189)–(2.191) or other estimations, such as

$$
\begin{aligned}
S_L &= \min(u_L, u_R) - \max\left(c_f(\mathbf{U}_L), c_f(\mathbf{U}_R)\right), \\
S_R &= \max(u_L, u_R) + \max\left(c_f(\mathbf{U}_L), c_f(\mathbf{U}_R)\right)
\end{aligned}
$$

In summary, the flux of HLLD solver reads as follows

$$
\mathbf{f}_{\mathrm{HLLD}}(\mathbf{U}_L, \mathbf{U}_R) = \begin{cases} \mathbf{f}_L & \text{if} S_L > 0 \\ \mathbf{f}_L^* = \mathbf{f}_L + S_L(\mathbf{U}_L^* - \mathbf{U}_L) & \text{if } S_L \le 0 < S_L^* \\ \mathbf{f}_L^{**} = \mathbf{f}_L + S_L(\mathbf{U}_L^* - \mathbf{U}_L) + S_L^*(\mathbf{U}_L^{**} - \mathbf{U}_L^*) & \text{if } S_L^* \le 0 < S_M \\ \mathbf{f}_R^{**} = \mathbf{f}_R + S_R(\mathbf{U}_R^* - \mathbf{U}_R) + S_R^*(\mathbf{U}_R^{**} - \mathbf{U}_R^*) & \text{if } S_M \le 0 < S_R^* \\ \mathbf{f}_R^* = \mathbf{f}_R + S_R(\mathbf{U}_R^* - \mathbf{U}_R) & \text{if } S_R^* \le 0 \le S_R \\ \mathbf{f}_R & \text{if } S_R < 0 \end{cases}
\tag{2.215}
$$

It should be noted that in the derivation above, we have assumed that $B_x \neq 0$. In the case of $B_x = 0$, the four-state HLLD solver is naturally reduced to the two-state solver since $S_L^* = S_R^* = S_M$, and there is no need to estimate the intermediate states $\mathbf{U}_\kappa^{**}$. We have also assumed that B_x is continuous in the Riemann problem, that is $B_{x_L} = B_{x_R}$. For one-dimensional problem, this assumption is satisfied automatically. For multi-dimensional problem, a solution to allow the existence of the inequality $B_{x_L} \neq B_{x_R}$ is proposed by Guo et al. [177]. For applications splitting the magnetic field into a strong internal magnetic field and an external component, the HLLD solver needs some more efforts [177, 326]. Gomes et al. [168] implemented the four-state HLLD solver for the generalized Lagrange multiplier (GLM) MHD with a mixed hyperbolic-parabolic correction proposed by Dedner et al. [96] to deal with the divergence-free condition of the magnetic field.

2.8.2.4 HLLEM Riemann Solver

To further improve the resolution of the waves between S_L and S_R, an anti-diffusion term is added to the HLLE solver for the tangential discontinuity in the Euler case, which results in the so-called HLLEM solver [119, 120]. Extending this idea to MHD, Wesenberg [494] proposed an HLLEM solver by adding anti-diffusion terms for the slow waves, Alfvén waves and tangential discontinuity. The HLLEM is also derived by Dumbser and Balsara [116] in the path-conservation fashion for general conservative and non-conservative hyperbolic systems, including MHD. The multidimensional analogue of the HLLEM (or HLLI) is derived in Balsara and Nkonga [24] and Balsara et al. [28]. The HLLEM scheme attempts to formulate the Roe scheme as correction to HLL. In such correction, the computational effort is reduced, the adjustment of the viscosity on the acoustic waves can be easily applied by choosing the bounding speeds S_L and S_R, the sonic point glitch can be avoided, and the failure of the standard Roe scheme for strong rarefaction waves can be healed. The construction runs as follows. For the sake of simplicity, assume that $\tilde{\lambda}_1 \leq \tilde{\lambda}_2 \leq \cdots \leq \tilde{\lambda}_m$ are the eigenvalues of the Roe matrix $|\tilde{\mathbf{A}}| = \tilde{\mathbf{R}}|\tilde{\mathbf{\Lambda}}|\tilde{\mathbf{L}}$. Thus, we can choose $S_L = \min(\tilde{\lambda}_1, 0)$ and $S_R = \max(\tilde{\lambda}_m, 0)$. With this setting, the numerical flux of HLLEM scheme can be written as

$$\mathbf{f}_{\text{HLLEM}} = \mathbf{f}_{\text{HLL}} - \varphi \frac{S_R S_L}{S_R - S_L} \tilde{\mathbf{R}}_* \tilde{\boldsymbol{\delta}}_* \tilde{\mathbf{L}}_* \left(\mathbf{U}_R - \mathbf{U}_L\right)$$

where the diagonal matrix $\tilde{\boldsymbol{\delta}}_* = \mathbf{I} - \frac{\tilde{\mathbf{\Lambda}}_* - |\tilde{\mathbf{\Lambda}}_*|}{2S_L} - \frac{\tilde{\mathbf{\Lambda}}_* + |\tilde{\mathbf{\Lambda}}_*|}{2S_R}$ with $\tilde{\mathbf{\Lambda}}_*$, $\tilde{\mathbf{R}}_*$, and $\tilde{\mathbf{L}}_*$ corresponding to linearly degenerate intermediate waves, and $\varphi \in [0, 1]$ is a flattener variable (e.g., [21, 116, 274]), which acts to identify regions of strong shocks within the computational domain and plays the role of switching the Riemann solver smoothly from HLL to HLLEM. This smooth transition technique was also used for carbuncle cure of HLLEM (HLLEMCC) in Kemm [239] for the Euler equations and in Kemm [241] for the shallow water equations. Since the goal of HLLEMCC is to prevent unphysical shear and entropy waves, its core is the computation of the residual in the Rankine–Hugoniot condition for these waves:

$$\tau = \mathbf{f}_R - \mathbf{f}_L - \tilde{v}_n(\mathbf{U}_R - \mathbf{U}_L)$$

with the Roe mean value for the normal velocity $\tilde{v}_n$. Now, take as the basic indicator the residual relative to the speed of sound $\tilde{a}$, respectively its Euclidean norm $\theta = \frac{\|\tau\|_2}{\tilde{a}}$, which vanishes for all shear and entropy waves. By introducing parameters $\alpha, \beta \in (0, 1)$, we can now define

$$\varphi(\theta, Fr_u) = \min\left\{1, \varepsilon\theta \max\left\{0, (1 - {Fr_u}^{\alpha})\right\}\right\}^{\beta}$$

for shallow water and

$$\varphi(\theta, M_u) = \min\left\{1, (\varepsilon\theta)^{\beta} \max\left\{0, (1 - {M_u}^{\alpha})\right\}\right\}$$

for the Euler equations, where Fr_u and M_u are the directional Froude and Mach numbers perpendicular to the cell face. A major advantage of this approach is that the modification can be applied to shear and entropy waves separately. As done before [15, 239, 241, 242], $\alpha = \beta = 0.33$ for both shallow water and gas dynamics, $\varepsilon = 1/100$ in the gas dynamics case and $\varepsilon = 10^{-3}$ for the shallow water equations.

Instead of using the residual in the Rankine–Hugoniot condition relative to the speed of sound, Kemm et al. (e.g., [242]) presented a modification of the numerical viscosity at shocks by employing the sum of the fluxes as weighting factor. By introducing the shock indicator

$$\theta_s = \varepsilon_s \left(1 - \frac{||\mathbf{f}_R - \mathbf{f}_L||_2}{||\mathbf{f}_R + \mathbf{f}_L||_2}\right)$$

with some small positive parameter ε_s, they modified the wave speeds of the Einfeldt choice Eq. (2.190) as follows

$$S_L = \min\left(u_L - c_f(\mathbf{U}_L), \theta_s\left(u_L - c_f(\mathbf{U}_L)\right) + (1-\theta_s)\left(u_{\text{Roe}} - c_f(\mathbf{U}_{\text{Roe}})\right), 0\right),$$
$$S_R = \max\left(u_R + c_f(\mathbf{U}_R), \theta_s\left(u_R + c_f(\mathbf{U}_R)\right) + (1-\theta_s)\left(u_{\text{Roe}} + c_f(\mathbf{U}_{\text{Roe}})\right), 0\right),$$

which results in a HLLEM-solver that does not enforce the Rankine–Hugoniot condition at steady shocks. In fact, the numerical viscosity is slightly increased as in the Osher–Solomon scheme.

2.8.3 *Some Remarks on the Properties of Riemann Solvers*

Numerical flux functions or Riemann solvers are evaluated based on an approximate solution of the Riemann problem at the cell face. They may be categorized as complete Riemann solvers and incomplete Riemann solvers. A complete Riemann solver is designed to resolve all waves present in a Riemann problem, While an incomplete Riemann solver will neglect some waves. Prominent examples of complete Riemann solvers are the Roe scheme (e.g., [389, 402]) and Osher–Solomon scheme (e.g., [139, 372]). While these solvers are preferable when complex wave structures as well as entropy and shear waves are expected, incomplete solvers are known to be robust in situations dominated by strong shocks. An example is the HLL-solver Harten et al. [192] whose construction is based only on the two outer waves of the Riemann problem. In Sect. 2.3, we have illustrated the derivation of Roe Riemann solver

$$\mathbf{f}_{\text{Roe}}(\mathbf{U}_L, \mathbf{U}_R) = \frac{1}{2}\left(\mathbf{f}(\mathbf{U}_L) + \mathbf{f}(\mathbf{U}_R) - \sum_{k=1}^{m} L_k(\mathbf{U}_R - \mathbf{U}_L)|\lambda_k|R_k\right)$$

in detail. Comparing with the complete Riemann solvers like Roe and Osher–Solomon, the wave structure of the Riemann fan is simplified for incomplete solvers in the context of MHD, such as the LLF, HLL, HLLC, and HLLD solvers. The degree of simplification reflects the ability to capture and spread discontinuities over few or more computational zones and it is ultimately responsible for the amount of numerical dissipation of the numerical scheme. In spite of its simplicity and robustness, the HLL solver is not able to resolve the isolated discontinuities and, as a result, is quite dissipative because the Riemann fan is constituted by only one intermediate state. For the Euler equations, the HLLC solver is a complete Riemann solver, but for MHD it does not contain the complete wave structure. The derivation of stable and mathematically consistent HLLC solvers for MHD has been the subject of investigation, and a number of inconsistencies have been highlighted (e.g., see the analysis in

Sect. 4.2 of Miyoshi and Kusano [324]). It is demonstrated by Miyoshi and Kusano [324] through several numerical tests that the HLLD solver is accurate enough in comparison with the Roe solver, but it is still widely believed that the Roe solver is the most accurate Riemann solver for MHD [317, 448, 474].

One may wonder why we should use the more diffusive and less accurate solvers, like HLL, as we have already got the accurate Roe solver in hand. First, the accuracy and robustness of the solvers are hard to be satisfied simultaneously. It has been found that the Roe solver may generate negative density and pressure due to the linearization of the Riemann problem. At the same time, the HLL solver is found to be positive preserving if S_L and S_R are bounded by certain conditions, and therefore more robust than the Roe solver [120, 225]. Second, more accurate solver would be computationally more expensive. By numerical tests [324] shows that the ratios of CPU time τ are $(\tau^{\mathrm{HLL}} : \tau^{\mathrm{HLLD}} : \tau^{\mathrm{Roe}}) = (1 : 1.35 : 2.67)$. Third, for a particular problem, the less accurate solver may be adequate to catch the discontinuities that is expected to appear in the solution [283]. Therefore, the choice of the Riemann solver is problem dependent, and one should consider the trade off between accuracy, robustness and computational efficiency.

2.8.4 The Hybrid Riemann Solver

The classical Riemann solvers we have introduced above, including HLL, HLLC, Roe (for both Euler and MHD) and HLLD (for MHD only), have been successfully applied in studies of a variety of problems (e.g., [271, 448, 469, 475]). In most situations, they work well. However, the Roe and HLLC solver for Euler equations, which are able to resolve the contact discontinuity and less diffusive, may suffer from carbuncle and odd-even decoupling phenomena that are called numerical shock instability (e.g., [73, 425]). Similar phenomena have also been found in MHD simulation utilizing Roe solver [186]. On the other hand, it is well known that the HLL solver and LLF solver are free from shock instabilities, although they lack the ability to resolve contact discontinuity for Euler equations and are more diffusive than Roe and HLLC. Of course, a solver that has both the contact resolving ability and shock stability would be desired. However, researches have demonstrated contradiction on these two capabilities (e.g., [123, 174, 374, 401]).

One method to cure the shock instability phenomenon is the hybrid Riemann solver, which is based on the classical Riemann solvers and may combine the advantages of different classical solvers. Certain instabilities can be controlled by changing the flavor of the dissipation mechanism according to the flow field properties rather than increasing the absolute level of dissipation. Therefore, it is straightforward and practical to develop a hybrid solver for improving both the accuracy and stability. The pioneer work of Quirk [394] uses a pressure gradient to detect a shock, and then switch the flux from the Roe to the HLL to avoid the carbuncle. Although very effective, the resulting algorithm does not seem to have gained popularity, apparently because of its empirical nature in the shock-sensor and also the discontinuous switching between two different fluxes. As mentioned by Zhang et al. [521], the discontinuous flux transition could be affecting the convergence ratio of the hybrid scheme. In the following, we will introduce some efforts toward the development of the hybrid solvers. In Sect. 2.8.4.1, the grid-aligned hybrid solver that bends two solvers continuously are introduced. It worth noting that this kind of hybrid method has also been utilized as one of the measures to keep density and pressure positive during the simulation (e.g., [78, 501]). In Sect. 2.8.4.2, we introduce a more elegant hybrid strategy that uses the idea about rotated Riemann solver and applies two different Riemann solvers in two different directions.

Some studies employing the essence of hybrid solver are motivated by other purposes. These studies are carried out not to solve the instability problem of the less diffusive solver, but to lower the dissipation of the HLL solver without increasing the computational expense significantly. The HLLX and HLLXω solver proposed by these studies are introduced in Sect. 2.8.4.3.

Most of the Riemann solvers can be written in the general form

$$\mathbf{f}(\mathbf{U}_L, \mathbf{U}_R) = \frac{1}{2}\left(\mathbf{f}(\mathbf{U}_L) + \mathbf{f}(\mathbf{U}_R)\right) - \frac{1}{2}\mathbf{D}(\mathbf{U}_L, \mathbf{U}_R)\left(\mathbf{U}_R - \mathbf{U}_L\right) \tag{2.216}$$

The first term is the physical flux consisting an average of $\mathbf{f}(\mathbf{U}_L)$ and $\mathbf{f}(\mathbf{U}_R)$. The second term is a diffusion term that is proportional to $(\mathbf{U}_R - \mathbf{U}_L)$ and makes the solver stable and oscillation free. The form of the numerical flux function $\mathbf{f}(\mathbf{U}_L, \mathbf{U}_R)$ is determined by the dissipation matrix $\mathbf{D}$. Here we present the dissipation matrices of several classical Riemann solvers.

- Lax–Friedrichs:

$$\mathbf{D}_{\text{LF}} = \frac{\Delta x}{\Delta t}\mathbf{I}. \tag{2.217}$$

- Rusanov or local Lax–Friedrichs:

$$\mathbf{D}_{\text{LLF}} = \max\{\bar{\lambda}(\mathbf{U}_L), \bar{\lambda}(\mathbf{U}_R)\}\mathbf{I} \tag{2.218}$$

$\bar{\lambda}(\mathbf{U}) = \max\{|\lambda_{\min}(\mathbf{U})|, |\lambda_{\max}(\mathbf{U})|\}$. For MHD, $\lambda_{\min} = u - c_f$, $\lambda_{\max} = u + c_f$.

- HLL:

$$\mathbf{D}_{\text{HLL}} = \frac{|\lambda_L|\lambda_R - |\lambda_R|\lambda_L}{\lambda_R - \lambda_L}\mathbf{I} + \frac{|\lambda_R| - |\lambda_L|}{\lambda_R - \lambda_L}\mathbf{A} \tag{2.219}$$

with $\lambda_L = S_L$, $\lambda_R = S_R$. The value of S_L and S_R may be found in Eqs. (2.189), (2.190) or (2.191). Here, $\mathbf{A} = \frac{\partial \mathbf{f}}{\partial \mathbf{U}}$ is the Jacobian matrix of $\mathbf{f}$.

- Lax–Wendroff (LW):

$$\mathbf{D}_{\text{LW}} = \frac{\Delta t}{\Delta x}\mathbf{A}^2. \tag{2.220}$$

- Roe:

$$\mathbf{D}_{\text{Roe}} = |\mathbf{A}|. \tag{2.221}$$

with $|\mathbf{A}|$ as defined in Eq. (2.53).

Based on these classical solvers, hybrid Riemann solvers have been developed. The dissipation matrix of the hybrid Riemann solver is constructed using weighted combination of classical Riemann solvers' dissipation matrices. The dissipation become controllable for the hybrid Riemann solver, and the right amount dissipation may be added to meet the solver designer's goal.

2.8.4.1 Hybrid Roe-LLF Riemann Solver

In order to deal with the carbuncle effect and improve accuracy, Chandrashekar [73] and Derigs et al. [107] proposed a hybrid Riemann solver, which bends the Roe and the local Lax–Friedrichs (LLF) solver continuously. The local Lax–Friedrichs solver is responsible for suppressing the carbuncle phenomenon, while the more accurate Roe solver is used in regions without large pressure jumps to enable the hybrid solver to track smooth parts of the solutions with higher accuracy. The dissipation matrix of the hybrid Riemann solver is

$$\mathbf{D}_{\text{hybrid}} = \omega \mathbf{D}_{\text{LLF}} + (1-\omega)\mathbf{D}_{\text{Roe}}. \tag{2.222}$$

The parameter $\omega \in [0, 1]$ is obtained using a local pressure jump indicator. With Eq. (2.222), the numerical flux function of the hybrid Riemann solver may be given as

$$\mathbf{f}_{\text{hybrid}}(\mathbf{U}_L, \mathbf{U}_R) = \frac{1}{2}(\mathbf{f}(\mathbf{U}_L) + \mathbf{f}(\mathbf{U}_R)) - \frac{1}{2}\mathbf{D}_{\text{hybrid}}(\mathbf{U}_L, \mathbf{U}_R)(\mathbf{U}_R - \mathbf{U}_L) \tag{2.223}$$

By setting the indicator ω as

$$\omega = \left|\frac{p_L - p_R}{p_L + p_R}\right|^{\frac{1}{2}}$$

the hybrid Riemann solver (2.223) is established in the framework of the entropy conservative/stable scheme for compressible Euler equations [73] and MHD equations [107].

Alternatively, for hypersonic flow simulation with high-order methods, Wang et al. [488] set ω as

$$\omega = 0.5 + 0.5\text{sign}(\nabla p - \alpha)\left(1 - \exp\left(-100|\nabla p - \alpha|\right)\right)$$
$$\nabla p = \frac{|p_L - p_R|}{\min(p_L, p_R)}$$

where α is a threshold parameter related to the shock strength. In Wang et al. [488], α is set to be 0.15 to ensure that the LLF flux function is active near a weak shock. The calculation of ∇p that takes the multidimensional effects into consideration is presented in Eqs. (14)–(16) of Wang et al. [488].

From the design of the jump indicator, the scheme is dominated by less dissipative Roe scheme for smooth parts of the flow (but also near e.g., contact discontinuities), while the more dissipative LLF is utilized near strong shocks.

2.8.4.2 The Rotated-Hybrid Riemann Solver

In the standard finite volume method, no matter how we solve the Riemann problem, the direction for the one-dimensional problem is fully fixed by the mesh geometry. Evidently, this grid-aligned normal may not always be appropriate for upwinding. Thus, rotated Riemann solver approaches have been proposed in order to choose freely the direction of the one-dimensional Riemann problem for improving shock-capturing accuracy (see Levy et al. [272] and references therein). Subsequent developments aim to gain both robustness and accuracy, and various rotated Riemann solvers are designed in order to both suppress carbuncles and keep low numerical diffusion (e.g., [206, 354, 397, 423, 425]).

A rotated Riemann solver applies two Riemann solvers in two different directions. Usually, the solver uses a more-dissipative and carbuncle-free solver to the velocity difference vector direction or the direction normal to shocks, while employs less dissipative solver such as the Roe solver in the other direction to prevent the resulting solver unacceptably dissipative. In this way, the rotated Riemann solver becomes carbuncle-free normal to shocks and less dissipative across shear layers. What's more, the rotated solver mechanism would introduce extra dissipation to suppress the carbuncle instability even using the same Roe solver in the two directions, as shown by Ren [397].

Following the literature (e.g., [206, 354, 397, 423, 425]), we learn that the rotated Riemann solver for the inviscid flux is based on the decomposition of the outward unit vector, $\mathbf{n}$, normal to the cell interface into two orthogonal directions, i.e.,

$$\mathbf{n} = \alpha_1 \mathbf{n}_1 + \alpha_2 \mathbf{n}_2, \quad \mathbf{n}_1 \cdot \mathbf{n}_2 = 0, \tag{2.224}$$

where $\mathbf{n}_1$ and $\mathbf{n}_2$ are two unit vectors, i.e., $|\mathbf{n}_1| = |\mathbf{n}_2| = 1$, and $\alpha_1 = \mathbf{n}_1 \cdot \mathbf{n}, \alpha_2 = \mathbf{n}_2 \cdot \mathbf{n}$, as shown in Fig. 2.11. In order to keep the same left and right states in both directions, the vectors $\mathbf{n}_1$ and $\mathbf{n}_2$ are chosen to make $\alpha_1 \geq 0$ and $\alpha_2 \geq 0$. The decomposition of $\mathbf{n}$ into two vectors $\mathbf{n}_1$ and $\mathbf{n}_2$ is a geometrical procedure, but physical feature of the flow problem should be taken into account to determine how to choose $\mathbf{n}_1$ and $\mathbf{n}_2$. Thus, to completely define the rotated-hybrid solvers, the normal direction $\mathbf{n}_1$ must be determined at every interface by being aligned with some physically meaningful direction, say, the velocity difference vector (normal to shocks and parallel to shears) taken over two adjacent cells, and $\mathbf{n}_2$ orthogonal to it. According to Ren [397], the vector $\mathbf{n}_1$ is defined by the following formula:

$$\mathbf{n}_1 = \begin{cases} \mathbf{n}, & \text{if } |\Delta\mathbf{v}| \leq \epsilon \\ \frac{\Delta\mathbf{v}}{|\Delta\mathbf{v}|}, & \text{otherwise} \end{cases} \tag{2.225}$$

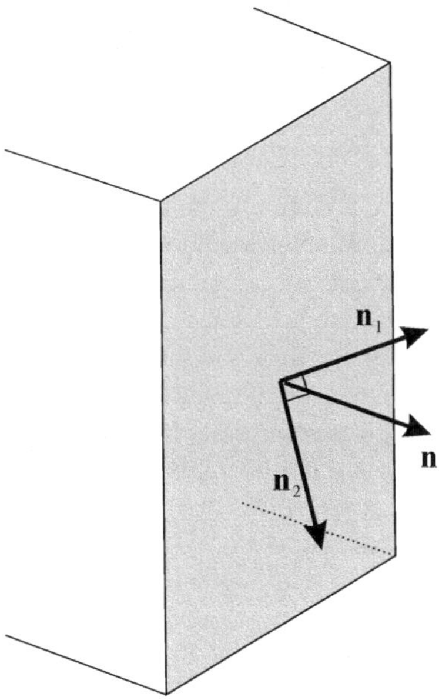

Fig. 2.11 Decomposition of the face normal

where ϵ is a small positive number, $\Delta\mathbf{v}$ denotes the velocity difference vector between two adjacent cells, i.e., $\Delta\mathbf{v} = (\Delta u, \Delta v, \Delta w) = (u_R - u_L, v_R - v_L, w_R - u_L)$, and $|\Delta\mathbf{v}| = \sqrt{(\Delta u)^2 + (\Delta v)^2 + (\Delta w)^2}$. This choice of $\mathbf{n}_1$ is used to select the direction normal to a shock or parallel to a shear if they exist (e.g., [206, 272, 354, 397, 423, 425]). Another direction $\mathbf{n}_2$ is defined, to be perpendicular to $\mathbf{n}_1$, by

$$\mathbf{n}_2 = \frac{(\mathbf{n}_1 \times \mathbf{n}) \times \mathbf{n}_1}{|(\mathbf{n}_1 \times \mathbf{n}) \times \mathbf{n}_1|}$$

so as to make sure that $\mathbf{n}_1$ and $\mathbf{n}_2$ are in the same state domain. Besides, in order to reduce the diffusion induced by the rotated-hybrid flux on the shear layer, Nishikawa and Kitamura [354] suggested to use a cell-tangent instead of normal $\mathbf{n}$, i.e., the direction $\mathbf{n}_1$ is determined as

$$\mathbf{n}_1 = \begin{cases} \frac{\Delta\mathbf{v}}{|\Delta\mathbf{v}|}, \text{ if } |\Delta\mathbf{v}| > \epsilon \\ \mathbf{n}_\perp, \text{otherwise} \end{cases} \tag{2.226}$$

where $\mathbf{n}_\perp$ denotes the tangential direction to the geometric face normal.

Let's recall that the rotational invariant of the MHD equations make it possible to write the flux at cell interface as $\mathbf{F}(\mathbf{U}) \cdot \mathbf{n} = (\mathbf{f}(\mathbf{U}), \mathbf{g}(\mathbf{U}), \mathbf{h}(\mathbf{U})) \cdot \mathbf{n} = \mathbf{T}^{-1}\mathbf{f}(\mathbf{TU})$, in which the unit vector $\mathbf{n}$ from Eq. (2.224) is substituted to obtain the following formula

$$\mathbf{F} \cdot \mathbf{n} = \alpha_1 \mathbf{F} \cdot \mathbf{n}_1 + \alpha_2 \mathbf{F} \cdot \mathbf{n}_2 = \alpha_1 \mathbf{T}_1^{-1}\mathbf{f}(\mathbf{T}_1\mathbf{U}) + \alpha_2 \mathbf{T}_2^{-1}\mathbf{f}(\mathbf{T}_2\mathbf{U}) \tag{2.227}$$

with $\mathbf{T}_p$ being the corresponding rotation matrix associated with $\mathbf{n}_p, p = 1, 2$. We can further rewrite Eq. (2.227) as follows:

$$\mathbf{F} \cdot \mathbf{n} = \alpha_1 \mathbf{f}(\mathbf{n}_1) + \alpha_2 \mathbf{f}(\mathbf{n}_2) \tag{2.228}$$

where $\mathbf{f}(\mathbf{n}_1) = \mathbf{T}_1^{-1}\mathbf{f}(\mathbf{T}_1\mathbf{U})$ and $\mathbf{f}(\mathbf{n}_2) = \mathbf{T}_2^{-1}\mathbf{f}(\mathbf{T}_2\mathbf{U})$, and $\mathbf{f}(\mathbf{T}_1\mathbf{U})$ and $\mathbf{f}(\mathbf{T}_2\mathbf{U})$ are the corresponding face fluxes approximately obtained from the one-dimensional Riemann solvers applied respectively in $\mathbf{n}_1$ and $\mathbf{n}_2$ directions as done for Eq. (2.71).

If there is a shock or a shear wave at the cell interface, the shock will propagate in the direction of $\mathbf{n}_1$ and the shear wave will move in the direction of $\mathbf{n}_2$. Therefore, when constructing the rotated Riemann solver, the more diffusive solver is employed in $\mathbf{n}_1$ to stabilize the shock while the less diffusive solver is applied in $\mathbf{n}_2$ to promote accuracy. For example, selecting the more diffusive Rusanov flux in $\mathbf{n}_1$ and the less diffusive Roe flux in $\mathbf{n}_2$, we get a new hybrid flux function based on the rotated-hybrid Riemann solver

$$\mathbf{F} \cdot \mathbf{n} = \alpha_1 \mathbf{f}_{\text{Rusanov}}(\mathbf{n}_1) + \alpha_2 \mathbf{f}_{\text{Roe}}(\mathbf{n}_2) \tag{2.229}$$

which achieves a hybrid numerical approximation to the numerical flux $\mathbf{F} \cdot \mathbf{n}$ in the direction normal to the interface. Here, $\mathbf{f}_{\text{Rusanov}}(\mathbf{n}_1)$ and $\mathbf{f}_{\text{Roe}}(\mathbf{n}_2)$ are one-dimensional Rusanov solver and Roe solver applied in $\mathbf{n}_1$ and $\mathbf{n}_2$ directions respectively, following a similar procedure as implemented for the one-dimensional Riemann solver in $\mathbf{n}$ direction.

In this way, the rotated Riemann solver approach generates hybrid fluxes in a simple and predigested way like a single Riemann solver, and does not require any problem-dependent tuning parameters. To date, there exist the currently rotated-hybrid flux functions available below.

(1) Rotated-hybrid Roe-Roe flux of $\mathbf{f}_{\text{Roe}}(\mathbf{n}_1)$ and $\mathbf{f}_{\text{Roe}}(\mathbf{n}_2)$ with $\mathbf{n}_1$ defined by Eq. (2.225) [397];

(2) Rotated-hybrid Rusanov-Roe flux of $\mathbf{f}_{\text{Rusanov}}(\mathbf{n}_1)$ and $\mathbf{f}_{\text{Roe}}(\mathbf{n}_2)$ with $\mathbf{n}_1$ defined by Eq. (2.226) [354];

(3) Rotated-hybrid HLL-Roe flux of $\mathbf{f}_{\text{HLL}}(\mathbf{n}_1)$ and $\mathbf{f}_{\text{Roe}}(\mathbf{n}_2)$ with $\mathbf{n}_1$ defined by Eq. (2.226) [354];

(4) Rotated-hybrid HLLC-HLLC flux of $\mathbf{f}_{\text{HLLC}}(\mathbf{n}_1)$ and $\mathbf{f}_{\text{HLLC}}(\mathbf{n}_2)$ with $\mathbf{n}_1$ defined by Eq. (2.225) [206];

(5) Rotated hybrid HLL-LDE flux of $\mathbf{f}_{\text{HLL}}(\mathbf{n}_1)$ and $\mathbf{f}_{\text{LDE}}(\mathbf{n}_2)$ with $\mathbf{n}_1$ defined by Eq. (2.226) [423]. Here, the LDE schemes not introduced explicitly here is a low diffusion version of the E-CUPS schemes [518]. The E-CUSP schemes use the total energy in the convective vector, which is a member of the convective upwind and split pressure (CUSP) family schemes [424].

Wu et al. [500] suggested an even simple hybrid method to avoid the rotated procedure of flux function and make direct combination of different fluxes. Their proposed numerical flux reads as follows

$$\mathbf{F} \cdot \mathbf{n} = \beta_1 \mathbf{f}_{\text{A}}(\mathbf{n}) + \beta_2 \mathbf{f}_{\text{B}}(\mathbf{n})$$

where $\mathbf{f}_A(\mathbf{n})$ can take a high resolution scheme such as Roe and $\mathbf{f}_B(\mathbf{n})$ is a dissipative less wave flux as HLL. β_1 and β_2 depending on decomposition of $\mathbf{n}$ are

$$\beta_1 = \alpha_1/(\alpha_1 + \alpha_2), \beta_2 = \alpha_2/(\alpha_1 + \alpha_2)$$

where α_1 and α_2 are defined by (2.224) with $\mathbf{n}_1$ given by Eq. (2.225). In order to reduce dissipation furthermore, Wu et al. [500] and Huang et al. [206] proposed another kind of hybrid HLLC-HLL flux by combining $\beta_1\mathbf{f}_{\text{HLLC}}(\mathbf{n}) + \beta_2\mathbf{f}_{\text{HLL}}(\mathbf{n})$ with the weights $\beta_1 = \frac{1}{2} + \frac{1}{2}\alpha_1/(\alpha_1 + \alpha_2), \beta_2 = \frac{1}{2}\alpha_2/(\alpha_1 + \alpha_2)$. According to Huang et al. [206], this hybrid HLLC-HLL solver is very robust in many severe benchmark problems, however, its disadvantage is that it is more dissipative than the rotated-hybrid HLLC-HLLC solver. For such kind of hybrid schemes, a matrix-based stability analysis was carried out by Shen et al. [425]. By combining the Roe with HLL flux in different directions and different flux components, they compared some hybrid schemes to illustrate different mechanisms in controlling shock instability.

For the choice of small positive parameter ϵ in Eq. (2.225), Ren [397] use $\epsilon = (10^{-3} \sim 10^{-5})u_{\text{ref}}$ with u_{ref} a reference velocity magnitude. With this choice of ϵ [397] found that the condition $|\Delta v| \leq \epsilon$ is always satisfied for his tests except in the regions containing shock waves and strong compression/rarefaction waves where the velocity-differences are large. Huang et al. [206] set $\epsilon = 10^{-4}U^*$ with U^* the free stream velocity or velocity at inlet of the flow field. The definition of $\mathbf{n}_1$ in Eq. (2.226) is utilized by Nishikawa and Kitamura [354] and Shen et al. [423]. $\epsilon = 10^{-6}$ is taken in the normalized Navier–Stokes equations by Shen et al. [423] in their tests. Nishikawa and Kitamura [354] set $\epsilon = 10^{-12}U_{\text{ref}}$ (simply a very small number), where U_{ref} is a reference velocity magnitude, using the definition of $\mathbf{n}_1$ (2.226) for all their test problems, and demonstrated that their rotated fluxes are fully capable without problem-dependent tuning of ϵ. Nishikawa and Kitamura [354] also concluded that once such an extremely small value is chosen for ϵ, the two choices, Eqs. (2.225) and (2.226), would not make any big difference in the numerical results because when $|\Delta\mathbf{v}| < \epsilon$ is true, the left and right states will be almost identical, implying that the choice of flux functions is insignificant.

Note that $\mathbf{n}_1$ defined in Eq. (2.226) differs from that in Eq. (2.225) in the alternative direction for a small velocity difference, i.e., the cell-face tangent instead of the normal, and this effectively turns off the dissipative solver (since $\alpha_1 = \mathbf{n} \cdot \mathbf{n}_1 = \mathbf{n} \cdot \mathbf{n}_\perp = 0$) and recovers the grid-aligned Roe flux for a smoothly varying streamwise velocity field. Thus, the rotated-hybrid fluxes, constructed according to the decomposition given by Eq. (2.226), do smoothly switch between two different fluxes. With the definition in Eq. (2.226), the rotated-hybrid fluxes, such as the Rusanov-Roe flux and the HLL-Roe flux, can be expressed as follows:

$$\mathbf{F} \cdot \mathbf{n} = \begin{cases} \alpha_1\mathbf{f}_{\text{Rusanov/HLL}}(\mathbf{n}_1) + \alpha_2\mathbf{f}_{\text{Roe}}(\mathbf{n}_2), \text{ if } |\Delta\mathbf{v}| > \epsilon \\ \mathbf{f}_{\text{Roe}}(\mathbf{n}), \text{ otherwise} \end{cases} \tag{2.230}$$

Clearly, even if the rotated-hybrid solver in Eq. (2.230) is activated (when $|\Delta \mathbf{v}| > \epsilon$), it introduces the Rusanov/ HLL flux only if $\alpha_1 \neq 0$ and only by a factor of α_1 (≤ 1). That is, the vector quantity $\alpha_1 \mathbf{n}_1$ is non-zero only at discontinuities and strongly varying flow fields, and therefore it could serve as a directional feature detector for anisotropic mesh adaptation. Furthermore, the flux becomes completely identical to the Rusanov/HLL flux only if $\alpha_1 = 1$, i.e., the velocity difference vector is perfectly aligned with the interface normal (e.g., a grid-aligned normal shock). This implies that the rotated-hybrid Rusanov-Roe and HLL-Roe fluxes are generally much less dissipative than the Rusanov/HLL fluxes themselves. Darwis et al. [90] carried out a comparative study of shock capturing schemes by using Roe, rotated-hybrid HLL-Roe and Rusanov flux functions, and concluded that: (a) Roe scheme has the property of good capturing contact discontinuity so that this property causes the Roe scheme to suffer shock instability in multidimensional problem; (b) Rusanov scheme is a dissipative scheme, which causes it to produce very diffusive solution and thus be unable to capture contact discontinuity; (c) Rotated-hybrid HLL-Roe scheme has both properties of Roe and Rusanov scheme, which makes rotated-hybrid HLL-Roe scheme capable in suppressing the shock instability and at the same time avoid diffusive solution. Zhang et al. [520] evaluated the relation between the rotated upwind-differencing direction and the performance of rotated upwind schemes by theoretical and numerical investigations with three commonly used upwind schemes: Roe scheme, AUSM+ (Advection Upwind Splitting Method; [284]) scheme and Van Leer scheme [268]. The theoretical analysis shows that the rotation procedure changes the dissipative terms of each upwind scheme, and the dissipative density term in the numerical flux function of rotated upwind schemes introduces extra dissipation. The numerical results show that the carbuncle phenomenon can be eliminated by rotated upwind schemes, and the contact resolving capability of rotated upwind schemes is deteriorated by the rotation procedure. Zhang et al. [520] defined a novel rotation strategy of the rotation angle by a pressure weight function in order to tune the dissipative effects adaptively.

According to Levy et al. [272], there are other possible choices of $\mathbf{n}_1$ such as the flow direction, pressure-gradient direction, or velocity-magnitude-gradient direction. However, none of these directional choices can remove the shock instability completely, as indicated by numerical tests in Sect. 4 of Ren [397]. Also, Ren [397] deemed that the use of the velocity-difference vector as an upwind direction is of physical significance, and can effectively eliminate the shock instabilities, at least, for his numerical tests.

These existing studies demonstrate that the resulting rotated-hybrid flux functions are not only robust for nonlinear shock instability, but also accurate for resolving shear layers. They are also directly applicable to both structured and unstructured grids. Furthermore, they are particularly simple in that implementing them into an existing code (with a Roe-type flux already built in, for instance) is an extremely simple task.

Following the same spirit (e.g., [354, 397]) as above, any two Riemann solvers can be rotated-hybridized, if we are able to determine some physically meaningful upwind direction by selecting a directional feature detector. Of course, the choice of flux formulations should also be better based on the consideration of the flow physics.

The flexibility of decomposing the interface normal into $\mathbf{n}_1$ and $\mathbf{n}_2$ and selecting the flux functions in each direction enables us to optimize the design of rotated-hybrid numerical schemes for solving hyperbolic conservation laws. By the way, if we do not consider the cost of computing two flux functions per interface, we can hybridize any two flux functions in the form of Eq. (2.228) configured for each individual interface to create a new rotated-hybrid flux that might work expectedly better. Even further, we can hybridize each component of the flux function independently with different solvers. However, this optimization of rotated-hybrid fluxes in the context of flow characteristics is a challenging task.

Exploring various combinations in MHD Riemann solvers is promising. The rotated-hybrid scheme is applied [140] to 3D MHD equations in the finite-volume frame. Similar to Eq. (2.227), rotated-hybrid MHD scheme hybridizes HLL family along direction $\mathbf{n}_1$ and Roe flux along direction $\mathbf{n}_2$

$$\mathbf{F}\cdot\mathbf{n} = \alpha_1 \mathbf{f}_{\text{HLL family}}(\mathbf{n}_1) + \alpha_2 \mathbf{f}_{\text{Roe}}(\mathbf{n}_2)$$

Benchmark problems, such as Orszag-Tang MHD vortex problem and numerical study of steady-state coronal structures, validate and demonstrate the robustness of the rotated-hybrid MHD scheme.

2.8.4.3 HLLX and HLLXω Solver

In order to reduce the dissipation of the HLL solver, the HLLX solver is developed [414, 415]. As argued by Schmidtmann and Torrilhon [414], while the slow wave is better resolved by HLLX solver, the computational expense of HLLX is not significantly more expensive than HLL. In deriving the HLLX solver, the dimensionless scalar dissipation function $d(\nu)$ is used to describe the dissipative property of the dissipation matrix $\mathbf{D}$. Here, $\nu = \lambda\frac{\Delta t}{\Delta x}$ is the dimensionless characteristic speed. The dimensionless scalar dissipation functions corresponding to the dissipation matrices in Eqs. (2.217)–(2.221) are as follows:

- Lax–Friedrichs:

$$d_{\text{LF}} = 1$$

- Rusanov or local Lax–Friedrichs:

$$d_{\text{LLF}} = \max\left(|\nu_{\min}|, |\nu_{\max}|\right)$$

  where $\nu_{\min}$ and $\nu_{\max}$ are the (non-dimensionalized) globally fastest left and right traveling wave speeds, i.e.,

$$\nu_{\min} = \frac{\Delta t}{\Delta x}\min(\lambda_{\min}(\mathbf{U}_L), \lambda_{\min}(\mathbf{U}_R)) \quad \nu_{\max} = \frac{\Delta t}{\Delta x}\max(\lambda_{\max}(\mathbf{U}_L), \lambda_{\max}(\mathbf{U}_R))$$

- HLL:

$$d_{\text{HLL}} = \frac{|\nu_L|\nu_R - |\nu_R|\nu_L}{\nu_R - \nu_L} + \frac{|\nu_R| - |\nu_L|}{\nu_R - \nu_L}\nu$$

where $\nu_L = \frac{\Delta t}{\Delta x}S_L$ and $\nu_R = \frac{\Delta t}{\Delta x}S_R$.

- Lax–Wendroff:

$$d_{\text{LW}} = \nu^2$$

- Roe:

$$d_{\text{Roe}} = |\nu|$$

HLLX solver is established on the basis of Lax–Friedrichs, HLL and Lax–Wendroff solvers. Its dissipation function has the following form

$$d_{\text{HLLX}} = \alpha_0 d_{\text{LF}} + \alpha_1 d_{\text{HLL}} + \alpha_2 d_{\text{LW}},$$

where the coefficients are given by

$$\begin{aligned}
\alpha_0 &= \alpha|\nu_{\min} \cdot \nu_{\max}|, \\
\alpha_1 &= 1 - \alpha(|\nu_{\max}| + |\nu_{\min}|), \\
\alpha_2 &= \alpha, \\
\alpha &= \frac{\nu_{\max} - \nu_{\min} - ||\nu_{\max}| - |\nu_{\min}||}{(\nu_{\max} - \nu_{\min})^2}
\end{aligned}$$

An even simpler but not as demonstrative way of defining the dissipation matrix is

$$\mathbf{D}_{\text{HLLX}} = \mathbf{D}_{\text{HLL}} + \alpha\frac{\Delta t}{\Delta x}(\mathbf{A} - \lambda_{\min}\mathbf{I})(\mathbf{A} - \lambda_{\max}\mathbf{I})$$

Hence, the corresponding numerical flux function is

$$\mathbf{f}_{\text{HLLX}}(\mathbf{U}_L, \mathbf{U}_R) = \mathbf{f}_{\text{HLL}} - \frac{\alpha}{2}\frac{\Delta t}{\Delta x}(\mathbf{A} - \lambda_{\min}\mathbf{I})(\mathbf{A} - \lambda_{\max}\mathbf{I})(\mathbf{U}_R - \mathbf{U}_L) \tag{2.231}$$

For the numerical flux function Eq. (2.231), it's necessary to know the flux Jacobian matrix $\mathbf{A}$. For MHD, $\mathbf{A}$ may be calculated by the explicit expression, like Eq. (2.50) with the Powell source term included. Another choice is utilizing a Jacobian-free implementation, as illustrated in Schmidtmann and Torrilhon [414]. In Eq. (2.231), or more general Eq. (2.216), the dissipation matrix is always multiplied by the state difference $\Delta\mathbf{U} = \mathbf{U}_R - \mathbf{U}_L$, so that the Jacobian matrix always appears as a matrix vector multiplication with $\Delta\mathbf{U}$. We make full use of this fact by using the finite difference formulation $\mathbf{A}\Delta\mathbf{U} = \frac{\partial \mathbf{f}}{\partial \mathbf{U}}\big|_{\mathbf{U}=\overline{\mathbf{U}}}\Delta\mathbf{U} = \lim_{\varepsilon\to 0}(\mathbf{f}(\overline{\mathbf{U}} + \varepsilon\Delta\mathbf{U}) - \mathbf{f}(\bar{\mathbf{U}}))/\varepsilon$ with $\overline{\mathbf{U}} = 0.5(\mathbf{U}_L + \mathbf{U}_R)$. Assuming $\varepsilon \ll 1$, we get

$$\mathbf{A}\Delta\mathbf{U} = \frac{\mathbf{f}(\overline{\mathbf{U}} + \varepsilon\Delta\mathbf{U}) - \mathbf{f}(\overline{\mathbf{U}})}{\varepsilon},$$

$$\mathbf{A}^2\Delta\mathbf{U} = \frac{\mathbf{f}(\overline{\mathbf{U}} + \varepsilon\frac{\mathbf{f}(\overline{\mathbf{U}}+\varepsilon\Delta\mathbf{U})-\mathbf{f}(\overline{\mathbf{U}})}{\varepsilon}) - \mathbf{f}(\overline{\mathbf{U}})}{\varepsilon}.$$

For larger systems of conservation law whose Jacobian is difficult to compute, this approach is particularly useful (e.g., [209, 253]).

To further reduce the dissipation of the HLLX solver, HLLXω solver is developed [414, 415]. In order to derive the HLLXω solver, the HLLω solver is first constructed with dissipation function

$$d_{\mathrm{HLL}\omega} = b_0 + b_1\nu,$$

The corresponding dissipation matrix of HLLω is

$$\mathbf{D}_{\mathrm{HLL}\omega} = b_0\frac{\Delta x}{\Delta t}\mathbf{I} + b_1\mathbf{A},$$

where the coefficients $b_0 = b_0(\omega)$ and $b_1 = b_1(\omega)$ are given by

$$b_0(\omega) = \frac{\nu_{\max}(\omega\nu_{\min}^2 + (1-\omega)|\nu_{\min}|) - \nu_{\min}(\omega\nu_{\max}^2 + (1-\omega)|\nu_{\max}|)}{\nu_{\max} - \nu_{\min}},$$

$$b_1(\omega) = \frac{(1-\omega)(|\nu_{\max}| - |\nu_{\min}|) + \omega(\nu_{\max}^2 - \nu_{\min}^2)}{\nu_{\max} - \nu_{\min}}.$$

Then the dissipation function of HLLXω is given as a weighted linear combination of the dissipation functions of Lax–Friedrichs (LF), HLLω, and Lax–Wendroff (LW) solver

$$d_{\mathrm{HLLX}\omega} = \beta_0 d_{\mathrm{LF}} + \beta_1 d_{\mathrm{HLL}\omega} + \beta_2 d_{\mathrm{LW}},$$

with

$$\beta_0(\omega) = \beta(\omega)\frac{(1-\omega)|\nu_{\min}\nu_{\max}|}{(1-\omega) + \omega(|\nu_{\min}| + |\nu_{\max}|)},$$

$$\beta_1(\omega) = 1 - \beta(\omega)\left(\frac{1-\omega}{|\nu_{\min}| + |\nu_{\max}|} + \omega\right)^{-1},$$

$$\beta_2(\omega) = \beta(\omega),$$

and

$$\beta(\omega) = \omega + (1-\omega)\frac{\nu_{\max} - \nu_{\min} - |(|\nu_{\max}| - |\nu_{\min}|)|}{(\nu_{\max} - \nu_{\min})^2}.$$

Here, $\omega \in [0, 1]$ is a tunable parameter which describes the amount of dissipation reduction. For $\omega = 0$, the HLLXω solver is reduced to the HLLX solver, which determines the monotone behavior of the solver. For $\omega = 1$, the HLLXω is reduced to the non-monotone LW solver. If we define $\bar{\mathbf{f}} = (\mathbf{f}(\mathbf{U}_L) + \mathbf{f}(\mathbf{U}_R))/2$, the HLLX$\omega$ numerical flux function can be formulated

$$\mathbf{f}_{\mathrm{HLLX}\omega}(\mathbf{U}_L, \mathbf{U}_R) = \bar{\mathbf{f}} + \beta_0 \left(\mathbf{f}_{\mathrm{LF}} - \bar{\mathbf{f}}\right) + \beta_1 \left(\mathbf{f}_{\mathrm{HLL}\omega} - \bar{\mathbf{f}}\right) + \beta_2 \left(\mathbf{f}_{\mathrm{LW}} - \bar{\mathbf{f}}\right)$$

which provides the following dissipation matrix

$$\mathbf{D}_{\mathrm{HLLX}\omega} = \mathbf{D}_{\mathrm{HLL}\omega} + \beta \frac{\Delta t}{\Delta x}(\mathbf{A} - \lambda_{\min}\mathbf{I})(\mathbf{A} - \lambda_{\max}\mathbf{I})$$

Therefore, the HLLXω numerical flux function can also be written as

$$\mathbf{f}_{\mathrm{HLLX}\omega}(\mathbf{U}_L, \mathbf{U}_R) = \mathbf{f}_{\mathrm{HLL}\omega} - \frac{\beta}{2}\frac{\Delta t}{\Delta x}(\mathbf{A} - \lambda_{\min}\mathbf{I})(\mathbf{A} - \lambda_{\max}\mathbf{I})(\mathbf{U}_R - \mathbf{U}_L)$$

It is remarked that the choice of ω remains problem-dependent. In Schmidtmann and Torrilhon [414], it is believed that $\omega \leq 0.5$ is a good choice. In Schmidtmann and Winters [415], $\omega = 0.925$ is used in their numerical tests.

2.9 Newton–Krylov Methods

As illustrated in Sect. 2.5.5, the semi-discrete form Eq. (2.89) of the governing equations is repeated here as a coupled set of non-linear ordinary differential equations

$$\frac{\partial \mathbf{U}}{\partial t} = -\mathbf{R}(\mathbf{U}) \tag{2.232}$$

which can be discretized by a variety of time stepping methods. Some explicit time stepping methods introduced in Sect. 2.5.5 are frequently used. Usually, explicit time integration methods are conditionally stable, and the time step is limited by the CFL condition. Meanwhile, explicit schemes and their boundary conditions are easy to implement, vectorize and parallelize, and require only limited memory storage. When simulating some stiff problems, the time step limited by the CFL condition may be orders of magnitude smaller than the time step required by accuracy. In this situation, the explicit time stepping methods are inefficient. Especially for large-scale steady state problems, the rate of convergence may slow down dramatically.

However, implicit time stepping methods allow for time step much larger than that limited by the CFL condition and make it possible for modelers to select the time step based entirely on the desired accuracy of the solution. Generally, implicit methods require greater computational resources, but a sufficiently large time step can pay off the expense of the implicit method and result in a more efficient time stepping method overall.

The commonly used implicit time discretization may be the following

$$\frac{\mathbf{U}^{n+1} + (\beta - 2)\mathbf{U}^n + (1 - \beta)\mathbf{U}^{n-1}}{\beta \Delta t} + \mathbf{R}(\mathbf{U}^{n+1}) \equiv \mathbf{R}^*(\mathbf{U}^{n+1}) = 0 \tag{2.233}$$

Here, β is a tunable parameter, Δt is the time step, and $\mathbf{U}^{n+1}$ is the solution of $\mathbf{U}$ at time level $n+1$. For $\beta = 2/3$, Eq. (2.233) is the so-called second order backward differentiation formula (BDF2), which is widely used by time-accuracy simulation. To compute stationary flows, the first-order accurate implicit Euler method with $\beta = 1$ is chosen since time accuracy is not required to reach steady state in the computation. When using BDF2, the first order backward-Euler method can be used when $\mathbf{U}^{n-1}$ is not available.

Solving Eq. (2.233) is equivalent to finding the root $\mathbf{U}^{n+1}$ of non-linear equations $\mathbf{R}^*(\mathbf{U^{n+1}}) = 0$, that is,

$$\mathbf{R}^*(\mathbf{U^{n+1}}) = \frac{\mathbf{U}^{n+1} + (\beta - 2)\mathbf{U}^n + (1-\beta)\mathbf{U}^{n-1}}{\beta \Delta t} + \mathbf{R}(\mathbf{U}^{n+1}) = 0 \tag{2.234}$$

which by application of Newton's method leads to the following linear system of equations

$$\frac{\partial \mathbf{R}^*}{\partial \mathbf{U}^{n+1}}(\mathbf{U}^{n+1,k+1} - \mathbf{U}^{n+1,k}) = -\mathbf{R}^*(\mathbf{U}^{n+1,k})$$

Setting $\Delta \mathbf{U}^{n+1,k} = (\mathbf{U}^{n+1,k+1} - \mathbf{U}^{n+1,k})$, the above equation becomes

$$\left[\frac{\mathbf{I}}{\beta \Delta t} + \frac{\partial \mathbf{R}}{\partial \mathbf{U}}\right] \Delta \mathbf{U}^{n+1,k} = \mathbf{J}^* \Delta \mathbf{U}^{n+1,k} = -\mathbf{R}^*(\mathbf{U}^{n+1,k}) \tag{2.235}$$

The term $\frac{\partial \mathbf{R}}{\partial \mathbf{U}} = \frac{\partial \mathbf{R}_{\text{inv}}}{\partial \mathbf{U}} - \frac{\partial \mathbf{R}_{\text{diff}}}{\partial \mathbf{U}} - \frac{\partial \mathbf{R}_s}{\partial \mathbf{U}}$ involves symbolically the Jacobian matrice of inviscid flux vectors $\mathbf{R}_{\text{inv}}$ and viscous flux vectors $\mathbf{R}_{\text{diff}}$, or even source vectors $\mathbf{R}_s$. In order to obtain the quadratic convergence of Newton's method, the linearization of the numerical flux function must be virtually exact. Unfavorably, explicit formation of the Jacobian matrix resulting from the exact linearization of any second-order numerical flux functions for inviscid fluxes can require excessive storage and is extremely expensive, let alone more than second-order numerical flux functions. In order to reduce the number of nonzero entries in the matrix and to simplify the linearization, some applications use only a first-order linearization of the numerical fluxes. This leads to a graph of the sparse matrix $\frac{\partial \mathbf{R}}{\partial \mathbf{U}}$ being identical to the graph of the supporting cells.

Equation (2.235) must be solved at each Newton iteration level k to obtain $\mathbf{U}^{n+1,k+1}$, the improved approximation for $\mathbf{U}^{n+1}$. This iterative procedure is repeated until an appropriate norm of the solution residual $\mathbf{R}^*$ is sufficiently small, i.e.,

$$\|\mathbf{R}^*(\mathbf{U}^{n+1,k+1})\|_2 < \epsilon \|\mathbf{R}^*(\mathbf{U}^{n+1,k})\|_2$$

where ϵ is the Newton convergence tolerance. The value of ϵ is problem-dependent. For steady-state problem, typically an ϵ in the range $\epsilon \approx 10^{-8} \sim 10^{-10}$ is appropriate to ensure overall solution convergence. For unsteady problems, where the convergence in each time step is required, a larger ϵ in the range of $10^{-1} \sim 10^{-2}$ can be used. The convergence speed of Newton's method can be improved by a proper initial guess, $\mathbf{U}^{n+1,0}$. An obvious choice is to set $\mathbf{U}$ of the previous time step as the initial guess, i.e., $\mathbf{U}^{n+1,0} = \mathbf{U}^n$. Improved initial guess can be obtained with polynomial extrapolation by using previous solution information. In Tóth [473], instead of solving Eq. (2.235) in a repeated way, Eq. (2.235) is solved only once. This single Newton iteration is more efficient for unsteady applications, as pointed out by Tóth [473].

Equation (2.235) can be rewritten as a system of linear algebraic equations,

$$\mathbf{A}\boldsymbol{\chi} = \mathbf{b}$$

where $\mathbf{A} = \mathbf{J}^*$ is the matrix of coefficients of elements a_{ij}, $\boldsymbol{\chi} = \Delta\mathbf{U}^{n+1,k}$ the vector of unknown variables χ_i, and $\mathbf{b} = -\mathbf{R}^*(\mathbf{U}^{n+1,k})$ the vector with elements b_i. Generally speaking, each row in the matrix $\mathbf{A}$ is an equation defined over one element of the computational domain, and the non-zero coefficients are those related to the neighbors of that element. The coefficient a_{ij} measures the strength of the link between the control volume average χ_i in a cell i and those in its neighbors. As a cell is connected to only few neighbors, with their number depending on the connectivity of the elements in the discretized domain, $\mathbf{A}$ is typically a large, sparse, and generally asymmetric matrix. That is, many of the coefficients of the resulting $\mathbf{A}$ matrix are zeros and its non-zero coefficients are a very small fraction of the matrix. For a structured grid system, the matrix $\mathbf{A}$ will be banded with all non-zero elements aligned along few diagonals.

In order to solve this linear system, an iteration method is usually used, and allowed to be halted after a certain reduction of $\|\mathbf{b} - \mathbf{A}\boldsymbol{\chi}\|_2$. The employment of the iteration method to solve the linear system leads to an overall procedure of Newton's method with iterations within iterations: the inner loop iterations involve solving the linear system (2.235) at each Newton iteration level k, and the outer loop is associated with the solution of the non-linear problem (2.234) by using Newton's method for each time step n. When the inner loop iterations are not fully converged at each Newton iteration level, the Newton's method is referred to as inexact. It is demonstrated by Dembo et al. [103] that even if the solution of the linear system (2.235) is inexact, rapid convergence of Newton's method can still be achieved. For inexact method, the following convergence criteria for each inner loop iteration is used

$$\|\mathbf{R}^*(\mathbf{U}^{n+1,k}) + \mathbf{J}^*\Delta\mathbf{U}^{n+1,k}\|_2 \leq \xi \|\mathbf{R}^*(\mathbf{U}^{n+1,k})\|_2$$

where ξ represents the convergence tolerance for the iterative solution of the linear system and is typically in the range $0.01 \sim 0.5$. There may exist a tradeoff between the effort of solving the linear system (2.235) and the required number of outer loop iteration. Too large a value for ξ leads to less iterations in the inner loops but more iterations in the outer loop, and vice versa. Several strategies for optimizing the total computational work with a variable ξ have been given by Stanley [122].

Although solving Eq. (2.235) directly is an option, the vast computational cost makes such an approach prohibitive. Equation (2.235) represents a system of linear simultaneous algebraic equations and needs to be solved at each time step. The most widely used methods to solve this linear system are approximate factorization and iterative solution methods. As one of approximate factorization methods, the lower-upper symmetric Gauss-Seidel (LU-SGS) method (e.g., [224, 294, 295, 512]) is attractive because of its good stability properties and competitive computational cost in comparison to explicit methods. Although the LU-SGS method is more efficient than its explicit counterpart, a significant number of time steps are still required to achieve the steady-state solution, due to the nature of the approximation factorization schemes. One way to speed up the convergence is to use iterative methods that offer a more economical approach for the solution of large sparse system. A particularly suitable class of iterative techniques is the so-called Krylov-subspace method. Several implementations of Krylov method have been proposed, such as the Bi-Conjugate Gradient Stabilised (Bi-CGSTAB) method, the Transpose-Free Quasi-Minimum Residual (TFQMR) method and the Conjugate Gradient Squared (CGS) method. However, it is widely believed that the most successful Krylov subspace method may be the Generalized Minimal Residual (GMRES) method. For the simulation of MHD problem, a variety of implicit time stepping methods have been implemented by using the Newton–Krylov-GMRES methods (e.g., [69, 209, 297, 398, 473]).

2.9.1 GMRES Realization

To solve Eq. (2.235) at time level $n + 1$ and Newton iteration level k, the GMRES method looks for an approximate solution of $\Delta \mathbf{U}^{n+1,k}$ at every interaction m

$$\Delta \mathbf{U}_m = \Delta \mathbf{U}_0 + \sum_{l=0}^{m-1} \beta_l (\mathbf{J}^*)^l \mathbf{r}_0$$

Here, $\Delta \mathbf{U}_m$ is required to belong to the subspace $\mathcal{K}_m$ spanned by the Krylov vectors, $\{\mathbf{r}_0, \mathbf{J}^*\mathbf{r}_0, (\mathbf{J}^*)^2\mathbf{r}_0, \ldots, (\mathbf{J}^*)^{m-1}\mathbf{r}_0\}$. Since the GMRES iteration is performed at fixed $n + 1$ and k, the time level index $n + 1$ and the Newton iteration level index k have been dropped from $\Delta \mathbf{U}_m$. The initial linear residual, $\mathbf{r}_0$, is defined by $\mathbf{r}_0 = -\mathbf{R}^* - \mathbf{J}^* \Delta \mathbf{U}_0$. Here, $\Delta \mathbf{U}_0$ is the initial guess, which is typically set to be zero [253]. The scalars β_l are determined to minimize $\|\mathbf{J}^* \Delta \mathbf{U}_m + \mathbf{R}^*\|_2$ in a least-squares sense. The parameter m is the number of search directions, which specifies the dimension of the Krylov subspace. As m increases, the storage requirements grow linearly with m and the computational load increases quadratically. To control memory and limit computational cost, the value of m typically lies between $10 \sim 40$. If this is not sufficient for a converged solution $\Delta \mathbf{U}_m$, the GMRES method has to be restarted. $\Delta \mathbf{U}_m$ obtained from the previous GMRES iteration will be used as the initial guess for the next GMRES iteration. The GMRES method requires the action of the Jacobian only in the form of matrix-vector products. Indeed, the Jacobian $\mathbf{J}^*$ itself does not need to be provided explicitly. Instead, the matrix-vector product is approximated by

$$\mathbf{J}^* \boldsymbol{\nu} = \frac{\boldsymbol{\nu}}{\beta \Delta t} + \frac{\mathbf{R}(\mathbf{U} + \epsilon \boldsymbol{\nu}) - \mathbf{R}(\mathbf{U})}{\epsilon} \tag{2.236}$$

Here, $\boldsymbol{\nu}$ is an intermediate vector involved in the GMRES iteration. By using this approximation, the linear system can be solved without forming or storing the true Jacobian, and therefore the whole method is referred to as Jacobian-free.

The finite difference approach for $\mathbf{R}$ in Eq. (2.236), also called matrix-free method, is very sensitive to the choice of the parameter ϵ. For large ϵ, the finite difference approximation is poorly approximated, and Newton's method does not converge for high CFL numbers. For too small ϵ, the result of the finite difference is contaminated by floating-point roundoff error. There are several theories about the computation of ϵ, but it is not always possible to find a appropriate ϵ because the accuracy of the finite differences also depends on the function $\mathbf{R}$. The problem of determining a suitable value for ϵ persists when higher-order approximations are employed (e.g., [51, 253]). ϵ_0 is typically taken as the square root of machine accuracy to balance truncation and round-off effects. For instance, $\epsilon = \epsilon_0/||\boldsymbol{\nu}||_2$, in which $\epsilon_0 \approx 10^{-6} \sim 10^{-8}$ is commonly used (e.g., [139, 339, 340, 342, 361, 365]).

A crucial ingredient of Krylov subspace methods is the evaluation of the product of the Jacobian matrix of the spatial operator, e.g., fluxes, and a Krylov vector $\boldsymbol{\nu}$, by using a finite difference scheme for the inviscid function $\mathbf{R}_{\text{inv}}$,

$$\frac{\partial \mathbf{R}_{\text{inv}}}{\partial \mathbf{U}} \cdot \boldsymbol{\nu} = \lim_{\epsilon \to 0} \frac{\mathbf{R}_{\text{inv}}(\mathbf{U} + \epsilon \boldsymbol{\nu}) - \mathbf{R}_{\text{inv}}(\mathbf{U})}{\epsilon} \approx \frac{\mathbf{R}_{\text{inv}}(\mathbf{U} + \epsilon \boldsymbol{\nu}) - \mathbf{R}_{\text{inv}}(\mathbf{U})}{\epsilon} \tag{2.237}$$

When the inviscid residual $\mathbf{R}_{\text{inv}} = \frac{1}{V_i} \sum_{j=1}^{N_{\text{faces}}} (\mathbf{F}_{\text{inv}} \cdot \mathbf{n})_{ij} A_{ij}$ are functions of the reconstructed variables on the left and the right side of a cell surface, the matrix-vector product with the Jacobian of these fluxes can be transformed to

$$\frac{\partial (\mathbf{F}_{\text{inv}} \cdot \mathbf{n})_{ij}}{\partial \mathbf{U}} \cdot \boldsymbol{\nu} = \frac{\partial (\mathbf{F}_{\text{inv}} \cdot \mathbf{n})_{ij}}{\partial \mathbf{U}_L} \frac{\partial \mathbf{U}_L}{\partial \mathbf{U}} \cdot \boldsymbol{\nu} + \frac{\partial (\mathbf{F}_{\text{inv}} \cdot \mathbf{n})_{ij}}{\partial \mathbf{U}_R} \frac{\partial \mathbf{U}_R}{\partial \mathbf{U}} \cdot \boldsymbol{\nu} \tag{2.238}$$

where $\mathbf{n}$ is the face normal. The matrices $\frac{\partial (\mathbf{F}_{\text{inv}} \cdot \mathbf{n})_{ij}}{\partial \mathbf{U}_L}$ and $\frac{\partial (\mathbf{F}_{\text{inv}} \cdot \mathbf{n})_{ij}}{\partial \mathbf{U}_R}$ can be exactly obtained by differentiating the flux function with respect to its first argument (that is, the "left" state solution vector) and its second argument (that is, the "right" state solution vector). The matrix-vector products $\frac{\partial \mathbf{U}_L}{\partial \mathbf{U}} \cdot \boldsymbol{\nu}$ and $\frac{\partial \mathbf{U}_R}{\partial \mathbf{U}} \cdot \boldsymbol{\nu}$ can be computed by the finite difference approach just described: $\frac{\partial \mathbf{U}_L}{\partial \mathbf{U}} \cdot \boldsymbol{\nu} \approx \frac{\mathbf{U}_L(\mathbf{U}+\epsilon \boldsymbol{\nu}) - \mathbf{U}_L(\mathbf{U})}{\epsilon}$ and $\frac{\partial \mathbf{U}_R}{\partial \mathbf{U}} \cdot \boldsymbol{\nu} \approx \frac{\mathbf{U}_R(\mathbf{U}+\epsilon \boldsymbol{\nu}) - \mathbf{U}_R(\mathbf{U})}{\epsilon}$. Since Eq. (2.238) employs the functions $\mathbf{U}_L(\mathbf{U})$ and $\mathbf{U}_R(\mathbf{U})$ for the finite differences, which are slightly better scaled than $\mathbf{R}_{\text{inv}}(\mathbf{U})$, this scheme is more stable than the finite difference approach (2.237) for the function $\mathbf{R}_{\text{inv}}(\mathbf{U})$ (e.g., [365]).

In the practical implementation of GMRES method, instead of using the basis of $\{\mathbf{r}_0, \mathbf{J}^*\mathbf{r}_0, (\mathbf{J}^*)^2\mathbf{r}_0, \ldots, (\mathbf{J}^*)^{m-1}\mathbf{r}_0\}$, an orthonormal basis $\{\mathbf{v}_1, \mathbf{v}_2, \ldots, \mathbf{v}_m\}$ of the Krylov subspace $\mathcal{K}_m$ is looked for. The Arnoldi iteration is used to find these orthonormal basis vectors. With this orthonormal basis, the approximated solution of $\Delta \mathbf{U}_m \in \mathcal{K}_m$ can be expressed as

$$\Delta \mathbf{U}_m = \Delta \mathbf{U}_0 + \mathbf{V}_m \mathbf{y}_m$$

Here, $\mathbf{V}_m$ is a matrix whose columns are $\mathbf{v}_1, \mathbf{v}_2, \ldots, \mathbf{v}_m$, and $\mathbf{y}_m$ is an appropriate vector.

The Arnoldi process makes use of an $(m+1) \times m$ upper Hessenberg matrix $\tilde{\mathbf{H}}_m$ and the partial similarity transform

$$\mathbf{J}^* \mathbf{Q}_m = \mathbf{Q}_{m+1} \tilde{\mathbf{H}}_m$$

The minimization of the residual $\|\mathbf{J}^* \Delta \mathbf{U}_m + \mathbf{R}^*\|_2$ is then equivalent to that of $\|\tilde{\mathbf{H}}_m \mathbf{y}_n - \beta \mathbf{e}_1\|_2$, where $\beta = \| - \mathbf{R}^* - \mathbf{J}^* \Delta \mathbf{U}_0\|_2$ and $\mathbf{e}_1 = (1, 0, 0, \ldots, 0)^T$. This procedure is outlined in **Algorithm** 2.9.1. For a more comprehensive description of the GMRES method, one may refer to Saad [409].

Algorithm 1 GMRES

1: $\mathbf{r}_0 = -\mathbf{R}^* - \mathbf{J}^* \Delta \mathbf{U}_0$, $\beta = \|\mathbf{r}_0\|_2$, $\mathbf{v}_1 = \mathbf{r}_0/\beta$
2: **for** j=1,2, . . ., m **do**
3: $\quad \mathbf{w}_j = \mathbf{J}^* \mathbf{v}_j$
4: $\quad$ **for** i=1,2, . . .,j **do**
5: $\qquad h_{ij} = (\mathbf{w}_j, \mathbf{v}_i)$
6: $\qquad \mathbf{w}_j = \mathbf{w}_j - h_{ij} \mathbf{v}_i$
7: $\quad$ **end for**
8: $\quad h_{j+1,j} = \|\mathbf{w}_j\|_2$
9: $\quad \mathbf{v}_{j+1} = \mathbf{w}_j / h_{j+1,j}$
10: **end for**
11: Define the $(m+1) \times m$ upper Hessenberg matrix $\tilde{\mathbf{H}}_m = \{h_{i,j}\}_{1 \le i \le m+1, 1 \le j \le m}$, and $\mathbf{V}_m = (\mathbf{v}_1, \mathbf{v}_2, \ldots, \mathbf{v}_m)$.
12: Compute $\mathbf{y}_m$ that minimizes $\|\tilde{\mathbf{H}}_m \mathbf{y}_m - \beta \mathbf{e}_1\|_2$ and $\Delta \mathbf{U}_m = \Delta \mathbf{U}_0 + \mathbf{V}_m \mathbf{y}_m$.
13: If converged stop, else set $\Delta \mathbf{U}_0 = \Delta \mathbf{U}_m$ and GoTo 1.

2.9.2 Row Scaling of the Linear Problem $\mathbf{A}\chi = \mathbf{b}$

The GMRES iterative solution method is sensitive to scaling of the linear equations. In order to obtain a more robust and efficient solver, row scaling strategy can be applied to scale each equation of the linearized system $\mathbf{A}\chi = \mathbf{b}$ such that their coefficients are all of similar magnitude. Such techniques have been used in the literature (e.g., [298, 361]).

Row scaling can be accomplished by scaling the conserved variable $\mathbf{U}$ and solution residual $\mathbf{R}$ vectors within each computational cell such that $\mathbf{U}' = \mathcal{Y}_U \mathbf{U}$ and $\mathbf{R}' = \mathcal{Y}_R \mathbf{R}$ with $\mathcal{Y}_U$ and $\mathcal{Y}_R$ diagonal matrices consisting of the scaling coefficients. Applying this scaling procedure transforms equation $\mathbf{A}\chi = \mathbf{b}$ into the following prescaled linear problem

$$(\mathcal{Y}_R \mathbf{J}^* \mathcal{Y}_U^{-1})(\mathcal{Y}_U \chi) = -\mathcal{Y}_R \mathbf{R}^*$$

which is then solved in place of the original system Eq. (2.235) to obtain $\mathcal{Y}_U \chi$, with the preconditioned GMRES method. For example, the diagonal scaling matrices applied to each computational cell may take the form

$$\mathcal{Y}_{U,i} = \text{diag}\left(\frac{1}{\rho'}, \frac{1}{\rho' a'}, \frac{1}{\rho' a'}, \frac{1}{\rho' a'}, \frac{1}{|\mathbf{B}'|}, \frac{1}{|\mathbf{B}'|}, \frac{1}{|\mathbf{B}'|}, \frac{1}{\rho' a'^2}\right)$$

$$\mathcal{Y}_{R,i} = \text{diag}\left(\frac{1}{\rho' a'}, \frac{1}{\rho' a'^2}, \frac{1}{\rho' a'^2}, \frac{1}{\rho' a'^2}, \frac{1}{|\mathbf{B}'| a'}, \frac{1}{|\mathbf{B}'| a'}, \frac{1}{|\mathbf{B}'| a'}, \frac{1}{\rho' a'^3}\right)$$

where ρ' and a' are reference pressure and sound speed respectively. $|\mathbf{B}'|$ is the reference total magnetic strength $\sqrt{B_x'^2 + B_y'^2 + B_z'^2}$. When the scaling is applied to the conserved variables in each cell, $\mathbf{U}$, as defined previously by $\mathbf{U} = (\rho, \rho u, \rho v, \rho w, B_x, B_y, B_z, e)^T$, the scaled solution vector, $\mathbf{U}'$, becomes

$$\mathbf{U}' = \left(\frac{\rho}{\rho'}, \frac{\rho u}{\rho' a'}, \frac{\rho v}{\rho' a'}, \frac{\rho w}{\rho' a'}, \frac{B_x}{|\mathbf{B}'|}, \frac{B_y}{|\mathbf{B}'|}, \frac{B_z}{|\mathbf{B}'|}, \frac{e}{\rho' a'^2}\right)^T$$

and the corresponding scaled solution residual, $\mathbf{R}'$, has the form

$$\mathbf{R}' = \left(\frac{R_1}{\rho' a'}, \frac{R_2}{\rho' a'^2}, \frac{R_3}{\rho' a'^2}, \frac{R_4}{\rho' a'^2}, \frac{R_5}{|\mathbf{B}'| a'}, \frac{R_6}{|\mathbf{B}'| a'}, \frac{R_7}{|\mathbf{B}'| a'}, \frac{R_8}{\rho' a'^3}\right)^T$$

with $\mathbf{R} = (R_1, R_2, \ldots, R_8)^T$. Alternatively, e can be scaled by $\rho'(a'^2 + (v_A')^2)$ instead of $\rho' a'^2$, while R_8 by $\rho'(a'^2 + (v_A')^2)a'$ instead of $\rho' a'^3$. Primed variables can take their local values at the previous time step or iteration level. Of course, the initial values can also be used as such primed variables.

2.9.3 Preconditioner

GMRES minimizes the norm of the computed residual vector over the subspace spanned by a certain number of orthogonal search directions. It is well known that the speed of convergence of an iterative algorithm for a linear system depends on the condition number of the matrix $\mathbf{A}$. GMRES works best when the eigenvalues of matrix $\mathbf{A}$ are clustered. The easiest and the most common way to improve the efficiency and robustness of GMRES is to use preconditioning to attempt to cluster the eigenvalues at a single value.

The efficiency of Krylov-subspace methods depends on a good preconditioner. The purpose of preconditioning is to cluster the eigenvalues of the matrix $\mathbf{J}^*$ and thus to reduce the number of GMRES iterations. Usually, right preconditioning of the form $(\mathbf{J}^*\mathbf{M}^{-1})(\mathbf{M}\Delta\mathbf{U}) = -\mathbf{R}^*$ and left preconditioning of the form $(\mathbf{P}^{-1}\mathbf{J}^*)\Delta\mathbf{U} = -\mathbf{P}^{-1}\mathbf{R}^*$ are used, where $\mathbf{M}$ and $\mathbf{P}$ are the right and left preconditioning matrices respectively. In comparison to left preconditioning, right preconditioning is often preferred since it does not affect the norm of the residual. Using the right preconditioning, one first solve

$$(\mathbf{J}^*\mathbf{M}^{-1})\mathbf{Z} = -\mathbf{R}^* \tag{2.239}$$

for $\mathbf{Z}$, and then solve

$$\mathbf{M}\Delta\mathbf{U} = \mathbf{Z}$$

for $\Delta\mathbf{U}$. The preconditioning matrix can be generated from the factorization of $\mathbf{J}^*$. Although occasionally the preconditioner matrix might be entailed explicitly, the preconditioned GMRES method is conventionally said to be "Jacobian free" due to the simplicity of the preconditioning matrix.

In order to fulfill the purpose of preconditioning, preconditioner $\mathbf{M}$ should be as close as possible to $\mathbf{J}^*$, i.e., $\mathbf{M} \approx \mathbf{J}^*$. On the other hand, $\mathbf{M}$ should be computationally more inexpensive to invert than $\mathbf{J}^*$. For a successful implementation of preconditioning, an optimal tradeoff between the convergence acceleration of Krylov method and the time spent for inverting the preconditioning matrix should be found. Among various methods of preconditioning (see Knoll and Keyes [253] and reference therein), the Incomplete Lower Upper (ILU) factorization method is commonly used.

By using the ILU factorization method, $\mathbf{M}$, is approximated by $\mathbf{M} = \mathcal{L}\mathcal{U} \approx \mathbf{J}^*$, and Jacobian matrix $\mathbf{J}^*$ is calculated in Sect. 2.9.4. $\mathcal{L}$ and $\mathcal{U}$ are the sparse lower and upper triangular matrices derived from the ILU(f) factorization of $\mathbf{J}^*$, where the level of fill-in, f, determines the accuracy of this approximation. The simplest choice is zero fill-in, i.e., $f = 0$. By employing ILU(0), $\mathcal{L}$ will share the same structure (non-zero pattern) as the lower part of $\mathbf{J}^*$, and $\mathcal{U}$ has the same structure as the upper part of $\mathbf{J}^*$. Despite its simplicity, the accuracy of ILU(0) may be insufficient to yield an adequate convergence rate. With higher fill-in level, i.e., $f > 0$, more non-zero entries are retained in $\mathcal{L}$ and $\mathcal{U}$, and thus a more accurate approximation of $\mathbf{J}^*$ is achieved. However, this is at the cost of greater computational work and storage. For the detail of the ILU factorization algorithm, one may refer to Saad [409] and Moukalled et al. [330].

An alternative choice of the factorization is the Block ILU (BILU) decomposition, which exploits the block structure of the Jacobian $\mathbf{J}^*$. In this approach, $\mathbf{J}^*$ is still approximated by $\mathcal{L}\mathcal{U} \approx \mathbf{J}^*$, but the elimination process is performed on a block level rather than on a point level [14, 483].

In the description above, an intermediate variable $\mathbf{Z}$ is introduced for the preconditioned GMRES method. However, $\mathbf{Z}$ never needs to be invoked explicitly. For the initial residual, we have $\mathbf{r}_0 = -\mathbf{R}^* - \mathbf{J}^*\mathbf{M}^{-1}\mathbf{Z}_0 = -\mathbf{R}^* - \mathbf{J}^*\Delta\mathbf{U}_0$ and $\mathbf{Z}_0$ is not needed. As seen in **Algorithm** 2.9.3, once this initial residual is computed, all subsequent Krylov subspace vectors can be computed without referring to $\mathbf{Z}$.

At the end, the approximate solution of $\mathbf{Z}$ is given by $\mathbf{Z}_m = \mathbf{Z}_0 + \mathbf{V}_m\mathbf{y}_m$. Since $\mathbf{Z}_0 = \mathbf{M}\Delta\mathbf{U}_0$, multiplying $\mathbf{Z}_m$ by $\mathbf{M}^{-1}$ yields the desired approximation of $\Delta\mathbf{U}$, i.e., $\Delta\mathbf{U}_m = \Delta\mathbf{U}_0 + \mathbf{M}^{-1}\mathbf{V}_m\mathbf{y}_m$.

Algorithm 2 Right-Preconditioned GMRES

$\mathbf{r}_0 = -\mathbf{R}^* - \mathbf{J}^*\Delta\mathbf{U}_0$, $\beta = \|\mathbf{r}_0\|_2$, $\mathbf{v}_1 = \mathbf{r}_0/\beta$
for j=1,2,. . .,m **do**
  $\mathbf{w}_j = \mathbf{J}^*\mathbf{M}^{-1}\mathbf{v}_j$
  for i=1,2, . . .,j **do**
    $h_{ij} = (\mathbf{w}_j, \mathbf{v}_i)$
    $\mathbf{w}_j = \mathbf{w}_j - h_{ij}\mathbf{v}_i$
  end for
  $h_{j+1,j} = \|\mathbf{w}_j\|_2$
  $\mathbf{v}_{j+1} = \mathbf{w}_j/h_{j+1,j}$
end for
Define the $(m+1)\times m$ upper Hessenberg matrix $\tilde{\mathbf{H}}_m = \{h_{i,j}\}_{1\leq i\leq m+1, 1\leq j\leq m}$, and $\mathbf{V}_m = (\mathbf{v}_1, \mathbf{v}_2, \ldots, \mathbf{v}_m)$
Compute $\mathbf{y}_m$ that minimizes $\|\tilde{\mathbf{H}}_m\mathbf{y}_m - \beta\mathbf{e}_1\|_2$ and $\Delta\mathbf{U}_m = \Delta\mathbf{U}_0 + \mathbf{M}^{-1}\mathbf{V}_m\mathbf{y}_m$.
If converged stop, else set $\Delta\mathbf{U}_0 = \Delta\mathbf{U}_m$ and GoTo 1.

It should be noted that in **Algorithm** 2.9.3, explicit form of $\mathbf{M}^{-1}$ is not required. When calculating $\mathbf{J}^*\mathbf{M}^{-1}\mathbf{v}_j$, the linear problem $\mathbf{M}\mathbf{p} = \mathbf{v}_j$ is first solved, and then $\mathbf{J}^*\mathbf{M}^{-1}\mathbf{v}_j = \mathbf{J}^*\mathbf{p}$ is obtained by using Eq. (2.236). If $\mathbf{M}$ is generated by the ILU factorization, to obtain $\mathbf{p}$ by solving $\mathbf{M}\mathbf{p} = \mathbf{v}_j$ is an easy job.

Regrettably, the ILU preconditioning requires that elements of the Jacobian matrix $\mathbf{J}^*$ be calculated and stored. In order to deal with this problem, some authors suggests to employ the GMRES method with a lower-upper symmetric Gauss-Seidel (LU-SGS) preconditioner (e.g., [295, 528]). The storage of the approximate Jacobian matrix can be completely avoided by approximating the Jacobian with numerical flux under the help of matrix-free techniques. GMRES method with LU-SGS preconditioner is usually achieved with left preconditioning, i.e., $(\mathbf{P}^{-1}\mathbf{J}^*)\Delta\mathbf{U} = -\mathbf{P}^{-1}\mathbf{R}^*$. The procedure of such implementation is listed in **Algorithm** 2.9.3.

Algorithm 3 Left-Preconditioned GMRES

$\mathbf{r}_0 = -\mathbf{P}^{-1}(\mathbf{R}^* + \mathbf{J}^*\Delta\mathbf{U}_0)$, $\beta = \|\mathbf{r}_0\|_2$, $\mathbf{v}_1 = \mathbf{r}_0/\beta$
for j=1,2,. . .,m **do**
  $\mathbf{w}_j = \mathbf{P}^{-1}\mathbf{J}^*\mathbf{v}_j$
  for i=1,2,. . .,j **do**
    $h_{ij} = (\mathbf{w}_j, \mathbf{v}_i)$
    $\mathbf{w}_j = \mathbf{w}_j - h_{ij}\mathbf{v}_i$
  end for
  $h_{j+1,j} = \|\mathbf{w}_j\|_2$
  $\mathbf{v}_{j+1} = \mathbf{w}_j/h_{j+1,j}$
end for
Define the $(m+1)\times m$ upper Hessenberg matrix $\tilde{\mathbf{H}}_m = \{h_{i,j}\}_{1\leq i\leq m+1, 1\leq j\leq m}$, and $\mathbf{V}_m = (\mathbf{v}_1, \mathbf{v}_2, \ldots, \mathbf{v}_m)$
Compute $\mathbf{y}_m$ that minimizes $\|\tilde{\mathbf{H}}_m\mathbf{y}_m - \beta\mathbf{e}_1\|_2$ and $\Delta\mathbf{U}_m = \Delta\mathbf{U}_0 + \mathbf{V}_m\mathbf{y}_m$.
If converged stop, else set $\Delta\mathbf{U}_0 = \Delta\mathbf{U}_m$ and GoTo 1.

From **Algorithm** 2.9.3, it may be seen that linear problems $\mathbf{Pr}_0 = -(\mathbf{R}^* + \mathbf{J}^*\Delta\mathbf{U}_0)$ and $\mathbf{Pw}_j = \mathbf{J}^*\mathbf{v}_j$ need to be solved to obtain $\mathbf{r}_0$ and $\mathbf{w}_j$. By employing LU-SGS method, the preconditioning matrix $\mathbf{P}$ is based on the factorization of the Jacobian matrix $\mathbf{J}^*$, i.e., $\mathbf{P} = (\mathbf{D} + \mathcal{L})\mathbf{D}^{-1}(\mathbf{D} + \mathcal{U})$, with $\mathbf{J}^* = \mathcal{L} + \mathbf{D} + \mathcal{U}$. Here, $\mathcal{L}$ consists only of terms in the strictly lower triangular matrix of $\mathbf{J}^*$, $\mathcal{U}$ of terms in the strictly upper triangular matrix of $\mathbf{J}^*$ and $\mathbf{D}$ of diagonal terms of $\mathbf{J}^*$. Each linear problem is solved by a forward sweep and a backward sweep. For an example, for $\mathbf{Pr}_0 = -(\mathbf{R}^* + \mathbf{J}^*\Delta\mathbf{U}_0) = \mathbf{r}'_0$, The forward step of the LU-SGS method is

$$\mathbf{Dr}_0^* = \mathbf{r}'_0 - \mathcal{L}\mathbf{r}_0^*$$

and the backward sweep is

$$\mathbf{Dr}_0 = \mathbf{Dr}_0^* - \mathcal{U}\mathbf{r}_0$$

The forward sweep and backward sweep are accomplished along $i + j = $ const. lines in 2D case and on $i + j + k = $ const. planes in 3D case. In this way, the off-diagonal terms involved in the $\mathcal{L}$ and $\mathcal{U}$ terms become known from the previous part of this sweep. Terms like $\mathcal{L}\mathbf{r}_0^*$ and $\mathcal{U}\mathbf{r}_0$ can be evaluated by matrix-free method of Eq. (2.236) so that the preconditioned matrix $\mathbf{P}$ is not computed and stored explicitly.

In contrast, there is another kind of the local-preconditioning methods, which is designed to identify and isolate the distinct physical phenomena, and then to precondition each physics implicitly. The physical source of stiffness is directly handled while other features are not affected. Generally, preconditioning consists of a matrix multiplying the time derivative term with the effect of reducing the disparity of the eigenvalues of associated Jacobian matrix, thus removing the stiffness of the equations. This kind of preconditioner can be categorized as "physics-based preconditioners" (e.g., [181, 282, 353, 356, 376, 380, 396, 480]). Physics-based preconditioners for the MHD system can be found in a variety of literatures (e.g., [7, 69, 71, 72, 95, 358, 399, 400]).

2.9.4 Approximate Jacobian Matrix

From Eq. (2.235), the key point to obtain the approximate Jacobian matrix $\mathbf{J}^* = \frac{\mathbf{I}}{\beta\Delta t} + \frac{\partial\mathbf{R}}{\partial\mathbf{U}}$ is to calculate the term $\frac{\partial\mathbf{R}}{\partial\mathbf{U}}$. Recall that the discretized residual term $\mathbf{R}$ can be expressed as $\mathbf{R} = \mathbf{R}_{\text{inv}} - \mathbf{R}_{\text{diff}} - \mathbf{R}_s$ with $\mathbf{R}_{\text{inv}}$ representing convective residual, $\mathbf{R}_{\text{diff}}$ the diffusion residual, and $\mathbf{R}_s$ the source residual. Calculation of the approximate Jacobian $\frac{\partial\mathbf{R}}{\partial\mathbf{U}}$ involves those of the inviscid Jacobian matrix $\frac{\partial\mathbf{R}_{\text{inv}}}{\partial\mathbf{U}}$, the viscous Jacobian $\frac{\partial\mathbf{R}_{\text{diff}}}{\partial\mathbf{U}}$, and the source Jacobian $\frac{\partial\mathbf{R}_s}{\partial\mathbf{U}}$.

Approximate Inviscid Jacobian

When the rotational invariance property of the flux function is utilized in the solver, the frame of reference is first rotated to a local frame in the $\hat{x}$-direction, the inviscid flux function at the interface is evaluated and then the frame of reference is rotated back to the original orientation, Sect. 2.5. In this case, the numerical flux across cell face can be written as

$$(\mathbf{F}_{\text{inv}} \cdot \mathbf{n})_{LR} = (\mathbf{F}_{\text{inv}} \cdot \mathbf{n})_{ij} = \mathbf{T}^{-1}\boldsymbol{\phi}_{\text{inv}}(\mathbf{TU}_L, \mathbf{TU}_R)$$

where $\boldsymbol{\phi}_{\text{inv}}$ can be any numerical flux function defined in Sect. 2.8, and $\mathbf{T}$ is a rotation matrix that rotates the momentum vector leaving the mass and energy fluxes unchanged. The matrix $\mathbf{T}^{-1}$ is the inverse of $\mathbf{T}$ which is also equal to the transpose of $\mathbf{T}$ since $\mathbf{T}$ is orthogonal.

By definition of the Riemann problem, the numerical flux $\boldsymbol{\phi}_{\text{inv}}(\mathbf{TU}_L, \mathbf{TU}_R)$ across a cell face is a function of the left and right states. Thus by the chain rule the Jacobian of $\boldsymbol{\phi}_{\text{inv}}$ with respect to $\mathbf{U}$ is

$$\begin{aligned}\frac{\partial\boldsymbol{\phi}_{\text{inv}}(\mathbf{TU}_L,\mathbf{TU}_R)}{\partial\mathbf{U}} &= \frac{\partial\boldsymbol{\phi}_{\text{inv}}(\mathbf{TU}_L,\mathbf{TU}_R)}{\partial(\mathbf{TU}_L)}\frac{\partial(\mathbf{TU}_L)}{\partial\mathbf{U}_L}\frac{\partial\mathbf{U}_L}{\partial\mathbf{U}} + \frac{\partial\boldsymbol{\phi}_{\text{inv}}(\mathbf{TU}_L,\mathbf{TU}_R)}{\partial(\mathbf{TU}_R)}\frac{\partial(\mathbf{TU}_R)}{\partial\mathbf{U}_R}\frac{\partial\mathbf{U}_R}{\partial\mathbf{U}} \\ &= \frac{\partial\boldsymbol{\phi}_{\text{inv}}(\mathbf{TU}_L,\mathbf{TU}_R)}{\partial(\mathbf{TU}_L)}\mathbf{T}\frac{\partial\mathbf{U}_L}{\partial\mathbf{U}} + \frac{\partial\boldsymbol{\phi}_{\text{inv}}(\mathbf{TU}_L,\mathbf{TU}_R)}{\partial(\mathbf{TU}_R)}\mathbf{T}\frac{\partial\mathbf{U}_R}{\partial\mathbf{U}}\end{aligned} \tag{2.240}$$

As the inviscid flux $\boldsymbol{\phi}_{\text{inv}}$, is computed using an Godunov-type upwind scheme, Sect. 2.5. Alternatively, the variables on the left and the right side of a cell surface $\mathbf{U}_L$ and $\mathbf{U}_R$ can be considered as functions of the reconstructed primitive variables $\mathbf{W}_L$ and $\mathbf{W}_R$. Applying the chain rule to the Jacobian of the inviscid fluxes $(\mathbf{F}_{\text{inv}} \cdot \mathbf{n})_{ij}$, we compute all Jacobians exactly

$$\begin{aligned}\frac{\partial\boldsymbol{\phi}_{\text{inv}}(\mathbf{TU}_L,\mathbf{TU}_R)}{\partial\mathbf{U}} &= \frac{\partial\boldsymbol{\phi}_{\text{inv}}(\mathbf{TU}_L,\mathbf{TU}_R)}{\partial(\mathbf{TU}_L)}\frac{\partial(\mathbf{TU}_L)}{\partial\mathbf{U}_L}\frac{\partial\mathbf{U}_L}{\partial\mathbf{W}_L}\frac{\partial\mathbf{W}_L}{\partial\mathbf{W}}\frac{\partial\mathbf{W}}{\partial\mathbf{U}} + \frac{\partial\boldsymbol{\phi}_{\text{inv}}(\mathbf{TU}_L,\mathbf{TU}_R)}{\partial(\mathbf{TU}_R)}\frac{\partial(\mathbf{TU}_R)}{\partial\mathbf{U}_R}\frac{\partial\mathbf{U}_R}{\partial\mathbf{W}_R}\frac{\partial\mathbf{W}_R}{\partial\mathbf{W}}\frac{\partial\mathbf{W}}{\partial\mathbf{U}} \\ &= \frac{\partial\boldsymbol{\phi}_{\text{inv}}(\mathbf{TU}_L,\mathbf{TU}_R)}{\partial(\mathbf{TU}_L)}\mathbf{T}\frac{\partial\mathbf{U}_L}{\partial\mathbf{W}_L}\frac{\partial\mathbf{W}_L}{\partial\mathbf{W}}\frac{\partial\mathbf{W}}{\partial\mathbf{U}} + \frac{\partial\boldsymbol{\phi}_{\text{inv}}(\mathbf{TU}_L,\mathbf{TU}_R)}{\partial(\mathbf{TU}_R)}\mathbf{T}\frac{\partial\mathbf{U}_R}{\partial\mathbf{W}_R}\frac{\partial\mathbf{W}_R}{\partial\mathbf{W}}\frac{\partial\mathbf{W}}{\partial\mathbf{U}}\end{aligned}$$

In practice, the Jacobian is usually approximated using just the first order terms of the upwind discretization procedure. Such first order approximation Jacobian is believed to provide a relatively inexpensive yet effective preconditioner (e.g., [51, 59, 295, 339, 340, 361]).

In the finite volume method, we need only the Jacobian of the inviscid residual $\mathbf{R}_{\text{inv}}$ with respect to the variables of cell i and then for each face, add this Jacobian to the Jacobian of the neighbouring cells with respect to the variables of cell i. For the cell i with the direct neighbors of j (Fig. 2.1), the inviscid residual $\mathbf{R}_{\text{inv}}$ can be recast in the following form

$$\mathbf{R}_{\text{inv}} = \frac{1}{V_i}\sum_{j=1}^{N_{\text{faces}}}(\mathbf{F}_{\text{inv}} \cdot \mathbf{n})_{ij}A_{ij} = \frac{1}{V_i}\sum_{j=1}^{N_{\text{faces}}}\mathbf{T}^{-1}\boldsymbol{\phi}_{\text{inv}}(\mathbf{TU}_L, \mathbf{TU}_R)A_{ij}$$

$$\mathbf{U}_L = \mathbf{U}_i, \mathbf{U}_R = \mathbf{U}_j$$

where $\mathbf{T}$ is the rotation matrix for each face and A_{ij} is the face area. In order to take derivative of the residual function respect to the solution vector of $\mathbf{U}$ at control volume i and its neighbors j's, we only consider the first neighbors as the Jacobian matrix is being computed to the first order of accuracy. Then, following Eq. (2.240), the row i entries of the inviscid Jacobian matrix $\frac{\partial \mathbf{R}_{\mathrm{inv}}}{\partial \mathbf{U}}$ can be written as

$$\begin{aligned} \frac{\partial \mathbf{R}_{\mathrm{inv}}}{\partial \mathbf{U}_j} &= \frac{1}{V_i}\mathbf{T}^{-1}\frac{\partial \boldsymbol{\phi}_{\mathrm{inv}}(\mathbf{TU}_i,\mathbf{TU}_j)}{\partial \mathbf{U}_j}A_{ij} = \frac{1}{V_i}\mathbf{T}^{-1}\frac{\partial \boldsymbol{\phi}_{\mathrm{inv}}(\mathbf{TU}_i,\mathbf{TU}_j)}{\partial (\mathbf{TU}_j)}\mathbf{T}A_{ij}, j \in \{V_j\}_i \\ \frac{\partial \mathbf{R}_{\mathrm{inv}}}{\partial \mathbf{U}_i} &= \frac{1}{V_i}\sum_{j=1}^{N_{\mathrm{faces}}}\mathbf{T}^{-1}\frac{\partial \boldsymbol{\phi}_{\mathrm{inv}}(\mathbf{TU}_i,\mathbf{TU}_j)}{\partial \mathbf{U}_i}A_{ij} = \frac{1}{V_i}\sum_{j=1}^{N_{\mathrm{faces}}}\mathbf{T}^{-1}\frac{\partial \boldsymbol{\phi}_{\mathrm{inv}}(\mathbf{TU}_i,\mathbf{TU}_j)}{\partial \mathbf{TU}_i}\mathbf{T}A_{ij} \end{aligned} \tag{2.241}$$

If the fluxes at cell faces are calculated based on Roe flux function, for example, for the face between cell i and its neighbor, cell j, the flux function can be expressed by

$$\boldsymbol{\phi}_{\mathrm{inv}}\left(\mathbf{U}_i, \mathbf{U}_j\right) = \frac{1}{2}\left(\mathbf{f}(\mathbf{U}_i) + \mathbf{f}(\mathbf{U}_j)\right) - \frac{1}{2}|\mathbf{A}|_{ij}\left(\mathbf{U}_j - \mathbf{U}_i\right) \tag{2.242}$$

Therefore, by ignoring changes in $|\mathbf{A}|_{ij}$ matrix, the flux function derivative terms in Eq. (2.241) can be simply computed according to Eq. (2.242)

$$\begin{aligned} \frac{\partial \boldsymbol{\phi}_{\mathrm{inv}}(\mathbf{TU}_i, \mathbf{TU}_j)}{\partial (\mathbf{TU}_i)} &= \left(\frac{\partial \boldsymbol{\phi}_{\mathrm{inv}}(\mathbf{U}_i, \mathbf{U}_j)}{\partial \mathbf{U}_i}\right)\Bigg|_{(\mathbf{U}_i,\mathbf{U}_j)=(\mathbf{TU}_i,\mathbf{TU}_j)} \\ &= \frac{1}{2}\left(\frac{\partial \mathbf{f}(\mathbf{U}_i)}{\partial \mathbf{U}_i} + |\mathbf{A}|_{ij}\right)\Bigg|_{(\mathbf{U}_i,\mathbf{U}_j)=(\mathbf{TU}_i,\mathbf{TU}_j)}, \\ \frac{\partial \boldsymbol{\phi}_{\mathrm{inv}}(\mathbf{TU}_i, \mathbf{TU}_j)}{\partial (\mathbf{TU}_j)} &= \left(\frac{\partial \boldsymbol{\phi}_{\mathrm{inv}}(\mathbf{U}_i, \mathbf{U}_j)}{\partial \mathbf{U}_j}\right)\Bigg|_{(\mathbf{U}_i,\mathbf{U}_j)=(\mathbf{TU}_i,\mathbf{TU}_j)} \\ &= \frac{1}{2}\left(\frac{\partial \mathbf{f}(\mathbf{U}_j)}{\partial \mathbf{U}_j} - |\mathbf{A}|_{ij}\right)\Bigg|_{(\mathbf{U}_i,\mathbf{U}_j)=(\mathbf{TU}_i,\mathbf{TU}_j)} \end{aligned} \tag{2.243}$$

We note that the data used to compute $\frac{\partial \mathbf{R}_{\mathrm{inv}}}{\partial \mathbf{U}}$ would ordinarily be the control volume averages $\mathbf{U}_i$ and $\mathbf{U}_j$. These approximations in the linearization process can cause the failure of the quadratic convergence of Newton's method due to the mismatch and inconsistency between the right- and left-hand sides in Eq. (2.235). Although the number of time steps (Newton iterations, if Δt tends to infinity) may increase, the cost per time step is significantly reduced when solving the resulting linear system, since it takes less CPU time to compute the Jacobian matrix and improves the conditioning of the simplified Jacobian matrix. However, it is suggested [339, 340] that using the reconstructed values in each control volume at the Gauss point produces a more effective preconditioning matrix; Nejat and Ollivier-Gooch [339, 340] used linear reconstruction data in computing the Jacobian. Using this more accurate data presents no additional cost, since the reconstruction has already been computed. Also, with this approach the effects of boundary conditions can be included in the approximate Jacobian by computing the Jacobian of the boundary flux with respect to the solution in the interior of the domain. In this case, only the $\frac{\partial \mathbf{R}_{\mathrm{inv}}}{\partial \mathbf{U}_i}$ term in the Jacobian needs to be computed. The above equations are written for an interior face which has a left cell and a right cell adjacent to it. At boundary cells, only the diagonal contribution from the left equation of Eq. (2.243) is used. The addition on the diagonal tends to stabilize the matrix solve calculations.

The implicit operator matrix in the left hand of Eq. (2.235), i.e., the Jacobian matrix $\mathbf{J}^*$, is a large and sparse block-septadiagonal matrix. The basic constituent unit of this matrix is 8×8 matrix block for MHD equations or 5×5 matrix block for Euler and Navier–Stokes equations. Suppose the simulation is carried out in a structured grid with N_x, N_y, and N_z grid cells in each dimension. For each cell i, its corresponding grid index are I, J, K where $I \in [1, N_x], J \in [1, N_y]$ and $K \in [1, N_z]$. Variable vector $\Delta\mathbf{U}_{I,J,K}$ of each grid cell is sorted to form the global vector $\Delta\mathbf{U}$ so that the index I runs fastest, then index J and index K runs slowest. In this way, the I_Gth element of the global vector $\Delta\mathbf{U}$ is $\Delta\mathbf{U}_{I,J,K}$, where $I_G = I + J \times N_x + K \times N_x \times N_y$. The non-zero elements in the I_Gth row of the Jacobian matrix $\mathbf{J}^*$ are associated with the cells that are closely associated to the I_Gth cell being updated. In other words, If update of physical values in the I_Gth cell depends on physical value in the J_Gth cell, non-zero elements will appear in the I_Gth row and J_Gth column. Since the solution of $\Delta\mathbf{U}_{I,J,K}$ always depends on the physical values in its own grid cell, elements at the position (I_G, I_G), i.e., the diagonal elements, are always non-zero. Any non-zero elements with $J_G < I_G$ are situated in the strictly lower part of matrix $\mathbf{J}^*$, while those with $J_G > I_G$ are in the strictly upper part of matrix $\mathbf{J}^*$. In what follows, a visual representation of $\frac{\mathbf{I}}{\beta\Delta t} + \frac{\partial\mathbf{R}_{\text{inv}}}{\partial\mathbf{U}}$ associated with the first order inviscid flux is given by

$$\begin{pmatrix}
* & * & \cdot & * & \cdot & * & \cdot & \cdot & \cdot & \cdot & \cdot & \cdot & \cdot \\
* & * & * & \cdot & * & \cdot & * & \cdot & \cdot & \cdot & \cdot & \cdot & \cdot \\
\cdot & * & * & * & \cdot & * & \cdot & * & \cdot & \cdot & \cdot & \cdot & \cdot \\
* & \cdot & * & * & * & \cdot & * & \cdot & * & \cdot & \cdot & \cdot & \cdot \\
\cdot & * & \cdot & * & * & * & \cdot & * & \cdot & * & \cdot & \cdot & \cdot \\
* & \cdot & * & \cdot & * & * & * & \cdot & * & \cdot & * & \cdot & \cdot \\
\cdot & \mathcal{A} & \cdot & \mathcal{B} & \cdot & \mathcal{C} & \mathcal{D} & \mathcal{E} & \cdot & \mathcal{F} & \cdot & \mathcal{G} & \cdot \\
\cdot & \cdot & * & \cdot & * & \cdot & * & * & * & \cdot & * & \cdot & * \\
\cdot & \cdot & \cdot & * & \cdot & * & \cdot & * & * & * & \cdot & * & \cdot \\
\cdot & \cdot & \cdot & \cdot & * & \cdot & * & \cdot & * & * & * & \cdot & * \\
\cdot & \cdot & \cdot & \cdot & \cdot & * & \cdot & * & \cdot & * & * & * & \cdot \\
\cdot & \cdot & \cdot & \cdot & \cdot & \cdot & * & \cdot & * & \cdot & * & * & * \\
\cdot & \cdot & \cdot & \cdot & \cdot & \cdot & \cdot & * & \cdot & * & \cdot & * & *
\end{pmatrix}$$

In the above matrix, zero elements are represent by '·' and non-zero elements are represent by '$*$'. Characters $\mathcal{A}$, $\mathcal{B}$, $\mathcal{C}$, $\mathcal{D}$, $\mathcal{E}$, $\mathcal{F}$, $\mathcal{G}$ are non-zero elements associated with cell i (with 3D cell index I, J, K), which are expressed explicitly as an example. The diagonal element $\mathcal{D} = \frac{\mathbf{I}}{\beta\Delta t} + \frac{1}{V_{I,J,K}} \sum_j \mathbf{T}^{-1} \frac{\partial\boldsymbol{\phi}_{\text{inv}}(\mathbf{TU}_{I,J,K}, \mathbf{TU}_j)}{\partial\mathbf{TU}_{I,J,K}} \mathbf{T}A^j_{I,J,K}$ with $j \in \{(I+1, J, K),\ (I-1, J, K), (I, J+1, K), (I, J-1, K), (I, J, K+1), (I, J, K-1)\}$. The off-diagonal elements $\mathcal{A} = \frac{1}{V_{I,J,K}} \mathbf{T}^{-1} \frac{\partial\boldsymbol{\phi}_{\text{inv}}(\mathbf{TU}_{I,J,K}, \mathbf{TU}_{I,J,K-1})}{\partial(\mathbf{TU}_{I,J,K-1})} \mathbf{T}A^{I,J,K-1}_{I,J,K}$, $\mathcal{B} = \frac{1}{V_{I,J,K}} \mathbf{T}^{-1} \frac{\partial\boldsymbol{\phi}_{\text{inv}}(\mathbf{TU}_{I,J,K}, \mathbf{TU}_{I,J-1,K})}{\partial(\mathbf{TU}_{I,J-1,K})} \mathbf{T}A^{I,J-1,K}_{I,J,K}$, $\mathcal{C} = \frac{1}{V_{I,J,K}} \mathbf{T}^{-1} \frac{\partial\boldsymbol{\phi}_{\text{inv}}(\mathbf{TU}_{I,J,K}, \mathbf{TU}_{I-1,J,K})}{\partial(\mathbf{TU}_{I-1,J,K})} \mathbf{T}A^{I-1,J,K}_{I,J,K}$, $\mathcal{E} = \frac{1}{V_{I,J,K}} \mathbf{T}^{-1} \frac{\partial\boldsymbol{\phi}_{\text{inv}}(\mathbf{TU}_{I,J,K}, \mathbf{TU}_{I+1,J,K})}{\partial(\mathbf{TU}_{I+1,J,K})} \mathbf{T}A^{I+1,J,K}_{I,J,K}$, $\mathcal{E} = \frac{1}{V_{I,J,K}} \mathbf{T}^{-1} \frac{\partial\boldsymbol{\phi}_{\text{inv}}(\mathbf{TU}_{I,J,K}, \mathbf{TU}_{I,J+1,K})}{\partial(\mathbf{TU}_{I,J+1,K})} \mathbf{T}A^{I,J+1,K}_{I,J,K}$, $\mathcal{G} = \frac{1}{V_{I,J,K}} \mathbf{T}^{-1} \frac{\partial\boldsymbol{\phi}_{\text{inv}}(\mathbf{TU}_{I,J,K}, \mathbf{TU}_{I,J,K+1})}{\partial(\mathbf{TU}_{I,J,K+1})} \mathbf{T}A^{I,J,K+1}_{I,J,K}$. Here, $A^{I+1,J,K}_{I,J,K}$ is the area of the interface between the cells (I, J, K) and $(I+1, J, K)$, and other A symbols with subscripts and superscripts indexed are defined similarly. Note that elements $\mathcal{B}$ and $\mathcal{D}$ are separated by $N_x - 1$ elements, and so do $\mathcal{F}$ and $\mathcal{D}$. Elements $\mathcal{A}$ and $\mathcal{D}$ are separated by $N_x \times N_y - 1$ elements, and so do $\mathcal{G}$ and $\mathcal{D}$. When the contribution from inviscid flux are taken into consideration, or higher-order inviscid flux are used, there would be more non-zero elements entering the Jacobian matrix, whose positions can be determined using the criteria introduced above.

If the change in solution is relatively small (start-up phase), the approximation in Eq. (2.243) with ignorance of changes in $|\mathbf{A}|$ works reasonably well. But, for very large changes in solution, keeping $|\mathbf{A}|$ constant can adversely affect the quality of preconditioning matrix, resulting in the degradation of the Newton-GMRES convergence rate. Since taking analytical derivative of $|\mathbf{A}|$ is quite challenging and expensive (e.g., [30, 31, 50, 339]), the following finite difference perturbation for inviscid Jacobian calculation

$$\frac{\partial \mathbf{R}_{\text{inv}}}{\partial (\mathbf{U})_k} = \frac{\mathbf{R}_{\text{inv}}(\mathbf{U} + \epsilon \boldsymbol{\alpha}_k) - \mathbf{R}_{\text{inv}}(\mathbf{U})}{\epsilon}$$

is a choice (e.g., [338, 339]), in which ϵ is the root square of machine accuracy and $\boldsymbol{\alpha}_k$ is a vector whose kth element is one and the rest of the elements are zero; $(\mathbf{U})_k$ is the kth component of the solution variable vector $\mathbf{U}$. Due to finite difference approach for $\mathbf{R}_{\text{inv}}$, Jacobian computation cost would be increased; but overall CPU-Time is not affected considerably since such procedure involves only one Jacobian computation per GMRES outer iteration in Newton-GMRES phase and small number of Newton iterations. As concluded by Nejat and Ollivier-Gooch [338], overall convergence rate is increased considerably by reducing number of outer iterations, as this approach leads to more accurate preconditioner matrix. Asgharzadeh and Borazjani [13] developed a computationally inexpensive analytical Jacobian for Newton–Krylov method to solve unsteady incompressible Navier–Stokes momentum equations on staggered overset-curvilinear grids with immersed boundaries, and used the analytical Jacobian to form a preconditioner for matrix-free method in order to improve its performance.

Approximate Viscous Jacobian and Approximate Source Jacobian

For high-order schemes of the flux function from the formulation of K-exact polynomial reconstruction and the rotational invariance property, the calculation of approximate viscous Jacobian is repeated again. For the viscous operator, the term $\mathbf{F}_{\text{diff}} \cdot \mathbf{n}$ in the equation for the viscous residual $\mathbf{R}_{\text{diff}} = \frac{1}{V_i} \sum_{j=1}^{N_{\text{faces}}} (\mathbf{F}_{\text{diff}} \cdot \mathbf{n})_{ij} A_{ij}$ is given by $\mathbf{F}_{\text{diff}} \cdot \mathbf{n} = \mathbf{f}_{\text{diff}} n_x + \mathbf{g}_{\text{diff}} n_y + \mathbf{h}_{\text{diff}} n_z$. In the evaluation of the viscous residual, the gradient values of cell-centered conserved variables, $\mathbf{U}$, as well as the corresponding cell-centered primitive variables, $\mathbf{W}$, may be used. These gradients on a given cell interface, can be evaluated by the methods described in the later part of Sect. 2.2, or for high-order schemes, by $\frac{1}{2}\left(\nabla \mathbf{W}_L^K + \nabla \mathbf{W}_R^K\right)$ via a K-exact least-squares approach, as shown in the later part of Sect. 2.5. Let $\mathbf{E}_f = (\mathbf{W}_f, \nabla \mathbf{W}_f)$ the extended solution vector $(\mathbf{W}, \nabla \mathbf{W})$ taking values at the interface f. Thus, $\frac{\partial \mathbf{F}_{\text{diff}} \cdot \mathbf{n}}{\partial \mathbf{U}}$ is performed through the chain rule:

$$\frac{\partial \mathbf{F}_{\text{diff}} \cdot \mathbf{n}}{\partial \mathbf{U}} = \frac{\partial \mathbf{f}_{\text{diff}}}{\partial \mathbf{E}_f} \frac{\partial \mathbf{E}_f}{\partial \mathbf{W}} \frac{\partial \mathbf{W}}{\partial \mathbf{U}} n_x + \frac{\partial \mathbf{g}_{\text{diff}}}{\partial \mathbf{E}_f} \frac{\partial \mathbf{E}_f}{\partial \mathbf{W}} \frac{\partial \mathbf{W}}{\partial \mathbf{U}} n_y + \frac{\partial \mathbf{h}_{\text{diff}}}{\partial \mathbf{E}_f} \frac{\partial \mathbf{E}_f}{\partial \mathbf{W}} \frac{\partial \mathbf{W}}{\partial \mathbf{U}} n_z$$

Here, $\frac{\partial \mathbf{W}}{\partial \mathbf{U}}$ is the standard Jacobian of the primitive variables with respect to the conserved variables and is readily available. The Jacobian, $\frac{\partial \mathbf{E}_f}{\partial \mathbf{W}}$, is due to geometry only, for it only depends on the interpolation used in the viscous flux evaluation described in the later part of Sect. 2.2. For high-order schemes, the cell face value $\mathbf{E}_f = \left(\frac{1}{2}\left(\mathbf{W}_L^K + \mathbf{W}_R^K\right), \frac{1}{2}\left(\nabla \mathbf{W}_L^K + \nabla \mathbf{W}_R^K\right)\right)_f$, see the later part of Sect. 2.5.

The Jacobian components, related to the source terms, are straightforward as they only rely on the solution values for local cell i and its neighborhoods. Thus, the contribution to the residual Jacobian can be calculated simply as $\frac{\partial \mathbf{R}_s}{\partial \mathbf{U}}$.

2.9.5 *Krylov–Schwarz Preconditioner*

In parallel computation, an additive Krylov–Schwarz global preconditioner is usually used in conjunction with the ILU local preconditioning (e.g., [253, 410, 411, 464]). The employment of additive Krylov–Schwarz preconditioner is favored by its perfect parallelism. The preconditioning is introduced through a subdomain-by-subdomain basis

$$\mathbf{M}^{-1} = \sum_{d=1}^{N_{\text{domain}}} \mathbf{B}_d^T \mathbf{M}_d^{-1} \mathbf{B}_d$$

Here, $\mathbf{B}_d$ is the global-to-local restriction operator for the dth subdomain, which gathers data for the subdomain from the global vector. N_{domain} is the total number of subdomains covering the computational domain. The sub-preconditioner $\mathbf{M}_d$ is constructed and inverted regardless of other subdomains. Afterwards, the global preconditioning matrix is formed by the local-to-global prolongation operator $\mathbf{B}_d^T$. An introductory description of Krylov–Schwarz methods may be found in Keyes [247]. Subdomain overlapping is permitted in Newton-Schwarz preconditioning. In general, the greater the overlap, the better the condition number of $\mathbf{J}^*\mathbf{M}^{-1}$ and the fewer the number of iterations. However, this is achieved at the cost of redundant computational work in the overlap region. Therefore, the overall computational effort may not be reduced by overlapping subdomains. The subpreconditioner, or the local preconditioner for the dth subdomain, $\mathbf{M}_d$, is approximated with ILU or BILU factorizations by $\mathbf{M}_d = \mathcal{L}_d \mathcal{U}_d \approx \mathbf{J}_d^*$, where $\mathbf{J}_d^*$ is the Jacobian of dth subdomain.

2.9.6 *Steady State Convergence*

To solve the steady state problem, we need to seek the root of the nonlinear equations $\mathbf{R} = 0$, where $\mathbf{R}$ is the right-hand-side of Eq. (2.232). The solution of $\mathbf{R} = 0$ may be found by taking an infinite time step, i.e., $\Delta t \to \infty$ and employing the Newton iteration formulated in Eq. (2.235). Newton method generally converges quadratically in the vicinity of the solution. Overall solver robustness depends critically on finding an approximate solution near enough to the steady-state solution that Newton iteration will succeed, while efficiency considerations necessitate the start-up procedure by which such an approximate solution can be computed as cheap as possible. Since the initial condition far from the steady state may cause a breakdown of the Newton iteration, a start-up algorithm is usually required in order to ensure the global convergence of the Newton method.

One widely used approach is the so-called Switched Evolution Relaxation (SER) technique [331]. By using SER, the temporal derivation term is retained and the first order backward-Euler method is used for time marching. This is realized by setting $\beta = 1$ in Eqs. (2.233)–(2.235). The time step is varied, starting from a finite-value, gradually increasing and becoming large enough as the steady solution is achieved. The time step may be calculated in the same way as presented for the explicit scheme but with a variable CFL number σ. Starting from a small value, σ is increased correspondingly to the reduction of the L_2-norm of the residual, according to $\sigma^{n+1} = \sigma^n \frac{\|\mathbf{R}(\mathbf{U}^{n-1})\|}{\|\mathbf{R}(\mathbf{U}^n)\|}$ [45] or $\sigma^{n+1} = \min(\sigma^n \frac{\|\mathbf{R}(\mathbf{U}^{n-1})\|}{\|\mathbf{R}(\mathbf{U}^n)\|}, 100)$ [528] that is limited to ensure stability.

Bramkamp et al. [51] proposed a start-up strategy, which is based on the switch between first-order and second-order accurate Jacobian-vector products. The first-order accurate Jacobian-vector product $\mathbf{J}^*_{\text{low}} \Delta\mathbf{U}$ is employed during the early iterations. When the relative residual of the density first drops below a prescribed threshold ν, $\mathbf{J}^*_{\text{low}} \Delta\mathbf{U}$ is switched to second-order accurate Jacobian-vector product $\mathbf{J}^*_{\text{high}} \Delta\mathbf{U}$. For time step n, let R_ρ denote the relative residual of the density ρ, i.e., $R_\rho = \frac{\|\mathbf{R}_{\rho,n}\|_1}{\|\mathbf{R}_{\rho,0}\|_1}$. With n_ν the first time step satisfying $R_\rho \le \nu$ the actual implementation is as follows

$$\mathbf{J}^*_{\text{low}} \Delta\mathbf{U} = -\mathbf{R}^*_{\text{high}}(\mathbf{U}), \text{ if } n \le n_\nu; \; \mathbf{J}^*_{\text{high}} \Delta\mathbf{U} = -\mathbf{R}^*_{\text{high}}(\mathbf{U}), \text{ if } n > n_\nu \quad (2.244)$$

To be specific, $\mathbf{R}_{\text{high}}(\mathbf{U})$ represents the higher-order spatial discretization used for the simulation. The Jacobian-vector product $\mathbf{J}^*_{\text{high}} \Delta\mathbf{U}$ is a matrix-free, consistently second-order spatial accurate implementation, which includes the reconstruction and limiting process, i.e., $\mathbf{J}^*_{\text{high}} \Delta\mathbf{U} = -\mathbf{R}^*_{\text{high}}(\mathbf{U})$. Whereas, $\mathbf{J}^*_{\text{low}}$ is based on a first-order accurate spatial discretization, such that only direct neighbouring cells are involved in the linearization. Bramkamp et al. [51] employed automatic differentiation to evaluate the matrix-vector products during the iterative solution of the system $\mathbf{J}^*_{\text{high}} \Delta\mathbf{U} = -\mathbf{R}^*_{\text{high}}(\mathbf{U})$. In Eq. (2.244), both linear systems are solved using the Bi-CGSTAB algorithm [487], combined with left preconditioning, which is based on a point-block-ILU(f) method. For problems tested, Bramkamp et al. [51] showed that the convergence of the Newton–Krylov solver by this hybrid strategy between first-order and second-order accurate matrix-vector products can significantly be improved, compared to standard first-order approximations.

Other authors suggest to switch after an a-priori prescribed number of time steps (e.g., [339, 340, 365]). As convergence performance and stability of the Newton-

GMRES technique are quite sensitive to the start-up, these authors performed several implicit iterations before switching to Newton–Krylov solver. Equation (2.233) with $\beta = 1$ is used for implicit pre-iterations. In this situation, for the left hand side of Eq. (2.235) the approximate first order Jacobian can be used, while the residuals in the right hand side are being computed to the right order of accuracy. Lower-upper symmetric-Gauss-Seidel method (LU-SGS) algorithm is qualified for solving a linear solver for these pre-iterations. Then, Newton–Krylov solver is switched; at this stage the $\frac{1}{\Delta t}$ term in Eq. (2.235) is removed (i.e., taking infinite time step). GMRES is used to solve the resultant linear system at each Newton iteration.

References

1. Abgrall R, Karni S (2010) A comment on the computation of non-conservative products. J Comput Phys 229(8):2759–2763. https://doi.org/10.1016/j.jcp.2009.12.015
2. Abramowitz M (1965) Handbook of mathematical functions, with formulas, graphs, and mathematical tables. Dover Publications, New York
3. Adams MF, Samtaney R, Brandt A (2010) Toward textbook multigrid efficiency for fully implicit resistive magnetohydrodynamics. J Comput Phys 229(18):6208–6219. https://doi.org/10.1016/j.jcp.2010.04.024
4. Ahn HT, Kallinderis Y (2006) Strongly coupled flow/structure interactions with a geometrically conservative ALE scheme on general hybrid meshes. J Comput Phys 219(2):671–696. https://doi.org/10.1016/j.jcp.2006.04.011
5. Ahn HT, Shashkov M (2007) Multi-material interface reconstruction on generalized polyhedral meshes. J Comput Phys 226(2):2096–2132. https://doi.org/10.1016/j.jcp.2007.06.033
6. Altmann C (2011) An explicit discontinuous Galerkin scheme with divergence cleaning for magnetohydrodynamics. Springer, Berlin, pp 357–364
7. Amari T, Luciani JF, Joly P (1999) A preconditioned semi-implicit method for magnetohydrodynamics equations. SIAM J Sci Comput 21(3):970–986. https://doi.org/10.1137/S1064827596304824
8. Anderson E, Bai Z, Bischof C, Blackford L, Demmel J, Dongarra J, Du Croz J, Greenbaum A, Hammarling S, McKenney A, Sorensen D (1999) LAPACK users' guide, 3rd edn. Society for industrial and applied mathematics, Philadelphia. https://doi.org/10.1137/1.9780898719604
9. Angot P, Dolejší V, Feistauer M, Felcman J (1998) Analysis of a combined barycentric finite volume-nonconforming finite element method for nonlinear convection-diffusion problems. Appl Math 43(4):263–310. https://doi.org/10.1023/A:1023217905340
10. Artebrant R, Schroll HJ (2006) Limiter-free third order logarithmic reconstruction. SIAM J Sci Comput 28(1):359–381. https://doi.org/10.1137/040620187
11. Aschwanden MJ (2004) Physics of the solar corona. Praxis Publishing Ltd, Chichester, An introduction
12. Aschwanden MJ (2019) New millennium solar physics. Springer, Cham
13. Asgharzadeh H, Borazjani I (2017) A Newton-Krylov method with an approximate analytical Jacobian for implicit solution of Navier-Stokes equations on staggered overset-curvilinear grids with immersed boundaries. J Comput Phys 331:227–256. https://doi.org/10.1016/j.jcp.2016.11.033
14. Axelsson O, Lindskog G (1986) On the eigenvalue distribution of a class of preconditioning methods. Numer Math 48(5):479–498. https://doi.org/10.1007/BF01389447

15. Bader G, Kemm F (2014) The carbuncle phenomenon in shallow water simulations. In: The second international conference on computational science and engineering (ICCSE-2014). Ho Chi Minh City, Vietnam
16. Balay S, Gropp WD, McInnes LC, Smith BF (1997) Efficient management of parallelism in object oriented numerical software libraries. In: Arge E, Bruaset AM, Langtangen HP (eds) Modern software tools in scientific computing. Birkhäuser Press, New York, pp 163–202
17. Balay S, Abhyankar S, Adams MF, Brown J, Brune P, Buschelman K, Dalcin L, Eijkhout V, Gropp WD, Kaushik D, Knepley MG, McInnes LC, Rupp K, Smith BF, Zampini S, Zhang H, Zhang H (2016) PETSc web page. http://www.mcs.anl.gov/petsc
18. Balsara DS (1998) Linearized formulation of the Riemann problem for adiabatic and isothermal magnetohydrodynamics. Astrophys J Suppl Ser 116(1):119. http://stacks.iop.org/0067-0049/116/i=1/a=119
19. Balsara DS (2010) Multidimensional HLLE Riemann solver: application to Euler and magnetohydrodynamic flows. J Comput Phys 229(6):1970–1993
20. Balsara DS (2012) A two-dimensional HLLC Riemann solver for conservation laws: application to Euler and magnetohydrodynamic flows. J Comput Phys 231:7476–7503. https://doi.org/10.1016/j.jcp.2011.12.025
21. Balsara DS (2012) Self-adjusting, positivity preserving high order schemes for hydrodynamics and magnetohydrodynamics. J Comput Phys 231(22):7504–7517. https://doi.org/10.1016/j.jcp.2012.01.032
22. Balsara DS (2015) Three dimensional HLL Riemann solver for conservation laws on structured meshes; application to Euler and magnetohydrodynamic flows. J Comput Phys 295:1–23. https://doi.org/10.1016/j.jcp.2015.03.056
23. Balsara DS, Kim J (2004) A comparison between divergence-cleaning and staggered-mesh formulations for numerical magnetohydrodynamics. Astrophys J 602(2):1079. http://stacks.iop.org/0004-637X/602/i=2/a=1079
24. Balsara DS, Nkonga B (2017) Multidimensional Riemann problem with self-similar internal structure-part III-A multidimensional analogue of the HLLI Riemann solver for conservative hyperbolic systems. J Comput Phys 346:25–48. https://doi.org/10.1016/j.jcp.2017.05.038
25. Balsara DS, Spicer DS (1999) A staggered mesh algorithm using high order godunov fluxes to ensure solenoidal magnetic fields in magnetohydrodynamic simulations. J Comput Phys 149(2):270–292. https://doi.org/10.1006/jcph.1998.6153
26. Balsara DS, Dumbser M, Abgrall R (2014) Multidimensional HLLC Riemann solver for unstructured meshes-with application to Euler and MHD flows. J Comput Phys 261:172–208. https://doi.org/10.1016/j.jcp.2013.12.029
27. Balsara DS, Garain S, Shu CW (2016) An efficient class of WENO schemes with adaptive order. J Comput Phys 326(Supplement C):780–804. https://doi.org/10.1016/j.jcp.2016.09.009
28. Balsara DS, Li J, Montecinos GI (2018) An efficient, second order accurate, universal generalized Riemann problem solver based on the HLLI Riemann solver. J Comput Phys 375:1238–1269. https://doi.org/10.1016/j.jcp.2018.09.018
29. Barmin A, Kulikovskiy A, Pogorelov N (1996) Shock-capturing approach and nonevolutionary solutions in magnetohydrodynamics. J Comput Phys 126(1):77–90. https://doi.org/10.1006/jcph.1996.0121
30. Barth TJ (1987) Analysis of implicit local linearization techniques for upwind and TVD algorithms. In: 25th AIAA Aerospace Sciences Meeting, Reno, NV, USA. https://doi.org/10.2514/6.1987-595
31. Barth TJ (1988) Implicit linearization procedures for upwind algorithms. Springer, Berlin, pp 1449–1456
32. Barth TJ (1992) Aspects of unstructured grids and finite-volume solvers for the Euler and Navier-Stokes equations. In: AGARD, special course on unstructured grid methods for advection dominated flows (SEE N92-27671 18-34), p 61

33. Barth TJ (1993) A 3-D least-squares upwind Euler solver for unstructured meshes. In: Napolitano M, Sabetta F (eds) Numerical methods in fluid dynamics. Lecture notes in physics, vol 414. Springer, Berlin, pp 240–244
34. Barth TJ (1993) Recent developments in high order k-exact reconstruction on unstructured mesh. AIAA paper 93-0668
35. Barth TJ (1994) Aspects of unstructured grids and finite-volume solvers for the Euler and Navier-Stokes equations. In: Proceedings of the VKI lecture series on computational. Fluid Dynamics. VKI-LS 1994-05, March 1994. 64
36. Barth TJ, Frederickson PO (1990) Higher order solution of the Euler equations on unstructured grids using quadratic reconstruction. In: Aerospace sciences meetings, American Institute of Aeronautics and Astronautics, Reno, NV, U.S.A. https://doi.org/10.2514/6.1990-1310.2514/6.1990-13
37. Barth TJ, Jespersen DC (1989) The design and application of upwind schemes on unstructured meshes. AIAA paper 89-0366
38. Bartschat K (2013) Computational methods for electron-atom collisions in plasma applications. J Phys D Appl Phys 46:334004. https://doi.org/10.1088/0022-3727/46/33/334004
39. Batten P, Clarke N, Lambert C, Causon DM (1997) On the choice of wavespeeds for the HLLC Riemann solver. SIAM J Sci Comput 18(6):1553–1570. https://doi.org/10.1137/S1064827593260140
40. Bermúdez A, Dervieux A, Desideri JA, Vázquez M (1998) Upwind schemes for the two-dimensional shallow water equations with variable depth using unstructured meshes. Comput Methods Appl Mech Eng 155(1):49–72. https://doi.org/10.1016/S0045-7825(97)85625-3
41. Billett SJ, Toro EF (1998) Unsplit WAF-type schemes for three dimensional hyperbolic conservation laws. Springer, Dordrecht, pp 75–124
42. Birdsall C, Langdon A (1991) Plasma physics via computer simulation. CRC Press, Boca Raton. https://doi.org/10.1201/9781315275048
43. Biskamp D (1997) Nonlinear magnetohydrodynamics. Cambridge University Press, Cambridge
44. Bittencourt JA (2004) Fundamentals of plasma physics. Springer, New York. https://doi.org/10.1007/978-1-4757-4030-1
45. Blazek J (2005) Computational fluid dynamics: principles and applications, 2nd edn. Elsevier, Amsterdam
46. Blom DS, Birken P, Bijl H, Kessels F, Meister A, van Zuijlen AH (2016) A comparison of Rosenbrock and Esdirk methods combined with iterative solvers for unsteady compressible flows. Adv Comput Math 42(6):1401–1426. https://doi.org/10.1007/s10444-016-9468-x
47. Boom PD, Zingg DW (2018) Optimization of high-order diagonally-implicit Runge-Kutta methods. J Comput Phys 371:168–191. https://doi.org/10.1016/j.jcp.2018.05.020
48. Brackbill JU, Barnes DC (1980) The effect of $\nabla \cdot \mathbf{B}$ on the numerical solution of the magnetohydrodynamic equations. J Comput Phys 35:426–430. https://doi.org/10.1016/0021-9991(80)90079-0
49. Braginskii SI (1965) Transport processes in a plasma. Rev Plasma Phys 1:205–311
50. Bramkamp F, Bucker H, Rasch A (2006) Using exact Jacobians in an implicit Newton-Krylov method. Comput Fluids 35(10):1063–1073. https://doi.org/10.1016/j.compfluid.2005.10.003
51. Bramkamp FD, Pollul B, Rasch A, Schieffer G (2014) Matrix-free second-order methods in implicit time integration for compressible flows using automatic differentiation. IJCSE 9:484–498
52. Brio M, Wu CC (1988) An upwind differencing scheme for the equations of ideal magnetohydrodynamics. J Comput Phys 75:400–422. https://doi.org/10.1016/0021-9991(88)90120-9
53. Burden RL, Faires JD (2016) Numerical analysis, 10th edn. Brooks/Cole Publishing Co., Baltimore

54. Burg C (2005) Higher order variable extrapolation for unstructured finite volume RANS flow solvers. In: Fluid dynamics and co-located conferences, American Institute of Aeronautics and Astronautics, Toronto, Ontario, Canada. https://doi.org/10.2514/6.2005-499910.2514/6.2005-4999
55. Burlaga LF (1995) Interplanetary magnetohydrodynamics, vol 3. Oxford University Press, New York
56. Butcher JC (2008) Numerical methods for ordinary differential equations, 2nd edn. Wiley Ltd., The Atrium, Southern Gate, Chichester, West Sussex PO19 8SQ, England. https://doi.org/10.1002/9780470753767
57. Butcher JC (2016) Numerical methods for ordinary differential equations, 3rd edn. Wiley Ltd., The Atrium, Southern Gate, Chichester, West Sussex PO19 8SQ, England. https://doi.org/10.1002/9781119121534
58. Čada M, Torrilhon M, (2009) Compact third-order limiter functions for finite volume methods. J Comput Phys 228(11):4118–4145. https://doi.org/10.1016/j.jcp.2009.02.020
59. Cai XC, Gropp WD, Keyes DE, Melvin RG, Young DP (1998) Parallel Newton-Krylov-Schwarz algorithms for the transonic full potential equation. SIAM J Sci Comput 19(1):246–265. https://doi.org/10.1137/S1064827596304046
60. Caplan RM, Mikić Z, Linker JA, Lionello R (2017) Advancing parabolic operators in thermodynamic MHD models: explicit super time-stepping versus implicit schemes with Krylov solvers. J. Phys.: Conf. Ser. 837(012):016. https://doi.org/10.1088/1742-6596/837/1/012016
61. Cargo P, Gallice G (1997) Roe matrices for ideal MHD and systematic construction of roe matrices for systems of conservation laws. J Comput Phys 136:446–466. https://doi.org/10.1006/jcph.1997.5773
62. Cary A, Dorgan A, Mani M (2009) Towards accurate flow predictions using unstructured meshes. In: Proceedings of the nineteenth AIAA computational fluid dynamics conference, American Institute of Aeronautics and Astronautics, AIAA paper 2009–3650. https://doi.org/10.2514/6.2009-3650
63. Castro MJ, LeFloch PG, Munoz-Ruiz ML, Parés C (2008) Why many theories of shock waves are necessary: convergence error in formally path-consistent schemes. J Comput Phys 227(17):8107–8129. https://doi.org/10.1016/j.jcp.2008.05.012
64. Castro MJ, Fernández-Nieto ED, Morales de Luna T, Narbona-Reina G, Parés C (2013) A HLLC scheme for nonconservative hyperbolic problems. Application to turbidity currents with sediment transport. ESAIM: M2AN 47(1):1–32. https://doi.org/10.1051/m2an/2012017
65. Castro MJ, Gallardo JM, Marquina A (2016) Approximate Osher-Solomon schemes for hyperbolic systems. Appl Math Comput 272:Part 2, 347–368
66. Cea L, French JR, Vázquez-Cendón ME (2006) Numerical modelling of tidal flows in complex estuaries including turbulence: an unstructured finite volume solver and experimental validation. Int J Numer Methods Eng 67(13):1909–1932. https://doi.org/10.1002/nme.1702
67. Cea L, Puertas J, Vázquez-Cendón ME (2007) Depth averaged modelling of turbulent shallow water flow with wet-dry fronts. Arch Comput Methods Eng 14(3):303–341. https://doi.org/10.1007/s11831-007-9009-3
68. Chacón L (2004) A non-staggered, conservative, $\nabla \cdot \mathbf{B} = 0$, finite-volume scheme for 3D implicit extended magnetohydrodynamics in curvilinear geometries. Comput Phys Commun 163(3):143–171. https://doi.org/10.1016/j.cpc.2004.08.005
69. Chacón L (2008) An optimal, parallel, fully implicit Newton-Krylov solver for three-dimensional viscoresistive magnetohydrodynamics. Phys Plasmas 15(5):056,103. https://doi.org/10.1063/1.2838244
70. Chacón L, Chen G (2016) A curvilinear, fully implicit, conservative electromagnetic PIC algorithm in multiple dimensions. J Comput Phys 316:578–597. https://doi.org/10.1016/j.jcp.2016.03.070
71. Chacón L, Stanier A (2016) A scalable, fully implicit algorithm for the reduced two-field low-β extended MHD model. J Comput Phys 326:763–772. https://doi.org/10.1016/j.jcp.2016.09.007

72. Chacón L, Knoll D, Finn J (2002) An implicit, nonlinear reduced resistive MHD solver. J Comput Phys 178(1):15–36. https://doi.org/10.1006/jcph.2002.7015
73. Chandrashekar P (2013) Kinetic energy preserving and entropy stable finite volume schemes for compressible Euler and Navier-Stokes equations. Commun Comput Phys 14(05):1252–1286
74. Charest MRJ, Groth CPT, Gauthier PQ (2015) A high-order central ENO finite-volume scheme for three-dimensional low-speed viscous flows on unstructured mesh. Commun Comput Phys 17:615–656. https://doi.org/10.4208/cicp.091013.281114a
75. Chen Y, Tóth G, Cassak P, Jia X, Gombosi TI, Slavin JA, Markidis S, Peng IB, Jordanova VK, Henderson MG (2017) Global three-dimensional simulation of Earth's dayside reconnection using a two-way coupled magnetohydrodynamics with embedded particle-in-cell model: initial results. J Geophys Res (Space Phys) 122(A10):10,318–10,335. https://doi.org/10.1002/2017JA024186
76. Choi H, Liu JG (1998) The reconstruction of upwind fluxes for conservation laws: its behavior in dynamic and steady state calculations. J Comput Phys 144(2):237–256. https://doi.org/10.1006/jcph.1998.5970
77. Christlieb AJ, Liu Y, Tang Q, Xu Z (2015) Positivity-preserving finite difference weighted ENO schemes with constrained transport for ideal magnetohydrodynamic equations. SIAM J Sci Comput 37(4):A1825–A1845. https://doi.org/10.1137/140971208
78. Christlieb AJ, Feng X, Seal DC, Tang Q (2016) A high-order positivity-preserving single-stage single-step method for the ideal magnetohydrodynamic equations. J Comput Phys 316:218–242. https://doi.org/10.1016/j.jcp.2016.04.016
79. Cimino A, Krause G, Elaskar S, Costa A (2016) Characteristic boundary conditions for magnetohydrodynamics: the Brio-Wu shock tube. Comput Fluids 127:194–210. https://doi.org/10.1016/j.compfluid.2016.01.001
80. Clain S, Diot S, Loubere R (2011) A high-order finite volume method for systems of conservation laws–Multi-dimensional Optimal Order Detection (MOOD). J Comput Phys 230(10):4028–4050. https://doi.org/10.1016/j.jcp.2011.02.026
81. Coirier WJ, Powell KG (1995) Adaptively refined Euler and Navier-Stokes solutions with a Cartesian-cell based scheme. In: South JC Jr, Thomas JL, Vanrosendale J (eds) ICASE/LaRC workshop on adaptive grid methods, pp 153–161
82. Colella P, Sekora MD (2008) A limiter for PPM that preserves accuracy at smooth extrema. J Comput Phys 227(15):7069–7076. https://doi.org/10.1016/j.jcp.2008.03.034
83. Colella P, Woodward PR (1984) The piecewise parabolic method (PPM) for gas-dynamical simulations. J Comput Phys 54(1):174–201. https://doi.org/10.1016/0021-9991(84)90143-8
84. Colonius T (2004) Modeling artificial boundary conditions for compressible flow. Annu Rev Fluid Mech 36(1):315–345. https://doi.org/10.1146/annurev.fluid.36.050802.121930
85. Correa CD, Hero R, Ma K (2011) A comparison of gradient estimation methods for volume rendering on unstructured meshes. IEEE Trans Vis Comput Graph 17(3):305–319. https://doi.org/10.1109/TVCG.2009.105
86. Cranmer SR, Winebarger AR (2019) The properties of the solar corona and its connection to the solar wind. arXiv:1811.00461
87. Cristofolini A (2016) Magnetohydrodynamics equations. In: Plasma modeling, 2053-2563. IOP Publishing, Bristol, pp 7–1 to 7–26. https://doi.org/10.1088/978-0-7503-1200-4ch7
88. Dai W, Woodward PR (1994) An approximate Riemann solver for ideal magnetohydrodynamics. J Comput Phys 111(2):354–372
89. Daldorff LKS, Tóth G, Gombosi TI, Lapenta G, Amaya J, Markidis S, Brackbill JU (2014) Two-way coupling of a global Hall magnetohydrodynamics model with a local implicit particle-in-cell model. J Comput Phys 268:236–254. https://doi.org/10.1016/j.jcp.2014.03.009
90. Darwis M, Abdullah K, Mohammed AN (2017) Comparative study of Roe, RHLL and Rusanov fluxes for shock-capturing schemes. IOP Conf Ser: Mater Sci Eng 243(1):012,007. http://stacks.iop.org/1757-899X/243/i=1/a=012007

91. Darwish M, Moukalled F (2003) TVD schemes for unstructured grids. Int J Heat Mass Transf 46(4):599–611. https://doi.org/10.1016/S0017-9310(02)00330-7
92. Davis P, Rabinowitz P (1956) Abscissas and weights for Gaussian quadratures of high order. J Res Nat Bur Stand 56(1):35–37
93. Davis SF (1988) Simplified second-order Godunov-type methods. SIAM J Sci Stat Comput 9(3):445–473. https://doi.org/10.1137/0909030
94. De Sterck H, Low BC, Poedts S (1999) Characteristic analysis of a complex two-dimensional magnetohydrodynamic bow shock flow with steady compound shocks. Phys Plasmas 6:954–969. https://doi.org/10.1063/1.873336
95. De Sterck H, Csik A, Abeele DV, Poedts S, Deconinck H (2001) Stationary two-dimensional magnetohydrodynamic flows with shocks: characteristic analysis and grid convergence study. J Comput Phys 166(1):28–62. https://doi.org/10.1006/jcph.2000.6640
96. Dedner A, Kemm F, Kröner D, Munz CD, Schnitzer T, Wesenberg M (2002) Hyperbolic divergence cleaning for the MHD equations. J Comput Phys 175:645–673. https://doi.org/10.1006/jcph.2001.6961
97. Dedner A, Kröner D, Rohde C, Wesenberg M (2003) Efficient divergence cleaning in three-dimensional MHD simulations. Springer, Berlin, pp 323–334
98. Dedner A, Rohde C, Wesenberg M (2003) A new approach to divergence cleaning in magnetohydrodynamic simulations. Springer, Berlin, pp 509–518
99. Dedner A, Kröner D, Rohde C, Wesenberg M (2005) Analysis and numerics for conservation laws. Chapter, Radiation magnetohydrodynamics: analysis for model problems and efficient 3D-simulations for the full system. Springer, Berlin, pp 163–202
100. Deka M, Brahmachary S, Thirumalaisamy R, Dalal A, Natarajan G (2018) A new Green-Gauss reconstruction on unstructured meshes. Part I: gradient reconstruction. J Comput Phys. https://doi.org/10.1016/j.jcp.2018.10.023
101. Delanaye M, Essers JA (1997) Finite volume scheme with quadratic reconstruction on unstructured adaptive meshes applied to turbomachinery flows. J Turbomach 119(2):263–269. https://doi.org/10.1115/1.2841109
102. Dellar PJ (2001) A note on magnetic monopoles and the one-dimensional MHD Riemann problem. J Comput Phys 172:392–398. https://doi.org/10.1006/jcph.2001.6815
103. Dembo RS, Eisenstat SC, Steihaug T (1982) Inexact newton methods. SIAM J Numer Anal 19(2):400–408. https://doi.org/10.1137/0719025
104. Dement DC, Ruffin SM (2018) Higher order cell centered finite volume schemes for unstructured Cartesian grids. In: AIAA SciTech forum, American Institute of Aeronautics and Astronautics, Kissimmee, Florida. https://doi.org/10.2514/6.2018-130510.2514/6.2018-1305
105. Demmel J (1997) Applied numerical linear algebra. Society for industrial and applied mathematics, Philadelphia. https://doi.org/10.1137/1.9781611971446
106. Deng X, Mao M, Tu G, Zhang H, Zhang Y (2012) High-order and high accurate CFD methods and their applications for complex grid problems. Commun Comput Phys 11(4):1081–1102. https://doi.org/10.4208/cicp.100510.150511s
107. Derigs D, Winters AR, Gassner GJ, Walch S (2016) A novel high-order, entropy stable, 3D AMR MHD solver with guaranteed positive pressure. J Comput Phys 317:223–256. https://doi.org/10.1016/j.jcp.2016.04.048
108. Derigs D, Gassner GJ, Walch S, Winters AR (2018) Entropy stable finite volume approximations for ideal magnetohydrodynamics. Jahresbericht der Deutschen Mathematiker-Vereinigung 120(3):153–219. https://doi.org/10.1365/s13291-018-0178-9
109. Derigs D, Winters AR, Gassner GJ, Walch S, Bohm M (2018) Ideal GLM-MHD: about the entropy consistent nine-wave magnetic field divergence diminishing ideal magnetohydrodynamics equations. J Comput Phys 364:420–467. https://doi.org/10.1016/j.jcp.2018.03.002

110. Diskin B, Thomas JL (2011) Comparison of node-centered and cell-centered unstructured finite-volume discretizations: inviscid fluxes. AIAA J 49(4):836–854. https://doi.org/10.2514/1.J050897
111. Diskin B, Thomas JL (2012) Comparison of node-centered and cell-centered unstructured finite-volume discretizations: inviscid fluxes. AIAA J 51(1):277–277. https://doi.org/10.2514/1.J051870
112. Diskin B, Thomas JL, Nielsen EJ, Nishikawa H, White JA (2010) Comparison of node-centered and cell-centered unstructured finite-volume discretizations: viscous fluxes. AIAA J 48(7):1326–1338. https://doi.org/10.2514/1.44940
113. Do S, Ha Y, Kang M, Kim CH (2016) Application of a multi-dimensional limiting process to central-upwind schemes for solving hyperbolic systems of conservation laws. J Sci Comput 69(1):274–291. https://doi.org/10.1007/s10915-016-0193-x
114. Dolejší V, Feistauer M, Felcman J, Klikova A (2002) Error estimates for barycentric finite volumes combined with nonconforming finite elements applied to nonlinear convection-diffusion problems. Appl Math 47(4):301–340. https://doi.org/10.1023/A:1021701705932
115. Du Z, Li J (2018) A Hermite WENO reconstruction for fourth order temporal accurate schemes based on the GRP solver for hyperbolic conservation laws. J Comput Phys 355:385–396. https://doi.org/10.1016/j.jcp.2017.11.023
116. Dumbser M, Balsara DS (2016) A new efficient formulation of the HLLEM Riemann solver for general conservative and non-conservative hyperbolic systems. J Comput Phys 304:275–319. https://doi.org/10.1016/j.jcp.2015.10.014
117. Dumbser M, Toro EF (2011) A simple extension of the Osher Riemann solver to non-conservative hyperbolic systems. J Sci Comput 48(1):70–88
118. Dumbser M, Toro EF (2011) On universal Osher-type schemes for general nonlinear hyperbolic conservation laws. Commun Comput Phys 10:635–671. https://doi.org/10.4208/cicp.170610.021210a
119. Einfeldt B (1988) On Godunov-type methods for gas dynamics. SIAM J Numer Anal 25(2):294–318. https://doi.org/10.1137/0725021
120. Einfeldt B, Munz C, Roe P, Sjogreen B (1991) On Godunov-type methods near low densities. J Comput Phys 92(2):273–295. https://doi.org/10.1016/0021-9991(91)90211-3
121. Einkemmer L (2016) A resistive magnetohydrodynamics solver using modern C++ and the boost library. Comput Phys Commun 206:69–77. https://doi.org/10.1016/j.cpc.2016.04.015
122. Eisenstat SC, Walker HF (1994) Globally convergent inexact Newton methods. SIAM J Optim 4(2):393–422. https://doi.org/10.1137/0804022
123. Elling V (2009) The carbuncle phenomenon is incurable. Acta Math Sci 29(6):1647–1656. https://doi.org/10.1016/S0252-9602(10)60007-0
124. Engquist B, Osher S (1980) Stable and entropy satisfying approximations for transonic flow calculations. Math Comput 34(149):45–75. https://doi.org/10.1090/S0025-5718-1980-0551290-1
125. Evans CR, Hawley JF (1988) Simulation of magnetohydrodynamic flows - A constrained transport method. Astrophys J 332:659–677. https://doi.org/10.1086/166684
126. Eymard R, Herbin R, Gallouët T (2010) Finite volume method. Scholarpedia 5(6):9835. https://doi.org/10.4249/scholarpedia.9835
127. Falle SAEG (2010) Computational magnetohydrodynamics. In: Blockley R, Shyy W (eds) Encyclopedia of aerospace engineering. Wiley Ltd, Hoboken, pp 1–7. https://doi.org/10.1002/9780470686652.eae067
128. Falle SAEG, Komissarov SS, Joarder P (1998) A multidimensional upwind scheme for magnetohydrodynamics. Mon Not R Astron Soc 297:265–277. https://doi.org/10.1046/j.1365-8711.1998.01506.x
129. Farhat C, Lanteri S (1994) Simulation of compressible viscous flows on a variety of MPPS: computational algorithms for unstructured dynamic meshes and performance results. Comput Methods Appl Mech Eng 119(1):35–60. https://doi.org/10.1016/0045-7825(94)00075-1

130. Farre C, Perez-Segarra CD, Soria M, Oliva A (2006) Analysis of different numerical schemes for the resolution of convection-diffusion equations using finite-volume methods on three-dimensional unstructured grids. Part II: numerical analysis. Numer Heat Trans, Part B: Fund 49(4):351–375. https://doi.org/10.1080/10407790500314996
131. Feistauer M, Felcman J, Lukáčová-Medvid'ová M (1995) Combined finite element-finite volume solution of compressible flow. J Comput Appl Math 63(1):179–199. https://doi.org/10.1016/0377-0427(95)00051-8
132. Felippa CA (2004) A compendium of FEM integration formulas for symbolic work. Eng Comput 21(8):867–890. https://doi.org/10.1108/02644400410554362
133. Felker KG, Stone JM (2018) A fourth-order accurate finite volume method for ideal MHD via upwind constrained transport. J Comput Phys 375:1365–1400. https://doi.org/10.1016/j.jcp.2018.08.025
134. Feng XS, Yang LP, Xiang CQ, Wu ST, Zhou YF, Zhong DK (2010) Three-dimensional solar wind modeling from the Sun to Earth by a SIP-CESE MHD model with a six-component grid. Astrophys J 723:300
135. Feng XS, Zhang SH, Xiang CQ, Yang LP, Jiang CW, Wu ST (2011) A hybrid solar wind model of the CESE+HLL method with a Yin-Yang overset grid and an AMR grid. Astrophys J 734:50
136. Feng XS, Xiang CQ, Zhong DK, F ZY, Yang LP, Ma XP, (2014) SIP-CESE MHD model of solar wind with adaptive mesh refinement of hexahedral meshes. Comput Phys Commun 185(7):1965–1980. https://doi.org/10.1016/j.cpc.2014.03.027
137. Feng XS, Zhang M, Zhou YF (2014) A new three-dimensional solar wind model in spherical coordinates with a six-component grid. Astrophys J Suppl Ser 214(1):6. http://stacks.iop.org/0067-0049/214/i=1/a=6
138. Feng XS, Ma XP, Xiang CQ (2015) Data-driven modeling of the solar wind from 1 R_s to 1 AU. J Geophys Res 120(12):10,159–10,174, 2015JA021911
139. Feng XS, Li CX, Xiang CQ, Zhang M, Li HC, Wei FS (2017) Data-driven modeling of the solar corona by a new three-dimensional path-conservative Osher-Solomon MHD model. Astrophys J Suppl Ser 233(1):10. http://stacks.iop.org/0067-0049/233/i=1/a=10
140. Feng XS, Liu XJ, Xiang CQ, Li HC, Wei FS (2019) A new MHD model with a rotated-hybrid scheme and solenoidality-preserving approach. Astrophys J 871(2):226. https://doi.org/10.3847/1538-4357/aafacf
141. Ferracina L, Spijker M (2008) Strong stability of singly-diagonally-implicit Runge-Kutta methods. Appl Numer Math 58(11):1675–1686. https://doi.org/10.1016/j.apnum.2007.10.004
142. Folkner D, Katz A, Sankaran V (2014) Design and verification methodology of boundary conditions for finite volume schemes. Comput Fluids 96(Supplement C):264–275. https://doi.org/10.1016/j.compfluid.2014.03.028
143. Freidberg JP (1987) Ideal magnetohydrodynamics. Springer, New York
144. Freret L, Groth C, De Sterck H (2017) A parallel high-order CENO finite-volume scheme with AMR for three-dimensional ideal MHD flows. In: Bittencourt ML, Dumont NA, Hesthaven JS (eds) Spectral and high order methods for partial differential equations ICOSAHOM 2016. Springer International Publishing, Cham, pp 343–355
145. Freret L, Groth C, Nguyen T, De Sterck H (2017) High-order finite-volume scheme with anisotropic adaptive mesh refinement: efficient inexact Newton method for steady three-dimensional flows, AIAA 2017-3297. Denver, Colorado. https://doi.org/10.2514/6.2017-3297
146. Freret L, Ivan L, De Sterck H, Groth CP (2017) A high-order finite-volume method with anisotropic AMR for ideal MHD flows. In: AIAA 2017–0845, AIAA SciTech forum, American Institute of Aeronautics and Astronautics, Grapevine, Texas. https://doi.org/10.2514/6.2017-084510.2514/6.2017-0845
147. Freret L, Ivan L, De Sterck H, Groth CPT (2019) High-order finite-volume method with block-based AMR for magnetohydrodynamics flows. J Sci Comput 79:176–208. https://doi.org/10.1007/s10915-018-0844-1

148. Friedrich O (1998) Weighted essentially non-oscillatory schemes for the interpolation of mean values on unstructured grids. J Comput Phys 144(1):194–212. https://doi.org/10.1006/jcph.1998.5988
149. Frink N (1994) Recent progress toward a three-dimensional unstructured Navier-Stokes flow solver. In: Aerospace sciences meetings, American Institute of Aeronautics and Astronautics, Reno, NV, U.S.A. https://doi.org/10.2514/6.1994-61
150. Frink N, Parikh P, Pirzadeh S (1991) A fast upwind solver for the Euler equations on three-dimensional unstructured meshes. In: Aerospace sciences meetings, American Institute of Aeronautics and Astronautics, Reno, NV. https://doi.org/10.2514/6.1991-10210.2514/6.1991-102
151. Frink NT (1992) Upwind scheme for solving the Euler equations on unstructured tetrahedral meshes. AIAA J 30:70–77. https://doi.org/10.2514/3.10884
152. Fu L (2019) A low-dissipation finite-volume method based on a new TENO shock-capturing scheme. Comput Phys Commun 235:25–39. https://doi.org/10.1016/j.cpc.2018.10.009
153. Fu L, Hu XY, Adams NA (2016) A family of high-order targeted ENO schemes for compressible-fluid simulations. J Comput Phys 305:333–359. https://doi.org/10.1016/j.jcp.2015.10.037
154. Fu L, Hu XY, Adams NA (2017) Targeted ENO schemes with tailored resolution property for hyperbolic conservation laws. J Comput Phys 349:97–121. https://doi.org/10.1016/j.jcp.2017.07.054
155. Fu L, Hu XY, Adams NA (2018) A new class of adaptive high-order targeted ENO schemes for hyperbolic conservation laws. J Comput Phys 374:724–751. https://doi.org/10.1016/j.jcp.2018.07.043
156. Fu ZF, Hu YQ (1995) Numerical simulation of space plasma (in Chinese). Anhui Science and Technology Press, Hefei
157. Fuchs F, McMurry A, Mishra S, Risebro N, Waagan K (2010) High order well-balanced finite volume schemes for simulating wave propagation in stratified magnetic atmospheres. J Comput Phys 229(11):4033–4058. https://doi.org/10.1016/j.jcp.2010.01.038
158. Fuchs FG, Mishra S, Risebro NH (2009) Splitting based finite volume schemes for ideal MHD equations. J Comput Phys 228:641–660. https://doi.org/10.1016/j.jcp.2008.09.027
159. Gardiner TA, Stone JM (2005) An unsplit Godunov method for ideal MHD via constrained transport. J Comput Phys 205(2):509–539. https://doi.org/10.1016/j.jcp.2004.11.016
160. Gardiner TA, Stone JM (2008) An unsplit Godunov method for ideal MHD via constrained transport in three dimensions. J Comput Phys 227(8):4123–4141. https://doi.org/10.1016/j.jcp.2007.12.017
161. Gassner G, Lorcher F, Munz CD (2007) A contribution to the construction of diffusion fluxes for finite volume and discontinuous Galerkin schemes. J Comput Phys 224(2):1049–1063. https://doi.org/10.1016/j.jcp.2006.11.004
162. Gerlinger P (2012) Multi-dimensional limiting for high-order schemes including turbulence and combustion. J Comput Phys 231(5):2199–2228. https://doi.org/10.1016/j.jcp.2011.10.024
163. Ghidaglia J, Pascal F (2002) On boundary conditions for multidimensional hyperbolic systems of conservation laws in the finite volume framework. Technical report, CMLA, ENS de Cachan, 2002. http://www.cmla.ens-cachan.fr/Utilisateurs/jmg/rapport.zip
164. Ghidaglia J, Pascal F (2005) The normal flux method at the boundary for multidimensional finite volume approximations in CFD. Eur J Mech B-Fluids 24(1):1–17
165. Godunov SK (1959) A difference method for numerical calculation of discontinuous solutions of the equations of hydrodynamics. Mat Sb (NS) 47(89)(3):271–306
166. Godunov SK (1972) Symmetric form of the equations of magnetohydrodynamics. Num Meth Mech Contin Medium 1:26–34
167. Golub GH, van Loan CF (1996) Matrix computations. Johns Hopkins University Press, Baltimore
168. Gomes AKF, Domingues MO, Schneider K, Mendes O, Deiterding R (2015) An adaptive multiresolution method for ideal magnetohydrodynamics using divergence cleaning with

parabolic-hyperbolic correction. Appl Numer Math 95:199–213. https://doi.org/10.1016/j.apnum.2015.01.007
169. González-Avilés JJ, Cruz-Osorio A, Lora-Clavijo FD, Guzmán FS (2015) Newtonian CAFE: a new ideal MHD code to study the solar atmosphere. Mon Not R Astron Soc 454:1871–1885. https://doi.org/10.1093/mnras/stv2013
170. Gottlieb S (2005) On high order strong stability preserving Runge-Kutta and multi step time discretizations. J Sci Comput 25(1):105–128. https://doi.org/10.1007/BF02728985
171. Gottlieb S, Shu CW, Tadmor E (2001) Strong stability-preserving high-order time discretization methods. SIAM Rev 43(1):89–112. https://doi.org/10.1137/S003614450036757X
172. Grappin R, Léorat J, Buttighoffer A (2000) Alfvén wave propagation in the high solar corona. Astron Astrophys 362:342–358
173. Gresho PM (1991) Incompressible fluid dynamics: some fundamental formulation issues. Annu Rev Fluid Mech 23(1):413–453. https://doi.org/10.1146/annurev.fl.23.010191.002213
174. Gressier J, Moschetta JM (2000) Robustness versus accuracy in shock-wave computations. Int J Numer Methods Fluids 33:313–332. https://doi.org/10.1002/1097-0363(20000615)33:3<313::AID-FLD7>3.3.CO;2-5
175. Groth CPT, De Zeeuw DL, Gombosi TI, Powell KG (2000) Global three-dimensional MHD simulation of a space weather event: CME formation, interplanetary propagation, and interaction with the magnetosphere. J Geophys Res 105:25,053–25,078. https://doi.org/10.1029/2000JA900093
176. Guo X (2015) An extended HLLC Riemann solver for the magneto-hydrodynamics including strong internal magnetic field. J Comput Phys 290:352–363. https://doi.org/10.1016/j.jcp.2015.02.048
177. Guo X, Florinski V, Wang C (2016) The HLLD Riemann solver based on magnetic field decomposition method for the numerical simulation of magneto-hydrodynamics. J Comput Phys 327:543–552. https://doi.org/10.1016/j.jcp.2016.09.057
178. Gurski KF (2004) An HLLC-type approximate Riemann solver for ideal magnetohydrodynamics. SIAM J Sci Comput 25(6):2165–2187. https://doi.org/10.1137/S1064827502407962
179. Gustafsson B (1975) The convergence rate for difference approximations to mixed initial boundary value problems. Math Comput 29(130):396–406. http://www.jstor.org/stable/2005559
180. Guzik SM, Gao X, Owen LD, McCorquodale P, Colella P (2015) A freestream-preserving fourth-order finite-volume method in mapped coordinates with adaptive-mesh refinement. Comput Fluids 123:202–217. https://doi.org/10.1016/j.compfluid.2015.10.001
181. Haack J, Jin S, Liu JG (2012) An all-speed asymptotic-preserving method for the isentropic Euler and Navier-Stokes equations. Commun Comput Phys 12:955–980. https://doi.org/10.4208/cicp.250910.131011a
182. Haider F, Brenner P, Courbet B, Croisille JP (2011) Efficient implementation of high order reconstruction in finite volume methods. In: Fořt J, Fürst J, Halama J, Herbin R, Hubert F (eds) Finite volumes for complex applications VI. Problem and perspectives, Springer, Berlin, pp 553–560
183. Haider F, Brenner P, Courbet B, Croisille JP (2014) Parallel implementation of k-exact finite volume reconstruction on unstructured grids. In: Abgrall R, Beaugendre H, Congedo PM, Dobrzynski C, Perrier V, Ricchiuto M (eds) High order nonlinear numerical schemes for evolutionary PDEs. Springer International Publishing, Cham, pp 59–75
184. Hairer E, Wanner G (2010) Solving ordinary differential equations II: stiff and differential-algebraic problems. Springer series in computational mathematics. Springer, Berlin. https://doi.org/10.1017/CBO9780511543241
185. Han SH, Lee J, Kim KH (2009) Accurate and robust pressure weight advection upstream splitting method for magnetohydrodynamics equations. AIAA J 47(4)
186. Hanawa T, Mikami H, Matsumoto T (2008) Improving shock irregularities based on the characteristics of the MHD equations. J Comput Phys 227:7952–7976. https://doi.org/10.1016/j.jcp.2008.05.006

187. Hans Goedbloed JP, Poedts S (2004) Principles of magnetohydrodynamics: with applications to laboratory and astrophysical plasmas. Cambridge University Press, Cambridge
188. Hans Goedbloed JP, Keppens R, Poedts S (2010) Advanced magnetohydrodynamics: with applications to laboratory and astrophysical plasmas. Cambridge University Press, Cambridge
189. Hans Goedbloed JP, Keppens R, Poedts S (2019) Magnetohydrodynamics of laboratory and astrophysical plasmas. Cambridge University Press, Cambridge
190. Harten A (1983) High resolution schemes for hyperbolic conservation laws. J Comput Phys 49(3):357–393. https://doi.org/10.1016/0021-9991(83)90136-5
191. Harten A, Hyman JM (1983) Self adjusting grid methods for one-dimensional hyperbolic conservation laws. J Comput Phys 50(2):235–269. https://doi.org/10.1016/0021-9991(83)90066-9
192. Harten A, Lax PD, van Leer B (1983) On upstream differencing and Godunov-type schemes for hyperbolic conservation laws. SIAM Rev 25(1):35–61
193. Harten A, Engquist B, Osher S, Chakravarthy SR (1987) Uniformly high order accurate essentially non-oscillatory schemes, III. J Comput Phys 71(2):231–303. https://doi.org/10.1016/0021-9991(87)90031-3
194. Haselbacher A, Blazek J (2000) Accurate and efficient discretization of Navier-Stokes equations on mixed grids. AIAA J 38:2094–2102. https://doi.org/10.2514/2.871
195. Hejranfar K, Azampour MH (2009) A dual-time implicit upwind scheme for computing three-dimensional unsteady compressible flows using unstructured moving grids. Springer, Berlin, pp 301–306
196. Hejranfar K, Kamali-Moghadam R (2012) Preconditioned characteristic boundary conditions for solution of the preconditioned Euler equations at low mach number flows. J Comput Phys 231(12):4384–4402. https://doi.org/10.1016/j.jcp.2012.01.040
197. Hejranfar K, Parseh K (2017) Application of a preconditioned high-order accurate artificial compressibility-based incompressible flow solver in wide range of reynolds numbers. Int J Numer Methods Fluids 86(1):46–77. https://doi.org/10.1002/fld.4407
198. Hejranfar K, Parseh K (2017) Preconditioned characteristic boundary conditions based on artificial compressibility method for solution of incompressible flows. J Comput Phys 345(Supplement C):543–564. https://doi.org/10.1016/j.jcp.2017.05.014
199. Helzel C, Rossmanith JA, Taetz B (2011) An unstaggered constrained transport method for the 3D ideal magnetohydrodynamic equations. J Comput Phys 230(10):3803–3829. https://doi.org/10.1016/j.jcp.2011.02.009
200. Holmes D, Connell S (1989) Solution of the 2D Navier-Stokes equations on unstructured adaptive grids. In: Fluid dynamics and co-located conferences, American Institute of Aeronautics and Astronautics, Buffalo, NY. https://doi.org/10.2514/6.1989-193210.2514/6.1989-1932
201. Hopkins PF (2016) A constrained-gradient method to control divergence errors in numerical MHD. Mon Not R Astron Soc 462:576–587. https://doi.org/10.1093/mnras/stw1578
202. Hopkins PF, Raives MJ (2015) Accurate, meshless methods for magnetohydrodynamics. Mon Not R Astron Soc 455:51–88. https://doi.org/10.1093/mnras/stv2180
203. Hu F (2018) The 6th-order weighted ENO schemes for hyperbolic conservation laws. Comput Fluids 174:34–45. https://doi.org/10.1016/j.compfluid.2018.07.008
204. Hu WR (1987) Cosmical magnetohydrodynamics (in Chinese). Science Press, Beijing
205. Huang C, Chen LL (2018) A simple smoothness indicator for the WENO scheme with adaptive order. J Comput Phys 352:498–515. https://doi.org/10.1016/j.jcp.2017.10.005
206. Huang K, Wu H, Yu H, Yan D (2011) Cures for numerical shock instability in HLLC solver. Int J Numer Methods Fluids 65(9):1026–1038. https://doi.org/10.1002/fld.2217
207. Innocenti ME, Johnson A, Markidis S, Amaya J, Deca J, Olshevsky V, Lapenta G (2017) Progress towards physics-based space weather forecasting with exascale computing. Adv Eng Softw 111:3–17. Advances in high performance computing: on the path to exascale software. https://doi.org/10.1016/j.advengsoft.2016.06.011
208. Ivan L (2011) Development of high-order CENO finite-volume schemes with block-based adaptive mesh refinement. Phd thesis, University of Toronto, Canada. http://hdl.handle.net/1807/29759

209. Ivan L, De Sterck H (2011) Three-dimensional MHD on cubed-sphere grids: parallel solution-adaptive simulation framework. In: AIAA 2011-3382 20th AIAA computational fluid dynamics conference, 27–30 June 2011, Honolulu, Hawaii
210. Ivan L, Groth CP (2014) High-order solution-adaptive central essentially non-oscillatory (CENO) method for viscous flows. J Comput Phys 257(Part A):830–862. https://doi.org/10.1016/j.jcp.2013.09.045
211. Ivan L, Groth CPT (2007) High-order central ENO finite-volume scheme with adaptive mesh refinement. AIAA paper 2007-4323
212. Ivan L, Susanto A, De Sterck H, Northrup SA, Groth CPT (2012) High-order central ENO finite-volume scheme for MHD on three-dimensional cubed-sphere grids. In: Proceedings of the seventh international conference on computational fluid dynamics (ICCFD7), vol 3382. Hawaii, pp 1325–1342
213. Ivan L, De Sterck H, Northrup SA, Groth CPT (2013) Multi-dimensional finite-volume scheme for hyperbolic conservation laws on three-dimensional solution-adaptive cubed-sphere grids. J Comput Phys 255:205–227
214. Ivan L, Sterck HD, Susanto A, Groth C (2015) High-order central ENO finite-volume scheme for hyperbolic conservation laws on three-dimensional cubed-sphere grids. J Comput Phys 282:157–182
215. Iyer PS (2015) Discrete roughness effects on high-speed boundary layers. PhD thesis, University of Minnesota
216. Mavriplis DJ (1997) Adaptive meshing techniques for viscous flow calculations on mixed element unstructured meshes. 34
217. Jalali A (2017) An adaptive higher-order unstructured finite volume solver for turbulent compressible flows. PhD thesis, University of British Columbia. https://doi.org/10.14288/1.0340789
218. Jalali A, Ollivier-Gooch C (2017) Higher-order unstructured finite volume RANs solution of turbulent compressible flows. Comput Fluids 143:32–47. https://doi.org/10.1016/j.compfluid.2016.11.004
219. Jalali A, Ollivier Gooch CF (2013) Higher-order finite volume solution reconstruction on highly anisotropic meshes. In: Fluid dynamics and co-located conferences, American Institute of Aeronautics and Astronautics, San Diego, CA. https://doi.org/10.2514/6.2013-256510.2514/6.2013-2565
220. Jalali A, Sharbatdar M, Ollivier-Gooch C (2014) Accuracy analysis of unstructured finite volume discretization schemes for diffusive fluxes. Comput Fluids 101:220–232. https://doi.org/10.1016/j.compfluid.2014.06.008
221. Jalali A, Sharbatdar M, Ollivier-Gooch C (2016) An efficient implicit unstructured finite volume solver for generalised newtonian fluids. Int J Comput Fluid Dyn 30(3):201–217. https://doi.org/10.1080/10618562.2016.1188202
222. Jameson A (2017) Evaluation of fully implicit Runge-Kutta schemes for unsteady flow calculations. J Sci Comput 73(2):819–852. https://doi.org/10.1007/s10915-017-0476-x
223. Jameson A, Vassberg JC (2001) A vertex-centroid (V-C) scheme for the gas-dynamics equations. In: Satofuka N (ed) Computational fluid dynamics 2000. Springer, Berlin, pp 37–52
224. Jameson A, Yoon S (1987) Lower-upper implicit schemes with multiple grids for the Euler equations. AIAA J 25:929–935. https://doi.org/10.2514/3.9724
225. Janhunen P (2000) A positive conservative method for magnetohydrodynamics based on HLL and Roe methods. J Comput Phys 160:649–661. https://doi.org/10.1006/jcph.2000.6479
226. Jardin S (2010) Computational methods in plasma physics. Chapman & Hall/CRC, New York, Computational science series. https://doi.org/10.1201/EBK1439810958
227. Jardin S (2012) Review of implicit methods for the magnetohydrodynamic description of magnetically confined plasmas. J Comput Phys 231(3):822–838. https://doi.org/10.1016/j.jcp.2010.12.025

228. Jawahar P, Kamath H (2000) A high-resolution procedure for Euler and Navier-Stokes computations on unstructured grids. J Comput Phys 164(1):165–203. https://doi.org/10.1006/jcph.2000.6596
229. Jeffrey A, Taniuti T (1964) Non-linear wave propagation with applications to physics and magnetohydrodynamics. Academic, New York
230. Jiang GS, Shu CW (1996) Efficient implementation of weighted ENO schemes. J Comput Phys 126(1):202–228. https://doi.org/10.1006/jcph.1996.0130
231. Jiang GS, Wu CC (1999) A high-order WENO finite difference scheme for the equations of ideal magnetohydrodynamics. J Comput Phys 150:561–594. https://doi.org/10.1006/jcph.1999.6207
232. Jiang RL, Fang C, Chen PF (2012) A new MHD code with adaptive mesh refinement and parallelization for astrophysics. Comput Phys Commun 183(8):1617–1633
233. John V, Rang J (2010) Adaptive time step control for the incompressible Navier-Stokes equations. Comput Methods Appl Mech Eng 199(9):514–524. https://doi.org/10.1016/j.cma.2009.10.005
234. Jones OS, Shumlak U, Eberhardt D (1997) An implicit scheme for nonideal magnetohydrodynamics. J Comput Phys 130(2):231–242. https://doi.org/10.1006/jcph.1996.5580
235. Kallinderis Y, Ahn H (2005) Incompressible Navier-Stokes method with general hybrid meshes. J Comput Phys 210(1):75–108. https://doi.org/10.1016/j.jcp.2005.04.002
236. Kang HM, Kim KH, Lee DH (2010) A new approach of a limiting process for multi-dimensional flows. J Comput Phys 229(19):7102–7128. https://doi.org/10.1016/j.jcp.2010.06.001
237. Katz A, Sankaran V (2012) An efficient correction method to obtain a formally third-order accurate flow solver for node-centered unstructured grids. J Sci Comput 51(2):375–393. https://doi.org/10.1007/s10915-011-9515-1
238. Katz A, Sankaran V (2012) High aspect ratio grid effects on the accuracy of Navier-Stokes solutions on unstructured meshes. Comput Fluids 65:66–79. Sixth international conference on computational fluid dynamics (ICCFD6). https://doi.org/10.1016/j.compfluid.2012.02.012
239. Kemm F (2008) A carbuncle free Roe-type solver for the Euler equations. In: Benzoni-Gavage S, Serre D (eds) Hyperbolic problems: theory, numerics, applications. Springer, Berlin, pp 601–608
240. Kemm F (2011) A comparative study of TVD-limiters-well-known limiters and an introduction of new ones. Int J Numer Methods Fluids 67(4):404–440. https://onlinelibrary.wiley.com/doi/abs/10.1002/fld.2357
241. Kemm F (2014) A note on the carbuncle phenomenon in shallow water simulations. Z Angew Math Mech 94(6):516–521. https://doi.org/10.1002/zamm.201200176
242. Kemm F (2018) Heuristical and numerical considerations for the carbuncle phenomenon. Appl Math Comput 320:596–613. https://doi.org/10.1016/j.amc.2017.09.014
243. Kennedy CA, Carpenter MH (2019) Higher-order additive Runge-Kutta schemes for ordinary differential equations. Appl Numer Math 136:183–205. https://doi.org/10.1016/j.apnum.2018.10.007
244. Keppens R, Tóth G, Botchev MA, van der Ploeg A (1999) Implicit and semi-implicit schemes: algorithms. Int J Numer Methods Fluids 30:335–352. https://doi.org/10.1002/(SICI)1097-0363(19990615)30:3<335::AID-FLD923>3.3.CO;2-L
245. Keppens R, Nool M, Toth G, Goedbloed J (2003) Adaptive mesh refinement for conservative systems: multi-dimensional efficiency evaluation. Comput Phys Commun 153(3):317–339. https://doi.org/10.1016/S0010-4655(03)00139-5
246. Keppens R, Meliani Z, van Marle A, Delmont P, Vlasis A, van der Holst B (2012) Parallel, grid-adaptive approaches for relativistic hydro and magnetohydrodynamics. J Comput Phys 231(3):718–744. https://doi.org/10.1016/j.jcp.2011.01.020

247. Keyes DE (1995) Aerodynamic applications of Newton-Krylov-Schwarz solvers. In: Deshpande SM, Desai SS, Narasimha R (eds) Numerical methods in fluid dynamics. Lecture notes in physics, vol 453. Springer, Berlin, pp 1–20. https://doi.org/10.1007/3-540-59280-6_95
248. Kifonidis K, Müller E (2012) On multigrid solution of the implicit equations of hydrodynamics-experiments for the compressible Euler equations in general coordinates. Astron Astrophys 544:A47. https://doi.org/10.1051/0004-6361/201116979
249. Kim KH, Kim C (2005) Accurate, efficient and monotonic numerical methods for multi-dimensional compressible flows: Part II: multi-dimensional limiting process. J Comput Phys 208(2):57,015. https://doi.org/10.1016/j.jcp.2005.02.022
250. Kim SD, Lee BJ, Lee HJ, Jeung IS (2009) Robust HLLC Riemann solver with weighted average flux scheme for strong shock. J Comput Phys 228:7634–7642. https://doi.org/10.1016/j.jcp.2009.07.006
251. Kim SE (2003) A multi-dimensional linear reconstruction scheme for arbitrary unstructured mesh. In: Fluid dynamics and co-located conferences, American Institute of Aeronautics and Astronautics, Orlando, Florida. https://doi.org/10.2514/6.2003-399010.2514/6.2003-3990
252. Kitamura K, Shima E (2012) Simple and parameter-free second slope limiter for unstructured grid aerodynamic simulations. AIAA J 50:1415–1426. https://doi.org/10.2514/1.J051269
253. Knoll DA, Keyes DE (2004) Jacobian-free Newton-Krylov methods: a survey of approaches and applications. J Comput Phys 193:357–397. https://doi.org/10.1016/j.jcp.2003.08.010
254. Koldoba AV, Romanova MM, Ustyugova GV, Lovelace RVE (2002) Three-dimensional magnetohydrodynamic simulations of accretion to an inclined rotator: the "cubed sphere" method. Astrophys J Lett 576(1):L53. http://stacks.iop.org/1538-4357/576/i=1/a=L53
255. Kriel A (2017) Error analysis of flux limiter schemes at extrema. J Comput Phys 328:371–386. https://doi.org/10.1016/j.jcp.2016.10.024
256. Lahooti M, Pishevar A (2012) A new fourth order central WENO method for 3D hyperbolic conservation laws. Appl Math Comput 218(20):10,258–10,270. https://doi.org/10.1016/j.amc.2012.04.003
257. Lal SA, Reji R, Santhosh K (2012) Evaluation of diffusive flux across faces of arbitrary shaped finite volume cells. Comput Fluids 57:225–236. https://doi.org/10.1016/j.compfluid.2012.01.001
258. Landi S, Velli M, Einaudi G (2005) Alfvén waves and shock wave formation at an X-point magnetic field configuration. Astrophys J 624(1):392. http://stacks.iop.org/0004-637X/624/i=1/a=392
259. Langdon AB (1992) On enforcing gauss' law in electromagnetic particle-in-cell codes. Comput Phys Commun 70(3):447–450. https://doi.org/10.1016/0010-4655(92)90105-8
260. Lax PD (1957) Hyperbolic systems of conservation laws II. Commun Pure Appl Math 10(4):537–566. https://doi.org/10.1002/cpa.3160100406
261. Ledentsov L, Somov B (2015) MHD discontinuities in solar flares: continuous transitions and plasma heating. Adv Space Res 56(12):2779–2792. Advances in solar physics. https://doi.org/10.1016/j.asr.2015.08.005
262. Ledentsov LS, Somov BV (2013) Continuous transitions between discontinuous magnetohydrodynamic flows of plasma and its heating. J Exp Theor Phys 117(6):1164–1172. https://doi.org/10.1134/S1063776113140148
263. Lee D, Deane AE (2009) An unsplit staggered mesh scheme for multidimensional magnetohydrodynamics. J Comput Phys 228:952–975. https://doi.org/10.1016/j.jcp.2008.08.026
264. Lee D, Faller H, Reyes A (2017) The piecewise cubic method (PCM) for computational fluid dynamics. J Comput Phys 341:230–257. https://doi.org/10.1016/j.jcp.2017.04.004
265. Lee E, Ahn HT, Luo H (2018) Cell-centered high-order hyperbolic finite volume method for diffusion equation on unstructured grids. J Comput Phys 355:464–491. https://doi.org/10.1016/j.jcp.2017.10.051
266. Lee X, Lee C (2013) An implicit algorithm based on iterative modified approximate factorization method coupling with characteristic boundary conditions for solving subsonic viscous flows. Sci China Phys, Mech Astron 56(6):1187–1208. https://doi.org/10.1007/s11433-013-5076-7

267. van Leer B (1979) Towards the ultimate conservative difference scheme. V. A second-order sequel to Godunov's method. J Comput Phys 32(1):101–136. https://doi.org/10.1016/0021-9991(79)90145-1
268. van Leer B (1982) Flux-vector splitting for the Euler equations. In: Krause E (ed) Eighth international conference on numerical methods in fluid dynamics. Springer, Berlin, pp 507–512
269. Leer BV (1977) Towards the ultimate conservative difference scheme III. Upstream-centered finite-difference schemes for ideal compressible flow. J Comput Phys 23(3):263–275. https://doi.org/10.1016/0021-9991(77)90094-8
270. Leibinger J, Dumbser M, Iben U, Wayand I (2016) A path-conservative Osher-type scheme for axially symmetric compressible flows in flexible visco-elastic tubes. Appl Numer Math 105:47–63. https://doi.org/10.1016/j.apnum.2016.02.001
271. LeVeque RJ (2004) Finite volume methods for hyperbolic problems. Cambridge University Press, Cambridge
272. Levy DW, Powell KG, van Leer B (1993) Use of a rotated Riemann solver for the two-dimensional Euler equations. J Comput Phys 106(2):201–214. https://doi.org/10.1016/S0021-9991(83)71103-4
273. Li BQ (2006) Discontinuous finite elements in fluid dynamics and heat transfer. Springer Science & Business Media, Berlin
274. Li CX, Feng XS, Xiang CQ, Zhang M, Li HC, Wei FS (2018) Solar coronal modeling by path-conservative HLLEM Riemann solver. Astrophys J 867(1):42. https://doi.org/10.3847/1538-4357/aae200
275. Li J, Du Z (2016) A two-stage fourth order time-accurate discretization for Lax–Wendroff type flow solvers I. Hyperbolic conservation laws. SIAM J Sci Comput 38(5):A3046–A3069. https://doi.org/10.1137/15M1052512
276. Li L, Liu X, Lou J, Luo H, Nishikawa H, Ren Y (2018) A discontinuous galerkin method based on variational reconstruction for compressible flows on arbitrary grids. In: AIAA SciTech forum, (AIAA 2018-0831), American Institute of Aeronautics and Astronautics, Kissimmee, Florida. https://doi.org/10.2514/6.2018-083110.2514/6.2018-0831
277. Li S (2005) An HLLC Riemann solver for magneto-hydrodynamics. J Comput Phys 203(1):344–357. https://doi.org/10.1016/j.jcp.2004.08.020
278. Li W (2014) Efficient implementation of high-order accurate numerical methods on unstructured grids. Springer, Berlin
279. Li W, Ren YX (2012) High-order k-exact WENO finite volume schemes for solving gas dynamic Euler equations on unstructured grids. Int J Numer Methods Fluids 70(6):742–763. https://doi.org/10.1002/fld.2710
280. Li W, Ren YX (2012) The multi-dimensional limiters for solving hyperbolic conservation laws on unstructured grids II. Extension to high order finite volume schemes. J Comput Phys 231(11):4053–4077. https://doi.org/10.1016/j.jcp.2012.01.029
281. Li W, Ren YX, Lei G, Luo H (2011) The multi-dimensional limiters for solving hyperbolic conservation laws on unstructured grids. J Comput Phys 230(21):7775–7795. https://doi.org/10.1016/j.jcp.2011.06.018
282. Li XS, Gu CW (2013) Mechanism of Roe-type schemes for all-speed flows and its application. Comput Fluids 86:56–70. https://doi.org/10.1016/j.compfluid.2013.07.004
283. Linde TJ (1998) A three-dimensional adaptive multifluid MHD model of the heliosphere. PhD thesis, The University of Michigan
284. Liou MS (1996) A sequel to AUSM: AUSM+. J Comput Phys 129(2):364–382. https://doi.org/10.1006/jcph.1996.0256
285. Liu H, Jiao X (2016) WLS-ENO: weighted-least-squares based essentially non-oscillatory schemes for finite volume methods on unstructured meshes. J Comput Phys 314:749–773. https://doi.org/10.1016/j.jcp.2016.03.039
286. Liu H, Yan J (2009) The direct discontinuous Galerkin (DDG) methods for diffusion problems. SIAM J Numer Anal 47(1):675–698. https://doi.org/10.1137/080720255

287. Liu X, Lou J, Li L, Luo H, Nishikawa H, Ren Y (2017) A compact high order finite volume method based on variational reconstruction for compressible flows on arbitrary grids. In: AIAA aviation forum, (AIAA 2017-3097), American Institute of Aeronautics and Astronautics, Denver, Colorado. https://doi.org/10.2514/6.2017-309710.2514/6.2017-3097
288. Liu XD, Osher S, Chan T (1994) Weighted essentially non-oscillatory schemes. J Comput Phys 115(1):200–212. https://doi.org/10.1006/jcph.1994.1187
289. Liu Y, Zhang W (2017) Accuracy preserving limiter for the high-order finite volume method on unstructured grids. Comput Fluids 149(Supplement C):88–99. https://doi.org/10.1016/j.compfluid.2017.03.008
290. Liu Y, Zhang W, Li C (2017) A novel multi-dimensional limiter for high-order finite volume methods on unstructured grids. Commun Comput Phys 22(5):1385–1412. https://doi.org/10.4208/cicp.OA-2017-0039
291. Londrillo P, del Zanna L (2004) On the divergence-free condition in Godunov-type schemes for ideal magnetohydrodynamics: the upwind constrained transport method. J Comput Phys 195:17–48. https://doi.org/10.1016/j.jcp.2003.09.016
292. Lorcher F, Gassner G, Munz CD (2008) An explicit discontinuous Galerkin scheme with local time-stepping for general unsteady diffusion equations. J Comput Phys 227(11):5649–5670. https://doi.org/10.1016/j.jcp.2008.02.015
293. Lu P, Oliveira M, Liu C (2010) High-order compact scheme for boundary points. Int J Comput Math 87(8):1795–1819. https://doi.org/10.1080/00207160802506512
294. Luo H, Baum JD, Löhner R (1998) A fast, matrix-free implicit method for compressible flows on unstructured grids. Springer, Berlin, pp 73–78. https://doi.org/10.1007/BFb0106564
295. Luo H, Baum JD, Löhner R (1998) A fast, matrix-free implicit method for compressible flows on unstructured grids. J Comput Phys 146(2):664–690. https://doi.org/10.1006/jcph.1998.6076
296. Luo H, Baum JD, Lohner R (2007) A Hermite WENO-based limiter for discontinuous Galerkin method on unstructured grids. J Comput Phys 225(1):686–713. https://doi.org/10.1016/j.jcp.2006.12.017
297. Lütjens H, Luciani JF (2010) XTOR-2F: a fully implicit Newton-Krylov solver applied to nonlinear 3D extended MHD in tokamaks. J Comput Phys 229:8130–8143. https://doi.org/10.1016/j.jcp.2010.07.013
298. Ma XP (2016) Data-driven MHD simulation fo time-dependent solar wind. PhD thesis, University of Chinese Academy of Sciences
299. Mandal J, Rao SP (2011) High resolution finite volume computations on unstructured grids using solution dependent weighted least squares gradients. Comput Fluids 44(1):23–31. https://doi.org/10.1016/j.compfluid.2010.11.021
300. Mandal J, Subramanian J (2008) On the link between weighted least-squares and limiters used in higher-order reconstructions for finite volume computations of hyperbolic equations. Appl Numer Math 58(5):705–725. https://doi.org/10.1016/j.apnum.2007.02.003
301. Mandal JC, Rao S, Subramanian J (2008) High-resolution finite volume computations using a novel weighted least-squares formulation. Int J Numer Methods Fluids 56(8):1425–1431. https://doi.org/10.1002/fld.1707
302. Marder B (1987) A method for incorporating Gauss' law into electromagnetic PIC codes. J Comput Phys 68(1):48
303. Markidis S, Lapenta G, Rizwan-uddin (2010) Multi-scale simulations of plasma with IPIC3d. Math Comput Simul 80(7):1509–1519. Multiscale modeling of moving interfaces in materials. https://doi.org/10.1016/j.matcom.2009.08.038
304. Marquina A (1994) Local piecewise hyperbolic reconstruction of numerical fluxes for nonlinear scalar conservation laws. SIAM J Sci Comput 15(4):892–915. https://doi.org/10.1137/0915054
305. Massa F, Noventa G, Lorini M, Bassi F, Ghidoni A (2018) High-order linearly implicit two-step peer schemes for the discontinuous Galerkin solution of the incompressible Navier-Stokes equations. Comput Fluids 162:55–71. https://doi.org/10.1016/j.compfluid.2017.12.003

306. Mathur SR, Murthy JY (1997) A pressure-based method for unstructured meshes. Numer Heat Trans Part B: Fund 31(2):195–215
307. Mavriplis DJ (1997) Unstructured grid techniques. Annu Rev Fluid Mech 29(1):473–514. https://doi.org/10.1146/annurev.fluid.29.1.473
308. Mavriplis DJ (2003) Revisiting the least-squares procedure for gradient reconstruction on unstructured meshes. In: Fluid dynamics and co-located conferences, American Institute of Aeronautics and Astronautics, Orlando, Florida. https://doi.org/10.2514/6.2003-398610.2514/6.2003-3986
309. Mavriplis DJ (2008) Unstructured-mesh discretizations and solvers for computational aerodynamics. AIAA J 46:1281–1298. https://doi.org/10.2514/1.34681
310. McCorquodale P, Colella P (2011) A high-order finite-volume method for conservation laws on locally refined grids. Commun Appl Math Comput Sci 6(1). https://doi.org/10.2140/camcos.2011.6.1
311. Merkin VG, Lionello R, Lyon JG, Linker J, Török T, Downs C (2016) Coupling of coronal and heliospheric magnetohydrodynamic models: solution comparisons and verification. Astrophys J 831(1):23. http://stacks.iop.org/0004-637X/831/i=1/a=23
312. Michalak C (2009) Efficient high-order accurate unstructured finite-volume algorithms for viscous and inviscid compressible flows. Phd thesis, University of British Columbia. https://doi.org/10.14288/1.0067108
313. Michalak C, Ollivier-Gooch C (2009) Accuracy preserving limiter for the high-order accurate solution of the Euler equations. J Comput Phys 228(23):8693–8711. https://doi.org/10.1016/j.jcp.2009.08.021
314. Michalak K, Ollivier-Gooch C (2007) Matrix-explicit GMRES for a higher-order accurate inviscid compressible flow solver. In: Fluid dynamics and co-located conferences, American Institute of Aeronautics and Astronautics. https://doi.org/10.2514/6.2007-394310.2514/6.2007-3943
315. Michel AN, Herget CJ (2007) Finite-dimensional vector spaces and matrices. Birkhäuser, Boston, pp 124–262. https://doi.org/10.1007/978-0-8176-4707-0_4
316. Mignone A (2014) High-order conservative reconstruction schemes for finite volume methods in cylindrical and spherical coordinates. J Comput Phys 270(Supplement C):784–814. https://doi.org/10.1016/j.jcp.2014.04.001
317. Mignone A (2017) MHD modeling: aims, usage, scales assessed, caveats, codes. Springer International Publishing, Cham, pp 187–213
318. Mignone A, Tzeferacos P (2010) A second-order unsplit Godunov scheme for cell-centered MHD: the CTU-GLM scheme. J Comput Phys 229(6):2117–2138
319. Mignone A, Tzeferacos P, Bodo G (2010) High-order conservative finite difference GLM-MHD schemes for cell-centered MHD. J Comput Phys 229(17):5896–5920. https://doi.org/10.1016/j.jcp.2010.04.013
320. Mignone A, Zanni C, Tzeferacos P, van Straalen B, Colella P, Bodo G (2012) The PLUTO code for adaptive mesh computations in astrophysical fluid dynamics. Astrophys J Suppl Ser 198:7. https://doi.org/10.1088/0067-0049/198/1/7
321. Miranda-Aranguren S, Aloy MA, Rembiasz T (2018) An HLLC Riemann solver for resistive relativistic magnetohydrodynamics. Mon Not R Astron Soc 476:3837–3860. https://doi.org/10.1093/mnras/sty419
322. Mitchell CR, Walters RW (1993) K-exact reconstruction for the Navier-Stokes equations on arbitrary grids. In: AIAA, aerospace sciences meeting and exhibit, 31st, Reno, NV, Jan 11–14, 1993, p 17
323. Miyoshi T, Kusano K (2001) A global MHD simulation of the Jovian magnetosphere interacting with/without the interplanetary magnetic field. J Geophys Res: Space Phys 106(A6):10,723–10,742. https://doi.org/10.1029/2000JA900153
324. Miyoshi T, Kusano K (2005) A multi-state HLL approximate Riemann solver for ideal magnetohydrodynamics. J Comput Phys 208(1):315–344. https://doi.org/10.1016/j.jcp.2005.02.017

325. Miyoshi T, Kusano K (2011) A comparative study of divergence-cleaning techniques for multi-dimensional MHD schemes. Plasma Fusion Res 6(2401):124. https://doi.org/10.1585/pfr.6.2401124
326. Miyoshi T, Terada N, Matsumoto Y, Fukazawa K, Umeda T, Kusano K (2010) The HLLD approximate Riemann solver for magnetospheric simulation. IEEE Trans Plasma Sci 38:2236–2242. https://doi.org/10.1109/TPS.2010.2057451
327. Mocz P, Pakmor R, Springel V, Vogelsberger M, Marinacci F, Hernquist L (2016) A moving mesh unstaggered constrained transport scheme for magnetohydrodynamics. Mon Not R Astron Soc 463(1):477–488. https://doi.org/10.1093/mnras/stw2004
328. Mohsen Karimian SA, Straatman AG (2006) Discretization and parallel performance of an unstructured finite volume Navier-Stokes solver. Int J Numer Methods Fluids 52(6):591–615. https://doi.org/10.1002/fld.1189
329. Moukalled F, Mangani L, Darwish M (2015) The finite volume method in computational fluid dynamics: an introduction with OpenFOAM and Matlab. Springer International Publishing, Cham. https://www.springer.com/gp/book/9783319168739
330. Moukalled F, Mangani L, Darwish M (2016) Solving the system of algebraic equations. Springer International Publishing, Cham, pp 303–364. https://doi.org/10.1007/978-3-319-16874-6_10
331. Mulder WA, van Leer B (1985) Experiments with implicit upwind methods for the Euler equations. J Comput Phys 59:232–246. https://doi.org/10.1016/0021-9991(85)90144-5
332. Munoz P, Jain N, Kilian P, Buchner J (2018) A new hybrid code (CHIEF) implementing the inertial electron fluid equation without approximation. Comput Phys Commun 224:245–264. https://doi.org/10.1016/j.cpc.2017.10.012
333. Muppidi S, Mahesh K (2008) Direct numerical simulation of passive scalar transport in transverse jets. J Fluid Mech 598:335–360. https://doi.org/10.1017/S0022112007000055
334. Murawski K (2011) Numerical solutions of magnetohydrodynamic equations. Bull Pol Acad Sci Tech Sci 59:219–226
335. Nakashima Y, Watanabe N, Nishikawa H (2016) Hyperbolic Navier-Stokes solver for three-dimensional flows. In: AIAA SciTech forum, American Institute of Aeronautics and Astronautics, San Diego, California. https://doi.org/10.2514/6.2016-110110.2514/6.2016-1101
336. Navarro A, Lora-Clavijo FD, González GA (2017) Magnus: a new resistive MHD code with heat flow terms. Astrophys J 844:57. https://doi.org/10.3847/1538-4357/aa7a13
337. Nazari F, Mohammadian A, Charron M (2015) High-order low-dissipation low-dispersion diagonally implicit Runge-Kutta schemes. J Comput Phys 286:38–48. https://doi.org/10.1016/j.jcp.2015.01.020
338. Nejat A, Ollivier-Gooch C (2006) A high-order accurate unstructured Newton-Krylov solver for inviscid compressible flows. In: Fluid dynamics and co-located conferences, American Institute of Aeronautics and Astronautics, San Francisco, California. https://doi.org/10.2514/6.2006-371110.2514/6.2006-3711
339. Nejat A, Ollivier-Gooch C (2008) Effect of discretization order on preconditioning and convergence of a high-order unstructured Newton-GMRES solver for the Euler equations. J Comput Phys 227(4):2366–2386. https://doi.org/10.1016/j.jcp.2007.10.024
340. Nejat A, Ollivier-Gooch C (2008) A high-order accurate unstructured finite volume Newton-Krylov algorithm for inviscid compressible flows. J Comput Phys 227(4):2582–2609. https://doi.org/10.1016/j.jcp.2007.11.011
341. Nguyen NT, Dumbser M (2015) A path-conservative finite volume scheme for compressible multi-phase flows with surface tension. Appl Math Comput 271:959–978
342. Nielsen EJ, Anderson WK, Walters RW, Keyes DE (1995) Application of Newton-Krylov methodology to a three-dimensional unstructured Euler code. AIAA Paper 95–1733–CP, AIAA
343. Nishikawa H (2007) A first-order system approach for diffusion equation. I: second-order residual-distribution schemes. J Comput Phys 227(1):315–352. https://doi.org/10.1016/j.jcp.2007.07.029

344. Nishikawa H (2010) Beyond interface gradient: a general principle for constructing diffusion schemes. In: Fluid dynamics and co-located conferences, American Institute of Aeronautics and Astronautics, Chicago, Illinois. https://doi.org/10.2514/6.2010-509310.2514/6.2010-5093
345. Nishikawa H (2011) New-generation hyperbolic Navier-Stokes schemes: O(1/h) speed-up and accurate viscous/heat fluxes. In: Fluid dynamics and co-located conferences, American Institute of Aeronautics and Astronautics, Honolulu, Hawaii. https://doi.org/10.2514/6.2011-304310.2514/6.2011-3043
346. Nishikawa H (2011) Robust and accurate viscous discretization via upwind scheme-I: basic principle. Comput Fluids 49(1):62–86. https://doi.org/10.1016/j.compfluid.2011.04.014
347. Nishikawa H (2011c) Two ways to extend diffusion schemes to Navier-Stokes schemes: gradient formula or upwind flux. In: Fluid dynamics and co-located conferences, American Institute of Aeronautics and Astronautics. https://doi.org/10.2514/6.2011-304410.2514/6.2011-3044
348. Nishikawa H (2012) Divergence formulation of source term. J Comput Phys 231(19):6393–6400. https://doi.org/10.1016/j.jcp.2012.05.032
349. Nishikawa H (2014) First-, second-, and third-order finite-volume schemes for diffusion. J Comput Phys 256:791–805. https://doi.org/10.1016/j.jcp.2013.09.024
350. Nishikawa H (2014b) First, second, and third order finite-volume schemes for advection-diffusion. J Comput Phys 273:287–309. https://doi.org/10.1016/j.jcp.2014.05.021
351. Nishikawa H (2016) On viscous scheme for cell-centered method. https://doi.org/10.13140/RG.2.1.1697.4965
352. Nishikawa H (2018) From hyperbolic diffusion scheme to gradient method: implicit green-gauss gradients for unstructured grids. J Comput Phys 372:126–160. https://doi.org/10.1016/j.jcp.2018.06.019
353. Nishikawa H (2018) On hyperbolic method for diffusion with discontinuous coefficients. J Comput Phys 367:102–108. https://doi.org/10.1016/j.jcp.2018.04.027
354. Nishikawa H, Kitamura K (2008) Very simple, carbuncle-free, boundary-layer-resolving, rotated-hybrid Riemann solvers. J Comput Phys 227(4):2560–2581. https://doi.org/10.1016/j.jcp.2007.11.003
355. Nishikawa H, Liu Y (2017) Accuracy-preserving source term quadrature for third-order edge-based discretization. J Comput Phys 344:595–622. https://doi.org/10.1016/j.jcp.2017.04.075
356. Nishikawa H, Liu Y (2018) Hyperbolic advection-diffusion schemes for high-Reynolds-number boundary-layer problems. J Comput Phys 352:23–51. https://doi.org/10.1016/j.jcp.2017.09.039
357. Nishikawa H, Nakashima Y (2018) Dimensional scaling and numerical similarity in hyperbolic method for diffusion. J Comput Phys 355:121–143. https://doi.org/10.1016/j.jcp.2017.11.008
358. Nishikawa H, Roe P, Suzuki Y, van Leer B (2003) A general theory of local preconditioning and its application to 2D ideal MHD equations. In: Fluid dynamics and co-located conferences, AIAA 2003–3704, American Institute of Aeronautics and Astronautics, Orlando, Florida. https://doi.org/10.2514/6.2003-370410.2514/6.2003-3704
359. Nishikawa K, Wakatani M (2000) Ideal magnetohydrodynamics. Springer, Berlin, pp 173–204. https://doi.org/10.1007/978-3-662-04078-2_10
360. Noelle S (1999) A comparison of third and second order accurate finite volume schemes for the two-dimensional compressible Euler equations. In: Jeltsch R, Fey M (eds) Hyperbolic problems: theory. Numerics, applications. Birkhäuser, Basel, pp 757–766
361. Northrup SA (2014) A parallel implicit adaptive mesh refinement algorithm for predicting unsteady fully-compressible reactive flows. PhD thesis, University of Toronto
362. Núñez-de la Rosa J, Munz CD (2016) XTROEM-FV: a new code for computational astrophysics based on very high order finite-volume methods - I. Magnetohydrodynamics. Mon Not R Astron Soc 455:3458–3479. https://doi.org/10.1093/mnras/stv2531
363. Ofman L (2010) Wave modeling of the solar wind. Living Rev Sol Phys 7(1):4 https://doi.org/10.12942/lrsp-2010-4

364. Okong'o N, Bellan J (2002) Consistent boundary conditions for multicomponent real gas mixtures based on characteristic waves. J Comput Phys 176(2):330–344. https://doi.org/10.1006/jcph.2002.6990
365. Olawsky F, Infed F, Auweter-Kurtz M (2004) Preconditioned Newton-Method for computing supersonic and hypersonic nonequilibrium flows. J Spacecr Rocket 41:907–914. https://doi.org/10.2514/1.4010
366. Oliveira DM (2017) Magnetohydrodynamic shocks in the interplanetary space: a theoretical review. Braz J Phys 47(1):81–95. https://doi.org/10.1007/s13538-016-0472-x
367. Ollivier-Gooch C, Altena MV (2002) A high-order-accurate unstructured mesh finite-volume scheme for the advection-diffusion equation. J Comput Phys 181(2):729–752. https://doi.org/10.1006/jcph.2002.7159
368. Ollivier-Gooch C, Nejat A, Michalak K (2009) Obtaining and verifying high-order unstructured finite volume solutions to the Euler equations. AIAA J 47:2105–2120. https://doi.org/10.2514/1.40585
369. Ollivier-Gooch CF (1997) High-order ENO schemes for unstructured meshes based on least-squares reconstruction. In: Conference: 35. Aerospace sciences meeting and exhibit, Reno, NV (United States), 6–9 Jan 1997. https://doi.org/10.2514/6.1997-540
370. Ollivier-Gooch CF (1997) Quasi-ENO schemes for unstructured meshes based on unlimited data-dependent least-squares reconstruction. J Comput Phys 133:6–17. https://doi.org/10.1006/jcph.1996.5584
371. Onofri M, Primavera L, Malara F, Londrillo P (2007) A compressible magnetohydrodynamic numerical code with time-dependent boundary conditions in cylindrical geometry. J Comput Phys 226(2):1874–1890. https://doi.org/10.1016/j.jcp.2007.06.015
372. Osher S, Solomon F (1982) Upwind difference schemes for hyperbolic conservation laws. Math Comput 38:339–V374
373. Pan J, Wang Q, Zhang Y, Ren Y (2018) High-order compact finite volume methods on unstructured grids with adaptive mesh refinement for solving inviscid and viscous flows. Chin J Aeronaut 31(9):1829–1841. https://doi.org/10.1016/j.cja.2018.06.018
374. Pandolfi M, D'Ambrosio D (2001) Numerical instabilities in upwind methods: analysis and cures for the "Carbuncle" phenomenon. J Comput Phys 166:271–301. https://doi.org/10.1006/jcph.2000.6652
375. Papalexandris M, Leonard A, Dimotakis P (2002) Unsplit algorithms for multidimensional systems of hyperbolic conservation laws with source terms. Comput Math Appl 44(1):25–49. https://doi.org/10.1016/S0898-1221(02)00128-1
376. Park H, Nourgaliev RR, Martineau RC, Knoll DA (2009) On physics-based preconditioning of the Navier-Stokes equations. J Comput Phys 228:9131–9146. https://doi.org/10.1016/j.jcp.2009.09.015
377. Park JS, Kim C (2012) Multi-dimensional limiting process for finite volume methods on unstructured grids. Comput Fluids 65:8–24. https://doi.org/10.1016/j.compfluid.2012.04.015
378. Park JS, Yoon SH, Kim C (2010) Multi-dimensional limiting process for hyperbolic conservation laws on unstructured grids. J Comput Phys 229(3):788–812. https://doi.org/10.1016/j.jcp.2009.10.011
379. Parker EN (1979) Cosmical magnetic fields: their origin and their activity. Clarendon Press, Oxford
380. Pelanti M (2017) Low Mach number preconditioning techniques for Roe-type and HLLC-type methods for a two-phase compressible flow model. Appl Math Comput 310:112–133. https://doi.org/10.1016/j.amc.2017.04.014
381. Perez-Segarra CD, Farre C, Cadafalch J, Oliva A (2006) Analysis of different numerical schemes for the resolution of convection-diffusion equations using finite-volume methods on three-dimensional unstructured grids. Part I: discretization schemes. Numer Heat Trans, Part B: Fund 49(4):333–350. https://doi.org/10.1080/10407790500314947

382. Pogorelov NV, Matsuda T (2000) Nonevolutionary MHD shocks in the solar wind and interstellar medium interaction. Astron Astrophys 354:697–702
383. Poinsot T, Lele S (1992) Boundary conditions for direct simulations of compressible viscous flows. J Comput Phys 101(1):104–129. https://doi.org/10.1016/0021-9991(92)90046-2
384. Polovin RV (1961) Shock waves in magnetohydrodynamics. Soviet Physics Uspekhi 3(5):677. http://stacks.iop.org/0038-5670/3/i=5/a=R02
385. Polovin RV, Cherkasova KP (1966) Magnetohydrodynamic waves. Sov Phys Uspekhi 9(2):278. http://stacks.iop.org/0038-5670/9/i=2/a=A08
386. Popov MV, Ustyugov SD (2007) Piecewise parabolic method on local stencil for gasdynamic simulations. Comput Math Math Phys 47(12):1970–1989. https://doi.org/10.1134/S0965542507120081
387. Porth O, Olivares H, Mizuno Y, Younsi Z, Rezzolla L, Moscibrodzka M, Falcke H, Kramer M (2017) The black hole accretion code. Comput Astrophys Cosmol 4(1):1. https://doi.org/10.1186/s40668-017-0020-2
388. Powell KG (1994) Approximate Riemann solver for magnetohydrodynamics (that works in more than one dimension). Technical report, Institute for Computer Applications in Science and Engineering, NASA Langley Research Center
389. Powell KG, Roe PL, Linde TJ, Gombosi TI, de Zeeuw DL (1999) A solution-adaptive upwind scheme for ideal magnetohydrodynamics. J Comput Phys 154:284–309. https://doi.org/10.1006/jcph.1999.6299
390. Powell KG, Toth G, De Zeeuw DL, Roe PL, Gombosi TI, Stout QF (2001) Development and validation of solution-adaptive, parallel schemes for compressible plasmas. In: AIAA-2001-2525, AIAA computational fluid dynamics conference, 15th, Anaheim, CA, June 11–14, 2001
391. Priest E, Forbes T (2000) Magnetic reconnection. Cambridge University Press, Cambridge
392. Pucci F, Onofri M, Malara F (2014) Evolution of magnetohydrodynamic waves in low layers of a coronal hole. Astrophys J 796(1):43. http://stacks.iop.org/0004-637X/796/i=1/a=43
393. Qiu J, Shu CW (2005) Hermite WENO schemes and their application as limiters for Runge-Kutta discontinuous Galerkin method II: two dimensional case. Comput Fluids 34(6):642–663. https://doi.org/10.1016/j.compfluid.2004.05.005
394. Quirk JJ (1994) A contribution to the great Riemann solver debate. Int J Numer Methods Fluids 18(6):555–574
395. Rausch RD, Batina JT, Yang HTY (1992) Spatial adaptation of unstructured meshes for unsteady aerodynamic flow computations. AIAA J 30(5):1243–1251. https://doi.org/10.2514/3.11057
396. Reisner J, Wyszogrodzki A, Mousseau V, Knoll D (2003) An efficient physics-based preconditioner for the fully implicit solution of small-scale thermally driven atmospheric flows. J Comput Phys 189:30–44. https://doi.org/10.1016/S0021-9991(03)00198-0
397. Ren YX (2003) A robust shock-capturing scheme based on rotated Riemann solvers. Comput Fluids 32(10):1379–1403. https://doi.org/10.1016/S0045-7930(02)00114-7
398. Reynolds DR, Samtaney R, Woodward CS (2006) A fully implicit numerical method for single-fluid resistive magnetohydrodynamics. J Comput Phys 219(1):144–162. https://doi.org/10.1016/j.jcp.2006.03.022
399. Reynolds DR, Samtaney R, Woodward CS (2010) Operator-based preconditioning of stiff hyperbolic systems. SIAM J Sci Comput 32(1):150–170. https://doi.org/10.1137/080713331
400. Reynolds DR, Samtaney R, Tiedeman HC (2012) A fully implicit Newton-Krylov-Schwarz method for tokamak magnetohydrodynamics: Jacobian construction and preconditioner formulation. Comput Sci Discov 5(1):014003
401. Robinet JC, Gressier J, Casalis G, Moschetta JM (2000) Shock wave instability and the carbuncle phenomenon: same intrinsic origin? J Fluid Mech 417:237–263. https://doi.org/10.1017/S0022112000001129
402. Roe PL (1981) Approximate Riemann solvers, parameter vectors, and difference schemes. J Comput Phys 43:357. https://doi.org/10.1016/0021-9991(81)90128-5

403. Roe PL (1985) Some contributions to the modelling of discontinuous flows. In: Lee RL, Sani RL, Shih TM, Gresho PM (eds) Large-scale computations in fluid mechanics, pp 163–193
404. Roe PL, Balsara DS (1996) Notes on the eigensystem of magnetohydrodynamics. SIAM J Appl Math 56:57–67. https://doi.org/10.1137/S003613999427084X
405. Rossmanith JA (2006) An unstaggered, high-resolution constrained transport method for magnetohydrodynamic flows. SIAM J Sci Comput 28:1766–1797. https://doi.org/10.1137/050627022
406. Rusanov VV (1961) The calculation of the interaction of non-stationary shock waves with barriers. Zh Vychisl Mat Mat Fiz 1:267–279. https://doi.org/http://mi.mathnet.ru/zvmmf7987
407. Ryu D, Jones TW (1995) Numerical magetohydrodynamics in astrophysics: algorithm and tests for one-dimensional flow. Astrophys J 442:228–258. https://doi.org/10.1086/175437
408. Ryu D, Miniati F, Jones TW, Frank A (1998) A divergence-free upwind code for multidimensional magnetohydrodynamic flows. Astrophys J 509:244–255. https://doi.org/10.1086/306481
409. Saad Y (2003) Iterative methods for sparse linear systems. Society for industrial and applied mathematics, 2nd edn. Philadelphia. https://doi.org/10.1137/1.9780898718003
410. Saad Y, Schultz MH (1986) GMRES: a generalized minimal residual algorithm for solving nonsymmetric linear systems. SIAM J Sci Stat Comput 7(3):856–869. https://doi.org/10.1137/0907058
411. Saad Y, van der Vorst HA (2000) Iterative solution of linear systems in the 20th century. J Comput Appl Math 123:1–33. https://doi.org/10.1016/S0377-0427(00)00412-X
412. Sánchez-Linares C, de Luna TM, Castro Díaz MJ (2016) A HLLC scheme for Ripa model. Appl Math Comput 272. Part 2:369–384. https://doi.org/10.1016/j.amc.2015.05.137
413. Sawyer SE (2012) Analysis and implementation of a high-order reconstruction algorithm for an unstructured finite volume flow solver. PhD thesis, University of Tennessee at Chattanooga
414. Schmidtmann B, Torrilhon M (2017) A hybrid Riemann solver for large hyperbolic systems of conservation laws. SIAM J Sci Comput 39(6):A2911–A2934. https://doi.org/10.1137/16M108567X
415. Schmidtmann B, Winters AR (2017) Hybrid entropy stable HLL-type Riemann solvers for hyperbolic conservation laws. J Comput Phys 330(Supplement C):566–570. https://doi.org/10.1016/j.jcp.2016.10.034
416. Schmidtmann B, Seibold B, Torrilhon M (2016) Relations between WENO3 and third-order limiting in finite volume methods. J Sci Comput 68(2):624–652. https://doi.org/10.1007/s10915-015-0151-z
417. Scott B (2005) An introduction to magnetohydrodynamics (MHD), or magnetic fluid dynamics. Springer, Berlin, pp 51–74. https://doi.org/10.1007/11360360_3
418. Sejekan CB, Ollivier-Gooch CF (2016) Improving finite-volume diffusive fluxes through better reconstruction. Comput Fluids 139:216–232. 13th USNCCM international symposium of high-order methods for computational fluid dynamics - a special issue dedicated to the 60th birthday of Professor David Kopriva. https://doi.org/10.1016/j.compfluid.2016.08.002
419. Sekora M, Colella P (2009) Extremum-preserving limiters for MUSCL and PPM. ArXiv e-prints arXiv:0903.4200
420. Selmin V (1993) The node-centred finite volume approach: bridge between finite differences and finite elements. Comput Methods Appl Mech Eng 102(1):107–138. https://doi.org/10.1016/0045-7825(93)90143-L
421. Serna S (2009) A characteristic-based nonconvex entropy-fix upwind scheme for the ideal magnetohydrodynamic equations. J Comput Phys 228(11):4232–4247. https://doi.org/10.1016/j.jcp.2009.03.001
422. Shang J (2001) Recent research in magneto-aerodynamics. Prog Aerosp Sci 37(1):1–20. https://doi.org/10.1016/S0376-0421(00)00015-4
423. Shen Y, Zha G, Huerta MA (2011) Rotated hybrid low diffusion ECUSP-HLL scheme and its applications to hypersonic flows. In: 20th AIAA computational fluid dynamics conference, AIAA 2011-3545, 27–30 June 2011, Honolulu, Hawaii. https://doi.org/10.2514/6.2011-3545

424. Shen Y, Zha G, Huerta MA (2012) E-CUSP scheme for the equations of ideal magnetohydrodynamics with high order WENO Scheme. J Comput Phys 231(19):6233–6247. https://doi.org/10.1016/j.jcp.2012.04.015
425. Shen Z, Yan W, Yuan G (2014) A stability analysis of hybrid schemes to cure shock instability. Commun Comput Phys 15:1320–1342. https://doi.org/10.4208/cicp.210513.091013a
426. Shen Z, Yan W, Yuan G (2016) A robust HLLC-type Riemann solver for strong shock. J Comput Phys 309(Supplement C):185–206. https://doi.org/10.1016/j.jcp.2016.01.001
427. Shepard D (1968) A two-dimensional interpolation function for irregularly-spaced data. In: Proceedings of the 1968 23rd ACM national conference ACM '68. ACM, New York, pp 517–524. https://doi.org/10.1145/800186.810616
428. Shi J, Hu C, Shu CW (2002) A technique of treating negative weights in WENO schemes. J Comput Phys 175(1):108–127. https://doi.org/10.1006/jcph.2001.6892
429. Shima E, Kitamura K, Fujimoto K (2010) New gradient calculation method for MUSCL Type CFD schemes in arbitrary polyhedra. In: Aerospace sciences meetings, American Institute of Aeronautics and Astronautics, Orlando, Florida. https://doi.org/10.2514/6.2010-108110.2514/6.2010-1081
430. Shima E, Kitamura K, Haga T (2013) Green-gauss/weighted-least-squares hybrid gradient reconstruction for arbitrary polyhedra unstructured grids. AIAA J 51(11):2740–2747. https://doi.org/10.2514/1.J052095
431. Shiota D, Kusano K, Miyoshi T, Nishikawa N, Shibata K (2008) A quantitative MHD study of the relation among arcade shearing, flux rope formation, and eruption due to the tearing instability. J Geophys Res: Space Phys 113(A3):A03S05. https://doi.org/10.1029/2007JA012516
432. Shiota D, Kusano K, Miyoshi T, Shibata K (2010) Magnetohydrodynamic modeling for a formation process of coronal mass ejections: interaction between an ejecting flux rope and an ambient field. Astrophys J 718(2):1305. http://stacks.iop.org/0004-637X/718/i=2/a=1305
433. Shiota D, Kataoka R, Miyoshi Y, Hara T, Tao C, Masunaga K, Futaana Y, Terada N (2014) Inner heliosphere MHD modeling system applicable to space weather forecasting for the other planets. Space Weather 12(4):187–204. https://doi.org/10.1002/2013SW000989
434. Shu CW (1987) TVB uniformly high-order schemes for conservation laws. Math Comp 49(179):105–121. https://doi.org/10.1090/S0025-5718-1987-0890256-5
435. Shu CW (1998) Essentially non-oscillatory and weighted essentially non-oscillatory schemes for hyperbolic conservation laws. Springer, Berlin, pp 325–432. https://doi.org/10.1007/BFb0096355
436. Shu CW (2009) High order weighted essentially nonoscillatory schemes for convection dominated problems. SIAM Rev 51(1):82–126. https://doi.org/10.1137/070679065
437. Shu CW (2016) High order WENO and DG methods for time-dependent convection-dominated PDEs: a brief survey of several recent developments. J Comput Phys 316(Supplement C):598–613. https://doi.org/10.1016/j.jcp.2016.04.030
438. Shu CW, Osher S (1988) Efficient implementation of essentially non-oscillatory shock-capturing schemes. J Comput Phys 77(2):439–471. https://doi.org/10.1016/0021-9991(88)90177-5
439. Shu CW, Osher S (1989) Efficient implementation of essentially non-oscillatory shock-capturing schemes, II. J Comput Phys 83(1):32–78. https://doi.org/10.1016/0021-9991(89)90222-2
440. Sitaraman H, Raja L (2013) A matrix free implicit scheme for solution of resistive magnetohydrodynamics equations on unstructured grids. J Comput Phys 251:364–382. https://doi.org/10.1016/j.jcp.2013.06.003
441. Skála J, Bárta M (2012) LSFEM implementation of MHD numerical solver. Appl Math 3(11A):1842–1850. https://doi.org/10.4236/am.2012.331250
442. Skála J, Bárta M, Varady M (2011) MHD modelling of multi-scale magnetic reconnection using the finite element method. CentL Eur Astrophys Bull 35:195–204
443. Skála J, Baruffa F, Büchner J, Rampp M (2015) The 3D MHD code GOEMHD3 for astrophysical plasmas with large Reynolds numbers. Code description, verification, and computational performance. Astron Astrophys 580:A48. https://doi.org/10.1051/0004-6361/201425274

444. Smith B (2011) PETSc (portable, extensible toolkit for scientific computation). Springer, Boston, pp 1530–1539
445. Somov BV (2012) Evolutionarity of MHD discontinuities. Springer, New York, pp 363–386
446. Sozer E, Brehm C, Kiris CC (2014) Gradient calculation methods on arbitrary polyhedral unstructured meshes for cell-centered CFD solvers. In: AIAA SciTech forum, American Institute of Aeronautics and Astronautics, National Harbor, Maryland. https://doi.org/10.2514/6.2014-144010.2514/6.2014-1440
447. Spekreijse S (1987) Multigrid solution of monotone second-order discretizations of hyperbolic conservation laws. Math Comput 49(179):135–155. https://doi.org/10.1090/S0025-5718-1987-0890258-9
448. Stone JM, Gardiner TA, Teuben P, Hawley JF, Simon JB (2008) Athena: a new code for astrophysical MHD. Astrophys J Suppl Ser 178:137
449. Strang G (2016) Introduction to linear algebra, 5th edn. Wellesley-Cambridge Press, Box 812060 Wellesley MA 02482. http://math.mit.edu/~gs/linearalgebra/
450. Subramaniam V, Raja LL (2017) Magnetohydrodynamic simulation study of plasma jets and plasma-surface contact in coaxial plasma accelerators. Phys Plasmas 24(6):062,507. https://doi.org/10.1063/1.4985320
451. Sun MT, Wu ST, Dryer M (1995) On the time-dependent numerical boundary conditions of magnetohydrodynamic flows. J Comput Phys 116:330–342. https://doi.org/10.1006/jcph.1995.1031
452. Suresh A (2000) Positivity-preserving schemes in multidimensions. SIAM J Sci Comput 22(4):1184–1198. https://doi.org/10.1137/S1064827599360443
453. Suresh A, Huynh H (1997) Accurate monotonicity-preserving schemes with Runge-Kutta time stepping. J Comput Phys 136(1):83–99. https://doi.org/10.1006/jcph.1997.5745
454. Susanto A, Ivan L, Sterck HD, Groth C (2013) High-order central ENO finite-volume scheme for ideal MHD. J Comput Phys 250:141–164. https://doi.org/10.1016/j.jcp.2013.04.040
455. Sweby PK (1984) High resolution schemes using flux limiters for hyperbolic conservation laws. SIAM J Numer Anal 21(5):995–1011. https://doi.org/10.1137/0721062
456. Syrakos A, Varchanis S, Dimakopoulos Y, Goulas A, Tsamopoulos J (2017) A critical analysis of some popular methods for the discretisation of the gradient operator in finite volume methods. Phys Fluids 29(12):127,103. https://doi.org/10.1063/1.4997682
457. Takahashi K, Yamada S (2014) On the existence and non-uniqueness of solutions of Riemann problems in ideal magnetohydrodynamics. In: Pogorelov NV, Audit E, Zank GP (eds) 8th International conference of numerical modeling of space plasma flows (ASTRONUM 2013). Astronomical society of the pacific conference series, vol 488, p 261
458. Takahashi K, Yamada S (2013) Regular and non-regular solutions of the Riemann problem in ideal magnetohydrodynamics. J Plasma Phys 79:335–356. https://doi.org/10.1017/S0022377812001006
459. Takahashi K, Yamada S (2014) Exact Riemann solver for ideal magnetohydrodynamics that can handle all types of intermediate shocks and switch-on/off waves. J Plasma Phys 80:255–287. https://doi.org/10.1017/S0022377813001268
460. Tanaka T (1994) Finite volume TVD scheme on an unstructured grid system for three-dimensional MHD simulation of inhomogeneous systems including strong background potential fields. J Comput Phys 111:381–390. https://doi.org/10.1006/jcph.1994.1071
461. Tanaka T (1995) Generation mechanisms for magnetosphere-ionosphere current systems deduced from a three-dimensional MHD simulation of the solar wind-magnetosphere-ionosphere coupling processes. J Geophys Res 100(A7):12,057–12,074
462. Thomas JL, Diskin B, Nishikawa H (2011) A critical study of agglomerated multigrid methods for diffusion on highly-stretched grids. Comput Fluids 41(1):82–93. Implicit solutions of Navier-Stokes equations. Special issue dedicated to Drs. W.R. Briley and H. McDonald. https://doi.org/10.1016/j.compfluid.2010.09.023
463. Thompson KW (1990) Time-dependent boundary conditions for hyperbolic systems, ii. J Comput Phys 89(2):439–461. https://doi.org/10.1016/0021-9991(90)90152-Q

464. Tidriri MD (2001) Development and study of Newton-Krylov-Schwarz algorithms. Int J Comput Fluid Dyn 15(2):115–126. https://doi.org/10.1080/10618560108970022
465. Titarev V, Toro E (2004) Finite-volume WENO schemes for three-dimensional conservation laws. J Comput Phys 201(1):238–260. https://doi.org/10.1016/j.jcp.2004.05.015
466. Toro EF (1999) Methods for multi-dimensional PDEs. Springer, Berlin, pp 539–579
467. Toro EF (2016) Chapter 2 - The Riemann problem: solvers and numerical fluxes. In: Abgrall R, Shu CW (eds) Handbook of numerical methods for hyperbolic problems. Handbook of numerical analysis, vol 17, Elsevier, Amsterdam, pp 19–54. https://doi.org/10.1016/bs.hna.2016.09.015
468. Toro EF, Spruce M, Speares W (1994) Restoration of the contact surface in the HLL-Riemann solver. Shock Waves 4:25–34. https://doi.org/10.1007/BF01414629
469. Toro T (2009) Riemann solvers and numerical methods for fluid dynamics: a practical introduction. Springer, Berlin
470. Torrilhon M (2003) Uniqueness conditions for Riemann problems of ideal magnetohydrodynamics. J Plasma Phys 69:253–276. https://doi.org/10.1017/S0022377803002186
471. Tóth G (2000) The $\nabla \cdot \mathbf{B} = 0$ constraint in shock-capturing magnetohydrodynamics codes. J Comput Phys 161(2):605–652. https://doi.org/10.1006/jcph.2000.6519
472. Tóth G, Odstrčil D (1996) Comparison of some flux corrected transport and total variation diminishing numerical schemes for hydrodynamic and magnetohydrodynamic problems. J Comput Phys 128(1):82–100. https://doi.org/10.1006/jcph.1996.0197
473. Tóth G, de Zeeuw DL, Gombosi TI, Powell KG (2006) A parallel explicit/implicit time stepping scheme on block-adaptive grids. J Comput Phys 217:722
474. Tóth G, Meng X, Gombosi TI, Ridley AJ (2011) Reducing numerical diffusion in magnetospheric simulations. J Geophys Res (Space Phys) 116:A07211. https://doi.org/10.1029/2010JA016370
475. Tóth G et al (2012) Adaptive numerical algorithms in space weather modeling. J Comput Phys 231:870
476. Tsoutsanis P (2018) Extended bounds limiter for high-order finite-volume schemes on unstructured meshes. J Comput Phys 362:69–94. https://doi.org/10.1016/j.jcp.2018.02.009
477. Tsoutsanis P, Titarev V, Drikakis D (2011) WENO schemes on arbitrary mixed-element unstructured meshes in three space dimensions. J Comput Phys 230(4):1585–1601. https://doi.org/10.1016/j.jcp.2010.11.023
478. Tu CY (1988) Solar terrestrial space physics volume 1 (in Chinese). Science Press, Beijing
479. Tu CY (1988) Solar terrestrial space physics volume 2 (in Chinese). Science Press, Beijing
480. Turkel E (1993) Review of preconditioning methods for fluid dynamics. Appl Numer Math 12(1):257–284. https://doi.org/10.1016/0168-9274(93)90122-8
481. van Albada GD, van Leer B, Roberts WW Jr (1982) A comparative study of computational methods in cosmic gas dynamics. Astron Astrophys 108:76–84
482. van der Holst B, Keppens R (2007) Hybrid block-AMR in Cartesian and curvilinear coordinates: MHD applications. J Comput Phys 226(1):925
483. van der Ploeg A, Keppens R, Tóth G (1997) Block incomplete LU-preconditioners for implicit solution of advection dominated problems. In: Hertzberger B, Sloot P (eds) High-performance computing and networking. Springer, Berlin, pp 421–430
484. Venkatakrishnan V (1993) On the accuracy of limiters and convergence to steady state solutions. AIAA paper 93-0880
485. Venkatakrishnan V (1995) Convergence to steady state solutions of the Euler equations on unstructured grids with limiters. J Comput Phys 118(1):120–130. https://doi.org/10.1006/jcph.1995.1084
486. Vinokur M (1989) An analysis of finite-difference and finite-volume formulations of conservation laws. J Comput Phys 81(1):1–52. https://doi.org/10.1016/0021-9991(89)90063-6
487. van der Vorst HA (1992) Bi-CGSTAB: a fast and smoothly converging variant of Bi-CG for the solution of nonsymmetric linear systems. SIAM J Sci Stat Comput 13(2):631–644. https://doi.org/10.1137/0913035

488. Wang D, Deng X, Wang G, Dong Y (2016) Developing a hybrid flux function suitable for hypersonic flow simulation with high-order methods. Int J Numer Methods Fluids 81:309–327. https://doi.org/10.1002/fld.4186
489. Wang Q, Ren YX, Li W (2016) Compact high order finite volume method on unstructured grids II: extension to two-dimensional Euler equations. J Comput Phys 314(Supplement C):883–908. https://doi.org/10.1016/j.jcp.2016.03.048
490. Wang Q, Ren YX, Pan J, Li W (2017) Compact high order finite volume method on unstructured grids III: variational reconstruction. J Comput Phys 337:1–26. https://doi.org/10.1016/j.jcp.2017.02.031
491. Wang Z (1998) A Quadtree-based adaptive Cartesian/Quad grid flow solver for Navier-Stokes equations. Comput Fluids 27(4):529–549. https://doi.org/10.1016/S0045-7930(97)00070-4
492. Wang Z, Fidkowski K, Abgrall R, Bassi F, Caraeni D, Cary A, Deconinck H, Hartmann R, Hillewaert K, Huynh H, Kroll N, May G, Persson PO, van Leer B, Visbal M (2013) High-order CFD methods: current status and perspective. Int J Numer Methods Fluids 72(8):811–845. https://doi.org/10.1002/fld.3767
493. Weiss JM, Maruszewski JP, Smith WA (1999) Implicit solution of preconditioned Navier-Stokes equations using algebraic multigrid. AIAA J 37:29–36. https://doi.org/10.2514/2.689
494. Wesenberg M (2002) Efficient MHD Riemann solvers for simulations on unstructured triangular grids. J Numer Math 10(2):37–71. https://doi.org/10.1515/JNMA.2002.37
495. White D (1997) Theory of tokamak plasmas. Cambridge University Press, Amsterdam. https://doi.org/10.1016/C2009-0-07544-8
496. Winters AR, Gassner GJ (2016) Affordable, entropy conserving and entropy stable flux functions for the ideal MHD equations. J Comput Phys 304:72–108. https://doi.org/10.1016/j.jcp.2015.09.055
497. Winters AR, Derigs D, Gassner GJ, Walch S (2017) A uniquely defined entropy stable matrix dissipation operator for high mach number ideal MHD and compressible Euler simulations. J Comput Phys 332:274–289. https://doi.org/10.1016/j.jcp.2016.12.006
498. Woodward P, Colella P (1984) The numerical simulation of two-dimensional fluid flow with strong shocks. J Comput Phys 54:115–173. https://doi.org/10.1016/0021-9991(84)90142-6
499. Wu CC (1990) Formation, structure, and stability of MHD intermediate shocks. J Geophys Res 95:8149–8175. https://doi.org/10.1029/JA095iA06p08149
500. Wu H, Shen L, Shen Z (2010) A hybrid numerical method to cure numerical shock instability. Commun Comput Phys
501. Xiong T, Qiu JM, Xu ZF (2016) Parametrized positivity preserving flux limiters for the high order finite difference WENO scheme solving compressible Euler equations. J Sci Comput 67(3):1066–1088. https://doi.org/10.1007/s10915-015-0118-0
502. Xisto CM, Páscoa JC, Oliveira PJ (2013) A pressure-based method with AUSM-type fluxes for MHD flows at arbitrary mach numbers. Int J Numer Methods Fluids pp 1165–1182. https://doi.org/10.1002/fld.3781
503. Yagi M, Seki K, Matsumoto Y (2009) Development of a magnetohydrodynamic simulation code satisfying the solenoidal magnetic field condition. Comput Phys Commun 180(9):1550–1557. https://doi.org/10.1016/j.cpc.2009.04.010
504. Yalim M, Abeele DV, Lani A, Quintino T, Deconinck H (2011) A finite volume implicit time integration method for solving the equations of ideal magnetohydrodynamics for the hyperbolic divergence cleaning approach. J Comput Phys 230(15):6136–6154. https://doi.org/10.1016/j.jcp.2011.04.020
505. Yalim MS (2008) An artificial compressibility analogy approach for compressible ideal MHD. PhD thesis, Université Libre de Bruxelles
506. Yang H, Chen Z, Przekwas A, Dudley J (2015) A high-order CFD method using successive differentiation. J Comput Phys 281:690–707. https://doi.org/10.1016/j.jcp.2014.10.046
507. Yang HQ, Harris RE (2016) Development of vertex-centered high-order schemes and implementation in FUN3D. AIAA J 54:3742–3760. https://doi.org/10.2514/1.J054561
508. Yang HQ, Harris RE, Yang SS (2018) Large eddy simulation of transition flow using high-order vertex-centered U-MUSCL schemes implemented in FUN3D. In: AIAA SciTech forum,

American Institute of Aeronautics and Astronautics, Kissimmee, Florida. https://doi.org/10.2514/6.2018-130310.2514/6.2018-1303

509. Yang LP, Feng XS, Xiang CQ, Liu Y, Zhao XP, Wu ST (2012) Time-dependent MHD modeling of the global solar corona for year 2007: driven by daily-updated magnetic field synoptic data. J Geophys Res 117(A16):A08110
510. Yee HC, Sjögreen B (2006) Efficient low dissipative high order schemes for multiscale MHD flows, II: minimization of $\nabla \cdot B$ numerical error. J Sci Comput 29:115–164. https://doi.org/10.1007/s10915-005-9004-5
511. Yee K (1966) Numerical solution of initial boundary value problems involving Maxwell's equations in isotropic media. IEEE Trans Antennas Propag 14(3):302–307. https://doi.org/10.1109/TAP.1966.1138693
512. Yoon S, Jameson A (1988) Lower-upper symmetric-Gauss-Seidel method for the Euler and Navier-Stokes equations. AIAA J 26:1025. https://doi.org/10.2514/3.10007
513. Yoon SH, Kim C, Kim KH (2008) Multi-dimensional limiting process for three-dimensional flow physics analyses. J Comput Phys 227(12):6001–6043. https://doi.org/10.1016/j.jcp.2008.02.012
514. Yuan KY, Huang YS, Yang HT, Pian T (1994) The inverse mapping and distortion measures for 8-node hexahedral isoparametric elements. Comput Mech 14(2):189–199
515. Zachary AL, Colellaz P (1992) A higher-order Godunov method for the equations of ideal magnetohydrodynamics. J Comput Phys 99(2):341–347. https://doi.org/10.1016/0021-9991(92)90213-I
516. Zachary AL, Malagoli A, Colella P (1994) A higher-order Godunov method for multidimensional ideal magnetohydrodynamics. SIAM J Sci Comput 15(2):263–284
517. Zanotti O, Dumbser M (2016) Efficient conservative ader schemes based on WENO reconstruction and space-time predictor in primitive variables. Comput Astrophys Cosmol 3(1):1. https://doi.org/10.1186/s40668-015-0014-x
518. Zha GC, Shen YQ, Wang BY (2008) Calculation of transonic flows using WENO method with a low diffusion E-CUSP upwind scheme. Aerospace sciences meetings, AIAA. American Institute of Aeronautics and Astronautics, Reno, Nevada, pp 2008–745
519. Zhang F (2017) A vertex-weighted-least-squares gradient reconstruction. ArXiv e-prints arXiv:1702.04518
520. Zhang F, Liu J, Chen B, Zhong W (2016) Evaluation of rotated upwind schemes for contact discontinuity and strong shock. Comput Fluids 134–135:11–22. https://doi.org/10.1016/j.compfluid.2016.05.010
521. Zhang F, Liu J, Chen B, Zhong W (2017) A robust low-dissipation AUSM-family scheme for numerical shock stability on unstructured grids. Int J Numer Methods Fluids 84:135–151. https://doi.org/10.1002/fld.4341
522. Zhang F, Liu J, Chen B (2018) Modified multi-dimensional limiting process with enhanced shock stability on unstructured grids. Comput Fluids 161:171–188. https://doi.org/10.1016/j.compfluid.2017.11.019
523. Zhang M, Feng XS (2015) Implicit dual-time stepping method for a solar wind model in spherical coordinates. Comput Fluids 115:115–123. https://doi.org/10.1016/j.compfluid.2015.03.020
524. Zhang M, Feng XS (2016) A comparative study of divergence cleaning methods of magnetic field in the solar coronal numerical simulation. Front Astron Space Sci 3(6). https://doi.org/10.3389/fspas.2016.00006
525. Zhang S (2005) Invertible Jacobian for hexahedral finite elements. Part 1. Bijectivity. Preprint available at https://www.math.udel.edu/~szhang/research/p/bijective1.ps
526. Zhang YT, Shu CW (2016) Chapter 5 - ENO and WENO schemes. In: Abgrall R, Shu CW (eds) Handbook of numerical methods for hyperbolic problems. Handbook of numerical analysis, vol 17, Elsevier, Amsterdam, pp 103–122. https://doi.org/10.1016/bs.hna.2016.09.009
527. Zhao Z, Zhu J, Chen Y, Qiu J (2019) A new hybrid WENO scheme for hyperbolic conservation laws. Comput Fluids 179:422–436. https://doi.org/10.1016/j.compfluid.2018.10.024

528. Zhou D, Lu ZL, Guo TQ (2018) Improvement of computational efficiency of circular function-based gas kinetic scheme by using Jacobian-free Newton-Krylov method. Comput Fluids 161:121–135. https://doi.org/10.1016/j.compfluid.2017.11.017
529. Zhu J, Shu CW (2018) A new type of multi-resolution WENO schemes with increasingly higher order of accuracy. J Comput Phys 375:659–683. https://doi.org/10.1016/j.jcp.2018.09.003
530. Ziegler U (2008) The NIRVANA code: parallel computational MHD with adaptive mesh refinement. Comput Phys Commun 179(4):227–244. https://doi.org/10.1016/j.cpc.2008.02.017
531. Ziegler U (2011) A semi-discrete central scheme for magnetohydrodynamics on orthogonal-curvilinear grids. J Comput Phys 230(4):1035–1063. https://doi.org/10.1016/j.jcp.2010.10.022
532. Zlatev Z, Dimov I, Faragó I, Georgiev K, Havasi Á (2017) Stability of the Richardson Extrapolation combined with some implicit Runge-Kutta methods. J Comput Appl Math 310:224–240. https://doi.org/10.1016/j.cam.2016.03.018

Chapter 3
A Finite Volume MHD Code in Spherical Coordinates for Background Solar Wind

Abstract A second-order Godunov-type finite volume method (FVM) to advance the equations of single-fluid solar wind plasma magnetohydrodynamics (MHD) in time has been implemented into a numerical code. This code operates on a three-dimensional (3D) spherical shell with both non-staggered and staggered grids on the overlapping grid system with hexahedral cells of quadrilateral frustum type. By merging geometrical factors in spherical coordinates into the reformulation of fluxes, flux evaluation is made easy to achieve, and thus many numerical schemes with the total variation diminishing (TVD) slope limiters and approximate Roe solvers intended for Cartesian case can follow in the present context of spherical grid described here. At the same time, alternative strategies to ensure a solenoidal magnetic field, such as projection Poisson (PP) solver, hyperbolic divergence cleaning (HDC) method derived from generalized Lagrange multiplier (GLM) formulation of MHD system and constrained transport (CT) method, are employed. In this chapter, an FVM is described exemplarily on a six-component composite grid system by using a minmod limiter for oscillation control. Additionally, an implicit dual time-stepping technique is demonstrated to model the steady state solar wind ambient. Being of second order in space and time, this model is written in FORTRAN language with Message Passing Interface (MPI) parallelization, and validated in modeling the large-scale structure of solar wind from the Sun to Earth process (hereafter called Sun-to-Earth Process MHD model, also STEP-MHD model for brief). To demonstrate the suitability of our code for the simulation of solar wind ambient from the Sun to Earth, selected results from Carrington rotations (CR) during different solar activity phases are presented to show its capability of producing structured solar wind in agreement with observations.

3.1 Introduction

The Sun can be taken as a sphere as a first approximation. Thus, the spherical shell, say from the solar surface at one solar radius (R_s) to the Earth at 215 R_s or beyond, is a natural choice of computational domain in solar-terrestrial space for solar wind modeling. The colatitude θ ranges from $\theta = 0$ to $\theta = \pi$ corresponding to the north

© Springer Nature Singapore Pte Ltd. 2020

X. Feng, *Magnetohydrodynamic Modeling of the Solar Corona and Heliosphere*, Atmosphere, Earth, Ocean & Space,
https://doi.org/10.1007/978-981-13-9081-4_3

and south poles, respectively. The azimuth ϕ covers $\phi = 0$ to $\phi = 2\pi$. For a spherical shell simulation domain like solar corona and heliosphere, the spherical grid might be a natural choice instead of Cartesian grid. The Cartesian grid filling a sphere of certain radius R takes $8R^3$ volume to cover the whole sphere, while spherical grid takes $4\pi R^3/3$ volume, which is 48% less than Cartesian grid. By choosing logarithmically spaced radial grid one automatically achieves very high spatial resolution at small radii while the total number of radial grid zones remains fairly small compared to a grid in Cartesian coordinates. For an example, the radial grid spacing [72] is fixed to be 0.01 R_s between 1.01 and 1.1 R_s, proportional to $\ln r$ within 7 R_s and proportional to $\sqrt{r}$ in more distance region. With angular spacing of $\pi/64$ in both colatitudinal and azimuthal directions, it takes $256 \times 64 \times 128$ cells to cover the simulated volume between the inner boundary sphere at 1.01 R_s and the outer boundary sphere at 250 R_s. The total grid number is approximately 2.0×10^6 for 3D spherical shell with the finest resolution to be $0.01R_s$ near the solar surface at radial direction. If one uses a cube of Cartesian uniform grid, one needs a grid of the order of 10^{14} grid zones to achieve the same resolution, whose memory and calculation requirement is beyond current supercomputing power.

A fundamental grid system adopted for solar wind modeling is the latitude-longitude grid under spherical shell geometry, especially suited for numerical schemes for the governing equations in the spherical polar coordinates [47, 71, 113, 133, 163, 179, 181, 183, 195], since its orthogonality is expected from numerical reasons. The grid mesh of the latitude-longitude grid, uniform when it is seen in the computational space of the spherical surface $S = \{(\theta^6, \phi), 0 \leq \theta \leq \pi, 0 \leq \phi \leq 2\pi\}$, is in fact non-uniform when observed in the real space. This is a feature common to the grid mesh in the spherical shell. Even if the numerical schemes are implemented in Cartesian or cylinder coordinates [30, 49, 67, 78, 99, 130, 171–174], the computational domain used in solar wind modeling has to approximate the spherical shell geometry.

Intuitively, the Sun's geometry, obviously spherical shaped, would suggest the use of spherical coordinates (r, θ, ϕ), especially since the radial convergence of lines of constant θ, ϕ can entail the extra advantage of enhanced spatial resolution near the Sun's surface. But, in such spherical computational domain, as it has been long well known, there exist two crucial numerical problems: the coordinate singularity and the grid convergence near the poles. In the latitude-longitude grid the existence of the coordinate singularity arises on the north pole $\theta = 0$ and the south pole $\theta = \pi$, which by using the L'Hospital's rule near the poles lead to three different forms for three local regions of the spherical surface S: $\nabla = \left(\frac{\partial}{\partial r}, \frac{\partial}{r\partial\theta}, \frac{1}{r\sin\theta}\frac{\partial}{\partial\phi}\right)$ for the local region $(\theta, \phi), 0 < \theta < \pi, 0 \leq \phi \leq 2\pi$ without the poles, $\nabla = \left(\frac{\partial}{\partial r}, \frac{\partial}{r\partial\theta}, \frac{1}{r}\frac{\partial^2}{\partial\theta\partial\phi}\right)$ for the north pole and $\nabla = \left(\frac{\partial}{\partial r}, \frac{\partial}{r\partial\theta}, -\frac{1}{r}\frac{\partial^2}{\partial\theta\partial\phi}\right)$ for the south pole. This observation prompts us to devise the numerical scheme for the MHD equations in three parts: low latitude-longitude region without the poles, the north and south poles, which result in coding three different subroutines for numerical computation of discretized equations. This may destroy the hyperbolicity of our hyperbolic system such as MHD equations and uniform difference schemes can not be directly applicable. In addition, this also increases the complexity of the computer programs.

The grid convergence near the poles is another problem in the latitude-longitude grid of spherical coordinates. The grid convergence requires a strict limitation on the time marching step Δt in time integration schemes. As $\theta \to 0$ or π near the poles, the metric factor $1/(r \sin \theta)$ (of the ϕ–component of the spherical gradient operator ∇) becomes extremely huge with a constant grid spacing $\Delta \phi$ in the longitudinal direction. In practice, the grid convergence brings us more difficulties than the coordinate singularity due to both the grid redundancy and the severe time step constraint (especially for large-scale numerical simulations with fine grids) imposed by the Courant–Friedrichs–Levy (CFL) stability condition.

For spherical coordinates, this means that enhancing spatial resolution at small r, however desired for physical reasons, would yield Δt much smaller than what the CFL criterion would require for most parts of the computational domain so that such small time-step seems impractical to be realized in numerical study. While adaptive mesh refinement (AMR) can, in principle, be used with spherical coordinates, its advantage is over-compensated by the fact that the convergence of grid spacing implies unacceptably low CFL numbers.

In the former pioneering works, full spherical grid has been used in solar wind study by developing 3D code for the governing MHD equations in spherical coordinates (r, θ, ϕ). To say a few, MAS model (MHD-Around-a-Sphere model, MAS model) [113, 125, 126, 146–149], COIN-MHD model (Corona interplanetary MHD model, COIN model) [45–47, 162, 163], Usmanov's model [179, 182, 183], Hayashi's model [71–74], and Feng's model [58]. For comparative analyses of other MHD models for solar wind, the reader may refer to the comprehensive survey by Feng et al. [51, 55] and Wu and Dryer [197].

Since it is not possible to construct a single grid to cover the entire 4π surface of a sphere, which is orthogonal, and at the same time, free of coordinate singularity or grid convergence, the decomposition of the spherical surface into subregions, patches or components, enables us to cover each component by a grid system that is individually orthogonal and singularity-free. The decomposition of the computational domain generates internal border or boundary between the components. There are two different approaches to handle the internal boundaries. One is the patched grid method [142] and the other is the overset grid method [23] or composite overlapping grid [169]. In the patched grid approach, the subdomains contact one another without any overlap on their borders. In the overlapping grid method, on the other hand, the subdomains partially overlap one another on their borders.

Such grids have been introduced in numerical MHD simulations. Ivan et al. [81–84] simulated the hyperbolic conservation laws of both the Euler and ideal MHD equations in a parallel block-based adaptive simulation framework of the cubed-sphere mesh (Fig. 3.1), which is proposed by Ronchi et al. [151] by dividing a spherical surface into six non-overlapping identical components (Fig. 3.1b), with the intersection of two sets of angularly equidistant great circles defining a gridded region, and each of the arcs of great circles being associated to either a vertical or horizontal coordinate line (Fig. 3.1a). The Yin-Yang grid with two identi-

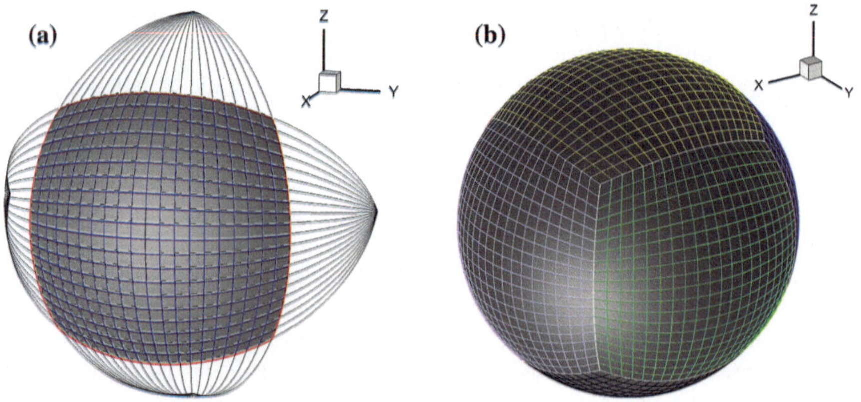

Fig. 3.1 One of the six meshes constructed from two sets of angularly equidistant great circles (**a**), and view of the spherical surface after the procedure outlined in (**a**) has been repeated to produce each of the six component meshes (**b**)

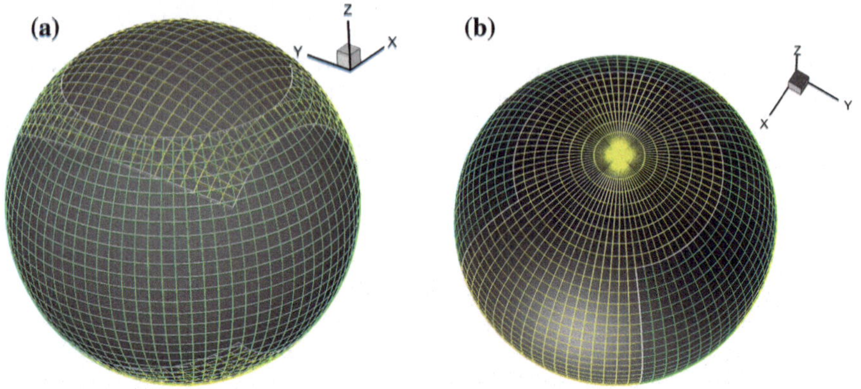

Fig. 3.2 An axis-free Yin-Yang grid configuration plotted on a spherical surface (**a**). Both the Yin (green) and Yang (yellow) grid are the low latitude part of the normal spherical polar grid and are identical in geometry. The Yang grid is obtained from the Yin grid (green marked by white edge) (**b**) by two rotations, and vice versa

cal overlapping grid components (Fig. 3.2), introduced by Kageyama and Sato [91] and Kageyama [90], has also been used by Feng et al. [52, 53], Shiota et al. [164, 165] and Shiota and Kataoka [166] in solar corona and solar wind study. Usmanov [181] and Usmanov et al. [184] suggested using a composite grid consisting of three overlapping spherical grids (Fig. 3.3). The first one is a spherical grid with a limited extension in latitude, and the other two components are designed to cover the polar regions in both hemispheres. The polar grids are fragments of a spherical coordinate system with its polar axis rotated by 90° and located in the equatorial plane of the first grid. The six-component grid (Fig. 3.4) system with six identical components

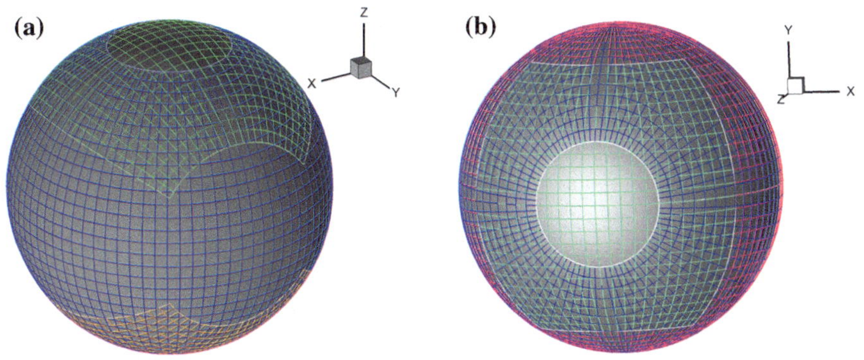

Fig. 3.3 Usmanov's overlapping grid with three components: three Composite spherical grid composed of a "main" spherical grid with excluded polar regions and two polar "patches", which are fragments of a spherical grid rotated by 90° relative to the main grid. One polar patch is shown in (**b**)

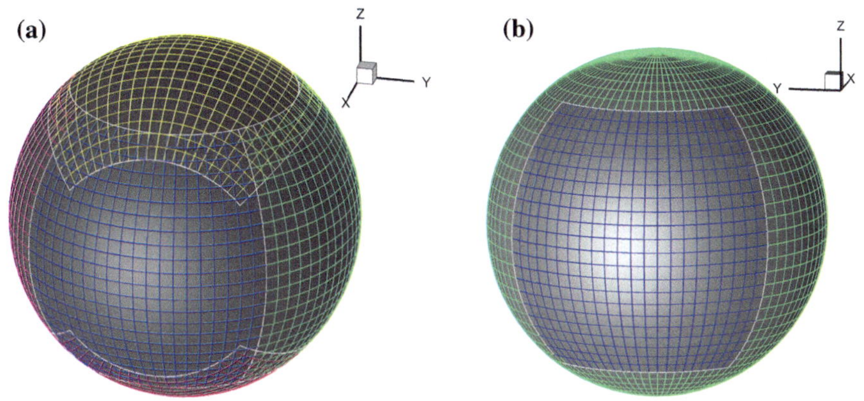

Fig. 3.4 Basic six-component grid: **a** a partition of a sphere into six identical components with partial overlaps and **b** each component grid with a rectangle in the (θ, ϕ) space

enveloping a spherical surface is introduced by Feng et al. [50] to the solar wind MHD modeling.

Although full spherical grid with composite overlapping components has no lateral boundaries, the composite overlapping grids have a disadvantage in common (e.g., [24, 138, 196]). The interpolation between the neighboring grid components does not ensure conservation of conserved quantities even if the finite-volume method employed on each grid component is conservative. Nonconservative and interpolation errors occur when a flow across the internal boundary is present. Although some conservative interpolations can be established [24, 138], the simple non-conservative interpolation like bicubic or Lagrange interpolation has been found to work well in practice for these overlapping composite grid mentioned above [50, 52–54, 57, 58, 75, 184, 193].

In such composite overlapping grid, particularly those with identical components, the transformation of coordinates and vector components between the neighboring components are straightforward and symmetric, thus making easy and straightforward to implement the grid in a 3D code already existing in employing finite volume method. Particularly, these merits apply naturally to a code employing finite difference scheme in spherical polar coordinates. Theoretically, the composite mesh grid with several components is a natural parallel domain decomposition of the whole sphere.

The aim of this chapter is to establish a general finite volume formulation for the solar wind plasma governing equations in spherical coordinates, in the frame work of hexahedral cells in such composite overlapping grid system. First of all, performance and efficiency of overlapping grid are given in Sect. 3.2. Then, the governing equations in spherical coordinates are written out in Sect. 3.3. Thirdly, a general formulation for the realization of finite volume method in these overlapping grids is described in Sect. 3.4. Fourthly, divergence cleaning procedures such as Poisson projection (PP), hyperbolic divergence cleaning (HDC) and constrained transport (CT) methods used in the STEP-MHD code for reducing $\nabla \cdot \mathbf{B}$ are outlined in Sect. 3.5. Fifthly, numerical scheme formulation and time integration of PP, HDC and CT are specified in Sects. 3.6.1 and 3.6.2. Sixthly, grid partition from the Sun to Earth and initial-boundary value setup used in STEP-MHD model are given in Sect. 3.7. Seventhly, numerical results are shown in Sect. 3.8. Finally, concluding remarks are made.

3.2 Performance and Efficiency of Overlapping Grid

The preponderance of overlapping spherical grid is further enhanced by the performance gain due to the increased size of the CFL time step, the fewer computational zones required to cover the full sphere, and thus less computational time. This part discusses the efficiency of these overlapping grids mentioned in the introduction. For the convenience of comparison, the same grid spacing is used.

Suppose each grid component of overlapping grid is spatially discretized into N_r spherical shells, where each shell is discritized by a latitude-longitude grid of N_θ and N_ϕ zones in θ and ϕ directions, respectively. The coordinate center of each grid zone is given by $r_i = r_{i-1} + \Delta r_i$ $(1 \leq i < N_r)$, $\theta_j = \theta_{\min} + j\Delta\theta - \frac{\Delta\theta}{2}$ $(1 < j < N_\theta)$, $\phi_k = \phi_{\min} + k\Delta\phi - \frac{\Delta\phi}{2}$ $(1 < k < N_\phi)$, where $r_0, \theta_{\min}, \phi_{\min}$ are minimum coordinate values in the r, θ, ϕ directions of each grid component and $\Delta\theta = \frac{\theta_{\max}-\theta_{\min}}{N_\theta}$, $\Delta\phi = \frac{\phi_{\max}-\phi_{\min}}{N_\phi}$.

In what follows, zone number and zone size of those grids are compared. Suppose $\Delta\theta = \Delta\phi = \delta$. To cover the whole 4π area , the zone number required by a grid is

$$\sum_{\ell=1}^{N_c}[(\theta_{max}^{(\ell)} - \theta_{min}^{(\ell)})/\delta] \times [(\phi_{max}^{(\ell)} - \phi_{min}^{(\ell)})/\delta] \tag{3.1}$$

where ℓ is the index of grid component and N_c is the total number of components. The conventional spherical grid could be viewed as a 1-component grid, that is, $N_c = 1$, $\theta_{\max}^{(1)} = \pi$, $\theta_{\min}^{(1)} = 0$, $\phi_{\max}^{(1)} = 2\pi$, $\phi_{\min}^{(1)} = 0$ for a conventional spherical grid.

The six component grid proposed by Feng et al. [50, 54, 58] consists of six identical component meshes to envelope a spherical surface with partial overlap on their boundaries. Each component grid is a low-latitude spherical mesh defined by $\theta_{\max}^{(\ell)} = 3\pi/4 + \delta$, $\theta_{\min}^{(\ell)} = \pi/4 - \delta$, $\phi_{\max}^{(\ell)} = 5\pi/4 + \delta$, $\phi_{\min}^{(\ell)} = 3\pi/4 - \delta$ for $\ell = 1, 2, \ldots, 6$.

The Yin-Yang grid, proposed by Kageyama and Sato [91] and Kageyama [90] and used by Feng et al. [52, 53], Shiota et al. [164, 165], and Shiota and Kataoka [166], is composed of two identical component grids combined in a complementary way to cover a spherical surface with partial overlap on their boundaries. Each component grid is a low-latitude spherical mesh with coordinate range of $\theta_{\max}^{(\ell)} = 3\pi/4 + \delta$, $\theta_{\min}^{(\ell)} = \pi/4 - \delta$, $\phi_{\max}^{(\ell)} = 7\pi/4 + \delta$, $\phi_{\min}^{(\ell)} = \pi/4 - \delta$ for $\ell = 1, 2$.

Usmanov [181] proposed a overlapping grid, which consists of a main low latitude component and two similar polar components. The coordinate range of the low latitude component is $\theta_{\max}^{(1)} = 138° + \delta/2$, $\theta_{\min}^{(1)} = 42° - \delta/2$, $\phi_{\max}^{(1)} = 360° + \delta$, $\phi_{\min}^{(1)} = 0° - \delta$. For the two polar components, the polar axis lies in the equatorial plane of the first coordinate system. The ranges of the coordinates are as follows: $\theta_{\max}^{(2)} = 144° + \delta/2$, $\theta_{\min}^{(2)} = 36° - \delta/2$, $\phi_{\max}^{(2)} = 144° + \delta/2$, $\phi_{\min}^{(2)} = 36° - \delta/2$ and $\theta_{\max}^{(3)} = 144° + \delta/2$, $\theta_{\min}^{(3)} = 36° - \delta/2$, $\phi_{\max}^{(3)} = 324° + \delta/2$, $\phi_{\min}^{(3)} = 216° - \delta/2$. Note that Usmanov et al. [184] changed the latitude extension of the main grid as $\theta_{\max}^{(1)} = 157.5° + \delta/2$, and $\theta_{\min}^{(1)} = 22.5° - \delta/2$.

The cubed-sphere [82, 83, 151] is a patched six component grid without overlapping area overset grid that covers a spherical surface with six component grids corresponding to six faces of a cube, in which each arc of great circles is a part of either a vertical or horizontal coordinate line, and the intersection of two sets of angularly equidistant great circles demarks a gridded region. Each component's range, defined by the equiangular coordinates η and ξ, can be confined by $\eta_{\max}^{(\ell)} = \pi/4$, $\eta_{\min}^{(\ell)} = -\pi/4$, $\xi_{\max}^{(\ell)} = \pi/4$, $\xi_{\min}^{(\ell)} = -\pi/4$ for $\ell = 1, 2, \ldots, 6$. If the grid spacing is angularly equidistant, Formula (3.1) is still able to calculate the total zone number if one takes η as θ and ξ as ϕ. Comparing to six component grid, which also cover the whole spherical shell by six grid components with each components taking an area of $\frac{\sqrt{2}}{2}\pi r^2$, each component of cubed-sphere occupies an area of $\frac{2}{3}\pi r^2$, which is 6% less. This is due to the fact that each component of six-component grid can be defined by the same coordinate system, however, components in cubed-sphere are defined differently. If each component of cubed-sphere is measured as a patch in the full sphere of the same spherical coordinate, one will find that the polar coordinates of the top edge lie approximately between 45° and 54.7°, while that of the bottom edge falls roughly between 125.3° and 135°. The azimuthal coordinate of the left and right edge is constant, which is −45° and 45° respectively.

Table 3.1 Zone number reduction

Grid type	0.5° (%)	1° (%)	2° (%)
Six component	23.3	21.6	18.2
Yin-Yang	23.8	22.8	20.5
Usmanov 1996	9.9	9.1	7.6
Usmanov 2012	–12.3	–13.2	–14.8
Cubed-sphere	25	25	25

By using Formula (3.1) to calculate the total zone number of a spherical shell for each overlapping grid, it is found that the use of overlapping grid reduces zone number required to cover the whole spherical region. Table 3.1 lists zone number reduction by each grid in percentage of conventional spherical grid's zone number. From Table 3.1, it is clear that the reduction increases with resolution for overlapping grid, since the number of extra grid for overlapping takes a minor part as resolution increases. Grid reduction of Yin-Yang and six-component grids are almost the same, while that of Usmanov grid achieves less reduction. The reduction of cube sphere does not change with resolution, since there is no overlapping grid in cubed-sphere. Usmanov 2012's grid [184] suffers from negative reduction, which means that it needs more zones than the conventional spherical grid. The gain depends only weakly on angular resolution and is problem independent.

By assuming zone centered variables, then one can define the smallest zone size of the original spherical grid as $d_\phi^{\mathrm{sph}} = r_0\delta\sin(\delta/2)$. For Yin-Yang and six-component grids, $d_\phi^{\mathrm{YY}} = d_\phi^{\mathrm{Six}} = r_0\delta\sin(\pi/4 - \delta/2)$. For Usmanov's grid, $d_\phi^{\mathrm{U}} = r_0\delta\sin(\theta_{\min} + \delta/2)$, where $\theta_{\min} = \min\,(\theta_{\min}^{(1)}, \theta_{\min}^{(2)}, \theta_{\min}^{(3)})$. According to Ronchi et al. [151], the smallest zone size of cube sphere is $d_\phi^{\mathrm{Cube}} = r_0\delta\frac{\sqrt{2}}{2}$. For each overlapping grid, the enlargement of the smallest zone size in azimuthal direction, as opposed to the conventional spherical grid, can be characterized by the ratio $\frac{d_\phi^{\mathrm{overlapping}}}{d_\phi^{\mathrm{sph}}}$ according to Wongwathanarat et al. [193], in which they made some comments on these aspects for the conventional spherical grid and Yin-Yang grid.

Table 3.2 presents the enlargement ratio of these different overlapping grids. In time dependent MHD equation, the time step is restricted by CFL condition $dt = \mathrm{CFL}/\left(\frac{C_r}{d_r} + \frac{C_\theta}{d_\theta} + \frac{C_\phi}{d_\phi}\right)$, where d_r, d_θ and d_ϕ are the minimum grid spacing in radial, polar and azimuthal directions, and C_r, C_θ, C_ϕ are the maximum characteristic speed in each direction. It may be seen from this equation that larger grid spacing would result in larger time step. For simulation of certain physical time, larger time step means fewer computation rounds and less computation time. Although the time step gain factor would not as big as those listed in Table 3.2 since d_r and d_θ are not affected, the employment of overlapping grid is expected to significantly accelerate computation due to grid number reduction and time step gain.

Table 3.2 Enlargement Ratio

Grid type	0.5°	1°	2°
Six component	161	80	39
Yin-Yang	161	80	39
Usmanov 1996	134	67	33
Usmanov 2012	86	42	21
Cubed-sphere	162	81	40

A spherical grid with overlapping structure provides important advantages compared with a Cartesian grid. On one hand, the length of a grid cell in the radial direction r and in the two angular directions perpendicular to it, θ and ϕ, can be modified independently. In such a way, whenever needed, the size of the cell can be subdivided in (r, θ, ϕ) directions so as to make small in the r direction and in the θ and ϕ directions at any spherical surface of any radial distance. Each cell in the spherical grids is connected by their faces with 6 neighboring cells and by the nodes with 26 neighboring cells. Thus, the spherical grid can be considered as an orthogonal mesh associated with spherical coordinates. This means that all cell elements of such an orthogonal lattice (the face areas, the cell volumes, and the edge lengths) have the properties of a spherical cell. On the other hand, boundary conditions near the sphere-shaped Sun become more straightforward to implement in the Sun-to-Earth computational domain of the solar wind modeling. The inner boundary is located at the solar surface with a spherical $r = 1R_s$ and, therefore, grid cells can be exactly either below or above the inner boundary. However, cubic cells near the Sun are partially inside and partially outside of the inner boundary, which makes the implementation of the inner boundary condition not straightforward and may result in numerical artifacts in simulations at low altitudes [50, 99].

Due to these advantages of composite grid systems, they have been employed in solar wind modeling [44, 50, 52–54, 57, 58, 83, 181, 184], with modern numerical schemes such as central weighted essentially non-oscillatory (CWENO), conservative element and solution element (CESE) and the Harten-Lax-van Leer(HLL) family schemes. The following sections will introduce a general semi-discrete central finite volume formulation for the solar wind plasma MHD equations in spherical coordinates, in the frame work of hexahedral cells in such composite overlapping grid system.

3.3 Conservative Form of Solar Wind MHD System

The time-dependent MHD equations of the total solar wind system in the frame of reference corotating with the Sun can be written conservatively in spherical coordinates as follows:

$$\frac{\partial \mathbf{U}}{\partial t} + \frac{1}{r^2}\frac{\partial r^2 \mathbf{F}_r}{\partial r} + \frac{1}{r\sin\theta}\frac{\partial \sin\theta \mathbf{F}_\theta}{\partial \theta} + \frac{1}{r\sin\theta}\frac{\partial \mathbf{F}_\phi}{\partial \phi} = \mathbf{S} \tag{3.2}$$

where

$$\mathbf{U} = \left(\rho, \rho v_r, \rho v_\theta, \rho v_\phi r\sin\theta, e, B_r, B_\theta, B_\phi\right)^T$$

$$\mathbf{F}_r = \left\{\begin{matrix} \rho v_r \\ p + \rho v_r^2 + \frac{-B_r^2 + B_\theta^2 + B_\phi^2}{2\mu} \\ \rho v_r v_\theta - \frac{B_r B_\theta}{\mu} \\ \left(\rho v_r v_\phi - \frac{B_r B_\phi}{\mu}\right) r\sin\theta \\ \rho v_r\left(\frac{\gamma p}{(\gamma-1)\rho} + \frac{1}{2}|\mathbf{v}|^2\right) + \frac{B_\theta}{\mu}(v_r B_\theta - v_\theta B_r) - \frac{B_\phi}{\mu}(v_\phi B_r - v_r B_\phi) \\ 0 \\ v_r B_\theta - v_\theta B_r \\ v_r B_\phi - v_\phi B_r \end{matrix}\right\}$$

$$\mathbf{F}_\theta = \left\{\begin{matrix} \rho v_\theta \\ \rho v_r v_\theta - \frac{B_r B_\theta}{\mu} \\ p + \rho v_\theta^2 + \frac{B_r^2 - B_\theta^2 + B_\phi^2}{2\mu} \\ \left(\rho v_\theta v_\phi - \frac{B_\theta B_\phi}{\mu}\right) r\sin\theta \\ \rho v_\theta\left(\frac{\gamma p}{(\gamma-1)\rho} + \frac{1}{2}|\mathbf{v}|^2\right) - \frac{B_r}{\mu}(v_r B_\theta - v_\theta B_r) + \frac{B_\phi}{\mu}(v_\theta B_\phi - v_\phi B_\theta) \\ v_\theta B_r - v_r B_\theta \\ 0 \\ v_\theta B_\phi - v_\phi B_\theta \end{matrix}\right\}$$

$$\mathbf{F}_\phi = \left\{\begin{matrix} \rho v_\phi \\ \rho v_r v_\phi - \frac{B_r B_\phi}{\mu} \\ \rho v_\theta v_\phi - \frac{B_\theta B_\phi}{\mu} \\ \left(\rho v_\phi^2 + p + \frac{B_r^2 + B_\theta^2 - B_\phi^2}{2\mu}\right) r\sin\theta \\ \rho v_\phi\left(\frac{\gamma p}{(\gamma-1)\rho} + \frac{1}{2}|\mathbf{v}|^2\right) + \frac{B_r}{\mu}(v_\phi B_r - v_r B_\phi) - \frac{B_\theta}{\mu}(v_\theta B_\phi - B_\theta v_\phi) \\ v_\phi B_r - v_r B_\phi \\ v_\phi B_\theta - v_\theta B_\phi \\ 0 \end{matrix}\right\}$$

$$\mathbf{S} = \mathbf{S}_1 + \mathbf{S}_2 + \mathbf{S}_3$$

with

$$\mathbf{S}_1 = \begin{pmatrix} 0 \\ \rho\frac{v_\theta^2+v_\phi^2}{r} + \left(\frac{2p}{r} + \frac{|\mathbf{B}|^2}{r\mu}\right) - \frac{B_\theta^2+B_\phi^2}{r\mu} \\ S_{1,3} \\ 0 \\ 0 \\ 0 \\ \frac{1}{r}(B_\theta v_r - B_r v_\theta) \\ \frac{1}{r}(v_r B_\phi - v_\phi B_r) + (B_\phi v_\theta - B_\theta v_\phi)\frac{\cot\theta}{r} \end{pmatrix}$$

$$\mathbf{S}_2 = \begin{pmatrix} 0 \\ -\rho\frac{GM_s}{r^2} + \rho\Omega\sin\theta(2v_\phi + \Omega r\sin\theta) + S_M \\ \rho\Omega\cos\theta(2v_\phi + \Omega r\sin\theta) \\ -2\rho\Omega(v_\theta\cos\theta + v_r\sin\theta)r\sin\theta \\ \rho v_r(-\frac{GM_s}{r^2} + \Omega^2 r\sin^2\theta) + \rho v_\theta\Omega^2 r\sin\theta\cos\theta + S_E + v_r S_M \\ 0 \\ 0 \\ 0 \end{pmatrix}$$

$$\mathbf{S}_3 = -\begin{pmatrix} 0 \\ \frac{B_r}{\mu} \\ \frac{B_\theta}{\mu} \\ \frac{B_\phi}{\mu} r\sin\theta \\ \frac{B_r v_r + B_\theta v_\theta + B_\phi v_\phi}{\mu} \\ v_r \\ v_\theta \\ v_\phi \end{pmatrix} \nabla\cdot\mathbf{B}$$

$$S_{1,3} = \left(p + \frac{|\mathbf{B}|^2}{2\mu}\right)\frac{\cot\theta}{r} + \frac{1}{r}\left(\frac{B_\theta B_r}{\mu} - \rho v_r v_\theta\right) + \left(\rho v_\phi^2 - \frac{B_\phi^2}{\mu}\right)\frac{\cot\theta}{r}$$

In the above equations, the dependent variables $\rho, \mathbf{v} = (v_r, v_\theta, v_\phi), \mathbf{B} = (B_r, B_\theta, B_\phi)$, and p are the plasma density, the flow velocity in the frame rotating with the Sun, the magnetic field, and the thermal pressure, respectively. M_S is the solar mass, Ω is the solar angular velocity, G the gravitational constant, γ is the polytropic index, t is the time, and r is the heliocentric distance. $e - \frac{p}{\gamma-1} + \rho\mathbf{v}^2/2 + \frac{\mathbf{B}^2}{2\mu}$ stands for the total energy consisting of the kinetic energy density, thermal energy density, and magnetic energy density written in terms of $\mathbf{B}$. $\mu = 4\pi \times 10^{-7}$ H/m is the magnetic permeability, $p = \Re\rho T$ is the thermal energy with T the bulk plasma temperature and $\Re$ the gas constant (1.653×10^{-2} km^2/s^2 K). Here, the flux vector functions $\mathbf{F}_r, \mathbf{F}_\theta, \mathbf{F}_\phi$ are extended by the Lorentz force and with a geometric source term $\mathbf{S} = \mathbf{S}_1 + \mathbf{S}_2 + \mathbf{S}_3$ accounting for the curvilinear structure of spherical geometry, extra terms of Coriolis, centrifugal and gravity forces, and the Powell source term [141].

In the momentum and energy source terms, S_M and S_E are proxy responsible for coronal heating and solar wind acceleration. In order to reflect the magnetic field topology in the heating and acceleration of solar wind [50, 130], following Feng et al. [50] the source terms S_M and S_E are given as follows:

$$S_M = M(\frac{r}{R_s} - 1)\exp(-\frac{r}{L_M}) \tag{3.3}$$

$$S_E = Q_1 \exp(-\frac{r}{L_{Q_1}}) + Q_2(\frac{r}{R_s} - 1)\exp(-\frac{r}{L_{Q_2}}) + \nabla \cdot \left(\kappa T^{\frac{5}{2}} \frac{\nabla T \cdot \mathbf{B}}{\mathbf{B}^2}\mathbf{B}\right) \tag{3.4}$$

where $Q_2 = Q_0 C_a, M = M_0 C_a$, $C_a = C'_a / max(C'_a)$ with $C'_a = \frac{(5.8-1.6e^{[1-(\theta_b/8.5)^3]})^{3.5}}{(1+f_s)^{2/7}}$, M_0, Q_0, and Q_1 are constants to be further specified in Sect. 3.7 later. Here, L_{Q_1}, L_{Q_2}, L_M are taken to be 1 R_s, $f_s = (\frac{R_s}{R_{ss}})^2 \frac{B_{R_s}}{B_{R_{ss}}}$ is the magnetic expansion factor, with B_{R_s} and $B_{R_{ss}}$ being magnetic field strength at the solar surface and at the heliocentric distance $R_{ss} = 2.5\ \text{R}_s$. While, θ_b denotes the minimum angular separation from an open magnetic field foot point to its nearest coronal hole boundary. The factor C'_a has been used [1, 135] to empirically study the solar wind. A Spitzer type thermal conduction has been added in the third term of S_E and $\kappa = 5\kappa_p = 1.6 \times 10^{-12}\ \text{Jm}^{-1}\text{s}^{-1}\text{K}^{-7/2}$ is chosen according to Eirik et al. [41].

Since angular momentum conservation is particularly important in circumstances where redistribution of angular momentum occurs, the azimuthal component of momentum equation for $\rho v_\phi r \sin\theta$ can be put into flux-conservation form, which is discussed in Sect. 3.4.3.

The primitive variables $\rho, \mathbf{v}, p, \mathbf{B}, r, t$ in these equations are normalized by the characteristic values $\rho_s, a_s, \rho_s a_s^2, \sqrt{\mu \rho_s a_s^2}, R_s$, and R_s/a_s, where ρ_s and a_s are the density and acoustic wave speed at the solar surface. The solar rotation is considered in the present study with angular velocity $\Omega = 2\pi/26.5$ radian day^{-1} (normalized by a_s/R_s in simulations).

In the low β plasma regions, magnetic energy is larger than pressure. When the pressure p is derived by subtracting the kinetic energy and magnetic energy from the total e, the discretization errors can be large enough to generate negative pressure. In order to avoid such unacceptable physical situation in the MHD flows, people usually algebraically subtract the potential magnetic field $\mathbf{B}_0$ from the conservative variables (e.g., [50, 66, 130, 174]) such that the total magnetic field $\mathbf{B}$ is split into the potential magnetic field $\mathbf{B}_0$ and its perturbed component $\mathbf{B}_1$ such as $\mathbf{B} = \mathbf{B}_0 + \mathbf{B}_1$. $e_1 = \frac{p}{\gamma-1} + \rho \mathbf{v}^2/2 + \frac{\mathbf{B}_1^2}{2\mu}$ stands for the modified total energy density consisting of the kinetic energy density, thermal energy density, and modified magnetic energy density written in $\mathbf{B}_1$. $\mathbf{U} = \left(\rho, \rho v_r, \rho v_\theta, \rho v_\phi r \sin\theta, e_1, B_{1r}, B_{1\theta}, B_{1\phi}\right)^T$ is sued for the conservative variables. For $\mathbf{B}_0$ is constant in time and force free, many terms about $\mathbf{B}_0$ in the right hand side can vanish.

3.4 Finite Volume Method for the Conservative MHD

This section is devoted to the description of finite volume method for the conservative MHD Eq. (3.2) in the frame work of hexahedral cells in such composite overlapping grid system. First of all, grid system is introduced.

3.4.1 *Geometric Elements of a Hexahedral Cell in Spherical Shell Domain*

Figure 3.5 displays a typical cell geometry in spherical coordinates. The grid consists of the spherical cells and each of them, in turn, consists of the faces (six), the nodes (eight), and the cell center (one). The faces are located on the layers of $r =$ const., $\theta =$ const., and $\phi =$ const. Each cell of a spherical grid is unique: The areas of the faces, the length of edges, and the cell volume are different and depend on the cell position on the grid. The cell walls in the r and θ directions are curved while the cell walls in ϕ direction are straight surfaces as shown in Fig. 3.5.

The spherical shell computational domain can be partitioned into the sliding cells $V_{i,j,k} = [r_{im}, r_{ip}] \times [\theta_{jm}, \theta_{jp}] \times [\phi_{km}, \phi_{kp}]$ as displayed in Fig. 3.5 of spacings $\Delta r(i)$, $\Delta\theta(j)$, $\Delta\phi(k)$ and marked by integer indices i, j, k. The bounding faces of the cell are marked by $im = i - 1/2(jm = j - 1/2, km = k - 1/2)$ and $ip = i + 1/2(jp = j + 1/2, kp = k + 1/2)$, where $i = 1, \ldots, N_r, j = 1, \ldots, N_\theta$, $k = 1, \ldots, N_\phi$. Then, the grid mesh reads as follows: $r_{im} = r_{(i-1)p}, r_{ip} = r_{im} + \Delta r(i), \theta_{jm} = \theta_{(j-1)p}, \theta_{jp} = \theta_{jm} + \Delta\theta(j), \phi_{km} = \phi_{(k-1)p}, \phi_{kp} = \phi_{km} + \Delta\phi(k)$. The

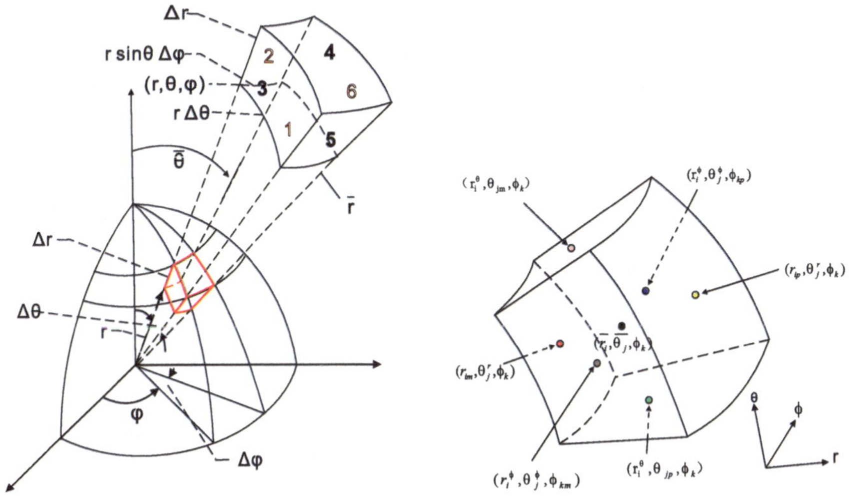

Fig. 3.5 Cell geometry in spherical coordinates

coordinates of the geometrical center of a cell $V_{i,j,k}$ are then given by (r_i, θ_j, ϕ_k) with $r_i = \frac{r_{ip}+r_{im}}{2}$, $\theta_j = \frac{\theta_{jp}+\theta_{jm}}{2}$, and $\phi_k = \frac{\phi_{kp}+\phi_{km}}{2}$. The coordinates of the cell's six geometrical face centers are $(r_{im}, \theta_j, \phi_k)$ and $(r_{ip}, \theta_j, \phi_k)$ on its r constant surfaces (Faces 1 and 4), $(r_i, \theta_{jm}, \phi_k)$ and $(r_i, \theta_{jp}, \phi_k)$ on its θ constant surfaces (Faces 2 and 5), and $(r_i, \theta_j, \phi_{km})$ and $(r_i, \theta_j, \phi_{kp})$ on its ϕ constant surfaces (Faces 3 and 6). The volume of this cell will be denoted by $V_{i,j,k} = \frac{1}{3}(r_{ip}^3 - r_{im}^3)(-\cos\theta_{jp} + \cos\theta_{jm})\Delta\phi(k)$.

The $i + 1/2$ surface area of the cell in r-direction (i.e. an r_{ip} constant surface) is: $S_{r,ip,j,k} = r_{ip}^2(\cos\theta_{jm} - \cos\theta_{jp})(\phi_{kp} - \phi_{km}) = r_{ip}^2(\cos\theta_{jm} - \cos\theta_{jp})\Delta\phi(k)$ and similar for $i - 1/2$ ($S_{r,im,j,k} = r_{im}^2(-\cos\theta_{jp} + \cos\theta_{jm})\Delta\phi(k)$). The $j + 1/2$ surface area of the cell in θ-direction (i.e. a θ_{jp} constant surface) is: $S_{\theta,i,jp,k} = \frac{1}{2}(r_{ip}^2 - r_{im}^2)\sin\theta_{jp}$ $(\phi_{kp} - \phi_{km}) = \frac{1}{2}(r_{ip}^2 - r_{im}^2)\sin\theta_{jp}\Delta\phi(k)$ and similar for $j - 1/2$ ($S_{\theta,i,jm,k} = \frac{1}{2}(r_{ip}^2 - r_{im}^2)\sin\theta_{jm}\Delta\phi(k)$). The $k + 1/2$ surface area of the cell in ϕ-direction (i.e. a ϕ_{kp} constant surface) is: $S_{\phi,i,j,kp} = \frac{1}{2}(r_{ip}^2 - r_{im}^2)(\theta_{jp} - \theta_{jm}) = \frac{1}{2}(r_{ip}^2 - r_{im}^2)\Delta\theta(j)$ and identical for $k - 1/2$ ($S_{\phi,i,j,km} = S_{\phi,i,j,kp}$).

The cell's volume-averaged coordinates are defined by $(\bar{r}_i, \bar{\theta}_j, \phi_k)$ with

$$\bar{r}_i \equiv \frac{1}{V_{i,j,k}}\int_{V_{i,j,k}} r dV = \frac{3(r_{ip}^4 - r_{im}^4)}{4(r_{ip}^3 - r_{im}^3)},$$

$$\bar{\theta}_j \equiv \frac{1}{V_{i,j,k}}\int_{V_{i,j,k}} \theta dV = \frac{\sin\theta_{jp} - \sin\theta_{jm} - \left(\theta_{jp}\cos\theta_{jp} - \theta_{jm}\cos\theta_{jm}\right)}{-\cos\theta_{jp} + \cos\theta_{jm}}$$

The area-averaged r-coordinates in θ- and ϕ-directions are $r_i^\theta = r_i^\phi \equiv \frac{1}{S_{\theta,i,j\pm\frac{1}{2},k}}$ $\int_{S_{\theta,i,j\pm\frac{1}{2},k}} rdS'_\theta = \frac{2(r_{ip}^3 - r_{im}^3)}{3(r_{ip}^2 - r_{im}^2)}$, and the area-averaged θ-coordinates in r- and ϕ-directions are $\theta_j^r \equiv \frac{1}{S_{r,i\pm\frac{1}{2},j,k}}\int_{S_{r,i\pm\frac{1}{2},j,k}} \theta dS'_r = \bar{\theta}_j$ and $\theta_j^\phi = \frac{\theta_{jp}+\theta_{jm}}{2}$. The coordinates of face centers are then $(r_{im}, \theta_j^r, \phi_k)$, $(r_{ip}, \theta_j^r, \phi_k)$, $(r_i^\theta, \theta_{jm}, \phi_k)$, $(r_i^\theta, \theta_{jp}, \phi_k)$, $(r_i^\phi, \theta_j^\phi, \phi_{km})$, $(r_i^\phi, \theta_j^\phi, \phi_{kp})$. It should be noted that the symbols for the cell and its volume, surface and its area are used indistinguishably.

3.4.2 Finite Volume Method

Carrying out the volume integral over the sliding element $V_{i,j,k}$, the set of coupled ordinary differential equations (ODEs) resulting from the application of a finite-volume formulation to Eq. (3.2) for cell (i, j, k) is given by

$$\frac{d\overline{\mathbf{U}}}{dt} = \frac{d}{dt}\frac{1}{V_{i,j,k}}\int_{V_{i,j,k}} \mathbf{U} dV = -\frac{1}{V_{i,j,k}}\sum_{l=1}^{6}(\mathbf{F}\cdot\mathbf{n}S_l) + \overline{\mathbf{S}}_{i,j,k} = \mathcal{R}_{i,j,k}(\overline{\mathbf{U}}, \overline{\mathbf{B}}) \quad (3.5)$$

where $\mathbf{n}$ and S_l are the unit outward normal vector and the area of cell-face l, respectively. The volume average for conservative variables $\mathbf{U}$ is

$$\overline{\mathbf{U}}_{i,j,k} = \frac{1}{V_{i,j,k}} \int_{V_{i,j,k}} \mathbf{U} dV \approx \mathbf{U}(\overline{r}_i, \overline{\theta}_j, \phi_k) = \mathbf{U}_{\overline{i},\overline{j},k}, \overline{\mathbf{S}}_{i,j,k} = \frac{1}{V_{i,j,k}} \int_{V_{i,j,k}} \mathbf{S} dV$$
$$= \mathbf{S}(\overline{\mathbf{U}}_{i,j,k}, \overline{i}, \overline{j}, k)$$

which are approximate to the centroid point value at position $(\overline{r}_i, \overline{\theta}_j, \phi_k)$.

Here,

$$\mathbf{F}\cdot\mathbf{n} = \begin{Bmatrix} -\mathbf{F}_r & \text{on face } 1 \\ \mathbf{F}_r & \text{on face } 4 \\ -\mathbf{F}_\theta & \text{on face } 2 \\ \mathbf{F}_\theta & \text{on face } 5 \\ -\mathbf{F}_\phi & \text{on face } 3 \\ \mathbf{F}_\phi & \text{on face } 6 \end{Bmatrix}, \quad \mathbf{F}\cdot\mathbf{n}S_l = \begin{Bmatrix} -(\mathbf{F}_r S_r)_{im,j,k} & \text{on face } \mathrm{l}=1 \\ (\mathbf{F}_r S_r)_{ip,j,k} & \text{on face } \mathrm{l}=4 \\ -(\mathbf{F}_\theta S_\theta)_{i,jm,k} & \text{on face } \mathrm{l}=2 \\ (\mathbf{F}_\theta S_\theta)_{i,jp,k} & \text{on face } \mathrm{l}=5 \\ -(\mathbf{F}_\phi S_\phi)_{i,j,km} & \text{on face } \mathrm{l}=3 \\ (\mathbf{F}_\phi S_\phi)_{i,j,kp} & \text{on face } \mathrm{l}=6 \end{Bmatrix}$$

The numerical fluxes, $\mathbf{F}\cdot\mathbf{n}$, at the quadrature point of each l face of cell (i, j, k) are determined from the solution of a Riemann problem. Given the left and right interface solution states, $\mathbf{U}_L$ and $\mathbf{U}_R$, an upwind numerical flux is evaluated by solving a Riemann problem in the direction defined by the normal to the face. The initial solution states, $\mathbf{U}_L$ and $\mathbf{U}_R$, are determined by performing the multi-dimensional limited piecewise linear solution reconstruction proposed by Barth and Jespersen [12], in conjunction with slope limiters (e.g., [12, 18, 93, 98, 132, 157, 186]). The least-squares solution of the linear reconstruction, based on a symmetric three-dimensional supporting stencil which generally includes 27 cells in total, is a natural choice with high-order accuracy [81–84, 170]. In particular, one can refer to Mignone [120], where high-order reconstruction schemes for the solution of hyperbolic conservation laws in orthogonal-curvilinear coordinates are formulated in the finite volume approach by employing a piecewise polynomial approximation to the zone-average values to reconstruct left and right interface states from within a computational zone to arbitrary order of accuracy via inverting a Vandermonde-like linear system of equations with spatially varying coefficients.

The numerical procedure above results in a second-order upwind finite-volume scheme. The average value of the source term $\overline{\mathbf{S}}_{i,j,k}$ at $(\overline{r}_i, \overline{\theta}_j, \phi_k)$ is computed with second-order accuracy by estimating the quantity $\mathbf{S}$ based on the appropriate cell-averaged solution variables (i.e., the value of the reconstructed polynomial at the cell centroid). In particular, the average value of $\nabla\cdot\mathbf{B}$ in the Powell source term $\mathbf{S}_3$ of cell (i, j, k) is obtained by the following discretization for $\nabla\cdot\mathbf{B}$:

$$\overline{(\nabla\cdot\mathbf{B})}_{i,j,k} = \frac{1}{V_{i,j,k}} \sum_{l=1}^{6} (\mathbf{B}_l \cdot \mathbf{n}S_l)_{i,j,k}$$

where $\mathbf{B}_l$ is an interface magnetic field computed as the arithmetic mean of the left and right reconstructed values, that is $\mathbf{B}_l = (\mathbf{B}_L + \mathbf{B}_R)/2$.

Usually, the value of the conserved solution vector $\mathbf{U}\,|_{ip,j,k}$ at the cell interface (Face (ip, j, k), for instance) can be evaluated by averaging the left and the right reconstructed solution states, $\mathbf{U}\,|_{ip,j,k} = \frac{1}{2}(\mathbf{U}_L + \mathbf{U}_R)$. However, the evaluation of the gradients for the conserved variables at the cell interface, such as the heat conduct term in S_E, $\nabla\mathbf{U}\,|_{ip,j,k}$ requires some additional work. Such cell-face gradients can be estimated by using the formula proposed by Mathur and Murthy [119]

$$\nabla\mathbf{U}\,|_{ip,j,k} = \frac{(\overline{\mathbf{U}}_{i+1,j,k} - \overline{\mathbf{U}}_{i,j,k})}{\Delta s}\frac{\mathbf{n}}{\mathbf{n}\cdot\mathbf{e}_s} + \left(\overline{\nabla\mathbf{U}}\,|_{ip,j,k} - \overline{\nabla\mathbf{U}}\,|_{ip,j,k}\cdot\mathbf{e}_s\frac{\mathbf{n}}{\mathbf{n}\cdot\mathbf{e}_s}\right) \tag{3.6}$$

where $\overline{\nabla\mathbf{U}}\,|_{ip,j,k}$ is the weighted average of the cell centered gradient at the cell interface given by

$$\overline{\nabla\mathbf{U}}\,|_{ip,j,k} = \alpha(\nabla\mathbf{U})_{i,j,k} + (1-\alpha)(\nabla\mathbf{U})_{i+1,j,k}$$

Δs is the distance between two centroids, unit vectors, $\mathbf{n}$, and $\mathbf{e}_s$ are the face norm, and the distance vector from the cell centroid to its neighbour's as shown in Fig. 3.6. The weighting factor, $\alpha = \frac{V_{i,j,k}}{V_{i+1,j,k}+V_{i,j,k}}$, is defined by cell volume ratios. The solution gradient $(\nabla\mathbf{U})_{i,j,k}$ at the centroid can be reconstructed by using either Green-Gauss or Least-Squares approach, which results in a second order spatial discretization.

For convenience of later use in the CT formulation, the area average for magnetic field $B_r(B_\theta, B_\phi)$ in face $r(\theta, \phi)$ is defined as follows:

Fig. 3.6 3D Cell face gradient reconstruction

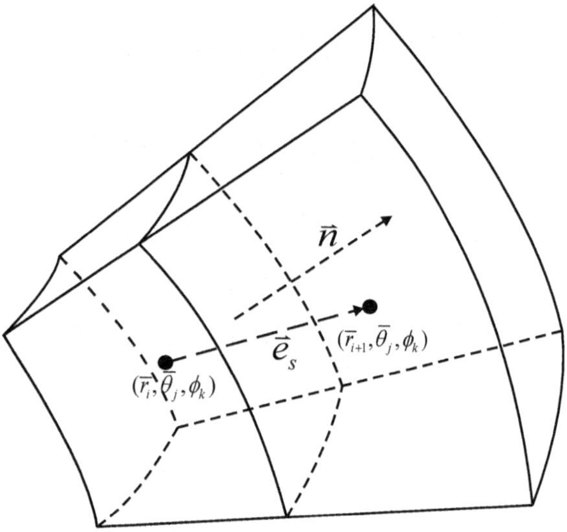

$$
\begin{aligned}
\langle B_r\rangle^r_{im,j,k} &= \frac{1}{S_{r,im,j,k}}\int_{S_{r,im,j,k}} B_r dS' \approx B_r(r_{im},\theta^r_j,\phi_k) = B_{r,im,j^r,k}\\
\langle B_\theta\rangle^\theta_{i,jm,k} &= \frac{1}{S_{\theta,i,jm,k}}\int_{S_{\theta,i,jm,k}} B_\theta dS' \approx B_\theta(r^\theta_i,\theta_{jm},\phi_k) = B_{\theta,i^\theta,jm,k}\\
\langle B_\phi\rangle^\phi_{i,j,km} &= \frac{1}{S_{\phi,i,j,km}}\int_{S_{\phi,i,j,km}} B_\phi dS' \approx B_\phi(r^\phi_i,\theta^\phi_j,\phi_{km}) = B_{\phi,i^\phi,j^\phi,km}
\end{aligned}
\tag{3.7}
$$

Here, the area average for B_r (B_θ, B_ϕ) in face r (θ, ϕ) is approximate to point value at position $(r_{im},\theta^r_j,\phi_k)$ $\left((r^\theta_i,\theta_{jm},\phi_k),(r^\phi_i,\theta^\phi_j,\phi_{km})\right)$.

3.4.3 Some Specifics in Polar Coordinates

As it is known, compared to Cartesian coordinate system, the conservation laws of vector quantities in polar coordinates yield extra terms in the equations as if an additional force is present. As it can be seen, the usual conservative form for ρv_ϕ in the ϕ momentum equation [47, 112, 182] reads:

$$
\begin{aligned}
&\frac{\partial \rho v_\phi}{\partial t} + \frac{1}{r^2}\frac{\partial r^2\left(\rho v_r v_\phi - B_r B_\phi/\mu\right)}{\partial r} + \frac{1}{r\sin\theta}\frac{\partial \sin\theta\left(\rho v_\theta v_\phi - B_\theta B_\phi/\mu\right)}{\partial\theta}\\
&+\frac{1}{r\sin\theta}\frac{\partial\left(\rho v_\phi^2 + p + (B_r^2 + B_\theta^2 - B_\phi^2)/(2\mu)\right)}{\partial\phi} = S_{1,4} - 2\rho\Omega(v_\theta\cos\theta + v_r\sin\theta) - \frac{B_\phi}{\mu}\nabla\cdot\mathbf{B}
\end{aligned}
\tag{3.8}
$$

where the extra terms $S_{1,4} = \frac{\cot\theta}{r}\left(\frac{B_\theta B_\phi}{\mu} - \rho v_\theta v_\phi\right) - \frac{1}{r}\left(\rho v_r v_\phi - \frac{B_r B_\phi}{\mu}\right)$ come from geometrical curvature in spherical coordinates. Here, reducing or avoiding such source terms in the spatial discretizations, one can replace the ϕ momentum equation with the following angular momentum equation [57, 100]. The ϕ component of the momentum equation, when written out in an angular momentum conserving form, becomes

$$
\begin{aligned}
&\frac{\partial \rho v_\phi r\sin\theta}{\partial t} + \frac{1}{r^2}\frac{\partial r^2\left(\rho v_r v_\phi - B_r B_\phi/\mu\right) r\sin\theta}{\partial r} + \frac{1}{r\sin\theta}\frac{\partial \sin\theta\left(\rho v_\theta v_\phi - B_\theta B_\phi/\mu\right) r\sin\theta}{\partial\theta}\\
&\quad+\frac{1}{r\sin\theta}\frac{\partial\left(\rho v_\phi^2 + p + (B_r^2 + B_\theta^2 - B_\phi^2)/(2\mu)\right) r\sin\theta}{\partial\phi}\\
&\quad= -2\rho\Omega(v_\theta\cos\theta + v_r\sin\theta) r\sin\theta - \frac{B_\phi}{\mu} r\sin\theta\nabla\cdot\mathbf{B} \equiv S_{2,4} + S_{3,4}
\end{aligned}
\tag{3.9}
$$

In this way, the extra terms $S_{1,4}$ in Eq. (3.8) have been completely incorporated into the momentum flux difference in Eq. (3.9).

In addition, the extra source terms in $\mathbf{S}_1$ for ρv_r, ρv_θ, B_r and B_ϕ equations should be also treated carefully. In order to clearly take a look at how these terms in polar coordinates can be formulated, introduce Δ_r, Δ_θ, and Δ_ϕ, as well as S_r, S_θ and S_ϕ such that Eq. (3.5) for polar coordinates can be written as:

$$V_{i,j,k}\frac{\partial \overline{\mathbf{U}}_{i,j,k}}{\partial t} = -(\Delta_r \mathbf{F}_r S_r)_{i,j,k} - (\Delta_\theta \mathbf{F}_\theta S_\theta)_{i,j,k} - (\Delta_\phi \mathbf{F}_\phi S_\phi)_{i,j,k} + \overline{\mathbf{S}}_{i,j,k} V_{i,j,k} \tag{3.10}$$

Here $(\Delta_r \mathbf{F}_r S_r)_{i,j,k} = \mathbf{F}_{r,ip,j,k} S_{r,ip,j,k} - \mathbf{F}_{r,im,j,k} S_{r,im,j,k}$ denotes the difference of the argument $\mathbf{F}_r S_r$ between ip and im interfaces, the same for Δ_θ and Δ_ϕ. Here $S_{r,ip,j,k}$, $S_{\theta,i,jp,k}$, $S_{\phi,i,j,kp}$ denote ip, jp, kp surfaces of cell (i, j, k) and $V_{i,j,k}$ is the volume of cell (i, j, k).

Notice that the terms like $\frac{w}{r}$ and $\frac{\cot\theta}{r}w$ in $\mathbf{S}_1$ for the r-direction momentum equation, the θ-direction momentum equation and B_θ, B_ϕ equations are generated when $\frac{\partial w}{\partial r}$ and $\frac{1}{r}\frac{\partial w}{\partial \theta}$ are converted into $\frac{1}{r^2}\frac{\partial r^2 w}{\partial r}$ and $\frac{1}{r\sin\theta}\frac{\partial \sin\theta w}{\partial \theta}$, respectively. w stands for a scalar function consisting of primitive variables ρ, v_r, v_θ, v_ϕ, B_r, B_θ, v_ϕ and p. In order to preserve discrete consistency, one should compute such source terms following the flux differences in r and θ directions by replacing $\frac{w}{r}V_{i,j,k}$ with $\frac{1}{2}w\Delta_r S_r = \frac{1}{2}(S_{r,ip,j,k} - S_{r,im,j,k})w_{i,j,k}$ and $\frac{\cot\theta}{r}wV_{i,j,k}$ with $w\Delta_\theta S_\theta = (S_{\theta,i,jp,k} - S_{\theta,i,jm,k})w_{i,j,k}$. That is, the source terms $\left(\frac{2p}{r} + \frac{|\mathbf{B}|^2}{r\mu}\right)$ and $\left(p + \frac{|\mathbf{B}|^2}{2\mu}\right)\frac{\cot\theta}{r}$ in $S_{1,2}$ and $S_{1,3}$ can be replaced with $\frac{1}{V_{i,j,k}}\overline{\left(p + \frac{|\mathbf{B}|^2}{2\mu}\right)}\Delta_r S_r$ and $\frac{1}{V_{i,j,k}}\overline{\left(p + \frac{|\mathbf{B}|^2}{2\mu}\right)}\Delta_\theta S_\theta$. In this way, $w(r, \theta, \phi) =$ constant case, although seldom occurs, can be identified in a uniform way in both flux differencing and the source term discretization. Otherwise, one can easily find that the case $p(r, \theta, \phi) =$ constant and $(v_r, v_\theta, v_\phi) = \mathbf{0}$ can not be recognized by the solver as a static solution to the pure Euler equations, that is, Eq. (3.2) without consideration of magnetic field, solar rotation, solar gravity and heating source terms. The reason why this is the case is because the term $2\frac{p}{r}$ arises due to the differences in the surface areas of the cell interfaces at $i - \frac{1}{2}$ and $i + \frac{1}{2}$ that are used in the $\Delta_r(\mathbf{F}_r S_r)$ term. So by compensating this with exactly the same but opposite one gets an exact cancelation, and thereby a constant pressure will be recognized as such.

Besides the consideration above, in practice one can also discretize the source terms $S_{1,2}$, $S_{1,3}$, $S_{1,7}$, $S_{1,8}$ as follows:

$$V_{i,j,k}\overline{S}_{1,2} = \int_{V_{i,j,k}} \frac{1}{r}\left[\rho(v_\theta^2 + v_\phi^2) + 2p + B_r^2/\mu\right] dV$$

$$= \left[\frac{1}{2}\overline{\left(\rho\left(v_\theta^2 + v_\phi^2\right) + \frac{B_r^2}{\mu}\right)}_{i,j,k} + \overline{p}_{i,j,k}\right]\Delta_r S_r,$$

$$V_{i,j,k}\overline{S}_{1,3} = \int_{V_{i,j,k}} \left(p + \rho v_\phi^2 + \frac{B_r^2 + B_\theta^2 - B_\phi^2}{2\mu}\right)\frac{\cot\theta}{r} dV + \int_{V_{i,j,k}} \frac{1}{r}\left(\frac{B_\theta B_r}{\mu} - \rho v_r v_\theta\right) dV$$

$$= \overline{\left(p + \rho v_\phi^2 + \frac{B_r^2 + B_\theta^2 - B_\phi^2}{2\mu}\right)}_{i,j,k} \Delta_\theta S_\theta + \frac{1}{2}\overline{\left(\frac{B_\theta B_r}{\mu} - \rho v_r v_\theta\right)}_{i,j,k} \Delta_r S_r$$

$$V_{i,j,k}\overline{S}_{1,7} = \int_{V_{i,j,k}} \frac{1}{r}\left(B_\theta v_r - B_r v_\theta\right) dV = \frac{1}{2}\overline{(B_\theta v_r - B_r v_\theta)}_{i,j,k} \Delta_r S_r$$

$$V_{i,j,k}\overline{S}_{1,8} = \int_{V_{i,j,k}} \frac{1}{r}\left(v_r B_\phi - v_\phi B_r\right) dV + \int_{V_{i,j,k}} \left(B_\phi v_\theta - B_\theta v_\phi\right) \frac{\cot\theta}{r} dV$$

$$= \frac{1}{2}\overline{\left(v_r B_\phi - v_\phi B_r\right)}_{i,j,k} \Delta_r S_r + \overline{\left(B_\phi v_\theta - B_\theta v_\phi\right)}_{i,j,k} \Delta_\theta S_\theta$$

From the flux reformulation involved in Eq. (3.10), it is easy to know that the Roe linearized [141], HLL family [68, 70, 76, 97, 102, 111, 123, 124, 127, 128, 219] approximate Riemann solvers, and a monotonic upstream scheme for conservation laws (MUSCL) [71, 130, 174] can be theoretically used in the computational studies. It is essentially true for us to use any of the numerical schemes developed for general conservation laws, with some multi-dimensional reconstruction procedure. On the overlapping area between two components, solution values at the ghost cell centers of one component grid are determined by the standard Lagrange interpolation from its neighboring component [50, 57, 58]. Note that this seems to reasonably work on any composite overlapping component grids mentioned in the introduction. At the end of this chapter, an appendix is devoted to the formulation of Roe and HLL schemes in some details.

In the solar wind MHD simulation, since $\nabla \cdot \mathbf{B}$ error is always a menace, reducing or avoiding it is a positive thing to do in the spatial discretizations. In what follows, three divergence-free cleaning procedures are given out as supplementary to the numerical discretization.

3.5 Divergence-Free Cleaning Procedure

In order to maintain a divergence-free magnetic field or at least reduce the impact of the errors on the numerical solution, divergence-cleaning methods and divergence-free methods have been proposed for simulating compressible MHD flows.

The divergence-cleaning methods correct the magnetic field for numerically ensuring $\nabla \cdot \mathbf{B} = 0$ or propagate any nonzero divergence out of the computational domain. The projection scheme solves an additional Poisson equation after each time step to correct errors of $\nabla \cdot \mathbf{B}$ and ensure $\nabla \cdot \mathbf{B} = 0$ numerically, and thus can be combined with any numerical scheme including high-order upwind-biased schemes. The projection scheme has been originally proposed by Brackbill [178], and later used widely [7, 32, 50, 99, 152, 178, 208]. The projection scheme is accurate if accuracy in Poisson solver is sufficient, but expensive, especially combined with AMR.

According to the concept of propagating any nonzero divergence out of the computational domain, 8-wave scheme [37, 86, 141] is proposed to modify MHD equations for non-zero divergence to be advected. The resultant non-conservative MHD equations with the source terms $\nabla \cdot \mathbf{B}(0, \mathbf{B}, \mathbf{v} \cdot \mathbf{B}, \mathbf{v})^T$ can be solved by any numerical method, and the robustness and accuracy are improved although the solution does not satisfy the discrete divergence-free magnetic field. Since the 8-wave formulation for MHD equations allow for the numerical errors of $\nabla \cdot \mathbf{B}$ to convect with the flow, the errors do not accumulate at a fixed location. However, $\nabla \cdot \mathbf{B}$ may not be very small. Although this simple, fast and robust approach tries to minimize the solenoidal errors, the use of a non-conservative form of the governing equations does not satisfy discrete conservation of momentum, total energy, and magnetic field, thus possibly producing incorrect shock jump conditions and speeds [175]. With the generalized Lagrange multiplier (GLM) formulation for MHD equations, by adding a scalar field coupled to $\nabla \cdot \mathbf{B}$ to advect and diffuse $\nabla \cdot \mathbf{B}$ away, Dedner et al. [35] proposed the hyperbolic/parabolic divergence cleaning method that transports the divergence errors to the boundaries of computational domain and also damps the errors at the same time [35, 48, 69, 82, 88, 99, 121, 122, 170, 176, 187, 202, 206].

In fully conservative form, the hyperbolic/parabolic cleaning method is simple and fast, and more effective than the 8-wave formulation. A limited case of the hyperbolic/parabolic divergence cleaning method leads to the diffusive control term $\nabla(\nu \nabla \cdot \mathbf{B})$ [52, 54, 58, 94, 118, 185] to deal with the divergence of the magnetic field [35, 118, 121], where usually $\nu = \frac{C}{\Delta t}(\frac{1}{\Delta x^2} + \frac{1}{\Delta y^2} + \frac{1}{\Delta z^2})^{-1}$, with Δx, Δy, and Δz being grid spacings in Cartesian coordinates [52, 54, 58, 94] with C a constant. In order to simulate compressible MHD turbulent flows with shock waves accurately, Kawai [92] proposed an artificial magnetic resistivity method by developing a divergence-free numerical method with the following three requirements simultaneously: (1) high-order accuracy, non-dissipative, and low dispersion error in smooth regions to resolve turbulence accurately, (2) the capability of robustly capturing numerical discontinuities, and (3) the ability of inherently maintaining the discrete form of a divergence-free magnetic field numerically. All these methods are easy to be applied to AMR grids (e.g., [52, 54, 58, 88, 94, 185]).

The divergence-free methods inherently maintain the discrete form of a divergence-free magnetic field numerically. The CT methods [40, 43] rely on a staggered representation of the magnetic field combined with a particular finite differencing discretization, and numerically maintain $\nabla \cdot \mathbf{B}^{n+1} = \nabla \cdot \mathbf{B}^n$ (where the superscript n represents the time step). Therefore, the CT schemes maintain a particular discrete form of the magnetic field divergence around the machine round-off error as long as the initial and boundary conditions are compatible with the constraints [87]. Furthermore, the staggered-grid-based approach has been combined [3, 10, 33, 34, 154] with conventional low-order (i.e., second-order or lower) Godunov-type finite-volume methods for the simulations of compressible MHD flows with shock waves. Tóth [175] proposed central difference scheme for induction equation to conserve $\nabla \cdot \mathbf{B}$ by combining co-located-grid-based CT methods with the conventional Godunov-type schemes. The divergence-free second-order Godunov-type directionally-unsplit schemes based on the staggered-grid formulation

[63, 64, 101], and the divergence-free genuinely multidimensional Riemann solvers [5, 6] have been suggested. Central finite-volume third- and fourth-order schemes on overlapping grid for both the volume-averaged variables and the face averaged magnetic field have been proposed by Li [103, 104], where the divergence-free condition of the magnetic field is preserved by the fourth-order divergence-free reconstruction and the constrained transport method. Ziegler [218–220] described a Godunov type central scheme for the equations of magnetohydrodynamics on orthogonal-curvilinear grids within a finite-volume framework by an adaptive mesh refinement implementation, with the divergence-free condition of the magnetic field as a built-in property of the scheme by virtue of a constrained-transport ansatz for the induction equation. The staggered-grid-based approaches are also extended to higher-order (i.e., third-order or higher) WENO [4, 9, 11] and ENO [117] type reconstruction procedures. Thus, the divergence-free methods are also promising approaches for simulating MHD flows with the conventional second-order or lower Godunov-type schemes or high-order-accurate divergence-free ENO and WENO schemes. Skinner and Ostriker [167] described the implementation of the Athena MHD code for cylindrical coordinates of the higher-order spatial reconstruction and characteristic evolution steps as well as the finite-volume and CT updates.

For comparative study of divergence-cleaning techniques and divergence-free methods, the reader may refer to related works [2, 129, 175, 211]. There seems no religious guide about what kind of divergence-controller is used, which depends on the problem investigated [58].

To maintain the $\nabla \cdot \mathbf{B} = 0$ constraint, the projection scheme, and hyperbolic divergence cleaning approach and the CT method are employed in STEP-MHD code, respectively.

3.5.1 Poisson Projection (PP) Method

In the PP scheme, the magnetic field $\mathbf{B}^*$ provided by the base scheme in the new time step $n+1$ is projected to a divergence free $\mathbf{B}^{n+1}$ field by the following procedure. With the help of solution Φ to a Poisson equation

$$\nabla^2 \Phi = \nabla \cdot \mathbf{B}^* \tag{3.11}$$

$$\mathbf{B}^{n+1} = \mathbf{B}^* - \nabla \Phi \tag{3.12}$$

so as to force the numerical divergence of $\mathbf{B}^{n+1} = \mathbf{B}^* - \nabla \Phi$ exactly zero. Obviously, this correction does not affect the current density $J = \nabla \times \mathbf{B}^{n+1} = \nabla \times \mathbf{B}^*$. The numerical divergence of $\mathbf{B}^{n+1}$ can be exactly zero if the $\nabla^2 \Phi$ in Eq. (3.11) is evaluated as a divergence of the gradient with the same difference operators as used for calculating $\nabla \cdot \mathbf{B}^*$. For such purpose, the evaluation of $\nabla \cdot \mathbf{B}^*$ at the volume center reads as follows

$$\overline{(\nabla \cdot \mathbf{B}^*)}_{i,j,k} = \frac{1}{V_{i,j,k}} \sum_{l=1}^{6} \left(\mathbf{B}_l^* \cdot \mathbf{n}S_l\right)_{i,j,k}$$

and that of scalar function Φ's gradient $\nabla\Phi$ follows from Green-Gauss theorem

$$\overline{(\nabla\Phi)}_{i,j,k} = \frac{1}{V_{i,j,k}} \sum_{l=1}^{6} \Phi_l \mathbf{n}S_l$$

In this context, for $\nabla \cdot \nabla\Phi$ one has

$$\overline{(\nabla \cdot \nabla\Phi)}_{i,j,k} = \frac{1}{V_{i,j,k}} \sum_{l=1}^{6} ((\nabla\Phi)_l \cdot \mathbf{n}S_l)_{i,j,k}$$

where the gradient value $(\nabla\Phi)_l$ at the interface can be estimated from formula (3.6).

In order to solve Eq. (3.11), a pseudo-time derivative is introduced to the equation [71]

$$\frac{\partial\Phi}{\partial\tau} = \nabla \cdot (\nabla\Phi) - \nabla \cdot \mathbf{B}^*$$

which can be solved by adopting a first-order backward finite difference scheme for the pseudo-time derivative with the pseudo-time step $\Delta\tau$ ($<\Delta t$). If one wants to obtain an accurate transient solution, the pseudo-time (sub-iterations) must get converged at each physical time step. But this is too costly to make the sub-iteration procedure performed until convergence to machine precision. Besides setting up convergence criterion $\frac{|\Phi^{(m+1)}-\Phi^{(m)}|}{\Delta\tau} \leq 10^{-6}$ of the pseudo-time (sub-iterations), maximal user-defined sub-iterations are also set up to avoid infinite iterations. Here, instead of solving fully, our code calculates several or 10 iteration steps of the relaxation equation.

Since the total energy e is an independent unknown in our method, after the implementation of the projection, the only change may be the magnetic energy, which results in the variation or even the negative occurrence of the thermal energy or specifically the temperature. After the projection step, in order to avoid numerical difficulty with the possible occurrence of negative thermal pressure, the total energy in our simulation is adjusted as

$$e^{n+1} = e^* + \left(\left|\mathbf{B}^{n+1}\right|^2 - \left|\mathbf{B}^*\right|^2\right)/(2\mu)$$

for the sake of numerical robustness, while the thermal energy and temperature obtained by the base scheme are kept unchanged. The same procedure has been stated [10, 50] to preserve thermal energy at the expense of energy conservation.

3.5.2 Hyperbolic Divergence Cleaning (HDC) Approach

In the HDC approach, an additional scalar field ψ is introduced to the MHD system by coupling the divergence constraint to Faraday's law according to

$$\begin{aligned} &\frac{\partial \mathbf{B}}{\partial t} - \nabla \times (\mathbf{v} \times \mathbf{B}) + \nabla \psi = 0, \\ &\frac{\partial \psi}{\partial t} + c_h^2 \nabla \cdot \mathbf{B} = -\frac{c_h^2}{c_p^2} \psi \end{aligned} \tag{3.13}$$

where c_h is the (constant) speed at which divergence errors are propagated while c_p is a constant controlling the rate at which monopoles are damped. The remaining equations in Eq. (3.2) are not changed and the conservative character is not lost. Now, Eq. (3.2) contain nine equations associated with variables $\mathbf{U} = \left(\rho, \rho v_r, \rho v_\theta, \rho v_\phi r \sin\theta, e, B_r, B_\theta, B_\phi, \psi\right)^T$, where the fluxes for magnetic field have following forms $F_{r,6} = \psi$, $F_{\theta,7} = \psi$, $F_{\phi,8} = \psi$, $S_{1,6} = \frac{2\psi}{r}$, $S_{1,7} = \frac{1}{r}(B_\theta v_r - B_r v_\theta) + \frac{\psi}{r}\cot\theta$, $S_{1,9} = -\frac{c_h^2}{c_p^2}\psi$. The evaluation for the source term $S_{1,6}$ follows that described in Sect. 3.4.3. The c_h is often chosen to be the largest of all MHD eigenvalues in the whole domain $c_h = \max_{i,j,k}\left(|v_r| + c_{fr}, |v_\theta| + c_{f\theta}, |v_\phi| + c_{f\phi}\right)$, or equivalently $c_h = \frac{\mathrm{CFL}}{\Delta t}\Delta h$, with $\Delta h = \min_{i,j,k}\left(\Delta r(i), r(i)\Delta\theta(j), r(i)\sin\theta(j)\Delta\phi(k)\right)$ and the CFL Courant factor. As for the choice of c_p, we set $\alpha = \frac{c_h}{c_p^2}\Delta h$ and $\alpha = 0.1$ in STEP-MHD code. This corresponds to $c_p = \sqrt{-\frac{\Delta t}{\ln c_d} c_h^2}$ with $c_d = \exp(-\alpha \mathrm{CFL})$. This kind of choice is used in a similar way [35, 69, 88, 121, 122, 170, 187, 202], where extensive numerical testing has shown that divergence errors are minimized when the parameter α or c_d lies in the range (0, 1). Initially, ψ is set to 0. At the inner boundary and outer boundary, ψ is fixed. The HDC method Eq. (3.13) is first derived from the GLM formulation of MHD system by Dedner et al. [35].

3.5.3 Constrained Transport (CT) Method

In order to use the CT method, the governing MHD Equations (3.2) are divided into two coupled subsystems of equations: a fluid part and a magnetic induction part. The fluid part for the vector $\mathbf{V} = \left(\rho, \rho v_r, \rho v_\theta, \rho v_\phi r \sin\theta, e\right)^T$ can be derived from the first five arguments's equations for $\mathbf{U}$ in Eq. (3.2)

$$\frac{\partial \mathbf{V}}{\partial t} + \frac{1}{r^2}\frac{\partial r^2 \mathbf{G}_r}{\partial r} + \frac{1}{r\sin\theta}\frac{\partial \sin\theta \mathbf{G}_\theta}{\partial \theta} + \frac{1}{r\sin\theta}\frac{\partial \mathbf{G}_\phi}{\partial \phi} = \mathbf{Q} \tag{3.14}$$

where $\mathbf{G}_r, \mathbf{G}_\theta, \mathbf{G}_\phi$, and $\mathbf{Q}$ are the first five arguments in $\mathbf{F}_r, \mathbf{F}_\theta, \mathbf{F}_\phi$, and $\mathbf{S}_1 + \mathbf{S}_2$.

The magnetic equations expressed in the form of Faraday's law read as follows:

$$
\begin{aligned}
&\frac{\partial B_r}{\partial t}+\frac{1}{r\sin\theta}\frac{\partial}{\partial\theta}(\sin\theta E_\phi)-\frac{1}{r\sin\theta}\frac{\partial}{\partial\phi}E_\theta=0\\
&\frac{\partial B_\theta}{\partial t}-\frac{1}{r}\frac{\partial}{\partial r}(rE_\phi)+\frac{1}{r\sin\theta}\frac{\partial}{\partial\phi}E_r=0\\
&\frac{\partial B_\phi}{\partial t}+\frac{1}{r}\frac{\partial}{\partial r}(rE_\theta)-\frac{1}{r}\frac{\partial}{\partial\theta}E_r=0\\
&E_r=(v_\phi B_\theta-v_\theta B_\phi),\, E_\phi=(v_\theta B_r-v_r B_\theta),\, E_\theta=(v_r B_\phi-v_\phi B_r)
\end{aligned}
\tag{3.15}
$$

The first subsystem (3.14) is the extended Euler system for the vector $\mathbf{V}=(\rho,\rho v_r,\rho v_\theta,r\sin\theta\rho v_\phi,e)^T$ of primary hydrodynamical variables with the corresponding flux vector functions $\mathbf{G}_r,\mathbf{G}_\theta,\mathbf{G}_\phi$ extended by the Lorentz force and with a geometric source term $\mathbf{Q}$ accounting for the curvilinear structure of spherical geometry, and source terms of Coriolis, centrifugal and gravity forces. We use the finite-volume discretization of Eq. (3.14), and a finite-area integration of Faraday's Eq. (3.15) over the surfaces rather than the volume of the cell. In the appendix, the discretization formulations for Eqs. (3.14) and (3.15) are presented separately.

The benefit for solving the equations of Faraday's law above without expanding the derivatives is that it enables us to preserve $\frac{\partial}{\partial t}(\nabla\cdot\mathbf{B})=0$ to round-off error, in contrast to truncation error. Usually, for finite-difference calculations on even the smallest meshes possible, the truncation errors of an algorithm (due to the inaccuracy of the scheme) are typically around 10^{-6} per step), which are generally tens 8–10 orders of magnitude greater than round-off errors caused by finite machine precision (for 64-bit computations). When the variations of $\mathbf{B}$ are expressed by using an equation in the form of Faraday's law, then the perturbation to $\mathbf{B}$, $\delta\mathbf{B}$, calculated at each time step, is equal to the curl of the electric field. Thus, one is able to formulate the divergence and curl of operator such that the analytic vector identity $\nabla\cdot\nabla\times=0$ is maintained to round-off error. Then $\nabla\cdot\delta\mathbf{B}=0$ to round-off error, hence $\frac{\partial}{\partial t}(\nabla\cdot\mathbf{B})=0$ to round-off error. This trick has been used for the MHD system in spherical coordinates (e.g., [57, 112, 219]).

3.5.4 Positivity Consideration

It is well known that the pressure may numerically become negative due to numerical errors caused by strong magnetic field. Although the use of magnetic field decomposition and $e_1=\frac{p}{\gamma-1}+\rho\mathbf{v}^2/2+\frac{\mathbf{B}_1^2}{2\mu}$ may mitigate such occurrence of negative pressure to some extent, for further insuring positivity the entropy equation

$$
\begin{aligned}
\frac{\partial\rho s}{\partial t}+\nabla\cdot(\rho s\mathbf{v})&=\frac{\partial\rho s}{\partial t}+\frac{1}{r^2}\frac{\partial(r^2\rho s v_r)}{\partial r}+\frac{1}{r\sin\theta}\frac{\partial(\sin\theta\rho s v_\theta)}{\partial\theta}+\frac{1}{r\sin\theta}\frac{\partial(\rho s v_\phi)}{\partial\phi}\\
&=-(\gamma-1)S_E/\rho^{\gamma-1}
\end{aligned}
$$

is optionally used instead of the total energy equation in low β and smooth regions without shocks [10, 57]. Here, the entropy density s of the ideal gas is in particular given by $s = \frac{p}{\rho^\gamma}$. For detailed discussions, refer to Feng et al. [57].

To avoid the occurrence of negative pressure, researchers sometimes designed switches to exclude low kinematic pressure, and the regions where strong magnetosonic shocks or mildly compressive motions (also strong compressive motions) may take place, such that entropy or pressure equation is optionally used instead of the energy equation in low β and smooth regions without shocks [10, 58]. Here, in case of the offending cells at risk of negative pressure, instead of solving the total energy equation one can use the internal energy $\epsilon = \frac{p}{\gamma-1}$ equation [10, 17, 50, 57, 58, 106, 153]

$$\frac{\partial \epsilon}{\partial t} + \nabla \cdot (\epsilon \mathbf{v}) = -p\nabla \cdot \mathbf{v} + S_E \tag{3.16}$$

in those unsafe regions such as $\beta < 0.01$ or $\epsilon/e < 0.05$ [50, 57]. Here, the integration of gradient term $p\nabla \cdot \mathbf{v}$ in the right-hand side is $\overline{(p\nabla \cdot \mathbf{v})}_{i,j,k} = \frac{1}{V_{i,j,k}} \int_{V_{i,j,k}} p\left(\frac{1}{r^2}\frac{\partial (r^2 v_r)}{\partial r} + \frac{1}{r\sin\theta}\frac{\partial (\sin\theta v_\theta)}{\partial \theta} + \frac{1}{r\sin\theta}\frac{\partial v_\phi}{\partial \phi}\right) dV = \frac{\overline{p}_{i,j,k}}{V_{i,j,k}}\left(\Delta_r(v_r S_r) + \Delta_\theta(v_\theta S_\theta) + \Delta_\phi(v_\phi S_\phi)\right)$.

It is remarked that besides the positivity-preserving switch method Eq. (3.16), some other positivity-preserving schemes for hyperbolic conservation laws are available. The last part of the appendix presents the description of the parametrized maximum-principle-preserving (MPP) flux-corrected limiting procedure implemented in STEP-MHD model after a brief introduction of the existing positivity-preserving methods.

In what follows, for the sake of convenience, the discretizations for Eq. (3.2) combined with the corresponding divergence constraint methods of Poisson projection (PP), hyperbolic divergence cleaning (HDC), and the constrained transport (CT) methods will be referred to as PP, HDC and CT hereafter. Also, these abbreviations stand for the associated governing equations if no misunderstanding occurs.

As stated above, our STEP code accommodates other numerical schemes such as Roe and HLL types described in the appendix. However, the following two sections briefly describe the numerical scheme formulation and time integration for PP, HDC and CT schemes by following an example of the FV discretization of Ziegler [219], and of course present the numerical results obtained by this HLL solver in the subsequent sections.

3.6 Numerical Scheme Formulation and Time Integration

This section presents the numerical scheme formulation and time integration with Poisson projection, hyperbolic divergence cleaning, and the constrained transport methods. For details about linear reconstruction, please refer to the appendix.

3.6.1 Scheme Formula with Poisson Projection and Hyperbolic Divergence Cleaning Methods

For spatial discretization of PP and HDC, a semi-discrete Godunov-type central scheme is used by strictly following those of Ziegler [219] with the FV discretization, where a second order accurate linear ansatz reconstruction is adopted, and in the linear reconstruction the derivative terms at volume-averaged coordinates are approximated by using minmod limiter for oscillation control. After knowing all the discretized forms of fluxes and source terms, in a cell with geometrical center (i, j, k), the semi-discrete for PP and HDC follows from Eqs. (3.5) or (3.10):

$$\frac{d\overline{\mathbf{U}}}{dt} = -\frac{3\left(r_{ip}^2\mathbf{F}^n_{r,ip,j,k} - r_{im}^2\mathbf{F}^n_{r,im,j,k}\right)}{\left(r_{ip}^3 - r_{im}^3\right)} - \frac{\sin\theta_{jp}\mathbf{F}^n_{\theta,i,jp,k} - \sin\theta_{jm}\mathbf{F}^n_{\theta,i,jm,k}}{r_i^{\theta}\left(\cos\theta_{jm} - \cos\theta_{jp}\right)} - \frac{\Delta\theta(j)}{r_i^{\phi}\left(\cos\theta_{jm} - \cos\theta_{jp}\right)}\frac{\mathbf{F}^n_{\phi,i,j,kp} - \mathbf{F}^n_{\phi,i,j,km}}{\Delta\phi(k)} + \overline{\mathbf{S}}_{i,j,k} \tag{3.17}$$

In a cell with geometrical center (i, j, k), $\overline{\mathbf{U}} = \mathbf{U}(\overline{r}_i, \overline{\theta}_j, \phi_k)$ denotes volume average for conservative variables $\mathbf{U}$. For PP and HDC, $\mathbf{U} = \left(\rho, \rho v_r, \rho v_\theta, \rho v_\phi r\sin\theta, e, B_r, B_\theta, B_\phi\right)^T$ and $\mathbf{U} = \left(\rho, \rho v_r, \rho v_\theta, \rho v_\phi r\sin\theta, e, B_r, B_\theta, B_\phi, \psi\right)^T$, respectively. The full system of ODEs (3.17) is integrated with a second-order Runge–Kutta scheme. Denoting the righthand side of Eq. (3.17) as $\mathcal{R}[\overline{\mathbf{U}}]$, then

$$\overline{\mathbf{U}}^{*} = \overline{\mathbf{U}}^{n} + \Delta t\mathcal{R}\left[\overline{\mathbf{U}}^{n}\right], \overline{\mathbf{U}}^{n+1} = \frac{1}{2}\overline{\mathbf{U}}^{n} + \frac{1}{2}\left(\overline{\mathbf{U}}^{*} + \Delta t\mathcal{R}[\overline{\mathbf{U}}^{*}]\right) \tag{3.18}$$

where time step Δt is constrained by the Courant–Friedrichs–Lewy stability condition:

$$\Delta t = \mathrm{CFL}/\max\left(\sqrt{\left(\frac{|v_r| + c_{fr}}{\Delta r}\right)^2 + \left(\frac{|v_\theta| + c_{f\theta}}{r\Delta\theta}\right)^2 + \left(\frac{|v_\phi| + c_{f\phi}}{r\sin\theta\Delta\phi}\right)^2}\right) \tag{3.19}$$

Here, CFL, v_r, v_θ, and v_ϕ are the Courant factor, the flow velocities in radial, colatitude and azimuthal directions, and $c_s = \sqrt{\frac{\gamma p}{\rho}}$, $c_A = \sqrt{\frac{B_r^2+B_\theta^2+B_\phi^2}{\mu\rho}}$ are the local sound and Alfvénic speeds. Here, c_{fr}, $c_{f\theta}$, and $c_{f\phi}$ are the local fast magnetosonic speeds in (r, θ, ϕ) directions, defined respectively by $c_{fr} = \frac{1}{\sqrt{2}}\sqrt{c_s^2 + c_A^2 + \left(\left(c_s^2 + c_A^2\right)^2 - 4c_s^2\frac{B_r^2}{\mu\rho}\right)^{\frac{1}{2}}}$, $c_{f\theta} = \frac{1}{\sqrt{2}}\sqrt{c_s^2 + c_A^2 + \left(\left(c_s^2 + c_A^2\right)^2 - 4c_s^2\frac{B_\theta^2}{\mu\rho}\right)^{\frac{1}{2}}}$, and $c_{f\phi} = \frac{1}{\sqrt{2}}\sqrt{c_s^2 + c_A^2 + \left(\left(c_s^2 + c_A^2\right)^2 - 4c_s^2\frac{B_\phi^2}{\mu\rho}\right)^{\frac{1}{2}}}$.

In the following runs of this chapter, CFL $= 0.5$ is employed. It is also obvious that when $\theta = 0, \pi$, the denominator in the right-hand side of Eq. (3.19) becomes potentially large which restricts the time step to be very small. This is one of the disadvantages to use the spherical coordinates instead of the Cartesian grid. Composite grid system can avoid polar singularities and convergence, thus, to some extent, enlarging the time step.

3.6.2 Scheme Formula with Constrained Transport Method

The CT implementation strictly follows that of Ziegler [219] by using the FV discretization for Eq. (3.14) with the same reconstruction as done for Eq. (3.17), and by averaging Eqs. (3.15) over facial areas to get semi-integral form of magnetic induction equations. For consistency, the reconstruction of second order accuracy for magnetic field is employed at cell face centers by using minmod limiter. The semi-discrete for the fluid part equation (3.14), in a cell with geometrical center (i, j, k), is constructed by following Eq. (3.23) of Ziegler [219]:

$$\frac{d\overline{\mathbf{V}}}{dt} = -\frac{3\left(r_{ip}^2\mathbf{G}^n_{r,ip,j,k} - r_{im}^2\mathbf{G}^n_{r,im,j,k}\right)}{\left(r_{ip}^3 - r_{im}^3\right)} - \frac{\sin\theta_{jp}\mathbf{G}^n_{\theta,i,jp,k} - \sin\theta_{jm}\mathbf{G}^n_{\theta,i,jm,k}}{r_i^\theta\left(\cos\theta_{jm} - \cos\theta_{jp}\right)} - \frac{\Delta\theta(j)}{r_i^\phi\left(\cos\theta_{jm} - \cos\theta_{jp}\right)}\frac{\mathbf{G}^n_{\phi,i,j,kp} - \mathbf{G}^n_{\phi,i,j,km}}{\Delta\phi(k)} + \overline{\mathbf{Q}}_{i,j,k} \tag{3.20}$$

Following Eq. (3.24) in Ziegler [219], by averaging Eqs. (3.15) over facial areas of a cell with geometrical center (i, j, k), one can obtain the semi-discrete form as follows

$$\begin{aligned}
\frac{dB_{r,im,j^r,k}}{dt} &= -\frac{\sin\theta_{jp}E^*_{\phi,im,jp,k} - \sin\theta_{jm}E^*_{\phi,im,jm,k}}{r_{im}\left(\cos\theta_{jm} - \cos\theta_{jp}\right)} \\
&\quad + \frac{\Delta\theta(j)}{r_{im}\left(\cos\theta_{jm} - \cos\theta_{jp}\right)}\frac{E^*_{\theta,im,j,kp} - E^*_{\theta,im,j,km}}{\Delta\phi(k)} \\
\frac{dB_{\theta,i^\theta,jm,k}}{dt} &= -\frac{2}{\sin\theta_{jm}\left(r_{im} + r_{ip}\right)}\frac{E^*_{r,i,jm,kp} - E^*_{r,i,jm,km}}{\Delta\phi(k)} \\
&\quad + \frac{2\left(r_{ip}E^*_{\phi,ip,jm,k} - r_{im}E^*_{\phi,im,jm,k}\right)}{r_{ip}^2 - r_{im}^2} \\
\frac{dB_{\phi,i^\phi,j^\phi,km}}{dt} &= -\frac{2\left(r_{ip}E^*_{\theta,ip,j,km} - r_{im}E^*_{\theta,im,j,km}\right)}{r_{ip}^2 - r_{im}^2} + \frac{2}{r_{im} + r_{ip}}\frac{E^*_{r,i,jp,km} - E^*_{r,i,jm,km}}{\Delta\theta(j)}
\end{aligned} \tag{3.21}$$

For concrete expressions of formula in Eqs. (3.20) and (3.21), refer to the appendix. Having discussed the numerical algorithm for calculating the spatial discretizations at a given cell (i, j, k), time-integration of Eqs. (3.20) and (3.21) can follow easily from that of Eq. (3.18) [58].

However, dual-time method [210] is used for time integration of Eqs. (3.20)–(3.21). After moving the discretized fluxes to the right hand sides of the governing Eqs. (3.20)–(3.21), and denoting them with their corresponding source terms as $\mathcal{R}_{\mathbf{V}}\left[\overline{\mathbf{V}},\widehat{\mathbf{B}}\right]$ and $\mathcal{R}_{\mathbf{B}}[\overline{\mathbf{V}},\widehat{\mathbf{B}}]$ respectively, the full MHD equations can be advanced implicitly

$$\overline{\mathbf{V}}^{*}=\overline{\mathbf{V}}^{n}+\delta t\mathcal{R}_{\mathbf{V}}\left[\overline{\mathbf{V}}^{*},\widehat{\mathbf{B}}^{n}\right],\quad \widehat{\mathbf{B}}^{*}=\widehat{\mathbf{B}}^{n}+\delta t\mathcal{R}_{\mathbf{B}}\left[\overline{\mathbf{V}}^{n},\widehat{\mathbf{B}}^{*}\right] \tag{3.22}$$

$$\overline{\mathbf{V}}^{**}=\overline{\mathbf{V}}^{*}+\delta t\mathcal{R}_{\mathbf{V}}\left[\overline{\mathbf{V}}^{**},\widehat{\mathbf{B}}^{*}\right],\quad \widehat{\mathbf{B}}^{**}=\widehat{\mathbf{B}}^{*}+\delta t\mathcal{R}_{\mathbf{B}}\left[\overline{\mathbf{V}}^{*},\widehat{\mathbf{B}}^{**}\right] \tag{3.23}$$

$$\overline{\mathbf{V}}^{n+1}=\frac{1}{2}\overline{\mathbf{V}}^{n}+\frac{1}{2}\overline{\mathbf{V}}^{**},\quad \widehat{\mathbf{B}}^{n+1}=\frac{1}{2}\widehat{\mathbf{B}}^{n}+\frac{1}{2}\widehat{\mathbf{B}}^{**}$$

In a cell with geometrical center (i,j,k), $\overline{\mathbf{V}}=\mathbf{V}(\overline{r}_i,\overline{\theta}_j,\phi_k)$ denotes volume average for conservative variables $\mathbf{V}$, and $\widehat{\mathbf{B}}=\left(B_r(r_{im},\theta_j^r,\phi_k),B_\theta(r_i^\theta,\theta_{jm},\phi_k),B_\phi(r_i^\phi,\theta_j^\phi,\phi_{km})\right)$ denotes area average for magnetic field with B_r (B_θ, B_ϕ) in face r (θ,ϕ) given by Eq. (3.7). The spatial reconstruction for the magnetic field is based on the staggered magnetic field components instead of cell-centered magnetic field. Thus, magnetic field is only updated at the face center. The face-valued magnetic fields $\widehat{\mathbf{B}}$ stored on the faces are only averaged to the zone center value $\overline{\mathbf{B}}_{i,j,k}$ when required. In this formulation, δt will be the physical time step with an integer times that of explicit time step Δt constrained by the usual CFL stability criterion.

To solve the equations accurately in time, a dual-time algorithm [15, 210] can be used to modify the implicit form Eqs. (3.22)–(3.23) by introducing a pseudo-time derivative $\frac{\partial}{\partial\tau}$. We adopt a first-order backward finite difference scheme for the pseudo-time derivative, take Eqs. (3.22) for example, then the new discretized equations become

$$\frac{\Delta\overline{\mathbf{V}}_{i,j,k}^{*,m}}{\Delta\tau}+\frac{\overline{\mathbf{V}}_{i,j,k}^{*,m+1}-\overline{\mathbf{V}}_{i,j,k}^{n}}{\delta t}=\mathcal{R}_{\mathbf{V}}^{m+1} \tag{3.24}$$

$$\frac{\Delta\widehat{\mathbf{B}}_{i,j,k}^{*,m}}{\Delta\tau}+\frac{\widehat{\mathbf{B}}_{i,j,k}^{*,m+1}-\widehat{\mathbf{B}}_{i,j,k}^{n}}{\delta t}=\mathcal{R}_{\mathbf{B}}^{m+1} \tag{3.25}$$

where $\Delta\overline{\mathbf{V}}_{i,j,k}^{*,m}=\overline{\mathbf{V}}_{i,j,k}^{*,m+1}-\overline{\mathbf{V}}_{i,j,k}^{*,m}$, $\Delta\widehat{\mathbf{B}}_{i,j,k}^{*,m}=\widehat{\mathbf{B}}_{i,j,k}^{*,m+1}-\widehat{\mathbf{B}}_{i,j,k}^{*,m}$, $\Delta\tau$ is the pseudo-time step size, and n is the physical time level with m being the pseudo-time level (the number of sub-iterations). The algorithm essentially consists of performing sub-iterations for every physical time step.

Evidently, Eqs. (3.24)–(3.25) approaches the exact nonlinear representation of the desired implicit discretization in physical time, when the derivative, $\frac{\partial(\overline{\mathbf{V}},\widehat{\mathbf{B}})}{\partial\tau}=0$. It is noted that the presence of the discretized physical time derivative represents a sink (rather than a source) in the pseudotime formulation, such that it acts to accelerate the approach to steady conditions in pseudotime. The pseudotime iteration may be expected to converge more and more rapidly as the magnitude of the physical time derivative increases. The dual-time evolution to the final nonlinear value ensures that any truncation errors or factorization terms used in the implicit time step are evaluated nonlinearly, thereby precluding linearization or factorization errors. This nonlinear solution procedure enables larger physical time steps, especially in highly stretched regions, thereby largely offsetting the costs of the iteration at each time level [96].

Using a Taylor expansion for the right hand side of Eqs. (3.24)–(3.25), one can get

$$\left[\left(\frac{1}{\Delta\tau}+\frac{1}{\delta t}\right)I-\left(\frac{\partial\mathcal{R}_{\mathbf{V}}}{\partial\overline{\mathbf{V}}}\right)^{m}\right]\Delta\overline{\mathbf{V}}^{*,m}_{i,j,k}=\mathcal{R}^{m}_{\mathbf{V}}-\frac{\overline{\mathbf{V}}^{*,m}_{i,j,k}-\overline{\mathbf{V}}^{n}_{i,j,k}}{\delta t} \tag{3.26}$$

$$\left[\left(\frac{1}{\Delta\tau}+\frac{1}{\delta t}\right)I-\left(\frac{\partial\mathcal{R}_{\mathbf{B}}}{\partial\widehat{\mathbf{B}}}\right)^{m}\right]\Delta\widehat{\mathbf{B}}^{*,m}_{i,j,k}=\mathcal{R}^{m}_{\mathbf{B}}-\frac{\widehat{\mathbf{B}}^{*,m}_{i,j,k}-\widehat{\mathbf{B}}^{n}_{i,j,k}}{\delta t} \tag{3.27}$$

where $\Delta\overline{\mathbf{V}}^{*,m}_{i,j,k}=\overline{\mathbf{V}}^{*,m+1}_{i,j,k}-\overline{\mathbf{V}}^{*,m}_{i,j,k}$, $\Delta\widehat{\mathbf{B}}^{*,m}_{i,j,k}=\widehat{\mathbf{B}}^{*,m+1}_{i,j,k}-\widehat{\mathbf{B}}^{*,m}_{i,j,k}$. Initially, $\overline{\mathbf{V}}^{*,0}_{i,j,k}=\overline{\mathbf{V}}^{n}_{i,j,k}$, $\widehat{\mathbf{B}}^{*,0}_{i,j,k}=\widehat{\mathbf{B}}^{n}_{i,j,k}$. As pointed out by Jameson [85], the simulation scheme will become unstable if pseudo-time step size $\Delta\tau$ is larger than the physical time step size δt. Here, the pseudo-time step size $\Delta\tau$ is constrained by $\Delta\tau=\min\left(\frac{\mathrm{CFL_{sub}}\,V_{i,j,k}}{\lambda^A+\lambda^B+\lambda^C},\frac{2\delta t}{3}\right)$, where $\mathrm{CFL_{sub}}=0.3$ is the CFL number for pseudo-time step, $\lambda^A=|v_r|+c_{fr}(\lambda^B=|v_\theta|+c_{f\theta},\lambda^C=|v_\phi|+c_{f\phi})$ is the maximum speed in $r(\theta,\phi)$ direction respectively, defined after Eq. (3.19).

Equations (3.26)–(3.27) are solved using the unfactored line Gauss-Seidel iteration [21, 207]. As pointed out by Yuan [207], this method can reach very large pseudo-time step since no factorization error is introduced. If we want to obtain an accurate transient solution, the pseudo-time (sub-iterations) must get converged at each physical time step. But, there is no well-established theory existing on the number of sub-iterations or residual convergence to obtain a specified physical-time accuracy while minimizing the computational cost per time step. Also, this is too costly to make the sub-iteration procedure performed until convergence to machine precision [77]. In the study of steady state solar corona for CR 2048, Zhang and Feng [210] investigated the influence of different CFL conditions on the radial speed profiles along a heliocentric distance. In the dual time stepping scheme, they found that the slight oscillation with different CFL can be introduced by the history of the inflow through the solar surface, which will cause a slight oscillation of the radial speed within the inner corona around the equator without changing the general structure of the problem. Also, the sub-iterations in dual time stepping scheme can cause some oscillation. Increasing the sub-iterations will decrease the oscillation, but add computational cost. Similarly, near large gradient regions, increasing the pseudo-time maximum iteration and refining grid can better enhance the resolution of the numerical solution.

Here, besides setting up convergence criterion $\left|\frac{\left(\Delta\overline{\mathbf{V}}^{*,m}_{i,j,k},\Delta\widehat{\mathbf{B}}^{*,m}_{i,j,k}\right)}{\Delta\tau}\right|\le 10^{-6}$, maximal sub-iterations 15 are also prescribed to avoid infinite iterations, as similarly done by Zhang and Feng [210], which is implemented in different parallel computational domain along the radial direction to ensure the time accuracy and at the same time improve the computational efficiency. The physical solution at new time level is updated: $\left(\overline{\mathbf{V}}^{*}_{i,j,k}, \widehat{\mathbf{B}}^{*}_{i,j,k}\right) = \left(\overline{\mathbf{V}}^{*,m+1}_{i,j,k}, \widehat{\mathbf{B}}^{*,m+1}_{i,j,k}\right)$.

It should be mentioned that this dual-time algorithm is also applicable to the numerical schemes with Poisson projection and hyperbolic divergence cleaning methods provided by Eq. (3.17) although it is described for the numerical scheme with the constrained transport method in Eqs. (3.20) and (3.21).

3.7 Grid Partition from the Sun to Earth and Initial-Boundary Value Setup

From the above arguments, it is evident that the aforementioned formulation holds for the overlapping grid with hexahedral cells. From Sect. 3.2, there exist eight weak singularities at the corners of the six identical sectors of the cubed-sphere grid [151] although it is far more uniform than a polar grid in terms of shape and size of the cells. Except negative reduction of grid number compared with a full spherical grid, the three components of the overlapping spherical meshes [181] are not identical, which is less convenient to directly achieve efficient load balancing in paralleling components. Relatively speaking, the overlapping grids such as the Yin-Yang and six-component overset grid with identical grid components, own the same metric on each component and the component grids can be transformed into each other by coordinate transformation, and thus the parallelization can be naturally and efficiently implemented in both the radial direction and components.

This chapter presents the model in the framework of six-component grids with hexahedral cells to envelope a spherical surface with partial overlap on their boundaries (Fig. 3.4). That is, each component grid of six-component grids is a low-latitude spherical mesh confined by the same region $(\frac{\pi}{4} - \delta \le \theta \le \frac{3\pi}{4} + \delta) \cap (\frac{3\pi}{4} - \delta \le \phi \le \frac{5\pi}{4} + \delta)$ but in different spherical coordinates [44, 50, 53, 54, 57]. Here, δ is proportionally dependent on the grid spacing entailed for the minimum overlapping area and is taken to be $2\Delta\theta$.

In the solar wind simulation of spherical-shell computational domain, composite grids with identical components seem to be a better choice in parallel implementation with the consideration of load balance. Here, the parallel implementation in the whole computational domain from 1 R_s to 230 R_s is realized by domain decomposition of six-component grids and radial direction partition. Due to the radial expansion, the solar wind plasma MHD governing equations are made to be discretized on highly non-uniform grids in order to resolve steep gradients in the solution. That is, in order to mitigate this discrete or geometrical stiffness caused by disparate mesh cell widths, the following grid partitions for each component are used: For 1–25 R_s, $N_\theta = N_\phi =$

42, $\Delta r(i) = 0.01\,R_s$ if $r(i) < 1.1\,R_s$; $\Delta r(i) = \min(A \times \log_{10}(r(i-1)), \Delta\theta \times r(i-1))$ with $A = 0.01/\log_{10}(1.09)$ if $r(i) < 3.5\,R_s$; $\Delta r(i) = \Delta\theta \times r(i-1)$ if $r(i) > 3.5\,R_s$. For 25–75 R_s, $N_\theta = N_\phi = 64$ and $\Delta r(i) = \Delta\theta \times r(i-1)$. For 75–166 R_s, $N_\theta = N_\phi = 96$ and $\Delta r(i) = \Delta\theta \times r(i-1)$. For 166–230 R_s, $N_\theta = N_\phi = 130$ and $\Delta r(i) = \Delta\theta \times r(i-1)$. To conform to the notations defined in Sect. 3.4.1, $r_{1m} = 1R_S$, $\theta_{1m} = \frac{\pi}{4}$, and $\phi_{1m} = \frac{3\pi}{4}$. This kind of grid partitions has been similarly done by Feng et al. [50, 57].

The line-of-sight photospheric magnetic data from the Wilcox Solar Observatory (WSO) is used with the potential field model to produce an initial 3D global magnetic field in the computational domain, while the initial distributions of plasma density ρ, pressure p, and the velocity $\mathbf{v}$ are obtained by Parker's solar wind flow [80, 137] with the temperature $T_s = 1.3 \times 10^6$ K and number density $\rho_s = 1.5 \times 10^8\ \mathrm{cm}^{-3}$ on the solar surface.

Here, the inner boundary at 1 R_s is fixed for simplicity. Since the outer boundary in interplanetary space is supersonic/super-Alfvénic region, the solar wind parameters at the outer boundary are obtained by simple extrapolation. The horizontal boundary or internal border values in (θ, ϕ) directions of each component grid are determined by interpolation from the neighbor stencils lying in its neighboring component grid, with the second-order Lagrange polynomial interpolation being derived by the corresponding boundary values in the stencils [50, 57, 58]. Evidently, the horizontal boundary or internal border values in (θ, ϕ) directions of each component grid can be simply obtained from the reconstructed values by the newly updated ones in the neighboring components.

From the Sun to Earth, the plasma density, the Alfvén velocity, interplanetary magnetic field, the plasma $\beta = \frac{2\mu p}{|\mathbf{B}|^2}$, and spatial grid size vary over orders of magnitude. This admits a large variation of the CFL stability limit from the corona to interplanetary space. Thus, time accuracy stability limitations on the time step in fine grid parts of the computational domain decrease the performance of numerical methods and it may become much more difficult to achieve efficient load balancing. The formulation presented in this chapter uses multitime stepping proposed by Feng et al. [50] by taking different time steps in different grid parts: 1–3.5 R_s, 3.5–10 R_s, 10–25 R_s, 25–75 R_s, 75–166 R_s, 166–230 R_s. This multitime stepping method somehow avoids the necessity of taking a single time step in the entire computational domain determined by the numerical CFL stability conditions (such as on the smallest grid cells near the Sun), to some extent can reduce the CFL number disparity that often produces excessive numerical dissipation and save computation time.

The heating and momentum addition coefficients M_0, Q_0, Q_1 in Eqs. (3.3) and (3.4) for PP and HDC models are given as $3.5 \times 10^{-13}\ \mathrm{Nm}^{-3}$, $1.18 \times 10^{-7}\ \mathrm{Jm}^{-3}\mathrm{s}^{-1}$, and $7.5 \times 10^{-9}\ \mathrm{Jm}^{-3}\mathrm{s}^{-1}$, while M_0, Q_0, Q_1 in Eqs. (3.3) and (3.4) for CT model are given as $7.9 \times 10^{-14}\ \mathrm{Nm}^{-3}$, $1.18 \times 10^{-7}\mathrm{Jm}^{-3}\mathrm{s}^{-1}$, and $7.5 \times 10^{-9}\ \mathrm{Jm}^{-3}\mathrm{s}^{-1}$. It is noted that different coefficients used in the acceleration source terms are aimed to see how the momentum addition would affect the solar wind speed.

3.8 Numerical Results

This section presents the numerical MHD results of the large-scale background structure from the solar surface to 1 AU for CR 2060 by PP and HDC schemes with time-integration Eqs. (3.18) of (3.17). The solar wind background study for CR 2064 from CT scheme with time-integration Eq. (3.18) of Eqs. (3.20) and (3.21) has been presented in our previous paper [58]. Indeed we can achieve the expected numerical results by the CT scheme for CR 2060, and thus omit it for save of space. Instead, as an exemplary application of dual-time integration in solar wind, we present the numerical results of global solar corona for four CRs: CR 1915 (solar minimum), CR 1930 (rising phase), CR 1965 (solar maximum) and CR 2030 (declining phase) from CT scheme from Eqs. (3.26) and (3.27). Although we only choose CT scheme for the test, there is no doubt that the dual-time integration can also be used for PP and HDC.

3.8.1 Numerical Results from the Sun to Earth from Poisson Projection and Hyperbolic Divergence Cleaning

This part presents the numerical results of CR 2060 for the steady state solar wind and compare the MHD results at 1 AU with the observation data.

Figure 3.7 displays the radial magnetic fields from the PP, HDC and PFSS models. From the figure, it can be seen that the three models obtain nearly the same magnetic neutral lines (MNLs). All MNLs from PP, HDC, and PFSS are fairly wavy and undulate between 25°N and 12°S. The MNLs' northernmost and southernmost points are located at Longitudes 150° and 240°. In addition, all the three models achieve much similar distribution patterns for the radial magnetic field strength, which is 0.2 gauss at north pole and 0.15 gauss at south pole.

Figures 3.8 and 3.9 show the PP and HDC model results for the magnetic field lines, radial speed v_r and number density N on two different meridional planes at $\phi = 180° - 0°$ and $\phi = 270° - 90°$ from 1–20 R_s, where arrowheads on the black lines stand for the direction of magnetic field. The two figures demonstrate that in the two meridian planes, both the PP and HDC models give almost the identical

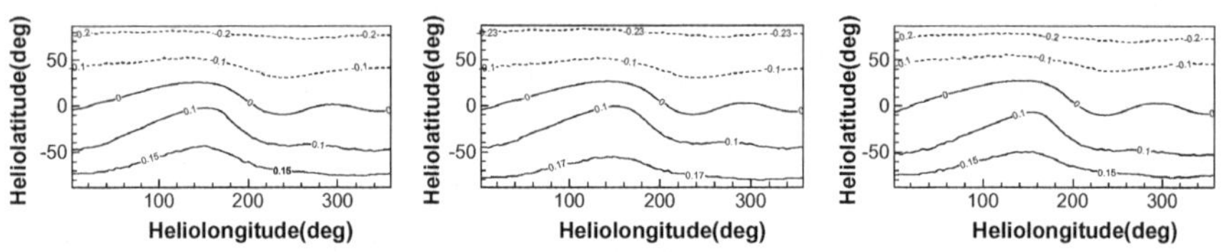

Fig. 3.7 The radial magnetic field with Gauss unit at 2.5 R_s for the PP model results (left column), HDC model results (middle column) and PFSS (right column)

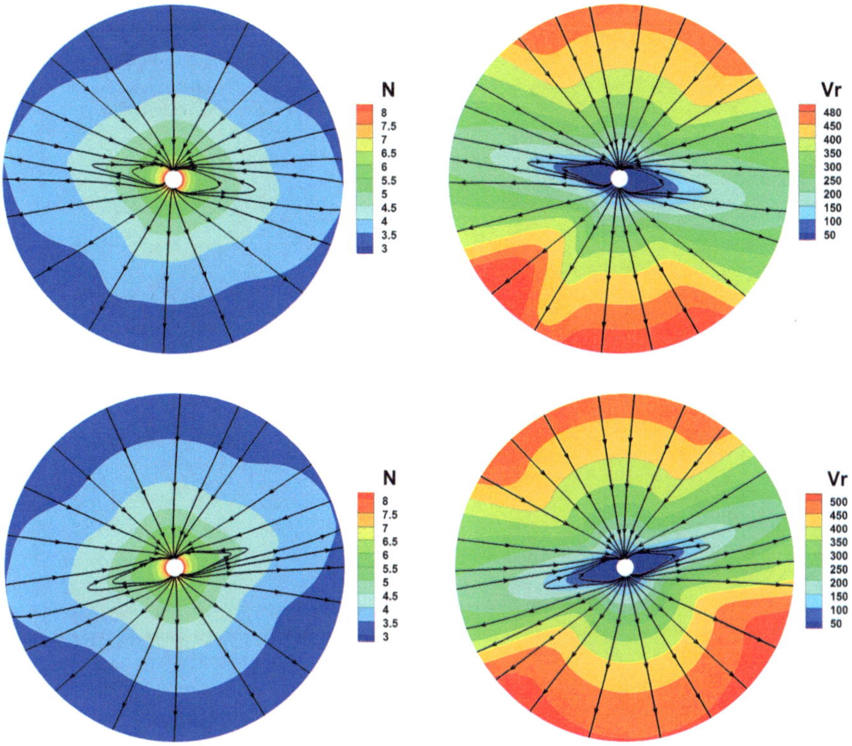

Fig. 3.8 The PP model results for magnetic field lines, radial speed v_r (km/s) and number density $N(\log_{10}/\text{cm}^3)$ on the meridional plane of $\phi = 180° - 0°$ (top), $\phi = 270° - 90°$ (bottom) from 1–20 R_s

magnetic topologies and distributions for corona plasma parameters. The high-speed low-density (HSLD) plasma flow occupies in the middle and high latitudes while the low-speed high-density (LSHD) plasma flow in the low latitudinal regions.

Figures 3.10 and 3.11 display the number density, radial speed and temperature maps for the PP and HDC MHD steady-state solution at 2.5 R_s, 20 R_s, 215 R_s, respectively. By comparing Figs. 3.10 and 3.11, one can infer that both the PP and HDC simulations present highly similar pictures in solar corona and interplanetary space. The LSHD solar wind occupies the low-latitudinal regions surrounding the radial extension of the MNL and the HSLD solar wind prevails the latitudes above 50°. The latitudinal widths of the LSHD solar wind for the PP and HDC models' results are about 55°, which is much wider than that of the LSHD solar wind for the 1996 solar minimum [131]. In addition, the wavy LSHD region near the Sun gradually evolves into a zone containing some folds due to the interaction between the streams of different speeds. The small troughs at Longitudes 240° and 350° on the sphere of 2.5 solar radii shift to 170° and 240° at 1 AU and become more evident. Quantitatively speaking, there are some slightly differences between the simulated

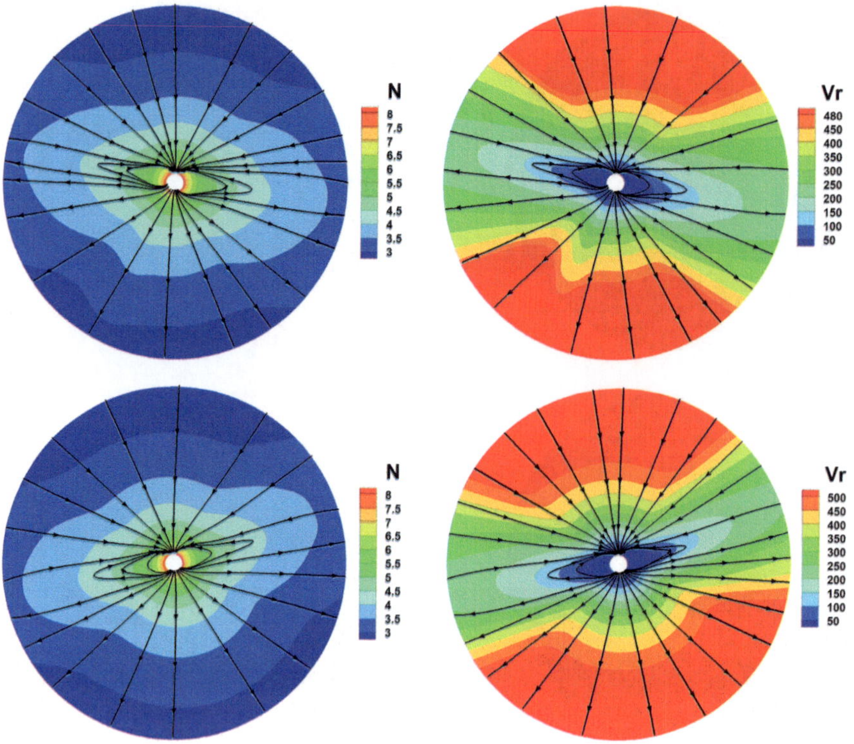

Fig. 3.9 The HDC model results for magnetic field lines, radial speed v_r (km/s) and number density $N\,(\log_{10}/\mathrm{cm}^3)$ on the meridional plane of $\phi = 180° - 0°$ (top), $\phi = 270° - 90°$ (bottom) from 1–20 R_s

results from the PP and HDC models. The proton density, bulk speed, and temperature for the LSHD region near the Earth's orbit are 20–22 cm^{-3}, 330–400 km s^{-1}, and 1.5–2 $\times$ 10^5 K from the PP model and 16–18 cm^{-3}, 360–400 km s^{-1}, and 1.8–2 $\times$ 10^5 K from the HDC model. The corresponding parameters for the HSLD region are 8–10 cm^{-3}, 650–700 km s^{-1}, and 3–3.2 $\times$ 10^5 K from the PP model and 4–8 cm^{-3}, 700–800 km s^{-1}, and 4 $\times$ 10^5 K from the HDC model. Therefore, the HDC model achieves a little sparser and faster solar wind in the HSLD regions.

Figure 3.12 is the variation profiles of number density N and radial speed v_r from 1 R_s to 20 R_s with different latitudes $\theta = 174°$ (solid line) and $\theta = 90°$ (dashed line) at the same longitude $\phi = 0°$, where $\theta = 174°$ corresponds to the HSLD region while $\theta = 90°$ corresponds to the LSHD region. Both modeled results show that the solar wind is accelerated monotonically outward from near the Sun. In the LSHD region, the plasma bulk speed increases from 50–60 km s^{-1} at 5 solar radii to about 220 km s^{-1} at 20 solar radii, which agrees with the observations provided by the Large Angle and Spectrometric Coronagraph (LASCO) instrument on the Solar and Heliospheric Observatory (SOHO) [159]. While the plasma flows in the HSLD region

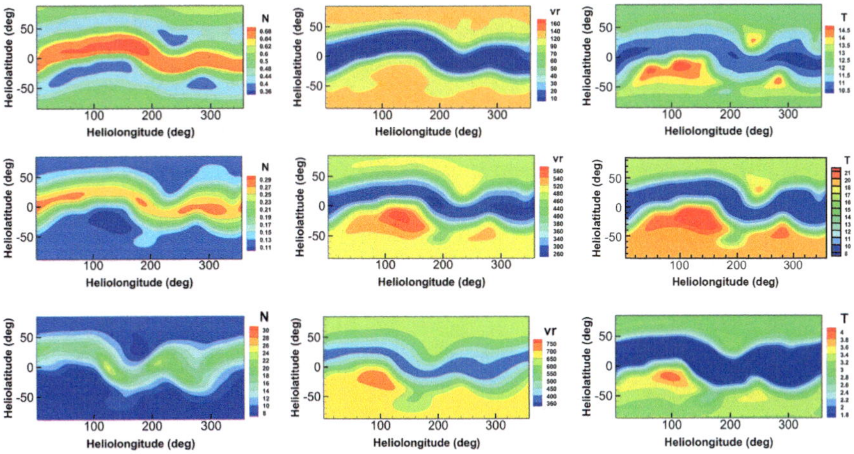

Fig. 3.10 The PP model results on the different surfaces at 2.5 R_s (top row), 20 R_s (middle row), 215 R_s (last row). The left column denotes number density N with unit 10^6 cm^{-3}, 10^4 cm^{-3}, cm^{-3} from top to bottom, the middle column denotes radial speed v_r with unit km/s, the last column denotes temperature T with units 10^5 K

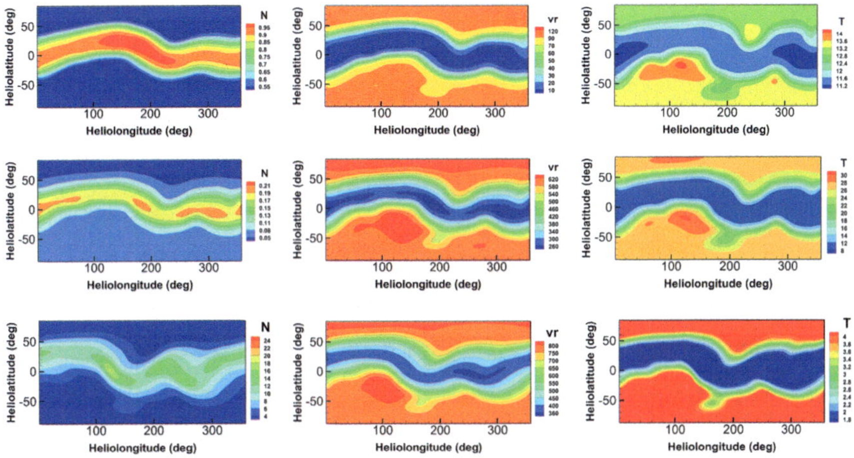

Fig. 3.11 The HDC model results on the different surfaces at 2.5 R_s (top row), 20 R_s (middle row), 215 R_s (last row). The left column denotes number density N with unit 10^6 cm^{-3}, 10^4 cm^{-3}, cm^{-3} from top to bottom, the middle column denotes radial speed v_r with unit km/s, the last column denotes temperature T with unit 10^5 K

is accelerated to 500 km s^{-1} at 10 solar radii and to 620 km s^{-1} at 20 solar radii, and the major acceleration of the solar wind occurs within a heliocentric distance of about 10 solar radii, which is in agreement with Cranmer [31].

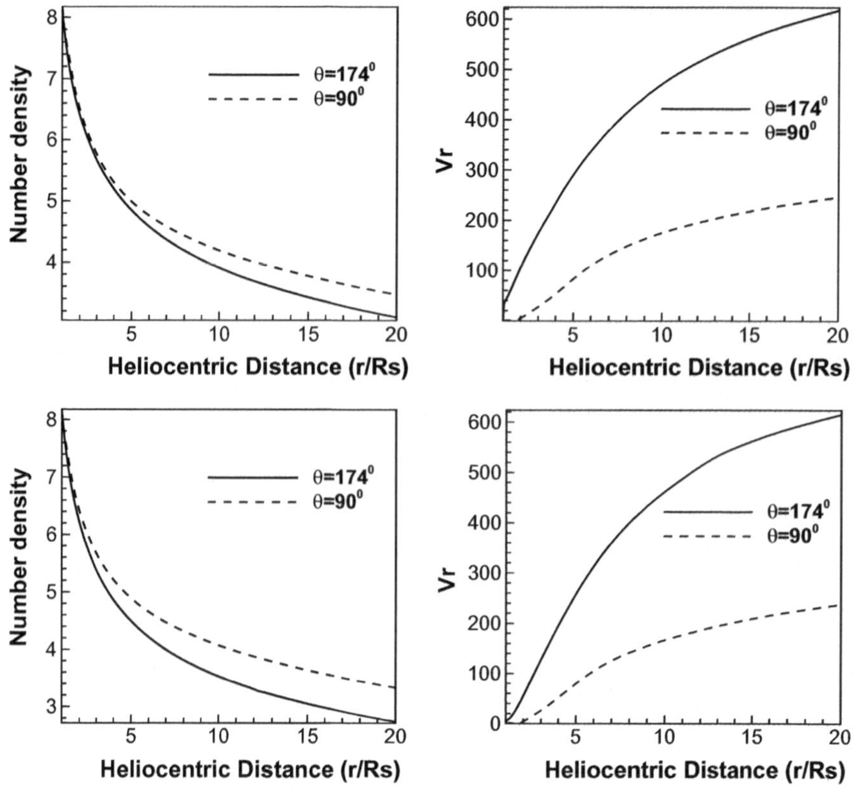

Fig. 3.12 The number density $N\,(\log_{10}/\mathrm{cm}^3)$ distribution (left) and radial speed v_r (km/s) profiles (right) along heliocentric distance with different latitudes $\theta = 174°$ and $\theta = 90°$ at the same longitude $\phi = 0°$. The top row displays PP MHD results and the bottom row does HDC MHD results

To quantitatively show how $\nabla \cdot \mathbf{B}$ evolves, the error for $\nabla \cdot \mathbf{B}$ is defined as Error $= \frac{\Sigma_{k=1}^{M} \int_{V_k} |(\nabla \cdot \mathbf{B})| dV}{\Sigma_{k=1}^{M} \int_{V_k} dV}$ [50], where M is the total number of cells in the computational domain and V_k is the kth sliding volume cell involved with the mesh grids. Figure 3.13 gives the temporal evolution of the Errors for $\nabla \cdot \mathbf{B}_1$ and $\nabla \cdot \mathbf{B}$ plotted in the computational domain. From Fig. 3.13 one can see that the use of PP and HDC approaches lead to the Errors for $\nabla \cdot \mathbf{B}_1$ and $\nabla \cdot \mathbf{B}$ around 10^{-5}. In practice, all the Errors stays the same after 25 h and no obvious large error appears after a long running time.

Figure 3.14 shows the PP and the HDC MHD results for the magnetic fields and radial speed on the equatorial plane from 20 R_s to 215 R_s, the color contours represent the radial solar wind speed, and the arrowheads denote the direction of of the radial magnetic field. Both the modeled results demonstrate two magnetic sectors on the equatorial plane and the modeled sector boundaries are almost coincident. Besides, the fast solar wind (>500 km s^{-1}) emanates nearly the same regions.

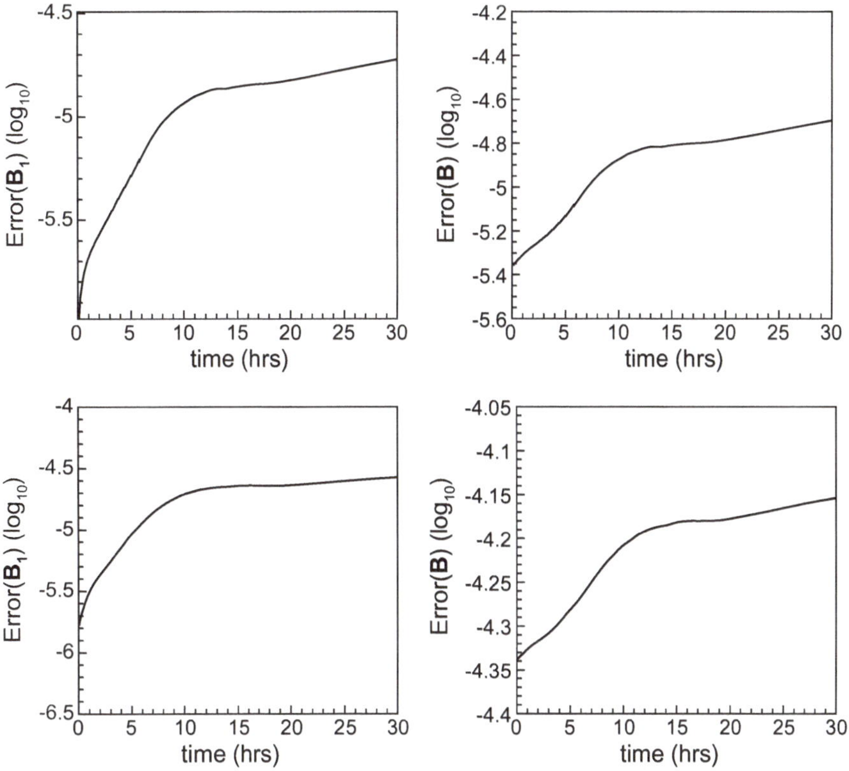

Fig. 3.13 Temporal evolution of error for $\nabla \cdot \mathbf{B}_1$ and $\nabla \cdot \mathbf{B}$ in the calculation with PP approach (top) and HDC approach (bottom)

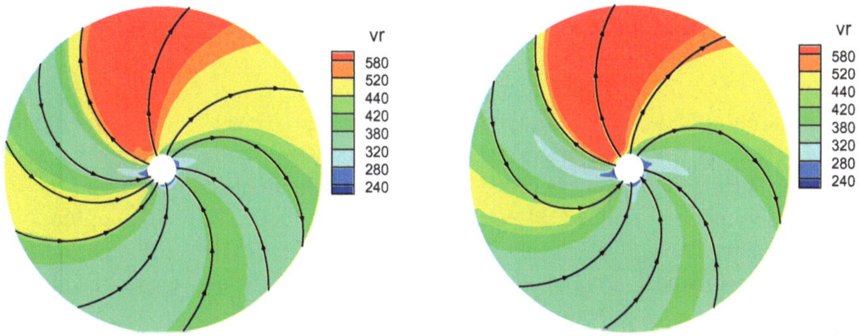

Fig. 3.14 The PP (left) and HDC (right) results for the magnetic fields, radial speed v_r (km/s) on the equatorial plane from 20 R_s to 215 R_s, where the color contours represent the radial solar wind speed, and streamlines denote the magnetic field lines

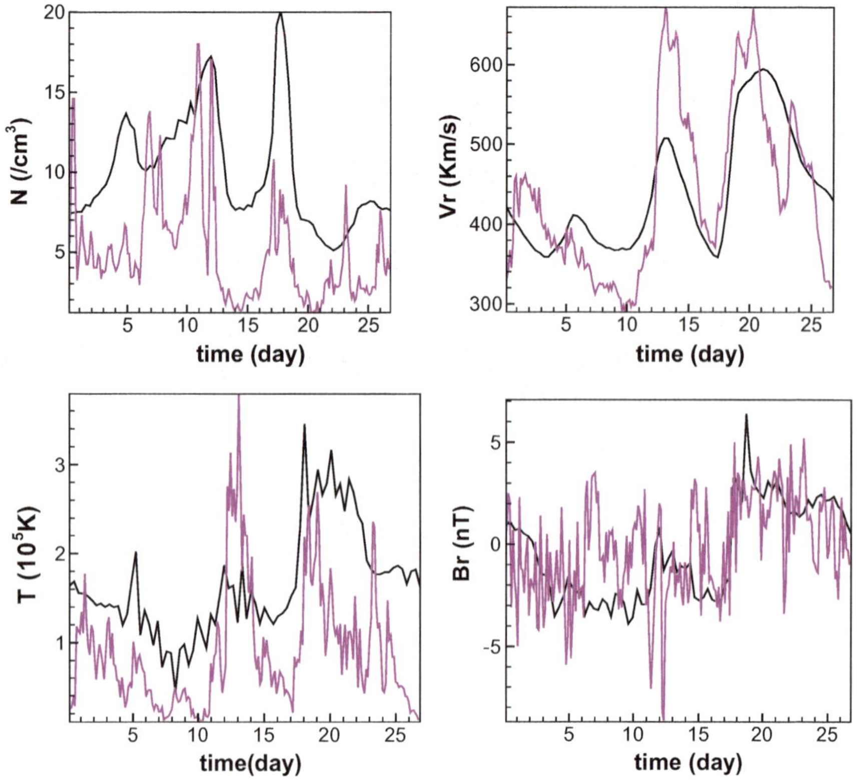

Fig. 3.15 Comparisons of the PP MHD results (black line) and the observational data (red line) obtained by OMNI: the first row is number density $N(\text{cm}^{-3})$, radial speed v_r (km/s), and the second row is temperature $T(10^5$ K) and magnetic field B_r (nT)

Figures 3.15 and 3.16 exhibit the comparisons among the PP MHD results, HDC MHD results and the OMNI observational data. The two figures reveal that both the PP and HDC models achieve the resembling temporal variation trends for the solar wind parameters at the Earth's orbit. By comparing the modeled results and the in-situ measurements from the OMNI data, one can find that both models capture two fast streams centering Days 14 and 21 with reasonable accuracy although the first modeled velocity peak is about 150 km s^{-1} lower than the measurement. In addition, both simulations give higher proton number density, slightly higher temperature, and weaker radial magnetic field than the observation. Overall, the simulations basically reproduce the observations.

To obtain the cusp magnetic configuration above the streamer and a thin current sheet, volumetric heating is necessary. Heating and momentum sources are important for achieving the typical observed characteristics of HSLD plasma flow in the hole and LSHD plasma flow in the streamer. However, the heating and momentum addition terms in this model are added ad hoc in the exponentially decaying form rather than

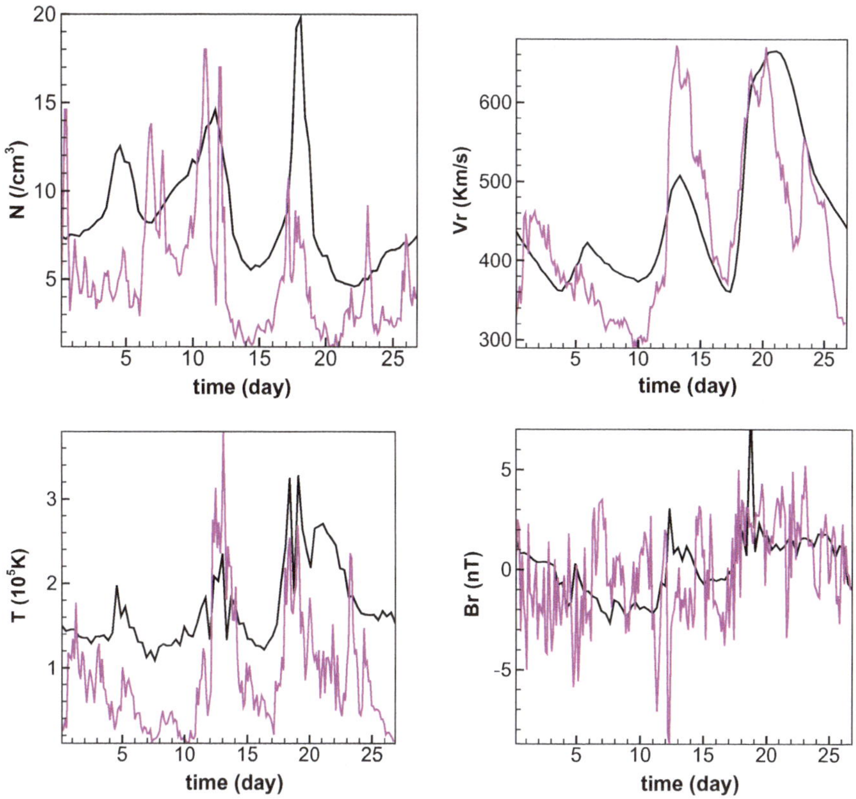

Fig. 3.16 Comparisons of the HDC MHD results (black line) and the observational data (red line) obtained by OMNI: the first row is number density $N\,(\mathrm{cm}^{-3})$, radial speed v_r (km/s), and the second row is temperature $T\,(10^5\ \mathrm{K})$ and magnetic field B_r (nT)

physically specified, although such source terms (e.g., [49, 50, 67, 130, 189]) are empirically suggested. Also we can improve our numerical model to better simulate the realistic global corona by incorporating, for example, a transition region at the base, since the inner boundary condition should be determined through the balance between radiative losses, thermal conduction, and heating in the transition region between the corona and the chromosphere [114].

3.8.2 Numerical Results of the Solar Corona from Constrained Transport

Figure 3.17 displays the contours of the β distribution on solar surface for CR 1915 (solar minimum), CR 1930 (rising phase), CR 1965 (solar maximum) and CR 2030

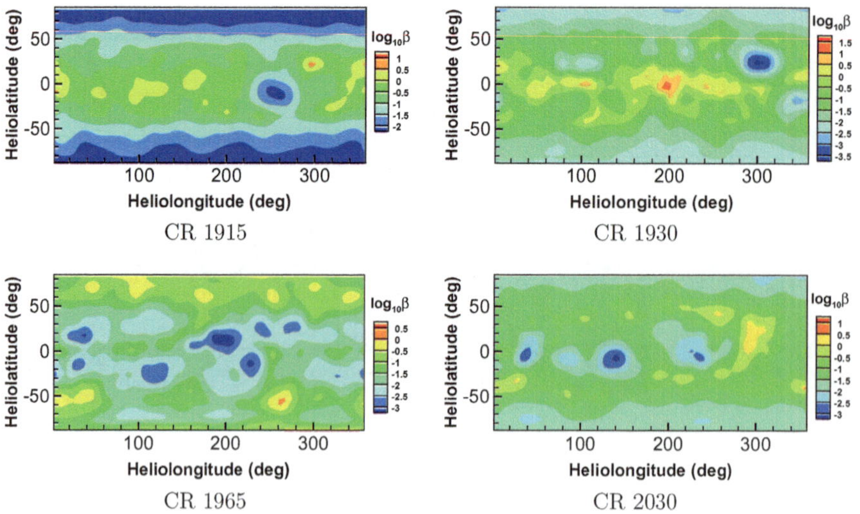

Fig. 3.17 Plasma β distributions on the solar surface: the top row displays the β distribution for CR 1915 (left) and CR 1930 (right), and the bottom row does for CR 1965 (left) and CR 2030 (right)

(declining phase). The figure reveals that the magnetic field near both poles is stronger than that in the low-latitude regions during CR 1915. From CR 1915 to CR 1965, the simulated magnetic field grows weaker in the polar regions and becomes stronger in low-latitude regions. Comparing the panels of the β distribution for CRs 1965 and 2030, one can find that the polar magnetic field strengthens and the low latitudinal magnetic field weakens. The changing trend is consistent with the supersynoptic charts of the solar magnetic field created from daily observational data available at http://obs.astro.ucla.edu/torsional.html. This implies that the smoothing or filtering method to preprocess the observed magnetic field used in this chapter is proper and retains the essential characteristics of the photospheric magnetic field measurements.

Figure 3.18 displays the radial magnetic field at 2.5 R_s for the four CRs using the CT and PFSS models. Comparing the results from the CT and PFSS models, one can find that both models give almost identical topologies of the MNLs and the distribution contours of the radial magnetic field at 2.5 solar radii. One of the differences between their results lies in that the CT model achieves a slightly stronger magnetic field in some regions. For CR 1915, the MNL is very flat and is confined to a narrow latitudinal zone ranging between 5°S and 15°N. There is only a shallow tilt structure extending from 240° to 290°. For CR 1930, the MNL is much wavy and the warp structure from 280° to 330° is relatively steep. The heliolatitudes of the northward hump and the southward trough are 40°N and 20°S. In CR 1965, the MNL exhibits very large inclination and its southernmost and northernmost excursions are as high as 55°S and 60°N. During CR 2030, the MNL is characteristic of a two-peak and two-valley structure. The two peaks are located at 0° and 210° and the two valleys

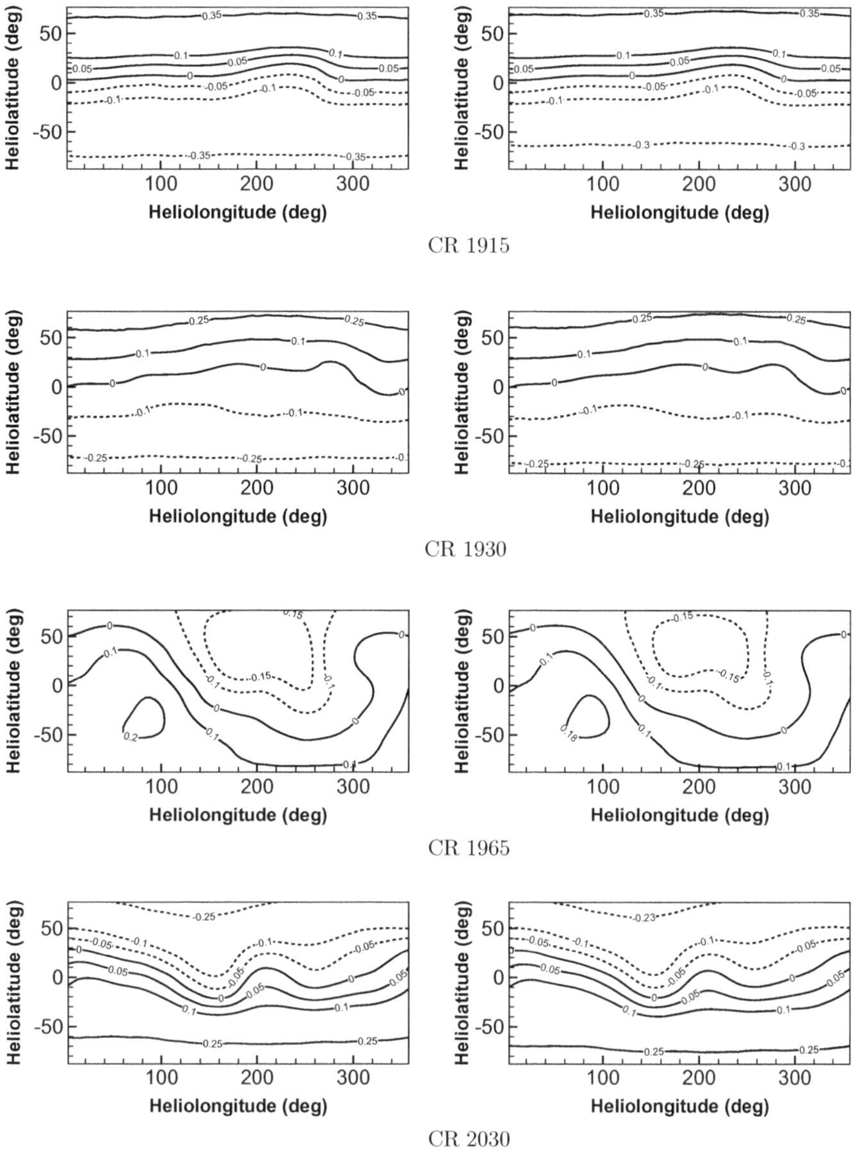

Fig. 3.18 CT model results for CRs 1915 (1st row), 1930 (2nd row), 1965 (3rd row) and 2030 (4th row) with $\delta t = 100\Delta t$: the radial magnetic field with unit Gauss at 2.5 R_s for MHD results (left) and PFSS (right)

at 160° and 250°. Generally speaking, the CT model captures the large-scale features of the MNLs corresponding to four solar activity phases.

Figures 3.19 and 3.20 display number density, radial speed and temperature maps for CRs 1915, 1930, 1965, and 2030 on the surfaces at 2.5 R_s, and 20 R_s. The

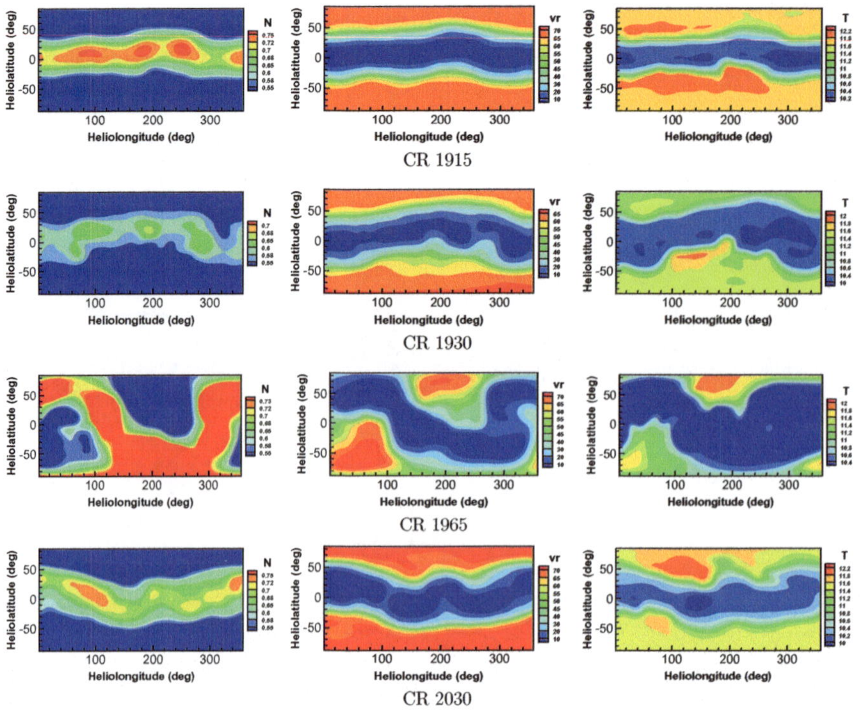

Fig. 3.19 CT model results for CRs 1915 (1st row), 1930 (2nd row), 1965 (3rd row) and 2030 (4th row) with $\delta t = 100\Delta t$: number density N with unit 10^6 cm^{-3} (left column); radial speed v_r with unit km/s (middle column) and temperature T with unit 10^5 K (right column) on the surface at 2.5 R_s

number density decreases with heliocentric distance while radial speed increases. The flow around the radial extension of MNLs has high density, low speed and low temperature, while at the open field region the inverse changing trends can be seen. The most striking feature in both figures is the LSHD plasma flow distributions varying from different solar activity phases. In CR 1915 near solar minimum, the LSHD plasma flow surrounds the MNL and is confined to about the latitudinal width of 50°. At the same time, the HSLD plasma flow dominates the middle- and high-latitude regions. In CR 1930 of solar rising phase, the LSHD plasma flow even appears at 65°N and 50°S, and latitudinal width is much wider than that in CR 1915 and amounts to 70°. For CR 1965 near solar maximum, the LSHD plasma flow is present at any latitude and dominates the global solar wind except small patches of HSLD plasma flow near both poles. For CR 2030 of the decline phase, the HSLD plasma flow occupies a large portion of high-latitude areas and intrudes into very latitudes. The stream interfaces of the LSHD and HSLD plasma flows begin to form at 20 solar radii. From Figs. 3.3 and 3.4, one can also find the effect of solar rotation. The hump of the LSHD plasma flow moves from 230° at 2.5 solar radii to 220° at 20

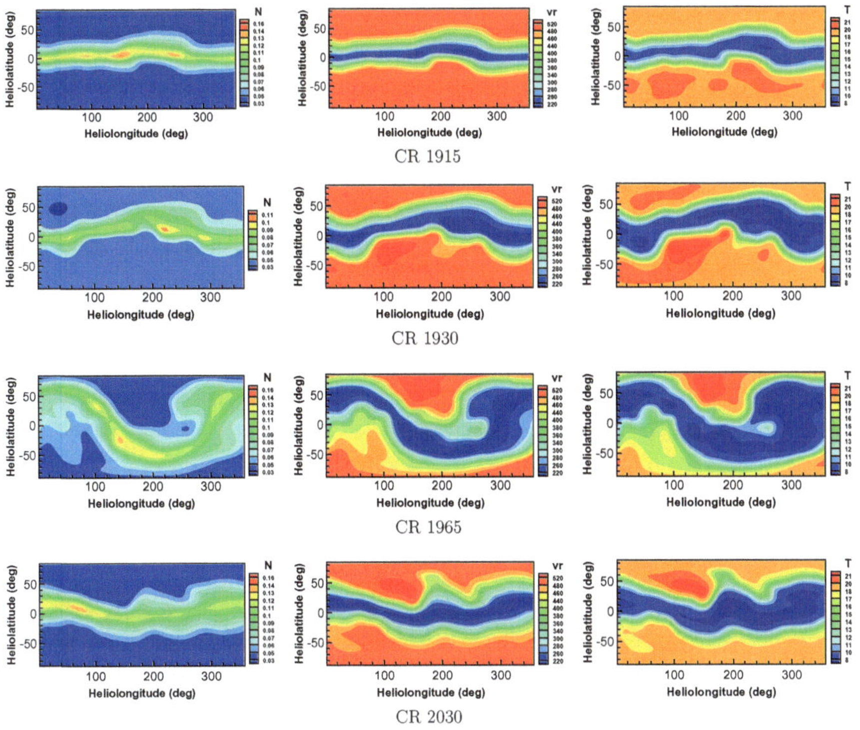

Fig. 3.20 CT model results for CRs 1915 (1st row), 1930 (2nd row), 1965 (3rd row) and 2030 (4th row) with $\delta t = 100\Delta t$: number density N with unit 10^4 cm^{-3} (left column); radial speed v_r with unit km/s (middle column) and temperature T with unit 10^5 K (right column) on the surface at 20 R$_s$

solar radii. The intrusion of the LSHD into the HSLD plasma flow in CR 2030 shifts from 170° at 2.5 solar radii to 160° at 20 solar radii and the interface also steepens because the HSLD solar wind catches up with the LSHD solar wind emitted earlier. Similar phenomena are also present in CR 1965.

Figure 3.21 presents the CT model results for CRs 1915, 1930, 1965, and 2030, the total speed v, the magnetic field lines, and Alfvénic surface on the meridional planes of $\phi = 180° - 0°$ and $\phi = 270° - 90°$ from 1–20 R$_s$. The Aflvénic Mach number, M_A, is defined as $\frac{|\mathbf{v}|\sqrt{\mu\rho}}{B}$ and the Aflvénic surface is plotted by directly connecting the adjacent Aflvénic critical points where $M_A = 1$. According to Hundhausen [80], M_A increases from a very small value (~0.1 or even smaller) near the solar surface to about 9 at 1 AU and any solar wind solution must pass Aflvénic critical points. The Aflvénic surface is fundamental to the magnetic topology of the solar corona and heliosphere by marking the causal disconnection of individual packets of plasma and magnetic flux from the Sun itself. On one hand, the Aflvénic surface that separates the super-Aflvénic solar wind from the sub-Aflvénic coronal expansion is usually used as the inner boundary for various heliospheric MHD models (see e.g., [134]);

on the other hand, the Aflvénic surface can be detectable remotely via motions in the visible corona [36]. Several recent works suggest a broad range of possible distances for the Aflvénic surface from 6 to 30 Rs. DeForest et al. [36] used the observed inbound compressive waves in the outer solar corona to set a strong lower limit on the location of the Aflvénic surface, which was 15 Rs in the streamer belt and 12 Rs in the coronal hole. Goelzer et al. [65] computed the Aflvénic critical points using OMNI2 data and found them varying systematically from 14 to 38 solar radii with the sunspot number. Schwadron et al. [156] and Smith et al. [168] treated the Aflvénic surface in the context of the heliospheric flux balance to derive it at 10–15 R_s. Zhao and Hoeksema [216] estimated the Aflvénic surface and placed its outer limit at 10–14 solar radii (R_s) around solar minimum. Wang et al. [190], Sheeley and Wang [160] and Sheeley et al. [161] observed isolated inbound retracting loops in streamers, and since never observing such a feature beyond 6 R_s, they attributed this lack to a low Aflvénic surface near 6 R_s.

From Fig. 3.21, it follows that the wispy Alfvénic surface implies that the solar wind is not spherical and higher velocities are achieved at open field regions. Meanwhile, the magnetic stress dominates the interaction in the inner corona below the Alfvénic surface, and around the minimum phase of solar activity the coronal magnetic field is largely dipole-like. The gradient of radial field strength in latitude, with stronger field in the open field region, bends stream lines from the radial direction toward the streamer belts. However, in the corona beyond the Alfvénic surface, it is the inertia of radially accelerated plasma flow that dominates the magnetic field-plasma interaction. Figure 3.21 also shows that the Alfvénic critical points are not spherical-symmetrically distributed, and the lowest (highest) Alfvénic critical point occurs in the LSHD plasma flow of streamer belt (the HSLD plasma flow of the open field region). This is in agreement with Pneuman and Kopp [140] and Zhao and Hoeksema [216].

Our numerical MHD results determine the location of the Alfvénic surface from the interaction of the solar wind flow and magnetic field. The Alfvénic surfaces for these four CRs derived from the MHD results on the meridional planes are roughly from about 8.5 to 12.5 R_s, 7.5 to 14 R_s, 7.5 to 15 R_s, and 7.5 to 13 R_s for CR 1915 (solar minimum), CR 1930 (rising phase), CR 1965 (solar maximum) and CR 2030 (declining phase), respectively. It is obvious that the upper limit for Aflvénic surface is higher for CR 1965 in solar maximum, the lowest one is for CR 1915 in solar minimum, However, the upper limit for Aflvénic surface for CR 1930 (rising phase) and CR 2030 (declining phase) lies in between. This implies that the upper limit for Aflvénic surface will move to high altitude in the open field region due to strong magnetic field. Thus our simulated results for the locations of the Aflvénic surfaces are consistent with the previous studies (see, e.g., [36, 65, 160, 161, 216]).

Figure 3.22 shows the steady-state MHD results of the radial speed v_r profiles along heliocentric distance at $\theta = 174°$, $\theta = 95°$ (CRs 1915 and 1930), $\theta = 45°$ (CR 1965) and $\theta = 80°$ (CR 2030) with $\phi = 0°$. In this figure, $\theta = 95°, 45°$, and $80°$ correspond to the vicinity of the streamer belt or the helioshperic current sheet (HCS) region (close to the equator near the solar minimum), while $\theta = 174°$ corresponds to the open field region. From the modeled results, one can find that in the open field

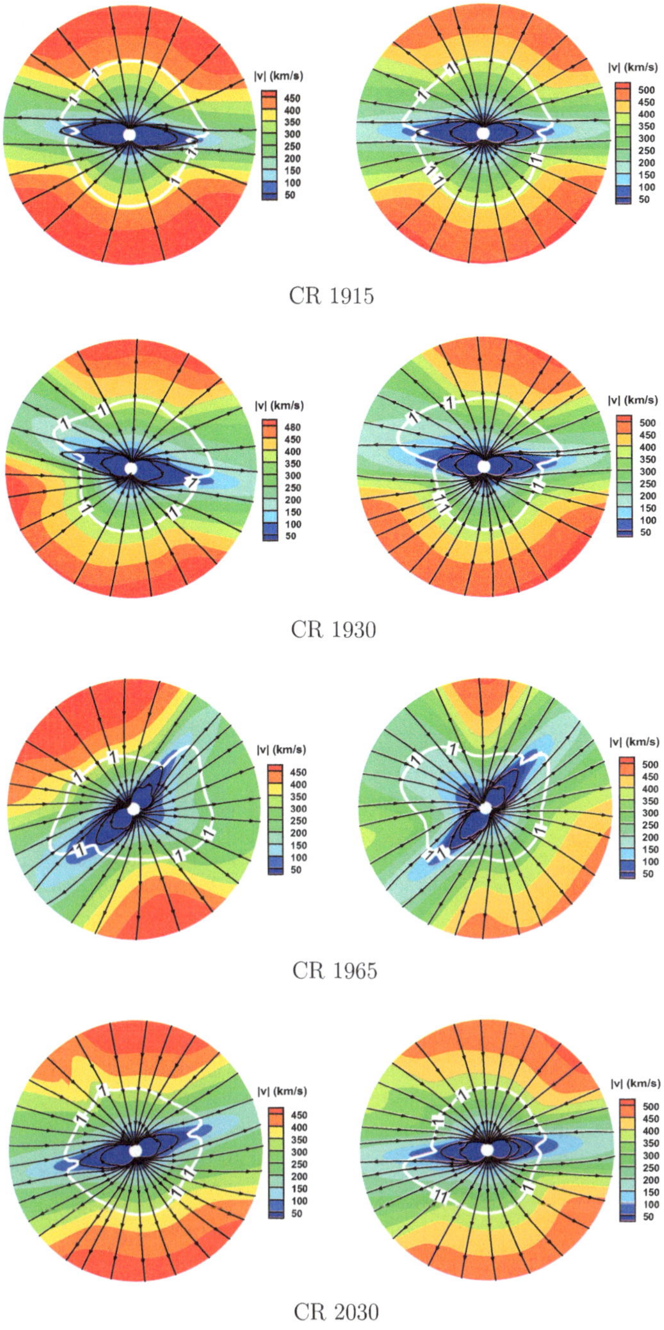

Fig. 3.21 CT model results for CRs 1915 (1st row), 1930 (2nd row), 1965 (3rd row) and 2030 (4th row) with $\delta t = 100\Delta t$: the total speed v (km/s), the magnetic field lines, and Alfvénic surface (white line) on the meridional plane of $\phi = 180° - 0°$ (left column), $\phi = 270° - 90°$ (right column) from 1–20 R_s

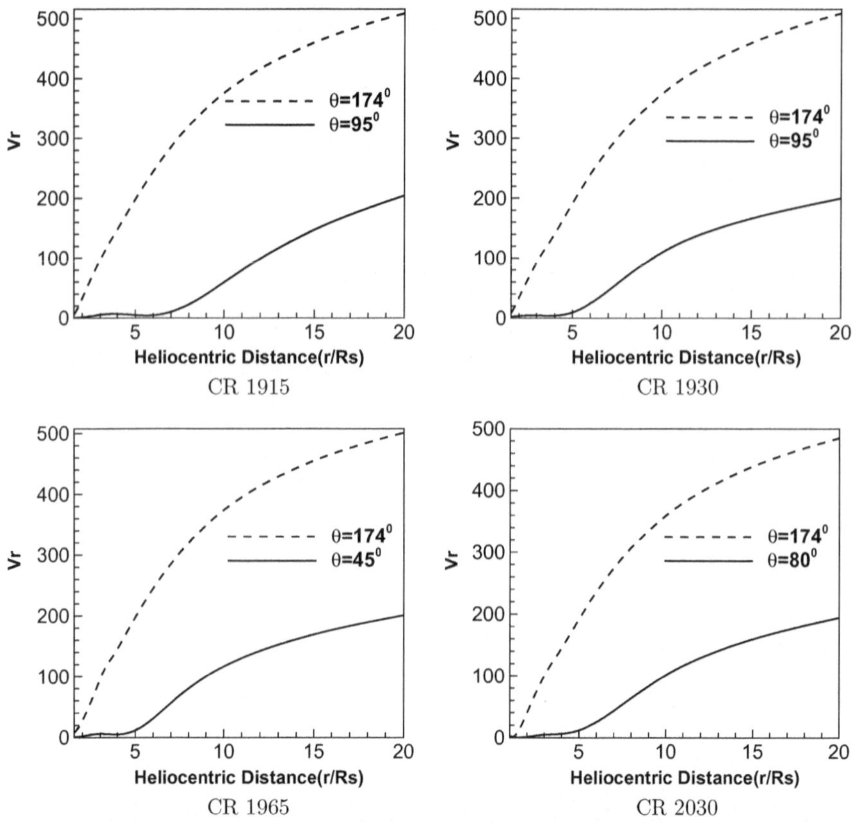

Fig. 3.22 The radial speed v_r (km/s) profiles with different latitudes at the same longitude $\phi = 0°$: the top row displays CR 1915 (left) and CR 1930 (right), and the bottom row gives CR 1965 (left) and CR 2030 (right)

region the speed is larger and stands for the HSLD solar wind, in the HCS region the speed is smaller and stands for the LSHD solar wind. In the LSHD region, the plasma bulk speed increases from 50–60 km s^{-1} at 5 solar radii to about 220 km s^{-1} at 20 solar radii and the plasma flows in the HSLD region is accelerated to 500 km s^{-1} at 20 solar radii. It should be noted here that the speed in the HDLS region is slightly lower than that in the results of PP and HDC because different coefficients in the acceleration source terms are used. Decreasing the momentum addition would cause the radial velocity curve at the pole to become low, by comparison between Figs. 3.22 and 3.12, as conformed to Wang et al. [189]. The large plasma β meridional gradient at the streamer boundary can maintain a high density inside the streamer and the radial extension of the external streamer stalk, as shown by Figs. 3.17, 3.21 and 3.22.

Figure 3.23 gives the temporal evolution of the Errors for $\nabla \cdot \mathbf{B}_1$ and $\nabla \cdot \mathbf{B}$ plotted in the computational domain. From this figure one can see that the Error for $\nabla \cdot \mathbf{B}_1$ is around 10^{-7} and that for $\nabla \cdot \mathbf{B}$ about 10^{-5} for these four Carrington rotations. Additionally, this Error stays almost the same after 7 h and no obvious large error

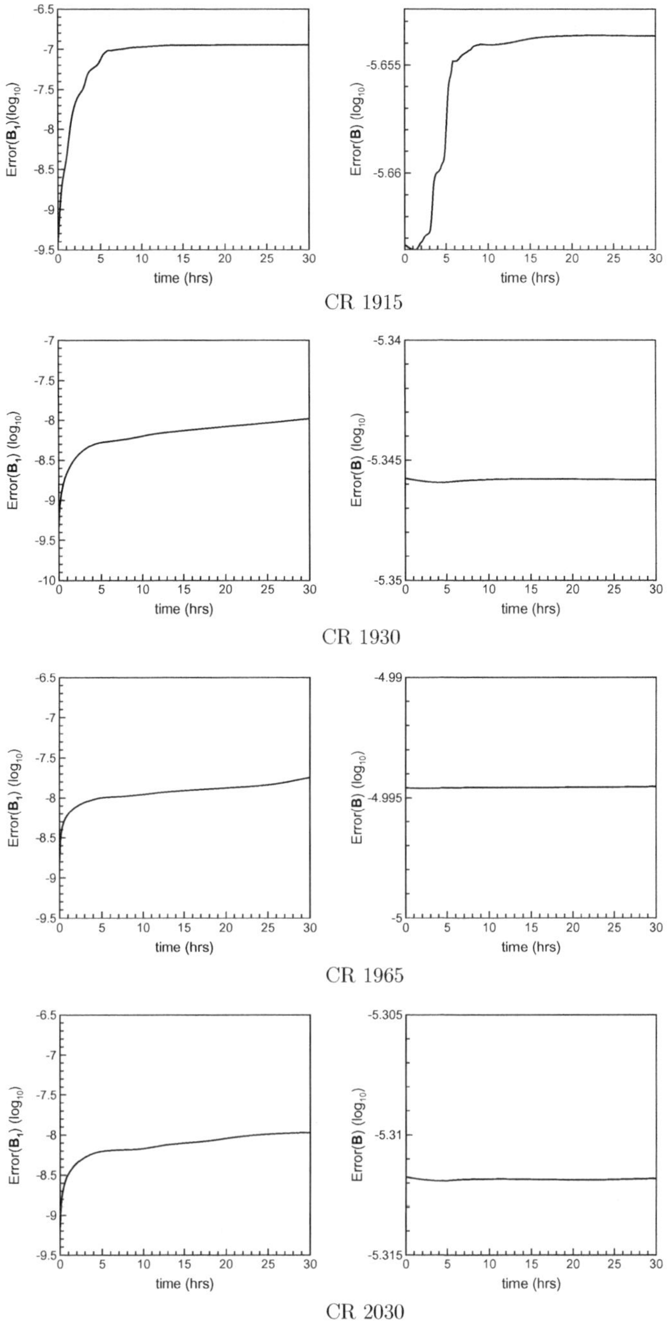

Fig. 3.23 CT model results for CRs 1915 (1st row), 1930 (2nd row), 1965 (3rd row) and 2030 (4th row) with $\delta t = 100\Delta t$: temporal evolution of Error for $\nabla \cdot \mathbf{B}_1$ and $\nabla \cdot \mathbf{B}$ in the calculation

appears after a long running time. This verifies that the numerical error for magnetic field divergence can be maintained to be acceptable during dual-time stepping.

3.9 Conclusions and Discussions

A 3D STEP-MHD model is established by employing the FV scheme in spherical geometry with a six-component grid for the solar wind plasma MHD system from the Sun to Earth. In the numerical scheme design, geometrical factors can be merged into the reformulation of fluxes such that flux evaluation is easily achieved, and many numerical schemes designed for Cartesian case can follow almost in the same way as done in the present context of spherical grid described here. In fact, the STEP code includes other numerical schemes such as Roe and HLL types for the solar wind plasma with eight-wave and generalized Lagrange multiplier formulated MHD equations given in the appendix. In order to maintain the solenoidal condition on the magnetic field for MHD calculations, three typical divergence cleaning approaches (PP, HDC, and CT methods) are included. In the CT formulation, two coupled subsystems of fluid part and magnetic field part consisting of the solar wind governing MHD equations, are integrated individually in time integration. In addition, dual time technique is exemplarily carried out for the CT method and the physical time step can be enlarged compared to that of explicit time step. Due to the enormous disparity in the fluid and electromagnetic frequencies, some form of implicit dual time integration is necessary for the MHD simulation. It is hoped that the present dual time technique may apply to the solar coronal magnetic field MHD reconstruction for active regions with the strong magnetic field, as evidenced from its capability of modeling small plasma β case.

The numerical formulation in principle applies to the composite overlapping grids with hexahedral cells of quadrilateral frustum type, particularly those with identical components or patches. Except for avoiding the problems of the coordinate singularity and the grid convergence near the poles in spherical polar coordinates, the use of the composite overlapping grids can facilitate the parallelization in $r-$, $\theta-$ and $\phi-$directions, with the efficient load balancing enhanced by the adoption of the multiple time-stepping method in the radial direction. Besides its performance gain, the composite mesh can also make easy to program the governing MHD equations since they have the same form in each component except for the source terms such as solar rotation and heating terms. On the other hand, the boundary points of each component grid are situated among the inner points of the neighboring component grids. Thus, after updating the solution at the inner points of all the grids, the boundary conditions for each component grid can be easily assigned using the values interpolated from the known values at the inner points of the overlapping grids. That is, one only needs to impose numerical boundary conditions at the boundary circles of the spherical shell computational domain; the boundary grids (overlapping grids) between components becomes "internal boundary", or one can say that there have no lateral boundaries for latitudinal and longitudinal directions. The simple non-conservative interpolation like bicubic or Lagrange interpolation has been found to perform well in practice for these overlapping composite grid, although nonconservative interpolation errors occur with presence of a flow across the internal boundary. A better choice is to determine the horizontal boundary or internal border values in $\theta-$ and $\phi-$directions of each component grid from the reconstructed values by the newly updated ones in the neighboring components.

As for the pressure positivity, we have used Eq. (3.16) similar to that of Feng et al. [58] and a switching method as done by Feng et al. [57] to find that both methods perform with no difference. Only for very strong magnetic field regions such as $\beta < 0.001$, these methods are in practice active to recover the positivity, and otherwise the schemes work well without them. As for numerical scheme's divergence preservation, three typical methods of magnetic field divergence-controller are employed in our code. Numerical tests show that they all effectively confine the divergence errors to the truncation errors. The PP solves a Poisson equation to remove $\nabla \cdot \mathbf{B}$ after each time step, which can be very accurate depending on how to solve the Poisson equation but computationally expensive. The multigrid Poisson solver may provide speedup [50]. The HDC couples a scalar field ψ to $\nabla \cdot \mathbf{B}$ to advect and diffuse $\nabla \cdot \mathbf{B}$ away, which is fully conservative, simple and fast. Although a little complicated, the CT, employing staggered (face centered) grid for the magnetic field to better constrain $\nabla \cdot \mathbf{B}$, is fast and accurate, even in the dual-time procedure. From our tests, the HDC procedure can confine $\nabla \cdot \mathbf{B}$ to almost machine error for the time-dependent part $\mathbf{B}_1$ (as mentioned above, we use the splitting form of magnetic field $\mathbf{B} = \mathbf{B}_1 + \mathbf{B}_0$ with time-dependent part $\mathbf{B}_1$ and time-independent part $\mathbf{B}_0$). In addition, all three models can retain the error of $\nabla \cdot \mathbf{B}$ as low as 10^{-5}. In fact, the magnetic divergence is the character of the evolution equation for the magnetic flux which analytically preserves the divergence during the evolution. That is, magnetic field with divergence-free from the initial conditions will be frozen during the calculation. Thus, the requirement $\nabla \cdot \mathbf{B} = 0$ is a problem of the initial conditions, but the discretization of these initial conditions may introduce divergence errors. This implies special demands on the discrete initial conditions to make its discrete divergence vanish, which is in general not easily obtained [87]. Overall, due to the dynamically evolving field of computational MHD, there is no point to be "religious" about these attributes, such as strict positivity and $\nabla \cdot \mathbf{B} = 0$. What kind of these methods is effective can be problem-dependent.

The developed STEP-MHD model is validated by using the PP and HDC models to obtain the large-scale structure of solar corona and solar wind for CR 2060 and employing the CT method to achieve the quasi-steady state for the solar corona during CRs 1915, 1930, 1965 and 2030. The simulated results from the PP and HDC models are almost the same for the topology of the MNL at 2.5 solar radii, the global distribution for plasma parameters at 2.5, 20, and 215 solar radii, and the temporal profiles of solar wind parameters at 1 AU. Both modeled results for CR 2060 show the domination of the HSLD solar wind at the latitudes higher than 50°, the longitudinal shifts of the peaks and troughs of the LSHD regions during the outward propagation, two magnetic sectors on the equatorial plane, and the interception of two fast streams at 1 AU on Days 14 and 21. The coronal solutions from the CT model for CRs 1915, 1930, 1965 and 2030 capture many coronal features of the corresponding solar activity phases. The simulated MNLs are flat in CR 1915, warped in CR 1930, highly inclined in CR 1965, and double peaked in CR 2030. The modeled latitudinal widths for the LSHD plasma flow are 50°, 70°, 110°, and 80° for CRs 1915, 1930, 1965 and 2030 respectively. The modeled Alfvénic surfaces range from 7.5 to 15 solar radii and are abruptly concave in the MNLs.

Although the STEP model can achieve many observed properties of solar corona and solar wind, the results for the present numerical model could not support a fully realistic model of the corona and solar wind simply because the physical description is incomplete. To attempt a "completely" physical description would involve the numerical solution to a very complex set of equations and invoking processes at the Sun, which are completely unknown (e.g., the sources of the momentum and heating terms). The precise description for the transition region, chromosphere, and photosphere is needed to understand the coupling processes among these regions, the resulting coronal heating mechanism, and the coronal dynamics. Besides, the achievement of a canonical solar wind model could utilize future solar observations and theoretical understanding to more narrowly and analytically define these terms, or characterize and quantify the key physical processes/mechanism to more physically integrate realistic coronal heating modules into 3D MHD codes.

In the computational aspect, the future development of the present STEP-MHD model will focus on some additional computing techniques, such as the preconditioning method to stiff hyperbolic systems arising in magnetohydrodynamics [144, 145, 155], time-derivation preconditioning [107, 177], dual time plus multigrid [96] or adaptive mesh refinement implementation in context of curvilinear coordinates [53, 54, 57, 110, 185, 218, 220] or graphic processing unit (GPU) accelerated technique [56].

Appendix: Discretizations, Reconstructions and MPP Flux-corrected Method

This appendix presents the used numerical techniques in some details for the discretized formulation of the solar wind MHD governing equations.

The STEP code advances the variables at the centroid point $(\overline{r}_i, \overline{\theta}_j, \phi_k)$ by constructing the semi-discretization for the solar wind MHD governing Eqs. (3.2) in a cell with geometrical center (i, j, k):

$$\begin{aligned}\frac{d\overline{\mathbf{U}}}{dt} = &-\frac{3\left(r_{ip}^2\mathbf{F}^n_{r,ip,j,k} - r_{im}^2\mathbf{F}^n_{r,im,j,k}\right)}{\left(r_{ip}^3 - r_{im}^3\right)} - \frac{\sin\theta_{jp}\mathbf{F}^n_{\theta,i,jp,k} - \sin\theta_{jm}\mathbf{F}^n_{,\theta,i,jm,k}}{r_i^\theta\left(\cos\theta_{jm} - \cos\theta_{jp}\right)} \\ &-\frac{\Delta\theta(j)}{r_i^\phi\left(\cos\theta_{jm} - \cos\theta_{jp}\right)}\frac{\mathbf{F}^n_{\phi,i,j,kp} - \mathbf{F}^n_{\phi,i,j,km}}{\Delta\phi(k)} + \overline{\mathbf{S}}_{i,j,k}\end{aligned} \tag{A.1}$$

where $\mathbf{U} = \left(\rho, \rho v_r, \rho v_\theta, \rho v_\phi r \sin\theta, e, B_r, B_\theta, B_\phi\right)^T$ or $\left(\rho, \rho v_r, \rho v_\theta, \rho v_\phi r \sin\theta, e, B_r, B_\theta, B_\phi, \psi\right)^T$ with primitive variables $\mathbf{W} = \left(\rho, v_r, v_\theta, v_\phi, p, B_r, B_\theta, B_\phi\right)^T$ or $\left(\rho, v_r, v_\theta, v_\phi, p, B_r, B_\theta, B_\phi, \psi\right)^T$.

The STEP code uses Roe and HLL solvers with PP, HDC and CT methods to keep the magnetic field divergence constraint. Flux formulations are displayed below.

Roe Type Flux with Powell's Eight-wave Formulation

According to the papers [71, 72, 74, 141], the Roe flux form is given as follows:

$$\begin{aligned}\mathbf{F}^n_{r,im,j,k} = &\frac{1}{2}\left(\mathbf{F}_r\left(\widetilde{\mathbf{U}}_{[i-1,j,k]}\left(r_{im}, \theta^r_j, \phi_k\right)\right) + \mathbf{F}_r\left(\widetilde{\mathbf{U}}_{[i,j,k]}\left(r_{im}, \theta^r_j, \phi_k\right)\right)\right) + \\ &\frac{1}{2}\left(\sum_J \mathbf{R}_{rJlm}\left|\lambda_{rJlm}\right|\left[\mathbf{L}_{rJlm}\left(\widetilde{\mathbf{U}}_{[i-1,j,k]}\left(r_{im}, \theta^r_j, \phi_k\right) - \widetilde{\mathbf{U}}_{[i,j,k]}\left(r_{im}, \theta^r_j, \phi_k\right)\right)\right]\right)\end{aligned} \tag{A.2}$$

$$\begin{aligned}\mathbf{F}^n_{r,ip,j,k} = &\frac{1}{2}\left(\mathbf{F}_r\left(\widetilde{\mathbf{U}}_{[i,j,k]}\left(r_{ip}, \theta^r_j, \phi_k\right)\right) + \mathbf{F}_r\left(\widetilde{\mathbf{U}}_{[i+1,j,k]}\left(r_{ip}, \theta^r_j, \phi_k\right)\right)\right) + \\ &\frac{1}{2}\left(\sum_J \mathbf{R}_{rJlp}\left|\lambda_{rJlp}\right|\left[\mathbf{L}_{rJlp}\left(\widetilde{\mathbf{U}}_{[i,j,k]}\left(r_{ip}, \theta^r_j, \phi_k\right) - \widetilde{\mathbf{U}}_{[i+1,j,k]}\left(r_{ip}, \theta^r_j, \phi_k\right)\right)\right]\right)\end{aligned} \tag{A.3}$$

$$\begin{aligned}\mathbf{F}^n_{\theta,i,jm,k} = &\frac{1}{2}\left(\mathbf{F}_\theta\left(\widetilde{\mathbf{U}}_{[i,j-1,k]}\left(r^\theta_i, \theta_{jm}, \phi_k\right)\right) + \mathbf{F}_\theta\left(\widetilde{\mathbf{U}}_{[i,j,k]}\left(r^\theta_i, \theta_{jm}, \phi_k\right)\right)\right) + \\ &\frac{1}{2}\left(\sum_J \mathbf{R}_{\theta Jlm}\left|\lambda_{\theta Jlm}\right|\left[\mathbf{L}_{\theta Jlm}\left(\widetilde{\mathbf{U}}_{[i,j-1,k]}\left(r^\theta_i, \theta_{jm}, \phi_k\right) - \widetilde{\mathbf{U}}_{[i,j,k]}\left(r^\theta_i, \theta_{jm}, \phi_k\right)\right)\right]\right)\end{aligned} \tag{A.4}$$

$$\begin{aligned}\mathbf{F}^n_{\theta,i,jp,k} = &\frac{1}{2}\left(\mathbf{F}_\theta\left(\widetilde{\mathbf{U}}_{[i,j,k]}\left(r^\theta_i, \theta_{jp}, \phi_k\right)\right) + \mathbf{F}_\theta\left(\widetilde{\mathbf{U}}_{[i,j+1,k]}\left(r^\theta_i, \theta_{jp}, \phi_k\right)\right)\right) + \\ &\frac{1}{2}\left(\sum_J \mathbf{R}_{\theta Jlp}\left|\lambda_{\theta Jlp}\right|\left[\mathbf{L}_{\theta Jlp}\left(\widetilde{\mathbf{U}}_{[i,j,k]}\left(r^\theta_i, \theta_{jp}, \phi_k\right) - \widetilde{\mathbf{U}}_{[i,j+1,k]}\left(r^\theta_i, \theta_{jp}, \phi_k\right)\right)\right]\right)\end{aligned} \tag{A.5}$$

$$\begin{aligned}\mathbf{F}^n_{\phi,i,j,km} = &\frac{1}{2}\left(\mathbf{F}_\phi\left(\widetilde{\mathbf{U}}_{[i,j,k-1]}\left(r^\phi_i, \theta^\phi_j, \phi_{km}\right)\right) + \mathbf{F}_\phi\left(\widetilde{\mathbf{U}}_{[i,j,k]}\left(r^\phi_i, \theta^\phi_j, \phi_{km}\right)\right)\right) + \\ &\frac{1}{2}\left(\sum_J \mathbf{R}_{\phi Jlm}\left|\lambda_{\phi Jlm}\right|\left[\mathbf{L}_{\phi Jlm}\left(\widetilde{\mathbf{U}}_{[i,j,k-1]}\left(r^\phi_i, \theta^\phi_j, \phi_{km}\right) - \widetilde{\mathbf{U}}_{[i,j,k]}\left(r^\phi_i, \theta^\phi_j, \phi_{km}\right)\right)\right]\right)\end{aligned} \tag{A.6}$$

$$\mathbf{F}^n_{\phi,i,j,kp} = \frac{1}{2}\left(\mathbf{F}_\phi\left(\widetilde{\mathbf{U}}_{[i,j,k]}\left(r_i^\phi, \theta_j^\phi, \phi_{kp}\right)\right) + \mathbf{F}_\phi\left(\widetilde{\mathbf{U}}_{[i,j,k+1]}\left(r_i^\phi, \theta_j^\phi, \phi_{kp}\right)\right)\right) + \frac{1}{2}\left(\sum_J \mathbf{R}_{\phi Jlp}\left|\lambda_{\phi Jlp}\right|\left[\mathbf{L}_{\phi Jlp}\left(\widetilde{\mathbf{U}}_{[i,j,k]}\left(r_i^\phi, \theta_j^\phi, \phi_{kp}\right) - \widetilde{\mathbf{U}}_{[i,j,k+1]}\left(r_i^\phi, \theta_j^\phi, \phi_{kp}\right)\right)\right]\right) \tag{A.7}$$

Be care of the subscripts. For instance, the subscript $rJlm$ $(rJlp)$ represents the r direction, Jth component, the arithmetic average at the interface $l = r_{im}$ $(r_{ip}$ $)$ of the variable between the left and right sides. $\widetilde{\mathbf{U}}_{[i,j,k]}$ denotes a second-order accurate linear reconstruction [219] on cell with geometrical center (i, j, k):

$$\begin{aligned}
\widetilde{\mathbf{U}}_{[i,j,k]} &= \overline{\mathbf{U}}_{i,j,k} + (\widetilde{\partial_r \mathbf{U}})_{i,j,k}(r - \overline{r}_i) + (\widetilde{\partial_\theta \mathbf{U}})_{i,j,k}(\theta - \overline{\theta}_j) + (\widetilde{\partial_\phi \mathbf{U}})_{i,j,k}(\phi - \phi_k) \\
(\widetilde{\partial_r \mathbf{U}})_{i,j,k} &= \text{minmod}\left(\frac{\overline{\mathbf{U}}_{i+1,j,k} - \overline{\mathbf{U}}_{i,j,k}}{\overline{r}_{i+1} - \overline{r}_i}, \frac{\overline{\mathbf{U}}_{i,j,k} - \overline{\mathbf{U}}_{i-1,j,k}}{\overline{r}_i - \overline{r}_{i-1}}\right) \\
(\widetilde{\partial_\theta \mathbf{U}})_{i,j,k} &= \text{minmod}\left(\frac{\overline{\mathbf{U}}_{i,j+1,k} - \overline{\mathbf{U}}_{i,j,k}}{\overline{\theta}_{j+1} - \overline{\theta}_j}, \frac{\overline{\mathbf{U}}_{i,j,k} - \overline{\mathbf{U}}_{i,j-1,k}}{\overline{\theta}_j - \overline{\theta}_{j-1}}\right) \\
(\widetilde{\partial_\phi \mathbf{U}})_{i,j,k} &= \text{minmod}\left(\frac{\overline{\mathbf{U}}_{i,j,k+1} - \overline{\mathbf{U}}_{i,j,k}}{\phi_{k+1} - \phi_k}, \frac{\overline{\mathbf{U}}_{i,j,k} - \overline{\mathbf{U}}_{i,j,k-1}}{\phi_k - \phi_{k-1}}\right)
\end{aligned} \tag{A.8}$$

For the Powell's eight wave formulation of MHD equations [141], the Jacobian matrices $\mathbf{A} = \frac{\partial \mathbf{F}_r}{\partial \mathbf{W}}$, $\mathbf{B} = \frac{\partial \mathbf{F}_\theta}{\partial \mathbf{W}}$, and $\mathbf{C} = \frac{\partial \mathbf{F}_\phi}{\partial \mathbf{W}}$ corresponding to the primitive variables in r, θ, and ϕ directions are:

$$\mathbf{A} = \begin{pmatrix}
v_r & \rho & 0 & 0 & 0 & 0 & 0 & 0 \\
0 & v_r & 0 & 0 & \frac{1}{\rho} & 0 & \frac{B_\theta}{\mu\rho} & \frac{B_\phi}{\mu\rho} \\
0 & 0 & v_r & 0 & 0 & 0 & -\frac{B_r}{\mu\rho} & 0 \\
0 & 0 & 0 & v_r & 0 & 0 & 0 & -\frac{B_r}{\mu\rho} \\
0 & \gamma p & 0 & 0 & v_r & 0 & 0 & 0 \\
0 & 0 & 0 & 0 & 0 & v_r & 0 & 0 \\
0 & B_\theta & -B_r & 0 & 0 & 0 & v_r & 0 \\
0 & B_\phi & 0 & -B_r & 0 & 0 & 0 & v_r
\end{pmatrix}$$

$$\mathbf{B} = \begin{pmatrix}
v_\theta & 0 & \rho & 0 & 0 & 0 & 0 & 0 \\
0 & v_\theta & 0 & 0 & 0 & -\frac{B_\theta}{\mu\rho} & 0 & 0 \\
0 & 0 & v_\theta & 0 & \frac{1}{\rho} & \frac{B_r}{\mu\rho} & 0 & \frac{B_\phi}{\mu\rho} \\
0 & 0 & 0 & v_\theta & 0 & 0 & 0 & -\frac{B_\theta}{\mu\rho} \\
0 & 0 & \gamma p & 0 & v_\theta & 0 & 0 & 0 \\
0 & -B_\theta & B_r & 0 & 0 & v_\theta & 0 & 0 \\
0 & 0 & 0 & 0 & 0 & 0 & v_\theta & 0 \\
0 & 0 & B_\phi & -B_\theta & 0 & 0 & 0 & v_\theta
\end{pmatrix}$$

$$
\mathbf{C} = \begin{pmatrix}
v_\phi & 0 & 0 & \rho & 0 & 0 & 0 & 0 \\
0 & v_\phi & 0 & 0 & 0 & -\frac{B_\phi}{\mu\rho} & 0 & 0 \\
0 & 0 & v_\phi & 0 & 0 & 0 & -\frac{B_\phi}{\mu\rho} & 0 \\
0 & 0 & 0 & v_\phi & \frac{1}{\rho} & \frac{B_r}{\mu\rho} & \frac{B_\theta}{\mu\rho} & 0 \\
0 & 0 & 0 & \gamma p & v_\phi & 0 & 0 & 0 \\
0 & -B_\phi & 0 & B_r & 0 & v_\phi & 0 & 0 \\
0 & 0 & -B_\phi & B_\theta & 0 & 0 & v_\phi & 0 \\
0 & 0 & 0 & 0 & 0 & 0 & 0 & v_\phi
\end{pmatrix}
$$

In fact, the Jacobian of each direction may be transferred to each other.

$$
\mathbf{T}_{AB}\left(\mathbf{A}\left(\mathbf{T}_{AB}\mathbf{W}\right)\right)\mathbf{T}_{AB}^{-1} = \mathbf{T}_{AB}\begin{pmatrix}
v_\theta & \rho & 0 & 0 & 0 & 0 & 0 & 0 \\
0 & v_\theta & 0 & 0 & \frac{1}{\rho} & 0 & \frac{B_r}{\mu\rho} & \frac{B_\phi}{\mu\rho} \\
0 & 0 & v_\theta & 0 & 0 & 0 & -\frac{B_\theta}{\mu\rho} & 0 \\
0 & 0 & 0 & v_\theta & 0 & 0 & 0 & -\frac{B_\theta}{\mu\rho} \\
0 & \gamma p & 0 & 0 & v_\theta & 0 & 0 & 0 \\
0 & 0 & 0 & 0 & 0 & v_\theta & 0 & 0 \\
0 & B_r & -B_\theta & 0 & 0 & 0 & v_\theta & 0 \\
0 & B_\phi & 0 & -B_\theta & 0 & 0 & 0 & v_\theta
\end{pmatrix}\mathbf{T}_{AB}^{-1} = \mathbf{B}\left(\mathbf{W}\right)
$$

where

$$
\mathbf{T}_{AB} = \mathbf{T}_{AB}^{-1} = \begin{pmatrix}
1&0&0&0&0&0&0&0\\
0&0&1&0&0&0&0&0\\
0&1&0&0&0&0&0&0\\
0&0&0&1&0&0&0&0\\
0&0&0&0&1&0&0&0\\
0&0&0&0&0&0&1&0\\
0&0&0&0&0&1&0&0\\
0&0&0&0&0&0&0&1
\end{pmatrix}
$$

Similarly, one can also obtain the Jacobian matrix $\mathbf{C}\left(\mathbf{W}\right) = \mathbf{T}_{AC}\left(\mathbf{A}\left(\mathbf{T}_{AC}\mathbf{W}\right)\right)\mathbf{T}_{AC}^{-1}$, and

$$
\mathbf{T}_{AC} = \mathbf{T}_{AC}^{-1} = \begin{pmatrix}
1&0&0&0&0&0&0&0\\
0&0&0&1&0&0&0&0\\
0&0&1&0&0&0&0&0\\
0&1&0&0&0&0&0&0\\
0&0&0&0&1&0&0&0\\
0&0&0&0&0&0&0&1\\
0&0&0&0&0&0&1&0\\
0&0&0&0&0&1&0&0
\end{pmatrix}
$$

The corresponding eigenvalues and eigenvectors of $\mathbf{B}\left(\mathbf{W}\right)$ and $\mathbf{C}\left(\mathbf{W}\right)$ can be generated as follows:

$$
\lambda_{\theta_J} = \lambda_{r_J}\left(\mathbf{T}_{AB}\mathbf{W}\right), \mathbf{r}_{\theta_J} = \mathbf{T}_{AB}\left(\mathbf{r}_{r_J}\left(\mathbf{T}_{AB}\mathbf{W}\right)\right), \mathbf{l}_{\theta_J} = \left(\mathbf{l}_{r_J}\left(\mathbf{T}_{AB}\mathbf{W}\right)\right)\mathbf{T}_{AB}^{-1}
$$

$$\lambda_{\phi J} = \lambda_{rJ}\left(\mathbf{T}_{AC}\mathbf{W}\right), \mathbf{r}_{\phi J} = \mathbf{T}_{AC}\left(\mathbf{r}_{rJ}\left(\mathbf{T}_{AC}\mathbf{W}\right)\right), \mathbf{l}_{\phi J} = \left(\mathbf{l}_{rJ}\left(\mathbf{T}_{AC}\mathbf{W}\right)\right)\mathbf{T}_{AC}^{-1}$$

For simplicity, here is only listed the eigensystem corresponding to r direction.

$$\lambda_{rJlm} = \lambda_{rJ}\left(\mathbf{U}_{lm}\right) = \lambda_{rJ}\left(\frac{1}{2}\left(\widetilde{\mathbf{U}}_{[i-1,j,k]}(r_{im}, \theta_j^r, \phi_k) + \widetilde{\mathbf{U}}_{[i,j,k]}(r_{im}, \theta_j^r, \phi_k)\right)\right) \quad (A.9)$$

The corresponding eigenvectors $\mathbf{L}_{rJlm}$ and $\mathbf{R}_{rJlm}$ are as follows:

$$\mathbf{L}_{rJlm} = \left(\mathbf{l}_{rJ}\frac{\partial \mathbf{W}}{\partial \mathbf{U}}\right)(\mathbf{U}_{lm}) = \left(\mathbf{l}_{rJ}\frac{\partial \mathbf{W}}{\partial \mathbf{U}}\right)\left(\frac{1}{2}\left(\widetilde{\mathbf{U}}_{[i-1,j,k]}(r_{im}, \theta_j^r, \phi_k) + \widetilde{\mathbf{U}}_{[i,j,k]}(r_{im}, \theta_j^r, \phi_k)\right)\right)$$
$$\mathbf{R}_{rJlm} = \left(\frac{\partial \mathbf{U}}{\partial \mathbf{W}}\mathbf{r}_{rJ}\right)(\mathbf{U}_{lm}) = \left(\frac{\partial \mathbf{U}}{\partial \mathbf{W}}\mathbf{r}_{rJ}\right)\left(\frac{1}{2}\left(\widetilde{\mathbf{U}}_{[i-1,j,k]}(r_{im}, \theta_j^r, \phi_k) + \widetilde{\mathbf{U}}_{[i,j,k]}(r_{im}, \theta_j^r, \phi_k)\right)\right) \quad (A.10)$$

$$\frac{\partial \mathbf{W}}{\partial \mathbf{U}} = \begin{pmatrix} 1 & 0 & 0 & 0 & 0 & 0 & 0 & 0 \\ -\frac{v_r}{\rho} & \frac{1}{\rho} & 0 & 0 & 0 & 0 & 0 & 0 \\ -\frac{v_\theta}{\rho} & 0 & \frac{1}{\rho} & 0 & 0 & 0 & 0 & 0 \\ -\frac{v_\phi}{\rho} & 0 & 0 & \frac{1}{\rho r \sin\theta} & 0 & 0 & 0 & 0 \\ \alpha\zeta & -v_r\zeta & -v_\theta\zeta & -\frac{v_\phi\zeta}{r\sin\theta} & \zeta & -\frac{B_r\zeta}{\mu} & -\frac{B_\theta\zeta}{\mu} & -\frac{B_\phi\zeta}{\mu} \\ 0 & 0 & 0 & 0 & 0 & 1 & 0 & 0 \\ 0 & 0 & 0 & 0 & 0 & 0 & 1 & 0 \\ 0 & 0 & 0 & 0 & 0 & 0 & 0 & 1 \end{pmatrix} \quad (A.11)$$

$$\frac{\partial \mathbf{U}}{\partial \mathbf{W}} = \begin{pmatrix} 1 & 0 & 0 & 0 & 0 & 0 & 0 & 0 \\ v_r & \rho & 0 & 0 & 0 & 0 & 0 & 0 \\ v_\theta & 0 & \rho & 0 & 0 & 0 & 0 & 0 \\ v_\phi r\sin\theta & 0 & 0 & \rho r\sin\theta & 0 & 0 & 0 & 0 \\ \alpha & \rho v_r & \rho v_\theta & \rho v_\phi & \frac{1}{\zeta} & \frac{B_r}{\mu} & \frac{B_\theta}{\mu} & \frac{B_\phi}{\mu} \\ 0 & 0 & 0 & 0 & 0 & 1 & 0 & 0 \\ 0 & 0 & 0 & 0 & 0 & 0 & 1 & 0 \\ 0 & 0 & 0 & 0 & 0 & 0 & 0 & 1 \end{pmatrix} \quad (A.12)$$

where $\alpha = \frac{1}{2}(v_r^2 + v_\theta^2 + v_\phi^2)$ and $\zeta = \gamma - 1$.

For convenience, denote $c_{fr}, c_{sr}, c_{f\theta}, c_{s\theta}, c_{f\phi}, c_{s\phi}$ the fast and slow magnetosonic speeds in (r, θ, ϕ) directions

$$c_{fr} = \frac{1}{\sqrt{2}}\sqrt{c_s^2 + c_A^2 + \left(\left(c_s^2 + c_A^2\right)^2 - 4c_s^2\frac{B_r^2}{\mu\rho}\right)^{\frac{1}{2}}}, c_{sr} = \frac{1}{\sqrt{2}}\sqrt{c_s^2 + c_A^2 - \left(\left(c_s^2 + c_A^2\right)^2 - 4c_s^2\frac{B_r^2}{\mu\rho}\right)^{\frac{1}{2}}}$$

$$c_{f\theta} = \frac{1}{\sqrt{2}}\sqrt{c_s^2 + c_A^2 + \left(\left(c_s^2 + c_A^2\right)^2 - 4c_s^2\frac{B_\theta^2}{\mu\rho}\right)^{\frac{1}{2}}}, c_{s\theta} = \frac{1}{\sqrt{2}}\sqrt{c_s^2 + c_A^2 - \left(\left(c_s^2 + c_A^2\right)^2 - 4c_s^2\frac{B_\theta^2}{\mu\rho}\right)^{\frac{1}{2}}}$$

$$c_{f\phi} = \frac{1}{\sqrt{2}}\sqrt{c_s^2 + c_A^2 + \left(\left(c_s^2 + c_A^2\right)^2 - 4c_s^2\frac{B_\phi^2}{\mu\rho}\right)^{\frac{1}{2}}}, c_{s\phi} = \frac{1}{\sqrt{2}}\sqrt{c_s^2 + c_A^2 - \left(\left(c_s^2 + c_A^2\right)^2 - 4c_s^2\frac{B_\phi^2}{\mu\rho}\right)^{\frac{1}{2}}}$$

where $c_s = \sqrt{\frac{\gamma p}{\rho}}$ and $c_A = \sqrt{\frac{B_r^2+B_\theta^2+B_\phi^2}{\mu\rho}}$ are the sound and total Alfvénic speeds, respectively. $c_{ax} = \frac{B_x}{\sqrt{\mu\rho}}$ is the Alfvénic speed in x-direction with $x = r, \theta, \phi$.

The eigenvectors associated with those eigenvalues are:

$$\lambda_{r1} = v_r - c_{fr}$$

$$\mathbf{r}_{r1} = (\rho\alpha_f, -c_{fr}\alpha_f, c_{sr}\alpha_s\beta_\theta M, c_{sr}\alpha_s\beta_\phi M, \alpha_f\gamma p, 0, \alpha_s\sqrt{\mu\rho}c_s\beta_\theta, \alpha_s\sqrt{\mu\rho}c_s\beta_\phi)^T$$

$$\mathbf{l}_{r1} = (0, -\frac{c_{fr}\alpha_f}{2c_s^2}, \frac{c_{sr}\alpha_s\beta_\theta M}{2c_s^2}, \frac{c_{sr}\alpha_s\beta_\phi M}{2c_s^2}, \frac{\alpha_f}{2\rho c_s^2}, 0, \frac{\alpha_s\beta_\theta}{2\sqrt{\mu\rho}c_s}, \frac{\alpha_s\beta_\phi}{2\sqrt{\mu\rho}c_s})$$

$$\lambda_{r2} = v_r - |c_{ar}|$$

$$\mathbf{r}_{r2} = (0, 0, -\frac{\beta_\phi}{\sqrt{2}}, \frac{\beta_\theta}{\sqrt{2}}, 0, 0, -\beta_\phi M\sqrt{\frac{\mu\rho}{2}}, \beta_\theta M\sqrt{\frac{\mu\rho}{2}})^T$$

$$\mathbf{l}_{r2} = (0, 0, -\frac{\beta_\phi}{\sqrt{2}}, \frac{\beta_\theta}{\sqrt{2}}, 0, 0, -\frac{\beta_\phi M}{\sqrt{2\mu\rho}}, \frac{\beta_\theta M}{\sqrt{2\mu\rho}})$$

$$\lambda_{r3} = v_r - c_{sr}$$

$$\mathbf{r}_{r3} = (\rho\alpha_s, -c_{rs}\alpha_s, -c_{fr}\alpha_f\beta_\theta M, -c_{fr}\alpha_f\beta_\phi M, \alpha_s\gamma p, 0, -\alpha_f\sqrt{\mu\rho}c_s\beta_\theta, -\alpha_f\sqrt{\mu\rho}c_s\beta_\phi)^T$$

$$\mathbf{l}_{r3} = (0, -\frac{c_{sr}\alpha_s}{2c_s^2}, -\frac{c_{fr}\alpha_f\beta_\theta M}{2c_s^2}, -\frac{c_{fr}\alpha_f\beta_\phi M}{2\alpha^2}, \frac{\alpha_s}{2\rho c_s^2}, 0, -\frac{\alpha_f\beta_\theta}{2\sqrt{\mu\rho}c_s}, -\frac{\alpha_f\beta_\phi}{2\sqrt{\mu\rho}c_s})$$

$$\lambda_{r4} = v_r, \quad \mathbf{r}_{r4} = (1, 0, 0, 0, 0, 0, 0, 0)^T, \quad \mathbf{l}_{r4} = (1, 0, 0, 0, -\frac{1}{c_s^2}, 0, 0, 0)$$

$$\lambda_{r5} = v_r, \quad \mathbf{r}_{r5} = (0, 0, 0, 0, 0, 1, 0, 0)^T, \quad \mathbf{l}_{r5} = (0, 0, 0, 0, 0, 1, 0, 0)$$

$$\lambda_{r6} = v_r + c_{sr}$$

$$\mathbf{r}_{r6} = (\rho\alpha_s, c_{sr}\alpha_s, c_{fr}\alpha_f\beta_\theta M, c_{fr}\alpha_f\beta_\phi M, \alpha_s\gamma p, 0, -\alpha_f\sqrt{\mu\rho}c_s\beta_\theta, -\alpha_f\sqrt{\mu\rho}c_s\beta_\phi)^T$$

$$\mathbf{l}_{r6} = (0, \frac{c_{sr}\alpha_s}{2c_s^2}, \frac{c_{fr}\alpha_f\beta_\theta M}{2c_s^2}, \frac{c_{fr}\alpha_f\beta_\phi M}{2c_s^2}, \frac{\alpha_s}{2\rho c_s^2}, 0, -\frac{\alpha_f\beta_\theta}{2\sqrt{\mu\rho}c_s}, -\frac{\alpha_f\beta_\phi}{2\sqrt{\mu\rho}c_s})$$

$$\lambda_{r7} = v_r + |c_{ar}|$$

$$\mathbf{r}_{r7} = (0, 0, -\frac{\beta_\phi}{\sqrt{2}}, \frac{\beta_\theta}{\sqrt{2}}, 0, 0, \beta_\phi M\sqrt{\frac{\mu\rho}{2}}, -\beta_\theta M\sqrt{\frac{\mu\rho}{2}})^T$$

$$\mathbf{l}_{r7} = (0, 0, -\frac{\beta_\phi}{\sqrt{2}}, \frac{\beta_\theta}{\sqrt{2}}, 0, 0, \frac{\beta_\phi M}{\sqrt{2\mu\rho}}, -\frac{\beta_\theta M}{\sqrt{2\mu\rho}})$$

$$\lambda_{r8} = v_r + c_{fr}$$

$$\mathbf{r}_{r8} = (\rho\alpha_f, c_{fr}\alpha_f, -c_{sr}\alpha_s\beta_\theta M, -c_{sr}\alpha_s\beta_\phi M, \alpha_f\gamma p, 0, \alpha_s\sqrt{\mu\rho}c_s\beta_\theta, \alpha_s\sqrt{\mu\rho}c_s\beta_\phi)^T$$

$$\mathbf{l}_{r8} = (0, \frac{c_{fr}\alpha_f}{2c_s^2}, -\frac{c_{sr}\alpha_s\beta_\theta M}{2c_s^2}, -\frac{c_{sr}\alpha_s\beta_\phi M}{2c_s^2}, \frac{\alpha_f}{2\rho c_s^2}, 0, \frac{\alpha_s\beta_\theta}{2\sqrt{\mu\rho}c_s}, \frac{\alpha_s\beta_\phi}{2\sqrt{\mu\rho}c_s})$$

where

$$M = sign(B_r), \alpha_f^2 = \frac{c_s^2 - c_{sr}^2}{c_{fr}^2 - c_{sr}^2}, \alpha_s^2 = \frac{c_{fr}^2 - c_s^2}{c_{fr}^2 - c_{sr}^2}, \beta_\theta = \frac{B_\theta}{\sqrt{B_\theta^2 + B_\phi^2}}, \beta_\phi = \frac{B_\phi}{\sqrt{B_\theta^2 + B_\phi^2}}$$

The eigenvectors given above are orthogonal, and, since α_f, α_s, β_θ, and β_ϕ all lie between zero and one, the eigenvectors are all well-formed, once these four parameters are defined. The only difficulties in defining these occur when $B_\theta^2 + B_\phi^2 = 0$, in which case β_θ and β_ϕ are ill-defined, and when $B_\theta^2 + B_\phi^2 = 0$, and $B_r^2 = \mu\rho c_s^2$, in which case α_f and α_s are ill-defined. The first case is fairly trivial; β_θ and β_ϕ represent direction cosines for the tangential component of the **B**-field, and in the case of a zero component, it is only important to choose so that $\beta_\theta^2 + \beta_\phi^2 = 1$ [141] or $\beta_\theta = \beta_\phi = \frac{1}{\sqrt{2}}$ [16, 158]. An approach for the case in which α_f and α_s are ill-defined has been outlined in detail by Roe and Balsara [150]. Indeed, they have shown in Roe and Balsara [150] that multiple solutions in this case for the linearized Riemann problem does not affect the value for the interface flux.

In case of $B_\theta^2 + B_\phi^2 = 0$ (say, $B_\theta^2 + B_\phi^2 < 10^{-10}$), one can set $\beta_\theta = \beta_\phi = \frac{1}{\sqrt{2}}$ as done by Brio and Wu [16]. The second case of $|c_{ar}| = c_s$ and $B_\theta^2 + B_\phi^2 = 0$ corresponds to the magnetosonic case, in which $c_{fr}^2 = c_{sr}^2 = c_s^2$, and $\alpha_f^2 + \alpha_s^2 = 1$. In order to remove the singularities in eigenvectors, in case of $|c_{ar}| - c_s = 0, B_\theta^2 + B_\phi^2 = 0$ (say, $||c_{ar}| - c_s| < 10^{-10}$ and $B_\theta^2 + B_\phi^2 < 10^{-10}$), one can follow the formulation of Case V suggested by Roe and Balsara [150] or directly set $\alpha_f^2 = \alpha_s^2 = \frac{1}{2}$ [158]. Thus, one will have well-defined orthogonal eigenvectors.

Roe Type Flux in GLM Form

Using the GLM [35], the divergence constraint is coupled with the conservation laws by introducing a new variable ψ. Now, the coupling part can be written in the following form

$$\begin{aligned}
&\frac{\partial \mathbf{B}}{\partial t} + \nabla \cdot (\mathbf{vB} - \mathbf{Bv} + \psi I) = 0 \\
&\frac{\partial \psi}{\partial t} + \nabla \cdot \left(c_h^2 \mathbf{B}\right) = -\frac{c_h^2}{c_p^2}\psi
\end{aligned} \tag{A.13}$$

Thus, the governing equations involve the ninth variable in $\mathbf{U} = \left(\rho, \rho v_r, \rho v_\theta, \rho v_\phi r \sin\theta, e, B_r, B_\theta, B_\phi, \psi\right)^T$, with primitive variables $\mathbf{W} = \left(\rho, v_r, v_\theta, v_\phi, p, B_r, B_\theta, B_\phi, \psi\right)^T$. In the GLM approach, divergence errors are propagated to the domain boundaries at finite speed c_h and damped at a rate given by c_h^2/c_p^2. The parameter c_h is the constant advection velocity of ψ. In order to avoid bringing additional constraint on the time step, it is evaluated as the fastest eigenvalue in the whole computation domain $c_h = \max_{i,j,k}\left(|v_r| + c_{fr}, |v_\theta| + c_{f\theta}, |v_\phi| + c_{f\phi}\right)$. The choice of constant c_p will be given below. In the GLM formulation, computing the flux across the interface in each direction can follow from decoupling the equations of normal magnetic field and ψ from the other seven equations [35, 121, 122, 170]. That is, when solving a one dimensional Riemann problem at a zone interface (say the r direction), these additional waves are decoupled from the remaining ones and are described by the following 2×2 linear hyperbolic system without the source term:

$$\begin{cases}
\frac{\partial B_r}{\partial t} + \frac{1}{r^2}\frac{\partial r^2 \psi}{\partial r} = 0 \\
\frac{\partial \psi}{\partial t} + \frac{1}{r^2}\frac{\partial}{\partial r}\left(c_h^2 r^2 B_r\right) = 0
\end{cases} \tag{A.14}$$

For a generic pair of left and right input states (B_{rL}, ψ_L) and (B_{rR}, ψ_R), the Godunov flux at the interface l of the system (A.14) can be computed exactly [35, 121, 122, 170] as

$$F_{r,B_r} = \psi_l = \frac{\psi_L + \psi_R}{2} - \frac{c_h(B_{rR} - B_{rL})}{2} \tag{A.15}$$

$$F_{r,\psi} = c_h^2 B_{rl} = c_h^2\left(\frac{B_{rL} + B_{rR}}{2} - \frac{\psi_R - \psi_L}{2c_h}\right) \tag{A.16}$$

This allows us to carry out the solution of the 2×2 linear Riemann problem separately before using any standard 7-wave Riemann solver for the one-dimensional MHD equations. Similarly, one can obtain the flux corresponding to the θ and ϕ directions. Specifically, the flux of the magnetic field and ψ components at the cell interfaces are as follows:

$$\begin{aligned}
F^n_{r,B_r,im,j,k} = &\frac{1}{2}\left[\widetilde{\psi}_{[i-1,j,k]}(r_{im}, \theta_j^r, \phi_k) + \widetilde{\psi}_{[i,j,k]}(r_{im}, \theta_j^r, \phi_k)\right] + \\
&\frac{1}{2}c_h\left(\widetilde{B}_{r[i-1,j,k]}(r_{im}, \theta_j^r, \phi_k) - \widetilde{B}_{r[i,j,k]}(r_{im}, \theta_j^r, \phi_k)\right) \\
F^n_{r,\psi,im,j,k} = &c_h^2\left\{\frac{1}{2}\left(\widetilde{B}_{r[i-1,j,k]}(r_{im}, \theta_j^r, \phi_k) + \widetilde{B}_{r[i,j,k]}(r_{im}, \theta_j^r, \phi_k)\right) + \right.
\end{aligned}$$

$$
\begin{aligned}
&\frac{1}{2c_h}\left(\widetilde{\psi}_{[i-1,j,k]}(r_{im},\theta_j^r,\phi_k)-\widetilde{\psi}_{[i,j,k]}(r_{im},\theta_j^r,\phi_k)\right)\Bigg\} \\
F^n_{\theta,B_\theta,i,jm,k} =\ &\frac{1}{2}\left[\widetilde{\psi}_{[i,j-1,k]}(r_i^\theta,\theta_{jm},\phi_k)+\widetilde{\psi}_{[i,j,k]}(r_i^\theta,\theta_{jm},\phi_k)\right]+ \\
&\frac{1}{2}c_h\left(\widetilde{B}_{\theta[i,j-1,k]}(r_i^\theta,\theta_{jm},\phi_k)-\widetilde{B}_{\theta[i,j,k]}(r_i^\theta,\theta_{jm},\phi_k)\right) \\
F^n_{\theta,\psi,i,jm,k} =\ &c_h^2\left\{\frac{1}{2}\left(\widetilde{B}_{\theta[i,j-1,k]}(r_i^\theta,\theta_{jm},\phi_k)+\widetilde{B}_{\theta[i,j,k]}(r_i^\theta,\theta_{jm},\phi_k)\right)+\right. \\
&\left.\frac{1}{2c_h}\left(\widetilde{\psi}_{[i,j-1,k]}(r_i^\theta,\theta_{jm},\phi_k)-\widetilde{\psi}_{[i,j,k]}(r_i^\theta,\theta_{jm},\phi_k)\right)\right\} \\
F^n_{\phi,B_\phi,i,j,km} =\ &\frac{1}{2}\left\{\widetilde{\psi}_{[i,j,k-1]}(r_i^\phi,\theta_j^\phi,\phi_{km})+\widetilde{\psi}_{[i,j,k]}(r_i^\phi,\theta_j^\phi,\phi_{km})\right\} \\
&+\frac{1}{2}c_h\left(\widetilde{B}_{\phi[i,j,k-1]}(r_i^\phi,\theta_j^\phi,\phi_{km})-\widetilde{B}_{\phi[i,j,k]}(r_i^\phi,\theta_j^\phi,\phi_{km})\right) \\
F^n_{\phi,\psi,i,j,km} =\ &c_h^2\left\{\frac{1}{2}\left(\widetilde{B}_{\phi[i,j,k-1]}(r_i^\phi,\theta_j^\phi,\phi_{km})+\widetilde{B}_{\phi[i,j,k]}(r_i^\phi,\theta_j^\phi,\phi_{km})\right)+\right. \\
&\left.\frac{1}{2c_h}\left(\widetilde{\psi}_{[i,j,k-1]}(r_i^\phi,\theta_j^\phi,\phi_{km})-\widetilde{\psi}_{[i,j,k]}(r_i^\phi,\theta_j^\phi,\phi_{km})\right)\right\}
\end{aligned}
\tag{A.17}
$$

Besides the fluxes listed above, other flux components not given here are still computed using those in Eqs. (A.2)–(A.7). For convenience, one can rewrite the fluxes for GLM system here by setting $\mathbf{F}^{n*}_{r,im,j,k}\equiv\left(F^{n*}_{rq,im,j,k}\right)^T$, $\mathbf{F}^{n*}_{r,ip,j,k}\equiv\left(F^{n*}_{rq,ip,j,k}\right)^T$, $\mathbf{F}^{n*}_{\theta,i,jm,k}\equiv\left(F^{n*}_{\theta q,i,jm,k}\right)^T$, $\mathbf{F}^{n*}_{\theta,i,jp,k}\equiv\left(F^{n*}_{\theta q,i,jp,k}\right)^T$, $\mathbf{F}^{n*}_{\phi,i,j,km}\equiv\left(F^{n*}_{\phi q,i,j,km}\right)^T$ and $\mathbf{F}^{n*}_{\phi,i,j,kp}\equiv\left(\mathrm{F}^{n*}_{\phi q,i,j,kp}\right)^T$, where $q=1,2,\ldots,8,9$:

$$
\begin{aligned}
\mathbf{F}^{n*}_{r,im,j,k} =\ &\frac{1}{2}\left(\mathbf{F}_r\left(\widetilde{\mathbf{U}}_{[i-1,j,k]}\left(r_{im},\theta_j^r,\phi_k\right)\right)+\mathbf{F}_r\left(\widetilde{\mathbf{U}}_{[i,j,k]}\left(r_{im},\theta_j^r,\phi_k\right)\right)\right)+ \\
&\frac{1}{2}\left(\sum_J\mathbf{R}_{rJlm}\left|\lambda_{rJlm}\right|\left[\mathbf{L}_{rJlm}\left(\widetilde{\mathbf{U}}_{[i-1,j,k]}\left(r_{im},\theta_j^r,\phi_k\right)-\widetilde{\mathbf{U}}_{[i,j,k]}\left(r_{im},\theta_j^r,\phi_k\right)\right)\right]\right) \\
\mathbf{F}^{n*}_{r,ip,j,k} =\ &\frac{1}{2}\left(\mathbf{F}_r\left(\widetilde{\mathbf{U}}_{[i,j,k]}\left(r_{ip},\theta_j^r,\phi_k\right)\right)+\mathbf{F}_r\left(\widetilde{\mathbf{U}}_{[i+1,j,k]}\left(r_{ip},\theta_j^r,\phi_k\right)\right)\right)+ \\
&\frac{1}{2}\left(\sum_J\mathbf{R}_{rJlp}\left|\lambda_{rJlp}\right|\left[\mathbf{L}_{rJlp}\left(\widetilde{\mathbf{U}}_{[i,j,k]}\left(r_{ip},\theta_j^r,\phi_k\right)-\widetilde{\mathbf{U}}_{[i+1,j,k]}\left(r_{ip},\theta_j^r,\phi_k\right)\right)\right]\right) \\
\mathbf{F}^{n*}_{\theta,i,jm,k} =\ &\frac{1}{2}\left(\mathbf{F}_\theta\left(\widetilde{\mathbf{U}}_{[i,j-1,k]}\left(r_i^\theta,\theta_{jm},\phi_k\right)\right)+\mathbf{F}_\theta\left(\widetilde{\mathbf{U}}_{[i,j,k]}\left(r_i^\theta,\theta_{jm},\phi_k\right)\right)\right)+ \\
&\frac{1}{2}\left(\sum_J\mathbf{R}_{\theta Jlm}\left|\lambda_{\theta Jlm}\right|\left[\mathbf{L}_{\theta Jlm}\left(\widetilde{\mathbf{U}}_{[i,j-1,k]}\left(r_i^\theta,\theta_{jm},\phi_k\right)-\widetilde{\mathbf{U}}_{[i,j,k]}\left(r_i^\theta,\theta_{jm},\phi_k\right)\right)\right]\right) \\
\mathbf{F}^{n*}_{\theta,i,jp,k} =\ &\frac{1}{2}\left(\mathbf{F}_\theta\left(\widetilde{\mathbf{U}}_{[i,j,k]}\left(r_i^\theta,\theta_{jp},\phi_k\right)\right)+\mathbf{F}_\theta\left(\widetilde{\mathbf{U}}_{[i,j+1,k]}\left(r_i^\theta,\theta_{jp},\phi_k\right)\right)\right)+
\end{aligned}
$$

$$\frac{1}{2}\left(\sum_{J}\mathbf{R}_{\theta Jlp}\left|\lambda_{\theta Jlp}\right|\left[\mathbf{L}_{\theta Jlp}\left(\widetilde{\mathbf{U}}_{[i,j,k]}\left(r_i^{\theta},\theta_{jp},\phi_k\right)-\widetilde{\mathbf{U}}_{[i,j+1,k]}\left(r_i^{\theta},\theta_{jp},\phi_k\right)\right)\right]\right)$$

$$\mathbf{F}^{n*}_{\phi,i,j,km}=\frac{1}{2}\left(\mathbf{F}_{\phi}\left(\widetilde{\mathbf{U}}_{[i,j,k-1]}\left(r_i^{\phi},\theta_j^{\phi},\phi_{km}\right)\right)+\mathbf{F}_{\phi}\left(\widetilde{\mathbf{U}}_{[i,j,k]}\left(r_i^{\phi},\theta_j^{\phi},\phi_{km}\right)\right)\right)+$$
$$\frac{1}{2}\left(\sum_{J}\mathbf{R}_{\phi Jlm}\left|\lambda_{\phi Jlm}\right|\left[\mathbf{L}_{\phi Jlm}\left(\widetilde{\mathbf{U}}_{[i,j,k-1]}\left(r_i^{\phi},\theta_j^{\phi},\phi_{km}\right)-\widetilde{\mathbf{U}}_{[i,j,k]}\left(r_i^{\phi},\theta_j^{\phi},\phi_{km}\right)\right)\right]\right)$$

$$\mathbf{F}^{n*}_{\phi,i,j,kp}=\frac{1}{2}\left(\mathbf{F}_{\phi}\left(\widetilde{\mathbf{U}}_{[i,j,k]}\left(r_i^{\phi},\theta_j^{\phi},\phi_{kp}\right)\right)+\mathbf{F}_{\phi}\left(\widetilde{\mathbf{U}}_{[i,j,k+1]}\left(r_i^{\phi},\theta_j^{\phi},\phi_{kp}\right)\right)\right)+$$
$$\frac{1}{2}\left(\sum_{J}\mathbf{R}_{\phi Jlp}\left|\lambda_{\phi Jlp}\right|\left[\mathbf{L}_{\phi Jlp}\left(\widetilde{\mathbf{U}}_{[i,j,k]}\left(r_i^{\phi},\theta_j^{\phi},\phi_{kp}\right)-\widetilde{\mathbf{U}}_{[i,j,k+1]}\left(r_i^{\phi},\theta_j^{\phi},\phi_{kp}\right)\right)\right]\right) \tag{A.18}$$

where $\mathbf{U}=\left(\rho,\rho v_r,\rho v_\theta,\rho v_\phi r\sin\theta,e,B_r,B_\theta,B_\phi,\psi\right)^T$, and the summation in J is carried out for all eigenvalues except $\pm c_h$, with the eigenvalues corresponding to r direction being $\lambda_{r1}=v_r-c_{fr}$, $\lambda_{r2}=v_r-|c_{ar}|$, $\lambda_{r3}=v_r-c_{sr}$, $\lambda_{r4}=v_r$, $\lambda_{r5}=v_r+c_{sr}$, $\lambda_{r6}=v_r+|c_{ar}|$, $\lambda_{r7}=v_r+c_{fr}$, and the corresponding eigenvectors $\mathbf{l}_{rJ}$ and $\mathbf{r}_{rJ}$, with respect to primitive variables $\mathbf{W}=\left(\rho,v_r,v_\theta,v_\phi,p,B_r,B_\theta,B_\phi,\psi\right)^T$. That is, seven of the nine eigenvalues of the GLM system are the same as the Powell's eight-wave form Powell et al. [141], with two different eigenvalues $\pm c_h$. The corresponding eigenvectors for these λ_J's are almost the same but presently with the ninth component being zero associated with ψ, except that the eigenvectors corresponding to the eigenvalues $\pm c_h$ are complex, however, those eigenvectors associated with the eigenvalues $\pm c_h$ are not used actually as seen from Eqs. (A.17). The eigenvalues on the interfaces are expressed in the same way as in Eq. (A.9) by $\lambda_{rJlm}=\lambda_{rJ}\left(\mathbf{U}_{lm}\right)=\lambda_{rJ}\left(\frac{1}{2}\left(\widetilde{\mathbf{U}}_{[i-1,j,k]}(r_{im},\theta_j^r,\phi_k)+\widetilde{\mathbf{U}}_{[i,j,k]}(r_{im},\theta_j^r,\phi_k)\right)\right)$, while the corresponding eigenvectors $\mathbf{L}_{rJlm}$ and $\mathbf{R}_{rJlm}$ corresponding to the conservative variables $\mathbf{U}=\left(\rho,\rho v_r,\rho v_\theta,\rho v_\phi r\sin\theta,e,B_r,B_\theta,B_\phi,\psi\right)^T$ can be calculated by following Eq. (A.10): $\mathbf{L}_{rJlm}=\left(\mathbf{l}_{rJ}\frac{\partial\mathbf{W}}{\partial\mathbf{U}}\right)\left(\mathbf{U}_{lm}\right)$, $\mathbf{R}_{rJlm}=\left(\frac{\partial\mathbf{U}}{\partial\mathbf{W}}\mathbf{r}_{rJ}\right)\left(\mathbf{U}_{lm}\right)$. Note that $\frac{\partial\mathbf{W}}{\partial\mathbf{U}}$ and $\frac{\partial\mathbf{U}}{\partial\mathbf{W}}$ are 9×9 matrices, whose left-upper blocks are the same as those in Eqs. (A.11) and (A.12), and right-bottom blocks have only one element 1.

Combining the reconstructed fluxes in Eqs. (A.17) and (A.18), one can obtain the final fluxes $\mathbf{F}^n_{r,im,j,k}$, $\mathbf{F}^n_{r,ip,j,k}$, $\mathbf{F}^n_{\theta,i,jm,k}$, $\mathbf{F}^n_{\theta,i,jp,k}$, $\mathbf{F}^n_{\phi,i,j,km}$ and $\mathbf{F}^n_{\phi,i,j,kp}$ for the GLM system:

$$\mathbf{F}^n_{r,im,j,k}=\left(F^{n*}_{r1,im,j,k},F^{n*}_{r2,im,j,k},F^{n*}_{r3,im,j,k},F^{n*}_{r4,im,j,k},F^{n*}_{r5,im,j,k},F^{n}_{r,B_r,im,j,k},\right.$$
$$\left.F^{n*}_{r7,im,j,k},F^{n*}_{r8,im,j,k},F^{n}_{r,\psi,im,j,k}\right)^T$$

$$\mathbf{F}^n_{r,ip,j,k}=\left(F^{n*}_{r1,ip,j,k},F^{n*}_{r2,ip,j,k},F^{n*}_{r3,ip,j,k},F^{n*}_{r4,ip,j,k},F^{n*}_{r5,ip,j,k},F^{n}_{r,B_r,ip,j,k},F^{n*}_{r7,ip,j,k},\right.$$
$$\left.F^{n*}_{r8,ip,j,k},F^{n}_{r,\psi,ip,j,k}\right)^T$$

$$\mathbf{F}^n_{\theta,i,jm,k} = \left(F^{n*}_{\theta1,i,jm,k}, F^{n*}_{\theta2,i,jm,k}, F^{n*}_{\theta3,i,jm,k}, F^{n*}_{\theta4,i,jm,k}, F^{n*}_{\theta5,i,jm,k}, F^{n*}_{\theta6,i,jm,k}, F^{n}_{\theta,B_\theta,i,jm,k}, F^{n*}_{r\theta8,i,jm,k}, F^{n}_{\theta,\psi,i,jm,k}\right)^T$$

$$\mathbf{F}^n_{\theta,i,jp,k} = \left(F^{n*}_{\theta1,i,jp,k}, F^{n*}_{\theta2,i,jp,k}, F^{n*}_{\theta3,i,jp,k}, F^{n*}_{\theta4,i,jp,k}, F^{n*}_{\theta5,i,jp,k}, F^{n*}_{\theta6,i,jp,k}, F^{n}_{\theta,B_\theta,i,jp,k}, F^{n*}_{r\theta8,i,jp,k}, F^{n}_{\theta,\psi,i,jp,k}\right)^T$$

$$\mathbf{F}^n_{\phi,i,j,km} = \left(F^{n*}_{\phi1,i,j,km}, F^{n*}_{\phi2,i,j,km}, F^{n*}_{\phi3,i,j,km}, F^{n*}_{\phi4,i,j,km}, F^{n*}_{\phi5,i,j,km}, F^{n*}_{\phi6,i,j,km}, F^{n*}_{\phi7,i,j,km}, F^{n}_{\phi,B_\phi,i,j,km}, F^{n}_{\phi,\psi,i,j,km}\right)^T$$

$$\mathbf{F}^n_{\phi,i,j,kp} = \left(F^{n*}_{\phi1,i,j,kp}, F^{n*}_{\phi2,i,j,kp}, F^{n*}_{\phi3,i,j,kp}, F^{n*}_{\phi4,i,j,kp}, F^{n*}_{\phi5,i,j,kp}, F^{n*}_{\phi6,i,j,kp}, F^{n*}_{\phi7,i,j,kp}, F^{n}_{\phi,B_\phi,i,j,kp}, F^{n}_{\phi,\psi,i,j,kp}\right)^T \quad \text{(A.19)}$$

It should be noted that the interface values ψ_l and B_{rl} from Eqs. (A.15) and (A.16) are used as more as possible for flux calculations. According to Harten et al. [70, 141], if a so-called "entropy fix" is not applied to Roe's scheme, expansion shocks can be permitted. The entropy fix is applied to the magnetosonic waves to bound those eigenvalues away from zero when the flow is expanding. This is done by replacing $|\lambda_{Jl}|$ with $|\lambda^*_{Jl}|$ in the discretized flux formulations above. Note that the replacement is implemented for the values of $|\lambda_{Jl}|$ $(v_r - c_{fr}, v_r - c_{sr}, v_r + c_{sr}, v_r + c_{fr})$ corresponding to the magnetoacoustic waves only, with $|\lambda^*_{Jl}|$ given by

$$|\lambda^*_{Jl}| = \begin{cases} |\lambda_{Jl}| & |\lambda_{Jl}| \geq \frac{\delta\lambda_J}{2} \\ \frac{\lambda^2_{Jl}}{\delta\lambda_J} + \frac{\delta\lambda_J}{4} & |\lambda_{Jl}| < \frac{\delta\lambda_J}{2} \end{cases} \quad \text{(A.20)}$$

where

$$\delta\lambda_J = \max\left(4\left(\lambda_{JR} - \lambda_{JL}\right), 0\right)$$

Another entropy correction formula suggested by Kermani and Plett [95] runs as follows:

$$|\lambda^*_{Jl}| = \begin{cases} \frac{\lambda^2_{Jl}+\epsilon^2}{2\epsilon} & |\lambda_{Jl}| < \epsilon \\ |\lambda_{Jl}| & |\lambda_{Jl}| \geq \epsilon \end{cases} \quad \text{(A.21)}$$

where ϵ has two choices with $\epsilon_1 = 2\max\left[0, \left(\lambda_{JR} - \lambda_{JL}\right)\right]$ and $\epsilon_2 = 4\max\left[0, \left(\lambda_{Jl} - \lambda_{JL}\right), \left(\lambda_{Jl} - \lambda_{JR}\right)\right]$. All these entropy fixing methods in Eqs. (A.20) and (A.21) are employed in our code.

Up to now, we are ready to solve Eq. (A.1) with the fluxes defined by Eq. (A.19). Finally, supplemented with the initial condition $\bar{\psi}^{n*}_{i,j,k}$ given by the output of the most recent step, one can solve the initial value problem obtained by the last equation of (A.14) without the $\nabla \cdot \mathbf{B}$ term

$$\frac{\partial \psi}{\partial t} = -\frac{c_h^2}{c_p^2}\psi \tag{A.22}$$

by integrating exactly for a time increment Δt_n, yielding

$$\psi_{i,j,k}^{n+1} = \exp(-\Delta t_n \frac{c_h^2}{c_p^2})\bar{\psi}_{i,j,k}^{n*} \tag{A.23}$$

The constant c_p^2 has the dimension of length squared over time and thus can be regarded as a diffusion coefficient. Dedner et al. [35] chose $c_p = \sqrt{0.18 c_h}$, depending on their numerical experiment. However, this definition is incomplete, since c_p^2/c_h has the dimension of length and thus it is not a dimensionless quantity. Thus, another choice defined by the dimensionless parameter $\alpha = \frac{\Delta h c_h}{c_p^2} \in [0, 1]$ is suggested [121, 122, 170], and can be regarded as the ratio of the diffusive and advective time scales, i.e., $\alpha = \frac{\Delta t_d}{\Delta t_a}$, where $\Delta t_d = \Delta h^2/c_p^2$ and $\Delta t_a = \Delta h/c_h$. Δh is the minimum grid spacing $\left(\min_{i,j,k}(\Delta r(i), r(i)\Delta\theta(j), r(i)\sin\theta(j)\Delta\phi(k))\right)$. The third choice $c_p = \sqrt{-\frac{\Delta t}{\ln c_d} c_h^2}$ with $c_d \in (0, 1)$ has been used [35, 69, 88, 202], where $c_h = \frac{\text{CFL}}{\Delta t}\Delta h$. If one chooses $c_d = \exp(-\alpha\text{CFL})$, then the later two choices becomes the same. In STEP-MHD code $\alpha = 0.1$, ψ is set to be zero initially and ψ is fixed at the inner boundary and outer boundary.

HLL Type Flux

$$\begin{aligned}
\mathbf{F}_{r,im,j,k}^{n} = &-\frac{a_{im,j^r,k}^{+} a_{im,j^r,k}^{-}}{a_{im,j^r,k}^{+} - a_{im,j^r,k}^{-}} \left(\widetilde{\mathbf{U}}_{[i-1,j,k]}(r_{im}, \theta_j^r, \phi_k) - \widetilde{\mathbf{U}}_{[i,j,k]}(r_{im}, \theta_j^r, \phi_k)\right) \\
&+ \frac{a_{im,j^r,k}^{+} \mathbf{F}_r \left(\widetilde{\mathbf{U}}_{[i-1,j,k]}(r_{im}, \theta_j^r, \phi_k), \widetilde{\mathbf{B}}_{[i-1,j,k]}(r_{im}, \theta_j^r, \phi_k)\right)}{a_{im,j^r,k}^{+} - a_{im,j^r,k}^{-}} \\
&- \frac{a_{im,j^r,k}^{-} \mathbf{F}_r \left(\widetilde{\mathbf{U}}_{[i,j,k]}(r_{im}, \theta_j^r, \phi_k), \widetilde{\mathbf{B}}_{[i,j,k]}(r_{im}, \theta_j^r, \phi_k)\right)}{a_{im,j^r,k}^{+} - a_{im,j^r,k}^{-}}
\end{aligned}$$

$$\begin{aligned}
\mathbf{F}_{\theta,i,jm,k}^{n} = &-\frac{b_{i^\theta,jm,k}^{+} b_{i^\theta,jm,k}^{-}}{b_{i^\theta,jm,k}^{+} - b_{i^\theta,jm,k}^{-}} \left(\widetilde{\mathbf{U}}_{[i,j-1,k]}(r_i^\theta, \theta_{jm}, \phi_k) - \widetilde{\mathbf{U}}_{[i,j,k]}(r_i^\theta, \theta_{jm}, \phi_k)\right) \\
&+ \frac{b_{i^\theta,jm,k}^{+} \mathbf{F}_\theta \left(\widetilde{\mathbf{U}}_{[i,j-1,k]}(r_i^\theta, \theta_{jm}, \phi_k), \widetilde{\mathbf{B}}_{[i,j-1,k]}(r_i^\theta, \theta_{jm}, \phi_k)\right)}{b_{i^\theta,jm,k}^{+} - b_{i^\theta,jm,k}^{-}}
\end{aligned}$$

$$- \frac{b^{-}_{i^\theta, jm,k} \mathbf{F}_\theta \left(\widetilde{\mathbf{U}}_{[i,j,k]}(r_i^\theta, \theta_{jm}, \phi_k), \widetilde{\mathbf{B}}_{[i,j,k]}(r_i^\theta, \theta_{jm}, \phi_k)\right)}{b^{+}_{i^\theta, jm,k} - b^{-}_{i^\theta, jm,k}}$$

$$\begin{aligned}
\mathbf{F}^n_{\phi,i,j,km} = &- \frac{c^{+}_{i^\phi, j^\phi, km} c^{-}_{i^\phi, j^\phi, km}}{c^{+}_{i^\phi, j^\phi, km} - c^{-}_{i^\phi, j^\phi, km}} \left(\widetilde{\mathbf{U}}_{[i,j,k-1]}(r_i^\phi, \theta_j^\phi, \phi_{km}) - \widetilde{\mathbf{U}}_{[i,j,k]}(r_i^\phi, \theta_j^\phi, \phi_{km})\right) \\
&+ \frac{c^{+}_{i^\phi, j^\phi, km} \mathbf{F}_\phi \left(\widetilde{\mathbf{U}}_{[i,j,k-1]}(r_i^\phi, \theta_j^\phi, \phi_{km}), \widetilde{\mathbf{B}}_{[i,j,k-1]}(r_i^\phi, \theta_j^\phi, \phi_{km})\right)}{c^{+}_{i^\phi, j^\phi, km} - c^{-}_{i^\phi, j^\phi, km}} \\
&- \frac{c^{-}_{i^\phi, j^\phi, km} \mathbf{F}_\phi \left(\widetilde{\mathbf{U}}_{[i,j,k]}(r_i^\phi, \theta_j^\phi, \phi_{km}), \widetilde{\mathbf{B}}_{[i,j,k]}(r_i^\phi, \theta_j^\phi, \phi_{km})\right)}{c^{+}_{i^\phi, j^\phi, km} - c^{-}_{i^\phi, j^\phi, km}}
\end{aligned}$$

Parameters $a^{+}_{im,j^r,k}, a^{-}_{im,j^r,k}, b^{+}_{i^\theta,jm,k}, b^{-}_{i^\theta,jm,k}, c^{+}_{i^\phi,j^\phi,km}, c^{-}_{i^\phi,j^\phi,km}$ involved in the flux functions are defined by the maximum and minimum speed on the centers of six faces

$$\begin{aligned}
a^{+}_{im,j^r,k} &= \max \left\{ (\widetilde{v}_r + \widetilde{c}_{fr})_{[i-1,j,k]}(r_{im}, \theta_j^r, \phi_k), (\widetilde{v}_r + \widetilde{c}_{fr})_{[i,j,k]}(r_{im}, \theta_j^r, \phi_k), 0 \right\} \\
a^{-}_{im,j^r,k} &= \min \left\{ (\widetilde{v}_r - \widetilde{c}_{fr})_{[i-1,j,k]}(r_{im}, \theta_j^r, \phi_k), (\widetilde{v}_r - \widetilde{c}_{fr})_{[i,j,k]}(r_{im}, \theta_j^r, \phi_k), 0 \right\} \\
b^{+}_{i^\theta,jm,k} &= \max \left\{ (\widetilde{v}_\theta + \widetilde{c}_{f\theta})_{[i,j-1,k]}(r_i^\theta, \theta_{jm}, \phi_k), (\widetilde{v}_\theta + \widetilde{c}_{f\theta})_{[i,j,k]}(r_i^\theta, \theta_{jm}, \phi_k), 0 \right\} \\
b^{-}_{i^\theta,jm,k} &= \min \left\{ (\widetilde{v}_\theta - \widetilde{c}_{f\theta})_{[i,j-1,k]}(r_i^\theta, \theta_{jm}, \phi_k), (\widetilde{v}_\theta - \widetilde{c}_{f\theta})_{[i,j,k]}(r_i^\theta, \theta_{jm}, \phi_k), 0 \right\} \\
c^{+}_{i^\phi,j^\phi,km} &= \max \left\{ (\widetilde{v}_\phi + \widetilde{c}_{f\phi})_{[i,j,k-1]}(r_i^\phi, \theta_j^\phi, \phi_{km}), (\widetilde{v}_\phi + \widetilde{c}_{f\phi})_{[i,j,k]}(r_i^\phi, \theta_j^\phi, \phi_{km}), 0 \right\} \\
c^{-}_{i^\phi,j^\phi,km} &= \min \left\{ (\widetilde{v}_\phi - \widetilde{c}_{f\phi})_{[i,j,k-1]}(r_i^\phi, \theta_j^\phi, \phi_{km}), (\widetilde{v}_\phi - \widetilde{c}_{f\phi})_{[i,j,k]}(r_i^\phi, \theta_j^\phi, \phi_{km}), 0 \right\}
\end{aligned} \tag{A.24}$$

CT Formulation

The CT method advances the variables $\mathbf{V} = \left(\rho, \rho v_r, \rho v_\theta, \rho v_\phi r \sin\theta, e\right)^T$ at the centroid point $(\overline{r}_i, \overline{\theta}_j, \phi_k)$ and $\mathbf{B} = \left(B_r, B_\theta, B_\phi\right)$ at the face center. In the derivation of numerical scheme with CT, the governing MHD equations are divided into two coupled subsystems of equations: a fluid part Eq. (3.14) and a magnetic induction part Eq. (3.15). The numerical formulation of fluid part Eq. (3.14) for the vector $\mathbf{V} = \left(\rho, \rho v_r, \rho v_\theta, \rho v_\phi r \sin\theta, e\right)^T$ can be derived from the first five arguments's equations for $\mathbf{U}$ in Eq. (3.2), where $\mathbf{G}_r$, $\mathbf{G}_\theta$, $\mathbf{G}_\phi$, $\mathbf{Q}$ are the first five arguments in $\mathbf{F}_r$, $\mathbf{F}_\theta$, $\mathbf{F}_\phi$, and $\mathbf{S}_1 + \mathbf{S}_2$. Hence, the discretized formulation (3.20) for $\mathbf{V}$ follows from those of the first five arguments associated with Eq. (A.1).

Following Eq. (3.24) in Ziegler [219], by averaging Eqs. (3.15) over facial areas of a cell with geometrical center (i, j, k), one can obtain the semi-discrete form Eq. (3.21) separately as follows

$$\frac{dB_{r,im,j^r,k}}{dt} = -\frac{\sin\theta_{jp}E^*_{\phi,im,jp,k} - \sin\theta_{jm}E^*_{\phi,im,jm,k}}{r_{im}(\cos\theta_{jm} - \cos\theta_{jp})} + \frac{\Delta\theta(j)}{r_{im}(\cos\theta_{jm} - \cos\theta_{jp})}\frac{E^*_{\theta,im,j,kp} - E^*_{\theta,im,j,km}}{\Delta\phi(k)}$$

$$\frac{dB_{\theta,i^\theta,jm,k}}{dt} = -\frac{2}{(r_{im} + r_{ip})\sin\theta_{jm}}\frac{E^*_{r,i,jm,kp} - E^*_{r,i,jm,km}}{\Delta\phi(k)} + \frac{2\left(r_{ip}E^*_{\phi,ip,jm,k} - r_{im}E^*_{\phi,im,jm,k}\right)}{r_{ip}^2 - r_{im}^2}$$

$$\frac{dB_{\phi,i^\phi,j^\phi,km}}{dt} = -\frac{2\left(r_{ip}E^*_{\theta,ip,j,km} - r_{im}E^*_{\theta,im,j,km}\right)}{r_{ip}^2 - r_{im}^2} + \frac{2}{r_{im} + r_{ip}}\frac{E^*_{r,i,jp,km} - E^*_{r,i,jm,km}}{\Delta\theta(j)}$$

Here, $E^*_{r,i,jm,km}$, $E^*_{\theta,im,j,km}$, $E^*_{\phi,im,jm,k}$ are the reconstructed values of the electric fields at cell edge $(r_i, \theta_{jm}, \phi_{km})$, $(r_{im}, \theta_j, \phi_{km})$, $(r_{im}, \theta_{jm}, \phi_k)$, defined respectively by

$$E^*_{r,i,jm,km} = -\frac{b^-_{i,jm,km}b^+_{i,jm,km}}{b^+_{i,jm,km} - b^-_{i,jm,km}}\left(\widetilde{B}_{\phi,i,jm+,km} - \widetilde{B}_{\phi,i,jm-,km}\right) + \frac{c^-_{i,jm,km}c^+_{i,jm,km}}{c^+_{i,jm,km} - c^-_{i,jm,km}}$$
$$\times\left(\widetilde{B}_{\theta,i,jm,km+} - \widetilde{B}_{\theta,i,jm,km-}\right) - \frac{1}{(b^+_{i,jm,km} - b^-_{i,jm,km})(c^+_{i,jm,km} - c^-_{i,jm,km})}$$
$$\times\Big(b^+_{i,jm,km}c^-_{i,jm,km}\widetilde{E}_{r,i,jm-,km+} - b^-_{i,jm,km}c^-_{i,jm,km}\widetilde{E}_{r,i,jm+,km+}$$
$$- b^+_{i,jm,km}c^+_{i,jm,km}\widetilde{E}_{r,i,jm-,km-} + b^-_{i,jm,km}c^+_{i,jm,km}\widetilde{E}_{r,i,jm+,km-}\Big)$$

$$E^*_{\theta,im,j,km} = \frac{a^-_{im,j,km}a^+_{im,j,km}}{a^+_{im,j,km} - a^-_{im,j,km}}\left(\widetilde{B}_{\phi,im+,j,km} - \widetilde{B}_{\phi,im-,j,km}\right) - \frac{c^-_{im,j,km}c^+_{im,j,km}}{c^+_{im,j,km} - c^-_{im,j,km}}$$
$$\times\left(\widetilde{B}_{r,im,j,km+} - \widetilde{B}_{r,im,j,km-}\right) - \frac{1}{(a^+_{im,j,km} - a^-_{im,j,km})(c^+_{im,j,km} - c^-_{im,j,km})}$$
$$\times\Big(a^+_{im,j,km}c^-_{im,j,km}\widetilde{E}_{\theta,im-,j,km+} - a^-_{im,j,km}c^-_{im,j,km}\widetilde{E}_{\theta,im+,j,km+}$$
$$- a^+_{im,j,km}c^+_{im,j,km}\widetilde{E}_{\theta,im-,j,km-} + a^-_{im,j,km}c^+_{im,j,km}\widetilde{E}_{\theta,im+,j,km-}\Big)$$

$$
\begin{aligned}
E^*_{\phi,im,jm,k} = & -\frac{a^-_{im,jm,k}a^+_{im,jm,k}}{a^+_{im,jm,k}-a^-_{im,jm,k}}\left(\widetilde{B}_{\theta,im+,jm,k}-\widetilde{B}_{\theta,im-,jm,k}\right)+\frac{b^-_{im,jm,k}b^+_{im,jm,k}}{b^+_{im,jm,k}-b^-_{im,jm,k}} \\
& \times\left(\widetilde{B}_{r,im,jm+,k}-\widetilde{B}_{r,im,jm-,k}\right)-\frac{1}{(b^+_{im,jm,k}-b^-_{im,jm,k})(a^+_{im,jm,k}-a^-_{im,jm,k})} \\
& \times\left(a^+_{im,jm,k}b^-_{im,jm,k}\widetilde{E}_{\phi,im-,jm+,k}-a^-_{im,jm,k}b^-_{im,jm,k}\widetilde{E}_{\phi,im+,jm+,k}\right. \\
& \left.-a^+_{im,jm,k}b^+_{im,jm,k}\widetilde{E}_{\phi,im-,jm-,k}+a^-_{im,jm,k}b^+_{im,jm,k}\widetilde{E}_{\phi,im+,jm-,k}\right)
\end{aligned}
$$

where

$$
\begin{aligned}
\widetilde{E}_{r,i,jm+,km+} &= -\left(\widetilde{v}_\theta\widetilde{B}_\phi-\widetilde{v}_\phi\widetilde{B}_\theta\right)_{[i,j,k]}(r_i,\theta_{jm},\phi_{km}) \\
\widetilde{E}_{r,i,jm-,km+} &= -\left(\widetilde{v}_\theta\widetilde{B}_\phi-\widetilde{v}_\phi\widetilde{B}_\theta\right)_{[i,j-1,k]}(r_i,\theta_{jm},\phi_{km}) \\
\widetilde{E}_{r,i,jm+,km-} &= -\left(\widetilde{v}_\theta\widetilde{B}_\phi-\widetilde{v}_\phi\widetilde{B}_\theta\right)_{[i,j,k-1]}(r_i,\theta_{jm},\phi_{km}) \\
\widetilde{E}_{r,i,jm-,km-} &= -\left(\widetilde{v}_\theta\widetilde{B}_\phi-\widetilde{v}_\phi\widetilde{B}_\theta\right)_{[i,j-1,k-1]}(r_i,\theta_{jm},\phi_{km}) \\
\widetilde{B}_{r[i,j,k]}(r_{im},\theta_{jm},\phi_k) &= \widetilde{B}_{1r[i,j,k]}(r_{im},\theta_{jm},\phi_k)+\widetilde{B}_{0r[i,j,k]}(r_{im},\theta_{jm},\phi_k) \\
\widetilde{B}_{r,im,j,km+} &= \widetilde{B}_{r[i,j,k]}(r_{im},\theta_j,\phi_{km}),\ \widetilde{B}_{r,im,j,km-}=\widetilde{B}_{r[i,j,k-1]}(r_{im},\theta_j,\phi_{km}) \\
\widetilde{B}_{r,im,jm+,k} &= \widetilde{B}_{r[i,j,k]}(r_{im},\theta_{jm},\phi_k),\ \widetilde{B}_{r,im,jm-,k}=\widetilde{B}_{r[i,j-1,k]}(r_{im},\theta_{jm},\phi_k)
\end{aligned}
$$

The formulation in θ and ϕ directions is similar to that in the r direction, which is omitted here. It is noted that, $(\widetilde{v}_r,\widetilde{v}_\theta,\widetilde{v}_\phi)$ is defined by Eq. (A.8) as above, while $(\widetilde{B}_r,\widetilde{B}_\theta,\widetilde{B}_\phi)$ is decided by the second-order magnetic field construction proposed in Sect. 3.2.4 by Ziegler [219]

$$
\begin{aligned}
\widetilde{B}_{r[i,j,k]}(\mathbf{r}) = & \frac{1}{2}\left(B_{r,ip,j^r,k}+B_{r,im,j^r,k}\right)+\left(B_{r,ip,j^r,k}-B_{r,im,j^r,k}\right)\frac{r-r_i}{\Delta r(i)} \\
& +\left[\widetilde{(\partial_\theta B_r)}_{ip,j,k}\frac{r-r_{im}}{\Delta r(i)}+\widetilde{(\partial_\theta B_r)}_{im,j,k}\frac{r_{ip}-r}{\Delta r(i)}\right](\theta-\theta^r_j) \\
& +\left[\widetilde{(\partial_\phi B_r)}_{ip,j,k}\frac{r-r_{im}}{\Delta r(i)}+\widetilde{(\partial_\phi B_r)}_{im,j,k}\frac{r_{ip}-r}{\Delta r(i)}\right](\phi-\phi_k)
\end{aligned}
$$

$$
\begin{aligned}
\widetilde{B}_{\theta[i,j,k]}(\mathbf{r}) = & \frac{1}{2}\left(B_{\theta,i^\theta,jp,k}+B_{\theta,i^\theta,jm,k}\right)+\left(B_{\theta,i^\theta,jp,k}-B_{\theta,i^\theta,jm,k}\right)\frac{\theta-\theta_j}{\Delta\theta(j)} \\
& +\left[\widetilde{(\partial_r B_\theta)}_{i,jp,k}\frac{\theta-\theta_{jm}}{\Delta\theta(j)}+\widetilde{(\partial_r B_\theta)}_{i,jm,k}\frac{\theta_{jp}-\theta}{\Delta\theta(j)}\right](r-r^\theta_i) \\
& +\left[\widetilde{(\partial_\phi B_\theta)}_{i,jp,k}\frac{\theta-\theta_{jm}}{\Delta\theta(j)}+\widetilde{(\partial_\phi B_\theta)}_{i,jm,k}\frac{\theta_{jp}-\theta}{\Delta\theta(j)}\right](\phi-\phi_k)
\end{aligned}
$$

$$\widetilde{B}_{\phi[i,j,k]}(\mathbf{r}) = \frac{1}{2}\left(B_{\phi,i^\phi,j^\phi,kp} + B_{\phi,i^\phi,j^\phi,km}\right) + \left(B_{\phi,i^\phi,j^\phi,kp} - B_{\phi,i^\phi,j^\phi,km}\right)\frac{\phi-\phi_k}{\Delta\phi(k)}$$
$$+ \left[\widetilde{(\partial_r B_\phi)}_{i,j,kp}\frac{\phi-\phi_{km}}{\Delta\phi(k)} + \widetilde{(\partial_r B_\phi)}_{i,j,km}\frac{\phi_{kp}-\phi}{\Delta\phi(k)}\right](r - r_i^\phi)$$
$$+ \left[\widetilde{(\partial_\theta B_\phi)}_{i,j,kp}\frac{\phi-\phi_{km}}{\Delta\phi(k)} + \widetilde{(\partial_\theta B_\phi)}_{i,j,km}\frac{\phi_{kp}-\phi}{\Delta\phi(k)}\right](\theta - \theta_j^\phi)$$

where $\widetilde{(\partial_\theta B_r)}_{i\pm\frac{1}{2},j,k}$ and other cross derivative terms are slope-limited approximations to the exact derivatives at cell faces. For example, using the minmod limiter,

$$\widetilde{(\partial_\theta B_r)}_{i\pm\frac{1}{2},j,k} = \text{minmod}\left\{\frac{B_{r,i\pm\frac{1}{2},(j+1)^r,k} - B_{r,i\pm\frac{1}{2},j^r,k}}{\theta_{j+1}^r - \theta_j^r}, \frac{B_{r,i\pm\frac{1}{2},j^r,k} - B_{r,i\pm\frac{1}{2},(j-1)^r,k}}{\theta_j^r - \theta_{j-1}^r}\right\}$$
$$\widetilde{(\partial_\phi B_r)}_{i\pm\frac{1}{2},j,k} = \text{minmod}\left\{\frac{B_{r,i\pm\frac{1}{2},j^r,k+1} - B_{r,i\pm\frac{1}{2},j^r,k}}{\phi_{k+1} - \phi_k}, \frac{B_{r,i\pm\frac{1}{2},j^r,k} - B_{r,i\pm\frac{1}{2},j^r,k-1}}{\phi_k - \phi_{k-1}}\right\}$$
$$\widetilde{(\partial_r B_\theta)}_{i,j\pm\frac{1}{2},k} = \text{minmod}\left\{\frac{B_{\theta,(i+1)^\theta,j\pm\frac{1}{2},k} - B_{\theta,i^\theta,j\pm\frac{1}{2},k}}{r_{i+1}^\theta - r_i^\theta}, \frac{B_{\theta,i^\theta,j\pm\frac{1}{2},k} - B_{\theta,(i-1)^\theta,j\pm\frac{1}{2},k}}{r_i^\theta - r_{i-1}^\theta}\right\}$$
$$\widetilde{(\partial_\phi B_\theta)}_{i,j\pm\frac{1}{2},k} = \text{minmod}\left\{\frac{B_{\theta,i^\theta,j\pm\frac{1}{2},k+1} - B_{\theta,i^\theta,j\pm\frac{1}{2},k}}{\phi_{k+1} - \phi_k}, \frac{B_{\theta,i^\theta,j\pm\frac{1}{2},k} - B_{\theta,i^\theta,j\pm\frac{1}{2},k-1}}{\phi_k - \phi_{k-1}}\right\}$$
$$\widetilde{(\partial_r B_\phi)}_{i,j,k\pm\frac{1}{2}} = \text{minmod}\left\{\frac{B_{\phi,(i+1)^\phi,j^\phi,k\pm\frac{1}{2}} - B_{\phi,i^\phi,j^\phi,k\pm\frac{1}{2}}}{r_{i+1}^\phi - r_i^\phi}, \frac{B_{\phi,i^\phi,j^\phi,k\pm\frac{1}{2}} - B_{\phi,(i-1)^\phi,j^\phi,k\pm\frac{1}{2}}}{r_i^\phi - r_{i-1}^\phi}\right\}$$
$$\widetilde{(\partial_\theta B_\phi)}_{i,j,k\pm\frac{1}{2}} = \text{minmod}\left\{\frac{B_{\phi,i^\phi,(j+1)^\phi,k\pm\frac{1}{2}} - B_{\phi,i^\phi,j^\phi,k\pm\frac{1}{2}}}{\theta_{j+1}^\phi - \theta_j^\phi}, \frac{B_{\phi,i^\phi,j^\phi,k\pm\frac{1}{2}} - B_{\phi,i^\phi,(j-1)^\phi,k\pm\frac{1}{2}}}{\theta_j^\phi - \theta_{j-1}^\phi}\right\}$$

As claimed by Ziegler [219], the above magnetic field reconstruction meets the following three features: (1) magnetic flux conservation

$$\langle B_r\rangle_{i\pm\frac{1}{2},j,k}^r = \frac{1}{S_{r,i\pm\frac{1}{2},j,k}}\int_{S_{r,i\pm\frac{1}{2},j,k}}\widetilde{B}_{r[i,j,k]}(i\pm\frac{1}{2},\theta',\phi')dS' \approx B_r(r_{i\pm\frac{1}{2}},\theta_j^r,\phi_k)$$
$$= B_{r,i\pm\frac{1}{2},j^r,k}$$
$$\langle B_\theta\rangle_{i,j\pm\frac{1}{2},k}^\theta = \frac{1}{S_{\theta,i,j\pm\frac{1}{2},k}}\int_{S_{\theta,i,j\pm\frac{1}{2},k}}\widetilde{B}_{\theta[i,j,k]}(r',j\pm\frac{1}{2},\phi')dS' \approx B_\theta(r_i^\theta,\theta_{j\pm\frac{1}{2}},\phi_k)$$
$$= B_{\theta,i^\theta,j\pm\frac{1}{2},k}$$
$$\langle B_\phi\rangle_{i,j,k\pm\frac{1}{2}}^\phi = \frac{1}{S_{\phi,i,j,k\pm\frac{1}{2}}}\int_{S_{\phi,i,j,k\pm\frac{1}{2}}}\widetilde{B}_{\phi[i,j,k]}(r',\theta',k\pm\frac{1}{2})dS' \approx B_\phi(r_i^\phi,\theta_j^\phi,\phi_{k\pm\frac{1}{2}})$$
$$= B_{\phi,i^\phi,j^\phi,k\pm\frac{1}{2}}$$

(2) continuity of $\widetilde{B}_r$ ($\widetilde{B}_\theta$, $\widetilde{B}_\phi$) in the r (θ, ϕ)-coordinate, i.e., $\widetilde{B}_{r[i,j,k]}(r_{im},\theta,\phi) = \widetilde{B}_{r[i-1,j,k]}(r_{im},\theta,\phi)$, $\widetilde{B}_{\theta[i,j,k]}(r,\theta_{jm},\phi) = \widetilde{B}_{\theta[i,j-1,k]}(r,\theta_{jm},\phi)$, and $\widetilde{B}_{\phi[i,j,k]}(r,\theta,\phi_{km}) = \widetilde{B}_{\phi[i,j,k-1]}(r,\theta,\phi_{km})$. Due to the continuity property of magnetic field components the longitudinal derivatives $\partial_r B_r$, $\partial_\theta B_\theta$, $\partial_\phi B_\phi$ can be approximated in a straight-

forward manner by using the facial magnetic field values without any slope limitation needed.

(3) Solenoidality of $\widetilde{\mathbf{B}}_{[i,jk]}$ on the discretization scale, i.e., $\left(\overline{\nabla \cdot \widetilde{\mathbf{B}}_{[i,jk]}}\right)_{i,j,k} = 0$. It is noted that the application of Gauss' law to $\left(\overline{\nabla \cdot \widetilde{\mathbf{B}}_{[i,jk]}}\right)_{i,j,k} = 0$ yields $\left[\Delta S_r \langle \widetilde{B}_{r[i,j,k]}\rangle^r\right]_{im}^{ip} + \left[\Delta S_\theta \langle \widetilde{B}_{\theta[i,j,k]}\rangle^\theta\right]_{jm}^{jp} + \left[\Delta S_\phi \langle \widetilde{B}_{\phi[i,j,k]}\rangle^\phi\right]_{km}^{kp} = 0$, which follows from the magnetic flux conservation and continuity stated here. Thus, one can infer that $\left(\overline{\nabla \cdot \widetilde{\mathbf{B}}_{[i,jk]}}\right)_{i,j,k} = \left(\overline{\nabla \cdot \mathbf{B}}\right)_{i,j,k}$ keeps its initial zero value on the discretization scale in the computation, since the initial magnetic field is cellwisely divergence-free and $\frac{d}{dt}\left(\overline{\nabla \cdot \mathbf{B}}\right)_{i,j,k} = 0$ by the present construction of spatial discretization. However, this method can not claim the more strict condition of pointwise solenoidality.

Similar to the above formulation given by Eq. (A.24), $a^+_{im,j,k}$, $a^-_{im,j,k}$, $b^+_{i,jm,k}$, $b^-_{i,jm,k}$, $c^+_{i,j,km}$, and $c^-_{i,j,km}$ are defined by the maximum and minimum speeds on the geometrical center of six faces

$$
\begin{aligned}
a^+_{im,j,k} &= \max\left\{(\widetilde{v}_r + \widetilde{c}_{fr})_{[i-1,j,k]}(r_{im}, \theta_j, \phi_k),\ (\widetilde{v}_r + \widetilde{c}_{fr})_{[i,j,k]}(r_{im}, \theta_j, \phi_k), 0\right\} \\
a^-_{im,j,k} &= \min\left\{(\widetilde{v}_r - \widetilde{c}_{fr})_{[i-1,j,k]}(r_{im}, \theta_j, \phi_k),\ (\widetilde{v}_r - \widetilde{c}_{fr})_{[i,j,k]}(r_{im}, \theta_j, \phi_k), 0\right\} \\
b^+_{i,jm,k} &= \max\left\{(\widetilde{v}_\theta + \widetilde{c}_{f\theta})_{[i,j-1,k]}(r_i, \theta_{jm}, \phi_k),\ (\widetilde{v}_\theta + \widetilde{c}_{f\theta})_{[i,j,k]}(r_i, \theta_{jm}, \phi_k), 0\right\} \\
b^-_{i,jm,k} &= \min\left\{(\widetilde{v}_\theta - \widetilde{c}_{f\theta})_{[i,j-1,k]}(r_i, \theta_{jm}, \phi_k),\ (\widetilde{v}_\theta - \widetilde{c}_{f\theta})_{[i,j,k]}(r_i, \theta_{jm}, \phi_k), 0\right\} \\
c^+_{i,j,km} &= \max\left\{(\widetilde{v}_\phi + \widetilde{c}_{f\phi})_{[i,j,k-1]}(r_i, \theta_j, \phi_{km}),\ (\widetilde{v}_\phi + \widetilde{c}_{f\phi})_{[i,j,k]}(r_i, \theta_j, \phi_{km}), 0\right\} \\
c^-_{i,j,km} &= \min\left\{(\widetilde{v}_\phi - \widetilde{c}_{f\phi})_{[i,j,k-1]}(r_i, \theta_j, \phi_{km}),\ (\widetilde{v}_\phi - \widetilde{c}_{f\phi})_{[i,j,k]}(r_i, \theta_j, \phi_{km}), 0\right\}
\end{aligned}
$$

In the same way, $a^+_{im,j-1,k}$, $a^-_{im,j-1,k}$, $a^+_{im,j,k-1}$, $a^-_{im,j,k-1}$, $b^+_{i-1,jm,k}$, $b^-_{i-1,jm,k}$, $b^+_{i,jm,k-1}$, $b^-_{i,jm,k-1}$, $c^+_{i-1,j,km}$, $c^-_{i-1,j,km}$, $c^+_{i,j-1,km}$, $c^-_{i,j-1,km}$ can be given. By using these maximum and minimum speeds on the geometrical center of six faces, the maximum and minimum speeds in the edges involved in $E^*_{r,i,jm,km}$, $E^*_{\theta,im,j,km}$, $E^*_{\phi,im,jm,k}$ can be defined as follows

$$
\begin{aligned}
a^+_{im,jm,k} &= \max\left\{a^+_{im,j,k}, a^+_{im,j-1,k}\right\}, \quad a^-_{im,jm,k} = \min\left\{a^-_{im,j,k}, a^-_{im,j-1,k}\right\} \\
a^+_{im,j,km} &= \max\left\{a^+_{im,j,k}, a^+_{im,j,k-1}\right\}, \quad a^-_{im,j,km} = \min\left\{a^-_{im,j,k}, a^-_{im,j,k-1}\right\} \\
b^+_{im,jm,k} &= \max\left\{b^+_{i,jm,k}, b^+_{i-1,jm,k}\right\}, \quad b^-_{im,jm,k} = \min\left\{b^-_{i,jm,k}, b^-_{i-1,jm,k}\right\} \\
b^+_{i,jm,km} &= \max\left\{b^+_{i,jm,k}, b^+_{i,jm,k-1}\right\}, \quad b^-_{i,jm,km} = \min\left\{b^-_{i,jm,k}, b^-_{i,jm,k-1}\right\} \\
c^+_{im,j,km} &= \max\left\{c^+_{i,j,km}, c^+_{i-1,j,km}\right\}, \quad c^-_{im,j,km} = \min\left\{c^-_{i,j,km}, c^-_{i-1,j,km}\right\} \\
c^+_{i,jm,km} &= \max\left\{c^+_{i,j,km}, c^+_{i,j-1,km}\right\}, \quad c^-_{i,jm,km} = \min\left\{c^-_{i,j,km}, c^-_{i,j-1,km}\right\}
\end{aligned}
$$

MPP Flux-corrected Method

Positivity-preserving stands for the capability to preserve the positivity of the determinative physical constraints for the solution to be positive within physically-permitted bounds. In ideal MHD, density ρ and pressure p should be positive, but one can encounter the negativeness of density and pressure when simulating problems with low plasma β, high Mach number, or much large magnetic energy compared with internal energy. Even with divergence-free methods, such positivity property is not always satisfied by approximated solutions. When negative density or/and pressure occur, numerical instability may develop so as to cause the breakdown of numerical simulations. Therefore, guaranteeing such determinative properties is a must under consideration in a numerical design.

The positivity-preserving approach introduced in Sect. 3.5.4 is usually the so-called "dual energy" formulation. According to some switching conditions based on the ratio ϵ/e between the internal energy ϵ and the total energy e or the magnitude of plasma β (usually in small internal energy and/or low plasma β regions), the entropy equation or pressure equation or the internal energy density equation is used to replace the conservative energy equation [10, 17, 20, 50, 57, 58, 106, 153, 176, 184]. Evidently, this kind of dual energy method is sensitive to the "switching conditions" besides its computational overhead. For some comments about the use of conservative versus nonconservative forms of the energy equations, the reader may refer to Sect. 2.7 of Usmanov et al. [180].

Recent years have witnessed some progresses in constructing conservative positivity-preserving schemes for hyperbolic conservation laws. Specifically, some efforts work on the positivity-preserving in Riemann solvers or the evolved solution polynomials in a finite volume or discontinuous Galerkin (DG) method for both hydrodynamic and MHD flows [19, 22, 42, 60, 86, 105, 116, 128, 139, 198, 199, 204, 205, 209, 209, 212–215, 217]. As demonstrated formerly (e.g., [13, 14, 139, 188]), the positivity property may fail to hold even with positivity-preserving numerical fluxes considered, positivity can be retained by further restricting the reconstructed states (e.g., [8, 59, 61, 62, 188]).

Also, parametrized MPP flux-corrected limiting procedure is an effective measure to preserve the positivity of density and the pressure for finite difference, finite volume and DG methods. Hu et al. [79] suggested a flux cut-off limiter for finite difference WENO schemes to maintain the positivity of density and pressure for the compressible Euler system. It should be mentioned that the MPP flux limiter [108, 203] or later improvement through the parametrized positivity-preserving flux limiter [201] has been developed for the compressible Euler system or hyperbolic conservation laws. Parent [136] suggested a class of flux-limited schemes for conservation laws of both high-resolution and positivity-preserving property through the use of the Steger-Warming method and TVD flux limiters. Christlieb et al. [29] developed a class of high order positivity-preserving finite difference WENO methods for the ideal MHD equations under the CT framework. Moreover, Wu [194] designed a high-order, density and pressure positivity-preserving methods for the general relativistic hydrodynamic (GRHD) equations with a general equation of state and carried it out at every step of the high-order strong stability preserving (SSP) time integration methods. For more details about the positivity-preserving limiters or MPP flux-corrected methods, see the existing literature (e.g., [26, 28, 29, 89, 109, 194, 200, 201, 203, 204, 209]) and references therein.

The positivity-preserving flux limiting approach is to seek a convex combination of the first-order monotone flux with the high-order flux, in the hope of that such combination can achieve both positivity-preserving property and high-order accuracy under certain conditions, e.g. some mild time step constraint. Popularly, positivity-preserving of the first-order Lax-Friedrichs scheme plays a pivotal role in the parametrized positivity-preserving flux-corrected limiting procedure. As pointed out by Zhang and Shu [212–214] (also Remark 2.4 of Zhang and Shu [212]), the local Lax-Friedrichs flux preserves positivity of both density and pressure. Xiong et al. [200, 201] showed that the global maximum principle can be preserved while the high order accuracy of the base scheme is maintained, when applied in incompressible flows and the compressible Euler equations, respectively. Through error analysis on one-dimensional non-uniform meshes, Christlieb et al. [28] stated that the proposed MPP flux limiting schemes can maintain high order of accuracy. Cheng et al. [22, 22] proved that the simple first-order local Lax-Friedrichs scheme coupled with forward Euler time discretization is positivity-preserving for one-dimensional MHD under the restriction CFL ≤ 0.5. Christlieb et al. [26, 29] also demonstrated that the corrected flux maintains the designed order of accuracy of the high-order flux for ideal MHD equations. Moreover, the overall scheme can also take the same CFL constraint of 0.5 as that of the local Lax-Friedrichs scheme [29].

The following is devoted to the MPP flux limiting technique for a second order positivity-preserving Godunov finite volume HLL method implemented for the solar wind plasma MHD equations. Based on the underlying first order building block of positivity-preserving Lax-Friedrichs, and the CT or GLM framework, the MPP solver described here is built with slightly extra CFL constraints due to spherical geometrical factor. In practice, the construction of the following positivity-preserving formulation abides by the framework of Christlieb et al. [29].

In order to establish positivity-preserving HLL scheme, Eqs. (3.17) or Eq. (A.1) with HLL type flux defined above can be used as the base scheme, where the second-order accurate linear reconstruction Eq. (A.8) is employed. Then, the variables at the centroid point $(\overline{r}_i, \overline{\theta}_j, \phi_k)$ can be advanced with a second-order Runge–Kutta scheme in Eq. (3.18). For clarity, rewrite these two stages explicitly

$$\overline{\mathbf{U}}^* = \overline{\mathbf{U}}^n + \Delta t \mathcal{R}_{\mathbf{U}}[\overline{\mathbf{U}}^n] \tag{A.25}$$

$$\overline{\mathbf{U}}^{**} = \overline{\mathbf{U}}^* + \Delta t \mathcal{R}_{\mathbf{U}}[\overline{\mathbf{U}}^*] \tag{A.26}$$

$$\overline{\mathbf{U}}^{n+1} = \frac{1}{2}\overline{\mathbf{U}}^n + \frac{1}{2}\overline{\mathbf{U}}^{**} = \overline{\mathbf{U}}^n + \frac{\Delta t}{2}\left(\mathcal{R}_{\mathbf{U}}[\overline{\mathbf{U}}^n] + \mathcal{R}_{\mathbf{U}}[\overline{\mathbf{U}}^*]\right) \tag{A.27}$$

By writing out Eq. (A.26) one has

$$\overline{\mathbf{U}}^{**} = \overline{\mathbf{U}}^{*} - \Delta t \left[\frac{3\left(r_{ip}^2 \mathbf{F}^{*}_{r,ip,j,k} - r_{im}^2 \mathbf{F}^{*}_{r,im,j,k}\right)}{\left(r_{ip}^3 - r_{im}^3\right)} + \frac{\sin\theta_{jp}\mathbf{F}^{*}_{\theta,i,jp,k} - \sin\theta_{jm}\mathbf{F}^{*}_{\theta,i,jm,k}}{r_i^{\theta}\left(\cos\theta_{jm} - \cos\theta_{jp}\right)} \right.$$
$$\left. + \frac{\Delta\theta(j)}{r_i^{\phi}\left(\cos\theta_{jm} - \cos\theta_{jp}\right)} \frac{\mathbf{F}^{*}_{\phi,i,j,kp} - \mathbf{F}^{*}_{\phi,i,j,km}}{\Delta\phi(k)} \right]\Bigg|_{\overline{\mathbf{U}}^{*}} + \Delta t \overline{\mathbf{S}}^{*}_{i,j,k}\Big|_{\overline{\mathbf{U}}^{*}}$$

where $\mathbf{F}^{*}$ is the flux evaluated at $\overline{\mathbf{U}}^{*}$ and $\mathbf{S}^{*}$ takes its value at $\overline{\mathbf{U}}^{*}$. Then substitute $\overline{\mathbf{U}}^{**}$ into Eq. (A.27) to obtain

$$\overline{\mathbf{U}}^{n+1} = \overline{\mathbf{U}}^{n} - \Delta t \left[\frac{3\left(r_{ip}^2 \mathbf{F}^{rk}_{r,ip,j,k} - r_{im}^2 \mathbf{F}^{rk}_{r,im,j,k}\right)}{\left(r_{ip}^3 - r_{im}^3\right)} + \frac{\sin\theta_{jp}\mathbf{F}^{rk}_{\theta,i,jp,k} - \sin\theta_{jm}\mathbf{F}^{rk}_{\theta,i,jm,k}}{r_i^{\theta}\left(\cos\theta_{jm} - \cos\theta_{jp}\right)} \right.$$
$$\left. + \frac{\Delta\theta(j)}{r_i^{\phi}\left(\cos\theta_{jm} - \cos\theta_{jp}\right)} \frac{\mathbf{F}^{rk}_{\phi,i,j,kp} - \mathbf{F}^{rk}_{\phi,i,j,km}}{\Delta\phi(k)} \right] + \Delta t \mathbf{S}^{rk}_{i,j,k} \tag{A.28}$$

where $\mathbf{F}^{rk}_{r,ip,j,k} = \frac{1}{2}(\mathbf{F}^{n}_{r,ip,j,k} + \mathbf{F}^{*}_{r,ip,j,k})$, $\mathbf{F}^{rk}_{\theta,i,jp,k} = \frac{1}{2}(\mathbf{F}^{n}_{\theta,i,jp,k} + \mathbf{F}^{*}_{\theta,i,jp,k})$, $\mathbf{F}^{rk}_{\phi,i,j,kp} = \frac{1}{2}$ $(\mathbf{F}^{n}_{\phi,i,j,kp} + \mathbf{F}^{*}_{\phi,i,j,kp})$, $\mathbf{S}^{rk}_{i,j,k} = \frac{1}{2}\left[\overline{\mathbf{S}}_{i,j,k} + \overline{\mathbf{S}}^{*}_{i,j,k}\right]$. Here, rk stands for Runge–Kutta integration method with two stages.

Before proceeding to state 3D case of positivity-preserving formulation, it is convenient to denote fluxes $\mathbf{F}_r$, $\mathbf{F}_\theta$, $\mathbf{F}_\phi$ as $\mathbf{f}$, $\mathbf{g}$, $\mathbf{h}$ when used in the first-order Lax-Friedrichs scheme. By combining the fluxes $\mathbf{f}$, $\mathbf{g}$, $\mathbf{h}$ in the first-order Lax-Friedrichs scheme and those in our HLL type scheme, one can configure the following fluxes

$$\begin{cases} \hat{\mathbf{F}}_{ip} = \omega_{ip,j,k}\left(\mathbf{F}^{rk}_{r,ip,j,k} - \mathbf{f}^{n}_{ip,j,k}\right) + \mathbf{f}^{n}_{ip,j,k} \\ \hat{\mathbf{F}}_{im} = \omega_{im,j,k}\left(\mathbf{F}^{rk}_{r,im,j,k} - \mathbf{f}^{n}_{im,j,k}\right) + \mathbf{f}^{n}_{im,j,k} \\ \hat{\mathbf{F}}_{jp} = \omega_{i,jp,k}\left(\mathbf{F}^{rk}_{\theta,i,jp,k} - \mathbf{g}^{n}_{i,jp,k}\right) + \mathbf{g}^{n}_{i,jp,k} \\ \hat{\mathbf{F}}_{jm} = \omega_{i,jm,k}\left(\mathbf{F}^{rk}_{\theta,i,jm,k} - \mathbf{g}^{n}_{i,jm,k}\right) + \mathbf{g}^{n}_{i,jm,k} \\ \hat{\mathbf{F}}_{kp} = \omega_{i,j,kp}\left(\mathbf{F}^{rk}_{\phi,i,j,kp} - \mathbf{h}^{n}_{i,j,kp}\right) + \mathbf{h}^{n}_{i,j,kp} \\ \hat{\mathbf{F}}_{km} = \omega_{i,j,km}\left(\mathbf{F}^{rk}_{\phi,i,j,km} - \mathbf{h}^{n}_{i,j,km}\right) + \mathbf{h}^{n}_{i,j,km} \end{cases} \tag{A.29}$$

By replacing $\mathbf{F}^{rk}$ in Eq. (A.28) with $\hat{\mathbf{F}}$ in Eqs. (A.29), (A.28) becomes

$$\overline{\mathbf{U}}^{n+1}(\omega) = \overline{\mathbf{U}}^{n} - \Delta t \left[\frac{3\left(r_{ip}^2 \hat{\mathbf{F}}_{ip} - r_{im}^2 \hat{\mathbf{F}}_{im}\right)}{\left(r_{ip}^3 - r_{im}^3\right)} + \frac{\sin\theta_{jp}\hat{\mathbf{F}}_{jp} - \sin\theta_{jm}\hat{\mathbf{F}}_{jm}}{r_i^{\theta}\left(\cos\theta_{jm} - \cos\theta_{jp}\right)} + \right.$$
$$\left. \frac{\Delta\theta(j)}{r_i^{\phi}\left(\cos\theta_{jm} - \cos\theta_{jp}\right)} \frac{\hat{\mathbf{F}}_{kp} - \hat{\mathbf{F}}_{km}}{\Delta\phi(k)} \right] + \Delta t \mathbf{S}^{rk}_{i,j,k} \tag{A.30}$$

where $\overline{\mathbf{U}}^{n+1}(\omega)$ stands for the solution obtained from Eq. (A.30) with the modified fluxes in Eq. (A.29) and $\omega = (\omega_{im,j,k}, \omega_{ip,j,k}, \omega_{i,jm,k}, \omega_{i,jp,k}, \omega_{i,j,km}, \omega_{i,j,kp})$. At this time, $\hat{\mathbf{F}}$ in Eq. (A.29) completes the flux modification of $\mathbf{F}^{rk}$ in Eq. (A.28). This procedure modifies the fluxes only at the final state (A.27), and is in practice favorable due to the time cost consideration. It should be mentioned that this kind of flux modification can be carried out in both time integration stages (A.25) and (A.26). Then to guarantee the positivity, one needs to find suitable $\omega_{ip,j,k}, \omega_{im,j,k}, \omega_{i,jp,k}, \omega_{i,jm,k}, \omega_{i,j,kp}, \omega_{i,j,km}$ such that the solutions satisfy

$$\begin{cases} \rho_{i,j,k}^{n+1}(\omega) \geq \epsilon_\rho \\ p_{i,j,k}^{n+1}(\omega) \geq \epsilon_p \end{cases}$$

where

$$\begin{cases} \epsilon_\rho = \min\limits_{i,j,k} \left({}^{\mathrm{LLF}}\rho_{i,j,k}^{n+1}, 10^{-13} \right) \\ \epsilon_p = \min\limits_{i,j,k} \left({}^{\mathrm{LLF}}p_{i,j,k}^{n+1}, 10^{-13} \right) \end{cases}$$

with ${}^{\mathrm{LLF}}\rho_{i,j,k}^{n+1}$, ${}^{\mathrm{LLF}}p_{i,j,k}^{n+1}$ obtained from the 1st order Lax-Fredrichs scheme.

To facilitate the description, the flux component associated with density equation is labeled with upper left superscript ρ. Then

$$ {}^{\mathrm{LLF}}\rho^{n+1} = \overline{\rho}^n - \Delta t \left[\frac{3\left(r_{ip}^2\, {}^{\rho}f_{ip,j,k}^n - r_{im}^2\, {}^{\rho}f_{im,j,k}^n\right)}{\left(r_{ip}^3 - r_{im}^3\right)} + \frac{\sin\theta_{jp}\, {}^{\rho}g_{i,jp,k}^n - \sin\theta_{jm}\, {}^{\rho}g_{i,jm,k}^n}{r_i^\theta\left(\cos\theta_{jm} - \cos\theta_{jp}\right)} + \frac{\Delta\theta(j)}{r_i^\phi\left(\cos\theta_{jm} - \cos\theta_{jp}\right)} \frac{{}^{\rho}h_{i,j,kp}^n - {}^{\rho}h_{i,j,km}^n}{\Delta\phi(k)} \right] $$

$$ \overline{\rho}^{n+1}(\omega) = \overline{\rho}^n - \Delta t \left[\frac{3\left(r_{ip}^2\, {}^{\rho}\hat{F}_{ip} - r_{im}^2\, {}^{\rho}\hat{F}_{im}\right)}{\left(r_{ip}^3 - r_{im}^3\right)} + \frac{\sin\theta_{jp}\, {}^{\rho}\hat{F}_{jp} - \sin\theta_{jm}\, {}^{\rho}\hat{F}_{jm}}{r_i^\theta\left(\cos\theta_{jm} - \cos\theta_{jp}\right)} + \frac{\Delta\theta(j)}{r_i^\phi\left(\cos\theta_{jm} - \cos\theta_{jp}\right)} \frac{{}^{\rho}\hat{F}_{kp} - {}^{\rho}\hat{F}_{km}}{\Delta\phi(k)} \right] \tag{A.31} $$

To preserve positive density, it is necessary to find the upper bounds $\Omega_{im,j,k}^\rho, \Omega_{ip,j,k}^\rho, \Omega_{i,jm,k}^\rho, \Omega_{i,jp,k}^\rho, \Omega_{i,j,km}^\rho, \Omega_{i,j,kp}^\rho$ of the limiting parameters $\omega_{im,j,k}, \omega_{ip,j,k}, \omega_{i,jm,k}, \omega_{i,jp,k}, \omega_{i,j,km}, \omega_{i,j,kp}$ such that any combination $(\omega_{im,j,k}, \omega_{ip,j,k}, \omega_{i,jm,k}, \omega_{i,jp,k}, \omega_{i,j,km}, \omega_{i,j,kp})$ belongs to S_ρ, where

$$S_\rho = \left[0,\, \Omega^{\rho}_{im,j,k}\right] \times \left[0,\, \Omega^{\rho}_{ip,j,k}\right] \times \left[0,\, \Omega^{\rho}_{i,jm,k}\right] \times \left[0,\, \Omega^{\rho}_{i,jp,k}\right] \times \left[0,\, \Omega^{\rho}_{i,j,km}\right] \times \left[0,\, \Omega^{\rho}_{i,j,kp}\right]$$

The requirement of positivity preserving of density implies the following inequality

$$\overline{\rho}^{n+1}(\omega) = \overline{\rho}^{n} - \Delta t \left[\frac{3\left(r_{ip}^{2}\,{}^{\rho}\hat{F}_{ip} - r_{im}^{2}\,{}^{\rho}\hat{F}_{im}\right)}{\left(r_{ip}^{3} - r_{im}^{3}\right)} + \frac{\sin\theta_{jp}\,{}^{\rho}\hat{F}_{jp} - \sin\theta_{jm}\,{}^{\rho}\hat{F}_{jm}}{r_i^{\theta}\left(\cos\theta_{jm} - \cos\theta_{jp}\right)} + \frac{\Delta\theta(j)}{r_i^{\phi}\left(\cos\theta_{jm} - \cos\theta_{jp}\right)} \frac{{}^{\rho}\hat{F}_{kp} - {}^{\rho}\hat{F}_{km}}{\Delta\phi(k)} \right] \geq \epsilon_\rho \tag{A.32}$$

with fluxes given by Eq. (A.29), i.e.,

$$\begin{cases} {}^{\rho}\hat{F}_{ip} = \omega_{ip,j,k}\left({}^{\rho}F^{rk}_{r,ip,j,k} - {}^{\rho}f^{n}_{ip,j,k}\right) + {}^{\rho}f^{n}_{ip,j,k} \\ {}^{\rho}\hat{F}_{im} = \omega_{im,j,k}\left({}^{\rho}F^{rk}_{r,im,j,k} - {}^{\rho}f^{n}_{im,j,k}\right) + {}^{\rho}f^{n}_{im,j,k} \\ {}^{\rho}\hat{F}_{jp} = \omega_{i,jp,k}\left({}^{\rho}F^{rk}_{\theta,i,jp,k} - {}^{\rho}g^{n}_{i,jp,k}\right) + {}^{\rho}g^{n}_{i,jp,k} \\ {}^{\rho}\hat{F}_{jm} = \omega_{i,jm,k}\left({}^{\rho}F^{rk}_{\theta,i,jm,k} - {}^{\rho}g^{n}_{i,jm,k}\right) + {}^{\rho}g^{n}_{i,jm,k} \\ {}^{\rho}\hat{F}_{kp} = \omega_{i,j,kp}\left({}^{\rho}F^{rk}_{\phi,i,j,kp} - {}^{\rho}h^{n}_{i,j,kp}\right) + {}^{\rho}h^{n}_{i,j,kp} \\ {}^{\rho}\hat{F}_{km} = \omega_{i,j,km}\left({}^{\rho}F^{rk}_{\phi,i,j,km} - {}^{\rho}h^{n}_{i,j,km}\right) + {}^{\rho}h^{n}_{i,j,km} \end{cases}$$

Then Eq. (A.32) becomes

$$\left(\omega_{im,j,k}\delta f_{im,j,k} + \omega_{ip,j,k}\delta f_{ip,j,k}\right) + \left(\omega_{i,jm,k}\delta f_{i,jm,k} + \omega_{i,jp,k}\delta f_{i,jp,k}\right) + \left(\omega_{i,j,km}\delta f_{i,j,km} + \omega_{i,j,kp}\delta f_{i,j,kp}\right) \geq \epsilon_\rho - {}^{\mathrm{LLF}}\rho^{n+1}_{i,j,k} \tag{A.33}$$

where

$$\begin{cases} \delta f_{im,j,k} = \dfrac{3\Delta t r_{im}^{2}}{r_{ip}^{3} - r_{im}^{3}}\left({}^{\rho}F^{rk}_{r,im,j,k} - {}^{\rho}f^{n}_{im,j,k}\right) \\ \delta f_{i,jm,k} = \dfrac{\Delta t \sin\theta_{jm}}{r_i^{\theta}\left(\cos\theta_{jm} - \cos\theta_{jp}\right)}\left({}^{\rho}F^{rk}_{\theta,i,jm,k} - {}^{\rho}g^{n}_{i,jm,k}\right) \\ \delta f_{i,j,km} = \dfrac{\Delta t \Delta\theta(j)}{r_i^{\phi}\left(\cos\theta_{jm} - \cos\theta_{jp}\right)\Delta\phi(k)}\left({}^{\rho}F^{rk}_{\phi,i,j,km} - {}^{\rho}h^{n}_{i,j,km}\right) \end{cases}$$

$$\begin{cases} \delta f_{ip,j,k} = -\dfrac{3\Delta t r_{ip}^2}{r_{ip}^3 - r_{im}^3}\left({}^{\rho}F^{rk}_{r,ip,j,k} - {}^{\rho}f^{n}_{ip,j,k}\right) \\ \delta f_{i,jp,k} = -\dfrac{\Delta t \sin\theta_{jp}}{r_i^{\theta}\left(\cos\theta_{jm} - \cos\theta_{jp}\right)}\left({}^{\rho}F^{rk}_{\theta,i,jp,k} - {}^{\rho}g^{n}_{i,jp,k}\right) \\ \delta f_{i,j,kp} = -\dfrac{\Delta t \Delta\theta(j)}{r_i^{\phi}\left(\cos\theta_{jm} - \cos\theta_{jp}\right)\Delta\phi(k)}\left({}^{\rho}F^{rk}_{\phi,i,j,kp} - {}^{\rho}h^{n}_{i,j,kp}\right) \end{cases}$$

Due to the positivity-preserving property of the 1th-order LLF scheme (proved below), ${}^{\mathrm{LLF}}\rho^{n+1}_{i,j,k} \geq \epsilon_\rho$, the right hand side $\epsilon_\rho - {}^{\mathrm{LLF}}\rho^{n+1}_{i,j,k} \leq 0$. Then we decouple the inequalities (A.33) based on the sign of the six defined items: $\delta f_{im,j,k}$, $\delta f_{ip,j,k}$, $\delta f_{i,jm,k}$, $\delta f_{i,jm,k}$, $\delta f_{i,j,km}$, $\delta f_{i,j,km}$, and define the upper bounds by limiting parameters as follows:

$$\Omega^{\rho}_{\mathrm{index}} = \begin{cases} 1 & \text{If } \delta f_{\mathrm{index}} \geq 0 \\ \min\left(1, \dfrac{\epsilon_\rho - {}^{\mathrm{LLF}}\rho^{n+1}_{i,j,k}}{\sum\limits_{\delta f_{\mathrm{index}}<0} \delta f_{\mathrm{index}}}\right) & \text{If } \delta f_{\mathrm{index}} < 0 \end{cases}$$

where the subscript "index" belongs to the index set $\{\{im, j, k\}, \{ip, j, k\}, \{i, jm, k\}, \{i, jp, k\}, \{i, j, km\}, \{i, j, kp\}\}$, and $\delta f_{\mathrm{index}} \in \left\{\delta f_{im,j,k}, \delta f_{ip,j,k}, \delta f_{i,jm,k}, \delta f_{i,jp,k}, \delta f_{i,j,km}, \delta f_{i,j,kp}\right\}$. Now we denote the sixty four vertices of

$$S_\rho = \left[0, \Omega^{\rho}_{im,j,k}\right] \times \left[0, \Omega^{\rho}_{ip,j,k}\right] \times \left[0, \Omega^{\rho}_{i,jm,k}\right] \times \left[0, \Omega^{\rho}_{i,jp,k}\right] \times \left[0, \Omega^{\rho}_{i,j,km}\right] \times \left[0, \Omega^{\rho}_{i,j,kp}\right]$$

to be $\mathcal{A}^{k_1,k_2,k_3,k_4,k_5,k_6}$, that is,

$$\mathcal{A}^{k_1,k_2,k_3,k_4,k_5,k_6} = \left(k_1\Omega^{\rho}_{im,j,k}, k_2\Omega^{\rho}_{ip,j,k}, k_3\Omega^{\rho}_{i,jm,k}, k_4\Omega^{\rho}_{i,jp,k}, k_5\Omega^{\rho}_{i,j,km}, k_6\Omega^{\rho}_{i,j,kp}\right)$$

with k_1, k_2, k_3, k_4, k_5, k_6 being 0 or 1. At present, for any set $\omega \in S_\rho$, $\overline{\rho}^{n+1}(\omega)$ obtained from Eq. (A.31) is obviously positive.

Under the condition of preserving positive density, it is ready to describe how to preserve the positivity of the pressure/internal energy. It is known that the pressure can be derived from the energy equation as follows:

$$p = (\gamma - 1)\left(e - \frac{1}{2}\frac{(\rho\mathbf{v})^2}{\rho} - \frac{1}{2\mu}\mathbf{B}^2\right)$$

Next a strategy is proposed to obtain positive pressure. More specifically, a subset S_p of S_ρ is identified such that $p^{n+1}_{i,j,k}$ is positive, i.e.,

$$S_p = \left\{ \omega = \left(\omega_{im,j,k}, \omega_{ip,j,k}, \omega_{i,jm,k}, \omega_{i,jp,k}, \omega_{i,j,km}, \omega_{i,j,kp} \right) \in S_\rho : p_{i,j,k}^{n+1}(\omega) \geq \epsilon_p \right\}$$

In this case, the vertices of S_p consist of

$$\mathcal{B}^{k_1,k_2,k_3,k_4,k_5,k_6} = r^{k_1,k_2,k_3,k_4,k_5,k_6} \mathcal{A}^{k_1,k_2,k_3,k_4,k_5,k_6} \tag{A.34}$$

and satisfy the following inequality:

$$p_{i,j,k}^{n+1}(r^{k_1,k_2,k_3,k_4,k_5,k_6} \mathcal{A}^{k_1,k_2,k_3,k_4,k_5,k_6}) \geq \epsilon_p \tag{A.35}$$

From (A.34) one can obtain $\mathcal{B}^{k_1,k_2,k_3,k_4,k_5,k_6}$ if $r^{k_1,k_2,k_3,k_4,k_5,k_6}$ is defined as below:

1. If $p_{i,j,k}^{n+1}(\mathcal{A}^{k_1,k_2,k_3,k_4,k_5,k_6}) \geq \epsilon_p$, then

$$r^{k_1,k_2,k_3,k_4,k_5,k_6} = 1$$

2. If $p_{i,j,k}^{n+1}(\mathcal{A}^{k_1,k_2,k_3,k_4,k_5,k_6}) < \epsilon_p$, one can obtain $r^{k_1,k_2,k_3,k_4,k_5,k_6}$ by solving the cubic equation (A.35) for the unknown variable $r \in [0, 1)$. Thanks to the concave property of pressure as a function of $\mathbf{U}$, we can get r in a simple way:

$$p(\alpha \mathbf{U}_1 + (1-\alpha)\mathbf{U}_2) \geq \alpha p(\mathbf{U}_1) + (1-\alpha) p(\mathbf{U}_2) \quad \text{if} \quad \rho(\mathbf{U}_1) > 0, \rho(\mathbf{U}_2) > 0$$

for $\alpha \in [0, 1]$. Evidently it follows that

$$\begin{aligned} p(\mathbf{U}(\alpha\omega_1 + (1-\alpha)\omega_2)) = p(\alpha \mathbf{U}(\omega_1) + (1-\alpha)\mathbf{U}(\omega_2)) \geq \alpha p(\mathbf{U}(\omega_1)) \\ + (1-\alpha) p(\mathbf{U}(\omega_2)) \end{aligned}$$

That is,

$$p(\alpha\omega_1 + (1-\alpha)\omega_2) \geq \alpha p(\omega_1) + (1-\alpha) p(\omega_2) \quad \text{if} \quad \rho(\omega_1) > 0, \rho(\omega_2) > 0 \tag{A.36}$$

where $\mathbf{U}(\omega)$, $\rho(\omega)$ and $p(\omega)$ are used to denote the solutions obtained from Eq. (A.30) with

$$\omega = \left(\omega_{im,j,k}, \omega_{ip,j,k}, \omega_{i,jm,k}, \omega_{i,jp,k}, \omega_{i,j,km}, \omega_{i,j,kp} \right)$$

If taking $\omega_1 = \mathcal{A}^{k_1,k_2,k_3,k_4,k_5,k_6}((k_1, k_2, k_3, k_4, k_5, k_6) \neq \mathbf{0})$, $\omega_2 = \mathcal{A}^{\mathbf{0}} = (0, 0, 0, 0, 0, 0)$, and $\alpha = r^{k_1,k_2,k_3,k_4,k_5,k_6}$, then $p\lfloor \alpha\omega_1 + (1-\alpha)\omega_2 \rfloor = p\left[r^{k_1,k_2,k_3,k_4,k_5,k_6} \mathcal{A}^{k_1,k_2,k_3,k_4,k_5,k_6} + (1 - r^{k_1,k_2,k_3,k_4,k_5,k_6}) \mathcal{A}^{\mathbf{0}} \right] = p\left(r^{k_1,k_2,k_3,k_4,k_5,k_6} \mathcal{A}^{k_1,k_2,k_3,k_4,k_5,k_6} \right)$. Thus, Eq. (A.36) becomes

$$\begin{aligned} & p\left(r^{k_1,k_2,k_3,k_4,k_5,k_6} \mathcal{A}^{k_1,k_2,k_3,k_4,k_5,k_6} \right) \\ & \geq r^{k_1,k_2,k_3,k_4,k_5,k_6} p(\mathcal{A}^{k_1,k_2,k_3,k_4,k_5,k_6}) + (1 - r^{k_1,k_2,k_3,k_4,k_5,k_6}) p(\mathcal{A}^{\mathbf{0}}) \end{aligned}$$

When the right-hand side of the above inequality $\geq \epsilon_p$ is assumed, it follows that

$$\begin{aligned}&p\left(r^{k_1,k_2,k_3,k_4,k_5,k_6}\mathcal{A}^{k_1,k_2,k_3,k_4,k_5,k_6}\right)\\&\geq r^{k_1,k_2,k_3,k_4,k_5,k_6}p(\mathcal{A}^{k_1,k_2,k_3,k_4,k_5,k_6})+(1-r^{k_1,k_2,k_3,k_4,k_5,k_6})p(\mathcal{A}^{\mathbf{0}})\geq\epsilon_p\end{aligned} \tag{A.37}$$

Since

$$\begin{cases}p(\mathcal{A}^{\mathbf{0}})\geq\epsilon_p\\p(\mathcal{A}^{k_1,k_2,k_3,k_4,k_5,k_6})<\epsilon_p\end{cases}$$

from (A.37) we get

$$r^{k_1,k_2,k_3,k_4,k_5,k_6}\leq\frac{p(\mathcal{A}^{\mathbf{0}})-\epsilon_p}{p(\mathcal{A}^{\mathbf{0}})-p(\mathcal{A}^{k_1,k_2,k_3,k_4,k_5,k_6})}\in[0,1)$$

or equivalently

$$r^{k_1,k_2,k_3,k_4,k_5,k_6}\leq\frac{\epsilon_p-p_{i,j,k}^{n+1}(\mathcal{A}^{0,0,0,0,0,0})}{p_{i,j,k}^{n+1}(\mathcal{A}^{k_1,k_2,k_3,k_4,k_5,k_6})-p_{i,j,k}^{n+1}(\mathcal{A}^{0,0,0,0,0,0})}$$

We can embark a rectangle subset $S_{\rho,p}$ inside S_p

$$\begin{aligned}S_{\rho,p}=&\left[0,\,\Omega_{im,j,k}^{\rho,p}\right]\times\left[0,\,\Omega_{ip,j,k}^{\rho,p}\right]\times\left[0,\,\Omega_{i,jm,k}^{\rho,p}\right]\times\left[0,\,\Omega_{i,jp,k}^{\rho,p}\right]\\&\times\left[0,\,\Omega_{i,j,km}^{\rho,p}\right]\times\left[0,\,\Omega_{i,j,kp}^{\rho,p}\right]\subset S_p,\end{aligned}$$

where

$$\begin{cases}\Omega_{im,j,k}^{\rho,p}=\min\limits_{k_{2,3,4,5,6}=0,1}\left(\mathcal{B}^{1,k_2,k_3,k_4,k_5,k_6}\right)\\\Omega_{ip,j,k}^{\rho,p}=\min\limits_{k_{1,3,4,5,6}=0,1}\left(\mathcal{B}^{k_1,1,k_3,k_4,k_5,k_6}\right)\\\Omega_{i,jm,k}^{\rho,p}=\min\limits_{k_{1,2,4,5,6}=0,1}\left(\mathcal{B}^{k_1,k_2,1,k_3,k_5,k_6}\right)\\\Omega_{i,jp,k}^{\rho,p}=\min\limits_{k_{1,2,3,5,6}=0,1}\left(\mathcal{B}^{k_1,k_2,k_3,1,k_5,k_6}\right)\\\Omega_{i,j,km}^{\rho,p}=\min\limits_{k_{1,2,3,4,6}=0,1}\left(\mathcal{B}^{k_1,k_2,k_3,k_4,1,k_6}\right)\\\Omega_{i,j,kp}^{\rho,p}=\min\limits_{k_{1,2,3,4,5}=0,1}\left(\mathcal{B}^{k_1,k_2,k_3,k_4,k_5,1}\right)\end{cases}$$

Finally, one can set

$$\begin{cases} \omega_{ip,j,k} = \min\left(\Omega^{\rho,p}_{ip,j,k}, \Omega^{\rho,p}_{(i+1)m,j,k}\right) \\ \omega_{im,j,k} = \min\left(\Omega^{\rho,p}_{im,j,k}, \Omega^{\rho,p}_{(i-1)p,j,k}\right) \\ \omega_{i,jp,k} = \min\left(\Omega^{\rho,p}_{i,jp,k}, \Omega^{\rho,p}_{i,(j+1)m,k}\right) \\ \omega_{i,jm,k} = \min\left(\Omega^{\rho,p}_{i,jm,k}, \Omega^{\rho,p}_{i,(j-1)p,k}\right) \\ \omega_{i,j,kp} = \min\left(\Omega^{\rho,p}_{i,j,kp}, \Omega^{\rho,p}_{i,j,(k+1)m}\right) \\ \omega_{i,j,km} = \min\left(\Omega^{\rho,p}_{i,j,km}, \Omega^{\rho,p}_{i,j,(k-1)p}\right) \end{cases} \tag{A.38}$$

With the above limiting bounds $\omega_{ip,j,k}, \omega_{im,j,k}, \omega_{i,jp,k}, \omega_{i,jm,k}, \omega_{i,j,kp}, \omega_{i,j,km}$ given by Eq. (A.38), one can restart to update $\overline{\mathbf{U}}^{n+1}$ from Eq. (A.30). This finishes the positivity-preserving procedure.

Since the first-order local Lax-Friedrichs scheme serves as the building block for the above MPP flux-corrected limiting formulation, we are left to prove positivity-preserving for the local Lax-Friedrichs scheme. In the above context, Lax-Friedrichs scheme can be written as follows:

$$\frac{d\overline{\mathbf{U}}}{dt} = -\frac{3\left(r_{ip}^2 \mathbf{f}^n_{ip,j,k} - r_{im}^2 \mathbf{f}^n_{im,j,k}\right)}{\left(r_{ip}^3 - r_{im}^3\right)} - \frac{\sin\theta_{jp}\mathbf{g}^n_{i,jp,k} - \sin\theta_{jm}\mathbf{g}^n_{i,jm,k}}{r_i^\theta\left(\cos\theta_{jm} - \cos\theta_{jp}\right)} - \frac{\Delta\theta(j)}{r_i^\phi\left(\cos\theta_{jm} - \cos\theta_{jp}\right)}\frac{\mathbf{h}^n_{i,j,kp} - \mathbf{h}^n_{i,j,km}}{\Delta\phi(k)} + \overline{\mathbf{S}}_{i,j,k}$$

where

$$\mathbf{f}^n_{im,j,k} = \frac{1}{2}\lambda_{im}\left(\overline{\mathbf{U}}_{i-1,j,k} - \overline{\mathbf{U}}_{i,j,k}\right) + \frac{1}{2}\left[\mathbf{f}\left(\overline{\mathbf{U}}_{i-1,j,k}\right) + \mathbf{f}\left(\overline{\mathbf{U}}_{i,j,k}\right)\right]$$

$$\mathbf{g}^n_{i,jm,k} = \frac{1}{2}\lambda_{jm}\left(\overline{\mathbf{U}}_{i,j-1,k} - \overline{\mathbf{U}}_{i,j,k}\right) + \frac{1}{2}\left[\mathbf{g}\left(\overline{\mathbf{U}}_{i,j-1,k}\right) + \mathbf{g}\left(\overline{\mathbf{U}}_{i,j,k}\right)\right]$$

$$\mathbf{h}^n_{i,j,km} = \frac{1}{2}\lambda_{km}\left(\overline{\mathbf{U}}_{i,j,k-1} - \overline{\mathbf{U}}_{i,j,k}\right) + \frac{1}{2}\left[\mathbf{h}\left(\overline{\mathbf{U}}_{i,j,k-1}\right) + \mathbf{h}\left(\overline{\mathbf{U}}_{i,j,k}\right)\right]$$

with

$$\lambda_{im} = \max\left\{(\overline{v}_r + \overline{c}_{fr})_{i-1,j,k}, (\overline{v}_r + \overline{c}_{fr})_{i,j,k}\right\},$$

$$\lambda_{ip} = \max\left\{(\overline{v}_r + \overline{c}_{fr})_{i,j,k}, (\overline{v}_r + \overline{c}_{fr})_{i+1,j,k}\right\}$$

$$\lambda_{jm} = \max\left\{(\overline{v}_\theta + \overline{c}_{f\theta})_{i,j-1,k}, (\overline{v}_\theta + \overline{c}_{f\theta})_{i,j,k}\right\},$$

$$\lambda_{jp} = \max\left\{(\overline{v}_\theta + \overline{c}_{f\theta})_{i,j,k}, (\overline{v}_\theta + \overline{c}_{f\theta})_{i,j+1,k}\right\}$$

$$\lambda_{km} = \max\left\{(\overline{v}_\phi + \overline{c}_{f\phi})_{i,j,k-1}, (\overline{v}_\phi + \overline{c}_{f\phi})_{i,j,k}\right\},$$

$$\lambda_{kp} = \max\left\{(\overline{v}_\phi + \overline{c}_{f\phi})_{i,j,k}, (\overline{v}_\phi + \overline{c}_{f\phi})_{i,j,k+1}\right\}$$

For simplicity, by setting

$$\begin{cases} \mathbf{f}_i^n = \mathbf{f}\left(\overline{\mathbf{U}}_{i,j,k}\right), & \mathbf{g}_j^n = \mathbf{g}\left(\overline{\mathbf{U}}_{i,j,k}\right), \quad \mathbf{h}_k^n = \mathbf{h}\left(\overline{\mathbf{U}}_{i,j,k}\right) \\ \mathbf{U}_i^n = \overline{\mathbf{U}}_{i,j,k}, & \mathbf{U}_j^n = \overline{\mathbf{U}}_{i,j,k}, \quad \mathbf{U}_k^n = \overline{\mathbf{U}}_{i,j,k} \end{cases}$$

then the first order local Lax-Friedrichs scheme with a forward Euler time-integration method reads

$$\begin{aligned} \overline{\mathbf{U}}_{i,j,k}^{n+1} = \overline{\mathbf{U}}_{i,j,k}^{n} &- \frac{3r_{ip}^2 \Delta t}{2\left(r_{ip}^3 - r_{im}^3\right)} \left[\mathbf{f}_{i+1}^n + \mathbf{f}_i^n - \lambda_{ip}(\mathbf{U}_{i+1}^n - \mathbf{U}_i^n)\right] \\ &+ \frac{3r_{im}^2 \Delta t}{2\left(r_{ip}^3 - r_{im}^3\right)} \left[\mathbf{f}_i^n + \mathbf{f}_{i-1}^n - \lambda_{im}(\mathbf{U}_i^n - \mathbf{U}_{i-1}^n)\right] \\ &- \frac{\Delta t \sin\theta_{jp}}{2r_i^\theta\left(\cos\theta_{jm} - \cos\theta_{jp}\right)} \left[\mathbf{g}_{j+1}^n + \mathbf{g}_j^n - \lambda_{jp}(\mathbf{U}_{j+1}^n - \mathbf{U}_j^n)\right] \\ &+ \frac{\Delta t \sin\theta_{jm}}{2r_i^\theta\left(\cos\theta_{jm} - \cos\theta_{jp}\right)} \left[\mathbf{g}_j^n + \mathbf{g}_{j-1}^n - \lambda_{jm}(\mathbf{U}_j^n - \mathbf{U}_{j-1}^n)\right] \\ &- \frac{\Delta t \Delta\theta(j)}{2r_i^\phi\left(\cos\theta_{jm} - \cos\theta_{jp}\right)\Delta\phi(k)} \left[\mathbf{h}_{k+1}^n + \mathbf{h}_k^n - \lambda_{kp}(\mathbf{U}_{k+1}^n - \mathbf{U}_k^n)\right] \\ &+ \frac{\Delta t \Delta\theta(j)}{2r_i^\phi\left(\cos\theta_{jm} - \cos\theta_{jp}\right)\Delta\phi(k)} \left[\mathbf{h}_k^n + \mathbf{h}_{k-1}^n - \lambda_{km}(\mathbf{U}_k^n - \mathbf{U}_{k-1}^n)\right] \end{aligned} \tag{A.39}$$

where Δt is the time marching step to be determined. Now, one can rewrite (A.39) as follows

$$\begin{aligned} \overline{\mathbf{U}}_{i,j,k}^{n+1} = \overline{\mathbf{U}}_{i,j,k}^{n} &\left(1 - \frac{3(\lambda_{ip} r_{ip}^2 + \lambda_{im} r_{im}^2)\Delta t}{2\left(r_{ip}^3 - r_{im}^3\right)} - \frac{\Delta t(\lambda_{jp}\sin\theta_{jp} + \lambda_{jm}\sin\theta_{jm})}{2r_i^\theta\left(\cos\theta_{jm} - \cos\theta_{jp}\right)}\right. \\ &\left. - \frac{\Delta t \Delta\theta(j)(\lambda_{kp} + \lambda_{km})}{2r_i^\phi\left(\cos\theta_{jm} - \cos\theta_{jp}\right)\Delta\phi(k)}\right) \\ &+ \frac{3(r_{im}^2 - r_{ip}^2)\Delta t}{2\left(r_{ip}^3 - r_{im}^3\right)}\mathbf{f}_i^n + \frac{\Delta t(\sin\theta_{jm} - \sin\theta_{jp})}{2r_i^\theta\left(\cos\theta_{jm} - \cos\theta_{jp}\right)}\mathbf{g}_j^n \\ &+ \frac{3r_{im}^2 \Delta t}{2\left(r_{ip}^3 - r_{im}^3\right)}\left(\mathbf{f}_{i-1}^n + \lambda_{im}\mathbf{U}_{i-1}^n\right) - \frac{3r_{ip}^2 \Delta t}{2\left(r_{ip}^3 - r_{im}^3\right)}\left(\mathbf{f}_{i+1}^n - \lambda_{ip}\mathbf{U}_{i+1}^n\right) \end{aligned}$$

$$
\begin{aligned}
&+\frac{\Delta t\sin\theta_{jm}}{2r_i^{\theta}\left(\cos\theta_{jm}-\cos\theta_{jp}\right)}\left(\mathbf{g}_{j-1}^n+\lambda_{jm}\mathbf{U}_{j-1}^n\right)\\
&-\frac{\Delta t\sin\theta_{jp}}{2r_i^{\theta}\left(\cos\theta_{jm}-\cos\theta_{jp}\right)}\left(\mathbf{g}_{j+1}^n-\lambda_{jp}\mathbf{U}_{j+1}^n\right)\\
&+\frac{\Delta t\Delta\theta(j)}{2r_i^{\phi}\left(\cos\theta_{jm}-\cos\theta_{jp}\right)\Delta\phi(k)}\left(\mathbf{h}_{k-1}^n+\lambda_{km}\mathbf{U}_{k-1}^n\right)\\
&-\frac{\Delta t\Delta\theta(j)}{2r_i^{\phi}\left(\cos\theta_{jm}-\cos\theta_{jp}\right)\Delta\phi(k)}\left(\mathbf{h}_{k+1}^n-\lambda_{kp}\mathbf{U}_{k+1}^n\right)
\end{aligned}
\tag{A.40}
$$

with $\Delta\phi(k)=\phi_{kp}-\phi_{km}$, $\Delta\theta(j)=\theta_{jp}-\theta_{jm}$. For density ρ, it follows from (A.40) that

$$
\begin{aligned}
\overline{\rho}_{i,j,k}^{n+1}=&\overline{\rho}_{i,j,k}^{n}\left(1-\frac{3(\lambda_{ip}r_{ip}^2+\lambda_{im}r_{im}^2)\Delta t}{2\left(r_{ip}^3-r_{im}^3\right)}-\frac{\Delta t(\lambda_{jp}\sin\theta_{jp}+\lambda_{jm}\sin\theta_{jm})}{2r_i^{\theta}\left(\cos\theta_{jm}-\cos\theta_{jp}\right)}\right.\\
&\left.-\frac{\Delta t\Delta\theta(j)(\lambda_{kp}+\lambda_{km})}{2r_i^{\phi}\left(\cos\theta_{jm}-\cos\theta_{jp}\right)\Delta\phi(k)}\right)\\
&+\frac{3(r_{im}^2-r_{ip}^2)\Delta t}{2\left(r_{ip}^3-r_{im}^3\right)}(\rho v_r)_i^n+\frac{\Delta t(\sin\theta_{jm}-\sin\theta_{jp})}{2r_i^{\theta}\left(\cos\theta_{jm}-\cos\theta_{jp}\right)}(v_\theta\rho)_j^n\\
&+\frac{3r_{im}^2\Delta t}{2\left(r_{ip}^3-r_{im}^3\right)}\left((v_r\rho)_{i-1}^n+\lambda_{im}\rho_{i-1}^n\right)-\frac{3r_{ip}^2\Delta t}{2\left(r_{ip}^3-r_{im}^3\right)}\left((v_r\rho)_{i+1}^n-\lambda_{ip}\rho_{i+1}^n\right)\\
&+\frac{\Delta t\sin\theta_{jm}}{2r_i^{\theta}\left(\cos\theta_{jm}-\cos\theta_{jp}\right)}\left((v_\theta\rho)_{j-1}^n+\lambda_{jm}\rho_{j-1}^n\right)-\frac{\Delta t\sin\theta_{jp}}{2r_i^{\theta}\left(\cos\theta_{jm}-\cos\theta_{jp}\right)}\\
&\left((v_\theta\rho)_{j+1}^n-\lambda_{jp}\rho_{j+1}^n\right)\\
&+\frac{\Delta t\Delta\theta(j)}{2r_i^{\phi}\left(\cos\theta_{jm}-\cos\theta_{jp}\right)\Delta\phi(k)}\left((v_\phi\rho)_{k-1}^n+\lambda_{km}\rho_{k-1}^n\right)\\
&-\frac{\Delta t\Delta\theta(j)}{2r_i^{\phi}\left(\cos\theta_{jm}-\cos\theta_{jp}\right)\Delta\phi(k)}\left((v_\phi\rho)_{k+1}^n-\lambda_{kp}\rho_{k+1}^n\right)
\end{aligned}
\tag{A.41}
$$

Since

$$
\begin{aligned}
\lambda_i&=\max\left\{(\overline{v}_r+\overline{c}_{fr})_{i-1,j,k},(\overline{v}_r+\overline{c}_{fr})_{i,j,k},(\overline{v}_r+\overline{c}_{fr})_{i+1,j,k}\right\}\\
&=\max(\lambda_{im},\lambda_{ip})>\max((v_r)_{i+1},(v_r)_i,(v_r)_{i-1})\\
\lambda_j&=\max\left\{(\overline{v}_\theta+\overline{c}_{f\theta})_{i,j-1,k},(\overline{v}_\theta+\overline{c}_{f\theta})_{i,j,k},(\overline{v}_\theta+\overline{c}_{f\theta})_{i,j+1,k}\right\}\\
&=\max(\lambda_{jm},\lambda_{jp})>\max((v_\theta)_{j+1},(v_\theta)_j,(v_\theta)_{j-1})\\
\lambda_k&=\max\left\{(\overline{v}_\phi+\overline{c}_{f\phi})_{i,j,k-1},(\overline{v}_\phi+\overline{c}_{f\phi})_{i,j,k},\overline{v}_\phi+\overline{c}_{f\phi})_{i,j,k+1}\right\}\\
&=\max(\lambda_{km},\lambda_{kp})>\max((v_\phi)_{k+1},(v_\phi)_k,(v_\phi)_{k-1})
\end{aligned}
\tag{A.42}
$$

there follow

$$\begin{cases}(\rho v_r)^n_{i-1}+\lambda_{im}\rho^n_{i-1}>0\\(\rho v_r)^n_i-\lambda_{im}\rho^n_i<0\\(\rho v_r)^n_{i+1}-\lambda_{ip}\rho^n_{i+1}<0\end{cases}\qquad\begin{cases}(\rho v_\theta)^n_{j-1}+\lambda_{jm}\rho^n_{j-1}>0\\(\rho v_\theta)^n_j-\lambda_{jm}\rho^n_j<0\\(\rho v_\theta)^n_{j+1}-\lambda_{jp}\rho^n_{j+1}<0\end{cases}$$
$$\begin{cases}(\rho v_\phi)^n_{k-1}+\lambda_{km}\rho^n_{k-1}>0\\(\rho v_\phi)^n_k-\lambda_{km}\rho^n_k<0\\(\rho v_\phi)^n_{k+1}-\lambda_{kp}\rho^n_{k+1}<0\end{cases}\tag{A.43}$$

Considering (A.41), (A.42) and (A.43), one obtains

$$\begin{aligned}\overline{\rho}^{n+1}_{i,j,k}\geq{}&\overline{\rho}^n_{i,j,k}\left(1-\frac{3(\lambda_{ip}r^2_{ip}+\lambda_{im}r^2_{im})\Delta t}{2\left(r^3_{ip}-r^3_{im}\right)}-\frac{\Delta t(\lambda_{jp}\sin\theta_{jp}+\lambda_{jm}\sin\theta_{jm})}{2r^\theta_i\left(\cos\theta_{jm}-\cos\theta_{jp}\right)}\right.\\&\left.-\frac{\Delta t\Delta\theta(j)(\lambda_{kp}+\lambda_{km})}{2r^\phi_i\left(\cos\theta_{jm}-\cos\theta_{jp}\right)\Delta\phi(k)}\right)+\frac{3(r^2_{im}-r^2_{ip})\Delta t}{2\left(r^3_{ip}-r^3_{im}\right)}\lambda_{im}(\rho)^n_i\\&-\frac{\Delta t}{2r^\theta_i}\left|\frac{(\sin\theta_{jm}-\sin\theta_{jp})}{(\cos\theta_{jm}-\cos\theta_{jp})}\right|\lambda_{jm}(\rho)^n_j\\\geq{}&\overline{\rho}^n_{i,j,k}\left(1-\frac{3(\lambda_{ip}+\lambda_{im})r^2_{ip}\Delta t}{2\left(r^3_{ip}-r^3_{im}\right)}-\frac{\Delta t}{2r^\theta_i}\left(\frac{(\lambda_{jp}\sin\theta_{jp}+\lambda_{jm}\sin\theta_{jm})}{(\cos\theta_{jm}-\cos\theta_{jp})}\right.\right.\\&\left.\left.+\left|\frac{(\sin\theta_{jm}-\sin\theta_{jp})}{(\cos\theta_{jm}-\cos\theta_{jp})}\right|\lambda_{jm}\right)-\frac{\Delta t\Delta\theta(j)(\lambda_{kp}+\lambda_{km})}{2r^\phi_i\left(\cos\theta_{jm}-\cos\theta_{jp}\right)\Delta\phi(k)}\right)\\\geq{}&\overline{\rho}^n_{i,j,k}\left(1-\frac{3r^2_{ip}\Delta t}{\left(r^3_{ip}-r^3_{im}\right)}\lambda_i-\frac{\Delta t}{2r^\theta_i}\left(\cot\frac{\Delta\theta(j)}{2}+\left|\cot\theta_j\right|\right)\lambda_j\right.\\&\left.-\frac{\Delta t\Delta\theta(j)}{2r^\phi_i\sin\theta_j\sin\frac{\Delta\theta(j)}{2}\Delta\phi(k)}\lambda_k\right)\\={}&\overline{\rho}^n_{i,j,k}\left(1-\frac{3r^2_{ip}\Delta t}{\left(r^3_{ip}-r^3_{im}\right)}\lambda_i-\frac{\Delta t}{2r^\theta_i\Delta\theta(j)}\left[\left(\cot\frac{\Delta\theta(j)}{2}+\left|\cot\theta_j\right|\right)\Delta\theta(j)\right]\lambda_j\right.\\&\left.-\frac{\Delta t\Delta\theta(j)}{2r^\phi_i\sin\theta_j\sin\frac{\Delta\theta(j)}{2}\Delta\phi(k)}\lambda_k\right)\end{aligned}\tag{A.44}$$

If $\overline{\rho}^n_{i,j,k}>0$ and

$$\left(1-\frac{3r_{ip}^2\Delta t}{\left(r_{ip}^3-r_{im}^3\right)}\lambda_i-\frac{\Delta t}{2r_i^\theta\Delta\theta(j)}\left[\left(\cot\frac{\Delta\theta(j)}{2}+\left|\cot\theta_j\right|\right)\Delta\theta(j)\right]\lambda_j\right.$$
$$\left.-\frac{\Delta t\Delta\theta(j)}{2r_i^\phi\sin\theta_j\sin\frac{\Delta\theta(j)}{2}\Delta\phi(k)}\lambda_k\right)>0$$

then $\overline{\rho}_{i,j,k}^{n+1}>0$ holds. If setting $\Delta t=\mathrm{CFL}\frac{\min\left(\Delta r(i),r_i^\theta\Delta\theta(j),r_i^\phi\sin\theta_j\Delta\phi(k)\right)}{\max\left(\lambda_i,\lambda_j,\lambda_k\right)}(\equiv\mathrm{AA})$, inequality (A.44) can be further enhanced by

$$1-\left(\frac{3r_{ip}^2\Delta r(i)}{\left(r_{ip}^3-r_{im}^3\right)}+\frac{1}{2}\left[\left(\cot\frac{\Delta\theta(j)}{2}+\left|\cot\theta_j\right|\right)\Delta\theta(j)\right]+\frac{\Delta\theta(j)}{2\sin\frac{\Delta\theta(j)}{2}}\right)\mathrm{CFL}>0 \tag{A.45}$$

Or equivalently $\Delta t=\min(\mathrm{AA},\mathrm{BB})$ with

$$\mathrm{BB}=\frac{1}{\max\left(\frac{3r_{ip}^2}{\left(r_{ip}^3-r_{im}^3\right)}\lambda_i+\frac{1}{2r_i^\theta\Delta\theta(j)}\left[\left(\cot\frac{\Delta\theta(j)}{2}+\left|\cot\theta_j\right|\right)\Delta\theta(j)\right]\lambda_j+\frac{\Delta\theta(j)}{2r_i^\phi\sin\theta_j\sin\frac{\Delta\theta(j)}{2}\Delta\phi(k)}\lambda_k\right)}$$

Certainly, this constraint of Δt will affect the original CFL number. Since $\frac{\pi}{4}\le\theta_j\le\frac{3\pi}{4}$ in the grid mesh of computational region, $\left|\cot\theta_j\right|\le 1$ does hold. Obviously, $\frac{\Delta\theta(j)}{2}\cot\frac{\Delta\theta(j)}{2}\approx 1$ and $\frac{\frac{\Delta\theta(j)}{2}}{\sin\frac{\Delta\theta(j)}{2}}\approx 1$ for small $\Delta\theta(j)$. In these cases, one has the following approximations

$$\frac{3r_{ip}^2}{\left(r_{ip}^3-r_{im}^3\right)}=\frac{1}{\Delta r(i)}\frac{3r_{ip}^2}{\left(r_{ip}^2+r_{ip}r_{im}+r_{im}^2\right)}\approx\frac{1}{\Delta r(i)}$$
$$\frac{1}{2r_i^\theta\Delta\theta(j)}\left[\left(\cot\frac{\Delta\theta(j)}{2}+\left|\cot\theta_j\right|\right)\Delta\theta(j)\right]\approx\frac{1}{r_i^\theta\Delta\theta(j)}\frac{2+\Delta\theta(j)}{2}$$
$$\frac{\Delta\theta(j)}{2r_i^\phi\sin\theta_j\sin\frac{\Delta\theta(j)}{2}\Delta\phi(k)}\approx\frac{\frac{\Delta\theta(j)}{2}}{\sin\frac{\Delta\theta(j)}{2}}\frac{1}{r_i^\phi\sin\theta_j\Delta\phi(k)}\approx\frac{1}{r_i^\phi\sin\theta_j\Delta\phi(k)}$$

From (A.45), it follows that $1-(3+\frac{\Delta\theta(j)}{2})\mathrm{CFL}>0$. That is, the original CFL is reduced by a factor $2+\frac{2+\Delta\theta(j)}{2}\approx 3$, which is due to the geometrical factor in spherical coordinates.

It is noted that the above formulations have used the following algebraic identities:

$$\sin\theta_{jm}+\sin\theta_{jp}=2\sin\frac{\theta_{jp}+\theta_{jm}}{2}\cos\frac{\theta_{jp}-\theta_{jm}}{2}=2\sin\theta_j\cos\frac{\Delta\theta(j)}{2}$$
$$\sin\theta_{jp}-\sin\theta_{jm}=2\cos\frac{\theta_{jp}+\theta_{jm}}{2}\sin\frac{\theta_{jp}-\theta_{jm}}{2}=2\cos\theta_j\sin\frac{\Delta\theta(j)}{2}$$

$$\cos\theta_{jm} - \cos\theta_{jp} = 2\sin\frac{\theta_{jp}+\theta_{jm}}{2}\sin\frac{\theta_{jp}-\theta_{jm}}{2} = 2\sin\theta_j\sin\frac{\Delta\theta(j)}{2}$$

$$\frac{\Delta t(\sin\theta_{jm} - \sin\theta_{jp})}{2r_i^{\theta}\left(\cos\theta_{jm} - \cos\theta_{jp}\right)} = \frac{\Delta t(-2\cos\theta_j\sin\frac{\Delta\theta(j)}{2})}{2r_i^{\theta}\left(2\sin\theta_j\sin\frac{\Delta\theta(j)}{2}\right)} = -\frac{\Delta t}{2r_i^{\theta}\Delta\theta(j)}\Delta\theta(j)\cot\theta_j$$

$$\frac{\Delta t(\sin\theta_{jm} + \sin\theta_{jp})}{2r_i^{\theta}\left(\cos\theta_{jm} - \cos\theta_{jp}\right)} = \frac{\Delta t(2\sin\theta_j\cos\frac{\Delta\theta(j)}{2})}{2r_i^{\theta}\left(2\sin\theta_j\sin\frac{\Delta\theta(j)}{2}\right)} = \frac{\Delta t}{2r_i^{\theta}\Delta\theta(j)}\Delta\theta(j)\cot\frac{\Delta\theta(j)}{2}$$

$$\frac{\sin\theta_{jm}}{2r_i^{\theta}\left(\cos\theta_{jm} - \cos\theta_{jp}\right)} = \frac{\sin\theta_{jm}}{4r_i^{\theta}\sin\theta_j\sin\frac{\Delta\theta(j)}{2}}$$

$$\frac{\sin\theta_{jp}}{2r_i^{\theta}\left(\cos\theta_{jm} - \cos\theta_{jp}\right)} = \frac{\sin\theta_{jp}}{4r_i^{\theta}\sin\theta_j\sin\frac{\Delta\theta(j)}{2}}$$

In the implementation, the correction steps can be carried out only at the end of each time step [28, 29, 200, 201, 203] instead of each stage of RK methods [25, 27, 194]. Christlieb et al. [29] show that the differences between the two operations are negligible through their numerical results by SSP-RK3 time integration. If there exist negative density or/and pressure in the intermediate stages of the RK method, the speed of sound can be calculated by $\sqrt{\gamma\frac{|p|}{|\rho|}}$ with the absolute values of density and pressure [29, 201]. As commented before [200, 201, 203], the flux limiter can be applied only at the final RK stage such that the implementation complexity is significantly reduced. They also claimed that if the MPP flux limiter is applied at each of the intermediate stage of RK method, due to the influence of the special cancellation of RK, high order temporal accuracy could be lost. However, Christlieb et al. [27] showed that the correction steps carried out only at the end of each time step may result in accumulation of the divergence error, with the low-storage 10-stage SSP-RK4 method for their numerical simulations. Thus, the operation that the correction steps are only implemented at the end of each time step is not recommended for RK methods with large stage numbers.

By the way, Wu [194] studied the high-order accurate density and pressure positivity preserving schemes under the Courant-Friedrichs-Lewy condition for the relativistic magnetohydrodynamics (RMHD) and reveals for the first time that the discrete divergence-free condition is closely connected with the density and pressure positivity-preserving property in theory. His formulation of the convexity, scaling invariance and Lax-Friedrichs splitting property can be hopefully used as the building block for high-order accuracy design of the MPP flux-corrected limiting procedure of MHD system. Moreover, a first order finite volume Lax-Friedrichs type scheme ((69) of Ranocha [143]) update procedure using an explicit Euler step in time, where the flux of the first order finite volume scheme is reformulated by logarithmic mean, is proved to be positivity-preserving. By numerical experiments, Ranocha [143] investigated robustness properties of entropy stable numerical fluxes and positivity preservation for compressible Euler equations via several entropy conservative fluxes enhanced with local Lax-Friedrichs type dissipation operators. This technique, combined with the entropy consistent schemes [38, 39, 115, 191, 192] for MHD equations, is expected to provide us entropy-stable positivity-preserving schemes for MHD system.

References

1. Arge CN, Odstrčil D, Pizzo VJ, Mayer LR (2003) Improved method for specifying solar wind speed near the Sun. In: Velli M, Bruno R, Malara F, Bucci B (eds) Solar Wind Ten, American institute of physics conference series, vol 679, pp 190–193
2. Balsara DS, Kim J (2004) A comparison between divergence-cleaning and staggered-mesh formulations for numerical magnetohydrodynamics. Astrophys J 602:1079–1090
3. Balsara DS (2004) Second-order-accurate schemes for magnetohydrodynamics with divergence-free reconstruction. Astrophys J Suppl Ser 151:149–184
4. Balsara DS (2009) Divergence-free reconstruction of magnetic fields and WENO schemes for magnetohydrodynamics. J Comput Phys 228:5040–5056
5. Balsara DS (2010) Multidimensional HLLE Riemann solver: application to Euler and magnetohydrodynamic flows. J Comput Phys 229:1970–1993
6. Balsara DS (2012a) A two-dimensional HLLC Riemann solver for conservation laws: application to Euler and magnetohydrodynamic flows. J Comput Phys 231:7476–7503
7. Balsara DS (1998) Total variation diminishing scheme for adiabatic and isothermal magnetohydrodynamics. Astrophys J Suppl Ser 116(1):133
8. Balsara DS (2012b) Self-adjusting, positivity preserving high order schemes for hydrodynamics and magnetohydrodynamics. J Comput Phys 231(22):7504–7517
9. Balsara DS, Shu CW (2000) Monotonicity preserving weighted essentially non-oscillatory schemes with increasingly high order of accuracy. J Comput Phys 160(2):405–452
10. Balsara DS, Spicer D (1999) Maintaining pressure positivity in magnetohydrodynamic simulations. J Comput Phys 148(1):133–148
11. Balsara DS, Meyer C, Dumbser M, Du H, Xu Z (2013) Efficient implementation of ADER schemes for Euler and magnetohydrodynamical flows on structured meshes - Speed comparisons with Runge-Kutta methods. J Comput Phys 235:934–969
12. Barth TJ, Jespersen DC (1989) The design and application of upwind schemes on unstructured meshes. AIAA Paper 89–0366
13. Berthon C (2005) Stability of the MUSCL schemes for the Euler equations. Commun Math Sci 3(2):133–157. https://projecteuclid.org:443/euclid.cms/1118778272
14. Berthon C (2006) Why the MUSCL-Hancock scheme is L^1-stable. Numerische Mathematik 104(1):27–46. https://doi.org/10.1007/s00211-006-0007-4
15. Blazek J (2005) Chapter 10-Consistency, accuracy and stability. In: Blazek J (ed) Computational fluid dynamics: principles and applications, 2nd edn. Elsevier Science, Oxford, pp 351–371
16. Brio M, Wu C (1988) An upwind differencing scheme for the equations of ideal magnetohydrodynamics. J Comput Phys 75(2):400–422
17. Bryan GL, Norman ML, Stone JM, Cen R, Ostriker JP (1995) A piecewise parabolic method for cosmological hydrodynamics. Comput Phys Commun 89:149–168. https://doi.org/10.1016/0010-4655(94)00191-4
18. Čada M, Torrilhon M (2009) Compact third-order limiter functions for finite volume methods. J Comput Phys 228(11):4118–4145. https://doi.org/10.1016/j.jcp.2009.02.020

19. Calgaro C, Creusé E, Goudon T, Penel Y (2013) Positivity-preserving schemes for Euler equations: sharp and practical CFL conditions. J Comput Phys 234:417–438. https://doi.org/10.1016/j.jcp.2012.09.040
20. Cargill PJ, Schmidt J, Spicer DS, Zalesak ST (2000) Magnetic structure of overexpanding coronal mass ejections: numerical models. J Geophys Res 105:7509–7520
21. Chen XY, Zha GC (2006) Implicit application of non-reflective boundary conditions for Navier-Stokes equations in generalized coordinates. Int J Numer Methods Fluids 50(7):767–793
22. Cheng Y, Li F, Qiu J, Xu L (2013) Positivity-preserving DG and central DG methods for ideal MHD equations. J Comput Phys 238:255–280
23. Chesshire G, Henshaw WD (1990) Composite overlapping meshes for the solution of partial differential equations. J Comput Phys 90(1):1–64
24. Chesshire G, Henshaw WD (1994) A scheme for conservative interpolation on overlapping grids. SIAM J Sci Comput 15(4):819–845
25. Christlieb AJ, Feng X, Jiang Y, Tang Q (2018) A high-order finite difference WENO scheme for ideal magnetohydrodynamics on curvilinear meshes. SIAM J Sci Comput 40(4):A2631–A2666. https://doi.org/10.1137/17M115757X
26. Christlieb AJ, Feng X, Seal DC, Tang Q (2016) A high-order positivity-preserving single-stage single-step method for the ideal magnetohydrodynamic equations. J Comput Phys 316(Supplement C):218–242, https://doi.org/10.1016/j.jcp.2016.04.016
27. Christlieb AJ, Rossmanith JA, Tang Q (2014) Finite difference weighted essentially non-oscillatory schemes with constrained transport for ideal magnetohydrodynamics. J Comput Phys 268(302):325. https://doi.org/10.1016/j.jcp.2014.03.001
28. Christlieb AJ, Liu Y, Tang Q, Xu Z (2015a) High order parametrized maximum-principle-preserving and positivity-preserving WENO schemes on unstructured meshes. J Comput Phys 281:334–351
29. Christlieb AJ, Liu Y, Tang Q, Xu Z (2015b) Positivity-preserving finite difference weighted ENO schemes with constrained transport for ideal magnetohydrodynamic equations. SIAM J Sci Comput 37(4):A1825–A1845
30. Cohen O, Sokolov IV, Roussev II, Gombosi TI (2008) Validation of a synoptic solar wind model. J Geophys Res: Space Phys 113(A3):A03104. https://doi.org/10.1029/2007JA012797, https://agupubs.onlinelibrary.wiley.com
31. Cranmer SR (2002) Coronal holes and the high-speed solar wind. Space Sci Rev 101:229–294
32. Crockett RK, Colella P, Fisher RT, Klein RI, McKee CF (2005) An unsplit, cell-centered Godunov method for ideal MHD. J Comput Phys 203:422–448
33. Dai W, Woodward PR (1998a) On the divergence-free condition and conservation laws in numerical simulations for supersonic magnetohydrodynamical flows. Astrophys J 494(1):317
34. Dai W, Woodward PR (1998b) A simple finite difference scheme for multidimensional magnetohydrodynamical equations. J Comput Phys 142(2):331–369
35. Dedner A, Kemm F, Kröner D, Munz CD, Schnitzer T, Wesenberg M (2002) Hyperbolic divergence cleaning for the MHD equations. J Comput Phys 175(2):645–673
36. DeForest CE, Howard TA, McComas DJ (2014) Inbound waves in the solar corona: a direct indicator of Alfvén surface location. Astrophys J 787(2):124
37. Dellar PJ (2001) A note on magnetic monopoles and the one dimensional MHD Riemann problem. J Comput Phys 172:392–398. https://doi.org/10.1006/jcph.2001.6815
38. Derigs D, Winters AR, Gassner GJ, Walch S (2016) A novel high-order, entropy stable, 3D AMR MHD solver with guaranteed positive pressure. J Comput Phys 317:223–256. https://doi.org/10.1016/j.jcp.2016.04.048
39. Derigs D, Winters AR, Gassner GJ, Walch S (2017) A novel averaging technique for discrete entropy-stable dissipation operators for ideal MHD. J Comput Phys 330:624–632. https://doi.org/10.1016/j.jcp.2016.10.055

40. DeVore CR (1991) Flux-corrected transport techniques for multidimensional compressible magnetohydrodynamics. J Comput Phys 92(1):142–160
41. Eirik E, Leer E, Holzer TE (2003) Two-dimensional magnetohydrodynamic models of the solar corona: mass loss from the streamer belt. Astrophys J 589(2):1040
42. Endeve E, Hauck CD, Xing Y, Mezzacappa A (2015) Bound-preserving discontinuous Galerkin methods for conservative phase space advection in curvilinear coordinates. J Comput Phys 287(Supplement C):151–183. https://doi.org/10.1016/j.jcp.2015.02.005
43. Evans CR, Hawley JF (1988) Simulation of magnetohydrodynamic flows - a constrained transport method. Astrophys J 332:659–677
44. Feng XS, Yang LP, Xiang CQ, Liu Y, Zhao XP, Wu ST (2012c) Numerical study of the global corona for CR 2055 driven by daily updated synoptic magnetic field. In: Astronomical society of the pacific conference series, vol 459, p 202
45. Feng XS, Wu ST, Fan QL, Wei FS (2002) A class of TVD type combined numerical scheme for MHD equations and its application to MHD numerical simulations (in Chinese). Chin J Space Sci 22:300–308
46. Feng XS, Wu ST, Wei FS, Fan QL (2003) A class of TVD type combined numerical scheme for MHD equations with a survey about numerical methods in solar wind simulations. Space Sci Rev 107:43–53
47. Feng XS, Xiang CQ, Zhong DK, Fan QL (2005) A comparative study on 3-D solar wind structure observed by Ulysses and MHD simulation. Chin Sci Bull 50(8):672–678
48. Feng XS, Zhou YF, Hu YQ (2006) A 3rd order WENO GLM-MHD scheme for magnetic reconnection. Chin J Space Sci 26(1):1
49. Feng XS, Zhou YF, Wu ST (2007) A novel numerical implementation for solar wind modeling by the modified conservation element/solution element method. Astrophys J 655:1110–1126
50. Feng XS, Yang LP, Xiang C, Wu ST, Zhou YF, Zhong DK (2010) Three-dimensional solar wind modeling from the Sun to Earth by a SIP-CESE MHD model with a six-component grid. Astrophys J 723:300–319
51. Feng XS, Xiang CQ, Zhong DK (2011a) The state-of-art of three-dimensional numerical study for corona-interplanetary process of solar storms (in Chinese). Sci Sin-Terrae 41:1–28
52. Feng XS, Zhang SH, Xiang CQ, Yang LP, Jiang CW, Wu ST (2011b) A hybrid solar wind model of the CESE+HLL method with a Yin-Yang overset grid and an AMR grid. Astrophys J 734:50
53. Feng XS, Jiang CW, Xiang CQ, Zhao XP, Wu ST (2012a) A data-driven model for the global coronal evolution. Astrophys J 758(1):62
54. Feng XS, Yang LP, Xiang CQ, Jiang CW, Ma XP, Wu ST, Zhong DK, Zhou YF (2012b) Validation of the 3D AMR SIP-CESE solar wind model for four Carrington rotations. Sol Phys 279(1):207–229
55. Feng XS, Xiang CQ, Zhong DK (2013a) Numerical study of interplanetary solar storms (in Chinese). Sci Sin-Terrae 43:912–933
56. Feng XS, Zhong DK, Xiang CQ, Zhang Y (2013b) GPU-accelerated computing of three-dimensional solar wind background. Sci China Earth Sci 56(11):1864–1880
57. Feng XS, Xiang CQ, Zhong DK, Zhou YF, Yang LP, Ma XP (2014a) SIP-CESE MHD model of solar wind with adaptive mesh refinement of hexahedral meshes. Comput Phys Commun 185(7):1965–1980
58. Feng XS, Zhang M, Zhou YF (2014b) A new three-dimensional solar wind model in spherical coordinates with a six-component grid. Astrophys J Suppl Ser 214(1):6
59. Feng XS, Li CX, Xiang CQ, Zhang M, Li HC, Wei FS (2017) Data-driven modeling of the solar corona by a new three-dimensional path-conservative Osher-Solomon MHD model. Astrophys J Suppl Ser 233:10
60. Fogang F, Tchuen G, Burtschell Y, Woafo P (2015) An extension of AUFSR scheme for the ideal magnetohydrodynamics equations. Comput Fluids 114:297–313. https://doi.org/10.1016/j.compfluid.2015.02.003

61. Fuchs FG, McMurry AD, Mishra S, Risebro NH, Waagan K (2010) Finite volume methods for wave propagation in stratified magneto-atmospheres. Commun Comput Phys 7(3):1–30
62. Fuchs FG, McMurry AD, Mishra S, Risebro NH, Waagan K (2011) Approximate Riemann solvers and robust high-order finite volume schemes for multi-dimensional ideal MHD equations. Commun Comput Phys 9:324–362. https://doi.org/10.4208/cicp.171109.070510a
63. Gardiner TA, Stone JM (2005) An unsplit Godunov method for ideal MHD via constrained transport. J Comput Phys 205(2):509–539
64. Gardiner TA, Stone JM (2008) An unsplit Godunov method for ideal MHD via constrained transport in three dimensions. J Comput Phys 227(8):4123–4141
65. Goelzer ML, Schwadron NA, Smith CW (2014) An analysis of Alfvén radius based on sunspot number from 1749 to today. J Geophys Res: Space Phys 119(1):115–120
66. Gombosi TI, Zeeuw DL, Powell KG, Ridley AJ, Sokolov IV, Stout QF, Tóth G (2003) Adaptive mesh refinement for global magnetohydrodynamic simulation. In: Büchner J, Scholer M, Dum CT (eds) Space plasma simulation, vol 615. Lecture notes in physics. Springer, Berlin Heidelberg, pp 247–274
67. Groth CPT, De Zeeuw DL, Gombosi TI, Powell KG (2000) Global three-dimensional MHD simulation of a space weather event: CME formation, interplanetary propagation, and interaction with the magnetosphere. J Geophys Res: Space Phys 105:25053–25078
68. Gurski K (2004) An HLLC-type approximate Riemann solver for ideal magnetohydrodynamics. SIAM J Sci Comput 25(6):2165–2187
69. Han SH, Lee JI, Kim KH (2009) Accurate and robust pressure weight advection upstream splitting method for magnetohydrodynamics equations. AIAA J 47:970–981
70. Harten A, Lax PD, Leer Bv (1983) On upstream differencing and Godunov-type schemes for hyperbolic conservation laws. SIAM Rev 25(1):35–61
71. Hayashi K (2005) Magnetohydrodynamic simulations of the solar corona and solar wind using a boundary treatment to limit solar wind mass flux. Astrophys J Suppl Ser 161:480–494
72. Hayashi K (2013) An MHD simulation model of time-dependent global solar corona with temporally varying solar-surface magnetic field maps. J Geophys Res: Space Phys 118(11):6889–6906
73. Hayashi K, Benevolenskaya E, Hoeksema T, Liu Y, Zhao XP (2006) Three-dimensional magnetohydrodynamic simulation of a global solar corona using a temperature distribution map obtained from SOHO EIT measurements. Astrophys J Lett 636:L165–L168
74. Hayashi K, Zhao XP, Liu Y (2008) MHD simulations of the global solar corona around the Halloween event in 2003 using the synchronic frame format of the solar photospheric magnetic field. J Geophys Res: Space Phys 113(A7):A07104
75. Henshaw WD (2010) Adaptive Mesh and Overlapping Grid Methods. Wiley, New York, pp 623–630
76. Honkkila V, Janhunen P (2007) HLLC solver for ideal relativistic MHD. J Comput Phys 223(2):643–656
77. Housman JA, Barad MF, Kiris CC, Kwak D (2011) Space-time convergence analysis of a dual-time stepping method for simulating ignition overpressure waves. Computational fluid dynamics 2010. Springer, Berlin, pp 645–652
78. Hu YQ, Feng XS, Wu ST, Song WB (2008) Three-dimensional MHD modeling of the global corona throughout solar cycle 23. J Geophys Res: Space Phys 113(A3):A03106
79. Hu XY, Adams NA, Shu CW (2013) Positivity-preserving method for high-order conservative schemes solving compressible Euler equations. J Comput Phys 242:169–80
80. Hundhausen AJ (1972) Coronal expansion and solar wind. Springer, New York
81. Ivan L, De Sterck H, Northrup SA, Groth CPT (2013) Multi-dimensional finite-volume scheme for hyperbolic conservation laws on three-dimensional solution-adaptive cubed-sphere grids. J Comput Phys 255:205–227
82. Ivan L, Sterck HD, Susanto A, Groth C (2015) High-order central ENO finite-volume scheme for hyperbolic conservation laws on three-dimensional cubed-sphere grids. J Comput Phys 282:157–182

83. Ivan L, De Sterck H, Northrup SA, Groth CPT (2011) Three-dimensional MHD on cubed-sphere grids: parallel solution-adaptive simulation framework. In: 20th AIAA CFD conference, vol 3382, pp 1325–1342
84. Ivan L, Susanto A, De Sterck H, Northrup SA, Groth CPT (2012) High-order central ENO finite-volume scheme for MHD on three-dimensional cubed-sphere grids. In: Proceedings of the seventh international conference on computational fluid dynamics (ICCFD7), Hawaii, 2012, vol 3382, pp 1325–1342
85. Jameson A (1991) Time dependent calculations using multigrid, with applications to unsteady flows past airfoils and wings. AIAA 91–1596
86. Janhunen P (2000) A positive conservative method for magnetohydrodynamics based on HLL and Roe methods. J Comput Phys 160(2):649–661
87. Jeltsch R, Torrilhon M (2005) Solenoidal discrete initialization for magnetohydrodynamics. In: Bock H, Phu H, Kostina E, Rannacher R (eds) Modeling, Simulation and Optimization of Complex Processes. Springer, Berlin Heidelberg, pp 235–253
88. Jiang RL, Fang C, Chen PF (2012) A new MHD code with adaptive mesh refinement and parallelization for astrophysics. Comput Phys Commun 183(8):1617–1633
89. Jiang Y, Xu Z (2013) Parametrized maximum principle preserving limiter for finite difference WENO schemes solving convection-dominated diffusion equations. SIAM J Sci Comput 35(6):A2524–A2553. https://doi.org/10.1137/130924937
90. Kageyama A (2005) Dissection of a sphere and Yin-Yang grids. J Earth Simulator 3:20–28
91. Kageyama A, Sato T (2004) Yin-Yang grid: an overset grid in spherical geometry. Geochem, Geophys, Geosystems 5(9):Q09005. https://doi.org/10.1029/2004GC000734
92. Kawai S (2013) Divergence-free-preserving high-order schemes for magnetohydrodynamics: an artificial magnetic resistivity method. J Comput Phys 251:292–318
93. Kemm F (2011) A comparative study of TVD-limitersąłwell-known limiters and an introduction of new ones. Int J Numer Methods Fluids 67(4):404–440
94. Keppens R, Meliani Z, van Marle A, Delmont P, Vlasis A, van der Holst B (2012) Parallel, grid-adaptive approaches for relativistic hydro and magnetohydrodynamics. J Comput Phys 231(3):718–744
95. Kermani MJ, Plett EG (2001) Modified entropy correction formula for the Roe scheme. AIAA 2001–0083
96. Kifonidis K, Müller E (2012) On multigrid solution of the implicit equations of hydrodynamics-experiments for the compressible Euler equations in general coordinates. Astron Astrophys 544:A47
97. Kim J, Balsara DS (2014) A stable HLLC Riemann solver for relativistic magnetohydrodynamics. J Comput Phys 270:634–639
98. Kitamura K, Shima E (2012) Simple and parameter-free second slope limiter for unstructured grid aerodynamic simulations. AIAA J 50(6):1415–1426
99. Kleimann J, Kopp A, Fichtner H, Grauer R (2009) A novel code for numerical 3-D MHD studies of CME expansion. Ann Geophys 27(3):989–1004
100. Kley W (1998) On the treatment of the Coriolis force in computational astrophysics. Astron Astrophys 338:L37–L41
101. Lee D, Deane AE (2009) An unsplit staggered mesh scheme for multidimensional magnetohydrodynamics. J Comput Phys 228(4):952–975
102. Li S (2005) An HLLC Riemann solver for magneto-hydrodynamics. J Comput Phys 203:344–357
103. Li ST (2008) High order central scheme on overlapping cells for magneto-hydrodynamic flows with and without constrained transport method. J Comput Phys 227(15):7368–7393
104. Li ST (2010) A fourth-order divergence-free method for MHD flows. J Comput Phys 229(20):7893–7910
105. Li F, Xu L (2012) Arbitrary order exactly divergence-free central discontinuous Galerkin methods for ideal MHD equations. J Comput Phys 231(6):2655–2675
106. Li ST, Li H, Cen R (2008) CosmoMHD: a cosmological magnetohydrodynamics code. Astrophys J Suppl Ser 174(1):1

107. Li D, Merkle C, Scott WM, Keefer D, Moeller T, Rhodes R (2011) Hyperbolic algorithm for coupled plasma/electromagnetic fields including conduction and displacement currents. AIAA J 49:909–920
108. Liang C, Xu Z (2014a) Parametrized maximum principle preserving flux limiters for high order schemes solving multi-dimensional scalar hyperbolic conservation laws. J Sci Comput 58(1):41–60
109. Liang C, Xu Z (2014b) Parametrized maximum principle preserving flux limiters for high order schemes solving multi-dimensional scalar hyperbolic conservation laws. J Sci Comput 58(1):41–60. https://doi.org/10.1007/s10915-013-9724-x
110. Li S, Li H (2003) A modern code for solving magneto-hydrodynamic or hydrodynamic equations. Technical report LA-UR-03-8925, Los Alamos National Lab
111. Linde T (2002) A practical, general-purpose, two-state HLL Riemann solver for hyperbolic conservation laws. Int J Numer Methods Fluids 40(3–4):391–402
112. Linker JA, Kivelson MG, Walker RJ (1991) A three-dimensional MHD simulation of plasma flow past io. J Geophys Res Space Phys 96(A12):21037–21053
113. Linker JA, Mikić Z, Biesecker DA, Forsyth RJ, Gibson SE, Lazarus AJ, Lecinski A, Riley P, Szabo A, Thompson BJ (1999) Magnetohydrodynamic modeling of the solar corona during whole sun month. J Geophys Res Space Phys 104(A5):9809–9830
114. Lionello R, Linker JA, Mikić Z (2001) Including the transition region in models of the large-scale solar corona. Astrophys J 546:542–551
115. Liu Y, Shu CW, Zhang M (2018) Entropy stable high order discontinuous Galerkin methods for ideal compressible MHD on structured meshes. J Comput Phys 354:163–178. https://doi.org/10.1016/j.jcp.2017.10.043
116. Lohmann C, Kuzmin D (2016) Synchronized flux limiting for gas dynamics variables. J Comput Phys 326:973–990. https://doi.org/10.1016/j.jcp.2016.09.025
117. Londrillo P, Zanna LD (2004) On the divergence-free condition in Godunov-type schemes for ideal magnetohydrodynamics: the upwind constrained transport method. J Comput Phys 195(1):17–48
118. Marder B (1987) A method for incorporating Gauss' law into electromagnetic PIC codes. J Comput Phys 68(1):48
119. Mathur SR, Murthy JY (1997) A pressure-based method for unstructured meshes. Numer Heat Transf, Part B: Fundam 31(2):195–215
120. Mignone A (2014) High-order conservative reconstruction schemes for finite volume methods in cylindrical and spherical coordinates. J Comput Phys 270:784–814
121. Mignone A, Tzeferacos P (2010) A second-order unsplit Godunov scheme for cell-centered MHD: the CTU-GLM scheme. J Comput Phys 229(6):2117–2138
122. Mignone A, Tzeferacos P, Bodo G (2010) High-order conservative finite difference GLM-MHD schemes for cell-centered MHD. J Comput Phys 229(17):5896–5920
123. Mignone A, Bodo G (2006) An HLLC Riemann solver for relativistic flows-II. Magnetohydrodynamics. Mon Not R Astron Soc 368:1040–1054
124. Mignone A, Ugliano M, Bodo G (2009) A five-wave Harten-Lax-van Leer Riemann solver for relativistic magnetohydrodynamics. Mon Not R Astron Soc 393:1141–1156. https://doi.org/10.1111/j.1365-2966.2008.14221.x
125. Mikić Z, Linker JA (1994) Disruption of coronal magnetic field arcades. Astrophys J 430:898–912
126. Mikić Z, Linker JA, Schnack DD, Lionello R, Tarditi A (1999) Magnetohydrodynamic modeling of the global solar corona. Phys Plasmas 6(5):2217–2224
127. Miyoshi T, Kusano K (2005) A multi-state HLL approximate Riemann solver for ideal magnetohydrodynamics. J Comput Phys 208:315–344
128. Miyoshi T, Terada N, Matsumoto Y, Fukazawa K, Umeda T, Kusano K (2010) The HLLD approximate Riemann solver for magnetospheric simulation. IEEE Trans Plasma Sci 38:2236–2242

129. Miyoshi T, Kusano K (2011) A comparative study of divergence-cleaning techniques for multi-dimensional MHD schemes. Plasma Fusion Res 6:2401124–2401124. https://doi.org/10.1585/pfr.6.2401124
130. Nakamizo A, Tanaka T, Kubo Y, Kamei S, Shimazu H, Shinagawa H (2009) Development of the 3-D MHD model of the solar corona-solar wind combining system. J Geophys Res: Space Phys 114(A07):A07109. https://doi.org/10.1029/2008JA013844
131. Neugebauer M (1999) The three-dimensional solar wind at solar activity minimum. Rev Geophys 37(1):107–126
132. Norman MR, Semazzi FHM, Nair RD (2009) Conservative cascade interpolation on the sphere: an intercomparison of various non-oscillatory reconstructions. Q J R Meteorol Soc 135(640):795–805. https://doi.org/10.1002/qj.402
133. Odstrčil D, Linker JA, Lionello R, Mikic Z, Riley P, Pizzo VJ, Luhmann JG (2002) Merging of coronal and heliospheric numerical two-dimensional MHD models. J Geophys Res 107:1493
134. Odstrčil D, Riley P, Zhao XP (2004) Numerical simulation of the 12 May 1997 interplanetary CME event. J Geophys Res: Space Phys 109(A2):A02116
135. Owens MJ, Spence HE, McGregor S, Hughes WJ, Quinn JM, Arge CN, Riley P, Linker J, Odstrčil D (2008) Metrics for solar wind prediction models: comparison of empirical, hybrid, and physics-based schemes with 8 years of L1 observations. Space Weather 6(8):S08001
136. Parent B (2012) Positivity-preserving high-resolution schemes for systems of conservation laws. J. Comput. Phys. 231(1):173–189
137. Parker EN (1963) Interplanetary dynamical processes. Interscience Publishers, New York, p 1963
138. Peng XD, Xiao F, Takahashi K (2006) Conservative constraint for a quasi-uniform overset grid on the sphere. Q J R Meteorol Soc 132(616):979–996
139. Perthame B, Shu CW (1996) On positivity preserving finite volume schemes for Euler equations. Numer Math 73(1):119–130. https://doi.org/10.1007/s002110050187
140. Pneuman GW, Kopp RA (1971) Gas-magnetic field interactions in the solar corona. Sol Phys 18:258–270
141. Powell KG, Roe PL, Linde TJ, Gombosi TI, De Zeeuw DL (1999) A solution-adaptive upwind scheme for ideal magnetohydrodynamics. J Comput Phys 154(2):284–309
142. Rai MM (1986) A conservative treatment of zonal boundaries for Euler equation calculations. J Comput Phys 62(2):472–503
143. Ranocha H (2018) Comparison of some entropy conservative numerical fluxes for the Euler equations. J Sci Comput 76:216–242. https://doi.org/10.1007/s10915-017-0618-1
144. Reynolds DR, Samtaney R, Tiedeman HC (2012) A fully implicit Newton-Krylov-Schwarz method for tokamak magnetohydrodynamics: Jacobian construction and preconditioner formulation. Comput Sci Discov 5(1):014003
145. Reynolds D, Samtaney R, Woodward C (2010) Operator-based preconditioning of stiff hyperbolic systems. SIAM J Sci Comput 32(1):150–170
146. Riley P, Lionello R, Linker JA, Mikić Z, Luhmann J, Wijaya J (2011) Global MHD modeling of the solar corona and inner heliosphere for the whole heliosphere interval. Sol Phys 274(1–2):361–377
147. Riley P, Linker JA, Lionello R, Mikić Z (2012) Corotating interaction regions during the recent solar minimum: the power and limitations of global MHD modeling. J Atmos Sol-Terr Phys 83:1–10
148. Riley P, Linker JA, Mikić Z (2013) On the application of ensemble modeling techniques to improve ambient solar wind models. J Geophys Res: Space Phys 118(2):600–607
149. Riley P, Linker JA, Mikić Z (2001) An empirically-driven global MHD model of the solar corona and inner heliosphere. J Geophys Res: Space Phys 106(A8):15,889–15,901
150. Roe P, Balsara D (1996) Notes on the eigensystem of magnetohydrodynamics. SIAM J Appl Math 56(1):57–67
151. Ronchi C, Iacono R, Paolucci PS (1996) The "cubed sphere": a new method for the solution of partial differential equations in spherical geometry. J Comput Phys 124(1):93–114

152. Ryu D, Jones TW (1995) Numerical magetohydrodynamics in astrophysics: algorithm and tests for one-dimensional flow. Astrophys J 442:228–258
153. Ryu D, Ostriker JP, Kang H, Cen R (1993) A cosmological hydrodynamic code based on the total variation diminishing scheme. Astrophys J 414:1–19
154. Ryu D, Miniati F, Jones TW, Frank A (1998) A divergence-free upwind code for multidimensional magnetohydrodynamic flows. Astrophys J 509:244–255
155. Samtaney R (2012) Implicit numerical methods for magnetohydrodynamics. In: Zheng L (ed) Topics in magnetohydrodynamics. Shanghai, InTech China, pp 59–84
156. Schwadron NA, Connick DE, Smith CW (2010) Magnetic flux balance in the heliosphere. Astrophys J Lett 722(2):L132
157. Serna S (2006) A class of extended limiters applied to piecewise hyperbolic methods. SIAM J Sci Comput 28(1):123–140. https://doi.org/10.1137/040611811
158. Serna S (2009) A characteristic-based nonconvex entropy-fix upwind scheme for the ideal magnetohydrodynamic equations. J Comput Phys 228(11):4232–4247
159. Sheeley NR, Wang YM, Hawley SH, Brueckner GE, Dere KP, Howard RA, Koomen MJ, Korendyke CM, Michels DJ, Paswaters SE, Socker DG, St Cyr OC, Wang D, Lamy PL, Llebaria A, Schwenn R, Simnett GM, Plunkett S, Biesecker DA (1997) Measurements of flow speeds in the corona between 2 and 30 R_{sun}. The Astrophysical Journal 484:472–478
160. Sheeley NRJ, Wang YM (2001) Coronal inflows and sector magnetism. Astrophys J Lett 562(1):L107
161. Sheeley NRJ, Warren HP, Wang YM (2004) The origin of postflare loops. Astrophys J 616(2):1224
162. Shen F, Feng XS, Wu ST, Xiang CQ, Song WB (2011) Three-dimensional MHD simulation of the evolution of the April 2000 CME event and its induced shocks using a magnetized plasma blob model. J Geophys Res: Space Phys 116:A04102
163. Shen F, Feng XS, Wu ST, Xiang CQ (2007) Three-dimensional MHD simulation of CMEs in three-dimensional background solar wind with the self-consistent structure on the source surface as input: numerical simulation of the January 1997 Sun-Earth connection event. J Geophys Res: Space Phys 112(A6):A06109. https://doi.org/10.1029/2006JA012164
164. Shiota D, Kusano K, Miyoshi T, Shibata K (2010) Magnetohydrodynamic modeling for a formation process of coronal mass ejections: interaction between an ejecting flux rope and an ambient field. Astrophys J 718(2):1305
165. Shiota D, Kataoka R, Miyoshi Y, Hara T, Tao C, Masunaga K, Futaana Y, Terada N (2014) Inner heliosphere MHD modeling system applicable to space weather forecasting for the other planets. Space Weather 12(4):187–204
166. Shiota D, Kataoka R (2016) Magnetohydrodynamic simulation of interplanetary propagation of multiple coronal mass ejections with internal magnetic flux rope (SUSANOO-CME). Space Weather 14(2):56–75
167. Skinner MA, Ostriker EC (2010) The Athena astrophysical magnetohydrodynamics code in cylindrical geometry. Astrophys J Suppl Ser 188(1):290, http://stacks.iop.org/0067-0049/188/i=1/a=290
168. Smith CW, Schwadron NA, DeForest CE (2013) Decline and recovery of the interplanetary magnetic field during the protracted solar minimum. Astrophys J 775(1):59
169. Steger J, Benek J (1987) On the use of composite grid schemes in computational aerodynamics. Comput Methods Appl Mech Eng 64:301–320
170. Susanto A, Ivan L, De Sterck H, Groth CPT (2013) High-order central ENO finite-volume scheme for ideal MHD. J Comput Phys 250:141–164
171. Tanaka T (1995) Generation mechanisms for magnetosphere-ionosphere current systems deduced from a three-dimensional MHD simulation of the solar wind-magnetosphere-ionosphere coupling processes. J Geophys Res 100(A7):12,057–12,074
172. Tanaka T (1999) Configuration of the magnetosphere-ionosphere convection system under northward IMF conditions with nonzero IMF By. J Geophys Res 104(A7):14,683–14,690
173. Tanaka T (2000) The state transition model of the substorm onset. J Geophys Res 105(A9):21,081–21,096

174. Tanaka T (1994) Finite volume TVD scheme on an unstructured grid system for three-dimensional MHD simulation of inhomogeneous systems including strong background potential fields. J Comput Phys 111:381–390
175. Tóth G (2000) The $\nabla \cdot \mathbf{B} = 0$ constraint in shock-capturing magnetohydrodynamics codes. Journal of Computational Physics 161(2):605–652
176. Tóth G, van der Holst B, Sokolov IV, De Zeeuw DL, Gombosi TI, Fang F, Manchester WB, Meng X, Najib D, Powell KG, Stout QF, Glocer A, Ma YJ, Opher M (2012) Adaptive numerical algorithms in space weather modeling. J Comput Phys 231(3):870–903
177. Turkel E (1999) Preconditioning techniques in computational fluid dynamics. Annu Rev Fluid Mech 31:385–416
178. Uand Brackbill JU, Barnes DC (1980) The effect of nonzero $\nabla \cdot \mathbf{B}$ on the numerical solution of the magnetohydrodynamic equations. J Comput Phys 35(3):426–430
179. Usmanov AV, Goldstein ML (2003) A tilted-dipole MHD model of the solar corona and solar wind. J Geophys Res: Space Phys 108(A9):1354. https://doi.org/10.1029/2002JA009777
180. Usmanov AV, Matthaeus WH, Goldstein ML, Chhiber R (2018) The steady global corona and solar wind: a three-dimensional MHD simulation with turbulence transport and heating. Astrophys J 865(1):25, http://stacks.iop.org/0004-637X/865/i=1/a=25
181. Usmanov AV (1996) A global 3-D MHD solar wind model with Alfvén waves. In: AIP Conference Proceedings, vol 382, p 141
182. Usmanov AV (1993) A global numerical 3-D MHD model of the solar wind. Sol Phys 146(2):377–396
183. Usmanov AV, Dryer M (1995) A global 3-D simulation of interplanetary dynamics in June 1991. Sol Phys 159:347–370
184. Usmanov AV, Goldstein ML, Matthaeus WH (2012) Three-dimensional magnetohydrodynamic modeling of the solar wind including pickup protons and turbulence transport. Astrophys J 754(1):40
185. van der Holst B, Keppens R (2007) Hybrid block-AMR in cartesian and curvilinear coordinates: MHD applications. J Comput Phys 226(1):925–946
186. Venkatakrishnan V (1993) On the accuracy of limiters and convergence to steady state solutions. AIAA Paper 93–0880
187. Vides J, Audit E, Guillard H, Nkonga B (2013) Divergence-free MHD simulations with the HERACLES code. In: ESAIM: Proceedings 43:180–194
188. Waagan K (2009) A positive MUSCL-Hancock scheme for ideal magnetohydrodynamics. J Comput Phys 228(23):8609–8626
189. Wang AH, Wu ST, Suess ST, Poletto G (1998) Global model of the corona with heat and momentum addition. J Geophys Res: Space Phys 103(A2):1913–1922
190. Wang YM, Sheeley NR, Howard RA, Cyr OCS, Simnett GM (1999) Coronagraph observations of inflows during high solar activity. Geophys Res Lett 26(9):1203–1206
191. Winters AR, Derigs D, Gassner GJ, Walch S (2017) A uniquely defined entropy stable matrix dissipation operator for high mach number ideal MHD and compressible Euler simulations. J Comput Phys 332:274–289. https://doi.org/10.1016/j.jcp.2016.12.006
192. Winters AR, Gassner GJ (2016) Affordable, entropy conserving and entropy stable flux functions for the ideal MHD equations. J Comput Phys 304:72–108. https://doi.org/10.1016/j.jcp.2015.09.055
193. Wongwathanarat A, Hammer NJ, Müller E (2010) An axis-free overset grid in spherical polar coordinates for simulating 3D self-gravitating flows. Astron Astrophys 514:A48
194. Wu K (2017) Design of provably physical-constraint-preserving methods for general relativistic hydrodynamics. Phys Rev D 95(103):001. https://doi.org/10.1103/PhysRevD.95.103001
195. Wu ST, Guo WP, Michels DJ, Burlaga LF (1999) MHD description of the dynamical relationships between a flux rope, streamer, coronal mass ejection, and magnetic cloud: an analysis of the January 1997 Sun-Earth connection event. J Geophys Res 104:14,789–14,802
196. Wu ZN, Xu SS, Gao B, Zhuang LS (2007) Review of numerical computation of compressible flows with artificial interfaces. Comput Fluids 36(10):1657

197. Wu ST, Dryer M (2015) Comparative analyses of current three-dimensional numerical solar wind models. Sci China Earth Sci 58:839–858
198. Wu K, Tang H (2017) Admissible states and physical-constraints-preserving schemes for relativistic magnetohydrodynamic equations. Math Model Methods Appl Sci 27(10):1871–1928. https://doi.org/10.1142/S0218202517500348
199. Xing Y, Zhang X, Shu CW (2010) Positivity-preserving high order well-balanced discontinuous Galerkin methods for the shallow water equations. Adv Water Resour 33(12):1476–1493. https://doi.org/10.1016/j.advwatres.2010.08.005
200. Xiong T, Qiu JM, Xu Z (2013) A parametrized maximum principle preserving flux limiter for finite difference RK-WENO schemes with applications in incompressible flows. J Comput Phys 252:310–331
201. Xiong T, Qiu JM, Xu Z (2016) Parametrized positivity preserving flux limiters for the high order finite difference WENO scheme solving compressible Euler equations. J Sci Comput 67(3):1066–1088. https://doi.org/10.1007/s10915-015-0118-0
202. Xisto CM, Páscoa JC, Oliveira PJ (2013) A pressure-based method with AUSM-type fluxes for MHD flows at arbitrary mach numbers. Int J Numer Methods Fluids 72:1165–1182
203. Xu Z (2014) Parametrized maximum principle preserving flux limiters for high order schemes solving hyperbolic conservation laws: one-dimensional scalar problem. Math Comp 83:2213–2238
204. Xu Z, Zhang X (2017) Chapter 4 - Bound-preserving high-order schemes. In: Abgrall R, Shu CW (eds) Handbook of numerical methods for hyperbolic problems, handbook of numerical analysis, vol 18, Elsevier, pp 81–102, https://doi.org/10.1016/bs.hna.2016.08.002
205. Yakovlev S, Xu L, Li F (2013) Locally divergence-free central discontinuous Galerkin methods for ideal MHD equations. J Comput Phys 4(1–2):80–91
206. Yalim MS, Abeele DV, Lani A, Quintino T, Deconinck H (2011) A finite volume implicit time integration method for solving the equations of ideal magnetohydrodynamics for the hyperbolic divergence cleaning approach. J Comput Phys 230(15):6136–6154
207. Yuan L (2002) Comparison of implicit multigrid schemes for three-dimensional incompressible flows. J Comput Phys 177:134–155
208. Zachary A, Malagoli A, Colella P (1994) A higher-order godunov method for multidimensional ideal magnetohydrodynamics. SIAM J Sci Comput 15(2):263–284
209. Zhang X (2017) On positivity-preserving high order discontinuous Galerkin schemes for compressible Navier Stokes equations. Journal of Computational Physics 328(Supplement C):301–343, https://doi.org/10.1016/j.jcp.2016.10.002
210. Zhang M, Feng XS (2015) Implicit dual-time stepping method for a solar wind model in spherical coordinates. Comput Fluids 115:115–123
211. Zhang M, Feng XS (2016) A comparative study of divergence cleaning methods of magnetic field in the solar coronal numerical simulation. Front Astron Space Sci 3(6):1–14
212. Zhang X, Shu CW (2010b) On positivity-preserving high order discontinuous Galerkin schemes for compressible Euler equations on rectangular meshes. J Comput Phys 229(23):8918–8934
213. Zhang X, Shu CW (2010a) On maximum-principle-satisfying high order schemes for scalar conservation laws. J Comput Phys 229(9):3091–3120. https://doi.org/10.1016/j.jcp.2009.12.030
214. Zhang X, Shu CW (2011) Positivity-preserving high order discontinuous Galerkin schemes for compressible Euler equations with source terms. J Comput Phys 230(4):1238–1248. https://doi.org/10.1016/j.jcp.2010.10.036
215. Zhang X, Xia Y, Shu CW (2012) Maximum-principle-satisfying and positivity-preserving high order discontinuoud Galerkin schemes for conservation laws on triangular meshes. J Sci Comput 50(1):29–62. https://doi.org/10.1007/s10915-011-9472-8
216. Zhao XP, Hoeksema JT (2010) The magnetic field at the inner boundary of the heliosphere around solar minimum. Sol Phys 266:379–390
217. Zhao X, Yang Y, Seyler CE (2014) A positivity-preserving semi-implicit discontinuous Galerkin scheme for solving extended magnetohydrodynamics equations. J Comput Phys 278:400–415

218. Ziegler U (2005) A solution-adaptive central-constraint transport scheme for magnetohydrodynamics. Comput Phys Commun 170:153–174
219. Ziegler U (2011) A semi-discrete central scheme for magnetohydrodynamics on orthogonal-curvilinear grids. J Comput Phys 230:1035–1063
220. Ziegler U (2012) Block-structured adaptive mesh refinement on curvilinear-orthogonal grids. SIAM J Sci Comput 34(3):C102–C121

Chapter 4
3D SIP-CESE MHD Model on Triangular Prism Grids

Abstract This chapter describes the 3D Solar-Interplanetary space-time Conservation Element and Solution Element (SIP-CESE) MHD model and its application on the solar wind study. SIP-CESE MHD model is established on pentahedral cells with each cell a triangular prism composed of two triangular bases and three rectangular sides. The basic principle as well as the implementation detail of the CESE numerical scheme are presented. Two examples are given to illustrate the validity and capacity of modeling solar wind: (i) two-dimensional coronal dynamical structure with multipole magnetic fields and (ii) three-dimensional coronal dynamical structure, using measured solar surface magnetic fields and the empirical values of the plasma properties on the solar surface as the initial conditions for the set of MHD equations, and then the relaxation method to achieve a quasi-steady state. These examples show that the MHD model possesses the ability to model the Sun-Earth environment. Finally, the evolution of the large-scale coronal magnetic structure during solar cycle 23 is investigated by the SIP-CESE MHD model.

4.1 Introduction

The MHD system used to describe the solar wind plasmas is a non-strictly hyperbolic system that is composed of eight equations coupled by inviscid compressible fluid dynamics and magnetic induction to take into account the non-linear interaction between the fluid and magnetic fields. Differing from the Euler equations with the unique sound wave mode, MHD equations have multiple modes of waves, i.e., entropy wave, Alfvén wave, fast magnetosonic and slow magnetosonic waves. This circumstance makes the eigenvalues and eigenvectors of the MHD system more complex than those of equations governing gas dynamics. Although the MHD system and the Euler equations are both non-linear systems consisting of conservation laws, many modern numerical schemes used in ordinary fluid dynamics cannot be directly transplanted to solve the MHD system. Solutions of the compressible MHD equations are characterized by complicated nonlinear wave structure and admit strong shocks and contact discontinuities. However, the ideal MHD equations exhibit solution degeneracies of a type that do not arise in conventional gas dynamics and, as

© Springer Nature Singapore Pte Ltd. 2020

X. Feng, *Magnetohydrodynamic Modeling of the Solar Corona and Heliosphere*, Atmosphere, Earth, Ocean & Space,
https://doi.org/10.1007/978-981-13-9081-4_4

they are normally written, have the added constraint of zero divergence of the magnetic field imposed by the condition that there are no observed magnetic monopoles. For many space and astrophysical plasma applications, such as the solar wind simulation, the solutions may also exhibit rapid transitions between a wide range of highly different physical regimes, including transitions from low speed subsonic to hypersonic and hyperAlfénic regimes, from high-density high-temperature to low-density low-temperature regimes, and from strongly magnetized (low-β) to weakly magnetized (high-β) regimes, where β is the ratio of thermal to magnetic pressures. To understand the space plasma phenomena associated with space weather, for many decades space scientists have been attracted to numerical study of the solar wind and its delicate interaction with the highly variable solar magnetic field. Thus, as seen from Chap. 1, global MHD models have a wide application in the solar-terrestrial research field, including numerical study of the solar wind structure and CME propagation. The prime objective of this chapter is to introduce a specific finite space-time volume integration method to 3D MHD equations in Cartesian coordinates, namely Space-Time Conservation Element and Solution Element (CESE) method, and its application to the solar wind study.

In a series of papers [9, 11–13, 63, 75, 80–83], Chang and his co-workers have successfully developed the CESE method in solving linear and nonlinear convection-diffusion equations in one, two, and three spatial dimensions. Different from the routine finite volume methods as introduced in Chap. 2, the principal idea of the CESE method is treating space and time as one entity in calculating flux balance, which is the key difference between the CESE method and traditional numerical methods. The second difference between the CESE method and other numerical methods is that the spatial-derivatives in the CESE method are considered as solving variables. Note, the physical parameters are defined to have smooth profiles inside an SE, while between SEs or in CEs, they may be discontinuous. That is the typical reason why the CESE method can capture sharp discontinuity within a few grid points.

The CESE method has been used to study flows with moving and steady shock, acoustic waves, complex vertical flows, detonations, shock/acoustics/vortices interactions, dam-break flows, etc. For a complete review of the development of CESE method, one may refer to Wang [62]. In two papers [77, 78], Zhang et al. applied a version of CESE method to the simulation of one-dimensional (1D) and two-dimensional (2D) ideal MHD equations. The shock tube MHD problem and the well known Orszag and Tangąs MHD vortex problem are used to validate their application. Feng et al. [15], Hu and Feng [25] applied the CESE method to 2.5-dimensional resistive MHD equations in Cartesian coordinates, with the purpose of modeling the magnetic reconnection study. For higher order implementation of MHD equations, refer to Yang et al. [72, 73]. In all these studies, the numerical results agree favorably with those obtained by using the higher order modern upwind schemes, but the CESE method has much simpler logics and implementation. However, applying the CESE method to model the solar wind in the corona is not straight forward, and it is necessary to reformulate the grid mesh suitable to the solar wind simulation. The present chapter introduces the CESE method for solar wind modeling on triangular prism grids in a spherical geometry. This chapter may be considered as a consolidation from the work of Feng et al. [16] and Hu et al. [27].

The reader is assumed to be familiar with some basic concepts of numerical methods for partial differential equations in general. For the essential background, the reader may consult Chap. 2 or virtually any standard book on numerical methods for differential equations (e.g., [6, 29, 37, 39, 53]).

4.2 Model Equations for Solar Wind

The most commonly used model to describe the dynamic behavior of solar wind and solar magnetic field in the corona and interplanetary space is the MHD model. Considering the effect of gravitational force, solar rotation and corona heating, the MHD equations can be written as:

$$\frac{\partial \mathbf{U}}{\partial t} + \frac{\partial \mathbf{F(U)}}{\partial x} + \frac{\partial \mathbf{G(U)}}{\partial y} + \frac{\partial \mathbf{H(U)}}{\partial z} = \eta \tag{4.1}$$

where

$$\mathbf{U} = (\rho, \rho u, \rho v, \rho w, e, B_x, B_y, B_z)^T = (u_1, u_2, u_3, u_4, u_5, u_6, u_7, u_8)^T,$$

$$\mathbf{F(U)} = \begin{pmatrix} \rho u \\ \rho u^2 + p_0 - {B_x}^2 \\ \rho uv - B_x B_y \\ \rho uw - B_x B_z \\ (e + p_0)u - B_x(uB_x + vB_y + wB_z) \\ 0 \\ uB_y - vB_x \\ uB_z - wB_x \end{pmatrix}$$

$$= (f_1, f_2, f_3, f_4, f_5, f_6, f_7, f_8)^T,$$

$$\mathbf{G(U)} = \begin{pmatrix} \rho v \\ \rho vu - B_y B_x \\ \rho v^2 + p_0 - {B_y}^2 \\ \rho vw - B_y B_z \\ (e + p_0)v - B_y(uB_x + vB_y + wB_z) \\ vB_x - uB_y \\ 0 \\ vB_z - wB_y \end{pmatrix}$$

$$= (g_1, g_2, g_3, g_4, g_5, g_6, g_7, g_8)^T,$$

$$\mathbf{H}(\mathbf{U}) = \begin{pmatrix} \rho w \\ \rho u w - B_x B_z \\ \rho v w - B_y B_z \\ \rho w^2 + p_0 - {B_z}^2 \\ (e + p_0) w - B_z(u B_x + v B_y + w B_z) \\ w B_x - u B_z \\ w B_y - v B_z \\ 0 \end{pmatrix}$$

$$= (h_1, h_2, h_3, h_4, h_5, h_6, h_7, h_8)^T,$$

and

$$\eta = \left(0, \rho\left[\mathbf{g} - \mathbf{\Omega} \times (\mathbf{\Omega} \times \mathbf{r}) - 2\mathbf{\Omega} \times \mathbf{u}\right], \rho\mathbf{u} \cdot [\mathbf{g} - \mathbf{\Omega} \times (\mathbf{\Omega} \times \mathbf{r})] + Q_e, \mathbf{0}\right)^T - \nabla \cdot \mathbf{B}\,(0, \mathbf{B}, \mathbf{u} \cdot \mathbf{B}, \mathbf{u})^T$$

Here, ρ is density, $\mathbf{u} = (u, v, w)$ is velocity vector and its components in x, y, z directions, p is gas pressure and $\mathbf{B}$ is magnetic-field vector. $\mathbf{g} = -GM_s\mathbf{r}/r^3$ is the solar gravitational acceleration, where G is the solar gravitational constant at the solar surface and M_s the solar mass. The solar angular speed $\mathbf{\Omega}$ is considered with angular velocity $\omega = |\mathbf{\Omega}| = 2\pi/27.2753$ radian day^{-1}. A factor of $1/\sqrt{\mu_0}$ has been absorbed into the definition of $\mathbf{B}$, with μ_0 being the magnetic permeability. The specific total energy $e = p/(\gamma - 1) + \rho\left(u^2 + v^2 + w^2\right)/2 + \left({B_x}^2 + {B_y}^2 + {B_z}^2\right)/2$ for a perfect gas with ratio of specific heat γ. The total pressure is $p_0 = p + \left({B_x}^2 + {B_y}^2 + {B_z}^2\right)/2$. Q_e is the volume heating rate which is described in Sect. 4.4.1. The source terms proportional to $\nabla \cdot \mathbf{B}$ have been added to the momentum, induction and energy equations to avoid accumulation of errors which are associated with high values of $\nabla \cdot \mathbf{B}$. The system was proposed by Powell et al. [40] based on the pioneer work of Godunov [19]. This modified form of the equations allows the derivation of an eight-wave approximate Riemann solver that can be used to construct an upwind solution scheme for multi-dimensional flows, and to some extent, mitigates the $\nabla \cdot \mathbf{B}$ error [40]. The ideal MHD system written in pure divergence form without the source term proportional to $\nabla \cdot \mathbf{B}$, while conservative, is not Gallilean invariant, has a zero eigenvalue in the Jacobian matrix, and is not symmetrizable [19]. For convenience, ρ, $\mathbf{u}$, p, $\mathbf{B}$, $\mathbf{r}$, t, $\mathbf{\Omega}$ and $\mathbf{g}$ are normalized by the characteristic values ρ_s, a_s, $\rho_s a_s^2$, $\sqrt{\rho_s a_s^2}$, R_s, R_s/a_s, a_s/R_s and a_s^2/R_s, where ρ_s and a_s are the density and sound speed at the solar surface of solar radius R_s.

4.3 Modified Space-Time Conservation Element and Solution Element Method

This section describes space-time conservation element and solution element method for Eq. (4.1). Trivially, Eq. (4.1) can be rewritten as

$$\frac{\partial u_m}{\partial t} + \frac{\partial f_m}{\partial x} + \frac{\partial g_m}{\partial y} + \frac{\partial h_m}{\partial z} = \eta_m \tag{4.2}$$

where $m = 1, 2, \ldots, 8$. Let (x, y, z, t) be the coordinates of a four-dimensional Euclidean space E_4. By using Gauss' divergence theorem in E_4, it can be shown that Eq. (4.2) is equivalent to the following integral equations:

$$\oint_{S(V)} \mathbf{q}_m \cdot d\mathbf{S} = \int_V \eta_m dV \tag{4.3}$$

Here $\mathbf{q}_m = (u_m, f_m, g_m, h_m)$, $S(V)$ is the boundary of an arbitrary space-time region V in E_4, and $\mathbf{q}_m \cdot d\mathbf{S}$ is the space-time flux $\mathbf{q}_m$ leaving the region V through the surface element $d\mathbf{S}$.

In the following, a brief discussion of this extended CESE method is provided for the spherical geometry suitable to solar-terrestrial simulation. For details refer to Feng et al. [16] and Hu et al. [27].

4.3.1 Conservation Element and Solution Element in Spherical Geometry

Let's consider a simple algorithm to construct a spherical grid on which to approximate the solar wind MHD solution described in Sect. 4.2. Such a grid can be constructed by starting with an ordinary octahedron inscribed inside the spherical surface $r = 1\ R_s$. Each face of the octahedron is subdivided into four new faces by bisecting the edges. Next, the new vertices are projected onto the spherical surface. Continuing this process, further partitioning the spherical triangles and projecting onto the sphere until the desired resolution is reached, the spherical surface $r = 1\ R_s$ is divided into non-overlapped triangles and any two neighboring triangles share a common side. Figure 4.1a shows the grid structure on the spherical inner boundary. Meanwhile, expand the spherical surface at different radial distances by defining the radial variation (here, $r(1) = 1\ R_s, r(i) = r(i-1) + \text{step} \times r(i-1)$ with step $= \frac{\pi}{60}$). Then three-dimensional spatial computational domain is divided into non-overlapped convex pentahedrons and any two neighboring pentahedrons share a common surface. Each pentahedral cell is a triangular prism composed of two triangular bases and three rectangular sides. Figure 4.1b shows a part of pentahedrons and the method of constructing a 3D grid structure by extending the position of spherically allocated grids outward from the inner boundary.

After the establishment of 3D grid structure, we define conservation element and solution element as follows. In Fig. 4.2, Q (marked by a circle) is the centroid of a typical pentahedron $B_1B_2B_3B_4B_5B_6$. Each of the central pentahedron's five neighboring pentahedrons is arbitrarily assigned an identification index $\ell = 1, 2, \ldots, 5$. Also, the centroid of the ℓth neighbor will be denoted by A_ℓ. As an example, the

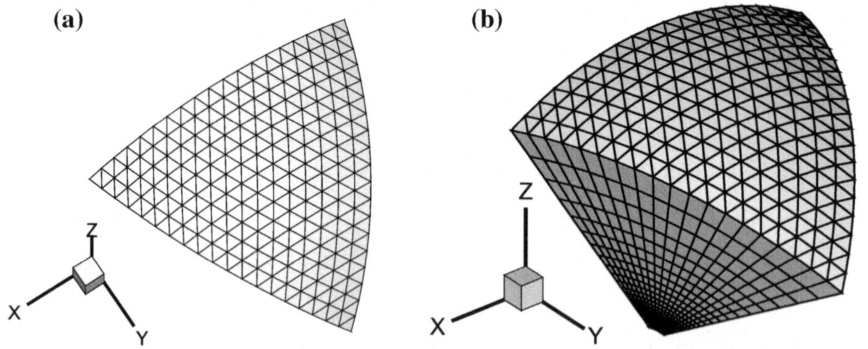

Fig. 4.1 Part of grid structure for the 3D CESE calculation: **a** at solar surface and **b** in the corona. Taken from Fig. 1 of Feng et al. [16] with permissions of the authors and the AAS

Fig. 4.2 Representative grid points in the x-y-z space. Taken from Fig. 2 of Feng et al. [16] with permissions of the authors and the AAS

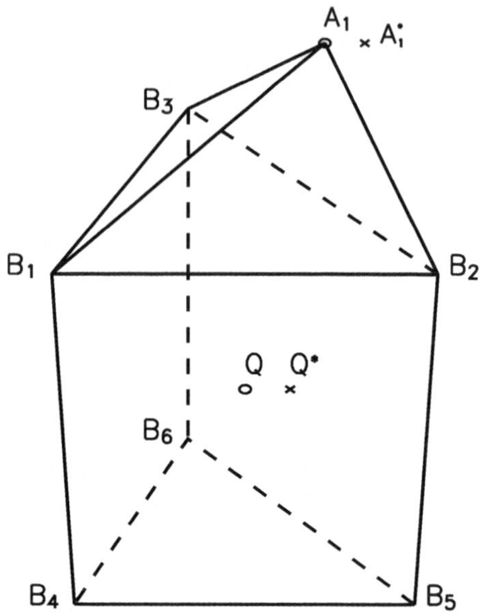

central pentahedron and its 1st, 5th neighbor is separated by the triangle $B_1B_2B_3$, $B_4B_5B_6$, respectively. The neighboring pentahedron with the index 2, 3 and 4 is referred to as the pentahedron which shares the quadrilateral $B_1B_2B_5B_4$, $B_2B_3B_6B_5$ and $B_3B_1B_4B_6$, respectively.

Given any $\ell = 1, 2, \ldots, 5$, a triangle is formed by the point A_ℓ and the two end points of any of the edges of the interface that separates the central pentahedron and its ℓth neighbor. Each of the triangles so formed with the same ℓ is arbitrarily assigned an index k_ℓ, and denoted by $\triangle(k_\ell, \ell)$, where k_ℓ is 1, 2, 3 when $\ell = 1, 5$, and k_ℓ is 1, 2, 3, 4 when $\ell = 2, 3, 4$. As an example, $\triangle A_1B_1B_2$, $\triangle A_1B_2B_3$, and

$\triangle A_1B_1B_3$ depicted by Fig. 4.2 have the same $\ell = 1$. Therefore they may be denoted by $\triangle(1, 1)$, $\triangle(2, 1)$ and $\triangle(3, 1)$, respectively.

Point Q and the two end points (say points B_1 and B_2) of any of the nine edges of the central pentahedron form a triangle. Each of nine triangles so formed is arbitrarily assigned an index $j = 1, 2, \ldots, 9$ and denoted by $\triangle(j)$.

Points A_1, B_1, A_2, B_2, A_3, B_3, A_4, B_4, A_5, B_5 and B_6 form a 18-faced polyhedron $B_1B_2B_3B_4B_5B_6A_1A_2A_3A_4A_5$. The centroid of the 18-faced polyhedron is referred to as the solution point associated with point Q. Note that the above 18-faced polyhedron hereafter is denoted by $V(18)$ and the centroid of $V(18)$ is denoted by Q^* and marked by a cross in Fig. 4.2.

Given any $\ell = 1, 2, 3, 4, 5$, points Q and A_ℓ and the vertices of the interface that separates the central pentahedron and its ℓth neighbor are the vertices of a polyhedron. This polyhedron hereafter is denoted by $V5(\ell)$. As an example, Q, A_1, B_1, B_2 and B_3 form a polyhedron denoted by $V5(1)$.

In the space-time computational domain, we assume that $t = n\Delta t$ at the nth time level ($n = 0, \frac{1}{2}, 1, \frac{3}{2}, \ldots$) with the marching time step Δt limited by the Courant-Friedrichs-Lewy (CFL) stability condition. Also, for a given $n > 0$, let Q, Q', and Q'' (not shown), respectively, be the points on the nth, $(n-\frac{1}{2})$th, and $(n+\frac{1}{2})$th time levels with point Q being their common spatial projection. Other space-time mesh points such as (i) Q^* and Q'^*, (ii) B_k, B'_k, and B''_k, k = 1, 2, 3, 4, 5, 6; and (iii) A_ℓ, A^*_ℓ, A'_ℓ, and A'^*_ℓ, $\ell = 1, 2, 3, 4, 5$, are defined similarly. Because geometric objects in E_4 generally are difficult to visualize, they will be described analytically in the following discussions.

A "plane" in E_4, by definition, is a subspace of E_4 defined by a linear equation i.e.,

$$a_1x + a_2y + a_3z + a_4t = a_0, \; ((a_1)^2 + (a_2)^2 + (a_3)^2 + (a_4)^2 \neq 0) \tag{4.4}$$

where a_k, $k = 0, 1, 2, 3, 4$, are constants. As a result, a hyperplane in E_4 is a three-dimensional subspace. Note that a hyperplane segment, by definition, is a bounded region of a hyperplane. A hyperplane of type I, denoted by $\Gamma(V; t_c)$, is formed by all the points (x, y, z, t) that satisfy the conditions $t = t_c$ and $(x, y, z) \in V$, where t_c is a constant and V denotes a 3D spatial region.

In contrast, a hyperplane segment of type II, denoted by $\Gamma(S; t_-, t_+)$, is formed by all the points (x, y, z, t) that satisfy the conditions $(x, y, z) \in S$ and $t_- \leq t \leq t_+$, where S denotes a spatial plane segment and t_- and $t_+ (t_- < t_+)$ are constants. Note that every point (x, y, z) on the spatial plane segment S satisfies a linear equation of the form

$$c_1x + c_2y + c_3z = c_0, \; ((c_1)^2 + (c_2)^2 + (c_3)^2 \neq 0)$$

where c_k, $k = 0, 1, 2, 3$, are constants. Thus every point (x, y, z, t) on $\Gamma(S; t_-, t_+)$ also satisfies a special form of Eq. (4.4).

In addition to the above two types of hyperplanes, we shall also consider "hyper-cylinder" in E_4. A hypercylinder, denoted by $\Lambda(V; t_-, t_+)$, is formed by all the points

(x, y, z, t) that satisfy the conditions $(x, y, z) \in V$ and $t_- \leq t \leq t_+$, where V is a 3D spatial region and t_- and $t_+ (t_- < t_+)$ are constants.

With the above preliminaries, SE(Q^*), the solution element of point Q^*, i.e., the point that lies on the nth time level and has Q^* as its spatial projection, is defined to be the union of $\Gamma(V(18); t^n)$ and $\Gamma(\Delta(j); t^{n-1/2}, t^{n+1/2})$, $j = 1, 2, \ldots, 9$, and their immediate neighborhoods. Moreover, the five Basic Conservation Elements(BCEs) of point Q, denoted by CE$_\ell(Q)$, $\ell = 1, 2, \ldots, 5$, are defined to be the hypercylinders $\Lambda(V5(\ell); t^{n-1/2}, t^n)$, $\ell = 1, 2, \ldots, 5$, respectively. In addition, the Compound Conservation Element (CCE) of point Q, denoted by CE(Q), is defined to be $\Lambda(V(18); t^{n-1/2}, t^n)$, i.e., the union of the above five BCEs.

We begin with the following preliminaries:

(a) Let Γ be a hyperplane segment lying within SE(Q^*). Let A be the area of Γ, (x_c, y_c, z_c, t_c) be the coordinates of the centroid of Γ, and $\mathbf{n}$ be a unit vector normal to Γ. Under the approximation that the flow variables can be represented by the first-order Taylor series inside each SE, it can be shown that

$$\int_\Gamma \mathbf{q}_m^* \cdot d\mathbf{S} = \mathbf{q}_m^*(x_c, y_c, z_c, t_c; Q^*) \cdot A\mathbf{n}$$

where $d\mathbf{S} = d\sigma\,\mathbf{n}$ with $d\sigma$ being the area of a surface element on Γ, and $\mathbf{q}_m^*(x_c, y_c, z_c, t_c; Q^*)$ is defined by Eq. (4.8).

(b) Let V denote the volume of $V(18)$. Because the unit outward normal vector (outward from the interior of CE(Q) of this face is $(0, 0, 0, 1)$), its surface vector (i.e., the unit outward normal vector multiplied by the area) is $(0, 0, 0, V)$.

(c) Let V^ℓ and (x^ℓ, y^ℓ, z^ℓ), respectively, denote the volume and the spatial coordinates of the centroid of any $V5(\ell)$. Then the surface vector, and the coordinates of the centroid of $\Gamma(V5(\ell); t^{n-1/2})$, respectively, are $(0, 0, 0, -V^\ell)$ and $(x^\ell, y^\ell, z^\ell, t^{n-1/2})$.

(d) Let $S_{k_\ell}^\ell$, $(n_{k_\ell x}^\ell, n_{k_\ell y}^\ell, n_{k_\ell z}^\ell)$, and $(x_{k_\ell}^\ell, y_{k_\ell}^\ell, z_{k_\ell}^\ell)$, respectively, denote the area, the spatial unit outward normal, and the coordinates of the centroid of any $\Delta(k_\ell, \ell)$. Then the surface vector, and the coordinates of the centroid of the side of face $\Gamma(\Delta(k_\ell, \ell); t^{n-1/2}, t^n)$, respectively, are $(\Delta t/2)S_{k_\ell}^\ell(n_{k_\ell x}^\ell, n_{k_\ell y}^\ell, n_{k_\ell z}^\ell, 0)$ and $(x_{k_\ell}^\ell, y_{k_\ell}^\ell, z_{k_\ell}^\ell, t^n - \Delta t/4)$.

By using the information presented in items (a), (b), (c) and (d), the flux of $\mathbf{q}_m^*$ leaving the other faces of CE(Q) can be evaluated in terms of the independent marching variables at points $A_\ell'^*$, $\ell = 1, 2, 3, 4, 5$.

4.3.2 Calculation of $(u_m)_{Q^*}$

This section is devoted to the calculation of $(u_m)_{Q^*}$ by integrating MHD equations on conservation element and solution element defined above. Following Chang's original approach [9], inside each SE the flow variables are assumed smooth, and are represented by the first-order Taylor series. For any $(x, y, z, t) \in$ SE(Q^*), $u_m(x, y, z, t)$, $f_m(x, y, z, t)$, $g_m(x, y, z, t)$ and $h_m(x, y, z, t)$ of Eq. (4.2) are approximated by

$$u_m^*(x, y, z, t; Q^*) = (u_m)_{Q^*} + (u_{mx})_{Q^*}(x - x_{Q^*}) + (u_{my})_{Q^*}(y - y_{Q^*}) + (u_{mz})_{Q^*}(z - z_{Q^*}) + (u_{mt})_{Q^*}(t - t^n) \quad (4.5)$$

$$\mathcal{F}_m^*(x, y, z, t; Q^*) = (\mathcal{F}_m)_{Q^*} + (\mathcal{F}_{mx})_{Q^*}(x - x_{Q^*}) + (\mathcal{F}_{my})_{Q^*}(y - y_{Q^*}) + (\mathcal{F}_{mz})_{Q^*}(z - z_{Q^*}) + (\mathcal{F}_{mt})_{Q^*}(t - t^n) \quad (4.6)$$

$$(\mathcal{F}_{m\xi})_{Q^*} = \sum_{\ell=1}^{8} (\mathcal{F}_{m\ell})_{Q^*}(u_{\ell\xi})_{Q^*}, \mathcal{F}_{m\ell} = \frac{\partial \mathcal{F}_m}{\partial u_\ell} \quad (4.7)$$

for $m = 1, 2, \ldots, 8$. Here, u_{mx}, u_{my}, u_{mz}, and u_{mt} represent the first order derivative of u_m with respect to x, y, z and t, respectively. Such a notation also apply to the subscript of $\mathcal{F}_m$. $\mathcal{F}_m$ stands for f_m, g_m or h_m, and ξ denotes x, y, z or t, and $(\mathcal{F}_{m\ell})_{8\times 8}$ denotes the Jacobian matrix of $\mathcal{F}$, which is given in appendix. $(x_{Q^*}, y_{Q^*}, z_{Q^*}, t^n)$ is the space-time coordinate of point Q^*. Following the definition of $\mathbf{q}_m$, we define

$$\mathbf{q}_m^*(x, y, z, t; Q^*) = (f_m^*(x, y, z, t; Q^*), g_m^*(x, y, z, t; Q^*), h_m^*(x, y, z, t; Q^*), u_m^*(x, y, z, t; Q^*)) \quad (4.8)$$

Applying Eq. (4.2), we have

$$(u_{mt})_{Q^*} = \eta_m - (f_{mx})_{Q^*} - (g_{my})_{Q^*} - (h_{mz})_{Q^*}$$
$$= \eta_m - \sum_{\ell=1}^{8} \left[(f_{m\ell})_{Q^*}(u_{\ell x})_{Q^*} + (g_{m\ell})_{Q^*}(u_{\ell y})_{Q^*} + (h_{m\ell})_{Q^*}(u_{\ell z})_{Q^*} \right] \quad (4.9)$$

Thus, the space-time flux conservation, Eq. (4.3), can be approximated by its discrete counterpart on the conservation element:

$$\oint_{S(CE)} \mathbf{q}_m^* \cdot d\mathbf{S} = \int_{CE} \eta_m^* dV \quad (4.10)$$

i.e., the total flux of $\mathbf{q}_m^*$ leaving CE(Q) is equal to the volume integrate of source terms. Then, with the aid of the above preliminaries, substituting Eqs. (4.5) and (4.9) into Eq. (4.10), we obtain the following equation

$$(u_m)_{Q^*}^n - \frac{\Delta t}{2}\eta(u_m)_{Q^*}^n = \left(\sum_{\ell=1}^{5} R_m^\ell \right) / V \quad (4.11)$$

which is the discretized form of the governing equations (4.1) or (4.2) for the solar wind MHD equations. Here

$$R_m^\ell = V^\ell \left[(u_m)_{A_\ell^{\prime *}}^{n-1/2} + (x^\ell - x_{A_\ell^{\prime *}})(u_{mx})_{A_\ell^{\prime *}}^{n-1/2} + (y^\ell - y_{A_\ell^{\prime *}})(u_{my})_{A_\ell^{\prime *}}^{n-1/2} + (z^\ell - z_{A_\ell^{\prime *}})(u_{mz})_{A_\ell^{\prime *}}^{n-1/2} \right]$$
$$- \frac{\Delta t}{2} \sum_{k_\ell} S_{k_\ell}^\ell \left\{ n_{k_\ell x}^\ell \left[(f_m)_{A_\ell^{\prime *}}^{n-1/2} + (x_{k_\ell}^\ell - x_{A_\ell^{\prime *}})(f_{mx})_{A_\ell^{\prime *}}^{n-1/2} + \frac{\Delta t}{4}(f_{mt})_{A_\ell^{\prime *}}^{n-1/2} \right.\right.$$
$$\left.\left. + (y_{k_\ell}^\ell - y_{A_\ell^{\prime *}})(f_{my})_{A_\ell^{\prime *}}^{n-1/2} + (z_{k_\ell}^\ell - z_{A_\ell^{\prime *}})(f_{mz})_{A_\ell^{\prime *}}^{n-1/2} \right] \right\}$$

$$-\frac{\Delta t}{2}\sum_{k_\ell} S^\ell_{k_\ell}\left\{n^\ell_{k_\ell y}\left[(g_m)^{n-1/2}_{A'^*_\ell} + (x^\ell_{k_\ell} - x_{A'^*_\ell})(g_{mx})^{n-1/2}_{A'^*_\ell} + \frac{\Delta t}{4}(g_{mt})^{n-1/2}_{A'^*_\ell}\right.\right.$$
$$\left.\left.+(y^\ell_{k_\ell} - y_{A'^*_\ell})(g_{my})^{n-1/2}_{A'^*_\ell} + (z^\ell_{k_\ell} - z_{A'^*_\ell})(g_{mz})^{n-1/2}_{A'^*_\ell}\right]\right\}$$
$$-\frac{\Delta t}{2}\sum_{k_\ell} S^\ell_{k_\ell}\left\{n^\ell_{k_\ell z}\left[(h_m)^{n-1/2}_{A'^*_\ell} + (x^\ell_{k_\ell} - x_{A'^*_\ell})(h_{mx})^{n-1/2}_{A'^*_\ell} + \frac{\Delta t}{4}(h_{mt})^{n-1/2}_{A'^*_\ell}\right.\right.$$
$$\left.\left.+(y^\ell_{k_\ell} - y_{A'^*_\ell})(h_{my})^{n-1/2}_{A'^*_\ell} + (z^\ell_{k_\ell} - z_{A'^*_\ell})(h_{mz})^{n-1/2}_{A'^*_\ell}\right]\right\} \quad (4.12)$$

where $\ell = 1, 2, 3, 4$ and 5, for flux conservation contributed from five neighboring CEs. $k_\ell = 1, 2, 3$ when $\ell = 1, 5$, and $k_\ell = 1, 2, 3, 4$ when $\ell = 2, 3, 4$. By solving Eq. (4.11), we can get $(u_m)_{Q^*}$.

At this time, solving the governing equations (4.1) or (4.2) for the solar wind model is equivalent to finding the solutions of (4.11). Due to the source term, Eq. (4.11) is a nonlinear equation for $(u_m)_{Q^*}$. Given the values of the marching variables at the $t^{n-1/2}$ time level, the right hand side of (4.11), i.e., (4.12), can be explicitly calculated. To update $(u_m)_{Q^*}$ by solving Eq. (4.11), the Newton's method is used. The initial condition for the iterations is calculated by assuming null source term.

4.3.3 Solution of Nonlinear Equation (4.11)

The treatment is identical to the one used in Yu and Chang [75]. As the source term $\eta(\mathbf{U})$ itself is a function of the unknown $\mathbf{U}$ at the new time level, a local iterative procedure is needed to determine $\mathbf{U}$. The discretized integral equation (4.11) reduces to the form

$$\mathbf{U} - \frac{\Delta t}{2}\eta(\mathbf{U}) = \mathbf{U}_H$$

where $\mathbf{U}_H$ is the local homogeneous solution (i.e. the solution for $\eta = 0$ locally). Note that $\mathbf{U}_H$ only depends on the solution at the previous time step, i.e., $\mathbf{U}_H$ is obtained using explicit formulas given by Eqs. (4.11) and (4.12). A Newtonian iteration procedure to determine $\mathbf{U}$ is then

$$\mathbf{U}^{(i+1)} = \mathbf{U}^{(i)} - \left(\frac{\partial \boldsymbol{\Phi}}{\partial \mathbf{U}^{(\mathrm{i})}}\right)^{-1}[\boldsymbol{\Phi}(\mathbf{U}^{(\mathrm{i})}) - \mathbf{U}_H] \quad (4.13)$$

where i is the iteration number and

$$\boldsymbol{\Phi}(\mathbf{U}) = \mathbf{U} - \frac{\Delta t}{2}\eta(\mathbf{U})$$

with

$$\eta = \left(0, \rho\left[\mathbf{g} - \boldsymbol{\Omega}\times(\boldsymbol{\Omega}\times\mathbf{r}) - 2\boldsymbol{\Omega}\times\mathbf{u}\right], \rho\mathbf{u}\cdot[\mathbf{g} - \boldsymbol{\Omega}\times(\boldsymbol{\Omega}\times\mathbf{r})] + Q_e, \mathbf{0}\right)^T - (\nabla\cdot\mathbf{B})^{n-\frac{1}{2}}_{Q^*}(0, \mathbf{B}, \mathbf{u}\cdot\mathbf{B}, \mathbf{u})^T$$

That is, $(\nabla \cdot \mathbf{B})_{Q^*}^{n-\frac{1}{2}}$ in the source term of η takes the value explicitly at the previous time step. Normally, $\mathbf{U}$ at the previous time step is a good initial guess $\mathbf{U}^{(0)}$, and the procedure takes about 3–4 iterations to convergence. $\frac{\partial \boldsymbol{\Phi}}{\partial \mathbf{U}}$ is calculated in Appendix.

4.3.4 Evaluation of $(u_{mx})_{Q^*}$, $(u_{my})_{Q^*}$ and $(u_{mz})_{Q^*}$

To proceed, a central difference type reconstruction approach and a weighted average are employed to calculate $(u_{mx})_{Q^*}$, $(u_{my})_{Q^*}$ and $(u_{mz})_{Q^*}$. First, we perform a spatial translation of the polyhedron $A_1^*A_2^*A_3^*A_4^*A_5^*$ so that the centroid of the resulting new polyhedron $A_1^oA_2^oA_3^oA_4^oA_5^o$ coincides with Q^*. Let the centroid of the polyhedron $A_1^*A_2^*A_3^*A_4^*A_5^*$ and its spatial coordinates be denoted by A^* and $(x_{A^*}, y_{A^*}, z_{A^*})$, respectively, and let $\delta x = x_{Q^*} - x_{A^*}$, $\delta y = y_{Q^*} - y_{A^*}$ and $\delta z = z_{Q^*} - z_{A^*}$. Then $(x_{A_\ell^o}, y_{A_\ell^o}, z_{A_\ell^o})$, the spatial coordinates of A_ℓ^o, $\ell = 1, 2, \ldots, 5$, are given by

$$x_{A_\ell^o} = x_{A_\ell^*} + \delta x, \quad y_{A_\ell^o} = y_{A_\ell^*} + \delta y \text{ and } z_{A_\ell^o} = z_{A_\ell^*} + \delta z.$$

As a preliminary for the following discussions, for $m = 1, 2, \ldots, 8$ and $\ell = 1, 2, \ldots, 5$, let's define

$$(u_m)_{A_\ell^o} \stackrel{\text{def}}{=} u_m^*(x_{A_\ell^o}, y_{A_\ell^o}, z_{A_\ell^o}, t^n; A'^*_\ell),$$

$$\delta u_m^\ell \stackrel{\text{def}}{=} (u_m)_{A_\ell^o} - (u_m)_{Q^*},$$

and

$$\delta x_\ell \stackrel{\text{def}}{=} x_{A_\ell^o} - x_{Q^*}, \quad \delta y_\ell \stackrel{\text{def}}{=} y_{A_\ell^o} - y_{Q^*}, \quad \delta z_\ell \stackrel{\text{def}}{=} z_{A_\ell^o} - z_{Q^*}.$$

Next consider the vertex B_1 depicted in Fig. 4.2. This vertex is the common vertex of the central pentahedron and three of its neighbors. As an example, let the identification indices ℓ of these three neighbors be 1, 2, and 3. Then, for any $m = 1, 2, \ldots, 8$, consider the four points in the $x - y - z - u_m$ space with the coordinates $(x_{Q^*}, y_{Q^*}, z_{Q^*}, (u_m)_{Q^*})$ and $(x_{A_\ell^o}, y_{A_\ell^o}, z_{A_\ell^o}, (u_m)_{A_\ell^o})$, $\ell = 1, 2, 3$. It can be shown that the values of $\partial u_m/\partial x$, $\partial u_m/\partial y$ and $\partial u_m/\partial z$ on the hyperplane that intercepts the above four points are given by

$$(u_{mx}^{(1)})_{Q^*} \stackrel{\text{def}}{=} \Delta_x/\Delta, \quad (u_{my}^{(1)})_{Q^*} \stackrel{\text{def}}{=} \Delta_y/\Delta, \quad (u_{mz}^{(1)})_{Q^*} \stackrel{\text{def}}{=} \Delta_z/\Delta \ (\Delta \neq 0), \tag{4.14}$$

where

$$\Delta \stackrel{\text{def}}{=} \begin{vmatrix} \delta x_1 & \delta y_1 & \delta z_1 \\ \delta x_2 & \delta y_2 & \delta z_2 \\ \delta x_3 & \delta y_3 & \delta z_3 \end{vmatrix}$$

and

$$\Delta_x \stackrel{\text{def}}{=} \begin{vmatrix} \delta u_m^1 & \delta y_1 & \delta z_1 \\ \delta u_m^2 & \delta y_2 & \delta z_2 \\ \delta u_m^3 & \delta y_3 & \delta z_3 \end{vmatrix}, \quad \Delta_y \stackrel{\text{def}}{=} \begin{vmatrix} \delta x_1 & \delta u_m^1 & \delta z_1 \\ \delta x_2 & \delta u_m^2 & \delta z_2 \\ \delta x_3 & \delta u_m^3 & \delta z_3 \end{vmatrix}, \quad \Delta_z \stackrel{\text{def}}{=} \begin{vmatrix} \delta x_1 & \delta y_1 & \delta u_m^1 \\ \delta x_2 & \delta y_2 & \delta u_m^2 \\ \delta x_3 & \delta y_3 & \delta u_m^3 \end{vmatrix}.$$

Note that $\Delta = 0$ if and only if the spatial projections of A_1^o, A_2^o, A_3^o and Q^* are coplanar. In the same procedure, we can obtain $(u_{mx}^{(k)})_{Q^*}$, $(u_{my}^{(k)})_{Q^*}$ and $(u_{mz}^{(k)})_{Q^*}$ for each vertex B_k in Fig. 4.2 ($k = 2, 3, \ldots, 6$).

With the above preliminaries, for each $m = 1, 2, \ldots, 8$, $(u_{mx})_{Q^*}$, $(u_{my})_{Q^*}$, and $(u_{mz})_{Q^*}$ may be evaluated by

$$(u_{mx})_{Q^*} = \frac{1}{6}\sum_{k=1}^{6}(u_{mx}^{(k)})_{Q^*}, \quad (u_{my})_{Q^*} = \frac{1}{6}\sum_{k=1}^{6}(u_{my}^{(k)})_{Q^*}, \quad (u_{mz})_{Q^*} = \frac{1}{6}\sum_{k=1}^{6}(u_{mz}^{(k)})_{Q^*}. \tag{4.15}$$

Alternatively, for a flow with steep gradients or discontinuities, the simple averages in Eq. (4.15) may be replaced by weighted averages to suppress numerical oscillations near a discontinuity, i.e.,

$$(u_{mx})_{Q^*} = \begin{cases} 0, & \text{if } \theta_{mk} = 0, \quad k = 1, 2, \ldots, 6, \\ \sum_{k=1}^{6}[(W_m^{(k)})^\alpha (u_{mx}^{(k)})_{Q^*}] / \sum_{k=1}^{6}(W_m^{(k)})^\alpha, & \text{otherwise}, \end{cases} \tag{4.16a}$$

$$(u_{my})_{Q^*} = \begin{cases} 0, & \text{if } \theta_{mk} = 0, \quad k = 1, 2, \ldots, 6, \\ \sum_{k=1}^{6}[(W_m^{(k)})^\alpha (u_{my}^{(k)})_{Q^*}] / \sum_{k=1}^{6}(W_m^{(k)})^\alpha, & \text{otherwise}, \end{cases} \tag{4.16b}$$

and

$$(u_{mz})_{Q^*} = \begin{cases} 0, & \text{if } \theta_{mk} = 0, \quad k = 1, 2, \ldots, 6, \\ \sum_{k=1}^{6}[(W_m^{(k)})^\alpha (u_{mz}^{(k)})_{Q^*}] / \sum_{k=1}^{6}(W_m^{(k)})^\alpha, & \text{otherwise}. \end{cases} \tag{4.16c}$$

Here $\alpha \geq 0$ is an adjustable constant (usually $\alpha = 1$ or $\alpha = 2$),

$$\theta_{mk} \stackrel{\text{def}}{=} \sqrt{[(u_{mx}^{(k)})_{Q^*}]^2 + [(u_{my}^{(k)})_{Q^*}]^2 + [(u_{mz}^{(k)})_{Q^*}]^2}, \tag{4.17}$$

and for each k, $W_m^{(k)}$ is the product of θ_{m1}, θ_{m2}, ..., θ_{m6} excluding θ_{mk}. To avoid dividing by zero, in practice a small positive number such as 10^{-60} is added to the denominators that appear in Eqs. (4.16a)–(4.16c). The parameter α controls the amount of numerical diffusion added to the scheme at the vicinity of strong gradient. The method is stable with $\alpha > 1$ and the solution generated by this method tends to become more smeared as the value of α increases.

In this section, the grid system and the CESE solver are formulated for the MHD solar wind model by discritizing MHD system on pentahedral cells with triangular prism structure in a spherical geometry. Parallelization can be easily realized for this kind of grid after specifying a "parallel" axis. In such computational domain of spherical shell for solar wind modeling, this parallel axis is naturally chosen to be the radial direction. The parallel axis is used to decompose the computational domain and to define the overlap data areas. In the CESE scheme, the most serious load in calculations comes from those for grid points, numerical flux, Eqs. (4.5), (4.9), (4.12), (4.14), (4.16) and the calculation of (4.13). These calculations are done sequentially in subroutines on a two-dimensional redundant data area, after copying 3D dependent variables from a distributed data area to two-dimensional redundant data areas. The implementation of SIP-CESE MHD model on hexahedral cells with six-component grid is presented in Chap. 5.

4.4 Numerical Simulation of Steady-State Solar Wind

In order to illustrate the numerical scheme, two examples are given for steady-state coronal atmospheric problems. The first experiment is the computation for steady-state solar wind using multiple magnetic field as an input. The second one is to apply this method to construct a realistic coronal structure using the measured photospheric magnetic field as the lower boundary conditions. We choose the adjustable constant $\alpha = 2$ in Eqs. (4.16a), (4.16b) and (4.16c) for these two numerical tests.

In order to assess the $\nabla \cdot \mathbf{B} = 0$ constraint numerically, as done in Zhang and Yu [77] the following error measurement:

$$Error = \frac{\sum_{k=1}^{M} \int\limits_{V(18)} |(\nabla \cdot \mathbf{B})_k| dV}{\sum_{k=1}^{M} \int\limits_{V(18)} dV} \tag{4.18}$$

is defined, where M is the number of mesh nodes in the computational domain. Here $(\nabla \cdot \mathbf{B})_k$ is calculated by Gauss's divergence theorem:

$$(\nabla \cdot \mathbf{B})_k = \frac{\int\limits_{V(18)} \nabla \cdot \mathbf{B} d\sigma}{\int\limits_{V(18)} d\sigma} = \frac{\oint\limits_{S(V(18))} \mathbf{B} \cdot d\mathbf{s}}{\int\limits_{V(18)} d\sigma} \tag{4.19}$$

where $V(18)$, as defined before, is a polyhedron in the three-dimensional spatial region. Eq. (4.19) is an volume average of $\nabla \cdot \mathbf{B}$ over a spatial cell, i.e., polyhedron $V(18)$. Here $(\nabla \cdot \mathbf{B})_k$ can be also simply calculated by

$$(\nabla \cdot \mathbf{B})_k = (u_{6x})_{Q^*} + (u_{7y})_{Q^*} + (u_{8z})_{Q^*}$$

since all these first-order derivatives are available at every time step. If meshes are congruent, Eq. (4.18) becomes

$$Error = \frac{\sum_{k=1}^{M} |(\nabla \cdot \mathbf{B})_k|}{M}$$

The steady-state is considered to be reached when the relative change in the conservative variables u_i from one time level to the next drops below 10^{-10}. We use a normalized measure defined by

$$\delta U = \sqrt{\frac{1}{8}\sum_{i=1}^{8} \frac{\sum_{Q^*}(u_i^{n+1/2} - u_i^n)^2}{\sum_{Q^*}(u_i^n)^2}}$$

Since smooth solutions are searched, the numerical schemes can easily achieve such accuracy in the steady-state solutions.

4.4.1 Numerical Simulation of Steady-State Solar Wind Using Multiple Magnetic Field as Input

This part presents the simulated steady-state corona by using multiple magnetic field as an input. To initialize the MHD code, a multipole magnetic field and Parker's solar wind solution [28] are used as the initial condition for Eq. (4.2). The multipole magnetic field is generated by a linear combination of dipole and octupole magnetic field as follows:

$$B_r = aB_r^{(2)} + bB_r^{(5)},\ B_\theta = aB_\theta^{(2)} + bB_\theta^{(5)}$$

where $B_r^{(2)} = 2\frac{R_s^3 \cos\theta}{r^3}$, $B_\theta^{(2)} = \frac{R_s^3 \sin\theta}{r^3}$, $B_r^{(5)} = R_s^5 \frac{4(5\cos^2\theta - 3)}{3r^5} \cos\theta$ and $B_\theta^{(5)} = R_s^5 \frac{(5\cos^2\theta - 1)}{r^5} \sin\theta$. Coefficients a and b can be chosen such that the magnetic field strength of the combined magnetic field at the pole and magnetic equator are 8.4 and 2.2 G, respectively.

The corresponding parameters involved in the MHD equations for this numerical calculation are $R_s = 6.96 \times 10^8$ m (solar radius), $\Re = 1.653 \times 10^4\ \mathrm{m^2/sec^2K}$ (gas constant), $n = 2 \times 10^{14}/\mathrm{m}^3$ (proton number density at $r = 1\ R_s$), $T = 1.85 \times 10^6$ K (temperature at $r = 1\ R_s$), and $\gamma = 1.05$ (the polytropic index).

In the MHD model, a volumetric energy input is needed in order to mimic the effects of energy absorption above the transition region, thermal conduction, and radiative losses as well as coronal heating. The heating function has the following form,

$$Q_e = \rho q_0 \left(T_0 - \gamma\frac{p}{\rho}\right) \exp\left[-\frac{(r - R_s)^2}{\sigma_0^2}\right]$$

where the target temperature, T_0, is 3.6×10^6 K at the pole, and 1.85×10^6 K at the equator. The heating function is defined with $q_0 = 10^6$ erg/s/g/K, where r is the heliocentric distance.

In order to obtain the steady state solar wind solution, boundary conditions are required at the inner boundary of the solution domain, defined by the solar surface at $r = 1R_s$, and outer boundaries, defined by the outer edges of the computational domain. The inner boundary conditions at the solar surface $r = 1R_s$ play a crucial role in obtaining physically relevant solutions [26, 30, 61, 70, 71]. The procedure for prescribing boundary conditions at the solar surface dependent on the local flow conditions has been successfully used, with plasma permitted to freely leave the sun and no "backflow" being allowed along the characteristic curve, which is called non-reflecting boundary condition. At the inner boundary, using near characteristic theory of partial differential equations, these authors employed the time-dependent characteristic boundary conditions for the plasma flow parameters density, velocity, temperature (pressure) and the magnetic field. These studies show that their non-reflecting boundary conditions works well for solar wind simulations.

It is known that numerical treatments to achieve non-reflecting boundary condition in conventional computational fluid dynamics methods have been developed based on theorems of the partial differential equation, and they could be categorized into the following three groups: (i) applying the method of characteristics to the discretized equations, (ii) the use of the buffer zone or a perfectly matched layer, and (iii) applying asymptotic analytical solution at the far field. In the setting of the CESE method, the non-reflecting boundary condition treatment is based on flux conservation near the computational boundary and letting the flux from the interior domain to the boundary CE smoothly exit to the exterior of the domain. Because surface element of each CE allows flow information to be propagated into the future, the numerical implementation of this flux-based method is extremely simple. Some works [7, 14, 33] have provided detailed discussions of various implementations of the above principle, including enforcing a back pressure.

In order to prescribe boundary conditions at the solar surface to be dependent on local flow conditions such that plasma is permitted to freely leave the reservoir, but no "backflow" is allowed, the following boundary conditions are adopted: for $u_r > 0$,

$$\rho = 1, p = 1/\gamma, \nabla \cdot (\rho \mathbf{u}) = 0, \mathbf{B} = \mathbf{B}_0$$

and for $u_r < 0$,

$$\frac{\partial \rho}{\partial r} = 0, \frac{\partial p}{\partial r} = 0, \mathbf{u} = 0, \mathbf{B} = \mathbf{B}_0$$

depending on whether the flow is toward or away from the solar surface. It should be mentioned that this kind of inner boundary boundary condition for primitive flow variables is similar to that used by Groth et al. [20].

At the outer (side) boundaries of the computational domain, the solar wind flow is essentially super-fast (hence super-Alfvénic). Simple zero-gradient (Neuman-type) and constant extrapolation boundary conditions are therefore appropriate and are used to specify the plasma properties at the outer boundary.

After the setup of initial and boundary conditions for Eq. (4.2), we run the numerical code and allow it to reach a new equilibrium state after $\delta U \sim O(10^{-10})$ (about

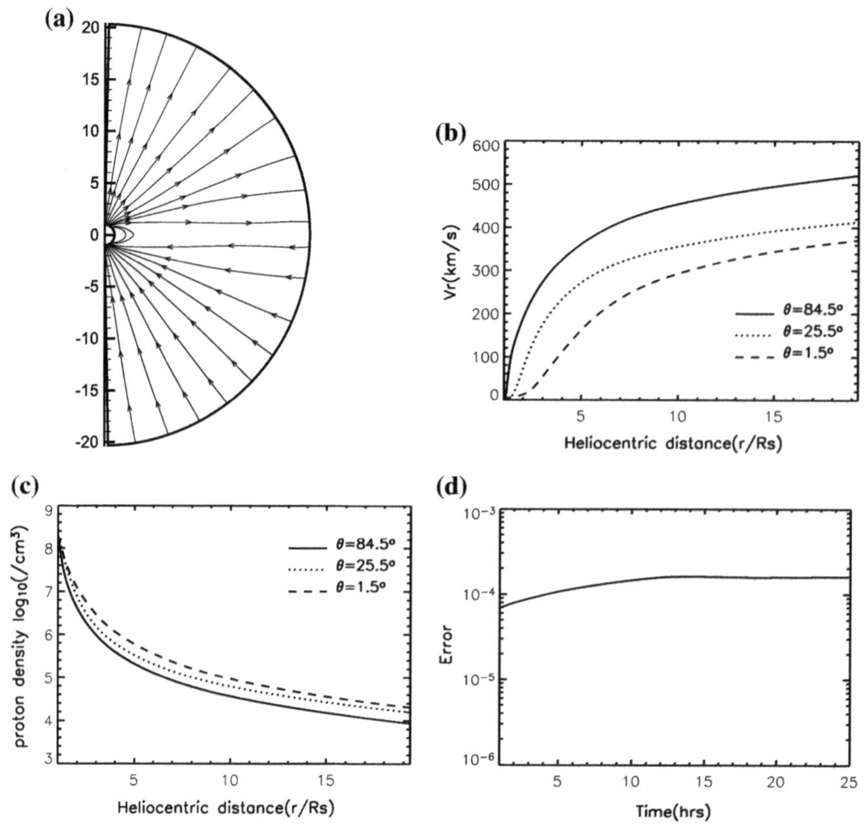

Fig. 4.3 Steady state solar wind with initial multipolar magnetic fields as input: **a** magnetic field topology; **b** the radial profile of velocity; **c** the radial profile of proton number density; and **d** error for $\nabla \cdot \mathbf{B}$. Taken from Fig. 3 of Feng et al. [16] with permissions of the authors and the AAS

90000 seconds) via relaxation technique [51]. Figure 4.3 depicts a two-dimensional (2D) cut through the 3D steady-state model in the meridional plane close to the sun. This solution represents the typical characteristics of a two-dimensional cut for coronal streamer structure with solar wind flow; these characteristics include: (a) magnetic topology, (b) the radial velocity vs heliocentric distance at various latitudinal angles, and (c) proton density vs heliocentric distance at various latitudinal angles. From these results, it is recognized that the proton number density decreases with increasing of heliocentric distance, becoming smaller toward the polar region. This picture agrees with the observed magnetic field topology of a coronal streamer at solar minimum [45]. This configuration is due to the interaction between the magnetic field and solar wind. Also, the radial velocity is higher at the polar region and increases in the radial direction. All these characteristics are consistent with previous solutions [51, 60, 76]. The simulated density structure qualitatively agrees both with the fact that coronal streamer centers are formed around the solar equator, and with the MK III observations of coronal polarized brightness near solar minimum [2]. The radial profile of bulk speed is consistent with measurements of flow speeds in the corona between 2 and 22 R_s [45]. Due to the heating source term, the radial velocity becomes higher than those obtained by Zhang and Wei [76] without heating.

4.4.2 3D Simulation for Coronal Structures Using Measured Photospheric Magnetic Field as Input

This section presents realistic 3D solar coronal atmospheric solutions. This numerical calculation is similar to the 3D case described in the previous section, but the prescribed initial magnetic field is based on measurements of the solar surface together with a potential field model. The measured photospheric magnetic field for Carrington rotation 1922 from the Wilcox Solar Observatory at Stanford University (http://wso.stanford.edu/synoptic.html) is used as the initial boundary condition to deduce a three-dimensional global potential magnetic field as shown in Fig. 4.4a and the projected magnetic field lines for different meridional planes shown in Fig. 4.4b and the corresponding contours of the radial component of the magnetic field at 2.5 R_s and 20 R_s as shown in Fig. 4.4c. This three-dimensional potential magnetic field, together with the assumed plasma properties (number density and temperature) and Parker's solar wind solution, is used as input to the numerical code to allow the numerical code to arrive at a quasi-equilibrium solution via the relaxation method. The same volumetric heating as in the above section is used here.

Figure 4.5 shows (a) the three-dimensional magnetic field topology in the northern hemisphere and (b) the corresponding field lines projected on two different meridional planes, which are the same as those shown in Fig. 4.4b. By examining both the initial (Fig. 4.4) and final relaxation solutions (Fig. 4.5), one immediately recognizes that the relaxation solution has revealed the new physics. Figure 4.6 shows the contours of radial magnetic field strength at the surfaces of 2.5 R_s and 20 R_s, respectively. Comparison of the results shown in Figs. 4.4c and 4.6 shows significant differences between the potential field approximation and the MHD solution, in that the MHD simulation exhibits more structure, even though the basic characteristics are similar. For example, the neutral line for the MHD solution shows significant magnetic shear, while in the case of the potential field approximation no magnetic shear is seen. It also could be noted that the strength of the magnetic field is decreasing with the increasing of height, but less than the potential field model which is proportional to $\frac{1}{r^2}$. Riley et al. [41] compared the steady-state magnetic field configuration of the solar corona calculated by global solar magnetohydrodynamic and potential field source surface model. The result presented here is in consistent with their result. The cusp-like features of the streamer belt appear in the MHD solution, which merge into the heliosphere current sheet. However, it is absent in the potential field solution.

Figure 4.7 shows the contour plots of the number density at the surfaces of 2.5 R_s and 20 R_s. These values in Fig. 4.7 are of the same order of magnitude as those

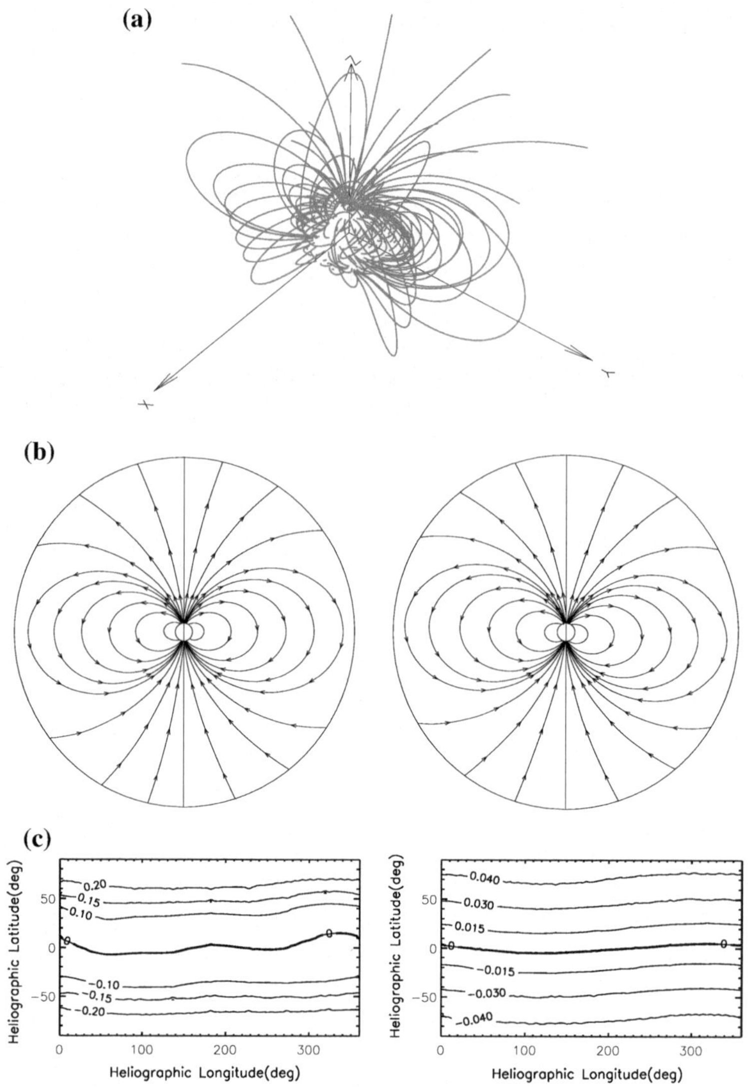

Fig. 4.4 Magnetic field profiles based on the measured line-of-sight magnetic field at WSO for Carrington Rotation 1922 by using the potential field model: **a** Three-dimensional representation of the magnetic field line in the northern hemisphere; **b** Magnetic field line projected on the meridian plane at $\phi = 198°–18°$ (left) and at $\phi = 247.5°–67.5°$ (right), from 1 R_s to 12 R_s; **c** The radial magnetic field at 2.5 R_s (Unit: Gauss) and 20 R_s (Unit: 10^{-6}T), respectively. Taken from Fig. 4 of Feng et al. [16] with permissions of the authors and the AAS

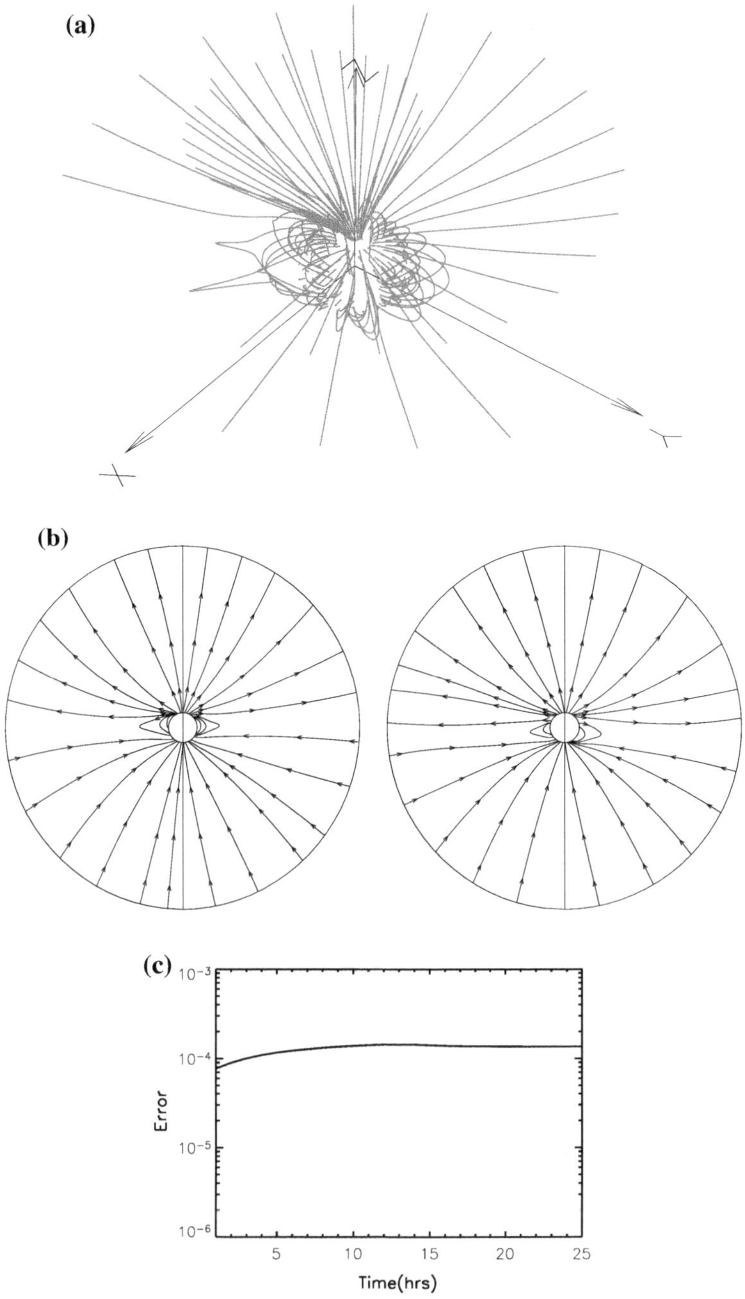

Fig. 4.5 **a** Three-dimensional numerical result of magnetic field of north hemi-sphere; **b** Computed magnetic field lines from 1 R_s to 12 R_s projected on the meridional plane at $\phi = 198°–18°$ (left) and $\phi = 247.5°–67.5°$ (right), respectively; **c** Error for $\nabla \cdot \mathbf{B}$. Taken from Fig. 5 of Feng et al. [16] with permissions of the authors and the AAS

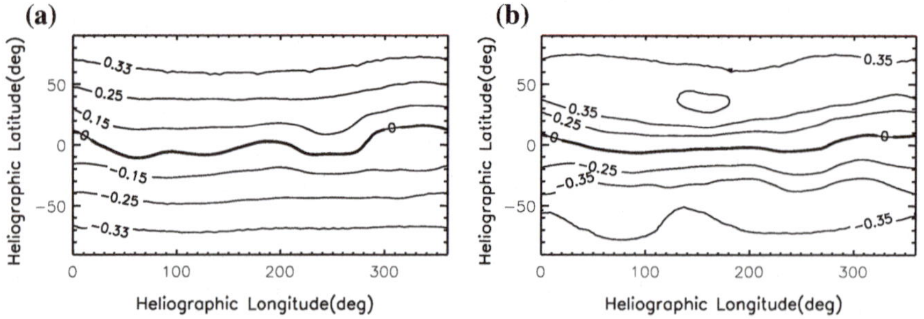

Fig. 4.6 Contours of radial magnetic field from MHD simulation for Carrington rotation 1922 **a** at 2.5 R_s (unit: Gauss); and **b** at 20 R_s (unit: 10^{-6}T), where the heavy black line indicates the neutral line. Taken from Fig. 6 of Feng et al. [16] with permissions of the authors and the AAS

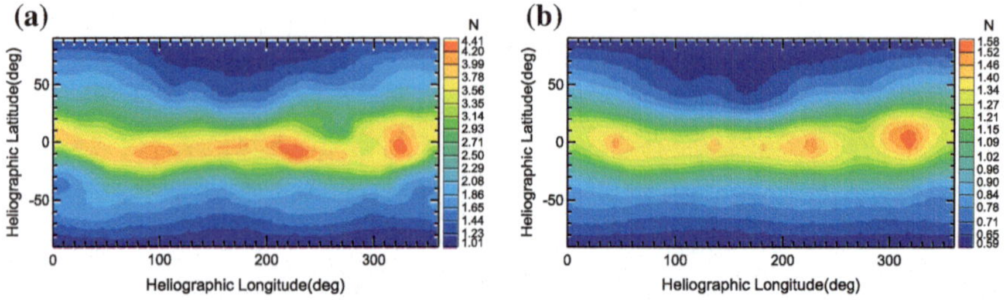

Fig. 4.7 Computed number density distributions on the surface **a** at 2.5 R_s (unit: 10^6/cm^3) and **b** at 20 R_s (unit: 10^4/cm^3). Taken from Fig. 7 of Feng et al. [16] with permissions of the authors and the AAS

previous results shown in Fig. 1 of Wei et al. [69] for the averaged distribution of the solar plasma number density (N) on the surface at 2.5 solar radii (R_s) estimated from K coronal brightness during Carrington rotations 1733–1742 in 1983.

Figures 4.8a and b show the distribution of radial velocity v_r on the surface at 2.5 R_s and 20 R_s, respectively. The numerical result at 2.5 R_s here is consistent with the averaged distribution of solar wind speed on the surface at 2.5 R_s calculated from the mapping of interplanetary scintillation data during Carrington rotations 1733–1742 in 1983; and also with results using data for Carrington rotation 1742 in 1983 in a special MHD method by Wei et al. [69]. By examining Figs. 4.7 and 4.8 together, we notice that the low density regions have higher radial speed, which indicates that there exists a north polar coronal hole, a south polar coronal hole and an equatorial structured distribution as a function of heliographic longitude. Figure 4.9a displays the radial velocity profiles along heliocentric distance at two different latitudes: $\theta = 66.5°$ and $-5.5°$ at the same longitude $\phi = 54°$, which corresponds, respectively, to the open field and current sheet regions. The corresponding radial density distribution of these two locations is shown in Fig. 4.9b. We notice that the velocity is consistent with the profile of low solar wind speed [45]. It should be further noted that the same color shown in Fig. 4.8 represents different velocities at two different heights, shown

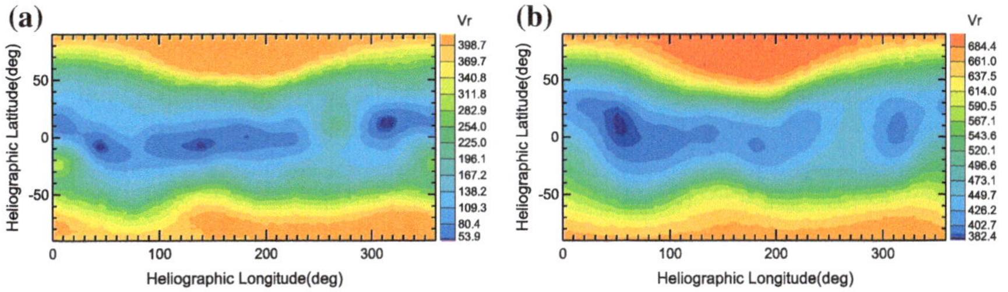

Fig. 4.8 Contours of radial velocity v_r on the surface **a** at 2.5 R_s and **b** at 20 R_s (unit: km/s). Taken from Fig. 8 of Feng et al. [16] with permissions of the authors and the AAS

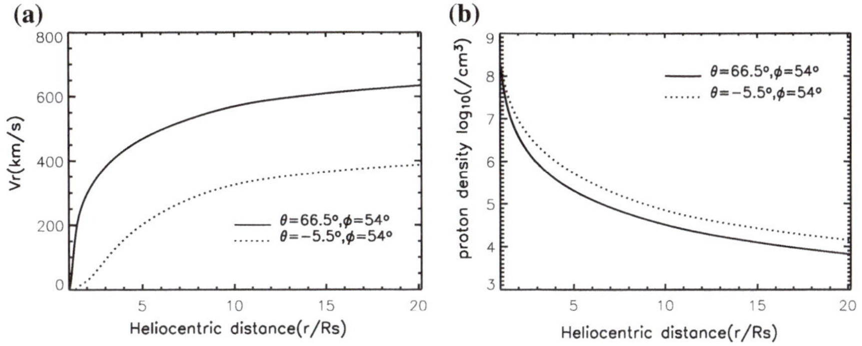

Fig. 4.9 **a** The radial velocity profiles vs heliocentric distance at two location with different latitudes $\theta = 66.5^\circ$ and $\theta = -5.5^\circ$ at the same Carrington longitude $\phi = 54^\circ$ and **b** the corresponding number density distributions. Taken from Fig. 9 of Feng et al. [16] with permissions of the authors and the AAS

by the legends on the right; the higher heights have higher velocities. This shows the expansion of the solar wind which is consistent with the result shown in Fig. 4.9a.

Although the polytropic law consideration combined with a volumetric heating may approximate the effects of the unknown heating processes for simulating the ambient solar wind, and the modeled result matches many features of the corona (as shown above), it is not able to better reproduce the fast streams in coronal holes and slow streams in streamers [35] because the present model's energy equation is too simplified.

4.5 Global Corona Throughout Solar Cycle 23

This section presents the evolution of the large-scale coronal magnetic structure for fifteen Carrington Rotations (CRs) spanning solar cycle 23 by the 3D SIP-CESE MHD model. We compare the location of the HCS as determined by the MHD model with that from the potential field source surface (PFSS) model to show the commonness and difference of these two models at different phases of the solar cycle. Also, the modeled results are compared with the observed white-light intensity pattern to illustrate the value of the simulations in interpreting white-light observations.

The solar magnetic field plays a crucial role in coronal and interplanetary physics, determining the structure of the solar corona and inner heliosphere [31]. However, there is a fundamental difficulty in understanding coronal and interplanetary structures: it is still impossible to obtain direct measurement of the weak coronal magnetic field. One principal approach for obtaining the global magnetic structure of the solar corona is the PFSS model, which was first developed independently by both Altschuler and Newkirk [44] and Schatten et al. [1]. This approach assumes that the coronal field is approximately potential and the field at a certain outer radial boundary (for instance, source surface) is completely radial, then the coronal magnetic field configuration can be calculated from the line-of-sight component of the magnetic field measured in the photosphere. The PFSS model has been successfully applied to the study of a wide range of solar and heliospheric topics. Another approach for studying these topics is the use of global magnetohydrodynamic (MHD) models (e.g., [20, 31, 32, 36, 55–58]). In principle, MHD modeling is a more powerful approach for studying the solar corona, which can self-consistently describe the coupled interaction between magnetic, plasma, and solar gravity forces.

The structures of the solar magnetic field vary systematically over a period of about 11 years in relation to the solar activity cycle, and four phases of a solar cycle can be identified: solar minimum, the rising phase, solar maximum, and the declining phase. Figure 4.10 summarizes several solar parameters measured through the solar cycle. The monthly averaged and smoothed sunspot number (SSN) are shown as a function of time in black, while the computed WSO tilt angle of the HCS is shown in blue. Fifteen Carrington Rotations are selected for the present study, as shown in Fig. 4.10, which spanned the period from June 1996 to December 2006. The two successive selected Carrington rotations are separated by 9$\sim$11 Carrington Rotations and can be broadly grouped into three categories: (1) the first five rotations that belong to solar minimum and the rising phase (CR 1910, CR 1919, CR 1928, CR 1937, and CR 1946), shown in red, during the early portion of solar cycle 23; (2) the second five rotations that surround solar maximum (CR 1955, CR 1964, CR 1974, CR 1984, and CR 1995), in green, during the middle portion; (3) the last five rotations that account for the declining phase and approach the solar minimum (CR 2006, CR 2017, CR 2028, CR 2039, and CR 2050), in blue, during the ending portion. This categorization is mirrored in the monthly smoothed sunspot number, which started with the monthly SSN at 8.0 in May 1996, peaked at 120.8 in April 2000, had a second, smaller peak at 115.5 in November 2001, and then followed a longer decay to 12.7 in November 2006.

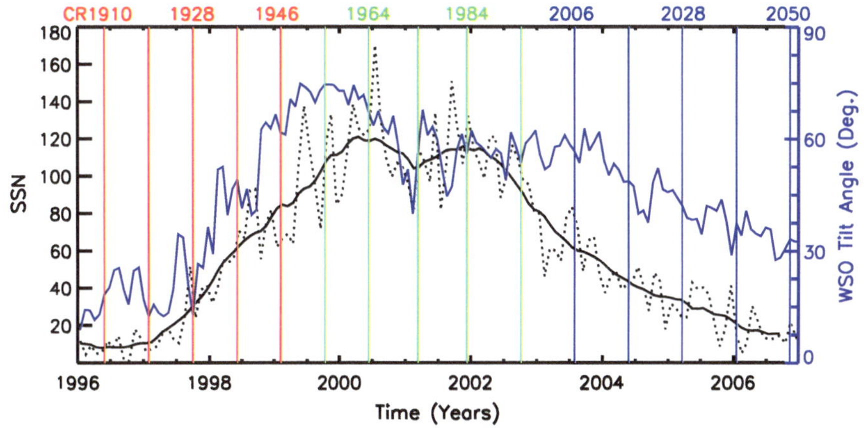

Fig. 4.10 The monthly averaged (dotted black line) and smoothed sunspot number (SSN, solid black line) are shown as a function of time; The computed WSO (Wilcox Solar Observatory) tilt angle of the HCS (solid blue line) is also plotted; Hu et al. [27] investigated fifteen Carrington Rotations, which are colored according to their associated phases in the solar cycle. Reprinted from Fig. 1 of Hu et al. [27] with permission of the John Wiley & Sons, Inc

4.5.1 Evolution of the HCS Topology and Coronal Magnetic Field Configuration

One of the principal advantages of the three-dimensional MHD model is that the global coronal structure can be directly obtained from the simulation results. In the following, we show that the evolution of the HCS topology and coronal magnetic field configuration, which are predicted by SIP-CESE MHD model through solar cycle 23.

In Fig. 4.11, the upper two rows show the HCS topology and the three-dimensional coronal field structure of the first five Carrington Rotations. We can easily see that, during the first three Carrington Rotations (CR 1910, CR 1919, and CR 1928), which are typical periods around solar minimum, the HCS can be approximately described as a flat plane with little spatial variability. The latitudinal excursion of HCS is very limited, and the HCS is nearly in the heliospheric equatorial plane. During these periods, the coronal magnetic field is dominated by a dipole pattern with an axis along the Sun's rotation axis. The large, closed loops connecting the northern and southern mid-latitudes produce the cusp-like features of the helmet streamers, which merge into the HCS at its top. At the same time, large coronal holes, appearing as bundles of open field lines, dominate the higher latitudes. As time goes into the following two Carrington Rotations (CR 1937 and CR 1946), the maximum latitudinal extent of the HCS expands rapidly to higher latitudes, while the coronal magnetic field shows a mixed dipole/quadrupole configuration with increasing tilt angle of the dipole. Especially in CR 1937, the HCS is highly warped and a large

Fig. 4.11 Evolution of the HCS topology and coronal magnetic field configuration throughout solar cycle 23. Five Carrington Rotations are classified into one group according to their associated phases in the solar cycle (see the context for the details). The HCSs are shown above as the isosurfaces of $B_r = 0$ and the magnetic field configurations are shown below, respectively, for each Carrington Rotation; Field lines have been assigned as black for open field lines and blue for close field lines; however, the same foot points at the photosphere are used in tracing the field lines for each Carrington Rotation. Reprinted from Fig. 2 of Hu et al. [27] with permission of the John Wiley & Sons, Inc

north-south asymmetry appears about the heliospheric equator, which is due to the rapid growth of the quadrupole strength of the coronal field [18].

The middle two rows in Fig. 4.11 illustrate the evolution of the coronal magnetic structure during the second five Carrington Rotations. As solar activity enhances significantly around solar maximum, the HCS can become nearly vertical to the heliospheric equator and displays its greatest complexity. In fact, in CR 1955, the HCS shows a duplex structure with an additional "tube-like" current sheet detached from the main HCS. During this period, the solar magnetic field undergoes its polarity reversal process [68], when the polar fields weaken and ultimately reverse polarity. The dominance of the polar fields gradually disappeared, and small coronal holes having opposite polarities may be present at mid- or low-latitudes. Occasionally, unipolar regions develop that are surrounded by a circular neutral contour, which could represent a "tube-like" current sheet detached from the main HCS. In fact, Hoeksema et al. [23] have shown that multiple current sheets probably exist in the two or three years near solar maximum. Then, during CR 1964, the HCS recovers its typical topology but still extends to high latitude and almost crosses over the poles. Around solar maximum, the global coronal field is dominated by higher order components rather than dipole, with streamers scattered at all latitudes including polar regions. The total numbers of open field lines reduce significantly compared with those around solar minimum, because the same foot points at the photosphere are used in tracing the field lines for every Carrington Rotation, indicating that the area of the open flux regions rooted in scattered coronal holes decreases obviously at solar maximum. In fact, the percentage of the open field area is shown in Fig. 4.12 as a function of solar cycle 23. The result shows that when the solar activity increases, the field becomes more closed, and more open field lines exhibit at solar minimum. Quantitatively, it can be easily seen that the percentage of the Sun's surface covered by open field decreases from $\sim$30% at solar minimum to $\sim$14% at solar maximum. This finding is consistent with a previous analysis for the long-term variations of the Sun's open magnetic flux [66]. After the maximum, solar activity shows a slight decrease as reflected by the monthly smoothed sunspot number in Fig. 4.10. In CR 1974, well after solar maximum, a moderate warped HCS presents itself with small inclination with respect to the heliospheric equator, when the dipole field regains its strength. However, by the end of 2001, solar activity increases again and, accordingly, the HCS regrows to high heliographic latitudes, looking like a tilted but relatively flat plane during CR 1984. This observation indicates that the coronal magnetic field is dominated by a tilted dipole pattern with relatively small quadrupole or higher order components. Afterwards, solar cycle 23 starts its longer declining phase in CR 1995.

The last two rows in Fig. 4.11 show the results of the last five Carrington Rotations. Although the solar cycle deeply enters into its declining phase during CR 2006 and CR 2017, the inclination of the HCS remains large with respect to the heliospheric equatorial plane. During these periods, a tilted dipole pattern still dominates the global coronal field with the tilt angle's becoming smaller and smaller. On the way to solar minimum, the HCS may also present large spatial variability. As in CR 2028, the HCS is highly warped and displays a two-fold structure, which can be attributed to the reinforcement of the higher order component of the coronal field, as shown

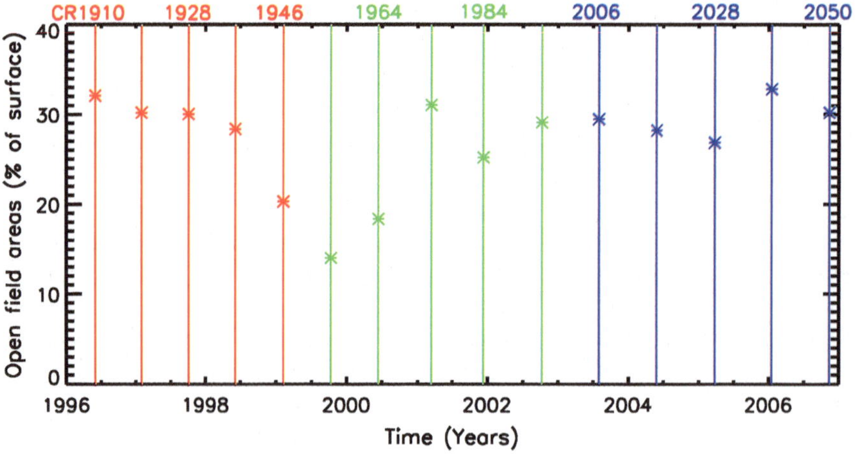

Fig. 4.12 Percentages of open field areas as a function of time during solar cycle 23. Reprinted from Fig. 3 of Hu et al. [27] with permission of the John Wiley & Sons, Inc

in the magnetic field configuration. However, the trend towards solar minimum is unavoidable. As time enters into CR 2039 and CR 2050, the HCS, slightly warped, returns to the near heliospheric equator. The dipole field is restored with its axis aligned with the Sun's rotation axis, while long-lived streamers (regions with blue closed field lines) and large polar coronal holes (regions with black open field lines) can persist over several solar rotations. These characters all indicate that solar cycle 23 is closing and the newly solar minimum epoch is coming up.

4.5.2 Comparisons with the PFSS Model Results and Observations

The PFSS model has been successfully applied to predict the coronal magnetic structure (see Luhmann [34] and references therein). It may be instructive to compare SIP-CESE MHD model results with those of the PFSS model. In fact, previous studies [38, 41] have compared some aspects of the MHD and PFSS results at specified solar activity phases. This section makes similar comparisons of these results covering the period of the whole of solar cycle 23. The PFSS model used here is almost identical to that used by Luhmann [34], in which the observed photospheric field is radial [64], $r = 2.5R_s$ is used as the source surface, and 13 orders of the harmonic coefficients are included to calculate the coronal magnetic field. On the other hand, since the beginning of 1996, the Large Angle and Spectrometric Coronagraph (LASCO) experiments on board the Solar and Heliospheric Observatory (SOHO) have recorded large numbers of high-resolution coronal images, providing an unprecedented opportunity to learn about the solar corona. It is now widely believed that the observed white-light bright structures in the outer corona are closely related to the neutral line surfaces of the magnetic field and coincide in location with the HCS [65, 67]. Thus, observations by SOHO/LASCO give us a criterion to verify the MHD results.

In Fig. 4.13, the left column shows the neutral lines on the source surface calculated from both the MHD and PFSS models. Comparisons between the two model results suggest that in general, the MHD result is in close agreement with that of the PFSS model for predicting the HCS. The neutral lines have almost the same shape and location for each Carrington Rotation (except for CR 1955), and the topology of the HCS evolves in a qualitatively similar manner during the course of the solar cycle. However, there are also several notable disagreements. In CR 1955, the PFSS neutral line displays a distinct four-sector structure with a different magnetic polarity at each pole, while the MHD neutral line shows a long switchback with an additional circular neutral line, and both poles have the same polarity. This disagreement may result from inaccurate polar field estimation due to the observation constraints for polar regions (above $\sim 70°$), which has significant effect on determining the neutral line topology [22]. In the PFSS model, the polar field strengths are determined based on annual statistical analysis [52, 84], which unavoidably may introduce ambiguous errors. In particular, at this specified period, the solar magnetic field undergoes its polarity reversal process [68], which makes the polar field more difficult to determine. We find that other PFSS models [43], which did not include the polar regions, obtained a similar double neutral line topology as the MHD model. In addition, there are also several local discrepancies between the MHD and PFSS neutral lines, where the MHD results show more distortions, especially in low activity intervals. This may seem somewhat surprising; however, Riley et al. [41] have proved that the source surface is less spherical at solar minimum than at solar maximum, which indicates the basic assumption of the PFSS model (that a spherical source surface exists) is more difficult to meet under low solar activity condition. The nonpotential effect associated with active regions may be another factor, which can be partially included in the MHD approach but not in the PFSS model. Under low activity condition, few long-lived active regions may be present, which can locally alter the location of the HCS above them remarkably [17, 65]. However around solar maximum, more active regions may scatter at all latitudes, where the alterations due to the nonpotential effects of these active regions may be offset in an average sense [41].

The middle and right columns in Fig. 4.13 show the Carrington format maps of the observed white-light streamer structures at the east and west limbs, respectively. These maps are constructed by extracting from successive LASCO C2 images a slice of east or west limb data at $r = 2.5\ R_s$ and assembling the strips into time-reversed sequences. White bars at the top of this map indicate missing data; the vertical streaks represent coronal mass ejections. The differences between the east and west limb maps may reflect the effect of the Sun's axial tilt relative to the sky plane, as well as intrinsic changes in the streamers occurring during half a solar rotation. Wang et al. [67] suggested that one origin of the observed white-light streamer structures is the large-scale plasma sheet located near the HCS. The streamer's highest intensities occur where the latitude of the current sheet varies least with longitude; at these locations, the plasma sheet is oriented edge-on in the sky plane, and the number of

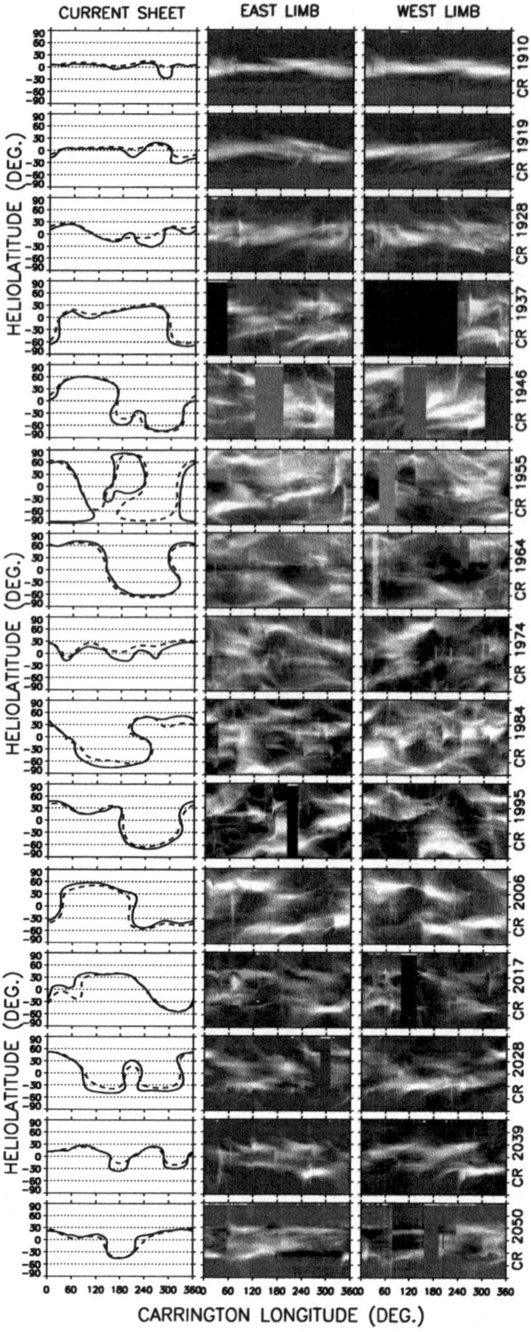

Fig. 4.13 Shape of the source surface neutral lines (left column) calculated from the MHD (solid line) and PFSS model (dashed line), and Carrington format maps of the white-light streamer structures at the east (middle column) and west limb (right column) constructed from successive LASCO C2 images. Reprinted from Fig. 4 of Hu et al. [27] with permission of the John Wiley & Sons, Inc

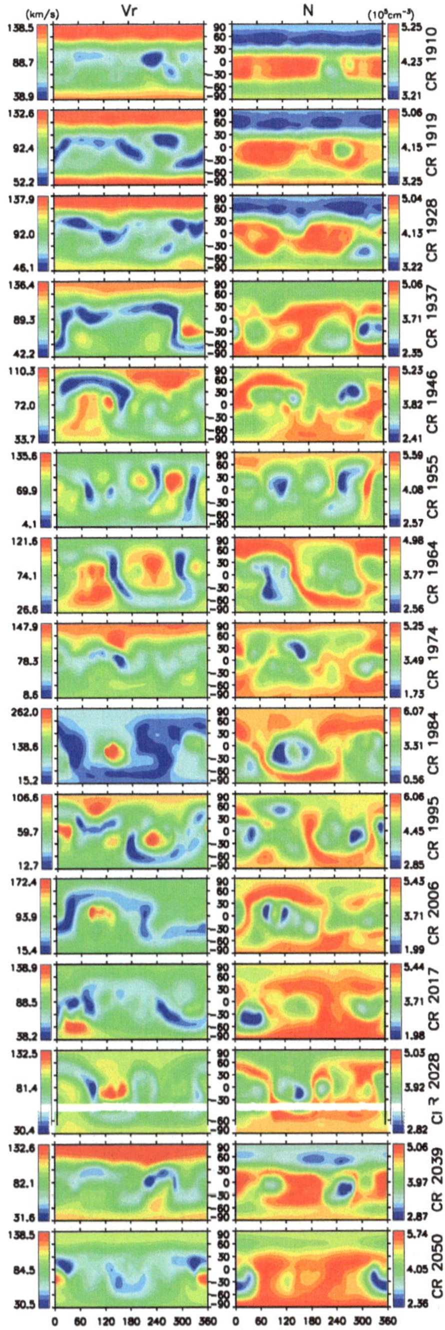

Fig. 4.14 Radial velocity and number density distributions on the surface at 2.5 R_S. Reprinted from Fig. 5 of Hu et al. [27] with permission of the John Wiley & Sons, Inc

electrons in the line of sight reaches its maximum. From Fig. 4.13, we can see that to a first approximation the locations of the HCS calculated from the MHD model do show good overall agreement with the bright structures in the observed intensity pattern, especially near solar minimum (CR 1910–CR 1928) and well after solar maximum (CR 1984–CR 2050). Around solar maximum (CR 1937–CR 1974), the discrepancies between the location of the HCS and the white-light observation are a little more obvious, while the coronal streamer structures display more complexity and temporal variation (see the notable differences between the east and west limbs). As it is known, the corona is dynamic. To partially address this, the high time-cadence data such as daily updated and hourly synoptic map from the observed magnetograms may be incorporated into the present model by data-driven procedures. This topic will be presented in Chap. 7.

To further compare the MHD results with the observations, Fig. 4.14 shows the radial velocity and number density distributions on the surface at 2.5 R_s. By examining these results together with the location of the HCS shown in Fig. 4.13, it is recognized that the flow speed (density) is lower (higher) in the neighborhood of the HCS and higher (lower) at the polar regions, which is qualitatively in agreement with the observed intensity pattern. Also, the values of the number density and radial speed, respectively, have the same order of magnitude as those shown in Figs. 1 and 2 of Wei et al. [69] for the averaged distribution of the solar plasma number density and solar wind speed on the surface at 2.5 R_s, where the plasma density is estimated from K coronal brightness and the solar wind speed are calculated from the mapping of interplanetary scintillation data during CRs 1733–1742. In addition, we can observe the solar cycle variations. When solar activity increases, the high speed stream spreads to the lower latitude while the density changes reversely. During the solar minimum, the higher speed is confined to the poles and a low speed is near the equator, but the density does vary almost inversely. However, despite these qualitative similarities, the model results have important quantitative differences from the real corona. In particular, the density contrast between open and closed field regions is smaller than observed, and the fast solar wind speed in the simulation is too low compared to measurements. These inaccuracies may be twofold: one can be related to the simple adiabatic energy equation assumed, and another can be due to the simplified input of using Parker 1D solar wind model; Improvements can be achieved by taking into account of coronal heating/acceleration mechanism and usage of self-consistent boundary value with more constraints by observations.

4.6 Conclusions and Discussions

This chapter establishes the 3D SIP-CESE MHD model on pentahedral cells with each cell a triangular prism composed of two triangular bases and three rectangular sides, and describes its application of the MHD model to study solar wind plasma. The ability of the CESE method is further extended in subsequent chapters to discuss the CESE-MHD model and its applications to background solar wind, coronal mass ejections, coronal non-linear force-free magnetic field extrapolations and coronal magnetic field evolutions and eruptions.

After the proposal of the second-order CESE method [9], high-order versions of CESE have been further developed for the compressible Euler and Navier–Stokes equations (e.g., [3–5, 8, 10, 46, 48, 59]). These works paid attention to the developmental effort for high-fidelity multi-dimensional, unstructured-mesh flow solvers using the CESE framework with focuses on: (1) time-accurate local time stepping, (2) higher-order accuracy, and (3) hybrid or unstructured meshes. Unlike other existing explicit high-order methods for unstructured meshes, the high-order CESE framework maintains a CFL condition of one for arbitrarily high-order formulations while retaining the same compact stencil as its second-order counterpart. In CESE schemes, the spatial derivatives are not reconstructed based on cell averages as in conventional finite-volume schemes. Instead, they are updated by independent schemes. Different CESE schemes share an identical time-marching scheme for solution vector $\mathbf{U}$, but they have different time-marching schemes for the spatial derivatives [48, 49].

Shen et al. [48] implemented the scheme on unstructured meshes consisting of hybrid elements and derived the corresponding high-order schemes. To improve the resolution of CESE schemes for contact discontinuities, Shen et al. [47, 49] proposed upwind CESE schemes by introducing characteristic-based fluxes into the framework of the CESE method. They coupled the upwind CESE scheme with Harten, Lax and van Leer (HLL) [21], contact restoring Harten, Lax and van Leer (HLLC) [54] and Roe [42] Riemann solvers. However, neither the central CESE scheme nor the upwind CESE scheme can preserve the positivity of densities and pressures. To further settle this problem, Shen et al. [50] constructed a maximum-principle-satisfying CESE scheme via the use of the simple and efficient limiter proposed by Zhang and Shu [79], which, in general, can be applied to both central and upwind CESE schemes. This scheme was successfully used [50] in the simulation of multicomponent flows to preserve the positivity of volume fractions. However, the positivity of the density and pressure was not completely guaranteed for the compressible Navier–Stokes equations under some extreme conditions. Hu and Khoo [24] suggested an effective slope limiter for the one-dimensional CESE Euler solver using a kinetic energy fix technique. Shen and Parsani [46] developed positivity-preserving CESE schemes for solving the compressible Euler and Navier–Stokes equations on unstructured hybrid meshes while maintaining the conservativeness and accuracy of baseline CESE schemes with local Lax-Friedrichs flux. The schemes require an a posteriori limiter that can be applied to both central and upwind CESE schemes. The proposed approach, as pointed out by Shen and Parsani [46], is robust and accurate in low density flows, for strong shocks, high Mach number flows and complex viscous flows on different types of meshes.

Yang et al. [72] established a high-order space-time CESE method with a compact stencil for MHD equations. Yang et al. [73] extended the upwind CESE scheme to multidimensional MHD in general curvilinear coordinates by transforming the MHD equations from the physical domain (general curvilinear coordinates) to the computational domain (rectangular coordinates). This upwind CESE MHD scheme also paves the way to achieve a flexible combination of the baseline CESE and any upwind finite volume method [74]. Benchmark tests [73, 74] have shown remarkable advantages of the upwind CESE MHD scheme over the original CESE in capturing discontinuities and maintaining robust property has shown at no extra computational cost.

Appendix

In this appendix, the Jacobian matrices for the fluxes **F**, **G**, **H**, and $\boldsymbol{\Phi}$ are calculated [16].

$$\frac{\partial \mathbf{F}}{\partial \mathbf{U}} = \begin{pmatrix} 0 & 1 & 0 & 0 & 0 & 0 & 0 & 0 \\ -u^2 + (\gamma-1)\frac{u^2+v^2+w^2}{2} & (3-\gamma)u & (1-\gamma)v & (1-\gamma)w & \gamma-1 & -\gamma B_x & (2-\gamma)B_y & (2-\gamma)B_z \\ -uv & v & u & 0 & 0 & -B_y & -B_x & 0 \\ -uw & w & 0 & u & 0 & -B_z & 0 & -B_x \\ F_1 & F_2 & F_3 & F_4 & F_5 & F_6 & F_7 & F_8 \\ 0 & 0 & 0 & 0 & 0 & 0 & 0 & 0 \\ -\frac{uB_y - vB_x}{\rho} & \frac{B_y}{\rho} & -\frac{B_x}{\rho} & 0 & 0 & -v & u & 0 \\ -\frac{uB_z - wB_x}{\rho} & \frac{B_z}{\rho} & 0 & -\frac{B_x}{\rho} & 0 & -w & 0 & u \end{pmatrix}$$

where

$$F_1 = -\frac{\gamma e u}{\rho} + (\gamma-1)u\left(u^2+v^2+w^2\right) + \frac{\gamma-2}{2}u\frac{B_y^2+B_z^2}{\rho} + \frac{\gamma}{2}u\frac{B_x^2}{\rho} + B_x\frac{vB_y+wB_z}{\rho}$$

$$F_2 = \gamma\frac{e}{\rho} + \frac{3}{2}(1-\gamma)u^2 + \frac{1-\gamma}{2}\left(v^2+w^2\right) - \frac{\gamma-2}{2}\frac{B_y^2+B_z^2}{\rho} - \frac{\gamma}{2}\frac{B_x^2}{\rho}$$

$$F_3 = (1-\gamma)uv - \frac{B_xB_y}{\rho}$$

$$F_4 = (1-\gamma)uw - \frac{B_xB_z}{\rho},\ F_5 = \gamma u$$

$$F_6 = -\gamma u B_x - (vB_y + wB_z)$$

$$F_7 = (2-\gamma)uB_y - vB_x$$

$$F_8 = (2-\gamma)uB_z - wB_x$$

$$\frac{\partial \mathbf{G}}{\partial \mathbf{U}} = \begin{pmatrix} 0 & 0 & 1 & 0 & 0 & 0 & 0 & 0 \\ -uv & v & u & 0 & 0 & -B_y & -B_x & 0 \\ -v^2 + (\gamma-1)\frac{u^2+v^2+w^2}{2} & (1-\gamma)u & (3-\gamma)v & (1-\gamma)w & \gamma-1 & (2-\gamma)B_x & -\gamma B_y & (2-\gamma)B_z \\ -vw & 0 & w & v & 0 & 0 & -B_z & -B_y \\ G_1 & G_2 & G_3 & G_4 & G_5 & G_6 & G_7 & G_8 \\ -\frac{vB_x-uB_y}{\rho} & -\frac{B_y}{\rho} & \frac{B_x}{\rho} & 0 & 0 & v & -u & 0 \\ 0 & 0 & 0 & 0 & 0 & 0 & 0 & 0 \\ -\frac{vB_z-wB_y}{\rho} & 0 & \frac{B_z}{\rho} & -\frac{B_y}{\rho} & 0 & 0 & -w & v \end{pmatrix}$$

where

$$G_1 = -\frac{\gamma ev}{\rho} + (\gamma-1)v\left(u^2+v^2+w^2\right) + \frac{\gamma-2}{2}v\frac{B_x^2+B_z^2}{\rho} + \frac{\gamma}{2}v\frac{B_y^2}{\rho} + B_y\frac{uB_x+wB_z}{\rho}$$

$$G_2 = (1-\gamma)vu - \frac{B_yB_x}{\rho}$$

$$G_3 = \gamma\frac{e}{\rho} + \frac{3}{2}(1-\gamma)v^2 + \frac{1-\gamma}{2}\left(u^2+w^2\right) - \frac{\gamma-2}{2}\frac{B_x^2+B_z^2}{\rho} - \frac{\gamma}{2}\frac{B_y^2}{\rho}$$

$$G_4 = (1-\gamma)vw - \frac{B_yB_z}{\rho}, G_5 = \gamma v$$

$$G_6 = (2-\gamma)vB_x - uB_y$$

$$G_7 = -\gamma vB_y - (uB_x + wB_z)$$

$$G_8 = (2-\gamma)vB_z - wB_y$$

$$\frac{\partial \mathbf{H}}{\partial \mathbf{U}} = \begin{pmatrix} 0 & 0 & 0 & 1 & 0 & 0 & 0 & 0 \\ -uw & 0 & w & u & 0 & -B_z & 0 & -B_x \\ -vw & 0 & w & v & 0 & 0 & -B_z & -B_y \\ -w^2 + (\gamma-1)\frac{u^2+v^2+w^2}{2} & (1-\gamma)u & (1-\gamma)v & (3-\gamma)w & \gamma-1 & (2-\gamma)B_x & (2-\gamma)B_y & -\gamma B_z \\ H_1 & H_2 & H_3 & H_4 & H_5 & H_6 & H_7 & H_8 \\ -\frac{wB_x-uB_z}{\rho} & -\frac{B_z}{\rho} & 0 & \frac{B_x}{\rho} & 0 & w & 0 & -u \\ -\frac{wB_y-vB_z}{\rho} & 0 & -\frac{B_z}{\rho} & \frac{B_y}{\rho} & 0 & 0 & w & -v \\ 0 & 0 & 0 & 0 & 0 & 0 & 0 & 0 \end{pmatrix}$$

where

$$H_1 = -\frac{\gamma ew}{\rho} + (\gamma-1)w\left(u^2+v^2+w^2\right) + \frac{\gamma-2}{2}w\frac{B_x^2+B_y^2}{\rho} + \frac{\gamma}{2}w\frac{B_z^2}{\rho} + B_z\frac{uB_x+vB_y}{\rho}$$

$$H_2 = (1-\gamma)wu - \frac{B_zB_x}{\rho}$$

$$H_3 = (1-\gamma)wv - \frac{B_z B_y}{\rho}$$

$$H_4 = \gamma\frac{e}{\rho} + \frac{3}{2}(1-\gamma)w^2 + \frac{1-\gamma}{2}\left(u^2+v^2\right) - \frac{\gamma-2}{2}\frac{B_x^2+B_y^2}{\rho} - \frac{\gamma}{2}\frac{B_z^2}{\rho}$$

$$H_5 = \gamma w$$

$$H_6 = (2-\gamma)wB_x - uB_z$$

$$H_7 = (2-\gamma)wB_y - vB_z$$

$$H_8 = -\gamma wB_z - (uB_x + vB_y)$$

$$\frac{\partial \boldsymbol{\Phi}}{\partial \mathbf{U}} = \mathbf{I} - \frac{\Delta t}{2}\frac{\partial \eta}{\partial \mathbf{U}} \equiv \begin{pmatrix} 1 & 0 & 0 & 0 & 0 & 0 & 0 & 0 \\ a_{21} & 1 & a_{23} & 0 & 0 & a_{26} & 0 & 0 \\ a_{31} & a_{32} & 1 & 0 & 0 & 0 & a_{37} & 0 \\ a_{41} & 0 & 0 & 1 & 0 & 0 & 0 & a_{48} \\ a_{51} & a_{52} & a_{53} & a_{54} & a_{55} & a_{56} & a_{57} & a_{58} \\ a_{61} & a_{62} & 0 & 0 & 0 & 1 & 0 & 0 \\ a_{71} & 0 & a_{73} & 0 & 0 & 0 & 1 & 0 \\ a_{81} & 0 & 0 & a_{84} & 0 & 0 & 0 & 1 \end{pmatrix}$$

where

$$a_{21} = -a_{\text{gra}}x\frac{\Delta t}{2} - \omega^2 x\frac{\Delta t}{2},\ a_{23} = -2\omega\frac{\Delta t}{2},\ a_{31} = -a_{\text{gra}}y\frac{\Delta t}{2} - \omega^2 y\frac{\Delta t}{2},\ a_{32} = 2\omega\frac{\Delta t}{2}$$

$$a_{41} = -a_{\text{gra}}z\frac{\Delta t}{2},\ a_{26} = a_{37} = a_{48} = (\nabla\cdot B)_{Q^*}^{n-1/2}\frac{\Delta t}{2}$$

$$a_{51} = -(\gamma-1)q_e\left(T_0 - \gamma(\gamma-1)\frac{u^2+v^2+w^2}{2}\right)\frac{\Delta t}{2} - (\nabla\cdot B)_{Q^*}^{n-1/2}\frac{uB_x+vB_y+wB_z}{\rho}\frac{\Delta t}{2}$$

$$a_{52} = -\left(a_{\text{gra}}x + \gamma(\gamma-1)^2 q_e u\right)\frac{\Delta t}{2} - \omega^2 x\frac{\Delta t}{2} + (\nabla\cdot B)_{Q^*}^{n-1/2}\frac{B_x}{\rho}\frac{\Delta t}{2}$$

$$a_{53} = -\left(a_{\text{gra}}y + \gamma(\gamma-1)^2 q_e v\right)\frac{\Delta t}{2} - \omega^2 y\frac{\Delta t}{2} + (\nabla\cdot B)_{Q^*}^{n-1/2}\frac{B_y}{\rho}\frac{\Delta t}{2}$$

$$a_{54} = -\left(a_{\text{gra}}z + \gamma(\gamma-1)^2 q_e w\right)\frac{\Delta t}{2} + (\nabla\cdot B)_{Q^*}^{n-1/2}\frac{B_z}{\rho}\frac{\Delta t}{2}$$

$$a_{55} = 1 + \gamma(\gamma-1)^2 q_e\frac{\Delta t}{2}$$

$$a_{56} = -\gamma(\gamma - 1)^2 q_e B_x \frac{\Delta t}{2} + \left((\nabla \cdot B)^{n-1/2}_{Q^*} u\right) \frac{\Delta t}{2}$$

$$a_{57} = -\gamma(\gamma - 1)^2 q_e B_y \frac{\Delta t}{2} + \left((\nabla \cdot B)^{n-1/2}_{Q^*} v\right) \frac{\Delta t}{2}$$

$$a_{58} = -\gamma(\gamma - 1)^2 q_e B_z \frac{\Delta t}{2} + \left((\nabla \cdot B)^{n-1/2}_{Q^*} w\right) \frac{\Delta t}{2}$$

$$a_{61} = -\frac{(\nabla \cdot B)^{n-1/2}_{Q^*} u}{\rho} \frac{\Delta t}{2}, a_{62} = \frac{(\nabla \cdot B)^{n-1/2}_{Q^*}}{\rho} \frac{\Delta t}{2}$$

$$a_{71} = -\frac{(\nabla \cdot B)^{n-1/2}_{Q^*} v}{\rho} \frac{\Delta t}{2}, a_{73} = \frac{(\nabla \cdot B)^{n-1/2}_{Q^*}}{\rho} \frac{\Delta t}{2}$$

$$a_{81} = -\frac{(\nabla \cdot B)^{n-1/2}_{Q^*} w}{\rho} \frac{\Delta t}{2}, a_{84} = \frac{(\nabla \cdot B)^{n-1/2}_{Q^*}}{\rho} \frac{\Delta t}{2}$$

with $a_{\text{gra}} = -\frac{GM_s}{r^3}$ and $q_e = q_0 \exp\left[-\frac{(r-R_s)^2}{\sigma_0^2}\right]$. Then, $\left(\frac{\partial \boldsymbol{\Phi}}{\partial \mathbf{U}}\right)^{-1}$ can be easily obtained by Mathematica.

References

1. Altschuler MD, Newkirk G (1969) Magnetic fields and the structure of the solar corona. I: methods of calculating coronal fields. Solar Phys 9:131–149. https://doi.org/10.1007/BF00145734
2. Biesecker DA, Thompson BJ, Gibson SE, Alexander D, Fludra A, Gopalswamy N, Hoeksema JT, Lecinski A, Strachan L (1999) Synoptic sun during the first whole sun month campaign: August 10 to September 8, 1996. J Geophys Res 104:9679–9690. https://doi.org/10.1029/1998JA900056
3. Bilyeu D, Yu STJ (2013) A 3D Unstructured-mesh Euler solver based on the fourth-order CESE method, American Institute of Aeronautics and Astronautics, San Diego, CA. Fluid dynamics and co-located conferences. AIAA, p 3071.https://doi.org/10.2514/6.2013-307110.2514/6.2013-3071
4. Bilyeu D, Chen YY, Yu STJ (2011) High-order CESE methods for the Euler equations. American Institute of Aeronautics and Astronautics, Orlando, Florida. Aerospace sciences meetings. AIAA, p 298. https://doi.org/10.2514/6.2011-29810.2514/6.2011-298
5. Bilyeu D, Chen YY, Yu ST (2012) A two-dimensional fourth-order CESE method for the Euler equations on triangular unstructured meshes. American Institute of Aeronautics and Astronautics, Nashville, Tennessee. Aerospace sciences meetings. AIAA, p 0293. https://doi.org/10.2514/6.2012-29310.2514/6.2012-293
6. Blazek J (2005) Computational fluid dynamics: principles and applications, 2nd edn. Elsevier, Amsterdam
7. Chang CL (2006) Time-accurate, unstructured mesh Navier-Stokes computations with the CESE method. AIAA, P 4780
8. Chang CL, Venkatachari BS, Cheng G (2013) Time-accurate local time stepping and high-order space time cese methods for multi-dimensional flows using unstructured meshes, American

Institute of Aeronautics and Astronautics, San Diego, CA. Fluid dynamics and co-located conferences. AIAA, P 3069. https://doi.org/10.2514/6.2013-306910.2514/6.2013-3069
9. Chang SC (1995) The method of space-time conservation element and solution element-a new approach for solving the Navier-Stokes and Euler equations. J Comput Phys 119:295
10. Chang SC (2010) A new approach for constructing highly stable high order CESE schemes. American Institute of Aeronautics and Astronautics, Orlando, Florida. Aerospace sciences meetings. AIAA, p 543. https://doi.org/10.2514/6.2010-54310.2514/6.2010-543
11. Chang SC, Himansu A (1997) Application of the space-time conservation element and solution element method to one-dimensional advection-diffusion problems. NASA TM-1997-206307
12. Chang SC, Himansu A, Loh CY, Wang XY, Yu ST, Jorgenson PCE (1997) Robust and simple non-reflecting boundary conditions for the space-time conservation element and solution element method. NASA TM-1997-2077
13. Chang SC, Wang XY, Chow CY (1999) The space-time conservation element and solution element method: a new high-resolution and genuinely multidimensional paradigm for solving conservation laws. J Comput Phys 156:89–136. https://doi.org/10.1006/jcph.1999.6354
14. Chang SC, Himansu A, Loh CY, Wang XY, Yu ST (2003) Robust and simple non-reflecting boundary conditions for the Euler equations-a new approach based on the space-time CE/SE method. NASA TM-2003-212495
15. Feng XS, Hu YQ, Wei FS (2006) Modeling the resistive MHD by the CESE method. Solar Phys 235:235–257. https://doi.org/10.1007/s11207-006-0040-6
16. Feng XS, Zhou YF, Wu ST (2007) A novel numerical implementation for solar wind modeling by the modified conservation element/solution element method. Astrophys J 655:1110
17. Forsyth RJ, Balogh A, Smith EJ, Gosling JT (1997) Ulysses observations of the northward extension of the heliospheric current sheet. Geophys Res Lett 24:3101–3104. https://doi.org/10.1029/97GL03099
18. Girish TE, Prabhakaran Nayar SR (1988) North-south asymmetry in the heliospheric current sheet and the IMF sector structure. Solar Phys 116:369–376. https://doi.org/10.1007/BF00157484
19. Godunov SK (1959) A difference method for numerical calculation of discontinuous solutions of the equations of hydrodynamics. Mat Sb 47(3):271–306
20. Groth CPT, De Zeeuw DL, Gombosi TI, Powell KG (2000) Global three-dimensional MHD simulation of a space weather event: CME formation, interplanetary propagation, and interaction with the magnetosphere. J Geophys Res 105:25053–25078. https://doi.org/10.1029/2000JA900093
21. Harten A, Lax P, Leer B (1983) On upstream differencing and Godunov-type schemes for hyperbolic conservation laws. SIAM Rev 25(1):35–61
22. Hoeksema JT, Wilcox JM, Scherrer PH (1982) Structure of the heliospheric current sheet in the early portion of sunspot cycle 21. J Geophys Res 87:10331–10338, https://doi.org/10.1029/JA087iA12p10331
23. Hoeksema JT, Wilcox JM, Scherrer PH (1983) The structure of the heliospheric current sheet: 1978–1982. J Geophys Res 88:9910–9918. https://doi.org/10.1029/JA088iA12p09910
24. Hu X, Khoo B (2004) Kinetic energy fix for low internal energy flows. J Comput Phys 193(1):243–259. https://doi.org/10.1016/j.jcp.2003.08.007
25. Hu YQ, Feng XS (2006) Numerical study for the bursty nature of spontaneous fast reconnection. Solar Phys 238:329–345. https://doi.org/10.1007/s11207-006-0212-4
26. Hu YQ, Wu ST (1984) A full-implicit-continuous-eulerian (FICE) scheme for multidimensional transient magnetohydrodynamic (MHD) flows. J Comput Phys 55:33–64. https://doi.org/10.1016/0021-9991(84)90014-7
27. Hu YQ, Feng XS, Wu ST, Song WB (2008) Three-dimensional MHD modeling of the global corona throughout solar cycle 23. J Geophys Res 113(A12):A03106
28. Hundhausen AJ (1972) Coronal expansion and solar wind. Springer, New York
29. Leveque RJ (2002) Finite volume methods for hyperbolic problems. Cambridge University Press, Cambridge

30. Li JQ, Wei FS, Feng XS (2001) Numerical simulation of asymmetric corona with multi-streamer structures. Geophys Res Lett 28(7):1359–1362. https://doi.org/10.1029/2000GL012062
31. Linker JA, Mikić Z, Biesecker DA, Forsyth RJ, Gibson SE, Lazarus AJ, Lecinski A, Riley P, Szabo A, Thompson BJ (1999) Magnetohydrodynamic modeling of the solar corona during whole sun month. J Geophys Res 104:9809–9830. https://doi.org/10.1029/1998JA900159
32. Lionello R, Linker JA, Mikić Z (2009) Multispectral emission of the sun during the first whole sun month: magnetohydrodynamic simulations. Astrophys J 690:902–912. https://doi.org/10.1088/0004-637X/690/1/902
33. Loh CY (2003) On a non-reflecting boundary condition for hyperbolic conservation laws. AIAA, p 3975
34. Luhmann JG, Li Y, Arge CN, Gazis PR, Ulrich R (2002) Solar cycle changes in coronal holes and space weather cycles. J Geophys Res (Space Phys) 107:1154. https://doi.org/10.1029/2001JA007550
35. McComas DJ, Bame SJ, Barraclough BL, Feldman WC, Funsten HO, Gosling JT, Riley P, Skoug R, Balogh A, Forsyth R, Goldstein BE, Neugebauer M (1998) Ulysses' return to the slow solar wind. Geophys Res Lett 25:1–4. https://doi.org/10.1029/97GL03444
36. Mikić Z, Linker JA, Schnack DD, Lionello R, Tarditi A (1999) Magnetohydrodynamic modeling of the global solar corona. Phys Plasmas 6:2217–2224. https://doi.org/10.1063/1.873474
37. Mitchell AR, Griffiths DF (1980) The finite difference method in partial differential equations. A Wiley-Interscience Publication, Chichester, Wiley
38. Neugebauer M, Forsyth RJ, Galvin AB, Harvey KL, Hoeksema JT, Lazarus AJ, Lepping RP, Linker JA, Mikic Z, Steinberg JT, von Steiger R, Wang YM, Wimmer-Schweingruber RF (1998) Spatial structure of the solar wind and comparisons with solar data and models. J Geophys Res 103:14587–14600. https://doi.org/10.1029/98JA00798
39. Pletcher RH, Tannehill JC, Anderson DA (2012) Computational fluid mechanics and heat transfer, 3rd edn. CRC Press, London
40. Powell KG, Roe PL, Linde TJ, Gombosi TI, De Zeeuw DL (1999) A solution-adaptive upwind scheme for ideal magnetohydrodynamics. J Comput Phys 154:284–309. https://doi.org/10.1006/jcph.1999.6299
41. Riley P, Linker JA, Mikić Z, Lionello R, Ledvina SA, Luhmann JG (2006) A comparison between global solar magnetohydrodynamic and potential field source surface model results. Astrophys J 653:1510–1516. https://doi.org/10.1086/508565
42. Roe P (1981) Approximate Riemann solvers, parameter vectors, and difference schemes. J Comput Phys 43(2):357–372. https://doi.org/10.1016/0021-9991(81)90128-5
43. Sanderson TR, Appourchaux T, Hoeksema JT, Harvey KL (2003) Observations of the sun's magnetic field during the recent solar maximum. J Geophys Res (Space Phys) 108:1035. https://doi.org/10.1029/2002JA009388
44. Schatten KH, Wilcox JM, Ness NF (1969) A model of interplanetary and coronal magnetic fields. Solar Phys 6:442–455. https://doi.org/10.1007/BF00146478
45. Sheeley NR, Wang YM, Hawley SH, Brueckner GE, Dere KP, Howard RA, Koomen MJ, Korendyke CM, Michels DJ, Paswaters SE, Socker DG, St Cyr OC, Wang D, Lamy PL, Llebaria A, Schwenn R, Simnett GM, Plunkett S, Biesecker DA (1997) Measurements of flow speeds in the corona between 2 and 30 $R_{\odot}$. Astrophys J 484:472–478
46. Shen H, Parsani M (2018) Positivity-preserving CE/SE schemes forsolving thecompressible euler and navierąvstokes equations on hybrid unstructuredmeshes. Comput Phys Commun 232:165–176. https://doi.org/10.1016/j.cpc.2018.05.011
47. Shen H, Wen CY (2016) A characteristic space-time conservation element and solution element method for conservation laws ii. multidimensional extension. J Comput Phys 305:775–792. https://doi.org/10.1016/j.jcp.2015.11.017
48. Shen H, Wen CY, Liu K, Zhang D (2015a) Robust high-order space-time conservative schemes for solving conservation laws on hybrid meshes. J Comput Phys 281:375–402. https://doi.org/10.1016/j.jcp.2014.10.023
49. Shen H, Wen CY, Zhang DL (2015b) A characteristic space-time conservation element and solution element method for conservation laws. J Comput Phys 288:101–118. https://doi.org/10.1016/j.jcp.2015.02.018

50. Shen H, Wen CY, Parsani M, Shu CW (2017) Maximum-principle-satisfying space-time conservation element and solution element scheme applied to compressible multifluids. J Comput Phys 330:668–692. https://doi.org/10.1016/j.jcp.2016.10.036
51. Steinolfson RS, Suess ST, Wu ST (1982) The steady global corona. Astrophys J 255:730–742. https://doi.org/10.1086/159872
52. Svalgaard L, Duvall TL Jr, Scherrer PH (1978) The strength of the Sun's polar fields. Solar Phys 58:225–239. https://doi.org/10.1007/BF00157268
53. Toro EF (2009) Riemann solvers and numerical methods for fluid dynamics. Springer, Berlin
54. Toro EF, Spruce M, Speares W (1994) Restoration of the contact surface in the HLL-Riemann solver. Shock Waves 4(1):25–34. https://doi.org/10.1007/BF01414629
55. Tóth G, van der Holst B, Sokolov IV, De Zeeuw DL, Gombosi TI, Fang F, Manchester WB, Meng X, Najib D, Powell KG, Stout QF, Glocer A, Ma YJ, Opher M (2012) Adaptive numerical algorithms in space weather modeling. J Comput Phys 231:870–903. https://doi.org/10.1016/j.jcp.2011.02.006
56. Usmanov AV (1996) A global 3-D MHD solar wind model with Alfvén waves. In: Winterhalter D, Gosling JT, Habbal SR, Kurth WS, Neugebauer M (eds) American Institute of Physics Conference Series, vol 382, pp 141–144, https://doi.org/10.1063/1.51468
57. Usmanov AV, Goldstein ML (2003) A tilted-dipole MHD model of the solar corona and solar wind. J Geophys Res (Space Physics) 108:1354. https://doi.org/10.1029/2002JA009777
58. Usmanov AV, Goldstein ML, Besser BP, Fritzer JM (2000) A global MHD solar wind model with WKB Alfvén waves: comparison with Ulysses data. J Geophys Res 105:12675–12696. https://doi.org/10.1029/1999JA000233
59. Venkatachari BS, Cheng G, Chang SC (2009) Development of a higher order CESE scheme for transient viscous flows, American Institute of Aeronautics and Astronautics, San Antonio, Texas. Fluid dynamics and co-located conferences. AIAA, p 3984. https://doi.org/10.2514/6.2009-398410.2514/6.2009-3984
60. Wang AH, Wu ST, Suess ST, Poletto G (1993) A two-dimensional MHD global coronal model - Steady-state streamers. Solar Phys 147:55–71. https://doi.org/10.1007/BF00675487
61. Wang AH, Wu ST, Suess ST (1913) Poletto G (1998) Global model of the corona with heat and momentum addition. J Geophys Res 103:1913–1922. https://doi.org/10.1029/97JA01770
62. Wang XY (2015) A Summary of the space-time consevation element and solution element (CESE) method. NASA TM-2015-218743
63. Wang XY, Chang SC (1999) A 2D non-splitting unstructured triangular mesh euler solver based on the space-time conservation element and solution element method. Comput Fluid Dyn J 8(2):309–325
64. Wang YM, Sheeley NR Jr (1992) On potential field models of the solar corona. Astrophys J 392:310–319. https://doi.org/10.1086/171430
65. Wang YM, Sheeley NR Jr, Howard RA, Kraemer JR, Rich NB, Andrews MD, Brueckner GE, Dere KP, Koomen MJ, Korendyke CM, Michels DJ, Moses JD, Paswaters SE, Socker DG, Wang D, Lamy PL, Llebaria A, Vibert D, Schwenn R, Simnett GM (1997) Origin and evolution of coronal streamer structure during the 1996 minimum activity phase. Astrophys J 485:875–889
66. Wang YM, Lean J, Sheeley NR Jr (2000a) The long-term variation of the Sun's open magnetic flux. Geophys Res Lett 27:505–508. https://doi.org/10.1029/1999GL010744
67. Wang YM, Sheeley NR Jr, Rich NB (2000b) Evolution of coronal streamer structure during the rising phase of solar cycle 23. Geophys Res Lett 27:149–152. https://doi.org/10.1029/1999GL010698
68. Wang YM, Sheeley NR, Andrews MD (2002) Polarity reversal of the solar magnetic field during cycle 23. J Geophys Res (Space Physics) 107:1465. https://doi.org/10.1029/2002JA009463
69. Wei FS, Feng XS, Cai HC, Zhou QJ (2003) Global distribution of coronal mass outputs and its relation to solar magnetic field structures. J Geophys Res (Space Physics) 108:1238. https://doi.org/10.1029/2002JA009439
70. Wu ST, Wang AH, Guo WP (1996) Numerical boundary conditions for solar magnetic hydrodynamic flows using the method of characteristics. Astrophys Space Sci 243:149. https://doi.org/10.1007/BF00644045

71. Wu ST, Wang AH, Liu Y, Hoeksema JT (2006) Data-driven magnetohydrodynamic model for active region evolution. Astrophys J 652:800–811. https://doi.org/10.1086/507864
72. Yang Y, Feng XS, Jiang CW (2017) A high-order CESE scheme with a new divergence-free method for MHD numerical simulation. J Comput Phys 349:561–581. https://doi.org/10.1016/j.jcp.2017.08.019
73. Yang Y, Feng XS, Jiang CW (2018a) An upwind CESE scheme for 2D and 3D MHD numerical simulation in general curvilinear coordinates. J Comput Phys 371:850–869. https://doi.org/10.1016/j.jcp.2018.05.014
74. Yang Y, Feng XS, Jiang CW, Manchester WB (2018b) A study of the emergence of flux rope from the solar convection zone into the atmosphere by using a novel numerical method. Phys Plasmas 25(9):090702. https://doi.org/10.1063/1.5044646
75. Yu ST, Chang SC (1997) Treatment of stiff source terms in conservation laws by the method of space-time conservation element/solution element. AIAA paper, P 0435
76. Zhang JH, Wei FS (1993) A time-dependent solution of transonic, trans-Alfvén MHD flow in the solar meridian. Sci China Math 36(1):83. https://doi.org/10.1360/ya1993-36-1-83
77. Zhang MJ, Yu ST (2003) Calculation of the ideal MHD equations by the CESE method without special treatment for the divergence-free constraint of magnetic field. AIAA paper, p 0324
78. Zhang MJ, Lin SC, Yu ST (2002) Application of the CE/SE method to the ideal MHD equations. AIAA paper, p 3888
79. Zhang X, Shu CW (2010) On maximum-principle-satisfying high order schemes for scalar conservation laws. J Comput Phys 229(9):3091–3120. https://doi.org/10.1016/j.jcp.2009.12.030
80. Zhang ZC, Yu ST (1999) Shock capturing without Riemann solver-A modified space-time CE/SE method for conservation laws. AIAA paper, p 0904
81. Zhang ZC, Yu ST, Chang SC, Himansu A, Jorgenson PCE (1999) A modified space-time conservation element and solution element method for Euler and Navier-Stokes equations. AIAA paper, p 3277
82. Zhang ZC, Yu ST, Wang XY, Chang SC, Himansu A, Jorgenson PCE (2000) The CE/SE method for Navier-Stokes equations using unstructured meshes for flow at all speeds. AIAA paper, p 393
83. Zhang ZC, Yu STJ, Chang SC (2002) A space-time conservation element and solution element method for solving the two- and three-dimensional unsteady Euler equations using quadrilateral and hexahedral meshes. J Comput Phys 175:168–199. https://doi.org/10.1006/jcph.2001.6934
84. Zhao XP, Hoeksema JT, Liu Y, Scherrer PH (2006) Success rate of predicting the heliospheric magnetic field polarity with Michelson Doppler Imager (MDI) synoptic charts. J Geophys Res (Space Physics) 111:A10108. https://doi.org/10.1029/2005JA011576

Chapter 5
3D SIP-CESE MHD Model on Six-Component Overset Grid System

Abstract This chapter introduces the implementation of the SIP-CESE MHD model in six component overset grid system with the aim of mitigating the problem of singularity and mesh convergence near the poles. The SIP-CESE MHD model introduced in Chap. 4 is applied to solar wind simulation with this grid system. New numerical features are also added, including: (1) The $\nabla \cdot \mathbf{B}$ constraint error cleaning procedure via an easy-to-use fast multigrid Poisson solver, (2) The Courant-number-insensitive method that reduces the numerical viscosity without generating any instability, (3) The time-integration by multiple time stepping, (4) The time-dependent boundary condition at the subsonic region by limiting the mass flux escaping through the solar surface. In order to produce fast and slow plasma streams of the solar wind, the volumetric heating and momentum source terms are included by considering the topological effect of the magnetic field expansion factor f_s and the minimum angular distance θ_b (at the photosphere) between an open field foot point and its nearest coronal hole boundary. The simulation result for the 3D steady-state background solar wind during Carrington rotation (CR) 1911 from the Sun to Earth is presented. The good agreements between numerical result and observations, such as the Large Angle and Spectrometric Coronagraph (LASCO) aboard the Solar Heliospheric Observatory satellite (SOHO) in the corona and *Wind* observations at 1 AU, demonstrated the efficiency, accuracy and the ability to reasonably produce the structured solar wind of the model.

5.1 Introduction

The spherical shell is a key computational domain in both computational geophysics and solar-terrestrial physics [36, 55, 64, 65, 76, 98, 117, 130–132, 148, 152]. Simulations in these fields such as geodynamo, mantle convection, global circulation of the atmosphere, space weather modeling from the Sun to Earth are all performed in the spherical shell geometry, such as, (r, θ, ϕ) with radius r between some intervals, if defined in the spherical coordinates, for example, $r_0 \le r \le r_1$, $0 \le \theta \le \pi$, and $0 \le \phi \le 2\pi$.

© Springer Nature Singapore Pte Ltd. 2020

X. Feng, *Magnetohydrodynamic Modeling of the Solar Corona and Heliosphere*, Atmosphere, Earth, Ocean & Space,
https://doi.org/10.1007/978-981-13-9081-4_5

However, there are drawbacks for implementing a conventional spherical grid. One drawback is the existence of coordinate singularity at both poles $\theta = 0, \pi$, where the third component derivative in the gradient operator $\nabla = (\nabla_r, \nabla_\theta, \nabla_\phi)$ becomes $\pm\frac{1}{r}\frac{\partial^2}{\partial\theta\partial\phi}$ [64, 65, 75] by using L'Hôpital's rule. Thus, in the polar region, spatial discretizations for the governing equations in spherical coordinates require special attention [75]. Another is the pole convergence problem implied by the fact that the spherical polar grid comprises of grid lines converging towards the north and south poles. In time dependent simulations, this would put a severe restriction on the time step size due to the Courant–Friedrichs–Levy (CFL) condition. In addition, lateral boundary conditions, which have to be imposed at the coordinate symmetry axis ($\theta = 0, \pi$), contaminate the simulations by causing undesired numerical artifacts near the axis.

The problems of the coordinate singularity and the grid convergence can be avoided by using Cartesian coordinates, which can have equal cell spacing everywhere and thus do save computing time on the larger cells. Many modern numerical schemes are usually established in Cartesian coordinates and can be directly used in solar wind modeling without any further modification. For these reasons, Cartesian coordinates (x, y, z) have also been widely used [52, 68, 134], for which numerics are faster, simpler (esp. with respect to multi-dimensional extension), and more stable. This is especially true for an MHD code built within a framework that allows for Cartesian AMR (e.g., [52, 68, 134]).

Evidently, the solar surface, described by a sphere of unit radius located inside the computational domain, can not be consistent with any of the Cartesian coordinate surfaces. Polyhedron-splitting method seems very efficient in generating mesh grid on the spherical surface of unit-sphere. In 3D solar wind simulation, regular octahedron (a Platonic solid composed of eight equilateral triangles, four of which meet at each vertex) and dodecahedron (a Platonic solid composed of 12 regular pentagonal faces, 20 vertices, and 30 edges, with three faces meeting at each vertex) splitting grids can be constructed by starting with a regular octahedron [38, 62] and dodecahedron [94] inscribed inside the unit spherical surface. These methods divide the sphere homogeneously into triangles of the same size with no directional preference such that grid singularities can be avoided and thus the computation stability can be effectively achieved. In this way, by piling up the triangle mesh along the heliocentric distance direction, a 3D grid system can be generated, which is unstructured in (θ, ϕ) directions and structured in radial direction. In the computational domain of spherical shell, the "parallel" axis is selected to be the radial direction. This 3D grid system can be efficiently suitable to the finite volume method-like numerical schemes, such as those successfully used formerly [38, 62, 94]. The grid mesh generated by polyhedron-splitting method is of sphere-surface body-fitting that can enable us to easily implement the inner boundary conditions. But, its obvious disadvantage may be its difficulty of paralleling in "(θ, ϕ)" directions.

The general mesh generation technique [54] seems a powerful tool to fit the spherical shell-shaped computational domain of solar wind modeling from the Sun to Earth. The advantages regarding mesh generation involve fully automated mesh generation based on the triangulated geometry, cell refinement and cell coarsening, automatic solution-adaptive mesh refinement/coarsening at run time, ideally shaped cells in the interior of the computational domain. However, the general mesh generation may seriously deform the cells in body-fitted grids resulting in loss of accuracy and numerical errors due to the quality of the triangulation. Typically, this method is also tedious, and is not able to accurately yield curved 3D boundaries. The grid system from this method may also meet parallel problem in "(θ, ϕ)" directions. Automatic mesh generation, a relatively new field, seems promising, but at present it has no sound criteria for obtaining the high quality mesh distribution.

In order to overcome these problems mentioned above, overset grids [20, 64, 65, 91, 120, 130, 152] may be one of the best choices. Among these overset grids, the Yin-Yang grid has been applied successfully in simulations of geoscience [64, 65, 152]. The cubed sphere [113], which constitutes a spherical surface with six nonoverlapping components, is another overset grid used for the numerical study of the mantle convection, thermal convection and shallow water [19, 21, 58]. Based on the concept of overset grid [20], Henshaw and Schwendeman [57] proposed a numerical solution approach for nonlinear evolutionary partial differential equations in complex 3D domains, where the domains are represented by overlapping structured grids, and block-structured adaptive mesh refinement (AMR) is used to locally increase the spatial resolution. Usmanov [130] also introduced a composite mesh in his 3D solar wind modeling. His composite mesh consists of three overlapping spherical meshes. The first one is the usual spherical mesh with a limited extension in latitude ($42 \le \theta_1 \le 138°, 0 \le \phi_1 \le 360°$). The polar axis of the mesh is directed along the solar rotation axis. Two other meshes are introduced to cover the polar regions in both hemispheres. These meshes are fragments of spherical coordinates ($36 \le \theta_{2,3} \le 144°, 26 \le \phi_2 \le 144°$, and $216 \le \phi_3 \le 324°$) with the polar axis along $(r, 90°, 90°)$ laying in the equatorial plane of the first coordinate system. Motivated by the Yin-Yang grid [64, 65, 152], the cubed sphere grid [113], and the successful application of polyhedron-splitting grid [38, 62, 94], a six-component grid system for solar wind MHD modeling [40, 42, 150] has been proposed for three-dimensional SIP-CESE MHD model. These overset grid system and the evaluation of their performance and efficiency have been detailed in Chap. 3. The preponderance of the overset grid is further enhanced by the performance gain due to the increased size of the CFL time step, the fewer computational zones required to cover the full sphere, and thus less computational time.

The present chapter introduces the construction of the six component grid, the boundary treatment at internal border between two grid components and the implementation of the SIP-CESE MHD model on six-component grid in detail. This chapter is organized as follows. (i) The governing MHD equations with the volumetric heating consideration are described. (ii) The six-component grid system and transformation between two components are introduced. (iii) Numerical features added to SIP-CESE MHD model are introduced, including multiple time stepping method, Courant-number insensitive method and multigrid method for Poisson solver to assure the solenoidal condition (i.e., $\nabla \cdot \mathbf{B}$) to be satisfied. (iv) Initial and boundary conditions are specified. (v) Numerical results for steady-state solar wind structure of CR 1911 are displayed. (vi) Conclusions and discussions are made. It should be noted that the content of this chapter was originally published in Feng et al. [40].

5.2 Model Description

5.2.1 Governing Equations

The solar wind evolution is governed by the modified MHD equations. By splitting the magnetic field into a time-independent potential magnetic field $\mathbf{B}_0$ and a deviation $\mathbf{B}_1$ [100, 125], we can write the MHD system governing solar-interplanetary plasma as

$$
\begin{aligned}
&\frac{\partial \rho}{\partial t} + \nabla \cdot \rho \mathbf{u} = 0\\
&\frac{\partial \rho \mathbf{u}}{\partial t} + \nabla \cdot \left[\rho \mathbf{u}\mathbf{u} + \mathbf{I}\left(p + \frac{1}{2}\mathbf{B}_1^2 + \mathbf{B}_1 \cdot \mathbf{B}_0 \right) - \mathbf{B}_1\mathbf{B}_1 - \mathbf{B}_1\mathbf{B}_0 - \mathbf{B}_0\mathbf{B}_1 \right] =\\
&\mathbf{j}_0 \times \mathbf{B}_0 + \rho \left[\mathbf{g} - \boldsymbol{\Omega} \times (\boldsymbol{\Omega} \times \mathbf{r}) \right] - 2\rho \boldsymbol{\Omega} \times \mathbf{u} + \mathbf{S}_m,\\
&\frac{\partial e_1}{\partial t} + \nabla \cdot \left[\mathbf{u}\left(e_1 + p + \frac{1}{2}\mathbf{B}_1^2 + \mathbf{B}_1 \cdot \mathbf{B}_0 \right) - (\mathbf{u} \cdot \mathbf{B}_1)(\mathbf{B}_1 + \mathbf{B}_0) \right] =\\
&- \mathbf{B}_1 \cdot \frac{\partial \mathbf{B}_0}{\partial t} + \mathbf{E} \cdot \mathbf{j}_0 + \rho \mathbf{u} \cdot [\mathbf{g} - \boldsymbol{\Omega} \times (\boldsymbol{\Omega} \times \mathbf{r})] + Q_e + \mathbf{u} \cdot \mathbf{S}_m,\\
&\frac{\partial \mathbf{B}_1}{\partial t} + \nabla \cdot (\mathbf{u}\mathbf{B} - \mathbf{B}\mathbf{u}) = -\frac{\partial \mathbf{B}_0}{\partial t}
\end{aligned}
\tag{5.1}
$$

where

$$
\mathbf{E} = \mathbf{u} \times \mathbf{B}, \ \ \mathbf{j}_0 = \nabla \times \mathbf{B}_0, \ \text{ and } e_1 = \frac{p}{\gamma - 1} + \frac{1}{2}\rho \mathbf{u}^2 + \frac{1}{2}\mathbf{B}_1^2
$$

the latter of which corresponds to the modified total energy density consisting of the kinetic, thermal, and magnetic energy density (written in terms of $\mathbf{B}_1$). A factor of $1/\sqrt{\mu_0}$ is included in the definition of $\mathbf{B}$, with μ_0 being the magnetic permeability. Since $\mathbf{B}_0$ is constant in time, many terms about $\mathbf{B}_0$ on the right hand side will vanish.

Here, ρ is the mass density, $\mathbf{u} = (u, v, w)$ is velocity vector and its components in x, y, and z directions, p is the thermal pressure, and $\mathbf{B} = \mathbf{B}_0 + \mathbf{B}_1$ denotes the total magnetic field consisting of the time-independent potential magnetic field $\mathbf{B}_0$ and its time-dependent derived part $\mathbf{B}_1$. t and $\mathbf{r}$ are time and position vector originating at the center of the Sun, $\mathbf{g} = -GM_s/r^3 \cdot \mathbf{r}$ is the solar gravitational force, G, M_s, $\boldsymbol{\Omega}$ and γ are the gravity constant, solar mass, solar angular speed and the ratio of specific heats. ρ, $\mathbf{u}$, p, $\mathbf{B}$, $\mathbf{r}$, t, and $\mathbf{g}$ are normalized by the characteristic values ρ_s, a_s, $\rho_s a_s^2$, $\sqrt{\rho_s a_s^2}$, R_s, R_s/a_s, and a_s^2/R_s, where ρ_s and a_s are the density and sound speed at the solar surface. The solar rotation is considered in the study with angular velocity $|\boldsymbol{\Omega}| = 2\pi/27.2753$ radian day^{-1} (here normalized by a_s/R_s). For γ, similar to that of Wu et al. [148], a variable polytropic index is used

$$\gamma = \begin{cases} 1.05, r/R_s \leq 5 \\ 1.05 + 0.03(r/R_s - 5), 5 \leq r/R_s \leq 22 \\ 1.56, \ r/R_s > 22 \end{cases}$$

$\mathbf{S}_m$ and Q_e stand for the momentum and energy source terms, which are responsible for acceleration and heating of the solar wind and will be described later.

The solar wind evolution is calculated in a reference frame of heliographic coordinates corotating with the Sun. For this reference coordinate system (r, θ, ϕ) stands for the position of a point in solar-interplanetary space and (x, y, z) is used to express its corresponding Cartesian coordinates. Sometimes, the analysis of computational results is carried out in the coordinate system at rest to remove these rotational effects.

5.2.2 *Determination of the Momentum Source Term* $\mathbf{S}_m$ *and the Energy Source Term* Q_e

The solar wind heating/acceleration has been debated by scientists for long [5, 6]. Since the creative work by Pneuman and Kopp [106], by empirically describing the source terms for the deposition of energy and/or momentum into the solar wind, many works have been devoted to the numerical study of producing the solar steady state conditions (e.g., [36, 38, 52, 68, 85, 88, 94, 122, 123, 129, 140, 148, 159]).

To mimic the effects of energy absorption above the transition region—thermal conduction, radiative losses, coronal heating and solar wind acceleration—these models usually use the spatial profiles as the source terms for the deposition of the energy or momentum by exponentials in radial distance, such as by ad hoc volumetric heating function of exponential decaying form originated with Hartle and Barnes [53]. Thermal conduction term of Spitzer type $\nabla\left(\xi T^{\frac{5}{2}} \frac{\nabla T \cdot \mathbf{B}}{B^2}\right) \cdot \mathbf{B}$ for collisionless regimes has also been considered in heating source term by some authors (e.g., [33, 72, 77, 122, 140]).

In some models, the energy and momentum exchange between the solar wind plasma and large-scale Alfvén wave [13, 31, 63, 132, 133] has been considered for heating and accelerating the solar wind. For instance, Usmanov et al. [133] and Usmanov and Goldstein [132] used a short-wavelength Wentzel–Kramers–Brillouin (WKB) approximation of the Alfvén wave turbulence to mimic the energy deposit for solar wind heating and acceleration.

The well-known Wang–Sheeley–Arge (WSA) model [2, 3], a relationship between the flux tube expansion factor f_s and satellite observations of the solar wind speed, has been used to drive the solar wind background simulation. The WSA model has been used in the kinematic HAFv.2 model [49, 118] to obtain the solar wind background. Furthermore, Roussev et al. [114] and Cohen et al. [22] used the WSA model as an input to present an improved 3D MHD study for the solar wind, where the processes of turbulent heating in the solar wind are characterized by a varied polytropic index. Using this method, Cohen et al. [22, 23] are able to well reproduce the fast and slow plasma streams of the solar wind.

In order to empirically specify the energy and momentum source term to take into account the magnetic field topology effects, we follow the work of Nakamizo et al. [94] of the expansion factor "f_s" in the volumetric heating function and momentum source term. Thus, the volumetric heating function and momentum source term are

$$Q_e = Q_1 \exp\left(-\frac{r}{L_{Q_1}}\right) + Q_2\left(\frac{r}{R_s} - 1\right)\exp\left(-\frac{r}{L_{Q_2}}\right) + \nabla\left(\xi T^{\frac{5}{2}}\frac{\nabla T\cdot \mathbf{B}}{\mathbf{B}^2}\right)\cdot \mathbf{B},$$
$$\mathbf{S}_m = M\left(\frac{r}{R_s} - 1\right)\exp\left(-\frac{r}{L_M}\right)\mathbf{r}/r \tag{5.2}$$

Here, r is the heliocentric distance, Q_1, Q_2 and L_{Q_1}, L_{Q_2} are the intensity and decay height of heating, and T is the temperature, M and L_M are the intensity and the decay height of the momentum addition. A Spitzer type thermal conduction has been added in the third term of Q_e and $\xi = 5\varepsilon_p = 1.6 \times 10^{-12}\,\mathrm{J\,m^{-1}\,s^{-1}\,K^{-7/2}}$ is choosen according to Endeve et al. [33].

However, different from that of Nakamizo et al. [94], formula (5.2) considers both the topology of the magnetic field expansion factor f_s and the minimum angular separation θ_b between an open field foot point and its nearest coronal hole boundary. Specifically, $Q_2 = Q_0 C_a$, $M = M_0 C_a$, where $C_a = C_a'/\max(C_a')$ with $C_a' = \frac{\left(5.8-1.6e^{[1-(\theta_b/8.5)^3]}\right)^{3.5}}{(1+f_s)^{2/7}}$. Some empirical functions with various free parameters, similar to the above formula C_a', has been employed to derive the solar wind speed near the Sun [3, 101]. Here, the constant values of Q_1, Q_0 and M_0 are $1.5 \times 10^{-9}\,\mathrm{J\,m^{-3}\,s^{-1}}$, $1.18 \times 10^{-7}\,\mathrm{J\,m^{-3}\,s^{-1}}$ and $7.9 \times 10^{-14}\,\mathrm{N\,m^{-3}}$, respectively. L_{Q_1}, L_{Q_2} and L_M are set to be 1 R_s.

The idea is motivated by the observation that there exists a close relationship between the solar wind velocity and the inverse of the expansion factor [71, 142] and the angular distance θ_b distinguishes the high speed solar winds from the low speed solar wind (high speed stream originating from the center of an open field region has large θ_b and low speed stream from the hole boundary has a small θ_b). This idea has been adopted to estimate solar wind terminal speed by empirical relations, such as the WSA model [84] and the Distance from the Coronal Hole Boundary (DCHB) model [111]. f_s and θ_b are usually determined from coronal magnetic field extrapolated by potential field source surface (PFSS) model [62, 79, 157]. The value of f_s is defined to be $f_s = \left(\frac{R_s}{R_{ss}}\right)^2 \frac{B_{R_s}(\theta_s,\phi_s)}{B_{R_{ss}}(\theta_{ss},\phi_{ss})}$ where R_s and $R_{ss} = 2.5\,R_s$ the solar surface radius and the source surface radius, (θ_s, ϕ_s) and (θ_{ss}, ϕ_{ss}) the foot point coordinates of the field line at the solar surface and the source surface, and $B_{R_s}(\theta_s, \phi_s)$ and $B_{R_{ss}}(\theta_{ss}, \phi_{ss})$ the foot point magnetic field strength at the solar surface and the source surface.

Once the distribution of the magnetic field in the solar corona is known, we can trace the magnetic field lines using the following equations:

$$\frac{dr}{ds}=\frac{B_r}{B},\ \frac{rd\theta}{ds}=\frac{B_\theta}{B},\ \frac{r\sin\theta d\phi}{ds}=\frac{B_\phi}{B} \tag{5.3}$$

where ds is a segment along the field line. Starting from the source surface, we use a numerical procedure, such as the Euler method or the Runge–Kutta method, to trace open magnetic field lines from $(R_{ss}, \theta_{ss}, \phi_{ss})$. We find the corresponding foot point of an open field line (R_s, θ_s, ϕ_s) at the solar surface to obtain the flux tube expansion factor f_s and minimum angular separation θ_b between open field foot point and its nearest coronal hole boundary. For convenience, we denote such obtained f_s and θ_b by $f_s(\theta_{ss}, \phi_{ss})$ and $\theta_b(\theta_{ss}, \phi_{ss})$.

In the heating source term we allow f_s and θ_b to have a 3D distribution from the solar surface at 1 R_s to the source surface at R_{ss}. For a given point $O = (r, \theta, \phi)$ in the open field region from the solar surface to the source surface, $f_s(r, \theta, \phi)$ and $\theta_b(r, \theta, \phi)$ are obtained by the following procedures. We first trace the magnetic field line from a grid point O to the source surface to find the foot point coordinate $(\theta_{ss}^O, \phi_{ss}^O)$ of that field line. Then, we calculate $f_s(\theta_{ss}^O, \phi_{ss}^O)$ and $\theta_b(\theta_{ss}^O, \phi_{ss}^O)$ using the method introduced just above. $f_s(r, \theta, \phi)$ and $\theta_b(r, \theta, \phi)$ are set to be equal to $f_s(\theta_{ss}^O, \phi_{ss}^O)$ and $\theta_b(\theta_{ss}^O, \phi_{ss}^O)$. If point O lies in a closed filed region, $f_s(r, \theta, \phi)$ is set to a large value (e.g., 1000) and $\theta_b(r, \theta, \phi)$ is set to zero. Repeating this procedure for each grid point, we will have three dimensional distribution of f_s and θ_b depending on (r, θ, ϕ) in the region $1 \le r \le R_{ss}$ for the heating source terms. Beyond R_{ss}, f_s and θ_b take their distributions on the source surface.

5.3 Gridding System

5.3.1 Six Component Mesh Grid System in the Computational Domain of the Spherical Shell

Here is introduced a composite mesh that consists of six identical component meshes to envelope a spherical surface with partial overlap on their boundaries (Fig. 5.1). Each component grid is a low-latitude spherical mesh, which is defined in the spherical coordinates by

$$\left(\frac{\pi}{4}-\delta\le\theta\le\frac{3\pi}{4}+\delta\right)\cap\left(\frac{3\pi}{4}-\delta\le\phi\le\frac{5\pi}{4}+\delta\right) \tag{5.4}$$

where δ is proportionally dependent on the grid spacing, entailed for minimum overlapping area. Each component is confined by the same region as that in Eq. (5.4) but in different spherical coordinates. We define the red-colored spherical mesh in Fig. 5.1

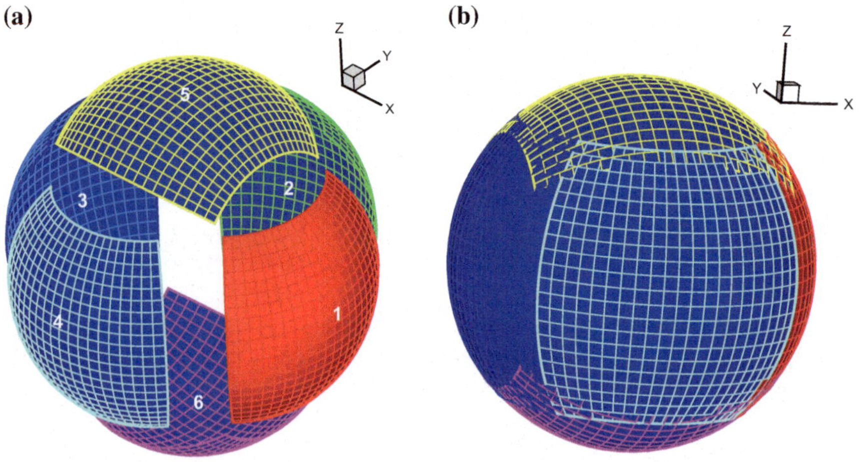

Fig. 5.1 Basic six-component grid: **a** a spherical overset grid consisting of six component grids and **b** a partition of a sphere into six identical components with partial overlaps. Each component grid is a rectangle in the (θ, ϕ) space. Taken from Fig. 1 of Feng et al. [40] with permissions of the authors and the AAS

as the reference heliographic or Cartesian coordinates. The relations between the reference heliographic coordinates C_1 and other coordinates $C_j (j = 2, 3, 4, 5, 6)$ are denoted in the Cartesian coordinates by $(x_2, y_2, z_2) = (y_1, -x_1, z_1)$, $(x_3, y_3, z_3) = (-x_1, -y_1, z_1)$, $(x_4, y_4, z_4) = (-y_1, x_1, z_1)$, $(x_5, y_5, z_5) = (z_1, y_1, -x_1)$, $(x_6, y_6, z_6) = (-z_1, y_1, x_1)$, where (x_1, y_1, z_1) is the Cartesian coordinates of the red-colored part associated with reference Cartesian coordinates; (x_2, y_2, z_2), (x_3, y_3, z_3), (x_4, y_4, z_4), (x_5, y_5, z_5), and (x_6, y_6, z_6) denote those of green-colored, blue-colored, cyan-colored, yellow-colored, and purple-colored parts, respectively, as labeled in Fig. 5.1. From these Cartesian coordinates' transforms, we can easily get their relationships in their associated spherical coordinates. For spatial discretization, we define mesh point on each component $C_\ell (\ell = 1, \ldots, 6)$ as

$$\theta_j^\ell = \theta_{\min} + (j-1)\Delta\theta, j = 0, 1, \ldots, N_\theta + 1$$

$$\phi_k^\ell = \phi_{\min} + (k-1)\Delta\phi, k = 0, 1, \ldots, N_\phi + 1$$

and

$$\Delta\theta = (\theta_{\max} - \theta_{\min})/(N_\theta - 1)$$

$$\Delta\phi = (\phi_{\max} - \phi_{\min})/(N_\phi - 1)$$

where N_θ and N_ϕ are the mesh numbers of the latitude and longitude, respectively. $\theta_{\min} = \frac{\pi}{4}, \theta_{\max} = \frac{3\pi}{4}, \phi_{\min} = \frac{3\pi}{4}, \phi_{\max} = \frac{5\pi}{4}$. For three-dimensional six-component

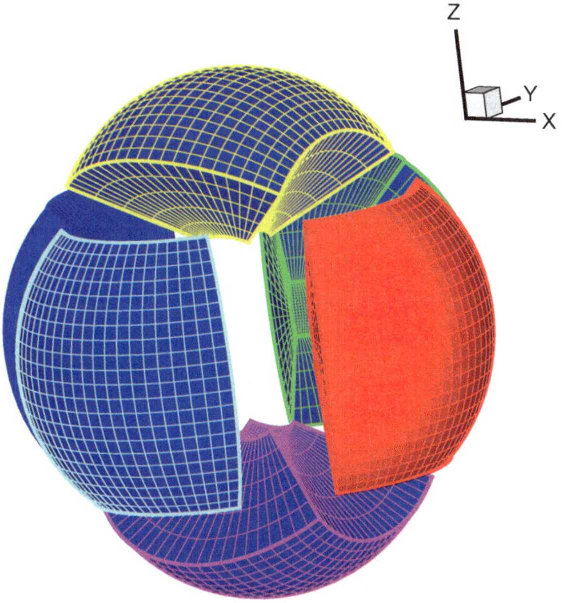

Fig. 5.2 3D six-component grid for spherical shell geometry, by stacking the six-component grids (Fig. 5.1) in r direction. Taken from Fig. 2 of Feng et al. [40] with permissions of the authors and the AAS

mesh, the construction is straightforward. The mesh grid in r-direction as shown in Fig. 5.2 is set up by stacking the basic six component grids shown in Fig. 5.1 in r direction by: $r(1) = 1R_s$, $r(i+1) = r(i) + \Delta r(i)$, where $i = 1, \ldots, N_r$. The grid spacing on the spherical surface is quasi-uniform, and the ratio of the minimum grid spacing over maximum grid spacing is approximately 0.707. For the spherical surface, the overlapping region between component grids is at least 2 grid sizes.

The parallel implementation in the whole computational domain of the simulated region is realized by basing domain decomposition of six component grids on the spherical surface and radial direction partition. In order to enhance the resolution in the "sphere" and radial direction, we devise 6 parts among radial parallel decomposition: 1–3.5 R_s, 3.5–10 R_s, 10–25 R_s, 25–75 R_s, 75–166 R_s, 166–247 R_s with the facility of multiple time-stepping integration (to be given in Sect. 5.4.5 below), and the following grid partitions are employed: for 1–25 R_s, $N_\phi = N_\theta = 2 \times 2^5 - 1$, $\Delta r(i) = 0.01\,R_s$ if $r(i) < 1.1\,R_s$; $\Delta r(i) = \min(A \times \lg(r(i-1)), \Delta\theta \times r(i-1))$ with $A = 0.01/\lg(1.09)$ if $r(i) < 3.5\,R_s$; and $\Delta r(i) = \Delta\theta \times r(i-1)$ if $r(i) > 3.5\,R_s$. For 25–75 R_s, $N_\phi = N_\theta = 3 \times 2^5 - 1$ and $\Delta r(i) = \Delta\theta \times r(i-1)$. For 75–166 R_s, $N_\phi = N_\theta = 2 \times 2^6 - 1$ and $\Delta r(i) = \Delta\theta \times r(i-1)$. For 166–247 R_s, $N_\phi = N_\theta = 3 \times 2^6 - 1$ and $\Delta r(i) = \Delta\theta \times r(i-1)$. Hereafter, such defined subdomains will be called C_ℓ^q, where $\ell, q = 1, \ldots, 6$. The innermost region is set at the solar surface, and the outermost region corresponds to the sphere with its radius 247 R_s. As a result, for 1-25 R_s, the angular resolution is about 1.4° in the latitudinal and longitudinal directions; for 25–75 R_s, the angular resolution is about 0.95° in the latitudinal and longitudinal directions; for 75–166 R_s, the angular resolution is about 0.71° in the latitudinal and longitudinal Directions; for 166–247 R_s, the angular resolution is about 0.48° in the latitudinal and longitudinal directions, and a spatial resolution in the radial direction gradually varies from 0.01 R_s at the inner boundary on the solar surface to 1.8 R_s near 1 AU at 215 R_s.

It is evident that the six-component composite mesh has the following advantages: (1) that the decomposition of the spherical shell owns six identical regions with the same metric and the component grids can be transformed into each other by coordinate transformation such that we can make efficient and concise programs, (2) that each component grid is just a regular low latitude part of latitude-longitude grid such that various program resources in the spherical or Cartesian coordinates grid can be used directly, (3) that the boundary or internal border value in the overlapping area can be easily determined by an interpolation from its neighbor component grids according to the related geometrical positions of component grids, (4) that the parallelization can be efficiently implemented both in radial direction and "(θ, ϕ)" directions, (5) that the six-component composite mesh is of sphere-surface body-fitting that enables us to easily implement the inner boundary conditions at the solar surface prescribed for solar wind modeling. These properties have been shared with by Yin-Yang grid [65] and cubed sphere grid [113].

In code programming, we can directly code the governing equations in each component grid as they are formulated. Meanwhile, we can take the advantages of using various resources available, such as mathematical tools and program libraries in the Cartesian or spherical coordinates. For the parallel computing, it is natural to divide the computational task into six parts, that is, we can realize the parallel in "(θ, ϕ)" directions. Because the component grids are identical, a balance of the computational loads between the processors can be perfectly achieved. When we have 6N processors, we decompose the radial direction of each component grid into N blocks for N processors each. As done before [38], the domain decomposition in radial direction is simple and straightforward in each component grid since it is a rectangular box in the computational space of (r, θ, ϕ).

5.3.2 Vector Transformation Formulae Between Field Vectors Among Six-Component Grids

As Fig. 5.1 indicates, if $\mathbf{u}_j = (u_j, v_j, w_j)$ is used to express a velocity vector defined on the jth part $C_j (j = 1, 2, 3, 4, 5, 6)$ of 6-component composite grid, then we will have the following relations among the adjacent parts $\mathbf{u}_j = (v_i, -u_i, w_i)$, where $(i, j) = (1, 2), (2, 3), (3, 4)$, and $(4, 1)$. $\mathbf{u}_j = (-v_i, u_i, w_i)$, where $(i, j) = (2, 1), (3, 2), (4, 3)$, and $(1, 4)$. $\mathbf{u}_j = (w_i, v_i, -u_i)$, where $(i, j) = (1, 5)$ and $(6, 1)$. $\mathbf{u}_j = (-w_i, v_i, u_i)$, where $(i, j) = (5, 1)$ and $(1, 6)$. $\mathbf{u}_5 = (w_4, -u_4, -v_4)$, $\mathbf{u}_4 = (-v_5, -w_5, u_5)$, $\mathbf{u}_6 = (-w_2, u_2, -v_2)$, $\mathbf{u}_2 = (v_6, -w_6, -u_6)$, $\mathbf{u}_5 = (w_2, u_2, v_2)$, and $\mathbf{u}_2 = (v_5, w_5, u_5)$. $\mathbf{u}_j = (w_i, -v_i, u_i)$ with $(i, j) = (3, 5)$ and $(5, 3)$. $\mathbf{u}_j = (-w_i, -v_i, -u_i)$ with $(i, j) = (3, 6)$ and $(6, 3)$. $\mathbf{u}_6 = (-w_4, -u_4, v_4)$, and $\mathbf{u}_4 = (-v_6, w_6, -u_6)$.

For other five components to component 1, we have the following relations: $\mathbf{u}_1 = (-v_2, u_2, w_2)$, $\mathbf{u}_1 = (-u_3, -v_3, w_3)$, $\mathbf{u}_1 = (v_4, -u_4, w_4)$, $\mathbf{u}_1 = (-w_5, v_5, u_5)$, and $\mathbf{u}_1 = (w_6, v_6, -u_6)$. For vector transformation from component 1 to other five components we know that $\mathbf{u}_2 = (v_1, -u_1, w_1)$, $\mathbf{u}_3 = (-u_1, -v_1, w_1)$, $\mathbf{u}_4 = (-v_1, u_1, w_1)$, $\mathbf{u}_5 = (w_1, v_1, -u_1)$, and $\mathbf{u}_6 = (-w_1, v_1, u_1)$. For magnetic field vector, we have the same formula.

5.4 CESE Solver Modified Space-Time Conservation Element and Solution Element Method

The governing equations (5.1) can be written as:

$$\frac{\partial \mathbf{U}}{\partial t} + \frac{\partial \mathbf{F}}{\partial x} + \frac{\partial \mathbf{G}}{\partial y} + \frac{\partial \mathbf{H}}{\partial z} = \boldsymbol{\eta} \tag{5.5}$$

whose mth ($m = 1, 2, \ldots, 8$) component is

$$\frac{\partial u_m}{\partial t} + \frac{\partial f_m}{\partial x} + \frac{\partial g_m}{\partial y} + \frac{\partial h_m}{\partial z} (\equiv \nabla \cdot \mathbf{q}_m) = \eta_m \tag{5.6}$$

where $m = 1, 2, \ldots, 8$. Here, $\mathbf{q}_m = (f_m, g_m, h_m, u_m)$, u_m, f_m, g_m, h_m and η_m are the mth component of $\mathbf{U}, \mathbf{F}, \mathbf{G}, \mathbf{H}$ and $\boldsymbol{\eta}$, respectively. Let (x, y, z, t) be the coordinates of a four-dimensional Euclidean space E_4 and $\nabla = (\partial_x, \partial_y, \partial_z, \partial_t)$ stands for the usual gradient in four dimensional Euclidean space. By using Gauss' divergence theorem in E_4, it can be shown that Eq. (5.6) is equivalent to the following integral equations:

$$\oint_{S(V)} \mathbf{q}_m \cdot d\mathbf{S} = \int_V \eta_m dV \tag{5.7}$$

Here $S(V)$ is the boundary of an arbitrary space-time region V in E_4, and $\mathbf{q}_m \cdot d\mathbf{S}$ is the space-time flux $\mathbf{q}_m$ leaving the region V through the surface element $d\mathbf{S}$. Equation (5.7) states that the total space-time flux of $\mathbf{q}_m$ leaving V through its boundary $S(V)$ is equal to the integration of the source term η_m over the space-time region V.

Based on the above integral in four dimensional Euclidean space, the space-time conservation element and solution element (CESE) method has been introduced in Chap. 4 for solar wind modeling. Different substantially from other well-established methods, the CESE is not an incremental improvement of a previously existing computational fluid dynamics one. The CESE method's nontraditional characteristics contain a unified treatment of space and time, the introduction of conservation element and solution element, and a novel shock capturing strategy without using Riemann solvers. Note that conservation elements are nonoverlapping space-time subdomains introduced such that (i) the computational domain is the union of these subdomains and (ii) space-time flux conservation can be ensured over each of them and also over the union of any combination of them. In contrast, each solution element is a space-time subdomain where any physical flux vector is approximated using simple smooth functions. Usually, a conservation element does not coincide with a solution element. This method is distinguished by the simplicity of its conceptual basis–a unified treatment of flux conservation in space and time.

It should be noted that one of the most important features of the CESE method is to employ an integral form of space-time flux conservation as the cornerstone for its numerical discretization. Instead, the conventional finite-volume methods like those introduced in Chap. 2 are derived based on Reynolds transport theorem [139], in which space and time are treated separately. By Reynolds transport theorem [139], using Gauss's law in three-dimensional space, the integration form of Eq. (5.6) can be written as

$$\frac{\partial}{\partial t}\int\limits_{V} u_m dV + \oint\limits_{S(V)} (f_m n_x + g_m n_y + h_m n_z) ds = \int\limits_{V} \eta_m dV \tag{5.8}$$

where dV and ds are the volume and surface element of a fixed spatial control volume V, $\mathbf{n} = (n_x, n_y, n_z)$ is a unit vector normal to the surface of the control volume. Alternatively, integrate Eq. (5.8) with respect to time from t_1 to t_2 to arrive at

$$\int\limits_{V} u_m dV \Bigg|_{t_1}^{t_2} = \int_{t_1}^{t_2} \left(-\oint\limits_{S(V)} (f_m n_x + g_m n_y + h_m n_z) ds + \int\limits_{V} \eta_m dV \right) \tag{5.9}$$

As such, the conservation laws state that the rate of change of the total amount of a substance contained in a fixed spatial control volume V is equal to the combination of the following two factors: (i) the flux of that substance across the boundary $S(V)$ of the control volume V, and (ii) the integration of the source term over the fixed spatial domain. The conventional finite volume methods concentrate on the evaluation of the right side of Eq. (5.9). The left hand side of Eq. (5.9) is usually discretized by a finite difference method such as the Runge–Kutta methods. Due to the above two factors, the fractional step (or splitting) methods resort to various measurements of numerically segregating the two factors and treating them in a sequential manner. The numerical fluxes, $(f_m n_x + g_m n_y + h_m n_z)$, at the quadrature point of each face of cell V are determined from the solution of a Riemann problem. Given the left and right interface solution states, $\mathbf{U}_L$ and $\mathbf{U}_R$, an upwind numerical flux is evaluated by solving a Riemann problem in the direction defined by the normal to the face. This is, the finite volume method evaluates the change of the conservative values of all control volumes by integrating their interface or edge fluxes. From the point view of programming it consists of two steps, calculating fluxes at every interface and gathering interface fluxes for every control volume. Therefore, the data structure is designed as having two primary arrays, one for control volumes, another for faces, with a bi-directional reference between control volumes and faces.

As pointed out by Yu and Chang [153], and Zhang et al. [156], this separate treatment of space and time prescribes a restriction on the space-time geometry of finite volumes and, as a result, classical Riemann problems arise naturally in the calculation of flux evaluation across an interface. In contrast, due to its unified treatment of space and time, CESE's flux conservation formulation allows a choice of the space-time geometry of CEs that avoids solving Riemann problems. This fundamental difference between the conventional finite-volume methods and the space-time flux conservation formulation unique to the CESE method has been clarified by Yu and Chang [153], and Zhang et al. [156].

In Chap. 4, we described the implementation of 3D SIP-CESE MHD solar wind model on the pentahedral mesh. Here, SIP-CESE MHD model will be implemented on six component grid system of hexahedral meshes introduced above, by showing how to use the CESE method to calculate $\mathbf{U} = (u_1, u_2, \ldots, u_8)$ and its derivatives at a new time level if these values at previous time step are given, and pointing out other numerical features that help improve the accuracy and efficiency of SIP-CESE MHD model. It should be noted that except for the shape of the mesh, the numerical implementation can almost proceed as what has been done in Chap. 4. However, for the reader's convenience and self-consistency, the complete implementation procedure is presented in some details below.

5.4.1 CEs and SEs in Six-Component Grid System

After the establishment of 3D spatial grid structure, conservation element and solution element are defined in space-time domain. Q (marked by a circle) is the centroid of a typical hexahedron $B_1B_2B_3B_4B_5B_6B_7B_8$ in Fig. 5.3a. Each of the central hexahedron's six neighboring hexahedrons is arbitrarily assigned an identification index

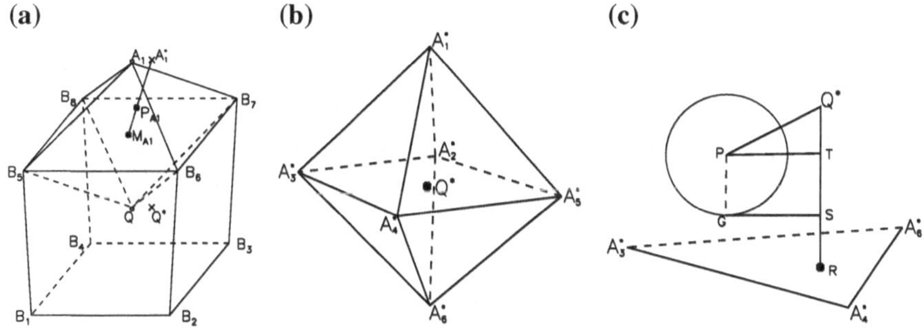

Fig. 5.3 Schematic illustration of CNIS method. Taken from Fig. 3 of Feng et al. [40] with permissions of the authors and the AAS

$\ell = 1, 2, \ldots, 6$. Also, the centroid of the ℓth neighbor will be denoted by A_ℓ. As an example, the central hexahedron $B_1B_2B_3B_4B_5B_6B_7B_8$ and its 1st neighbor is separated by the quadrilateral $B_5B_6B_7B_8$.

Given any $\ell = 1, 2, \ldots, 6$, a triangle is formed by the point A_ℓ and the two end points of any of the edges of the interface that separates the central hexahedron and its ℓth neighbor. Each of the triangles so formed with the same ℓ is arbitrarily assigned an index $k_\ell = 1, 2, 3, 4$, and denoted by $\triangle(k_\ell, \ell)$. As an example, $\triangle A_1B_5B_6$, $\triangle A_1B_6B_7$, $\triangle A_1B_7B_8$, and $\triangle A_1B_8B_5$ depicted by Fig. 5.3a have the same $\ell = 1$. Therefore they may be denoted by $\triangle(1, 1)$, $\triangle(2, 1)$, $\triangle(3, 1)$, and $\triangle(4, 1)$ respectively.

Point Q and the two end points (say points B_5 and B_6) of any of the twelve edges of the central hexahedron form a triangle. Each of twelve triangles so formed is arbitrarily assigned an index $j = 1, 2, \ldots, 12$ and denoted by $\triangle(j)$.

The points A_1, B_1, A_2, B_2, A_3, B_3, A_4, B_4, A_5, B_5, A_6, B_6, B_7 and B_8 form a 24-faced polyhedron $B_1B_2B_3B_4B_5B_6B_7B_8A_1A_2A_3A_4A_5A_6$. The centroid of the 24-faced polyhedron is referred to as the solution point associated with point Q. Note that the above 24-faced polyhedron hereafter is denoted by $V(24)$ and the centroid of $V(24)$ is denoted by Q^* and marked by a cross in Fig. 5.3a.

Given any $\ell = 1, 2, \ldots, 6$, points Q and A_ℓ and the vertices of the quadrilateral interface that separates the central hexahedron and its ℓth neighbor are the vertices of a octahedron. This octahedron hereafter is denoted by $V8(\ell)$. As an example, Q, A_1, B_5, B_6, B_7 and B_8 form a octahedron denoted by $V8(1)$.

In the space-time computational domain, we assume that $t = n\Delta t$ at the nth time level ($n = 0, \frac{1}{2}, 2, \frac{3}{2}, \ldots$), with the marching time step Δt constrained by the Courant–Friedrichs–Lewy (CFL) stability condition. Also, for a given $n > 0$, let Q, Q', and Q'' (not shown), respectively, be the points on the nth, $(n - \frac{1}{2})$th, and $(n + \frac{1}{2})$th time levels with point Q being their common spatial projection. Other space-time mesh points such as (i) Q'^* and Q''^*, (ii) B_k, B'_k, and B''_k, $k = 1, 2, \ldots, 8$; and (iii) A_ℓ, A^*_ℓ, A'_ℓ, and A'^*_ℓ, $\ell = 1, 2, \ldots, 6$, are defined similarly. Because geometric objects in E_4 generally are difficult to visualize, they will be described analytically in the following discussions.

A "plane" in E_4, by definition, is a subspace of E_4 defined by a linear equation i.e.,

$$a_1x + a_2y + a_3z + a_4t = a_0, \ ((a_1)^2 + (a_2)^2 + (a_3)^2 + (a_4)^2 \neq 0) \tag{5.10}$$

where a_k, $k = 0, 1, 2, 3, 4$, are constants. As a result, a hyperplane in E_4 is a three-dimensional subspace. Note that a hyperplane segment, by definition, is a bounded region of a hyperplane. A hyperplane of type I, denoted by $\Gamma(V; t_c)$, is formed by all the points (x, y, z, t) that satisfy the conditions $t = t_c$ and $(x, y, z) \in V$, where t_c is a constant and V denotes a 3D spatial region.

In contrast, a hyperplane segment of type II, denoted by $\Gamma(S; t_-, t_+)$, is formed by all the points (x, y, z, t) that satisfy the conditions $(x, y, z) \in S$ and $t_- \leq t \leq t_+$, where S denotes a spatial plane segment and t_- and $t_+ (t_- < t_+)$ are constants. Note that every point (x, y, z) on the spatial plane segment S satisfies a linear equation of the form

$$c_1 x + c_2 y + c_3 z = c_0, \ ((c_1)^2 + (c_2)^2 + (c_3)^2 \neq 0)$$

where $c_k, k = 0, 1, 2, 3$, are constants. Thus every point (x, y, z, t) on $\Gamma(S; t_-, t_+)$ also satisfies a special form of Eq. (5.10).

In addition to the above two types of hyperplanes, we shall also consider "hypercylinder" in E_4. A hypercylinder, denoted by $\Lambda(V; t_-, t_+)$, is formed by all the points (x, y, z, t) that satisfy the conditions $(x, y, z) \in V$ and $t_- \leq t \leq t_+$, where V is a 3D spatial region and t_- and $t_+ (t_- < t_+)$ are constants.

With the above preliminaries, SE(Q^*), the solution element of point Q^*, i.e., the point that lies on the nth time level and has Q^* as its spatial projection, is defined to be the union of $\Gamma(V(24); t^n)$ and $\Gamma(\Delta(j); t^{n-1/2}, t^{n+1/2})$, $j = 1, 2, \ldots, 12$, and their immediate neighborhoods. Moreover, the six Basic Conservation Elements (BCEs) of point Q, denoted by CE$_\ell(Q)$, $\ell = 1, 2, \ldots, 6$, are defined to be the hypercylinders $\Lambda(V8(\ell); t^{n-1/2}, t^n)$, $\ell = 1, 2, \ldots, 6$, respectively. In addition, the Compound Conservation Element (CCE) of point Q, denoted by CE(Q), is defined to be $\Lambda(V(24); t^{n-1/2}, t^n)$, i.e., the union of the above six BCEs.

We proceed with the following preliminaries:

(a) Let Γ be a hyperplane segment lying within SE(Q^*). Let A be the area of Γ, (x_c, y_c, z_c, t_c) be the coordinates of the centroid of Γ, and $\mathbf{n}$ be a unit vector normal to Γ. Then it can be shown that

$$\int_\Gamma \mathbf{q}_m^* \cdot d\mathbf{S} = \mathbf{q}_m^*(x_c, y_c, z_c, t_c; Q^*) \cdot A\mathbf{n}$$

where $d\mathbf{S} = d\sigma \mathbf{n}$ with $d\sigma$ being the area of a surface element on Γ. The definition of $\mathbf{q}_m^*(x_c, y_c, z_c, t_c; Q^*)$ is given below.

(b) Let V denote the volume of $V(24)$. Because the unit outward normal vector (outward from the interior of CE(Q)) of this face is $(0, 0, 0, 1)$, its surface vector (i.e., the unit outward normal vector multiplied by the area) is $(0, 0, 0, V)$.

(c) Let V^ℓ and (x^ℓ, y^ℓ, z^ℓ), respectively, denote the volume and the spatial coordinates of the centroid of any $V8(\ell)$. Then the surface vector, and the coordinates of the centroid of $\Gamma(V8(\ell); t^{n-1/2})$, respectively, are $(0, 0, 0, -V^\ell)$ and $(x^\ell, y^\ell, z^\ell, t^{n-1/2})$.

(d) Let $S_{k_\ell}^\ell$, $(n_{k_\ell x}^\ell, n_{k_\ell y}^\ell, n_{k_\ell z}^\ell)$, and $(x_{k_\ell}^\ell, y_{k_\ell}^\ell, z_{k_\ell}^\ell)$, respectively, denote the area, the spatial unit outward normal, and the coordinates of the centroid of any $\Delta(k_\ell, \ell)$. Then the surface vector, and the coordinates of the centroid of the side of face $\Gamma(\Delta(k_\ell, \ell); t^{n-1/2}, t^n)$, respectively, are $(\Delta t/2) S_{k_\ell}^\ell (n_{k_\ell x}^\ell, n_{k_\ell y}^\ell, n_{k_\ell z}^\ell, 0)$ and $(x_{k_\ell}^\ell, y_{k_\ell}^\ell, z_{k_\ell}^\ell, t^n - \Delta t/4)$.

By using the information presented in items (a), (b), (c) and (d), the flux of $\mathbf{q}_m^*$ leaving the other faces of CE(Q) can be evaluated in terms of the independent marching variables at points $A_\ell'^*$, $\ell = 1, 2, \ldots, 6$.

5.4.2 Calculation of $(u_m)_{Q^*}$

This section is devoted to the calculation of $(u_m)_{Q^*}$, the numerical approximation of the value of u_m at point Q^*, by solving MHD equations on conservation element and solution element defined above.

Following Chang's original approach [16], inside each SE the flow variables are assumed smooth, and are represented by the first-order Taylor series. For any $(x, y, z, t) \in \mathrm{SE}(Q^*)$, $u_m(x, y, z, t)$, $f_m(x, y, z, t)$, $g_m(x, y, z, t)$ and $h_m(x, y, z, t)$ are approximated by the following formulae for $u_m^*(x, y, z, t; Q^*)$, $f_m^*(x, y, z, t; Q^*)$, $g_m^*(x, y, z, t; Q^*)$ and $h_m^*(x, y, z, t; Q^*)$:

$$u_m^*(x, y, z, t; Q^*) = (u_m)_{Q^*} + (u_{mx})_{Q^*}(x - x_{Q^*}) + (u_{my})_{Q^*}(y - y_{Q^*}) + (u_{mz})_{Q^*}(z - z_{Q^*}) + (u_{mt})_{Q^*}(t - t^n) \tag{5.11}$$

$$\mathcal{F}_m^*(x, y, z, t; Q^*) = (\mathcal{F}_m)_{Q^*} + (\mathcal{F}_{mx})_{Q^*}(x - x_{Q^*}) + (\mathcal{F}_{my})_{Q^*}(y - y_{Q^*}) + (\mathcal{F}_{mz})_{Q^*}(z - z_{Q^*}) + (\mathcal{F}_{mt})_{Q^*}(t - t^n) \tag{5.12}$$

$$u_{m\xi} = \frac{\partial u_m}{\partial \xi}, \mathcal{F}_{m\xi} = \sum_{\ell=1}^{8} \mathcal{F}_{m\ell} u_{\ell\xi}, \mathcal{F}_{m\ell} = \frac{\partial \mathcal{F}_m}{\partial u_\ell}, \tag{5.13}$$

$$\mathbf{q}_m^*(x, y, z, t; Q^*) \equiv (f_m^*(x, y, z, t; Q^*), g_m^*(x, y, z, t; Q^*), h_m^*(x, y, z, t; Q^*), u_m^*(x, y, z, t; Q^*)) \tag{5.14}$$

for $m = 1, 2, \ldots, 8$. Here, $\mathcal{F}_m$ stands for f_m, g_m or h_m, and ξ denotes x, y, z or t. $(\mathcal{F}_{mx})_{Q^*}$ can be considered as the numerical approximation of the value of $\frac{\partial \mathcal{F}_m}{\partial x}$ at point Q^*, and other variables with subscript Q^* are defined similarly. $(\mathcal{F}_{m\ell})_{8\times 8}$ denotes the Jacobian matrix, which is given in Appendix of Chap. 4. $(x_{Q^*}, y_{Q^*}, z_{Q^*}, t^n)$ is the space-time coordinate of point Q^*. Moreover, we assume that, for any $(x, y, z, t) \in \mathrm{SE}(Q^*)$, and any $m = 1, \ldots, 8$,

$$\frac{\partial u_m^*(x, y, z, t; Q^*)}{\partial t} + \frac{\partial f_m^*(x, y, z, t; Q^*)}{\partial x} + \frac{\partial g_m^*(x, y, z, t; Q^*)}{\partial y} + \frac{\partial h_m^*(x, y, z, t; Q^*)}{\partial z} = \eta_m \tag{5.15}$$

Obviously, we have

$$(u_{mt})_{Q^*} = (\eta_m)_{Q^*} - (f_{mx})_{Q^*} - (g_{my})_{Q^*} - (h_{mz})_{Q^*}$$

$$= (\eta_m)_{Q^*} - \sum_{\ell=1}^{8} \left[(f_{m\ell})_{Q^*}(u_{\ell x})_{Q^*} + (g_{m\ell})_{Q^*}(u_{\ell y})_{Q^*} + (h_{m\ell})_{Q^*}(u_{\ell z})_{Q^*} \right] \tag{5.16}$$

Thus, the space-time flux conservation, Eq. (5.7), can be approximated by its discrete counterpart on conservation element:

$$\oint_{S(CE)} \mathbf{q}_m^* \cdot d\mathbf{S} = \int_{CE} \eta_m dV \tag{5.17}$$

i.e., the total flux of $\mathbf{q}_m^*$ leaving CE(Q) through its boundary equals to the integration of the source term over CE(Q). Then, with the aid of the above preliminaries, substituting Eqs. (5.11) and (5.16) into Eq. (5.17), we obtain the following equation

$$(u_m)_{Q^*}^n - \frac{\Delta t}{2}\eta(u_m)_{Q^*}^n = \left(\sum_{\ell=1}^{6} R_m^\ell\right) / V \tag{5.18}$$

which is the discretized form of the governing equations (5.1) or (5.6) for the solar wind model. Here, the marching time step Δt is limited by the Courant–Friedrichs–Lewy (CFL) stability condition and $(u_m)_{Q^*}^n$ stands for the value of $(u_m)_{Q^*}$ at nth time level. Meanwhile,

$$\begin{aligned}
R_m^\ell = {} & V^\ell\left[(u_m)_{A_\ell'^*}^{n-1/2} + (x^\ell - x_{A_\ell'^*})(u_{mx})_{A_\ell'^*}^{n-1/2} + (y^\ell - y_{A_\ell'^*})(u_{my})_{A_\ell'^*}^{n-1/2} + (z^\ell - z_{A_\ell'^*})(u_{mz})_{A_\ell'^*}^{n-1/2}\right] \\
& - \frac{\Delta t}{2}\sum_{k_\ell=1}^{4} S_{k_\ell}^\ell \left\{ n_{k_\ell x}^\ell \left[(f_m)_{A_\ell'^*}^{n-1/2} + (x_{k_\ell}^\ell - x_{A_\ell'^*})(f_{mx})_{A_\ell'^*}^{n-1/2} + \frac{\Delta t}{4}(f_{mt})_{A_\ell'^*}^{n-1/2} \right.\right. \\
& \left.\left. + (y_{k_\ell}^\ell - y_{A_\ell'^*})(f_{my})_{A_\ell'^*}^{n-1/2} + (z_{k_\ell}^\ell - z_{A_\ell'^*})(f_{mz})_{A_\ell'^*}^{n-1/2} \right]\right\} \\
& - \frac{\Delta t}{2}\sum_{k_\ell=1}^{4} S_{k_\ell}^\ell \left\{ n_{k_\ell y}^\ell \left[(g_m)_{A_\ell'^*}^{n-1/2} + (x_{k_\ell}^\ell - x_{A_\ell'^*})(g_{mx})_{A_\ell'^*}^{n-1/2} + \frac{\Delta t}{4}(g_{mt})_{A_\ell'^*}^{n-1/2} \right.\right. \\
& \left.\left. + (y_{k_\ell}^\ell - y_{A_\ell'^*})(g_{my})_{A_\ell'^*}^{n-1/2} + (z_{k_\ell}^\ell - z_{A_\ell'^*})(g_{mz})_{A_\ell'^*}^{n-1/2} \right]\right\} \\
& - \frac{\Delta t}{2}\sum_{k_\ell=1}^{4} S_{k_\ell}^\ell \left\{ n_{k_\ell z}^\ell \left[(h_m)_{A_\ell'^*}^{n-1/2} + (x_{k_\ell}^\ell - x_{A_\ell'^*})(h_{mx})_{A_\ell'^*}^{n-1/2} + \frac{\Delta t}{4}(h_{mt})_{A_\ell'^*}^{n-1/2} \right.\right. \\
& \left.\left. + (y_{k_\ell}^\ell - y_{A_\ell'^*})(h_{my})_{A_\ell'^*}^{n-1/2} + (z_{k_\ell}^\ell - z_{A_\ell'^*})(h_{mz})_{A_\ell'^*}^{n-1/2} \right]\right\}
\end{aligned} \tag{5.19}$$

where $\ell = 1, 2, \ldots, 6$, for flux conservation contributed from six neighboring CEs. Using Eqs. (5.18) and (5.19) given above, it can be seen clearly that, for the current 3D case, the only independent discrete variables associated with the space-time solution point Q^* are $(u_m)_{Q^*}$, $(u_{mx})_{Q^*}$, $(u_{my})_{Q^*}$, $(u_{mz})_{Q^*}$, $m = 1, \ldots, 8$. By solving Eq. (5.18), we can get $(u_m)_{Q^*}$.

At this time, solving the governing equations (5.1) or (5.6) for the solar wind model is equivalent to finding the solutions to (5.18). Due to the source term, Eq. (5.18) is a nonlinear equation for $(u_m)_{Q^*}$. Given the values of the solution variables at the $t^{n-1/2}$ time level, the right hand side of (5.18), i.e., (5.19), can be explicitly calculated. To calculate $(u_m)_{Q^*}$ by solving Eq. (5.18), the Newton's method is used. The initial condition for the iterations is calculated by assuming null source term.

Following SIP-CESE MHD method given by Feng et al. [38], solving the governing equations (5.1) leads us to the discretized integral equation $\boldsymbol{\Phi}(\mathbf{U})$ $\left(\equiv \mathbf{U} - \frac{\Delta t}{2}\boldsymbol{\eta}(\mathbf{U})\right)$ $= \mathbf{U}_H$ and its iteration procedure for the corresponding system of linear equations

$$\frac{\partial \boldsymbol{\Phi}(\mathbf{U})}{\partial \mathbf{U}^{(i)}}\left(\mathbf{U}^{(i+1)} - \mathbf{U}^{(i)}\right) = -(\boldsymbol{\Phi}(\mathbf{U}^{(i)}) - \mathbf{U}_H) \tag{5.20}$$

for the unknown $\mathbf{U}^{(i+1)} - \mathbf{U}^{(i)}$, with the iteration number $i = 0, 1, 2, \ldots$. Here, $\mathbf{U}_H$ is the local homogeneous solution (i.e. the solution for $\boldsymbol{\eta} = 0$ locally). Note that $\mathbf{U}_H$ only depends on the solution at the previous time step, i.e., $\mathbf{U}_H$ is obtained using explicit formulas (5.19). Rather than actually computing the inverse of this Jacobian matrix 8×8 Jacobian matrix $\frac{\partial \boldsymbol{\Phi}(\mathbf{U})}{\partial \mathbf{U}^{(i)}}$ like (17) of Feng et al. [38], one can directly solve (5.20) by "intel MKL" with Fortran library function (https://software.intel.com/en-us/articles/intel-math-kernel-library-documentation/) according to Press et al. [109]. Normally, $\mathbf{U}$ at the previous time step is a good initial guess $\mathbf{U}^{(0)}$, and the procedure takes about 3–4 iterations to convergence. It should be noted that $\frac{\partial \boldsymbol{\Phi}}{\partial \mathbf{U}}$ is calculated in Appendix of Chap. 4.

5.4.3 Time Marching for $(u_{mx})_{Q^*}$, $(u_{my})_{Q^*}$ and $(u_{mz})_{Q^*}$

We proceed to calculate $(u_{mx})_{Q^*}$, $(u_{my})_{Q^*}$ and $(u_{mz})_{Q^*}$ with a central difference type reconstruction approach [38, 141, 154–156]. First, we define a new polyhedron whose vortexes are $A_\ell^o(\ell = 1, 2, \ldots, 6)$. According to the definition of SEs, we can get the approximated u_m at A_ℓ^o by using the Taylor series expansion in space and time. By using these u_m and the solution of $(u_m)_{Q^*}$ at solution point Q^*, then we can calculate spatial derivatives $(u_{mx}^{(k)})_{Q^*}$, $(u_{my}^{(k)})_{Q^*}$ and $(u_{mz}^{(k)})_{Q^*}$, for $k = 1, 2, \ldots, 8$. Finally, by using simple average or re-weighting average, we can obtain the $(u_{mx})_{Q^*}$, $(u_{my})_{Q^*}$ and $(u_{mz})_{Q^*}$ at solution point Q^*. A finite-difference approach similar to that given in Chap. 4 is employed here to evaluate $(u_{mx})_{Q^*}$, $(u_{my})_{Q^*}$ and $(u_{mz})_{Q^*}$.

To be specific, we first perform a spatial translation of the polyhedron $A_1^*A_2^*A_3^*A_4^*A_5^*A_6^*$ so that the centroid of the resulting new polyhedron $A_1^oA_2^oA_3^oA_4^oA_5^oA_6^o$ coincides with Q^*. Let the centroid of the polyhedron $A_1^*A_2^*A_3^*A_4^*A_5^*A_6^*$ and its spatial coordinates be denoted by A^* and $(x_{A^*}, y_{A^*}, z_{A^*})$, respectively, and let $\delta x = x_{Q^*} - x_{A^*}$, $\delta y = y_{Q^*} - y_{A^*}$ and $\delta z = z_{Q^*} - z_{A^*}$. Then $(x_{A_\ell^o}, y_{A_\ell^o}, z_{A_\ell^o})$, the spatial coordinates of A_ℓ^o, $\ell = 1, 2, \ldots, 6$, are given by

$$x_{A_\ell^o} = x_{A_\ell^*} + \delta x, \quad y_{A_\ell^o} = y_{A_\ell^*} + \delta y \text{ and } z_{A_\ell^o} = z_{A_\ell^*} + \delta z.$$

As a preliminary for the following discussions, for $m = 1, 2, \ldots, 8$ and $\ell = 1, 2, \ldots, 6$, let's define

$$(u_m)_{A_\ell^o} \stackrel{\text{def}}{=} u_m^*(x_{A_\ell^o}, y_{A_\ell^o}, z_{A_\ell^o}, t^n; A_\ell'^*),$$

$$\delta u_m^\ell \stackrel{\text{def}}{=} (u_m)_{A_\ell^o} - (u_m)_{Q^*},$$

and

$$\delta x_\ell \stackrel{\text{def}}{=} x_{A_\ell^o} - x_{Q^*}, \quad \delta y_\ell \stackrel{\text{def}}{=} y_{A_\ell^o} - y_{Q^*}, \quad \delta z_\ell \stackrel{\text{def}}{=} z_{A_\ell^o} - z_{Q^*}.$$

Next consider the vertex B_5 depicted in Fig. 5.3a. This vertex is the common vertex of the central hexahedron and three of its neighbors. As an example, let the identification indices ℓ of these three neighbors be 1, 2, and 3. Then, for any $m = 1, 2, \ldots, 8$, consider the four points in the $x - y - z - u_m$ space with the coordinates $(x_{Q^*}, y_{Q^*}, z_{Q^*}, (u_m)_{Q^*})$ and $(x_{A_\ell^o}, y_{A_\ell^o}, z_{A_\ell^o}, (u_m)_{A_\ell^o})$, $\ell = 1, 2, 3$. It can be shown that the values of $\partial u_m/\partial x$, $\partial u_m/\partial y$ and $\partial u_m/\partial z$ on the hyperplane that intercepts the above four points are given by

$$(u_{mx}^{(1)})_{Q^*} \stackrel{\text{def}}{=} \Delta_x/\Delta, \quad (u_{my}^{(1)})_{Q^*} \stackrel{\text{def}}{=} \Delta_y/\Delta, \quad (u_{mz}^{(1)})_{Q^*} \stackrel{\text{def}}{=} \Delta_z/\Delta \ (\Delta \neq 0), \tag{5.21}$$

where

$$\Delta \stackrel{\text{def}}{=} \begin{vmatrix} \delta x_1 & \delta y_1 & \delta z_1 \\ \delta x_2 & \delta y_2 & \delta z_2 \\ \delta x_3 & \delta y_3 & \delta z_3 \end{vmatrix}$$

and

$$\Delta_x \stackrel{\text{def}}{=} \begin{vmatrix} \delta u_m^1 & \delta y_1 & \delta z_1 \\ \delta u_m^2 & \delta y_2 & \delta z_2 \\ \delta u_m^3 & \delta y_3 & \delta z_3 \end{vmatrix}, \quad \Delta_y \stackrel{\text{def}}{=} \begin{vmatrix} \delta x_1 & \delta u_m^1 & \delta z_1 \\ \delta x_2 & \delta u_m^2 & \delta z_2 \\ \delta x_3 & \delta u_m^3 & \delta z_3 \end{vmatrix}, \quad \Delta_z \stackrel{\text{def}}{=} \begin{vmatrix} \delta x_1 & \delta y_1 & \delta u_m^1 \\ \delta x_2 & \delta y_2 & \delta u_m^2 \\ \delta x_3 & \delta y_3 & \delta u_m^3 \end{vmatrix}.$$

Note that $\Delta = 0$ if and only if the spatial projections of A_1^o, A_2^o, A_3^o and Q^* are coplanar and that for each $k = 2, 3, \ldots, 8$, $(u_{mx}^{(k)})_{Q^*}$, $(u_{my}^{(k)})_{Q^*}$, and $(u_{mz}^{(k)})_{Q^*}$ are defined by the above definition procedure except that B_1 is replaced by B_k. In the same procedure, we can obtain $(u_{mx}^{(k)})_{Q^*}$, $(u_{my}^{(k)})_{Q^*}$ and $(u_{mz}^{(k)})_{Q^*}$ for each vertex B_k in Fig. 5.3a ($k = 2, 3, \ldots, 8$).

With the above preliminaries, for each $m = 1, 2, \ldots, 8$, $(u_{mx})_{Q^*}$, $(u_{my})_{Q^*}$, and $(u_{mz})_{Q^*}$ may be evaluated by

$$(u_{mx})_{Q^*} = \frac{1}{8}\sum_{k=1}^{8}(u_{mx}^{(k)})_{Q^*}, \quad (u_{my})_{Q^*} = \frac{1}{8}\sum_{k=1}^{8}(u_{my}^{(k)})_{Q^*}, \quad (u_{mz})_{Q^*} = \frac{1}{8}\sum_{k=1}^{8}(u_{mz}^{(k)})_{Q^*}. \tag{5.22}$$

Alternatively, for a flow with steep gradients or discontinuities, the simple averages in Eq. (5.22) may be replaced by weighted averages, i.e.,

$$(u_{mx})_{Q^*} = \begin{cases} 0, & \text{if } \theta_{mk} = 0, \ k = 1, 2, \ldots, 8, \\ \sum_{k=1}^{8}[(W_m^{(k)})^\alpha (u_{mx}^{(k)})_{Q^*}] / \sum_{k=1}^{8}(W_m^{(k)})^\alpha, & \text{otherwise,} \end{cases} \tag{5.23a}$$

$$(u_{my})_{Q^*} = \begin{cases} 0, & \text{if } \theta_{mk} = 0, \quad k = 1, 2, \ldots, 8, \\ \sum_{k=1}^{8} [(W_m^{(k)})^\alpha (u_{my}^{(k)})_{Q^*}] / \sum_{k=1}^{8} (W_m^{(k)})^\alpha, & \text{otherwise,} \end{cases} \tag{5.23b}$$

and

$$(u_{mz})_{Q^*} = \begin{cases} 0, & \text{if } \theta_{mk} = 0, \quad k = 1, 2, \ldots, 8, \\ \sum_{k=1}^{8} [(W_m^{(k)})^\alpha (u_{mz}^{(k)})_{Q^*}] / \sum_{k=1}^{8} (W_m^{(k)})^\alpha, & \text{otherwise.} \end{cases} \tag{5.23c}$$

Here $\alpha \geq 0$ is an adjustable constant (usually $\alpha = 1$ or $\alpha = 2$),

$$\theta_{mk} \stackrel{\text{def}}{=} \sqrt{[(u_{mx}^{(k)})_{Q^*}]^2 + [(u_{my}^{(k)})_{Q^*}]^2 + [(u_{mz}^{(k)})_{Q^*}]^2}, \tag{5.24}$$

and for each k, $W_m^{(k)}$ is the product of $\theta_{m1}, \theta_{m2}, \ldots, \theta_{m8}$ excluding θ_{mk}. Note that (i) to avoid dividing by zero, in practice a small positive number such as 10^{-60} is added to the denominators that appear in Eqs. (5.23a)–(5.23c); and (ii) Eqs. (5.23a)–(5.23c) reduce to Eq. (5.22) if $\alpha = 0$.

5.4.4 Courant Number Insensitive Method

As demonstrated by Chang and Wang (2003), the local grid Courant numbers (CFL) ν, generally affects the accuracy of the numerical solution computed from the CESE scheme by using a globally fixed time marching step, and a huge disparity in ν across the non-uniform mesh make the solutions highly dissipative in regions where $\nu << 1$. In solar wind simulations, the grid points are accumulated heavily near the Sun due to the spherical shell geometry of computational domain, and the plasma flow and magnetic field vary over many orders of magnitude in the solar-terrestrial space. Such distributions of mesh grids and physical parameters lead to a huge disparity in ν across the mesh. Thus, making the numerical solution insensitive to variation in ν is necessary to get rid of the excessive numerical dissipation caused by small ν and achieve a higher numerical accuracy. To this end, a multidimensional Courant number insensitive scheme (CNIS) for Euler solvers proposed by Chang and Wang [17] is employed to CESE-MHD model by re-designing the procedure for the calculation of the spatial derivatives $\mathbf{U}_{mx}$, $\mathbf{U}_{my}$ and $\mathbf{U}_{mz}$ as follows. As shown in Fig. 5.3a, assume that M_{A1} is the centroid of octahedron $QB_5B_6B_7B_8A_1$, and P_{A1} is a point on line segment $\overline{M_{A1}A_1^*}$ defined by

$$x_{P_{A1}} = x_{M_{A1}} + \nu(x_{A_1^*} - x_{M_{A1}})$$

$$y_{P_{A1}} = y_{M_{A1}} + \nu(y_{A_1^*} - y_{M_{A1}})$$

$$z_{P_{A1}} = z_{M_{A1}} + \nu(z_{A_1^*} - z_{M_{A1}})$$

where ν represents the grid CFL at Q^*. The corresponding other five points P_{A2}, P_{A3}, P_{A4}, P_{A5} and P_{A6} can be similarly given. The values at P_{A1}, P_{A2}, P_{A3}, P_{A4}, P_{A5} and P_{A6} are calculated from the Taylor expansion at A_1^*, A_2^*, A_3^*, A_4^*, A_5^* and A_6^*, respectively. In CNIS, the spatial derivatives $\mathbf{U}_{mx}$, $\mathbf{U}_{my}$ and $\mathbf{U}_{mz}$ at Q^* are obtained with weighted central difference type reconstruction approach by basing the values at P_{A1}, P_{A2}, P_{A3}, P_{A4}, P_{A5} and P_{A6}, instead of the values at A_1^*, A_2^*, A_3^*, A_4^*, A_5^* and A_6^* as done before [38].

From this procedure, it is easily seen that when $\nu \to 0$ (P_A approaches M_A), it will approach the original non-dissipative 'a' scheme as shown by Chang and Wang [17], while $\nu \to 1$ (P_A approaches A^*), it switches to the so-called α scheme as we used before [38]. By the introduction of P_A point, the numerical dissipative generated by small local CFL can be effectively reduced, which is helpful in yielding accurate solutions.

For calculating the local CFL number ν, a simplified 3D grid CFL calculation is used as proposed by Yen et al. [151]. The simplified 3D grid CFL, according to Yen et al. [151], is defined by enforcing the necessary condition of stability, namely, to require the analytical domain of dependence to be bounded by the numerical domain of dependence, associated with a solution point. Here, both domains of dependence are defined as follows: The numerical domain of dependence (Fig. 5.3b) is the octahedron by the neighbor solution points A_1^*, A_2^*, A_3^*, A_4^*, A_5^* and A_6^*. The analytical domain of dependence, as shown in Fig. 5.3c, is a sphere with a radius of $V_f \Delta t/2$ (V_f is the local speed of fast magnetosonic wave at Q^*) and with center at point P that is displaced by $(-u\Delta t/2, -v\Delta t/2, -w\Delta t/2)$ from point Q^*. To satisfy the necessary condition of stability, the sphere, the analytical domain of dependence, has to be bounded by the eight faces of the octahedron, the numerical domain of dependence. Referring to Fig. 5.3c, the stability requirement is to enforce the geometrical relation between the sphere and triangle $A_3^*A_4^*A_6^*$ by $\overline{Q^*T} + \overline{TS} < \overline{Q^*R}$ or $\nu_1 = (\overline{Q^*T} + \overline{TS})/\overline{Q^*R} < 1$, where the line segments Q^*R and PG are perpendicular to the triangle $A_3^*A_4^*A_6^*$, the line segment Q^*T is the projection of Q^*P on Q^*R, and the length of PG equals to the sphere radius $V_f \Delta t/2$, TS is its projection on Q^*R. The corresponding ν_2, ν_3, ν_4, ν_5, ν_6, ν_7 and ν_8 can be similarly obtained by applying the similar geometrical requirement to the other seven faces of the octahedron. The local CFL ν at the solution point Q^* is defined to be the maximum value among ν_1, ν_2, ν_3, ν_4, ν_5, ν_6, ν_7 and ν_8. It should be pointed out that like what had been done to the tetrahedron grid by Yen et al. [151], we calculate the simplified grid CFL for only half time-step marching instead of the formal grid CFL for a full Δt in order not to incur extra mesh management effort. The formal grid CFL for a full Δt involves the influencing cells to a solution point larger than the immediate neighboring cells, and the resulting grid CFL calculation is expensive and complex to apply for large-scale parallel computation application of solar wind modeling with composite six-component grids.

5.4.5 Multiple Time Stepping

The plasma density, the Alfvén velocity, interplanetary magnetic fields, the plasma β and spatial grid size vary over many orders of magnitude from the Sun to Earth. This implies a large variation of the Courant–Friedrichs–Lewy (CFL) stability limit from corona to interplanetary space. Typically, the MHD time step is 1–3 s in the corona by the CFL criterion, and it will be 100–300 s in interplanetary space. If applying one uniform time step in the entire solar-terrestrial domain, time accuracy stability limitations on the time step in fine grid parts of the computational domain decrease the performance of numerical methods and it may become much more difficult to achieve efficient load balancing. The multi-time-stepping method [82, 136] is to take different time steps in different parts of the grid with large spatial grid difference, and it avoids the necessity of taking a single time step in the entire computational domain determined by the numerical CFL stability conditions (such as on the smallest grid cells). The application of the multi-time-stepping method can reduce the Courant–Friedrichs–Lewy (CFL) number disparity that often produces excessive numerical dissipation and saves computation resource.

Here, according to Van der Ven et al. [136] and Chang et al. [18], a multiple time-stepping algorithm is implemented in the radial direction with the following spherical shell partitions: 1–3.5 R_s, 3.5–10 R_s, 10–25 R_s, 25–75 R_s, 75–166 R_s, and 166–247 R_s introduced before. Obviously, each fixed spherical shell partition q consists of six components. The time step $\Delta t_q^{\mathrm{CFL}}$ is the minimum of the usual time steps determined in C_ℓ^q with $\ell = 1, \ldots, 6$ by using the CFL stability condition with the formal Courant number 0.8. Then, the bulk time step Δt_q in subdomain C_ℓ^q for $q > 1$ is further constrained as follows: $\Delta t_{q+1} = M_q \Delta t_q$, $M_q = \mathrm{int}(\Delta t_{q+1}^{\mathrm{CFL}} / \Delta t_q)$ and $q = 1, 2, 3, 4, 5$, as shown in Fig. 5.4. For $q = 1$, $\Delta t_q = \Delta t_q^{\mathrm{CFL}}$.

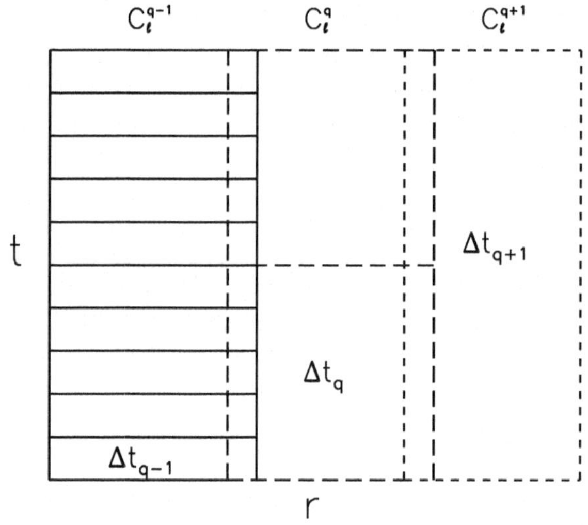

Fig. 5.4 Multiple time-stepping in the radial direction. Taken from Fig. 4 of Feng et al. [40] with permissions of the authors and the AAS

When the flow in C_ℓ^q travels M_q iterations of Δt_q, the flow in C_ℓ^{q+1} advances a single step of Δt_{q+1}. In this way, the flow in C_ℓ^q advances as many time steps of Δt_q as $M_q \times M_{q+1} \times \cdots \times M_5$ for $q \leq 5$ in order to fit in the synchronization time step Δt_s defined as the maximum of these bulk time steps ($\Delta t_s = \Delta t_6$ here). After each synchronization time level the bulk time step for each subdomain C_ℓ^q is determined again.

For the overlapping boundary interface between subdomain C_ℓ^q and subdomain C_ℓ^{q+1}, subdomain C_ℓ^q needs to advance M_q time steps of Δt_q and subdomain C_ℓ^{q+1} does one time step of Δt_{q+1} to arrive at the same time level. Hence, at the same time level the boundary values are obtained by simple cubic spline spatial interpolation, that is, the boundary values of C_ℓ^q are achieved by interpolating the associated C_ℓ^{q+1} values and vice versa. Otherwise, at every time step the qth subdomain's values at the boundary points (in subdomain C_ℓ^{q+1}) are determined by $U_{(q)}(t + m\Delta t_q) = U_{(q+1)}(t) + m\Delta t_q (U_t)_{(q+1)}(t)$, where $m = 1, \ldots, M_q - 1$ and $U_{(q+1)}(t)$, $(U_t)_{(q+1)}(t)$ stand for the values of U and its time derivative in subdomain C_ℓ^{q+1} at time t. However, in order to keep the space-time conservation of fluxes, when the flow in C_ℓ^{q+1} is advanced we need to replace the $(q+1)$th subdomain's flux at the interface between qth and $(q+1)$th subdomain $\text{flux}_{(q+1)}(t)\Delta t_{q+1}$ at its boundary points by $\left(\text{flux}_{(q)}(t) + \text{flux}_{(q)}(t + \Delta t_q) + \cdots + \text{flux}_{(q)}(t + (M_q - 1)\Delta t_q)\right)\Delta t_q$, where $\text{flux}_{(q)}(t + m\Delta t_q)$ is the flux at the interface between qth and $(q+1)$th subdomain in C_ℓ^q at time $t + m\Delta t_q$.

5.4.6 *Multigrid Method for Cleaning Magnetic Field Divergence*

The numerical error, caused by the nonzero divergence of the magnetic field $\nabla \cdot \mathbf{B}$, can influence the numerical stability and accuracy in multidimensional MHD simulations. In order to to keep $\nabla \cdot \mathbf{B}$ to an allowable numerical error, practioners of MHD simulations have devised some efficient methods, such as the constrained transport (CT) or CT/Central Difference (CD) approach [11, 29, 34, 70, 73, 78, 121, 127, 160], the reconstruction method [9, 10, 12, 73, 74, 78], the eight-wave formulation [108], the mixed hyperbolic/parabolic correction procedure with the so-called generalized Lagrange multiplier (GLM) method [30, 37, 48, 67, 68, 87] and the projection method [8, 10, 14, 28, 127, 128]. For detailed descriptions of magnetic field divergence cleaning methods, the reader may refer to Chap. 2.

Here, the projection method is chosen to eliminate $\nabla \cdot \mathbf{B}$, since it does not depends on specific differences and can achieve a better result compared to other methods [10, 127]. In this method, the $\mathbf{B}^*$ field provided by the base scheme in time step $n+1$ is projected to a divergence-free $\mathbf{B}^{n+1}$ field by the following procedure. Since magnetic field can be expressed by the sum of a curl and a gradient $\mathbf{B}^* = \nabla \times \mathbf{A} + \nabla \Phi$, we can arrive at a Poisson equation

$$\nabla^2 \Phi = \nabla \cdot \mathbf{B}^* \tag{5.25}$$

With the help of solution Φ to Eq. (5.25), the aim of the projection method is to correct the magnetic field to

$$\mathbf{B}^{n+1} = \mathbf{B}^* - \nabla \Phi \tag{5.26}$$

so as to force the numerical divergence of $\mathbf{B}^{n+1}$ exactly zero. Obviously, this correction does not affect the current density $J = \nabla \times \mathbf{B}^{n+1} = \nabla \times \mathbf{B}^*$. In order to solve the Poisson equation (5.25), Tóth [127] used a stabilized biconjugate gradient iterative solver, while Tanaka [126] suggested a conjugate residual method. Tóth [127] also showed that its order of accuracy cannot be worser than that of his base scheme or a CT/CD type discretization therein. But, the usual method for the Poisson equation is very time-consuming, especially for large-scale simulation of solar-terrestrial space. Thus, a fast solver for the Poisson equation is very necessary. Multigrid methods [61] are such highly efficient numerical techniques for this purpose, and can reduce low-frequency error components by changing grid size such that the global error disappears as a local error on coarse meshes, which is just suitable to such projection procedure.

The approximate projection operator [110] with the full multigrid (F-cycle) method [61] described below is employed to solve the Poisson equation (5.25). Usually, the approximate projection operator does not depress high frequency modes standing for a local decoupling of the magnetic field as occurred for the decoupling of the velocity field in numerical solutions of incompressible flows [1, 110]. For robust consideration, filtering is necessary, especially for long-term integrations. For this purpose, after carrying out the cleaning procedure through (5.26), it is also advisable to have a diffusive term of the form [28, 30]

$$\frac{\partial(\nabla \cdot \mathbf{B})}{\partial t} = \mu \nabla^2 (\nabla \cdot \mathbf{B})$$

act on the divergence of $\mathbf{B}(\equiv \mathbf{B}^{n+1})$. This may be described to eliminate a spatial derivative by pulling out a divergence operator

$$\frac{\partial \mathbf{B}}{\partial t} = \mu \nabla (\nabla \cdot \mathbf{B}) \tag{5.27}$$

where $\mu = C\frac{(\Delta x)^2}{\Delta t}$, Δx is the minimal grid spacing, Δt is the time step, and C a constant. A finite-difference approximation [28, 90, 110] may be used to achieve a simple, single-step filter. Here, Δt_q is used in μ for C_ℓ^q and constant C is taken to be 0.05. The application of this filter (5.27) is twofold: its decrease of the cell-centered divergence and effective damping of checkerboard modes [28, 30, 90, 110].

Thus, the use of (5.25)–(5.27) will conclude the magnetic field divergence cleaning procedure used. In what follows a F-cycle method is introduced for solving the

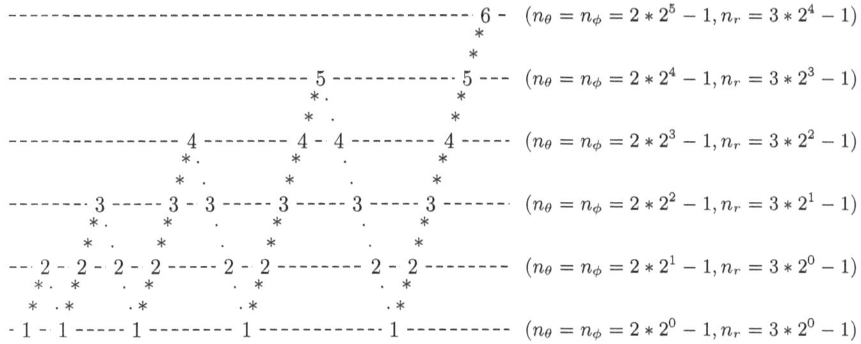

Fig. 5.5 Schematic diagram of the F-cycle iteration and grid levels on subdomain C_ℓ^1. Taken from Fig. 5 of Feng et al. [40] with permissions of the authors and the AAS

Poisson equation (5.25) by multi-cubic interpolation for both restriction and prolongation between grids of different resolutions.

In Fig. 5.5, numbers 1, 2, 3, 4, 5, 6 stand for grid levels from coarse to finest grids for subdomain C_ℓ^1. Level 1 is the coarsest grid and Level 6 is the finest grid that is also the grids employed in this chapter. Similar grid level charts follow for C_ℓ^q. In this figure, star '$*$' means prolongating the solution from coarse grid to fine grid, while dot '$\cdot$' denotes restricting the solution from fine grid to coarse grid. Figure 5.5 indicates one F-cycle by starting from Level 1 through prolongations and restrictions to Level 6: $1 \to 2 \to 1 \to 2 \to 3 \to 2 \to 1 \to 2 \to 3 \cdots \to 5 \to 6$. In each grid level, Gauss–Seidel method is used as smoother to reduce errors. In the program, two F-cycles have been run when physical time is less than 15 h, three F-cycles have been performed from 15 to 25 h, and four F-cycles after 25 h. Along with the time advancing, the error of divergence of magnetic field is accumulated, more F-cycle runs have to be used in order to achieve more accurate solution to Poisson equation and to further reduce the divergence of magnetic field, which will lead to the cost of more computational time.

In order to facilitate the F-cycle iteration in the spherical shell computational domain of solar-terrestrial space, Eq. (5.25) is solved on each part $C_\ell^q(\ell, q = 1, \ldots, 6)$ (introduced in section "gridding system"), conforming to the implementation of multiple time-stepping. In solving Eq. (5.25) by multigrid, we do not parallelize the multigrid method in each C_ℓ^q. If we use multi-processor to deal with the F-cycle method, the size of the meshes in each processor at the coarser levels of the multigrid cycle is quite limited (typically 8 or 16 cells). In this case, most of the CPU time will be then wasted in data exchanges between processors. Solving Eq. (5.25) on each C_ℓ^q will give us a proper time-cost/performance. In the present work, at each new time step in C_ℓ^q, $\nabla \cdot \mathbf{B}^*$ is sent to a certain process (assigned to subdomain C_ℓ^q in parallel implementation of the solar-terrestrial space) to finish the F-cycle iteration, then the obtained Φ through solving Eq. (5.25) is dispatched to every process associated with C_ℓ^q, on which the cleaning procedure (5.26) is done and then the filter (5.27) is used.

It should be mentioned that Neumann boundary condition is used only for the inner boundary at the solar surface, and for all other boundaries Dirichlet boundary condition is employed. Meanwhile, the code for this F-cycle is modified from free source codes at website https://www2.cisl.ucar.edu/resources/legacy/mudpack. In fact, such cleaning procedure is done for the time-dependent part of magnetic field $\mathbf{B}_1$.

Since the modified total energy e_1 is an independent unknown in the CESE method, after the implementation of the projection, the only change may be the magnetic energy, which results in the variation or even the negative occurrence of the thermal energy or specifically the temperature. In the extreme case of plasma $\beta << 1$, the occurrence of negative pressure may also appear. Thus, after the projection step, in order to avoid numerical difficulty with the possible occurrence of negative thermal pressure, the total energy in the simulation is adjusted as

$$e_1^{n+1} = e_1^* + (|\mathbf{B}_1^{n+1}|^2 - |\mathbf{B}_1^*|^2)/2 \tag{5.28}$$

for the sake of numerical robustness, while the thermal energy and temperature obtained by the base CESE scheme are kept unchanged. The same procedure has been stated [11, 29] for their flux-CT algorithm to preserve thermal energy at the expense of energy conservation.

In order to assess the $\nabla \cdot \mathbf{B} = 0$ constraint numerically, as done in Eq. (18) by Feng et al. [38] we define the following error measurement:

$$\text{Error} = \frac{\sum_{\text{k}=1}^{\text{M}} \int\limits_{\text{V}(24)} |(\nabla \cdot \mathbf{B})_{\text{k}}| \text{dV}}{\sum_{\text{k}=1}^{\text{M}} \int\limits_{\text{V}(24)} \text{dV}}$$

where M is the number of mesh nodes in the computational domain and $V(24)$, which is similar to $V(18)$ in Feng et al. [38], is a 24-faced polyhedron involved with mesh grid in the three-dimensional spatial region.

5.5 Boundary Conditions and Initial Conditions

For numerical simulation of solar-interplanetary space problems in the whole spherical shell between 1 R_s and 215 R_s or beyond, three types of boundary conditions are needed to be considered: horizontal boundary conditions at the six components' borders in overset parts, lower boundary conditions at $r = 1\ R_s$ and top boundary conditions at $r = 215\ R_s$ or beyond.

The boundary prescription on the inner or lower boundary surface is a challenge: the lower boundary at solar surface 1 R_s is situated at the subsonic/sub-Alfvénic region, thus the numerical inconsistency will appear unless the proper treatment is provided there. The projected normal characteristic method [93] imposes the boundary treatment satisfying both the underlying MHD equations and the prescribed boundary constraints. By combining the projected normal characteristic method and the solar wind mass flux limit derived by Neugebauer [96] from Ulysses, a modified version of this method has been developed [55, 56]. Such time-dependent inner boundary condition is used here by limiting the mass flux escaping through the solar surface, in order to get the mathematical and physical consistency for sub-Alfvénic fluid and produce a more realistic plasma conditions in the coronal hole and streamer.

Since the top boundary at 215 R_s or beyond is in the supersonic/super-Alfvénic region, the nonreflecting projected normal characteristic boundary condition is used here so as to give more stable simulation [147, 149].

For horizontal (θ, ϕ) boundary or internal border between two components, according to the general overset technique (e.g., [20]), interpolations are applied on the boundary of each component grid to set the boundary values, or internal boundary condition. That is, the horizontal boundary or internal border values of each component grid are determined by interpolation from the neighbor stencils lying in its neighboring component grid, with the interpolation coefficients being derived by the relative position of the boundary point in the stencils. As Fig. 5.6a shows, if a C_5's horizonal boundary (hb) point $Q(\theta^5_{hb}, \phi^5_{hb})$ lies in a neighboring component C_1, we first determine its corresponding coordinates in the C_1's grid, for example, (θ^1_0, ϕ^1_0), then obtain the values at $Q(\theta^1_0, \phi^1_0)$ by means of interpolations of the known values at C_1's grid points neighboring to $Q(\theta^1_0, \phi^1_0)$. To be specific, 16 grid points neighboring to $Q(\theta^1_0, \phi^1_0)$ are used for two-dimensional bi-cubic spline interpolation (Fig. 5.6b). First we obtain the values at the points A, B, C, and D along the θ direction by using one-dimensional cubic spline interpolation, then we interpolate the $Q(\theta^1_0, \phi^1_0)$ values along the ϕ direction with the computed values at A, B, C, and D. For scalar variables, the interpolation is simply direct. However, for vector fields, we

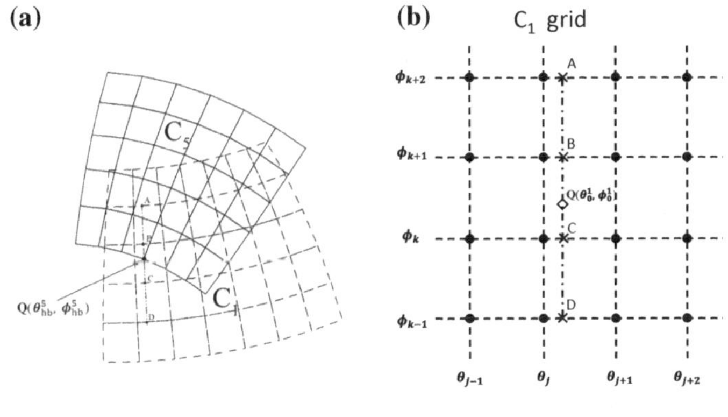

Fig. 5.6 Schematic diagram of **a** the boundary interpolation between C_5 and C_1 components and **b** the two-dimensional cubic spline interpolation on C_1 component. Taken from Fig. 6 of Feng et al. [40] with permissions of the authors and the AAS

need the rotation transformation between different coordinates, because the expressions of a vector in each component grid are different. For instance, transformation $\mathbf{u}(Q(\theta_{hb}^5, \phi_{hb}^5)) = \begin{pmatrix} 0 & 0 & 1 \\ 0 & 1 & 0 \\ -1 & 0 & 0 \end{pmatrix} \mathbf{u}(Q(\theta_0^1, \phi_0^1))$ is used for vector field $\mathbf{u}$. The boundary points in other component grids could be achieved similarly. After the values at these boundary points are known, we can calculate the whole component region by the numerical procedure. As the numerical tests later will show, using this interpolation to communicate data across boundaries does not degrade the numerical accuracy of the whole model.

In order to realistically generate the structured solar wind, we specify the inner initial boundary magnetic field with the observed line of sight photospheric magnetic field data. The observed photospheric magnetic field from the Wilcox Solar Observatory (WSO) at Stanford University is used to deduce a 3D global potential magnetic field as initial magnetic input. Parker's solar wind flow [104] provides the initial distributions of the plasma density ρ, gas pressure p, and the plasma velocity $\mathbf{u}$. Here, the initial solar surface temperature and number density are set to be 1.3×10^6 K and $1.5 \times 10^8\,\mathrm{cm}^{-3}$, respectively. Then, the code is run in time-relaxation manner until a steady-state equilibrium between flow and magnetic fields is achieved by satisfying some error criteria [38].

5.6 Numerical Results for Steady-State Solar Wind Structure of CR 1911

This section presents the 3D numerical results of structured solar wind from the Sun to Earth for CR 1911, which are obtained by executing the techniques introduced in the above sections. The model is run on a 456-core cluster (19 nodes with each node having 24 cores run at 2.4 GHz). With the use of 156 MPI processes for CR 1911, it takes about 40 h of wall time to obtain a steady state solution at the physical time 250 h.

Figure 5.7 shows the magnetic field lines, radial velocity and number density on two different meridional planes at $\phi = 180° - 0°$ and $\phi = 270° - 90°$. Figure 5.8 is the three-dimensional magnetic field distribution in the solar inner corona drawn by solid black lines while the arrowheads denote the direction of the magnetic field lines, and the color contours represent the number density distribution on the solar surface. The final relaxation solution has revealed the three-dimensional asymmetrical streamer resulting in an arcade structure.

From Figs. 5.7 and 5.8, it is seen that a significant dipole component dominates the large-scale coronal magnetic field, which is a typical characteristic of heliospheric magnetic field during periods of solar minimum. In these field configurations, magnetic field lines at high latitudes extend into interplanetary space to form coronal holes of high speed and low density. As a result, the plasma from both polar coro-

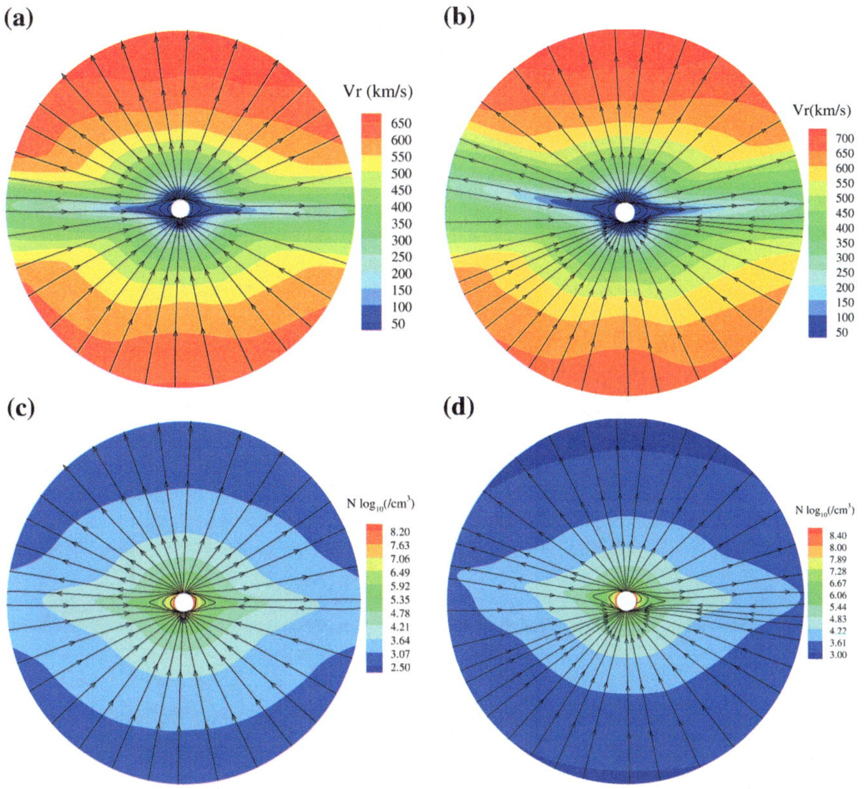

Fig. 5.7 The calculated steady solar coronal solution for computed magnetic fields, radial speed v_r and number density N on the meridional planes at $\phi = 180° - 0°$ (*left panel*) (**a**, **c**) and $\phi = 270° - 90°$ (*right panel*) (**b**, **d**), from 1 to 20 R_s. The color contours stand for the radial speed, number density and streamlines denote the magnetic field lines. Taken from Fig. 7 of Feng et al. [40] with permissions of the authors and the AAS

nal holes occupies almost all of the heliospheric latitudes beyond 5 R_s at which the magnetic field is essentially open everywhere. While, at lower latitudes around the equator, a helmet streamer stretched by the solar wind can be observed within about three solar radii. Above the steamer, we can see a thin current sheet between different magnetic polarities. This scenario of the helmet streamer-current sheet system is consistent with those depicted by Pneuman and Kopp [106] and Gosling et al. [51].

Due to the proper inner boundary treatment of combining the projected normal characteristic method and the mass flux limit, the model can indeed generate reasonable contrasts of the plasma density (also temperature and velocity omitted here) between the coronal hole and streamer as shown by Fig. 5.8. This point has been pointed out by Hayashi [55] and Hayashi et al. [56]. The fixed boundary condition will of course smears the density structure on the solar surface.

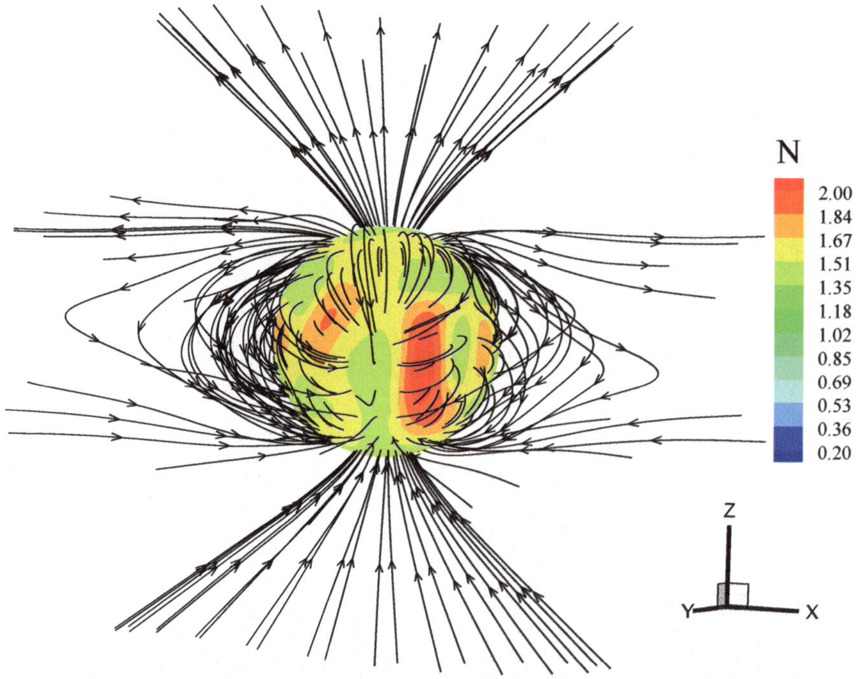

Fig. 5.8 3D representation of the solar coronal magnetic field drawn as solid black lines. The color contours represent the number density (unit: $10^8\,\mathrm{cm}^{-3}$) on the solar surface. Taken from Fig. 8 of Feng et al. [40] with permissions of the authors and the AAS

The coronal polarized brightness measurements are often used to diagnose the coronal density structure. The top panels in Fig. 5.9 present the synthesized polarized brightness images from 2.3 R_s to 6 R_s projected on the meridional planes with $\phi = 180° - 0°$ and $\phi = 270° - 90°$ respectively, which are computed from the 3D MHD simulation results. As it is known, CR 1911 starts on June 28 and ends on July 24, 1996. If the coronal structure is assumed not to change significantly within one CR, which is a reasonable hypothesis during the period of solar minimum, then the coronal polarized brightness images observed on July 5 and June 28 roughly correspond to the left top and right top panels in Fig. 5.9 respectively. Thus, for comparison two LASCO/SOHO observed polarized brightness images are given at the bottom panels of Fig. 5.9. From this figure, it is found that there exists a qualitative agreement between the simulation results and the observations. The most highlighted feature in the right column is the relatively broad latitudinal extent of bright structures in the east limb. This feature results from the projection effect of the high-density regions between $\phi = 230°$ and $\phi = 280°$, which can be seen clearly from the third subfigure in the left column of Fig. 5.10 and from the synoptic polarized brightness images in both the east and the west limbs displayed in Fig. 5.11. These bright structures are also the demonstration of the tilt of the magnetic field neutral line near this region. It

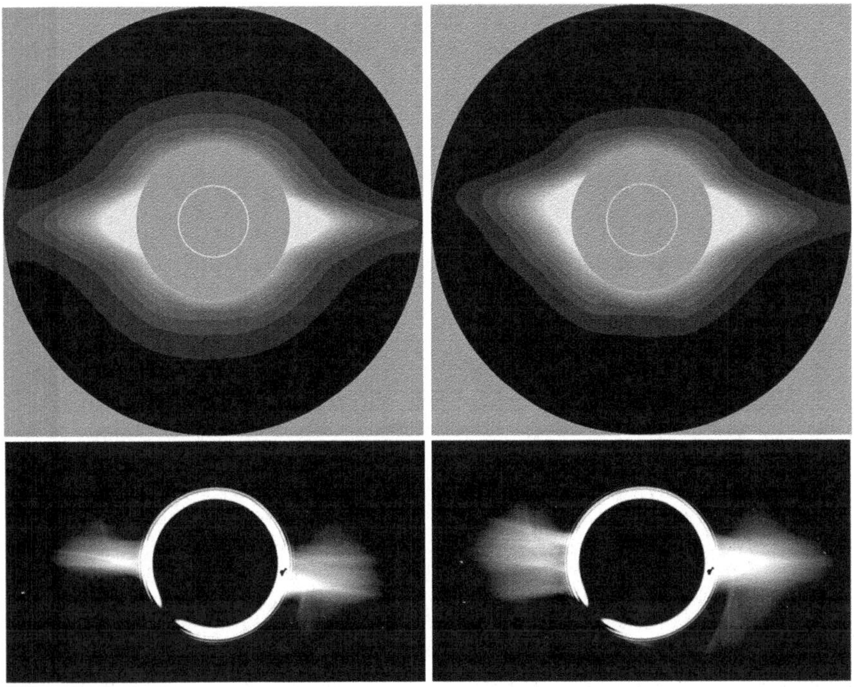

Fig. 5.9 The coronal polarized brightness images from 2.3 R_s to 6 R_s computed from the simulation (top panels) and observed by LASCO C2/ SOHO (bottom panels). The left top and right top panels are the views projected on the meridional planes with $\phi = 180^\circ - 0^\circ$ and $\phi = 270^\circ - 90^\circ$, respectively. The bottom panels are the observations made on 5 July and 28 June, respectively. Taken from Fig. 9 of Feng et al. [40] with permissions of the authors and the AAS

should be noted that there is a small latitudinal difference between the bright structures obtained from simulation and observation. The difference results from the fact that the observations made by SOHO are about 7° north from the solar equator and the synthesized images are viewed from 1 AU in the solar equator. The other reason for this difference is due to the imperfect extrapolation of the polar photospheric magnetic field [2]. In addition, the distributions of high-density streamers in Fig. 5.7 are consistent with the observation of LASCO C2 when considering the little difference mentioned above. In general, the simulation can well reproduce the observed coronal density structure.

Figure 5.10 displays the background solar wind synoptic maps for the MHD steady state solution on the different solar surfaces at 2.5 R_s, 20 R_s and 215 R_s. This configuration is due to the interaction between the magnetic field and solar wind flow field. Quantitatively, N and T decrease with helocentric distance while v_r increases. The flow in the polar regions is faster than that near the equatorial regions, and the heliospheric current sheet (HCS) region is surrounded by higher N. In the top panel of Fig. 5.10, the neutral lines for the MHD solution show significant magnetic shear,

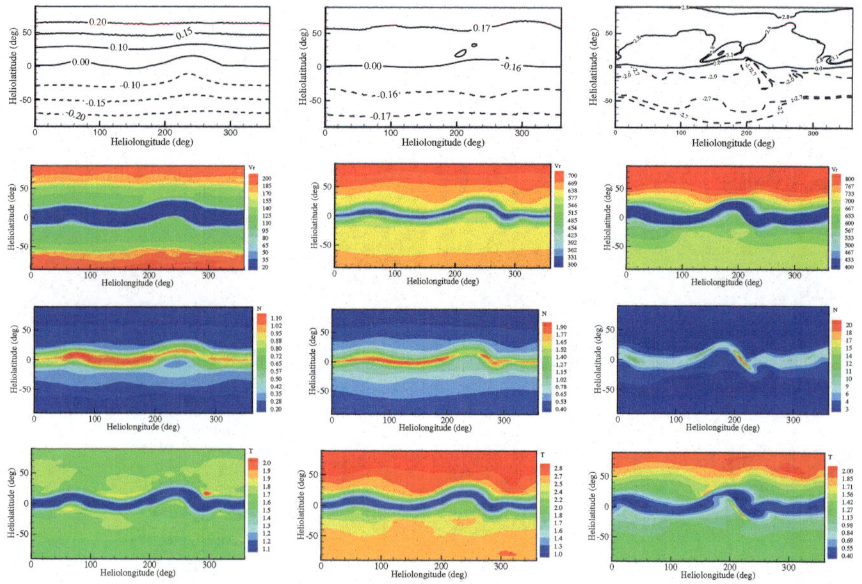

Fig. 5.10 Contours of MHD steady state solution on the different surfaces at 2.5 R_s (*left column*), 20 R_s (*middle column*) and 215 R_s (*right column*). The top panel denotes radial magnetic field B_r with unit: Gauss, 10^{-6} T, and nT from left to right; the second panel is the radial speedy v_r with unit: km s^{-1}; the third panel displays number density N with unit: 10^6 cm^{-3}, 10^4 cm^{-3} and cm^{-3} from left to right; the bottom panel stands for temperature T with unit: 10^6 K, 10^6 K and 10^5 K from left to right. Taken from Fig. 10 of Feng et al. [40] with permissions of the authors and the AAS

and the magnetic field strength falls off along with heliocentric distance, but much less than that of the potential field. The strong warped structure in the streamer belt, located between 230° and 280° longitudes and encompassed by low plasma pressure region with both low N and T, is spatially coincident with the extensions of the coronal hole boundaries (seen θ_b distribution below in Fig. 5.12a). By examining the radial magnetic field, velocity, number density and temperature distributions at 2.5 R_s and 20 R_s, it is recognized that the speed and density patterns around the HCS and in the polar regions are in agreement with the synoptic maps of LASCO C2 in Fig. 5.11. Also, the values of the number density and radial speed at 2.5 R_s and 20 R_s, respectively, are of the same magnitude as those shown in Figs. 7 and 8 of Feng et al. [38], Fig. 5 of Hu et al. [62], Figs. 1 and 2 of Wei et al. [146]. Additionally, the speeds at 2.5 R_s and 20 R_s are supported by the results derived from the observations by LASCO C2 and C3 [107, 144], by interplanetary scintillation [15] and by the Ulysses solar corona experiment [105].

From the distributions of plasmas and magnetic field at 215 R_s in the right column of Fig. 5.10, it is seen that the bimodal structure of the solar wind (a combination of uniform, unipolar, and tenuous fast wind, with relatively dense slow wind for the solar minimum period) and even the width of the slow wind band and the sharpness

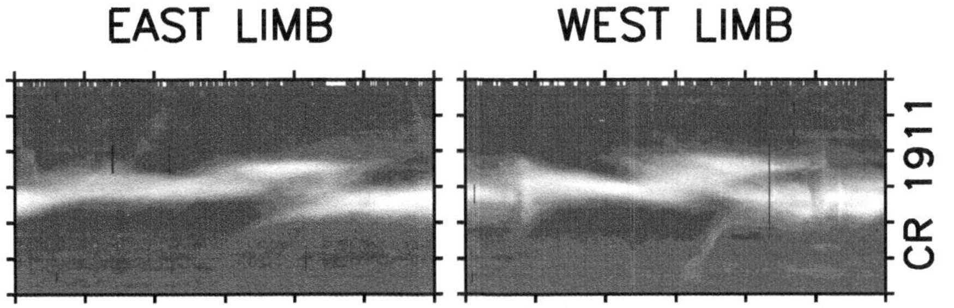

Fig. 5.11 Synoptic maps of the white-light polarized brightness at the east (*left*) and west (*right*) limbs from LASCO C2 images. Taken from Fig. 11 of Feng et al. [40] with permissions of the authors and the AAS

of velocity transitions are well reproduced in the model. The absence of a significant gradient of B_r with respect to heliolatitude in fast flows is also observed in this simulation.

By comparison with the four panels at 215 R_s in the right column of Fig. 5.10, the fast solar speed can be associated with the presence of an interaction region (high temperature, high density and intensive magnetic field). And the interaction region is approximately aligned with the Parker spiral direction. From the warped structure of the HCS between 230° and 280° at 2.5 R_s and beyond, the features of corotating interaction regions (CIR) can be seen obviously. At the southern flank of low-speed stream around the HCS, there is a compression region characteristic of steep gradients of density, temperature and radial magnetic field, which corresponds to the fast solar wind overtaking the slow solar wind. While, at the northern flank, there is a rarefaction region and the physical quantities vary smoothly. The rarefaction region is associated with the low-speed stream following the high-speed stream. The different features of the two regions can be seen more clearly from Fig. 5.16. In addition, it can also be seen that the maximum latitudinal width of slow wind flow region decreases approximately from 45° at 2.5 R_s to 25° at 215 R_s. This behavior is a consequence of magnetic pressure gradients.

Figures 5.12a, b compare the derived coronal hole boundaries mapped on the photosphere and the expansion factors "f_s" at 2.5 R_s with $\phi = 180°$ versus θ for the MHD model (solid line) and for the PFSS model (dash line). Following Linker et al. [76] and Riley et al. [112], by identifying the closed-field regions and the open-field regions through tracing magnetic field lines on the lower boundary, a map of the coronal hole boundaries obtained by the MHD model and by the PFSS model is presented in Fig. 5.12a. It is found that the coronal hole boundaries produced by the MHD model and the source-surface model are similar in this calculation. While slight differences exist, the two techniques have achieved the same qualitative characteristics for CR 1911. In particular, the polar coronal holes have the same topological configuration such as overall shape, span and approximately the same area. In Fig. 5.12b, the heliolatitude changes of expansion factors "f_s" at 2.5 R_s with $\phi = 180°$ for the MHD model (solid line) and for the PFSS model (dash line)

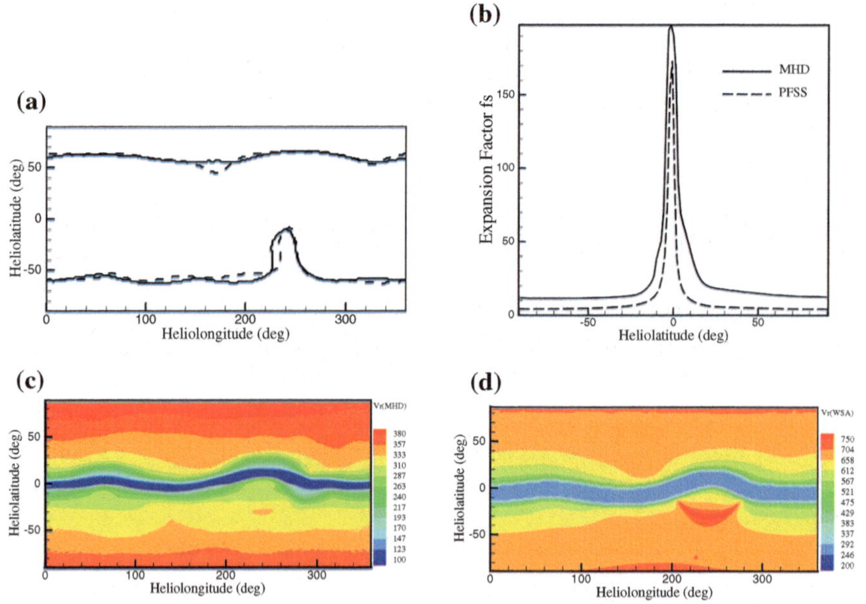

Fig. 5.12 **a** The derived boundary of coronal holes mapped on the photosphere and **b** the expansion factors "f_s" at 2.5 R_s with $\phi = 180°$ from MHD model (solid line) and PFSS model (dash line). The bottom panel displays the distributions of the calculated radial velocity v_r (unit: km s^{-1}) at 5 R_s from the steady state result by MHD model (**c**) and Wang–Sheeley relationship by PFSS model (**d**). Taken from Fig. 12 of Feng et al. [40] with permissions of the authors and the AAS

bear the same patterns, while the expansion factor from the MHD model is higher than that from the PFSS model. Figures 5.12c, d also display the distribution of the radial velocity v_r at 5 R_s from the MHD model and the WSA model, the latter of which is calculated by $v_r = 265 + \frac{1.5}{(1+f_s)^{2/7}}(5.8 - 1.6e^{[1-(\theta_b/7.5)^3]})^{3.5}$ with the help of θ_b and the expansion factor f_s [101]. By examining this figure, we observe that the expansion factor of the MHD model is higher than that of the PFSS model. This indicates that the MHD model will give a large expansion of magnetic field and thus present a relatively low speed, which is further confirmed by the radial speed variation in the bottom panel of Fig. 5.12. This can be interpreted physically in terms of the pressure excised by the plasma that further spreads out the magnetic field in the MHD simulation.

Figure 5.13 displays two cases of the heating coefficient dependence on f_s only and a combination of θ_b and f_s. Figures 5.13a, b are the distributions of the corresponding heating coefficients C_a and C_b. They are associated with just two cases of heating coefficients with θ_b and without θ_b if the exponential decaying factor is neglected from Q_e and $\mathbf{S}_m$. Figures 5.13c, d show the MHD steady radial speeds when the heating coefficients C_a and C_b are used in the volumetric heating source terms. From this figure, it is easily seen that θ_b plays an important role in the heating process. C_a acts in a much wider open field region than C_b, and C_a is larger than C_b. Correspondingly,

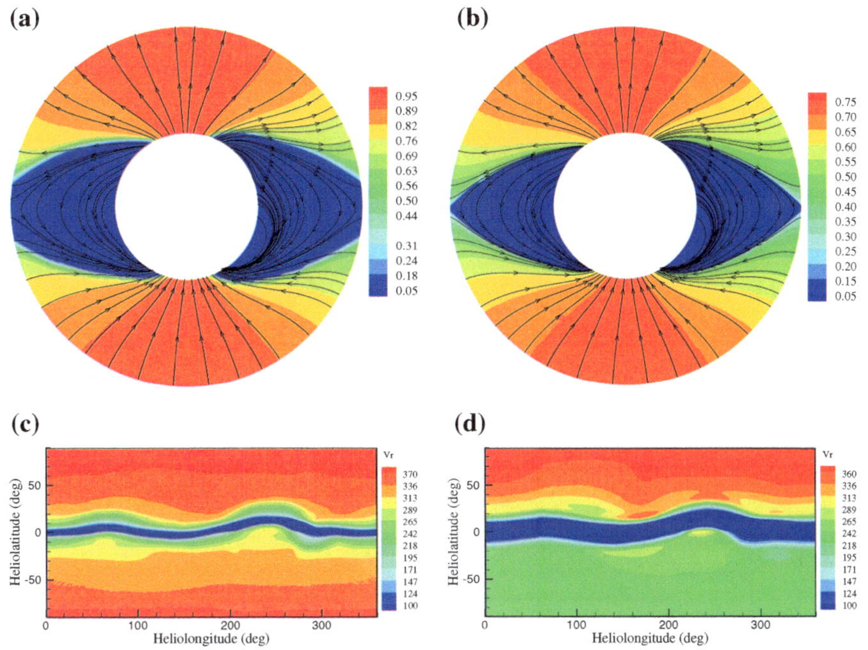

Fig. 5.13 The meridional distribution at $\phi = 180° - 0°$ of heating coefficient **a** C_a and **b** $C_b = \frac{1}{(1+f_s)^{2/7}}$. The distributions of the calculated MHD steady radial speeds v_r (unit: km s^{-1}) on the surface at 5 R_s with the heating coefficients respectively given by C_a (c) and C_b (d). Taken from Fig. 13 of Feng et al. [40] with permissions of the authors and the AAS

the distributions of the MHD steady radial speeds v_r on the surface at 5 R_s with the heating coefficients given by C_a and C_b have some differences such as a narrower low speed region and a higher speed if θ_b is considered.

Figure 5.14 shows radial variations in the flow speed v_r and number density N from 1 R_s to 20 R_s versus heliocentric distance at $\phi = 0°$ with $\theta = 60°$ by the solid line and $\theta = 2°$ by the dashed line, which correspond to the open field region and closed field region. Evidently, there exist a high speed and low density stream in the polar region ($\theta = 60°$, $\phi = 0°$). At 10 R_s and above, the flow is super-Alfvénic and supersonic nearly everywhere and changes very little. However, a low-speed and high-density flow is present in the current sheet region ($\theta = 2°$, $\phi = 0°$) and the velocity agrees with the profile of low solar wind speed [116]. The overall distributions of velocity and density are similar to those obtained by Feng et al. [38], but the fast stream here is much higher due to the addition of the heating and acceleration source terms.

It is clear from Fig. 5.14 that heating has much effect on the plasma distribution for an open field, but has little effect for a closed region. That is, by setting the energy and momentum intensity Q_2, M related to the expansion factor f_s and the minimum angular distance θ_b, the heating and acceleration mostly act on the coronal hole and

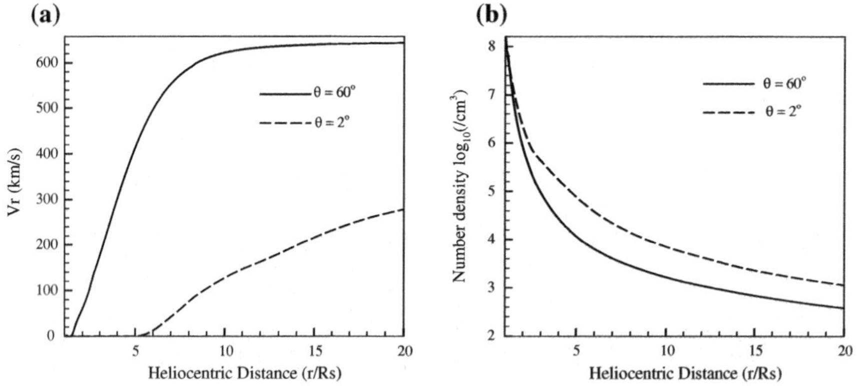

Fig. 5.14 (a) Radial velocity profiles along heliocentric distance at two locations with different latitudes $\theta = 60°$ and $\theta = 2°$ at the same longitude $\phi = 0°$ and (b) the corresponding number density distributions. Taken from Fig. 14 of Feng et al. [40] with permissions of the authors and the AAS

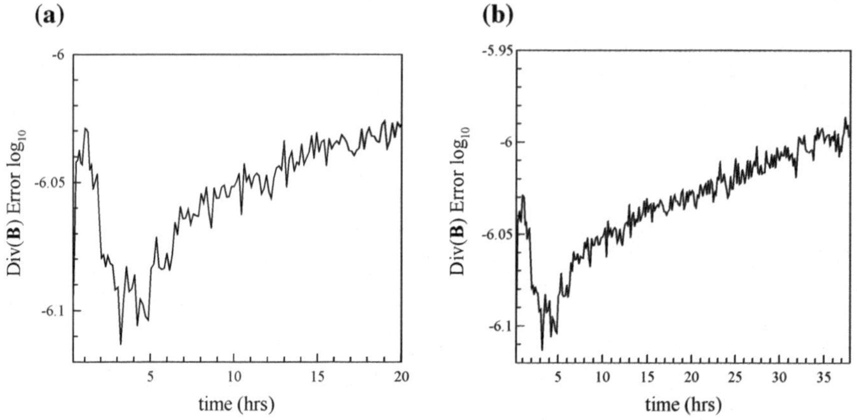

Fig. 5.15 Temporal evolution of Error for $\nabla \cdot \mathbf{B}$ during the calculation. Taken from Fig. 15 of Feng et al. [40] with permissions of the authors and the AAS

exert little influence on the closed region of the streamer, where the expansion factor goes to infinity.

Figure 5.15 for temporal evolution of Error for $\nabla \cdot \mathbf{B}$ defined in Sect. 5.4.6 is plotted in the region from 1 to 26 R_s, since the magnetic filed in this region is strong. From the evolution of $|\nabla \cdot \mathbf{B}|$ as shown in Fig. 5.15, the use of multigrid method can keep the divergence free condition in an error much smaller than that without any special treatment [38]. The estimation of Error for $|\nabla \cdot \mathbf{B}|$ is of about 10^{-6} and continues to be the same even after a long running time, and no obvious large error accumulation from $|\nabla \cdot \mathbf{B}|$ appears. In the run, the Error remains around 10^{-6} until the steady state is reached, although large-time marching sees a little increase as

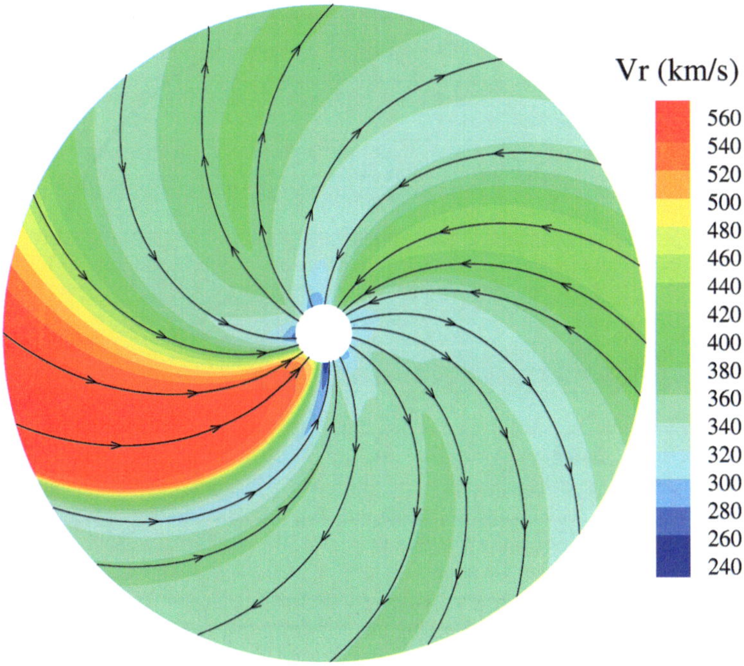

Fig. 5.16 The calculated MHD steady inner heliosphere solution in the solar equatorial plane from 20 R_s to 215 R_s. The color contours represent the radial solar wind speed, and streamlines denote the magnetic field lines. Taken from Fig. 16 of Feng et al. [40] with permissions of the authors and the AAS

shown by Fig. 5.15b. More F-cycles can greatly reduce this Error but cost much CPU time.

Figure 5.16 shows the interplanetary magnetic field (IMF) lines of the steady state together with the distribution of the radial solar wind speed in the solar equatorial plane. The arrowheads denote the direction of the magnetic field line. The solar wind stretches the IMF outward into Archimedean spirals due to the solar rotation and the IMF freezing-in effect. The spirals coil more tightly in slow speed flow than in high speed flow. For instance, calculated from Fig. 5.16, the angle between the radial direction and the IMF at 1 AU is 44° when the solar wind speed is 320 km s^{-1} and is 38° when the solar wind speed is 560 km s^{-1}. The conclusion that the IMF lines coil more tightly in the slow stream has been also pointed out by Odstrcil and Pizzo [97].

Figure 5.17 gives the two-state nature of a polar diagram of the computed solar wind speed at 1 AU as a function of heliolatitude. Flow speed profiles are displayed in the meridional plane ($\phi = 180° - 0°$). From Fig. 5.17, it can be clearly seen a transition from one solar wind region to another at the midlatitudes and the width of transition is to some extent determined by the thickness of the HCS. This simulation result is quite consistent with Ulysses speed observation when normalized to 1 AU [52, 83, 133].

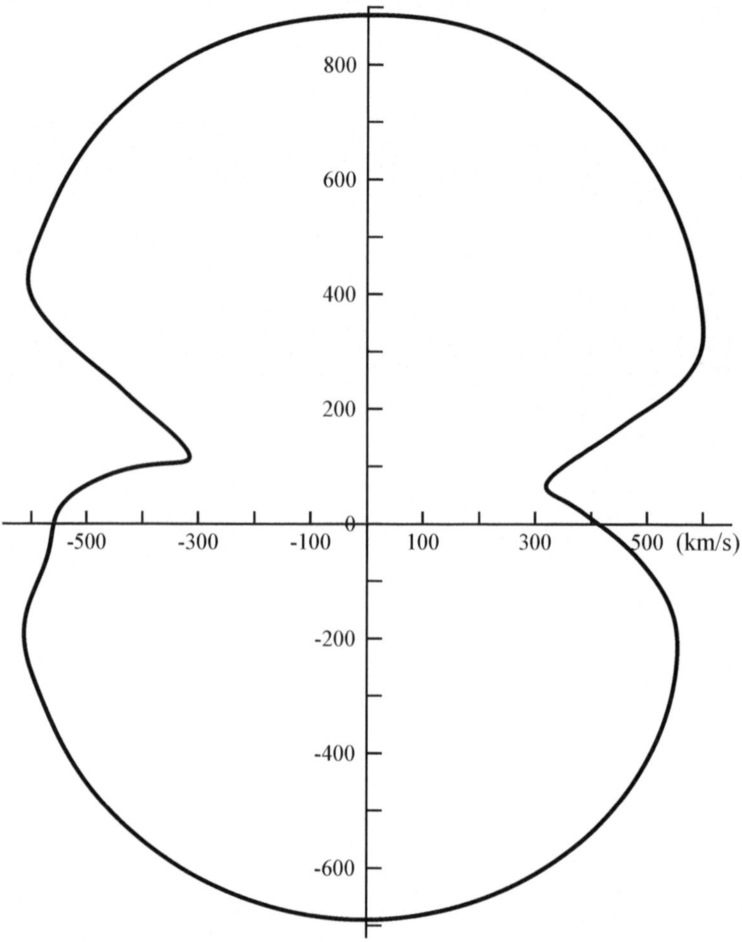

Fig. 5.17 The polar plot of computed solar wind speed in the meridional plane ($\phi = 180° - 0°$) at 215 R_s (1 AU). Taken from Fig. 17 of Feng et al. [40] with permissions of the authors and the AAS

Figure 5.18 exhibits an analogous comparison with the 1-h averaged WIND plasma data. WIND was located in the ecliptic plane near 1 AU ($212 \sim 213\ R_s$) and observed a more complex pattern of variations than those only partially reproduced by the MHD solution. In Fig. 5.18, there is an overall good agreement between the simulations and the observations. The simulation well reproduces the observed relative high-speed wind part ($>500\,\mathrm{km\,s^{-1}}$), together with their structure variations. From these comparisons, we can see that CIR-associated features, such as observed enhancements of density, temperature and magnetic field magnitude in front of the high-speed wind or at the leading edge of the high-speed wind, are well reproduced. The rate of density enhancement and the slope of the rising solar wind speed in the simulation are closer to those of the 1-h averaged data. However, some differences

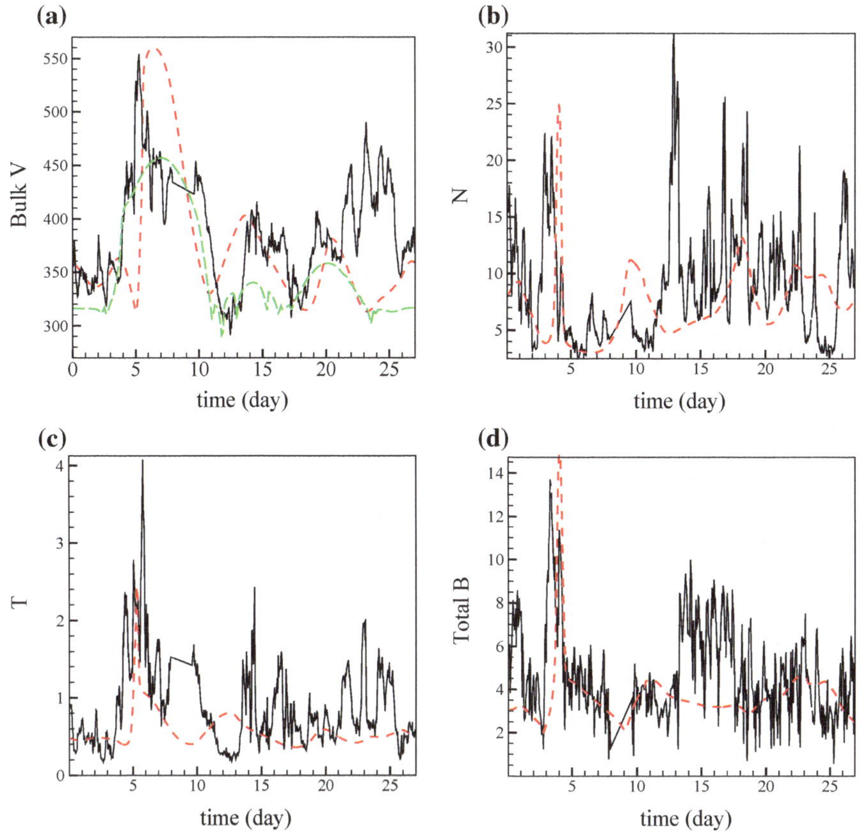

Fig. 5.18 Comparisons of the MHD results (dashed red line), the predicted speed from Wang–Sheeley model (dashed green line) and the observational data (solid black line) obtained by the WIND satellite: (a) bulk flow speed V (unit: km s^{-1}), (b) number density N (unit: cm^{-3}), (c) temperature T (unit: 10^5 K) and (d) total magnetic field magnitude B (unit: nT). Here data from *Wind* starts from June 30, 1996. Taken from Fig. 18 of Feng et al. [40] with permissions of the authors and the AAS

exist between the simulation results and the observational data. One difference is the arrival time of the fast wind. In the simulation, the fast wind is shifted late about 1–2 days. The second discrepancy is the structure of the solar wind speed in the last few days. The discrepancies in the last few days partially come from the uncertainties of the photospheric magnetic field measurements, the influence of differential rotation of the photosphere, and the inconsistency of the photospheric magnetic field at the beginning and ending days of a Carrington rotation due to its evolution.

By the way, the solar wind speed obtained by the MHD simulation is also closer to the observation than that derived by the Wang–Sheeley relationship $v_r = 267.5 + 410.0/f_s^{0.4}$ [2]. However, the advantage of MHD model over the WSA model is that the former can provide all the physical parameters everywhere in the computational domain, while the latter is used to predict the background solar wind speed and the interplanetary magnetic field (IMF) polarity at 1 AU (http://legacy-www.swpc.noaa.gov/ws/).

5.7 Conclusions and Discussions

The 3D SIP-CESE MHD Model [38, 39, 62, 158, 159] has been greatly improved from the consideration of the following aspects: grid system, CNIS method, time integration, magnetic field divergence cleaning procedure and volumetric heating. For grid system, here is introduced a composite mesh that consists of six identical component meshes to cover a spherical surface with partial overlap on their boundaries. This grid system can avoid mesh convergence and singularities at the pole regions. Meanwhile, the grid system allows easy-parallelling not only in "(θ, ϕ)" directions but also in radial direction. CNIS method can reduce high numerical dissipation in regions with small CFL numbers and thus keep the accuracy of the solution. In time integration, multiple time stepping method is used by dividing solar-terrestrial space into six subdomains. This method enhances the convergence stability and speeds up the calculation by easing CFL number disparity. In order to fix the magnetic field divergence error caused by numerical scheme, the F-cycle iteration can maintain the $\nabla \cdot \mathbf{B}$ error to an acceptable level of 10^{-6}, and it has a proper time-cost/performance about 40% CPU time in time step. For the acceleration of solar wind, volumetric heating source terms are considered by using a 3D distribution profile based on the expansion factor f_s and the angular distance θ_b. This heating is weakened at the location where the magnetic field is in over-radial expansion ($f_s > 1.0$), and also it is strengthened at the location where the magnetic field in under-radial expansion ($f_s < 1.0$). The use of angular distance θ_b can effectively distinguish the high speed solar wind from the low speed solar wind (high speed wind emanating from the center of a coronal hole has large θ_b and low speed wind from the hole boundary has a small θ_b).

Overall, CESE-MHD model can produce all the physical parameters everywhere within the computation domain. This may imply that the solar surface-global heliospheric structure connection can be predicted by the simulation up to a generally acceptable level, although there remain many unsatisfactory points as discussed here. As can be seen, CESE-MHD model validations for some other CRs show that the numerical results for the CRs during solar minimum are usually better than those for the CRs during solar maximum, when compared with observations in the solar corona and at 1 AU. This may be due to the fact that the initial input of magnetic field based on the potential field model can not take into account of the observed heliospheric open flux from active regions and CMEs by way of interchange reconnection frequently occurring at solar maximum [22]. Differences occur due to many factors such as the uncertainties in photospheric magnetic measurements (especially in polar regions), the imperfectness of the potential magnetic field approximation, the occurrence of some coronal mass ejections (CMEs) during CR 1911 (June 28–July

25, 1996) (http://cdaw.gsfc.nasa.gov/CME_list/UNIVERSAL/1996_07/univ1996_07.html), the shortage of more sound physical-based as well as observational data supported foundation of heating mechanism, and the neglect of interaction between solar wind and interplanetary discontinuities.

Although the heating consideration in this chapter gives favorable numerical results, volumetric heating alone cannot be the only acceleration process acting on the solar wind and other sources presently unclear are needed to act within the region between the lower corona and the source surface. Further characterizing and quantifying the key physical processes/mechanism will clarify an operational route to more physically integrate realistic coronal heating modules into 3D MHD codes. In solar wind modeling, coronal heating and the acceleration of the ambient solar wind is a challenging that modelers have been facing with. Although significant progress has been seen in the last few decades' study on coronal heating and the acceleration of the ambient solar wind (e.g., [27, 102, 103]), there does not exist a complete understanding of the relevant physical processes, nor does there exist a quantitatively precise census of which coronal structures contribute to specific types of solar wind. About solar wind acceleration, there still exist some fundamental questions to which answers are left open: What are the physical processes that dominate coronal heating? Where does most of the slow wind come from on the Sun? Do fast and slow wind streams originate from two separate mechanisms? The progress of these problems can be referred to the relevant studies and reviews (e.g., [4, 24, 26, 27, 60, 69, 161]). Most of the proposed physical models can be categorized into two broad paradigms [27], the wave/turbulence-driven (WTD) models and the reconnection/loop-opening (RLO) models. The former assumes that the convection-driven motions of the flux-tube footpoints generates wave-like fluctuations traveling up into the extended corona and the waves partially reflect back down toward the Sun, evolve into strong MHD turbulence, and dissipate gradually. These models can generate realistic fast and slow wind conditions with the wave amplitudes estimated in the corona and heliosphere [25, 59, 80, 81, 124, 137, 138, 143, 145]. The latter assumes the solar wind flux tubes are influenced by magnetic reconnection with continuously emerging closed-field regions. In these models, the mass, momentum, and energy of the solar wind are input from loops of varying properties in the low corona [7, 45, 46, 115]. In addition, many interplanetary physical processes, such as turbulence, stream-stream interactions, and Coulomb collisions, challenge us on tracing a fluid parcel at 1 AU back to the coronal source region.

In terms of 3D MHD solar wind modeling, modelers are not only confronted with the difficulty of no final solution to the problem of solar coronal heating (SCH) and the solar wind acceleration (SWA), but also with the challenge of how to translate these proposed theoretical models of SCH and SWA into the proper formulae employable in 3D MHD solar wind simulations. The challenge is so severer that most of the current 3D MHD solar wind models adopt the specially-designed volumetric heating and acceleration terms to achieve the observed solar corona and solar wind structures (e.g., [22, 32, 35, 40, 41, 43, 44, 52, 76, 88, 95, 112, 114, 133], such as those introduced in Sect. 5.2.2. On one hand, under current circumstances, such choice is reasonable to generate solar wind solution that compares well with observation. On the other hand, these volumetric heating and acceleration terms often contain non-self-consistent coronal heating physics or include several free parameters to be determined. Although the developed Alfvén Wave Solar Model (AWSoM) [86, 119, 135] extended the Alfvén wave turbulence description with a 3D solar corona/solar wind model that self-consistently incorporates low frequency Alfvén wave turbulence, it is still at a relatively early state of development and much work remains to be done, as pointed out by Gombosi et al. [50]. Mikić et al. [89] used a new heating model based on the dissipation of Alfvén waves, and a new energization mechanism to twist the magnetic field in filament channels, and predicted the corona for the 21 August 2017 total solar eclipse. The modeled results compared favourably with observations of the eclipse in white light and extreme ultraviolet, however the discrepancies between the model and observations are considered to arise from limitations in our ability to observe the Sun's magnetic field.

With a combination of in-situ and remote-sensing instruments and the highly elliptical inner-heliosphere operational orbit, Solar Orbiter, a European Space Agency (ESA) mission to explore the Sun in unprecedented detail, will provide critical remote-sensing and in-situ measurements to address the central question of heliophysics how the Sun creates and controls the heliosphere [92, 99]. The Parker Solar Probe launched in August 2018 will use Venus gravitational assists to gradually move the perihelion of the orbit from 35 solar radii (R_s) down to less than 10 R_s from the center of the Sun within about six years [47, 66]. The Parker Solar Probe will explore the atmosphere of the Sun and pass from the solar wind into the solar corona for the first time. Thus it will make unexpected discoveries, and likely improve fundamental understanding of the physics of the corona and solar wind.

References

1. Aprovitola A, Denaro FM (2007) A non-diffusive, divergence-free, finite volume-based double projection method on non-staggered grids. Int J Numer Methods Fluids 53(7):1127–1172
2. Arge CN, Pizzo VJ (2000) Improvement in the prediction of solar wind conditions using near-real time solar magnetic field updates. J Geophys Res 105:10,465–10,480. https://doi.org/10.1029/1999JA900262
3. Arge CN, Luhmann JG, Odstrčil D, Schrijver CJ, Li Y (2004) Stream structure and coronal sources of the solar wind during the May 12th, 1997 CME. J Atmos Sol Terr Phys 66(15–16):1295–1309
4. Aschwanden M (2006) Physics of the solar corona: an introduction with problems and solutions, 2nd edn. Springer, Berlin
5. Aschwanden MJ (2004) Physics of the solar corona. An introduction. Praxis Publishing Ltd, Chichester
6. Aschwanden MJ, Burlaga LF, Kaiser ML, Ng CK, Reames DV, Reiner MJ, Gombosi TI, Lugaz N, Manchester W, Roussev II, Zurbuchen TH, Farrugia CJ, Galvin AB, Lee MA, Linker JA, Mikić Z, Riley P, Alexander D, Sandman AW, Cook JW, Howard RA, Odstrčil D, Pizzo VJ, Kóta J, Liewer PC, Luhmann JG, Inhester B, Schwenn RW, Solanki SK, Vasyliunas VM, Wiegelmann T, Blush L, Bochsler P, Cairns IH, Robinson PA, Bothmer V, Kecskemety K, Llebaria A, Maksimovic M, Scholer M, Wimmer-Schweingruber RF (2008) Theoretical

modeling for the STEREO mission. Space Sci Rev 136:565–604. https://doi.org/10.1007/s11214-006-9027-8
7. Axford WI, McKenzie JF (1992) The origin of high speed solar wind streams. In: Marsch E, Schwenn R (eds) Solar wind seven. COSPAR colloquia series, vol 3. Pergamon Press, Oxford, pp 1–5
8. Balbás J, Tadmor E, Wu CC (2004) Non-oscillatory central schemes for one- and two-dimensional MHD equations: I. J Comput Phys 201:261–285. https://doi.org/10.1016/j.jcp.2004.05.020
9. Balsara DS (2009) Divergence-free reconstruction of magnetic fields and WENO schemes for magnetohydrodynamics. J Comput Phys 228:5040–5056. https://doi.org/10.1016/j.jcp.2009.03.038
10. Balsara DS, Kim J (2004) A comparison between divergence-cleaning and staggered-mesh formulations for numerical magnetohydrodynamics. Astrophys J 602:1079–1090. https://doi.org/10.1086/381051
11. Balsara DS, Spicer DS (1999) A staggered mesh algorithm using high order Godunov fluxes to ensure solenoidal magnetic fields in magnetohydrodynamic simulations. J Comput Phys 149:270–292. https://doi.org/10.1006/jcph.1998.6153
12. Balsara DS, Rumpf T, Dumbser M, Munz CD (2009) Efficient, high accuracy ADER-WENO schemes for hydrodynamics and divergence-free magnetohydrodynamics. J Comput Phys 228:2480–2516. https://doi.org/10.1016/j.jcp.2008.12.003
13. Barnes A (1992) Acceleration of the solar wind. Rev Geophys 30:43–55. https://doi.org/10.1029/91RG02816
14. Brackbill JU, Barnes DC (1980) The effect of nonzero $\nabla \cdot \mathbf{B}$ on the numerical solution of the magnetohydrodynamic equations. J Comput Phys 35(3):426–430
15. Breen AR, Thomasson P, Jordan CA, Tappin SJ, Fallows RA, Canals A, Moran PJ (2002) Slow and fast solar wind acceleration near solar maximum. Adv Space Res 30:433–436. https://doi.org/10.1016/S0273-1177(02)00339-3
16. Chang S (1995) The method of space-time conservation element and solution element–a new approach for solving the Navier-Stokes and Euler equations. J Comput Phys 119:295–324. https://doi.org/10.1006/jcph.1995.1137
17. Chang SC, Wang XY (2003) Multi-dimensional Courant number insensitive CE/SE Euler solvers for application involving highly non-uniform meshes. AIAA-2003-5280
18. Chang SC, Wu Y, Yang V, Wang XY (2005) Local time-stepping procedures for the space-time conservation element and solution element method. Int J Comput Fluid Dyn 19:359–380. https://doi.org/10.1080/10618560500092610
19. Chen CG, Xiao F (2008) Shallow water model on cubed-sphere by multi-moment finite volume method. J Comput Phys 227(10):5019–5044
20. Chesshire G, Henshaw WD (1990) Composite overlapping meshes for the solution of partial differential equations. J Comput Phys 90:1–64. https://doi.org/10.1016/0021-9991(90)90196-8
21. Choblet G (2005) Modelling thermal convection with large viscosity gradients in one block of the cubed sphere. J Comput Phys 205:269–291. https://doi.org/10.1016/j.jcp.2004.11.005
22. Cohen O, Sokolov IV, Roussev II, Arge CN, Manchester WB, Gombosi TI, Frazin RA, Park H, Butala MD, Kamalabadi F, Velli M (2007) A semiempirical magnetohydrodynamical model of the solar wind. Astrophys J Lett 654:L163–L166. https://doi.org/10.1086/511154
23. Cohen O, Sokolov IV, Roussev II, Gombosi TI (2008) Validation of a synoptic solar wind model. J Geophys Res 113(A12):A03104. https://doi.org/10.1029/2007JA012797
24. Cranmer SR (2012) Self-consistent models of the solar wind. Space Sci Rev 172:145–156. https://doi.org/10.1007/s11214-010-9674-7
25. Cranmer SR, van Ballegooijen AA, Edgar RJ (2007) Self-consistent coronal heating and solar wind acceleration from anisotropic magnetohydrodynamic turbulence. Astrophys J Suppl Ser 171:520–551. https://doi.org/10.1086/518001
26. Cranmer SR, Matthaeus WH, Breech BA, Kasper JC (2009) Empirical constraints on proton and electron heating in the fast solar wind. Astrophys J 702:1604–1614. https://doi.org/10.1088/0004-637X/702/2/1604

27. Cranmer SR, Gibson SE, Riley P (2017) Origins of the ambient solar wind: implications for space weather. Space Sci Rev 212(3):1345–1384. https://doi.org/10.1007/s11214-017-0416-y
28. Crockett RK, Colella P, Fisher RT, Klein RI, McKee CF (2005) An unsplit, cell-centered Godunov method for ideal MHD. J Comput Phys 203:422–448. https://doi.org/10.1016/j.jcp.2004.08.021
29. Cunningham AJ, Frank A, Varnière P, Mitran S, Jones TW (2009) Simulating magnetohydrodynamical flow with constrained transport and adaptive mesh refinement: algorithms and tests of the AstroBEAR code. Astrophys J Suppl Ser 182:519–542. https://doi.org/10.1088/0067-0049/182/2/519
30. Dedner A, Kemm F, Kröner D, Munz CD, Schnitzer T, Wesenberg M (2002) Hyperbolic divergence cleaning for the MHD equations. J Comput Phys 175:645–673. https://doi.org/10.1006/jcph.2001.6961
31. Dewar RL (1970) Interaction between hydromagnetic waves and a time-dependent, inhomogeneous medium. Phys Fluids 13:2710–2720. https://doi.org/10.1063/1.1692854
32. Downs C, Roussev II, van der Holst B, Lugaz N, Sokolov IV, Gombosi TI (2010) Toward a realistic thermodynamic magnetohydrodynamic model of the global solar corona. Astrophys J 712:1219–1231. https://doi.org/10.1088/0004-637X/712/2/1219
33. Endeve E, Leer E, Holzer TE (2003) Two-dimensional magnetohydrodynamic models of the solar corona: mass loss from the streamer belt. Astrophys J 589:1040–1053. https://doi.org/10.1086/374814
34. Evans CR, Hawley JF (1988) Simulation of magnetohydrodynamic flows - a constrained transport method. Astrophys J 332:659–677. https://doi.org/10.1086/166684
35. Evans RM, Opher M, Oran R, van der Holst B, Sokolov IV, Frazin R, Gombosi TI, Vásquez A (2012) Coronal heating by surface Alfvén wave damping: implementation in a global magnetohydrodynamics model of the solar wind. Astrophys J 756:155. https://doi.org/10.1088/0004-637X/756/2/155
36. Feng XS, Xiang CQ, Zhong DK, Fan QL (2005) A comparative study on 3-D solar wind structure observed by Ulysses and MHD simulation. Chin Sci Bull 50:672–678
37. Feng XS, Zhou YF, Hu YQ (2006) A 3rd order WENO GLM-MHD scheme for magnetic reconnection. Chin J Space Sci 26:1–7
38. Feng XS, Zhou YF, Wu ST (2007) A novel numerical implementation for solar wind modeling by the modified conservation element/solution element method. Astrophys J 655:1110–1126. https://doi.org/10.1086/510121
39. Feng XS, Zhang Y, Yang LP, Wu ST, Dryer M (2009) An operational method for shock arrival time prediction by one-dimensional CESE-HD solar wind model. J Geophys Res 114(A13):A10103
40. Feng XS, Yang LP, Xiang CQ, Wu ST, Zhou YF, Zhong DK (2010) Three-dimensional solar WIND modeling from the Sun to Earth by a SIP-CESE MHD model with a six-component grid. Astrophys J 723:300–319. https://doi.org/10.1088/0004-637X/723/1/300
41. Feng XS, Jiang CW, Xiang CQ, Zhao XP, Wu ST (2012) A data-driven model for the global coronal evolution. Astrophys J 758(1):62
42. Feng XS, Xiang CQ, Zhong DK, Zhou YF, Yang LP, Ma XP (2014) SIP-CESE MHD model of solar wind with adaptive mesh refinement of hexahedral meshes. Comput Phys Commun 185:1965–1980. https://doi.org/10.1016/j.cpc.2014.03.027
43. Feng XS, Ma XP, Xiang CQ (2015) Data-driven modeling of the solar wind from 1 Rs to 1 AU. J Geophys Res: Space Phys 120(12):10,159–10,174. https://doi.org/10.1002/2015JA021911
44. Feng XS, Li CX, Xiang CQ, Zhang M, Li HC, Wei FS (2017) Data-driven modeling of the solar corona by a new three-dimensional path-conservative Osher-Solomon MHD model. Astrophys J Suppl Ser 233(1):10. http://stacks.iop.org/0067-0049/233/i=1/a=10
45. Fisk LA (2003) Acceleration of the solar wind as a result of the reconnection of open magnetic flux with coronal loops. J Geophys Res (Space Phys) 108:1157. https://doi.org/10.1029/2002JA009284

46. Fisk LA, Schwadron NA, Zurbuchen TH (1999) Acceleration of the fast solar wind by the emergence of new magnetic flux. J Geophys Res 104:19,765–19,772. https://doi.org/10.1029/1999JA900256
47. Fox NJ, Velli MC, Bale SD, Decker R, Driesman A, Howard RA, Kasper JC, Kinnison J, Kusterer M, Lario D, Lockwood MK, McComas DJ, Raouafi NE, Szabo A (2016) The Solar Probe Plus mission: humanity's first visit to our Star. Space Sci Rev 204:7–48. https://doi.org/10.1007/s11214-015-0211-6
48. Fragile PC, Anninos P, Gustafson K, Murray SD (2005) Magnetohydrodynamic simulations of shock interactions with radiative clouds. Astrophys J 619:327–339. https://doi.org/10.1086/426313
49. Fry CD, Dryer M, Smith Z, Sun W, Deehr CS, Akasofu SI (2003) Forecasting solar wind structures and shock arrival times using an ensemble of models. J Geophys Res 108:1070. https://doi.org/10.1029/2002JA009474
50. Gombosi TI, van der Holst B, Manchester WB, Sokolov IV (2018) Extended MHD modeling of the steady solar corona and the solar wind. Living Rev Sol Phys 15(1):4. https://doi.org/10.1007/s41116-018-0014-4
51. Gosling JT, Asbridge JR, Bame SJ, Feldman WC, Borrini G, Hansen RT (1981) Coronal streamers in the solar wind at 1 AU. J Geophys Res 86:5438–5448. https://doi.org/10.1029/JA086iA07p05438
52. Groth CPT, De Zeeuw DL, Gombosi TI, Powell KG (2000) Global three-dimensional MHD simulation of a space weather event: CME formation, interplanetary propagation, and interaction with the magnetosphere. J Geophys Res 105:25,053–25,078. https://doi.org/10.1029/2000JA900093
53. Hartle RE, Barnes A (1970) Nonthermal heating in the two-fluid solar wind model. J Geophys Res 75:6915–6931. https://doi.org/10.1029/JA075i034p06915
54. Hartmann D, Meinke M, Schröder W (2008) An adaptive multilevel multigrid formulation for Cartesian hierarchical grid methods. Comput Fluids 37(9):1103–1125
55. Hayashi K (2005) Magnetohydrodynamic simulations of the solar corona and solar wind using a boundary treatment to limit solar wind mass flux. Astrophys J Suppl Ser 161:480. https://doi.org/10.1086/491791
56. Hayashi K, Benevolenskaya E, Hoeksema T, Liu Y, Zhao XP (2006) Three-dimensional magnetohydrodynamic simulation of a global solar corona using a temperature distribution map obtained from SOHO EIT measurements. Astrophys J Lett 636:L165–L168
57. Henshaw WD, Schwendeman DW (2008) Parallel computation of three-dimensional flows using overlapping grids with adaptive mesh refinement. J Comput Phys 227:7469–7502
58. Hernlund J, Tackley P (2003) Three-dimensional spherical shell convection at infinite Prandtl number using the 'cubed sphere' method. Comput Fluid Solid Mech 931–933
59. Hollweg JV (1986) Transition region, corona, and solar wind in coronal holes. J Geophys Res 91:4111–4125. https://doi.org/10.1029/JA091iA04p04111
60. Hollweg JV (2008) The solar wind: our current understanding and how we got here. J Astrophys Astron 29:217–237. https://doi.org/10.1007/s12036-008-0028-8
61. Holst M, Saied F (1993) Multigrid solution of the Poisson-Boltzmann equation. J Comput Chem 14(1):105–113
62. Hu YQ, Feng XS, Wu ST, Song WB (2008) Three-dimensional MHD modeling of the global corona throughout solar cycle 23. J Geophys Res 113(A12):A03106. https://doi.org/10.1029/2007JA012750
63. Jacques SA (1977) Momentum and energy transport by waves in the solar atmosphere and solar wind. Astrophys J 215:942–951. https://doi.org/10.1086/155430
64. Kageyama A (2005) Dissection of a sphere and "Yin-Yang grid" grids. J Earth Simulator 3:20–28
65. Kageyama A, Sato T (2004) "Yin-Yang grid": an overset grid in spherical geometry. Geochem Geophys Geosyst 5:Q09005. https://doi.org/10.1029/2004GC000734
66. Kasper JC, Abiad R, Austin G, Balat-Pichelin M, Bale SD, Belcher JW, Berg P, Bergner H, Berthomier M, Bookbinder J, Brodu E, Caldwell D, Case AW, Chandran BDG, Cheimets

P, Cirtain JW, Cranmer SR, Curtis DW, Daigneau P, Dalton G, Dasgupta B, DeTomaso D, Diaz-Aguado M, Djordjevic B, Donaskowski B, Effinger M, Florinski V, Fox N, Freeman M, Gallagher D, Gary SP, Gauron T, Gates R, Goldstein M, Golub L, Gordon DA, Gurnee R, Guth G, Halekas J, Hatch K, Heerikuisen J, Ho G, Hu Q, Johnson G, Jordan SP, Korreck KE, Larson D, Lazarus AJ, Li G, Livi R, Ludlam M, Maksimovic M, McFadden JP, Marchant W, Maruca BA, McComas DJ, Messina L, Mercer T, Park S, Peddie AM, Pogorelov N, Reinhart MJ, Richardson JD, Robinson M, Rosen I, Skoug RM, Slagle A, Steinberg JT, Stevens ML, Szabo A, Taylor ER, Tiu C, Turin P, Velli M, Webb G, Whittlesey P, Wright K, Wu ST, Zank G (2016) Solar wind electrons alphas and protons (SWEAP) investigation: design of the solar wind and coronal plasma instrument suite for Solar Probe Plus. Space Sci Rev 204:131–186. https://doi.org/10.1007/s11214-015-0206-3

67. Kataoka R, Ebisuzaki T, Kusano K, Shiota D, Inoue S, Yamamoto TT, Tokumaru M (2009) Three-dimensional MHD modeling of the solar wind structures associated with 13 December 2006 coronal mass ejection. J Geophys Res 114(A13):A10102. https://doi.org/10.1029/2009JA014167
68. Kleimann J, Kopp A, Fichtner H, Grauer R (2009) A novel code for numerical 3-D MHD studies of CME expansion. Ann Geophys 27(3):989–1004
69. Klimchuk JA (2006) On solving the coronal heating problem. Sol Phys 234:41–77. https://doi.org/10.1007/s11207-006-0055-z
70. Lee D, Deane AE (2009) An unsplit staggered mesh scheme for multidimensional magnetohydrodynamics. J Comput Phys 228:952–975. https://doi.org/10.1016/j.jcp.2008.08.026
71. Levine RH, Altschuler MD, Harvey JW (1977) Solar sources of the interplanetary magnetic field and solar wind. J Geophys Res 82:1061–1065. https://doi.org/10.1029/JA082i007p01061
72. Li B, Habbal SR, Li X, Mountford C (2005) Effect of the latitudinal distribution of temperature at the coronal base on the interplanetary magnetic field configuration and the solar wind flow. J Geophys Res 110(A9):A12112. https://doi.org/10.1029/2005JA011332
73. Li ST (2008) High order central scheme on overlapping cells for magneto-hydrodynamic flows with and without constrained transport method. J Comput Phys 227(15):7368–7393
74. Li ST, Li H (2004) A novel approach of divergence-free reconstruction for adaptive mesh refinement. J Comput Phys 199:1–15. https://doi.org/10.1016/j.jcp.2004.01.027
75. Linker JA, Kivelson MG, Walker RJ (1991) A three-dimensional MHD simulation of plasma flow past Io. J Geophys Res 96:21. https://doi.org/10.1029/91JA02132
76. Linker JA, Mikić Z, Biesecker DA, Forsyth RJ, Gibson SE, Lazarus AJ, Lecinski A, Riley P, Szabo A, Thompson BJ (1999) Magnetohydrodynamic modeling of the solar corona during whole Sun month. J Geophys Res 104:9809–9830. https://doi.org/10.1029/1998JA900159
77. Lionello R, Linker JA, Mikić Z (2001) Including the transition region in models of the large-scale solar corona. Astrophys J 546(1):542
78. Londrillo P, Zanna LD (2004) On the divergence-free condition in Godunov-type schemes for ideal magnetohydrodynamics: the upwind constrained transport method. J Comput Phys 195:17–48
79. Luhmann JG, Li Y, Arge CN, Gazis PR, Ulrich R (2002) Solar cycle changes in coronal holes and space weather cycles. J Geophys Res 107:1154. https://doi.org/10.1029/2001JA007550
80. Matsumoto T, Shibata K (2010) Nonlinear propagation of Alfvén waves driven by observed photospheric motions: application to the coronal heating and spicule formation. Astrophys J 710:1857–1867. https://doi.org/10.1088/0004-637X/710/2/1857
81. Matthaeus WH, Zank GP, Oughton S, Mullan DJ, Dmitruk P (1999) Coronal heating by magnetohydrodynamic turbulence driven by reflected low-frequency waves. Astrophys J Lett 523:L93–L96. https://doi.org/10.1086/312259
82. Maurits NM, van der Ven H, Veldman AEP (1998) Explicit multi-time stepping methods for convection-dominated flow problems. Comput Methods Appl Mech Eng 157(1–2):133–150
83. McComas DJ, Bame SJ, Barraclough BL, Feldman WC, Funsten HO, Gosling JT, Riley P, Skoug R, Balogh A, Forsyth R, Goldstein BE, Neugebauer M (1998) Ulysses' return to the slow solar wind. Geophys Res Lett 25:1–4. https://doi.org/10.1029/97GL03444

84. McGregor SL, Hughes WJ, Arge CN, Owens MJ, Odstrčil D (2011) The distribution of solar wind speeds during solar minimum: calibration for numerical solar wind modeling constraints on the source of the slow solar wind. J Geophys Res (Space Phys) 116:A03101. https://doi.org/10.1029/2010JA015881
85. McKenzie JF, Axford WI, Banaszkiewicz M (1997) The fast solar wind. Geophys Res Lett 24:2877–2880. https://doi.org/10.1029/97GL02097
86. Meng X, van der Holst B, Tóth G, Gombosi TI (2015) Alfvén wave solar model (AWSoM): proton temperature anisotropy and solar wind acceleration. Mon Not R Astron Soc 454(4):3697–3709. https://doi.org/10.1093/mnras/stv2249
87. Mignone A, Tzeferacos P, Bodo G (2010) High-order conservative finite difference GLM-MHD schemes for cell-centered MHD. J Comput Phys 229(17):5896–5920
88. Mikić Z, Linker JA, Schnack DD, Lionello R, Tarditi A (1999) Magnetohydrodynamic modeling of the global solar corona. Phys Plasmas 6:2217–2224. https://doi.org/10.1063/1.873474
89. Mikić Z, Downs C, Linker JA, Caplan RM, Mackay DH, Upton LA, Riley P, Lionello R, Török T, Titov VS, Wijaya J, Druckmüller M, Pasachoff JM, Carlos W (2018) Predicting the corona for the 21 August 2017 total solar eclipse. Nat Astron 2:913–921. https://doi.org/10.1038/s41550-018-0562-5
90. Miller GH, Colella P (2001) A high-order Eulerian Godunov method for elastic-plastic flow in solids. J Comput Phys 167(1):131–176
91. Mohan Rai M (1986) A conservative treatment of zonal boundaries for Euler equation calculations. J Comput Phys 62(2):472–503
92. Müller D, Marsden RG, St Cyr OC, Gilbert HR (2013) Solar orbiter. Exploring the Sun-heliosphere connection. Sol Phys 285:25–70. https://doi.org/10.1007/s11207-012-0085-7
93. Nakagawa Y, Hu YQ, Wu ST (1987) The method of projected characteristics for the evolution of magnetic arches. Astron Astrophys 179:354–370
94. Nakamizo A, Tanaka T, Kubo Y, Kamei S, Shimazu H, Shinagawa H (2009) Development of the 3-D MHD model of the solar corona-solar wind combining system. J Geophys Res 114(A13):7109. https://doi.org/10.1029/2008JA013844
95. Nakamizo A, Tanaka T, Kubo Y, Kamei S, Shimazu H, Shinagawa H (2009) Development of the 3-D MHD model of the solar corona-solar wind combining system. J Geophys Res (Space Phys) 114:A07109. https://doi.org/10.1029/2008JA013844
96. Neugebauer M (1999) The three-dimensional solar wind at solar activity minimum. Rev Geophys 37:107–126. https://doi.org/10.1029/1998RG900001
97. Odstrčil D, Pizzo VJ (1999) Distortion of the interplanetary magnetic field by three-dimensional propagation of coronal mass ejections in a structured solar wind. J Geophys Res 104(28):225. https://doi.org/10.1029/1999JA900319
98. Odstrčil D, Linker JA, Lionello R, Mikic Z, Riley P, Pizzo VJ, Luhmann JG (2002) Merging of coronal and heliospheric numerical two-dimensional MHD models. J Geophys Res 107:1493. https://doi.org/10.1029/2002JA009334
99. Ofman L, Viñas AF, Moya PS (2011) Hybrid models of solar wind plasma heating. Ann Geophys 29:1071–1079. https://doi.org/10.5194/angeo-29-1071-2011
100. Ogino T, Walker RJ (1984) A magnetohydrodynamic simulation of the bifurcation of tail lobes during intervals with a northward interplanetary magnetic field. Geophys Res Lett 11:1018–1021. https://doi.org/10.1029/GL011i010p01018
101. Owens MJ, Arge CN, Spence HE, Pembroke A (2005) An event-based approach to validating solar wind speed predictions: high-speed enhancements in the Wang-Sheeley-Arge model. J Geophys Res 110(A9):A12105. https://doi.org/10.1029/2005JA011343
102. Owens MJ, Lockwood M, Barnard LA (2017) Coronal mass ejections are not coherent magnetohydrodynamic structures. Sci Rep 7(1):4152. https://doi.org/10.1038/s41598-017-04546-3
103. Owens MJ, Lockwood M, Riley P (2017) Global solar wind variations over the last four centuries. Sci Rep 7:41548. https://doi.org/10.1038/srep41548
104. Parker EN (1963) Interplanetary dynamical processes. Interscience Publishers, New York

105. Pätzold M, Tsurutani BT, Bird MK (1997) An estimate of large-scale solar wind density and velocity profiles in a coronal hole and the coronal streamer belt. J Geophys Res 102(A11):24,151–24,160
106. Pneuman GW, Kopp RA (1971) Gas-magnetic field interactions in the solar corona. Sol Phys 18(2):258–270
107. Porfir'eva GA, Yakunina GV, Delone AB, Oreshina AV, Oreshina IV (2009) Coronal streamers on the Sun and their physical properties. J Phys Stud 13(2):2901–2906
108. Powell KG, Roe PL, Linde TJ, Gombosi TI, De Zeeuw DL (1999) A solution-adaptive upwind scheme for ideal magnetohydrodynamics. J Comput Phys 154(2):284–309
109. Press WH, Teukolsky SA, Vetterling WT, Flannery BP (1992) Numerical recipes in FORTRAN. The art of scientific computing, vol 1, 2nd edn. Cambridge University Press, Cambridge
110. Rider WJ (1998) Filtering non-solenoidal modes in numerical solutions of incompressible flows. Int J Numer Methods Fluids 28(5):789–814
111. Riley P, Linker JA, Mikić Z (2001) An empirically-driven global MHD model of the solar corona and inner heliosphere. J Geophys Res 106:15,889–15,902. https://doi.org/10.1029/2000JA000121
112. Riley P, Linker JA, Mikić Z, Lionello R, Ledvina SA, Luhmann JG (2006) A comparison between global solar magnetohydrodynamic and potential field source surface model results. Astrophys J 653:1510–1516
113. Ronchi C, Iacono R, Paolucci PS (1996) The cubed sphere: a new method for the solution of partial differential equations in spherical geometry. J Comput Phys 124(1):93–114
114. Roussev II, Gombosi TI, Sokolov IV, Velli M, Manchester IV, DeZeeuw DL, Liewer P, Tóth G, Luhmann J (2003) A three-dimensional model of the solar wind incorporating solar magnetogram observations. Astrophys J Lett 595:L57–L61
115. Schwadron NA, McComas DJ (2003) Solar wind scaling law. Astrophys J 599:1395–1403. https://doi.org/10.1086/379541
116. Sheeley NR, Wang YM, Hawley SH, Brueckner GE, Dere KP, Howard RA, Koomen MJ, Korendyke CM, Michels DJ, Paswaters SE et al (1997) Measurements of flow speeds in the corona between 2 and 30 R. Astrophys J 484:472–478
117. Shen F, Feng X, Wu ST, Xiang C (2007) Three-dimensional MHD simulation of CMEs in three-dimensional background solar wind with the self-consistent structure on the source surface as input: numerical simulation of the January 1997 Sun-Earth connection event. J Geophys Res 112(A11):A06109. https://doi.org/10.1029/2006JA012164
118. Smith ZK, Dryer M, McKenna-Lawlor SMP, Fry CD, Deehr CS, Sun W (2009) Operational validation of HAFv2's predictions of interplanetary shock arrivals at Earth: declining phase of solar cycle 23. J Geophys Res 114(A13):A05106. https://doi.org/10.1029/2008JA013836
119. Sokolov IV, van der Holst B, Oran R, Downs C, Roussev II, Jin M, Ward B, Manchester I, Evans RM, Gombosi TI (2013) Magnetohydrodynamic waves and coronal heating: unifying empirical and MHD turbulence models. Astrophys J 764(1):23. http://stacks.iop.org/0004-637X/764/i=1/a=23
120. Steger J, Benek J (1987) On the use of composite grid schemes in computational aerodynamics. Comput Methods Appl Mech Eng 64:301–320. https://doi.org/10.1016/0045-7825(87)90045-4
121. Stone JM, Gardiner T (2009) A simple unsplit Godunov method for multidimensional MHD. New Astron 14(2):139–148
122. Suess ST, Wang AH, Wu ST (1996) Volumetric heating in coronal streamers. J Geophys Res 101(A9):19,957–19,966
123. Suess ST, Wang AH, Wu ST, Poletto G, McComas DJ (1999) A two-fluid, MHD coronal model. J Geophys Res 104(A3):4697–4708
124. Suzuki TK, Inutsuka SI (2006) Solar winds driven by nonlinear low-frequency Alfvén waves from the photosphere: parametric study for fast/slow winds and disappearance of solar winds. J. Geophys. Res (Space Phys) 111:A06101. https://doi.org/10.1029/2005JA011502

125. Tanaka T (1994) Finite volume TVD scheme on an unstructured grid system for three-dimensional MHD simulation of inhomogeneous systems including strong background potential fields. J Comput Phys 111(2):381–389
126. Tanaka T (2003) Finite volume TVD schemes for magnetohydrodynamics on unstructured grids. In: Büchner J, Dum C, Scholer M (eds) Space plasma simulation. Lecture notes in physics, vol 615. Springer, Berlin, pp 275–295
127. Tóth G (2000) The $\nabla \cdot \mathbf{B} = 0$ constraint in shock-capturing magnetohydrodynamics codes. J Comput Phys 161:605–652. https://doi.org/10.1006/jcph.2000.6519
128. Tóth G, de Zeeuw DL, Gombosi TI, Powell KG (2006) A parallel explicit/implicit time stepping scheme on block-adaptive grids. J Comput Phys 217:722–758. https://doi.org/10.1016/j.jcp.2006.01.029
129. Usmanov AV (1993) A global numerical 3-D MHD model of the solar wind. Sol Phys 146(2):377–396
130. Usmanov AV (1996) A global 3-D MHD solar wind model with Alfvén waves. In: Winterhalter D, Gosling JT, Habbal SR, Kurth WS, Neugebauer M (eds) American institute of physics conference series, vol 382, pp 141–144. https://doi.org/10.1063/1.51468
131. Usmanov AV, Dryer M (1995) A global 3-D simulation of interplanetary dynamics in June 1991. Sol Phys 159(2):347–370
132. Usmanov AV, Goldstein ML (2003) A tilted-dipole MHD model of the solar corona and solar wind. J Geophys Res 108(A9):1354
133. Usmanov AV, Goldstein ML, Besser BP, Fritzer JM (2000) A global MHD solar wind model with WKB Alfvén waves: comparison with Ulysses data. J Geophys Res 105(A6):12,675–12,695
134. van der Holst B, Keppens R (2007) Hybrid block-AMR in Cartesian and curvilinear coordinates: MHD applications. J Comput Phys 226:925–946. https://doi.org/10.1016/j.jcp.2007.05.007
135. van der Holst B, Sokolov IV, Meng X, Jin M, Manchester WB IV, Tóth G, Gombosi TI (2014) Alfvén wave solar model (AWSoM): coronal heating. Astrophys J 782:81
136. Van der Ven H, Niemann-Tuitman BE, Veldman AEP (1997) An explicit multi-time-stepping algorithm for aerodynamic flows. J Comput Appl Math 82(1):423–431
137. Velli M, Grappin R, Mangeney A (1991) Waves from the Sun. Geophys Astrophys Fluid Dyn 62:101–121. https://doi.org/10.1080/03091929108229128
138. Verdini A, Velli M, Matthaeus WH, Oughton S, Dmitruk P (2010) A turbulence-driven model for heating and acceleration of the fast wind in coronal holes. Astrophys J Lett 708:L116–L120. https://doi.org/10.1088/2041-8205/708/2/L116
139. Vinokur M (1989) An analysis of finite-difference and finite-volume formulations of conservation laws. J Comput Phys 81(1):1–52
140. Wang AH, Wu ST, Suess ST, Poletto G (1998) Global model of the corona with heat and momentum addition. J Geophys Res 103(A2):1913–1922. https://doi.org/10.1029/97JA01770
141. Wang XY, Chang SC (1999) A 2d non-splitting unstructured triangular mesh Euler solver based on the space-time conservation element and solution element method. Comput Fluid Dyn J 8(2):309–325
142. Wang YM, Sheeley NR Jr (1990) Solar wind speed and coronal flux-tube expansion. Astrophys J 355:726–732. https://doi.org/10.1086/168805
143. Wang YM, Sheeley NR Jr (1991) Why fast solar wind originates from slowly expanding coronal flux tubes. Astrophys J Lett 372:L45–L48. https://doi.org/10.1086/186020
144. Wang YM, Sheeley NR Jr, Walters JH, Brueckner GE, Howard RA, Michels DJ, Lamy PL, Schwenn R, Simnett GM (1998) Origin of streamer material in the outer corona. Astrophys J Lett 498:L165. https://doi.org/10.1086/311321
145. Wang YM, Ko YK, Grappin R (2009) Slow solar wind from open regions with strong low-coronal heating. Astrophys J 691:760–769. https://doi.org/10.1088/0004-637X/691/1/760
146. Wei FS, Feng XS, Cai HC, Zhou QJ (2003) Global distribution of coronal mass outputs and its relation to solar magnetic field structures. J Geophys Res 108(A6):1238. https://doi.org/10.1029/2002JA009439

147. Wu ST, Wang AH, Guo WP (1996) Numerical boundary conditions for solar magnetic hydrodynamic flows using the method of characteristics. Astrophys Space Sci 243:149–153. https://doi.org/10.1007/BF00644045
148. Wu ST, Guo WP, Michels DJ, Burlaga LF (1999) MHD description of the dynamical relationships between a flux rope, streamer, coronal mass ejection, and magnetic cloud: an analysis of the January 1997 Sun-Earth connection event. J Geophys Res 104:14,789–14,802. https://doi.org/10.1029/1999JA900099
149. Wu ST, Wang AH, Liu Y, Hoeksema JT (2006) Data-driven magnetohydrodynamic model for active region evolution. Astrophys J 652:800–811. https://doi.org/10.1086/507864
150. Yang LP, Feng XS, Xiang CQ, Zhang SH, Wu ST (2011) Simulation of the unusual solar minimum with 3D SIP-CESE MHD model by comparison with multi-satellite observations. Sol Phys 271:91–110. https://doi.org/10.1007/s11207-011-9785-7
151. Yen JC, Duell EG, Martindale W (2006) CAA using 3D CESE method with a simplified Courant number insensitive scheme. AIAA-2006-2417
152. Yoshida M, Kageyama A (2004) Application of the Yin-Yang grid to a thermal convection of a Boussinesq fluid with infinite Prandtl number in a three-dimensional spherical shell. Geophys Res Lett 31(L12):609
153. Yu ST, Chang SC (1997) Treatments of stiff source terms in conservation laws by the method of space-time conservation element and solution element. AIAA-97-0435
154. Zhang ZC, Yu ST (1999) Shock capturing without Riemann solver—a modified space-time CE/SE method for conservation laws. AIAA-99-0904
155. Zhang ZC, Yu ST, Chang SC, Himansu A, Jorgenson PCE (1999) A modified space-time conservation element and solution element method for Euler and Navier-Stokes equations. AIAA-99-3277
156. Zhang ZC, Yu ST, Chang SC (2002) A space-time conservation element and solution element method for solving the two- and three-dimensional unsteady Euler equations using quadrilateral and hexahedral meshes. J Comput Phys 175(1):168–199
157. Zhao XP, Hoeksema JT, Liu Y, Scherrer PH (2006) Success rate of predicting the heliospheric magnetic field polarity with Michelson Doppler imager (MDI) synoptic charts. J Geophys Res 111(A10):A10108. https://doi.org/10.1029/2005JA011576
158. Zhou YF, Feng XS (2008) Numerical study of successive CMEs during November 4–5, 1998. Sci China E 51(10):1600–1610
159. Zhou YF, Feng XS, Wu ST (2008) Numerical simulation of the 12 May 1997 CME event. Chin Phys Lett 25(2):790–793
160. Ziegler U (2005) A solution-adaptive central-constraint transport scheme for magnetohydrodynamics. Comput Phys Commun 170:153–174. https://doi.org/10.1016/j.cpc.2005.04.002
161. Zurbuchen TH (2007) A new view of the coupling of the Sun and the heliosphere. Annu Rev Astron Astrophys 45(1):297–338

Chapter 6
AMR Implementation of 3D SIP-CESE MHD Model on Six-Component Overset Grid System

Abstract Coronal-heliospheric space is characterized by disparate temporal and spatial scales as well as by different relevant physics in different domains such as the corona and the inner heliosphere. The scalable, massively parallel, block-based, adaptive-mesh refinement (AMR) allows resolving such disparate spatial and temporal scales throughout the computational domain with even less cells but can generate a necessary resolution. This chapter is devoted to the adaptive mesh refinement (AMR) implementation of Solar-Interplanetary space-time conservation element and solution element (CESE) magnetohydrodynamic model (SIP-CESE MHD model) with the aid of the parallel AMR package PARAMESH. Two AMR realization strategies are employed: one uses a solution adaptive technique directly for the CESE solver in the six-component grid system introduced in Chap. 5, while the other is implemented for the CESE solver of associated partial differential equations (PDEs) in the reference space of curvilinear coordinates transformed from the original governing partial differential equations in the physical space. Under these two AMR implementations, tests are carried out for the solar wind background study, and numerical results are compared with the observations in the solar corona and in interplanetary space from the Solar and Heliospheric Observatory (SOHO) and spacecraft data from OMNI.

6.1 Introduction

As reviewed in Chap. 1, in quantitatively producing the solar wind structures, 3D global MHD models have been developed. As pointed out by space weather scientists (e.g., [4, 27, 32, 118]), high performance computational models require further improvements in order to make the real- or faster than real-time numerical prediction of adverse space weather events and their influence on the geospace environment.

Corona–interplanetary space involves the presence of different temporal and spatial scales on which solar wind plasma occurs throughout the vast solar–interplanetary space, and contains many critical features, such as discontinuities and heliospheric current sheet, that have spatial scales many orders of magnitude smaller than the system size. The numerical grids are either adapted to the small scale features in

© Springer Nature Singapore Pte Ltd. 2020

X. Feng, *Magnetohydrodynamic Modeling of the Solar Corona and Heliosphere*, Atmosphere, Earth, Ocean & Space,
https://doi.org/10.1007/978-981-13-9081-4_6

the system, or a brute force approach is used with as high numerical resolution as possible while fighting the limits of available computational power.

In order to solve numerical problems with multi-orders of spatial/temporal scales and minimize memory requirements and CPU time, the adaptive mesh refinement (AMR) technique has become a mature tool to refine or coarsen mesh grids with the expected spatial resolution according to spatial gradients of physical quantities or other specific criteria. AMR promises to be capable of better resolving the solution features with less cells, although it may generate a substantial computational overhead due to the remeshing, refluxing at coarse–fine interfaces, and guardcell filling.

To support the parallel implementation in numerically solving partial differential equations (PDEs), there have been a number of AMR software infrastructures available, such as AmrLib/BoxLib [92], Chombo [23], GrACE [84], PARAMESH [67, 68] and SAMRAI [43]. These softwares provide developers with an easy route to realize AMR with an existing code of fixed mesh. AMR technique has been successfully applied to the MHD problems with disparate spatial and temporal scales [63, 90, 102, 132], among which FLASH [41, 63], PLUTO [74], ATHENA [102], NIRVANA [132] and BATSRUS [90, 108, 109] developed for astrophysics and space weather modeling are several representatives using oct-tree adaptive meshes.

In Chap. 5, we have introduced the implementation of SIP-CESE MHD model on six-component grid. The objective of this chapter is to describe the application of scalable massively-parallel block-based adaptive mesh refinement (AMR) in six-component grid system (Fig. 5.1 of Chap. 5) for the three-dimensional (3D) SIP-CESE MHD model.

In order to realize the 3D SIP-CESE MHD model on six-component grids with the AMR capabilities for solar wind MHD simulations, our considerations run from the points of view of both the grid system with the solar surface sphere fitting and an easy-to-use AMR implementation. As it is well known, our numerical modeling requires the use of the spherical shell computational domain to describe the coronal-heliospheric space, which makes us have to treat the lower spherical boundary at the solar surface locating the center of the computational domain. At the same time, dealing with the spherical boundary properly is a critical issue as the Sun is the source of the activities in the corona and interplanetary space [30, 31, 55]. In such a spherical shell computational domain, how to take the advantage of the AMR implementation on a logically Cartesian structured mesh for a parallel code needs a delicate concern. Usually, such domain can be partitioned into rectangular or polyhedron mesh grids. If rectangular box is used, although we can easily implement the AMR with the help of some existing tools such as PARAMESH, CHOMBO, and NIRVANA, we can meet problems in boundary fitting and boundary condition implementation due to the fact that the spherical surface cannot be consistently described by any of rectangular box. To overcome this, the employment of body-fitted multi-block mesh and cutting-cell method can be a choice, which nevertheless brings the issues of heavy management of the data structure and difficult implementation of the lower spherical boundary conditions, especially for the projected characteristic method in need of the finite difference operation on the solar surface [31, 50, 123]. If polyhedron cell is used

[30, 78], the sphere and boundary condition can be easily fitted. But for polyhedron grid system, we can meet difficulties in AMR implementation, since there are no conveniently easy-to-use existing grid mesh tools or it is too complicated to be ready for use such as OVERTURE available at http://overtureframework.org/.

In six-component grid mesh, the surface of a sphere is made of six adjoining grid faces that cover the whole sphere with partial overlapping areas. The six-component grid faces lend themselves naturally to a multi-component mesh data structure and allow the discretization to be carried out with structured grids, thereby permitting easier implementation of competitive numerical schemes. The 3D six-component grid is obtained by stacking a sequence of concentric 2D spherical shell grids in the radial direction and forming six three-dimensional blocks, each of which is a quadrangular frustum pyramid. The quasi-uniform and self similar multi-component nature of the six-component grid makes it ideally suited for implementation on massively parallel architectures [31]. However, implementing a parallel-AMR approach on the six-component grid system implies a considerable undertaking because that most of the AMR software infrastructures are based on a logically Cartesian coordinate.

In this chapter, the AMR implementation is carried out with the help of PARAMESH, which is an open-source software package of Fortran 90 subroutines designed to ease the task of adding parallelization and dynamic, adaptive mesh refinement [67]. The same package has been used by the code FLASH and ATHENA in astrophysics community [67, 82], and is available on the website https://sourceforge.net/projects/paramesh/. Two AMR realization strategies are implemented: one is to employ a solution-adaptive technique directly for the CESE solver in the six-component grid system introduced in Chap. 5, while the other is implemented for the CESE solver of associated PDEs in the reference space of curvilinear coordinates transformed from the original governing PDEs in the physical space. We will address these two strategies in Sects. 6.4 and 6.5, respectively. Here, Strategy I is extracted from Feng et al. [39], and Strategy II is adapted from Feng et al. [35] and Jiang et al. [55]. The validation results are extracted from Feng et al. [35].

6.2 Governing Equations

The three-dimensional equations governing solar wind plasma can be described by the following MHD equations in Cartesian coordinates with diffusion terms:

$$\frac{\partial \mathbf{U}}{\partial t} + \frac{\partial \mathbf{F}}{\partial x} + \frac{\partial \mathbf{G}}{\partial y} + \frac{\partial \mathbf{H}}{\partial z} - \frac{\partial \mathbf{F}_\mu}{\partial x} - \frac{\partial \mathbf{G}_\mu}{\partial y} - \frac{\partial \mathbf{H}_\mu}{\partial z} = \mathbf{S} \tag{6.1}$$

where

$$\mathbf{U} = \left(\rho, \rho u, \rho v, \rho w, e, B_x, B_y, B_z\right)^T,$$

$$\mathbf{F} = \begin{pmatrix} \rho u \\ \rho u^2 + p_0 - B_x^2 \\ \rho uv - B_x B_y \\ \rho uw - B_x B_z \\ (e + p_0)u - (\mathbf{u} \cdot \mathbf{B}) B_x \\ 0 \\ u B_y - B_x v \\ u B_z - B_x w \end{pmatrix}, \mathbf{G} = \begin{pmatrix} \rho v \\ \rho vu - B_y B_x \\ \rho v^2 + p_0 - B_y^2 \\ \rho vw - B_y B_z \\ (e + p_0)v - (\mathbf{u} \cdot \mathbf{B}) B_y \\ v B_x - B_y u \\ 0 \\ v B_z - B_y w \end{pmatrix},$$

$$\mathbf{H} = \begin{pmatrix} \rho w \\ \rho wu - B_z B_x \\ \rho wv - B_z B_y \\ \rho w^2 + p_0 - B_z^2 \\ (e + p_0)w - (\mathbf{u} \cdot \mathbf{B}) B_z \\ w B_x - B_z u \\ w B_y - B_z v \\ 0 \end{pmatrix},$$

$$\mathbf{F}_\mu = \begin{pmatrix} 0 \\ \nu\rho \frac{\partial u}{\partial x} \\ \nu\rho \frac{\partial v}{\partial x} \\ \nu\rho \frac{\partial w}{\partial x} \\ 0 \\ \mu \nabla \cdot \mathbf{B} \\ \eta j_z \\ -\eta j_y \end{pmatrix}, \mathbf{G}_\mu = \begin{pmatrix} 0 \\ \nu\rho \frac{\partial u}{\partial y} \\ \nu\rho \frac{\partial v}{\partial y} \\ \nu\rho \frac{\partial w}{\partial y} \\ 0 \\ -\eta j_z \\ \mu \nabla \cdot \mathbf{B} \\ \eta j_x \end{pmatrix}, \mathbf{H}_\mu = \begin{pmatrix} 0 \\ \nu\rho \frac{\partial u}{\partial z} \\ \nu\rho \frac{\partial v}{\partial z} \\ \nu\rho \frac{\partial w}{\partial z} \\ 0 \\ \eta j_y \\ -\eta j_x \\ \mu \nabla \cdot \mathbf{B} \end{pmatrix},$$

$$\mathbf{S} = (0, \rho \mathbf{F}_0, \rho \mathbf{u} \cdot (\mathbf{F}_0 + 2\boldsymbol{\Omega} \times \mathbf{u}), \mathbf{0})^T - \nabla \cdot \mathbf{B}(0, \mathbf{B}, \mathbf{u} \cdot \mathbf{B}, \mathbf{u})^T + (0, \mathbf{S}_m, Q_e + \mathbf{u} \cdot \mathbf{S}_m, \mathbf{0})^T$$

with the total pressure $p_0 = p + \frac{1}{2}\mathbf{B}^2$. As usual, $e = \frac{p}{\gamma - 1} + \frac{1}{2}\rho\mathbf{u}^2 + \frac{1}{2}\mathbf{B}^2$ stands for the total energy density. Here, ρ, $\mathbf{u}$, p and $\mathbf{B}$ are the mass density, plasma velocity, gas pressure and magnetic field. γ is the specific heat ratio and is taken to be 1.5 here.

In Eq. (6.1), the external force $\mathbf{F}_0 = -\frac{GM_s}{r^3}\mathbf{r} - \boldsymbol{\Omega} \times (\boldsymbol{\Omega} \times \mathbf{r}) - 2\boldsymbol{\Omega} \times \mathbf{u}$ is the sum of solar gravity force and inertial force due to the corotating frame with the Sun. G, M_s, and $\boldsymbol{\Omega}$ are the gravity constant, solar mass and solar angular speed. Also, Powell's source terms $-\nabla \cdot \mathbf{B}(0, \mathbf{B}, \mathbf{u} \cdot \mathbf{B}, \mathbf{u})^T$ [90] and the diffusive control term $\nabla(\mu\nabla \cdot \mathbf{B})$ [33, 70, 110] have been added in the MHD equations (6.1), to deal with the divergence of magnetic field. Here, following Feng et al. [33], $\mu = \frac{1.3}{\Delta t}(\frac{1}{\Delta x^2} + \frac{1}{\Delta y^2} + \frac{1}{\Delta z^2})^{-1}$, where Δx, Δy, and Δz are grid spacings in Cartesian coordinates and Δt is the time step.

In Eq. (6.1), ν is the kinematic plasma viscosity and η the electrical resistivity. The model is resistive MHD when ν and η are finite, while it becomes ideal MHD when $\nu = 0$ and $\eta = 0$. Note that $\mathbf{F}_\mu$, $\mathbf{G}_\mu$ and $\mathbf{H}_\mu$ may contain any term associated with the first derivative of the MHD variables.

In the numerical implementation, following Tanaka [106] we split the full magnetic field vector $\mathbf{B}$ into the sum of a potential magnetic field $\mathbf{B}_0$ and a time-dependent deviation $\mathbf{B}_1$ (i.e., $\mathbf{B} = \mathbf{B}_0 + \mathbf{B}_1$), where $\mathbf{B}_1$ is calculated by the numerical scheme and $\mathbf{B}_0$ reconstructed with spherical harmonics based on the line-of-sight observations of the photospheric magnetic field. In the AMR implementation of this chapter, during the calculation the values of $\mathbf{B}_0$ on derefined or refined grid points are calculated from the spherical harmonics. The time-dependent part $\mathbf{B}_1$ is involved in the updating procedure. Solving for $\mathbf{B}_1$ is not only inherently more accurate than solving for the full magnetic field vector $\mathbf{B}$, but also helpful to avoid the negative pressure during the calculation. We will address this point in Sect. 6.3 in detail. In the validation tests, these numerical techniques enable us to maintain the $\nabla \cdot \mathbf{B}$ error to an acceptable level of 10^{-6}.

The primitive variables ρ, $\mathbf{u}$, p, $\mathbf{B}$, r, and t in these equations are normalized by the characteristic values ρ_s, a_s, $\rho_s a_s^2$, $\sqrt{\rho_s a_s^2}$, $\mathrm{R_s}$, and $\mathrm{R_s}/a_s$, where ρ_s and a_s are the density and ion-acoustic wave speed at the solar surface of one solar radius $\mathrm{R_s}$. The solar rotation is considered in the present study with angular velocity $|\mathbf{\Omega}| = 2\pi/27.2753\,\mathrm{rad\ day^{-1}}$ (normalized by $a_s/\mathrm{R_s}$ in simulations). A factor of $1/\sqrt{\mu_0}$ has been absorbed into the definition of $\mathbf{B}$, with μ_0 being the magnetic permeability.

In order to achieve the observed pattern of fast and slow solar winds through MHD simulations, researchers often add various kinds of heating terms into the MHD models. Here, the momentum-source term $\mathbf{S}_m$ in the momentum equations and the energy-source term Q_e in the energy equation are constructed from the flux tube expansion factor f_s and minimum angular separation θ_b between open field foot point and its nearest coronal hole boundary, with the help of Wang–Sheeley–Arge (WSA) model. The formula of the source terms and their underlying physics are illustrated in Sect. 5.2.2 of Chap. 5.

The solar wind evolution is calculated in coordinates corotating with the Sun. In this coordinate system we use (r, θ, ϕ) for the position of a point in solar-interplanetary space and (x, y, z) is used to express its corresponding Cartesian coordinates, and the Cartesian coordinates (x, y, z) and its corresponding spherical coordinates (r, θ, ϕ) will be referred to alternatively. Sometimes, the analysis of computational results is carried out in the coordinate system at rest in order to compare with the observations.

6.3 Positivity-Preserving Consideration in SIP-CESE MHD Model

Positivity-preservation is a crucial property that a flux discretization scheme should possess. Positivity-preservation stands for the capability to preserve the positivity of the determinative properties for the solution to be positive within physically-permitted bounds. In ideal MHD, density ρ and pressure p should be non-negative, but such positivity property is not always satisfied by approximated solutions. One can encounter this when simulating problems with low density, high Mach number, or much large magnetic energy compared with internal energy. When this occurs, numerical instability may develop and the simulation can break down. Therefore, a method guaranteeing positivity of the determinative properties has to be considered in a numerical design.

When the MHD equations are written in near conservative form, in order to derive the value of the pressure p one has to subtract off the kinetic energy $\frac{1}{2}\rho\mathbf{u}^2$ and magnetic energy $\frac{\mathbf{B}^2}{2}$ from the total energy $e = \frac{p}{\gamma-1} + \frac{1}{2}\rho\mathbf{u}^2 + \frac{\mathbf{B}^2}{2}$, which can be quite large relative to the value of the internal energy or pressure, especially in regions such as the solar corona with very small plasma $\beta = \frac{2p}{\mathbf{B}^2}$. Such case can lead to negativity problems that density and pressure lose their positivity (in a scheme density and pressure should remain positive under a suitable CFL-condition) (e.g., [11, 42, 47, 54, 112]). Thus discretization errors made in computing the total energy, the kinetic energy and the magnetic energy can be large enough to generate negative pressure. This results in an unacceptable physical situation in the computation of solar-terrestrial space plasma flows.

In order to maintain pressure positivity, Balsara and Spicer [11] used a hybrid scheme: both the conservative energy and the entropy equations are solved. Close to shock waves the energy equation is used to obtain the correct weak solution, at other places the more robust and positive entropy equation $\frac{\partial}{\partial t}\left(\frac{p}{\rho^{\gamma-1}}\right) + \nabla\cdot\left(\mathbf{u}\frac{p}{\rho^{\gamma-1}}\right) = 0$ can be used. That is, the conservative energy and entropy equations are switched to be solved according to some criteria of shock identifier. Similarly, instead of entropy equation, in regions of small plasma β and/or dynamic pressure, the pressure equation $\frac{\partial p}{\partial t} + \nabla\cdot(\mathbf{u}p) = -(\gamma-1)p\nabla\cdot\mathbf{u}$ is used to replace the conservative energy equation [31, 109]. That is, instead of solving the total energy equation we use the internal energy $\varepsilon = \frac{p}{\gamma-1}$ equation

$$\frac{\partial\varepsilon}{\partial t} + \nabla\cdot(\varepsilon\mathbf{u}) = -p\nabla\cdot\mathbf{u} + Q_{\mathrm{e}} \tag{6.2}$$

in those unsafe regions such as $\beta < 0.01$ or $\varepsilon/e_1 < 0.05$.

In MPI-AMRVAC (Message-Passing Interface-Adaptive Mesh Refinement Versatile Advection Code), Keppens et al. [57] suggested a positivity fixing strategy as follows: (1) identify all cells (within the same grid block) that represent physical states surrounding a faulty cell in a rectangular zone up to n_{pos} cells away; (2) convert those cells to primitive variables; and (3) for all but the magnetic field components, replace the faulty cell values by the average of surrounding physical state cells. Finally, conservative variables are reverted where needed.

It is helpful in positivity fix to use the splitting method by writing the full magnetic field vector $\mathbf{B}$ as the sum of a time-independent potential magnetic field $\mathbf{B}_0$ and a deviation $\mathbf{B}_1$ (i.e., $\mathbf{B} = \mathbf{B}_0 + \mathbf{B}_1$), where $\mathbf{B}_1$ is calculated by the numerical scheme.

After splitting, this problem can be mitigated by rewriting the energy equation in terms of the modified total energy density

$$e_1 = \frac{p}{\gamma - 1} + \frac{1}{2}\rho \mathbf{u}^2 + \frac{1}{2}\mathbf{B}_1^2 \tag{6.3}$$

As pointed out by Gombosi et al. [47], solving for the deviation $\mathbf{B}_1$ from the embedded field $\mathbf{B}_0$ is inherently more accurate than solving for the full magnetic field vector $\mathbf{B}$. This splitting approach was first suggested by Ogino and Walker [81], applied to the Godunov-type schemes by Tanaka [106] and Nakamizo et al. [78], the Roe-type approximate Riemann solver by Powell et al. [90] and Gombosi et al. [47], and the CESE scheme employed by Feng et al. [31, 33, 35].

The positivity-preserving procedure runs as follow: First solve the total energy equation of Eq. (6.1). If negative values occur, then switch to solve Eq. (6.2). Furthermore if there still exist negative values, switch to the positivity fixing strategy proposed by Keppens et al. [57] and finally rederive the total energy from Eq. (6.3). In the solar wind modeling, the radial speed should be kept to flow out. That is, v_r must be positive. The non-positiveness of v_r may occur in the near solar surface. In order to confirm the positiveness of v_r during the calculation, we also replace the faulty cell values by the average of surrounding physical state cells.

For other useful techniques in positivity-preserving, please refer to the works [10, 85] or the comments in Chap. 3.

6.4 AMR Strategy I

AMR strategy I uses the CESE solver in the six-component grid system introduced in Chap. 5 to achieve the solution-adaptive technique. A hexahedral grid structure for three dimensions has been established by Sun and Takayama [104], where every cell points to its six faces, and every face points to its two neighboring cells, as shown in Fig. 6.1. Every face is uniquely oriented such that its two neighboring cells can be identified by *left* and *right*, according to the orientation as shown in Fig. 6.1b. The six faces of each cell are ordered by the following rules: (1) the same orientation must be assigned to two opposite faces; (2) the orientations of three pairs of opposite faces must constitute a local right-handed coordinate frame in which six faces is unambiguously ordered. These definitions and neighboring information are sufficient to be required by establishing the CESE solver introduced in Chap. 5.

As pointed out by Sun and Takayama [104], these strict definitions in the data structure reduce the complexity of the unstructured adaptive solver without losing generality, and do not require additional memory to store the definitions, but store the neighboring cell information following the rules.

In the AMR strategy of Berger and Colella [13] and MacNeice et al. [67], a subgrid will be created in regions of its parent grid needing higher resolution, and each grid evolves as a separate initial boundary value problem. The whole grid hierarchy

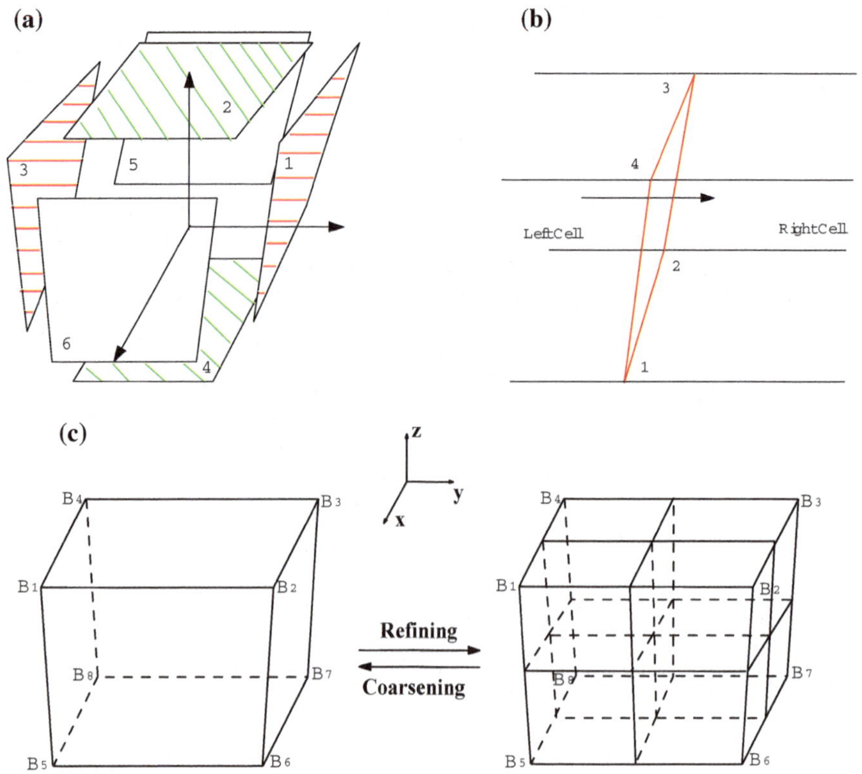

Fig. 6.1 AMR strategy for grid adaptation: **a** every cell stores its location and points to six faces; **b** every face stores the locations of its four vertices and points to two neighboring cells; and **c** a parent cell is divided into eight children. From Fig. 2 of Feng et al. [39]

evolves recursively, and organized in a tree structure. As usual, in the refinement procedure, the one-level-difference rule that no two neighboring cells differ by more than one refinement level is obeyed. That is, once a cell cannot be refined due to the level difference between itself and one or a few neighboring cells, the neighboring cells are refined no matter what values of their refine and coarse are. A cell to be refined is divided into eight children as shown in Fig. 6.1c.

The physical values at newly generated cells in the adaptation procedure are determined by those on the old mesh according to the method proposed by Sun and Takayama [104]. That is, in the refinement procedure the conservative variables of new children are linearly interpolated from those of their parent, while in coarsening procedure flow variables of a coarsened parent cell are the volume-weighted average of these of its deleted children. This kind of interpolation and the weighted average for new cells has been proven to preserve conservation [103].

With the above orientation or ordering of cells and interpolation during the refining and coarsening process, PARAMESH can be easily accommodated to manage the AMR grid system such that we can directly use most of the default operations provided by PARAMESH, such as the refining or coarsening of the blocks, prolongation, and restriction. Therefore, the data transferring and guard cell filling between blocks can be automatically accomplished by the PARAMESH package without large modification.

The ghost cell filling of the edges of each of the six components is achieved as follows. On the overlapping area between two components, solution values at the ghost cell centers of one component grid are determined by the standard Lagrange interpolation from its neighboring component. The grid layer buffer δ is suitably chosen for the overlapping region to perform such interpolation. Here, $\delta = 2\Delta\theta$ is chosen such that the overlapping region between two components contains at least two layers of grid cells of each other to guarantee that none of the neighboring component's ghostcell centers is used in the interpolation stencil. For instance, a ghost cell center G has coordinates (x_G, y_G, z_G) in its corresponding neighboring component, which does not necessarily coincide with any mesh cell center of the neighboring component, while the ghost cell center G's coordinates can be denoted by (x'_G, y'_G, z'_G) when seen as a point in the ghostcell associated component. The interpolated value u_m at the point $G(x_G, y_G, z_G)$ in the neighboring component is computed by $u_m(G) = \sum_{k=0}^{2}\sum_{j=0}^{2}\sum_{i=0}^{2} P_i^G(x)P_j^G(y)P_k^G(z)u_m(x_i, y_j, z_k)$, where $P_l^G(\xi)$ is the Lagrange interpolating polynomial $P_l^G(\xi) = \prod_{k=0,k\neq l}^{2} \frac{\xi_G-\xi_k}{\xi_l-\xi_k}$ with ξ being x, y or z and u_m the mth component of the solution vector $\mathbf{U}$. The twenty-seven (3^3) nodes with coordinates $(x_i, y_j, z_k)(i, j, k = 0, 1, 2)$ denote the inner mesh cell centers of the very neighboring component that are closest to G, and these values $u_m(x_i, y_j, z_k)(i, j, k = 0, 1, 2)$ are already known in the neighboring component. Afterwards, through the transformation between components, $\mathbf{U}(G) = (u_1(G), \ldots, u_8(G))$ at the point $G(x_G, y_G, z_G)$ is converted to $\mathbf{U}'(x'_G, y'_G, z'_G)$ in the ghostcell associated component, and $\mathbf{U}'(x'_G, y'_G, z'_G)$ is used to fill the value at the ghost cell center $G(x'_G, y'_G, z'_G)$. The transformation of velocity and magnetic field vectors in the solution vector $\mathbf{U}$ between different components has been provided in detail in Sect. 5.3.2 of Chap. 5. This kind of interpolation with the third order accuracy works well in practice [33–35] for the second order accurate CESE solver.

In solar wind modeling, the topology of helioshperic current sheet is an important structure and thus only the curl of the magnetic field is used as the refinement strategy to capture the current sheet. The standard deviation about zero $\tau_e = \sqrt{\frac{\sum_{i=1}^{N}\chi_{e_i}^2}{N}}$ is computed for $\chi_e = V^{0.5}\frac{|\nabla\times\mathbf{B}|}{|\mathbf{B}|+\epsilon\sqrt{p}}$ and $\epsilon = 10^{-10}$ to set the thresholds of refining and coarsening grid [25], with V being the spatial cell volume. When the maximum of this criterion in one block is greater than $\kappa\tau_e$, this block is flagged to be refined, while if the maximum in one block is less than $\varsigma\tau_e$, this block is flagged to be coarsened. κ and ς are selected according to different physical time (t_H in unit of hours) intervals during the code's running. That is, when $t_H < 20$, $\kappa = 3$ and $\varsigma = 0.1$; when $20 \leq t_H < 50$, $\kappa = 6$ and $\varsigma = 0.2$; when $50 \leq t_H < 100$, $\kappa = 10$ and $\varsigma = 0.3$; when $100 \leq t_H < 160$, $\kappa = 15$ and $\varsigma = 0.4$; when $t_H \geq 160$, $\kappa = 6$ and $\varsigma = 0.1$.

As demonstrated in Chap. 5, the multiple time-stepping algorithm is implemented in a six-component grid system with the radial direction decomposed into six subdomains: 1–10 R_s, 10–20 R_s, 20–50 R_s, 50–100 R_s, 100–170 R_s, 170–328 R_s. Each subdomain corresponds to one time step. We first calculate the smallest time step [$\Delta t_{\min}$] in the whole domain. Then, the smallest time step in each subdomain [Δt_{q}] is calculated. The time step for each block in each subdomain [Δt_{b}] is taken to be the same as $\Delta t_{\mathrm{b}} = 2^{\mathrm{int}(\log_2(\mathrm{t_q}/\mathrm{t_{min}}))} \times \Delta t_{\mathrm{q}}$. Finally, all of the blocks are advanced as done by Jiang et al. [55].

6.5 AMR Strategy II

Transforming the governing equations of conservation laws into general curvilinear coordinates has been used in constructing numerical algorithms for long (e.g., [16, 111]). In AMR strategy II, the governing equations (6.1) is transformed from physical space (x, y, z) to the reference space (ξ, η, ζ) by specifying the nonsingular mapping

$$x = x(\xi, \eta, \zeta);\ y = y(\xi, \eta, \zeta);\ z = z(\xi, \eta, \zeta) \tag{6.4}$$

As the MHD equations are solved in the reference space (ξ, η, ζ), the grid is still in rectangular coordinates that make the implementation of AMR much conveniently and straightforwardly. The CESE method employed by AMR strategy II for the MHD equations in general curvilinear coordinates has been proposed and validated for some MHD benchmark problems [55]. For the present section to be self-contained, in what follows we present the main points of curvilinear coordinate transformation from physical to reference coordinates from the context of solar-wind modeling.

We use six-component grid system in the physical space, each of which deals with the same grids, basic equations, and numerical tasks. Hence we do not have to distinguish them and we only need to describe the grid partition and the associated coordinate transform. Keep in mind that our physical space is Cartesian coordinates (x, y, z), its corresponding spherical coordinates (r, θ, ϕ) or solution variables defined in them, and we will cross-refer to them indistinguishably. In both θ and ϕ directions, the grid points are even spaced such that $\Delta\theta = \Delta\phi$. However, in the r direction, a new variable λ is introduced as a reference coordinate in reference space, which is exponentially related with r by $r = a^{\lambda}, a = 1.481$. In this way, the grid spacing Δr in the r direction is always around $r\Delta\theta$ and $r \sin\theta \Delta\phi$ ($\Delta\theta$ is already chosen to be equal to $\Delta\phi$) by choosing $\Delta\lambda = \log_a(1 + \Delta\theta)$, so that each grid cell is always approximately a cube. The $\sin\theta$ term varies from 1 to $1/\sqrt{2}$ within the 6 grid components, which shows that avoiding the singularity of the spherical grid indeed results in more regular cells. Thus, the curvilinear or reference coordinates (ξ, η, ζ) used in the CESE solver refer to (λ, θ, ϕ) here. Meanwhile, the grid cell in the reference space (ξ, η, ζ) is a rectangular box. In principle, any parameter $a > 1$ can be a choice. In order to make the cells more regular in the physical space, we only need to choose $\Delta\lambda = \log_a(1 + \Delta\theta)$ such that $\Delta r = a^{\lambda}(a^{\Delta\lambda} - 1) = r\Delta\theta$. We have no preference of selecting the parameter a, but the larger the parameter a, the denser the grid in the reference space, which makes no difference due to the AMR implementation. We have tested many cases for such a choice $a \in [1.2, 3]$, but have seen no difference in the numerical results. The choice $a = 1.481$ is just one case for presenting the simulated results here.

To be specific, the transformation between the coordinates of the reference space and the physical space for one component reads as follows:

$$\begin{cases} x_1 = a^\xi \sin\eta\cos\zeta \\ y_1 = a^\xi \sin\eta\sin\zeta \\ z_1 = a^\xi \cos\eta \end{cases}, \quad \begin{cases} \xi = \log_a\sqrt{x_1^2+y_1^2+z_1^2} \\ \eta = \arccos(z_1/\sqrt{x_1^2+y_1^2+z_1^2}) \\ \zeta = \arctan(y_1/x_1) \end{cases}. \tag{6.5}$$

which will be denoted by $x = x(\xi, \eta, \zeta), y = y(\xi, \eta, \zeta), z = z(\xi, \eta, \zeta)$ or $\xi = \xi(x, y, z), \eta = \eta(x, y, z), \zeta = \zeta(x, y, z)$. The Jacobian matrix of this nonsingular curvilinear coordinate transform $\mathbf{J}$ is written as:

$$\mathbf{J} = \frac{\partial(x, y, z)}{\partial(\xi, \eta, \zeta)} = \begin{pmatrix} a^\xi \sin\eta\cos\zeta\ln a & a^\xi\cos\eta\cos\zeta & -a^\xi\sin\eta\sin\zeta \\ a^\xi\sin\eta\sin\zeta\ln a & a^\xi\cos\eta\sin\zeta & a^\xi\sin\eta\cos\zeta \\ a^\xi\cos\eta\ln a & -a^\xi\sin\eta & 0 \end{pmatrix}, \tag{6.6}$$

and its determinant is $J = |\mathbf{J}| = a^{3\xi}\sin\eta\ln a$.

With these transformations (6.5), (6.6) and those of variables between components, we can get the corresponding transformation of coordinates between any two components or transformation of solution vector variables between the reference space and the physical space on an arbitrary component. The 3D MHD equations (6.1) of solar wind plasma in the reference coordinates (ξ, η, ζ) can read below:

$$\frac{\partial\hat{\mathbf{U}}}{\partial t} + \frac{\partial\hat{\mathbf{F}}}{\partial\xi} + \frac{\partial\hat{\mathbf{G}}}{\partial\eta} + \frac{\partial\hat{\mathbf{H}}}{\partial\zeta} - \frac{\partial\hat{\mathbf{F}}_\mu}{\partial\xi} - \frac{\partial\hat{\mathbf{G}}_\mu}{\partial\eta} - \frac{\partial\hat{\mathbf{H}}_\mu}{\partial\zeta} = \hat{\mathbf{S}}, \tag{6.7}$$

where $\hat{\mathbf{U}}, \hat{\mathbf{F}}, \hat{\mathbf{G}}, \hat{\mathbf{H}}, \hat{\mathbf{F}}_\mu, \hat{\mathbf{G}}_\mu, \hat{\mathbf{H}}_\mu, \hat{\mathbf{S}}$ are given in Eq. (21) by Jiang et al. [55].

Note that Eq. (6.7) is still written in conservation form, just like Eq. (6.1). As a result, the CESE solver for Eq. (6.7) in the reference coordinate space (ξ, η, ζ) can be derived in the same way as that for Eq. (6.1) according to Jiang et al. [55].

With the CESE solver for both Eqs. (6.1) and (6.7), we can easily take turns between the physical solution variables and the reference solution variables. If the initial inputs of the physical solution variables are given, we first use $\mathbf{F}, \mathbf{G}, \mathbf{H}, \mathbf{S}$ to calculate the fluxes $\hat{\mathbf{F}}, \hat{\mathbf{G}}, \hat{\mathbf{H}}, \hat{\mathbf{S}}$ and apply the CESE solver to Eq. (6.7) to obtain the reference solution variables $\hat{\mathbf{U}}$ and their first-order derivatives $(\hat{\mathbf{U}}_\xi, \hat{\mathbf{U}}_\eta, \hat{\mathbf{U}}_\zeta)$ at the new time step. Finally, we can recover the physical solution variables $\mathbf{U}$. Of course, we can reverse the above process from the reference solution variables to the physical solution variables.

It is evident that the only difference between the CESE solver in Cartesian coordinate and reference coordinate is the coordinate transformation **J**, and consequently the same solver applies to both the grid systems. Meanwhile, the present CESE solver in the reference space (ξ, η, ζ), formulated through the curvilinear transform from the physical space to the reference space, solves the transformed MHD equations with rectangular box cells in a logically Cartesian space, and the treatment of time iteration by integrating two half timesteps into one full timestep can be seen as a one-step update that leads to low-storage and makes the scheme suitable building blocks for adaptive mesh refinement calculations. These advantages greatly simplify the effective realization of numerical methods and allow us to easily carry out the AMR implementation with the help of PARAMESH.

Since the present AMR is realized in the reference space (ξ, η, ζ) and one component grid in the physical space corresponds to one associated reference component gird, the PARAMESH package decomposes every reference component (i.e., our computational space) into many blocks of the same size, and organizes all the blocks in the whole computational space (patched by the six reference components) into an oct-tree structure. In the implementation, we store both the solution variables and their first-order derivatives $(\hat{\mathbf{U}}, \hat{\mathbf{U}}_\xi, \hat{\mathbf{U}}_\eta, \hat{\mathbf{U}}_\zeta)$ at each solution point, which have the number of variables `nvf` $= 4 \times 8 = 32$. The block size is set to be the same as AMR strategy I. Again, PARAMESH manages all the rectangular blocks, and does not need to distinguish which group the blocks belong to, such that we can directly use most of the default operations provided by PARAMESH, such as the refining or coarsening of the blocks, prolongation, and restriction. Therefore, the data transferring and guard cell filling between blocks can be automatically accomplished by the package without large modification. In practice, the data exchange between these blocks in the same component is completed by PARAMESH. However, the data exchange at overlapping layers between different components adopt the same Lagrange interpolation as AMR strategy I. This, of course, will sacrifice the conservation at the component interfaces. The code is tested with dipole, multiple magnetic field and the observed magnetic field as input, and it is found that no spurious reflections are observed from the grid interface. This may be due to the fact that the conservative properties are more important to model shock waves either due to coronal mass ejections (CMEs) or corotating interaction regions (CIRs) than to model smooth solar wind.

The refinement strategy and multiple time-stepping algorithm employed by AMR strategy I also apply to AMR strategy II, and therefore are not repeated here.

6.6 Considerations of Second Order Accuracy AMR Schemes

Sufficient spatial resolution is one desirable characteristic for a numerical discretization scheme. It involves the capability to capture continuous and discontinuous waves with few nodes while not introducing spurious oscillations, and can be usually achieved through flux or slope limiters in a numerical scheme. High-order resolution approximations more than second order (for instance, the piecewise parabolic method (PPM) [22, 24] or corner transport upwind PPM (CTU-PPM) [44, 45] or 5th order weighted essentially non-oscillatory (WENO) [56] are effective in increasing the accuracy per unit mesh cell of finite difference methods under almost all circumstances. This also appears to be true for flows containing discontinuities such as shock waves. The nature of their effectiveness must include the increased cost incurred when invoking higher order approximations. The general belief is that the high-order numerical scheme will provide greater accuracy although the high-order method is complex and time consuming. The question to be considered is whether such improvement in accuracy is worth the additional cost. By taking account of the "balance" between accuracy and cost, we suggest that the MHD modeling from the Sun to Earth benefits from the second order scheme. In particular, in numerical study of solar-terrestrial physics problems, we prefer the AMR implementation of the 2nd order spatial resolution scheme without flux vector splitting. Here are given some comments on the advantages of using such kind of schemes.

On one hand, since formal order of accuracy is lost in a shock wave propagating in a coupled system of equations [69], the higher order methods do not accumulate the significant advantage available for smooth linear problems. In coupled hyperbolic problems with discontinuities, it was proven by Majda and Osher [69] that numerical methods with modest restrictions are subject to a loss of formal accuracy. The accuracy degenerates to first-order in all, but very special cases between characteristics emanating from a discontinuity. Later studies [17, 26] have revisited this issue, but without any evidence to modify Majda and Osher's result. Examples are found by Donat and Osher [26], where sub-cell resolution can overcome this problem. To some extent, in analogy, this problem can be reduced or mitigated by the AMR scheme because of the ability of the scheme to place grid points at the location in the flow where the largest truncation errors occur [13].

Greenough and Rider [48] and Rider and Kamm [94] carried out a quantitative comparison of numerical methods for the compressible Euler equations: fifth-order WENO, The piecewise parabolic method (PPM) and piecewise-linear method (PLM) of Godunov. In light of their conclusions, the higher-order method's advantage observed in the linear regime is not pronounced when nonlinearity is introduced. Similar work in this area indicates that the multi-dimensional results share similar phenomena that formally high order methods are comparable to the second-order Godunov method for nonlinear problems in multiple spatial dimensions [A.W. Cook, W.H. Cabot, J.A. Greenough, A comparison of high and low order schemes for shock-induced mixing, UCRL-JC-144109.]. These studies show that the impact of the grid spacing is more important than that of the difference accuracy on the performance of a numerical scheme.

On the other hand, the governing equations describing the solar wind plasma are often modified to include other terms such as the thermal conduction and the resistivity, divergence–cleaning procedure, which leads to the difficulty in globally guaranteeing the same high-order accuracy as the high-order scheme owns. As done in the numerical study of incompressible Navier–Stokes equations, simulations of turbulent channel flow indicate that the second-order central-difference scheme resolves the turbulent spectrum better than the high-order upwind schemes, and there is little evidence that the high-order treatment of the pressure equation adds sufficiently to the overall accuracy of the scheme even if fourth-order accurate central-difference approximations are used to construct high-order Laplacians in the pressure equation [105]. Also, AMR reflux and interpolation method may mitigate the overall high-order accuracy. Usually, higher-resolution numerical schemes need more ghostcells, which must be taken care of in AMR implementation in order to keep flux conservation. In particular, more ghostcells bring the challenging of the inner boundary determination at the solar surface [31]. So far, efforts to achieve better accuracy in numerical relativity have focused either on implementing second order accurate adaptive mesh refinement or on defining higher order accurate differences and update schemes. Lehner et al. [61] argued for the combination that adaptive gridding techniques provide to resolve fine scales (in addition to a more efficient use of resources) together with the higher accuracy furnished by higher order schemes when the solution is adequately resolved. But, to define a convenient higher order adaptive mesh refinement scheme should devise a novel approach to dealing with the grid boundaries for the use of high order operators within an adaptive framework.

Although many higher order accurate schemes such as WENO and PPM have existed for a long time, implementing them to the AMR framework is still in its infancy. Li and Hyman [62] combined high order finite difference WENO with AMR, where the use of a piecewise linear prolongation in space reduces the scheme to second order accuracy. Shen et al. [99] proposed a finite difference AMR-WENO method for hyperbolic conservation laws by combining the AMR framework with the high order finite difference WENO method in space and the total variation diminishing (TVD) Runge–Kutta (RK) method in time (WENO-RK) by a high order coupling. These AMR-WENO studies showed a significant improvement of the fifth order AMR-WENO over the third order one, not only in accuracy for smooth problems, but also in its ability in resolving complicated solution structures, due to the very low numerical diffusion of high order schemes. At the same time, it is found that designing a robust AMR-WENO scheme that is both conservative and high order (higher than second order) is a difficult task, due to the mass inconsistency of coarse and fine grid solutions at the initial stage in a finite difference scheme. In realizing such high order coupling, the following major roadblocks exist: (1) it is difficult to maintain high order accuracy across several levels of grids. In the AMR algorithm, the fine mesh solutions are interpolated from the coarse mesh solutions. Solutions at boundary points are needed not only at the current time but also at intermediate fine grid time steps (subcycles) as well as RK sub-stages. The same temporal and spatial accuracy are required in the data prolongation procedures as those of the base integrator. (2) introducing a high order accurate scheme in the AMR setting makes it harder to maintain local mass conservation. It is critical to conserve the properties of solutions when the mesh resolution changes. According to Berger and Colella [13], a flux correction step was introduced to preserve the mass conservation for the traditional finite volume AMR method. However, the space-time conservative CESE formulation differs from previous techniques because of global and local flux conservation in a space-time domain without resorting to interpolation or extrapolation, and no flux correction was needed [19, 20]. (3) in a higher order accurate scheme, robust and inexpensive refinement criteria are difficult to find. (4) for the MHD equations, multi-dimensional AMR MHD computations must handle the additional complication of maintaining a solenoidal magnetic field [33, 35, 57, 109].

Although Berger and Colella [13] proposed an adaptive mesh refinement (AMR) scheme for hydrodynamics to conserve scalar quantities (e.g., mass, energy) and numerical fluxes, additional challenges are presented in physical systems satisfying the Stokes's law type of equation with the divergence free evolution of vector fields, such as velocity fields in incompressible hydrodynamics and magnetic fields in magnetohydrodynamics (MHD). Divergence control methods are highly application dependent. As pointed out by Tóth et al. [109], since the $\nabla \cdot \mathbf{B} = 0$ condition is a consequence of the initial condition only, the local time stepping conventionally used in AMR will lead to a steady state solution with non-zero $\nabla \cdot \mathbf{B}$. To mitigate this problem, modifications of the induction equation, such as the 8-wave scheme [90], the parabolic/hyperbolic cleaning or the generalized Lagrangian multiplier [124], and the projection scheme [15]) may be used so that the divergence free condition depends on the boundary conditions and not on the initial conditions. The constrained transport scheme [12, 28, 133], although popular, cannot be combined with the local time stepping towards steady state, because it relies on the initial condition to maintain the divergence-free magnetic field. Meanwhile, although the directionally splitting approach can be easily applied to any kind of grids and in any dimensions for hydrodynamic problems, it is unsuitable for MHD in that dimensionally split MHD algorithms cannot preserve the divergence-free constraint between each one-dimensional update [9, 44, 45], because it is impossible to enforce the divergence-free constraint between partial updates unless all three components of the magnetic field are updated together, which in turn violates the assumption crucial to the splitting method that each dimensional operator is independent and can be split from the others. As pointed out by Balsara [9], the directionally split algorithms are also unsuitable for AMR applications in which the fluxes and electric fields that are needed for flux correction or electric field correction steps should contain all the contributions from all the waves that arrive at the boundaries or the edges (e.g., [6–8]). In principle, the fluxes and fields that are obtained from directionally split algorithms do lose this property, while the unsplit algorithms do have this attribute. The CESE method itself does inherently satisfy the discrete divergence constraint of the magnetic field, however the numerical results from the application of the CESE method can be improved further by using a divergence cleaning procedure [29, 31, 33, 35, 64, 91, 109, 130, 131] such as the projection method, the constraint-transport method, the hyperbolic/parabolic cleaning method or the generalized Lagrangian multiplier, and eight-wave method.

A detailed comparison of AMR versus high order schemes is given by Jameson [53] for a range of different problems. It is believed that the AMR scheme will be advantageous if the AMR region is below a certain proportion of the entire domain; and that it is advantageous to have an AMR scheme with as high order as the regularity of the partial differential equation. As pointed out by Jameson [53], AMR methods can be very effective as long as two fundamental requirements are satisfied:

(1) The region of the domain needing refinement must be confined to a relatively small fraction of the computation domain. Roughly no more than 1/3 of the domain should be at the finest grid spacing. (2) The numerical order of the AMR scheme should be as close as possible to the numerical order of the computational data and should not exceed it. That is, if one has a flow that is essentially shock dominated, roughly of piecewise linear structure, and if the shocks never fill more than 1/3 of the domain, then low order AMR schemes can offer a large computational savings when compared to other methods of calculation. These two fundamental requirements are evidenced by the solar-terrestrial phenomena such as heliospheric current sheet (HCS), corotating interaction regions (CIRs) and coronal mass ejections (CMEs) or interplanetary shocks, which can never occupy more than 1/3 of the vast solar-terrestrial space.

It should be also mentioned that the degeneracy of eigenvalues in MHD system suggest us to avoid the flux-splitting schemes as possible as we can. The MHD equation system forms a non-strictly hyperbolic system. The eigenvalues also can reduce locally to coincide with each other when the magnetic field switches polarity. As a consequence, the analytic structure of the weak solution is unknown where degeneracy occurs. Nevertheless, most numerical approaches for solving this equation system adopt the characteristic-based or approximate Riemann [77, 89, 128] and total variation diminishing (TVD) schemes [106]. Roe [97] has pointed out that one could not expect to find a legitimate solution by solving differential equations in regions where the true solution is not differentiable. Therefore, the numerical methods built on the integral rather than differential conservative laws should be preferentially considered. In the CESE method, by treating space and time as one entity, the ideal MHD equations are formulated in a space-time integral form, and are solved by the space-time conservation element and solution element method. Contrast to the modern upwind methods, no reconstruction procedure or Riemann solver is needed in this approach. Its computational logic and operational count are much simpler and more efficient. Moreover, no special treatment has been employed to maintain the divergence-free condition for the magnetic field [29, 131]. Nevertheless, the divergence free constraint has been faithfully maintained in smooth region, while the magnitude of $|\nabla \cdot \mathbf{B}|$ is bounded in regions near shocks.

Near the computational domain boundaries, the accuracy of the spatial derivatives has to be decreased. Conventionally, centered differences have to be replaced by one-sided differences because grid points are available only on the interior side of the boundary. As pointed out by Poinsot and Lele [87], the overall accuracy of the scheme is not affected if the order of approximation near the boundary is equal to the scheme order minus one. Numerical validations [31, 33, 34, 36, 126] indicate that this is not a major difficulty.

Based on the above arguments or comments, the second order spatial resolution schemes is a flexible choice in AMR modeling for solar wind. To some extent, based on the space-time conservation integral method without using Riemann solver, the CESE MHD solver of second order accuracy in space-time domain is a reasonable candidate for MHD-based modeling solar-terrestrial physics problems.

6.7 Numerical Results

To validate the application of adaptive mesh refinement in six-component grid system for the 3D SIP-CESE MHD model, the simulation results for steady solar corona and interplanetary space are compared with the observation from multiple spacecraft. The study intervals are chosen to be CR 1967, CR 2009, CR 2060 and CR 2094, which represent the characteristics of the solar maximum, declining, minimum and rising phases in solar cycles (SCs) 23 and 24 respectively.

The calculations are performed between 1 R_s (i.e., the base of the corona) and 328 R_s. The initial solar-surface temperature and number density are 1.3×10^6 K and $2.0 \times 10^8\,\text{cm}^{-3}$. The solar corona and interplanetary space is modeled by ideal MHD, so that both ν and η are set to zero. We employ the projected-normal characteristic inner boundary condition combined with the mass flux escaping through the solar surface (e.g., [50]). Then, the code is initialized by using the potential magnetic field based on the line-of-sight measurements of the photospheric magnetic field from Wilcox Solar Observatory (WSO) at Stanford University for the CRs of interest and Parker solar-wind solution. Finally, the model is run in time-relaxation method until a quasi-steady state is achieved. In fact, the same boundary conditions and initial conditions given in Sect. 5 of Feng et al. [31] are observed here.

Initially, the computational domain in every reference component is divided into $14 \times 4 \times 4$ blocks with each block consisting of $6 \times 6 \times 6$ cells. These correspond to $N_\theta = N_\phi = 25$ and $\Delta\theta = \pi/48$ by defining gird points on each component in physical space as $\theta_j^\ell = \theta_{\min} + j\Delta\theta$, $j = 0, 1, \ldots, N_\theta + 1$, $\phi_k^\ell = \phi_{\min} + k\Delta\phi$, $k = 0, 1, \ldots, N_\phi + 1$ and $\Delta\theta = (\theta_{\max} - \theta_{\min})/(N_\theta - 1)$, $\Delta\phi = (\phi_{\max} - \phi_{\min})/(N_\phi - 1)$, where N_θ and N_ϕ are the mesh numbers of the latitude and longitude, respectively. $\theta_{\min} = \frac{\pi}{4}$, $\theta_{\max} = \frac{3\pi}{4}$, $\phi_{\min} = \frac{3\pi}{4}$, $\phi_{\max} = \frac{5\pi}{4}$. The innermost region is set on the solar surface at 1 R_s and the outermost region on the sphere at 328 R_s, such that each component of the physical grids is equivalent to $0 \le \xi \le 14.75$, $\pi/4 - \delta \le \eta \le 3\pi/4 + \delta$, and $3\pi/4 - \delta \le \zeta \le 5\pi/4 + \delta$ in reference space, with $\Delta\xi = \log_a(1 + \Delta\theta)$, and $\Delta\eta = \Delta\zeta = \delta = \Delta\theta = \frac{\pi}{48}$.

The grid throughout the simulation is refined to obtain a grid cell size of about 0.15 R_s near the current sheet within 20 R_s and it is about 0.7 R_s near 1 AU. The maximum grid cell size is 1.3 R_s in the corona and 7 R_s in the inner heliosphere.

The validation shows that both AMR strategies mentioned above obtain almost the same results. Thus, we will only present the results obtained from strategy II in the following.

6.7.1 Comparisons with the Observation Near the Sun

Figure 6.2 presents the synoptic maps of the coronal holes from both observations and simulations for CR 1967 (Column a), CR 2009 (Column b), CR 2060 (Column c), and CR 2094 (Column d). The first row is the synoptic observations on the solar surface

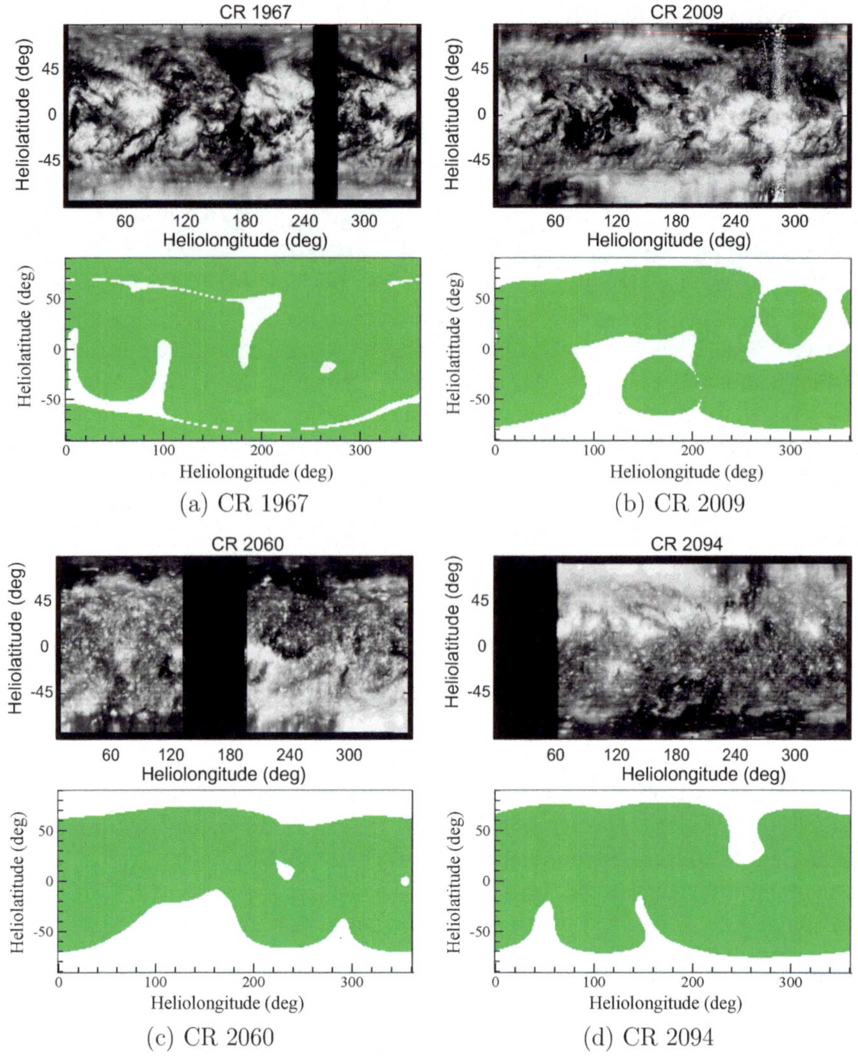

Fig. 6.2 Synoptic maps of the coronal holes at 1 R_s for CR 1967 (Column a), CR 2009 (Column b), CR 2060 (Column c), and CR 2094 (Column d). The first row is the SOHO/EIT 195 Å observation, in which the black vertical stripes represent data gaps. The second row presents the open-field regions and closed-field ones from the SIP-AMR-CESE MHD model, where the green areas are of closed magnetic field and the white regions are of open field. From Fig. 2 of Feng et al. [35]

at the wavelength of 195 Å taken with the Extreme ultraviolet Image Telescope (EIT) aboard the spacecraft of SOlar and Heliospheric Observatory (SOHO), where coronal holes appear as dark areas of weaker coronal emission due to lower density and temperature [49]. The second row shows the open-field and closed-field regions from the SIP-AMR-CESE MHD model.

It is generally accepted that the areas and the distributions of the coronal holes are different in different solar activity phases. Seen from the observation and the simulation in CR 1967, no polar coronal holes (PCHs) appear at either pole. In fact, this is one of the most significant features before the completion of the reversal of the polar field during solar maximum phases [14, 49, 113]. Only a few isolated coronal holes (ICHs) scatter around the Carrington longitudes of $\phi = 10°$, $100°$, $180°$ and $270°$ at the low and middle heliographic latitudes. We should note that the simulated locations of the ICHs are roughly in agreement with the observed ones but the total area of the ICHs from the simulation is a little smaller than that from measurements. For CR 2009 in the declining phase, the EIT observation and the calculation reveal that the PCHs tend to be asymmetric about both poles and that the two PCHs extend equatorward across the solar equator. The latitudinal median for the equatorial boundaries of PCHs excluding the extending holes is about 74°, and the lowest latitude is about $-10°$ for the northern extending hole and 10° for the southern one. In CR 2060, the Sun came into its minimum phase. The northern and southern PCHs are nearly symmetric about both poles. In addition, two small ICHs appear near the solar equator around $\phi = 230°$ and $350°$ from both the observation and simulation, which was rare during previous solar minima and was investigated thoroughly (e.g., [1, 107, 114, 125]). Additionally, the corona holes occupy the largest area in the four selected CRs, which agrees with that of Harvey and Recely [49]. The locations of the coronal hole boundaries are almost the same as those derived from the model of potential field source surface (PFSS) basing on the photospheric magnetograms of Mount Wilson Solar Observatory [66]. With the end of the extending solar minimum came the rising phase of solar cycle 24. In CR 2094, both PCHs from the simulation and observation shrink poleward rapidly and the simulated latitudinal median for the equatorial boundaries of PCHs moves poleward from about 63° in CR 2060 to 68° in CR 2094.

Associated with the differences in shapes and areas of the coronal holes on the solar surface among the four selected CRs, there also exist some corresponding distinctions near the Sun among them. Figure 6.3 shows the measured and computed synoptic maps at $2.5\,R_s$ for CR 1967 (Column a), CR 2009 (Column b), CR 2060 (Column c), and CR 2094 (Column d). The first and second rows are the white-light polarized brightness (pB) images at the east and west limbs from Large Angle Spectrometric COronograph (LASCO) C2 on board SOHO. Each image in the third row superimposes the isolines of the radial magnetic field on the contour map of the simulated number density N and the fourth row shows the contour maps of the radial speed v_r. In Row 4, the black solid lines denote the magnetic neutral lines (MNLs). The most significant differences among the four CRs shown in Fig. 6.3 include the topologies of the MNLs and the distributions of the low-density high-speed (LDHS) plasma flows. In the simulation for CR 1967, the MNL along the longitudes from 330° through 0° to 120° is almost coincident with the latitudinal circle of 60°N except that there is a dip around the longitude of 40° and the MNL along the longitudes from 160° to 300° lies in the latitude of 60°S. As a result, the MNL around the longitudes of 130° and 320° is nearly vertical and the heliosphere at low and middle latitudes is divided into two unipolar regions according to the IMF polarities. The

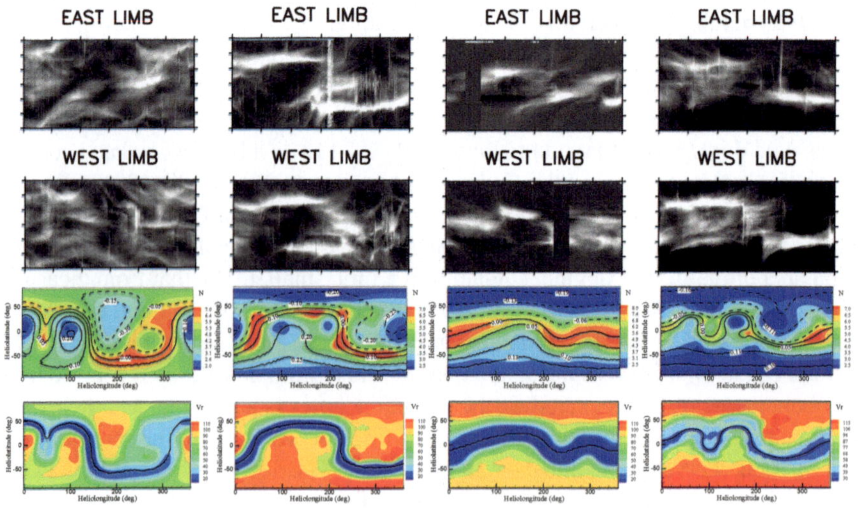

(a) CR 1967 (b) CR 2009 (c) CR 2060 (d) CR 2094

Fig. 6.3 Synoptic maps at 2.5 R_s for CR 1967 (Column a), CR 2009 (Column b), CR 2060 (Column c), and CR 2094 (Column d). The first and second rows are the white-light polarized brightness at the east and west limbs from SOHO/LASCO C2. The third row overlays the isolines of the radial magnetic field (unit: Gauss) on the contour map of the simulated number density N (unit: 10^5 cm^{-3}) and the fourth rows are the pseudo-color images of the radial speed v_r (unit: km s^{-1}). In Row 3, the dashed lines stand for inward radial magnetic field and the solid ones outward radial magnetic field. The black lines in the last row denote the magnetic neutral lines. From Fig. 3 of Feng et al. [35]

latitudes of the MNL from the simulation are a slightly smaller than those derived from Ulysses' crossings of the sector structure when it traveled from 80.2°S to 70°N [100, 101]. In addition, the LDHS regions are confined to the small areas in the neighborhood of the longitudes of 0°, 100° and 200° and the high-density low-speed regions spread widely, which is in agreement with Ulysses solar wind observations [71, 72]. Therefore, the bright features can be observed almost everywhere from the synoptic maps of pB measurements. However, the bright structures in the pB map recorded in CR 2009 are roughly distributed around the computed MNL, which lies at about 40°N from $\phi = 60°$ to 300° and 40°S from 240° to 20° and changes abruptly near the longitudes of 40° and 220°. It is obvious that the dark regions in the pB maps are consistent with the distribution of calculated LDHS plasma flows, which not only cover both poles, but also occupy the low- and middle-latitude regions around the longitudes of 0°, 100° and 200°. The latitudinal span of the high-density low-speed solar wind is compatible with the measurements of solar wind speed by Ulysses during this CR [73]. For CR 2060, the brightest streaks at both limbs in the pB maps from LASCO C2 are well organized around of the wavy MNL, which is confined to low latitudes near the solar equator and extends northward to 25°N at $\phi = 160°$ and southward to 15°S at $\phi = 230°$. Additionally, the percentage of the

LDHS regions is highest among the four CRs. As far as CR 2094 is concerned, the bright features in LASCO C2 observations can be seen at very broad latitudes due to the complex structure of the MNL, which displays a pattern of two peaks and two valleys seen from the simulation. The two peaks are located at $(\theta, \phi) = (35^\circ, 70^\circ)$ and $(35^\circ, 160^\circ)$. The narrow valley is located at $(-25^\circ, 100^\circ)$ and the wide one at $(-30^\circ, 260^\circ)$. The median of equatorial boundaries for the LDHS plasma flows from unipolar PCHs is about the latitude of 50° excluding the extending part of northern PCH in the wide trough.

Now we investigate the states of the solar wind from 2.5 to $6\,R_s$. Figure 6.4 displays the coronal observations and the simulated results for CR 1967 (Column a), CR 2009 (Column b), CR 2060 (Column c), and CR 2094 (Column d). The first four rows extract the computed results on the meridional plane at $\phi = 0^\circ - 180^\circ$ and some related observations. The first and second rows are the white light pB images from 2.3 to $6\,R_s$ recorded by LASCO C2/ SOHO and computed from the simulation, respectively. The LASCO C2 data are available from the webpage of http://sharpp.nrl.navy.mil/cgi-bin/swdbi/lasco/img_short/form. The third row is the simulated magnetic field topologies projected on the meridional planes from 1 to $6\,R_s$. The fourth row is the simulated radial solar wind speed on the meridional plane from 1 to $6\,R_s$, where the white quadrilaterals represent the blocks of different refinement levels showing that the regions near the current sheet are identified by the grid resolution. The last row exhibits the simulated three-dimensional current sheets from 1 to $6\,R_s$.

In CR 1967, the bright streaks extend at almost all latitudes in the images for both the numerical results and the observations and the dark area at southeast quadrant in the C2 observation results from the occulter's pylon of the telescope. The brightest structures extending farthest in CR 2009 just lie in the simulated streamers because most regions are covered with the LDHS plasma flows, which can be clearly seen from the simulation. For CR 2060, the current sheet is concentrated on the solar equator and of limited latitudinal excursion. The HCS during this period can be approximated as a flat plane, although the HCS is not as conventional as those in previous minima [107, 125] and slow solar winds span more latitudes so that the spacecraft near the Earth sampled fewer high solar wind flows [66]. Past studies have also revealed the different shapes of HCS in different solar activity phases. The HCS in the solar maximum extends to very high latitudes (e.g., [21]) and the HCS in the solar minimum is almost coincident with the solar equator (e.g., [21, 31, 52, 75, 96]). During CR 2094, we can find that the bright fan-like structures cover low and middle latitudes in both the observed and synthesized pB images, and that the current sheet fluctuates with relatively large amplitude and can not be described as a nearly flat plane any more, which is similar to the simulated HCS structure of CR 1934 by Mikić et al. [75]. We also notice that the simulated solar wind speeds at 6 solar radii are higher than those in van der Holst et al. [52], which probably results from the major acceleration region for the simulation is a little nearer to the Sun than that of the model by van der Holst et al. [52].

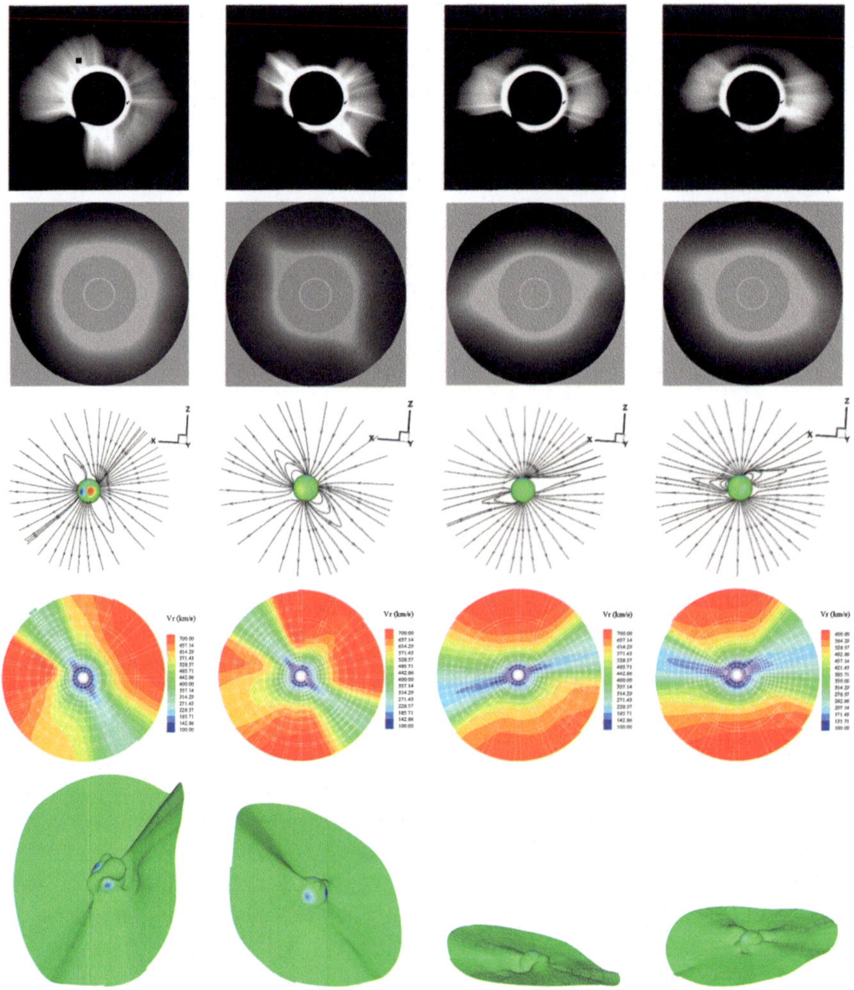

(a) CR 1967 (b) CR 2009 (c) CR 2060 (d) CR 2094

Fig. 6.4 The coronal observations and the simulation results on the meridional plane at $\phi = 0° - 180°$ for CR 1967 (Column a), CR 2009 (Column b), CR 2060 (Column c), and CR 2094 (Column d). The first and second rows are the white light pB images from 2.3 to 6 R_s recorded by LASCO C2/SOHO and computed from the simulation, respectively. The third row is the simulated magnetic field topology projected on the meridional planes from 1 to 6 R_s. The fourth row is the simulated radial solar wind speed on the meridional planes from 1 to 6 R_s, where the white quadrilaterals denote the grid blocks. The last row exhibits the simulated current sheets from 1 to 6 R_s. From Fig. 4 of Feng et al. [35]

6.7.2 Comparisons with the Observation in Interplanetary Space

After discussing the coronal states of the four CRs near the Sun, there follows the comparison between the simulated and observed results in interplanetary space.

The first row in Fig. 6.5 exhibits the composite images of the iso-lines for the radial speed v_r and the pseudo-color images of the mass flux density, F_{m}, escaping through the surface at $20\,\mathrm{R_s}$. Row 2 stands for the steady solutions in the solar equatorial plane for the selected four CRs and the white quadrilaterals imply that the blocks of higher refinement levels are clustering near the low speed region, which means that current sheet structure is well captured by the adaptive meshes. Row 3 and 4 show the simulated proton number densities N and v_r at $215\,\mathrm{R_s}$. Columns (a), (b), (c), and (d) correspond to the results for CRs 1967, 2009, 2060, and CR 2094. In those maps, the dashed green lines in the first row and the black solid lines in the third and last rows denote the magnetic neutral lines.

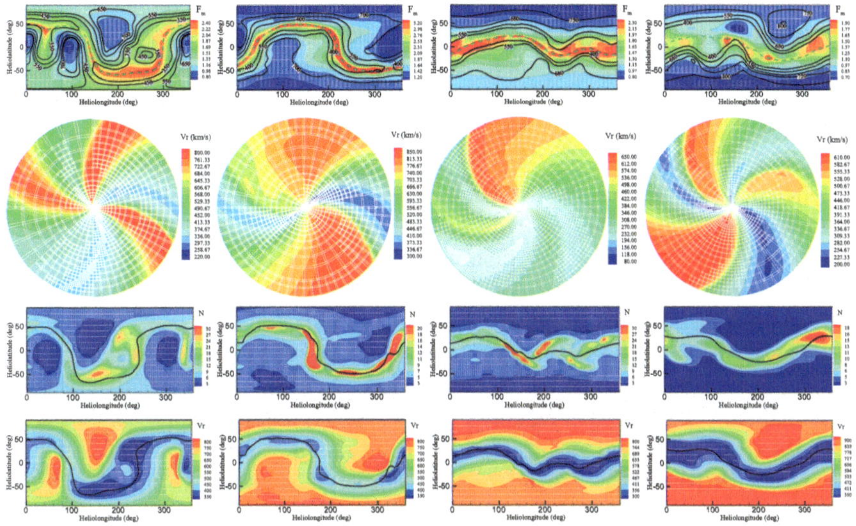

(a) CR 1967 (b) CR 2009 (c) CR 2060 (d) CR 2094

Fig. 6.5 The simulated steady solution in interplanetary space from the MHD model for CR 1967 (Column a), CR 2009 (Column b), CR 2060 (Column c), and CR 2094 (Column d). The isolines of the radial speed v_r (unit: $\mathrm{km\,s^{-1}}$) superimposed on the synoptic pseudo-color image of the mass flux density F_m (unit: $10^8\,\mathrm{km\,s^{-1}\,cm^{-3}\,R_s^2}$) at $20\,\mathrm{R_s}$ (Row 1), the pseudo-color images of the radial velocity in the solar equatorial plane (Row 2), the synoptic contours of the simulated proton number density N (unit: $\mathrm{cm^{-3}}$) (Row 3) and v_r (unit: $\mathrm{km\,s^{-1}}$) (Row 4) at $215\,\mathrm{R_s}$. The white quadrilaterals in Row 2 represent the grid blocks, where the blocks of higher refinement levels are associated with the resulting AMR grid intersected with the solar equator plane. In Rows 3 and 4, the black lines denote the magnetic neutral lines. From Fig. 5 of Feng et al. [35]

Comparing Figs. 6.3 and 6.5, we can see that for all the four CRs, the plasma flows of $v_r \geq 60\,\mathrm{km\,s^{-1}}$ at $2.5\,\mathrm{R_s}$ have been accelerated to the intermediate-speed solar winds (ISWs, $450 \leq v_r \leq 700\,\mathrm{km\,s^{-1}}$) or fast solar winds (FSWs, $v_r \geq 700\,\mathrm{km\,s^{-1}}$) of low mass flux density at $20\,\mathrm{R_s}$ and the slow solar winds (SSWs, $v_r \leq 450\,\mathrm{kms^{-1}}$) at $20\,\mathrm{R_s}$ of high density basically surrounding the HCS originate from the plasma flows of $v_r \leq 60\,\mathrm{km\,s^{-1}}$ at $2.5\,\mathrm{R_s}$. As a result, FSW covers the smallest area in CR 1967 and the largest in CR 2060. In addition, all SSW, ISW and FSW have experienced an acceleration of about $\sim$50–100 $\mathrm{km\,s^{-1}}$ from 20 to $215\,\mathrm{R_s}$. The radial variations of the acceleration for different types of solar winds beyond $2.5\,\mathrm{R_s}$ are consistent with those derived from both Ulysses solar corona experiment [86] and LASCO C2 and C3 measurements [88, 117]. The values of the mass flux densities and solar wind speeds are of the same magnitude as those given by Wei et al. [119], who analyzed the global distribution of coronal mass output at $2.5\,\mathrm{R_s}$ and its relation to solar magnetic field structures by using observational data from K-coronal brightness. Additionally, the results have also revealed the longitudinal deflections of the solar wind due to solar rotation and the freezing-in effect of interplanetary magnetic field (IMF) by comparing the shapes of HCSs at different solar radii and the distribution of solar wind speed on the solar equator plane. The spirals coil more tightly in slow-speed flow than those in high-speed flow as pointed out by Odstrčil and Pizzo [80]. The simulated results on the equatorial planes and at $215\,\mathrm{R_s}$ display the structure of corotating interaction regions (CIRs) when the rear ISSs and FSSs overtake the front SSWs, and at $215\,\mathrm{R_s}$ the CIRs are evident around $\phi = 160°$ and $340°$ in CR 2009, $\phi = 140°$ and $200°$ in CR 2060, and $\phi = 300°$ in CR 2094. It is interesting to notice that for CR 2094 the two-peak and two-valley pattern of the HCS near the Sun become less apparent and more regular near the Earth, due to the interaction between ISWs and SSWs during their outward propagation.

Figure 6.6 demonstrates the calculated MHD steady solutions at 1 AU for CR 1967 (Column a), CR 2009 (Column b), CR 2060 (Column c), and CR 2094 (Column d). The rows from top to bottom show the comparisons between the MHD results and the 1-h averaged OMNI data near the Earth for radial solar wind speed v_r, number density N, temperature T, and radial magnetic field B_r, respectively. Roughly speaking, the modeled steady solutions in Fig. 6.6 can capture the observed changing trends of the solar wind parameters in CR 1967, but all the three peaks of solar wind speed decline more slowly than that observed, and have a bit higher speeds and proton number densities. For CR 2009, the model achieves two humps in the speed profile, which agrees well with the in-situ measurements. However, it should be noted that the first simulated hump rises 1.5 days earlier. The simulation shows the best consistence with the measurements during CR 2060. The simulated maxima in the profiles of v_r, N, and T and their arrivals can match the observations reasonably well. The simulation gives a fairly good match in CR 2094 for the two intermediate-speed flows observed from Day 12 to 15 and from Day 17 to 20, but the first one arrives 2.5 days earlier. The relatively large discrepancy between the simulation and the measurements from Day 26 to 2 should come from the significant differences of the photospheric magnetic fields between the start and the end of the CR. Generally speaking, the accuracy of the simulated solar wind speeds is at the same level as

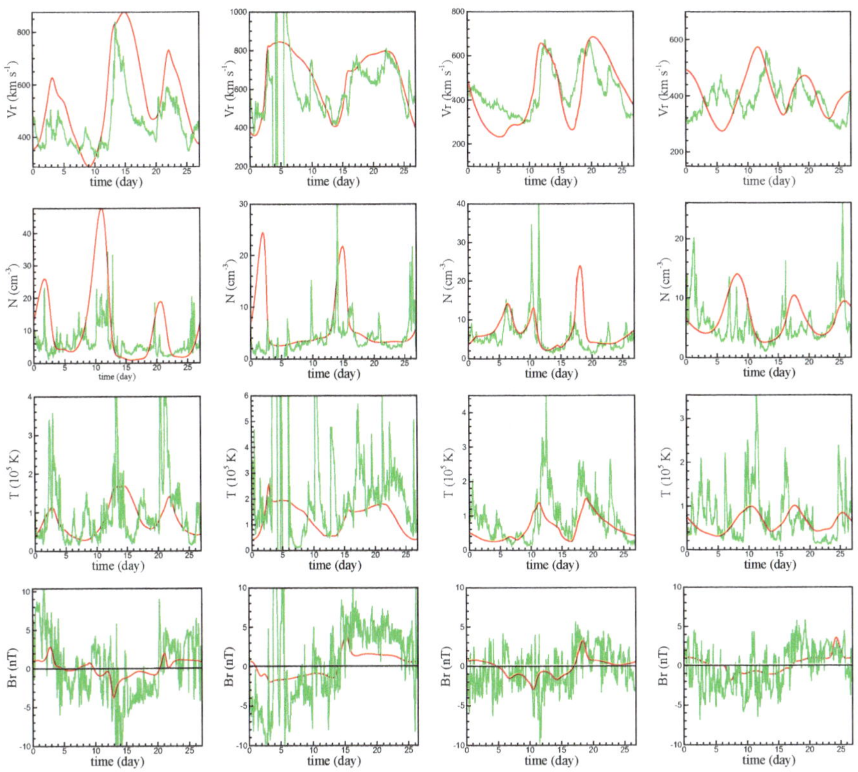

(a) CR 1967 (b) CR 2009 (c) CR 2060 (d) CR 2094

Fig. 6.6 The calculated MHD steady at 1 AU for CR 1967 (Column a), CR 2009 (Column b), CR 2060 (Column c), and CR 2094 (Column d). The first, second, third, fourth rows are the comparisons between the MHD results and the 1-h averaged OMNI data near 1 AU for radial solar wind speed v_r, number density N, temperature T, and radial magnetic field B_r, respectively, where the green lines denote the observations and the red lines represent the numerical results. From Fig. 6 of Feng et al. [35]

that for other models (e.g., [21, 52, 96]). From the plots of the radial magnetic field during the four CRs, we can see that the IMF polarities and their changes are captured by the simulations with fairly good accuracy. However, the simulated magnitudes are only one-third to one-half of those from OMNI data for all the four CRs, which exists in almost all the 3D MHD models of steady and transient solar winds by using the potential field as the initial state (e.g., [21, 52, 65, 96]).

From the above comparisons, we can see that the simulated results both near the Sun and in interplanetary space reproduce the observations much better for the solar minimum and declining phases than those for the solar maximum and rising phases, which may be resulted from the fact the photospheric magnetic fields around the solar maximum and the rising phase change so rapidly due to their evolution that the changes will cause much error if we assume they basically change little during a CR. To incorporate the daily updated solar magnetograms into the present model may partially improve the situation [3, 75]. In addition, the discrepancies between the simulations and the observations for the solar maximum and rising phases can partially be attributed to the contaminations by the increasing number of coronal mass ejections.

6.8 Conclusions and Discussions

This chapter carries out SIP-CESE MHD model on six-component grid by employing two different AMR strategies. In AMR strategy I, the solution-adaptive technique enables the direct application of the CESE solver introduced in previous chapters. In AMR strategy II, through the use of nonsingular transform from the physical space to the reference space, the quadrangular frustum pyramid cell in the spherical shell computational domain for solar wind modeling becomes the conventional rectangular box in the reference space (ξ, η, ζ), which can be seen as the usual Cartesian coordinate. Both strategies make it possible to directly utilize most of the default operations provided by PARAMESH. Data transferring and guard cell filling between blocks can be automatically accomplished by the PARAMESH package without large modification. As a consequence, we obtain a grid system that well fit the spherical surface of the Sun and an easy-to-use AMR implementation at the same time.

Decomposition of a spherical surface grid (θ, ϕ) into six identical components with piling up the radial direction, combined with a simple radial decomposition of the spherical shell computational domain, can be managed to yield very good load balancing since the processor workload and the communication load are very similar for each processor. However, this is not necessarily true for other grids, in particular AMR in Cartesian coordinates.

Numerical validations through CRs 1967, 2009, 2060, and 2094 for the solar maximum, declining, minimum and rising phases are carried out. The numerical simulations have reproduced a lot of features near the Sun during the corresponding solar activity phases. For CR 1967, no PCHs appear at either pole, and only a few ICHs scatter at the low and middle heliographic latitudes. The high-density low-speed regions can be observed everywhere. For CR 2009, LDHS plasma flows from the simulation are coincident with pB measurements from LASCO C2 and the latitudinal span of the high-density low-speed solar wind is compatible with Ulysses' measurements of solar wind speed. The numerical results in CR 2060 have some distinctions from the previous minima demonstrated by other researchers [107, 125]. The MNL in CR 2094 displays the structure of two peaks and two valleys, which is also consistent with the past results [51]. Many observed interplanetary structures have also been reproduced by the simulations. At 1 AU, the modeled steady solutions have captured the observed changing trends of the solar wind parameters for all the four CRs except that some peaks of solar wind speed decline more slowly and arrive no more than 3 days earlier than observed. In addition, the IMF polarities and their changes are captured by the simulated results with fairly good accuracy.

On the one hand, the model can reproduce most of the observed features among the four selected CRs, which are essentially determined by the photospheric magnetic fields, especially their dipole and quadrupole components [98] and the polar open fields [114]; on the other hand, it should be also noted that there exist some differences between the numerical results and observations, such as the magnitude of radial magnetic fields and arrival times of ISWs at the Earth. As discussed in Chap. 5, coronal heating and solar wind acceleration has much room for improvement. The uncertainties of photospheric (especially polar regions) magnetic measurements and imperfectness of the potential magnetic field are also important factors leading to the differences between the observations and numerical results. Although the CESE-MHD model is able to reproduce the locations of the coronal hole boundaries in EIT measurements, the magnitude of open flux tends to be lower compared with in-situ measurements at 1 AU, and the discrepancy becomes greater around solar maximum when the in-situ IMF can be almost twice as strong as our model. A lot of research has revealed that a significant portion of the heliospheric open magnetic flux originates from active regions and coronal mass ejections by means of interchange reconnections [21, 40, 83, 95, 127]. As pointed out by Mikić et al. [76], the discrepancies between the modeled results and observations arise from limitations in our ability to observe the Sun's magnetic field. In addition, the occurrence of some coronal mass ejections (CMEs) during the simulated period and the interaction between solar wind and interplanetary discontinuities, which have been neglected in our model, can also affect the variation trends of solar wind parameters.

Smarter AMR algorithms allow models to focus the computational effort at regions where the most spatial resolution is required. This capability enables the model of the ambient solar wind to generate more detailed fine structure in the lower corona and interplanetary space with reduced computational resource, and therefore improves the computational efficiency of the model. Besides AMR, there exists other advanced computational techniques that can be used by the solar wind model. Along with additional physical processes and more complex field topologies included, physics-based models become more expensive to run, which can lead to issues with accessibility, data transfer speeds, and guarantees of acceptable time to solution. For such data intensive computing, accelerated computing using Graphics Processing Unit (GPU) technology has become a viable tool. A number of developers of space weather models or space plasma simulations have already explored the use of GPUs (e.g., [5, 37, 38, 46, 59, 60, 116, 120–122]). However, these adopters have to re-write the computation-intensive part of their code using low-level APIs such as CUDA or OpenCL, which usually requires great human-power resources. The OpenACC 2.6 standard (https://www.openacc.org) offers an easy-to-use GPU-accelerated computations, while maintaining compatibility with existing MPI implementations. The present OpenACC probably paves the way for the modelers to modify these existing codes into GPU versions, just as currently being done by the Predictive Science Inc. (https://devblogs.nvidia.com/solar-storm-modeling-gpu-openacc/). Besides, dual-time stepping method [58, 129], and preconditioning technique (e.g., [2, 18, 79, 93, 115]) for the MHD system, which may accelerate and optimize algorithms themselves, can be further investigated for speeding up solar wind modeling.

Notes: Figure 6.1 and Sects. 6.3, 6.4 and 6.6 are reprinted from *Computer Physics Communications*, 185, Xueshang Feng, Changqing Xiang, Dingkun Zhong, Yufen Zhou, Liping Yang, Xiaopeng Ma, SIP-CESE MHD model of solar wind with adaptive mesh refinement of hexahedral meshes, 1965–1980. Copyright 2014, with permission from Elsevier. Figures 6.2, 6.3, 6.4, 6.5 and 6.6, Sects. 6.2 and 6.7, and part of Sect. 6.5 are reprinted by permission from Springer Nature Customer Service Centre GmbH: Springer *Solar Physics*, Validation of the 3D AMR SIP-CESE solar wind model for four Carrington rotations, Feng XS, Yang LP, Xiang CQ, Jiang CW, Ma X, Wu ST, Zhong DK, Zhou YF. Copyright 2012. Part of Sect. 6.5 is also reprinted by permission from Springer Nature Customer Service Centre GmbH: Springer *Solar Physics*, AMR Simulations of Magnetohydrodynamic Problems by the CESE Method in Curvilinear Coordinates, Chaowei Jiang, Xueshang Feng, Jian Zhang, Dingkun Zhong. Copyright 2010.

References

1. Abramenko V, Yurchyshyn V, Linker J, Mikić Z, Luhmann J, Lee CO (2010) Low-latitude coronal holes at the minimum of the 23rd solar cycle. Astrophys J 712:813
2. Amari T, Luciani JF, Joly P (1999) A preconditioned semi-implicit method for magnetohydrodynamics equations. SIAM J Sci Comput 21(3):970–986. https://doi.org/10.1137/S1064827596304824
3. Arge CN, Pizzo VJ (2000) Improvement in the prediction of solar wind conditions using near-real time solar magnetic field updates. J Geophys Res 105(10):465
4. Aschwanden MJ, Burlaga LF, Kaiser ML, Ng CK, Reames DV, Reiner MJ, Gombosi TI, Lugaz N, Manchester W, Roussev II, Zurbuchen TH, Farrugia CJ, Galvin AB, Lee MA, Linker JA, Mikić Z, Riley P, Alexander D, Sandman AW, Cook JW, Howard RA, Odstrčil D, Pizzo VJ, Kóta J, Liewer PC, Luhmann JG, Inhester B, Schwenn RW, Solanki SK, Vasyliunas VM, Wiegelmann T, Blush L, Bochsler P, Cairns IH, Robinson PA, Bothmer V, Kecskemety K, Llebaria A, Maksimovic M, Scholer M, Wimmer-Schweingruber RF (2008) Theoretical modeling for the STEREO mission. Space Sci Rev 136:565–604. https://doi.org/10.1007/s11214-006-9027-8
5. Asensio IA, Laguna AA, Aissa MH, Poedts S, Ozak N, Lani A (2019) A GPU-enabled implicit finite volume solver for the ideal two-fluid plasma model on unstructured grids. Comput Phys Commun 239:16–32. https://doi.org/10.1016/j.cpc.2019.01.019
6. Balsara DS (1998) Total variation diminishing scheme for adiabatic and isothermal magnetohydrodynamics. Astrophys J Suppl Ser 116(1):133
7. Balsara DS (2001) Divergence-free adaptive mesh refinement for magnetohydrodynamics. J Comput Phys 174:614–648. https://doi.org/10.1006/jcph.2001.6917
8. Balsara DS (2001) Total variation diminishing scheme for relativistic magnetohydrodynamics. Astrophys J Suppl Ser 132:83–101. https://doi.org/10.1086/318941
9. Balsara DS (2004) Second-order-accurate schemes for magnetohydrodynamics with divergence-free reconstruction. Astrophys J Suppl Ser 151:149–184
10. Balsara DS (2012) Self-adjusting, positivity preserving high order schemes for hydrodynamics and magnetohydrodynamics. J Comput Phys 231(22):7504–7517

11. Balsara DS, Spicer D (1999) Maintaining pressure positivity in magnetohydrodynamic simulations. J Comput Phys 148(1):133–148
12. Balsara DS, Spicer DS (1999) A staggered mesh algorithm using high order Godunov fluxes to ensure solenoidal magnetic fields in magnetohydrodynamic simulations. J Comput Phys 149(1):270–292
13. Berger MJ, Colella P (1989) Local adaptive mesh refinement for shock hydrodynamics. J Comput Phys 82:64–84. https://doi.org/10.1016/0021-9991(89)90035-1
14. Bilenko IA (2002) Coronal holes and the solar polar field reversal. Astron Astrophys 396:657
15. Brackbill JU, Barnes DC (1980) The effect of nonzero $\nabla \cdot \mathbf{B}$ on the numerical solution of the magnetohydrodynamic equations. J Comput Phys 35(3):426–430. https://doi.org/10.1016/0021-9991(80)90079-0
16. Bridges TJ (2008) Conservation laws in curvilinear coordinates: a short proof of Vinokur's theorem using differential forms. Appl Math Comput 202(2):882–885
17. Casper J, Carpenter MH (1998) Computational considerations for the simulation of shock-induced sound. SIAM J Sci Comput 19(3):813–828
18. Chacón L, Stanier A (2016) A scalable, fully implicit algorithm for the reduced two-field low-β extended MHD model. J Comput Phys 326:763–772. https://doi.org/10.1016/j.jcp.2016.09.007
19. Chang SC (1995) The method of space-time conservation element and solution element: a new approach for solving the Navier-Stokes and Euler equations. J Comput Phys 119(2):295–324. https://doi.org/10.1006/jcph.1995.1137
20. Chang SC, Wang XY, Chow CY (1999) The space-time conservation element and solution element method: a new high-resolution and genuinely multidimensional paradigm for solving conservation laws. J Comput Phys 156(1):89–136
21. Cohen O, Sokolov IV, Roussev II, Gombosi TI (2008) Validation of a synoptic solar wind model. J Geophys Res 113(A12):A03, 104
22. Colella P, Woodward PR (1984) The piecewise parabolic method (PPM) for gas-dynamical simulations. J Comput Phys 54(1):174–201
23. Colella P et al (2007) Chombo software package for AMR applications design document. Technical report, Applied Numerical Algorithms Group, NERSC Division, Lawrence Berkeley National Laboratory Berkeley, CA
24. Dai W, Woodward PR (1998) A simple finite difference scheme for multidimensional magnetohydrodynamical equations. J Comput Phys 142(2):331–369
25. De Zeeuw DL (1993) A quadtree-based adaptively-refined Cartesian-grid algorithm for solution of the Euler equations. PhD thesis, University of Michigan, Ann Arbor, MI, USA, UMI Order No. GAX94-09674
26. Donat R, Osher S (1990) Propagation of error into regions of smoothness for non-linear approximations to hyperbolic equations. Comput Methods Appl Mech Eng 80(1–3):59–64
27. Dryer M (2007) Space weather simulation in 3D MHD from the Sun to the Earth and beyond to 100 AU: a modeler's perspective of the present state of the art. Asian J Phys 16:97–121
28. Evans CR, Hawley JF (1988) Simulation of magnetohydrodynamic flows - a constrained transport method. Astrophys J 332:659–677. https://doi.org/10.1086/166684
29. Feng XS, Hu YQ, Wei FS (2006) Modeling the resistive MHD by the CESE method. Solar Phys 235:235–257
30. Feng XS, Zhou YF, Wu ST (2007) A novel numerical implementation for solar wind modeling by the modified conservation element/solution element method. Astrophys J 655:1110
31. Feng XS, Yang LP, Xiang CQ, Wu ST, Zhou YF, Zhong DK (2010) Three-dimensional solar wind modeling from the Sun to Earth by a SIP-CESE MHD model with a six-component grid. Astrophys J 723:300
32. Feng XS, Xiang CQ, Zhong DK (2011) The state-of-art of three-dimensional numerical study for corona-interplanetary process of solar storms (in chinese). Sci Sin-Terrae 41:1–28
33. Feng XS, Zhang SH, Xiang CQ, Yang LP, Jiang CW, Wu ST (2011) A hybrid solar wind model of the CESE+HLL method with a Yin-Yang overset grid and an AMR grid. Astrophys J 734:50

34. Feng XS, Jiang CW, Xiang CQ, Zhao XP, Wu ST (2012) A data-driven model for the global coronal evolution. Astrophys J 758(1):62
35. Feng XS, Yang LP, Xiang CQ, Jiang CW, Ma XP, Wu ST, Zhong DK, Zhou YF (2012) Validation of the 3D AMR SIP-CESE solar wind model for four Carrington rotations. Solar Phys 279:207–229
36. Feng XS, Yang LP, Xiang CQ, Liu Y, Zhao XP, Wu ST (2012) Numerical study of the global corona for CR 2055 driven by daily updated synoptic magnetic field. In: Pogorelov NV, Font JA, Audit E, Zank GP (eds) Numerical modeling of space plasma slows (ASTRONUM 2011), Astronomical society of the pacific conference series, vol 459, p 202
37. Feng XS, Zhong DK, Xiang CQ, Zhang Y (2013) GPU-accelerated computing of three-dimensional solar wind background. Sci China Earth Sci 56(11):1864–1880
38. Feng XS, Zhong DK, Xiang CQ, Zhang Y (2013) GPU computing in space weather modeling. In: Pogorelov NV, Audit E, Zank GP (eds) Numerical modeling of space plasma flows (ASTRONUM2012), Astronomical society of the pacific conference series, vol 474, p 131
39. Feng XS, Xiang CQ, Zhong DK, Zhou YF, Yang LP, Ma XP (2014) SIP-CESE MHD model of solar wind with adaptive mesh refinement of hexahedral meshes. Comput Phys Commun 185:1965–1980. https://doi.org/10.1016/j.cpc.2014.03.027
40. Fisk LA, Schwadron NA, Zurbuchen TH (1999) Acceleration of the fast solar wind by the emergence of new magnetic flux. J Geophys Res 104(19):765
41. Fryxell B (2000) Flash: an adaptive mesh hydrodynamics code for modeling astrophysical thermonuclear flashes. Astrophys J Suppl Ser 131:273
42. Fuchs FG, McMurry AD, Mishra S, Risebro NH, Waagan K (2010) Finite volume methods for wave propagation in stratified magneto-atmospheres. Commun Comput Phys 7(3):1–30
43. Garaizar X, Hornung R, Kohn S (1999) Structured adaptive mesh refinement applications infrastructure. Technical report, Lawrence Livermore National Laboratory
44. Gardiner TA, Stone JM (2005) An unsplit Godunov method for ideal MHD via constrained transport. J Comput Phys 205(2):509–539
45. Gardiner TA, Stone JM (2008) An unsplit Godunov method for ideal MHD via constrained transport in three dimensions. J Comput Phys 227(8):4123–4141. https://doi.org/10.1016/j.jcp.2007.12.017
46. Germaschewski K, Raeder J (2011) Using automated code generation to support high performance extended MHD integration in OpenGGCM. In: Pogorelov NV, Audit E, Zank GP (eds) 5th international conference of numerical modeling of space plasma flows (ASTRONUM 2010), Astronomical society of the pacific conference series, vol 444, p 197
47. Gombosi TI, de Zeeuw DL, Powell KG et al (2003) Adaptive mesh refinement for global magnetohydrodynamic simulation. In: Büchner J, Dum C, Scholer M (eds) Space plasma simulation. Lecture notes in physics, vol 615. Springer, Berlin, pp 247–274
48. Greenough JA, Rider WJ (2004) A quantitative comparison of numerical methods for the compressible Euler equations: fifth-order WENO and piecewise-linear Godunov. J Comput Phys 196:259–281
49. Harvey KL, Recely F (2002) Polar coronal holes during cycles 22 and 23. Solar Phys 211:31
50. Hayashi K (2005) Magnetohydrodynamic simulations of the solar corona and solar wind using a boundary treatment to limit solar wind mass flux. Astrophys J Suppl Ser 161:480
51. Hoeksema JT, Wilcox JM, Scherrer PH (1983) The structure of the heliospheric current sheet-1978-1982. J Geophys Res 88(A12):9910–9918. https://doi.org/10.1029/JA088iA12p09910
52. van der Holst B, Manchester WB, Frazin RA, Vásquez AM, Tóth G, Gombosi TI (2010) A data-driven, two-temperature solar wind model with Alfvén waves. Astrophys J 725(1):1373
53. Jameson L (2003) AMR vs high order schemes. J Sci Comput 18:1–24
54. Janhunen P (2000) A positive conservative method for magnetohydrodynamics based on HLL and Roe methods. J Comput Phys 160(2):649–661. https://doi.org/10.1006/jcph.2000.6479
55. Jiang CW, Feng XS, Zhang J, Zhong DK (2010) AMR simulations of magnetohydrodynamic problems by the CESE method in curvilinear coordinates. Solar Phys 267:463–491
56. Jiang GS, Wu CC (1999) A high-order WENO finite difference scheme for the equations of ideal magnetohydrodynamics. J Comput Phys 150:561–594

57. Keppens R, Meliani Z, van Marle A, Delmont P, Vlasis A, van der Holst B (2012) Parallel, grid-adaptive approaches for relativistic hydro and magnetohydrodynamics. J Comput Phys 231(3):718–744
58. Kifonidis K, Müller E (2012) On multigrid solution of the implicit equations of hydrodynamics. Experiments for the compressible Euler equations in general coordinates. Astron Astrophys 544:A47. https://doi.org/10.1051/0004-6361/201116979
59. Lani A, Yalim MS, Poedts S (2014) A GPU-enabled finite volume solver for global magnetospheric simulations on unstructured grids. Comput Phys Commun 185(10):2538–2557
60. Innocenti ME, Johnson A, Markidis S, Amaya J, Deca J, Olshevsky V, Lapenta G (2017)Progress towards physics-based space weather forecasting with exascale computing. Adv Eng Softw 111:3–17
61. Lehner L, Liebling SL, Reula O (2006) AMR, stability and higher accuracy. Class Quantum Gravity 23(16):S421
62. Li ST, Hyman JM (2003) Adaptive mesh refinement for finite difference WENO schemes. Technical report, LA-UR-03-8927, Los Alamos National Lab
63. Linde T (2002) MHD simulations with the FLASH code. In: APS March meeting abstracts, p F3.005
64. Liu YX, Ji Z, Feng XS, Zhou YF (2010) CESE method for resistive magnetohydrodynamics. Chin J Space Sci 30(3):211–220
65. Lugaz N, Downs C, Shibata K, Roussev II, Asai A, Gombosi TI (2011) Numerical investigation of a coronal mass ejection from an Anemone active region: reconnection and deflection of the 2005 August 22 eruption. Astrophys J 738(2):127
66. Luhmann JG, Lee CO, Li Y, Arge CN, Galvin AB, Simunac K, Russell CT, Howard RA, Petrie G (2009) Solar wind sources in the late declining phase of cycle 23: effects of the weak solar polar field on high speed streams. Solar Phys 256(1):285–305. https://doi.org/10.1007/s11207-009-9354-5
67. MacNeice P, Olson KM, Mobarry C, de Fainchtein R, Packer C (2000) PARAMESH: a parallel adaptive mesh refinement community toolkit. Comput Phys Commun 126:330–354
68. MacNeice P, Olson KM, Mobarry C, de Fainchtein R, Packer C (2011) PARAMESH V4.1: parallel adaptive mesh refinement. Astrophys Sour Code Libr. arXiv:1106.009
69. Majda A, Osher S (1977) Propagation of error into regions of smoothness for accurate difference approximations to hyperbolic equations. Commun Pure Appl Math 30(6):671–705
70. Marder B (1987) A method for incorporating Gauss' law into electromagnetic PIC codes. J Comput Phys 68(1):48–55
71. McComas DJ, Elliott HA, Gosling JT, Reisenfeld DB, Skoug RM, Goldstein BE, Neugebauer M, Balogh A (2002) Ulysses' second fast-latitude scan: complexity near solar maximum and the reformation of polar coronal holes. Geophys Res Lett 29(9):1290
72. McComas DJ, Elliott HA, Schwadron NA, Gosling JT, Skoug RM, Goldstein BE (2003) The three-dimensional solar wind around solar maximum. Geophys Res Lett 30(10):1517
73. McComas DJ, Elliott HA, Gosling JT, Skoug RM (2006) Ulysses observations of very different heliospheric structure during the declining phase of solar activity cycle 23. Geophys Res Lett 33:9102
74. Mignone A, Bodo G, Massaglia S, Matsakos T, Tesileanu O, Zanni C, Ferrari A (2007) PLUTO: a numerical code for computational astrophysics. Astrophys J Suppl Ser 170:228–242. https://doi.org/10.1086/513316
75. Mikić Z, Linker JA, Schnack DD, Lionello R, Tarditi A (1999) Magnetohydrodynamic modeling of the global solar corona. Phys Plasmas 6:2217–2224
76. Mikić Z, Downs C, Linker JA, Caplan RM, Mackay DH, Upton LA, Riley P, Lionello R, Török T, Titov VS, Wijaya J, Druckmüller M, Pasachoff JM, Carlos W (2018) Predicting the corona for the 21 August 2017 total solar eclipse. Nat Astron 2:913–921. https://doi.org/10.1038/s41550-018-0562-5
77. Myong RS, Roe PL (1998) On Godunov-type schemes for magnetohydrodynamics. J Comput Phys 147:545–567

78. Nakamizo A, Tanaka T, Kubo Y, Kamei S, Shimazu H, Shinagawa H (2009) Development of the 3-D MHD model of the solar corona-solar wind combining system. J Geophys Res (Space Physics) 114:A07109. https://doi.org/10.1029/2008JA013844
79. Nishikawa H, Roe P, Suzuki Y, van Leer B (2003) A general theory of local preconditioning and its application to 2D ideal MHD equations. In: 16th AIAA computational fluid dynamics conference, American Institute of Aeronautics and Astronautics, Orlando, Florida, Fluid dynamics and co-located conferences, AIAA 2003-3704. https://doi.org/10.2514/6.2003-370410.2514/6.2003-3704
80. Odstril D, Pizzo VJ (1999) Distortion of the interplanetary magnetic field by three-dimensional propagation of coronal mass ejections in a structured solar wind. J Geophys Res 104:28, 225
81. Ogino T, Walker RJ (1984) A magnetohydrodynamic simulation of the bifurcation of tail lobes during intervals with a northward interplanetary magnetic field. Geophys Res Lett 11(10):1018–1021. https://doi.org/10.1029/GL011i010p01018
82. Olson K (2006) PARAMESH: a parallel adaptive grid tool. In: James M, Gunther B, Nobuyuki S, Akin E, Anil D (eds) Parallel computational fluid dynamics 2005: theory and applications: proceedings of the parallel CFD conference. Elsevier, Amsterdam, pp 341–348
83. Owens MJ, Crooker NU (2006) Coronal mass ejections and magnetic flux buildup in the heliosphere. J Geophys Res 111(A10):A10104
84. Parashar M (2007) Grace grid adaptive computational engine. Technical report, Rutgers University
85. Parent B (2012) Positivity-preserving high-resolution schemes for systems of conservation laws. J Comput Phys 231(1):173–189
86. Pätzold M, Tsurutani BT, Bird MK (1997) An estimate of large-scale solar wind density and velocity profiles in a coronal hole and the coronal streamer belt. J Geophys Res 102(24):151
87. Poinsot TJ, Lele SK (1992) Boundary conditions for direct simulations of compressible viscous flows. J Comput Phys 101:104–129
88. Porfir'eva GA, Yakunina GV, Delone AB, Oreshina AV, Oreshina IV (2009) Coronal streamers on the Sun and their physical properties. J Phys Stud 13(2):2901–2905
89. Powell KG, Roe PL, Myong RS, Aee DD (1995) An upwind scheme for magnetohydrodynamics. AIAA 95-1704-CP, pp 661–671
90. Powell KG, Roe PL, Linde TJ, Gombosi TI, De Zeeuw DL (1999) A solution-adaptive upwind scheme for ideal magnetohydrodynamics. J Comput Phys 154:284–309. https://doi.org/10.1006/jcph.1999.6299
91. Qamar S, Mudasser S (2010) On the application of a variant CE/SE method for solving two-dimensional ideal MHD equations. Appl Numer Math 60(6):587–606
92. Rendleman CA, Beckner VE, Lijewski M, Crutchfield W, Bell JB (2000) Parallelization of structured, hierarchical adaptive mesh refinement algorithms. Comput Vis Sci 3:147
93. Reynolds DR, Samtaney R, Tiedeman HC (2012) A fully implicit Newton-Krylov-Schwarz method for tokamak magnetohydrodynamics: Jacobian construction and preconditioner formulation. Comput Sci Discov 5(1):014003. http://stacks.iop.org/1749-4699/5/i=1/a=014003
94. Rider W, Kamm J (2006) How effective are high-order approximations in shock-capturing methods? is there a law of diminishing returns? In: Groth C, Zingg DW (eds) Computational fluid dynamics 2004. Springer, Berlin, pp 401–405
95. Riley P (2007) An alternative interpretation of the relationship between the inferred open solar flux and the interplanetary magnetic field. Astrophys J Lett 667:L97
96. Riley P, Lionello R, Linker JA, Mikic Z, Luhmann J, Wijaya J (2011) Global MHD modeling of the solar corona and inner heliosphere for the whole heliosphere interval. Solar Phys 274:361–377
97. Roe PL (1986) Characteristic-based schemes for the Euler equations. Annu Rev Fluid Mech 18:337–365
98. Sanderson TR, Appourchaux T, Hoeksema JT, Harvey KL (2003) Observations of the Sun's magnetic field during the recent solar maximum. J Geophys Res 108(A1):1035. https://doi.org/10.1029/2002JA009388

99. Shen C, Qiu JM, Christlieb A (2011) Adaptive mesh refinement based on high order finite difference WENO scheme for multi-scale simulations. J Comput Phys 230:3780–3802
100. Smith EJ (2011) Solar cycle evolution of the heliospheric magnetic field: the Ulysses legacy. J Atmos Solar Terr Phys 73:277
101. Smith EJ, Marsden RG, Balogh A, Gloeckler G, Geiss J, McComas DJ, McKibben RB, MacDowall RJ, Lanzerotti LJ, Krupp N, Krueger H, Landgraf M (2003) The sun and heliosphere at solar maximum. Science 302(5648):1165–1169. https://doi.org/10.1126/science.1086295
102. Stone JM, Gardiner TA, Teuben P, Hawley JF, Simon JB (2008) Athena: a new code for astrophysical MHD. Astrophys J Suppl Ser 178:137–177. https://doi.org/10.1086/588755
103. Sun M (1998) Numerical and experimental studies of shock wave interaction with bodies. PhD thesis, Tohoku University
104. Sun M, Takayama K (2001) A solution-adaptive technique using unstructured hexahedral grids. AIAA-2001-2656
105. Tafti D (1996) Comparison of some upwind-biased high-order formulations with a second-order central-difference scheme for time integration of the incompressible Navier-Stokes equations. Comput Fluids 25(7):647–665
106. Tanaka T (1994) Finite volume TVD scheme on an unstructured grid system for three-dimensional MHD simulation of inhomogeneous systems including strong background potential fields. J Comput Phys 111:381–390. https://doi.org/10.1006/jcph.1994.1071
107. Tokumaru M, Kojima M, Fujiki K, Hayashi K (2009) Non-dipolar solar wind structure observed in the cycle 23/24 minimum. Geophys Res Lett 36:9101
108. Tóth G, Sokolov IV, Gombosi TI, Chesney DR, Clauer CR, De Zeeuw DL, Hansen KC, Kane KJ, Manchester WB, Oehmke RC, Powell KG, Ridley AJ, Roussev II, Stout QF, Volberg O, Wolf RA, Sazykin S, Chan A, Yu B, Kota J (2005) Space weather modeling framework: a new tool for the space science community. J Geophys Res Space Phys 110(A12):12, 226
109. Tóth G, van der Holst B, Sokolov IV, De Zeeuw DL, Gombosi TI, Fang F, Manchester WB, Meng X, Najib D, Powell KG, Stout QF, Glocer A, Ma YJ, Opher M (2012) Adaptive numerical algorithms in space weather modeling. J Comput Phys 231:870–903. https://doi.org/10.1016/j.jcp.2011.02.006
110. van der Holst B, Keppens R (2007) Hybrid block-AMR in Cartesian and curvilinear coordinates: MHD applications. J Comput Phys 226(1):925–946
111. Vinokur M (1974) Conservation equations of gasdynamics in curvilinear coordinate systems. J Comput Phys 14(2):105–125
112. Waagan K (2009) A positive MUSCL-Hancock scheme for ideal magnetohydrodynamics. J Comput Phys 228:8609–8626
113. Waldmeier M (1981) Cyclic variations of the polar coronal hole. Solar Phys 70:251
114. Wang Y, Robbrecht E, Sheeley NR (2009) On the weakening of the polar magnetic fields during solar cycle 23. Astrophys J 707:1372–1386. https://doi.org/10.1088/0004-637x/707/2/1372
115. Wang Y, Feng XS, Xiang CQ (2019) An effective matrix-free implicit scheme for the magnetohydrodynamic solar wind simulations. Comput Fluids 179:67–77. https://doi.org/10.1016/j.compfluid.2018.10.014
116. Wang Y, Feng XS, Zhou YF, Gan XB (2019) A multi-GPU finite volume solver for magnetohydrodynamics-based solar wind simulations. Comput Phys Commun 238:181–193. https://doi.org/10.1016/j.cpc.2018.12.003
117. Wang YM, Sheeley Jr NR, Walters JH, Brueckner GE, Howard RA, Michels DJ, Lamy PL, Schwenn R, Simnett GM (1998) Origin of streamer material in the outer corona. Astrophys J Lett 498(2):L165–L168. http://stacks.iop.org/1538-4357/498/i=2/a=L165
118. Watermann J, Wintoft P, Sanahuja B, Saiz E, Poedts S, Palmroth M, Milillo A, Metallinou FA, Jacobs C, Ganushkina NY, Daglis IA, Cid C, Cerrato Y, Balasis G, Aylward AD, Aran A (2009) Models of solar wind structures and their interaction with the Earth's space environment. Space Sci Rev 147(3):233–270. https://doi.org/10.1007/s11214-009-9494-9
119. Wei FS, Feng XS, Cai HC, Zhou QJ (2003) Global distribution of coronal mass outputs and its relation to solar magnetic field structures. J Geophys Res 108:1238

120. Wong HC, Wong UH, Feng X, Tang Z (2011) Efficient magnetohydrodynamic simulations on graphics processing units with CUDA. Comput Phys Commun 182(10):2132–2160
121. Wong UH, Aoki T, Wong HC (2014a) Efficient magnetohydrodynamic simulations on distributed multi-GPU systems using a novel GPU direct-MPI hybrid approach. Comput Phys Commun 185(7):1901–1913
122. Wong UH, Wong HC, Ma Y (2014b) Global magnetohydrodynamic simulations on multiple GPUs. Comput Phys Commun 185(1):144–152
123. Wu ST, Wang AH, Liu Y, Hoeksema JT (2006) Data-driven magnetohydrodynamic model for active region evolution. Astrophys J 652:800
124. Yalim M, Abeele DV, Lani A, Quintino T, Deconinck H (2011) A finite volume implicit time integration method for solving the equations of ideal magnetohydrodynamics for the hyperbolic divergence cleaning approach. J Comput Phys 230(15):6136–6154. https://doi.org/10.1016/j.jcp.2011.04.020
125. Yang LP, Feng XS, Xiang CQ, Zhang SH, Wu ST (2011) Simulation of the unusual solar minimum with 3D SIP-CESE MHD model by comparison with multi-satellite observations. Solar Phys 271:91–110
126. Yang LP, Feng XS, Xiang CQ, Liu Y, Zhao XP, Wu ST (2012) Time-dependent MHD modeling of the global solar corona for year 2007: driven by daily-updated magnetic field synoptic data. J Geophys Res 117(A16):A08110
127. Yeates AR, Mackay DH, van Ballegooijen AA, Constable JA (2010) A nonpotential model for the Sun's open magnetic flux. J Geophys Res 115(A14):A09, 112
128. Zachary AL, Colella P (1992) A higher-order Godunov method for the equations of ideal magnetohydrodynamics. J Comput Phys 99:341–347
129. Zhang M, Feng XS (2015) Implicit dual-time stepping method for a solar wind model in spherical coordinates. Comput Fluids 115:115–123. https://doi.org/10.1016/j.compfluid.2015.03.020
130. Zhang M, Blankson I, Chang SC, Lin SC, Yu STJ (2004) Solving magnetohydrodynamic equations without special treatment for divergence-free magnetic field. AIAA J 42:2605–2608
131. Zhang M, John Yu ST, Henry Lin SC, Chang SC, Blankson I (2006) Solving the MHD equations by the space time conservation element and solution element method. J Comput Phys 214:599–617
132. Ziegler U (2008) The nirvana code: parallel computational MHD with adaptive mesh refinement. Comput Phys Commun 179(4):227–244. https://doi.org/10.1016/j.cpc.2008.02.017
133. Ziegler U (2011) A semi-discrete central scheme for magnetohydrodynamics on orthogonal-curvilinear grids. J Comput Phys 230(4):1035–1063. https://doi.org/10.1016/j.jcp.2010.10.022

Chapter 7
Data-Driven MHD Modeling of Solar Wind

Abstract Solar wind is rather than of steady state, but physically dynamic corresponding to the solar rotation, solar mass flow and solar magnetic field evolution. In fact, solar wind is constantly time varying. Data-driven modeling of solar wind here means to use continuously time-varying solar observations as input to drive models to produce solar wind background and reproduce eruptive process of solar active region, instead of using one instantaneous cadence of observation as input. The latter is in contrast called data-constrained modeling. The idea of data-driven modeling is to let continuously observed data and MHD evolve to produce what is happening next. This chapter describes how to use the global photospheric field maps generated by the solar surface flux Transport (SFT) model for long-interval synoptic data products or time interpolation for short-interval synoptic data products as input to drive the CESE-MHD coronal and solar wind model.

7.1 Introduction

Physically speaking, photospheric magnetic field dominates solar coronal states, and thus controls the heliospheric large-scale solar wind structures [57, 63, 68]. Global estimates of the solar photospheric magnetic field distribution are critical for most numerical solar wind modeling efforts. The magnetic field in the photosphere is inferred by Zeeman-splitting measurements. Currently, the line-of-sight (LOS) component of the photospheric magnetic field can be obtained with the most reliability. Unfortunately, it is impossible to obtain the true instantaneous global measurement of the LOS photospheric magnetic field at a certain time because there is no direct photospheric magnetic field observation of the far-side of the Sun at present. However, over the course of a solar rotation, it is possible to generate a synoptic map from photospheric magnetograms, which may provide a general view of the global LOS magnetic flux distribution [39]. The conventional full-Carrington Rotation (full-CR) synoptic maps are constructed by the following procedures. First, the magnetograms are remapped into longitude, measured from the central meridian, and sin(Latitude). These remapped images are then shifted to the appropriate longitude in the Carrington frame according to its central meridian passing time, and merged together using a

© Springer Nature Singapore Pte Ltd. 2020

X. Feng, *Magnetohydrodynamic Modeling of the Solar Corona and Heliosphere*, Atmosphere, Earth, Ocean & Space,
https://doi.org/10.1007/978-981-13-9081-4_7

weighted mean for overlapping regions to form a full-surface picture of the solar magnetic field. The weighted factor is a function of longitudinal distance ϕ, like $\cos^4 \phi$, ensuring that measurements taken at a particular Carrington time contribute most to that Carrington longitude in the synoptic map. Synoptic maps from different observatories are constructed using magnetograms of different cadences. For example, the Wilcox Solar Observatory (WSO) uses daily magnetograms to construct synoptic maps [2]. Synoptic maps generated by the Michelson Doppler Imager (MDI) onboard the Solar and Heliospheric Observatory (SOHO) spacecraft utilize 96-min cadence magnetograms [87]. Synoptic maps provided by the Global Oscillation Network Group (GONG) uses one-minute magnetograms gathered from GONG's six sites' observation (https://gong.nso.edu/data/dmac_magmap/). Synoptic map from the Helioseismic and Magnetic Imager (HMI) onboard the Solar Dynamics Observatory (SDO) utilizes 720 s line-of-sight magnetograms [34]. Assuming that the magnetic fields are approximately radial at the photosphere, the LOS component synoptic map may be converted to radial component synoptic map. Most current MHD models for the solar wind from the Sun to Earth usually initialize the computation by using potential magnetic field based on those full-CR synoptic maps and Parker solar wind solution and then run the model in time-relaxation method until a steady-state equilibrium is achieved. To list a few, the WSO synoptic maps are employed by Feng et al. [16, 17], and Cohen et al. [8]. The MDI/SOHO synoptic maps are used by Hayashi et al. [32], Cohen et al. [9] and Riley et al. [79]. The HMI/SDO synoptic maps are used by van der Holst et al. [98] and Meng et al. [62], and the 3D MHD corona simulation module of Hayashi et al. [34] has become part of the HMI/SDO data pipeline. The GONG synoptic maps are accepted by a variety of models [20, 77, 97] and employed by intercomparison study between different models [43]. The model simulation results have an underlying sensitivity to the synoptic maps, but no definite conclusion has been made [27, 77, 78] about which data source would systematically produce better result.

Due to the LOS measurements only available from the ground stations or the instruments board on spacecraft orbiting near the ecliptic plane, there are data gaps in the polar regions of the synoptic maps, because the projection effect on the LOS measurements leads to too much noise of the data in polar regions and once a year the inclination of the elliptic plane with respect to the solar equatorial plane causes data from any one pole to be missing entirely at certain time. It is necessary to correct the problematic values in the polar region of the input synoptic map for both ground-based and spacecraft observation, especially for period near solar minimum when the photospheric polar fields are the main source of the interplanetary magnetic filed (IMF) flux. A variety of such polar field correction methods have been put forward. For example, Svalgaard et al. [88] analyzed the magnetograms at Wilcox Solar Observatory (WSO) and concluded that the magnetic field poleward 55° latitude was nearly radial and of the form $11.5\cos^8 \theta$ (where θ is colatitude) during 1976–1977. Wang and Sheeley [103] developed potential models of solar corona by using similar formulae to extrapolate the polar field. Arge and Pizzo [2] filled the polar field by fitting a second-order polynomial to the most reliable observations, which were made when the absolute values of the solar b angles were greater than 5°. Using the data from SOHO/Michelson Doppler Imager (MDI), Liu et al. [54] compared the performances of interpolating the polar fields with the following seven methods: one dimensional (1D) cubic spline interpolation method, the potential field method, the smoothed 1D cubic spline interpolation method, the method of Svalgaard et al. [88], the two dimensional (2D) temporal interpolation method, the 2D spatial interpolation method, and the flux-transport model based method [81]. Sun et al. [87] developed a technique by combining a two-dimensional spatial/temporal interpolation and a simple version of the flux transport model.

The full-CR synoptic maps are constructed to represent the global distribution of the photospheric magnetic field of the 27.2753 days' interval under the assumption that the photospheric magnetic field does not change significantly during one CR. However, it is well known that even during solar minimum, the photospheric field is always evolving and at solar maximum period, it can change drastically over a Carrington rotation. In order to give more realistic global photospheric magnetic field, the high-cadence synoptic map is developed such that the most recent observations are merged into the synoptic map whenever available (e.g., [2, 34, 73]). The advantage of such high-cadence map is that the part of the map directed towards Earth consists of the most recent magnetic observations available, especially east of the Sun's central meridian. For instance, the daily-updated synoptic map are routinely provided by WSO (http://wso.stanford.edu/forms/prsyn.html) and HMI/SDO (http://jsoc.stanford.edu/ajax/lookdata.html), while the hourly or 6-h updated synoptic map is provided by GONG (https://gong2.nso.edu/archive/patch.pl?menutype=z).

Global simulation of dynamic solar wind is one of the challenging problems in space weather modeling. The MHD modeling for the solar corona and solar wind is essentially time-independent, which integrate the MHD equations to obtain a steady-state solution by the time-relaxation method. The implicit assumption in the full-CR synoptic map-based MHD models is that the evolution of the field is quasi-steady and the evolution can be approximated by a sequence of steady states [48]. To validate the quasi-steady approximation, the photospheric magnetic fields are supposed to evolve slowly compared to the timescale for establishing a steady solution. However, as mentioned above, the photospheric field is always evolving. Also, the time history of the coronal field is neglected in calculating its steady state structure. Another disadvantage about the quasi-steady assumption is that MHD model results can not simultaneously responds to the dynamic or time-dependent evolution of the global corona and solar wind, which are rather than of steady state, but physically dynamic corresponding to the solar rotation, solar mass flow and solar magnetic field evolution, especially during solar eruptions [59, 94, 126]. The basic processes that produce the ambient solar wind or occur in it are controlled by intrinsically time-dependent processes [76]. The steady solar wind prescribed by the full-CR synoptic maps can basically capture some global structures, but they can hardly reproduce the dynamic features in the solar corona and heliosphere (e.g., [17, 92, 96, 97]).

In order to better determine background solar wind changes, researchers advanced their simulation using high-cadence synoptic maps of the photospheric magnetic field instead of the full-CR synoptic maps. Arge and Pizzo [2] input the daily-updated synoptic maps into the Wang-Sheele-Arge (WSA) model, and improved the prediction of the solar wind conditions with a continuous empirical function. Hayashi [31] applied the temporally varying solar radial magnetic field to the time-dependent solar wind MHD model, which can trace the evolution of the global solar coronal response to variations of the photospheric magnetic field. Intriligator et al. [42] combined the WSA model with the daily-updated synoptic maps to achieve the inner boundary at 0.1AU for the model of hybrid heliospheric modeling system with pickup protons (HHMS-PI). In addition, Hernández et al. [35] incorporated the data of far-side active regions derived from helioseismology into the daily-updated data.

It is well known that solar differential rotation causes the solar surface to spin at different rates as measured by motions of structures on the photosphere. However, the synoptic maps of photospheric magnetic field are generally based on the premise that the photosphere rotates rigidly at different latitudes, so they contain no information on the solar differential rotation. Lionello et al. [53] used a global 3D MHD model to study the effects of differential rotation on the coronal magnetic field by introducing an artificial differential rotation into a steady corona solution obtained from the time relaxation method. They identified examples of interchange reconnection and found other changes of the magnetic field topology due to the differential rotation. Zhao et al. [125] proposed a modified version of synoptic charts, named a "synchronic frame", in which the positions of the synoptic data are longitudinally shifted in accordance with the solar differential rotation. The synchronic frame data of photospheric magnetic field have been used by Hayashi et al. [33] in simulating the global solar corona around the Halloween event in 2003.

Another way to involve the effect of solar differential rotation into the evolution of synoptic maps is the surface flux transport (SFT) model. The SFT model [13, 105] describes the evolution of the distribution of magnetic flux (i.e., radial magnetic field) at the photosphere as a combined result of differential rotation, meridian flow and supergranular diffusion [58]. New magnetic flux emergence are added to the model either by empirical method [112] or using observation data [115]. Previous study has demonstrate its ability to match observed strength and distribution of photosphere radial magnetic field ([82] and reference therein). It is able to provide the Snapshots of the Entire-surface Photospheric Magnetic Flux (SEPMF). The Air Force Data Assimilative Photospheric Flux Transport (ADAPT) model developed by Arge et al. [4] incorporate a modified version of the Worden and Harvey's SFT model [115] with EnLS data assimilation method to provide input for WSA model. A latter version of ADAPT has been modified to assimilate helioseismic far-side active region data [5, 36]. Together with the coronal models, the SFT model has been successfully applied to describe a number of phenomena on the Sun [44, 81, 105, 109, 120, 121].

Meanwhile, modelers should be very cautious to utilize the daily-updated or higher-cadence synoptic maps to drive the solar wind simulations. These synoptic maps usually contain too much noise and very strong magnetic field in active regions. A proper smoothing procedure is needed to make the photospheric magnetic map friendly as input to MHD models. In addition, the synoptic maps can serve as the near-realistic inner boundary conditions for the radial magnetic field of solar corona. However, updating the radial magnetic field by employing the synoptic maps without special care probably leads to the physical inconsistency and the unphysical oscillations near the sub-Alfvénic solar surface boundary [64–66, 101]. In order to minimize these effects, the projected normal characteristic (PNC) method [64–67, 101] is used to prescribe the inner boundary conditions [18, 30, 31, 100, 118].

Besides the solar photospheric magnetic field, other data on the solar surface may also be used tentatively to prescribe the bottom boundary conditions in the MHD simulations. The temperature map derived from the SOHO/EIT as input has been tailored by Hayashi et al. [32] in the steady state corona model. Using the Solar Terrestrial Relations Observatory (STEREO) A and B spacecraft Extreme Ultraviolet Imagers (EUVI) images taken simultaneously, Vásquez et al. [99] applied the technique of Differential Emission Measure Tomography (DEMT) to produce the maps of the 3D extreme ultraviolet (EUV) emissivity, and of a 3D version of the standard DEM analysis but without projection effects, which allows in turn the derivation of 3D maps of the electron density and temperature [24]. Wang et al. [100] investigated the energy transport across the photosphere during CR 2009 through conducting a 3D MHD simulation, which was driven by the synoptic chart of the global transverse velocity measurements near the photosphere from the Global Oscillation Network Group (GONG) and the synoptic maps of the full-resolution LOS magnetic field on the photosphere from Synoptic Optical Long-term Investigations of the Sun (SOLIS). These derived maps of the electron density, temperature, and transverse velocity, of course, will be beneficial to the 3D MHD numerical study of the corona. However, these maps are obtained under the assumption that the coronal plasma parameters change little within a CR. Currently, these data cannot provide dynamic evolution information.

In this chapter, the data-driven modeling of the corona and solar wind is established by employing the AMR-CESE-MHD model introduced in Chap. 6. By data-driven modeling, we mean to use continuously time-varying solar observations as input, instead of using one instantaneous cadence of observation as input, the latter of which is called data-constrained modeling. In such data-driven effort, the code first achieves a steady-state equilibrium of solar wind using the synoptic map on the beginning date. Then, based on the simulated background steady-state solar wind, the CESE-MHD model is advanced with the changing bottom boundary conditions, which are determined from the PNC method and the continuously time-varying solar observations. Finally, one can achieve the temporally varying corona and solar wind structures in response to the changing photospheric magnetic field. Section 7.2 describes the method that self-consistently combines the PNC method and the photospheric synoptic maps evolved with time. Synoptic maps of different sources, cadences and data processing procedures are utilized to drive the corona and solar wind model. In Sect. 7.3, the daily-update synoptic maps from WSO are used. In Sect. 7.4, the SFT model initiated by Solar and Heliospheric Observatory (SOHO)/Michelson Doppler Imager (MDI) synoptic map is employed. In Sect. 7.5, results are obtained by using six-hour cadence synoptic maps provided by GONG. In what follows, the presentations in Sects. 7.3, 7.4 and 7.5 are extracted from Yang et al. [119], Feng et al. [18–20] respectively.

7.2 Projected Normal Characteristic Boundary Conditions

This section presents projected normal characteristic boundary conditions at the solar surface. For the basic equations governing the solar wind plasma, refer to Sect. 5.2 of Chap. 5. With the radial magnetic field B_r given by synoptic maps, we use the PNC method plus the plasma mass flux limit to prescribe the eight variables on the solar surface. Just as Yeh and Dryer [122] pointed out, the constraint, $(\nabla \times \mathbf{E})_r = -\frac{\partial B_r}{\partial t}$, should be satisfied on the boundary surface so that the sinusoidal condition, $\nabla \cdot \mathbf{B} = 0$, will be preserved. That is to say, when B_r on the solar surface is fixed in time, the tangential component of the electric field at the boundary, $\mathbf{E}_t$, is generally set to zero. However, when B_r varies with time, it is necessary to specify a non-zero $\mathbf{E}_t$ to make the flux evolution match the observed changes of the magnetic field [63] on the boundary surface. Furthermore, the monopolar component in each magnetogram is subtracted out as Linker et al. [49] did before being used in the computation. According to the generalized Helmholtz theorem in the classical electromagnetic theory [86], $\mathbf{E}_t$ can be separated into two parts: one is an irrotational component expressed by the gradient of a scalar function and the other part is a solenoidal field denoted by the rotation of a radial vector function [63, 86]: $\mathbf{E}_t = \nabla_t \Phi + \nabla_t \times (\Psi \mathbf{e}_r)$, where Φ and Ψ are arbitrary scalar functions of θ and ϕ, $\mathbf{e}_r$ stands for unit vector in radial direction and ∇_t indicates tangential derivative operator in the $\theta - \phi$ plane. Obviously, the first term represents the effects of the transverse magnetic field (i.e., the shear and the normal electric current), whereas the second term varies with the magnetic flux of B_r [63]. Therefore, we only use a nonzero Ψ to describe the tangential electric field on the solar surface because the line-of-sight magnetograms can only provide the radial component of the magnetic field. Combining Faraday's law of electromagnetic induction with $\mathbf{E}_t = \nabla_t \times (\Psi \mathbf{e}_r)$, we can get

$$\nabla_t^2 \Psi = \frac{\partial B_r}{\partial t} \tag{7.1}$$

where $\frac{\partial B_r}{\partial t}$ may be obtained from linear or higher order differencing of successive synoptic maps. Once we get $\frac{\partial B_r}{\partial t}$, we can solve Eq. (7.1) on the solar surface to obtain the potential Ψ, thus specifying the evolution of $\mathbf{E}_t$. Eq. (7.1) is solved on the solar surface by a direct fast Poisson solver [46].

From the ideal MHD condition

$$\mathbf{E}_t = -(\mathbf{u} \times \mathbf{B})_t \tag{7.2}$$

we can obtain two additional constraints of the bottom boundary conditions on the solar surface:

$$\frac{\partial E_\theta}{\partial t} = B_\phi \frac{\partial v_r}{\partial t} + v_r \frac{\partial B_\phi}{\partial t} - B_r \frac{\partial v_\phi}{\partial t} - v_\phi \frac{\partial B_r}{\partial t}, \tag{7.3}$$

and

$$\frac{\partial E_\phi}{\partial t} = B_r \frac{\partial v_\theta}{\partial t} + v_\theta \frac{\partial B_r}{\partial t} - B_\theta \frac{\partial v_r}{\partial t} - v_r \frac{\partial B_\theta}{\partial t}, \tag{7.4}$$

Another five bottom boundary conditions are needed to carefully specify the evolution of the remaining seven physical quantities (i.e., ρ, p, v_r, v_θ, v_ϕ, B_θ, B_ϕ), such that the incompatibility or inconsistentness in physics and mathematics can be avoided. The PNC method, originated by Nakagawa [64–67] and practically used by previous MHD studies [16, 29, 32, 100, 101, 116–118], can minimize the physical inconsistency and reduce the unphysical vibrations near the sub-Alfvénic solar surface boundary to an acceptable level.

In order to use the PNC method along the radial direction (the normal of the inner boundary surface), the governing MHD equations are cast in vector-matrix form under spherical coordinates [29, 100, 117, 118]

$$\frac{\partial \mathbf{W}}{\partial t} = -\mathbf{A}\frac{\partial \mathbf{W}}{\partial r} - \mathbf{B}\frac{\partial \mathbf{W}}{r\partial \theta} - \mathbf{C}\frac{\partial \mathbf{W}}{r\sin\theta\partial\phi} + \mathbf{S} \equiv -\mathbf{A}\frac{\partial \mathbf{W}}{\partial r} + \mathbf{S}_r \tag{7.5}$$

where $\mathbf{W}$ consists of the primary physical quantities such as $\mathbf{W} = (\rho,\ v_r,\ v_\theta,\ v_\phi,\ p,\ B_r,\ B_\theta,\ B_\phi)^T$. The detailed coefficient matrices $\mathbf{A}$, $\mathbf{B}$, $\mathbf{C}$ and source term $\mathbf{S}$ can be found in Wu et al. [118], and will not be repeated. The Jacobian matrix $\mathbf{A}$ can be expressed as $\mathbf{A} = \mathbf{R}\boldsymbol{\Lambda}\mathbf{L}$, where $\mathbf{R}$ and $\mathbf{L}$ are the matrix form of the normalized right and left eigenvectors of $\mathbf{A}$, respectively, which satisfy $\mathbf{RL} = \mathbf{LR} = \mathbf{I}$ and $\boldsymbol{\Lambda}$ is the diagonal matrix whose diagonal elements are the eigenvalues. For the right and left eigenvectors, please refer to Wu et al. [117, 118] and Hayashi [29]. Here, only the eigenvalues are given in a nondecreasing order,

$$\lambda_1 = v_r - v_f,\ \lambda_2 = v_r - v_A,\ \lambda_3 = v_r - v_s,\ \lambda_4 = v_r,$$

$$\lambda_5 = v_r,\ \lambda_6 = v_r + v_f,\ \lambda_7 = v_r + v_A,\ \lambda_8 = v_r + v_s,$$

where v_A is the Alfvén wave speed defined by $v_A = |B_r|/\sqrt{\rho}$, and $v_{f,s}$ are the fast/slow magneto-acoustic speeds given by

$$v_{f,s} = \sqrt{\frac{1}{2}\left(\frac{\gamma p + \mathbf{B}\cdot\mathbf{B}}{\rho} \pm \sqrt{\left(\frac{\gamma p + \mathbf{B}\cdot\mathbf{B}}{\rho}\right)^2 - 4\frac{\gamma p B_r^2}{\rho^2}}\right)}.$$

By multiplying the matrix $\mathbf{L}$ from the left, Eq. (7.5) becomes

$$\mathbf{L}\frac{\partial \mathbf{W}}{\partial t} = -\mathbf{L}\left(\mathbf{R}\boldsymbol{\Lambda}\mathbf{L}\frac{\partial \mathbf{W}}{\partial r} + \mathbf{S}_r\right) = -\boldsymbol{\Lambda}\mathbf{L}\frac{\partial \mathbf{W}}{\partial r} + \mathbf{L}\mathbf{S}_r \tag{7.6}$$

which gives a set of projected characteristic equations (i.e., compatibility equations). The mth row of Eq. (7.6) reads

$$\mathbf{L}_m \frac{\partial \mathbf{W}}{\partial t} = -\lambda_m \mathbf{L}_m \frac{\partial \mathbf{W}}{\partial r} + \mathbf{L}_m \mathbf{S}_r$$

which corresponds to the compatibility equation along the mth projected characteristic $dr/dt = \lambda_m$.

According to the PNC method [67, 117], the compatibility equations along the incoming waves are discarded because they do not carry any reasonable information for our physical purpose, but the information of the outgoing MHD waves is used to update the MHD variables on the boundary. In the implementation, we always suppose that $0 < v_r < v_s$ at the solar surface, just as Hayashi [29] did. In this situation, three negative eigenvalues (λ_1, λ_2, and λ_3) correspond to the outgoing waves, and thus we have three compatibility equations associated with λ_1, λ_2, and λ_3 to use in prescribing the physical quantities. Up to now, two remaining constraints are required in order to accomplish the boundary treatments. Like Hayashi [29], in order to stabilize the computation, the local mass flux flowing through the solar surface is limited by Ulysses observation [69] and v_r is assumed to be greater than a small value δv_r ($\delta v_r = 0.001$ km s^{-1}). According to the radial flow speed or solar surface mass flux $(\rho v_r)_c$ with the typical value of 5.0×10^{13} s^{-1} cm^{-2} [29], the two remaining constraints mentioned above are determined in the following three cases:

Case I. If $0 < \rho v_r < (\rho v_r)_c$, the plasma density and pressure are assumed to be fixed

$$\frac{\partial \rho}{\partial t} = 0, \quad \frac{\partial p}{\partial t} = 0 \tag{7.7}$$

Case II. If $v_r < \delta v_r$, v_r is reset to be δv_r, and at the same time the adiabatic enthalpy is fixed. That is,

$$v_r = \delta v_r, \quad \frac{\partial}{\partial t}(p/\rho^\gamma) = 0 \tag{7.8}$$

Case III. If $\rho v_r > (\rho v_r)_c$, the mass flux escaping through the solar surface is settled to be the critical value $(\rho v_r)_c$, and the adiabatic enthalpy is fixed.

$$\rho v_r = (\rho v_r)_c, \quad \frac{\partial}{\partial t}(p/\rho^\gamma) = 0 \tag{7.9}$$

The application of one of Eqs. (7.7), (7.8) and (7.9), with the help of $\frac{\partial B_r}{\partial t}$, Eqs. (7.2), and the three compatibility equations associated to λ_1, λ_2, and λ_3, enables us to specify the evolution of the eight physical quantities at the solar surface. Actually, Case I is first applied on the whole surface to get a temporary updated version of $(v_r)_t$. Then, if $(v_r)_t$ is less than δv_r, Case II is used to update the solution variables, and if $(v_r)_t$ is greater than $(\rho v_r)_c$, Case III is employed.

Obviously, $\frac{\partial v_\theta}{\partial t}$ and $\frac{\partial v_\phi}{\partial t}$ can be derived from Eqs. (7.3) and (7.4) by dividing B_r. If $|B_r|$ is very small, the division will cause unacceptable errors, and even make the code fail to proceed during the simulation. Thus, if $|B_r|$ is detected to be smaller than 0.1 G, the magnetic field will be calculated by PFSS model, and the velocities are set to be parallel to the magnetic field. An example of the regions affected are illustrated

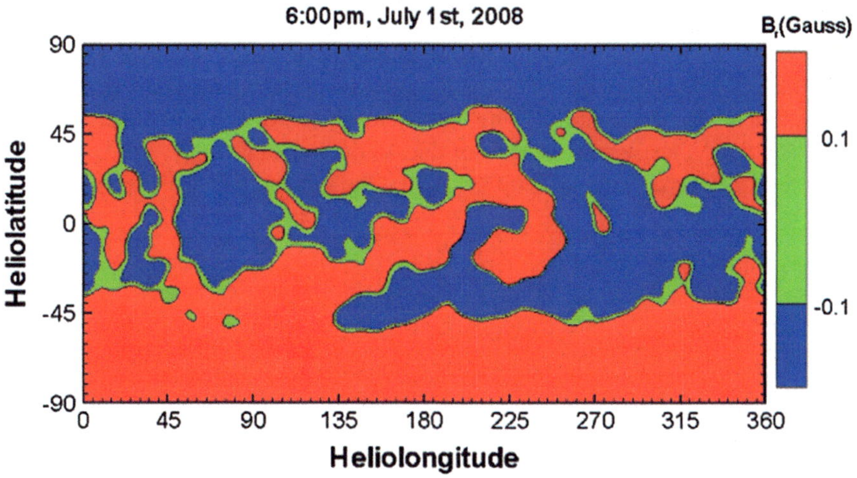

Fig. 7.1 Illustration of the regions where $|B_r| < 0.1$ G. The regions with green color are specially treated in the boundary determination instead of using the characteristic projection method. Reprinted from Fig. 1 of Feng et al. [20] with permission of the John Wiley & Sons, Inc.

in Fig. 7.1 with green color, from which we can see that only small parts of the solar surface are affected. In order to guarantee positive pressures and densities, if $\rho < \rho_{\min}$ or $p < p_{\min}$, $\frac{\partial \rho}{\partial t} = 0$ and $\frac{\partial p}{\partial t} = 0$ are strictly applied to limit their tendency to decrease. Here, $\rho_{\min}$ and $p_{\min}$ are chosen to be 1.5×10^7 cm^{-3} and 10^{-4} Pa.

7.3 Global Solar Corona for Year 2007 Modeling Driven by Daily-Updated Synoptic Maps

7.3.1 Data Processing

The daily-updated synoptic maps used to drive the evolution is provided by Stanford's Wilcox Solar Observatory (WSO). the daily-updated magnetic field synoptic charts are preprocessed before being input into the calculation. Due to the line saturation effects, a saturation factor of 1.8 is chosen to modify the observed magnetic field strength of magnetograms according to the calibration analysis of WSO and other solar observatories [41, 88]. To remove the projection effects, B_r is specified at the solar surface with the relationship $B_r = B_{\text{los}}/\cos\theta$, where θ is the heliographic latitude and B_{los} is the measured line-of-sight photospheric field, by supposing that the magnetic field is radial at the photosphere [56, 103]. At the same time, the polar magnetic fields are completed by the expression $B_r = B_p \cos^8\theta$, where B_p is the field strength at the very pole [88, 104]. Finally, the monopolar component in the data is subtracted out as Linker et al. [49] did.

We input a sequence of daily-updated magnetic field synoptic charts, in which the magnetic fields change at a rate corresponding to real time. Since the cadence of the daily-updated synoptic maps is about 24 h, the magnetic field data on the solar surface are prepared by simply taking the difference of the two time sequences of the maps and then linearly increasing with time to drive the evolution. The expression used to update the radical magnetic field B_r on the solar surface for next time step is given as

$$B_r(\theta, \phi)^{n+1} = B_r(\theta, \phi)^n + \triangle B_r(\theta, \phi)\Delta t \tag{7.10}$$

with

$$\triangle B_r(\theta, \phi) = \frac{\partial B_r(\theta, \phi)}{\partial t} = \frac{(B_r(\theta, \phi))_{m+1} - (B_r(\theta, \phi))_m}{24 \times 3600} \tag{7.11}$$

where Δt is the usual time step restricted by the Courant-Friedrichs-Lewy (CFL) condition, m denotes the mth daily-updated synoptic map and n is the nth time step. The method introduced in Sect. 7.2 is used to prescribe boundary condition.

7.3.2 Solution Implementation

The computational region used in this section extends from 1 to 27 R_s. Initially, the computational domain is divided into $8 \times 4 \times 4$ blocks with each block consisting of $6 \times 6 \times 6$ cells, and the blocks near the Sun are refined to have two refinement levels. To start the computation, the Parker's hydrodynamic isothermal solar wind solution [72] is used to give the initial values of the plasma density, ρ, gas pressure p, and the plasma velocity $\mathbf{u}$. At the same time, the potential magnetic field based on the daily-updated magnetic field synoptic data of the first day in Year 2007 is calculated in the corona as initial magnetic field input. Here, the initial temperature and number density on the solar surface are prescribed to be 1.3×10^6 K and 1.5×10^8 cm^{-3}, respectively. Then, we integrate the MHD equations in time by relaxation until a converged steady-state equilibrium is achieved. In what follows, the solar wind solution is updated by being subjected to the time-dependent solar surface boundary conditions (described in Sect. 7.2) driven by daily-updated magnetic field synoptic data to obtain the coronal responses to the surface magnetic field variations in Year 2007. During the calculation, a grid cell size of 0.015 R_s is achieved on the solar surface and at most five levels of grid refinement are obtained in the computational domain. The grid throughout the simulation is refined or derefined to obtain a grid cell size of about 0.21 R_s near the current sheet and a maximum grid cell size of about 2 R_s at the outer boundary.

7.3.3 *Numerical Results*

This section presents the evolution of the large-scale coronal structures during Year 2007 obtained from the MHD-DUSM simulation, and compares them with the available coronal and interplanetary observations. In Sects. 7.3.3.1, 7.3.3.2 and 7.3.3.3, the changing features of the coronal structures obtained by the 3D MHD-DUSM model are compared with the observations from the EUVI and the white-light measurements from the inner coronagraph (COR1). Both EUVI and COR1 are parts of the Sun Earth Connection Coronal and Heliospheric Investigation (SECCHI, Howard et al. [40]) instrument suite on board the spacecraft of the Solar Terrestrial Relations Observatory ahead of the Earth (STEREO A). In addition, the polarized brightness (pB) observations from the Large Angle Solar COronagraph (LASCO) C2 onboard SOHO are also used to check the MHD-DUSM model. In Sect. 7.3.3.4, a detailed comparison will be performed between the simulated results and the interplanetary measurements from both Ulysses and OMNI, which are mapped to 20 R_s from the interplanetary space, as done by Neugebauer et al. [68] and Linker et al. [49]. Section 7.3.3.5 displays some results from the MHD-MSM for CR 2061 with the same code in order to demonstrate the differences between these two MHD models. It should be noted that although the simulated results for the whole year are obtained, for clarity, the results from four CRs 2052, 2055, 2058, 2061, which evenly distributed within Year 2007, are analyzed in detail. The results of other CRs are presented in a supplementary way.

7.3.3.1 Comparison with Disc Images

Coronal holes are believed to be associated with open magnetic field regions and to be the source regions of fast solar wind [10], in which, due to the expansion of the solar wind, the plasma has lower density and temperature than in other coronal regions, such as the streamers. Thus, coronal holes are typically identified as dark regions in ultraviolet (UV) and X-rays radiations by their low emission intensity [49]. Here we assume that the observed coronal holes correspond to the open-field regions. In order to determine the boundaries between the open-field and closed-field regions for both the MHD-DUSM and PFSS results, we trace the 3D magnetic field lines to distinguish the regions of open field on the solar surface, from which the field lines reach the outer boundaries at 5 R_s for the MHD-DUSM model and at 2.5 R_s for the PFSS model [16].

Figure 7.2 displays the synoptic maps of coronal holes at 1 R_s from the SECCHI EUVI observations (Column 1), MHD-DUSM model (Column 2) and PFSS model (Column 3) for CRs 2052, 2055, 2058 and 2061. The synoptic images of EUVI observations are constructed by assembling the daily-observed disk images. In Columns 2 and 3, the modeled coronal holes (i.e., the open-field regions) are represented with black regions and the other regions (i.e., the closed-field regions) with white regions. For the purpose of comparison with the observations and PFSS results, one can select

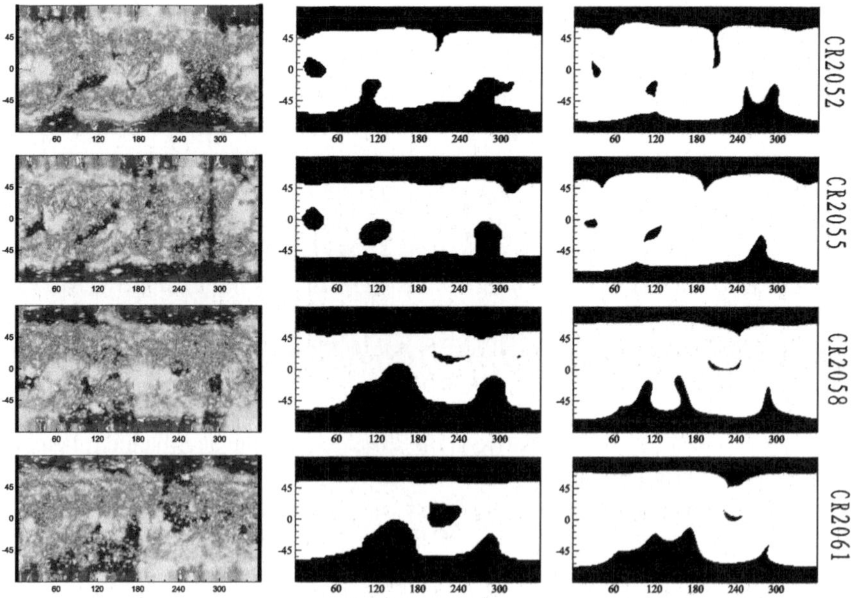

Fig. 7.2 Synoptic maps of the coronal holes at 1 R_s for CRs 2052, 2055, 2058 and 2061. The first column is the observation by STEREO-A SECCHI EUVI 195 Å. The second and third columns are the simulated coronal holes from the MHD-DUSM and PFSS models, respectively. Reprinted from Fig. 1 of Yang et al. [119] with permission of the John Wiley & Sons, Inc.

the MHD-DUSM result on the last day of each Carrington rotation (CR), in which the radial magnetic fields on the solar surface have been updated completely by the daily observed magnetograms.

Seen from both the observations and the simulations in Fig. 7.2, the most pronounced coronal feature during the whole year is the presence and persistence of the mid- and low-latitude coronal holes (MLCHs), which occupy a significant portion of the solar surface. In fact, Abramenko and Yurchyshyn [1] found that there were several isolated long-lived equatorial coronal holes dispersed on the solar surface during 2007, which was different from the observations in the previous solar minima. Now we examine the isolated low-latitude MLCH centered at about $\theta = -10°$ and $\phi = 25°$. The EUVI images show that the MLCH first appears in CR 2052 (although rather weak in the observation) and survives until CR 2058 for nearly 6 rotations. However, in both the MHD-DUSM and PFSS results, the life of the MLCH is shorter than the observation by about a CR, and its location is too far north. In addition, its size is too large in the MHD-DUSM simulation as compared to that in the EUVI image. These discrepancies are possibly attributed to the inaccurate polar fields caused by the imperfect interpolation of data gaps in the polar regions. In fact, some studies have revealed that polar fields have significant effects on the distributions and sizes of the coronal holes [112]. Another long-duration isolated MLCH from Columns 1 and 2 is located at $\theta = 0°$ and $\phi = 220° - 240°$, firstly detected in

CR 2057, and continuously observed throughout the end of this year. Abramenko and Yurchyshyn [1] analyzed this persistent MLCH in detail, and found that it survived almost two years, began to shrink gradually during 2009 May, and finally vanished in June.

In Figs. 7.2 and 7.11, it can be seen that both the EUVI observations and the MHD-DUSM model results reveal that polar coronal holes (PCHs) continuously change during the year of 2007 and appear different in shapes and sizes from one rotation to another. Basically speaking, the southern PCH is mainly characterized by two extended CHs and more evident than the northern one. The extended CH at about $\phi = 120°$ in CR 2052 appears at the start of the year 2007, then increases rapidly and remains a relatively large size until the end of this year. However, the other extended CH about $\phi = 240°$ behaves somewhat differently seen from the observations and the MHD-DUSM simulations. The EUVI observations show that it is very big at the first sight, then decreases, and disappears completely in CR 2064, whereas it extends more northward in the MHD-DUSM model results, especially in CR 2058. Sometimes, the PCHs are a little patchy and are obscured by bright features, which have not been reproduced by both the MHD-DUSM and PFSS models.

Comparing the modeled results with the observations, we can see that the MHD-DUSM has basically captured the features and evolution of the coronal holes. The shapes and distributions of the modeled open-field regions roughly agree with those of the observed CHs. However, there are some points worth noting. (i) The areas of both PCHs in the EUVI image are smaller than those in the MHD-DUSM results, but larger than those in the PFSS model; (ii) The shapes and sizes of extended PCHs are different among the EUVI observations, the PFSS and the MHD-DUSM results, and the MHD-DUSM results have much more extended PCHs than those from the PFSS model; (iii) More isolated MLCHs can be clearly seen in the EUVI measurements, although the MHD-DUSM model obviously captures the two largest and most stable isolated ones with one centered at about $\theta = -10°$ and $\phi = 30°$ and the other at $\theta = 25°$ and $\phi = 240°$. The PFSS result is not as pronounced as the MHD-DUSM calculation; (iv) The evolution of both the extended PCHs and the isolated MLCHs is more evident in the observations than those in the MHD-DUSM and PFSS models.

It is also instructive to examine the distributions of the number density, temperature, and radial speed on the solar surface with time-dependent inner boundary conditions described in Sect. 7.2. Figure 7.3 shows the synoptic maps of the simulated number density N (unit: 10^8 cm^{-3}), temperature T (unit: 10^6 K), and radial speed v_r (unit: km s^{-1}) at 1 R_s for CRs 2052, 2055, 2058 and 2061, where the black lines denote the magnetic neutral lines. As expected, with the time-dependent inner boundary condition, N, T and v_r vary reasonably on the solar surface with respect to the magnetic field topology obtained by the MHD-DUSM model. In the open-field regions, i.e., the coronal holes as shown Fig. 7.2, the plasma achieves higher speed, lower temperature and density. However, the nearly stagnant plasma in the closed field regions is of higher temperature and density. The typical density at the base of the coronal hole obtained from the MHD-DUSM model is about 1×10^8 cm^{-3}, the temperature is 2×10^6 K and the radial speed is 5 km s^{-1}. Near the magnetic neutral lines, they are 2×10^8 cm^{-3}, 2.5×10^6 K and 0.5 km s^{-1}, respectively. Thus the

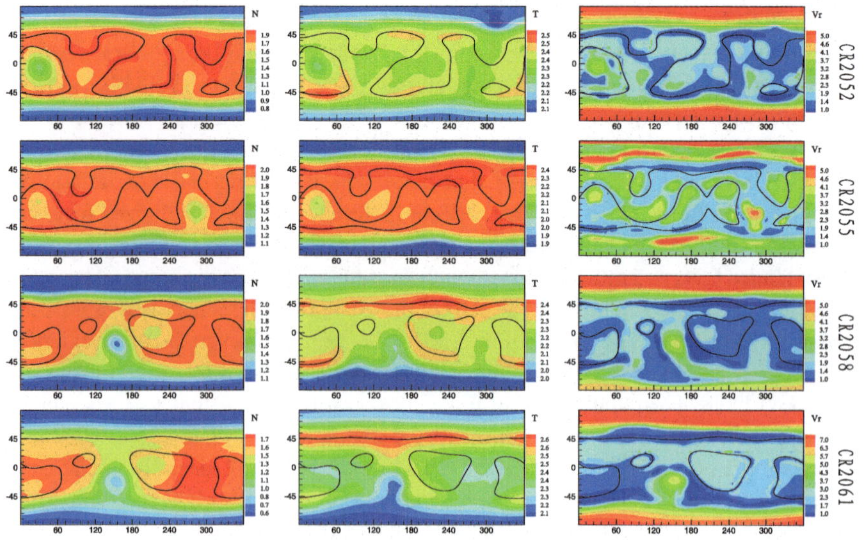

Fig. 7.3 Synoptic maps of the simulated number density N (unit: 10^8 cm^{-3}), temperature T (unit: 10^6 K), and radial speed v_r (unit: km s^{-1}) at 1 R_s for CRs 2052, 2055, 2058 and 2061, where the black lines denote the magnetic neutral lines. Reprinted from Fig. 2 of Yang et al. [119] with permission of the John Wiley & Sons, Inc.

typical ratios of the simulated parameters in the coronal hole to those in the streamer are about 1:2 for the proton number density, 1:1.3 for the temperature and 10:1 for the radial speed. These numerical values of temperature and density are in good agreement with those derived from the analysis of the reconstructed STEREO/EUVI local differential emission measure (LDEM) for different coronal features (e.g., [24, 99]). It should be also noted that from CR 2052 to 2064, the distributions of thin, cool and fast plasma are automatically adjusted to keep pace with the shapes and distributions of coronal holes, and the properties of the plasma density and radial velocity from the simulation are more sensitive to the associated magnetic field geometry than those of the temperature. Therefore, the model has indeed generated reasonable distributions of the plasma density, temperature and velocity on the solar surface due to the proper inner boundary treatments of combining the projected normal characteristic method and the mass flux limit with the evolution of radial magnetic field and the tangential component of the electric field.

7.3.3.2 Comparison with Synoptic pB Images

The synoptic pB images obtained by a coronagraph are predominantly produced by the line-of-sight integrated electron density of coronal structures with the photosphere occulted. Thus, the bright regions shown in white light images are usually interpreted as the dense streamers viewed on the limbs and the dark regions as the less

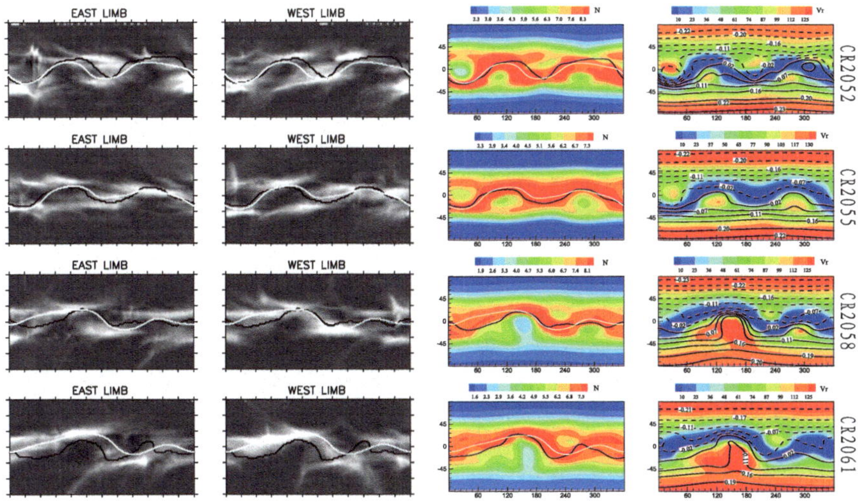

Fig. 7.4 Synoptic maps at 2.5 R_s for CRs 2052, 2055, 2058 and 2061. The first and second columns are the white-light pB observations at the east and west limbs from SOHO LASCO C2. The third column is the simulated number density N (unit: 10^5 cm $^{-3}$). In the first, second, and third columns, the black lines denote the magnetic neutral lines from the MHD-DUSM model and the white lines from the PFSS model. The fourth column overlays the isolines of the radial magnetic field (unit: Gauss) on the contour maps of the simulated radial speed v_r (unit: km s^{-1}). Reprinted from Fig. 3 of Yang et al. [119] with permission of the John Wiley & Sons, Inc.

dense coronal holes. As the physical characteristics of the solar corona are generally thought to be controlled by coronal magnetic field, white-light images also reflect the coronal magnetic field configurations. It is believed that the observed white-light bright structures observed farther than 2.5 R_s are coincident in locations of the current sheets and therefore are closely related to the surfaces where the radial magnetic field Br changes its sign (e.g., [41, 107]). In this section, the synoptic pB maps from the white-light observations of SOHO LASCO C2 and STEREO-A SECCHI COR1 are used to diagnose the 3D structures of the corona and to verify the MHD-DUSM results.

Figure 7.4 presents the synoptic maps at 2.5 R_s for CRs 2052, 2055, 2058 and 2061. The first and second columns are the observed images at the east limb from SOHO LASCO C2 and at the west limb from STEREO-A SECCHI COR1. The synoptic images are constructed by extracting a slice of east or west limb data centered at r = 2.5 R_s from each of a 27.3-day successive LASCO C2 or SECCHI COR1 images, and arranging the strips in time-reversed sequences, where the vertical streaks represent coronal mass ejections [110]. The third column is the simulated number density N (unit: 10^5 cm^{-3}). In the first, second, and third columns, the black solid lines denote the magnetic neutral lines from the MHD-DUSM model and the white lines from the PFSS model. The fourth column overlays the isolines of the radial magnetic field (unit: Gauss) on the contour maps of the simulated radial speed v_r (unit: km s^{-1}).

Figure 7.4 provides the current sheets of the four CRs calculated from both the MHD-DUSM and PFSS models are surrounded by the bright white-light structures at the east and west limbs. Most of the brightest structures lie where the latitudes of the current sheet vary least with longitude. Wang et al. [108] pointed out that at these locations, the plasma sheet was oriented edge-on in the sky plane and the number of electrons in the line of sight reached its maximum. For each rotation, a pronounced 4-sector structure of the current sheet, i.e., a rough double sinusoid, is located between the latitudes of $\pm 30°$ both in the observation and in the simulation at 2.5 R_s, which indicates that there is a strong quadrupole component originating in the observed photospheric field. Although the latitudes of the HCS along Carrington longitude range between 1° and 15°, the 4-sector structure is relatively stable throughout the whole year. During the evolution from CR 2052 to 2055, the current sheet reaches its maximum southward excursions at about Longitudes 60° and 200°, near which the current sheet bends sharply northward. From CR 2056 to 2060, the first maximum southward excursion of the current sheet at $\phi = 60°$ becomes flat gradually, and eventually disappears from CR 2061 to 2064. Whereas, the second maximum southward excursion at $\phi = 200°$ progressively increases in longitude and arrives at $\phi = 220°$ in CR 2064. The HCSs obtained from both MHD-DUSM and PFSS models have almost the same overall shape and location for each CR, and evolve in a similar manner during the year of 2007. However, there are also several local discrepancies between them. The HCSs from the MHD-DUSM model are more distorted and deflect more abruptly, compared with those from the PFSS model. In addition, the HCSs from the MHD-DUSM results displace a little southward relatively to those from the PFSS results. However, the bright structures in the white-light pB observations are more complicated than HCSs. Some very bright structures from observations and the high-density from the MHD-DUSM model are indeed far away from HCSs. The first region lies at 30°N in the first and last 30 degrees of longitudes during CRs 2052, 2055, and 2061. The other one is located at 30°S around Longitude 170° during CRs 2052 and 2055. These high-density regions possibly result from the pseudo-streamers [75, 111], i.e., unipolar dense regions far from the neutral lines.

Also, Fig. 7.4 presents the distributions of the simulated proton number density, radial speed and radial magnetic field on the surface at 2.5 R_s. By examining these results, we can recognize that in each rotation, the locations of the bright structures in the white-light pB images at the east and west limbs, which are identified as the streamer belts, are characterized by the lowest velocity and highest plasma density in the simulation, and the dark regions are coincident with the locations of the increased flow speed and decreased plasma density. The high-density low-speed flow spans 20 degrees in latitude around the magnetic neutral line, whereas the low-density high-speed (LDHS) flow mainly covers the middle and high latitudes, where the magnetic field changes slowly. Meanwhile, the LDHS flow also scatters in a few low-latitude regions, which are associated with the isolated coronal holes or the persistent isolated MLCHs by combining Fig. 7.4 with Figs. 7.2 and 7.3. These LDHS flows produce the dark regions near the solar equator in white-light images at the east and west limbs. It can be also found that the differences in shapes and areas of the coronal holes on the solar surface between different rotations are also reflected on the distributions of the number density, radial speed and radial magnetic field on the surface at 2.5 R_s, which demonstrates that the time-dependent solar surface boundary condition can drive both the evolution of coronal holes and the corresponding evolution of streamer belts beyond 2.5 R_s. Besides, we can see that the typical values of the simulated number density and radial speed are about 7.5×10^5 cm^{-3} and 25 km s^{-1} near the current sheet and about 2.0×10^5 cm^{-3} and 110 km s^{-1} at middle and high latitudes, which are in good agreement with the results on the surface at 2.5 R_s near the solar minimum period from the empirical model developed by Wei et al. [113]. Due to the presence of the coronal heating, the MHD-DUSM model basically yields the observed large contrasts in the number density and radial speed between open and closed field regions.

7.3.3.3 Comparison with Daily Observations

To further demonstrate that the MHD-DUSM solution can describe the specific coronal observations, here is presented the results corresponding to some daily measurements. Figure 7.5 shows the evolution of the observed and simulated streamers and coronal holes during the year 2007 at about 27-day intervals between January 22 and December 15, which are correspondent to Day 13 of the associated CRs. The first column is the composite coronal pB image from 1.15 to 6 R_s, where the data from 1.15 to 2.3 R_s comes from Mauna Loa Solar Observatory (MLSO) Mark-IV coronameter observation, and the outer fields of view from 2.3 to 6 R_s are from LASCO C2 observations. The second column displays the derived pB images from the simulation results on the same plane shown in the first column, in which the images are enhanced inside 2.3 R_s, from 2.3 to 4 R_s and outside 4 R_s, separately. The third column is the disk observations from STEREO-A SECCHI EUVI 195 Å for the same views as the coronal composite images in the first column, and the fifth column is the simulated 3D magnetic field lines from 1 to 2 R_s with open-field regions on the solar surface shown in black and closed-field regions at 1 R_s shown as gray. The last column is the 3D representation of the HCS.

Figure 7.5 shows that in each pB image, the bright, thin streamers are restricted to a narrow range of lower latitudes and the higher latitudes are dominated by the large, dark polar coronal holes. Almost every observed images show very bright long streamer-like structures above 3 R_s at the east and west limbs, although they seem wider in the simulation results. These streamer-like structures result from the emission of the classic helmet streamers close to the sky plane. Both the simulation and the observations reveal that these streamer-like structures do not extend radially outward from their foot points, but cover relatively large latitudes near the Sun. However, except the brightest and sharpest structures at both limbs, there are also diffusive and less bright radial structures outside 3 R_s, some of which disappear in some CRs and then reappear in other CRs. Seen from the Mark-IV coronagraph observation and the simulated results presented in Figs. 7.4 and 7.5, most of these diffusive structures result from the projection effects of the high-density structures off the plane of sky, which are associated with a single, warped current sheet encircling

Fig. 7.5 The coronal observations and simulated results on the meridional plane at $\phi = 90° - 270°$ for CRs 2052, 2055, 2058 and 2061. Columns 1 and 2 are the pB images obtained by the coronal composite observations and the MHD-DUSM results from 1.15 to 6 R_s, respectively. Column 3 is the STEREO-A SECCHI EUVI 195 Å observation on the solar disk for the same days as in Column 1, and column 4 is the simulated 3D magnetic field lines from 1 to 2 R_s with open-field regions shown in black and closed-field regions in gray. The last column is the 3-D representations of the HCSs. Reprinted from Fig. 4 of Yang et al. [119] with permission of the John Wiley & Sons, Inc.

the Sun as shown in the last row of Fig. 7.5. However, some of the diffusive structures may originate from the pseudo-streamers as discussed in Sect. 7.3.3.2. Besides, in both the Mark-IV and the LASCO C2 observations, there were sharp polar boundaries between the brightest streamers and the adjacent coronal hole regions. By contrast, the equatorward boundaries, where the lines of sight pass the streamer materials out of the plane of the sky, are relatively diffusive in appearance as demonstrated by Wang [102]. Although the coronal large-scale structure changes very slowly from one CR to another near the solar minimum, there are still some changing features of the streamers and coronal holes during this period. On the east limb, there are initially 2 streamer-like structures. They are almost at the equator on CR 2054 and then go back north and split in CR 2056. Whereas, the bright long streamer-like structure on the west limb of each image is clearly moving south during the year, which first appears at about $\theta = 20°$ in CR 2052, then drifts towards and crosses the solar equator, and finally reaches about $\theta = -30°$ in CR 2064.

As far as the coronal holes are concerned, the fourth column shows that there are only small-sized isolated MLCHs near the east disk from CR 2052 to 2056 and the simulation has produced some relatively larger ones. In CR 2057, the isolated MLCH first appears at the center of the disk, and persists till the end of this year although its shapes and sizes slightly change from one rotation to the next. Nevertheless, the extended southern coronal hole near the east limb behaves somewhat different, which first appears on the south hemisphere in CR 2058. Its size increases from CR 2059 to CR 2061, then decreases from CR 2062 on and becomes the largest in CR 2064. Comparisons of the simulated results with the observations evidently shows that the basic 3D structure of the streamer belt and coronal holes have been captured by the MHD-DUSM model. Their evolution shows a general consistency with the observations except that the extended southern coronal hole near the west limb obtained from the MHD-DUSM model appears about one CR later, which is possibly due to the longitudinal inaccuracy of the simulation during the coronal evolution.

7.3.3.4 Comparison with Interplanetary Measurements

Since there are no in-situ solar wind observations near the outer boundary in the simulation to verify the simulated results, the interplanetary measurements are mapped back to 20 R_s by using a ballistic approximation, in which the variations in longitude are computed from the time interval required for a plasma parcel traveling from 20 R_s to the spacecraft location with the in-situ measured solar wind velocity, just as done by Mikić et al. [63] and Linker et al. [49]. In order to test the simulated results at both high and low heliographic latitudes, the MHD-DUSM model predictions are compared with the mapped observations from both Ulysses and OMNI data. At the same time, the Wang-Sheeley-Arge (WSA) model [60, 61] is also employed to check the MHD-DUSM model.

Figure 7.6 displays the synoptic images at 20 R_s for CRs 2052, 2055, 2058 and 2061. In every CR, the left column is the simulated radial speed v_r (unit: km s^{-1}) from the MHD-DUSM model and the right column is the predicted radial speed from the WSA model [61]. The 1-h averaged Ulysses' observations are mapped back to 20 R_s, beyond which it is assumed that there is no changes for the radial speed. The mapped points are plotted versus heliographic latitude and Carrington longitude with the black-line bounded regions representing the distributions of the measured radial speed. In all these maps, the same contour levels for the radial speed are used as shown in the bottom. Figure 7.7 shows temporal profiles of the radial solar wind speed and the radial magnetic field polarities from the simulated results of the MHD-DUSM model (red dotted lines), the WSA model (blue dashed lines) and the mapped observational results from the OMNI data (black dashed lines) at 20 R_s for CRs 2052, 2055, 2058 and 2061.

As shown in Fig. 7.6, when Ulysses orbiting in the high latitudes (> 50°) far away from the current sheet, both the MHD-DUSM simulation and the WSA model modified by McGregor et al. [61] have well reproduced the observations of the solar wind speed, and the MHD-DUSM simulation has also predicted the presence of the intermediate speed solar wind at the middle latitudes during CR 2057 (Fig. 7.15). The MHD-DUSM model gives a relatively wider low speed solar wind than the modified

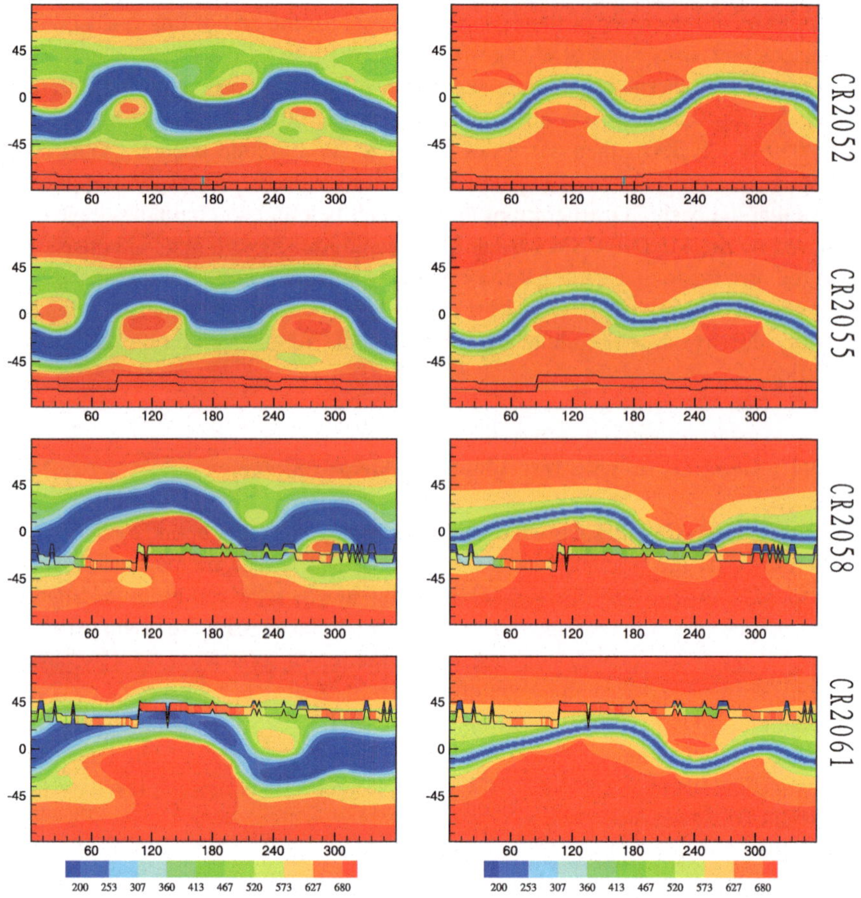

Fig. 7.6 Synoptic images of the mapped Ulysses' measurements and models' results at 20 R_s for CRs 2052, 2055, 2058 and 2061. In every CR, the left column is the simulated radial speed v_r (unit: km s^{-1}) from the MHD-DUSM model and the right column is the predicted radial speed from the WSA model [61]. The mapped points are plotted versus heliographic latitude and Carrington longitude at 20 R_s with the observations from Ulysses shown in black-line bounded regions. In all these maps, the same contour levels for the radial speed are used as shown in the bottom. Reprinted from Fig. 5 of Yang et al. [119] with permission of the John Wiley & Sons, Inc.

WSA model. However, both models have overestimated the coverage of the high speed solar wind from the southern polar coronal hole from Longitudes 50° to 170° for CRs 2058 and 2059. The MHD-DUSM simulation has also underestimated the solar wind speed along Latitude 40°N between Longitudes 30° and 230° in CR 2061.

Both the modeled achievements and the mapped measurements near Earth observe that the low-latitude regions are not dominated by the low-speed solar wind like the previous two minima [91], but interrupted by at least two broad persistent and pervasive intermediate or fast streams during 2007, although their peak value

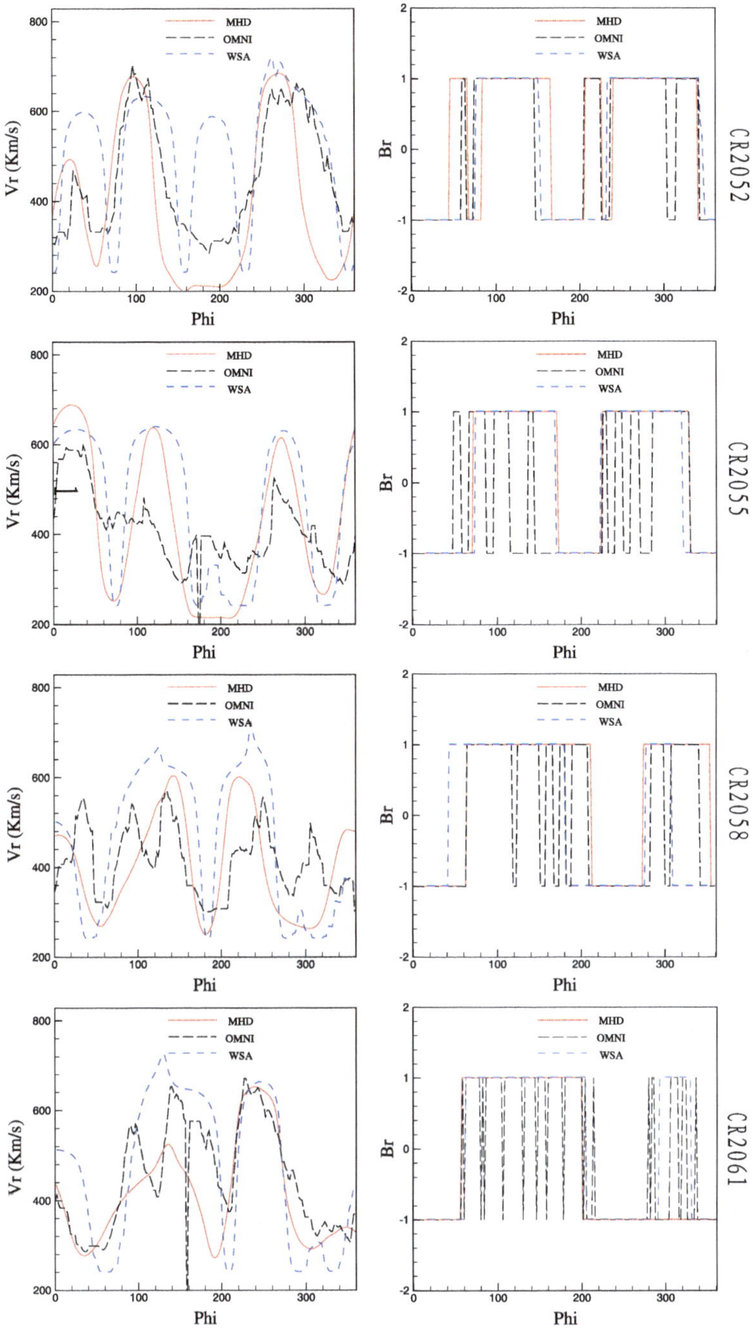
MHD
OMNI
WSA
Vr (Km/s)
Phi
Br
CR2052
CR2055
CR2058
CR2061

◀**Fig. 7.7** Comparisons between the mapped interplanetary measurements from OMNI data and the models' results for CRs 2052, 2055, 2058 and 2061. For each CR, the left column shows the temporal profiles of the radial speed v_r (unit: km s^{-1}) from the mapped observations (black dashed lines), the MHD-DUSM (red dotted lines), and the WSA formula (blue dashed lines). The right column presents the profiles of the radial magnetic field polarities obtained from the mapped observations (black dashed lines), the MHD-DUSM model (red solid lines), and the WSA formula (blue dashed lines), where "+1" stands for the radial magnetic field away from the Sun and "−1" towards the Sun. Reprinted from Fig. 6 of Yang et al. [119] with permission of the John Wiley & Sons, Inc.

($\sim$600 km s^{-1}) is somewhat lower than the flow speed of the common fast polar streams in the previous solar minimum. The phasing of the fast streams predicted by both the MHD-DUSM and WSA models is offset by less than two days from the observations. Meanwhile, the continual emergence of these equatorial fast wind leads to the significant increase of the averaged solar wind speed at the equator with the occurrence rate at about 600 km s^{-1}, which was almost comparable to that at about 400–700 km s^{-1} as pointed out by Tokumaru et al. [91]. Another important point revealed by the MHD-DUSM model and the observations is that the solar wind speed structures evolve considerably during the year 2007. The two broad fast streams are first discerned in CRs 2052–2053, and the first fast stream becomes less prominent in CR 2054. In CR 2055, the two broad fast streams almost disappear. The fast stream at about $\phi = 30°$ becomes more evident since CR 2056, but it only lasts to the end of CR 2059. The area of the first broad fast stream at about $\phi = 130°$ is over estimated by both the MHD-DUSM and WSA models, which becomes evident after CR 2060. For the fast stream at about $\phi = 220°$, the MHD-DUSM model have basically reproduced its evolution during 2007.

As far as the magnetic polarities in low latitudes are concerned, the MHD-DUSM model predicts them with a very satisfactory accuracy except for a few errors in CRs 2059 and 2060 due to the waves and perturbations occurring in the field, which leads the opposite polarity to be measured rather than the true field polarity [48]. Generally speaking, the sector structure of the magnetic field predicted by the MHD-DUSM model seems closer to the observation than that from the PFSS model. We can see that the number of different sectors predicted by the MHD-DUSM model approaches to that from the observations, despite that the crossing longitudes are not well aligned, with somewhat later or earlier than the modeled passage of the current sheet. Furthermore, the sector structures from the PFSS model are much more uniform than those from both the observation and the MHD-DUSM model.

In order to quantitatively demonstrate the predicting ability of the MHD-DUSM model, we compute the correlation coefficients between the radial speeds from the mapped interplanetary measurements and the results of the MHD-DUSM ($C_{V_r-\mathrm{MHD}}$) and WSA ($C_{V_r-\mathrm{WSA}}$) models in Table 7.1. Also, this table lists the hit ratios of the simulated radial magnetic field polarities from these two models, which are represented by $R_{B_r-\mathrm{MHD}}$ and $R_{B_r-\mathrm{WSA}}$. The table tells that both models give comparable and basically similar predicting results. There are 8 CRs whose correlation coefficients are greater than 50% for both of them. They both give relatively poor predictions from CRs 2055 to 2059. For CRs 2055 and 2056, the relatively poor results are

Table 7.1 Statistical analysis between the models' results and the mapped OMNI data. Reprinted from Table 1 of Yang et al. [119] with permission of the John Wiley & Sons, Inc

	C_{V_r}–MHD	C_{V_r}–WSA	R_{B_r}–MHD	R_{B_r}–WSA
CR 2052	0.745	0.578	0.817	0.845
CR 2053	0.578	0.306	0.879	0.842
CR 2054	0.694	0.512	0.770	0.743
CR 2055	0.345	0.360	0.665	0.680
CR 2056	0.402	0.599	0.765	0.648
CR 2057	0.339	0.401	0.641	0.674
CR 2058	0.384	0.374	0.802	0.715
CR 2059	0.339	0.434	0.773	0.780
CR 2060	0.828	0.680	0.768	0.807
CR 2061	0.745	0.710	0.867	0.839
CR 2062	0.651	0.538	0.846	0.822
CR 2063	0.794	0.760	0.907	0.653
CR 2064	0.861	0.811	0.763	0.742

probably attributed to a series of small-scale transients intercepted by multi spacecraft near the Earth during this period [45], which displaced the originally recurrent high-speed solar wind flow. The poor predictions for CRs 2057, 2058 and 2059 may result from the northward offset of the simulated low-speed solar wind band from the observations, which can be seen from Fig. 7.6 and is possibly associated with inaccurate interpolation of the polar fields. In addition, both models give much better prediction of the radial magnetic field polarities.

7.3.3.5 Comparison with the MHD-MSM Results

In order to demonstrate the differences between the results from the MHD-DUSM and the MHD-MSM models, here is presented the MHD-MSM results for CR 2061, which is chosen without any preference. Figure 7.8 displays the synoptic map of the simulated coronal holes at 1 R_s. Figure 7.9 is the synoptic maps of the simulated number density N and radial speed v_r, where the first row is N (unit: 10^8 cm^{-3}) (left column) and v_r (unit: km s^{-1}) (right column) at 1 R_s, and the second row is N (unit: 10^5 cm^{-3}) at 2.5 R_s (left column) and v_r (unit: km s^{-1}) at 20 R_s (right column), where the black-line bounded region denotes the mapped Ulysses' measurements. In Fig. 7.9, the black lines denote the magnetic neutral lines. Figure 7.10 shows a comparison for the simulated 3D magnetic field lines from 1 to 2 R_s between the MHD-DUSM (left column) and the MHD-MSM models (right column), where the contours represent B_r (unit: Gauss) on the solar surface. It should be mentioned that we select the same starting points at the photosphere to trace the field lines in order to make a direct comparison between these two MHD models.

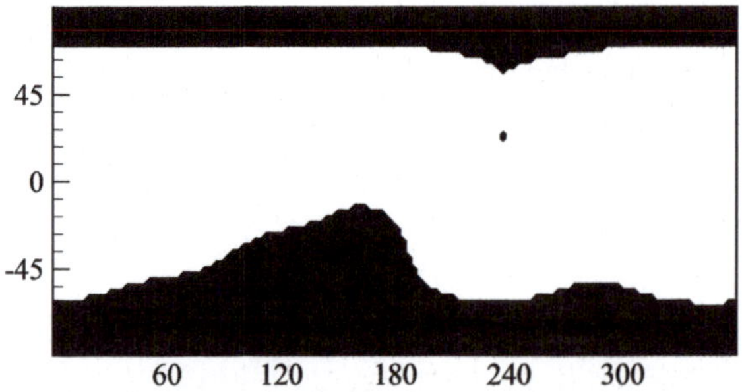

Fig. 7.8 The synoptic map of the simulated coronal holes at 1 R_s from the MHD-MSM model for CR 2061. Reprinted from Fig. 7 of Yang et al. [119] with permission of the John Wiley & Sons, Inc.

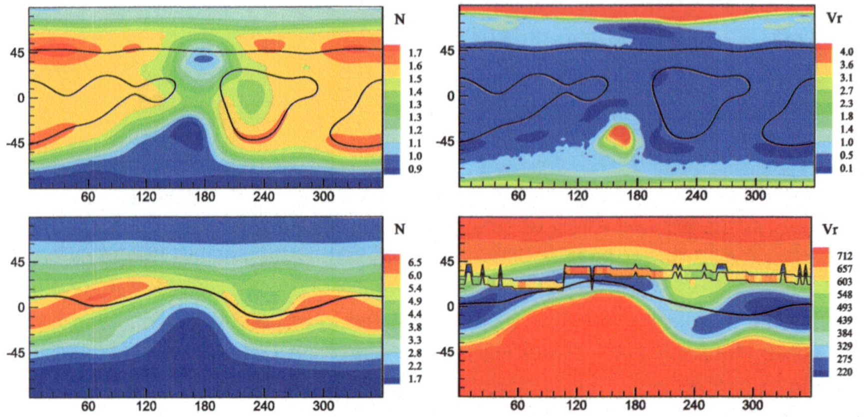

Fig. 7.9 Synoptic maps of the simulated number density N and radial speed v_r from the MHD-MSM model for CR 2061. The first row is N (unit: 10^8 cm^{-3}) (left column) and v_r (unit: km s^{-1}) (right column) at 1 R_s. The second row is N (unit: 10^5 cm^{-3}) at 2.5 R_s (left column) and v_r (unit: km s^{-1}) at 20 R_s (right column), where the black-line bounded region denotes the mapped Ulysses' measurements. In all these maps, the black lines denote the magnetic neutral lines. Reprinted from Fig. 8 of Yang et al. [119] with permission of the John Wiley & Sons, Inc.

The most conspicuous difference between Figs. 7.8 and 7.2 is the behavior of the MLCH. The long-duration isolated MLCH centering at about $\theta = 0°$ and $\phi = 220°$ is very small and almost disappears in the MHD-MSM simulation. Meanwhile, the extension of CHs in the southern hemisphere is small, but the extended CH in the northern hemisphere is closer to the EUVI observation than that in the MHD-DUSM model. From Fig. 7.9, we can see that the magnetic neutral line is relatively flat and the distributions of N and v_r on the solar surface are more regular and simpler than the results from the MHD-DUSM model. The densest regions at 2.5 R_s are highly concentrated near both sides of the magnetic neutral lines. In addition, the radial speed in the CH can roughly be comparable to that from the MHD-DUSM model. Meanwhile, the distribution of the simulated radial speed v_r at 20 R_s in Fig. 7.9 shows that the MHD-MSM overestimates the coverage of the high-speed solar

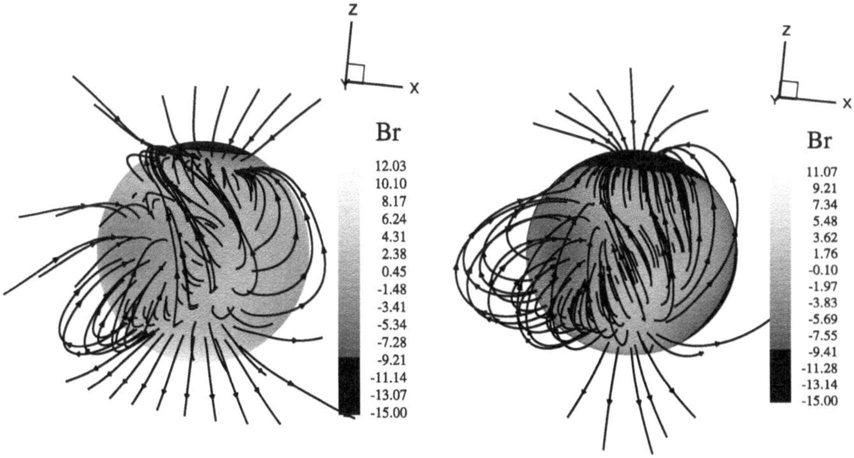

Fig. 7.10 3D representations of the simulated magnetic field field from 1 to 2 R_s for CR 2061. The left figure is from the MHD-DUSM model, and the right figure from the MHD-MSM model, where the contours represent B_r (unit: Gauss) on the solar surface. The same starting points at the photosphere are used in tracing the field lines in order to make direct comparisons between these two MHD models. Reprinted from Fig. 9 of Yang et al. [119] with permission of the John Wiley & Sons, Inc.

wind from the southern polar coronal hole, and underestimates the coverage of the intermediate-speed solar wind at the middle latitudes. The high-speed solar wind observed by Ulysses during CR 2061 is also missed by the MHD-MSM simulation. As to the magnetic polarities at low latitudes, the MHD-MSM model predicts the less accurate crossing longitudes of the current sheet, but basically gives the four-sector structure for the passage of Earth.

Figure 7.10 compares the magnetic field topologies obtained by both the MHD-DUSM and the MHD-MSM models. The adoption of daily-updated magnetograms has a significant effect on the organization of the magnetic field as compared to the MHD-MSM model. The most striking feature is that the MHD-DUSM model has captured the sheared field lines that are slightly aligned with the neutral line separating the different polarities. These sheared field lines indicates that the solar corona is far from the potential state. Whereas the MHD-MSM is closer to a potential solution. In addition, the connection of both large and small closed loops is also different for these two MHD models. The MHD-DUSM model displays more small-scale loops at the base of corona, and these small-scale loops become deformed and appear nonpotential. Downs et al. [14] found that inclusion of the additional energy transport terms of coronal heating, electron heat conduction, and optically thin radiative cooling into the governing MHD energy equation enabled the evolution of the magnetic field during the simulation to be no longer strictly potential.

As a supplementary, the simulated results for the remaining CRs 2053, 2054, 2056, 2057, 2059, 2060, 2062, 2063, and 2064 during year 2007 are presented in Figs. 7.11, 7.12, 7.13, 7.14, 7.15 and 7.16.

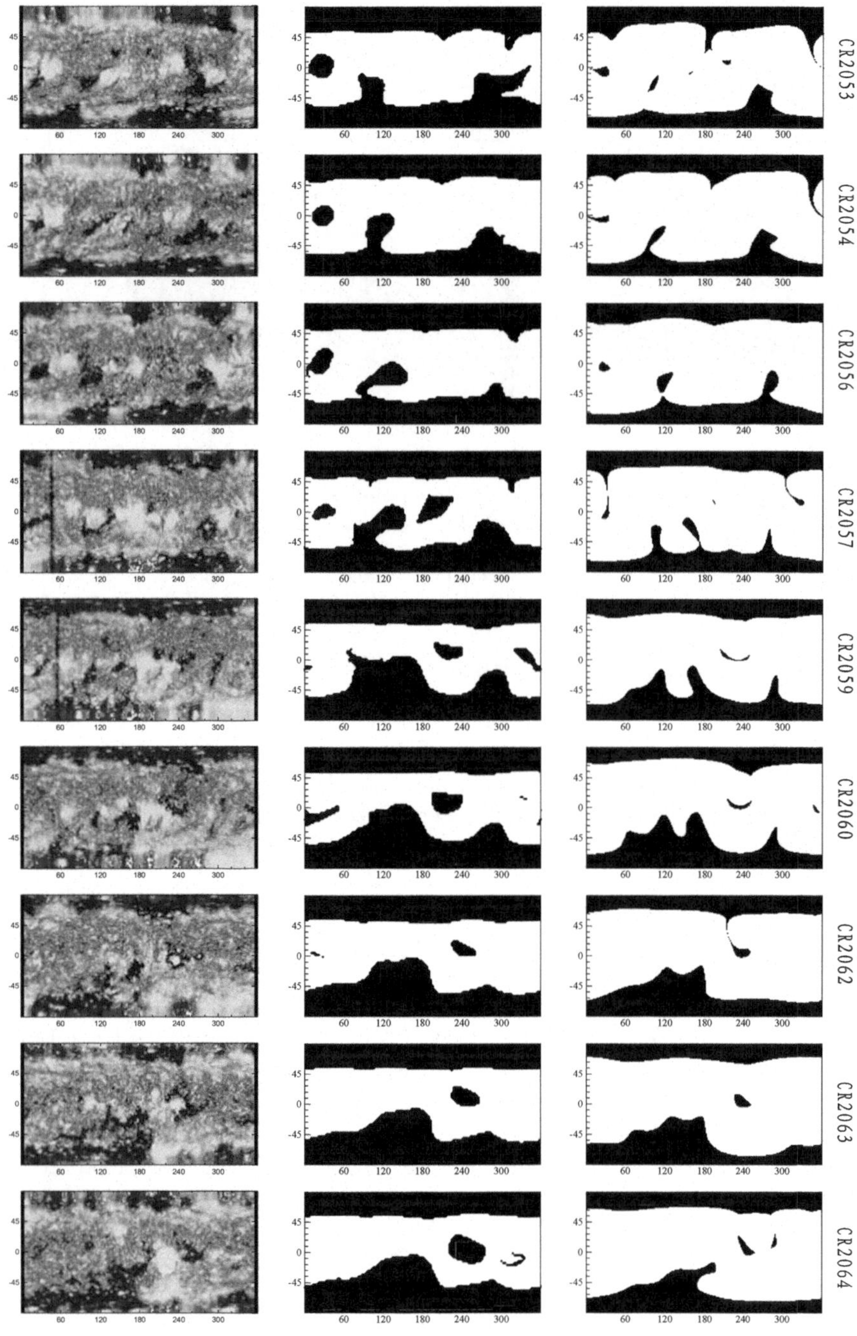

Fig. 7.11 The same as Fig. 7.2 but for CRs 2053, 2054, 2056, 2057, 2059, 2060, 2062, 2063, and 2064. Reprinted from Fig. B1 of Yang et al. [119] with permission of the John Wiley & Sons, Inc.

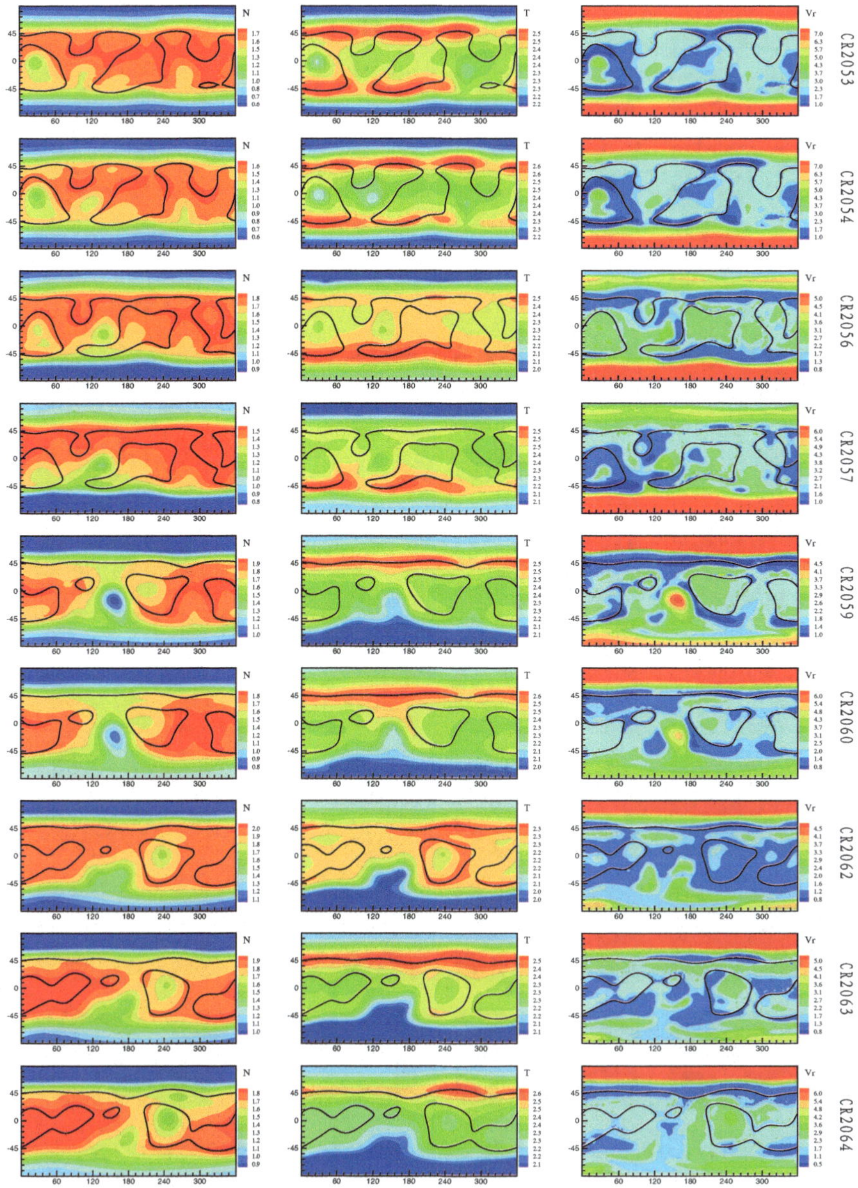

Fig. 7.12 The same as Fig. 7.3 but for CRs 2053, 2054, 2056, 2057, 2059, 2060, 2062, 2063, and 2064. Reprinted from Fig. B2 of Yang et al. [119] with permission of the John Wiley & Sons, Inc.

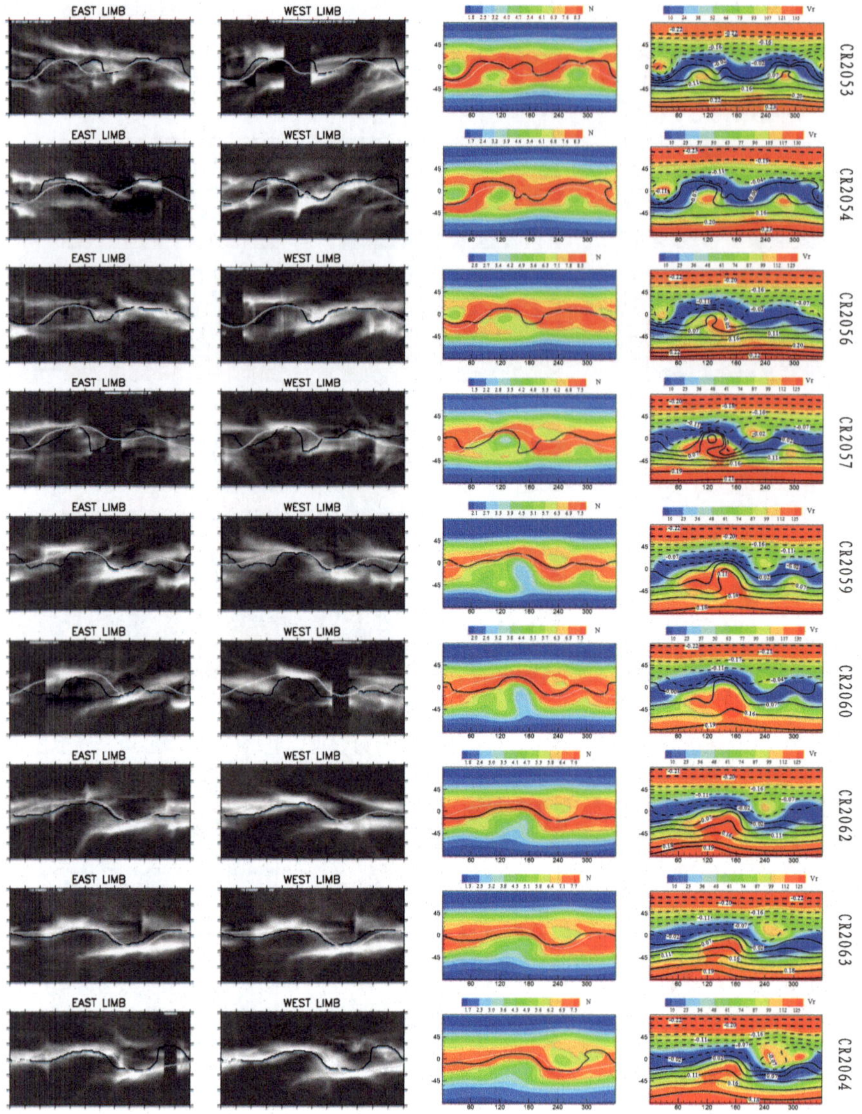

Fig. 7.13 The same as Fig. 7.4 but for CRs 2053, 2054, 2056, 2057, 2059, 2060, 2062, 2063, and 2064. Reprinted from Fig. B3 of Yang et al. [119] with permission of the John Wiley & Sons, Inc.

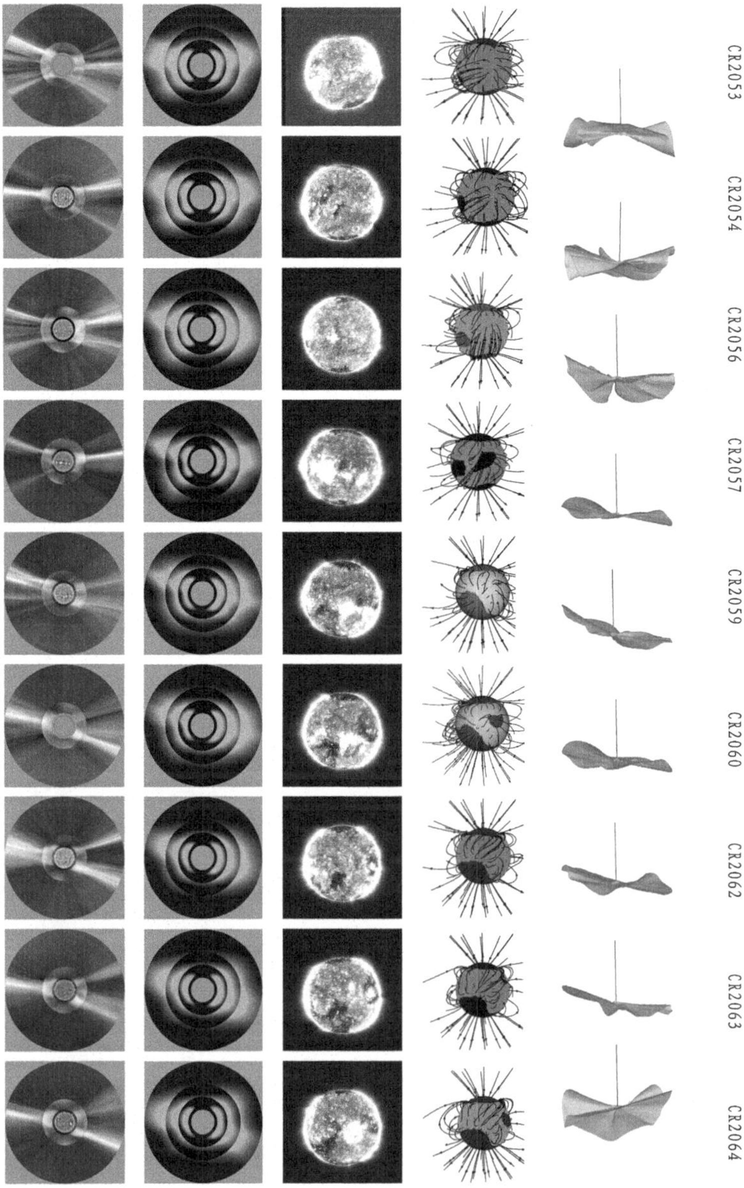

Fig. 7.14 The same as Fig. 7.5 but for CRs 2053, 2054, 2056, 2057, 2059, 2060, 2062, 2063, and 2064. Reprinted from Fig. B4 of Yang et al. [119] with permission of the John Wiley & Sons, Inc.

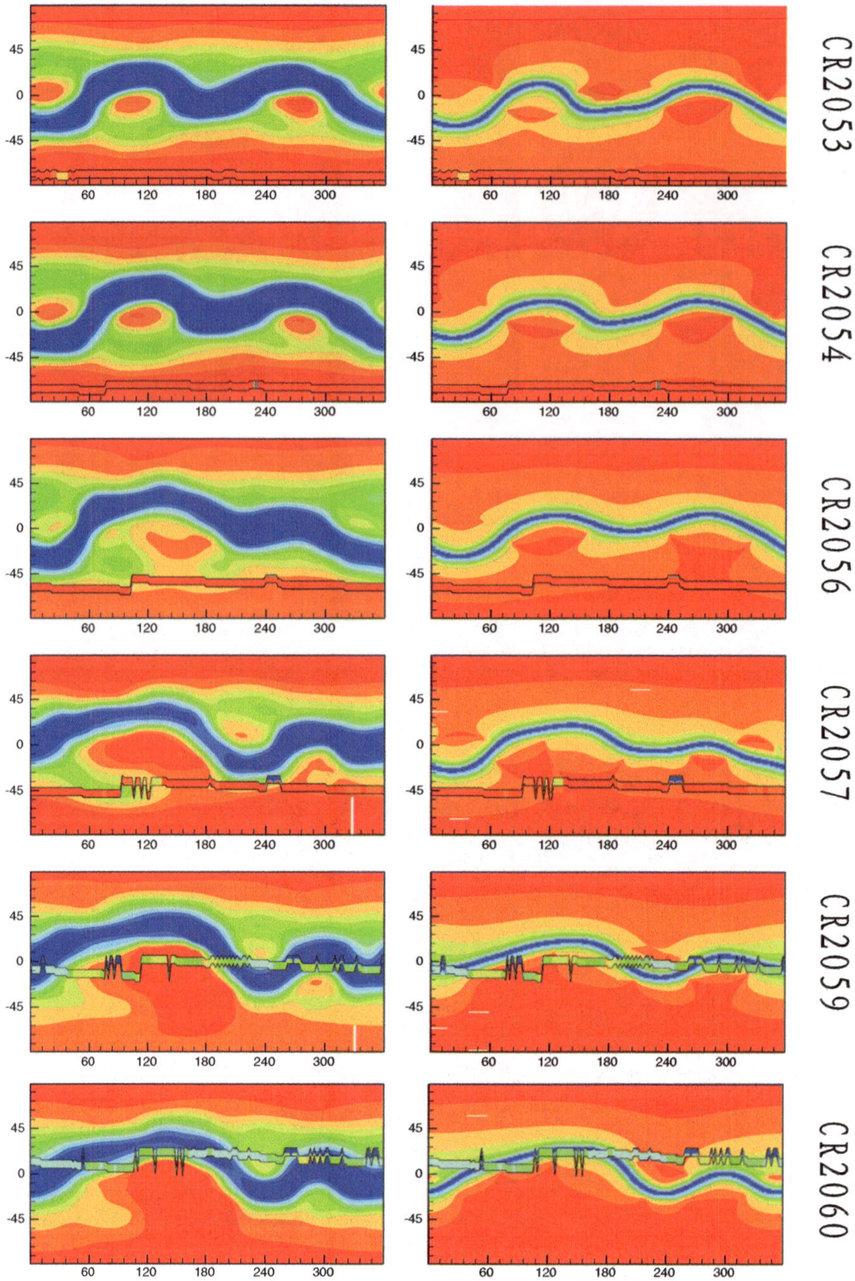

Fig. 7.15 The same as Fig. 7.6 but for CRs 2053, 2054, 2056, 2057, 2059, 2060, 2062, 2063, and 2064. Reprinted from Fig. B5 of Yang et al. [119] with permission of the John Wiley & Sons, Inc.

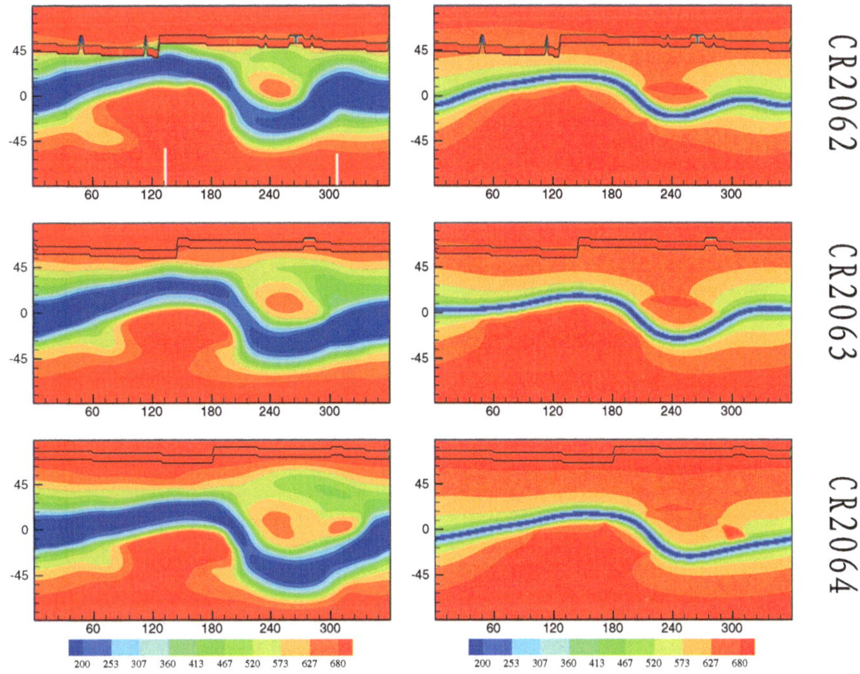

Fig. 7.15 (continued)

7.4 Corona Evolution Driven by SFT Model

7.4.1 The Surface Flux-Transport Model

The surface flux transport (SFT) model describes the radial component evolution of large-scale photospheric magnetic field with time by using the magnetic induction equation in spherical coordinates

$$\frac{\partial B_r}{\partial t} = -\Omega(\theta)\frac{\partial B_r}{\partial \phi} - \frac{1}{R\sin\theta}\frac{\partial}{\partial\theta}(v_\theta B_r \sin\theta) + \frac{\eta}{R^2}\left[\frac{1}{\sin\theta}\frac{\partial}{\partial\theta}\left(\sin\theta\frac{\partial B_r}{\partial\theta}\right) + \frac{1}{\sin^2\theta}\frac{\partial^2 B_r}{\partial\phi^2}\right] + \mathcal{S}, \tag{7.12}$$

where the first two terms in right-hand side of the equation describe the magnetic flux convection by the differential rotation and the meridional flow. Here the angular velocity of the differential rotation is given by Snodgrass [84]

$$\Omega(\theta) = 0.18 - 2.3\cos^2\theta - 1.62\cos^4\theta \text{ deg day}^{-1} \tag{7.13}$$

Fig. 7.16 The same as Fig. 7.7 but for CRs 2053, 2054, 2056, 2057, 2059, 2060, 2062, 2063, and 2064. Reprinted from Fig. B6 of Yang et al. [119] with permission of the John Wiley & Sons, Inc.

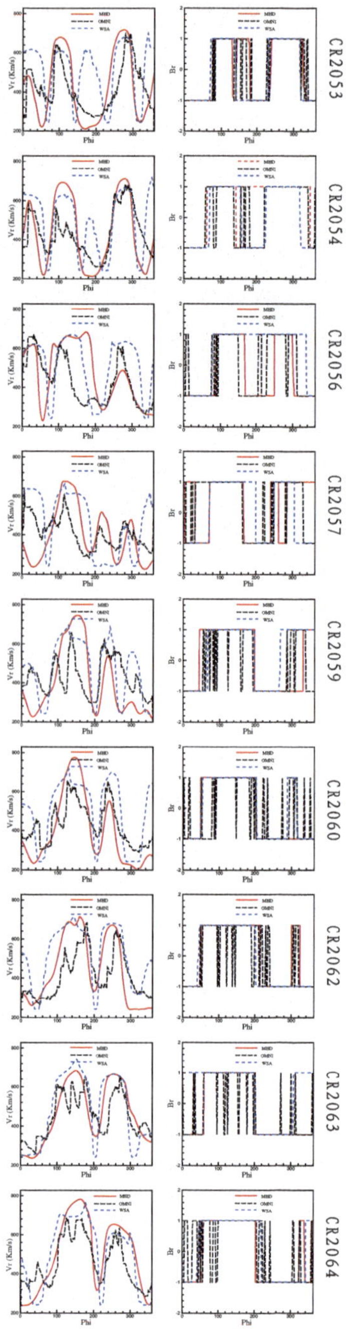

and for the meridional flow v_θ, we adopt the profile given by Yeates et al. [120]

$$v_\theta(\theta) = C \cos\left[\frac{\pi(\theta_{\max} + \theta_{\min} - 2\theta)}{2(\theta_{\max} - \theta_{\min})}\right] \cos\theta, \tag{7.14}$$

where $\theta_{\min}$ and $\theta_{\max}$ are the poleward boundaries and $C = -36$ m s^{-1} such that the maximum flow at mid-latitudes is 16 m s^{-1}.

The third term in the right-hand side of Eq. (7.12) is the diffusion effect that models the random walk of magnetic flux owing to the changing super-granular convection pattern [47]. The diffusivity is also adopted with the same value $\eta = 450$ km^2 s^{-1} as done by Yeates et al. [120]. The additional source term S can be used to describe emergence of new magnetic flux, which is important for long-term evolution with several months or years. For short-term evolution or without new sunspots, it is often neglected.

If discretized on the θ-ϕ plane, the SFT equation can be solved numerically by the finite-difference method under given the initial-boundary value condition, *i.e.*, the initial map of B_r and a proper boundary condition. The computational domain is the entire photosphere of $[0, \pi] \times [0, 2\pi]$ and uniformly divided with grid size $\Delta\theta = \Delta\phi = 1°$. A forward-time central-space (FTCS) scheme is used, which gives the numerical counterpart of Eq. (7.12) as

$$\frac{(B_r)_{i,j}^{n+1} - (B_r)_{i,j}^{n}}{\Delta t} = \frac{\partial (B_r)_{i,j}}{\partial t} = f((B_r)^n, \theta, \phi) \tag{7.15}$$

Here i, j and n are the indices for θ, ϕ and t, respectively and f denotes the numerical discertization of the right hand of Eq. (7.12) for short. The reason why such a simple FTCS scheme can be used is that the diffusion effect is far more sufficient for numerical stability. The time step Δt is chosen to satisfy the CFL condition

$$\Delta t = \text{CFL} \min\left(\frac{R\sin\theta\Delta\phi}{v_{\max}}, \frac{(R\sin\theta\Delta\phi)^2}{\eta}\right) \tag{7.16}$$

where CFL is the Courant number and $v_{\max}$ is the maximum advection speed. Considering the problem of grid convergence at both poles, in the practical computation, the above finite-difference method is only used in the domain of $5° \leq \theta \leq 175°$. Then we can use a time step in the order of minutes and we fix it to $\Delta t = 8$ min. While for the polar meshes, values are obtained by simply extrapolating and averaging from the solution value at $\theta = 5°$ or $175°$. That is reasonable since the measured magnetic field are least reliable at the poles. Also on the classic synoptic magnetogram that is usually provided in the format of pixels equally-spaced in longitude and sine latitude, the polar region $\theta \leq 5° (\geq 175°)$ is represented by nearly only one pixel in the latitude direction. In the ϕ direction, boundary conditions are periodic naturally.

To initialize the SFT simulation, we produce the SEPMF based on the SOHO/MDI synchronic frames of photospheric magnetic flux, which are available on-line

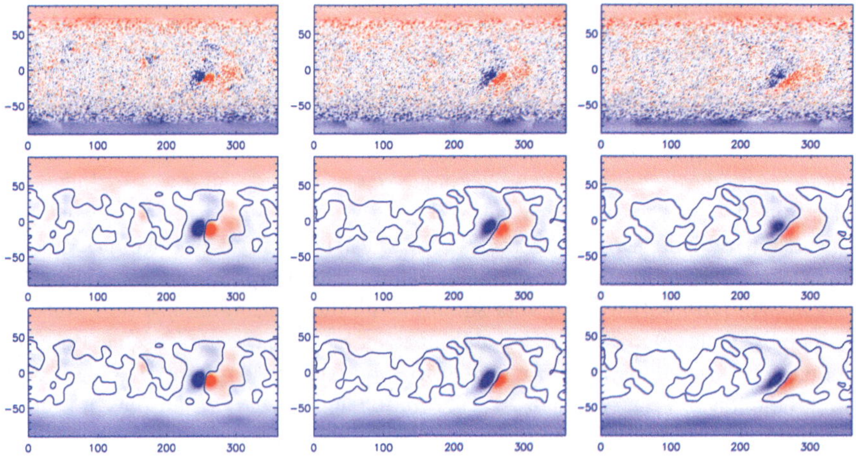

Fig. 7.17 Comparison between the SFT simulated SEPMFs at times of September 4, October 2 and October 29, 1996 (the three panels in the bottom row) and the observed MDI synchronic frames at the same times (The first two rows. The second row shows the smoothed version of the first row). Red and blue colors indicate positive and negative B_r with a saturation level of ± 20G. The zero contours of the observed smoothed synchronic frames and the simulated SEPMFs are also plotted for comparison of large-scale structures. Taken from Fig. 4 of Feng et al. [18] with permissions of the authors and the AAS

http://soi.stanford.edu/magnetic/synoptic/. For details, refer to Hayashi et al. [33] and Sun et al. [87].

For a preliminary study, here is simulated the evolution from the September 4 to October 29, 1996 (near the minimum of the solar activity cycle 23) with the source term $\mathcal{S} = 0$, since no new sunspot was observed during this time interval. The numerical results given in Fig. 7.17 clearly show the shearing and decaying of a typical bipolar active region (NOAA AR 07986 at latitude $\sim -10°$) due to the surface motion and diffusion. The SFT model reproduces well the evolution of both the active region and the background field. Note that the SEPMF input for numerical computation have been smoothed to remove the noises, which is done by 50 days of diffusion alone in the SFT model. Although many small-scale structures of the active region are smoothed out, the large-scale features with which we are concerned, remain well in the map. This simulated time-dependent map of B_r is employed as boundary input to drive the coronal model. Besides the radial magnetic field variation on the bottom boundary specified by the SFT model, the other seven constraints are determined by the PNC method introduced in Sect. 7.2.

7.4.2 Solution Implementation

The computational domain extends from the corona bottom to 35 R_s and initially, the grid spacing in both θ and ϕ directions is $2°$ globally. Grid blocks touching the bottom surface are refined with two levels to provide the same resolution (1°) with the SFT map. An MHD equilibrium on September 4, 1996 is employed as the initial input of the data-driven simulation. The equilibrium is obtained by the same method for calculating the quasi-steady corona, i.e., the MHD relaxation of a Parker solar-wind solution and a potential magnetic field. The Parker solution is specified by the solar-surface temperature 1.8 MK and proton number density 2.0×10^8 cm^{-3}, and the potential field is obtained using and the potential field is obtained using the synchronic frame on September 4, 1996, i.e., the initial map in the SFT model.

The numerical time step constraint Δt (i.e., CFL condition) with the grid size (~4 Mm) and Alfvén speed (~500 km s^{-1}) at the bottom is about several seconds. Using such time step to model months' evolution will cost tens of days with our present computational facility, which is unbearable for a preliminary study. Thus we accelerate the evolution artificially by inputting the SEPMF at a rate that is enhanced by 10 times compared to real time, as done by Lionello et al. [53]. As a result, the 8-min cadence SEPMF provided by the SFT model is regarded as 48 s in the corona model. We fix the time step of the bottom blocks to be 4 s. Hence it takes 12 steps to achieve the cadence of the SFT maps, and data needed at the in-between steps are provided by time-linear interpolation of two successive maps. We note that this artificially enhanced variation of the maps gives us credible results since the photospheric fields changed insignificantly in this chosen period. It should be also noted that we refer to Carrington longitude as longitude in this section for simplicity.

7.4.3 *Simulation Results*

This section presents the evolution of the corona from September 4, 1996 to October 29, 1996 by coupling the SFT and the coronal models.

7.4.3.1 Large-Scale Coronal Structures

Figure 7.18 compares the simulated results with observed coronal composite images from solar surface to 6 R_s. Panels (a)–(c) show the composite images observed on September 4, October 2, and October 29, 1996 from 1.15 to 2.30 R_s by HAO/MLSO Mark-IV coronameter and from 2.30 to 6.00 R_s by SOHO/LASCO-C2 coronagraph. The disk images were observed on the same dates by SOHO/EIT at He II 304 Å.

Figures 7.18d–f are the derived white-light polarized-brightness (pB) images from the simulation for the same dates of the observations. These images are also displayed from 1.15 to 6 R_s and enhanced inside and outside 2.3 R_s separately. The third row of Fig. 7.18 gives the magnetic field lines projected on the sky planes with the color contours of surface B_r, and the three panels in the bottom row show calculated entire surface distribution of number density N at 2.5 R_s on September 4, October 2, and October 29, 1996.

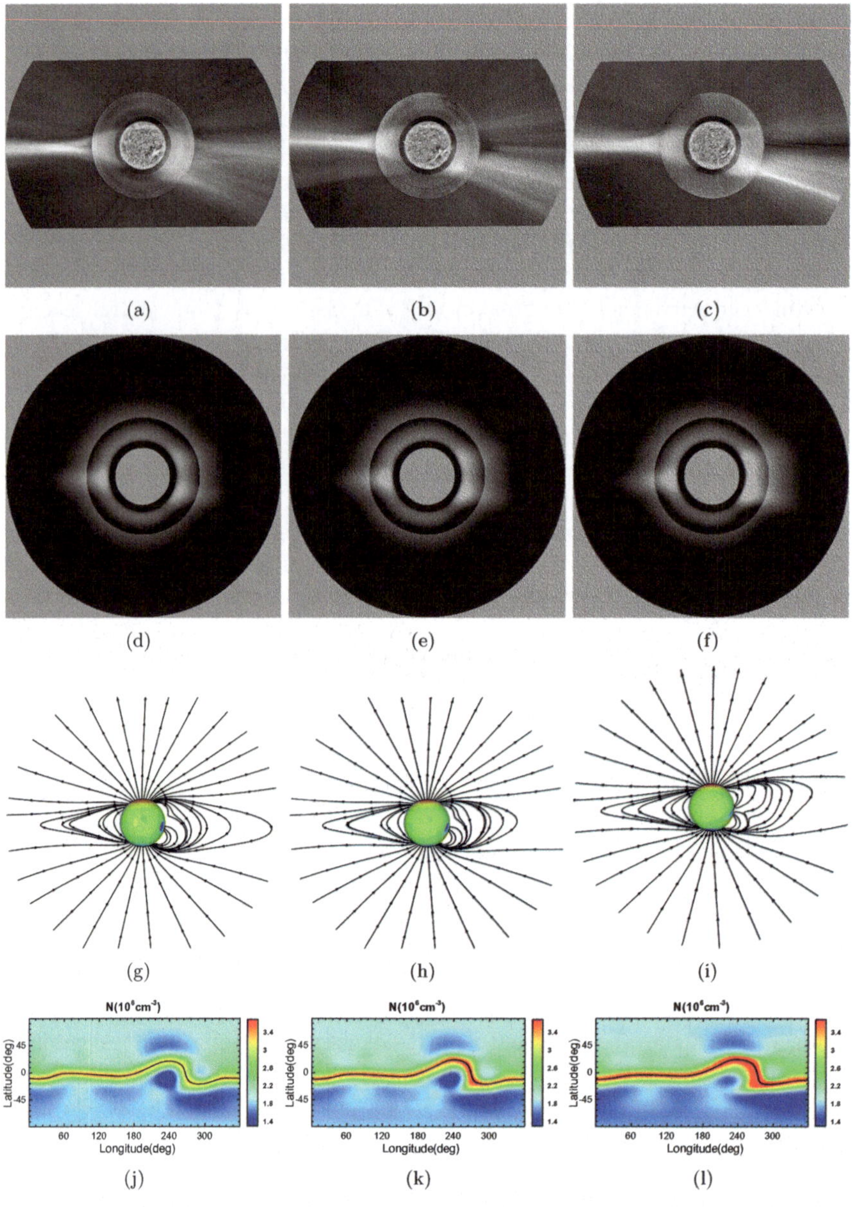

Fig. 7.18 Comparisons of the observed and simulated coronal structures. Panels **a–c** give the coronal composite images on September 4, October 2 and October 29. **d–f** are the simulated pB images and **g–i** are the projections of the magnetic field lines of the same dates; **j–l** are the pseudo-color global synchronic snapshots of number density N (unit: 10^6 cm^{-3}) at 2.5 R_s with black lines denoting the magnetic neutral lines. In this Figure, the observer on the Earth is roughly located at Longitude 180°. Taken from Fig. 5 of Feng et al. [18] with permissions of the authors and the AAS

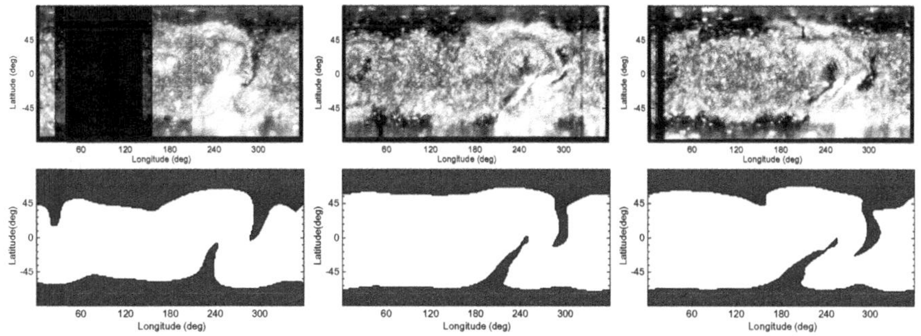

Fig. 7.19 Comparisons of the coronal-hole locations: SOHO/EIT 195 Å synoptic images (the first row) and the simulated coronal-hole maps on September 4 (CR 1913), October 2 (CR 1914) and October 29, 1996 (CR 1915). In the second row, dark regions denote the coronal holes. Taken from Fig. 6 of Feng et al. [18] with permissions of the authors and the AAS

In the simulated period, the structure of the coronal streamers changed very little, as shown in Fig. 7.17. The equatorial 'jet' that dominates the east limbs of Figs. 7.18a–f is a typical helmet streamer at solar minimum, showing the simple quasi-dipole magnetic topology. At the west limbs, the presence of multiple bright structures is due to the warp of magnetic field neutral line around the longitude of $\phi = 270°$ (see the neutral line plot in the last row). By a detailed examination one can see the gradual separating of the west limb streamers with time, and these features are well represented by the evolution of the magnetic field lines. Besides, Figs. 7.18j–l show that the number density in the streamer grows slightly, which is consistent with the pB images.

Figure 7.19 compares the simulated coronal holes with those from the SOHO/EIT 195 Å observations. The three panels in the top row are the EIT 195 Å Carrington synoptic maps of Carrington rotations 1913, 1914 and 1915. The dark vertical strips in this figure denote the observation gaps. On these maps, the dark regions are typically identified as the coronal holes with lower levels of emission due to lower plasma densities and temperatures. The coronal holes are believed to be associated with the regions of open magnetic field stretched by the solar wind. By tracing the 3D field lines in the numerical results, we produce the maps to identify open-field (black) and closed-field regions (white) on the Sun (the second row of the figure). In CR 1913 the most conspicuous observed feature is the elephant trunk-like northern extending coronal hole (ECH) approximately at the longitude of 270°, which is also captured by the MHD model. The simulation results show that the longitudinal width of the ECH at the latitude of 40° decreases from 25° in CR 1913 to 15° in CR 1915. Meanwhile, the longitudinal width of the ECH at the solar equator increases by 10°. It should be noted that the southern ECH also reaches the equator approximately at Longitude 240° with a long channel, and it becomes more and more inclined with time because of the differential rotation included in the SFT model. All these features of the CHs and their evolutions are clearly shown by both the EIT images and the simulated maps. Besides, a visible shrinking of the polar coronal holes during the simulated period from CR 1913 to CR 1915 can be seen in the modeled maps. This is probably due to the combining effect of the poleward meridional flows and the diffusive process in the SFT model. The observed images also show such a trend of evolution, even though not as pronounced as the simulation shows.

7.4.3.2 Magnetic Field in the Active Region

Figure 7.20 shows the magnetic evolution of AR 07896 by plotting the 3D field lines (black lines) over the contours of the radial field (pseudo-color image). The first row gives the MHD results, and the second row the potential field extrapolations of the corresponding SEPMF. From left to right, the time sequence is the same as that of Fig. 7.18.

At first glance of these results, one clearly recognizes that the field is getting weaker and the shear is growing stronger, which results from the surface diffusion and the surface flow of the differential rotation and meridional motion. In order to demonstrate the coronal responses to these photospheric physical processes, Fig. 7.21 shows the 3D magnetic field topologies of the same time sequence with the central meridian of 180° longitude. From this figure, we can find that the closed field lines originating from the edges of the active region extend outwards almost to about 10 R_s,

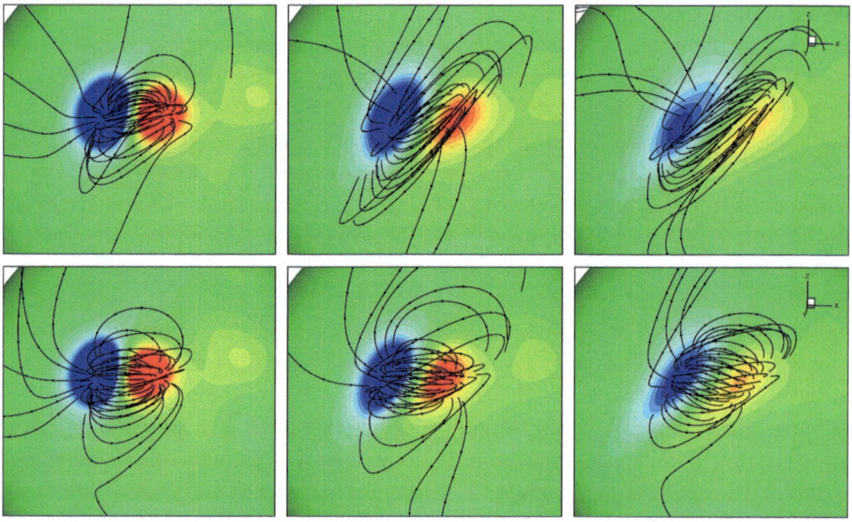

Fig. 7.20 Simulated magnetic evolution of the active region (the first row) and comparison with the potential field extrapolations. From left to right, they correspond to the results on September 4, October 2 and October 29, 1996. Black lines with arrows denote the magnetic field lines and their directions. Background image stands for the contours of the radial field. Red color indicates positive B_r and blue negative, with a saturation level of $\pm$ 20G. Taken from Fig. 7 of Feng et al. [18] with permissions of the authors and the AAS

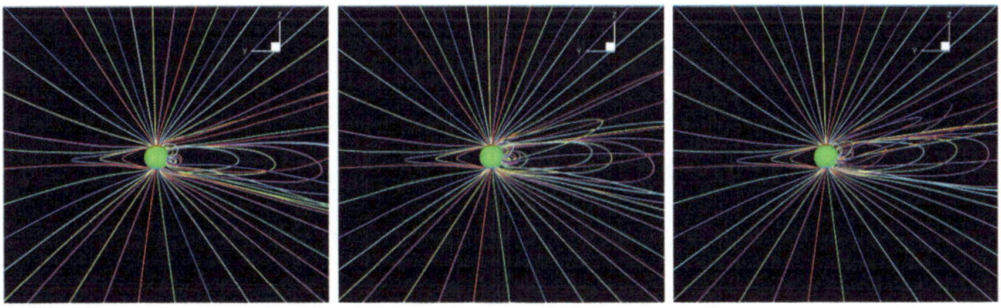

Fig. 7.21 3D views of the field lines in the numerical simulation of the same time sequence in Fig. 7.20 (left to right) with the central meridian located at 180° longitude. Note that this figure is plotted in the view different from Fig. 7.20 to show more clearly the 3D field lines associated with the AR. The field lines in these panels are rendered with different colors and traced from the same foot points at the coronal base. Taken from Fig. 8 of Feng et al. [18] with permissions of the authors and the AAS

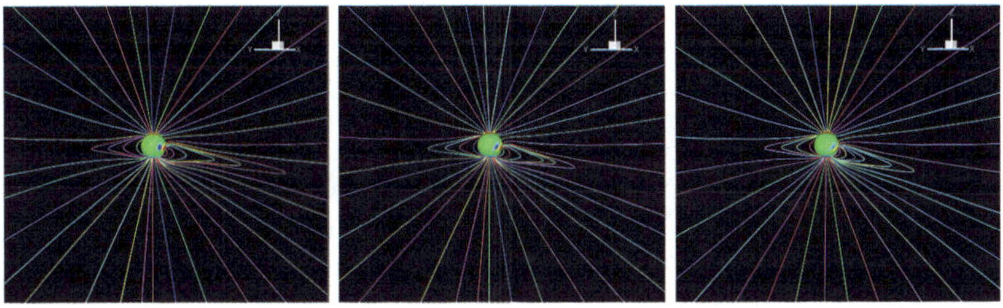

Fig. 7.22 3D views of the field lines in the numerical simulation on the same dates as those in Fig. 7.21, but the longitude of the central meridian is 220°. Taken from Fig. 9 of Feng et al. [18] with permissions of the authors and the AAS

the outer boundary of Fig. 7.21, which probably results from the large gradient of the magnetic force due to the strong magnetic field associated with the active region, the artificially heating added in the MHD model, and the enhanced differential rotation [57] included in the SFT model. Figure 7.22 presents 3D views of the simulated field lines on the same dates as those in Fig. 7.21, but the longitude of the central meridian is 220°. From this figure, we find that the closed magnetic field lines in the coronal streamers only extend outward to 5-7 R_s. We also note that the closed-field region lay within 5 R_s in Linker et al. [50], who explored the coronal evolution by introducing an artificial surface flow in a small region of the solar surface. The introduced artificial surface flow by Linker et al. [50] was far below the enhanced differential rotation used in the present paper. Therefore, the strong magnetic field in AR 07896 and the enhanced differential rotation should be two important reasons for the large radial extent of the closed-field regions displayed in Fig. 7.21. In addition, this phenomenon was also investigated in detail by Lionello et al. [53], who analyzed the coronal response to the enhanced photospheric differential rotation with the help of the resistive MHD model. The small loop structures near the Sun may be formed through the processes of the magnetic reconnection discussed in Lionello et al. [53]

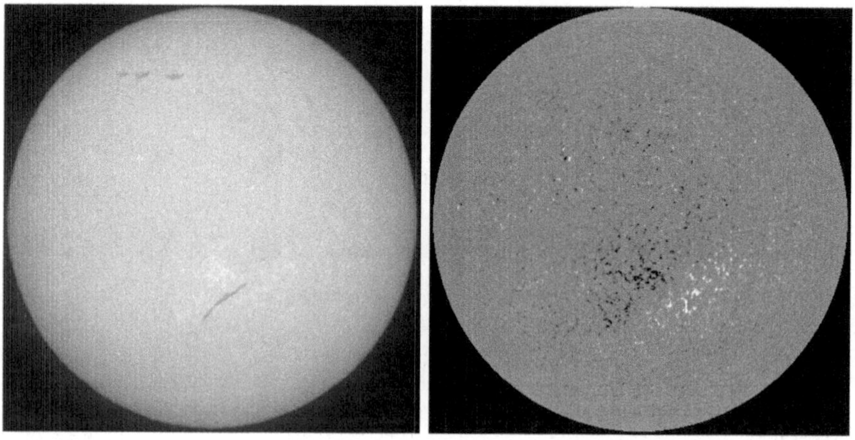

Fig. 7.23 Hα image from BBSO (left) and MDI full-disk magnetogram (right) on October 23, 1996. Taken from Fig. 10 of Feng et al. [18] with permissions of the authors and the AAS

due to the unavoidable numerical viscosity and the releasing plasmoids during the magnetic reconnection will probably become the near-Sun counterparts of the small-scale transient events observed in interplanetary space (e.g., [22, 45]). Meanwhile, the magnetic reconnection is also required in order to understand the rigid rotation of the coronal holes [106, 124]. However, the analysis in depth needs further investigation.

Secondly, the field lines gradually become aligned to the polarity neutral line (PNL) due to the shearing, especially on October 29, 1996 (CR 1915), some field lines are almost parallel to the PNL, which shows the growth of non-potential energy. This field configuration is also supported by observations. Figure 7.23 shows the Hα images of the solar disk from Big Bear Solar Observatory (BBSO) and the MDI full-disk magnetogram on October 23, 1996, when the active region is roughly at the central meridian. Note that above the active region a large-scale filament lies along the PNL. To support the filament, the field lines are commonly believed to be approximately along the long axis of the filament [55], although the fine-structure field could be much complicated with multi-twisted ropes [52]. Such continuous evolution of non-potential magnetic field configuration as well as the coronal plasma flows can hardly be reproduced by the static model such as potential field source surface model or nonlinear force-free models (e.g., [74, 76, 114]), which neglect the inertial effects and the gas pressure and incorporate no temporal history of the fields.

7.4.3.3 Solar Wind at 0.1 AU

This part shows the evolution of the solar wind parameters, e.g., the number density N, the radial velocity v_r and the magnetic flux B_r at roughly 0.1 AU (21.5 R_s) in Fig. 7.24. These near-Sun solar wind parameters are important and can be served as boundary conditions for a heliospheric model that is focused on the outer corona and

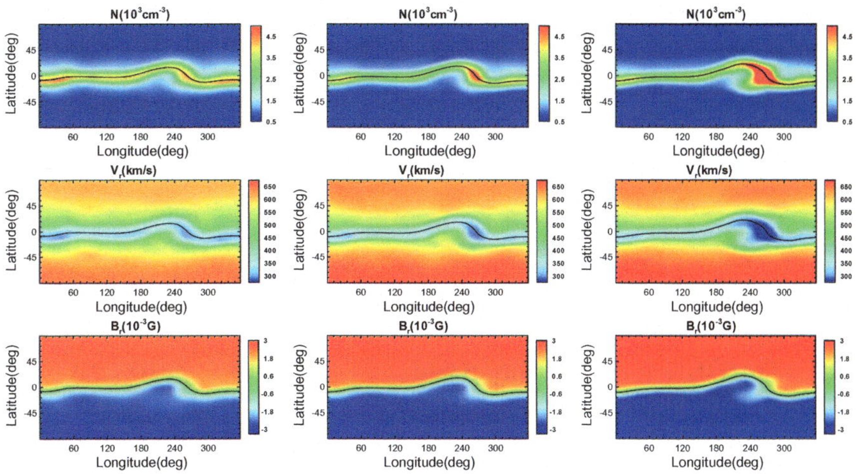

Fig. 7.24 The entire surface distribution of the solar wind parameters simulated at 0.1 AU on September 4, October 2 and October 29, 1996 from left to right. The first row gives the number density N, the second row for the radial velocity v_r, and the last row for the magnetic flux B_r. Taken from Fig. 11 of Feng et al. [18] with permissions of the authors and the AAS

interplanetary space (e.g., [12, 70, 71, 89]). Comparison shows that the neutral line warps around $\phi = 250^\circ$ with the slightly enhanced density, although the variation is small. The velocity evolution shows that the model successfully produces the bimodal solar wind structure with the high speed ~600 km/s and low speed ~300 km/s, which is consistent with previous simulations or empirical model for solar minimum [61]. Furthermore, we compare the numerical velocity distribution with those derived by the empirical Wang-Sheeley-Arge (WSA) model,

$$V(f_s, \theta_b) = 240 + 675C_a \text{ km/s} \tag{7.17}$$

where C_a is defined as

$$C_a = \left(1 - 0.8\exp\left(-(\theta_b/\phi)^\beta\right)\right)^3 / (1 + f_s)^{\frac{2}{9}}. \tag{7.18}$$

Here f_s is the areal expansion factor of the magnetic flux tubes and θ_b is the minimum angular separation at the photosphere between an open field foot point and its nearest coronal hole boundary, and the remaining parameters are $\phi = 2.8^\circ$ and $\beta = 1.25$. The same coefficient C_a has been used by McGregor et al. [61] to derive the WSA model at 0.1 AU. Figure 7.25 presents the results based on the synoptic maps of the observed magnetic field. Basically, the overall structures of the velocity obtained from the MHD simulation in Fig. 7.24 and the WSA model in Fig. 7.25 are roughly consistent. Differences between them are mainly twofold: the MHD results present a little lower value of the velocity of the fast solar wind flow by ~50 km/s than the empirical WSA model and at the same time gives a much narrower coronal streamer.

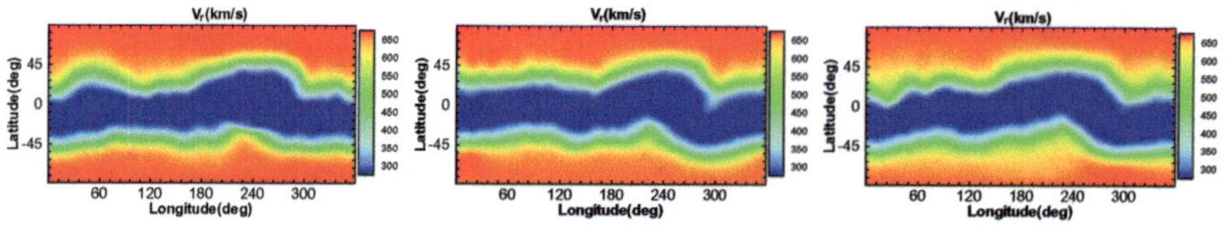

Fig. 7.25 The solar wind velocity at 0.1 AU, computed by the empirical WSA formula (7.17) of f_s and θ_b based on the Carrington synotic maps of the observed photospheric magnetic flux in CRs 1913, 1914, and 1915 centered at September 4, October 2 and October 29, 1996, respectively. Taken from Fig. 12 of Feng et al. [18] with permissions of the authors and the AAS

7.5 Data-Driven Modeling of Solar Wind from 1 R_s to 1 AU

7.5.1 Magnetic Synoptic Data Pre-processing

This section focuses attention on how to generate the time-varying magnetic field data to drive the model according to the standard magnetogram synoptic maps available from GONG at http://gong.nso.edu/data/magmap/QR/bqs/.

We choose the synoptic map data of photospheric magnetic field as the input to the SIP-CESE-MHD model based on the following considerations. Firstly, the GONG network is comprised of six solar observatories at different longitudes of the Earth to make 24-hour-a-day observations of the Sun [28, 37, 38], which enables the GONG network to provide six-hour-cadence synoptic maps in 2008. Secondly, it uses the Ni I line at 676.8 nm in the solar spectrum to measure the LOS photospheric field strength. Comparing with the observations using the Fe I 523.3 nm line (e.g., WSO), the data doesn't need to be corrected due to the saturation effects [3, 95]. Finally, polar fields not well observed have been represented by a cubic polynomial surface fit to the observed fields at neighboring latitudes.

The six-hourly-updated synoptic maps of photospheric magnetograms must be preprocessed before they are input into the SIP-CESE-MHD model. To this end, all the data points are synchronized to the observation time of the magnetograms at the central meridian by taking the solar differential rotation into consideration. For a data point at the Carrington longitude ϕ_0 and colatitude θ on the six-hourly-updated synoptic map, the new longitude ϕ on the corrected map is given by Yeates et al. [120]. The derived relationship between ϕ and ϕ_0 reads

$$\phi_0 = \frac{\Omega\,(\phi - \phi_{\rm ref})}{\Omega' + \Omega} + \phi_{\rm ref}.$$

Here, $\phi_{\rm ref}$ is the Carrington longitude of the central meridian on the six-hourly-updated synoptic map of GONG. Ω is the rotation rate of the Carrington frame with respect to an observer on the Earth, and has the value of 13.2 deg/day. Ω' is the angular velocity of solar differential rotation, for which we use the formula presented by Snodgrass [84],

$$\Omega' = 0.18 - 2.3\cos^2\theta - 1.62\cos^4\theta \text{ deg day}^{-1}$$

Then the magnetic field at (θ, ϕ) is interpolated from the original synoptic map according to the coordinates (θ, ϕ_0). It should be mentioned that the longitude ϕ may exceed the left or right edge of the original map, and thus we have to use three neighbouring synoptic maps to generate one corrected map.

In order to be consistent with the co-rotating frame used in SIP-CESE-MHD model, all the synoptic maps corrected above are adjusted by shifting the meridian at the Carrington longitude of 360° to their left edges, so that their initial Carrington longitudes are aligned. As we know, it is the photospheric magnetic field that the synoptic maps record. In order to accommodate the synoptic maps to the model's bottom boundary on the solar surface, a Gaussian-smooth procedure [31] is applied to reduce the noise, which makes the photospheric magnetic maps much easier to be dealt with by the MHD solver. Finally, the magnetic monopoles are removed by subtracting the excess global magnetic flux from the smoothed synoptic maps. For convenience of later use, we use the word "processed" to mean the six-hourly-updated synoptic map from GONG after the above preprocessing procedures.

The time steps of SIP-CESE-MHD model on the solar surface are much smaller than the 6-h cadence of synoptic maps. To fill the gaps between two neighbouring processed synoptic maps, a third-order Lagrange-interpolation over time is applied. For any specified time t, firstly find three consecutive maps $B_{r,1}$, $B_{r,2}$ and $B_{r,3}$ on the condition that their observation time t_1, t_2 and t_3 are nearest to t among all the processed maps. Then the radial magnetic field on the solar surface at time t can be interpolated by

$$B_r(t) = \frac{(t-t_2)(t-t_3)}{(t_1-t_2)(t_1-t_3)}B_{r,1} + \frac{(t-t_1)(t-t_3)}{(t_2-t_1)(t_2-t_3)}B_{r,2} + \frac{(t-t_1)(t-t_2)}{(t_3-t_1)(t_3-t_2)}B_{r,3}.$$

The interpolated synoptic maps converge to the processed ones every 6 h. As a result, this method doesn't lose time accuracy due to the long-term simulation.

7.5.2 Simulation Implementation

In the simulation, the computational domain extends from 1 R_s to 256 R_s, and each component consists of $14 \times 4 \times 4$ blocks. Each block is an $8 \times 8 \times 8$ cubic grid in a reference space (λ, θ, ϕ) where $\lambda = \ln r$ and (r, θ, ϕ) are the spherical coordinates. The grid near the current sheet is refined at most 4 times which makes the grid spacings become 1/16 of the original ones. The MHD variables are initiated with Parker solar wind solution and magnetic field calculated from the potential field. After adequate time of relaxation, the solution given by the SIP-CESE-MHD solver tends to be steady. Once the equilibrium state of solar wind is achieved, we start to apply time-dependent boundary conditions to investigate the dynamic corona and interplanetary structures in response to the time-evolving photospheric magnetic field.

7.5.3 Simulation Results

Based on the data-driven SIP-CESE-MHD model described above, we simulate the evolution of the solar corona and solar wind from July 1 to August 11, 2008 by using the six-hourly-updated synoptic maps from GONG as input. The initial solar wind is generated with the synoptic map of the photospheric magnetic field on July 1st. After the equilibrium state is achieved, the 164 consecutive preprocessed synoptic maps are applied to drive the SIP-CESE-MHD model. In the simulation, the parameters f_s and θ_b used in the volume heating coefficients are recalculated once a day. The calculations were performed on the TianHe-1 (A) supercomputer located at National Supercomputer Center in Tianjin, China (http://nscc-tj.gov.cn/en/resources/resources_1.asp#TH-1A). This piece of work was completed in 4×24 h of wall time on 8 computing nodes with 96 CPU cores and 96 GB RAM. In what follows, we display the modeled results and compare them with the solar coronal and interplanetary observations.

Figure 7.26 displays the synoptic maps of open field distributions at 1 Rs from the initial steady MHD result constrained by the synoptic map on July 1st (the top panel), the solution of the data-driven model result on July 8th (the middle panel), and the SOHO/EIT observations (http://lasco-www.nrl.navy.mil/carr_maps) during CR 2072 which lasted from July 6th to August 3rd (the bottom panel). The modeled coronal holes (i.e., the open-field regions) are shaded black.

From both the observations and the simulations in Fig. 7.26, we find that the northern polar coronal hole (PCH) occupies the largest area for the data-driven MHD model and the smallest area for the steady state MHD model. The median latitudes of the northern PCH's equatorward boundary from the data-driven MHD model, the steady-state MHD model and the EIT observation are 63°, 72° and 68°. For the southern PCH, the data-driven MHD model and the steady state MHD model give similar coverage, but the EIT observations cannot accurately record it because the spacecraft orbited 6° north of the solar equator. In addition, both the data-driven MHD model and the steady-state MHD model capture the equatorial coronal hole (ECH) centered around the longitude of 290°, but the size and shape from the former are more close to those from the EIT observations. It should be noted that compared with the EIT observations, the ECH from the data-driven and steady state models is a little northward. Abramenko and Yurchyshyn [1] investigated this coronal hole and found that it lasted almost two years and finally vanished in June, 2009. Although the data-driven model captures this ECH with reasonable accuracy, it fails to capture the small-sized ECHs centered on 90° and 190° in longitude.

Figure 7.27 shows the open field distributions from the data-driven MHD model and the EIT observations at 195 Å on the specified dates. These modeled images show relatively good agreements with the observations except that small-sized dark regions near the solar equator on July 16th are not well reproduced in the simulated image. In addition, the north PCH covers a little larger area in the modeled image on July 31st. It should be noted that some observed dark regions in Figs. 7.26 and 7.27 may be transient coronal holes associated with coronal mass ejections (CMEs)

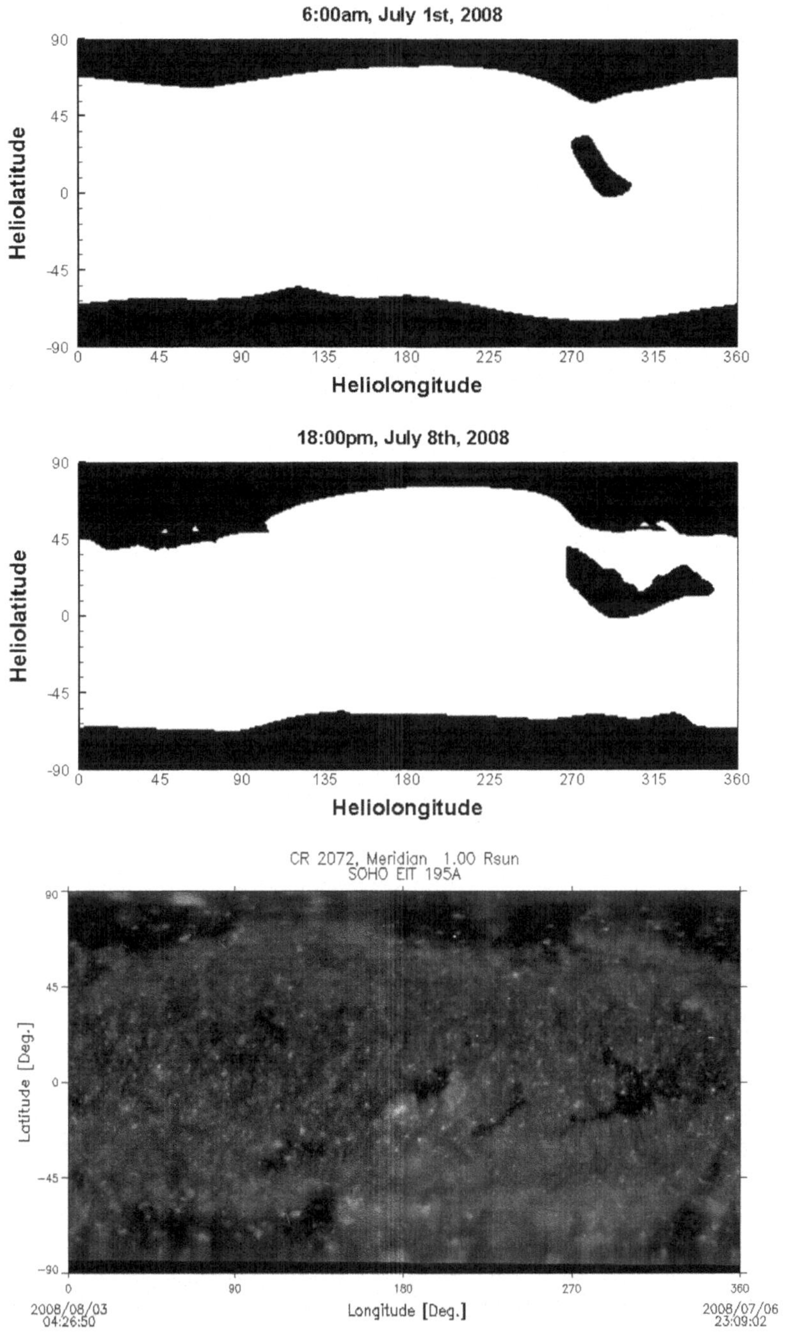

Fig. 7.26 The synoptic maps of the open and closed field distributions obtained from the steady-state MHD model (upper), the data-driven MHD model (middle), and the EIT observations at 195 Å (bottom). Reprinted from Fig. 2 of Feng et al. [20] with permission of the John Wiley & Sons, Inc.

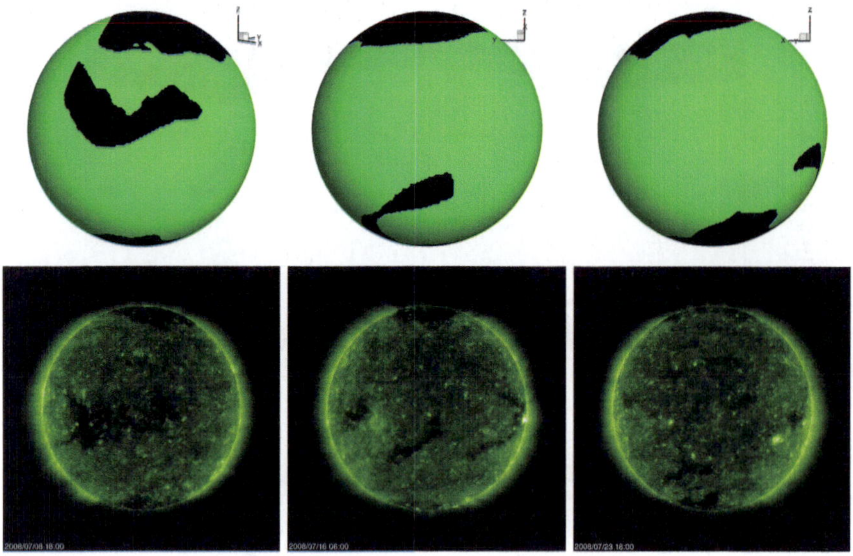

Fig. 7.27 The open and closed field distributions from the data-driven MHD model results (top row) and the EIT observations at 195 Å (bottom row) on July 8th (left column), July 16th (middle column), and July 23rd (right column). Reprinted from Fig. 3 of Feng et al. [20] with permission of the John Wiley & Sons, Inc.

because there were many CMEs remotely recorded in the STEREO data set available from http://sidc.oma.be/cactus. It should also be noted that CMEs are not the sole cause for the missing small-sized CHs in the simulated result because some of the effects that occur in the corona are caused by processes that only occur in the corona, such as reconnection or destabilization of structures following magnetic shearing (e.g., [15]). In addition, it is probably impossible to map CME transient features in detail.

Figure 7.28 shows the white-light polarized brightness (pB) images from the data-driven MHD model and from the LASCO/SOHO observations on three specified dates. It is well known that the bright regions shown in pB images are indicators of the dense streamers or pseudo-streamers viewed on the limbs and the dark regions to the coronal holes [6, 7, 23].

In Fig. 7.28, the bright streamers span a narrow range of lower latitudes and the dark PCHs dominate higher latitudes. Almost every observed image shows very bright long streamer-like structures at the east and west limbs, although the correspondent bright structures seem wider in the simulation results. In the image on July 16th, we expect the bright configurations at both limbs in the northern hemisphere result from the pseudo-streamers at Carrington longitudes 110° and 290° [80]. In order to clearly demonstrate the pseudo-streamer structures, we present the magnetic field lines near the Sun projected on the sky plane on July 17th in Fig. 7.29. Additionally, the other two wide bright structures near the solar equator are associ-

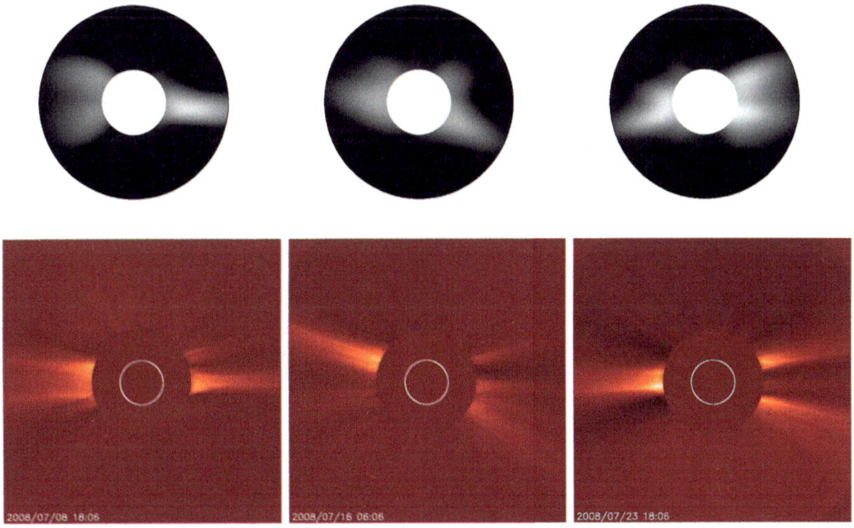

Fig. 7.28 The white-light polarized brightness images from the data-driven MHD model (top row) and from the LASCO/SOHO observations (bottom row) on July 8th (left column), July 16th (middle column), and July 23rd (right column). Reprinted from Fig. 4 of Feng et al. [20] with permission of the John Wiley & Sons, Inc.

ated with the abrupt deflect of HCS around Carrington longitude 270°, which can be seen clearly in Fig. 7.30.

Figure 7.30 presents the synoptic maps of the proton number density and radial bulk speed on July 1st and July 14th at 2.5 R_s. Comparing the results from the steady state MHD model on July 1st and the data-driven MHD model on July 14th, we find that both models achieve roughly similar global distributions of the solar wind parameters. The HCSs are basically coincident with the regions of high density and low speed, while the regions poleward of 45° are covered with the solar wind of high speed and lower density. However, a bulge centered at the longitude of 220° appears in the HCS profile in the data-driven model's result. As a result, the HCS becomes more tilted than that from the steady MHD model's result. A high-density region is also present near $(\theta, \phi) = (-20°, 290°)$ in the data-driven model's result.

In order to further validate the data-driven solar wind model, we plot the loci of the Alfvén surfaces (ASs) in the meridian planes of 270°–90° in Fig. 7.31. The figure demonstrates that the ASs lie between 9 and 12 R_s in the PCHs and between 4 and 15 R_s in the coronal streamers. This is consistent with the recent studies that suggest that the Alfvén surfaces are located at a wide range of possible heliocentric distances, from 6 to 30 R_s. Zhao and Hoeksema [123] determined the outer limit of AS in coronal streamers to be 10–14 R_s around solar minimum. Sheeley et al. [83] inferred that the AS in coronal streamers should be outside of 6 R_s. Goelzer et al. [26] employed a model of the interplanetary magnetic field to obtain the location of the AS from sunspot numbers and then placed the AS at 15 R_s during solar minimum and

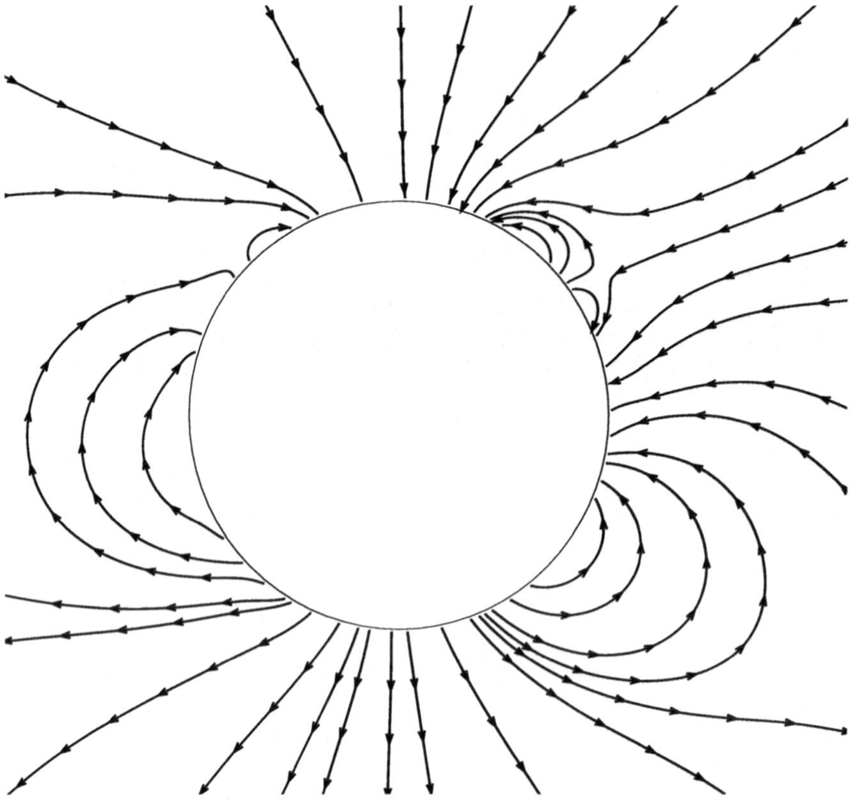

Fig. 7.29 The solar coronal magnetic field lines from the data-driven MHD model on July 17th, 2008. The field lines are projected on the meridian plane perpendicular to the Sun-Earth direction; the circle indicates the solar surface. Reprinted from Fig. 5 of Feng et al. [20] with permission of the John Wiley & Sons, Inc.

30 R_s during solar maximum. DeForest et al. [11] identified the AS by analyzing a sequence of coronagraph images from STEREO-A/COR2 and determined the AS's altitude to be beyond 15 R_s in the streamer belt and 12 R_s in the coronal hole. It should be noted that, in the figure, the sunward concaved structures on the ASs result from the minima of the local coronal magnetic field and the Alfvén speed.

Figure 7.32 exhibits the synoptic maps of the proton number density and radial bulk speed on July 1st (left column) and July 14th (right column) at 20 R_s. Comparing the results on July 1st and July 14th, we can find some perceptible changes of the solar wind between the results from both models, although the main structures are very similar including the basic shape of the heliospheric current sheet (HCS) and the global distributions of the plasma parameters. In this figure, the prominent differences between the results from the steady MHD model and the data-driven MHD model are the appearances of the northward bulge of the HCS around the longitude of 180°

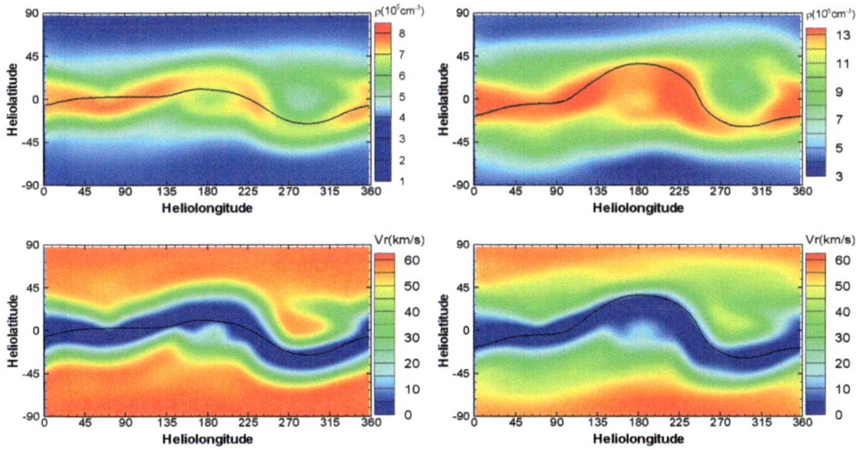

Fig. 7.30 Synoptic maps of the proton number density (top row) and radial bulk speed (bottom row) on July 1st (left column) and July 14th (right column) at 2.5 R_S. The black lines indicate the locations where $B_r = 0$. Reprinted from Fig. 6 of Feng et al. [20] with permission of the John Wiley & Sons, Inc.

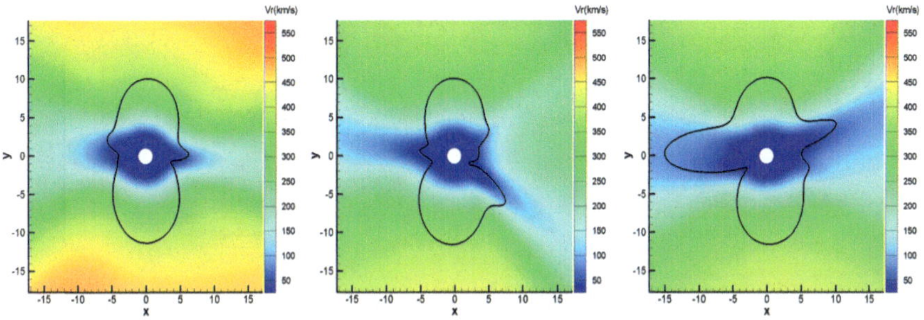

Fig. 7.31 Contours of solar wind radial bulk speeds in the 270°-90° meridian planes on July 8th (left), July 16th (middle), and July 23rd (right). The black lines indicate the locations where the Alfvénic Mach number is 1. Reprinted from Fig. 7 of Feng et al. [20] with permission of the John Wiley & Sons, Inc.

and the associated high-dense structure centered at $(\theta, \phi) = (30°, 180°)$ in the data-driven MHD model. Another significant change is the southward shift of the HCS near the longitude of 0°. Figure 7.33 presents the synoptic maps of the proton number density and radial bulk speed at 1 AU. This figure shows that the high-dense structure centered at $(\theta, \phi) = (30°, 180°)$ at 20 R_S shifts to $(\theta, \phi) = (30°, 90°)$. Comparing with Fig. 7.30, the hump of the HCS becomes wider and flatter and the trough centered at Longitude 200° becomes deeper and more apparent.

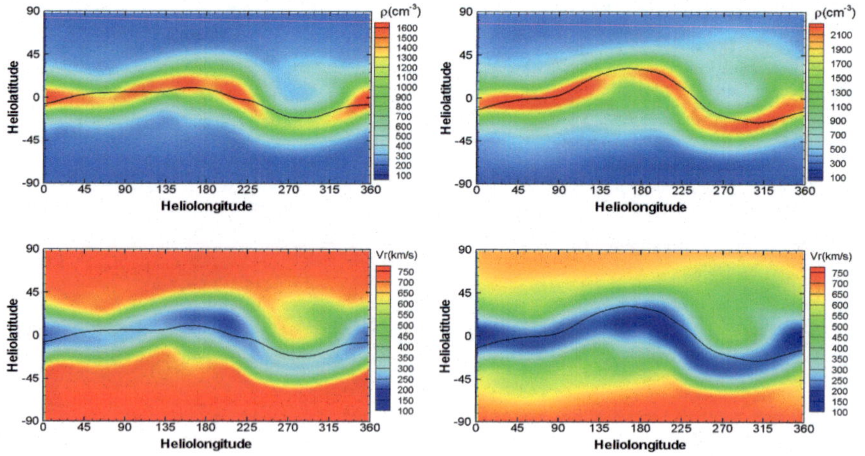

Fig. 7.32 Synoptic maps of the proton number density (top row) and radial bulk speed (bottom row) on July 1st (left column) and July 14th (right column) at 20 R_s. The black lines indicate the locations where $B_r = 0$. Reprinted from Fig. 8 of Feng et al. [20] with permission of the John Wiley & Sons, Inc.

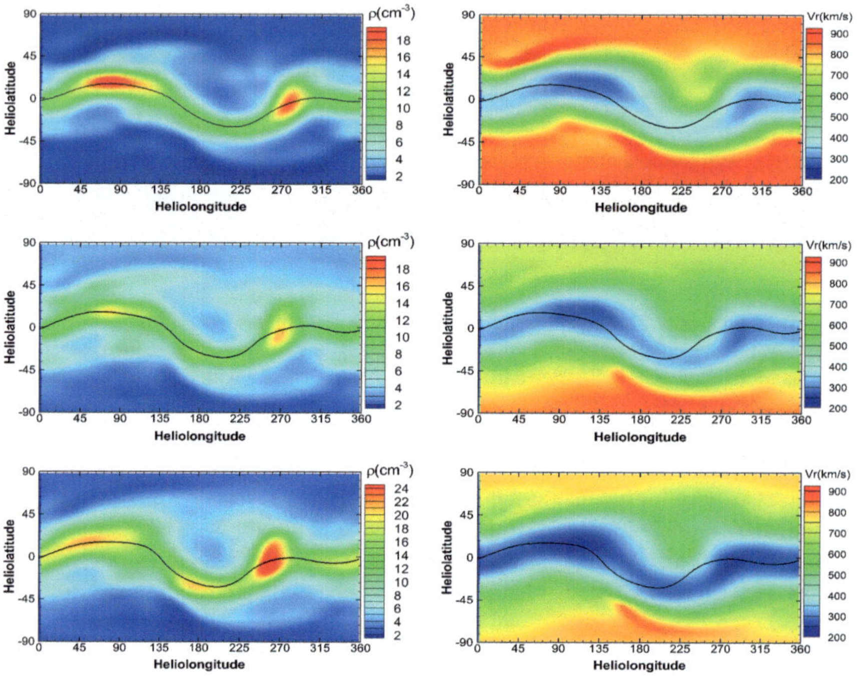

Fig. 7.33 Synoptic maps of the proton number density (left column), and radial bulk speed (right column) on July 1st (top row), July 14th (middle row), and July 23rd (bottom row) at 1 AU. The black lines indicate the locations where $B_r = 0$. Reprinted from Fig. 9 of Feng et al. [20] with permission of the John Wiley & Sons, Inc.

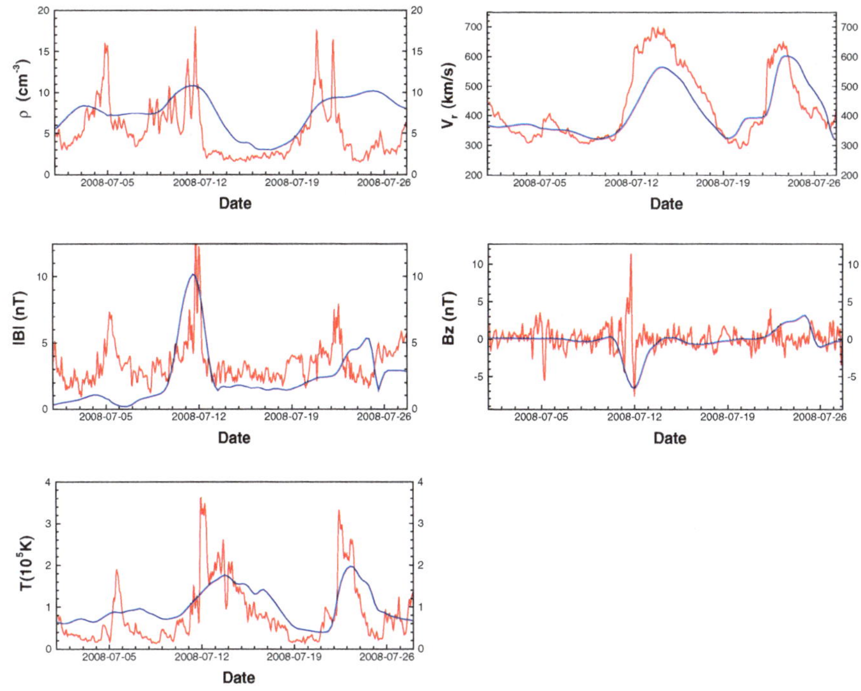

Fig. 7.34 Comparison between simulation results and OMNI data. The red lines are OMNI in-situ measurements, and the blue lines represent our simulation results. From top left to bottom right, the parameters are the proton number density, radial bulk speed, total magnetic field strength, and the temperature. Reprinted from Fig. 10 of Feng et al. [20] with permission of the John Wiley & Sons, Inc.

Figure 7.34 compares the simulation results from the data-driven MHD model and the OMNI data at 1 AU. The comparison shows that the result of the data-driven model roughly captures the main features of the solar wind observation. In the figure, both the observation and the data-driven model show similar variation trends of the radial bulk speed, total magnetic field strength and the temperature. Both the measurements and the data-driven model capture two high-speed streams with reasonable accuracy. The simulation obtains relatively higher density and lower solar wind temperature than the OMNI data. The modeled hump of the solar wind speed centered on July 14th is somewhat lower than the observation, but the simulated hump on July 23rd fits the OMNI data very well. Indeed, there are some discrepancies between the simulation and the observation. The total magnetic field strength from the simulated result is much lower than that observed for the most part of the investigated period, and the improvement of this situation requires the modelers' dedicated efforts.

7.6 Conclusions

This chapter discusses the implementation of data-driven MHD model to achieve the time-varying coronal and solar wind structure. The model is based on the AMR-CESE-MHD model introduced in Chap. 6. The PNC method is employed to self-consistently coupling the input photosphere magnetic field synoptic map data with the MHD model. Synoptic maps of different sources, cadences and data processing procedures are utilized to drive the model. The daily-update synoptic maps from WSO, the SFT model initiated by SOHO/MHI synoptic map and the GONG's synoptic maps with differential rotation corrections are utilized in Sect. 7.3, Sect. 7.4 and Sect. 7.5 respectively. The simulated results are compared with the remote observations near the Sun and the in-situ or mapped interplanetary measurement. These comparisons demonstrate the data-driven model is able to generate the general structures of the global corona, such as coronal streamers and pseudo-streamers, the locations of coronal holes, the structure of the current sheet, and their evolution. The treads of the background solar wind are roughly reproduced. In Sect. 7.3, the difference between the results obtained from quasi-steady model and data-driven model are compared. In Sect. 7.4, the nonpotential magnetic configuration overlaying the active region, which is supported by the observation of a quiescent filament in the H_α image, is partially reproduced by the model. In Sect. 7.5, the modeled heliocentric distances of the AS lie in the range that previous work presented. It should be noted that possible input data for the AMR-CESE-MHD model are not limited to what are used in this chapter. By the way, the AMR-CESE-MHD model can be possibly driven by a variety of observational data, including HMI vector magnetogram and horizontal velocity data.

As mentioned in the introduction, in the context of 3D solar wind modeling, there are at least two challenges to properly employ the measurements of photospheric magnetic fields. One challenge is no instantaneous observation available for the global photospheric magnetic fields. The other one is seasonally unobserved data gaps near the poles because measurements of the photospheric magnetic field are now only available from the ground stations or the instruments board on spacecraft near the Earth.

Associated with the two above difficulties and other unresolved reasons, underestimation of the interplanetary field flux is always present for all MHD simulations for the solar wind with the maps of photospheric magnetic field measurements as input and the coronal base as the model's lower boundary. Linker et al. [51] suggested that the deficiency of the open magnetic flux could be attributed to either the underestimation of the Sun's magnetic flux by typical observatory maps, or other sources of open magnetic flux other than the regions that are obviously dark in EUV and X-ray emission. Tsuneta et al. [93] estimated that the concentrated kilogauss patches near polar regions would give the equivalent flux of a 10 G field over the region 20° from the pole and pointed out that the polar magnetic flux could be significantly underestimated near solar minimum. Some studies indicated that an important portion of the slow wind might arise from interchange reconnection between closed and open fields (e.g., [21, 90]). Therefore, significant amounts of open field are originated from the regions that are iterspersed with closed-field regions and are not dark in EUV and X-ray emission. The improvement of the data-driven modeling depends on the development of new observation techniques and products. For an example, the Polarimetric and Helioseismic Imager (PHI) on board the Solar Orbiter (SO) spacecraft will provide the first measurement of photospheric vector magnetic field from a heliographic latitude higher than 7° [25]. The SO will approach the Sun to within 0.28 AU and reach a heliographic latitude of up to 33°. Once it begins to observe the Sun, the PHI/SO instrument will have a unique view of the solar poles, allowing not just more precise and exact measurements of the polar field than ever before, but also allowing us to follow the dynamics of individual magnetic features at high latitudes and to determine solar surface and sub-surface flows right up to the poles [85]. These measurements will surely shed light on the solution of the data-gap problem at the solar polar region and significantly improve solutions of 3D data-driven MHD models for the solar wind. At the same time, the state-of-the-art machine learning/data science methods can also be utilized to diagnose the problem of missing magnetic flux, which requires integrated collaborations from the communities of data providers, expert modelers, and computer and data scientists.

References

1. Abramenko V, Yurchyshyn V (2010) Intermittency and multifractality spectra of the magnetic field in solar active regions. Astrophys J 722:122–130
2. Arge CN, Pizzo VJ (2000) Improvement in the prediction of solar wind conditions using near-real time solar magnetic field updates. J Geophys Res Space Phys 105(A5):10,465–10,479
3. Arge CN, Hildner E, Pizzo VJ, Harvey JW (2002) Two solar cycles of nonincreasing magnetic flux. J Geophys Res Space Phys 107(A10):1319
4. Arge CN, Henney CJ, Koller J, Toussaint WA, Harvey JW, Young S (2011) Improving data drivers for coronal and solar wind models. In: Pogorelov NV, Audit E, Zank GP (eds) 5th international conference of numerical modeling of space plasma flows (ASTRONUM 2010). Astronomical society of the Pacific conference series, vol 444, p 99
5. Arge CN, Henney CJ, Hernandez IG, Toussaint WA, Koller J, Godinez HC (2013) Modeling the corona and solar wind using ADAPT maps that include far-side observations. AIP Conf Proc 1539(1):11–14
6. Blackwell DE, Petford AD (1966) Observations of the 1963 July 20 solar eclipse. I, spectroscopic separation of the F and K components of the solar corona at large distances from the Sun. Mon Not R Astron Soc 131:383
7. Blackwell DE, Petford AD (1966) Observations of the 1963 July 20 solar eclipse. II, the electron density in the solar corona in the region $5<R/R_0<16$ obtained from measurements of Fraunhofer line depth and the polarization of the F corona. Mon Not R Astron Soc 131:399
8. Cohen O, Sokolov IV, Roussev II, Arge CN, Manchester WB, Gombosi TI, Frazin RA, Park H, Butala MD, Kamalabadi F, Velli M (2007) A semiempirical magnetohydrodynamical model of the solar wind. Astrophys J Lett 654:L163–L166. https://doi.org/10.1086/511154
9. Cohen O, Sokolov IV, Roussev II, Gombosi TI (2008) Validation of a synoptic solar wind model. J Geophys Res Space Phys 113(A3):A03104
10. de Toma G (2010) Evolution of coronal holes and implications for high-speed solar wind during the minimum between cycles 23 and 24. Sol Phys 213–226: https://doi.org/10.1007/s11207-010-9677-2

11. DeForest CE, Howard TA, McComas DJ (2014) Inbound waves in the solar corona: a direct indicator of Alfvén surface location. Astrophys J 787:124
12. Detman T, Smith Z, Dryer M, Fry CD, Arge CN, Pizzo V (2006) A hybrid heliospheric modeling system: background solar wind. J Geophys Res (Space Phys) 111(A10):A07102
13. DeVore CR, Boris JP, Sheeley NR Jr (1984) The concentration of the large-scale solar magnetic field by a meridional surface flow. Sol Phys 92:1–14
14. Downs C, Roussev II, van der Holst B, Lugaz N, Sokolov IV, Gombosi TI (2010) Toward a realistic thermodynamic magnetohydrodynamic model of the global solar corona. Astrophys J 712:1219–1231. https://doi.org/10.1088/0004-637X/712/2/1219
15. Dunn T, Jackson BV, Hick PP, Buffington A, Zhao XP (2005) Comparative analyses of the CSSS calculation in the UCSD tomographic solar observations. Sol Phys 227:339–353
16. Feng XS, Yang LP, Xiang CQ, Wu ST, Zhou YF, Zhong DK (2010) Three-dimensional solar wind modeling from the Sun to Earth by a SIP-CESE MHD model with a six-component grid. Astrophys J 723(1):300
17. Feng XS, Zhang SH, Xiang CQ, Yang LP, Jiang CW, Wu ST (2011) A hybrid solar wind model of the CESE+HLL method with a Yin-Yang overset grid and an AMR grid. Astrophys J 734:50
18. Feng XS, Jiang CW, Xiang CQ, Zhao XP, Wu ST (2012) A data-driven model for the global coronal evolution. Astrophys J 758(1):62
19. Feng XS, Yang LP, Xiang CQ, Liu Y, Zhao XP, Wu ST (2012) Numerical study of the global corona for CR 2055 driven by daily-updated synoptic magnetic field. In: Astronomical society of the Pacific conference series, vol 459, p 202
20. Feng XS, Ma XP, Xiang CQ (2015) Data-driven modeling of the solar wind from 1 R_s to 1 AU. J Geophys Res (Space Phys) 120(A9):10. https://doi.org/10.1002/2015JA021911
21. Fisk LA, Schwadron NA, Zurbuchen TH (1998) On the slow solar wind. Space Sci Rev 86:51–60. https://doi.org/10.1023/A:1005015527146
22. Foullon C, Lavraud B, Luhmann JG, Farrugia CJ, Retinò A, Simunac KDC, Wardle NC, Galvin AB, Kucharek H, Owen CJ, Popecki M, Opitz A, Sauvaud JA (2011) Plasmoid releases in the heliospheric current sheet and associated coronal hole boundary layer evolution. Astrophys J 737:16
23. Frazin RA, Vásquez AM, Kamalabadi F, Park H (2007) Three-dimensional tomographic analysis of a high-cadence LASCO-C2 polarized brightness sequence. Astrophys J Lett 671:L201–L204
24. Frazin RA, Vásquez AM, Kamalabadi F (2009) Quantitative, three-dimensional analysis of the global corona with multi-spacecraft differential emission measure tomography. Astrophys J 701:547–560
25. Gandorfer A, Solanki SK, Woch J, Martínez Pillet V, Álvarez Herrero A, Appourchaux T (2011) The solar orbiter mission and its polarimetric and helioseismic imager (SO/PHI). GONG-SoHO 24: New Era Seismol Sun Sol Like Stars 271(1):012086. https://doi.org/10.1088/1742-6596/271/1/012086
26. Goelzer ML, Schwadron NA, Smith CW (2014) An analysis of Alfvén radius based on sunspot number from 1749 to today. J Geophys Res Space Phys 119(1):115–120
27. Gressl C, Veronig AM, Temmer M, Odstrčil D, Linker JA, Mikić Z, Riley P (2014) Comparative study of MHD modeling of the background solar wind. Sol Phys 289:1783–1801. https://doi.org/10.1007/s11207-013-0421-6
28. Harvey JW, Hill F, Hubbard RP, Kennedy JR, Leibacher JW, Pintar JA, Gilman PA, Noyes RW, Title AM, Toomre J, Ulrich RK, Bhatnagar A, Kennewell JA, Marquette W, Patron J, Saa O, Yasukawa E (1996) The global oscillation network group (GONG) project. Science 272:1284–1286
29. Hayashi K (2005) Magnetohydrodynamic simulations of the solar corona and solar wind using a boundary treatment to limit solar wind mass flux. Astrophys J Suppl Ser 161:480–494
30. Hayashi K (2012) An MHD simulation model of time-dependent co-rotating solar wind. J Geophys Res Space Phys 117(A8):A08105. https://doi.org/10.1029/2011JA017490

31. Hayashi K (2013) An MHD simulation model of time-dependent global solar corona with temporally varying solar-surface magnetic field maps. J Geophys Res Space Phys 118(11):6889–6906
32. Hayashi K, Benevolenskaya E, Hoeksema T, Liu Y, Zhao XP (2006) Three-dimensional magnetohydrodynamic simulation of a global solar corona using a temperature distribution map obtained from SOHO EIT measurements. Astrophys J Lett 636:L165–L168
33. Hayashi K, Zhao XP, Liu Y (2008) MHD simulations of the global solar corona around the Halloween event in 2003 using the synchronic frame format of the solar photospheric magnetic field. J Geophys Res Space Phys 113(A7):A07104
34. Hayashi K, Hoeksema JT, Liu Y, Bobra MG, Sun XD, Norton AA (2015) The helioseismic and magnetic imager (HMI) vector magnetic field pipeline: magnetohydrodynamics simulation module for the global solar corona. Sol Phys 290:1507–1529. https://doi.org/10.1007/s11207-015-0686-z
35. Hernández IG, Hill F, Lindsey C (2007) Calibration of seismic signatures of active regions on the far side of the Sun. Astrophys J 669(2):1382
36. Hickmann KS, Godinez HC, Henney CJ, Arge CN (2015) Data assimilation in the adapt photospheric flux transport model. Sol Phys 290(4):1105–1118
37. Hill F, Fischer G, Forgach S, Grier J, Leibacher JW, Jones HP, Jones PB, Kupke R, Stebbins RT, Clay DW (1994a) The global oscillation network group site survey, 2: results. Sol Phys 152:351–379
38. Hill F, Fischer G, Grier J, Leibacher JW, Jones HB, Jones PP, Kupke R, Stebbins RT (1994b) The global oscillation network group site survey, 1: data collection and analysis methods. Sol Phys 152:321–349
39. Hoeksema JT (1984) Structure and evolution of the large scale solar and heliospheric magnetic fields. PhD thesis, Stanford University, CA
40. Howard R, Moses J, Vourlidas A, Newmark J, Socker D, Plunkett S, Korendyke C, Cook J, Hurley A, Davila J, Thompson W, St Cyr O, Mentzell E, Mehalick K, Lemen J, Wuelser J, Duncan D, Tarbell T, Wolfson C, Moore A, Harrison R, Waltham N, Lang J, Davis C, Eyles C, Mapson-Menard H, Simnett G, Halain J, Defise J, Mazy E, Rochus P, Mercier R, Ravet M, Delmotte F, Auchere F, Delaboudiniere J, Bothmer V, Deutsch W, Wang D, Rich N, Cooper S, Stephens V, Maahs G, Baugh R, McMullin D, Carter T (2008) Sun Earth Connection Coronal and Heliospheric Investigation (SECCHI). Space Sci Rev 136:67–115. https://doi.org/10.1007/s11214-008-9341-4
41. Hu YQ, Feng XS, Wu ST, Song WB (2008) Three-dimensional MHD modeling of the global corona throughout solar cycle 23. J Geophys Res 113(A12):A03106. https://doi.org/10.1029/2007JA012750
42. Intriligator DS, Detman T, Gloecker G, Gloeckler C, Dryer M, Sun W, Intriligator J, Deehr C (2012) Pickup protons: comparisons using the three-dimensional MHD HHMS-PI model and Ulysses SWICS measurements. J Geophys Res Space Phys 117(A6):A06104
43. Jian LK, MacNeice PJ, Taktakishvili A, Odstrcil D, Jackson B, Yu HS, Riley P, Sokolov IV, Evans RM (2015) Validation for solar wind prediction at Earth: comparison of coronal and heliospheric models installed at the CCMC. Space Weather 13:316–338. https://doi.org/10.1002/2015SW001174
44. Jiang J, Cameron R, Schmitt D, Schüssler M (2010) Modeling the Sun's open magnetic flux and the heliospheric current sheet. Astrophys J 709:301–307. https://doi.org/10.1088/0004-637X/709/1/301
45. Kilpua EKJ, Luhmann JG, Gosling J, Li Y, Elliott H, Russell CT, Jian L, Galvin AB, Larson D, Schroeder P, Simunac K, Petrie G (2009) Small solar wind transients and their connection to the large-scale coronal structure. Sol Phys 256:327–344. https://doi.org/10.1007/s11207-009-9366-1
46. Lai MC, Wang WC (2002) Fast direct solvers for Poisson equation on 2D polar and spherical geometries. Numer Methods Partial Differ 18(1):56–68. https://doi.org/10.1002/num.1038
47. Leighton RB (1964) Transport of magnetic fields on the Sun. Astrophys J 140:1547

48. Lepri ST, Antiochos SK, Riley P, Zhao L, Zurbuchen TH (2008) Comparison of heliospheric in situ data with the quasi-steady solar wind models. Astrophys J 674:1158–1166. https://doi.org/10.1086/524347
49. Linker JA, Mikić Z, Biesecker DA, Forsyth RJ, Gibson SE, Lazarus AJ, Lecinski A, Riley P, Szabo A, Thompson BJ (1999) Magnetohydrodynamic modeling of the solar corona during Whole Sun Month. J Geophys Res 104:9809–9830. https://doi.org/10.1029/1998JA900159
50. Linker JA, Lionello R, Mikić Z, Titov VS, Antiochos SK (2011) The evolution of open magnetic flux driven by photospheric dynamics. Astrophys J 731:110
51. Linker JA, Caplan RM, Downs C, Riley P, Mikic Z, Lionello R, Henney CJ, Arge CN, Liu Y, Derosa ML, Yeates A, Owens MJ (2017) The open flux problem. Astrophys J 848:70. https://doi.org/10.3847/1538-4357/aa8a70
52. Lionello R, Mikić Z, Linker JA, Amari T (2002) Magnetic field topology in prominences. Astrophys J 581:718–725
53. Lionello R, Riley P, Linker JA, Mikić Z (2005) The effects of differential rotation on the magnetic structure of the solar corona: magnetohydrodynamic simulations. Astrophys J 625(1):463
54. Liu Y, Hoeksema JT, Zhao X, Larson RM (2007) MDI synoptic charts of magnetic field: interpolation of polar fields. In: American astronomical society meeting abstracts #210, American Astronomical Society, Washington, D.C., Bulletin of the American Astronomical Society, vol 39, p 129
55. Low BC (1996) Solar activity and the corona. Sol Phys 167:217–265
56. Luhmann JG, Li Y, Arge CN, Gazis PR, Ulrich R (2002) Solar cycle changes in coronal holes and space weather cycles. J Geophys Res 107:1154. https://doi.org/10.1029/2001JA007550
57. Mackay DH, van Ballegooijen AA (2006) Models of the large-scale corona. I. Formation, evolution, and liftoff of magnetic flux ropes. Astrophys J 641(1):577
58. Mackay DH, Yeates AR (2012) The Sun's global photospheric and coronal magnetic fields: observations and models. Living Rev Sol Phys 9:6. https://doi.org/10.12942/lrsp-2012-6
59. McComas DJ, Ebert RW, Elliott HA, Goldstein BE, Gosling JT, Schwadron NA, Skoug RM (2008) Weaker solar wind from the polar coronal holes and the whole Sun. Geophys Res Lett 35:L18103
60. McGregor SL, Hughes WJ, Arge CN, Odstrcil D, Schwadron NA (2011) The radial evolution of solar wind speeds. J Geophys Res 116:A03106. https://doi.org/10.1029/2010JA016006
61. McGregor SL, Hughes WJ, Arge CN, Owens MJ, Odstrcil D (2011) The distribution of solar wind speeds during solar minimum: calibration for numerical solar wind modeling constraints on the source of the slow solar wind. J Geophys Res 116(A3):A03101. https://doi.org/10.1029/2010JA015881
62. Meng X, van der Holst B, Tóth G, Gombosi TI (2015) Alfvén wave solar model (AWSoM): proton temperature anisotropy and solar wind acceleration. Mon Not R Astron Soc 454:3697–3709. https://doi.org/10.1093/mnras/stv2249
63. Mikić Z, Linker JA, Schnack DD, Lionello R, Tarditi A (1999) Magnetohydrodynamic modeling of the global solar corona. Phys Plasmas 6(5):2217–2224
64. Nakagawa Y (1980) Evolution of solar magnetic fields–a new approach to MHD initial-boundary value problems by the method of nearcharacteristics. Astrophys J 240:275–299
65. Nakagawa Y (1981) Evolution of magnetic field and atmospheric response. I—three-dimensional formulation by the method of projected characteristics. Astrophys J 247:707
66. Nakagawa Y (1981) Evolution of magnetic field and atmospheric responses. II—formulation of proper boundary equations. Astrophys J 247:719
67. Nakagawa Y, Hu YQ, Wu ST (1987) The method of projected characteristics for the evolution of magnetic arches. Astron Astrophys 179:354–370
68. Neugebauer M, Forsyth RJ, Galvin AB, Harvey KL, Hoeksema JT, Lazarus AJ, Lepping RP, Linker JA, Mikic Z, Steinberg JT, von Steiger R, Wang YM, Wimmer-Schweingruber RF (1998) Spatial structure of the solar wind and comparisons with solar data and models. J Geophys Res Space Phys 103(A7):14,587–14,599

69. Neugebauer M, Liewer PC, Smith EJ, Skoug RM, Zurbuchen TH (2002) Sources of the solar wind at solar activity maximum. J Geophys Res 107:1488. https://doi.org/10.1029/2001JA000306
70. Odstrcil D, Pizzo VJ, Linker JA, Riley P, Lionello R, Mikic Z (2004a) Initial coupling of coronal and heliospheric numerical magnetohydrodynamic codes. J Atmos Sol Terr Phys 66:1311–1320. https://doi.org/10.1016/j.jastp.2004.04.007
71. Odstrcil D, Riley P, Zhao XP (2004b) Numerical simulation of the 12 May 1997 interplanetary CME event. J Geophys Res (Space Phys) 109(A18):A02116. https://doi.org/10.1029/2003JA010135
72. Parker EN (1963) Interplanetary dynamical processes. Interscience Publishers, New York
73. Petrie GJ, Bolding J, Clark R, Donaldson-Hanna K, Harvey JW, Hill F, Toner C, Wentzel TM (2007) GONG synoptic magnetograms and coronal magnetic field modeling. In: AGU fall meeting abstracts SH01-2873
74. Riley P (2007) An alternative interpretation of the relationship between the inferred open solar flux and the interplanetary magnetic field. Astrophys J Lett 667:L97–L100
75. Riley P, Luhmann JG (2012) Interplanetary signatures of unipolar streamers and the origin of the slow solar wind. Sol Phys 277:355–373. https://doi.org/10.1007/s11207-011-9909-0
76. Riley P, Linker JA, Mikić Z, Lionello R, Ledvina SA, Luhmann JG (2006) A comparison between global solar magnetohydrodynamic and potential field source surface model results. Astrophys J 653:1510–1516
77. Riley P, Linker JA, Lionello R, Mikić Z (2012) Corotating interaction regions during the recent solar minimum: the power and limitations of global MHD modeling. J Atmos Sol Terr Phys 83:1–10
78. Riley P, Ben-Nun M, Linker JA, Mikić Z, Svalgaard L, Harvey J, Bertello L, Hoeksema T, Liu Y, Ulrich R (2014) A multi-observatory inter-comparison of line-of-sight synoptic solar magnetograms. Sol Phys 289:769–792. https://doi.org/10.1007/s11207-013-0353-1
79. Riley P, Linker JA, Arge CN (2015) On the role played by magnetic expansion factor in the prediction of solar wind speed. Space Weather 13:154–169. https://doi.org/10.1002/2014SW001144
80. Rušin V, Druckmüller M, Aniol P, Minarovjech M, Saniga M, Mikić Z, Linker JA, Lionello R, Riley P, Titov VS, (2010) Comparing eclipse observations of the 2008 August 1 solar corona with an MHD model prediction. Astron Astrophys 513:A45
81. Schrijver CJ, DeRosa ML, Title AM (2002) What is missing from our understanding of long-term solar and heliospheric activity? Astrophys J 577(2):1006
82. Sheeley NR Jr (2005) Surface evolution of the Sun's magnetic field: a historical review of the flux-transport mechanism. Living Rev Sol Phys 2:5. https://doi.org/10.12942/lrsp-2005-5
83. Sheeley NR Jr, Warren HP, Wang YM (2004) The origin of postflare loops. Astrophys J 616:1224–1231
84. Snodgrass HB (1983) Magnetic rotation of the solar photosphere. Astrophys J 270:288–299
85. Solanki SK, del Toro Iniesta JC, Woch J, Gandorfer A, Hirzberger J, Schmidt W, Appourchaux T, Alvarez-Herrero A (2015) The polarimetric and helioseismic imager for solar orbiter: SO/PHI. In: Nagendra KN, Bagnulo S, Centeno R, Jesús Martínez González M (eds) Polarimetry, IAU symposium, vol 305, pp 108–113. https://doi.org/10.1017/S1743921315004615
86. Song WM, Zhang XJ, Wang Y, Ren XY (1999) The perfect project theorem for a vector function on vector wave function space. Chin J Radio Sci 14(3):3
87. Sun X, Liu Y, Hoeksema J, Hayashi K, Zhao X (2011) A new method for polar field interpolation. Sol Phys 270:9–22
88. Svalgaard L, Duvall TL Jr, Scherrer PH (1978) The strength of the Sun's polar fields. Sol Phys 58:225–239
89. Taktakishvili A, Pulkkinen A, MacNeice P, Kuznetsova M, Hesse M, Odstrcil D (2011) Modeling of coronal mass ejections that caused particularly large geomagnetic storms using ENLIL heliosphere cone model. Space Weather 90:S06002
90. Titov VS, Mikić Z, Linker JA, Lionello R, Antiochos SK (2011) Magnetic topology of coronal hole linkages. Astrophys J 731:111. https://doi.org/10.1088/0004-637X/731/2/111

91. Tokumaru M, Kojima M, Fujiki K (2010) Solar cycle evolution of the solar wind speed distribution from 1985 to 2008. J Geophys Res 115(A4):A04102. https://doi.org/10.1029/2009JA014628
92. Tóth G, van der Holst B, Sokolov IV, Zeeuw DLD, Gombosi TI, Fang F, Manchester WB, Meng X, Najib D, Powell KG, Stout QF, Glocer A, Ma YJ, Opher M (2012) Adaptive numerical algorithms in space weather modeling. J Comput Phys 231(3):870–903
93. Tsuneta S, Ichimoto K, Katsukawa Y, Lites BW, Matsuzaki K, Nagata S, Orozco Suárez D, Shimizu T, Shimojo M, Shine RA, Suematsu Y, Suzuki TK, Tarbell TD, Title AM (2008) The magnetic landscape of the Sun's polar region. Astrophys J 688:1374–1381. https://doi.org/10.1086/592226
94. Tu CY, Marsch E (1995) MHD structures, waves and turbulence in the solar wind: observations and theories. Space Sci Rev 73:1–210
95. Ulrich RK (1992) Analysis of magnetic fluxtubes on the solar surface from observations at Mt. Wilson of A5250; A5233. In: Giampapa MS, Bookbinder JA (eds) Cool stars, stellar systems, and the sun. Astronomical society of the Pacific conference series, vol 26. American Astronomical Society, Washington, D.C., p 265
96. Usmanov AV, Goldstein ML (2006) A three-dimensional MHD solar wind model with pickup protons. J Geophys Res Space Phys 111(A7):A07101
97. van der Holst B, Manchester WB IV, Frazin RA, Vásquez AM, Tóth G, Gombosi TI (2010) A data-driven, two-temperature solar wind model with Alfvén waves. Astrophys J 725:1373–1383. https://doi.org/10.1088/0004-637X/725/1/1373
98. van der Holst B, Sokolov IV, Meng X, Jin M, Manchester WB IV, Tóth G, Gombosi TI (2014) Alfvén wave solar model (AWSoM): coronal heating. Astrophys J 782:81. https://doi.org/10.1088/0004-637X/782/2/81
99. Vásquez AM, Frazin RA, Manchester WB IV (2010) The solar minimum corona from differential emission measure tomography. Astrophys J 715:1352–1365
100. Wang AH, Wu ST, Tandberg-Hanssen E, Hill F (2011) Utilization of multiple measurements for global three-dimensional magnetohydrodynamic simulations. Astrophys J 732:19
101. Wang S, Hu YQ, Wu ST (1982) Full-implicit continuous Eulerian scheme in the spherical coordinates and its applications to solar phenomena. Sci Sin Ser Math Phys Tech Sci 25:1305–1318
102. Wang YM (1996) Nonradial coronal streamers. Astrophys J Lett 456:L119. https://doi.org/10.1086/309871
103. Wang YM, Sheeley NR (1992) On potential-field models of the solar corona. Astrophys J 392:310–319
104. Wang YM, Sheeley NR Jr (1988) The solar origin of long-term variations of the interplanetary magnetic field strength. J Geophys Res 93:11,227–11,236. https://doi.org/10.1029/JA093iA10p11227
105. Wang YM, Nash AG, Sheeley NR Jr (1989) Magnetic flux transport on the Sun. Science 245(4919):712–718
106. Wang YM, Hawley SH, Sheeley NR Jr (1996) The magnetic nature of coronal holes. Science 271:464–469. https://doi.org/10.1126/science.271.5248.464
107. Wang YM, Sheeley NR Jr, Howard RA, Kraemer JR, Rich NB, Andrews MD, Brueckner GE, Dere KP, Koomen MJ, Korendyke CM, Michels DJ, Moses JD, Paswaters SE, Socker DG, Wang D, Lamy PL, Llebaria A, Vibert D, Schwenn R, Simnett GM (1997) Origin and evolution of coronal streamer structure during the 1996 minimum activity phase. Astrophys J 485:875. https://doi.org/10.1086/304467
108. Wang YM, Sheeley NR, Rich NB (2000) Evolution of coronal streamer structure during the rising phase of solar cycle 23. Geophys Res Lett 27:149–152. https://doi.org/10.1029/1999GL010698
109. Wang YM, Sheeley NR Jr, Lean J (2002) Meridional flow and the solar cycle variation of the Sun's open magnetic flux. Astrophys J 580(2):1188
110. Wang YM, Sheeley NR, Andrews MD (2002) Polarity reversal of the solar magnetic field during cycle 23. J Geophys Res 107:1465. https://doi.org/10.1029/2002JA009463

111. Wang YM, Sheeley NR Jr, Rich NB (2007) Coronal pseudostreamers. Astrophys J 658:1340–1348. https://doi.org/10.1086/511416
112. Wang YM, Robbrecht E, Sheeley NR Jr (2009) On the weakening of the polar magnetic fields during solar cycle 23. Astrophys J 707:1372–1386. https://doi.org/10.1088/0004-637X/707/2/1372
113. Wei F, Feng X, Cai H, Zhou Q (2003) Global distribution of coronal mass outputs and its relation to solar magnetic field structures. J Geophys Res 108:1238. https://doi.org/10.1029/2002JA009439
114. Wiegelmann T (2008) Nonlinear force-free modeling of the solar coronal magnetic field. J Geophys Res 113(A12):A03S02
115. Worden J, Harvey J (2000) An evolving synoptic magnetic flux map and implications for the distribution of photospheric magnetic flux. Sol Phys 195:247–268. https://doi.org/10.1023/A:1005272502885
116. Wu ST, Wang JF (1987) Numerical tests of a modified full implicit continuous Eulerian (FICE) scheme with projected normal characteristic boundary conditions for MHD flows. Comput Methods Appl Mech Eng 64:267–282
117. Wu ST, Zheng H, Wang S, Thompson BJ, Plunkett SP, Zhao XP, Dryer M (2001) Three-dimensional numerical simulation of MHD waves observed by the extreme ultraviolet imaging telescope. J Geophys Res 106:25,089–25,102. https://doi.org/10.1029/2000JA000447
118. Wu ST, Wang AH, Liu Y, Hoeksema JT (2006) Data-driven magnetohydrodynamic model for active region evolution. Astrophys J 652:800–811
119. Yang LP, Feng XS, Xiang CQ, Liu Y, Zhao X, Wu ST (2012) Time-dependent MHD modeling of the global solar corona for year 2007: driven by daily-updated magnetic field synoptic data. J Geophys Res Space Phys 117(A8):8110
120. Yeates AR, Mackay DH, van Ballegooijen AA (2007) Modelling the global solar corona: filament chirality observations and surface simulations. Sol Phys 245:87–107
121. Yeates AR, Mackay DH, van Ballegooijen AA (2008) Modelling the global solar corona II: coronal evolution and filament chirality comparison. Sol Phys 247:103–121
122. Yeh T, Dryer M (1985) A constraint on boundary data for magnetic solenoidality in MHD calculations. Astrophys Space Sci 117:165–171. https://doi.org/10.1007/BF00660919
123. Zhao XP, Hoeksema JT (2010) The magnetic field at the inner boundary of the heliosphere around solar minimum. Sol Phys 266:379–390
124. Zhao XP, Hoeksema JT, Scherrer PH (1999) Changes of the boot-shaped coronal hole boundary during Whole Sun Month near sunspot minimum. J Geophys Res 104:9735–9752
125. Zhao XP, Scherrer PH, Hoeksema JT (2004) The effect of the differential rotation of photospheric magnetic features on synoptic frames of the photospheric magnetic field. In: AGU fall meeting abstracts, p B419
126. Zurbuchen TH, Richardson IG (2006) In-situ solar wind and magnetic field signatures of interplanetary coronal mass ejections. Space Sci Rev 123:31–43

Chapter 8
Coronal Force-Free Field Extrapolations

Abstract Solar corona is the source region of many explosive events, e.g., flares and CMEs, which play a leading role in causing adverse space weather. All the solar eruptions are energized from the magnetic fields, which is generated below the photosphere and transported into the upper atmosphere, i.e., the corona. Consequently, the coronal magnetic field holds the key to understanding the origin of the energetic events. Due to the absence of direct measurement, the three-dimensional magnetic field in the solar corona is usually "extrapolated" in numerical ways from the photosphere where a two-dimensional surface of magnetic vectors can be routinely measured. Up to the present, the nonlinear force-free field (NLFFF) model dominates the physical models for magnetic field extrapolation in the low corona. This chapter briefly reviews the present methods for NLFFF extrapolation with emphasis on the CESE–MHD–NLFFF code.

8.1 Introduction

The magnetic field plays a fundamental role in almost all of the physical processes in the Sun's corona, such as the formation of coronal loops and prominences (or filaments), the production of solar flares, filament eruptions, and coronal mass ejections, as well as the determination of the structure of solar wind [104]. However, it is very difficult to make a direct measurement of the coronal magnetic field from emissions of the extremely tenuous plasma of the corona by the traditional spectropolarimetric methods (using the Zeeman and Hanle effects). Works that have been done to measure the coronal fields using the radio and infrared wave bands (e.g., [33, 73]) can only give fragmentary and occasional data. Up to the present, the routine measurement of the Sun's magnetic field that we can rely on is restricted to only a single layer of solar atmospheric, i.e., the photosphere. This is rather unfortunate since the role of the magnetic field playing in the corona is much more important than that in the photosphere.

Idea of numerical modeling the three-dimensional (3D) coronal magnetic field, with boundary constraint by the photospheric magnetograms dates back to nearly a half century ago [3]. In that model, they made a crude assumption that the corona is

© Springer Nature Singapore Pte Ltd. 2020
X. Feng, *Magnetohydrodynamic Modeling of the Solar Corona and Heliosphere*, Atmosphere, Earth, Ocean & Space,
https://doi.org/10.1007/978-981-13-9081-4_8

free of electric current due to the limitation that only the line-of-sight (LoS) magnetogram was measured, from which no information of electric current can be obtained. By this, a potential field can be well constructed by using the Legendre polynomial expansion in the spherical geometry with the LoS magnetogram. With the ever-growing ability of observations, vector magnetogram, that is, both the LoS and transverse component of the photospheric magnetic field (and information of the electric current passing through the photosphere can be derived), can be measured in a rather high precision and resolution, and many coronal field reconstructions based on these modern data have been developed. Almost all the modern codes are based on a basic assumption that the corona is free of Lorentz force (i.e., the magnetic pressure force and the magnetic tension force cancel each other). This is because in most parts of the low corona (with a height less than ~400 Mm from the coronal base), particularly in the strongly-magnetized active regions, the plasma β (a ratio of gas pressure p to magnetic pressure $B^2/(2\mu_0)$, i.e., $\beta = 2\mu_0 p/B^2$) is extremely low ($\beta \ll 1$) [34]. Thus in the case of quasi-static equilibrium (the plasma flow velocity $v \ll v_{\rm A}$, where v_A is the Alfvén speed), the pressure gradient, solar gravity, and inertial force can all be neglected from the momentum equation of MHD if compared to the magnetic pressure force, and the only remained Lorentz force must be self-balanced, i.e., $\mathbf{J} \times \mathbf{B} = \mathbf{0}$, where $\mathbf{J}$ is the electric current density and $\mathbf{B}$ is the magnetic field. This means that the magnetic tension and pressure force balances each other. By using the Ampere's law $\mathbf{J} = \nabla \times \mathbf{B}$, the force-free equation can also written as

$$\nabla \times \mathbf{B} = \alpha \mathbf{B}, \tag{8.1}$$

where the scalar α is called the force-free factor. In the case of current-free, $\alpha = 0$, and a constant α in the full coronal volume results in a linear force-free field (LFFF) model. Generally, α varies spatially, and such a general model is so-called the nonlinear force-free field (NLFFF) model. However, an interesting fact about NLFFF model is that the α does not change along a specific field line, $\mathbf{B} \cdot \nabla\alpha = 0$, which can be easily proven by taking divergence of Eq. 8.1 and using the divergence-free condition for $\mathbf{B}$.

Current models for reconstructing the low coronal field is dominated by the NLFFF model. The fact that NLFFF model is superior over other much simpler force-free models, i.e., the potential field and the LFFF models has long been demonstrated from many observations. From coronal loops and vector magnetograms, it was clearly indicated that there is significant non-potentiality in many solar active regions [98, 99], in particular during its forming and growing phase as shearing and twisting motions at the photosphere usually stress the coronal field significantly. Furthermore, the force-free parameter α directly derived from the measured vector magnetogram, that is $\alpha = J_z/B_z = \nabla \times (B_x, B_y)/B_z$, is found to be a generally space-dependent function. Often in the core of AR, α deviates substantially from zero with different signs, while in the envelope region out of the core it is much closer to zero. Also in the studies of using field lines to forward fitting the coronal loops (e.g., [13]), different sets of observed coronal loops cannot be modeled by a single LFFF extrapolations but requires nonlinear solution consisting of totally different α. Consequently, the free magnetic energy and magnetic topology cannot be characterized appropriately or accurately from the potential field and LFFF models.

On the other hand, the MHD model (e.g., [61, 63, 135, 136]) is less commonly used than the NLFFF model due to reasons of twofold. Firstly, there is a lack of observed information of gas parameters such as the surface plasma density and velocity, which are critical boundary conditions for the full MHD simulations [1, 2, 120, 122]; secondly the numerical realization of full MHD simulation is greatly limited by the present computational capability. For instance, the Courant-Friedrichs-Lewy (CFL) condition puts rather severe restrictions on the size of the time step for most explicit schemes because densities in the corona are very low while magnetic field strengths in active regions can be quite high (~kG) which inescapablely results in a extremely high Alfvén speed [1]. The computational limitation of full MHD arises especially when applying to very high-resolution and large-field-view magnetograms currently available.

Solving the NLFFF equation directly, however, is proved to be a difficult task. As is known, the system is nonlinear intrinsically and even the existence and uniqueness of a solution for a given boundary condition are not proved theoretically; solutions have rarely been found in closed analytic form (e.g., [76]) and predominantly they can only be approximately realized by resorting to numerical ways [131]. Moreover, observation only provide a single layer of data on the bottom boundary (i.e., the photosphere), and even worse, on that layer the magnetic field is forced significantly by the dense plasma and thus conflicts with the fundamental force-free assumption. Besides the noise in the observation, measurement error and instrumental uncertainty (e.g., the well known 180° ambiguity of the transverse fields) are all rather unfavorable for practical computation. As a result, at present it is difficult to seek an exact force-free solution with the observation information fully satisfied. The best scenario is to find a balance between the force-free constraint and the deviation of the final solution from the real observation, i.e., to seek an approximately force-free solution that matches the photospheric field measurements as well as possible.

8.2 Numerical Ways of Solving the NLFFF

There are many different numerical codes that have been developed to implement the NLFFF extrapolations, which have been comprehensively described in a series of papers [9, 22, 25, 26, 41, 49, 81, 92, 100, 128, 131]. Here we only review the essentials of them. The most popular methods underlying the NLFFF codes are three types by including the MHD relaxation method [23, 50, 55, 56, 60, 85, 94, 114, 115, 140], the optimization method [48, 125–127, 130] and the Grad-Rubin method [5, 6, 38, 96, 123, 124], and all of them share a common approach of iteration.

In the MHD relaxation method, the NLFFF field is achieved by the time evolution to static equilibrium of the MHD equations and allow the boundary conditions to change to match the observed magnetogram. Zero-beta assumption, i.e., discarding the plasma pressure, is usually made and viscous or frictional force is adopted in the equation for a fast relaxation. In its most simplified form, the MHD relaxation method is also called as the magneto-frictional method [94] in which the MHD momentum equation is replaced with

$$\nu \mathbf{v} = \mathbf{J} \times \mathbf{B}, \tag{8.2}$$

meaning that the frictional force $\mathbf{F} = -\nu \mathbf{v}$ is balanced by the Lorentz force in the relaxation process. Thus the magnetic induction equation is the only one needed to be solved. The initial condition is given by a potential field computed from the vertical component of the magnetogram, and the boundary condition evolves sequentially until matching the vector magnetogram.

The optimization method [126] is much more straightforward. It transforms the solution of the NLFFF equation to a simply a minimization problem by seeking the minimum of a volume integral function of $\mathbf{B}$ defined with Lorentz force and magnetic field divergence as

$$L = \int_V [B^{-2}|(\nabla \times \mathbf{B}) \times \mathbf{B}|^2 + |\nabla \cdot \mathbf{B}|^2] dV. \tag{8.3}$$

For the given modeling volume V, it also starts from a potential field with lower boundary values replaced by the required vector magnetogram, and then evolves the integral function L to minimize the Lorentz force and divergence.

Both of the MHD relaxation and optimization method use directly the transverse components of the full vector magnetogram as the lower boundary conditions, while in the Grad-Rubin method the boundary condition is specified by the force-free factor α, which is derived by $\alpha = J_z/B_z$ from the vector magnetogram. In the iteration of the Grad-Rubin method, magnetic field is updated from $\mathbf{B}^n$ to $\mathbf{B}^{n+1}$ by first solving α

$$\mathbf{B}^n \cdot \nabla \alpha = 0; \tag{8.4}$$

for the modeling volume with boundary conditions derived from the vector magnetogram, then solving the Ampere's law

$$\nabla \times \mathbf{B}^{n+1} = \alpha \mathbf{B}^n. \tag{8.5}$$

Of course, $\mathbf{B}^{n+1}$ should also fulfill divergence-free condition and its vertical component at the lower boundary matches that of the vector magnetogram. When the iteration converges, the NLFFF solution is believed to be reached.

Other methods, which are less frequently used, include the upward integration method [87, 105, 134], the boundary element method [42, 43, 138, 139] and the force-free electrodynamics method [24]. In recent years, reconstruction models using forward fitting of observed coronal loops as the major constraint with only the LoS magnetogram are developed (e.g., [10, 13, 80]), which is promising to become a new step of NLFFF modelings.

In addition to the differences among these methods, the specific realizations (i.e., the codes) differ significantly in many other aspects from software to hardware, e.g., the mesh configuration, the numerical scheme and boundary conditions, the language of the code (i.e., IDL, C or Fortran), the hardware architecture and the degree of parallelization. As a consequence, these codes perform very differently from each other with the computational speed and extrapolation accuracy. Metcalf et al. [84] and Schrijver et al. [100] have carried out detailed comparisons of some representative codes using respectively the semi-analytic Low and Lou's force-free solutions [76] and a Sun-like test case constructed by van Ballegooijen et al. [119]. They show that although all the tested codes can achieve the reference solutions qualitatively, considerable differences exist in quantity. It is pointed out by their analysis that the way of implementation of the method plays the same important role as its underlying approach for causing such differences. The differences are much more significant and critical when applying these codes to real solar magnetograms. By computing 14 NLFFF models for the famous region AR 10930 from Hinode/SOT magnetogram around a major X-class flare with different codes and boundary conditions, Schrijver et al. [101] concluded that "the model fields differ markedly in geometry, energy content, and force-freeness". Such conclusion is confirmed by DeRosa et al. [25] when they used the codes for another region AR 10953 extrapolating even more (60) different NLFFF models. They also evaluated the mean misalignment angle of the field lines with the stereoscopically determined loops, which ranges from 24° to 44°. Most recently, DeRosa et al. [26] showed that spatial resolution also heavily influences the extrapolation results. In light of these inconsistences, it is recommended to use the NLFFF with cautions, especially in studies with quantitative analysis, and the imporvement for better agreement of the modeled field lines with corresponding coronal loops is required for a reliable NLFFF model.

8.3 The CESE–MHD–NLFFF Method

The CESE–MHD–NLFFF code [56] belongs to the class of MHD relaxation methods, and thus the basic approach is the same, using some kind of fictitious dissipation to drive the MHD system to an equilibrium in which all the forces can be neglected if compared to the Lorentz force and the boundary vector map is satisfied. In this code, a set of modified zero-β MHD equations is solved with frictional force and splitting of the total magnetic field as

$$
\begin{aligned}
\frac{\partial \rho \mathbf{v}}{\partial t} &= (\nabla \times \mathbf{B}_1) \times \mathbf{B} - \nu \rho \mathbf{v}, \\
\frac{\partial \mathbf{B}_1}{\partial t} &= \nabla \times (\mathbf{v} \times \mathbf{B}) + \nabla(\mu \nabla \cdot \mathbf{B}_1) - \mathbf{v} \nabla \cdot \mathbf{B}_1, \\
\frac{\partial \mathbf{B}_0}{\partial t} &= \mathbf{0}, \; \nabla \times \mathbf{B}_0 = \mathbf{0}, \; \nabla \cdot \mathbf{B}_0 = 0, \\
\rho &= |\mathbf{B}|^2, \;\; \mathbf{B} = \mathbf{B}_0 + \mathbf{B}_1,
\end{aligned}
\tag{8.6}
$$

where $\mathbf{B}_0$ is the potential-field component of the total field $\mathbf{B}$, i.e., a potential magnetic field constructed from only the normal component of the magnetogram. Solving the potential field is a trivial task which can implemented easily by Green's function method or similar ways. By the splitting of the total field, one only need to solve the non-potential field component $\mathbf{B}_1$. Such method was originally proposed in Tanaka [108] in modeling the global magnetosphere field of the Earth, which consists of a strong intrinsic potential field induced by the source of electric currents inside of the Earth, and it was proved that the method can improve the numerical accuracy. For the magnetic fields in the solar corona, such a splitting also has a similar physical meaning [91]. The potential component $\mathbf{B}_0$ arises from photospheric or sub-photospheric currents and can be regarded as invariant during a flare, whereas the non-potential component $\mathbf{B}_1$ arises from large-scale coronal currents (above the photosphere) and is the source of the flare energy. On the other hand, the accuracy improvement by splitting for extrapolation a field close to potential state should be substantial, whereas for an extremely non-potential field, the improvement might be limited. The function of the frictional force $-\nu\rho\mathbf{v}$ has been mentioned above, and the value of coefficient ν is given as $\nu = \Delta t/\Delta x^2$, according to the time step Δt and local grid size Δx. They also used a $\mu = 0.4\Delta x^2/\Delta t$ for maximizing the diffusive speed of the numerical magnetic monopole. Similar to other zero-β codes, a pseudo-plasma density $\rho = |\mathbf{B}|^2$ is used to equalize the Alfvén speed of the whole domain for the purposed of accelerating the relaxation in the weak field regions, e.g., for the higher coronal configurations.

The main advancement of this code over the other available ones is that a modern numerical scheme, namely, the AMR-CESE-MHD method [59], is utilized. This method was well-tested as a robust and versatile scheme for solving a wide range of MHD problems from data-driven simulation of active region evolutions [63] to the global solar wind modeling [29], as well as the global magnetosphere models [121]. For extrapolation of the coronal magnetic field, it shows power in realizations in different coordinate systems, including in local Cartesian box of a single AR, and in large-scale or globally, in Spherical wedges or shells. This is owing to a generalization of the original CESE scheme to a arbitrarily curvilinear coordinate system [59], thus any shape of the 3D physical domain expressed in (x, y, z) can be mapped to a reference space (ξ, η, ζ) in which the computational volume is a Cartesian box.

The ability of using the adaptive mesh refinement (AMR) and parallel computing techniques in the code is also a useful improvement in NLFFF extrapolations. As also suggested by Wiegelmann [128], parallelized AMR computing could be very efficient for mitigating the contradiction between the computational demand for extrapolating current (and even future) high-resolution and large field-of-view magnetograms and the computational resource limitations. The distribution of the coronal magnetic field is highly non-uniform in both transverse and vertical (or radial) directions even within a single AR, and the electric currents are also usually concentrated within small areas, e.g., in the core region of ARs. Thus it is natural to use a grid adaptively resolving the different regions of different gradient of the field. For another important issue, reconstruction that need to be performed with much larger field of view (to more entirely characterize the currents between ARs and high over the AR) and

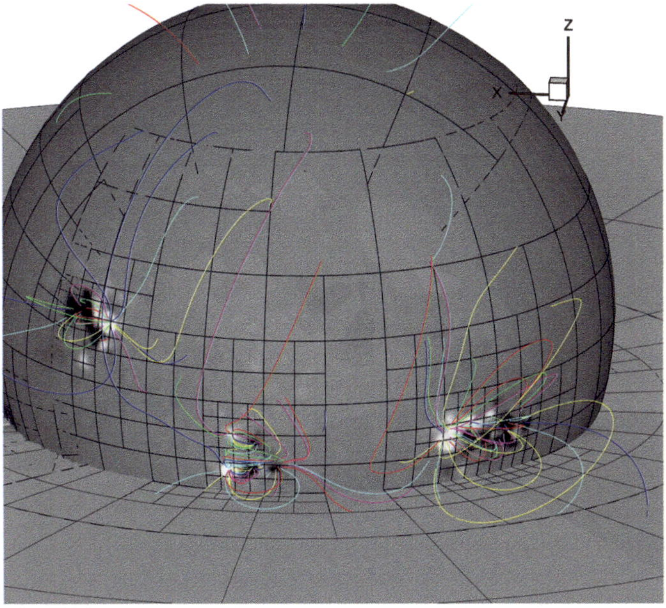

Fig. 8.1 A example of global extrapolation with AMR grid. The sphere represents the photosphere with grid structure outlined by the black lines (note that here each mesh cell stands for one grid block). The color lines represent the magnetic field lines and the grey image represents the radial field B_r. A slice of $z = -0.3$ is plotted to show the radial mesh structure. Note that the high-resolution blocks are clustered around the active regions. This figure is not for a real NLFFF result but an extrapolation of the potential field based on the synoptic LoS map of solar Carrington rotation 2029 obtained from SOHO/MDI. Mesh for the active regions are refined with three more grid levels than the background full-sphere grid (a block-AMR algorithm [90] is used and the ratio of resolutions between the grid levels is two). The grid structure can be dynamically adjusted during the MHD-relaxation process to capture the strong currents and important magnetic structures (e.g., flux ropes) by designing carefully the AMR refinement criteria. Taken from Fig. 10 of Jiang and Feng [62] with permissions of the authors and the AAS

with higher spatial resolution (to capture the fine critical structures such magnetic null point) is severely limited by the computational capability. Application with grid of about 500^3 pixels is almost the upper limit for computational capability of most recently developed or updated codes. This is rather unsatisfactory if considering extrapolation with the 4096×4096 SDO/HMI magnetograms. This issue can be resolved promisingly via combining the global extrapolation code with the adaptive mesh refinement, according to the intrinsic characteristics of the solar magnetic field in which the active regions represents only a small fraction of the whole surface. By the AMR technique, one can take focus on local corona, e.g., some active regions in the context of global extrapolation with the corresponding high-resolution vector magnetograms embedded in a low-resolution global vector or LoS map, as illustrated in Fig. 8.1.

8.4 Assessments

Assessments of the CESE–MHD–NLFFF code were performed through the reconstructions of a series of known solutions (in either analytic or numerical form) of force-free fields [55, 58]. One well-known solution is the half-analytic NLFFF model derived by Low and Lou [76], which has served as a standard benchmark for extrapolation codes [6, 42, 60, 100, 115, 125]. The original form of the solution is expressed in a spherical geometry and by arbitrarily positioning a rectangle box in the space of the analytical fields, one obtains a test case suited for extrapolation in Cartesian coordinates, in which one boundary of the box can be used to represent the bottom boundary condition for the extrapolation of the overlaying fields. Figure 8.2 shows a perspective visualization of a specific case of the Low and Lou [76]'s solution and the extrapolation by the CESE–MHD–NLFFF code in a rectangle box. As can be seen, this model is simply a configuration of sheared magnetic arcades with rather smooth distribution of the magnetic flux at the bottom surface. Almost all NLFFF codes can do this extrapolation with a satisfying accuracy [100], and it represents a elementary experiment for NLFFF codes. In spherical coordinates, the Low and Lou [76]'s model can also be used for the global coronal field extrapolation tests, as shown in Fig. 8.3.

An advanced experiment that has been used for NLFFF codes is a numerical approximately force-free solution of a magnetic flux rope [84], which is constructed by the so-called "magnetic flux rope insertion" technique [116, 118] within a realistic magnetic environment of a solar AR. Magnetic flux rope is believed to be a fundamental structure that accounts for magnetic free energy storage in the lower corona and solar eruptions [37, 68, 111]. However, such knowledge about the existence of coronal flux rope is in a large part derived from NLFFF extrapolations (e.g., [7, 18, 20, 39]). As a result, the reliability of recovering magnetic flux rope from extrapolations must be critically examined for all NLFFF codes. In the flux-rope insertion model of van Ballegooijen [116], they first inserted an S-shaped flux bundle into a potential field associated with AR 10814, and then let the unbalanced system relax to a equilibrium through the magneto-frictional approach (see Fig. 8.4a). As a more stringent and realistic test, it is possible to mimic the analysis of real observational data while still knowing the properties of the field to be modeled. As shown Fig. 8.4, this model field contains clearly a low-lying S-shaped flux rope surrounded by a near-potential field with a topological quasi-separatrix layer. Thus, it challenges the extrapolation code much more than the configuration of simple sheared arcedes in Low and Lou [76]'s model. Indeed, for this model, not all the NLFFF codes examined in the joint study of Metcalf et al. [84] passed the test, i.e., recovering the flux rope correctly, while the CESE–MHD–NLFFF code performed rather well with a flux rope clearly present in the extrapolation results (see Fig. 8.4b).

A range of parameters for critical comparison of the results for the above benchmarks by the CESE–MHD–NLFFF code with other available ones has been carried out by Jiang and Feng [55]. Their results (Table 8.1) show that the performance in extrapolation accuracy reaches the same level of the present best-performing code, i.e., the optimization code of Wiegelmann [126].

Another frequently used test for extrapolation codes is an analytic model of magnetic flux rope that is designed by Titov and Démoulin [110]. This flux rope model

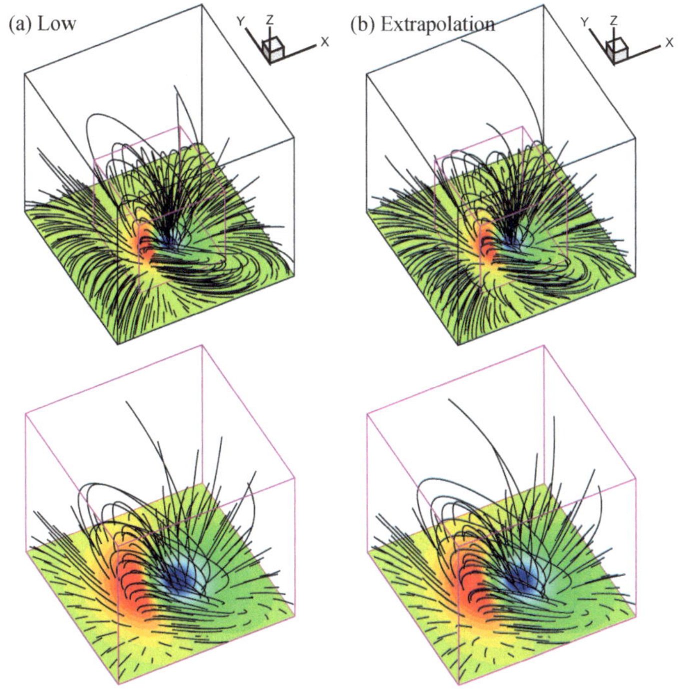

Fig. 8.2 3D magnetic field lines of the Low and Lou [76]'s NLFFF model in rectangle box and the extrapolation by the CESE–MHD–NLFFF code. This is the same case of CASE LL1 in Jiang and Feng [100] and Schrijver et al. [55]. The bottom surfaces of the 3D box show contour of B_z. The bottom row enlarges the central region outlined by the small cube in the top row. Taken from Fig. 4 of Jiang and Feng [55] with permissions of the authors and the AAS

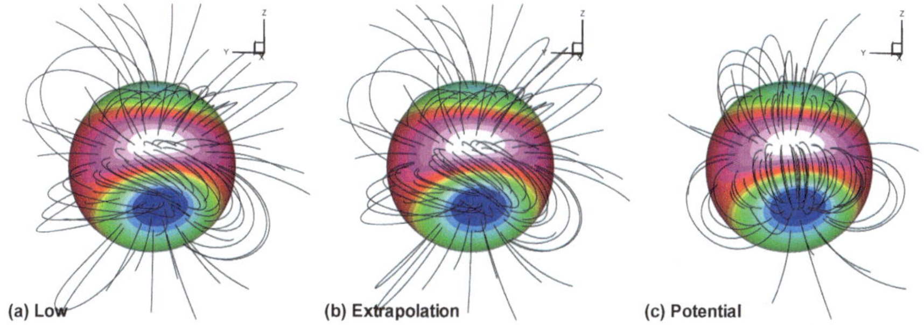

Fig. 8.3 3D magnetic field lines of the Low and Lou [76]'s NLFFF model in spherical geometry. The sphere represents contour of B_r on the photosphere surface. **a** Low and Lou's solution; **b** the extrapolation result by the CESE–MHD–NLFFF code [62]; **c** the potential field with the same B_r on the sphere. Taken from Fig. 5 of Jiang et al. [62] with permissions of the authors and the AAS

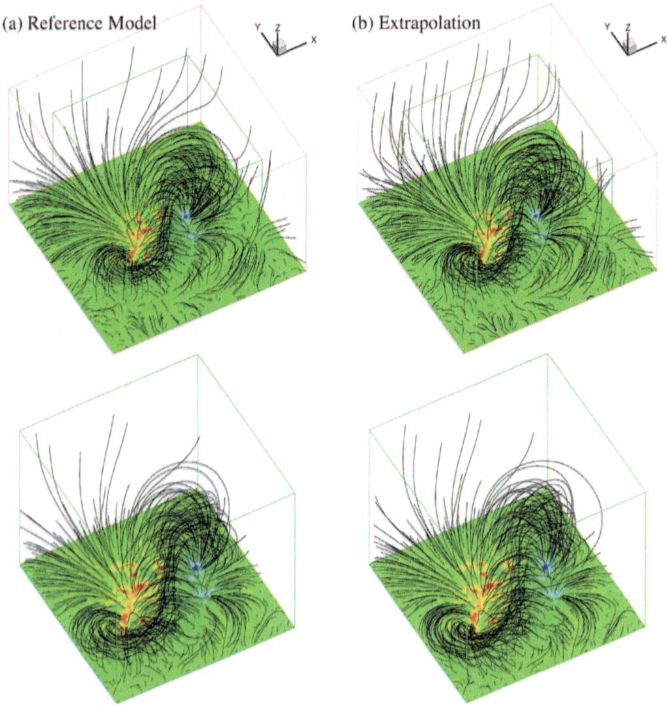

Fig. 8.4 Magnetic field lines of extrapolation test for the van Ballegooijen flux rope model. **a** the reference model, **b** the extrapolation field by the CESE–MHD–NLFFF code. The bottom row enlarges the central region outlined by the small cube in the top row. Taken from Fig. 10 of Jiang and Feng [55] with permissions of the authors and the AAS

is of considerable interest due to its relevance to the structures of solar active regions and eruptive magnetic field, which is demonstrated by many investigations [82, 95, 101, 109, 111, 112]. Basically, this model is constructed to simulate an bipolar active region containing a half-buried, toroidal current-carrying flux rope balanced (or confined) by the overlying arched near-potential field (as shown in Fig. 8.5). Excessive twist of the flux rope can trigger its kink instability [111], and a too fast decaying of the overlying field with height can trigger the torus instability of the system [68]. With different parameter settings, two different configurations, the one with a bald patch separatrix surface (BPSS), and the one without BPSS but with a hyperbolic flux tube (HFT), can be modeled, as two stages of partial and full emergence of flux rope, respectively. Figure 8.5 shows the comparison of the original model and the extrapolation by the CESE–MHD–NLFFF code, and Fig. 8.6 shows the reconstructed topological interfaces of BPSS and HFT. The reliability of using the CESE–MHD–NLFFF code to reconstruct these flux rope models and their related topology structures (e.g., BPSS and HFT) is validated by Jiang and Feng [58]. Basing on the present and all the previous tests (e.g., [55, 56]), confidence is gained

Table 8.1 The first-three best results of parameters for evaluating the extrapolation accuracy for the benchmark tests in the literature. The superscript $*$ denotes the reported results in Metcalf et al. [84] and Schrijver et al. [100], and $**$ denotes the reported results in Valori et al. [115], and $***$ denotes results in Jiang et al. [55]. Definition of these parameters can be found in Metcalf et al. [84] and Schrijver et al. [100]. The first four parameters quantify the relative errors of the extrapolation solution with the original one. For their results, the closer to unity, the better the extrapolation is. $E/E_{\rm pot}$ is the ratio of the total magnetic energy to the potential field energy. CWsin and $\langle |f_i| \rangle$ measures the force-freeness and divergence-freeness, respectively, and ideally they should be zero. Adapted from Tables 1 and 5 of Jiang and Feng [55] with permissions of the authors and the AAS

Model	$C_{\rm vec}$	$C_{\rm CS}$	$E'_{\rm n}$	$E'_{\rm m}$	$E/E_{\rm pot}$	CWsin	$\langle \vert f_i \vert \rangle$
For the Low and Lou model							
Low	1.000	1.000	1.000	1.000	1.242	0.014	0.94×10^{-4}
Wiegelmann*	1.00	1.00	0.97	0.96	1.26		
Jiang***	1.000	0.997	0.964	0.912	1.241	0.015	1.67×10^{-4}
Valori**	0.999	0.99	0.95	0.87	1.23	0.009	
Potential	0.858	0.869	0.498	0.443	1.000		
For the van Ballegooijen model							
Reference	1.000	1.000	1.000	1.000	1.343	0.107	1.02×10^{-4}
Wiegelmann*	1.00	0.99	0.89	0.73	1.34	0.11	
Jiang***	0.995	0.978	0.885	0.734	1.337	0.133	1.11×10^{-4}
Valori*	0.98	0.98	0.84	0.71	1.25	0.15	
Potential	0.852	0.952	0.687	0.665	1.000		

in applying this code to the realistic coronal field if the magnetogram fulfills the force-free constraints [57].

8.5 Preprocessing of Solar Vector Magnetogram

However, the vector magnetograms from observations are usually not force-free fields, and this is a major headache to the NLFFF modelers. Up to the present, the only routinely available magnetic fields are derived from measurements for the photosphere. At the photosphere the plasma β is usually significantly larger than what can be regarded as negilible, and the Lorentz force cannot be self-balanced and must be balanced by pressure force of the dense plasma, and thus are in conflict with the fundamental assumption of force-freeness. The force-free condition for the Lorentz force is only approxmately fulfilled at a height of roughly 400 km above the photosphere surface [34, 74, 83]. To mitigate such problem, Wiegelmann et al. [132] first suggested to modify the photospheric magnetograms to simulate the force-free

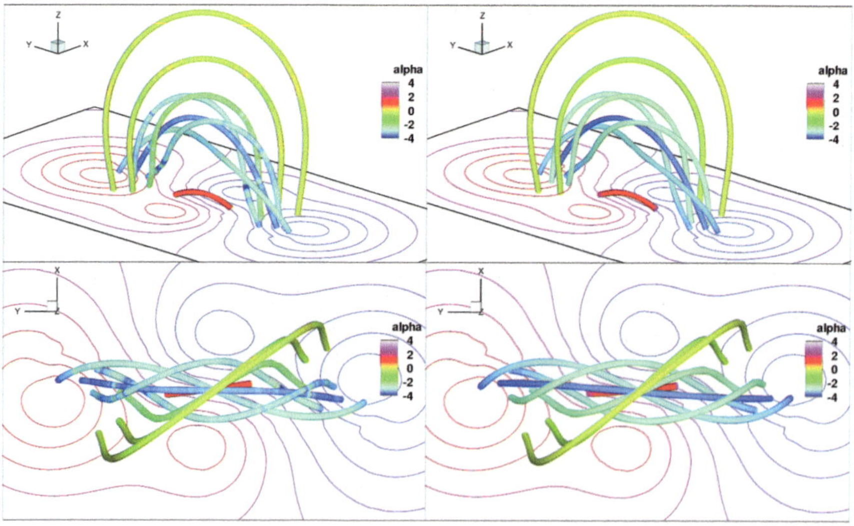

Fig. 8.5 Selected field lines for The High_HFT case: left are from the reference model and right from the extrapolation. Contour lines on the bottom are plotted for B_z. The field lines are color-coded by the value of force-free factor α. Taken from Fig. 3 of Jiang and Feng [58] with permission of the Research in Astronomy and Astrophysics (RAA)

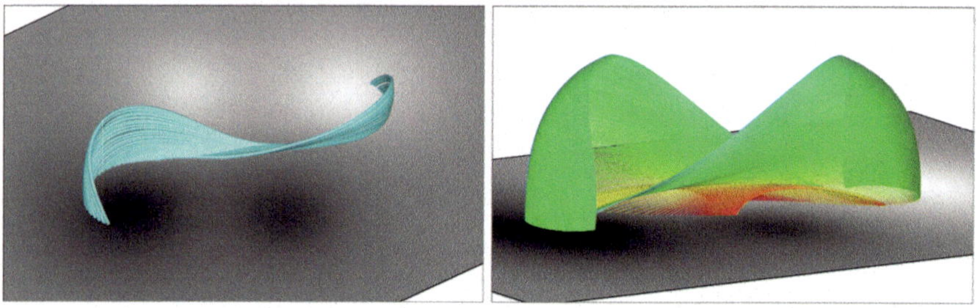

Fig. 8.6 The reconstructed BPSS (left) and HFT (right) illustrated with continuous set of field lines (colors of the lines are used for a good visualization of the structures). They characterize the interface layers that separates the flux rope from its ambient field. For details refer to From Jiang and Feng [58]. Taken from Fig. 5 of Jiang and Feng [58] with permission of the Research in Astronomy and Astrophysics (RAA)

chromospheric magnetograms, and such a practice is often called as 'preprocessing of solar vector magnetogram' [31, 32, 84, 130, 137], which has been routinely used for NLFFF extrapolation codes. The preprocessing is based on the assumption that since the interface between the photosphere and the bottom of the force-free domain is rather thin (say, about 400 km), especially if compared with the spatial scale of the coronal field (about tens of megameters), the basic field structures of the chromosphere should be very similar to those of the photosphere, except that (i) there must be some smoothness of the structures due to the fast expansion of field from the high-β to low-β regions and (ii) the very fine magnetic elements are just closed within this interface and thus show no signal in the force-free domain. Thanks to these reasons, modifications that need to be made on the photospheric field to mimic the force-free chromospheric field ought to be not significant and can hopefully be made within/around the error margins of the measurement.

The magnetogram preprocessing can be guided by the following conditions that must be fulfilled by a ideally force-free field map [4, 86].

$$\begin{aligned} \int_S B_z \, dx \, dy = 0, \quad F_x = \int_S B_x B_z \, dx \, dy = 0, \\ F_y = \int_S B_y B_z \, dx \, dy = 0, \quad F_z = \int_S E_B \, dx \, dy = 0, \\ T_x = \int_S y E_B \, dx \, dy = 0, \quad T_y = \int_S x E_B \, dx \, dy = 0, \\ T_z = \int_S (y B_x B_z - x B_y B_z) \, dx \, dy = 0 \end{aligned} \tag{8.7}$$

where $E_B = B_x^2 + B_y^2 - B_z^2$. These expressions are derived from the volume integrals of the total divergence, magnetic force and torque of an ideally force-free field by using Gauss' divergence theorem to transform them to the surface integrals. In the case of the coronal field, the full surface are usually restricted within the bottom magnetogram since the contribution from other boundaries can be neglected. Several preprocessing codes [31, 84, 132] have been developed and they follow the basic approach proposed by Wiegelmann et al. [132]. A functional L is designed by adding up the χ^2 deviations from the constraints of Eq. (8.7), the terms that control deviation from the raw data and the smoothness with different weights, e.g.,

$$L = \mu_1 L_1 + \mu_2 L_2 + \mu_3 L_3 + \mu_4 L_4 \tag{8.8}$$

where $\boldsymbol{\mu} = (\mu_1, \mu_2, \mu_3, \mu_4)$ is the weighting factor, $L_1 = F_x^2 + F_y^2 + F_z^2$, $L_2 = T_x^2 + T_y^2 + T_z^2$, $L_3 = \int_S |\mathbf{B} - \mathbf{B}_{\text{obs}}|^2 \, ds$, and L_4 measures the roughness of the data. Then the target magnetogram is searched by minimizing the functional L using an optimization method. Different algorithms of smoothing and optimization have been utilized and the results are more or less different, as shown in a comparison study by Fuhrmann et al. [32]. Also the differences can result from different choices of the weighting factors.

Following the same approach, Jiang and Feng [57] developed a preprocessing code with, again, the magnetic-field splitting form, in which the vector magnetogram is also split into a potential field part and a non-potential field part and the two parts are handled separately. Preprocessing of the potential part is simply performed by taking the data sliced at a plane about 400 km (approximately one pixel for an HMI magnetogram) above the photosphere from the 3D potential field which is extrapolated from the observed vertical field. Then the non-potential part is modified and smoothed by an optimization method to fulfill the constraints of total magnetic force-freeness and torque-freeness. One advantage of using such splitting way is that the preprocessing of the non-potential field part can be guided by the extents of force-freeness and smoothness of the preprocessed potential-field part. This is because in the practical computation, particular attention is needed to paid on to what extents the force is needed to be removed and the smoothing can be performed. In practical computation based on numerical discretization, an accurate satisfaction of force-free constraints is apparently not necessary. Also the extent of the smoothing for the data needs to be carefully determined, if we want to mimic the expansion of the magnetic field from the photosphere to some specific height above. Over-smoothing of the data may smear the basic structures while a too-limited smoothing cannot filter the small-scale noise sufficiently. A careful choice of the weighting factors $\boldsymbol{\mu}$ is required to deal with these problems. In Jiang and Feng [57] the values of force-freeness and smoothness calculated from the preprocessed potential-field part are used as a reference, and the target magnetogram is required to has the same level of force-freeness and smoothness as its potential part. It is found that these requirements can restrict well the free parameters, i.e., the weighted factors $\boldsymbol{\mu}$ in the optimization function.

However, we should note that Eq. (8.7) are only necessary conditions but not sufficient for an ideally force-free magnetogram, meaning that the magnetogram may still contains force even these conditions are satisfied. In other words, the preprocessed magnetogram is more suitable for NLFFF modeling than the raw data, rather than a completely consistent boundary condition.

8.6 Extrapolations from SDO/HMI Vector Magnetograms

The Helioseismic and Magnetic Imager (HMI) on board the *Solar Dynamics Observatory* (SDO) provides photospheric vector magnetograms with a high resolution both in space and time. It observes the full Sun with a 4K×4K CCD whose spatial sampling is 0.5 arcsec per pixel. Raw filtergrams are obtained at 6 different wavelengths and 6 polarization states in the Fe I 6173 Å absorption line, and are collected and converted to observable quantities (like Dopplergrams, continuum filtergrams, and line-of-sight and vector magnetograms) on a rapid time cadence. For the vector magnetic data, each set of filtergrams takes 135 s to be completed. To obtain vector magnetograms, Stokes parameters are first derived from filtergrams observed over a 12-min interval and then inverted through the Very Fast Inversion of the Stokes Vector [17]. The 180° azimuthal ambiguity in the transverse field is resolved by an improved version of the "minimum energy" algorithm [69]. Regions of interest with strong magnetic field are automatically identified near real time [113]. A detailed description on how the vector magnetograms are produced can be found on the website http://jsoc.stanford.edu/jsocwiki/VectorPaper and the HMI pipeline paper [44]. A special data product is provided for analysis of ARs, which is the Space-weather HMI Active Region Patches (SHARP) vector magnetogram 'hmi.sharp_cea_720s' [16]. The SHARP data tracks the whole life of specific ARs with cadence of 12 min, and it

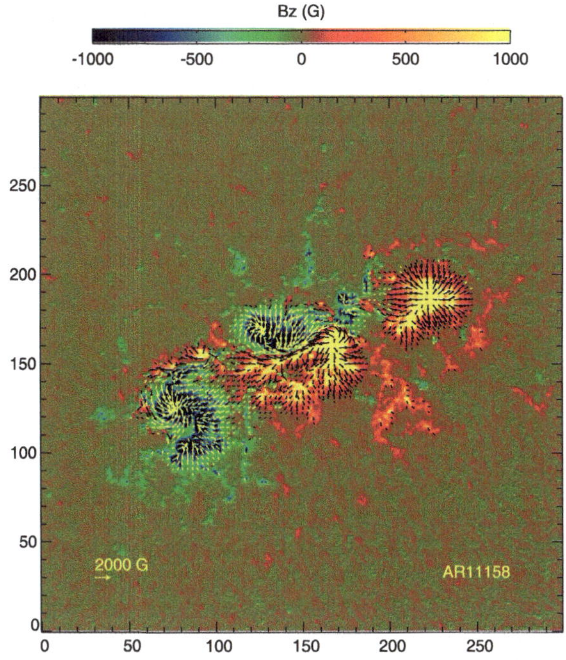

Fig. 8.7 Vector magnetograms for AR 11158 taken at 20:36 UT on 2011 February 14. The background shows the vertical components with saturation values of ±1000 G; the vectors represent the transverse fields (above 200 G). The length unit is arcsec, and the map as shown is rebined by 2 × 2 from the original SHARP data. Adapted from Fig. 2 of Jiang and Feng [56] with permissions of the authors and the AAS

provides two formats of spatial mapping, i.e., direct cutouts and Lambert cylindrical equal area projection (CEA) remapped images. The CEA re-mapping assumes a local Cartesian coordinate system centered on the tracked AR, thus it is well suited for extrapolations in rectangle box for ARs of typical sizes (say, spanning less than 20° on the solar spherical surface).

Extrapolation from HMI data was sampled for a well-studied region, AR 11158, which is also the first disambiguated vector field data released by the HMI project for science analysis. This AR has a complex, multipolar structure, exhibiting fast flux emergence with photospheric motion of shearing and rotation of the sunspots during its passage of the solar disk. In the period, it produced several major flares and coronal mass ejections (CMEs) including the first X-class flare of cycle 24 on 2012 February 15. The AR with its major flare events have been analyzed in many papers for different purposes, including the formation of CMEs and associated coronal waves, (e.g., [102, 107]), evidence of abrupt and permanent changes of photospheric magnetic fields after flare (e.g., [67, 75]), triggering of flares by magnetic twists and MHD simulation of eruptions (e.g., [51–53]), and magnetic topology relevant to flares (e.g., [106, 143]). In particular for modelling the coronal magnetic field, Wiegelmann et al. [133] used this AR for optimizing their extrapolation code with HMI data, and Aschwanden et al. [13] used it for presenting their NLFFF code using coronal loop fitting techniques. Figure 8.7 shows vector magnetogram of AR 11158 taken at 20:36 UT on 2011 February 14, several hours before its major X flare. This AR is well isolated from other ones and is almost flux-balanced, with the main

Table 8.2 Quality of the magnetogram of AR 11158 for NLFFF extrapolation. Adapted from Table 1 of Jiang and Feng [56] with permissions of the authors and the AAS

Data	ϵ_{flux}	ϵ_{force}	ϵ_{torque}	S_x	S_y	S_z
AR 11158						
Raw map	0.017	0.072	0.073	5.05E-03	6.12E-03	1.05E-03
Preprocessed map	0.019	0.002	0.002	7.83E-05	1.01E-04	8.74E-05
Numerical potential	0.019	0.001	0.002	8.78E-05	8.42E-05	8.74E-05

polarities concentrated in the central field of view (FoV). Strong shearing along the major PIL and twisting of the polarities can be clearly seen from the horizontal field, which all indicate a significant non-potentiality of the coronal field.

Preprocessing of the vector magnetogram reduces the parameters of measuring force-freeness of the map, i.e., ϵ_{force} and ϵ_{torque} by more than an order of magnitude (see Table 8.2), making the residual force around the level of numerical error (i.e., the parameters are close to those of the numerical potential field). Here the parameters in Table 8.2 are respectively defined by

$$\epsilon_{\text{flux}} = \frac{\int_S B_z \, dx \, dy}{\int_S |B_z| \, dx \, dy}, \quad \epsilon_{\text{force}} = \frac{|F_x| + |F_y| + |F_z|}{\int_S P_B \, dx \, dy},$$
$$\epsilon_{\text{torque}} = \frac{|T_x + |T_y| + |T_z|}{\int_S \sqrt{x^2 + y^2} P_B \, dx \, dy} \tag{8.9}$$

where $P_B = B_x^2 + B_y^2 + B_z^2$, and S_x, S_y, S_z measure the smoothness of the components B_m $(m = x, y, z)$, which are defined by Jiang and Feng [57]. As shown the smoothness of the preprocessed data is very close to those of their potential part, and is consistent among three components, which is unlike the raw data with very different smoothness for different components. Smoothing of the data can be clearly seen by comparing the raw and preprocessed magnetograms as shown in Fig. 8.8, and especially in the J_z map that shows that the random noise is suppressed effectively (but not totally).

The observed coronal loops in X-ray and EUV images give us a proxy of the magnetic field line geometry and are thus a good constraint for magnetic field model of the corona (e.g., [12, 80]). This is because in most part of the corona, the resistivity is sufficiently small that the plasma is "frozen" with the magnetic field, an interesting phenomenon in ideal MHD. As such, a reliable coronal field model should be able to reproduce the geometry of the loops by its 3D field lines. Figure 8.9 compares the extrapolated field with coronal loops observed by SDO/AIA in the wavelength of 171 Å, in which the relatively long loops are distinguished the best among all channels. Note that here the field lines are traced from the photosphere at locations selected roughly according to the footpoints of the visible bright loops, and the view

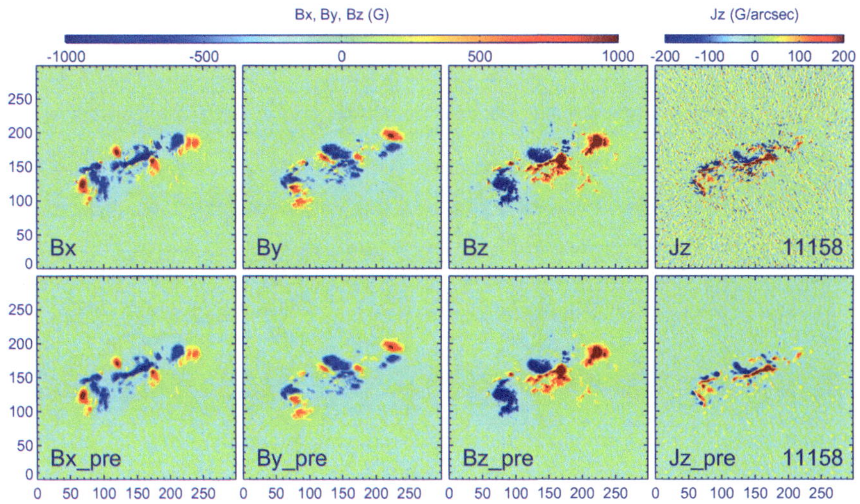

Fig. 8.8 Raw and preprossessed magnetogram for AR 11158, shown for each component and the vertical component of current density, which is derived by $J_z = (\nabla \times \mathbf{B})_z$ numerically with central difference. Refer to Jiang and Feng [56] for more details. Taken from Fig. 3 of Jiang and Feng [56] with permissions of the authors and the AAS

angle of the field lines is approximately aligned with the AIA image because the field is based on a local rectangle of which the bottom does not match the spherical surface of the Sun. Thus such comparison cannot be critical (see Sect. 8.7), although overall most of the extrapolated field lines resemble the loops. Some misalignments between the modeled field lines and the observation are also obvious, especially, the large loops near the north-west boundary of the FoV.

An important result of interest is that in the AR core, the field lines are strongly sheared along the major PIL and slightly twisted (as compared with the potential solution), indicating the existence of strong electric currents along the field lines. From an image of vertical integral of the current density in the extrapolation volume (see Fig. 8.10), one can see that the strong-current regions are localized in different small regions (see the boxes labeled as B and C). The corresponding field lines are also shown in the right panel of Fig. 8.10. There is strong twist of the field lines in region C, forming a magnetic flux rope. Such extrapolated field is consistent with observation studies that a filament related with a X-class flare and CME exists in the core region B [107], and there are small eruptions in region C due to flux emergence [106].

The quality of the extrapolation needs to be checked by computing the metrics as in the benchmark tests. The force-freeness of the extrapolation data is usually measured by a current-weighted sine metric (CWsin) defined by

$$\mathrm{CWsin} \equiv \frac{\int_V J\sigma dV}{\int_V JdV}; \quad \sigma = \frac{\mathbf{J} \times \mathbf{B}}{JB}, \tag{8.10}$$

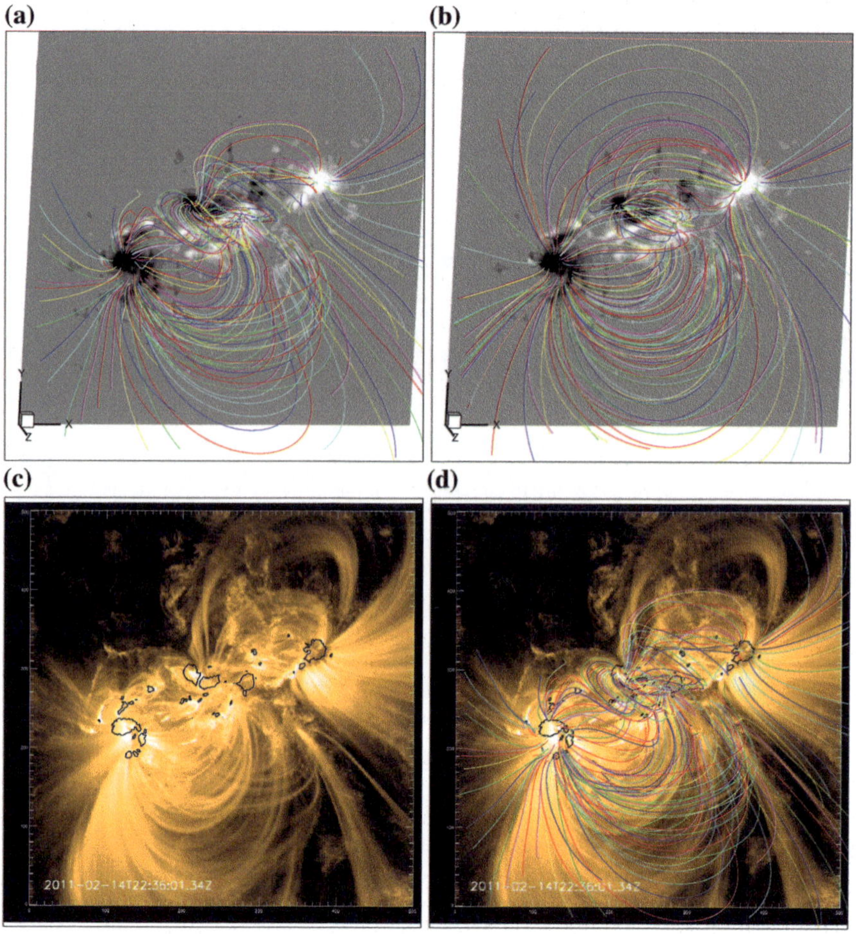

Fig. 8.9 Comparison of extrapolation field lines with AIA 171 Å loops for AR 11158: the NLFFF lines (**a**), the potential field lines (**b**), the AIA image (**c**) and NLFFF lines overlaying the AIA image (**d**). Contour lines for ±1000 G (the black curves) of LoS photospheric field are over-plotted on the AIA images, and for all the panels the field lines are traced from the same set of footpoints on the bottom surface. Taken from Fig. 5 of Jiang and Feng [56] with permissions of the authors and the AAS

where $B = |\mathbf{B}|$, $J = |\mathbf{J}|$ and V is the computational volume. Also a current-square-weighted sine metric ($\mathrm{C^2Wsin}$) can be similarly defined by

$$\mathrm{C^2Wsin} \equiv \frac{\int_V J^2 \sigma dV}{\int_V J^2 dV}, \tag{8.11}$$

with more weight on the strong-current regions. The divergence-freeness is measured by $\langle |f_i| \rangle$

Table 8.3 Results of the metrics for NLFFF extrapolation of AR 11158. Full region is extrapolation box of $[0, 300](x) \times [0, 300](y) \times [0, 150](z)$. Region A is $[53, 222] \times [104, 191] \times [0, 50]$; Region B is $[116, 181] \times [145, 175] \times [0, 30]$; Region C is $[62, 102] \times [132, 161] \times [0, 20]$. The FoVs of the regions are shown by the boxes in Fig. 8.10. The energy unit is 10^{32} erg. Taken from Table 2 of Jiang and Feng [56] with permissions of the authors and the AAS

Region	CWsin	C^{2}Wsin	$\langle\|f_i\|\rangle$	$E_{\rm tot}$	$E_{\rm pot}$	$E_{\rm free}$	$E_{\rm free}/E_{\rm pot}$ (%)	$E_{\rm free}/(E_{\rm free})_{\rm full}$ (%)
Raw								
Full	0.32	0.18	6.59E-04	10.9	8.97	1.93	22	100
A	0.18	0.11	1.25E-03	8.3	6.24	2.06	33	107
B	0.07	0.06	7.17E-04	3.24	2.09	1.15	55	60
C	0.16	0.11	1.64E-03	0.67	0.47	0.20	43	10
Preprocessed								
Full	0.30	0.14	6.18E-04	9.83	8.18	1.65	20	100
A	0.17	0.10	1.20E-03	7.54	5.61	1.93	34	117
B	0.06	0.05	7.20E-04	2.97	1.81	1.16	64	70
C	0.16	0.12	1.67E-03	0.59	0.42	0.17	40	10

$$\langle|f_i|\rangle = \frac{1}{V}\int_V \frac{\nabla\cdot\mathbf{B}}{6B/\Delta x}dV. \tag{8.12}$$

The energy contents, i.e., the total energy $E_{\rm tot}$, the potential energy $E_{\rm pot}$ and the free energy $E_{\rm free}$, are respectively,

$$E_{\rm tot} = \int_V \frac{B^2}{8\pi}dV, \quad E_{\rm pot} = \int_V \frac{B_{\rm pot}^2}{8\pi}dV, \quad E_{\rm free} = E_{\rm tot} - E_{\rm pot} \tag{8.13}$$

where $B_{\rm pot}$ is the potential field strength. Table 8.3 shows the results of the metrics for extrapolations from both the raw and preprocessed magnetograms, and the metrics for four different regions including the full extrapolation box and the subregions A, B, and C as outlined in Fig. 8.10.

The current-weighted mean sine of the full extrapolation volume is $\sim$0.3, thus the misalignment angle between the magnetic field and current is about 17° on average. Such value is much larger than those from benchmark tests [which are $\sim$0.1 (6°) or smaller, see Jiang and Feng [55]], but is comparable to previously reported results by other NLFFF codes on real solar magnetograms (e.g., the average CWsin by various NLFFF codes applied to AR 10930 [101] and AR 10953 [25] are 0.36 and 0.28, respectively). With such a large misalignment angle, the result seems to be far away from an exactly force-free solution that has a zero misalignment angle. However it should be noted that the metric CWsin may not be a good monitor for numerical solutions, which unavoidably have random numerical errors because of limited resolution. For example, CWsin is close to 1 even for a potential field solution

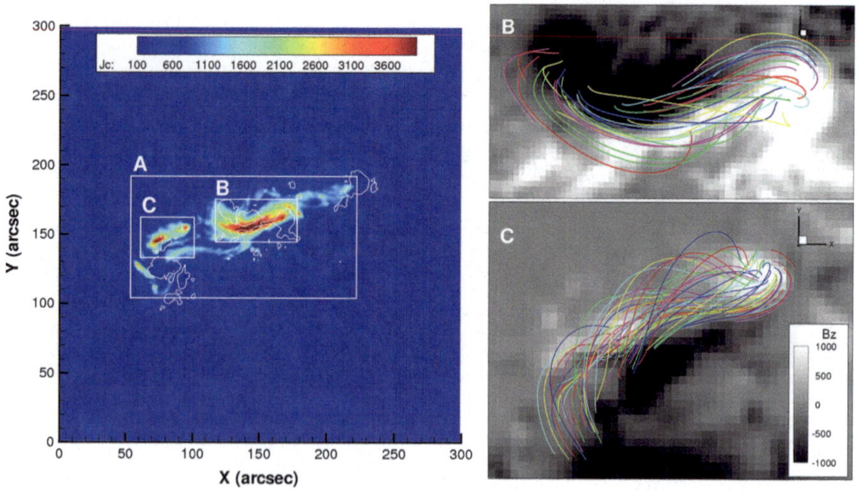

Fig. 8.10 Strong-current regions and their magnetic structures for AR 11158. Left panel is the image of the vertical integral of current density, i.e., $J_C = \int |\mathbf{J}|dz$ where the current is calculated by $\mathbf{J} = \nabla \times \mathbf{B}$ with a unit of G pixel^{-1}; Regions of strong current are outlines by the boxes A, B and C. Right panels show the magnetic structures of the strong-current regions of B and C. Taken from Fig. 6 of Jiang and Feng [56] with permissions of the authors and the AAS

computed by Green's function method or other numerical realization. The reason is that the numerical difference, used for computing the current $\mathbf{J} = \nabla \times \mathbf{B}$ from $\mathbf{B}$, gives small but finite currents, whose directions are randomly from 0° to 180°, thus an average of the full volume should give a misalignment angle of ~90°. The noise in the observation data, mainly in the weak field regions, is also a major source for the random numerical errors. In these regions, the actual magnetic elements are probably smaller than the observed or numerical pixel size, and the field directions generally exhibit a random pattern on the image. To reduce influence from such errors in computing the metric, one can either put larger weight of current (e.g., use C^2Wsin) or compute CWsin within the strong-current subregions only. As expected, the misalignment angle decreases significantly by measuring in this way. For the full region, C^2Wsin are only half of CWsin. For the subregions, CWsin are also only half or less for the full region, reaching the level of those from the benchmark tests [55]. In particular, the misalignment angle is only about 4° in subregion B, showing that the force-free assumption is well modeled.

Regarding the energy contents, it is interesting to note that the free energy of subregion A exceeds that of the full region, meaning that the free energy content in the full volume excluding subregion A is negative. This is, however, not surprising since any sub-volume energy content of the non-potential field may be lower than the potential energy (e.g., [61, 77]). Also the measurement error may result in this negative free energy since it is very small compared to the total free energy. No matter which case is true, it can be clearly seen that the spatial distribution of free energy is largely co-spatial with that of the current, since the subregion A contains most of the currents of the whole volume. This confirms that the free energy in the corona is actually stored by the current-carrying field (where non-potentiality is strong), but not necessarily in the magnetic flux concentrations.

Comparison of the results extrapolated from the raw and preprocessed magnetograms shows that the improvement by preprocessing is negligible. This is because the raw data already satisfies the boundary force-free conditions well. Also due to the smoothing, the result for the preprocessed data gives slightly lower energy contents than those for the raw data.

As the CESE–MHD–NLFFF is based on MHD relaxation approach, it is important to monitor the relaxation process to study whether the iteration converges, since there is no theory to guarantee this. The convergence process of the computations can be monitored by temporal evolution of several quantities, including the residual of field between two successive iterations

$$\mathrm{res}^n(\mathbf{B}) = \sqrt{\frac{1}{3}\sum_{\chi=x,y,z}\frac{\sum_i (B^n_{i\chi} - B^{n-1}_{i\chi})^2}{\sum_i (B^n_{i\chi})^2}} \tag{8.14}$$

(where n denotes the iteration step), the metric CWsin and the total energy content. As shown in Fig. 8.11, the system converges smoothly and fast. During the first 10 τ_A, the residual keeps increasing because the transverse field is inputted at the bottom continuously, which drives the system away from the initial potential field. After this driving process, the residual drops immediately, indicating a fast relaxation of the system. With about 40 τ_A (nearly 10000 iterations), the residual is already reduced to $\sim 10^{-5}$, and all the metrics and energy almost stagnate afterward. Thus the computations can actually be terminated once the residual is below 10^{-5}, which is consistent with previous studies for benchmark cases [55, 62]. It is also noteworthy that the convergence process is rather smooth, without any obvious oscillation or abrupt variation of the residual or the metrics, so the iteration is "safe". This is a good feature of the code over other iteration codes for extrapolation, e.g., the Valori et al. [115]'s magnetofrictional code or the Wheatland [124]'s Grad-Rubin-like code, which usually show strong oscillatory in the iteration or even fail to converge occasionally [25, 101].

8.7 Critical Comparison of Reconstructed Field with Coronal Loops

Various extrapolation codes exist but their results often drastically differ. Thus a critical comparison of the modeled magnetic field lines with the observed coronal loops is strongly required to establish the credibility of the model. Such quantitative measurements of the agreement between modeled field lines and observed coronal loop geometries can be carried out in different ways. In the ideal case, if we have the

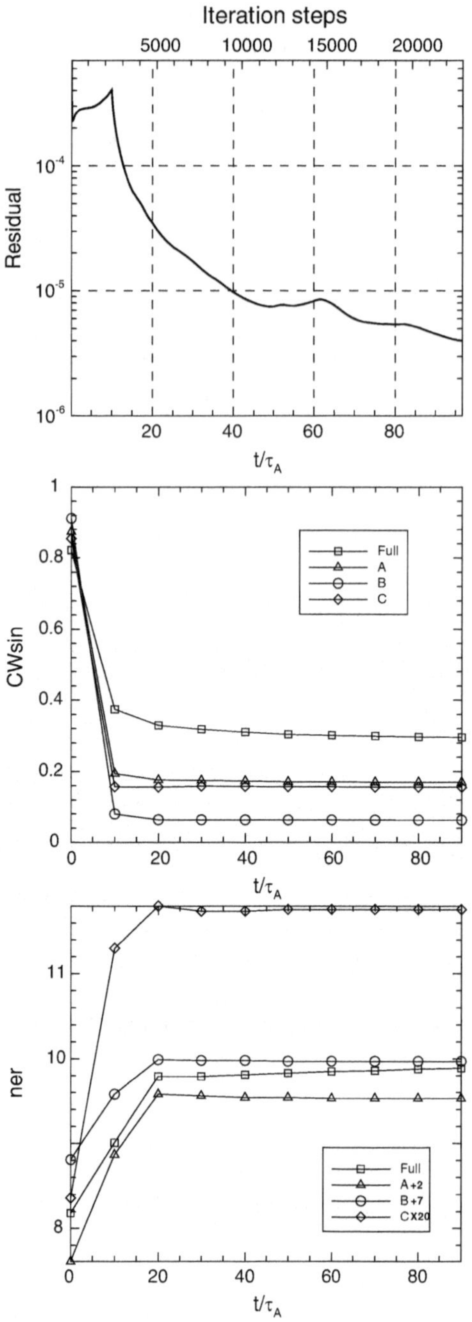

Fig. 8.11 Convergence of computations: temporal evolution of the residual, the metric CWsin and the total energy content in the iteration process of extrapolation for AR 11158. Note that for a better plot for the energy contents of different regions, their values are scaled properly as shown in the line legends. Adapted from Fig. 9 of Jiang and Feng [56] with permissions of the authors and the AAS

information of 3D trajectories of the coronal loops, we can directly calculate the misalignment angle between the field lines and the loops. For example, DeRosa et al. [25] assessed a variety of popular NLFFF codes for modeling AR 10953 by comparing the model field lines to 3D trajectories of coronal loops that are observed stereoscopically by the twin Solar Terrestrial Relations Observatory (STEREO) spacecraft [11]. It was found that the misalignment angle between the field lines and the loops amounts to rather large values of 24°–44°, and none of the examined NLFFF models improved significantly upon the value found for the potential field model. However, we note that their used loops are mostly at the periphery of the AR, which is often close to potential state. Another method employs a slight variation by measuring the distance between a loop and a projected field line [19, 72, 79]. The same idea of a "distance" has also been used in 3D, with STEREO data [129]. Without tracing the trajectories of coronal loops, Wiegelmann et al. [133] compared the model field lines with SDO/AIA images by calculating the sum of gradient of the AIA image intensity along each projected field lines, and the result is assumed to reach its minimum if all the field lines are co-aligned with the corresponding loops.

As in most cases one have only 2D images of coronal loops, Gary et al. [36] developed a method of deriving the 3D structure of observed 2D coronal loops independent of heliostereoscopy, and suggested that it can determine the matching of extrapolated magnetic fields with coronal loops. In that method, an automated loop recognition scheme [8, OCCULT-2] is first used to extract 2D loop structures from EUV images and then the extracted loops are fitted with 2D cubic Bézier splines that are based on 4 control points [35]. The 2D splines are further extended to 3D with the heights of all 4 control points set as free parameters. The heights are determined by minimizing the misalignment angles of the 3D splines with field lines of a given model of coronal magnetic field, and the resulting 3D splines are regarded as the trajectories of the corresponding coronal loops in 3D. So the image of these loops represents a 2D projection of these 3D structures. Naturally, different magnetic field models will result in different sets of misalignment angles, and a perfect matching model can presumably yield a misalignment angle of zero. Thus this method provides a powerful tool of comparing different magnetic-field models of the corona by using EUV images obtained from one single viewpoint.

Duan et al. [27] evaluated the CESE–MHD–NLFFF code employing Gary [36]'s method to compute the misalignment angles of the model field lines and identified loops. The target region used in the examination is AR 12158 prior to its major eruption on 2014 September 10, which exhibits a well-shaped sigmoidal core that indicates a non-potential field (Fig. 8.12). Three models are compared with coronal loops, one using the CESE–MHD–NLFFF code with the original SHARP vector magnetogram (NLFFF1), another with the preprocessed magnetogram (NLFFF2). In addition, they also compared the extrapolated field by an so-called non force-free extrapolation (NFFF) model [45–47], which provides an alternative way to NLFFF extrapolation. Duan et al. [27] applied the procedures of Gary [36] to two AIA channels, i.e., 94 Å and 171 Å, which represent two separate features of the AR field. The 94 Å shows loops at the AR core that are presumably associated with significantly non-potential magnetic field, and they are usually compact and short, while the

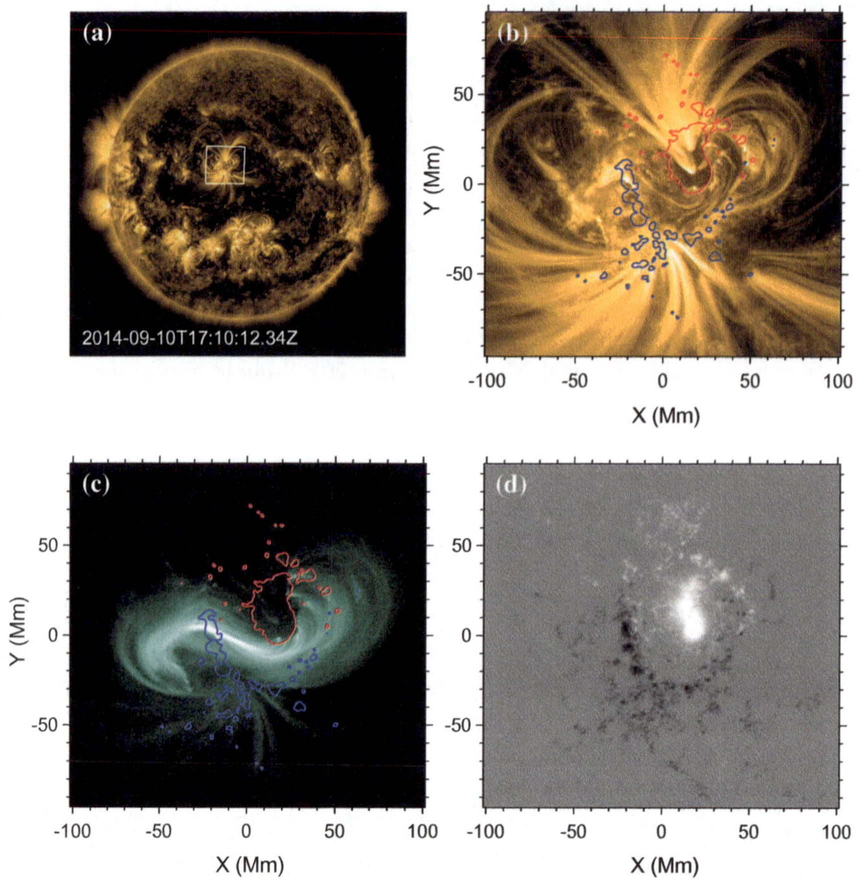

Fig. 8.12 Observations of AR 12158 at the time 17:10 UT on 2014 September 10. **a** AIA 171 Å full-disk image with location of AR 12158 marked by the box. **b** Close view of the boxed region. **c** Same as (**b**) but in AIA 94 Å channel. **d** SHARP CEA remapped LoS magnetogram. Taken from Fig. 1 of Duan et al. [27] with permissions of the authors and the AAS

171 Å loops correspond to the enveloping field that is often close to a potential state. Figures 8.13 and 8.14 give the results of the minimized misalignment angle for each traced loop in the two AIA channels. It is found that model NLFFF2 gives the best results in fitting to the AIA 94 Å loops as it has the smallest values for both the median and mean of the misalignment angles. For the AIA 171 Å loops, NLFFF2 also gives the smallest median value, while the smallest average value is given by the potential field model. Comparing the results for the two AIA channels, it is noticed that the fitting of the modeled field lines to loops is much better for the 94 Å than for 171 Å. In particular, in the best model (NLFFF2), the misalignment angles for the 94 Å loops, which have median value of $\sim 10°$ and mean value of $\sim 15°$, are less than half of those for the 171 Å loops ($\sim 30°$). For 94 Å loops, all the non-

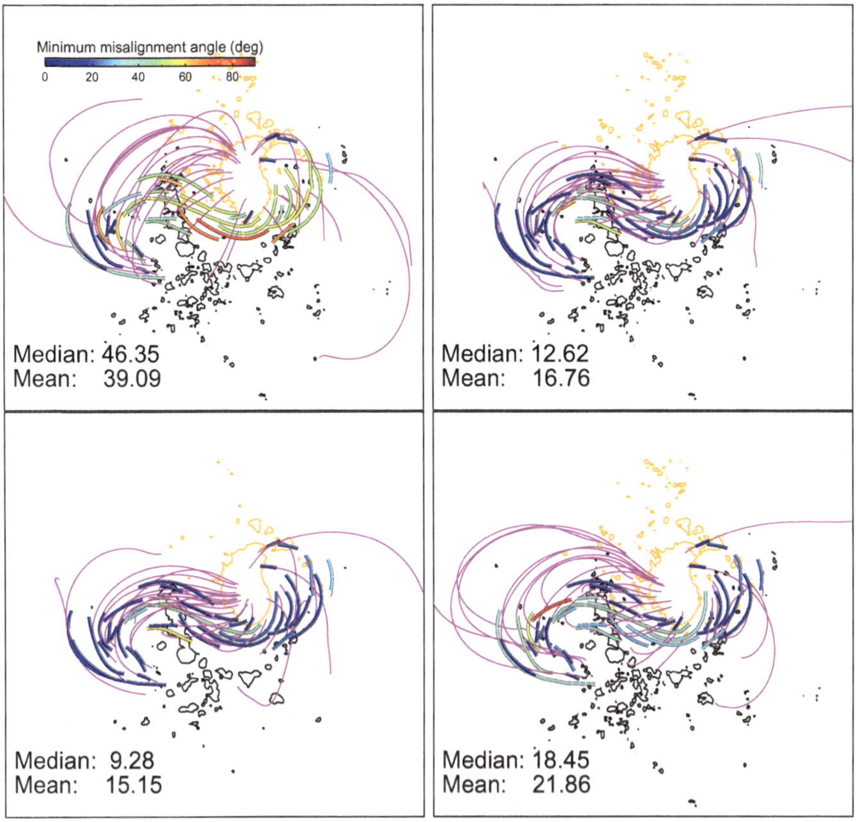

Fig. 8.13 Minimum misalignment angles between the magnetic field model and the Bézier splines (for AIA 94 Å loops) when extended to 3D. The four panels clockwise from the top left to the bottom left are results, respectively, for the potential magnetic field, the NLFFF1, the NFFF, and the NLFFF2 models. Each loop, as shown by the thick curves, is color-coded by the value of its misalignment angle ($0 - 90°$) as indicated by the colorbar in the first panel. Also a magnetic field line (shown by thin curves) is traced from the mid-point of each loop. The contour lines are plotted for B_z at -500 G (colored as black) and 500 G (yellow). For each model, the median and mean values of the misalignment angles for all the loops are shown. Taken from Fig. 5 of Duan et al. [27] with permissions of the authors and the AAS

potential models give significantly smaller value of misalignment than the potential field model, which confirms that these loops are associated with magnetic field of significant non-potentiality. Improvement from preprocessing of the magnetogram for NLFFF extrapolation is also confirmed by the decrease of the misalignment angles.

One more thing worth noting is that the misalignment angles of the extrapolated field lines with the AIA 94 Å loops are on the order of the results using coronal loop forward-fitting NLFFF code shown in Aschwanden [10], who found that the 3D misalignment angle amounts to an average value of $19° \pm 3°$ for their studied ARs. This

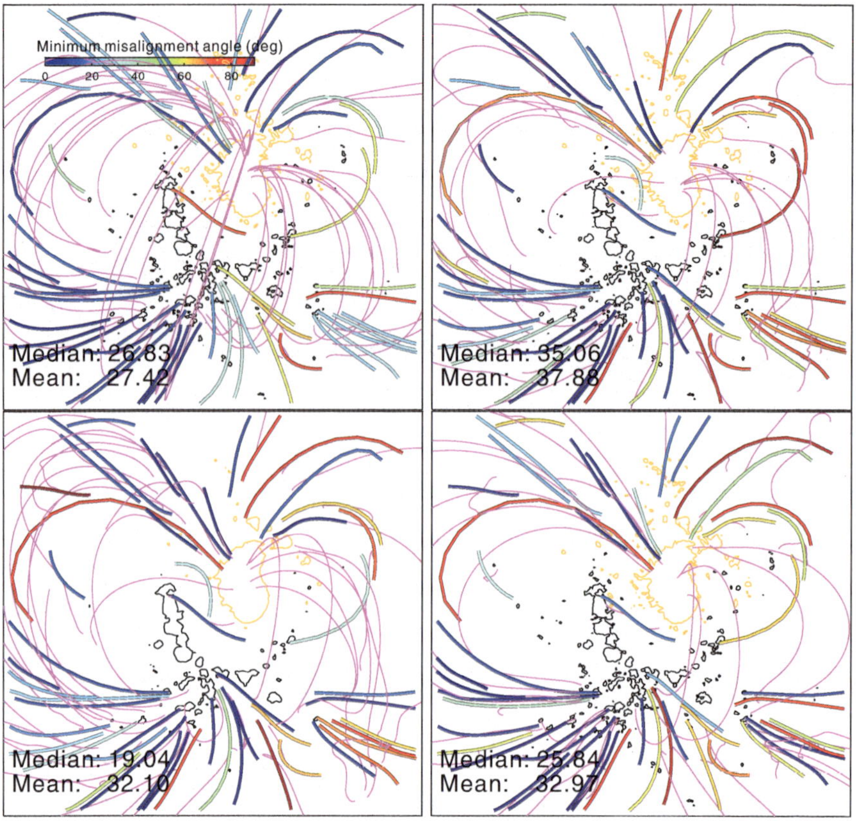

Fig. 8.14 Same as Fig. 8.13 but for AIA 171 Å loops. Taken from Fig. 6 of Duan et al. [27] with permissions of the authors and the AAS

suggests that the CESE–MHD–NLFFF code, even without using the information of coronal loops in constraining the magnetic field, performs as good as some coronal-loop forward-fitting models. However, for the 171 Å loops, the CESE–MHD–NLFFF extrapolation performs, at the best, only close to the potential field model (with misalignment of $\sim 30°$). Such value is comparable to the misalignment angles (24°–44°) as resulted from a comparison of various NLFFF models for AR 10953 by DeRosa [25], who also found that those NLFFF models perform no better than potential model. The NLFFF1 without preprocessing gives even larger misalignment angle than the potential model, and it is slightly improved from the preprocessing. The different values for 94 Å and 171 Å show that the CESE–MHD–NLFFF code can faithfully reconstruct the AR core field (which is significantly non-potential) than the envelope field (which is presumably close to potential). The large misalignment of the extrapolated field with the 171 Å loops is because the field lines corresponding to the long loops do not relax sufficiently during the MHD-relaxation process. Since in the code the relaxation speed is uniform and the relaxation time is thus proportional to the field line length, the long field lines need more time to relax than the short ones. Moreover, the long field lines often extend close to or reach side and top boundaries of the computational volume and are subject to the boundary effects, as all these boundaries are fixed with the initial potential field (however note that the computational volume is larger than the field-of-view of the magnetogram). Another factor that might be partially responsible for this effect is the departure from the Cartesian coordinates assumption further away from the core. Thus a further optimization of the relaxation process with realization of spherical geometry is needed for a better reconstruction of the enveloping field.

8.8 Applications

In the SDO era, there seems to be little doubt that magnetic flux rope plays an important role in solar eruptions. Many observations have suggested the existence of flux rope preceding eruptions [20, 21, 70, 71, 88, 93, 141]. From NLFFF extrapolations, flux ropes are usually also reproduced with the observed vector magnetograms [18, 20, 39, 40, 64–66]. However, how a flux rope is formed and what triggers its disruption are still under intense debates. On a large scale, flux ropes might either emerge, due to magnetic buoyancy, from the convection zone where the magnetic twist is generated [28, 78], or is created directly in the corona through photospheric shearing motion and flux cancellation [117]. On the triggers for eruption, resistive (i.e., reconnection) and ideal MHD mechanism (instabilities of flux rope) are usually invoked, and often conflict with each other [14, 30, 54, 97, 103]. Due to the insufficient information from direct observations, it is still unable to clarify the roles of these theoretical models played in the realistic cases. Thus recent investigations resort to NLFFF modeling of the realistic event, by which useful insight can be provided.

Since coronal *Sigmoid* is believed as a manifestation of magnetic flux rope (although their exact relationship is unclear), Jiang et al. [65] studied the formation and eruption of an compact sigmoid in AR 11283 by using the CESE–MHD–NLFFF code. They reconstructed a time sequence of magnetic fields following the quasi-static evolution of the corona. This event involves a variety of magnetic processes including flux emergence, flux cancellation, ideal MHD instability, and breakout reconnection, and with the NLFFF fields they attempted to discriminate the roles of these processes in the formation and triggering of the sigmoid eruption. As shown in Fig. 8.15, a weakly twisted flux rope is built up gradually from the initially sheared arcades. By topological analysis, it is found that below the flux rope exists a bald patch separatrix surface (BPSS), which consists of a group of field lines that get tangential to the photosphere surface at the PIL. This indicates that the flux rope body is attached to the photosphere, as can be seen in the vertical cross sections of the 3D field. It has been shown that slow reconnection can occurs at the BPSS below flux rope through flux cancellation [15], which builds up the flux rope. Study of the modeled flux rope suggests that the eruption is caused by the fact that the flux rope grows gradually until its axis runs into a torus instability (TI, [68]) domain. A strik-

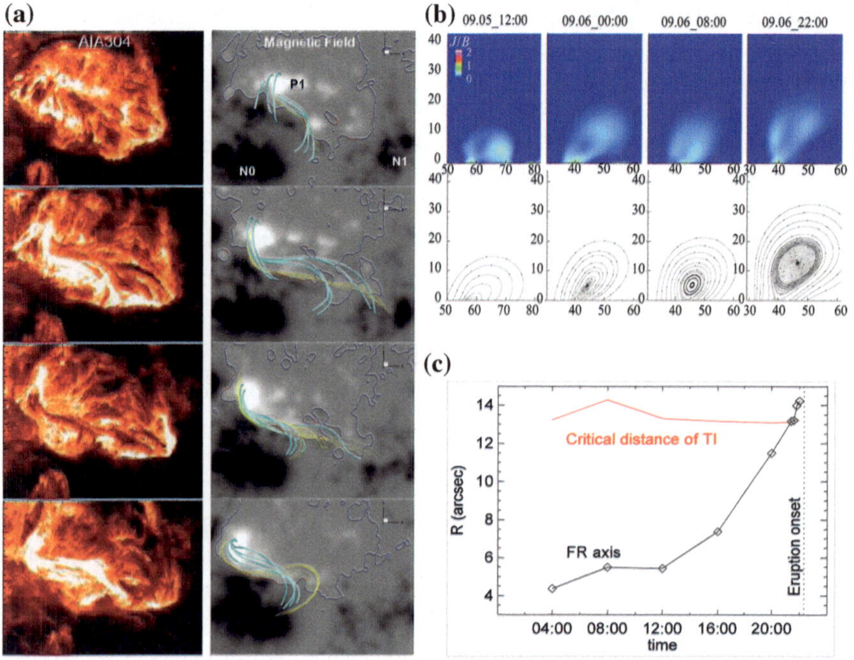

Fig. 8.15 Evolution of the core field leading to the flux rope in AR 11283 on 2011 September 6. **a** AIA-304 images (left) and sample coronal field lines (right), the blue contour lines indicate the PIL, and the red portions denote the BPs; field lines in yellow are traced through the BPs, forming the BPSS; the cyan lines represent some sample field lines closely above the BPSS. **b** Evolution of the central cross section of the FR. The top panels show currents normalized by the field strength, and the bottom panels show the field-line tracing of the poloidal flux of the FR, which forms helical field lines centered at the axis of the FR. **c** Time evolution for the distance from the BP to the FR axis, and the critical distance from the BP to the TI domain during the whole day of September 6. The vertical dashed line denotes the time of the eruption onset. Adapted from Figs. 8, 9, 15 of Jiang et al. [64] with permissions of the authors and the AAS

ingly good match of the TI theory and the eruption time is obtained, as the observed flare and eruption occurs immediately after the apex of rope axis entered into the TI domain (see Fig. 8.15c). The NLFFF solution also provides instructive information on the relationship of the observed sigmoid with the flux rope. Shortly before its eruption, a sound S-shaped EUV emission can be seen (Fig. 8.16) in the AIA 94 Å. By overlapping the magnetic field lines on the EUV image, it is found that the shape of the sigmoid is matched by the BPSS field lines rather than the main body of the flux rope. This offers a strong support that the enhanced current sheet forming along the sigmoid-shaped BPSS produces the enhanced emission of the EUV and SXR sigmoid [37, 82, 110].

In another application, Jiang et al. [65] showed a capability of the CESE–MHD–NLFFF code in extrapolation of a flux rope that supports a large-scale (length of ~300 Mm) intermediate filament between AR 11283 and a weak polarity region

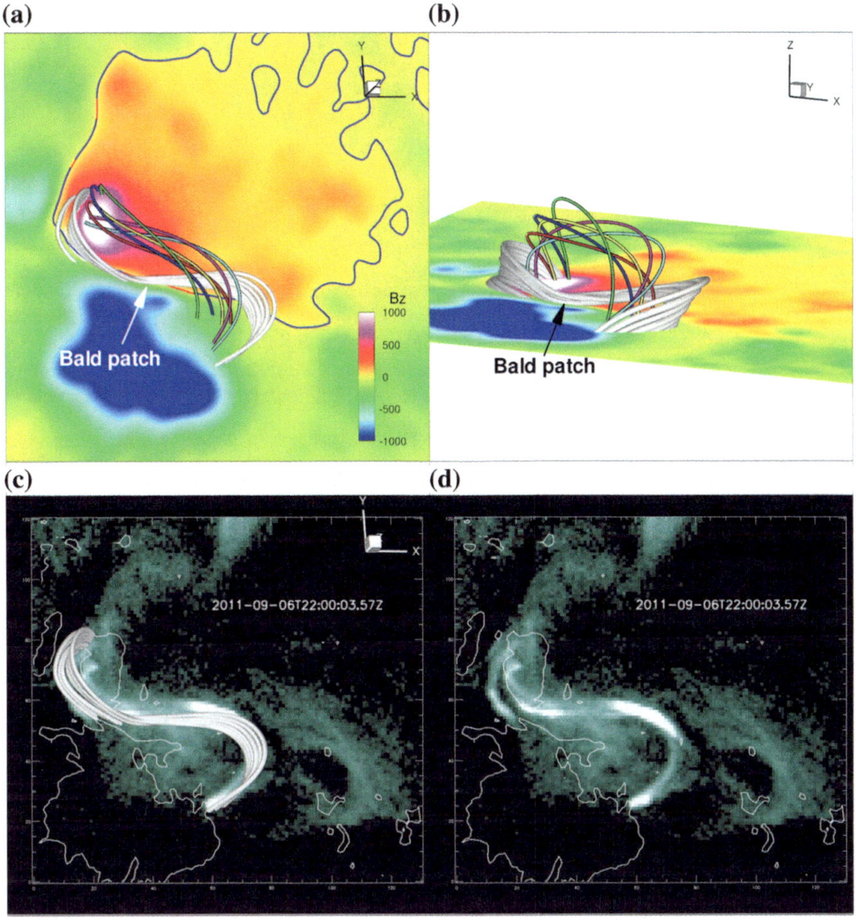

Fig. 8.16 Observation and NLFFF reconstruction of the sigmoid field at 22:00 UT on September 6. The white thick lines are the BPSS field lines. The inner body of the FR is shown by the color field lines. In particular the yellow line is the FR axis, which is traced through the center of the helical flux in the central cross section of the FR as shown in Fig. 8.15. The field lines are shown in both the SDO view (**a**) and a side view (**b**). Panels (**c**) and (**d**) are the AIA-94 image of the sigmoid and the BPSS field lines are overlaid on the AIA observation in panel (**c**). Contours of ±500 G for B_z are also overlaid on the AIA images. Taken from Fig. 10 of Jiang et al. [64] with permissions of the authors and the AAS

(Fig. 8.17). The result seems to be unique in the sense that almost all current NLFFF extrapolations of FRs are limited to relatively small-scale filaments that are close to sunspots and along main PILs (i.e., AR filament) with strong transverse field and magnetic shear, and the existence of an flux rope is usually predictable. In contrast, the intermediate filament lies along the weak-field region (photospheric field strength <100 G), where the PIL is very fragmented due to small parasitic polarities on both side of the PIL and the transverse field has a low value of signal-to-noise ratio. So,

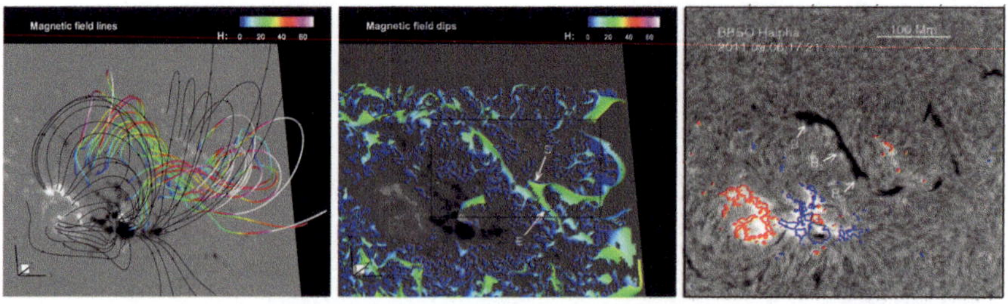

Fig. 8.17 Extrapolation of a magnetic flux rope associated with a large-scale intermediate filament. From left to right are modeled magnetic field lines, the simulated magnetic dips, and the observed filament structures. In the left two panels, the magnetogram of B_z is shown as the gray background and the color represents the height from the bottom (in units of Mm). In the right panel, the overlaid contours are shown for B_z (red for 200 G and blue for -200 G). Adapted from Fig. 1 of Jiang et al. [65] with permissions of the authors and the AAS

there seems to no hint of flux rope in the magnetogram, although a flux rope can be of prevision from the presence of the filament, which is commonly thought to be supported in magnetic dips of flux rope. By comparing the filament structure and the numerically reproduced magnetic dips of the extrapolated flux rope, one can see rather good agreements between them in morphology. In particular, some of the filament barbs are also reproduced. These results supports the theory of the flux-rope-dip model for filaments. Study of the stability of flux rope confirms that the filament is stably sustained because the flux rope is weakly twisted and strongly confined by the overlying closed arcades.

Zhang et al. [142] used the CESE–MHD–NLFFF code to substantiate their observations of several cases of EUV loops from "open" to closed during flares, which is predicted by the standard magnetic reconnection model for flares. As a particular example, they studied such transformation of loops in AR 11520 during the flare that took place on 2012 July 12. From SDO/AIA images, they found that "open" EUV loops rooted in a facula of the AR deflected to the post-flare loops and then merged with them while the flare ribbon approached the facula. Meanwhile, "open" EUV loops rooted in a pore disappeared from top to bottom as the flare ribbon swept over the pore. By contrasting the magnetic field as extrapolated for the pre-flare and post-flare coronae, they identified that the set of field lines corresponding to these flare loops apparently "open" prior to the flare become closed after it. This also demonstrates that CESE–MHD–NLFFF code is capable to sensitively capture the coronal magnetic configuration changes during flares.

8.9 Discussions

The NLFFF extrapolation is a commonly-used tool for studying solar activities associated with the coronal magnetic field, and it dominates the existing models that attempt to characterize the realistic environment of the solar atmosphere, especially, for solar eruption studies, which have unique importance for space weather researches. It has been demonstrated that the NLFFF modelings can shed important lights on the understanding the eruption mechanisms, by successfully recovering the twisted magnetic flux ropes, the magnetic null-point configurations, as well as other critical magnetic topological structures.

On the other hand, the NLFFF models have major limitations. The interaction of magnetic field with plasma is totally omitted in the NLFFF models, and thus they cannot be used to study any dynamics associated with the temporal evolution of the corona. Even for the quasi-static corona, the force-free assumption is challenged by Peter et al [89], who showed that β is of the order of a few percent to 10% or even more that is actually larger than what can be negligible. Also the photospheric magnetic field, the only data that is used for extrapolations, directly conflicts with the assumption of force-freeness. Advanced models should be of fully MHD, and a few of them has arisen, such as the data-driven MHD model described in the next chapter, which can provide a step forward in understanding the evolution of coronal magnetic fields by characterizing the true dynamics that leads to solar eruptions.

References

1. Abbett WP, Fisher GH (2003) A coupled model for the emergence of active region magnetic flux into the solar corona. Astrophys J 582:475–485. https://doi.org/10.1086/344613
2. Abbett WP, Mikić Z, Linker JA, McTiernan JM, Magara T, Fisher GH (2004) The photospheric boundary of Sun-to-Earth coupled models. J Atmos Sol-Terr Phys 66:1257–1270. https://doi.org/10.1016/j.jastp.2004.03.016
3. Altschuler MD, Newkirk G (1969) Magnetic fields and the structure of the solar corona. I: methods of calculating coronal fields. Solar Phys 9:131–149. https://doi.org/10.1007/BF00145734
4. Aly JJ (1989) On the reconstruction of the nonlinear force-free coronal magnetic field from boundary data. Solar Phys 120:19–48. https://doi.org/10.1007/BF00148533
5. Amari T, Boulmezaoud TZ, Mikić Z (1999) An iterative method for the reconstructionbreak of the solar coronal magnetic field. I. Method for regular solutions. Astron Astrophys 350:1051–1059
6. Amari T, Boulmezaoud TZ, Aly JJ (2006) Well posed reconstruction of the solar coronal magnetic field. Astron Astrophys 446:691–705. https://doi.org/10.1051/0004-6361:20054076
7. Amari T, Canou A, Aly JJ (2014) Characterizing and predicting the magnetic environment leading to solar eruptions. Nature 514(7523):465–469
8. Aschwanden M, De Pontieu B, Katrukha E (2013) Optimization of curvilinear tracing applied to solar physics and biophysics. Entropy 15:3007–3030. https://doi.org/10.3390/e15083007
9. Aschwanden MJ (2005) Physics of the solar corona. An introduction with problems and solutions, 2nd edn. Springer, New York
10. Aschwanden MJ (2013) Nonlinear force-free magnetic field fitting to coronal loops with and without stereoscopy. Astrophys J 763:115. https://doi.org/10.1088/0004-637X/763/2/115
11. Aschwanden MJ, Wulser JP, Nitta NV, Lemen JR (2008) First three-dimensional reconstructions of coronal loops with the STEREO A and B spacecraft. I. Geometry. Astrophys J 679(1):827 http://stacks.iop.org/0004-637X/679/i=1/a=827
12. Aschwanden MJ, Wuelser JP, Nitta NV, Lemen JR, DeRosa ML, Malanushenko A (2012) First three-dimensional reconstructions of coronal loops with the STEREO A+B spacecraft.

IV. Magnetic modeling with twisted force-free fields. Astrophys J 756:124. https://doi.org/10.1088/0004-637X/756/2/124
13. Aschwanden MJ, Sun X, Liu Y (2014) The magnetic field of active region 11158 during the 2011 February 12–17 flares: differences between photospheric extrapolation and coronal forward-fitting methods. Astrophy J 785(1):34 http://stacks.iop.org/0004-637X/785/i=1/a=34
14. Aulanier G (2014) The physical mechanisms that initiate and drive solar eruptions. In: Schmieder B, Malherbe JM, Wu ST (eds) IAU Symposium, IAU Symposium, vol 300., pp 184–196. https://doi.org/10.1017/S1743921313010958
15. Aulanier G, Török T, Démoulin P, DeLuca EE (2010) Formation of Torus-unstable flux ropes and electric currents in erupting sigmoids. Astrophys J 708:314–333. https://doi.org/10.1088/0004-637X/708/1/314
16. Bobra MG, Sun X, Hoeksema JT, Turmon M, Liu Y, Hayashi K, Barnes G, Leka KD (2014) The helioseismic and magnetic imager (HMI) Vector magnetic field pipeline: SHARPs - Space-Weather HMI active region patches. Solar Phys 289:3549–3578. https://doi.org/10.1007/s11207-014-0529-3
17. Borrero J, Tomczyk S, Kubo M, Socas-Navarro H, Schou J, Couvidat S, Bogart R (2011) Vfisv: very fast inversion of the stokes vector for the helioseismic and magnetic imager. Solar Phys 273:267–293. https://doi.org/10.1007/s11207-010-9515-6
18. Canou A, Amari T (2010) A twisted flux rope as the magnetic structure of a filament in active region 10953 observed by Hinode. Astrophys J 715:1566–1574. https://doi.org/10.1088/0004-637X/715/2/1566
19. Chae J, Moon YJ (2005) Magnetic twist of an extreme-ultraviolet coronal loop inferred from plasma constriction. Astrophys J 629:1110–1121. https://doi.org/10.1086/431660
20. Cheng X, Ding MD, Guo Y, Zhang J, Jing J, Wiegelmann T (2010) Re-flaring of a post-flare loop system driven by flux rope emergence and twisting. Astrophys J Lett 716:L68–L73. https://doi.org/10.1088/2041-8205/716/1/L68
21. Cheng X, Zhang J, Ding MD, Olmedo O, Sun XD, Guo Y, Liu Y (2013) Investigating two successive flux rope eruptions in a solar active region. Astrophys J Lett 769:L25. https://doi.org/10.1088/2041-8205/769/2/L25
22. Cheng X, Guo Y, Ding M (2017) Origin and structures of solar eruptions I: magnetic flux rope. Sci China Earth Sci 60(8):1383–1407. https://doi.org/10.1007/s11430-017-9074-6
23. Chodura R, Schlueter A (1981) A 3D code for MHD equilibrium and stability. J Comput Phys 41:68–88. https://doi.org/10.1016/0021-9991(81)90080-2
24. Contopoulos I, Kalapotharakos C, Georgoulis MK (2011) Nonlinear force-free reconstruction of the global solar magnetic field: methodology. Solar Phys 269:351–365. https://doi.org/10.1007/s11207-011-9713-x
25. DeRosa ML, Schrijver CJ, Barnes G, Leka KD, Lites BW, Aschwanden MJ, Amari T, Canou A, McTiernan JM, Régnier S, Thalmann JK, Valori G, Wheatland MS, Wiegelmann T, Cheung MCM, Conlon PA, Fuhrmann M, Inhester B, Tadesse T (2009) A critical assessment of nonlinear force-free field modeling of the solar corona for active region 10953. Astrophys J 696:1780–1791. https://doi.org/10.1088/0004-637X/696/2/1780
26. DeRosa ML, Wheatland MS, Leka KD, Barnes G, Amari T, Canou A, Gilchrist SA, Thalmann JK, Valori G, Wiegelmann T, Schrijver CJ, Malanushenko A, Sun X, Régnier S (2015) The influence of spatial resolution on nonlinear force-free modeling. Astrophys J 811:107. https://doi.org/10.1088/0004-637X/811/2/107
27. Duan AY, Jiang CW, Hu Q, Zhang H, Gary GA, Wu ST, Cao JB (2017) Comparison of two coronal magnetic field models to reconstruct a sigmoidal solar active region with coronal loops. Astrophys J 842(2):119. http://stacks.iop.org/0004-637X/842/i=2/a=119
28. Fan Y (2009) The emergence of a twisted flux tube into the solar atmosphere: sunspot rotations and the formation of a coronal flux rope. Astrophys J 697:1529–1542. https://doi.org/10.1088/0004-637X/697/2/1529
29. Feng XS, Yang LP, Xiang CQ, Jiang CW, Ma XP, Wu ST, Zhong DK, Zhou YF (2012) Validation of the 3D AMR SIP-CESE solar wind model for four carrington rotations. Solar Phys 279:207–229. https://doi.org/10.1007/s11207-012-9969-9

30. Forbes TG, Linker JA, Chen J, Cid C, Kóta J, Lee MA, Mann G, Mikić Z, Potgieter MS, Schmidt JM, Siscoe GL, Vainio R, Antiochos SK, Riley P (2006) CME theory and models. Space Sci Rev 123:251–302. https://doi.org/10.1007/s11214-006-9019-8
31. Fuhrmann M, Seehafer N, Valori G (2007) Preprocessing of solar vector magnetograms for force-free magnetic field extrapolation. Astron Astrophys 476:349–357. https://doi.org/10.1051/0004-6361:20078454
32. Fuhrmann M, Seehafer N, Valori G, Wiegelmann T (2011) A comparison of preprocessing methods for solar force-free magnetic field extrapolation. Astron Astrophys 526:A70. https://doi.org/10.1051/0004-6361/201015453
33. Gary DE, Hurford GJ (1994) Coronal temperature, density, and magnetic field maps of a solar active region using the owens valley solar array. Astrophys J 420:903–912. https://doi.org/10.1086/173614
34. Gary GA (2001) Plasma beta above a solar active region: rethinking the paradigm. Solar Phys 203:71–86
35. Gary GA, Hu Q, Lee JK (2014a) A rapid, manual method to map coronal-loop structures of an active region using cubic Bézier curves and its applications to misalignment angle analysis. Solar Phys 289:847–865. https://doi.org/10.1007/s11207-013-0359-8
36. Gary GA, Hu Q, Lee JK, Aschwanden MJ (2014b) Determining the 3D structure of the corona using vertical height constraints on observed active region loops. Solar Phys. 289:3703–3721. https://doi.org/10.1007/s11207-014-0548-0
37. Gibson SE, Fan Y, Török T, Kliem B (2006) The evolving sigmoid: evidence for magnetic flux ropes in the corona before, during, and after CMEs. Space Sci Rev 124:131–144. https://doi.org/10.1007/s11214-006-9101-2
38. Grad H, Rubin H (1958) Hydromagnetic equilibria and force-free fields. In: 2nd international conference on the peaceful uses of atomic energy, vol 31, p 386
39. Guo Y, Schmieder B, Démoulin P, Wiegelmann T, Aulanier G, Török T, Bommier V (2010) Coexisting flux rope and dipped arcade sections along one solar filament. Astrophys J 714:343–354. https://doi.org/10.1088/0004-637X/714/1/343
40. Guo Y, Ding MD, Cheng X, Zhao J, Pariat E (2013) Twist accumulation and topology structure of a solar magnetic flux rope. Astrophys J 779:157. https://doi.org/10.1088/0004-637X/779/2/157
41. Guo Y, Cheng X, Ding M (2017) Origin and structures of solar eruptions II: magnetic modeling. Sci China Earth Sci 60(8):1408–1439. https://doi.org/10.1007/s11430-017-9081-x
42. He H, Wang H (2008) Nonlinear force-free coronal magnetic field extrapolation scheme based on the direct boundary integral formulation. J Geophys Res 113(A12):A05S90. https://doi.org/10.1029/2007JA012441
43. He H, Wang H, Yan Y (2011) Nonlinear force-free field extrapolation of the coronal magnetic field using the data obtained by the Hinode satellite. J Geophys Res 116(A15):A01101. https://doi.org/10.1029/2010JA015610
44. Hoeksema JT, Liu Y, Hayashi K, Sun X, Schou J, Couvidat S, Norton A, Bobra M, Centeno R, Leka KD, Barnes G, Turmon M (2014) The helioseismic and magnetic imager (HMI) vector magnetic field pipeline: overview and performance. Solar Phys 289:3483–3530. https://doi.org/10.1007/s11207-014-0516-8
45. Hu Q, Dasgupta B (2008) An improved approach to non-force-free coronal magnetic field extrapolation. Solar Phys 247:87–101. https://doi.org/10.1007/s11207-007-9090-7
46. Hu Q, Dasgupta B, Choudhary DP, Buchner J (2008) A practical approach to coronal magnetic field extrapolation based on the principle of minimum dissipation rate. Astrophys J 679:848–853. https://doi.org/10.1086/587639
47. Hu Q, Dasgupta B, Derosa ML, Büchner J, Gary GA (2010) Non-force-free extrapolation of solar coronal magnetic field using vector magnetograms. J Atmos Sol-Terr Phys 72:219–223. https://doi.org/10.1016/j.jastp.2009.11.014
48. Inhester B, Wiegelmann T (2006) Nonlinear force-free magnetic field extrapolations: comparison of the grad rubin and wheatland sturrock roumeliotis algorithm. Solar Phys 235:201–221. https://doi.org/10.1007/s11207-006-0065-x

49. Inoue S (2016) Magnetohydrodynamics modeling of coronal magnetic field and solar eruptions based on the photospheric magnetic field. Prog Earth Planet Sci 3:19. https://doi.org/10.1186/s40645-016-0084-7
50. Inoue S, Kusano K, Magara T, Shiota D, Yamamoto TT (2011) Twist and connectivity of magnetic field lines in the solar active region NOAA 10930. Astrophys J 738:161. https://doi.org/10.1088/0004-637X/738/2/161
51. Inoue S, Hayashi K, Shiota D, Magara T, Choe GS (2013) Magnetic structure producing X- and M-class solar flares in solar active region 11158. Astrophys J 770:79. https://doi.org/10.1088/0004-637X/770/1/79
52. Inoue S, Hayashi K, Magara T, Choe GS, Park YD (2014) Magnetohydrodynamic Simulation of the X2.2 solar flare on 2011 February 15. I. Comparison with the observations. Astrophys J 788(2):182. http://stacks.iop.org/0004-637X/788/i=2/a=182
53. Inoue S, Hayashi K, Magara T, Choe GS, Park YD (2015) Magnetohydrodynamic simulation of the x2.2 solar flare on 2011 February 15. II. Dynamics connecting the solar flare and the coronal mass ejection. Astrophys J 803(2):73. http://stacks.iop.org/0004-637X/803/i=2/a=73
54. Janvier M, Aulanier G, Démoulin P (2015) From coronal observations to mhd simulations, the building blocks for 3D models of solar flares (invited review). Solar Phys 290:3425–3456. https://doi.org/10.1007/s11207-015-0710-3
55. Jiang CW, Feng XS (2012) A new implementation of the magnetohydrodynamics-relaxation method for nonlinear force-free field extrapolation in the solar corona. Astrophys J 749:135. https://iopscience.iop.org/article/10.1088/0004-637X/749/2/135/
56. Jiang CW, Feng XS (2013) Extrapolation of the solar coronal magnetic field from SDO/HMI magnetogram by a CESE-MHD-NLFFF code. Astrophys J 769:144. https://doi.org/10.1088/0004-637X/769/2/144
57. Jiang CW, Feng XS (2014) Preprocessing the photospheric vector magnetograms for an NLFFF extrapolation using a potential-field model and an optimization method. Solar Phys 289:63–77. https://doi.org/10.1007/s11207-013-0346-0
58. Jiang CW, Feng XS (2016) Testing a solar coronal magnetic field extrapolation code with the Titov-Démoulin magnetic flux rope model. Res Astron Astrophys 16(1):015. https://doi.org/10.1088/1674-4527/16/1/015
59. Jiang CW, Feng XS, Zhang J, Zhong DK (2010) AMR simulations of magnetohydrodynamic problems by the CESE method in curvilinear coordinates. Solar Phys 267:463–491
60. Jiang CW, Feng XS, Fan Y, Xiang C (2011) Reconstruction of the coronal magnetic field using the CESE-MHD method. Astrophys J 727:101. https://doi.org/10.1088/0004-637X/727/2/101
61. Jiang CW, Feng XS, Wu ST, Hu Q (2012a) Study of the three-dimensional coronal magnetic field of active region 11117 around the time of a confined flare using a data-driven CESE-MHD model. Astrophys J 759:85. https://doi.org/10.1088/0004-637X/759/2/85
62. Jiang CW, Feng XS, Xiang CQ (2012b) A new code for nonlinear force-free field extrapolation of the global corona. Astrophys J 755:62. https://doi.org/10.1088/0004-637X/755/1/62
63. Jiang CW, Feng XS, Wu ST, Hu Q (2013) Magnetohydrodynamic simulation of a sigmoid eruption of active region 11283. Astrophys J Lett 771(2):L30
64. Jiang CW, Wu ST, Feng XS, Hu Q (2014a) Formation and eruption of an active region sigmoid. I. A study by nonlinear force-free field modeling. Astrophys J 780(1):55
65. Jiang CW, Wu ST, Feng XS, Hu Q (2014b) Nonlinear force-free field extrapolation of a coronal magnetic flux rope supporting a large-scale solar filament from a photospheric vector magnetogram. Astrophys J Lett 786:L16. https://doi.org/10.1088/2041-8205/786/2/L16
66. Jing J, Tan C, Yuan Y, Wang B, Wiegelmann T, Xu Y, Wang H (2010) Free magnetic energy and flare productivity of active regions. Astrophys J 713:440–449. https://doi.org/10.1088/0004-637X/713/1/440
67. Jing J, Park SH, Liu C, Lee J, Wiegelmann T, Xu Y, Deng N, Wang H (2012) Evolution of relative magnetic helicity and current helicity in NOAA active region 11158. Astrophys J Lett 752:L9. https://doi.org/10.1088/2041-8205/752/1/L9

68. Kliem B, Török T (2006) Torus instability. Phys Rev Lett 96(25):255002. https://doi.org/10.1103/PhysRevLett.96.255002
69. Leka K, Barnes G, Crouch A, Metcalf T, Gary G, Jing J, Liu Y (2009) Resolving the 180 ambiguity in solar vector magnetic field data: evaluating the effects of noise, spatial resolution, and method assumptions. Solar Phys 260:83–108. https://doi.org/10.1007/s11207-009-9440-8
70. Li T, Zhang J (2013a) Fine-scale structures of flux ropes tracked by erupting material. Astrophys J Lett 770:L25. https://doi.org/10.1088/2041-8205/770/2/L25
71. Li T, Zhang J (2013b) Homologous flux ropes observed by the solar dynamics observatory atmospheric imaging assembly. Astrophys J Lett 778:L29. https://doi.org/10.1088/2041-8205/778/2/L29
72. Lim EK, Jeong H, Chae J, Moon YJ (2007) A check for consistency between different magnetic helicity measurements based on the helicity conservation principle. Astrophys J 656:1167–1172. https://doi.org/10.1086/510575
73. Lin H, Kuhn JR, Coulter R (2004) Coronal magnetic field measurements. Astrophys J Lett 613:L177–L180. https://doi.org/10.1086/425217
74. Liu S, Su JT, Zhang HQ, Deng YY, Gao Y, Yang X, Mao XJ (2012a) A statistical study on force-freeness of solar magnetic fields in the photosphere. Publications of the Astronomical Society of Australia (PASA) 30:e005. https://doi.org/10.1017/pasa.2012.005
75. Liu Y, Zhao J, Schuck PW (2012b) Horizontal flows in the photosphere and subphotosphere of two active regions. Solar Phys 195: https://doi.org/10.1007/s11207-012-0089-3
76. Low BC, Lou YQ (1990) Modeling solar force-free magnetic fields. Astrophys J 352:343–352. https://doi.org/10.1086/168541
77. Mackay DH, Green LM, van Ballegooijen A (2011) Modeling the dispersal of an active region: quantifying energy input into the corona. Astrophys J 729:97. https://doi.org/10.1088/0004-637X/729/2/97
78. Magara T (2004) A model for dynamic evolution of emerging magnetic fields in the sun. Astrophys J 605:480–492. https://doi.org/10.1086/382148
79. Malanushenko A, Yusuf MH, Longcope DW (2011) Direct measurements of magnetic twist in the solar corona. Astrophys J 736:97. https://doi.org/10.1088/0004-637X/736/2/97
80. Malanushenko A, Schrijver CJ, DeRosa ML, Wheatland MS, Gilchrist SA (2012) Guiding nonlinear force-free modeling using coronal observations: first results using a quasi-grad-rubin scheme. Astrophys J 756:153. https://doi.org/10.1088/0004-637X/756/2/153
81. McClymont AN, Jiao L, Mikic Z (1997) Problems and progress in computing three-dimensional coronal active region magnetic fields from boundary data. Solar Phys 174:191–218
82. McKenzie DE, Canfield RC (2008) Hinode XRT observations of a long-lasting coronal sigmoid. Astron Astrophys 481:L65–L68. https://doi.org/10.1051/0004-6361:20079035
83. Metcalf TR, Jiao L, McClymont AN, Canfield RC, Uitenbroek H (1995) Is the solar chromospheric magnetic field force-free? Astrophys J 439:474–481. https://doi.org/10.1086/175188
84. Metcalf TR, DeRosa ML, Schrijver CJ, Barnes G, van Ballegooijen AA, Wiegelmann T, Wheatland MS, Valori G, McTtiernan JM (2008) Nonlinear force-free modeling of coronal magnetic fields. II. Modeling a filament arcade and simulated chromospheric and photospheric vector fields. Solar Phys 247(2):269–299
85. Mikic Z, McClymont AN (1994) Deducing coronal magnetic fields from vector magnetograms. In: Balasubramaniam KS, Simon GW (eds) Solar active region evolution: comparing models with observations, astronomical society of the pacific conference series, vol 68, p 225
86. Molodenskii MM (1969) Integral properties of force-free fields. Sov Astron 12:585
87. Nakagawa Y (1974) Dynamics of the solar magnetic field. I. Method of examination of force-free magnetic fields. Astrophys J 190:437–440. https://doi.org/10.1086/152895
88. Patsourakos S, Vourlidas A, Stenborg G (2013) Direct evidence for a fast coronal mass ejection driven by the prior formation and subsequent destabilization of a magnetic flux rope. Astrophys J 764:125. https://doi.org/10.1088/0004-637X/764/2/125

89. Peter H, Warnecke J, Chitta LP, Cameron RH (2015) Limitations of force-free magnetic field extrapolations: revisiting basic assumptions. Astron Astrophys 584:A68. https://doi.org/10.1051/0004-6361/201527057
90. Powell KG, Roe PL, Linde TJ, Gombosi TI, de Zeeuw DL (1999) A solution-adaptive upwind scheme for ideal magnetohydrodynamics. J Comput Phys 154:284–309. https://doi.org/10.1006/jcph.1999.6299
91. Priest ER, Forbes TG (2002) The magnetic nature of solar flares. Astron Astrophys Rev 10:313–377. https://doi.org/10.1007/s001590100013
92. Régnier S (2013) Magnetic field extrapolations into the corona: success and future improvements. Solar Phys 288(2):481–505. https://doi.org/10.1007/s11207-013-0367-8
93. Régnier S, Walsh RW, Alexander CE (2011) A new look at a polar crown cavity as observed by SDO/AIA. Struct Dyn Astron Astrophys 533:L1. https://doi.org/10.1051/0004-6361/201117381
94. Roumeliotis G (1996) The "Stress-and-Relax" method for reconstructing the coronal magnetic field from vector magnetograph data. Astrophys J 473:1095. https://doi.org/10.1086/178219
95. Roussev II, Forbes TG, Gombosi TI, Sokolov IV, DeZeeuw DL, Birn J (2003) A three-dimensional flux rope model for coronal mass ejections based on a loss of equilibrium. Astrophys J Lett 588:L45–L48. https://doi.org/10.1086/375442
96. Sakurai T (1981) Calculation of force-free magnetic field with non-constant α. Solar Phys 69(2):343–359
97. Schmieder B, Aulanier G, Vršnak B (2015) Flare-CME models: an observational perspective (invited review). Solar Phys 290:3457–3486. https://doi.org/10.1007/s11207-015-0712-1
98. Schrijver CJ (2016) The nonpotentiality of coronae of solar active regions, the dynamics of the surface magnetic field, and the potential for large flares. Astrophys J 820(2):103
99. Schrijver CJ, De Rosa ML, Title AM, Metcalf TR (2005) The nonpotentiality of active-region coronae and the dynamics of the photospheric magnetic field. Astrophys J 628:501–513. https://doi.org/10.1086/430733
100. Schrijver CJ, De Rosa ML, Metcalf TR, Liu Y, McTiernan J, Régnier S, Valori G, Wheatland MS, Wiegelmann T (2006) Nonlinear force-free modeling of coronal magnetic fields Part I: a quantitative comparison of methods. Solar Phys 235:161–190. https://doi.org/10.1007/s11207-006-0068-7
101. Schrijver CJ, Elmore C, Kliem B, Török T, Title AM (2008) Observations and modeling of the early acceleration phase of erupting filaments involved in coronal mass ejections. Astrophys J 674:586–595. https://doi.org/10.1086/524294
102. Schrijver CJ, Aulanier G, Title AM, Pariat E, Delannée C (2011) The 2011 February 15 X2 flare, ribbons, coronal front, and mass ejection: interpreting the three-dimensional views from the solar dynamics observatory and STEREO guided by magnetohydrodynamic flux-rope modeling. Astrophys J 738:167. https://doi.org/10.1088/0004-637X/738/2/167
103. Shibata K, Magara T (2011) Solar flares: magnetohydrodynamic processes. Living Rev Sol Phys 8:6
104. Solanki SK, Inhester B, Schüssler M (2006) The solar magnetic field. Rep Prog Phys 69:563–668. https://doi.org/10.1088/0034-4885/69/3/R02
105. Song MT, Fang C, Tang YH, Wu ST, Zhang YA (2006) A new and fast way to reconstruct a nonlinear force-free field in the solar corona. Astrophys J 649:1084–1092. https://doi.org/10.1086/506249
106. Sun X, Hoeksema JT, Liu Y, Chen Q, Hayashi K (2012a) A non-radial eruption in a quadrupolar magnetic configuration with a coronal null. Astrophys J 757:149. https://doi.org/10.1088/0004-637X/757/2/149
107. Sun X, Hoeksema JT, Liu Y, Wiegelmann T, Hayashi K, Chen Q, Thalmann J (2012b) Evolution of magnetic field and energy in a major eruptive active region based on SDO/HMI observation. Astrophys J 748:77. https://doi.org/10.1088/0004-637X/748/2/77
108. Tanaka T (1994) Finite volume TVD scheme on an unstructured grid system for three-dimensional MHD simulation of inhomogeneous systems including strong background potential fields. J Comput Phys 111:381–390. https://doi.org/10.1006/jcph.1994.1071

109. Titov V, Török T, Mikic Z, Linker JA (2014) A method for embedding circular force-free flux ropes in potential magnetic fields. Astrophys J 790(2):163
110. Titov VS, Démoulin P (1999) Basic topology of twisted magnetic configurations in solar flares. Astron Astrophys 351:707–720
111. Török T, Kliem B (2005) Confined and ejective eruptions of kink-unstable flux ropes. Astrophys J Lett 630:L97–L100. https://doi.org/10.1086/462412
112. Török T, Kliem B, Titov VS (2004) Ideal kink instability of a magnetic loop equilibrium. Astron Astrophys 413:L27–L30. https://doi.org/10.1051/0004-6361:20031691
113. Turmon M, Jones HP, Malanushenko OV, Pap JM (2010) Statistical feature recognition for multidimensional solar imagery. Solar Phys 262:277–298. https://doi.org/10.1007/s11207-009-9490-y
114. Valori G, Kliem B, Keppens R (2005) Extrapolation of a nonlinear force-free field containing a highly twisted magnetic loop. Astron Astrophys 433:335–347. https://doi.org/10.1051/0004-6361:20042008
115. Valori G, Kliem B, Fuhrmann M (2007) Magnetofrictional extrapolations of low and Lou's force-free equilibria. Solar Phys 245:263–285. https://doi.org/10.1007/s11207-007-9046-y
116. van Ballegooijen AA (2004) Observations and modeling of a filament on the sun. Astrophys J 612:519–529. https://doi.org/10.1086/422512
117. van Ballegooijen AA, Martens PCH (1989) Formation and eruption of solar prominences. Astrophys J 343:971–984. https://doi.org/10.1086/167766
118. van Ballegooijen AA, Priest ER, Mackay DH (2000) Mean field model for the formation of filament channels on the sun. Astrophys J 539:983–994. https://doi.org/10.1086/309265
119. van Ballegooijen AA, Deluca EE, Squires K, Mackay DH (2007) Modeling magnetic flux ropes in the solar atmosphere. J Atmos Sol-Terr Phys 69:24–31. https://doi.org/10.1016/j.jastp.2006.06.007
120. Wang AH, Wu ST, Tandberg-Hanssen E, Hill F (2011) Utilization of multiple measurements for global three-dimensional magnetohydrodynamic simulations. Astrophys J 732:19. https://doi.org/10.1088/0004-637X/732/1/19
121. Wang J, Du A, Zhang Y, Zhang T, Ge Y (2015) Modeling the Earth's magnetosphere under the influence of solar wind with due northward IMF by the AMR-CESE-MHD model. Sci China Earth Sci 58(7):1235–1242
122. Welsch BT, Fisher GH, Abbett WP, Régnier S (2004) ILCT: recovering photospheric velocities from magnetograms by combining the induction equation with local correlation tracking. Astrophys J 610:1148–1156. https://doi.org/10.1086/421767
123. Wheatland MS (2004) Parallel construction of nonlinear force-free fields. Solar Phys 222:247–264. https://doi.org/10.1023/B:SOLA.0000043579.93988.6f
124. Wheatland MS (2006) A fast current-field iteration method for calculating nonlinear force-free fields. Solar Phys 238:29–39. https://doi.org/10.1007/s11207-006-0232-0
125. Wheatland MS, Sturrock PA, Roumeliotis G (2000) An optimization approach to reconstructing force-free fields. Astrophys J 540:1150–1155. https://doi.org/10.1086/309355
126. Wiegelmann T (2004) Optimization code with weighting function for the reconstruction of coronal magnetic fields. Solar Phys 219:87–108. https://doi.org/10.1023/B:SOLA.0000021799.39465.36
127. Wiegelmann T (2007) Computing nonlinear force-free coronal magnetic fields in spherical geometry. Solar Phys 240:227–239. https://doi.org/10.1007/s11207-006-0266-3
128. Wiegelmann T (2008) Nonlinear force-free modeling of the solar coronal magnetic field. J Geophys Res (Space Physics) 113:A03S02. https://doi.org/10.1029/2007JA012432
129. Wiegelmann T, Neukirch T (2002) Including stereoscopic information in the reconstruction of coronal magnetic fields. Solar Phys 208:233–251. https://doi.org/10.1023/A:1020537403934
130. Wiegelmann T, Neukirch T (2006) An optimization principle for the computation of MHD equilibria in the solar corona. Astron Astrophys 457:1053–1058. https://doi.org/10.1051/0004-6361:20065281
131. Wiegelmann T, Sakurai T (2012) Solar force-free magnetic fields. Living Rev Sol Phys 9:5

132. Wiegelmann T, Inhester B, Sakurai T (2006) Preprocessing of vector magnetograph data for a nonlinear force-free magnetic field reconstruction. Solar Phys 233:215–232. https://doi.org/10.1007/s11207-006-2092-z
133. Wiegelmann T, Thalmann JK, Inhester B, Tadesse T, Sun X, Hoeksema JT (2012) How should one optimize nonlinear force-free coronal magnetic field extrapolations from SDO/HMI vector magnetograms? Solar Phys 281:37–51. https://doi.org/10.1007/s11207-012-9966-z
134. Wu ST, Sun MT, Chang HM, Hagyard MJ, Gary GA (1990) On the numerical computation of nonlinear force-free magnetic fields. Astrophys J 362:698–708. https://doi.org/10.1086/169307
135. Wu ST, Wang AH, Liu Y, Hoeksema JT (2006) Data-driven magnetohydrodynamic model for active region evolution. Astrophys J 652:800–811. https://doi.org/10.1086/507864
136. Wu ST, Wang AH, Gary GA, Kucera A, Rybak J, Liu Y, Vrśnak B, Yurchyshyn V (2009) Analyses of magnetic field structures for active region 10720 using a data-driven 3D MHD model. Adv Space Res 44:46–53. https://doi.org/10.1016/j.asr.2009.03.020
137. Yamamoto TT, Kusano K (2012) Preprocessing magnetic fields with chromospheric longitudinal fields. Astrophys J 752:126. https://doi.org/10.1088/0004-637X/752/2/126
138. Yan Y, Li Z (2006) Direct boundary integral formulation for solar non-constant-α Force-free magnetic fields. Astrophys J 638:1162–1168. https://doi.org/10.1086/499064
139. Yan Y, Sakurai T (2000) New boundary integral equation representation for finite energy force-free magnetic fields in open space above the sun. Solar Phys 195:89–109
140. Yang WH, Sturrock PA, Antiochos SK (1986) Force-free magnetic fields-The magneto-frictional method. Astrophys J 309:383–391. https://doi.org/10.1086/164610
141. Zhang J, Cheng X, Ding MD (2012) Observation of an evolving magnetic flux rope before and during a solar eruption. Nat Commun 3:747. https://doi.org/10.1038/ncomms1753
142. Zhang J, Yang S, Li T, Zhang Y, Li L, Jiang C (2013) Magnetic reconnection: from "open" extreme-ultraviolet loops to closed post-flare ones observed by SDO. Astrophys J 776:57. https://doi.org/10.1088/0004-637X/776/1/57
143. Zhao J, Li H, Pariat E, Schmieder B, Guo Y, Wiegelmann T (2014) Temporal evolution of the magnetic topology of the noaa active region 11158. Astrophys J 787(1):88. http://stacks.iop.org/0004-637X/787/i=1/a=88

Chapter 9
Data-Driven MHD Modeling of Coronal Magnetic Evolutions and Eruptions

Abstract Solar corona is always in dynamical evolution as driven by the surface motions on the photosphere, which stress the coronal field by shearing, twisting and even braiding in a somewhat random way. Such processes inject continuously magnetic free energy and helicity into the corona, which, however, cannot be reserved forever. As a result, solar eruptions occur from time to time, like a volcanic eruption, which explosively releases the magnetic free energy into other forms like kinetic and thermal energies. As such dynamical process is not included in the force-free field extrapolation models, MHD model, in particular, constrained and even driven by the observed magnetograms, is required to follow the coronal dynamic evolution. Such data-driven MHD models have been recently developed and applied to study the solar active-region evolution as well as its eruption, which offers a new way for understanding sophisticatedly their underlying magnetic topology and mechanism. This chapter briefly presents data-driven CESE–MHD modeling of coronal magnetic evolutions and eruptions.

9.1 Introduction

Observations show that the magnetic field on the solar photosphere changes continuously due to magnetic emergence, cancellation, transport, and dispersal by surface motions such as differential rotation and meridional circulation. Although direct measurement of the coronal magnetic field is impossible, it is generally accepted that this three-dimensional (3D) field is evolving in response to or driven by the evolution of the photospheric field. Thus the solar corona also changes dynamically with diverse processes including the interaction of newly emerging structures with the pre-existing field, twisting and shearing of the magnetic field arcades, and magnetic reconnection. It is commonly believed that these processes can drive the field away from the potential or the force-free equilibrium, and may trigger solar eruptions such as flares and coronal mass ejections.

The macroscopic behavior of dynamic coronal evolution can be described self-consistently by MHD model. Typically, the evolution of the corona can be characterized by two different categories, quasi-static evolution and eruption. The quasi-static

© Springer Nature Singapore Pte Ltd. 2020
X. Feng, *Magnetohydrodynamic Modeling of the Solar Corona and Heliosphere*, Atmosphere, Earth, Ocean & Space,
https://doi.org/10.1007/978-981-13-9081-4_9

evolution often last for hours to days while the eruption occurs in a relatively shorter time scale in minutes to a few hours interrupting the quasi-static phase. The quasi-static evolution of corona is justified by the fact that the photospheric motions (with speed typically on the order of km s^{-1}) are far slower than the coronal reaction speed which is the Alfvén speed (typically on the order of Mm s^{-1}) in the corona. Thus, without eruptions, the corona can be assumed to be in successive MHD equilibrium because the photospheric driving motions are so slow that at any instant the coronal field has enough time to relax to a new equilibrium. Thus it is reasonable to calculate MHD equilibrium matching the observed magnetic field on the photosphere to represent the quasi-static state, and a time-sequence of such equilibrium match the time-sequence of magnetogram can be used to simulate the quasi-static evolution. The NLFFF extrapolation as described in Chap. 8 is actually a simplification of the MHD equilibrium by neglecting all the plasma effects (such as the inertia, the pressure and gravity force) in the MHD momentum equation due to the low-β nature in the corona. However, the low β condition is not satisfied everywhere in the solar atmosphere, for example, in the photosphere, and thus the observed data does not provide a consistent boundary condition for the model based on an exact force-free assumption, and some ad hoc preprocessing (to remove the force in the raw magnetogram) is usually done to prepare the vector magnetograms for the extrapolation codes. Furthermore, there are critical locations essential to the magnetic topology in the corona, for example, the magnetic null points and current sheets (CSs), and in these locations the plasma β is usually large because of the field strength is weaker than its surroundings. To maintain force balance, the plasma pressure might play an important role in these places, which are, however, excluded by the force-free assumption. The full MHD relaxation model, thus, provide a more realistic one than the NLFFF model, because it can cope with different regions of either force-free or non-force-free, and formation of CS is proved to be a natural process in MHD relaxation. The data-constrained MHD equilibrium model, which uses the photospheric magnetiogram as the lower boundary, provides a step further to study the quasi-static structure and evolution of the corona [19, 33, 87].

During solar eruptions, such equilibrium is strongly destroyed because the plasma speed could be on the order of Alfvén speed, and topological changes of the magnetic field (i.e. magnetic reconnection) happen. There are models [3, 25, 28, 34, 41] using the NLFFF reconstructed immediately preceding eruption (thus the unstable nature of the field has already well developed) as the initial condition for MHD simulation, which prove to be able to reproduce the fast dynamic phase of the erupting field (e.g., [34]). However, these kinds of simulations do not self-consistently show how the pre-eruptive field is formed and disrupted, and thus may not be used to identify the true triggering mechanism. Also, such kind of models might not be able to reproduce confined flare, which is not likely triggered by the large-scale instability of the pre-flare magnetic field. In such cases a self-consistent data-driven MHD model [35] is required to follow the full evolution of the corona all the way from an initial stable state to an eruption.

Currently, MHD model driven by data can be classified into three categories according to the aims of their application, (1) data-constrained MHD equilibrium,

(2) MHD modeling of eruptive processes initialized by NLFFF extrapolations, and (3) full data-driven MHD model of formation and eruption of eruptive structures.

9.2 Data-Constrained MHD Equilibrium

By an assumption that the dynamic evolution of the coronal magnetic field (in non-eruptive state) can be approximated by successive equilibria, data-constrained MHD models have been developed to obtain a time sequence of MHD equilibria, each of which is based on a vector magnetogram representing a snapshot of the photospheric magnetic field. Similar to the NLFFF codes, usually a potential field is calculated for the initial magnetic field, while different models of plasma are used. By relaxing an initial plasma and magnetic field to a MHD equilibrium in which the bottom boundary magnetic field matches the vector magnetogram, each MHD equilibrium is regarded as a snapshot of the quasi-static evolution of the corona. One of the applications of these models is to study the plasma flow in the photospheric layer and the derived parameters, such as the Poynting flux, the magnetic helicity flux and other derived parameters [19, 80, 89]. In these applications the plasma is configured as with typical values in the photosphere or lower atmosphere (plasma β on the order of unity), thus they can not be used to simulate the corona. In the original model of Wu et al. [87], the initial plasma is not even in an equilibrium. Specifically, in the horizontal direction, the plasma density at the bottom is proportional to the transverse magnetic field strength. All these models regard the bottom boundary as the photosphere and the boundary conditions are derived from the method of characteristics [87], which can include the observed vector magnetograms as part of the bottom boundary conditions in a self-consistent way to generate MHD flow. There also exist non-MHD methods of calculating the photospheric plasma flow, such as the feature tracking methods, e.g. the LCT and ILCT [85] and DAVE [67] methods, and all these methods uses time-sequence of continuously observed magnetograms as input. Wang et al. [80] compared the photospheric plasma flow from their data-constrained MHD model with the one computed by the ILCT method [85] for AR 8210. Remarkable differences are found between the two methods, naturally due to the different assumptions, and the inconsistence indicates that uncertainties exist in either the MHD model or the ILCT method or both. Fan et al. [19] used the same model of Wu et al. [87] with a different numerical scheme. They applied it to a simulated data produced by an idealized simulation of a magnetic flux rope emerging from below the photosphere [52], and found a good correspondence between the model derived Poynting flux with that from the original data.

Another more important and success application of the data-constrained MHD-equilibrium model is to study the coronal magnetic field. To this end, the plasma is initially given with typical values in the solar corona to fulfill the extremely low-β and highly rarefied (i.e. very large ratio of Alfvén speed to the plasma sound speed) condition. Jiang et al. [32] first tested their model, by using the same numerical CESE scheme as their extrapolation code CESE–MHD–NLFFF [30], to the Low and

Lou [49]'s semi-analytic force-free solutions. CESE–MHD–NLFFF model achieved rather high accuracy for recovering the original solution, and offered the best result comparably as reported in Schrijver et al. [65]. Such application is also followed by Fan [17], who studied the magnetic topological changes during a major flare in AR 10930 using Hinode/SOT vector magnetograms. In the following we review two examples of such applications of studying the coronal magnetic field and its association with solar flares.

9.2.1 A Magnetic Bald-Patch Flare in AR 11117

Jiang et al. [33] investigated the magnetic field of AR 11117 around the time of a small confined flare occurred on 2010 October 25, during which the magnetic configuration changes very little. The model reproduces the basic structures of the 3D magnetic field, as supported by the visual similarity between the field lines and the coronal loops observed by AIA, which shows that the coronal field can indeed be well characterized by the MHD equilibrium in most time (see Fig. 9.1). With the time-sequence of magnetic field, the energy budget for the flare can be calculated as a decrease of the free energy immediately before and after the flare time, which is an upper bound of the flare energy. The energy budget is computed as 1.7×10^{30} erg which seems to be adequate to power this minor flare, whose energization is about several percent of a typical X-class one.Such quantitative estimation, however, usually contains very large uncertainties, resulting from either measurement errors (systematic error and random noise) of the HMI magnetogram data or the model assumption of MHD quasi-static equilibrium. The latter is a much more basic problem than the data errors and is encountered by any extrapolation of the magnetic field with static (for example, the NLFFF model) or quasi-static models, because the assumption of a static state of the magnetic field is unjustified by the onset of the flare, which can make the magnetic field very dynamic.

Nevertheless, important information of the flare mechanism can be investigated by studying the pre-flare configuration, in particular, in the aspect of magnetic topology, and such topology analysis is proved to be rather robust. According to the magnetic squashing degree distribution at the lower boundary of the model, it was found that there exists a magnetic bald patch (BP) configuration in the flare site, which are places where magnetic field lines become tangent to the photosphere at the PIL [e.g. 46, 74, 86]. As shown in Fig. 9.2, the field line crosses the PIL from *negative* to *positive* polarity in such places, which is opposite to the normal case and we called it as BP field lines. Since the photosphere can be regarded as a line-tying boundary for the coronal field, a BP field line is thus very special because it is anchored at the BP in addition to its footpoints. As a result, the continuous set of BP field lines defines a separatrix surface (BPSS) of magnetic topology, across which the field-line linkage is discontinuous. This is rather important topological structure, since BP is the only place that can define magnetic separatrices besides the case that magnetic nulls are present in the corona. The field lines immediately above a BP are concave

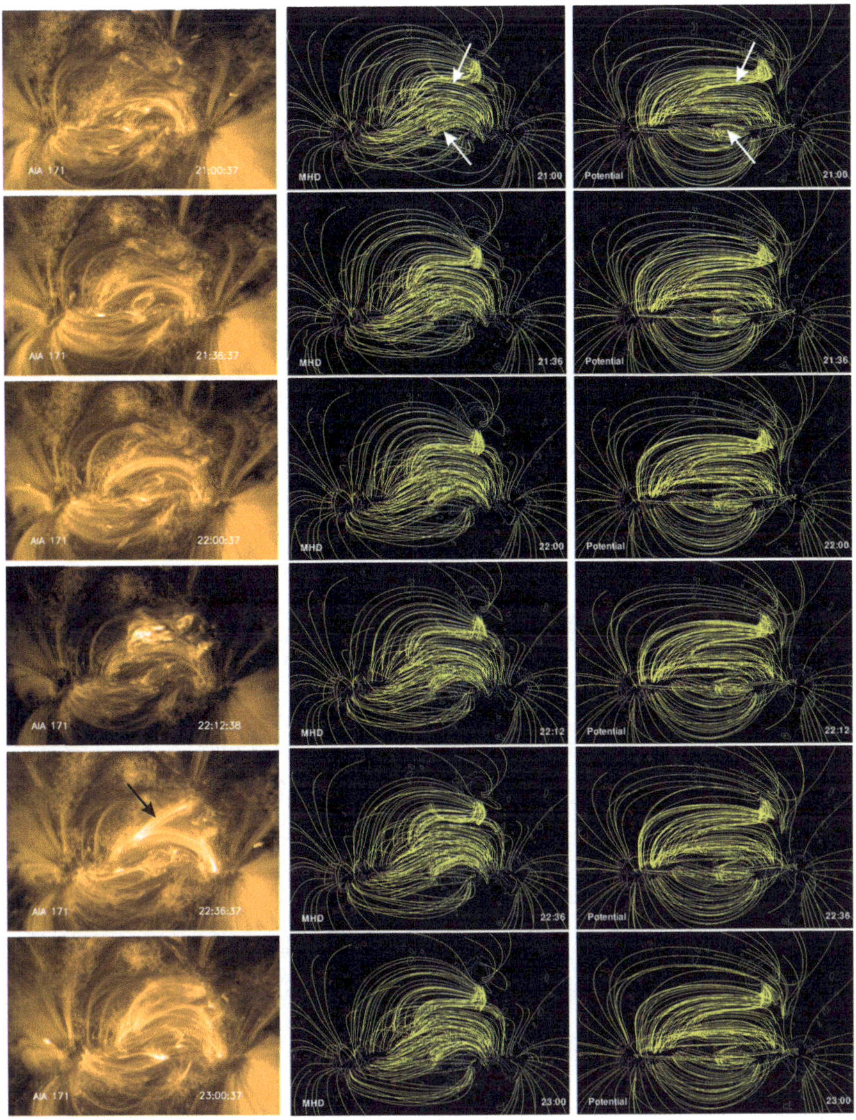

Fig. 9.1 Comparison of the data-constrained modeled field lines with *SDO*/AIA 171 Å image for AR 11117. The left column is the AIA images, the middle column is the selected magnetic field lines from the MHD results. Also the potential field lines are plotted in the last column. The field lines in all the panels are traced from the same footpoints on the photosphere and the color contours of photospheric B_z are plotted on the background. Taken from Fig. 6 of Jiang et al. [33] with permissions of the authors and the AAS

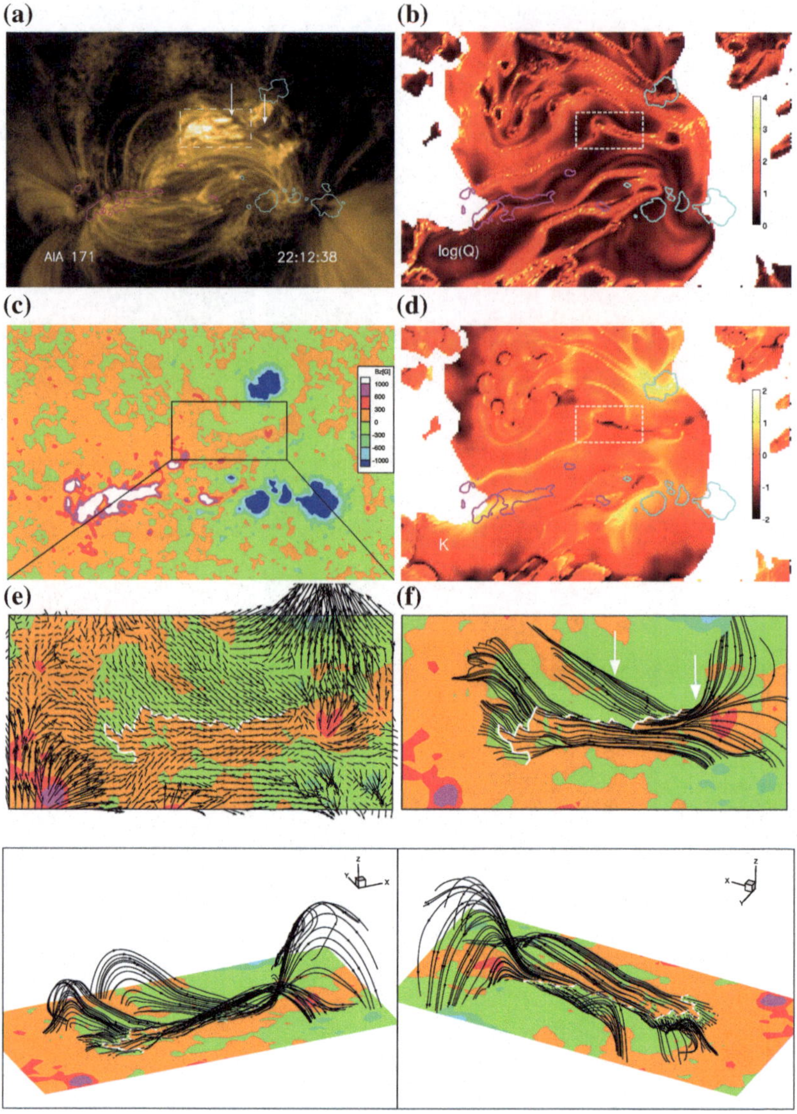

Fig. 9.2 **a** The AIA 171 image at 22:12, with the contour lines plotted for the LoS photospheric field at ±1000 G and the dashed box showing the location of flare with loops brightened. **b** The squashing factor Q in logarithmic scale; samely contours are plotted at ±1000 G for photospheric B_z and the dashed box outlines the flare location. **c** The B_z map with the flare location outlined by a black box which is enlarged in panel **e**. **d** Same as **b** but for the expansion-contraction degree K. **e** The vector magnetogram in the flare location with the thick white line denoting the bald patch (BP). Panel **f** gives some examples of the field lines (i.e. the BP separatrix field lines) that are tangent to the photosphere at the BP (the thick white line). The last two panels show different 3D views of the BP-separatrix field lines plotted in panel **f**. The BP is denoted by the thick white lines. The z–axis scale is doubled for a better view of the field lines. Adapted from Figs. 7 and 8 of of Jiang et al. [33] with permissions of the authors and the AAS

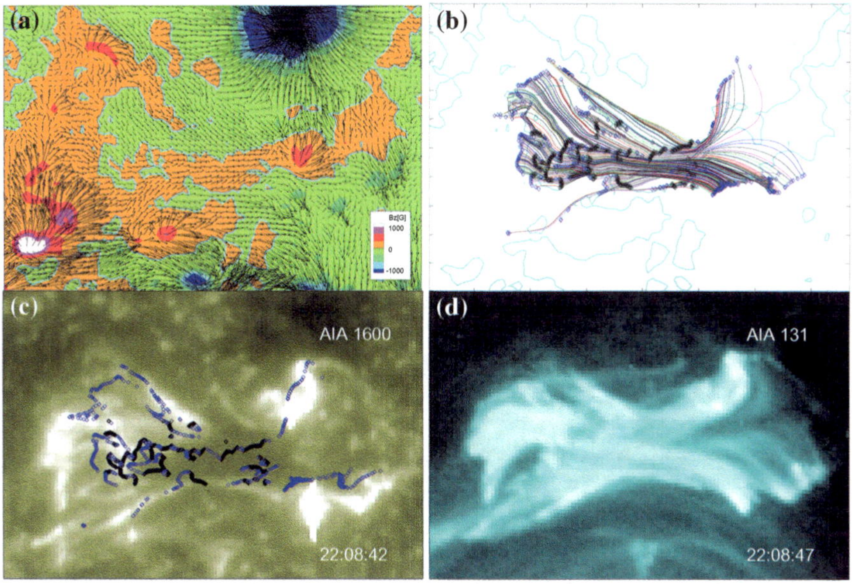

Fig. 9.3 Spatial relation between the flare site and the BPSS. **a** Magnetic vector fields at the photosphere taken by HMI at 22:12 UT; **b** Magnetic field lines form the BPSS. The small diamond shapes in black denote the BPs, and those in blue for the footpoints of the BPSS field lines. The thick cyan line represents the photospheric magnetic PILs. **c** Flare brightening patches observed by AIA in 1600 Å, and the overlaid are the BPs and BPSS footpoints. **d** Hot flaring loops observed by AIA in 131 Å. A nice agreement of the BPs and BPSS footpoints with the flare patches and the BPSS with the flaring loops can be seen. Taken from Fig. 3 of Jiang et al. [37] with permission of the Research in Astronomy and Astrophysics (RAA)

up (i.e. forming magnetic dips) and can hold material against solar gravity, which thus usually implies the existence of filament associated with BP [51, 73].

Further comparison of the flare observation with the modeled magnetic field demonstrates that the small flare is correlated with the magnetic BPSS configuration: first, the footpoints of the BPSS field lines are co-spatial with the flare ribbon that presents multiple brightening kernels (Fig. 9.3); secondly, the shape of the BPSS is in good agreement with the hot flaring loop observed in AIA-131; thirdly, a small filament is found to overlay the flare kernels, indicating that field lines there exhibit some dips to support this filament, while after the flare it disappeared with the appearance of a new set of coronal loops expanding after the flare, just in the same location of the missing mini-filament. It is also noted that before the flare there is an emergence of a small positive polarity under the west lobe of the BPSS, which could be the driver and trigger of reconnection in the BPSS. Thus, the accelerated particles there traced the field lines near the BP-separatrix and produced flare ribbons, and the field lines attached to the photosphere at the BP before the flare are lift up by the reconnection and expand upward, manifesting by the emergence of the post-flare loops and the disappearance of the mini-filament. Previous studies also show that the BP

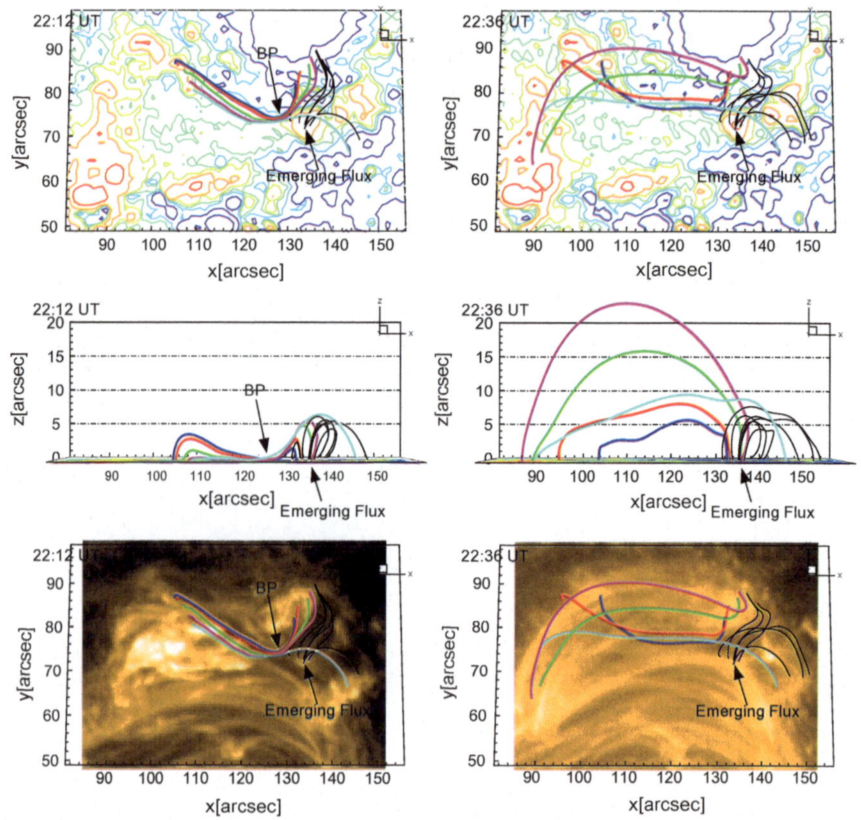

Fig. 9.4 Magnetic configuration change from pre-flare state (left panels) to post-flare state (right panels). The magnetic field lines are shown in SDO view angle (top) and side view (middle), and overlaid on AIA-171 images (bottom). A group of BPSS field lines are shown in thick colored lines. Their footpoints in the west (the right side) are fixed for the pre-flare and post-flare field lines. A group of field lines from the emerging polarity are shown in black color. The contour lines represent photospheric magnetogram B_z, red for positive and blue for negative. Nice agreement of the modeled field lines with the coronal loops can be seen. Taken from Fig. 6 of Jiang et al. [37] with permission of the Research in Astronomy and Astrophysics (RAA)

configuration is usually related with flux emergence, which triggers reconnection in the BPSS, for example in the formation of active region [55] and in the surges and arch filament systems [53]. Such evolution process is illustrated in Fig. 9.4, which shows the pre-flare and post-flare configuration calculated by the MHD model. The field lines rooted in the emerging flux expand upward (compared to the post-flare state with the pre-flare state) and push the overlaying BPSS, which continuously increases electric currents in the BPSS and free energy to the system. When the current sheet is enhanced enough for resistivity to be important, magnetic reconnection sets in at the BPSS, and the field lines confined around the BP by the photosphere plasma are lift up, expand as they relax, forming the post-flare loops. The field lines might also undergo slipping reconnection as they expand, with the footpoints not necessarily being in the same pre-flare locations.

9.2.2 Reconstruction of Large-Scale Pre-flare Current Sheet

In another application, Jiang et al. [38] shows that the model is able to reconstruct a large-scale current sheet (CS) existing prior to a flare, and the footpoints of the field lines traced from the CS clearly track the observed flare ribbons. This flare event is rather peculiar that exhibits unique X-shaped ribbons [44], thus such results are important to understand the flaring process.

According to the Parker theory [47, 48, 50, 56, 57], CSs spontaneously form in the evolution of coronal magnetic fields in response to slow photospheric footpoint motions, which is a basic MHD process. In other words, in general the coronal field contains CSs, which not only exist during flare/eruption but also form in the quasi-static evolution of the corona. This means that the presence of CSs should be ubiquitous in the corona, like the presence of flux ropes. To capture precisely the CS, an AMR is designed to automatically capture narrow layer with strong currents as well as resolve the strong magnetic field region (e.g. sunspot regions) with strong gradients, which is illustrated in Fig. 9.5. For capturing such intense current layer, the value of J/B is used to guide the refinement of the mesh. J/B is often shown to be a better indicator that can highlight CS-like distribution than the current density J itself [18, 21, 36]. This is because in numerical realization, the CS usually have both larger current density and weaker field than its neighborhoods. As can be seen in Fig. 9.5, the J/B value in the CS abruptly increases over its neighborhoods, and thus it is captured by the mesh points with highest resolution. During the calculation, any location with J/B becoming larger than $0.2/\Delta$ (where Δ is the local grid size) will be refined by a factor of two.

Using the data-constrained MHD model of the coronal equilibrium with a pre-flare magnetogram for AR 11967, Jiang et al. [38] revealed that there is a large-scale coronal CS responsible for a confined flare (see Fig. 9.6) recurring three times in the region. These homologous flares demonstrate a remarkable X shape (i.e. X-shaped ribbons and X-shaped brightening loops), which is rather unique. The flare site has a quadrupole magnetic configuration, and the X extends long "arms" and "legs" reaching to the four magnetic polarities. Also different from typical two-ribbon flares, these flare ribbons barely exhibit any movement during the whole flaring phase. There have been efforts to elaborate the magnetic topology behind such a unique flare by using potential-field and NLFFF extrapolations, but neither model can reproduce the observed mophorology [40, 43]. For example, based on a potential field extrapolation from the vertical component of the vector magnetogram, Liu et al. [43] found that there is a double-null magnetic topology that indeed defines a X-shaped magnetic skeleton, but obviously it does not match the geometry of the observed X-shaped flare ribbon. An NLFFF extrapolation produces a result even worse than the potential

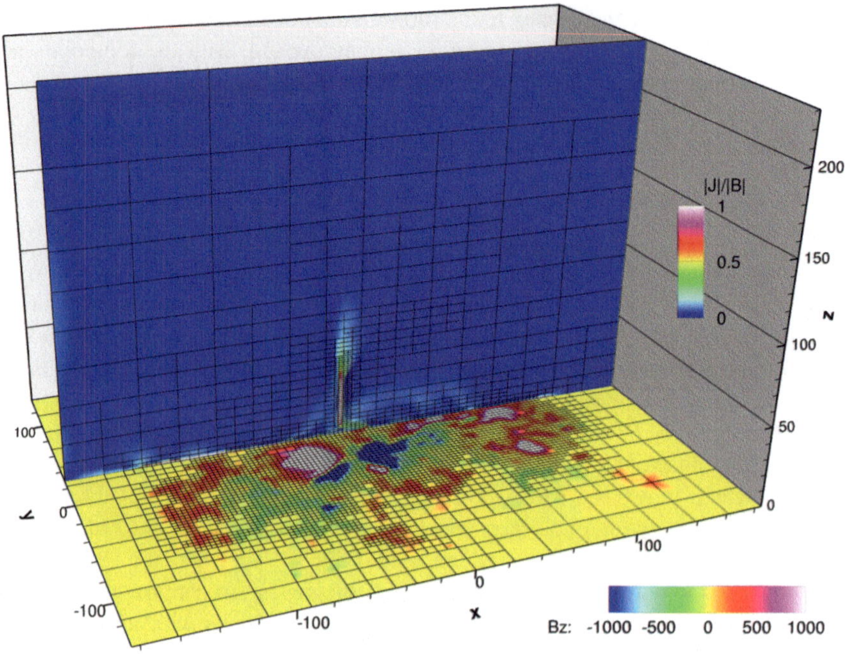

Fig. 9.5 Structure of the AMR grid. Six levels of refinement are used in this computation, and in each level the grid size is refined by a factor of 2. The lower slice of $z = 0$ is shown with the distribution of B_z on the photosphere and the vertical slice (pseudo-colored by J/B) is a cross section of the 3D volume. The grid lines show the structure of the grid blocks and each block consists of 8^3 cells which is omitted here. Unit of length is Mm. The highest resolution in $x - y$ plane is 0.36 Mm. Note that the resolution in vertical direction is higher by a factor of 2 than that in the horizontal direction. This is designed considering that the magnetic field often expands more strongly in the vertical direction than that in horizontal direction. Taken from Fig. 1 of Jiang et al. [38] with permissions of the authors and the AAS

field model in matching the flare topology, as also shown by Liu et al. [43]. On the other hand, the MHD model reveals a CS distinctly in the corona, which is located at the center of the quadrupolar configuration, and extends vertically all the way from the lower boundary of the model box to a height of 40 Mm (Fig. 9.7). Clearly this CS separates the field lines into two distinct connections, one connecting P1-N2, and the other connecting P2-N1 as shown. Thus the field lines traced from the CS naturally form an X-shaped configuration, and their footpoints on the bottom surface delineate a sound X shape. This X shape, with its center, arms and legs, almost coincides with the observed one in different AIA channels (see Fig. 9.8). Such a good match of the CS structure and the flare mophorology provides a strong evidence that reconnection at the CS produced the flare, since the locations of chromospheric flare ribbons are well recognized to correspond to the footpoint locations of those magnetic field lines

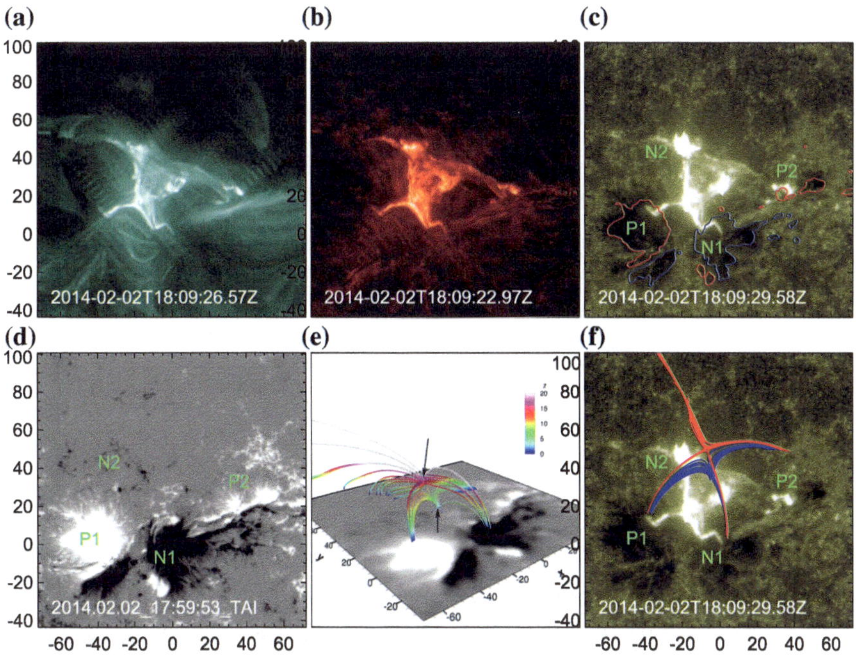

Fig. 9.6 SDO observations of the X-shaped M3.1 flare at 18:11 UT on 2014 February 2. **a**, **b** and **c** are snapshots of the flare at main phase in AIA 94 Å, 304 Å, and 1600 Å. **d** The photospheric magnetic flux distribution (magnetogram) observed by HMI at 10 min prior to the flare peak time. Four polarities associated with the flare are labeled as P1, N1, P2, and N2. Here P1 and N1 denote the major sunspots defining AR 11967; P2 refers to the positive flux region in the north of N1, including the plage region as well as the elongated positive polarity of strong field, while N2 denotes the plage region in the north of P1. Overlaid contours in **c** are photospheric magnetic field of 1000 G (red) and −1000 G (blue). **e** Magnetic field lines of a potential field extrapolation from the magnetogram. The two arrows mark the locations of two magnetic null points of the potential field. These two nulls are close to each other in the horizontal direction, one situating at altitude of ∼4 Mm and the other at ∼18 Mm. The field lines are traced in the neighborhood of the nulls, with the thick ones for the null point higher while the thin ones for the null lower. **f** The double-null related field lines overlaid on the AIA 1600 Å image. Lines of the higher (lower) null is colored as red (blue). The unit of length in all the panels is Mm. Taken from Fig. 2 of Jiang et al. [38] with permissions of the authors and the AAS

that undergo reconnection in the CS [60]. A further analysis based on calculating the magnetic squashing degree for the MHD model shows that the CS is formed at an HFT, because the X-shaped footpoints are co-spatial with two intersecting QSLs that define the HFT. This is an important result since HFT is predicted to be a place most readily for formation of flare CS.

The CS is rather robustly captured by the data-constrained MHD model [38] with different resolutions. Also, it is pointed out by Jiang et al. [38] that basically the CS's location is not sensitive to the resolution of grid, although the thickness of the

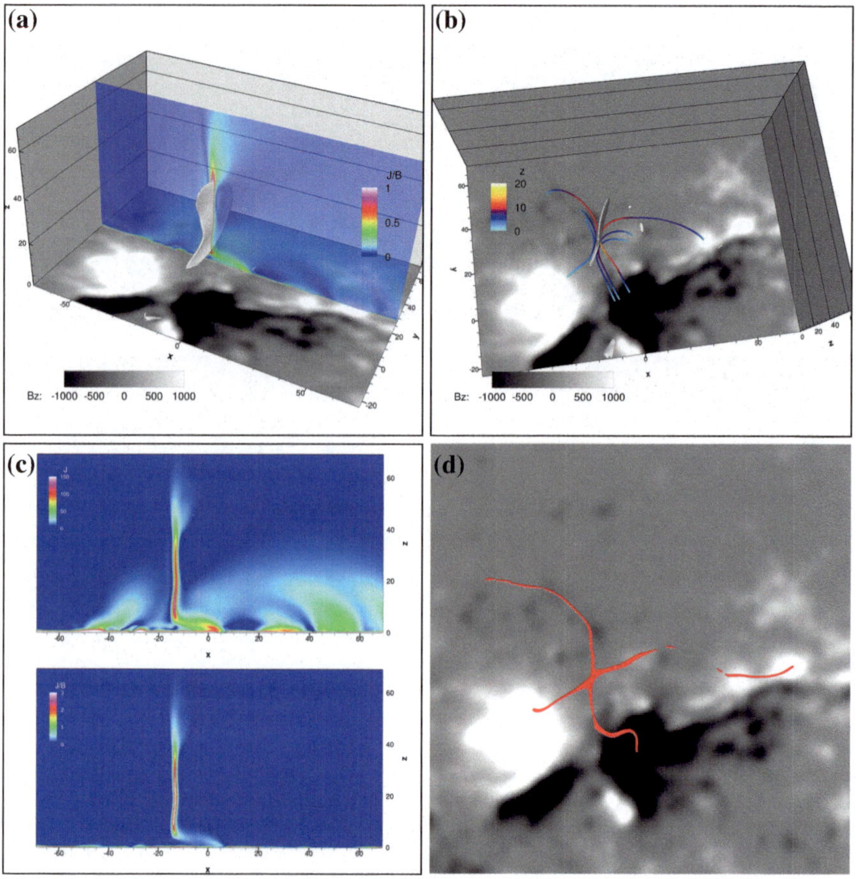

Fig. 9.7 Structure of the CS related with the X-shaped flare at 18:11 UT. **a** The 3D gray object is iso-surface of $J/B = 1\,\mathrm{Mm}^{-1}$ which is used to define the CS. A vertical cross section of the volume is shown with distribution of J/B. **b** Another view of the CS. The sampled field lines are traced from the two sides of the CS, and the color represents the height z. **c** Distribution of current density J and J/B on the vertical cross section shown in **a**. Here $J = |\nabla \times \mathbf{B}|$ and its unit is G Mm^{-1}. **d** The red dots are footpoints of all the closed field lines that are traced from the CS, and they are overlaid on the magnetogram. A total number of 2469 field lines are traced. The unit of length in all the panels is Mm. Taken from Fig. 3 of Jiang et al. [38] with permissions of the authors and the AAS

CS and its current intensity depend on the grid size. Presumably, the current-density distribution will approach to a δ function center at the CS, i.e. a magnetic tangential discontinuity, if the grid size approaches infinitesimal, but before that the microscopic behavior of plasma must be considered, which is beyond the scope of MHD.

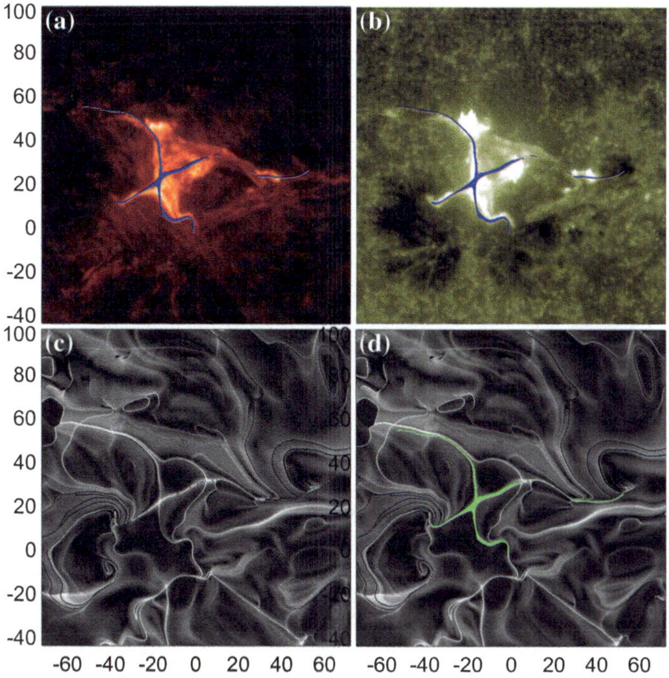

Fig. 9.8 Comparison of the modeled results with the observed flare ribbons. Panels **a** and **b** are respectively AIA 304 Å and 1600 Å images overlaid with the footpoints of magnetic field lines contacting the CS (in blue dots). **c** Map of magnetic squashing degree $\log Q$. The color ranging from black to white represents 0–5. **d** The same footpoints in **a** and **b** overlaid on the squashing-degree map. Taken from Fig. 4 of Jiang et al. [38] with permissions of the authors and the AAS

9.3 Combination of NLFFF Extrapolation and MHD Simulation

Since the NLFFF extrapolation provides a more realistic coronal magnetic field than the potential field, it is thus preferable to use the NLFFF solution as an initial condition of MHD simulation. This approach has been demonstrated to be very useful in simulation of eruptive process beginning from an unstable or marginally stable NLFFF extrapolation immediately prior to the eruption [3, 25, 34, 41, 59, 90]. A comprehensive review of this subject can be found in Inoue et al. [27]. In the first application, Jiang et al. [34] studied the eruption of a sigmoidal flux rope in AR 11283, which is found to possess a rather magnetic complex consisting of a spine-fan topology linking multiple polarities. Analysis of the pre-flare NLFFF model suggests that the flux rope already enters into the threshold of torus instability, by which the eruption is driven as demonstrated in the MHD simulation. The initiation and subsequent evolution of the eruptive flux rope are reproduced to resemble remarkably the corresponding EUV features, such as the sigmoidal emission, the filament ejection as well as the quasi-circular flare ribbons (see Fig. 9.9). This result has two-fold

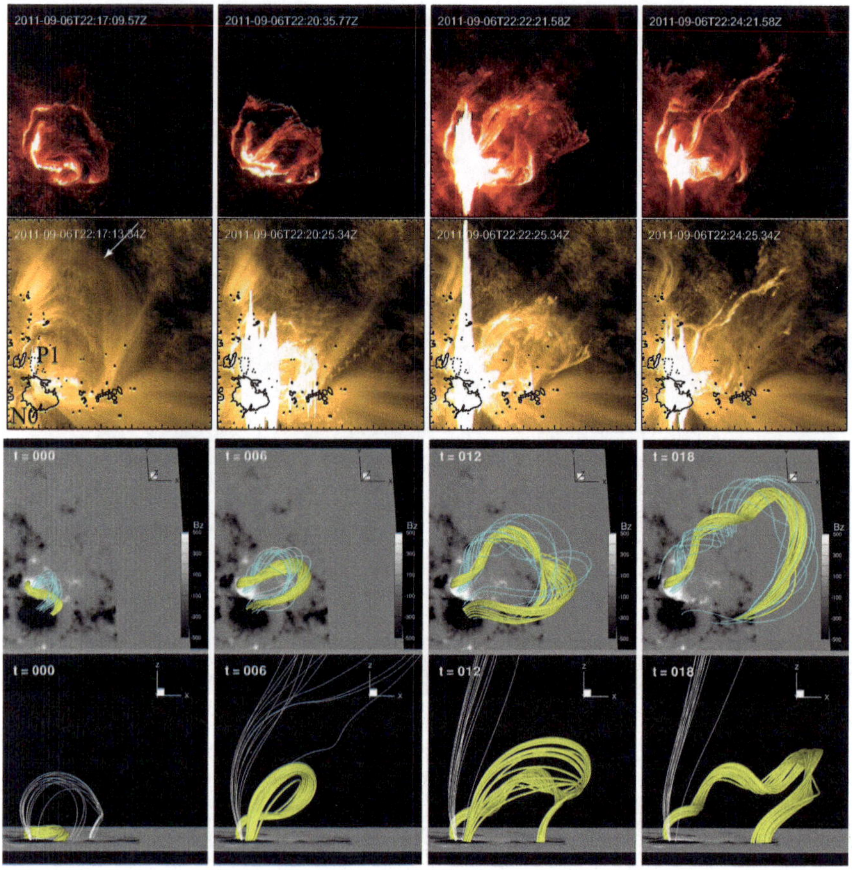

Fig. 9.9 Comparison of observation and simulation of the sigmoid eruption that is associated with a X2.1 flare occurring on 2011 September 6. The first two rows show the SDO/AIA observations in 304 and 171 Å, respectively. The contours overlying the images are the line-of-sight magnetic field with ±500 G. The last two rows show the evolution of magnetic field lines in the MHD simulation. Viewing angle in the upper row is aligned with the SDO/AIA observations, while the bottom row is side view from south. Two sets of the core field lines (cyan and yellow) are shown, which are traced from the same positive polarity at the bottom. Another set of lines that initially touch the null point are traced from the negative polarity to show their reconnection and opening. The bottom magnetogram of B_z is also shown. Adapted from Figs. 1 and 5 of Jiang et al. [34] with permissions of the authors and the AAS

significance. On one hand, it further evidenced the agreement of the NLFFF extrapolated field with the realistic corona; one the other hand, it shows the reliability of the MHD model in characterizing the dynamics in solar eruptions. As previous simulations of the initiation of solar eruptions were commonly carried out with idealized or hypothetical magnetic field configurations, the combination of NLFFF and MHD models provide a possibility of reproducing realistic and complex eruption process that is beyond the scope of idealized or theoretical models.

Similar way of simulating realistic eruptive events was then employed by other modelers [3, 25, 26, 41, 90]. Kliem et al. [41] studied the eruption on 2010 April 8, which is the first eruptive event witnessed by SDO. They used a data-constrained NLFFF model constructed with the techniques of flux rope insertion and magneto-frictional relaxation [77, 78], which is unstable, to initialize their zero-β simplified MHD modeling. It yields a very good agreement with a number of observed features, like the strongly inclined initial rise path and the close temporal association between the coronal mass ejection and the onset of flare reconnection. The same approach is used to model the process of magnetic twist release during a filament eruption [90]. Inoue et al. [25] investigated the eruption mechanism of an X2.2 flare in the well-known AR 11158 [69]. They first extrapolated an NLFFF model using SDO/HMI vector magnetogram observed two hours before the flare, and found this NLFFF is stable in MHD simulation. Thus, an enhanced anomalous resistivity was used to increase the magnetic twist through tether-cutting reconnection in those sheared arcades in the AR core, after which the quasi-equilibrium was broken and an eruption followed. Amari et al. [3] modeled the eruption process of a major flare and eruption in the AR 10930 by their NLFFF extrapolation code [1] and MHD simulation with Hinode/SOT vector magnetograms. A series of NLFFF extrapolations were performed to follow the quasi-stationary evolution of the pre-flare coronal field, and the eruption is simulated by a NLFFF input into the MHD model. However, similar to the work of Inoue et al. [25], as the input NLFFF model is extrapolated with data observed several hours prior to the eruption, it is still stable and shows no eruption in the MHD model. Further stress of the coronal field through artificially-specified converging flows on the boundary conditions are used to drive the flux rope to an unstable state until its eruption. The converging flows near the main PIL below the flux rope are often used to mimic flux cancellation, which is recognized to be a effective mechanism for buildup coronal flux rope and triggers flare [2, 42]. Very recently, using the same approach, Amari et al. [4] found a failed eruptive magnetic flux rope in the super AR 12192, in which the so-called "magnetic cage" (i.e. the overlying strapping field lines) plays a key role in the confined flares.

However, it should be noted that such combination of NLFFF and MHD models does not reproduce the eruption initiation in a self-consistent way in correspondence with the temporal evolution of the photospheric field. In other words, the evolution of the coronal field from a stable equilibrium to eruption matches seamlessly the variation in the photospheric magnetogram is still out of reach. For instance, in the simulations of Jiang et al., Kliem et al. [34, 41], their NLFFF models actually already deviate from force balance due to numerical residual Lorentz forces. Actually, almost all the available NLFFF codes actually generate non-force-free magnetic field data with residual Lorentz force that is often non-negligible. This is indicated by the misalignment of the current **J** and magnetic field vectors [65], which is usually measured by CWsin, a current-weighted sine of the angle between **J** and **B**. CWsin is typically in the range of 0.2 to 0.4 see, e.g. DeRosa et al. [13, 14], Schrijver et al. [66]. Another metric $E_{\nabla\times\mathbf{B}}$ measuring more directly the residual Lorentz force is defined as the average ratio of the force to sum of the magnitudes of magnetic tension and pressure forces [15]. These two metrics for the typical NLFFF extrapolation are respectively, Cwsin ~ 0.2 and $E_{\nabla\times\mathbf{B}} = 0.2$ [e.g. see the last column of Table 2 in 14]. Such residual force can instantly induce plasma motions in a low-beta and highly tenuous plasma environment. This motion provides a way of perturbing the system. If the system is significantly far away from unstable, such unbalance state will relax to a new MHD equilibrium as the induced motion can alter the magnetic field, which in turn generate restoring force to stabilize the system. Else if the system is unstable or not far away from unstable, the perturbation could grow and lead a drastic dynamical evolution of the system, i.e. eruption. On the other hand, the works of Inoue et al., Amari et al. [3, 25] begin simulation from a stable NLFFF and artificially modify the NLFFF by boundary flows or resistivity, which is used to bring the field to a eruption. The resulted boundary magnetic field, thus, does not match the observed magnetogram during the eruption.

9.4 Data-Driven MHD Model

The so-called DARE (data-driven active region evolution) model [35] enables us to follow the dynamical evolution of the coronal magnetic field driven directly by a time sequence of solar magnetograms. Actually, this is the first one of such kind of data-driven model capable of describing the full evolution of non-potential coronal magnetic field. The model numerically solves the full set of time-dependent 3D MHD equations with the bottom boundary conditions partially specified by the evolving magnetic field data as measured for a single layer, i.e. the photosphere. Similar to the NLFFF extrapolation, this model does not include the physics of the transition layer starting from the photosphere to the base of the corona, but simply putting the bottom at the coronal base. It proves to be a big challenge in both physical and numerical aspects to fully include the transition region in a model that attempts to study the large scale coronal evolution. Firstly, it is required to consider the still unknown mechanism of coronal heating to explain how the temperature increases steeply from thousands of degrees to millions. Even more, the ionization degree at the photosphere is extremely low, making the MHD model inappropriate [79]. Secondly, the computational overhead is huge in the case that the grid size should be as small as a few of kms to resolve the steep transition region while at the same time, the whole region should be as large as hundreds of Mms.

As the bottom boundary of the model is actually the coronal base, the magnetic field measured on the photosphere is used as a reasonable approximation of the field at the coronal base. The plasma thermodynamics are simplified by an adiabatic energy equation as the model focuses on the structure and evolution of the coronal magnetic field and its interaction with plasma, which dominates the basic dynamics in the corona. Although no explicit resistivity is included in the magnetic induction equation, magnetic reconnection is still allowed due to numerical diffusion if any current sheet (CS) forms and becomes thin enough with thickness close to the grid resolution [see, e.g. 16, 76]. A small kinematic viscosity ν is used with its value corresponding to the viscous diffusion time ($\tau_\nu = L^2/\nu$, where L is the unit length) as $\sim 10^2$ of the Alfvén time ($\tau_A = L/v_A$, where v_A is the Alfvén speed) in the strong field region. The plasma is initialized as in a hydrostatic, isothermal state $T = 10^6$ K (with sound speed $c_S = 128$ km s^{-1}) with solar gravity. It is configured to make the plasma β as small as 2×10^{-3} (the maximal v_A is 4 Mm s^{-1}) to mimic the coronal low-β (highly tenuous) condition [20]. Here the unit length L is set as 16 arcsec (or 11.5 Mm), doubling the length of a basic grid block (8 arcsec) used in the model, and the unit time is set as $\tau = L/c_S = 90$ s.

The model equation is solved using an advanced space-time high-accuracy scheme (AMR–CESE–MHD, refer to Jiang et al. [31]). The computational volume is sufficiently large to enclose the eruptive region of interest and its surrounding magnetic topology of relevance, and at the same time consists of a sufficiently small grid size of $\Delta = 360$ km (equal to 0.5 arcsec on the Sun) matching that of the HMI pixel. This is realized by a non-uniform mesh based on the magnetic flux distribution. The smallest grid is made around the flux emerging site, where the photospheric field changes most actively. Grid size is increased gradually to 4 arcsec near the side and top boundaries.

One data product of the SDO/HMI observations, the Space-weather HMI Active Region Patches (SHARP) vector magnetogram 'hmi.sharp_cea_720s' [7, 22], is most suitable for being used to drive the MHD model. With cadence of 12 min and spatial resolution of 1 arcsec, the SHARP data is adequate for simulation of relatively long-term evolution (hours to days) of eruptive AR magnetic structures from their origin to eruption. Furthermore, the SHARP data includes inverted magnetic field data, projected and re-mapped on the cylindrical equal area (CEA) Cartesian coordinate system centered on the tracked AR, which is well suited for simulations performed in the Cartesian coordinate system.

Smoothing is needed when data from observation is involved in a computing scheme based on numerical finite difference. Besides, the lower boundary of the MHD model represents the base of the corona rather than the photosphere and the magnetic structures should be broadened from the photosphere to the coronal base. Such broadening can be simulated by using Gauss smoothing of the photospheric data, as suggested by Yamamoto and Kusano [91], who compared the Gauss smoothed Hinode data with the chromospheric data observed by SOLIS and find that the correlation is above 0.95 for Gaussian σ of $2 \sim 5$ arcsec, and suggested to use Gaussian smoothing in this range when simulating the photospheric field broadening. It is thus recommended to use the lower bound value of σ (2 arcsec) to keep the details of the magnetic structures as much as one can. Additionally, the data also needs to be smoothed in time to remove short-term temporal oscillations and spikes due to bad pixels, which is handled with Gaussian window of $\sigma = 4 \times 12$ min in Jiang et al. [35]. An example of comparison of the data before and after being smoothed is shown in Fig. 9.10.

As being driven directly by data from observations, it is absolutely essential for the model to have inputted data with good quality and reliability. But the observed data is not perfect. Errors and uncertainties of the SHARP data come from several aspects that may affect numerical results. The first one is the random error which is, as estimated conservatively, about 200 G in weak field regions and 70 G where the

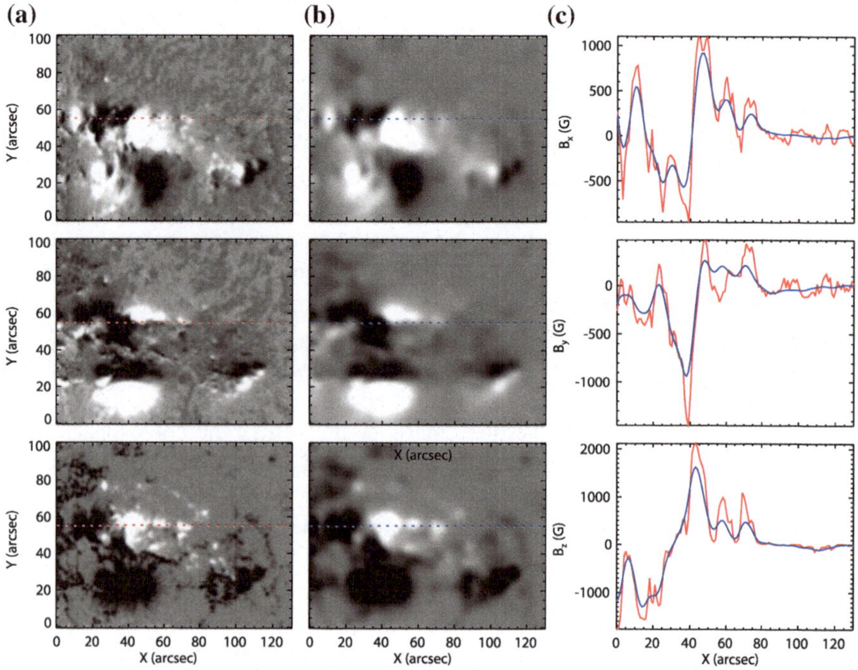

Fig. 9.10 Comparison of the original SHARP data before and after being smoothed. The data is sampled at the time of 00:00 UT on September 6, 2011. From top to bottom, panels are shown for the three components of the magnetic field B_x, B_y, and B_z, respectively. **a** The original map. **b** The smoothed map. Colors are saturated at white for value above 500 G and black for value below −500 G. **c** Plots of a random horizontal line, the location of which is shown in **a** and **b**. Red (blue) for data before (after) being smoothed. Taken from Supplementary Fig. 1 of Jiang et al. [35] licensed under the Creative Commons Attribution 4.0 International License (CC BY 4.0) available at https://creativecommons.org/licenses/by/4.0/

field is strong. Another one is the systematic errors due to the periodic variation of the SDO orbit, which results in daily temporal oscillations of the data. Also the data might lose its reliability at the flare time due to anomalous flare-related emissions. All these uncertainties should be kept in mind and carefully estimated with their possible influence in analyzing numerical results. Only the modeled coronal magnetic fields also agree reasonably with the corresponding AIA observations then can one reliably draw conclusions from the results.

9.4.1 Numerical Scheme and Boundary Conditions

To fully model the dynamic evolution is extremely computational expensive. When parallelized with a medium number (for example, a hundred) of CPUs (3 GHz), each time-step advancing of the computing code takes about 5 s. On the one hand, the explicit code requires that the time step (i.e. the size of iteration step in time) must be smaller than $\Delta / \max(v_A) \approx 0.1$ s due to the Courant–Friedrichs–Lewy stability condition [10]. Accordingly, to update in one hour of physics time needs about 50 h wall clock time. On the other hand, self-consistent modeling of the long-term energy buildup process leading to the eruption initiation requires the code to simulate a physics time interval of days. Thus, the whole evolution process would require months of wall clock computing time. To make the computation manageable, Jiang et al. [35] speed up the cadence of inputting the HMI data by 40 times. This is justified by the fact that the photospheric flow speed in accordance with the photospheric field evolution is about $0.1 \sim 1\,\mathrm{km\,s^{-1}}$ [45, 85]. So the evolution speed of the boundary field, even enhanced by a factor of 40, is still sufficiently small compared with the coronal Alfvén speed ($\sim$ Mm $\mathrm{s^{-1}}$), and the basic reaction of the coronal field to the bottom changes should not be affected in the non-eruptive time duration. As a result, one hour in the HMI data accounts for one τ in the simulation. When comparing the simulation with the EUV observations, such scaling also applies to the AIA data in the quasi-static evolution phase, but for the eruptive duration, in principle, time should be scaled according to the ratio of the realistic coronal Alfvén speed to the modeled one, for the purpose of giving a reasonable morphological similarity between simulations and observations from AIA for the eruption process.

The projected-characteristics method based on the wave-decomposition principle of the full MHD system [88] offers an approach for continuous inputting the boundary vector field to the model. It is proved that the method can naturally simulate the transferring of magnetic energy and helicity to the corona from below Wu et al. [88] by self-consistently calculating the surface flow field [80], which otherwise would have to be derived by local correlation tracking or similar techniques [67, 85]. Linear interpolate of the data in time is used to produce a data set with cadence matching the time step of the MHD model, because the cadence of the HMI data is 12 min.

9.4.2 Magnetic Flux Emergence Leading to an Eruption in AR 11283

The DARE–MHD model was first applied to simulate the overall process from the formation to the initiation of a solar eruptive flare in AR 11283 [35]. The event is characterized by a fast magnetic flux emergence of over three days from 2011 September 4 (day 1) to September 6 (day 3) that finally leads to an M-class eruptive flare on day 3. The time interval used for the data-driven simulation is chosen that there is no eruption or flare from the AR before the studied one, and during this interval the AR is not far away from the disk center, as shown in Fig. 9.11, which provides an uninterrupted window for measuring the changes of the photospheric magnetic field by the HMI. Furthermore, the initial state of the coronal field on day 1 is close to a potential field, thus one can see how the magnetic free energy is accumulated through the dynamic response of the coronal field to the photospheric field evolution.

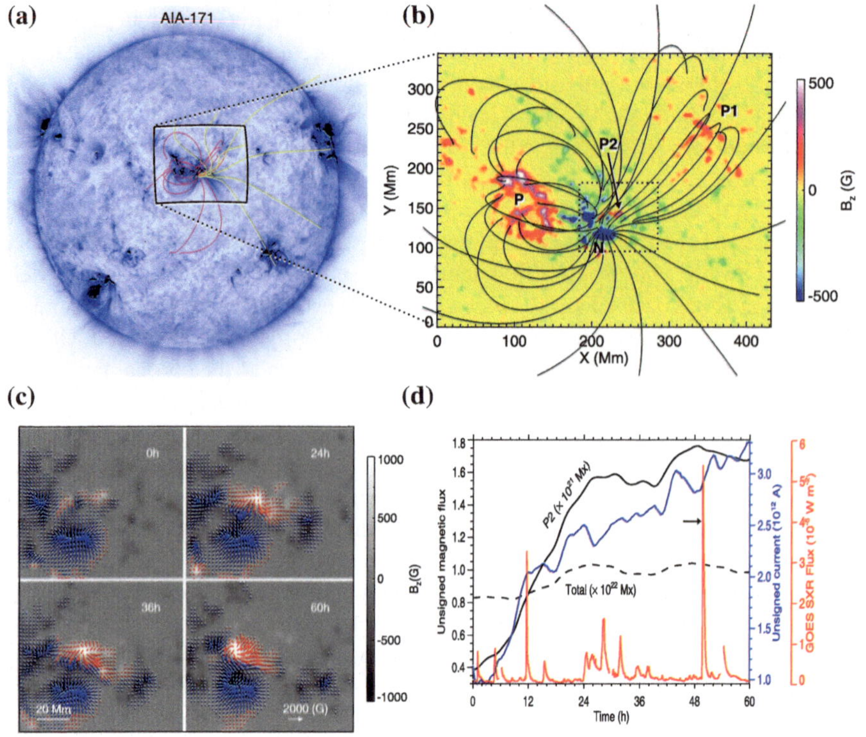

Fig. 9.11 **a** A full-disk SDO AIA 171 Å image of the Sun observed near the end of 2011 September 5 (day 2). Overlaid on the image are selected magnetic field lines of global potential-field-source-surface model with the pink (yellow) color denoting closed (open) flux. **b** Magnetic environment associated with the flux-emerging region denoted by the dashed box. The magnetic field lines as shown are calculated by the potential field model in a local Cartesian coordinate system. The background image is the map of photospheric magnetic flux with the main polarities labeled as N, P, P1, and P2, where P2 is the newly emerging one and surrounded by negative flux of N. **c** Evolution of magnetic flux distribution B_z and transverse magnetic components (B_x, B_y) shown by the arrows, which are colored as red (blue) for regions of positive (negative) flux. Transverse field less than 100 G is not shown. The field of view (FoV) for the selected region is displayed by the dashed box in panel **b**. Time starts from 00:00 UT on 2011 September 4. **d** Evolution of unsigned magnetic flux for the emerging positive polarity P2 (black solid line) and the whole region shown in panel **a** (black dashed line). The blue line shows unsigned electric current crossing the photospheric surface of P2, which indicates an increase of the non-potentiality of the coronal magnetic field. The GOES soft X-ray (SXR) flux is also shown as the red line. The arrow denotes the M5.3 flare produced by the FER, while the preceding flares recorded are not related with this region. Adapted from Figs. 1 and 2 of Jiang et al. [35] licensed under CC BY 4.0 available at https://creativecommons.org/licenses/by/4.0/

The magnetic configuration involved with the flare is rather complex (see Fig. 9.11), consisting of two major polarities and a remote one at slightly far away from the AR. Some part of the flux near the flare core region even opens to the interplanetary space, making the topology more complex. The evolution of the photospheric magnetic field

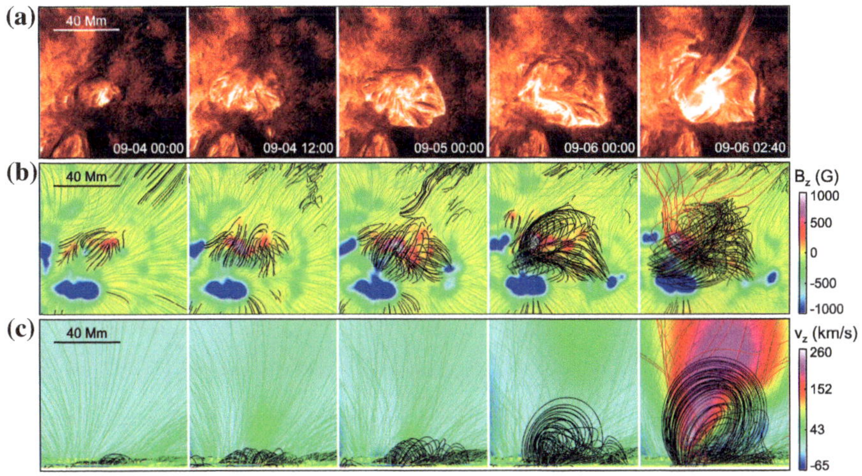

Fig. 9.12 Comparison of the simulated coronal magnetic field of the flux emerging region in AR 11283 with the SDO/AIA observations. **a** AIA 304 Å images at different times from the initial emergence to the eruption. **b** Top view of the corresponding magnetic field evolution at different times ($t = 0$, 12, 24, 48, and 57) from the MHD model. The field lines are traced from footpoints evenly distributed at the bottom surface, which is shown with the photospheric magnetic flux map. Field lines closed (open) in the box are colored black (green), while those becoming open from the closed during the eruption are colored red. **c** Side view of the magnetic field lines from south (i.e. the horizontal and vertical axes are x and z, respectively). The background shows a 2D central cross section of the 3D volume and its color indicates the value of vertical component of velocity. Taken from Fig. 5 of Jiang et al. [35] licensed under CC BY 4.0 available at https://creativecommons.org/licenses/by/4.0/

in the studied interval is dominated by a parasitic positive polarity (P2) emerging into N (Fig. 9.11). New flux is injected mainly in day 1, then followed by a fast shearing motion of P2 with respect to N. Interestingly, the flare emission consists of a quasi-circular ribbon [83] enclosing the newly-emerged polarity P2 and two small remote brightening patches outside of the circular ribbon, one at polarity P and the other at P1. The apparent path of these filament ejections is nearly linear without twist or rotation and it co-aligns well with that of the open flux from N, suggesting that the open flux might be involved with the eruption and provides a channel for the escape of the ejecta.

Since the initial field shows no signature of non-potentiality, Jiang et al. [35] simply started the simulation with a potential field extrapolation [63] from the vertical component of the photospheric field. The simulated magnetic field and its evolution show impressive resemblance in morphology to the AIA observation (Fig. 9.12). With the magnetic topology analysis, Jiang et al. [35] also confirms that the emergence of magnetic polarity P2 results in a topological separatrix like a closed dome separating the new emerging flux from the pre-existing one. As can be seen, the closed field lines with connection to the newly-emerging polarity are encircled within the separatrix, while outside of it are pre-existing field lines, and with the increasing of the new flux

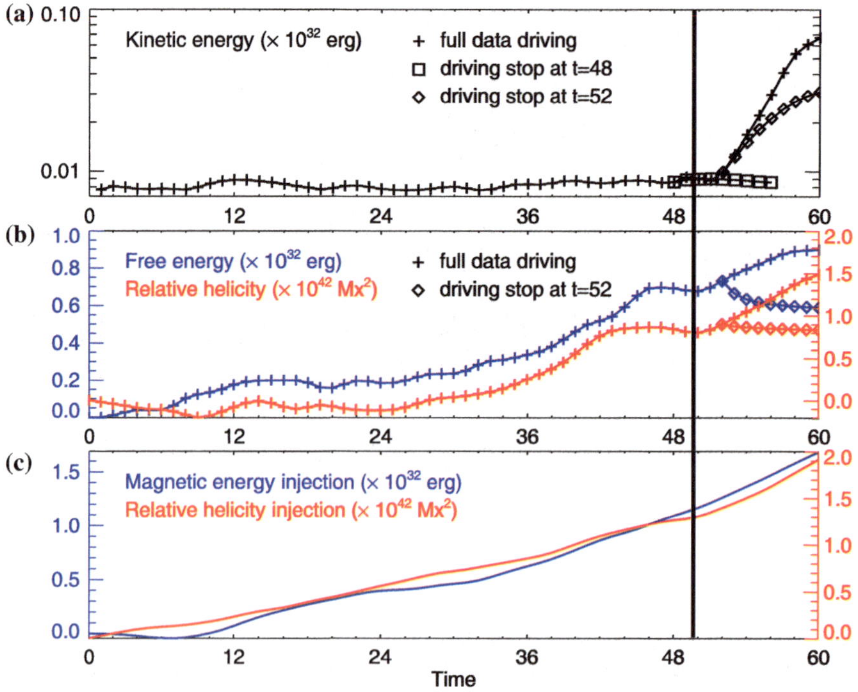

Fig. 9.13 Temporal evolution of energy and related quantities derived from the data-driven MHD simulation for the eruptive flare in AR 11283. **a** Total kinetic energy. In addition to the run with bottom driving supplied all the time, two experiments are carried out: one (the other) with driving ended at $t = 48$ ($t = 52$), shortly before (after) the eruption onset. **b** Free magnetic energy and relative magnetic helicity. **c** Estimated amount of injected magnetic energy and helicity through the bottom boundary. The vertical dashed line through the figures denotes the start time of the GOES flare. Taken from Fig. 4 of Jiang et al. [35] licensed under CC BY 4.0 available at https://creativecommons.org/licenses/by/4.0/

the separatrix expands in both area and height. Quiescent magnetic reconnection should occur at the separatrix for the successive replacement in the corona of the old flux with the new one [71]. Probably as a result of heating by such reconnection, the separatrix location is manifested in the EUV image (AIA 304 Å) as a bright kernel expanding with time (Fig. 9.12a). Such a distinct evolving feature is usually observed when new flux is emerging into a region of opposite polarity [54, 68, 71].

The growing shear of the coronal field is well recovered. As can be seen along the south part of photospheric polarity inversion line (PIL) separating P2 and N, the chromospheric filament threads (and the corresponding magnetic field lines) become more and more co-aligned with the PIL. Such stressing of the magnetic field corresponds to the continuous increasing of its free energy and relative helicity (Fig. 9.13). It is noted that magnetic flux rope is not produced in the model as the magnetic twist number of the core field is mostly less than a half turn.

One noteworthy success of the model is that the timing of the simulated eruption matches the observed flare time within less than 2 h, which provides a strong support that the model correctly captured the key transition of dynamics from pre-eruption to eruption. The onset of the simulated eruption is characterized by a rapid exponential growth of the system kinetic energy, whose evolution consists of two distinct phases (see Fig. 9.13a), a quasi-static evolution phase (from $t = 0$ to $t = 51$) and an eruption phase (after $t = 51$). In the first phase, the coronal MHD system evolves stably in response to the changing of the photospheric field. The kinetic energy keeps a rather low value (compared with the magnetic energy) without noticeable variation. On the other hand, there is a continuous injection of magnetic energy through the bottom boundary derived from the Poynting flux. Most of this added energy goes to the non-potential energy (i.e. the free energy), especially on the second day when the fast shearing motion of the emerging polarity commences. During this phase, the free magnetic energy, which can be used to power eruptions, is accumulated to an amount close to 10^{32} erg.

From the time $t_c = 51$, the kinetic energy begins to rapidly rise resembling an exponential growth, and within a short time interval from $t = 51$ to 60, it increases by about one order of magnitude. This clearly indicates that the system runs into a loss of quasi-equilibrium, i.e. a fast eruptive state, which is confirmed by tracking the evolution of magnetic field configuration (Fig. 9.12). Note that even through the eruption, the magnetic free energy keeps increasing due to the un-interrupted injection of energy into the volume. Additionally, we carried out two experimental runs of the model (Fig. 9.13a), the first (second) with the photosphere driving ended shortly before (after) t_c. In the first run the kinetic energy decreases eventually without any sign of eruption, while in the second run it evolves similarly as in the case of full-time driving, indicating that the eruption can only occur with the data driving supplied beyond the critical time point t_c. In the second run, the magnetic free energy drops as expected during the eruption (Fig. 9.13b). The released magnetic energy is on the order of 10^{31} erg, which is comparable with the energy budget for M-class flares [62].

Besides the free energy, the relative magnetic helicity is also an important indicator of the non-potentiality of the magnetic field by quantifying the magnetic twist and writhe [6]. Figure 9.13b shows that the relative helicity evolves in a similar way as the free energy because of the similar injection of helicity flux through the bottom boundary (Fig. 9.13c). Some observational and theoretical studies [12, 29] indicate that there is a threshold (0.25 ± 0.05) for the ratio of relative helicity to the square of total magnetic flux, and eruption seems to occur only when this threshold is exceeded. The estimated value of this ratio near the eruption is about 0.01, which is far below the aforementioned threshold. Such inconsistence might suggest that here the eruption is not directly driven by magnetic twist (or flux rope), and an analysis of the specific topology is required.

Jiang et al. [35] further provides an explanation from the point of view of magnetic topology on how the eruption is triggered and initiated. It is found that a critical change of the topology from a bald-patch type to a X-line type plays a key role. Such finding relies on calculation of the squashing degree (Q) of the field lines,

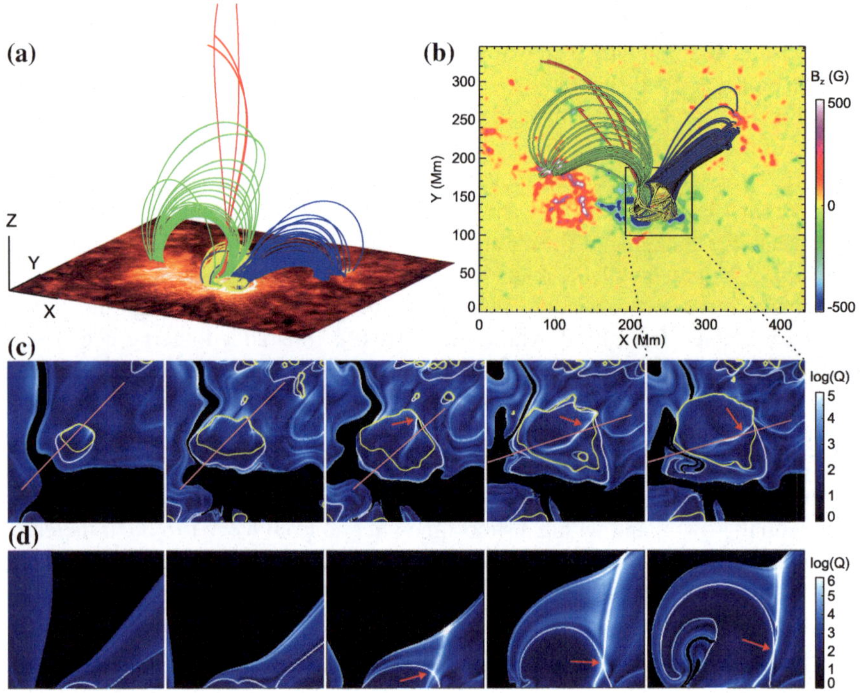

Fig. 9.14 Magnetic topology evolution of AR 11283. **a** Sampled field lines traced from topology separatrix at modeling time $t = 57$ (with different colors denoting different connectivity as shown). They represent the field lines undergoing reconnection during the flare. At the bottom is AIA 304 Å image near the flare peak time to show the flare ribbons. It can be seen that the locations of flare ribbons are matched well by those footpoints of the reconnecting field lines. **b** Top view of the field lines with the background image showing the photospheric magnetic flux. Extents shown in **a** and **b** are identical. **c** Magnetic squashing degree $\log(Q)$ at the bottom surface at different times ($t = 0$, 12, 24, 48, and 57). The separatrix is distinctly revealed by the quasi-circular narrow line with $\log(Q) > 5$. The black regions represent footpoints of open flux. The yellow lines are PILs. The arrows denote the newly-formed QSL. **d** Squashing degree maps for vertical cross sections whose locations are denoted by the oblique lines shown in panel **c**. The arrows denote the X-like configuration that is formed along with the new QSL. Taken from Fig. 6 of Jiang et al. [35] licensed under CC BY 4.0 available at https://creativecommons.org/licenses/by/4.0/

which is useful to locate the important topological structures like separatrices and quasi-separatrix layers (QSLs) [11, 75].

Specifically, Fig. 9.14 shows the evolution of the Q map. Initially the separatrix surface between the emerging flux and pre-existing one is simply a "bald-patch" type [74], as for the field lines that form the separatrix surface, each has one point touching tangentially with the bottom surface. All these points of tangency form a special part of the separatrix at the bottom surface where it coincides with the PIL (see Fig. 9.15a), and near there its vertical cross section demonstrates a U shape. With the growing of the newly-emerging flux system, part of its edge gets into contact with that of the open

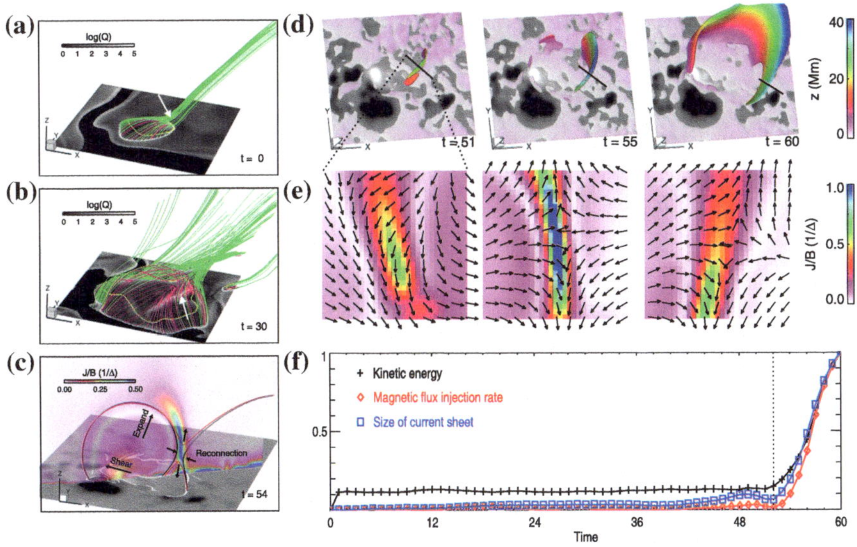

Fig. 9.15 Formation of a jet-like reconnection structure and current sheet development in the reconnection site. **a** Local magnetic topology at $t = 0$ for the newly-emerging polarity P2. Squashing degree map is shown on the bottom surface. Magnetic field lines that form the magnetic topology separatrix surface are traced from the circular line with $\log(Q) > 5$. The PIL at the bottom is shown by the yellow line. As indicated by the arrow, these field lines become tangential to the bottom surface at the locations where the PIL coincides with the separatrix. **b** Same as **a** but for $t = 30$, when the new QSL (marked by the arrow) has formed. The closed magnetic field within the separatrix is colored in red. **c** Illustration of the jet-like reconnection as eruption trigger mechanism (simulation time at $t = 54$). The transverse size of the bottom surface is $140 \times 120\,\mathrm{Mm}^2$. Field lines in black (red) denote magnetic flux before (after) reconnection. Large arrows denote the bottom shearing and the resulting expansion of the closed arcade. Small arrows indicate the inflow and outflow at the reconnection site. The bottom surface shows the map of B_z overlaid by the white lines showing the trace of separatrix and QSL ($\log(Q) > 5$). The vertical cross section false-colored by value of J/B shows distinctly a current sheet at the reconnection site. Note that the reconnecting field lines are not coplanar, thus the configuration is fully 3D. **d** 3D shape of the current sheet. It is defined as the region with $J/B > 1/(2\Delta)$ (where Δ is the minimal grid size in the model), which consists of intense current layers with width of $\sim \Delta$, thin enough for resistivity to take effect [21]. Its color denotes the height z from the bottom, and the bottom surface is shown with the photospheric magnetogram. **e** Flow directions at a vertical cross section of the current sheet, whose horizontal extent is denoted by the short line in **a**. Reconnection inflow and outflow can be clearly seen after $t = 51$. **f** Evolution of the size of the current sheet, magnetic flux injection rate (defined by $\oint_S |\mathbf{B}|\mathbf{v}d\mathbf{S}$ where $\mathbf{v}$ is plasma velocity and S is the full surface of the current sheet) compared with that of the kinetic energy. All are scaled by their values at $t = 60$. Adapted from Figs. 8 and 9 of Jiang et al. [35] licensed under CC BY 4.0 available at https://creativecommons.org/licenses/by/4.0/

flux (Figs. 9.14c, d, see the changes from $t = 0$ to 12), and the Q maps show that a new QSL is created after around $t = 24$ during the emerging process. The new QSL makes the topological surface as a mixed type of a bald-patch and an X-line configuration (see Fig. 9.15b), of which the vertical cross section appears as an X-shaped structure similar to the topology at a magnetic null point (Fig. 9.14d). The emergence of such

X-line structure provides a favorable configuration for reconnection in the corona. The later development of the shearing of the core field increases magnetic pressure and makes its overlying field expand towards the north in the environment of highly asymmetric magnetic flux distribution. This results in a jet-like configuration (as seen in Fig. 9.15c), in which reconnection can occur between the newly-emerged outer arcade (connecting P2 and N) with the side flux of much longer connection paths to polarities P and P1 and even some open flux. Once the reconnection sets in, a positive feedback is naturally established between the reconnection, which reduces the inward magnetic tension force that confines the flux below, and the consequent outward expansion of the closed arcades, which in turn enhances the reconnection. Such a mechanism is essentially in correspondence with the breakout eruption model [5].

Unlike the static extrapolation methods, the MHD model can self-consistently reproduce electric current sheets, which spontaneously form, and study of current sheets provides a direct pinpoint of reconnection. Jiang et al. [35] shows that a current sheet is built up at the X-line slightly before the eruption and grows impulsively, extending to almost the whole separatrix surface during the eruption (Fig. 9.15d). As a result, the global magnetic topology is fully involved in the reconnection (Fig. 9.14a, b), which provides a plausible explanation of why there forms the circular flare ribbon and additionally the two remote flaring patches. The linear ejection of filaments around the flare ribbon is most likely a result of the opening of the overlying magnetic field, which is reasonably shown by the model. As can be seen in Fig. 9.12, the field lines whose color changes from black to red denote the flux becoming open from closed configuration during the eruption. This is also reflected in the map of squashing degree that shows 'holes' corresponding to the open flux cutting into the closed circular separatrix.

To characterize how fast this reconnection occurs in the simulation, Jiang et al. [35] further estimate its size as well as the rate of magnetic flux injection into the current sheet (i.e. the reconnected magnetic flux). It is found that the rate of reconnection is temporally coupled with the acceleration of the plasma (Fig. 9.15f), clearly indicating the positive feedback between the reconnection and field expansion.

9.4.3 X-Class Confined Flares in Super AR 12192

Another noteworthy application of the DARE–MHD model is the study of the mechanism of a major X-class solar flare that occurred in the super AR 12192, the largest one of all ARs during the last 25 years. This region produced a series of X-class flares but none of them results in CME, and the strongest one in the series reaches X3.1, which sets a record in the flare energy for CMEless events [72] since the confined flares ever observed were predominantly below X-class [93]. A series of studies attempt to explain why these extremely powerful flares are confined [8, 27, 39, 70] according to the theory of torus instability, and they find that the overlying strapping field is significantly stronger than that for triggering torus instability. However, the

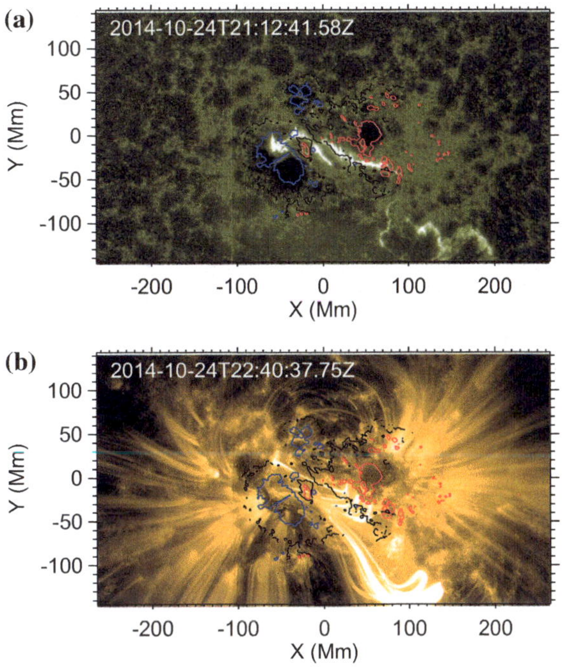

Fig. 9.16 Overview of the X3.1 flare occurring in AR 12192. **a** AIA 1600 Å image of the flare ribbon and **b** AIA 171 Å image of the post-flare loop. Both images are CEA re-mapped with field of view identical to that of the SHARP CEA magnetogram for this AR. Overlaid contours indicate the vertical component (B_z) of the photospheric magnetic field at the same time (red for 1000 G, blue for −1000 G, and black for the PIL with the total field strength above 200 G). Taken from Fig. 1 of Jiang et al. [36] with permissions of the authors and the AAS

reason why and how these flares occur, or more specifically, it is not clear from those studies how the magnetic fields underlying these non-eruptive flares in AR 12192 do evolve. When inspecting the SDO/AIA images one only see a series of brightening of coronal loops without much changes in their shape. The flare loops are seen to connect the boundaries of the strong magnetic polarities. Interestingly, there is also a set of rather long loops with remote connection to the southwest corner of the field of view as shown [e.g. 8, 70]. Chromospheric ribbons of the flare (Fig. 9.16a) consist of mainly two bands on both sides of the central part of the PIL. Distinct from typical two-ribbon flares, these two ribbons showed barely separation motion. As such, the coronal configuration changes are not easy to interpret from these EUV observations.

Jiang et al. [36] carried out a MHD–DARE simulation of the region starting from a NLFFF extrapolation of vector magnetogram on two days before the X3.1 flare. It is shown that in the core region a narrow layer of intense electric current extends vertically ($\sim$ 35 Mm) above along the central part of the PIL separating the two major polarities of the AR. Such an association of intense current layer with PIL of AR core seems to be common for flare-productive ARs [e.g. 70]. With the magnetic evolution driven by the photospheric magnetograms, the current layer was successively enhanced until it became so thin as a current sheet such that a tether-cutting reconnection between the sheared magnetic arcades was set in, which can be responsible for the flaring process, and after the reconnection, the current is significantly dissipated.

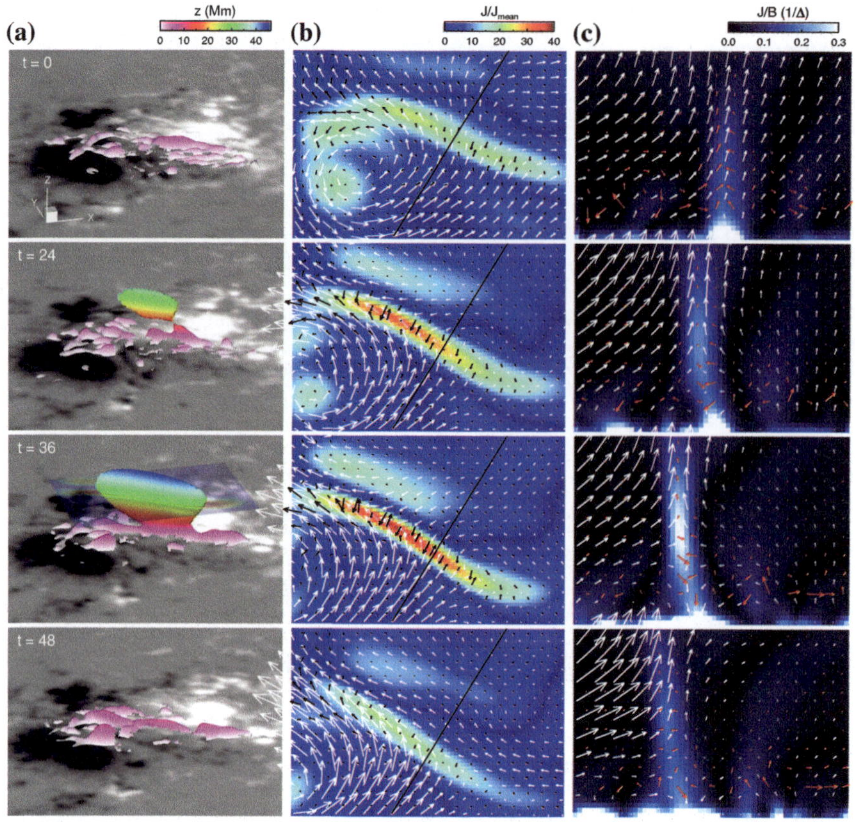

Fig. 9.17 Evolution of the electric current in the MHD model. **a** 3D shape of the CS (defined in the text) at four different times from the beginning to the end of the simulation. The color denotes the height from the bottom surface, which is shown by the photospheric B_z map. **b** A horizontal slice of the volume at $z = 28$ Mm. Its position is indicated by the transparent image shown at $t = 36$ of **a**. The color shows electric current density J (scaled by the average value J_{mean} at $t = 0$), and the arrows show plasma velocity vectors (white arrows) as well as Lorentz force vector (black arrows). The time is the same as **a** from top to bottom. **c** A vertical slice whose horizontal location is denoted by the black inclined line in **b**. Vertically, it extends from the bottom to $z = 57$ Mm. The color shows the value of J/B and the arrows represent the velocity (white) and Lorentz force (red). Note that in **b** and **c** each pixel represents a computational grid. *An animation for the 3D CS evolution and a horizontal slice is provided at* http://iopscience.iop.org/0004-637X/828/1/62/suppdata/apjaa3119f4_video.mp4, *in which the time step is* $\tau/5$. Taken from Fig. 4 of Jiang et al. [36] with permissions of the authors and the AAS

The formation and dissipation of the current sheet is depicted in Fig. 9.17 and quantitatively shown in Fig. 9.18. There is a successive formation, expansion and shrinkage of the current sheet as seen. At the beginning, the initial current layer is gradually squeezed from both two sides by the magnetic stress, which is shown by the plasma flows and the Lorentz force vectors around the current layer. Consequently

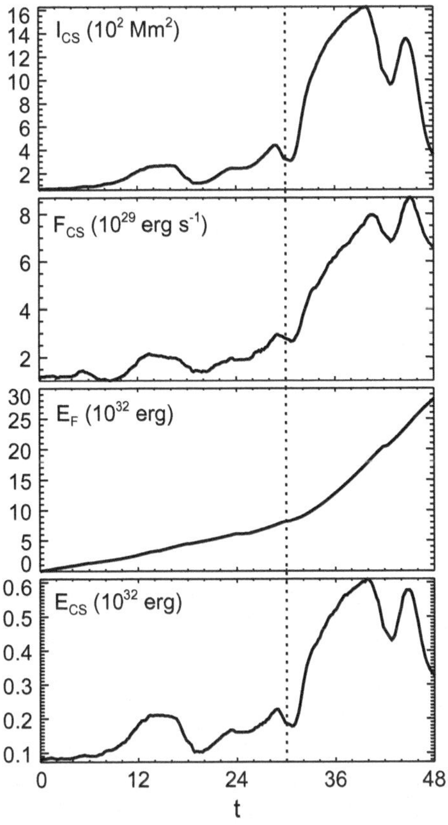

Fig. 9.18 Temporal evolution of different parameters associated with the CS which is defined by the volume V with $J/B > 0.2\Delta x$. **a** The intensity of the CS, defined by integrating J/B for the full volume V of the CS, which gives a value $I_{\rm CS} = \int J/BdV$ in unit of area. **b** Rate of magnetic energy injection into the CS, expressed by $F_{\rm CS} = -\int_S \mathbf{P}dS = -\int_V \nabla \cdot \mathbf{P}dV$, where $\mathbf{P} = \frac{1}{4\pi}\mathbf{B} \times (\mathbf{v} \times \mathbf{B})$ is the Poynting flux vector with **v** the plasma velocity, and S is the surface area of the CS volume V. **c** Total magnetic energy injected into the CS with time, i.e. $E_F = \int_0^t F_{\rm CS}dt$ at time t. **d** Total magnetic energy reserved in the CS volume, i.e. ($E_{\rm CS} = \int_V \frac{B^2}{8\pi}dV$), so the happening of reconnection can be indicated if the magnetic energy stored in this volume is less than those injected into it. The dashed line denotes $t = 30$. Taken from Fig. 5 of Jiang et al. [36] with permissions of the authors and the AAS

the current layer becomes thinner and more intense, leading to the increase of the value J/B and thus the formation of the CS, which is defined as $J/B > 0.2\Delta x$ [21, 36]. At the time of $t = 36$ (close to the peak stage of the CS development, see below), the CS extends from the bottom of the simulated domain up to heights of 50 Mm. With the slow evolution of the basic configuration, the CS also moves slowly from north to south, but overall its location is roughly the same in the whole period. Pattern of reconnection-like plasma flow, i.e. horizontal inflow at both sides of the CS and vertical outflow to up and down, can be seen (Figs. 9.17b, c, $t = 24$ and 36).

The quantitative measurement of the simulated CS shows more clearly that its evolution consists of two phases of distinct behaviors: in the first phase from $t = 0$ to approximately 30, it keeps a relatively small value (~ 200 Mm^2); in the second phase from $t = 30$ to the end of the simulation, it first increases impulsively and reaches the peak value of 1600 Mm^2 at nearly $t = 40$, and then decreases rapidly to a value similar to that in the first phase. The evolution of the rate of magnetic energy injected into the CS, $F_{\rm CS}$, shows a similar trend. In the later phase (i.e. from $t = 30$ to the end), the total magnetic energy input is about 10^{33} erg, while the magnetic energy in the CS volume is smaller by nearly two orders in magnitude and can be negligible. This indicates that the amount of the magnetic energy injected into the CS is mostly released via reconnection (converted into other forms), and the energy release rate can be approximated by $F_{\rm CS}$. Such an impulsive increase and decrease of the energy release rate in the CS can be interpreted as a simulated "flare" process, which releases magnetic energy of about 10^{33} erg by a rate of $\sim 10^{29}$ erg s^{-1} (the rate can reach an order higher in the peak time). Such amount of energy and conversion rate appears to be an order higher than typical-size ARs, e.g. AR 11158 [70]. On the other hand, the first phase ($t = 0$ to 30) can be regarded as a quasi-static evolution duration for which the time unit τ can be scaled as being one hour. This means that the simulated flare ($t = 30$) began 15 h ahead of the real X3.1 flare ($t \sim 45$), and the model fails to match the flare time. Such mismatch of the model with the reality is in a large part due to the over-simplification of the magnetic reconnection process, which is much more complex since it is related to the microscopic behavior of plasma. In the MHD model, the reconnection is simply resulted by numerical viscosity in the CS region, and its behavior depends on the numerical aspects of the model. For example, the thickness of the modeled CS and the numerical viscosity are often sensitive to the grid resolution (as well as the specific numerical scheme). A much thinner CS can develop if using a smaller grid size, and the onset of reconnection in the CS might be postponed. Namely, with a smaller grid size, a CS can sustain even stronger current and thus even larger gradient of magnetic field, which might needs more time to form. Further experiments using different grid resolutions will be required to quantify this effect.

Nevertheless, comparison with the AIA observations shows that the model almost reproduced exactly the location of the chromospheric flare ribbons, and the morphology of the reconnecting field lines and the simulated EUV image resembles well the flaring coronal loops (Fig. 9.19). The chromospheric flare ribbons are recognized to be an indicator of the footpoint locations of magnetic field lines that underwent reconnection during flares [60], because the non-thermal particles accelerated at the reconnection site travel down along these field lines toward the photosphere and collide with the dense chromosphere to cause enhanced heating [61]. By tracing magnetic field lines that are in contact or pass through the reconnecting CS during its peak phase (see Fig. 9.19a), Jiang et al. [36] locates the footpoints of these field lines that were undergoing reconnection or formed immediately after reconnection. It can be seen that the footpoints of these field lines, as mapped on the photosphere,

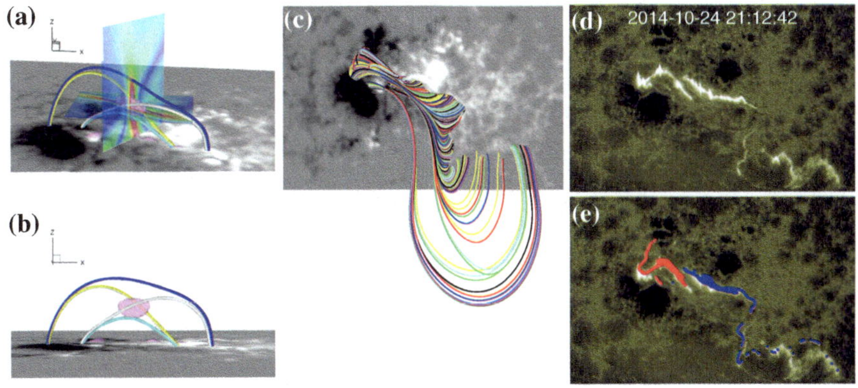

Fig. 9.19 Illustration of the tether-cutting reconnection process in the model. **a** Sampled field lines and horizontal and vertical cross sections of the reconnection site. The white and yellow curves represent the "before" flare field lines and blue and cyan curves show the new reconnected field lines. The pink colored area shows the iso-surface of $J/B = 0.3/\Delta x$, which is sandwiched between the pre-reconnection field lines at their crossing point. **b** Side view of the sampled field lines. **c** The reconnecting field lines (modeling time at $t = 40$). They are shown with different colors for better discrimination of each field lines. **d** AIA 1600 Å image of the chromospheric flare ribbons. **e** Comparison of the footpoints of reconnecting field lines in the model with the observed flare ribbons. Footpoints in positive (negative) magnetic flux are shown with blue (red) color. Adapted from Figs. 6, 8, and 9 of Jiang et al. [36] with permissions of the authors and the AAS

appear to form two curved narrow areas separated by the central PIL (Figs. 9.19a). A strikingly good match in both the location and the shape of these simulated footpoint areas with the observed chromospheric flare ribbons is evident (Figs. 9.19b,c), including even the relatively weak ribbon extending far in the southwest direction. Such a precise match of the flare ribbon outperforms the QSL method that has been proposed to locate the flare ribbon [64] from static extrapolations such as NLFFF. It should be noted that the QSLs are only possible sites for reconnection, while they cannot tell where the reconnection occurs specifically in a flare. Moreover, by inspecting the map of QSLs, one usually sees much more complex structures than those of target flare ribbons. Thus without knowing ribbon locations in advance, it is still problematic to identify from all the QSLs the particular flare-related one. The MHD model, on the other hand, can simulate the flare reconnection process in a self-consistent way and directly identify the reconnecting field lines, and thus is able to precisely reproduce the flare ribbon locations. Furthermore, Jiang et al. [36] produced a synthetic EUV image using a method similar to that used by Cheung and DeRosa [9], which presents a good morphological similarity to the AIA-94 images of the hot flaring loops (Fig. 9.20).

The MHD–DARE simulation provides direct explanations why the flare failed to erupt, in addition to analysis of the magnetic decay index [27, 39, 70]. As can be seen from a direct look of the field lines (see Fig. 9.19), the pre-flare magnetic arcades are twisted (or stressed) by a rather weak extent, which is consistent with a

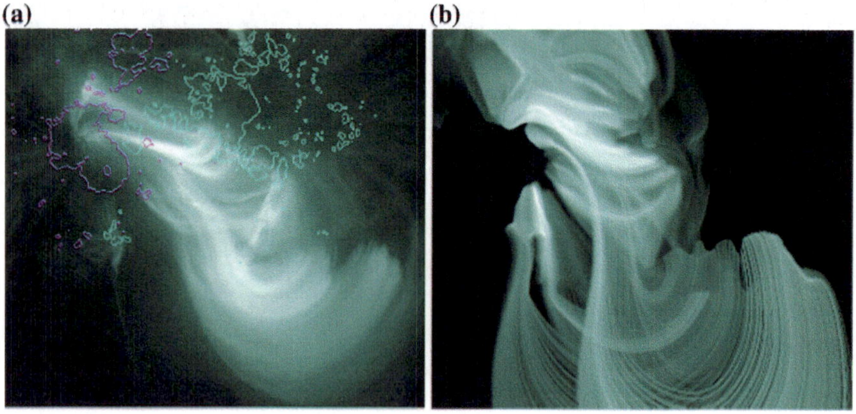

Fig. 9.20 Comparison of the AIA-94 image of the flare (**a**) with a synthetic EUV image of the flaring loops based on the modeled magnetic field and currents (**b**). The synthetic image is generated by first tracing a large number of field lines from their footpoints uniformly distributed at the bottom. Then each field line is assigned with a value of emission assumed to be J/BJ^2 at the peak value point of J/B along the field line, where J/B can be regarded as the dissipation rate of current, and by selecting its maximum, the emission from the reconnecting field lines is emphasized. Finally the total emission is obtained by integration through the volume along the line-of-sight (here simply along the z-axis), and forms the synthetic image. Adapted from Fig. 9 of Jiang et al. [36] with permissions of the authors and the AAS

quantitative study of the magnetic twist by Inoue et al. [27] for the same flare event using a NLFFF model. They found that the magnetic twists before the flare is mostly less than a half turn. Consequently, the sheared magnetic arcades do not show a well-formed two-J shape like those in many sigmoid ARs with eruptive events. Inoue et al. [27] also showed that after the flare, the magnetic twists are almost reserved without release. This is consistent with our modeled results that the post-reconnection field lines are still sheared arcades, without forming an escaping magnetic flux rope. In our simulation, the reconnected long field lines on top of the CS only expand slightly without propagating further to make the overlying arcades open. Thus, in the context of tether-cutting scenario, only the first-stage of tether-cutting occurred, while the second stage, the formation of a flux rope and reconnection below the flux rope, did not happen.

The MHD simulation also provides an insight in explaining another unusual fact of the flare that the enhancements of the horizontal field at the photosphere is significantly weaker [70], than in many other large flares [58, 81, 82, 84]. In the proposed coronal "implosion" mechanism [24], such enhancement of photospheric field is due to the downward contraction of the short magnetic arcades formed after the reconnection and their push on the photosphere. As found in the simulation of Jiang et al. [36], the reconnection site is rather high above the photosphere (see Fig. 9.19), the altitude of the reconnection point is $\sim$30 Mm), which is consistent with the result of Thalmann et al. [72] from the study of the limited variation of the flare-ribbon separations. Thus, the reconnected short field lines below the CS expanded still rather highly in the corona. So during its contraction, the effect of its push on the bottom might be reduced significantly by the strong corona field below. Thus the large altitude of the reconnection sites and the long extension of the reconnected arcades provide a plausible explanation for the weak "implosion" effect. Such explanation is yet to be confirmed by a further quantitative analysis of the MHD simulation data.

Comparative study of this flare with eruptive one of AR 11283 in Sect. 9.4.2 may offer insight into the different magnetic natures of eruptive and confined flares. For the eruptive event, the eruption onset time is matched by the model much better, with less than a time lag of 2τ during a whole simulated period of 60τ. Unlike this event, the simulated eruption in AR 11283 is due to the formation of a jet-like magnetic configuration that favors breakout-like reconnection [5]. So in that event, the critical condition causing eruption is the formation of a reconnection-favorable magnetic topology, which can be regarded as a macroscopic behavior. Whereas in AR 12192, the cause of its flare depends more on the triggering of the reconnection rather than the formation of a favorable topology, since the basic topology is simply a configuration of sheared arcades and it does not change for days. In other words, the cause of this flare relies more on the microscopic behavior of the plasma and thus is more subtle. The MHD model seems to be able to characterize well the macroscopic structures and evolutions, while it may not appropriately simulate the microscopic aspects of the plasma, and thus, the triggering of the reconnection. Such difference might be common and even fundamental between eruptive flares and confined flares, in particular, those events without noticeable changes of the coronal structures.

Finally, the reconnection-related quantities derived from the model should be taken with cautions because of the simplification of reconnection. A deeper understanding of the flare dynamics requires knowledge of the microscopic nature of magnetic reconnection. In addition, the quantitative results could suffer uncertainties from several aspects including HMI data and model settings. For example, the evolution speed of the photospheric magnetic field is increased in the model for saving computing time, and this might affect the coronal evolution, most likely at the time around the flare.

9.4.4 Formation and Eruption of a Small Flux Rope Driven by Flux Emergence

The DARE-MHD model was further applied to support the observations of a flux emergence event leading to flux rope formation and eruption [92]. Using high spatial and temporal data from New Solar Telescope (NST) and SDO, Yan et al. [92] observed the emergence of a small-scale flux rope near a large sunspot that triggers a M-class flare and a CME. It seems that the observations match many features predicted by theoretical simulations of emerging flux tubes from below the convection zone to the corona [23]. One footpoint of the observed flux rope roots in a newly formed mini-sunspot, and during the emergence process, clockwise rotation

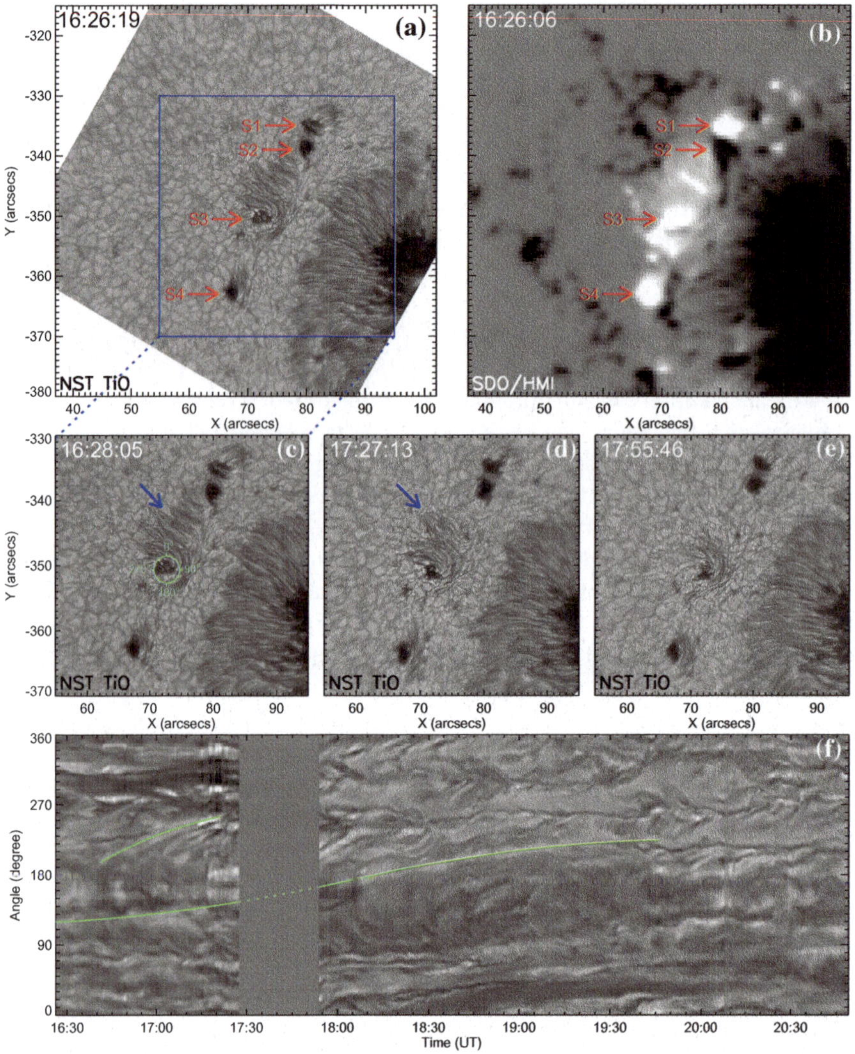

Fig. 9.21 Evolution of small satellite sunspots in active region NOAA 12403. **a** The NST TiO image showing the appearance of the small satellite sunspots indicated by the red arrows. **b** The corresponding HMI magnetogram. **c–e** The change of the small satellite sunspots in TiO images. The change of the penumbra between S2 and S3 is indicated by the blue arrows. The green circle is the position of the time slice. **f** Time-slice plot acquired along the green circles marked in panel **c**, and the green curved lines indicate the rotation of the penumbra filament around the center of the sunspot S3. Taken from Fig. 4 of Yan et al. [92] with permissions of the authors and the AAS

is clearly observed for the sunspots (see Fig. 9.21). The current density calculated from the transverse magnetic fields exhibits two concentrated regions, suggesting the existence of a volumetric channel of current in the corona. Further evidence

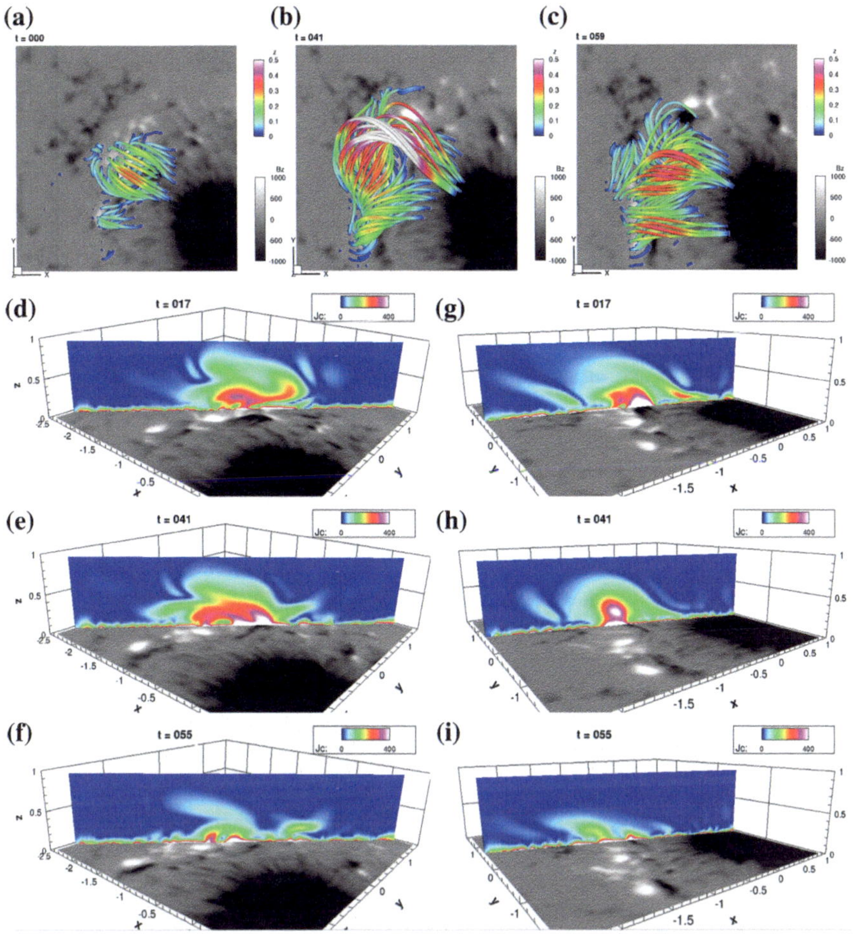

Fig. 9.22 Data-driven MHD simulation of flux rope emergence and eruption. **a**–**c**: The emergence of the flux rope; **d**–**f**: the current along the flux rope; **g**–**i**: the current perpendicular to the flux rope. Taken from Fig. 7 of Yan et al. [92] with permissions of the authors and the AAS

for the formation of such flux rope is the gradual increasing of the electric current, the Lorentz force, and transverse magnetic fields. The DARE simulation provides coronal magnetic field evolution matching these high resolution observations (Fig. 9.22). It is clear that the magnetic field emerges in the course of the simulation from t = 0 to 41, as driven by the evolution of the photospheric field. There are magnetic field lines with twist and shear during the emergence of the field, which constitute the flux rope. Furthermore, the current increases rapidly along the flux rope and the rising of the flux rope can also be seen from the cross-section of the current perpendicular to the axis of the flux rope. At around t = 55, both the twisted magnetic flux and the corresponding current density decreased quickly, indicating the eruption of the flux rope that releases some of the non-potential energy of the pre-eruption field, after which the magnetic structure is approximately potential. These observation and simulation confirm the existence of flux rope in the solar eruptions [90].

9.5 Discussions

The physics of solar eruptions is extremely complicated due to their underlying complex and elusive magnetic topology. While theoretical and idealized models are challenged by many new observations, the data-driven techniques provide a viable way of revealing the complex structure of solar active regions and their eruptions. Further accompanied with the advanced analysis tools such as the QSL method, the magnetic configuration can be studied in an unprecedentedly accuracy, and thus observation features can even be "reproduced", for example, formation of flux rope, locations of flare ribbons, hot flare loops as well as their dynamic evolutions. Based on these realistic reproductions of the coronal magnetic field and its dynamic evolution, the scenario of how solar eruptive structures are formed and initiated can be investigated much easier and faithfully than that obtained by using observations only or combined with static extrapolations.

Nevertheless, it should be noted that the current models are still over simplified in handling the plasma. For instance, the assumption of setting the model's lower boundary at the coronal base is rather crude because the input magnetograms are actually measured at the photosphere. To characterize more realistically the solar atmosphere, it is necessary to include a plasma with temperature stratified from the photosphere level (of $\sim$ 5000 K) to the corona ($\sim$ 1 MK) in the model. By modeling the solar atmosphere of the full domain from the thin layers of photosphere, chromosphere and transition region until to the corona, we can then describe self-consistently the behavior of the magnetic field in a highly stratified plasma with β from > 1 to $<< 1$. The thermodynamics should be also improved such as including thermal conduction, coronal heating source, as well as radiation loss, by which the resulted plasma emission can hopefully be directly compared with the EUV images.

References

1. Amari T, Aly JJ, Luciani JF, Boulmezaoud TZ, Mikić Z (1997) Reconstructing the solar coronal magnetic field as a force-free magnetic field. Sol Phys 174:129–149
2. Amari T, Luciani JF, Aly JJ, Mikic Z, Linker J (2003) Coronal mass ejection: initiation, magnetic helicity, and flux ropes I. Boundary motion-driven evolution. Astrophys J 585:1073–1086. https://doi.org/10.1086/345501
3. Amari T, Canou A, Aly JJ (2014) Characterizing and predicting the magnetic environment leading to solar eruptions. Nature 514(7523):465–469
4. Amari T, Canou A, Aly JJ, Delyon F, Alauzet F (2018) Magnetic cage and rope as the key for solar eruptions. Nature 554:211–215. https://doi.org/10.1038/nature24671

5. Antiochos SK, DeVore CR, Klimchuk JA (1999) A model for solar coronal mass ejections. Astrophys J 510:485–493. https://doi.org/10.1086/306563
6. Berger MA, Field GB (1984) The topological properties of magnetic helicity. J Fluid Mech 147:133–148. https://doi.org/10.1017/S0022112084002019
7. Bobra MG, Sun X, Hoeksema JT, Turmon M, Liu Y, Hayashi K, Barnes G, Leka KD (2014) The helioseismic and magnetic imager (HMI) vector magnetic field pipeline: SHARPs - space-weather HMI active region patches. Sol Phys 289:3549–3578. https://doi.org/10.1007/s11207-014-0529-3
8. Chen H, Zhang J, Ma S, Yang S, Li L, Huang X, Xiao J (2015) Confined flares in solar active region 12192 from 18 to 29 October 2014. Astrophys J Lett 808(1):L24
9. Cheung MCM, DeRosa ML (2012) A method for data-driven simulations of evolving solar active regions. Astrophys J 757:147. https://doi.org/10.1088/0004-637X/757/2/147
10. Courant R, Friedrichs K, Lewy H (1967) On the partial difference equations of mathematical physics. IBM J Res Dev 11(2):215–234
11. Demoulin P, Henoux JC, Priest ER, Mandrini CH (1996) Quasi-separatrix layers in solar flares I. Method. Astron Astrophys 308:643–655
12. Démoulin P, Mandrini C, van Driel-Gesztelyi L, Thompson B, Plunkett S, Kővári Z, Aulanier G, Young A (2002) What is the source of the magnetic helicity shed by CMEs? The long-term helicity budget of AR 7978. Astron Astrophys 382(2):650–665
13. DeRosa ML, Schrijver CJ, Barnes G, Leka KD, Lites BW, Aschwanden MJ, Amari T, Canou A, McTiernan JM, Régnier S, Thalmann JK, Valori G, Wheatland MS, Wiegelmann T, Cheung MCM, Conlon PA, Fuhrmann M, Inhester B, Tadesse T (2009) A critical assessment of nonlinear force-free field modeling of the solar corona for active region 10953. Astrophys J 696:1780–1791. https://doi.org/10.1088/0004-637X/696/2/1780
14. DeRosa ML, Wheatland MS, Leka KD, Barnes G, Amari T, Canou A, Gilchrist SA, Thalmann JK, Valori G, Wiegelmann T, Schrijver CJ, Malanushenko A, Sun X, Régnier S (2015) The Influence of spatial resolution on nonlinear force-free modeling. Astrophys J 811:107. https://doi.org/10.1088/0004-637X/811/2/107
15. Duan AY, Jiang CW, Hu Q, Zhang H, Gary GA, Wu ST, Cao JB (2017) Comparison of two coronal magnetic field models to reconstruct a sigmoidal solar active region with coronal loops. Astrophys J 842(2):119. http://stacks.iop.org/0004-637X/842/i=2/a=119
16. Fan Y (2010) On the eruption of coronal flux ropes. Astrophys J 719(1):728
17. Fan Y (2011) A magnetohydrodynamic model of the 13 December 2006 eruptive flare. Astrophys J 740:68. https://doi.org/10.1088/0004-637X/740/2/68
18. Fan Y, Gibson SE (2007) Onset of coronal mass ejections due to loss of confinement of coronal flux ropes. Astrophys J 668:1232–1245. https://doi.org/10.1086/521335
19. Fan YL, Wang HN, He H, Zhu XS (2011) Study of the poynting flux in active region 10930 using data-driven magnetohydrodynamic simulation. Astrophys J 737:39. https://doi.org/10.1088/0004-637X/737/1/39
20. Gary GA (2001) Plasma beta above a solar active region: rethinking the paradigm. Sol Phys 203:71–86
21. Gibson SE, Fan Y (2006) Coronal prominence structure and dynamics: a magnetic flux rope interpretation. J Geophys Res 111(A12):103. https://doi.org/10.1029/2006JA011871
22. Hoeksema JT, Liu Y, Hayashi K, Sun X, Schou J, Couvidat S, Norton A, Bobra M, Centeno R, Leka KD, Barnes G, Turmon M (2014) The helioseismic and magnetic imager (HMI) vector magnetic field pipeline: overview and performance. Sol Phys 289:3483–3530. https://doi.org/10.1007/s11207-014-0516-8
23. Hood AW, Archontis V, MacTaggart D (2012) 3D MHD flux emergence experiments: idealised models and coronal interactions. Sol Phys 278(1):3–31. https://doi.org/10.1007/s11207-011-9745-2
24. Hudson HS (2000) Implosions in coronal transients. Astrophys J Lett 531(1):L75. http://stacks.iop.org/1538-4357/531/i=1/a=L75
25. Inoue S, Hayashi K, Magara T, Choe GS, Park YD (2014) Magnetohydrodynamic simulation of the X2.2 solar flare on 15 February 2011. I. Comparison with the observations. Astrophys J 788(2):182. http://stacks.iop.org/0004-637X/788/i=2/a=182

26. Inoue S, Hayashi K, Magara T, Choe GS, Park YD (2015) Magnetohydrodynamic simulation of the x2.2 solar flare on 15 February 2011. II. Dynamics connecting the solar flare and the coronal mass ejection. Astrophys J 803(2):73. http://stacks.iop.org/0004-637X/803/i=2/a=73
27. Inoue S, Hayashi K, Kusano K (2016) Structure and stability of magnetic fields in solar active region 12192 based on the nonlinear force-free field modeling. Astrophys J 818:168. https://doi.org/10.3847/0004-637X/818/2/168
28. Inoue S, Shiota D, Bamba Y, Park SH (2018) Magnetohydrodynamic modeling of a solar eruption associated with an X9.3 flare observed in the active region 12673. Astrophys J 867(1):83. http://stacks.iop.org/0004-637X/867/i=1/a=83
29. Jacobs C, Poedts S, van der Holst B (2006) The effect of the solar wind on CME triggering by magnetic foot point shearing. Astron Astrophys 450:793–803. https://doi.org/10.1051/0004-6361:20054670
30. Jiang CW, Feng XS (2013) Extrapolation of the solar coronal magnetic field from SDO/HMI magnetogram by a CESE-MHD-NLFFF code. Astrophys J 769:144. https://doi.org/10.1088/0004-637X/769/2/144
31. Jiang CW, Feng XS, Zhang J, Zhong DK (2010) AMR simulations of magnetohydrodynamic problems by the CESE method in curvilinear coordinates. Sol Phys 267:463–491
32. Jiang CW, Feng XS, Fan Y, Xiang C (2011) Reconstruction of the coronal magnetic field using the CESE-MHD method. Astrophys J 727:101. https://doi.org/10.1088/0004-637X/727/2/101
33. Jiang CW, Feng XS, Wu ST, Hu Q (2012) Study of the three-dimensional coronal magnetic field of active region 11117 around the time of a confined flare using a data-driven CESE-MHD model. Astrophys J 759:85. https://doi.org/10.1088/0004-637X/759/2/85
34. Jiang CW, Feng XS, Wu ST, Hu Q (2013) Magnetohydrodynamic simulation of a sigmoid eruption of active region 11283. Astrophys J Lett 771(2):L30
35. Jiang CW, Wu ST, Feng XS, Hu Q (2016) Data-driven MHD simulation of a flux-emerging active region leading to solar eruption. Nature Comm 7(11):11522. https://doi.org/10.1038/ncomms11522
36. Jiang CW, Wu ST, Yurchyshyn VB, Wang H, Feng XS, Hu Q (2016) How did a major confined flare occur in super solar active region 12192. Astrophys J 828(1):62
37. Jiang CW, Feng XS, Wu ST, Hu Q (2017) A magnetic bald-patch flare in solar active region 11117. Res Astron Astrophys 17(9):93. https://doi.org/10.1088/16744527/17/9/93
38. Jiang CW, Yan X, Feng XS, Duan AY, Hu Q, Zuo P, Wang Y (2017) Reconstruction of a large-scale pre-flare coronal current sheet associated with a homologous X-shaped flare. Astrophys J 850(1):8. http://stacks.iop.org/0004-637X/850/i=1/a=8
39. Jing J, Xu Y, Lee J, Nitta NV, Liu C, Park SH, Wiegelmann T, Wang HM (2015) Comparison between the eruptive X2.2 flare on 15 February 2011 and confined X3.1 flare on 24 October 2014. Res Astron Astrophys 15(9):1537
40. Kawabata Y, Inoue S, Shimizu T (2017) Non-potential field formation in the X-shaped quadrupole magnetic field configuration. Astrophys J 842(2):106. http://stacks.iop.org/0004-637X/842/i=2/a=106
41. Kliem B, Su YN, van Ballegooijen AA, DeLuca EE (2013) Magnetohydrodynamic modeling of the solar eruption on 8 April 2010. Astrophys J 779:129. https://doi.org/10.1088/0004-637X/779/2/129
42. Linker JA, Mikić Z, Lionello R, Riley P, Amari T, Odstrcil D (2003) Flux cancellation and coronal mass ejections. Phys Plasmas 10:1971–1978. https://doi.org/10.1063/1.1563668
43. Liu R, Chen J, Wang Y, Liu K (2016) Investigating energetic X-shaped flares on the outskirts of a solar active region. Sci Rep 6(34):021
44. Liu R, Kliem B, Titov VS, Chen J, Wang Y, Wang H, Liu C, Xu Y, Wiegelmann T (2016) Structure, stability, and evolution of magnetic flux ropes from the perspective of magnetic twist. Astrophys J 818:148. https://doi.org/10.3847/0004-637X/818/2/148
45. Liu Y, Zhao J, Schuck PW (2013) Horizontal flows in the photosphere and subphotosphere of two active regions. Sol Phys 287:279–291. https://doi.org/10.1007/s11207-012-0089-3
46. Low BC (1992) Formation of electric-current sheets in the magnetostatic atmosphere. Astron Astrophys 253:311–317

47. Low BC (1996) The spontaneous formation of current-sheets in astrophysical magnetic fields. Springer, Dordrecht, Netherlands, pp 109–131
48. Low BC (2010) The Craig-Sneyd analytic solutions to the Parker problem. Sol Phys 266:277–291. https://doi.org/10.1007/s11207-010-9619-z
49. Low BC, Lou YQ (1990) Modeling solar force-free magnetic fields. Astrophys J 352:343–352. https://doi.org/10.1086/168541
50. Low BC, Petrie GJD (2005) The internal structures and dynamics of solar quiescent prominences. Astrophys J 626(1):551–562
51. Mackay DH, Karpen JT, Ballester JL, Schmieder B, Aulanier G (2010) Physics of solar prominences: II. Magnetic structure and dynamics. Space Sci Rev 151:333–399. https://doi.org/10.1007/s11214-010-9628-0
52. Manchester WB, Gombosi TI, Roussev I, Ridley A, De Zeeuw DL, Sokolov IV, Powell KG, Tóth G (2004) Modeling a space weather event from the Sun to the Earth: CME generation and interplanetary propagation. J Geophys Res 109(A18):2107. https://doi.org/10.1029/2003JA010150
53. Mandrini CH, Démoulin P, Schmieder B, Deng YY, Rudawy P (2002) The role of magnetic bald patches in surges and arch filament systems. Astron Astrophys 391:317–329. https://doi.org/10.1051/0004-6361:20020745
54. Mandrini CH, Schmieder B, Démoulin P, Guo Y, Cristiani GD (2014) Topological analysis of emerging bipole clusters producing violent solar events. Sol Phys 289:2041–2071. https://doi.org/10.1007/s11207-013-0458-6
55. Pariat E, Aulanier G, Schmieder B, Georgoulis MK, Rust DM, Bernasconi PN (2004) Resistive emergence of undulatory flux tubes. Astrophys J 614:1099–1112. https://doi.org/10.1086/423891
56. Parker EN (1972) Topological dissipation and the small-scale fields in turbulent gases. Astrophys J 174:499
57. Parker EN (1994) Spontaneous current sheets in magnetic fields: with applications to stellar X-rays. International series in astronomy and astrophysics, vol 1. Oxford University Press, New York
58. Petrie GJD (2012) The abrupt changes in the photospheric magnetic and lorentz force vectors during six major neutral-line flares. Astrophys J 759:50. https://doi.org/10.1088/0004-637X/759/1/50
59. Prasad A, Bhattacharyya R, Kumar S (2017) Magnetohydrodynamic modeling of solar coronal dynamics with an initial non-force-free magnetic field. Astrophys J 840:37. https://doi.org/10.3847/1538-4357/aa6c58
60. Qiu J (2009) Observational analysis of magnetic reconnection sequence. Astrophys J 692:1110–1124. https://doi.org/10.1088/0004-637X/692/2/1110
61. Reid HAS, Vilmer N, Aulanier G, Pariat E (2012) X-ray and ultraviolet investigation into the magnetic connectivity of a solar flare. Astron Astrophys 547:A52. https://doi.org/10.1051/0004-6361/201219562
62. Saint-Hilaire P, Benz AO (2005) Thermal and non-thermal energies of solar flares. A&A 435(2):743–752. https://doi.org/10.1051/0004-6361:20041918
63. Sakurai T (1989) Computational modeling of magnetic fields in solar active regions. Space Sci Rev 51:11–48. https://doi.org/10.1007/BF00226267
64. Savcheva A, Pariat E, McKillop S, McCauley P, Hanson E, Su Y, Werner E, DeLuca EE (2015) The relation between solar eruption topologies and observed flare features I. Flare ribbons. Astrophys J 810(2):96. http://stacks.iop.org/0004-637X/810/i=2/a=96
65. Schrijver CJ, De Rosa ML, Metcalf TR, Liu Y, McTiernan J, Régnier S, Valori G, Wheatland MS, Wiegelmann T (2006) Nonlinear force-free modeling of coronal magnetic fields Part I: a quantitative comparison of methods. Sol Phys 235:161–190. https://doi.org/10.1007/s11207-006-0068-7
66. Schrijver CJ, DeRosa ML, Metcalf T, Barnes G, Lites B, Tarbell T, McTiernan J, Valori G, Wiegelmann T, Wheatland MS, Amari T, Aulanier G, Démoulin P, Fuhrmann M, Kusano K, Régnier S, Thalmann JK (2008) Nonlinear force-free field modeling of a solar active region

around the time of a major flare and coronal mass ejection. Astrophys J 675:1637–1644. https://doi.org/10.1086/527413
67. Schuck PW (2008) Tracking vector magnetograms with the magnetic induction equation. Astrophys J 683(2):1134. http://stacks.iop.org/0004-637X/683/i=2/a=1134
68. Shibata K, Nakamura T, Matsumoto T, Otsuji K, Okamoto TJ, Nishizuka N, Kawate T, Watanabe H, Nagata S, UeNo S, Kitai R, Nozawa S, Tsuneta S, Suematsu Y, Ichimoto K, Shimizu T, Katsukawa Y, Tarbell TD, Berger TE, Lites BW, Shine RA, Title AM (2007) Chromospheric anemone jets as evidence of ubiquitous reconnection. Science 318:1591. https://doi.org/10.1126/science.1146708
69. Sun X, Hoeksema JT, Liu Y, Wiegelmann T, Hayashi K, Chen Q, Thalmann J (2012) Evolution of magnetic field and energy in a major eruptive active region based on SDO/HMI observation. Astrophys J 748:77. https://doi.org/10.1088/0004-637X/748/2/77
70. Sun X, Bobra MG, Hoeksema JT, Liu Y, Li Y, Shen C, Couvidat S, Norton AA, Fisher GH (2015) Why is the great solar active region 12192 flare-rich but CME-poor? Astrophys J Lett 804(2):L28
71. Tarr LA, Longcope DW, McKenzie DE, Yoshimura K (2014) Quiescent reconnection rate between emerging active regions and preexisting field, with associated heating: NOAA AR 11112. Sol Phys 289:3331–3349. https://doi.org/10.1007/s11207-013-0462-x
72. Thalmann JK, Su Y, Temmer M, Veronig AM (2015) The confined X-class flares of solar active region 2192. Astrophys J Lett 801:L23
73. Titov VS, Démoulin P (1999) Basic topology of twisted magnetic configurations in solar flares. Astron Astrophys 351:707–720
74. Titov VS, Priest ER, Demoulin P (1993) Conditions for the appearance of "bald patches" at the solar surface. Astron Astrophys 276:564
75. Titov VS, Hornig G, Démoulin P (2002) Theory of magnetic connectivity in the solar corona. J Geophys Res 107:1164. https://doi.org/10.1029/2001JA000278
76. Török T, Kliem B (2005) Confined and ejective eruptions of kink-unstable flux ropes. Astrophys J Lett 630:L97–L100. https://doi.org/10.1086/462412
77. van Ballegooijen AA (2004) Observations and modeling of a filament on the Sun. Astrophys J 612:519–529. https://doi.org/10.1086/422512
78. van Ballegooijen AA, Deluca EE, Squires K, Mackay DH (2007) Modeling magnetic flux ropes in the solar atmosphere. J Atmos Sol-Terr Phys 69:24–31. https://doi.org/10.1016/j.jastp.2006.06.007
79. Vranjes J, Poedts S, Pandey BP, de Pontieu B (2008) Energy flux of Alfvén waves in weakly ionized plasma. Astron Astrophys 478:553–558. https://doi.org/10.1051/0004-6361:20078274
80. Wang AH, Wu ST, Liu Y, Hathaway D (2008) Recovering photospheric velocities from vector magnetograms by using a three-dimensional, fully magnetohydrodynamic model. Astrophys J Lett 674:L57–L60. https://doi.org/10.1086/528844
81. Wang H (2006) Rapid changes of photospheric magnetic fields around flaring magnetic neutral lines. Astrophys J 649:490–497. https://doi.org/10.1086/506320
82. Wang H, Liu C (2010) Observational evidence of back reaction on the solar surface associated with coronal magnetic restructuring in solar eruptions. Astrophys J Lett 716:L195–L199. https://doi.org/10.1088/2041-8205/716/2/L195
83. Wang H, Liu C (2012) Circular ribbon flares and homologous jets. Astrophys J 760:101. https://doi.org/10.1088/0004-637X/760/2/101
84. Wang S, Liu C, Liu R, Deng N, Liu Y, Wang H (2012) Response of the photospheric magnetic field to the X2.2 flare on (15 February 2011) Astrophys J Lett 745:L17. https://doi.org/10.1088/2041-8205/745/2/L17
85. Welsch BT, Fisher GH, Abbett WP, Régnier S (2004) ILCT: recovering photospheric velocities from magnetograms by combining the induction equation with local correlation tracking. Astrophys J 610:1148–1156. https://doi.org/10.1086/421767
86. Wolfson R (1989) Current sheet formation in a sheared force-free-magnetic field. Astrophys J 344:471–477. https://doi.org/10.1086/167816

87. Wu ST, Zhang TX, Tandberg-Hanssen E, Liu Y, Feng X, Tan A (2005) Numerical magnetohydrodynamic experiments for testing the physical mechanisms of coronal mass ejections acceleration. Sol Phys 225:157–175. https://doi.org/10.1007/s11207-004-2568-7
88. Wu ST, Wang AH, Liu Y, Hoeksema JT (2006) Data-driven magnetohydrodynamic model for active region evolution. Astrophys J 652:800–811. https://doi.org/10.1086/507864
89. Wu ST, Wang AH, Gary GA, Kucera A, Rybak J, Liu Y, Vrśnak B, Yurchyshyn V (2009) Analyses of magnetic field structures for active region 10720 using a data-driven 3D MHD model. Adv Space Res 44:46–53. https://doi.org/10.1016/j.asr.2009.03.020
90. Xue Z, Yan XL, Cheng X, Yang L, Su Y, Kliem B, Zhang J, Liu Z, Bi Y, Xiang Y et al (2016) Observing the release of twist by magnetic reconnection in a solar filament eruption. Nat Commun 7(11):11837
91. Yamamoto TT, Kusano K (2012) Preprocessing magnetic fields with chromospheric longitudinal fields. Astrophys J 752:126. https://doi.org/10.1088/0004-637X/752/2/126
92. Yan XL, Jiang CW, Xue ZK, Wang JC, Priest ER, Yang LH, Kong DF, Cao WD, Ji HS (2017) The eruption of a small-scale emerging flux rope as the driver of an M-class flare and of a coronal mass ejection. Astrophys J 845(1):18. http://stacks.iop.org/0004-637X/845/i=1/a=18
93. Yashiro S, Akiyama S, Gopalswamy N, Howard RA (2006) Different power-law indices in the frequency distributions of flares with and without coronal mass ejections. Astrophys J Lett 650:L143–L146. https://doi.org/10.1086/508876

Chapter 10
Modeling Coronal Mass Ejections by CESE MHD Model

Abstract Coronal mass ejections (CMEs) consist of large-scale eruptions of magnetized plasma from the Sun, and they are considered to be the major drivers of adverse space weather disturbances that strongly affect our high-tech activities of humankind. Even the fastest CMEs require almost a day to arrive at Earth and initiate a geomagnetic storm. This allows, in principle, sufficient time to predict their impact. The geoeffectiveness of CMEs, that is, of the associated interplanetary CME (ICME) or magnetic cloud (MC), is primarily dependent on their Earth-side magnetic field direction (Bz), their velocity, and their associated ram pressure upon arrival at the magnetosphere. It is therefore highly expected to predict these parameters before an ICME (and the shock that potentially precedes it) arrives at Earth. A promising tool for such purpose is magnetohydrodynamic (MHD) numerical simulations. An accurate modeling of their onset and propagation up to 1 AU represents a key issue for more reliable space weather forecasts. To this end, a lot of CME-related models have been developed to describe their pre-eruption structures, their initiations, and their eruptions, and also the propagation of CMEs from the Sun to the Earth has been numerically investigated. In this chapter, after we briefly introduce the CME models so far, we numerically study the time-dependent evolution and propagation of the CME from the Sun to Earth using the 3D SIP-CESE MHD model introduced in Chap. 4, compare the simulation results with spacecraft observations and analyze in detail the CME's propagation characteristics.

10.1 Introduction

Coronal mass ejections (CMEs) are spectacular events involving the expulsion of significant amounts of solar material (10^{15-16} grams) into the heliosphere [61], with speeds observed from 20 to 2600 km/s. Associated with this eruption is energy release of 10^{31-32} ergs, which is presumably supplied by the magnetic field. The CME occurrence rate increases from $\sim$0.5 per day near solar minimum to $\sim$6 near solar maximum. They are responsible for non recurrent disturbances in the interplanetary medium, and their interactions with the Earth's magnetosphere cause severe geo-effective storms. Geomagnetic storms can perturb and interrupt electric power

© Springer Nature Singapore Pte Ltd. 2020

X. Feng, *Magnetohydrodynamic Modeling of the Solar Corona and Heliosphere*, Atmosphere, Earth, Ocean & Space,
https://doi.org/10.1007/978-981-13-9081-4_10

grids, cause radio blackouts and navigation system failures, and damage satellite systems [127, 134]. Therefore, the studies on CMEs are important for us not only to understand the spectacular phenomenon, but also to predict space weather.

Since the discovery of CMEs in the early 1970s, the spectacular phenomenon has attracted wide interest, and has been extensively investigated through observations, theoretical analysis, and numerical simulations. The space era has provided critical data that enable us to understand the properties of the solar wind and CMEs and to develop the capability of global quantitative modeling. The observations of the source regions of CMEs were obtained from extreme ultraviolet (EUV) spectrographic imagers, soft and hard X-ray imagers, white light coronagraph and white light heliospheric imagers on board the following spacecraft: the Orbiting Solar Observatory 7 (OSO-7) mission, Skylab, the Solar Maximum Mission (SMM) satellite, Yohkoh (Solar-A), SPARTAN 201, Solar and Heliospheric Observatory (SOHO), the Transition Region And Coronal Explorer (TRACE) satellite, the Ramaty High Energy Solar Spectroscopic Imager (RHESSI), Hinode (Solar-B), Solar Terrestrial Relations Observatory (STEREO) A and B, and the Solar Dynamics Observatory (SDO). Interplanetary signatures of CMEs came from the in situ measurements of magnetic field and solar space plasma instruments on board Mariner 2, Pioneer 6 and 9, Interplanetary Monitoring Platform 8 (IMP-8), Helios 1 and 2, Pioneer 10 and 11, Voyager 1 and 2, International Sun-Earth Explorer 3 (ISEE-3), Ulysses, WIND, Advanced Composition Explorer (ACE), and STEREO A and B. Ground based observation, like interplanetary scintillation (IPS), also contributes to our understanding of CMEs' interplanetary propagation. There are excellent reviews and book chapters published on observations of CMEs (e.g., [6, 10, 19, 58, 73, 75, 103]).

The study of the origin, the structure, and the propagation and evolution characteristics of these violent solar eruptive phenomena is essential for a deeper insight into space weather physics. It is well known that single spacecraft observation cannot directly describe the global structures of CMEs and their evolution in interplanetary space. The determination of CME's global structure and kinematics, requires in situ observations from multiple vantage points. Now, we have several spacecraft monitoring the Sun including SOHO [31] and STEREO [68]. In particular, after 2006, the NASA STEREO mission started to provide multi spacecraft in situ observations (e.g., [71, 111]), and image ICMEs in white light along the entire Sun-Earth line [37]. However, the observations can only provide the space weather conditions near the Sun and at 1 AU, and no routine observations for the solar wind are available to bridge the gap between the Sun and the Earth. Therefore, the real global structure of CMEs and the underlying physics governing CME from the initiation in the solar corona to propagation in the heliosphere are not well understood due to sparse observations. In order to realistically reflect the 3D characteristics of CME propagation and its interaction with the structured solar wind, as commented in the existing literature (e.g., [14, 22, 23, 32, 42, 50, 81, 122, 123, 167]), the numerical 3D MHD simulation has become one of the key tools in modeling space weather events. Moreover, the 3D MHD simulation results can help us to get a deeper insight into space weather physics aided by a better understanding of the propagation characteristics. This insight is crucial for better and more accurate space weather predictions.

Currently, to our knowledge there is no CME model self-consistently developed to fully explain all the observed features of solar eruptions and related phenomena. The basic pre-eruption configuration and the topological changes in the magnetic field that result in the conversion of a large fraction of the magnetic energy into kinetic energy are not yet well-known. Until now, a number of theoretical models have been proposed to interpret the initiation of CMEs. Here we only briefly mention three commonly-used scenarios: (i) The mass loading model, which starts the equilibrium within a prominence between magnetic tension and the gravity force of the mass above the structure. The slow accumulation of more mass merely causes the prominence to bend downwards under the increased weight of its load [43, 172]. Eruption can occur when some restructuring of the magnetic field causes part of the material to suddenly drain away, resulting in a loss of equilibrium; (ii) The flux cancelation model, which assumes that there is a threshold for the amount of flux cancelation that is defined as the mutual disappearance of magnetic fields of opposite polarity at the neutral line separating them [104] possibly due to submerging, annihilating, or being expelled upward. Below the threshold, a magnetic flux rope forms above the neutral line in 3D arcades and can support prominence material [79, 82]. Beyond the threshold, the configuration erupts [1–3], which transforms a significant fraction of the magnetic energy into kinetic energy; (iii) The breakout model, which describes the eruption of multipolar topology with a magnetic null [4, 5, 27, 28, 59, 60, 97, 98]. In the breakout model, a CME represents the catastrophic disruption of the force balance between the upward magnetic pressure of the highly-sheared filament-channel field and the downward tension of overlying quasi-potential field. When the strongly sheared fields near the neutral line cause the field lines to rise, slow reconnection at the null point transfers overlying flux in the central arcade to the neighboring arcades, eventually destabilizing the central arcade. These models idealize configurations and are primarily intended to investigate specific aspects of CMEs, such as their initiation mechanisms (e.g., [2, 3, 5, 9, 20, 38, 45, 69, 76, 108, 146, 147]). More detailed Reviews on CME initiation mechanisms can be found in a vast literature (e.g., [8, 19, 44, 46, 47, 54, 73, 74, 80, 89–91, 103, 107, 110, 173]).

In the numerical studies, many methods have been employed to mimic perturbations of CMEs when investigating the propagation of CMEs in interplanetary space with the help of 3D MHD models, although a choice of a reasonable representation of the 3D CME model is still debatable. Simple pressure/velocity pulses have been first applied to simulate the propagation of CME events [55, 115–117, 120, 130, 137, 162, 164–166]. The cone model [106, 109, 113, 170, 174] is often used to model CMEs in MHD simulations [29, 30, 105]. The model assumes that the CMEs propagate with constant angular widths and speeds. This assumption is founded on the facts that Large Angle and Spectrometric Coronagraph experiment (LASCO) observations of halo CMEs (e.g., [15, 126]) have revealed that the bulk velocity tends to be radial; and the angular width, $2\Delta\theta$, tends to remain constant as CME propagates through the corona. The cone model is currently used at the Community Coordinated Modeling Center (CCMC) for predicting arrival times of CMEs (e.g., [105]).

However, the preceding models lack details about the magnetic field carried by a CME. In order for the simulated CME ejecta to truly represent a magnetic cloud, one needs to incorporate magnetic field structure, controlled by the ambient external field at the location where CME is added, into the model. A simple but convenient way is to superimpose a spherical magnetic plasmoid or blob of high speed, high density, and high pressure on the ambient steady state solar wind [17, 18, 138, 139, 175–178]. Besides, various flux rope models have also been proposed to simulate CMEs in many studies, including the analytical [144] flux rope model (e.g., [36, 65, 88, 94, 102, 131, 132, 141, 148, 169]), the data-constrained Gibson-Low flux rope [49, 84] model [13, 14, 66, 67, 101], the flux rope model fitted with the graduated cylindrical shell (GCS: [142]) model [133] and the spheromak-type magnetic flux ropes [39, 70, 93, 140, 150, 151]. However, these models neglect the modeling of the eruption, and simplify initial conditions for the CME by inserting a magnetic flux rope that is not in magnetic equilibrium into the background corona. This "out-of-equilibrium" approach may not be well suited for modeling the frequently observed, Bz-relevant rotation of CMEs about their rise direction (e.g., [26, 143]), especially in cases where most of the rotation occurs low in the corona, while the ejected flux is still accelerating. In contrast, simulations that start from a stable pre-eruptive configuration that was constructed using several instances of the TDm coronal flux-rope model [145] produce CME properties such as the field strength, orientation, and speed self consistently [148]. This allows one to predict the properties and impact of a potential eruption before it has occurred, and it provides the means to assess the strength and impact of eruptions that a given source region on the Sun may produce.

Meanwhile, some authors have numerically studied the propagation characteristics for observed CME events by 3D MHD simulation, and the simulation results can reproduce time-dependent evolution of 3D structures and be used for comparison with observations by the Large Angle and Spectrometric Coronagraph experiment (LASCO), or STEREO, and WIND/ACE (e.g., [83, 94, 95, 101, 102, 120, 121, 137, 138, 163, 166, 168, 177]). For example, Wu et al. [168] described the evolution of the January 1997 CME event by 2.5D MHD model and compared the simulation results with SOHO and WIND data. Odstrčil et al. [120] simulated 12 May 1997 CME event and provided a global heliospheric picture of transient disturbance interacting with solar wind stream structures. By reproducing the shape of 24 November 2000 CMEs event as observed by the LASCO C3 coronograph, Lugaz et al. [94] investigated the three-dimensional aspects of the propagation for the CMEs event with the Space Weather Modeling Framework (SWMF), and compared synthetic images with observations from STEREO. Manchester et al. [102] numerically modeled 28 October 2003 CME event with SWMF and found excellent agreement with both the general morphology and the quantitative brightness of the model CME with LASCO observations. Shen et al. [138] investigated the dynamical interaction of April 2000 CME event with the background solar wind flow and the evolution of the physical parameters by 3D time-dependent MHD simulation. Lionello et al. [83] described the propagation of an ICME from the Sun to Earth. Wu et al. [166] studied the CME event on 15 March 2013 that produced a large solar energetic particle (SEP) event and severe geomagnetic activity by the 3D MHD simulation. Török et al. [148] presented a magnetohydrodynamic (MHD) simulation of 14 July 2000 "Bastille Day" eruption that produced a very strong geomagnetic storm. Borovikov et al. [14] developed a computational tool with the purpose of serving as an operational system in providing live information on the current potential threats posed by SEP based on the observations of the Sun, and discussed the fundamentals of magnetohydrodynamical simulations to be employed as a critical part of the desired forecasting system.

An understanding of the effects of CMEs on geo-space is essential for forecasting the geo-space weather. So an important topic for space weather forecast is to predict whether or not a CME can be geo-effective. In fact, previous researchers have shown that CMEs may not propagate radially from the corona to the interplanetary (IP) space (e.g., [17, 24, 25, 63, 149, 158, 179]). MacQueen et al. [99] found that 29 CMEs observed during the Skylab epoch underwent an average 2.2° equatorward deflection. Wei and Dryer [159] found that all of the flare-associated shock waves tended to propagate toward the low latitude region near the solar equator and the fastest propagation directions tended toward the heliospheric current sheet (HCS) by statistically studying 149 flare-shock events based on interplanetary scintillation (IPS) observations. The equatorward deflection propagation of the CME event during its outward propagation has been reported by Cremades and Bothmer [24]. They statistically study 124 structured CME events observed by LASCO during 1996–2002 and found that CMEs in solar minimum years (1996–1998) systematically deflected 20° to lower latitudes. CME deflections have become a very popular research topic since the NASA STEREO mission was successfully launched. For example, Kilpua et al. [72] studied two high-latitude CMEs on 2 November 2008 using observations from STEREO and WIND spacecraft. They showed that the first CME was strongly deflected to the equator and suggested that the relative contribution of deflecting CMEs to the near ecliptic CME rate could be significant at solar minimum conditions. Shen et al. [136] analyzed the obvious deflection of the 8 October 2007 CME and found that the deflection of this CME might be caused by a nonuniform distribution of the background magnetic field energy density. Bemporad et al. [12] and Zuccarello et al. [179] found that the 21 September 2009 CME showed a deflection toward the HCS during its propagation due to an imbalance in the magnetic pressure and tension forces. Based on the SOHO and STEREO observations, the longitudinal deflections of CMEs were also suggested by some researchers (e.g., [56, 85, 86, 96, 156, 157]). In the numerical simulation, Jacobs et al. [63], and van der Holst et al. [149] studied the influence of the background solar wind on the evolution properties of CMEs and the corresponding shocks by means of a 2.5D (axi-symmetric) ideal MHD model. Chané et al. [17, 18] quantified the effect of the initial magnetic polarity of CMEs on the evolution properties and geo-effectiveness of CMEs by using a 2.5D MHD simulations. They found that the CME from high latitude deflected toward the equator and hit the Earth when its initial CME magnetic field and the background wind magnetic field had the same polarity, while the same CME with an opposite polarity missed the Earth.

The main aim of this chapter is to present the numerical study of CME's propagation characteristics during its propagation from the solar corona to IP space. First, we simulate two typical CME events by 3D MHD model: (i) Sun-Earth connection halo CME event occurred on 4 November 1997 and (ii) the first fast CME observed by STEREO SECCHI/HI for the full Sun-Earth line occurred on 3 April 2010. We compare the simulation results with spacecraft observations and analyze in detail the several factors affecting the CME's geo-effectiveness. Second, we study the CME latitudinal deflection (LD) and the factors (the initial magnetic polarity, the parameters of CME model, and the background magnetic field) that influence such deflection based on the solar wind structures at solar minimum, during which the corona has a relatively simple configuration, and the current sheet is usually located near the equator.

10.2 Model Description

It is well known that the solar wind ambient is important in the propagation study of CMEs in interplanetary space. Therefore, we firstly simulate the solar wind ambient by using 3D SIP-CESE MHD model in Chap. 4 with the observation-based magnetic field and spherically symmetric Parker solar wind as the initial condition. In order to produce the realistic fast and slow solar wind structure, an additional coronal heating input is needed with general agreement, although the entire scenario for the coronal heating/solar wind acceleration is still unknown. The details and history of this subject have been debated by scientists for a long time (e.g., [7, 11, 21, 57, 78, 92]). Here, to mimic the effects of coronal heating/solar wind acceleration, a simple source function $Q_e = Q_0/f_s \exp(-r/L_Q)$ in the energy equation is employed by following the work of Nakamizo et al. [114]. Here, the constant Q_0 is set to be $1.2 \times 10^{-6}\,\mathrm{Jm^{-3}s^{-1}}$, and L_Q is set to be 0.9 R_s. f_s is the expansion factor defined as $f_s = (\frac{R_s}{r})^2 \frac{B_{R_s}}{B_r}$ where R_s and r are the solar radius and the distance from the solar center, and B_{R_s} and B_r are magnetic field strength at the solar surface and at r. In the simulation, the expansion factor is time-invariant during calculating the background solar wind. The involvement of expansion factor in the heating source term is encouraged by the fact that the solar wind speed is inversely correlated with the expansion rate of the magnetic flux tube in the corona [77]. Feng et al. [41] designed another kind of volumetric heating and momentum heating method according to the expansion factor and the angular separation between an open field foot point and its nearest coronal hole boundary. Here, the parametric source function is only added to the energy equation by involving the topological effect of the magnetic field expansion factor f_s to reproduce the fast and slow solar wind structure.

For the specific heat ratio γ, similar to that of Jacobs and Poedts [62] a variable polytropic index is used:

$$\gamma = \begin{cases} \gamma_0 + (\gamma_1 - \gamma_0)\sin^2\left(\frac{\pi}{2}\frac{r-r_1}{r_2-r_1}\right) & \text{if } r_1 \le r \le r_2, \\ \gamma_1 & \text{if } \quad r > r_2, \end{cases} \tag{10.1}$$

where the value of γ is smoothly varying between the value γ_0 close to the Sun, $\gamma_0 = 1.2$ and its value near 1AU, $\gamma_1 = 1.46$. The variation in γ takes place between the radial distances r_1 and r_2, where r_1 and r_2 are taken to be 1 R_s and 20 R_s respectively.

Since there is no general consensus yet on the triggering mechanism of solar eruptions, we instead assume that the CME has already been initiated. Here, a CME is mimicked by a 3D spherical plasmoid with high speed, high pressure and high plasma density initially taken to be out of equilibrium near the solar surface on the observed source location of the flare/CME. As such, the model does not address the question of how such an energetic CME is initiated, but describes the three-dimensional propagation and evolution of the CME event from the low corona to the interplanetary space, by providing the time variation of solar wind parameters and magnetic field, which generally are vital to geospace space weather forecasting.

Here, a high speed, high density and high pressure spherical plasmoid is employed to generate the CME. This kind of model is far from being a state-of-the-art CME initiation model with very complex features, but it gives us an easy way to study the evolution and propagation of the CME in the corona and heliospheric space. The density, temperature and radial velocity profiles of the initial perturbation are defined as follows:

$$\begin{aligned} \rho &= \rho_0 + \rho_{\max}(1 - a^2/a_{\text{cme}}^2) \\ T &= T_0 + T_{\max}(1 - a^2/a_{\text{cme}}^2) \\ v_r &= v_{r0} + v_{\max}(1 - a^2/a_{\text{cme}}^2) \end{aligned} \tag{10.2}$$

where a_{cme} is the radius of the plasmoid, a denotes the distance from the center of the plasmoid, and ρ_0, v_{r0} and T_0 are the density, radial velocity, and temperature of the background solar wind. $\rho_{\max}$, $v_{\max}$ and $T_{\max}$ are the maximum density, radial velocity and temperature added on top of the background solar wind, respectively. In order to match the observation data as the best fit as possible, the choices of these parameters are empirical.

Gibson and Fan [48] believed that the spheroidal magnetic field may be a natural product of a plasmoid ejected from a solar flare via reconnection, although the actual existence in the solar wind is under debate (e.g., [39, 152]). Whereas Wang et al. [154, 155] described certain existence evidence on plasmoid in the magnetotail. Here, the initial magnetic field of the spherical plasmoid is assumed to take the spheromak-type magnetic field form as the "classical" spheromak of Farrugia et al. [39] in local spherical coordinates $(r^\ell, \theta^\ell, \phi^\ell)$:

$$\begin{aligned} B_{r^\ell} &= (2B_0/\alpha r^\ell) j_1(\alpha r^\ell) \cos\theta^\ell \\ B_{\theta^\ell} &= -(B_0/\alpha r^\ell)[\sin(\alpha r^\ell) - j_1(\alpha r^\ell)] \sin\theta^\ell \\ B_{\phi^\ell} &= \pm B_0 j_1(\alpha r^\ell) \sin\theta^\ell \end{aligned} \tag{10.3}$$

where B_0 is a constant, and $\alpha = 4.493409458 a_{\text{cme}}^{-1}$ is the constant derived from the force-free condition of $\nabla \times \mathbf{B} = \alpha \mathbf{B}$ with the boundary condition of $B_{r^\ell} = 0$ at $r^\ell = a_{\text{cme}}$. $j_1(x) = x^{-2} \sin x - x^{-1} \cos x$ is the spherical Bessel function.

10.3 Sun-to-Earth Propagation of the 4 November 1997 CME Event

10.3.1 *Observational Properties of the CME Event*

The 4 November 1997 CME event was associated with the active region NOAA 8100, where a flare of class 2B/X2.1 began at 05:54 UT. The SOHO spacecraft recorded the CME between 05:52 and 06:08 UT. Dulk et al. [33] and Eto et al. [35] reported that the lift-off time of the CME occurred within a few minutes of the peak time of the impulsive X2/3B flare (S14W34). The CME was of halo type, with a reported angular width of 360°. From the LASCO/SOHO observations, Dulk et al. [33] obtained a speed about 750 km/s in the plane of sky. The CME-caused disturbance, having an average speed 646 km/s from 1 R_s to 1 AU, arrived at Earth and produced a strong shock at 22:10 UT on 6 November, about 65 h after the flare and CME lift-off. Several hours later, the magnetic cloud arrived at 1 AU. Figure 10.10 plots the solar wind bulk velocity, density, proton temperature, magnetic field, and magnetic field components in geocentric solar ecliptic [GSE] coordinates by dots for a period of 4 days from November 6 to 10. The shock front is indicated by the solid vertical line. The magnetic cloud, indicated by the two dotted vertical lines following the shock sheath, starts at November 7, 05:00 UT and ends at November 8, 12:00 UT. The figure shows a typical magnetic cloud: the magnetic field direction varies slowly, the magnetic field strength increases, and density, plasma proton temperature and thermal pressure decrease, i.e., the ejecta is a low-beta plasmoid. Moreover, the solar wind speed decreases as the ejecta passes.

In the following, the results of the 3D MHD numerical simulation are presented, including the solar wind ambient obtained from the method of time relaxation, the morphology and propagation of the CME, and the comparison between the numerical results and the in-situ data from WIND.

10.3.2 *Solar Wind Ambient*

The global steady-state solar wind for this event is obtained by the 3D SIP-CESE MHD model with Parker's 1D solar wind solution and measured photospheric magnetic fields as the initial values. The temperature and plasma density at the inner boundary are typically taken to be 1.5×10^6 K and 1.67×10^{-13} kgm^{-3} respectively. The computational domain covers $-90° \le \theta \le 90°$, $0° \le \phi \le 360°$, and 1 $R_s \le r \le 235$ R_s. Here, for the analysis of computational results, the corresponding simulated data are transformed to the inertial frame. The details of computing the background solar wind [40, 41] are omitted here.

Figure 10.1 shows the background solar wind solution for the experiment of the 4 November 1997 CME event at a solar coronal meridional plane (left-hand) and an inner heliosphere solar equatorial plane (right-hand) respectively. The color image

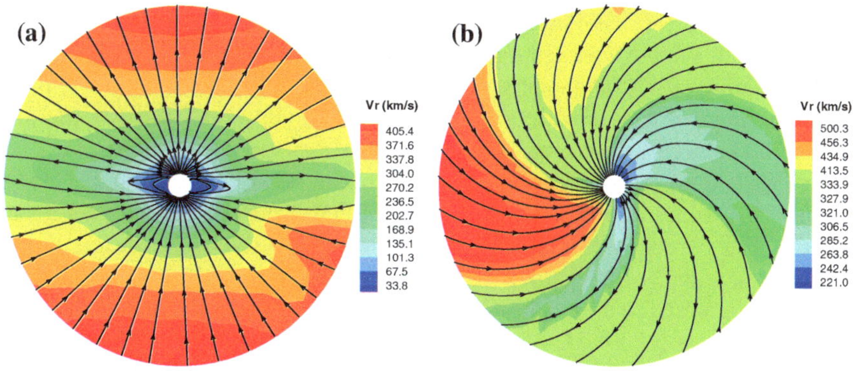

Fig. 10.1 Magnetic field and radial speed for the steady-state solar wind solution. **a** Solar coronal solution at $\phi = 180° - 0°$ from 1 R_s to 15 R_s and **b** inner heliosphere solution in the solar equatorial plane from 15 to 215 R_s. The color contours represent the radial speed and streamlines denote the magnetic field lines. Reprinted from Fig. 2 of Zhou et al. [177] with permission of the John Wiley & Sons, Inc

indicates the radial speed of the plasma while the magnetic field is represented by solid black lines. From Fig. 10.1a, we can see that the magnetic field and radial speed possess a typical characteristic of solar minimum. At lower latitudes around the equator, a helmet streamer with low radial speed is stretched by the solar wind. At high latitudes, the magnetic field is carried out with the solar wind to achieve an open configuration with high radial speed. Figure 10.1b shows that the IMF lines are stretched by the solar wind outward into Archimedean spirals due to the solar rotation and the IMF freezing-in effect.

Figure 10.2 shows the steady state background solar wind parameters at 1 AU. The results simultaneously reproduce the high- and low-temperature regions, the

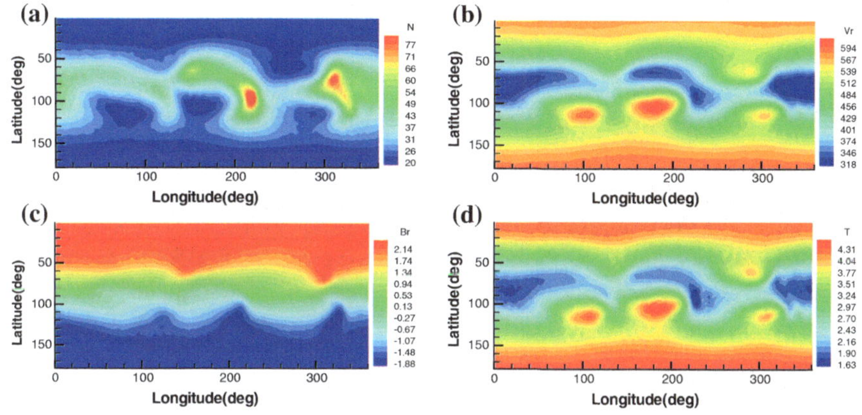

Fig. 10.2 The steady state solar wind parameters at 1 AU. **a** Number density (cm^{-3}) and **b** Radial speed (km/s). **c** Radial magnetic field (nT) and **d** Plasma temperature (10^5 K). Reprinted from Fig. 3 of Zhou et al. [177] with permission of the John Wiley & Sons, Inc

fast (low density) and slow (high density) solar wind structure in interplanetary space. From this figure, we can find that the radial speed profile is in agreement with the temperature profile, and opposite to the density profile. Inspection of Fig. 10.2b reveals an approximate bimodal outflow pattern with slow wind leaving the Sun around 350 km/s near the current sheet region and high-speed wind above 600 km/s in the high-latitude region. The variation in the background solar wind in this model results from coronal heating and flux expansion. But we find the radial speed is higher than the observed speed near the Earth, which may be caused by the low expansion factor owing to the contamination of six impulsive solar events in November 1997.

Then, superposed on the quiet background solar wind, a spherical plasmoid is used to mimic 4 November 1997 CME event. The center of the CME is initially placed at 1.8 R_s. The radius of the initial CME, a_{cme}, is 0.6 R_s. The value of $v_{\max}$ is set to be 300 km/s. $\rho_{\max}$ and $T_{\max}$ are assumed to be five and three that of the ambient solar wind at the center of the CME, respectively. The maximal magnetic field strength B_0 is 5.0 G. The CME is launched at S14W34 to conform to the location of the flare/CME. In order to find a better agreement with the in-situ IMF observations, we rotate the polar axis of the ejected spheromak toward the negative y-axis direction, so that the polar axis is inclined about 42° from the negative y-axis. During the simulation of the CME event, we have tested various angles, and found that the angle can best match the observed magnetic field. The introduction of the plasma plasmoid to the corona has added 5.77×10^{31} ergs of magnetic energy, 4.33×10^{30} ergs of kinetic and 5.55×10^{31} ergs of thermal energy. The total energy (magnetic+kinetic+thermal energy) increase is 1.175×10^{32} ergs. Here, the energy injected into the computation domain is estimated by $\Delta E = E_1 - E_0$, where E_1 (E_0) is the energy after (before) CME initiation. The magnetic, kinetic and thermal energy values are calculated using the expressions: $E_m = \int\limits_A \frac{B^2}{2} dA$, $E_k = \int\limits_A \frac{1}{2}\rho V^2 dA$, $E_t = \int\limits_A \frac{p}{\gamma-1} dA$, respectively. In the simulation, t = 0 is taken to be at 05:54 UT on 4 November 1997, when the flare associated with the halo CME was observed from AR 8100. Figure 10.3 shows the 3D schematic picture of the initial coronal magnetic field at t = 0.

10.3.3 Morphology and Propagation of the CME

Figure 10.4 compares the observed and synthetic LASCO C2 and EIT images for the ejection. The top two figures are the C2 and EIT running-difference images for the CME. The bottom row shows the corresponding synthetic images. By comparing the simulation results and observations, we can see clearly that the simulated white light image is in good agreement with the observed one, and that the simulated density enhancement compares well with the EIT intensity enhancement.

Figure 10.5 shows a 3D view of the CME at t = 1.0, 3.0, 5.0, and 10 h after the initiation. The false color image shows the velocity magnitude in the x–y equatorial plane. The magnetic field of the CME is colored to illustrate the velocity magnitude. The evolution of the CME on the meridional at $\phi = 210°$ is shown in Fig. 10.6 at

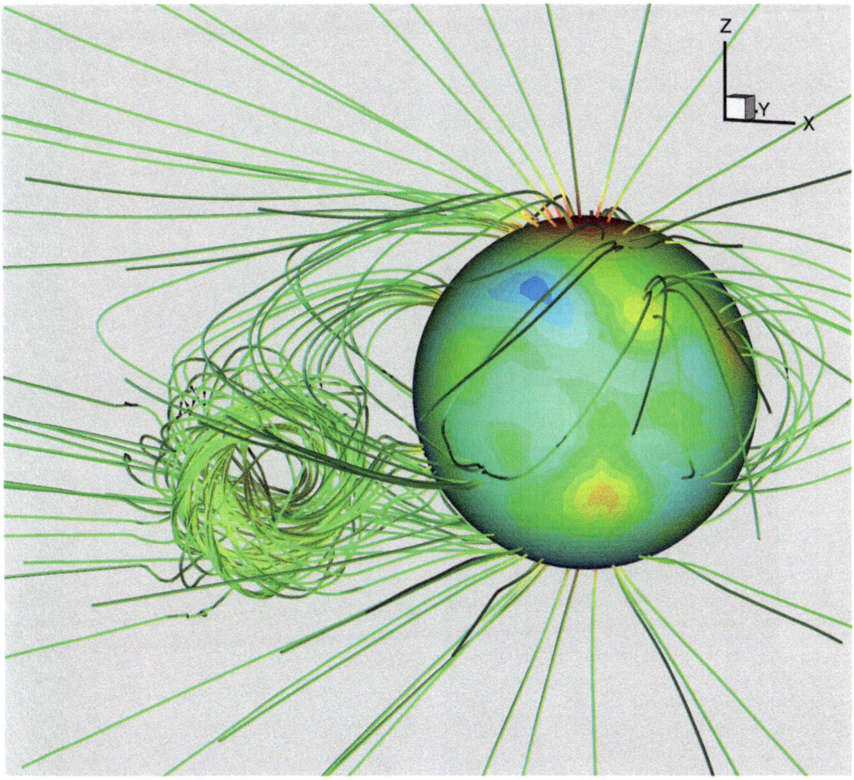

Fig. 10.3 Three dimensional view of the initial coronal magnetic field. Field lines of the CME are shown in color to illustrate the magnetic field strength. The color contours represent the radial magnetic field strength on the solar surface. Reprinted from Fig. 1 of [177] with permission of the John Wiley & Sons, Inc

four different instants (t = 1.0, 3.0, 5.0 and 10.0 h). This figure depicts the system in 2-D meridional slices (x–z plane) with false color images of the plasma velocity magnitude upon which solid black lines represent the magnetic field. From the two figures, we find that initially the magnetic field evolves nearly in self-similar manner. The plasmoid rapidly expands and is expelled from the corona since the injected CME has higher plasma density, speed, and magnetic pressure than the solar wind ambient. The radial and angular sizes of the CME increase with time. The plasma velocity magnitude shows that the evolution of the CME begins with rapid acceleration to a speed over 780 km/s at 7 R_s which almost agrees with the speed in the plane of sky from the LASCO observation, then undergoes a gradual deceleration to a speed of 710 km/s at 42 R_s. In this simulation, most of the magnetic field lines of CME are connected to the inner boundary from the beginning, as shown in Fig. 10.6, since the magnetic field of the spherical plasmoid is introduced by simple superposition on the ambient field.

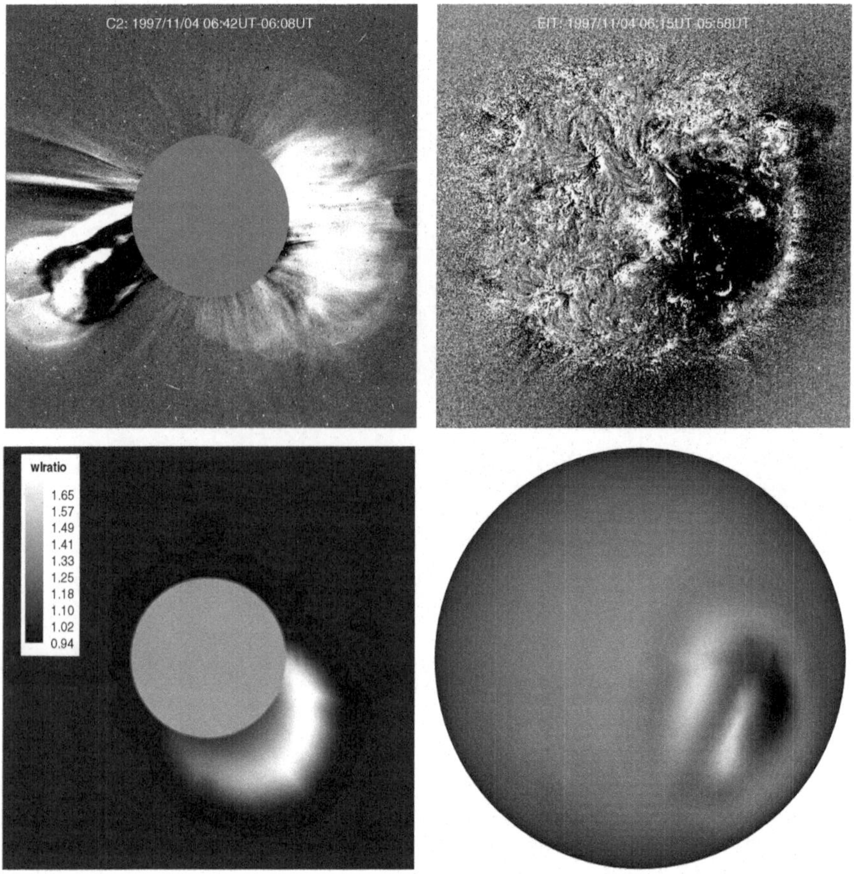

Fig. 10.4 Top: C2 and EIT difference images. Bottom: Synthetic white light running ratio image corresponding to C2 field-of-view (left), numerically simulated density-enhancement image on the solar disk (right). Reprinted from Fig. 4 of Zhou et al. [177] with permission of the John Wiley & Sons, Inc

Figure 10.7 shows the time profiles for the relative number density, velocity, relative temperature and magnetic field at selected solar radii r_i along $\theta = 0°$, $\phi = 210°$ (the top row) and $\phi = 180°$ (the bottom row). We find that the overall similarity (such as sharp rise and slow decay of the velocity's maxima, and the sharp jumps of the magnetic field, density and temperature) of the computed time profiles in the inner heliosphere ($r = 10\ R_s$, $20\ R_s$). The forward CME shock is well developed. It originates at the leading edge of the CME due to the collision of high-speed CME plasma with the background slow stream. The distribution of plasma parameters in the figure shows a significant expansion of the CME. As the CME propagates further into interplanetary space ($r = 50\ R_s$, $100\ R_s$), it continues to expand due to the pressure gradients, the difference between the speeds at the leading and trailing edges

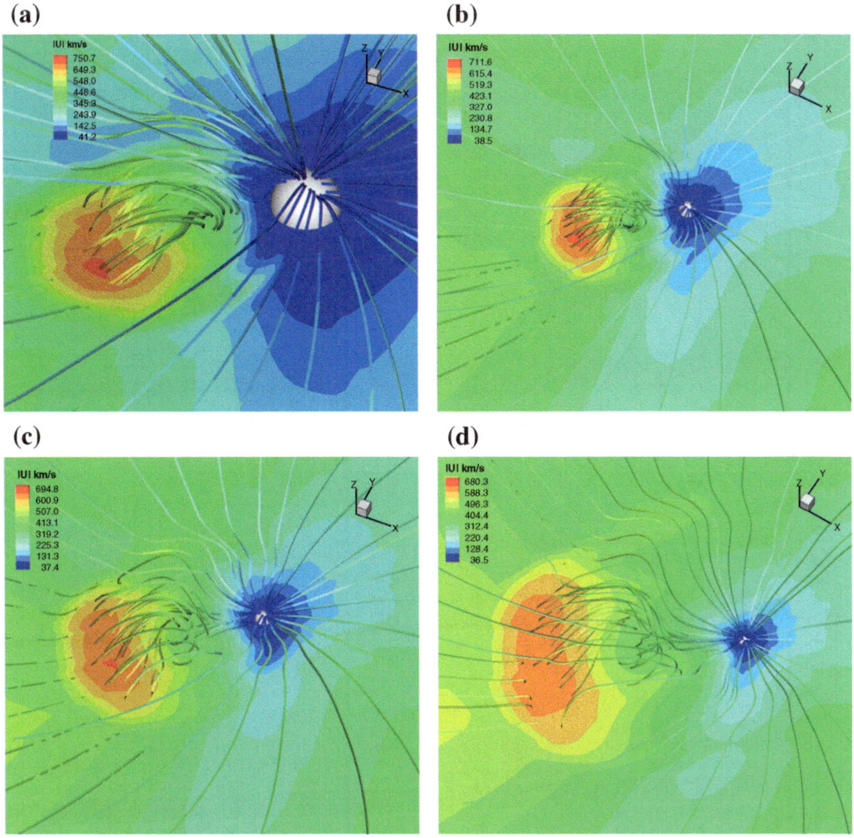

Fig. 10.5 A 3D representation of the CME is shown at **a** 1 h, **b** 3 h, **c** 5 h, **d** 10 h after initiation. The color code represents the velocity magnitude in the x-y equatorial plane. Field lines of the CME are shown in color to illustrate velocity magnitude. The Sun is shown with a white sphere. Reprinted from Fig. 5 of Zhou et al. [177] with permission of the John Wiley & Sons, Inc

decreases, and velocity's maxima also decrease. Simultaneously, we can find that the variation amplitude at $\phi = 180°$ is smaller than that at $\phi = 210°$, especially the difference of the magnetic field is more obvious. The arrival time of the disturbance at the same heliocentric distance at $\phi = 180°$ delays a few hours relative to that at $\phi = 210°$, and that the farther away from the Sun the disturbance is, the longer the time delay is.

Figure 10.8 shows the simulated global view of the relative density $((\rho - \rho_0)/\rho_0)$ (the top two rows) and the velocity (the bottom row) at 2.0° south, close to the Earth's latitude at different instants, where ρ is the density, and ρ_0 the density of the background solar wind. The top row shows the domain from 1 to 30 R_s within 5 h after the initiation, and the bottom two rows show views from 1 R_s to 1 AU after 20 h. From this figure, we find that the CME almost propagates along the Parker spiral. There is almost no pronounced eastward deflection of the CME during its

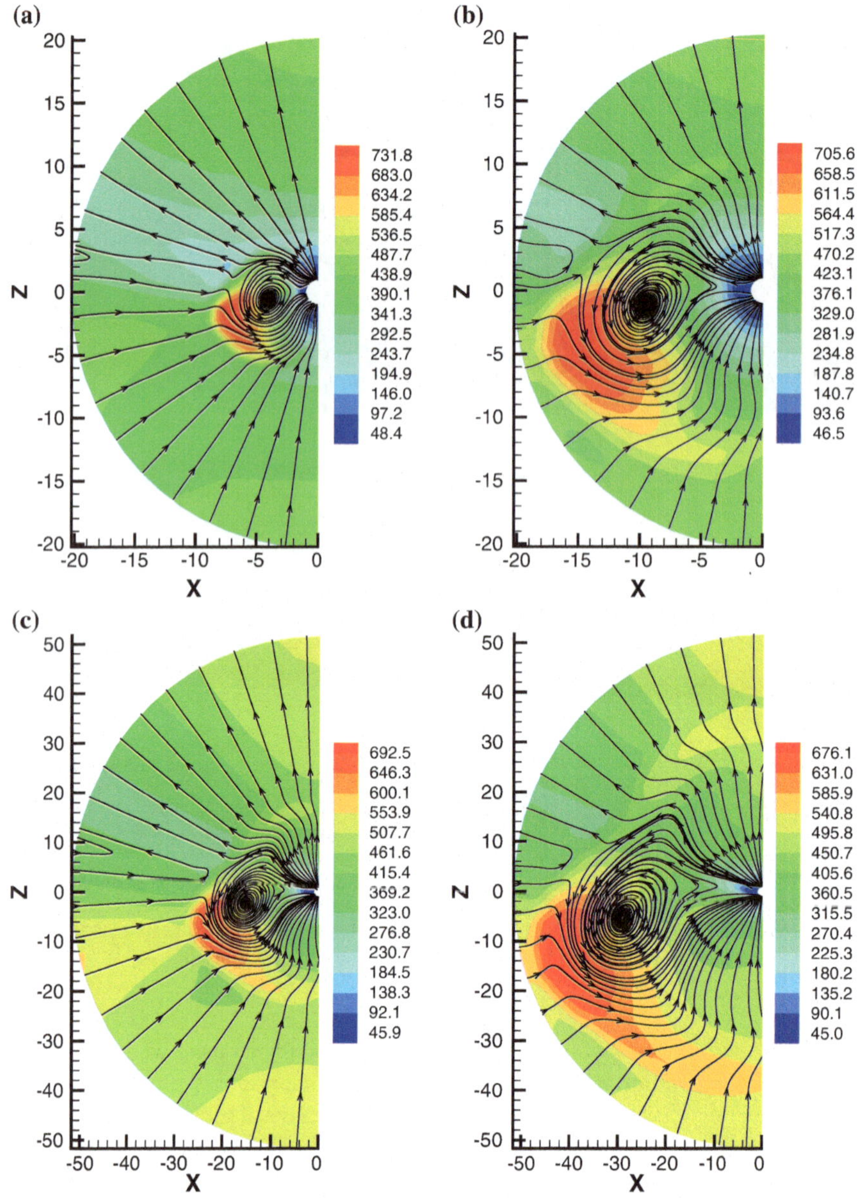

Fig. 10.6 Time evolution of the CME in the x–z meridional plane at **a** t = 1 h, **b** t = 3 h, **c** t = 5 h, **d** t = 10 h. Solid black lines display magnetic "streamlines" superimposed upon a false color image of the velocity magnitude. Reprinted from Fig. 6 of Zhou et al. [177] with permission of the John Wiley & Sons, Inc

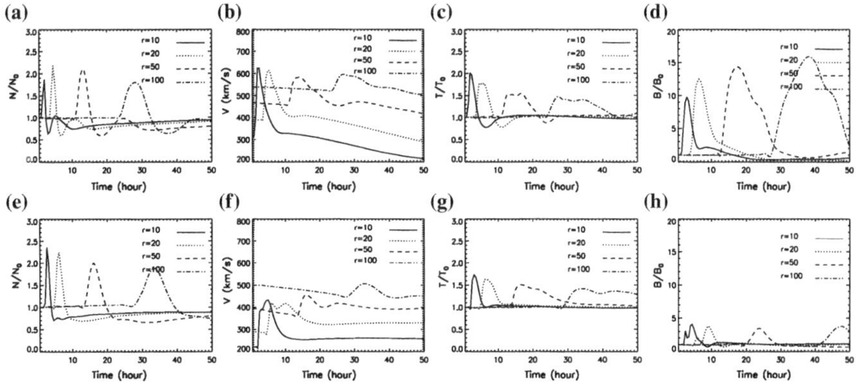

Fig. 10.7 Time profiles for the relative number density N/N$_0$, velocity V, relative temperature T/T$_0$, and relative magnetic field B/B$_0$, at selected solar radii along $\theta = 0°$, $\phi = 210°$ (top row) and $\phi = 180°$ (bottom row). Time is given in hours after the CME onset. Reprinted from Fig. 7 of Zhou et al. [177] with permission of the John Wiley & Sons, Inc

outward propagation because of the little speed difference between the CME and the background solar wind, which can be seen in the plots (g–i) of Fig. 10.8. This is consistent with the observational analyses of the deflection of CMEs' propagation in interplanetary space by Wang et al. [156]. Asymmetry of the CME is clearly seen in heliolongitude as it passes into non-uniform solar wind ambient, and the asymmetry of the CME in heliolatitude can also be seen clearly from Figs. 10.9a–c.

The evolution of the CME in interplanetary space at $\phi = 210°$ is displayed in Figs. 10.9 at t = 20, 30 and 50 h after the CME injection. The magnetic field is represented by solid black lines. The false color images of Figs. 10.9a–c indicate the relative density, while the false color image of Fig. 10.9d indicates the magnetic field strength.

The distribution of the relative density exhibits a concave-outward shape possible stemming either from intrinsic speed gradients within the CME or from the interactions with the structured ambient solar wind. The CME propagates slower near the streamer belt and faster in the open field region. Near the streamer belt, the CME plows most directly into the slow, dense streamer belt flow, its excess momentum is quickly dissipated, a compression is built up at its front, and the disturbance abruptly decelerates. By contrast, the open-field extensions of the CME are pulled rapidly outward by the fast flow toward the poles. Riley et al. [128] suggested that the observations were indicative of a strong velocity shear between Ulysses and Geotail spacecraft, with the CME being launched such that one part was in a high speed stream while the other part was in a slow stream.

The radial and meridional widths of the CME increase with distance. But the rate of radial expansion slowly decreases as the CME is expanding. The initially circular magnetic field gradually distorts owing to its interaction with a slower solar wind ambient. It expands latitudinally while being squeezed radially. Thus the initial

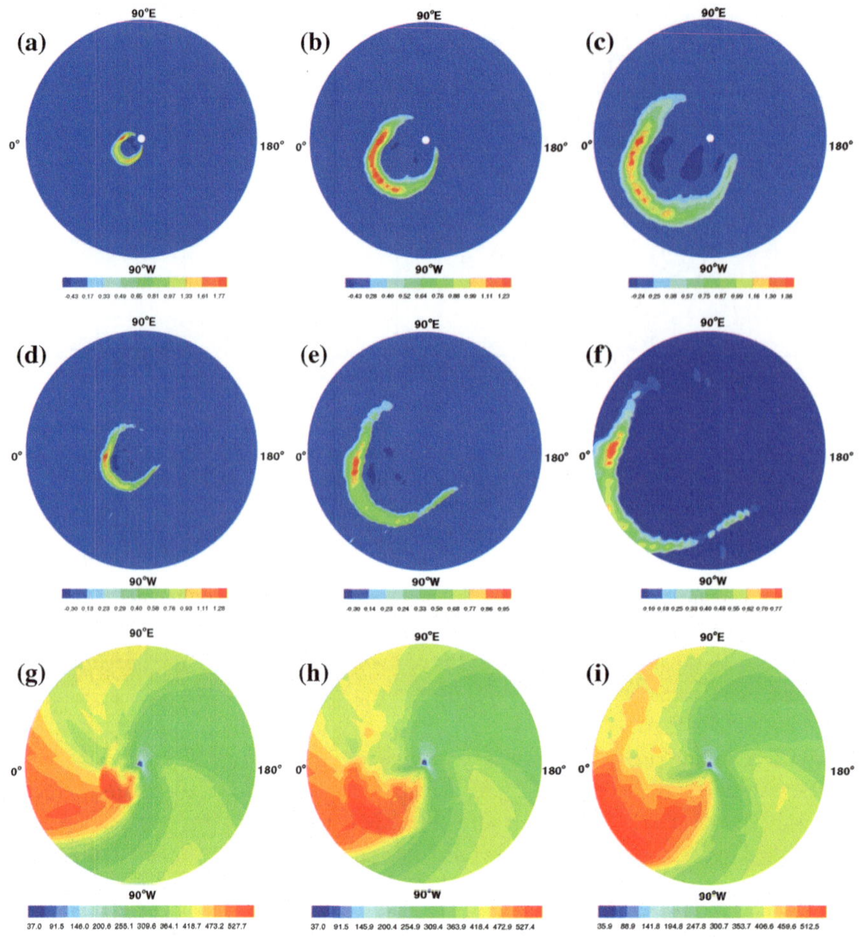

Fig. 10.8 The simulated profiles of the relative density (top two rows) and the velocity (bottom row) in the x–y azimuthal plane at **a** t = 1 h, **b** t = 3 h, **c** t = 5 h, **(d & g)** t = 20 h, **(e & h)** t = 40 h, **(f & i)** t = 60 h. Reprinted from Fig. 8 of Zhou et al. [177] with permission of the John Wiley & Sons, Inc

circular shape has evolved into an elliptical, lentil-like one and that while the leading edge is still curved, the trailing edge has been flattened. Such shape changes have also been observed in other numerical simulations [16, 119] due to the different hydrodynamic forces acting on each side. At t = 50 h, the numerical simulation shows that the trailing part of the magnetic field begins to flatten noticeably. As the CME expands, each part of its surface pushes against different plasma conditions. Since the pressure and density at the rear are higher than those at the front, the expansion will be inhibited at the rear relative to the front. This is also a feature of 1D models [53, 129]. While moving around the boundary from the rear to the sides,

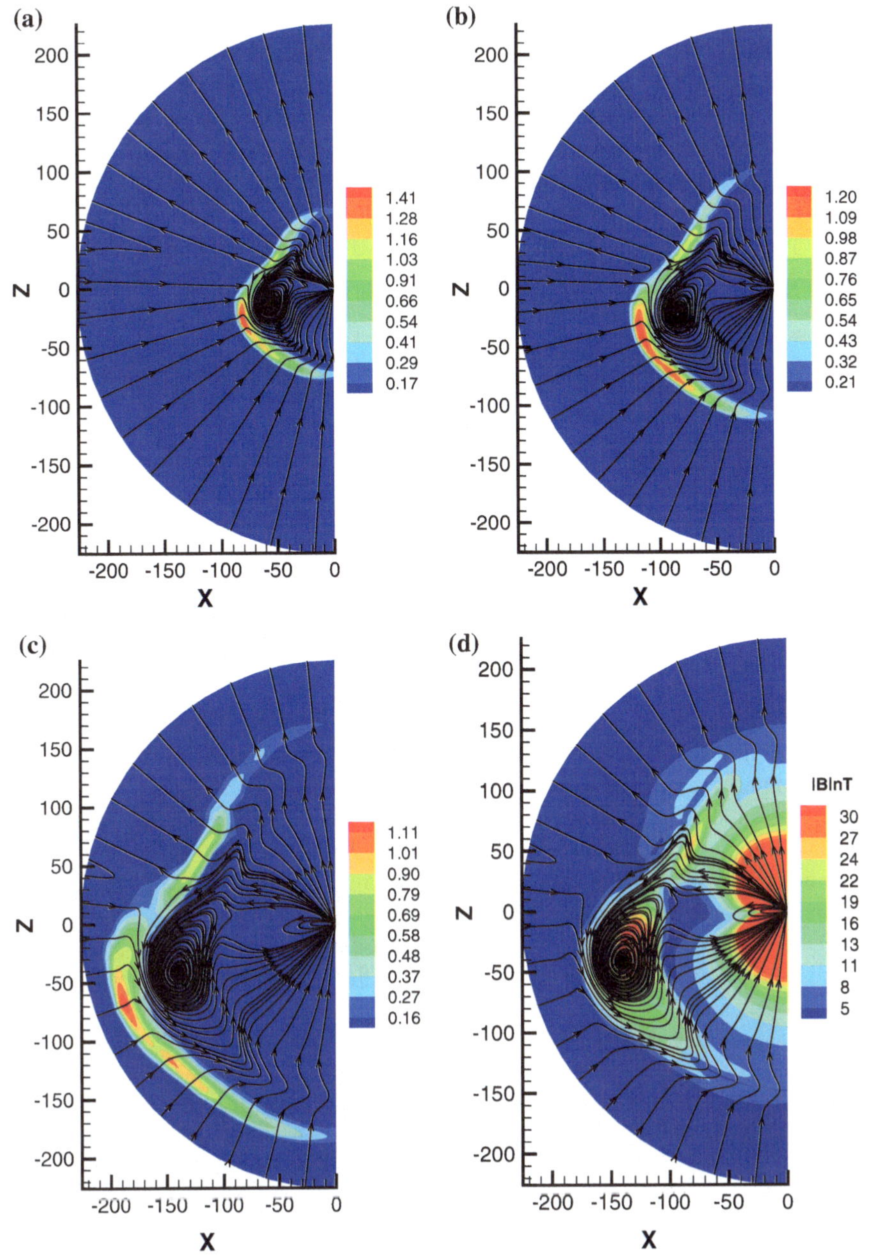

Fig. 10.9 Global view of the CME interaction with the background solar wind in interplanetary space at $\phi = 210°$ and different times: **a** t = 20 h, **b** t = 30 h, **c** and **d** t = 50 h. Solid black lines display magnetic "streamlines". Frames **a**, **b**, and **c** show the false color images of the relative density, while frame **d** shows the false color images of the magnetic field strength. Reprinted from Fig. 9 of Zhou et al. [177] with permission of the John Wiley & Sons, Inc

the expansion gradually encounters the regions of lower external plasma pressure and density so that the expansion becomes easier. Since the rear part of the CME expands by different distances while going around the boundary, this cumulative effect leads to the flattened shape. The front of the CME expands more easily than the side parts since it encounters a lower pressure and density.

The compression of the magnetic field is clearly seen on the north in Fig. 10.9d, which is most pronounced where the magnetic field of the CME moves through the dense plasma sheet, but it doesn't penetrate the streamer belt. However, the CME still moves southward until it reaches the boundary of the open field with high speed. This may be caused by the higher thermal and magnetic pressure at the faster solar wind region.

10.3.4 Comparison with In-Situ Data

To make a comparison with in-situ measurements, Fig. 10.10 simultaneously shows the computed plasma parameters at the Earth and its 30° westward and WIND observations for the same time interval. From this figure, we find that the simulated solar wind parameters at 30° westward of the Earth can better reproduce many basic structures: such as an increase in the magnetic field magnitude, the behaviors of three components of magnetic field, in particular southward B_z and the large-scale smooth magnetic field rotation. The discrepancy may result from the following factors such as the less accurate distribution of the initial disturbance, the inadequate expansion owing to the less realistic solar wind ambient, and almost no deflection of the CME due to the small speed difference between the CME and the solar wind ambient. For a better comparison with the WIND observation, we shift 4 h ahead to the time axis of simulated profiles and multiply the computed magnetic field by 2 just because the computed magnetic field is nearly half of that by observation.

From the figure, we can see that the simulated background velocity and density at the Earth is higher, the magnetic field is smaller than those by observation. The simulated plasma density is 3–4 times large, and the simulated velocity is almost 150 km s^{-1} greater than that by observation. The reason for this may be due to the single fluid model used, the volumetric heating assumption and other reasons as pointed out by Feng et al. [41]. In this simulation, the choice of such a heating may obtain a typical background solar wind characteristic of solar minimum due to the consideration of magnetic field topology. To further understand this problem, more observations of multiple spacecraft (such as SDO, STEREO, SOHO, ACE, WIND, or other future missions) will probably help us develop the ability of including physically realistic coronal heating modules into 3D MHD codes, improve the determination of the structure of the ambient solar wind, and further numerically characterize the 3D propagation of CMEs through the heliosphere.

From the computed B_z at 30° westward of the Earth shown in Fig. 10.10, we find that the field turns southward at ~18:00 UT on 6 November 1997 for nineteen hours, reaching a minimum of −6 nT. The magnetic field then smoothly rotates northward

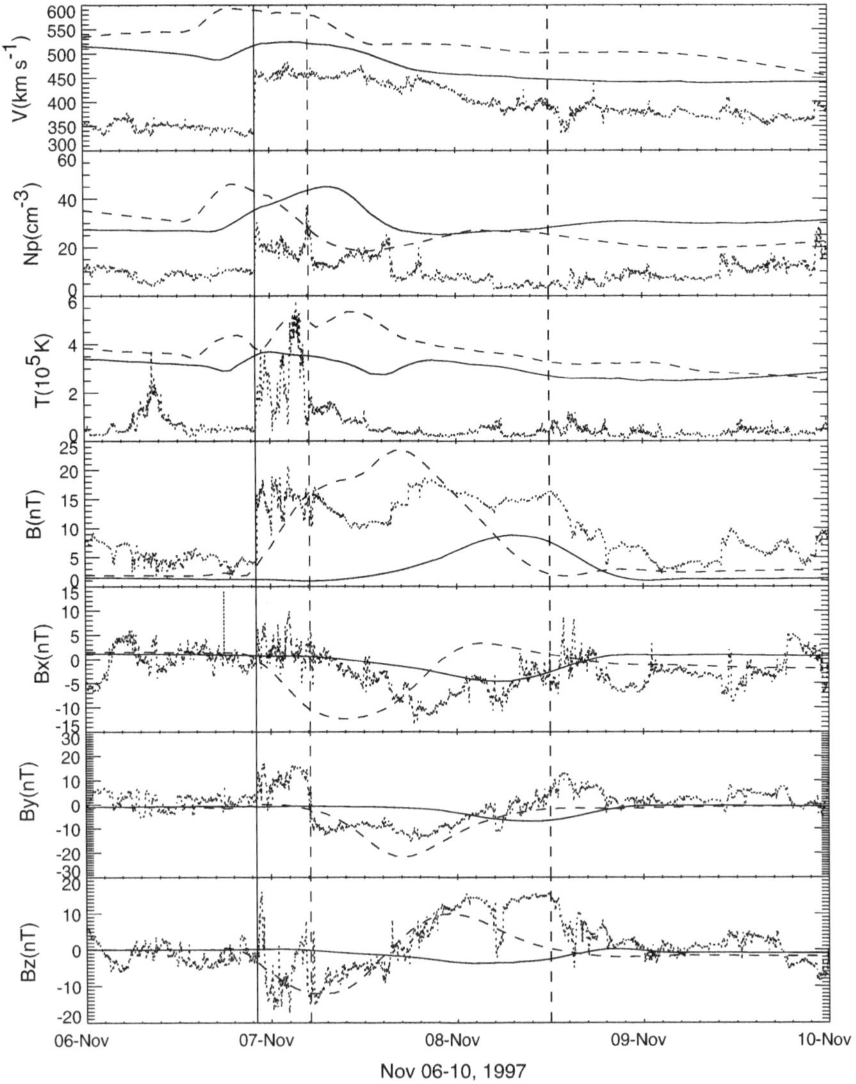

Fig. 10.10 Evolution of solar wind parameters at the Earth during the 04 Nov 1997 event. From top to bottom, shown are flow velocity, number density, temperature, magnetic field, and three components of magnetic field in GSE coordinates. The simulated results at the Earth are shown by solid lines. The WIND observations are shown by dots. The dashed lines show the solar wind profile computed at a position 30° westward of the Earth. Reprinted from Fig. 10 of Zhou et al. [177] with permission of the John Wiley & Sons, Inc

reaching a similar magnitude, lasting more than twenty hours. This magnetic field south-north rotation is mainly due to the passage through the helical field of the CME mentioned by Wu et al. [168]. The strong southward to northward turning of the IMF and prolonged southward IMF shown in this figure can produce a strong magnetospheric response. But the modeled magnetic field at 1 AU has nearly 50% magnetic magnitude of the observed magnetic cloud. The reason for this may be also due to the small background magnetic field caused by the imperfectness of potential field source surface model and the coarse grid resolution in interplanetary space.

10.4 Propagation and Evolution of the 3 April 2010 CME

10.4.1 Observational Properties of the CME Event

The 3 April 2010 event was the first fast (800 km/s) ICME including a magnetic cloud observed by both the STEREO/HI instruments and a near Earth spacecraft. At this time, the separation angles of STEREO A and B with Earth are 67° and 71°, respectively. The CME was an Earth-directed one, which was associated with a B7.4 X-ray flare recorded by GOES X-ray monitor from NOAA Active Region (AR) 11059 (S25W03). The flare was first seen in the Extreme Ultraviolet Imaging Telescope (EIT) on board SOHO between 09:14 UT and 10:00 UT associated with a faint EIT wave.

The STEREO observations suggest that the CME was traveling at a velocity of $\sim$1000 km/s close to the Sun, and that it gradually decelerated to $\sim$700 km/s when the CME impacted the Earth. Consistent with these measurements, in-situ instruments on board the NASA ACE spacecraft observed an interplanetary (IP) shock traveling with a solar wind velocity of $\sim$750 km/s at 7:51 UT on April 5. The IP shock was followed by an IP coronal mass ejection (ICME), which caused an extended geomagnetic storm with minimum Dst $= -72$ nT on April 5–7, 2010 [171]. The 3 April 2010 event has been studied by several authors [87, 112, 161, 171].

In the following, we perform a comprehensive study on the evolution and propagation of the 3 April 2010 CME/ shock by comparing the remote sensing observations of STEREO in the corona, the in situ observations of WIND, with the simulation results, and analyze several propagation characteristics of the CME.

10.4.2 Solar Wind Ambient

Similar to the above section, the quasi-steady solar wind is computed with the observed photospheric magnetic field and Parker's 1D solar wind solution [125] as the initial values. The temperature and plasma density at the inner boundary are

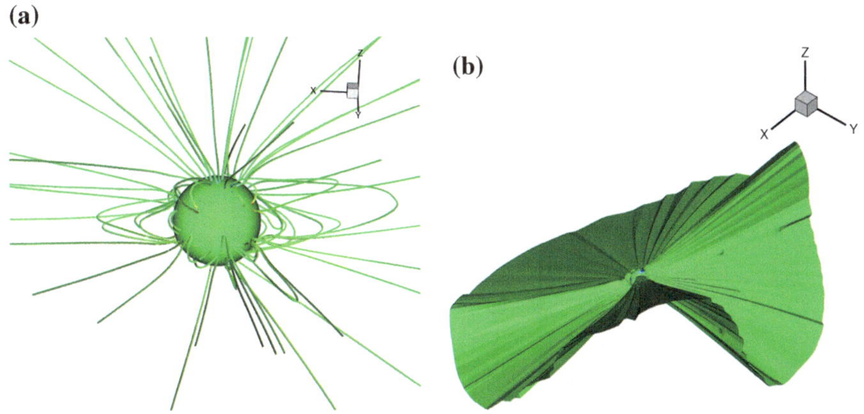

Fig. 10.11 **a** Three-dimensional steady-state magnetic field structure, **b** steady-state current sheet in the solar corona domain. Reprinted from Fig. 1 of Zhou and Feng [178] with permission of the John Wiley & Sons, Inc

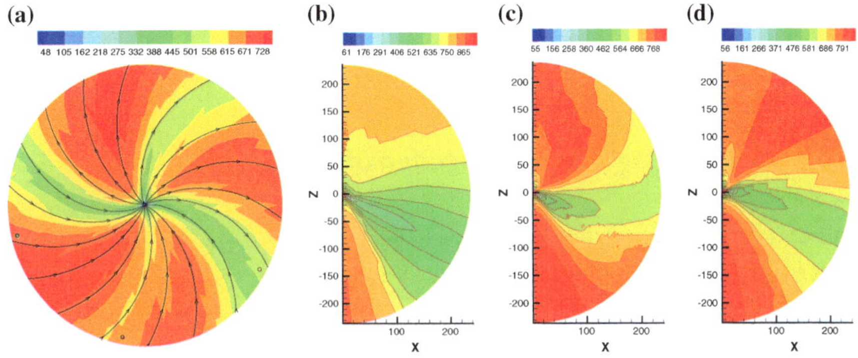

Fig. 10.12 **a** Steady-state solution in the solar equatorial plane. The black streamlines denote the magnetic field lines, the green, red and yellow circles represent STEREO A, Earth and STEREO B, respectively. **b–d** Steady-state solution at the meridional plane of E67, E0 and W71, respectively. The color contours represent the radial speed (km/s). Reprinted from Fig. 2 of Zhou and Feng [178] with permission of the John Wiley & Sons, Inc

typically taken to be 1.5×10^6 K and 0.85×10^{-13} kg m^{-3} respectively. The computational domain covers $-90^\circ \leq \theta \leq 90^\circ$, $0^\circ \leq \phi \leq 360^\circ$, and $1\ R_s \leq r \leq 235\ R_s$.

Figure 10.11 shows the three-dimensional coronal field and HCS structure. From the figure, we can easily see that the magnetic field and radial speed possess a typical characteristic of the rising phase. The coronal magnetic field shows a mixed dipole/quadrupole configuration with large tilt angle of the dipole. And the latitudinal extent of the HCS expands to higher latitudes, a north-south asymmetry appears around the heliospheric equator due to the quadrupole strength of the coronal field. Figure 10.12a shows the steady-state solution in the solar equatorial plane. The result shows that the magnetic field generates the sector structure in interplanetary space.

Figure 10.12b shows a cut through the ecliptic as well as along the meridional plane of E67, E0 and W71. The figure gives the information on how the solar wind speed is structured in IP space along the flank of the CME as viewed on STEREO A, Earth and STEREO B meridional planes. We can find that there are significant differences in the speed distribution.

In the simulation, the parameters of the CME model in Sect. 10.2 are empirically chosen to match the simulated properties of the ICME reasonable well with in situ measurements at 1 AU. The CME is launched at 2.0 R_s at t = 09:20 UT, when the CME is first observed by EIT in the lower corona. The radius of the initial CME, a_{cme}, is 0.8 R_s. Assume that $\rho_{\max}$ and $T_{\max}$ are six and five times larger than that of the solar wind ambient at the center of the CME, respectively. The value of $v_{\max}$ is set to be 500 km/s. The maximal magnetic field strength B_0 is 5.0 G. The CME is launched at S25W03 to conform to the location of the flare/CME. In order to find a better agreement with the in-situ IMF observations, we rotate the polar axis of the ejected spheromak toward the negative y-axis direction, so that the polar axis is inclined about 20° from the negative y-axis. The introduction of the plasmoid to the corona has added 6.93×10^{31} ergs of magnetic energy, 1.54×10^{30} ergs of kinetic and 3.68×10^{31} ergs of thermal energy. The total energy (magnetic+kinetic+thermal energy) increase is 1.08×10^{32} ergs.

10.4.3 Morphology and Propagation of the CME

A three dimensional view together with a radial velocity distribution in $y = 0$ plane at 10 h after the eruption of the CME is shown in Fig. 10.13a. The corresponding

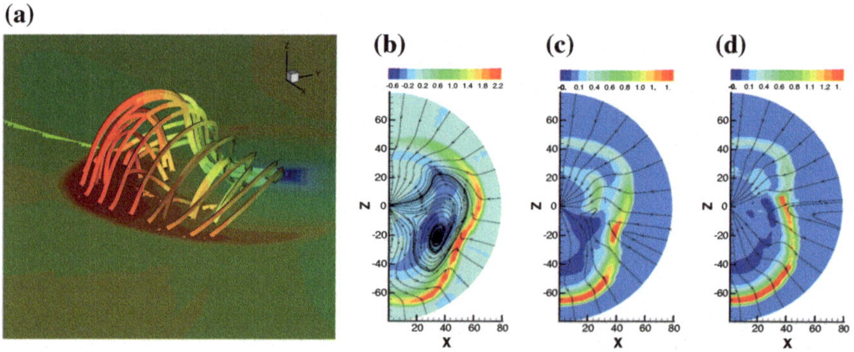

Fig. 10.13 **a** A 3D representation of the CME is shown at 10 h after initiation. The color code represents the velocity magnitude in the x–y equatorial plane. Field lines of the CME are shown in color to illustrate velocity magnitude. The Sun is shown with a green sphere. **b–d** Contour plots of the relative density distribution after 10.0 h at E67, E0 and W71 meridional planes, respectively. The black lines represent the magnetic field. The relative density is defined as follow: $\overline{\rho} = (\rho - \rho_{\text{wind}})/\rho_{\text{wind}}$. Reprinted from Fig. 3 of Zhou and Feng [178] with permission of the John Wiley & Sons, Inc

2-D meridional slices at E67, E0 and W71 with false color images of the relative density is depicted in Figs. 10.13b–d, respectively, upon which black lines represent the magnetic field. At this time, the CME front has reached 58 R_s. The maximum radial velocity of the CME is approximately 770 km/s. From Fig. 10.13b, we find that the front of the flux rope begins to flatten, since the flux rope is stalled by the dense plasma sheet around the HCS. This phenomenon has been found in earlier work [101, 177]. However, the flux rope doesn't expand to E67 and W71 meridional planes, as seen from Figs. 10.13c–d. The distribution of the relative density exhibits a concave-outward shape in Figs. 10.13b–d due to the HCS largely distorting the shock front. The large-scale concave feature of the shock front is consistent with previous simulations and observations [101, 118].

Figure 10.14 produces synthetic line-of-sight differential images of CME observed from STEREO A and B by integrating the simulated density and compare observed and synthetic images. The figures of the first and third columns are the CME running difference images for the CME from STEREO A and B, respectively. The second and forth columns show the corresponding synthetic images. The top two panels are from STEREO A and B COR1 and COR2, respectively. From the top two rows, it is found that the simulated images capture some features of the event such as the bright outer loop, but unfortunately miss the bright core of density structure, where three-part

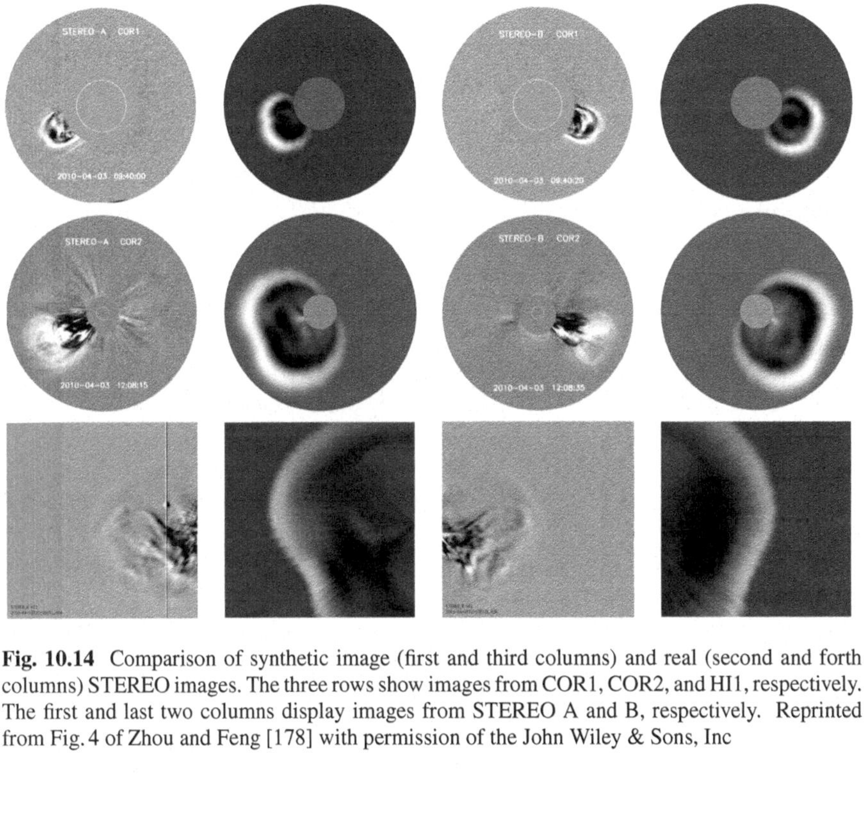

Fig. 10.14 Comparison of synthetic image (first and third columns) and real (second and forth columns) STEREO images. The three rows show images from COR1, COR2, and HI1, respectively. The first and last two columns display images from STEREO A and B, respectively. Reprinted from Fig. 4 of Zhou and Feng [178] with permission of the John Wiley & Sons, Inc

structure was observed in COR1 and COR2. This might be due to the simple CME model used here. In the bottom row, by comparing synthetic images of the CME in the heliosphere from the point of view of STEREO A and B with real images, a good agreement is seen between the overall shape and position of the CME in real and synthetic images of the CME observed. In particular, the concave-outward density structure is well reproduced. However, the size of CME in the synthetic image is larger than the real one. This might be due to an overexpansion of the simulated CME owing to its initiation setup.

10.4.4 Properties of the CME at 1 AU

Figure 10.15 shows the 3D structure of the distorted flux rope and its interaction with the solar wind at t = 43 h, when the front of the CME reaches 1 AU. Here, the front of the CME is shown as an iso-surface of $\overline{\rho} = 2\rho_{\text{wind}}$, and false color representations of velocity magnitude cover the iso-surface. The yellow sphere is positioned at the Earth and has a radius of 5 R_s. It is clear that the front of the CME already arrives at Earth at this time from the figure. The solar wind magnetic field in color illustrates the velocity magnitude. These lines fully characterize field line topology near 1 AU. This figure demonstrates that the magnetic field structure evolves to a crescent shape near 1 AU.

Figure 10.16 gives plasma parameters of the simulation results as a function of time at 1 AU (solid lines) and makes a comparison with in situ measurements of WIND observation between April 4 and 9 (dashed lines). From this figure, it is found that the simulation successfully reproduces the main features of ICME, for example, the similar curves of the plasma density and velocity, the behaviors of magnetic field: in particular an increase in the magnetic field magnitude, B_x and southward B_z, and the large-scale smooth magnetic field rotation.

From Fig. 10.16, one finds that the IP shock arrives at WIND on April 5 at 07:58 UT. The model predicts the shock arrival at Earth at about 05:50 UT on April 5, with an error of 2 h compared to the WIND shock arrival time (SAT). So the model provides a good prediction of the SAT for this event. The shock is followed by an ICME from April 5 at 12:00 UT to April 6 at 17:00 UT. The observed ICME has the low proton temperature, smooth and strong magnetic field, and declining speed profile. For the magnetic field, one can see that the simulated and measured profiles at 1 AU are very similar, in particular, for B_x and B_z components and the large-scale smooth magnetic field rotation. The computed B_z at 1 AU in Fig. 10.16 demonstrates that the field turns northward at $\sim$10:00 UT on April 5, 2010 for 17.5 h, reaching a maximum of 8 nT. Then the magnetic field smoothly rotates southward reaching a minimum of -3.3 nT, lasting more than 11 h. The north-south rotation of this magnetic field is mainly due to the passage through the helical field of the CME suggested by Wu et al. [168]. The magnetic field shown in Fig. 10.16 has a prolonged negative (southward) component spanning the ICME. A southward magnetic field can reconnect with geomagnetic fields and thus is a key factor in producing geomagnetic storms [34, 52]. But the

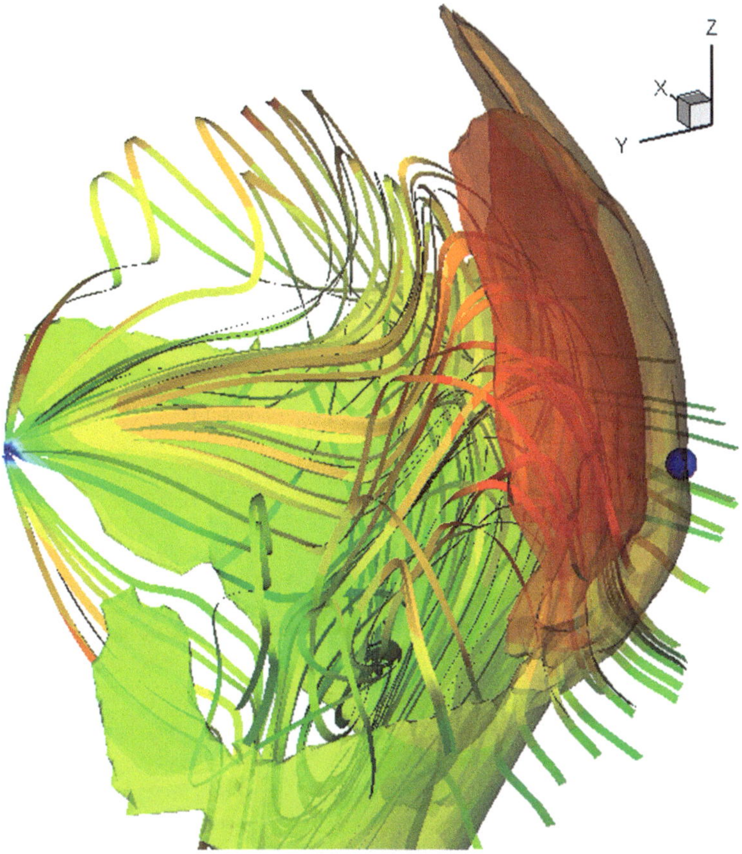

Fig. 10.15 A 3-D representation of the CME is shown at 43 h after initiation. The color code represents the velocity magnitude. A 3-D colored iso-surface of $\overline{\rho} = 2\rho_{\text{wind}}$ is drawn. The blue sphere shows the position of the Earth and has a radius of 5 R_s. Reprinted from Fig. 5 of Zhou and Feng [178] with permission of the John Wiley & Sons, Inc

simulated magnetic field at 1 AU is smaller than that of the observed magnetic cloud. This may be attributed to the small background magnetic field caused by the imperfectness of potential field source surface model and the coarse grid resolution in IP space.

10.4.5 Propagation Characteristics of the CME

The interplanetary observations suggest that the ICME encountered the Earth, while it missed the STEREO A and B spacecraft. In the following, we investigate several factors that affect the probability of the CME's encounter with the spacecraft, such

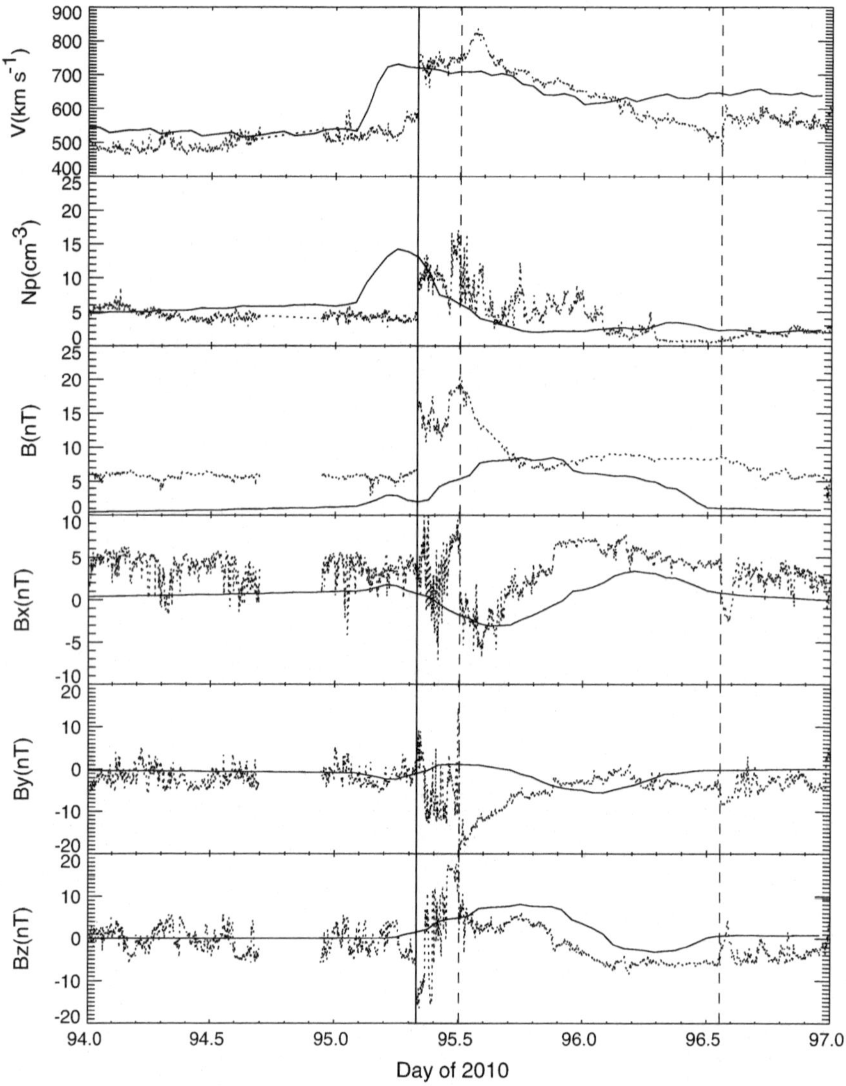

Fig. 10.16 Comparison between the in situ data obtained by the WIND spacecraft and the simulation during the 3 April 2010 event. From top to bottom, shown are flow velocity, number density, magnetic field, and three components of magnetic field in geocentric solar ecliptic (GSE) coordinates. The vertical solid line indicates the time of the ICME-driven shock. The vertical dashed line indicates the starting and ending time of the magnetic cloud. The simulation results at the Earth are shown by the solid lines. The WIND observations are shown by the dotted lines. Reprinted from Fig. 6 of Zhou and Feng [178] with permission of the John Wiley & Sons, Inc

as the speed, span angle, and the deflection, etc. This is also an interesting topic in predicting whether a CME will encounter the Earth.

With the help of simulation results, we can derive the CME trajectory from near the Sun all the way to 1 AU during its outward propagation. In order to quantitatively study the propagation trajectory of CME, the 3D centroid $\mathrm{PATH_{CME}}$ of CME is defined by the relative mass:

$$\mathrm{PATH_{CME}}(\mathrm{r},\theta,\phi)=\frac{\sum_{\mathrm{ijk}}(\mathrm{r_i},\theta_\mathrm{j},\phi_\mathrm{k})\bar{\rho}_{\mathrm{ijk}}\mathrm{r_i^2}\sin\theta_\mathrm{j}\mathrm{dr_i}\mathrm{d}\theta_\mathrm{j}\mathrm{d}\phi_\mathrm{k}}{\sum_{\mathrm{ijk}}\bar{\rho}_{\mathrm{ijk}}\mathrm{r_i^2}\sin\theta_\mathrm{j}\mathrm{dr_i}\mathrm{d}\theta_\mathrm{j}\mathrm{d}\phi_\mathrm{k}}$$

which includes CME central distance from the Sun, CME central latitude and CME central longitude. In the definition of $\mathrm{PATH_{CME}}$, $\overline{\rho}$ denotes the relative density (identified by the 20% enhancement of the density, i.e. $\overline{\rho}=1.2\rho_{\mathrm{wind}}$), and the indices i, j and k run over all grid points in the radial, meridional and azimuthal directions, respectively.

Figure 10.17 shows the propagation trajectory of $\mathrm{PATH_{CME}}$'s latitude and longitude with its distance. The dotted line in the figure shows the position angles of HCS at different heights. The black and red solid lines show the variation of $\mathrm{PATH_{CME}}$'s latitude and longitude away from the Sun-Earth line respectively. From the figure, we find that the CME shows a deflection of about 5° toward the solar equator during its outward propagation from the corona to about 20 R_s, this is due to the effect of the magnetic energy density gradient. Then the CME almost propagates along the HCS in the latitudinal direction in interplanetary space. And the result also displays that the CME does not deflect in longitudinal direction, since the $\mathrm{PATH_{CME}}$'s longitude away from the Sun-Earth line only shows about 2° variation during its outward propagation from the corona to about 1 AU.

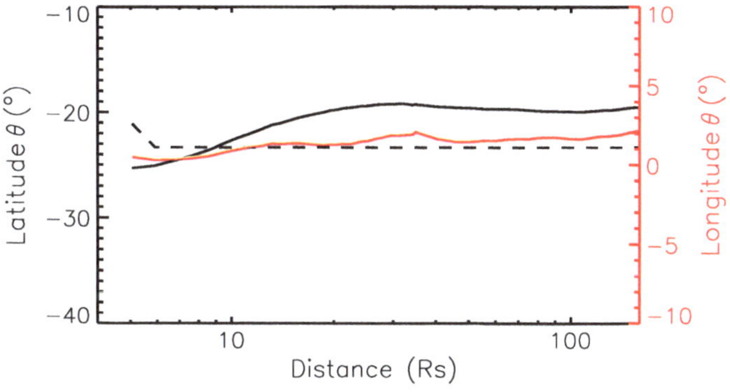

Fig. 10.17 Propagation trajectory of $\mathrm{PATH_{CME}}$'s latitude (black solid line) and longitude (red solid line). The dotted line shows the position angles of HCS at different heights. Reprinted from Fig. 7 of Zhou and Feng [178] with permission of the John Wiley & Sons, Inc

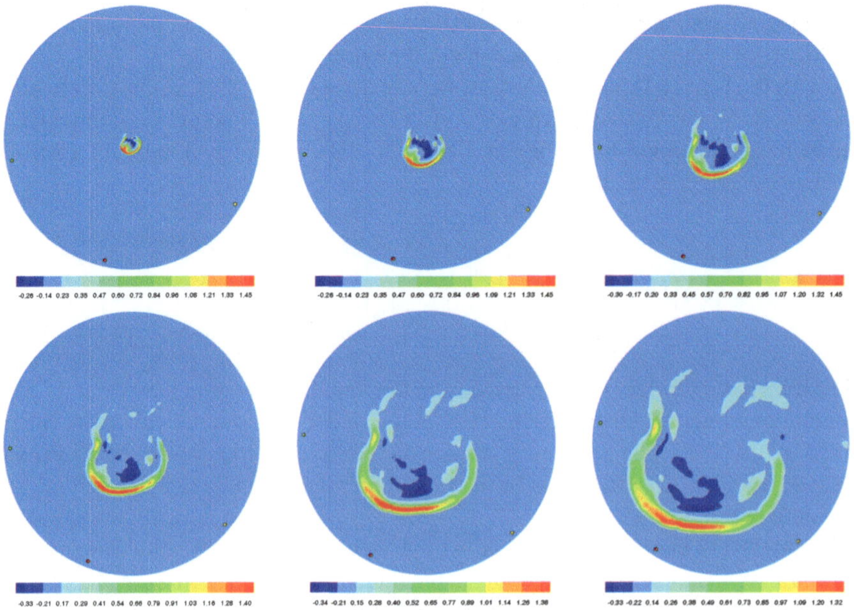

Fig. 10.18 The simulated profiles of the relative density in the x-y azimuthal plane at **a** 5 h, **b** 10 h, **c** 15 h, **d** 20 h, **e** 30 h, and **f** 40 h. The red circle shows the position of the Earth, the yellow and green circles show the position of STERO A and B, respectively. Reprinted from Fig. 8 of Zhou and Feng [178] with permission of the John Wiley & Sons, Inc

Figure 10.18 shows the simulated global view of the relative density in the equatorial plane at different instants. The red circle in the figure shows the position of the Earth. From this figure, we find that the CME almost propagates along the Sun-Earth line. There is almost no pronounced westward or eastward deflection of the CME during its outward propagation because of the little speed difference between the CME and the solar wind background. This is consistent with the observational analysis of the deflection of CMEs' propagation in interplanetary space by Wang et al. [156]. The angle between the initial CME radial direction and the IMF at 1AU calculated from the simulated solar wind background results is 36°, the Sun rotates about 33° from the east to west when the magnetic cloud center arrives at the Earth. This means that the CME almost moves outward along the Parker spiral interplanetary magnetic field. An asymmetry structure of the CME is clearly seen in heliolongitude as it passes into non-uniform solar wind ambient, and the asymmetry of the CME in heliolatitude can also be clearly seen from Fig. 10.18.

In the following, we study the evolution of the spread angle of the corresponding CME. Because the wing of the CME can encounter the spacecraft, even if the center of the CME misses it. If the Earth is near the spacecraft, this means that the wing of the CME can interact with the magnetosphere and may create a severe geomagnetic storm. We define the spread angle by means of the relative density contour plots

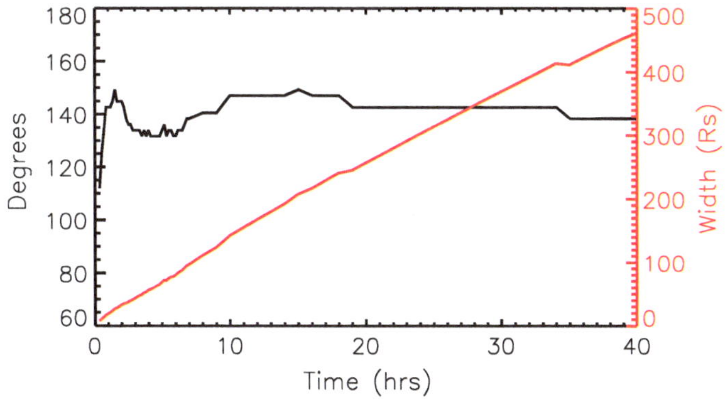

Fig. 10.19 The evolution of the spread angle (black) and meridional width (red) of the CME in time. Reprinted from Fig. 9 of Zhou and Feng [178] with permission of the John Wiley & Sons, Inc

with the relative density $\overline{\rho}$ being 1 (i.e. $\overline{\rho} = 2\rho_{\text{wind}}$). The spread angle of this contour measured from the solar center is defined as the spread angle of the CME. Though this definition is rather simple, at least it is based on mathematics and can reflect the propagation characteristics of the CME to some extent. Figure 10.19 shows the evolution of the spread angle (black line) and meridional width (red line) in time. The spread angle quickly evolves to 145° just 1.5 h after the CME initiation, at which time the CME center is at a distance of 10 R_s. The quick increase of spread angle during the first 1.5 h of the CME's propagation is clearly correlated with the rapid expansion due to the initial high speed, pressure, and plasma density plasmoid imposed on the solar wind ambient. Meanwhile, the CME's velocity rapidly decelerates, as seen from Fig. 10.20a. Though it has some small fluctuations for the first 10 h, the meridional width of the CME continues to expand, as seen from the red line in this figure. Afterwards, the spread angle almost remains constant. This means that the CME continues to expand and the meridional width of the CME increases during its propagating outward in IP space (see the red line in Fig. 10.19). The CME's expansion should continue even if it passes the spacecraft until a pressure equilibrium is established. This have been approved by previous works (e.g. [53, 153]). By studying the evolution of the radial width of the ICMEs (up to ~5 AU), these authors found that ICMEs expanded and the average width of ICMEs increased as traveling outward, since the internal pressure was generally larger than that of the surrounding wind.

One of the central points of space-weather forecasting is the prediction of ICME arrival at Earth. It is known that the CME velocity is a key factor affecting its arrival time. Figure 10.20 shows the CME's velocity along the Sun-Earth line vs time. From the figure, we find that the CME rapidly accelerates to about 1200 km/s just 0.5 h after the CME initiation, then the speed begins to decelerate to about 800 km/s until the CME arrives at the Earth. By comparison with the WIND observation data (marked by the cross in Fig. 10.20), one can find that the simulation result and observation

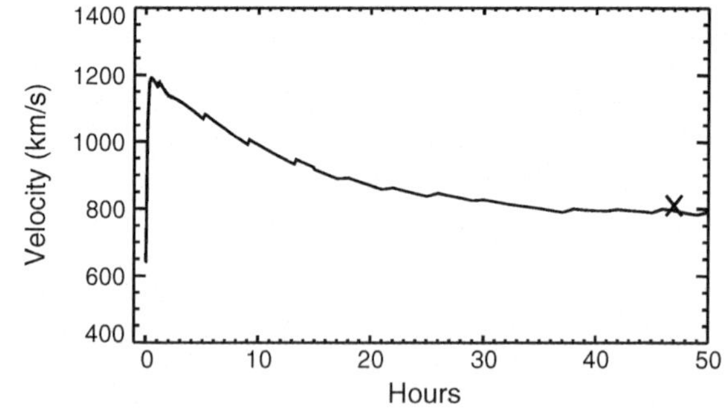

Fig. 10.20 Time-velocity profile along the Sun-Earth line, the cross marks the WIND observation data. Reprinted from Fig. 10 of Zhou and Feng [178] with permission of the John Wiley & Sons, Inc

result are almost the same. Xie et al. [171] gave three approximations of the velocity along the Sun-Earth line by observation data and approved that the Harmonic Mean method gave the smallest error, and the simulated result is similar to that by the Harmonic Mean method. This implies the effectiveness of the numerical result.

Figure 10.21a gives the velocity of the CME's center. From the figure, it is found that the speed of the CME's center rapidly accelerates to about 1600 km/s just several minutes after the CME initiation due to the imposed initial pressure imbalance. Then the speed quickly begins to decelerate to about 936 km/s at 4 h after the CME initiation, when the CME center already reaches 24 R_s in interplanetary space. The rapid deceleration with a relatively short time is associated with the interactions with the solar wind ambient due to the large drag by the background medium and the quick expansion of the CME (see Fig. 10.19). Figure 10.21b shows the velocity of the center as a function of distance as a solid line. We find that the variety of the velocity along heliocentric distance within 24 R_s is similar to the empirical relationship (dashed line in Fig. 10.21b) put forth by Sheeley et al. [135] and simulated by Manchester et al. [100]. Sheeley et al. [135] described the deceleration of fast halo CMEs observed with LASCO, by giving the following empirical relationship:

$$r(t) = r_0 + v_1 t + (1 - e^{-t/\tau})(v_0 - v_1)\tau$$

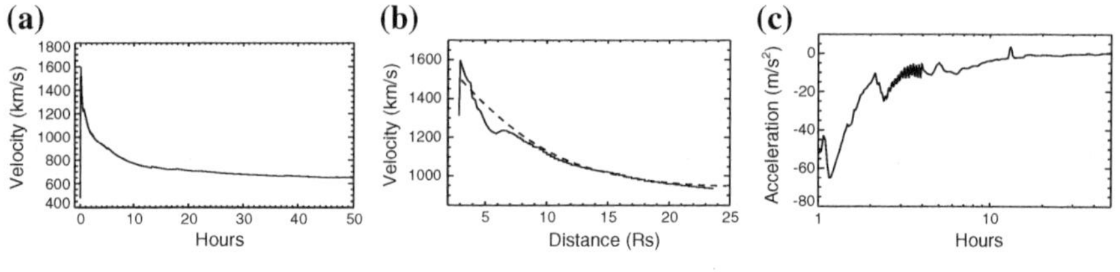

Fig. 10.21 Time-velocity profile (**a**), velocity-distance profile (**b**), and time-acceleration profile (**c**) along the center of CME. The dashed line in the middle figure shows an empirical relationship expressed by an exponential decay in velocity that closely approximate the evolution of the model CME [135]. Reprinted from Fig. 11 of [178] with permission of the John Wiley & Sons, Inc

$$v(t) = v_1 + (v_0 - v_1)e^{-t/\tau}$$

It is found that for $r_0 = 3.0\ R_s$, $v_0 = 1503$ km/s, and $v_1 = 936$ km/s and $\tau = 1.0$ hour, the above formula gives a close fit to the numerical result. After 4 h, the speed keeps on slowly decreasing to 658 km/s through the interaction with the solar wind background till the IP shock arrives at 1 AU. The result verifies that the velocity profile of the CME's center includes three phases: an impulsive acceleration near the corona, a faster deceleration phase and then a more gradual deceleration phase during the fast CME event propagation. In order to better demonstrate the variety of the velocity, the corresponding acceleration along the CME's center from 1 to 50 h is shown in Fig. 10.21c. We can find that the acceleration along the CME's center is about -55 m/s^2 at 1 h, and this value is very similar to that obtained by the acceleration $a(t) = -(v_0 - v_1)/\tau e^{-t/\tau}$ for the fast CME event in Sheeley et al. [135]. Then it rapidly rises to -10 m/s^2 at 4 h after the CME initiation, and slowly rises to about -0.5 m/s^2 when the shock arrives at 1 AU. During the time, the speed dropped from 936 to 658 km/s in about 39.5 h, the corresponding acceleration is -1.95 m/s^2, which is consistent with the IP acceleration $a = 2.193 - 0.0054u$ from the empirical shock arrival (ESA) model of Gopalswamy et al. [51], where u is the CME speed. For $u = 750$ km/s, $a = -1.857$ m/s^2.

10.5 Numerical Study of the Latitudinal Deflection of CME

This section analyzes and studies the latitudinal deflection (LD) of CME during its outward propagation, and quantifies the effect of the background magnetic field and the initiation parameters (such as the initial magnetic polarity and the parameters of the CME model) on the LD of CMEs. In order to study the deflection of CME in the latitudinal direction, we choose the steady-state solar wind during Carrington rotation 1922 as the background solar wind, in which the well-studied halo CME event on 12 May 1997 was observed, and the corona has a relatively simple configuration. Firstly, we compute the steady-state background solar wind, and try to mimic the 12 May 1997 halo CME event in the period. Then we study the LD of CME and the factors influencing it based on the same background field configuration.

10.5.1 Solar Wind Ambient

Figure 10.22a depicts a 2D cut of magnetic structure and velocity for the steady-state solar wind in the solar corona (SC) domain in the meridional plane. Figure 10.22b shows the three-dimensional heliospheric current sheet in the inner heliospheric (IH) domain. The color contours indicate the radial speed, and the magnetic field is represented by solid black lines. From the figures, we can see that the coronal field structure has a typical characteristic at solar minimum when the corona has a

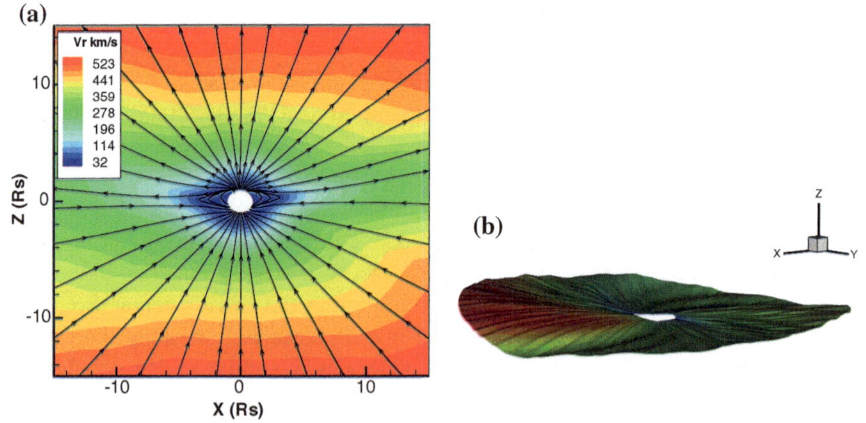

Fig. 10.22 **a** Magnetic field and radial speed for the steady-state solar wind solution in the SC domain in the solar-terrestrial meridian plane. **b** Steady-state current sheet in the IH domain. The color contours represent the radial speed. Reprinted from Fig. 1 of Zhou and Feng [175] with permission of the John Wiley & Sons, Inc

relatively simple configuration. The coronal magnetic field is dominated by a dipole pattern, and the HCS is nearly in the heliospheric equatorial plane, and the latitudinal excursion of HCS is very limited. A helmet streamer is formed at low latitude around the equator with low radial speed. An open configuration of the magnetic field is achieved at high latitudes with high radial speed. The steady-state solar wind reveals an approximate bimodal outflow pattern, slow-speed wind leaves the Sun around 350 km s^{-1} near the current sheet region, and high-speed wind leaves above 600 km s^{-1} in the high-latitude region.

In order to mimic the 12 May 1997 halo CME event, the parameters of the CME model in Sect. 10.2 are empirically chosen to try to match the WIND data. The CME's center is placed at 2.0 R_s, and the radius of the initial CME, a_{cme}, is 0.8 R_s. The value of $v_{\max}$ is set to be 600 km s^{-1}. $\rho_{\max}$ and $T_{\max}$ are set to be five times and three times the ambient solar wind value at the center of the CME, respectively. The maximal magnetic field strength B_0 is 5.0 G. The CME is initially centered at N21W07 to conform to the location of the observed flare/CME. For a better agreement with the IMF observed by spacecraft, the polar axis of the ejected spheromak is rotated toward the negative y-axis direction to make it incline about 5° from the negative y-axis. The initial plasma plasmoid has added to the corona 8.22×10^{24} J of magnetic energy, 3.89×10^{23} J of kinetic and 1.76×10^{24} J of thermal energy. The increase in total energy (magnetic+kinetic+thermal energy) is 1.04×10^{25} J. In the simulation, we take 04:35 UT on 12 May 1997, when EIT first observed the CME in the lower corona, to be the initial time of the eruption.

10.5.2 Analysis of the Propagation Trajectory of CMEs

This part studies the propagation trajectory of CMEs and analyzes the factors that influence the propagation trajectory of the corresponding CME, in particular the LD of CME during its outward propagation up to IP space. Figure 10.23 shows the evolution of CME on the solar-terrestrial meridian plane at four different instants (t = 1.0, 2.0, 5.0 and 10.0 h) after adding the force-free configuration, travelling from the corona to IP space. The red lines in the panels show the position angle of 12 May 1997 CME event observed, which is at the northern hemisphere 21°. The yellows lines are 60° north and south away from the observed position angle of CME. From the figure, we find that this CME shows an equatorward deflection.

Why does the CME deflect toward the low-latitude region near the solar equator (roughly symmetrical to the heliospheric current sheet)? We can explain the reason by applying the theoretical method proposed by Shen et al. [136] to the CME, in which the direction and magnitude of the deflection are well consistent with the gradient (with the conventional minus sign in front) of magnetic energy density, $< -\nabla\omega >$, where $\omega = \frac{B^2}{2\mu_0}$ is the magnetic energy density. The magnetic energy density distribution of the simulated steady-state solar wind at 3 R_s is shown in Fig. 10.24a. The asterisk sign in the figure denotes the associated source location of the halo CME. From this figure, it is seen that the distribution of the background magnetic energy density is nonuniform. The black dashed line in the figure shows the corresponding position of the HCS, which separates regions of the solar wind where the magnetic field points toward or away from the Sun. We can see that the magnetic energy density increases gradually above and below the HCS, and the direction of

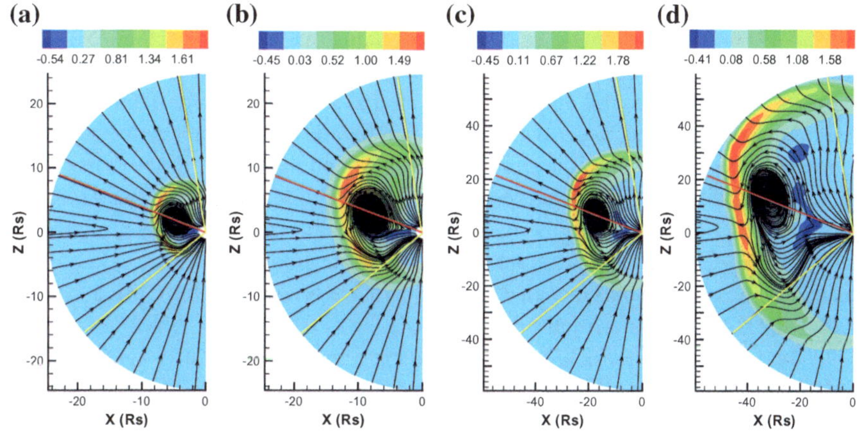

Fig. 10.23 Contour plots of the relative density distribution for the CME on the solar-terrestrial meridian plane after 1.0, 2.0, 5.0 and 10.0 h. The black lines represent the magnetic field. The relative density is defined as follow: $\overline{\rho} = (\rho - \rho_{\text{wind}})/\rho_{\text{wind}}$. Reprinted from Fig. 4 of Zhou and Feng [175] with permission of the John Wiley & Sons, Inc

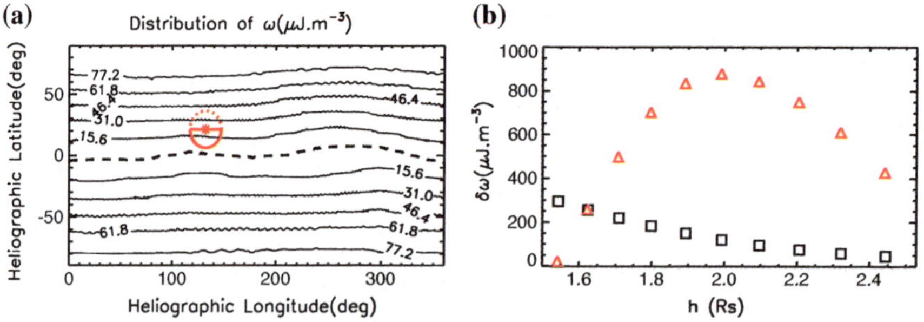

Fig. 10.24 **a** Distribution of the magnetic energy density ω at the height of 3 Rs. The dashed line shows the neutral line at this height. The red asterisk sign shows the position of this CME at the height. The regions enclosed by the red dotted line and the solid line show the selected upper and lower part, respectively. **b** The variation of $\delta\omega$ of the CME (red triangles) and the background magnetic energy density (black squares) with the height. Reprinted from Fig. 5 of Zhou and Feng [175] with permission of the John Wiley & Sons, Inc

the magnetic energy density gradient is toward the HCS, where the magnetic energy density is lowest. The magnetic energy density gradient makes the CME deflect to the region with lower magnetic-energy density. The figure mainly reflects the influence of this background magnetic energy density on CME deflection in the latitudinal direction. This magnetic configuration, due to the non-radial magnetic energy density gradient, will supply a force compelling the ejecta to move toward the low-latitude region.

Figure 10.24b analyzes the variation of $\delta\omega$ in the corona near the associated source location of the halo CME in, where $\delta\omega = \omega_U - \omega_L$, and ω_U and ω_L are the average energy density in the upper and the lower part of the simulated CME, respectively. The CME-perturbed magnetic field will act an equatorward force on the CME if $\delta\omega$ is positive, while a polar-ward force would act on the CME when $\delta\omega$ is negative. The regions enclosed by the red dotted line and the red solid line in Fig. 10.24a show the selected upper part and lower part in a 15° angle range. The squares in Fig. 10.24b show the variation of $\delta\omega$ of the simulated background magnetic energy density, and the red triangles show the variation of $\delta\omega$ after CME initiation. From the figure, we find that $\delta\omega$ of the CME and the background magnetic energy density are positive. $\delta\omega$ of the CME reaches maximum at 2 R_s, then it gradually decreases along the distance. The result means that an equatorward force is acting on the CME and could deflect the CME to a lower latitude.

In the following, the factors, which influence the deflection propagation of the CME, are investigated. We first analyze and quantify the effect of the initial flux rope magnetic field polarity on the evolutionary features of the CMEs in IP space. Depending on the parameter α, two different cases of CMEs simulations are carried out. In the above simulation, $\alpha > 0$, the magnetic field direction of the CME on the solar-terrestrial meridian plane is parallel to that of the background solar wind, which is referred as "parallel" CME. When $\alpha < 0$, the magnetic field direction of the

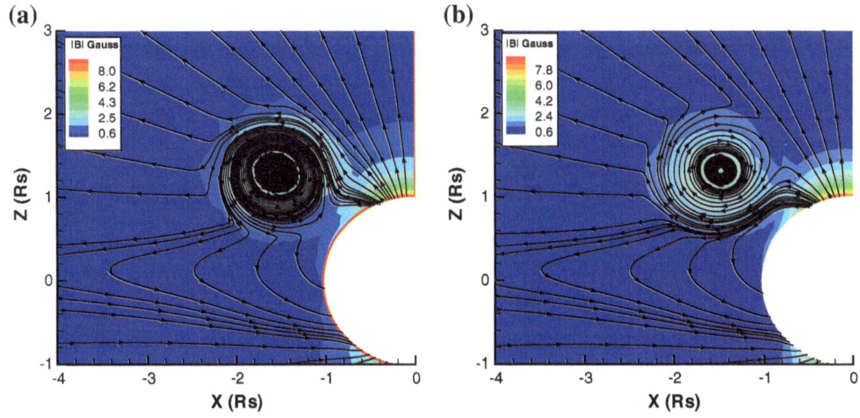

Fig. 10.25 magnetic field configuration in the y = 0 plane in Case A (**a**) ("parallel" CME) and Case B (**b**) ("anti-parallel" CME). The color contours represent the magnetic field strength. Reprinted from Fig. 6 of Zhou and Feng [175] with permission of the John Wiley & Sons, Inc

CME on the solar-terrestrial meridian plane is anti parallel to that of the background magnetic field, which is referred as "anti-parallel" CME. In order to better study the characteristics of the CME propagation, in particular the LD of CME, the "parallel" CME and "anti-parallel" CME are launched at higher latitude and the same longitude of the halo CME simulated above: such as N41W07. The two CMEs are labeled as Cases A and B, respectively. Figure 10.25 shows schematic pictures of the initial magnetic field on the meridional plane through the center for Case A (Fig. 10.25a) and Case B (Fig. 10.25b). In the numerical simulations, the CMEs have the same initial density, temperature, velocity and magnetic field strength.

Figure 10.26 shows the evolution of the CMEs on the solar-terrestrial meridian plane at four different instants (t = 1.0, 2.0, 5.0 and 10.0 h) for Cases A and B, respectively, traveling from the corona to IP space. The top row shows the result of Case A, the bottom row displays the result of Case B. The red lines in the figures show the position angle of CMEs which are launched at N41. From the figure, it is clear that the "parallel" CME shows an equatorward deflection. Because the HCS is located near the solar equator during this period, the "parallel" CME deflects to the HCS. The "anti-parallel" CME veers toward the pole, because the direction of the background magnetic field is opposite to the magnetic field of the CME, and a strong magnetic reconnection occurs. The bottom row of Fig. 10.26, shows that the magnetic flux of the flux rope is peeling off due to the reconnection with the background field as CME propagates outward. Finally, the flux rope no longer exists after most of the magnetic flux is reconnected.

In order to quantitatively study the evolution path and LD of CME, we define the 2D center and CME central latitude on the solar-terrestrial meridian plane by the relative mass:

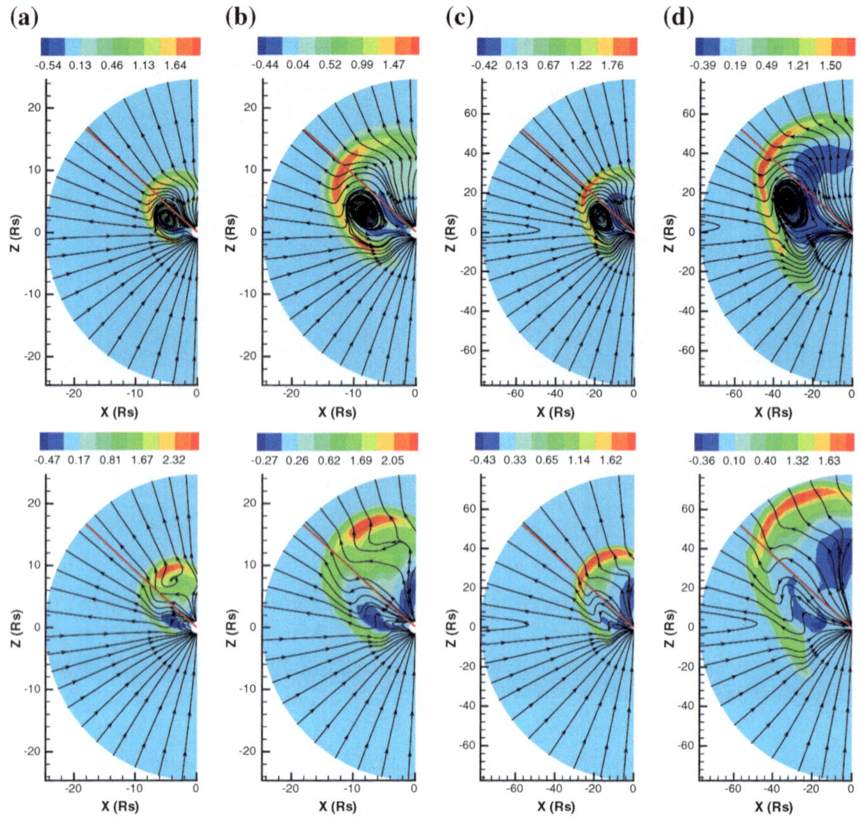

Fig. 10.26 Contour plots of the relative density distribution for the CME on the solar-terrestrial meridian plane after 1.0 (**a**), 2.0 (**b**), 5.0 (**c**) and 10.0 (**d**) h. The black lines represent the magnetic field. Above: "parallel" CME; below: "anti-parallel" CME. Reprinted from Fig. 7 of Zhou and Feng [175] with permission of the John Wiley & Sons, Inc

$$\mathrm{CORM}(r) = \frac{\sum\limits_{ij} \bar{\rho}_{ij} r_i^2 dr_i d\theta_j}{\sum\limits_{ij} \bar{\rho}_{ij} r_i dr_i d\theta_j}, \quad \mathrm{CORM}(\theta) = \frac{\sum\limits_{ij} \bar{\rho}_{ij} \theta_j r_i dr_i d\theta_j}{\sum\limits_{ij} \bar{\rho}_{ij} r_i dr_i d\theta_j}$$

where $\overline{\rho}$ denotes the relative density (identified by the 20% enhancement of the density, i.e. $\overline{\rho} = 1.2\rho_{\text{wind}}$), and the indices i and j run over all grid points in the radial and azimuthal directions on the solar-terrestrial meridian plane, respectively.

Figure 10.27 shows the propagation trajectory of CME central latitude on the solar-terrestrial meridian plane after 10 h for Cases A and B, respectively. Data are sampled every 20 min. Figure 10.27a shows the latitude-time profiles, and Fig. 10.27b shows the latitude-distance profiles. The squares in the figure show the position angles of HCS at different heights. The red diamonds show the propagation trajectory of the CME central latitude for Case A, and the blue diamonds show the result of Case B.

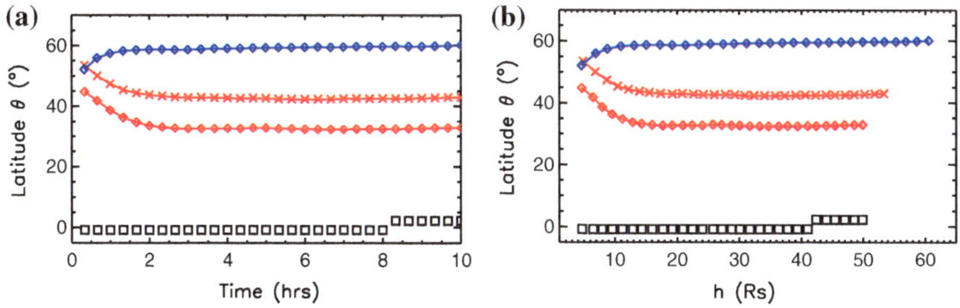

Fig. 10.27 The propagation trajectory of Case A (red diamonds), Case B (blue diamonds) and Case C (red crosses). The squares indicate position angles of HCS. Panel **a** to **b** show the latitude-time, and latitude-height plots, respectively. Reprinted from Fig. 8 of Zhou and Feng [175] with permission of the John Wiley & Sons, Inc

Fig. 10.28 Squares: The variation of $\delta\omega$ of the background magnetic energy density. Red triangle: The variation of $\delta\omega$ for Case A. Blue triangles: The variation of $\delta\omega$ with the minus sign for Case B. Reprinted from Fig. 9 of Zhou and Feng [175] with permission of the John Wiley & Sons, Inc

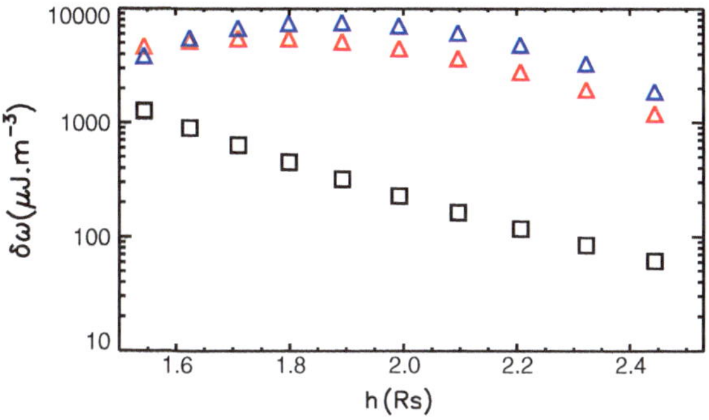

It can be seen that the "parallel" CME deflects to the HCS and the "anti-parallel" CME deflects toward the polar region at the beginning, then the CMEs propagate almost along the radial direction in IP space. This is similar to the scenario of Chané et al. [17]. From the figure, one can find that the "parallel" CME shows a deflection of about 8.5° toward the HCS during its outward propagation from the corona to IP space, and the "anti-parallel" CME shows a deflection of about 18° toward the pole.

The reason of these different configurations between Cases A and B is shown in Fig. 10.28, which shows the variation of $\delta\omega$ in the corona near the center of the CMEs. The squares in the figure show the variation of $\delta\omega$ of the background magnetic energy density, the red triangles in the figure show the variation of $\delta\omega$ after CME initiation for Case A, the blue triangles show the result of Case B with the minus sign in front. From the figure, it is found that $\delta\omega$ of the background magnetic energy density at N41W07 is larger than that of the background magnetic energy density at N21W07 (compare with Fig. 10.24). $\delta\omega$ of coronal magnetic energy density after CME initiation is positive for the "parallel" CME, but is negative for the "anti-parallel" CME. This indicates that an equatorward force is acting on the "parallel" CME, and it could be deflected to a lower latitude. A polar-ward force is acting on the "anti-parallel" CME, and it could be deflected to a higher latitude.

Also, here is simulated the CME labeled as Case C, which is launched from higher latitude at N61W07 and has the same polarity as that of Case A. Figure 10.27 shows the propagation trajectory of CME central latitude on the solar-terrestrial meridian plane after 10 h for Case C by using the red crosses. From the figure, it can be seen that Case C shows a deflection of about 18° toward the HCS. The deflection extent is larger than that of Case A, because the magnetic energy density gradient at higher latitudes is stronger. This result indicates that during solar minimum, even CMEs that are originated from high latitude can be easily deflected toward the equator and eventually result in geoeffective events. From the results of Fig. 10.27, it is found that the CME deflection almost occurs within the corona. In the corona, the magnetic field is dominant, the magnetic energy density gradient makes the CME deflect to the low magnetic-energy density region. The magnetic energy density gradient decreases quickly when a CME propagates outward. When the CME propagates into the interplanetary space, a weak magnetic energy density gradient is not sufficient to make the CME deflect.

In fact, we are most interested in the CME that deflects toward the equator. In the following, in order to study the effect of the four parameters of the CME model on the equatorward deflection of CME, a total of 8 cases of numerical experiments are performed by adjusting the four parameters of the CME model: decreasing the magnetic field, increasing the density and temperature, and decreasing the velocity, based on Case A (referred as Case 1 hereafter). Wherein Cases 2–5 are performed by adjusting one of the four parameters based on Case 1, and Cases 6–8 are performed by adjusting one of the three parameters except the magnetic field parameter based on Case 5. To obtain these results, the amplitudes of the four perturbations (B, ρ, T and V) are scaled by δB, $\delta\rho$, δT, and δV, respectively. Quantitatively, the values $\delta B = 5$ and $\delta B = 3$ mean that the maximal magnetic field strength B_0 is 5.0 G and 3.0 G, respectively. Similarly, $\delta V = 300$ (600) means that the value of v_{max} is set to be 300 (600) km s^{-1}. $\delta\rho = 5$ (10) corresponds to a density rise of a factor of 5 (10) from the background solar wind at the center of the CME; and $\delta T = 3$ (6) corresponds to a temperature rise of a factor of 3 (6). The experiments are carried out by using these various perturbations with the corresponding energy inputs shown in Table 10.1. Figure 10.29 shows the propagation trajectory of CME central latitude on the meridian plane after 10 h for the 8 cases in Table 10.1.

From the figure, one can find that Case 5 shows a deflection of about 15.5° toward the HCS, the deflection extent of Case 5 is clearly larger than that of Case 1. But the deflection extent of Cases 2–4 is almost the same as that of Case 1, and the difference between the deflection extent of Cases 6–8 and that of Case 5 is also very small. So it results that the magnetic field parameter of CME model has the greatest impact on the propagation trajectory of the CME, and that the smaller the magnetic field strength is, the larger the deflection extent is. When the magnetic field parameter is smaller, the ratio of the background magnetic field and the magnetic field parameter of CME model becomes larger. Then the influence of background magnetic energy density gradient on the equatorward deflection of CME is greater. However, the other three parameters do not have significant effect on the propagation trajectory of the CME.

Table 10.1 Physical characteristics of the perturbations and their corresponding energy contents for all 11 cases of the numerical experiments. Reprinted from Table 1 of Zhou and Feng [175] with permissions of the John Wiley & Sons, Inc

Case	δ_B	δ_ρ	δ_T	δ_V	Magnetic energy (J)	Kinetic energy (J)	Thermal energy (J)	Total energy (J)
1	5.0	5	3	600	8.13×10^{24}	2.78×10^{23}	1.06×10^{24}	9.47×10^{24}
2	5.0	10	3	600	8.13×10^{24}	4.77×10^{23}	1.85×10^{24}	1.05×10^{25}
3	5.0	5	6	600	8.13×10^{24}	2.78×10^{23}	1.79×10^{24}	1.01×10^{25}
4	5.0	5	3	300	8.13×10^{24}	8.72×10^{22}	1.06×10^{24}	9.28×10^{24}
5	3.0	5	3	600	2.94×10^{24}	2.78×10^{23}	1.06×10^{24}	4.28×10^{24}
6	3.0	10	3	600	2.94×10^{24}	4.77×10^{23}	1.85×10^{24}	5.27×10^{24}
7	3.0	5	6	600	2.94×10^{24}	2.78×10^{23}	1.79×10^{24}	5.01×10^{24}
8	3.0	5	3	300	2.94×10^{24}	8.72×10^{22}	1.06×10^{24}	4.09×10^{24}
9	5.0	10	3	600	8.13×10^{24}	3.11×10^{23}	1.15×10^{24}	9.58×10^{24}
10	5.0	10	3	600	8.13×10^{24}	4.05×10^{23}	1.81×10^{24}	1.03×10^{25}
11	5.0	10	3	600	8.13×10^{24}	2.50×10^{23}	8.93×10^{23}	9.21×10^{24}

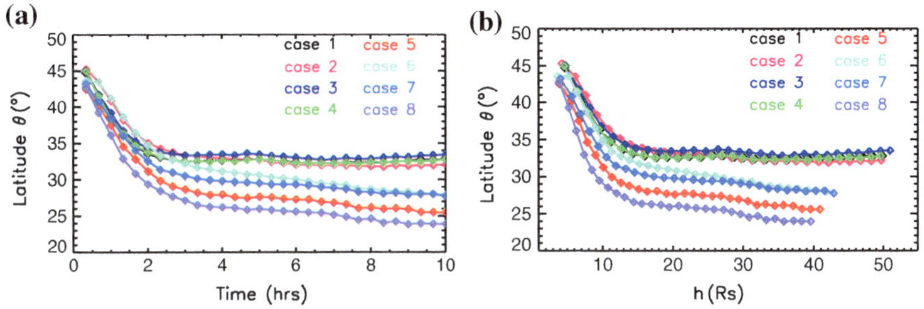

Fig. 10.29 The propagation trajectory for the 8 cases. Panel **a** and **b** show the latitude-time plot and latitude-height plot, respectively. Reprinted from Fig. 10 of Zhou and Feng [175] with permission of the John Wiley & Sons, Inc

Meanwhile, we study the effect of the background magnetic field on the propagation trajectory of the equatorward deflection of CME. For example, the background solar wind is enhanced with 2 times (Case 9) and 4 times (Case 10) the observed photospheric magnetic field. Based on the two types of background wind, we study the propagation trajectory of the CME, which has the same initial conditions as that of Case A, including the polarity. Figure 10.30 shows the propagation trajectory of the CME central latitude of the Cases 9–10 on the solar-terrestrial meridian plane. From the figure, it is found that Case 9 shows a deflection of about 17° toward the HCS, and the deflection extent of Case 10 is about 18°. By comparing the result of Case 1, we find that the background magnetic field substantially affects the propagation trajectory of the CMEs. It is clear that the larger the background magnetic field strength is, the larger deflection extent is. The magnetic field energy density gradient of Cases 9-10 is clearly much larger than that of Case 1 (compare Fig. 10.31 with

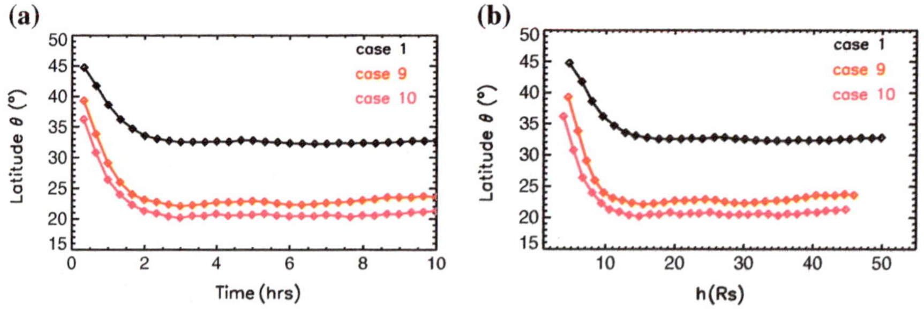

Fig. 10.30 The propagation trajectory for Case 1, Case 9 and Case 10. Panel **a** to **b** show the latitude-time, and latitude-height plots, respectively. Reprinted from Fig. 11 of Zhou and Feng [175] with permission of the John Wiley & Sons, Inc

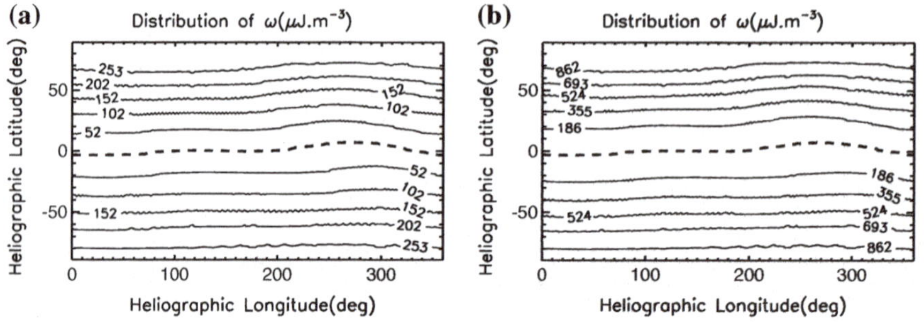

Fig. 10.31 Distribution of the magnetic-energy density ω at the height of 3 Rs for Case 9 and 10, respectively. The dashed lines show the neutral lines at this height. Reprinted from Fig. 12 of Zhou and Feng [175] with permission of the John Wiley & Sons, Inc

Fig. 10.24a). Similarly, Gui et al. [56] provided that there was a positive correlation between the deflection rate and the strength of the magnetic energy density gradient. Figure 10.32 shows the evolution of the CME on the solar-terrestrial meridian plane at four different instants (t = 1.0, 2.0, 5.0 and 10.0 h) for Case 9. From the figure, we can see that the flux rope deflects toward the HCS and finally gets into the streamer, and then propagates almost along it.

Furthermore, the deflection of all the above cases is studied except Case B, which is only one example that deflects toward the polar region. Figure 10.33a shows the correlation between the LD extent and average transit velocity for the 11 cases. Here, Case C is labeled as Case 11. The average transit velocity is calculated by simply employing the ratio of the distance between the simulated 2D CME center (say at R_2) at 10 h and its beginning (say at R_1) to 10 h, that is $(R_2 - R_1)/(10 \times 3600)$. Wherein, the green plus marks the data of Case 11. The correlation coefficient is only −0.355795 for Cases 1–11. From the figure, it is found that Cases 9–11 are far away from the majority of points. Without Cases 9–11, it is seen that there is a clear anti-correlation between the deflection extent and the CME average transit speed for Cases 1–8, which are based on the same background magnetic field and originated from the same source region. And the correlation coefficient increases to −0.987874

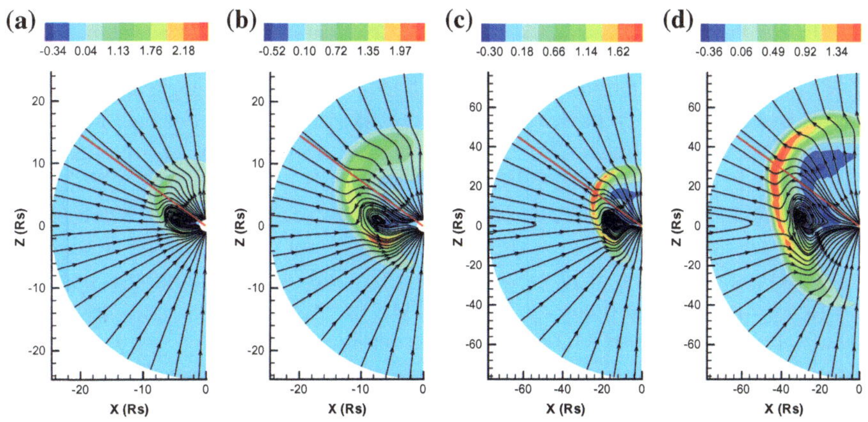

Fig. 10.32 The evolution of the CMEs on the solar-terrestrial meridian plane at four different instants (t = 1.0 (**a**), 2.0 (**b**), 5.0 (**c**) and 10.0 (**d**) h) for Case 9. The black lines represent the magnetic field. Reprinted from Fig. 13 of Zhou and Feng [175] with permission of the John Wiley & Sons, Inc

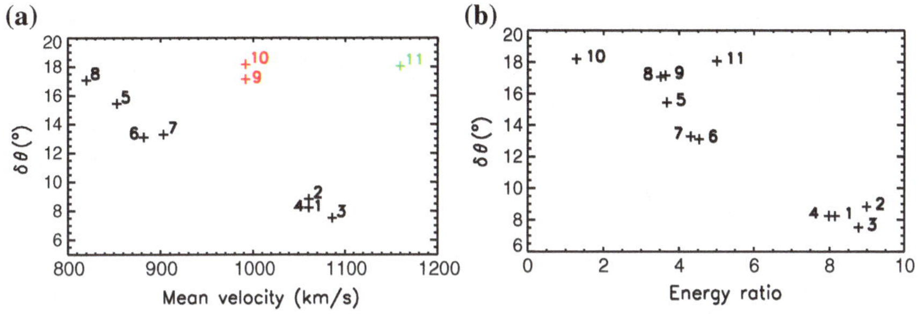

Fig. 10.33 Two panels show the plots between the LD extent versus (**a**) average transit velocity, (**b**) energy ratio. Reprinted from Fig. 14 of Zhou and Feng [175] with permission of the John Wiley & Sons, Inc

without Cases 9–11. It is noted that Gui et al [56] found a weak anti-correlation between the deflection rate and the speed. Figure 10.33b shows the relation between the LD extent and the energy ratio $E_{\mathrm{cme}}/E_{\mathrm{sw}}$ within the perturbed source region, where E is the total energy. Table 10.1 provides the corresponding energy of the eleven cases. Thus, this sample shows an anti-correlation between the deflection extent and the energy ratio. The correlation coefficient is about –0.914621.

10.6 Discussions

Currently, the observations available have not been able to track CMEs from the initiation in the solar corona to propagation in the heliosphere, to adequately constrain the 3D MHD space weather models. The CME initiation processes in most 3D MHD models are theoretical (e.g., see review paper by Chen [19], Green et al. [54] and references therein). Some pre-defined form of magnetic field structures and/or pressure/velocity pulses (e.g., [14, 160]) are added to the background to trigger CMEs in the model. While these methods are beneficial to investigate the fundamental physics of CME initiation. A more self-consistent way to initiate a CME, especially for the forecast purpose, seems to use a 3D MHD active region model to reconstruct a flux rope structure, and then input this flux rope structure to a 3D MHD global solar wind model to launch a CME. Such a method is expected not to rely on any empirical or educated CME initiation process. Jiang et al. [64] presented a self-consistent 3D MHD simulation directly driven by 3-day sequential data of surface vector magnetograms without any priori hypotheses. The study can realistically simulate the whole process from origin to onset of a solar eruption. An effort of such simulation has been carried out by Wu et al. [169] by a data-driven 3-D MHD active region evolution (MHD-DARE) model. Strictly speaking, however, in this study both the ambient solar wind and the flux rope are generated not in a data-driven but in a data-constrained manner. It is promising to consider how to couple the data-driven ambient simulation introduced in Chap. 7 and the data-driven active region simulation introduced in Chap. 9 to conduct a all-data-driven simulation of the Sun-Earth system, which would be a step forward in developing more self-consistent modelling for space weather research and prediction. Pagano et al. [124] proposed a technique for specifying the pre-eruptive plasma properties in the corona. Then, time dependent MHD simulations describe the ejection of magnetic flux rope to produce time dependent boundary conditions for the magnetic field and plasma at 4 R_s, and thus to provide interplanetary space weather forecasting models with accurate time dependent boundary conditions of erupting magnetic flux ropes in the upper solar corona. This technique is a viable one to improve predictive capability for the occurrence of events on the Sun relevant to space weather.

In the future, further basic research and development are encouraged to better understand the physics of CME origin that can be transitioned to enhanced forecasting capabilities. It can certainly be expected that major advancements in the observations will be encountered as further space missions off the Sun-Earth line and in the inner heliosphere gradually fill the gaps.

References

1. Amari T, Luciani JF, Mikić Z, Linker J (2000) A twisted flux rope model for coronal mass ejections and two-ribbon flares. Astrophys J Lett 529:L49–L52. https://doi.org/10.1086/312444
2. Amari T, Luciani JF, Aly JJ, Mikić Z, Linker J (2003) Coronal mass ejection: initiation, magnetic helicity, and flux ropes. I. Boundary motion-driven evolution. Astrophys J 585:1073–1086. https://doi.org/10.1086/345501
3. Amari T, Luciani JF, Aly JJ, Mikić Z, Linker J (2003) Coronal mass ejection: initiation, magnetic helicity, and flux ropes. II. Turbulent diffusion-driven evolution. Astrophys J 595:1231–1250. https://doi.org/10.1086/377444

4. Antiochos SK (1998) The magnetic topology of solar eruptions. Astrophys J Lett 502:L181–L184. https://doi.org/10.1086/311507
5. Antiochos SK, DeVore CR, Klimchuk JA (1999) A Model for Solar Coronal Mass Ejections. Astrophys. J. 510:485–493. https://doi.org/10.1086/306563
6. Aschwanden MJ (2019) New millennium solar physics. Springer Nature, Cham, Switzerland
7. Aschwanden MJ, Burlaga LF, Kaiser ML, Ng CK, Reames DV, Reiner MJ, Gombosi TI, Lugaz N, Manchester WB IV, Roussev II, Zurbuchen TH, Farrugia CJ, Galvin AB, Lee MA, Linker JA, Mikić Z, Riley P, Alexander D, Sandman AW, Cook JW, Howard RA, Odstrčil D, Pizzo VJ, Kóta J, Liewer PC, Luhmann JG, Inhester B, Schwenn RW, Solanki SK, Vasyliunas VM, Wiegelmann T, Blush L, Bochsler P, Cairns IH, Robinson PA, Bothmer V, Kecskemety K, Llebaria A, Maksimovic M, Scholer M, Wimmer-Schweingruber RF (2008) Theoretical modeling for the STEREO mission. Space Sci Rev 136:565–604. https://doi.org/10.1007/s11214-006-9027-8
8. Aulanier G (2013) The physical mechanisms that initiate and drive solar eruptions. Proc Int Astron Union 8(S300):184–196. https://doi.org/10.1017/S1743921313010958
9. Aulanier G, Török T, Démoulin P, DeLuca EE (2010) Formation of torus-unstable flux ropes and electric currents in erupting sigmoids. Astrophys J 708:314–333. https://doi.org/10.1088/0004-637X/708/1/314
10. Balogh A, Lanzerotti LJ, Suess ST (2008) The Heliosphere through the solar activity cycle. Springer-Praxis Books and Springer Science+Business Media. https://doi.org/10.1007/978-3-540-74302-6
11. Barnes A (1992) Acceleration of the solar wind. Rev Geophys 30(1):43–55. https://doi.org/10.1029/91RG02816
12. Bemporad A, Zuccarello FP, Jacobs C, Mierla M, Poedts S (2012) Study of multiple coronal mass ejections at solar minimum conditions. Solar Phys 281:223–236. https://doi.org/10.1007/s11207-012-9999-3
13. Borovikov D, Sokolov IV, Manchester WB IV, Jin M, Gombosi TI (2017) Eruptive event generator based on the Gibson-Low magnetic configuration. J Geophys Res (Space Phys) 122:7979–7984. https://doi.org/10.1002/2017JA024304
14. Borovikov D, Sokolov IV, Roussev II, Taktakishvili A, Gombosi TI (2018) Toward a quantitative model for simulation and forecast of solar energetic particle production during gradual events. I. Magnetohydrodynamic background coupled to the SEP model. Astrophys J 864:88. https://doi.org/10.3847/1538-4357/aad68d
15. Brueckner GE, Howard RA, Koomen MJ, Korendyke CM, Michels DJ, Moses JD, Socker DG, Dere KP, Lamy PL, Llebaria A, Bout MV, Schwenn R, Simnett GM, Bedford DK, Eyles CJ (1995) The large angle spectroscopic coronagraph (LASCO). Sol Phys 162(1):357–402. https://doi.org/10.1007/BF00733434
16. Cargill PJ, Schmidt J, Spicer DS, Zalesak ST (2000) Magnetic structure of overexpanding coronal mass ejections: numerical models. J Geophys Res 105:7509–7520. https://doi.org/10.1029/1999JA900479
17. Chané E, Jacobs C, van der Holst B, Poedts S, Kimpe D (2005) On the effect of the initial magnetic polarity and of the background wind on the evolution of CME shocks. Astron Astrophys 432:331–339. https://doi.org/10.1051/0004-6361:20042005
18. Chané E, van der Holst B, Jacobs C, Poedts S, Kimpe D (2006) Inverse and normal coronal mass ejections: evolution up to 1 AU. Astron Astrophys 447:727–733. https://doi.org/10.1051/0004-6361:20053802
19. Chen PF (2011) Coronal mass ejections: models and their observational basis. Living Rev Sol Phys 8:1. https://link.springer.com/article/10.12942/lrsp-2011-1
20. Chen PF, Shibata K (2000) An emerging flux trigger mechanism for coronal mass ejections. Astrophys J 545:524–531. https://doi.org/10.1086/317803
21. Cranmer SR (2002) Coronal holes and the high-speed solar wind. Space Sci Rev 101:229–294
22. Cranmer SR, Winebarger AR (2019) The properties of the solar corona and its connection to the solar wind. https://arxiv.org/abs/1811.00461

23. Cranmer SR, Gibson SE, Riley P (2017) Origins of the ambient solar wind: implications for space weather. Space Sci Rev 212(3):1345–1384. https://doi.org/10.1007/s11214-017-0416-y
24. Cremades H, Bothmer V (2004) On the three-dimensional configuration of coronal mass ejections. Astron Astrophys 422:307–322. https://doi.org/10.1051/0004-6361:20035776
25. Cremades H, Bothmer V, Tripathi D (2006) Properties of structured coronal mass ejections in solar cycle 23. Adv Space Res 38:461–465. https://doi.org/10.1016/j.asr.2005.01.095
26. Démoulin P (2008) A review of the quantitative links between CMEs and magnetic clouds. Ann Geophys 26(10):3113–3125. https://doi.org/10.5194/angeo-26-3113-2008
27. DeVore CR, Antiochos SK (2005) Magnetic free energies of breakout coronal mass ejections. Astrophys J 628:1031–1045. https://doi.org/10.1086/431141
28. DeVore CR, Antiochos SK (2008) Homologous confined filament eruptions via magnetic breakout. Astrophys J 680:740–756. https://doi.org/10.1086/588011
29. Dewey RM, Baker DN, Anderson BJ, Benna M, Johnson CL, Korth H, Gershman DJ, Ho GC, McClintock WE, Odstrčil D, Philpott LC, Raines JM, Schriver D, Slavin JA, Solomon SC, Winslow RM, Zurbuchen TH (2015) Improving solar wind modeling at Mercury: incorporating transient solar phenomena into the WSA-ENLIL model with the Cone extension. J Geophys Res: Space Phys 120(7):5667–5685
30. Dewey RM, Baker DN, Mays ML, Brain DA, Jakosky BM, Halekas JS, Connerney JEP, Odstrčil D, Luhmann JG, Lee CO (2016) Continuous solar wind forcing knowledge: providing continuous conditions at Mars with the WSA-ENLIL + Cone model. J Geophys Res (Space Phys) 121:6207–6222. https://doi.org/10.1002/2015JA021941
31. Domingo V, Fleck B, Poland A (1995) The SOHO mission: an overview. Sol Phys 162(1–2):1–37. https://doi.org/10.1007/BF00733425
32. Dryer M (2007) Space weather simulation in 3D MHD from the Sun to the Earth and beyond to 100 AU: a modeler's perspective of the present state of the art. Asia J Phys 16:97–121
33. Dulk GA, Leblanc Y, Bougeret JL (1999) Type II shock and CME from the corona to 1 AU. Geophys Res Lett 26:2331–2334. https://doi.org/10.1029/1999GL900454
34. Dungey JW (1961) Interplanetary magnetic field and the auroral zones. Phys Rev Lett 6:47–48. https://doi.org/10.1103/PhysRevLett.6.47
35. Eto S, Isobe H, Narukage N, Asai A, Morimoto T, Thompson B, Yashiro S, Wang T, Kitai R, Kurokawa H, Shibata K (2002) Relation between a Moreton wave and an EIT wave observed on 1997 November 4. Publ Astron Soc Jpn 54:481–491. https://doi.org/10.1093/pasj/54.3.481
36. Evans RM, Opher M, Gombosi TI (2011) Learning from the outer heliosphere: interplanetary coronal mass ejection sheath flows and the ejecta orientation in the lower corona. Astrophys J 728:41. https://doi.org/10.1088/0004-637X/728/1/41
37. Eyles C, Harrison R, Davis C, Waltham N, Shaughnessy B, Mapson-Menard H, Bewsher D, Crothers S, Davies J, Simnett G, Howard R, Moses J, Newmark J, Socker D, Halain JP, Defise JM, Mazy E, Rochus P (2009) The heliospheric imagers onboard the STEREO mission. Sol Phys 254(2):387–445. https://doi.org/10.1007/s11207-008-9299-0
38. Fan Y, Gibson SE (2003) flux tube into a preexisting coronal arcade. Astrophys J Lett 589(2):L105. http://stacks.iop.org/1538-4357/589/i=2/a=L105
39. Farrugia CJ, Osherovich VA, Burlaga LF (1995) Magnetic flux rope versus the spheromak as models for interplanetary magnetic clouds. J Geophys Res 100:12. https://doi.org/10.1029/95JA00272
40. Feng XS, Zhou YF, Wu ST (2007) A novel numerical implementation for solar wind modeling by the modified conservation element/solution element method. Astrophys J 655:1110–1126. https://doi.org/10.1086/510121
41. Feng XS, Yang LP, Xiang CQ, Wu ST, Zhou YF, Zhong DK (2010) Three-dimensional solar wind modeling from the Sun to Earth by a SIP-CESE MHD model with a six-component grid. Astrophys J 723:300
42. Feng XS, Xiang CQ, Zhong DK (2011) The state-of-art of three-dimensional numerical study for corona-interplanetary process of solar storms (in Chinese). Sci Sin-Terrae 41(6):1–28

43. Fong B, Low BC, Fan Y (2002) Quiescent solar prominences and magnetic-energy storage. Astrophys J 571:987–998. https://doi.org/10.1086/340070
44. Forbes T (2010) Models of coronal mass ejections and flares. In: Carolus J Schrijver, George L. Siscoe (eds) Heliophysics: space storms and radiation: causes and effects. Cambridge University Press, london, p 159
45. Forbes TG (1990) Numerical simulation of a catastrophe model for coronal mass ejections. J Geophys Res 95:11,919–11,931. https://doi.org/10.1029/JA095iA08p11919
46. Forbes TG (2000) A review on the genesis of coronal mass ejections. J Geophys Res 105:23,153–23,166. https://doi.org/10.1029/2000JA000005
47. Forbes TG, Linker JA, Chen J, Cid C, Kóta J, Lee MA, Mann G, Mikić Z, Potgieter MS, Schmidt JM, Siscoe GL, Vainio R, Antiochos SK, Riley P (2006) CME theory and models. Space Sci Rev 123:251–302. https://doi.org/10.1007/s11214-006-9019-8
48. Gibson SE, Fan Y (2008) Partially ejected flux ropes: implications for interplanetary coronal mass ejections. J Geophys Res (Space Phys) 113:A09103. https://doi.org/10.1029/2008JA013151
49. Gibson SE, Low BC (1998) A time-dependent three-dimensional magnetohydrodynamic model of the coronal mass ejection. Astrophys J 493:460–473
50. Gombosi TI, van der Holst B, Manchester WB IV, Sokolov IV (2018) Extended MHD modeling of the steady solar corona and the solar wind. Living Rev Sol Phys 15(1):4. https://doi.org/10.1007/s41116-018-0014-4
51. Gopalswamy N, Lara A, Manoharan PK, Howard RA (2005) An empirical model to predict the 1-AU arrival of interplanetary shocks. Adv Space Res 36:2289–2294. https://doi.org/10.1016/j.asr.2004.07.014
52. Gosling JT, McComas DJ, Phillips JL, Bame SJ (1991) Geomagnetic activity associated with Earth passage of interplanetary shock disturbances and coronal mass ejections. J Geophys Res 96:7831–7839. https://doi.org/10.1029/91JA00316
53. Gosling JT, Riley P, McComas DJ, Pizzo VJ (1998) Overexpanding coronal mass ejections at high heliographic latitudes–Observations and simulations. J Geophys Res 103:1941–1954. https://doi.org/10.1029/97JA01304
54. Green LM, Török T, Vršnak B, Manchester WB IV, Veronig A (2018) The origin, evolution and predictability of solar eruptions. Space Sci Rev 214:46. https://doi.org/10.1007/S11214-017-0462-5
55. Groth CPT, De Zeeuw DL, Gombosi TI, Powell KG (2000) Global three-dimensional MHD simulation of a space weather event: CME formation, interplanetary propagation, and interaction with the magnetosphere. J Geophys Res 105:25,053–25,078. https://doi.org/10.1029/2000JA900093
56. Gui B, Shen CL, Wang YM, Ye PB, Liu J, Wang S, Zhao X (2011) Quantitative analysis of CME deflections in the corona. Sol Phys 271:111–139. https://doi.org/10.1007/s11207-011-9791-9
57. Hollweg JV, Isenberg PA (2002) Generation of the fast solar wind: a review with emphasis on the resonant cyclotron interaction. J Geophys Res (Space Phys) 107:1147. https://doi.org/10.1029/2001JA000270
58. Howard TA, Tappin SJ (2009) Interplanetary coronal mass ejections observed in the heliosphere: 1 review of theory. Space Sci Rev 147:31–54. https://doi.org/10.1007/s11214-009-9542-5
59. Hu YQ (2002) Catastrophic behavior of coronal magnetic flux ropes in partially open magnetic fields. In: Wang H, Xu R (eds) Solar-terrestrial magnetic activity and space environment, COSPAR colloquia series, vol 14. Pergamon, pp 117–124. https://doi.org/10.1016/S0964-2749(02)80142-1
60. Hu YQ, Li GQ, Xing XY (2003) Equilibrium and catastrophe of coronal flux ropes in axisymmetrical magnetic field. J Geophys Res: Space Phys 108(A2):1072. https://doi.org/10.1029/2002JA009419
61. Hundhausen AJ (1987) The origin and propagation of coronal mass ejections (R). In: Pizzo VJ, Holzer T, Sime DG (eds) Sixth international solar wind conference, p 181

62. Jacobs C, Poedts S (2011) A polytropic model for the solar wind. Adv Space Res 48:1958–1966. https://doi.org/10.1016/j.asr.2011.08.015
63. Jacobs C, Poedts S, Van der Holst B, Chané E (2005) On the effect of the background wind on the evolution of interplanetary shock waves. Astron Astrophys 430:1099–1107. https://doi.org/10.1051/0004-6361:20041676
64. Jiang CW, Wu ST, Feng XS, Hu Q (2016) Data-driven magnetohydrodynamic modelling of a flux-emerging active region leading to solar eruption. Nat Commun 7:11522. https://doi.org/10.1038/ncomms11522
65. Jin M, Manchester WB IV, van der Holst B, Oran R, Sokolov I, Toth G, Liu Y, Sun XD, Gombosi TI (2013) Numerical simulations of coronal mass ejection on 2011 March 7: one-temperature and two-temperature model comparison. Astrophys J 773:50
66. Jin M, Manchester WB IV, van der Holst B, Sokolov I, Tóth G, Mullinix RE, Taktakishvili A, Chulaki A, Gombosi TI (2017) Data-constrained coronal mass ejections in a global magnetohydrodynamics model. Astrophys J 834:173. https://doi.org/10.3847/1538-4357/834/2/173
67. Jin M, Manchester WB IV, van der Holst B, Sokolov I, Tóth G, Vourlidas A, de Koning CA, Gombosi TI (2017) Chromosphere to 1 AU simulation of the 2011 March 7th event: a comprehensive study of coronal mass ejection propagation. Astrophys J 834:172. https://doi.org/10.3847/1538-4357/834/2/172
68. Kaiser M, Kucera T, Davila J, St Cyr O, Guhathakurta M, Christian E (2008) The stereo mission: an introduction. In: Russell C (ed) The STEREO mission. Springer, New York, pp 5–16. https://doi.org/10.1007/978-0-387-09649-0-2
69. Karpen JT, Antiochos SK, DeVore CR (2012) The mechanisms for the onset and explosive eruption of coronal mass ejections and eruptive flares. Astrophys J 760(1):81. http://stacks.iop.org/0004-637X/760/i=1/a=81
70. Kataoka R, Ebisuzaki T, Kusano K, Shiota D, Inoue S, Yamamoto TT, Tokumaru M (2009) Three-dimensional MHD modeling of the solar wind structures associated with 13 December 2006 coronal mass ejection. J Geophys Res (Space Phys) 114(A13):A10102. https://doi.org/10.1029/2009JA014167
71. Kilpua E, Liewer P, Farrugia C, Luhmann J, Möstl C, Li Y, Liu Y, Lynch B, Russell C, Vourlidas A, Acuna M, Galvin A, Larson D, Sauvaud J (2009) Multispacecraft observations of magnetic clouds and their solar origins between 19 and 23 May 2007. Sol Phys 254(2):325–344. https://doi.org/10.1007/s11207-008-9300-y
72. Kilpua EKJ, Pomoell J, Vourlidas A, Vainio R, Luhmann J, Li Y, Schroeder P, Galvin AB, Simunac K (2009) STEREO observations of interplanetary coronal mass ejections and prominence deflection during solar minimum period. Ann Geophys 27:4491–4503. https://doi.org/10.5194/angeo-27-4491-2009
73. Kleimann J (2012) 4 π models of CMEs and ICMEs (invited review). Sol Phys 281:353–367. https://doi.org/10.1007/s11207-012-9994-8
74. Klimchuk JA (2001) Theory of coronal mass ejections. Washington DC American geophysical union geophysical monograph series 125:p143
75. Koskinen HEJ, Baker DN, Balogh A, Gombosi T, Veronig A, von Steiger R (2017) Achievements and challenges in the science of space weather. Space Sci Rev 212:1137–1157. https://doi.org/10.1007/s11214-017-0390-4
76. Kusano K, Bamba Y, Yamamoto TT, Iida Y, Toriumi S, Asai A (2012) Magnetic field structures triggering solar flares and coronal mass ejections. Astrophys J 760(1):31. http://stacks.iop.org/0004-637X/760/i=1/a=31
77. Levine RH, Altschuler MD, Harvey JW (1977) Solar sources of the interplanetary magnetic field and solar wind. J Geophys Res 82:1061–1065. https://doi.org/10.1029/JA082i007p01061
78. Li X, Lu Q, Li B (2007) Ion pickup by finite amplitude parallel propagating Alfvén waves. Astrophys J Lett 661:L105–L108. https://doi.org/10.1086/518420
79. Linker JA, Lionello R, Mikić Z, Amari T (2001) Magnetohydrodynamic modeling of prominence formation within a helmet streamer. J Geophys Res 106:25,165–25,176. https://doi.org/10.1029/2000JA004020

80. Linker JA, Mikić Z, Lionello R, Riley P, Amari T, Odstrčil D (2003) Flux cancellation and coronal mass ejections. Phys. Plasmas 10:1971–1978. https://doi.org/10.1063/1.1563668
81. Linker JA, Caplan RM, Downs C, Riley P, Mikić Z, Lionello R, Henney CJ, Arge CN, Liu Y, Derosa ML, Yeates A, Owens MJ (2017) The open flux problem. Astrophys J 848:70. https://doi.org/10.3847/1538-4357/aa8a70
82. Lionello R, Mikić Z, Linker JA, Amari T (2002) Magnetic field topology in prominences. Astrophys J 581:718–725. https://doi.org/10.1086/344222
83. Lionello R, Downs C, Linker JA, Török T, Riley P, Mikić Z (2013) Magnetohydrodynamic simulations of interplanetary coronal mass ejections. Astrophys J 777(1):76. http://stacks.iop.org/0004-637X/777/i=1/a=76
84. Lites BW, Low BC, Martinez Pillet V, Seagraves P, Skumanich A, Frank ZA, Shine RA, Tsuneta S (1995) The possible ascent of a closed magnetic system through the photosphere. Astrophys J 446:877. https://doi.org/10.1086/175845
85. Liu Y, Davies JA, Luhmann JG, Vourlidas A, Bale SD, Lin RP (2010) Geometric triangulation of imaging observations to track coronal mass ejections continuously out to 1 AU. Astrophys J Lett 710:L82–L87. https://doi.org/10.1088/2041-8205/710/1/L82
86. Liu Y, Thernisien AF, Luhmann JG, Vourlidas A, Davies JA, Lin RP, Bale S (2010) Reconstructing CMEs with coordinated imaging and in situ observations: global structure, kinematics, and implications for space weather forecasting. AGU fall meeting abstracts, p B1861
87. Liu Y, Luhmann JG, Bale SD, Lin RP (2011) Solar source and heliospheric consequences of the 2010 April 3 coronal mass ejection: a comprehensive view. Astrophys J 734(2):84. http://stacks.iop.org/0004-637X/734/i=2/a=84
88. Loesch C, Opher M, Alves MV, Evans RM, Manchester WB IV (2011) Signatures of two distinct driving mechanisms in the evolution of coronal mass ejections in the lower corona. J Geophys Res (Space Phys) 116:A04106. https://doi.org/10.1029/2010JA015582
89. Low BC (1994) Magnetohydrodynamic processes in the solar corona: flares, coronal mass ejections, and magnetic helicity. Phys Plasmas 1:1684–1690. https://doi.org/10.1063/1.870671
90. Low BC (1996) Solar activity and the corona. Solar Phys 167:217–265. https://doi.org/10.1007/BF00146338
91. Low BC (2001) Coronal mass ejections, magnetic flux ropes, and solar magnetism. J Geophys Res 106:25,141–25,164. https://doi.org/10.1029/2000JA004015
92. Lu Q, Chen L (2009) Ion heating by a spectrum of obliquely propagating low-frequency Alfvén waves. Astrophys J 704(1):743. http://stacks.iop.org/0004-637X/704/i=1/a=743
93. Lugaz N, Manchester WB IV, Gombosi TI (2005) Numerical simulation of the interaction of two coronal mass ejections from Sun to Earth. Astrophys J 634:651–662. https://doi.org/10.1086/491782
94. Lugaz N, Manchester WB IV, Roussev II, Tóth G, Gombosi TI (2007) Numerical investigation of the homologous coronal mass ejection events from active region 9236. Astrophys J 659:788–800. https://doi.org/10.1086/512005
95. Lugaz N, Vourlidas A, Roussev II, Morgan H (2009) Solar–terrestrial simulation in the STEREO era: the 24–25 January 2007 eruptions. Sol Phys 256:269–284. https://doi.org/10.1007/s11207-009-9339-4
96. Lugaz N, Roussev II, Sokolov IV, Jacobs C (2010) Solar-terrestrial simulations of CMEs with a realistic initiation mechanism: case study for active region 10069. Twelfth international solar wind conference 1216:440–443. https://doi.org/10.1063/1.3395898
97. Lynch BJ, Antiochos SK, MacNeice PJ, Zurbuchen TH, Fisk LA (2004) Observable properties of the breakout model for coronal mass ejections. Astrophys J 617:589–599. https://doi.org/10.1086/424564
98. MacNeice P, Antiochos SK, Phillips A, Spicer DS, DeVore CR, Olson K (2004) A numerical study of the breakout model for coronal mass ejection initiation. Astrophys J 614:1028–1041. https://doi.org/10.1086/423887
99. MacQueen RM, Hundhausen AJ, Conover CW (1986) The propagation of coronal mass ejection transients. J Geophys Res 91:31–38. https://doi.org/10.1029/JA091iA01p00031

100. Manchester WB IV, Gombosi TI, Roussev I, de Zeeuw DL, Sokolov IV, Powell KG, Tóth G, Opher M (2004) Three-dimensional MHD simulation of a flux rope driven CME. J Geophys Res (Space Phys) 109:A01102. https://doi.org/10.1029/2002JA009672
101. Manchester WB IV, Gombosi TI, Roussev I, Ridley A, de Zeeuw DL, Sokolov IV, Powell KG, Tóth G (2004) Modeling a space weather event from the Sun to the Earth: CME generation and interplanetary propagation. J Geophys Res (Space Phys) 109:A02107. https://doi.org/10.1029/2003JA010150
102. Manchester WB IV, Vourlidas A, Tóth G, Lugaz N, Roussev II, Sokolov IV, Gombosi TI, De Zeeuw DL, Opher M (2008) Three-dimensional MHD simulation of the 2003 October 28 coronal mass ejection: comparison with LASCO coronagraph observations. Astrophys J 684:1448–1460. https://doi.org/10.1086/590231
103. Manchester WB IV, Kilpua EKJ, Liu YD, Lugaz N, Riley P, Török T, Vršnak B (2017) The physical process of CME/ICME evolution. Space Sci Rev 212:1159. https://doi.org/10.1007/S11214-017-0394-0
104. Martin SF, Livi SHB, Wang J (1985) The cancellation of magnetic flux. II - In a decaying active region. Aust J Phys 38:929–959. https://doi.org/10.1071/PH850929
105. Mays ML, Taktakishvili A, Pulkkinen A, MacNeice PJ, Rastätter L, Odstrčil D, Jian LK, Richardson IG, LaSota JA, Zheng Y, Kuznetsova MM (2015) Ensemble modeling of CMEs Using the WSA-ENLIL+Cone model. Sol Phys 290:1775–1814. https://doi.org/10.1007/s11207-015-0692-1
106. Michalek G (2006) An asymmetric cone model for halo coronal mass ejections. Sol Phys 237(1):101–118. https://doi.org/10.1007/s11207-006-0075-8
107. Mikić Z, Lee MA (2006) An introduction to theory and models of CMEs, shocks, and solar energetic particles. Space Sci Rev 123:57–80. https://doi.org/10.1007/s11214-006-9012-2
108. Mikić Z, Linker JA (1994) Disruption of coronal magnetic field arcades. Astrophys J 430:898–912. https://doi.org/10.1086/174460
109. Millward G, Biesecker D, Pizzo V, de Koning CA (2013) An operational software tool for the analysis of coronagraph images: determining CME parameters for input into the WSA-Enlil heliospheric model. Space Weather 11:57–68. https://doi.org/10.1002/swe.20024
110. Mittal N, Narain U (2010) Initiation of CMEs: a review. J Atmos Sol-Terr Phys 72:643–652. https://doi.org/10.1016/j.jastp.2010.03.011
111. Möstl C, Farrugia C, Biernat H, Leitner M, Kilpua E, Galvin A, Luhmann J (2009) Optimized Grad–Shafranov reconstruction of a magnetic cloud using STEREO-Wind observations. Sol Phys 256(1–2):427–441. https://doi.org/10.1007/s11207-009-9360-7
112. Möstl C, Temmer M, Rollett T, Farrugia CJ, Liu Y, Veronig AM, Leitner M, Galvin AB, Biernat HK (2010) STEREO and Wind observations of a fast ICME flank triggering a prolonged geomagnetic storm on 5–7 April 2010. Geophy Res Lett 37(24):L24103. https://doi.org/10.1029/2010GL045175
113. Na H, Moon YJ, Lee H (2017) Development of a full ice-cream CONE model for halo coronal mass ejections. Astrophys J 839:82. https://doi.org/10.3847/1538-4357/aa697c
114. Nakamizo A, Tanaka T, Kubo Y, Kamei S, Shimazu H, Shinagawa H (2009) Development of the 3-D MHD model of the solar corona-solar wind combining system. J Geophys Res (Space Phys) 114:A07109. https://doi.org/10.1029/2008JA013844
115. Odstrčil D, Pizzo VJ (1999) Three-dimensional propagation of CMEs in a structured solar wind flow: 1. CME launched within the streamer belt. J Geophys Res 104:483–492. https://doi.org/10.1029/1998JA900019
116. Odstrčil D, Pizzo VJ (1999) Three-dimensional propagation of coronal mass ejections in a structured solar wind flow 2. CME launched adjacent to the streamer belt. J Geophys Res 104:493–504. https://doi.org/10.1029/1998JA900038
117. Odstrčil D, Pizzo VJ (2009) Numerical heliospheric simulations as assisting tool for interpretation of observations by STEREO heliospheric imagers. Sol Phys 259:297–309. https://doi.org/10.1007/s11207-009-9449-z
118. Odstrčil D, Dryer M, Smith Z (1996) Propagation of an interplanetary shock along the heliospheric plasma sheet. J Geophys Res 101:19,973–19,986. https://doi.org/10.1029/96JA00479

119. Odstrčil D, Linker JA, Lionello R, Mikić Z, Riley P, Pizzo VJ, Luhmann JG (2002) Merging of coronal and heliospheric numerical two-dimensional MHD models. J Geophys Res (Space Phys) 107:1493. https://doi.org/10.1029/2002JA009334
120. Odstrčil D, Riley P, Zhao XP (2004) Numerical simulation of the 12 May 1997 interplanetary CME event. J Geophys Res (Space Phys) 109:A02116. https://doi.org/10.1029/2003JA010135
121. Odstrčil D, Pizzo VJ, Arge CN (2005) Propagation of the 12 May 1997 interplanetary coronal mass ejection in evolving solar wind structures. J Geophys Res (Space Phys) 110:A02106. https://doi.org/10.1029/2004JA010745
122. Owens MJ, Lockwood M, Barnard LA (2017) Coronal mass ejections are not coherent magnetohydrodynamic structures. Sci Rep 7(1):4152. https://doi.org/10.1038/s41598-017-04546-3
123. Owens MJ, Lockwood M, Riley P (2017) Global solar wind variations over the last four centuries. Sci Rep 7:41548. https://doi.org/10.1038/srep41548
124. Pagano P, Mackay DH, Yeates AR (2018) A new technique for observationally derived boundary conditions for space weather. J Space Weather Space Clim 8:A26. https://doi.org/10.1051/swsc/2018012
125. Parker EN (1963) Interplanetary dynamical processes. Interscience Publishers, New York
126. Plunkett SP, Thompson BJ, Howard RA, Michels DJ, St Cyr OC, Tappin SJ, Schwenn R (1997) Lamy PL (1998) Lasco observations of an Earth-directed coronal mass ejection on May 12. Geophys Res Lett 25(14):2477–2480. https://doi.org/10.1029/98GL50307
127. Pulkkinen T (2007) Space weather: terrestrial perspective. Living Rev Sol Phys 4:1 https://doi.org/10.12942/lrsp-2007-1
128. Riley P, Gosling JT, Pizzo VJ (1997) A two-dimensional simulation of the radial and latitudinal evolution of a solar wind disturbance driven by a fast, high-pressure coronal mass ejection. J Geophys Res 102:14,677–14,686. https://doi.org/10.1029/97JA01131
129. Riley P, Gosling JT, Pizzo VJ (2001) Investigation of the polytropic relationship between density and temperature within interplanetary coronal mass ejections using numerical simulations. J Geophys Res 106:8291–8300. https://doi.org/10.1029/2000JA000276
130. Riley P, Caplan RM, Giacalone J, Lario D, Liu Y (2016) Properties of the fast forward shock driven by the July 23 2012 extreme coronal mass ejection. Astrophys J 819:57. https://doi.org/10.3847/0004-637X/819/1/57
131. Roussev II, Sokolov IV, Forbes TG, Gombosi TI, Lee MA, Sakai JI (2004) A numerical model of a coronal mass ejection: shock development with implications for the acceleration of GeV protons. Astrophys J Lett 605:L73–L76
132. Schmidt JM, Cairns IH, Hillan DS (2013) Prediction of type II solar radio bursts by three-dimensional MHD coronal mass ejection and kinetic radio emission simulations. Astrophys J Lett 773:L30. https://doi.org/10.1088/2041-8205/773/2/L30
133. Schmidt JM, Cairns IH, Xie H, St Cyr OC, Gopalswamy N (2016) CME flux rope and shock identifications and locations: comparison of white light data, Graduated Cylindrical Shell model, and MHD simulations. J Geophys Res (Space Phys) 121:1886–1906. https://doi.org/10.1002/2015JA021805
134. Schwenn R (2006) Space weather: the solar perspective. Living Rev Sol Phys 3:2 https://doi.org/10.12942/lrsp-2006-2
135. Sheeley NR, Walters JH, Wang YM, Howard RA (1999) Continuous tracking of coronal outflows: two kinds of coronal mass ejections. J Geophys Res: Space Phys 104(A11):24,739–24,767. https://doi.org/10.1029/1999JA900308
136. Shen CL, Wang YM, Gui B, Ye PZ, Wang S (2011) Kinematic evolution of a slow CME in corona viewed by STEREO-B on 8 October 2007. Sol Phys 269:389–400. https://doi.org/10.1007/s11207-011-9715-8
137. Shen F, Feng XS, Wu ST, Xiang CQ (2007) Three-dimensional MHD simulation of CMEs in three-dimensional background solar wind with the self-consistent structure on the source surface as input: numerical simulation of the January 1997 Sun-Earth connection event. J Geophys Res (Space Phys) 112:A06109. https://doi.org/10.1029/2006JA012164

138. Shen F, Feng XS, Wu ST, Xiang CQ, Song WB (2011) Three-dimensional MHD simulation of the evolution of the April 2000 CME event and its induced shocks using a magnetized plasma blob model. J Geophys Res (Space Phys) 116:A04102. https://doi.org/10.1029/2010JA015809
139. Shen F, Shen CL, Zhang J, Hess P, Wang YM, Feng XS, Cheng HZ, Yang Y (2014) Evolution of the 12 July 2012 CME from the Sun to the Earth: data-constrained three-dimensional MHD simulations. J Geophys Res 119:7128–7141
140. Shiota D, Kataoka R (2016) Magnetohydrodynamic simulation of interplanetary propagation of multiple coronal mass ejections with internal magnetic flux rope (SUSANOO-CME). Space Weather 14(2):56–75
141. Shiota D, Kusano K, Miyoshi T, Shibata K (2010) Magnetohydrodynamic modeling for a formation process of coronal mass ejections: interaction between an ejecting flux rope and an ambient field. Astrophys J 718:1305–1314
142. Thernisien AFR, Howard RA, Vourlidas A (2006) Modeling of flux rope coronal mass ejections. Astrophys J 652:763–773. https://doi.org/10.1086/508254
143. Thompson WT, Kliem B, Török T (2012) 3d reconstruction of a rotating erupting prominence. Sol Phys 276(1):241–259. https://doi.org/10.1007/s11207-011-9868-5
144. Titov VS, Démoulin P (1999) Basic topology of twisted magnetic configurations in solar flares. Astron Astrophys 351:707–720
145. Titov VS, Török T, Mikić Z, Linker JA (2014) A method for embedding circular force-free flux ropes in potential magnetic fields. Astrophys J 790(2):163. http://stacks.iop.org/0004-637X/790/i=2/a=163
146. Török T, Kliem B (2005) Confined and ejective eruptions of kink-unstable flux ropes. Astrophys J Lett 630(1):L97. http://stacks.iop.org/1538-4357/630/i=1/a=L97
147. Török T, Kliem B (2007) Numerical simulations of fast and slow coronal mass ejections. Astron Nachr 328:743. https://doi.org/10.1002/asna.200710795
148. Török T, Downs C, Linker JA, Lionello R, Titov VS, Mikić Z, Riley P, Caplan RM, Wijaya J (2018) Sun-to-Earth MHD simulation of the 2000 July 14 "Bastille day" eruption. Astrophys J 856(1):75. http://stacks.iop.org/0004-637X/856/i=1/a=75
149. van der Holst B, Poedts S, Chané E, Jacobs C, Dubey G, Kimpe D (2005) Modelling of solar wind, CME initiation and CME propagation. Space Sci Rev 121:91–104. https://doi.org/10.1007/s11214-006-6541-7
150. Vandas M, Fischer S, Pelant P, Dryer M, Smith Z, Detman T (1997) Propagation of a spheromak 1. Some comparisons of cylindrical and spherical magnetic clouds. J Geophys Res 102:24,183–24,194. https://doi.org/10.1029/97JA02257
151. Vandas M, Fischer S, Dryer M, Smith Z, Detman T (1998) Propagation of a spheromak 2. Three-dimensional structure of a spheromak. J Geophys Res 103:23,717–23,726. https://doi.org/10.1029/98JA01902
152. Vandas M, Odstrčil D, Watari S (2002) Three-dimensional MHD simulation of a loop-like magnetic cloud in the solar wind. J Geophys Res (Space Phys) 107:1236. https://doi.org/10.1029/2001JA005068
153. Wang C, Du D, Richardson JD (2005) Characteristics of the interplanetary coronal mass ejections in the heliosphere between 0.3 and 5.4 AU. J Geophys Res (Space Phys) 110(A10):A10107. https://doi.org/10.1029/2005JA011198
154. Wang RS, Lu QM, Du AM, Wang S (2010) In situ observations of a secondary magnetic island in an ion diffusion region and associated energetic electrons. Phys Rev Lett 104(17):175003. https://doi.org/10.1103/PhysRevLett.104.175003
155. Wang RS, Lu QM, Li X, Huang C, Wang S (2010) Observations of energetic electrons up to 200 keV associated with a secondary island near the center of an ion diffusion region: a Cluster case study. J Geophys Res (Space Phys) 115(A14):A11201. https://doi.org/10.1029/2010JA015473
156. Wang YM, Shen CL, Wang S, Ye PZ (2004) Deflection of coronal mass ejection in the interplanetary medium. Sol Phys 222:329–343. https://doi.org/10.1023/B:SOLA.0000043576.21942.aa

157. Wang YM, Xue XH, Shen CL, Ye PZ, Wang S, Zhang J (2006) Impact of major coronal mass ejections on geospace during 2005 September 7–13. Astrophys J 646:625–633. https://doi.org/10.1086/504676
158. Wang YM, Chen C, Gui B, Shen CL, Ye PZ, Wang S (2011) Statistical study of coronal mass ejection source locations: understanding CMEs viewed in coronagraphs. J Geophys Res (Space Phys) 116:A04104. https://doi.org/10.1029/2010JA016101
159. Wei FS, Dryer M (1991) Propagation of solar flare-associated interplanetary shock waves in the heliospheric meridional plane. Sol Phys 132:373–394
160. Wold AM, Mays ML, Taktakishvili A, Jian LK, Odstrčil D, MacNeice P (2018) Verification of real-time WSA-ENLIL+Cone simulations of CME arrival-time at the CCMC from 2010 to 2016. J Space Weather Space Clim 8:A17. https://doi.org/10.1051/swsc/2018005
161. Wood BE, Wu CC, Howard RA, Socker DG, Rouillard AP (2011) Empirical reconstruction and numerical modeling of the first geoeffective coronal mass ejection of solar cycle 24. Astrophys J 729:70. https://doi.org/10.1088/0004-637X/729/1/70
162. Wood BE, Wu CC, Rouillard AP, Howard RA, Socker DG (2012) A coronal hole's effects on coronal mass ejection shock morphology in the inner heliosphere. Astrophys J 755:43. https://doi.org/10.1088/0004-637X/755/1/43
163. Wu CC, Fry CD, Dryer M, Wu ST, Thompson B, Liou K, Feng XS (2007) Three-dimensional global simulation of multiple ICMEs' interaction and propagation from the Sun to the heliosphere following the 25 28 October 2003 solar events. Adv Space Res 40:1827–1834. https://doi.org/10.1016/j.asr.2007.06.025
164. Wu CC, Fry CD, Wu ST, Dryer M, Liou K (2007) Three-dimensional global simulation of interplanetary coronal mass ejection propagation from the Sun to the heliosphere: solar event of 12 May 1997. J Geophys Res (Space Phys) 112:A09104. https://doi.org/10.1029/2006JA012211
165. Wu CC, Dryer M, Wu ST, Wood BE, Fry CD, Liou K, Plunkett S (2011) Global three-dimensional simulation of the interplanetary evolution of the observed geoeffective coronal mass ejection during the epoch 1–4 August 2010. J Geophys Res (Space Phys) 116(A15):A12103. https://doi.org/10.1029/2011JA016947
166. Wu CC, Liou K, Vourlidas A, Plunkett S, Dryer M, Wu ST, Mewaldt RA (2016) Global magnetohydrodynamic simulation of the 15 March 2013 coronal mass ejection event - interpretation of the 30–80 MeV proton flux. J Geophys Res (Space Phys) 121:56–76. https://doi.org/10.1002/2015JA021051
167. Wu ST, Dryer M (2015) Comparative analyses of current three-dimensional numerical solar wind models. Sci China Earth Sci 58(6):839–858. https://doi.org/10.1007/s11430-015-5062-1
168. Wu ST, Guo WP, Michels DJ, Burlaga LF (1999) MHD description of the dynamical relationships between a flux rope, streamer, coronal mass ejection, and magnetic cloud: an analysis of the January 1997 Sun-Earth connection event. J Geophys Res 104:14,789–14,802. https://doi.org/10.1029/1999JA900099
169. Wu ST, Zhou YF, Jiang CW, Feng XS, Wu CC, Hu Q (2016) A data-constrained three-dimensional magnetohydrodynamic simulation model for a coronal mass ejection initiation. J Geophys Res (Space Phys) 121:1009–1023. https://doi.org/10.1002/2015JA021615
170. Xie H, Ofman L, Lawrence G (2004) Cone model for halo CMEs: application to space weather forecasting. J Geophys Res (Space Phys) 109:A03109. https://doi.org/10.1029/2003JA010226
171. Xie H, Odstrčil D, Mays L, St Cyr OC, Gopalswamy N, Cremades H (2012) Understanding shock dynamics in the inner heliosphere with modeling and type II radio data: the 2010-04-03 event. J Geophys Res (Space Phys) 117(A4):A04105. http://dx.doi.org/10.1029/2011JA017304
172. Zhang M, Low BC (2004) Magnetic energy storage in the two hydromagnetic types of solar prominences. Astrophys J 600:1043–1051. https://doi.org/10.1086/379891
173. Zhang M, Low BC (2005) The hydromagnetic nature of solar coronal mass ejections. Ann Rev Astron Astroph 43:103–137. https://doi.org/10.1146/annurev.astro.43.072103.150602

174. Zhao XP, Plunkett SP, Liu W (2002) Determination of geometrical and kinematical properties of halo coronal mass ejections using the cone model. J Geophys Res (Space Phys) 107:1223
175. Zhou YF, Feng XS (2013) MHD numerical study of the latitudinal deflection of coronal mass ejection. J Geophys Res (Space Phys) 118:6007–6018. https://doi.org/10.1002/2013JA018976
176. Zhou YF, Feng XS (2017) Numerical study of the propagation characteristics of coronal mass ejections in a structured ambient solar wind. J Geophys Res (Space Phys) 122:1451–1462. https://doi.org/10.1002/2016JA023053
177. Zhou YF, Feng XS, Wu ST, Du D, Shen F, Xiang CQ (2012) Using a 3-D spherical plasmoid to interpret the Sun-to-Earth propagation of the 4 November 1997 coronal mass ejection event. J Geophys Res (Space Phys) 117:A01102. https://doi.org/10.1029/2010JA016380
178. Zhou YF, Feng XS, Zhao XH (2014) Using a 3-D MHD simulation to interpret propagation and evolution of a coronal mass ejection observed by multiple spacecraft: the 3 April 2010 event. J Geophys Res: Space Phys 119:9321–9333
179. Zuccarello FP, Bemporad A, Jacobs C, Mierla M, Poedts S, Zuccarello F (2012) The role of streamers in the deflection of coronal mass ejections: comparison between STEREO three-dimensional reconstructions and numerical simulations. Astrophys J 744:66. https://doi.org/10.1088/0004-637X/744/1/66

Chapter 11
Future Avenue

As we are approaching the end of this book, it is necessary to identify current challenges and potential directions to advance reliable predictive capability of the corona-heliosphere system, develop a research roadmap to tackle the scientific challenges of coupling observations and modeling with emerging data-science research to derive new knowledge from mass data, promote trans-disciplinary collaborations in the future, and find areas of convergence across disciplines. Topics discussed in this chapter include physical understanding of the coronal-heliospheric system, observational data availability and application, model metrics and validation, ensemble forecasting, data assimilation and machine learning. As space weather modeling and forecasting is an interdisciplinary field, it is difficult, if not impossible, to give a complete roadmap of future development. However, we still hope that the discussions in the following of this chapter would be of some help to inspire new research outcome in this field.

The Sun emits energy in two different forms. One is electromagnetic waves (including visible lights), which causes tropospheric disturbances, such as tornados and hurricanes. The other one is solar wind (being composed of charged particles), which causes geomagnetic disturbances and brilliant aurora, known as magnetic storms and substorms. The fields of space weather and terrestrial weather or meteorology have a lot in common [36, 48, 126, 128]. Both of them are dealing with the dynamics of gaseous media subject to instabilities that produce the disturbances recognized as weather. The adverse space weather and terrestrial weather condition are both caused by the travelling storm systems, such as CME for space weather and hurricanes for terrestrial weather. More than two centuries have seen the development of terrestrial weather both as a scientific discipline and an indispensable social infrastructure. Significant breakthroughs in terrestrial weather forecasting were achieved, especially in the latter 20th century. Comparing with terrestrial weather, space weather is a younger field, as space weather research began to carry out systematically only after 1957, when the first man-made satellite was launched. Since it is natural and advantageous for a younger field to look to the older field for

© Springer Nature Singapore Pte Ltd. 2020

X. Feng, *Magnetohydrodynamic Modeling of the Solar Corona and Heliosphere*, Atmosphere, Earth, Ocean & Space,
https://doi.org/10.1007/978-981-13-9081-4_11

guidance, the future avenue of space weather may be paved by the experience of terrestrial weather.

Compared with the maturity of terrestrial weather prediction, space weather forecasting is at a relatively nascent stage. There is still much room for gaining more comprehensive and realistic understanding of the physics of space weather. The solar corona is heated to temperatures of millions of degrees, and solar plasma flows out into interplanetary space at supersonic speeds. Fast-speed solar wind streams are believed to be connected to the central regions of large coronal holes, while slow-speed solar wind streams appear to come from a wide range of sources, including streamers, pseudostreamers, coronal loops, active regions, and coronal hole boundaries. Although significant progress has been seen in the last few decades' study of coronal heating and the acceleration of the ambient solar wind (e.g., [24, 25, 100, 101]), there does not exist a complete understanding of the relevant physical processes, nor does there exist a quantitatively precise census of which coronal structures contribute to specific types of solar wind. About solar wind acceleration, there still exist some fundamental questions to which answers are left open: What are the physical processes that dominate coronal heating? Where does most of the slow wind come from on the Sun? Do fast and slow wind streams originate from two separate mechanisms?

Complicating our understanding even more is the fact that processes such as turbulence, stream-stream interactions, and Coulomb collisions can make it difficult to unambiguously map a parcel measured at 1 AU back down to its coronal source. Up to now, there are no 3D simulations that self-consistently include shocks and turbulence to assess which mechanism dominates in the heating and acceleration (e.g., [41]). The developed Alfvén Wave Solar Model (AWSoM) [49, 82, 129] combined the Alfvén wave turbulence description with a three dimensional solar corona/solar wind model that self-consistently incorporates low frequency Alfvén wave turbulence. While this approach has the promise to explain the origin of both the fast and slow solar wind states, it is still at a relatively early state of development and much work remains to be done, as pointed out by Gombosi et al. [41].

As drivers of adverse space weather, eruptive flares, filament eruptions, coronal mass ejections and failed eruptions are commonly thought to be produced by the Sun due to a release of energy stored in the coronal magnetic field during its drastic reconfiguration. Green et al. [44] reviewed the observations and physical mechanisms behind this eruptive activity, with a view to making an assessment of the current capability of forecasting these events for space weather risk and impact mitigation.

From the observational perspective, the role of photospheric flows and magnetic environment should be more clearly understood in order to identify the evolutionary path that ultimately means a magnetic structure will erupt. Meanwhile, the energy required to power a CME (and any associated flare) cannot be supplied to the corona on the timescale of the dynamic event and instead it must be built up and stored in the coronal magnetic field in the hours or days beforehand. Time-dependent evolution models are required to understand the physical processes that create an eruptive magnetic field configuration and the associated currents, as well as the processes that trigger and drive the eruption. Time-dependent modeling could determine more realistic magnetic field configurations and their evolution. By continuously data-driven modeling, the evolution history of the magnetic field is retained, and the energy is allowed to accumulate and release. Developing the capability to conduct data driven simulations (e.g., [56]), relevant for active regions from their birth to their dispersal and decay into the background, will greatly aid this goal. Whilst a wealth of observations exist, more realistic models are encouraged in order to asses the full range of complexity of the solar atmosphere and the criteria that encapsulates the eruptive status of the magnetic field. Also, there has been significant interest in flying a satellite to the L5 point that lies 60° behind Earth. Such future mission with a magnetograph at the L5 point could significantly enrich data available in increasing accuracy of global nonpotential modeling [71, 105, 147].

Afterwards, we need to understand how structures originating in coronal eruptions can be connected to their far removed interplanetary counterparts. It remains a tantalizing mystery as to how these erupting magnetic fields evolve to form the complex structures as observed in the solar wind at Earth. CMEs traveling faster than the solar wind drive forward propagating shocks that can extend far beyond the ejecta, and whose geometry can be largely modified by the structure of the ambient solar wind (e.g., [74, 100]). This point deserves further attention of the modelers.

The fundamental nature of the solar wind magnetosphere interaction is driven by magnetic reconnection, therefore depending on the orientation of the solar wind magnetic field (e.g., [31]). During magnetospheric circulation, dayside and nightside reconnection do not typically behave as the simple picture originally described by Dungey [29] and in fact there is a process of loading and unloading. The unloading process drives strong, structured electrical currents in the ionosphere. Understanding this process requires knowledge of both the loading and unloading of energy in the magnetosphere at the magnetopause and in the magnetotail respectively. Thus, to improve magnetospheric space weather forecasting may involve the physical processes: Transition from L1 to the magnetopause and the magnetosheath; Solar wind magnetosphere coupling and magnetotail energy release, and questions: in what way the solar wind may evolve from L1 to the bow shock, and in what way it is processed in the magnetosheath; what is the physics that controls how solar wind plasma, momentum and energy all enter the magnetosphere? Where does this occur and on what timescale? and how do bursty bulk flows form in the near Earth magnetotail that is associated with particle injection and auroral activity. To explore the answers to these questions, it is necessary to examine the evolution of the solar wind from L1 to the magnetopause, the location, timing, and extent of reconnection on the magnetopause, the physics and temporal variability of reconnection, the nature of bursty bulk flows and their coupling to the inner magnetosphere, and the effects of magnetospheric preconditioning.

Numerical simulations have evolved to a state where a semi-realistic modeling of observed eruptions has become possible. State-of-the-art simulations employ measured magnetograms, a realistic treatment of the energy transfer in the corona, and propagate ICMEs to 1 AU or beyond (e.g., [41, 68, 74, 123]), and also utilize L1 measurements as the boundary conditions to perform magnetospheric space weather forecasting (e.g., [31]). In the future study, it is promising to consider how to conduct

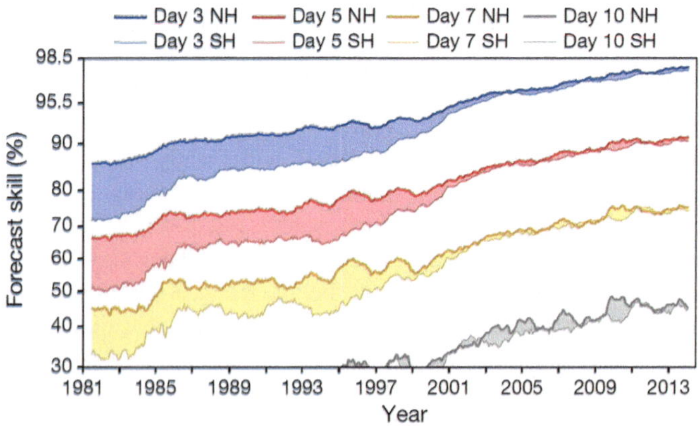

Fig. 11.1 Terrestrial weather forecast skill at three-, five-, seven- and ten-day ranges, computed over the extra-tropical northern hemisphere (NH) and southern hemisphere (SH). Forecast skill is measured by the correlation between the forecasts and the verifying analysis of the height of the 500-hPa level. Values bigger than 60% indicate useful forecasts, while those higher than 80% represent a high degree of accuracy. The convergence of the curves for NH and SH after 1999 indicates the revolution in exploiting satellite data through the use of variational data. Reprinted from Bauer et al. [8] by permission from Springer Nature Customer Service Centre GmbH: Nature Publishing AG *Nature*, The quiet revolution of numerical weather prediction, Peter Bauer, Alan Thorpe and Gilbert Brunet. Copyright 2015

a all-data-driven simulation of the Sun-Earth system by coupling the data-driven ambient simulation introduced in Chap. 7 and the data-driven active region simulation introduced in Chap. 9, which would be a step forward in developing more self-consistent modelling for space weather research and prediction.

In the historical scope, terrestrial weather forecast was revolutionized by the introduction of operational numerical weather prediction (NWP) in the 1950s [127]. Comparing the forecast with what actually occurs, an increasing weather forecast skill is seen since then. For an example, as shown in Fig. 11.1, the forecast skill in the range from 3 to 10 days ahead has been increasing by about one day per decade. Such a steady improvement of forecast skill may come from the fact that the numerical prediction models are simulations of nature itself. Improving the numerical models necessarily increase forecast skill. For space weather numerical models, the transition from research to operation has just began. In 2011, the Wang–Shelley–Arge–Enlil (WSA-Enlil) Cone model, which has been reviewed in Chap. 1, was transitioned into operation [106, 121]. It is commonly accepted that this is the first large-scale, physics-based space weather prediction model that come into service for daily space weather forecast. Nowadays, it is responsible to provide the 1–4- day forecast of solar wind near the Earth. A major lesson learned from the success of terrestrial NWP is that it is important to begin operational forecasting using physics-based numerical model as quick as possible [127]. With continuous feedback between researcher and forecaster, researchers would understand where to exert effort to achieve optimal operation effect. With this tight coupling, the increase of forecast ability will occur in a incremental way [128]. Therefore, much effort should be afforded to transition the sophisticated solar wind models reviewed in Chap. 1 from research to operation and to provide desirable services of space weather prediction [73].

In order to transform a MHD solar wind model from research to operation, two critical elements are metrics and model validation. As defined by Spence et al. [130], metrics are clearly specified and standardized quantifications of how well models predict the physical measurements especially important to terrestrial or space weather. In terrestrial NWP, by using metrics, the success and failure of forecasts is accurately known and pathways to improve predictive skill can be effectively tested. The improvement of forecast ability may be revealed by tracking metrics over time. Space weather may also benefit from metrics in such a way. Currently, two commonly used metrics for solar wind forecast are the mean-square errors and the correlation coefficients (e.g., [55, 94, 95, 97]). For events with significant space weather effects, such as SIRs, IMF sector boundaries, and most importantly, CMEs, metrics like hit/miss rates, correct/false alarms rates and arrival time error are used (e.g., [7, 55, 72, 114, 117, 142, 151]).

Validation refers to the broad assessment of a model's output through comparison with observations, whether statistically or via case study analysis [130]. Validation is an essential tool to assess and understand the strength and weakness of the model and guide modellers' path to eliminating or reducing the model's shortcomings. It is crucial to advancing the science in space weather models and to assessing our current forecasting ability. To carry out validation, metrics may be used as objective evaluation tools. For some MHD solar wind models, validation studies for a period of about one year have been implemented [45, 55]. Lee et al. [64], Li and Feng [66] and Wold et al. [152] have extended the time span of the validation to about three and seven years respectively. However, full-solar-cycle (i.e. 11-year) validation is still needed to demonstrate model's ability under different solar activity condition. Currently, most systematic validation studies of solar wind model are based on in-situ observations at the first Lagrange point L1 point. The continuous monitoring of the solar wind data at the L1 point, currently by STEREO, DSCOVR, ACE, SOHO, and Wind spacecraft, provides rich data to analyze the solar wind driving of the magnetosphere-ionosphere system or to provide the solar wind conditions driving space weather. In-situ data at the L1 point is continuously available into the foreseeable future (e.g., [120]). These missions can provide images of the solar wind that flows between the solar corona and spacecraft making in-situ measurements for scientists to directly connect processes imaged near the Sun with the subsequent effects measured in the solar wind. Meanwhile, metrics calculated at the L1 point can provide a consecutive record of model's ability in predicting near-Earth space weather condition.

Since the 1970s, many solar-wind remote-sensing observing tools have become available, such as sensitive coronagraphs, both ground-based and space-borne; heliospheric white-light imagers; X-ray imaging telescopes; low-frequency radio telescopes providing interplanetary scintillation (IPS) data; and space-borne kilometric-wave radio receivers. Determining the global properties of CMEs/ICMEs or more stable structures such as stream/co-rotating interaction regions (SIRs/CIRs), their morphology, and their motion, requires that these structures are tracked over an extended period of time during their outward propagation from the Sun. Heliospheric remote-sensing observations are of fundamental importance since they provide data on the solar-wind properties beyond the near-Sun region and those inaccessible to in-situ measurements.

However, validation based on those remote sensing observation is challenging. Although we may derive some physical properties of the corona by reconstruction techniques from remote sensing observation, the reconstruction itself is usually model-based and will bring in errors (e.g., [27, 84–87, 113]). At the same time, most reconstructed maps from coronal observations reproduce the morphological resemblances with corona images by spacecraft. A reliable validation method based on remote-sensing data requires efforts taken in the future.

Interplanetary scintillation (IPS) of compact radio sources at meter to centimeter wavelengths, such as Solar-Terrestrial Environment Laboratory (STELab) IPS observation, IPS observations with the Ooty Radio Telescope (ORT), the European Incoherent SCATter (EISCAT) IPS observations, serves as a useful ground-based method for investigating the solar wind [52–54, 137–140, 154]. Measurements of IPS have been used for many years to probe the solar wind throughout the inner heliosphere (e.g., [6, 10, 33, 53, 115, 140]). The greater sensitivity of IPS to small variations in electron density makes it the ideal complement to white-light imaging in investigating the solar wind disturbances. A more global coverage of IPS observations of the inner heliosphere with multiple IPS-capable systems, a better understanding of how interplanetary disturbances form and change throughout the inner heliosphere with the aid of the many models for heliospheric investigations should now be more easily achieved than ever before. For instance, the LOFAR4SW (the Low Frequency Radio Array (LOFAR) for Space Weather) design proposal will enable large-scale IPS monitoring projects generating unique data for the European (and worldwide) space weather research community [34, 35]. Also, polarized heliospheric imaging is a promising technology that should be developed for space weather monitoring and prediction, as pointed out by DeForest et al. [28]. Heliospheric imagers like SECCHI/STEREO can track coronal mass ejections (CMEs) as they cross the solar system, using sunlight scattered by electrons in the CME, which has been demonstrated to improve the forecasting of impact probability and arrival time for Earth-directed CMEs. Polarized imaging allows locating CMEs in three dimensions from a single vantage point. DeForest et al. [28] suggested deployment as an instrument on NOAA's Deep Space Climate Observatory follow-on near the Sun-Earth L1 Lagrange point, as a stand-alone constellation of smallsats in low Earth orbit, or as an instrument located at the Sun-Earth L5 Lagrange point.

Besides, Strugarek et al. [132] proffered an inner-heliosphere space mission, the Observatories of Solar Corona and Active Regions (OSCAR) mission. It is designed to identify the 3D structure of coronal loops and to study the trigger mechanisms of CMEs in solar Active Regions (ARs) as well as their evolution and propagation processes in the inner heliosphere. It also provides monitoring and forecasting of geo-effective CMEs and CIRs. Ritter et al. [118] also proposed a concept mission consisting of six spacecraf that are equally spaced in a heliocentric orbit at 0.72 AU.

The objective is to to increase space weather forecasting time by means of a near real-time information service based upon in-situ and remote measurements of the Earth-bounding CMEs' properties. What is more, several new concepts of space missions at L5 have also been put forward. The fourth and fifth Sun-Earth Lagrange points (L4 and L5) lead and trail Earth and are located at 60 degrees away from the Sun-Earth line. L5 point is an unparalleled advantageous location for space weather research and monitoring [143]. Gopalswamy et al. [43] put forward an L5 mission concept, the Earth-Affecting Solar Causes Observatory (EASCO) that aims to bridge the gaps in observational capabilities of current missions with key additional measurements such as radio, magnetograph and X-ray imaging. Trichas et al. [141] proposed the Carrington-L5 mission to to obtain solar magnetograms from L5 and Weinzierl et al. [147] explored its possible improvements on space weather forecast. Lavraud et al. [63] also proposed a concept for a small mission to the Sun-Earth L5 point, the investigation of solar-terrestrial activity and transients (INSTANT) mission. The INSTANT mission would determine coronal mass ejection (CME) kinematics with unparalleled accuracy and measure the coronal magnetic fields. Thomas et al. [135] proposed that space weather L5 monitor with in-situ plasma detectors should be located at the same heliographic latitude as the Earth so as to optimize the forecasting ability of the monitor and to reduce the chance of missing important events. Based on PFSS models, Petrie et al. [104] presented a new effort to combine the magnetogram data from multiple telescopes such as from Earth/L1, L3, L4, and L5 with different strengths or viewpoints that may become standard practice in future, leading to enhanced heliospheric models and space weather research. Such observations will fill known gaps in our scientific understanding of the three-dimensional, time-evolving Sun and heliosphere. Also, Gibson et al. [39] presented the science obtainable through extra Sun-Earth line vantages, including solar dynamo studies, solar atmospheric global connections, and solar wind evolution and transient interactions, and discussed the benefits from multi-vantage observations for space-weather monitoring, particularly from the far side of the Sun, at high latitudes including over the solar poles, or from near-quadrature angles relative to the Earth (e.g., the Sun-Earth L4 and L5 Lagrangian points). As pointed out by Gibson et al. [39], the extra Sun-Earth line vantages present unique opportunities for answering the outstanding science questions of heliophysics, for improving space-weather monitoring and prediction, and for revealing new discoveries about our Sun and solar system.

In-situ measurement gap for solar corona will be filled by NASA's Parker Solar Probe launched in 2018, which is the first spacecraft to fly into the low solar corona of $<10R_s$ [37, 144]. ESA's Solar Orbiter to be launched in the near future will not only fly to the Sun as close as 0.3 AU, but also provide observations in the solar wind out of the ecliptic plane [89]. Once the in-situ observations in the inner heliosphere are obtained from these two spacecraft, validation of MHD solar coronal models will be carried out more fully.

The future coordinated investigation of available ground and space based booming data paves us the way to gain the physics of the energy transport from the convective layers of the Sun into the heliosphere and triggers new theoretical approaches to the evolution of the heliospheric field. Substantially, observational advances offer progress in theory and modeling, with a broad range of models permitting plausible scenarios to explain coronal heating and solar wind acceleration, which is a challenging scientific question in solar physics. For instance, the planned coordinated observations within the framework of space missions, such as the Interhelioprobe, Solar Orbiter, Parker Solar Probe, Hinode, SDO, STEREO [60], will allow us to explore the structure and dynamics of magnetic fields in the polar regions of the Sun, to study the mechanisms of the solar dynamo and solar cycle, to gain a deeper insight into the process of corona heating and acceleration of the solar wind, and to get a response to a number of current pressing issues of heliophysics. Especially, novel and improved observations of the vector magnetic field in the extended solar atmosphere will provide constraints for the models under consideration, leading to our understanding the magnetic configuration of the pre-eruption structure from solar observations, the advances in modeling the magnetic configuration before the eruption, and realistic simulations of the eruption process. In this aspect, space missions, such as SOHO, SDO and Hinode, have already enriched our knowledge.

Additionally, network ground-based and multi-spacecraft validation will give us a more comprehensive understanding of the validity, strengths and weaknesses of our model. As new validation methodologies are continuously proposed by the terrestrial weather community (e.g., [46]), solar wind modelers can also make use of these new methodologies in a creative way to validate the model from an novel viewpoint in the future.

When reviewing the development of terrestrial NWP, Bauer et al. [8] concluded that the largest benefit in predictive skill has been obtained from ensemble forecasting, data assimilation and physical process representation. Ensemble forecasting (or ensemble modelling) is a numerical method that use multiple forecasts to generate a broad sample of the possible future states of a dynamical system [58, 77]. Individual forecasts that constitute an ensemble forecast are usually called ensemble members. In practice, ensemble members can be created by using different observational inputs, boundary conditions, initial conditions, physical processes and free parameters. Ensemble forecasting is able to quantify prediction uncertainties and determine forecast confidence. An ensemble contains additional information about possible model realizations and their probabilities, and thus provide a probabilistic forecast. The sensitivity of the forecast to the variation of model realization can also be revealed. For space weather purposes the dynamical system may be the solar atmosphere, interplanetary space in which solar disturbances propagate, Earth's magnetosphere, and/ or Earth's space atmosphere interaction region. Uncertainty quantification and ensemble techniques in space weather prediction have been underway (e.g., [1, 16, 22, 47, 55, 58, 78, 88, 90, 91, 99, 103, 108, 122]). In the future, the availability of multiple MHD models at the operational level is desirable to provide diversity of forecast methodology. The use of ensemble forecasting requires the ability to run predictions faster than real time.

The ensemble mean or a weighted mean can be employed as a single forecast, which should perform better than individual ensemble members. Ensemble methods are certainly useful for providing an improved forecast as well as an estimate of forecast uncertainty, and uncertainties will be utilized more frequently in future iterations of space weather predictions issued to end users. As such, the end users are able to make better decisions according to uncertainties since they can understand the trustworthiness, or robustness of the results.

In recent years, a variety of studies on the ensemble forecasting of CMEs have been carried out. Some of these studies focus on ensemble forecasting of one single CME event [21, 65], while other studies included more CME events in the ensemble [32, 77]. There are common conclusion reached by these studies. The reliability of ensemble CME-arrival predictions is heavily dependent on the initial distribution of CME input parameters, like initial speed, direction, width and mass of the CME. The solar wind background also play an important role in CME forecasting. Without using an ambient solar wind background that provides the realistic conditions, it is impossible to accurately forecast the CME arrival times and properties using even realistic CME fit parameters in the corona. All of these studies are carried out by using WSA-Enlil model. For other 3D MHD models, ensemble forecast is surely beneficial and should be carried out in the future. In Mays et al. [77], it is found that the ensemble members in their study did not sample a wide-enough spread in CME input parameters. For the ensemble forecasting of terrestrial NWP, determining ensemble members consistently and seamlessly so that the ensemble provides a good estimate of uncertainty across a wide range of prediction scales is also challenging. In the experience of terrestrial NWP, the input of mathematics and statical physics expertise are crucially important for further development of ensemble member generation method. In Pizzo et al. [107], it is argued that there is no evidence that the propagation of CME is chaotic, and therefore, CME forecasting calls for different tactics than employed for terrestrial NWP. This conclusion highlights that rather than duplicate the ensemble forecast methods from terrestrial NWP directly, we should adopt these methods in accordance with the characteristics of CME and solar wind forecasting.

In the ensemble forecasting of CME, the ambient solar wind is found to be highly sensitive to the input photospheric synoptic maps. This conclusion is confirmed by ensemble forecasts of ambient solar wind itself, with relatively small (i.e., $\sim$10) ensemble of MHD coronal models and only a single CR [67, 116]. Despite the fact that these solar wind ensemble forecasts are only initial proof-of-concept, the ensemble approach did provide a more reliable forecast than any individual realization, suggesting that this approach is worth pursuing. However, running the MHD solar wind model is computationally expensive. One obstacle to implement a larger ensemble of solar wind forecast with longer period is the limited computational resource. To mitigate this problem, Owens and Riley [98] employed a simple, computationally cheap, one-dimensional upwind scheme to propagate coronal solution to 1AU and produced a large ensemble (576) of time series at the base of the Sun-Earth line. They sampled the near-Sun solar wind conditions at a large range of positions around the sub-Earth point. This large ensemble of near-Earth solar wind speeds is used to construct a probabilistic solar wind forecast with a quantitative estimate of the forecast uncertainty. Such a way of sampling is based on the idea that the positional errors of large scale global solar wind structure, particularly in the location of the slow solar wind band, mean that a single deterministic forecast of the near-Earth solar wind speed does not fully exploit the near-Sun solar wind information. However, such an ensemble may actually be achieved by using a single set of global result, and may be used in future ensemble forecast study. In order to analyse their ensemble results, Owens and Riley [98] adopted the "cost/loss" analysis [48, 102] developed for meteorological applications, and provided a good example of effective usage of mature tool from terrestrial weather. It has been recognized by terrestrial weather forecaster that the error of the ensemble mean forecast contains systematic error and these systematic errors cause biases in the ensemble distribution. The ensemble distribution is shifted in its mean position and does not cover the true uncertainties. However, the bias can be corrected by statistical post processing algorithms (e.g., [26, 146]). Such bias correction can be considered in ensemble forecast studies of solar wind and CME in the future.

Murray [90] highlighted some examples of success stories in space weather ensemble methods. Nevertheless, the continuing emergence of ensemble techniques in many parts of the space weather system is encouraging. Further basic research and development with ensembles, particularly in areas such as solar eruptive forecasting, may proceed to offer useful insights that can be transitioned to operational forecasting as well as providing prediction errors to aid end-user understanding. This may ultimately aid space weather forecasts to begin to rival meteorological efforts in terms of accuracy. Since the ensemble would require a great deal of computational resources, it is important to develop optimized method for selecting ensemble members. This in turn requires an understanding of the sensitivity in the input data, boundary conditions, initial conditions and model parameterization.

Among many options for improving forecasts, data assimilation is also a promising choice, but this is challenging because of sparse temporal and spatial measurements in the solar-terrestrial space. The term "data assimilation" was coined in the late sixties by the meteorological community to denote a process that combines prior information of a system simulated in a numerical model with new observational information to obtain, as accurately as possible, a description of the evolution of the system (e.g., [134]). Observations enter model's forecast cycle through the process of data assimilation, which provide a safeguard against model error. Extensive use of data assimilation greatly improves forecast skill in numerical terrestrial weather prediction. Indeed, improvements in forecast skill in numerical terrestrial weather prediction models over the past 30 years have directly coincided with improvements in data assimilation schemes.

Although plenty of techniques have been developed by the meteorological community, whether they are applicable to current space weather models needs to be investigated. A simple application of proven techniques that work well in meteorology might not work for the space weather application, because models covering different domains may have different nature of sensitivity to the initial conditions and drivers. For space-weather forecasting, data assimilation is underdeveloped in space weather applications (e.g., [5, 14, 30, 38, 40, 50, 59, 61, 62, 81, 83, 124, 145, 148]). Data assimilation attempts have been made in three main domains: the photosphere, the solar wind and the ionosphere. As the number of observations of the ionosphere increases, various ionospheric data assimilation methods have been applied to the thermosphere-ionosphere system, such as 3DVar [14], 4DVar [145], Local Ensemble Transform Kalman Filter (LETKF) [30], Iterative Reinitialization, Driver Estimation and Assimilation (IRIDEA) technique [133] and transformative framework for data assimilation and calibration of the physical ionosphere-thermosphere models by using a Kalman filter [81]. Using The Medium Electron A (MEA) instrument aboard the Combined Release and Radiation Effects Satellite (CRRES) satellite data, and a radial diffusion model, Shprits et al. [124] showed that sparse CRRES MEA data is sufficient for reconstructing the electron phase space density (PSD), with high resolution in time and space using the Kalman filter. By using the Utah State University Global Assimilation of Ionospheric Measurements-Gauss Markov model and a Kalman filter as a basis for assimilating a diverse set of measurements, Gardner et al. [38] investigated the distribution of ionospheric plasma for storm-enhanced density (SED) structures in the midlatitude ionosphere over the continental United States during storm times over the continental United States during four storm periods: two strong, October 2003 and November 2003, and two moderate, August 2010 and August 2011. It was found that SED during strong storm was formed from the plasma in the northern equatorial anomaly crest, with the plasma in the SED channel lifting the closer it came to the high latitudes, while the moderate storm-enhanced density appeared to be unconnected to the northern anomaly crest but was rather produced locally in the SED channel, along with no corresponding increase in layer height associated with the SED evident in the model.

At present, due to relatively sparse data and technical challenges, data assimilation is particularly underdeveloped for the photosphere and solar wind (e.g., [5, 40, 50, 59, 61, 62, 148]). With the help of the Air Force Data Assimilative Photospheric Flux Transport (ADAPT) model [5], photospheric data assimilation uses observations of the magnetic field at the Sun's surface with physics-based temporal evolution to improve the inner boundary condition for coronal models. The improved representation of the corona can in turn be used to improve inner-boundary conditions for solar wind models. Merka et al. [83] presented the procedure and initial results of the data assimilation of solar wind observation in near-Earth space with Enlil model. However, no following paper of theirs on data assimilation is seen. Effort is also taken by Innocenti et al. [50] who applied the ensemble technique to an empirical solar wind forecasting model by using ACE observation at L1 point. Lang et al. [62] conducted a proof-of-concept study of the data assimilation in the solar wind. An artificially generated "true state" is compared with the data assimilation results. Pseudo observations are obtained from the "true state" at a fixed point at 0.6 AU. Their analyses indicate that the data assimilation result performs better than model-only result, especially for density and radial momentum variables. However, the improvement brought in by the data assimilation in each observation time step is quickly advected away from the observation location. This is because the model-only result immediately replaces the assimilated result in the time steps after assimilation due to the strong radial flow. This phenomenon implies the fact that the solar wind differ from the typical dynamics of the atmosphere, as the solar wind is strongly driven by the condition at the top of the corona. This difference should be considered in the future study of the solar wind data assimilation. In Lang et al. [62], the data assimilation is realized by EMPIRE (Employing Message Passing Interface for Researching Ensembles available at http://www.met.reading.ac.uk/~darc/empire) data assimilation system [12], which contains data assimilation codes that link to the model via a small number of MPI commands. Data assimilation libraries like EMPIRE, Data Assimilation Research Testbed (DART) [3] and Parallel Data Assimilation Framework (PDAF) [92] are highly optimized and not dependent on how specific model works. Minimal coding is needed to couple any solar wind model to these systems if they are employed. It should be noted that data assimilation requires the characterization of data quality as well as data robustness, often requiring careful processing of observational data for incorporation into models.

Another application of data assimilation method to the solar wind prediction is presented by Lang and Owens [61], where the first solar wind variational data assimilation scheme is developed by using a simple solar wind propagation model. This scheme enables solar wind observation far from the Sun, such as 1 AU, to update and improve the inner boundary conditions (about 0.1 AU) of interplanetary solar wind model. However, as pointed out by Lang and Owens [61], the requirement for an adjoint model used in this work means that the variational approach is not practical for a full MHD solar wind model. For data-assimilation with more complex MHD models, in order to map observational information from near-Earth space to the inner-boundary using solar wind observations that can then be used to drive more complex models, a hybrid data assimilation method, such as Ensemble-4DVar [2, 42], or a smoother-based data assimilation method, such as the Iterative Ensemble Kalman Smoother [11], is worthy of further study. Nevertheless, due to the improvement of the forecast skill, the technique proposed by Lang and Owens [61] for such a simple solar wind propagation model may provide an optimum set of boundary condition, which can then be used to drive MHD models to obtain better results in the future. By the way, driven by the applications of space weather modeling, there witnesses a significant development of numerical algorithms based on the hybrid Particle-In-Cell (PIC) ion and fluid electron model (e.g., [20, 23, 51, 76, 96, 131, 150]). PIC modeling has proceeded to model space weather events in the heliosphere and simulate space weather processes such as the conversion of magnetic energy to kinetic energy, and the interaction of the Earth space environment with incoming solar plasma (e.g., [51]).

The development of machine learning has been stimulated by the recent explosion of data from large-scale numerical simulations that can generate exabytes of data such as multiple runs needed for forecasts with the ensemble modeling approach, and from observations including imaging by spacecraft-borne imagers. There has been a substantial push within the space weather community to try machine learning techniques used by data scientists [79] to advance space weather forecasting efforts. Up to now, machine learning techniques have been widely applied to various topics of space weather, such as identification of coronal mass ejections (e.g., [69, 70, 93, 136]), classification and prediction of solar wind (e.g., [17, 153]) and prediction of disruptions to global navigation satellite systems signals at high latitudes (e.g., [80]).

Machine learning refers to a set of methods and algorithms [18] that (1) make predictions in time or space of a continuous quantity (regression); (2) assign a datum to a class within a prespecified set (classification); (3) assign a datum to a class within a set that is determined by the algorithm itself (clustering); (4) reduce the dimensionality of a dataset, by exposing relationships among variables; and (5) establish linear and nonlinear relationships and causalities among variables. In space weather forecasting, machine learning techniques can be used to automatically identify interesting regions like solar active regions, coronal holes, coronal mass ejections, and magnetic reconnection events, as well as to select features; to discover knowledge in studying causality and relationships within highly dimensional data, and to cluster similar events, with the aim of deepening our physical understanding; to provide forecasting capability of dealing with large class imbalances and/or significant data gaps to forecast important space weather events from a combination of solar images, solar wind, and geospace in situ data; to facilitate modeling that is somewhat different from forecasting and involves a higher level approach where the focus is on discovering the underlying physical and long-term behavior of the system; and also to discover the nonlinear map that describes the system evolution. Machine learning algorithms shed lights on two characteristics: being accurate and being robust against new data (i.e., being able to generalize). Most physics models, like solar wind MHD model, have the characteristic that every process that led to their development is completely intelligible by the human mind. This kind of model is often referred to as "white box" since each component is transparent. In contrast to white-box models of the "human mind", machine learning methods are often referred to as black-box, signifying that the mathematical structure and the relationships between variables are so complicated that it is often not useful to try to understand them, as long as they deliver the expected results. Although machine learning is a powerful tool, no physical information is involved or provided by the formula of the black box.

For an application based perspective, such a black-box model is satisfactory since promising results may be obtained from it. However, the black box provides or involves limited physical information, which makes it incomprehensible and completely useless. Besides the black-box and white-box model, the gray-box model, which integrates a first-principle model and a statistical model, has attracted researchers' attention by its capability: known linear/nonlinear phenomena can be embedded in the first-principle model, and an unknown relationship among variables can be embedded in the statistical model by extracting such a relationship from the data. In general, gray-box models are more accurate than simplified first-principle models, less complicated than physics-based MHD models, and more easily interpreted than statistical models. Within the framework of gray-box modeling, the machine learning technique can be also utilized in calibrating the assumptions or free parameters involved. Therefore, in the future study, the gray-box model may be a more reasonable choice for solar wind model development. For more detailed applications of machine learning to space weather forecasting, please refer to various chapters of the book entitled "Machine Learning Techniques for Space Weather" [19]. Also, Camporeale et al. [15] presented a review on the application of machine learning in space weather to a few areas, such as the forecasting of geomagnetic indices, of relativistic electrons at geosynchronous orbits, of solar flares occurrence, of coronal mass ejection propagation time, and of solar wind speed, introduced some machine learning concepts tailored to the space weather community, and pointed out a selected number of open challenges that the space weather community should undertake in the next decade. To learn more about machine learning, the interested reader is suggested to refer to Christoph Molnar's book entitled "Interpretable Machine Learning- A Guide for Making Black Box Models Explainable" (http://leanpub.com/interpretable-machine-learning), which is written to make machine learning models and their decisions interpretable for machine learning practitioners, data scientists and statisticians. A free HTML version of the book can be found at: https://christophm.github.io/interpretable-ml-book/.

Since useful data is usually "small" and expensive to acquire, enthusiasm about "big data" has set up a new paradigm of learning partial differential equations from small data. Such kind of methodology is employed to deal with the problem of learning, system identification, or data-driven discovery of partial differential equations (PDEs) (e.g., [9, 13, 109–112, 119]), by blending conservation laws, physical principles, and/or phenomenological behaviors expressed by partial differential equations with the data sets available. In particular, Raissi and Karniadakis [109] introduced hidden physics models, which are essentially data-efficient learning machines capable of leveraging the underlying laws of physics, expressed by time dependent and nonlinear partial differential equations, to extract patterns from high-dimensional data generated from experiments. Also, Raissi et al. [112] put forth a deep learning framework that enables the synergistic combination of mathematical models and data, by introducing an effective mechanism for regularizing the training of deep neural networks in small data regimes. Numerical solution to high-dimensional PDEs have been a longstanding computational challenge. Numerical discretization methods become infeasible in higher dimensions due to the explosion in the number of grid points and the demand for reduced time step size. Sirignano and Spiliopoulos [125] proposed to solve high-dimensional PDEs by approximating the solution with a deep neural network, which is trained to satisfy the differential operator, and initial-boundary conditions using stochastic gradient descent at randomly sampled spatial points so as to avoid the need to form a mesh (which is infeasible in higher dimensions) and instead convert the PDE problem into a machine learning problem. These works have investigated how to solve differential equations by using deep neural networks. The underlying idea is that a neural network constructs a nonlinear map between inputs and outputs such that a set of equations can be enforced to be very accurately satisfied on a given number of points in a computational domain. Both machine learning and standard methods can be understood as complementary to solve partial differential equations. Since coronal-heliospheric models introduced in this book are based on numerically solving 3D MHD equations, deep learning technique might become an promising alternative as well as complementary tool in space weather modeling. However, it is worth exploring under what situations such expensive physics-based simulations might be replaced or supplemented by a fast machine learning approximation.

In a word, machine learning, as an interdisciplinary field closely related with artificial intelligence, pattern recognition, data mining, statistics, probability theory, optimization, statistical physics, and theoretical computer science, is expected to learn or establish complex relationships or patterns from or between data and physical laws, and thus is promising to provide novel tools in making accurate space weather prediction decisions.

Today, among the many challenges facing space weather/space physics communities there is a growing need for an analysis and visualization package that can examine the results from numerical and empirical models as well as observational data. To such purpose, CISM-DX package, a suite of integrated data, models, and data and model explorers, is developed for research and education from the Center for Integrated Space Weather Modeling (CISM) Data and Model Explorer. It is capable of completing highly sophisticated visualizations of the results from a diverse set of models throughout the connected Sun-Earth system. Wiltberger et al. [149] presented a overview of the community-developed CISM-DX package for visualizing and analyzing space science model output and observational data. This package is being released under an open source licensing agreement to facilitate and encourage community use and contribution, while the core packages included in CISM-DX are available from http://lasp.colorado.edu/cism/CISM_DX.

Multispacecraft, multi-instrument missions are becoming the norm. With the advent of the Heliophysics/Geospace System Observatory (H/GSO), a complement of multi-spacecraft missions and ground-based observatories to study the space environment, data retrieval, analysis, and visualization of space physics data can be formidable. Optimal science yield from the constellation observatory system also demands efficient platform-independent data query, ingestion, joint analysis and visualization techniques. The Space Physics Environment Data Analysis System (SPEDAS), which grew out of the community's efforts to satisfy this growing need, has been placed under NASA support in recognition of its potential for optimal science analysis from past, current and future missions. SPEDAS, a grass-roots software development platform (http://www.spedas.org), is now officially supported by NASA Heliophysics as part of its data environment infrastructure. It serves more than a dozen space missions and ground observatories and can integrate the full complement of past and upcoming space physics missions with minimal resources, following clear, simple, and well-proven guidelines. SPEDAS can readily query and ingest all International Solar Terrestrial Physics (ISTP)-compatible products from the Space Physics Data Facility (SPDF), enabling access to a vast collection of historic and current mission data. The planned incorporation of Heliophysics Application Programmer's Interface (HAPI) standards will facilitate data ingestion from distributed datasets that adhere to these standards. As Heliophysics is moving towards constellation class missions, and reliance on large arrays of ground-based datasets is also increasing (SuperDARN: Super Dual Auroral Radar Network; TREX: Transition Region Explorer; ULTIMA: Ultra Large Terrestrial International Magnetic Array, etc.) this trend for efficient analysis keeps continuing. SPEDAS is currently Interactive Data Language (IDL)-based, efforts are underway for it to further work with python. Angelopoulos et al. [4] reviewed the SPEDAS development history, goals, current implementation and software management, interfaces with other organizations, resources and support structure available to the community, and future development plans. Developed by the Institute for Space-Earth Environmental Research,

Nagoya University, Japan visualization tool for three-dimensional plasma velocity distributions (ISEE_3D) is an interactive visualization tool for three-dimensional plasma velocity distribution functions [57]. As a plug-in for SPEDAS, ISEE_3D provides a variety of methods to visualize the distribution function of space plasma: scatter, volume, and isosurface modes. Written in IDL, the tool also has a wide range of functions, such as displaying magnetic field vectors and two-dimensional slices of distributions to facilitate extensive analysis. Currently, the visualization tool can be used for plasma data from Geotail/LEP (Geotail low-energy particle (LEP) experiment), MMS (magnetospheric multiscale), and THEMIS (Time History of Events and Macroscale Interactions during Substorms).

The space weather community has begun to use frontier methods such as data assimilation, machine learning, and ensemble modeling to advance current operational forecasting efforts [90]. Promising directions for the further development of the corona-interplanetary weather forecasting involve more than those mentioned above, and it is difficult to completely outline a roadmap. Advanced forecasting of corona-interplanetary weather requires prediction of near-Earth solar-wind conditions on the basis of remote solar observations. This is typically achieved using numerical magnetohydrodynamic models initiated by photospheric and solar coronal observations. The accuracy of such forecasts is being continually improved through better numerics, better determination of the boundary conditions and better representation of the underlying physical processes. As discussed by Bauer et al. [8] for the quiet revolution of numerical terrestrial weather prediction, both scientific and technological challenges are also met in numerical space weather prediction. Significant advances in space weather forecast skill will come from scientific and technological innovation in computing, the representation of physical processes in parameterizations, coupling of solar-terrestrial system components, the use of observations with advanced data assimilation algorithms, and the consistent description of uncertainties through ensemble methods and how they interact across multi-temporal spatial scales. Many research models typically focus only on specific areas of the Sun-Earth system and have been developed internationally; there is an urgent need for a scientific assessment of their suitability to operate within a coupled space weather forecast system. For a given model to be appropriate for operational forecasting, it is crucial to meet certain requirements: being robust, validated, capable of producing timely forecasts, able to couple to other models, and easy to facilitate maintenance and evolution from the view of good quality software engineering principles. Numerical space weather prediction entails combined scientific, observing and computational technology advances, and depends on more interdisciplinary research at the science-technology interface. Since severe space weather is a global threat that requires a coordinated global response, international collaboration within the United Nations (UN) and international coordination between member states are recommended for a future approach [75]. In a word, an integrated framework for space weather modeling that brings together the different but complementary approaches is very likely the most effective strategy in modeling and prediction of space weather. Observational studies go simultaneously with modelling, and improved forecasting techniques such as metrics, validation, ensemble forecasting and data assimilation are aiding the transition to operations. In the future, improved forecasts are likely to derive from three areas: ongoing scientific investigation using existing and future space missions; availability of multipoint operational real time data from space-based and ground-based observations; and improved modelling schemes and techniques. In each of these areas, prospects are bright and a significant improvement should be expected in our ability to forecast space weather over the coming years.

References

1. Amerstorfer T, Möstl C, Hess P, Temmer M, Mays ML, Reiss MA, Lowrance P, Bourdin PA (2018) Ensemble prediction of a halo coronal mass ejection using heliospheric imagers. Space Weather 16(7):784–801. https://doi.org/10.1029/2017SW001786
2. Amezcua J, Goodliff M, Van Leeuwen PJ (2017) A weak-constraint 4DEnsembleVar. Part I: formulation and simple model experiments. Tellus A: dynamic meteorology and oceanography 69(1):1271564. https://doi.org/10.1080/16000870.2016.1271564
3. Anderson J, Hoar T, Raeder K, Liu H, Collins N, Torn R, Avellano A (2009) The data assimilation research testbed: a community facility. Bull Am Meteorol Soc 90:1283. https://doi.org/10.1175/2009BAMS2618.1
4. Angelopoulos V, Cruce P, Drozdov A, Grimes EW, Hatzigeorgiu N, King DA, Larson D, Lewis JW, McTiernan JM, Roberts DA, Russell CL, Hori T, Kasahara Y, Kumamoto A, Matsuoka A, Miyashita Y, Miyoshi Y, Shinohara I, Teramoto M, Faden JB, Halford AJ, McCarthy M, Millan RM, Sample JG, Smith DM, Woodger LA, Masson A, Narock AA, Asamura K, Chang TF, Chiang CY, Kazama Y, Keika K, Matsuda S, Segawa T, Seki K, Shoji M, Tam SWY, Umemura N, Wang BJ, Wang SY, Redmon R, Rodriguez JV, Singer HJ, Vandegriff J, Abe S, Nose M, Shinbori A, Tanaka YM, UeNo S, Andersson L, Dunn P, Fowler C, Halekas JS, Hara T, Harada Y, Lee CO, Lillis R, Mitchell DL, Argall MR, Bromund K, Burch JL, Cohen IJ, Galloy M, Giles B, Jaynes AN, Le Contel O, Oka M, Phan TD, Walsh BM, Westlake J, Wilder FD, Bale SD, Livi R, Pulupa M, Whittlesey P, DeWolfe A, Harter B, Lucas E, Auster U, Bonnell JW, Cully CM, Donovan E, Ergun RE, Frey HU, Jackel B, Keiling A, Korth H, McFadden JP, Nishimura Y, Plaschke F, Robert P, Turner DL, Weygand JM, Candey RM, Johnson RC, Kovalick T, Liu MH, McGuire RE, Breneman A, Kersten K, Schroeder P (2019) The space physics environment data analysis system (SPEDAS). Space Sci Rev 215(1):9. https://doi.org/10.1007/s11214-018-0576-4
5. Arge CN, Henney CJ, Koller J, Compeau CR, Young S, MacKenzie D, Fay A, Harvey JW (2010) Air force data assimilative photospheric flux transport (ADAPT) model. Twelfth international solar wind conference 1216:343–346
6. Asai K, Kojima M, Tokumaru M, Yokobe A, Jackson BV, Hick PL, Manoharan PK (1998) Heliospheric tomography using interplanetary scintillation observations: 3. correlation between speed and electron density fluctuations in the solar wind. J Geophys Res Space Phys 103(A2):1991–2001. https://doi.org/10.1029/97JA02750
7. Austin HJ, Savani NP (2018) Skills for forecasting space weather. Weather 73(11):362–366. https://doi.org/10.1002/wea.3076
8. Bauer P, Thorpe A, Brunet G (2015) The quiet revolution of numerical weather prediction. Nature 525:47–55. https://doi.org/10.1038/nature14956
9. Berg J, Nyström K (2018) A unified deep artificial neural network approach to partial differential equations in complex geometries. Neurocomputing 317:28–41. https://doi.org/10.1016/j.neucom.2018.06.056

10. Bisi MM, Fallows RA, Breen AR, O'Neill IJ (2010) Interplanetary scintillation observations of stream interaction regions in the solar wind. Sol Phys 261:149–172. https://doi.org/10.1007/s11207-009-9471-1
11. Bocquet M, Sakov P (2014) An iterative ensemble Kalman smoother. Q J R Meteorol Soc 140:1521–1535. https://doi.org/10.1002/qj.2236
12. Browne PA, Wilson S (2015) A simple method for integrating a complex model into an ensemble data assimilation system using MPI. Environ Model Softw 68:122–128. https://doi.org/10.1016/j.envsoft.2015.02.003
13. Brunton SL, Proctor JL, Kutz JN (2016) Discovering governing equations from data by sparse identification of nonlinear dynamical systems. Proc Natl Acad Sci 113(15):3932–3937. https://doi.org/10.1073/pnas.1517384113
14. Bust GS, Mitchell CN (2008) History, current state, and future directions of ionospheric imaging. Rev Geophys 46(1):RG1003. https://doi.org/10.1029/2006RG000212
15. Camporeale E (2019) The challenge of machine learning in space weather nowcasting and forecasting. Accepted. https://doi.org/10.1029/2018SW002061
16. Camporeale E, Shprits Y, Chandorkar M, Drozdov A, Wing S (2016) On the propagation of uncertainties in radiation belt simulations. Space Weather 14(11):982–992. https://doi.org/10.1002/2016SW001494
17. Camporeale E, Caré A, Borovsky JE (2017) Classification of solar wind with machine learning. J Geophys Res Space Phys 122(11):10,910–10,920. https://doi.org/10.1002/2017JA024383
18. Camporeale E, Wing S, Johnson JR (2018) Introduction. In: Camporeale E, Wing S, Johnson JR (eds) Machine learning techniques for space weather, p xiii. Elsevier, Amsterdam. https://doi.org/10.1016/B978-0-12-811788-0.09987-X
19. Camporeale E, Wing S, Johnson JR (2018b) Machine learning techniques for space weather. Elsevier, Amsterdam
20. Cannon P, Angling M, Barclay L, Curry C, Dyer C, Edwards R, Greene G, Hapgood M, Horne R, Jackson D et al (2002) The hybrid multiscale simulation technology: an introduction with application to astrophysical and laboratory plasmas scientific computation. Springer, Berlin
21. Cash MD, Biesecker DA, Pizzo V, de Koning CA, Millward G, Arge CN, Henney CJ, Odstrčil D (2015) Ensemble modeling of the 23 July 2012 coronal mass ejection. Space Weather 13:611–625. https://doi.org/10.1002/2015SW001232
22. Cash MD, Biesecker DA, Pizzo V, Koning CA, Millward G, Arge CN, Henney CJ, Odstrčil D (2015) Ensemble modeling of the 23 July 2012 coronal mass ejection. Space Weather 13(10):611–625. https://doi.org/10.1002/2015SW001232
23. Chacón L, Chen G (2016) A curvilinear, fully implicit, conservative electromagnetic PIC algorithm in multiple dimensions. J Comput Phys 316:578–597. https://doi.org/10.1016/j.jcp.2016.03.070
24. Cranmer SR, Winebarger AR (2019) The properties of the solar corona and its connection to the solar wind. Ann Rev Astron Astrophys 57:1–30
25. Cranmer SR, Gibson SE, Riley P (2017) Origins of the ambient solar wind: implications for space weather. Space Sci Rev 212(3):1345–1384. https://doi.org/10.1007/s11214-017-0416-y
26. Cui B, Toth Z, Zhu Y, Hou D (2012) Bias correction for global ensemble forecast. Weather Forecast. 27:396–410. https://doi.org/10.1175/WAF-D-11-00011.1
27. de Patoul J, Foullon C, Riley P (2015) 3D electron density distributions in the solar corona during solar minima: assessment for more realistic solar wind modeling. Astrophys J 814:68. https://doi.org/10.1088/0004-637X/814/1/68
28. DeForest CE, Howard TA, Webb DF, Davies JA (2016) The utility of polarized heliospheric imaging for space weather monitoring. Space Weather 14:32–49. https://doi.org/10.1002/2015SW001286
29. Dungey JW (1961) Interplanetary magnetic field and the auroral zones. Phys Rev Lett 6:47–48. https://doi.org/10.1103/PhysRevLett.6.47

30. Durazo JA, Kostelich EJ, Mahalov A (2017) Local ensemble transform Kalman filter for ionospheric data assimilation: observation influence analysis during a geomagnetic storm event. J Geophys Res Space Phys 122(9):9652–9669. https://doi.org/10.1002/2017JA024274
31. Eastwood JP, Nakamura R, Turc L, Mejnertsen L, Hesse M (2017) The scientific foundations of forecasting magnetospheric space weather. Space Sci Rev 212(3):1221–1252. https://doi.org/10.1007/s11214-017-0399-8
32. Emmons D, Acebal A, Pulkkinen A, Taktakishvili A, MacNeice P, Odstrčil D (2013) Ensemble forecasting of coronal mass ejections using the WSA-ENLIL with CONED model. Space Weather 11:95–106. https://doi.org/10.1002/swe.20019
33. Fallows RA, Breen AR, Bisi MM, Jones RA, Wannberg G (2006) Dual-frequency interplanetary scintillation observations of the solar wind. Geophys Res Lett 33(11):L11106. https://doi.org/10.1029/2006GL025804
34. Fallows RA, Asgekar A, Bisi MM, Breen AR, ter-Veen S, (2013) The dynamic spectrum of interplanetary scintillation: first solar wind observations on LOFAR. Sol Phys 285:127–139. https://doi.org/10.1007/s11207-012-9989-5
35. Fallows RA, Bisi MM, Forte B, Ulich T, Konovalenko AA, Mann G, Vocks C (2016) Separating Nightside Interplanetary and Ionospheric Scintillation with LOFAR. Astrophys J Lett 828:L7. https://doi.org/10.3847/2041-8205/828/1/L7
36. Folini D (2018) Climate, weather, space weather: model development in an operational context. J Space Weather Space Clim 8:A32. https://doi.org/10.1051/swsc/2018021
37. Fox NJ, Velli MC, Bale SD, Decker R, Driesman A, Howard RA, Kasper JC, Kinnison J, Kusterer M, Lario D, Lockwood MK, McComas DJ, Raouafi NE, Szabo A (2016) The Solar Probe Plus mission: humanity's first visit to our star. Space Sci Rev 204:7–48. https://doi.org/10.1007/s11214-015-0211-6
38. Gardner LC, Schunk RW, Scherliess L, Eccles V, Basu S, Valladeres C (2018) Modeling the midlatitude ionosphere storm-enhanced density distribution with a data assimilation model. Space Weather 16:1539–1548. https://doi.org/10.1029/2018SW001882
39. Gibson SE, Vourlidas A, Hassler DM, Rachmeler LA, Thompson MJ, Newmark J, Velli M, Title A, McIntosh SW (2018) Solar physics from unconventional viewpoints. Front Astron Space Sci 5:32. https://doi.org/10.3389/fspas.2018.00032
40. Godinez HC, Yu Y, Lawrence E, Henderson MG, Larsen B, Jordanova VK (2016) Ring current pressure estimation with ram-scb using data assimilation and van Allen probe flux data. Geophysical Research Letters 43(23):11,948–11,956. https://doi.org/10.1002/2016GL071646
41. Gombosi TI, van der Holst B, Manchester WB, Sokolov IV (2018) Extended MHD modeling of the steady solar corona and the solar wind. Living Rev Sol Phys 15:4. https://doi.org/10.1007/s41116-018-0014-4
42. Goodliff M, Amezcua J, Van Leeuwen PJ (2017) A weak-constraint 4DEnsembleVar. Part II: experiments with larger models. Tellus A: dyn meteorol and ocean 69(1): 1271565. https://doi.org/10.1080/16000870.2016.1271565
43. Gopalswamy N, Davila JM, St Cyr OC, Sittler EC, Auchère F, Duvall TL, Hoeksema JT, Maksimovic M, MacDowall RJ, Szabo A, Collier MR (2011) Earth-affecting solar causes observatory (EASCO): a potential international living with a star mission from Sun-Earth L5. J Atmos Sol Terr Phys 73:658–663. https://doi.org/10.1016/j.jastp.2011.01.013
44. Green LM, Török T, Vršnak B, Manchester W, Veronig A (2018) The origin, early evolution and predictability of solar eruptions. Space Sci Rev 214(1):46. https://doi.org/10.1007/s11214-017-0462-5
45. Gressl C, Veronig AM, Temmer M, Odstrčil D, Linker JA, Mikić Z, Riley P (2014) Comparative study of MHD modeling of the background solar wind. Sol Phys 289:1783–1801. https://doi.org/10.1007/s11207-013-0421-6
46. Guan H, Zhu Y (2017) Development of Verification Methodology for Extreme Weather Forecasts. Weather Forecast 32:479–491. https://doi.org/10.1175/WAF-D-16-0123.1

47. Guerra JA, Pulkkinen A, Uritsky VM (2015) Ensemble forecasting of major solar flares: first results. Space Weather 13(10):626–642. https://doi.org/10.1002/2015SW001195
48. Henley EM, Pope ECD (2017) Cost-loss analysis of ensemble solar wind forecasting: space weather use of terrestrial weather tools. Space Weather 15:1562–1566. https://doi.org/10.1002/2017SW001758
49. van der Holst B, Sokolov IV, Meng X, Jin M, Manchester IWB, Tóth G, Gombosi TI (2014) Alfvén wave solar model (AWSoM): coronal heating. Astrophys J 782(2):81. http://stacks.iop.org/0004-637X/782/i=2/a=81
50. Innocenti ME, Lapenta G, VršNak B, Crespon F, Skandrani C, Temmer M, Veronig A, Bettarini L, Markidis S, Skender M (2011) Improved forecasts of solar wind parameters using the Kalman filter. Space Weather 9:S10005. https://doi.org/10.1029/2011SW000659
51. Innocenti ME, Johnson A, Markidis S, Amaya J, Deca J, Olshevsky V, Lapenta G (2017) Progress towards physics-based space weather forecasting with exascale computing. Adv Eng Softw 111:3–17. https://doi.org/10.1016/j.advengsoft.2016.06.011
52. Jackson BV, Hick PP, Buffington A, Yu HS, Bisi MM, Tokumaru M, Zhao X (2015a) A determination of the North-South heliospheric magnetic field component from inner corona closed-loop propagation. Astrophys J Lett 803:L1. https://doi.org/10.1088/2041-8205/803/1/L1
53. Jackson BV, Odstrčil D, Yu HS, Hick PP, Buffington A, Mejia-Ambriz JC, Kim J, Hong S, Kim Y, Han J, Tokumaru M (2015b) The UCSD kinematic IPS solar wind boundary and its use in the ENLIL 3-D MHD prediction model. Space Weather 13:104–115. https://doi.org/10.1002/2014SW001130
54. Jackson BV, Yu HS, Buffington A, Hick PP, Nishimura N, Nozaki N, Tokumaru M, Fujiki K, Hayashi K (2016) Exploration of solar photospheric magnetic field data sets using the UCSD tomography. Space Weather 14(12):1107–1124. https://doi.org/10.1002/2016SW001481
55. Jian LK, MacNeice PJ, Taktakishvili A, Odstrčil D, Jackson B, Yu HS, Riley P, Sokolov IV, Evans RM (2015) Validation for solar wind prediction at Earth: comparison of coronal and heliospheric models installed at the CCMC. Space Weather 13:316–338. https://doi.org/10.1002/2015SW001174
56. Jiang CW, Wu ST, Feng XS, Hu Q (2016) Data-driven magnetohydrodynamic modelling of a flux-emerging active region leading to solar eruption. Nat Commun 7:11522. https://doi.org/10.1038/ncomms11522
57. Keika K, Miyoshi Y, Machida S, Ieda A, Seki K, Hori T, Miyashita Y, Shoji M, Shinohara I, Angelopoulos V, Lewis JW, Flores A (2017) Visualization tool for three-dimensional plasma velocity distributions (isee_3d) as a plug-in for spedas. Earth Planets Space 69(1):170. https://doi.org/10.1186/s40623-017-0761-9
58. Knipp DJ (2016) Advances in space weather ensemble forecasting. Space Weather 14:52–53. https://doi.org/10.1002/2016SW001366
59. Koller J, Chen Y, Reeves GD, Friedel RHW, Cayton TE, Vrugt JA (2007) Identifying the radiation belt source region by data assimilation. J Geophys Res (Space Physics) 112:A06244. https://doi.org/10.1029/2006JA012196
60. Kuznetsov V (2015) Solar and heliospheric space missions. Adv Space Res 55(3):879–885. https://doi.org/10.1016/j.asr.2014.07.034
61. Lang M, Owens MJ (2019) A variational approach to data assimilation in the solar wind. Space Weather 17:59–83. https://doi.org/10.1029/2018SW001857
62. Lang M, Browne P, van Leeuwen PJ, Owens M (2017) Data assimilation in the solar wind: challenges and first results. Space Weather 15:1490–1510. https://doi.org/10.1002/2017SW001681
63. Lavraud B, Liu Y, Segura K, He J, Qin G, Temmer M, Vial JC, Xiong M, Davies JA, Rouillard AP, Pinto R, Auchère F, Harrison RA, Eyles C, Gan W, Lamy P, Xia L, Eastwood JP, Kong L, Wang J, Wimmer-Schweingruber RF, Zhang S, Zong Q, Soucek J, An J, Prech L, Zhang A, Rochus P, Bothmer V, Janvier M, Maksimovic M, Escoubet CP, Kilpua EKJ, Tappin J, Vainio R, Poedts S, Dunlop MW, Savani N, Gopalswamy N, Bale SD, Li G, Howard T, DeForest C, Webb D, Lugaz N, Fuselier SA, Dalmasse K, Tallineau J, Vranken D, Fernández JG (2016) A

small mission concept to the Sun-Earth Lagrangian L5 point for innovative solar, heliospheric and space weather science. J Atmos Sol Terr Phys 146:171–185. https://doi.org/10.1016/j.jastp.2016.06.004
64. Lee CO, Luhmann JG, Odstrčil D, MacNeice PJ, de Pater I, Riley P, Arge CN (2009) The solar wind at 1 AU during the declining phase of solar cycle 23: comparison of 3D numerical model results with observations. Sol Phys 254:155–183. https://doi.org/10.1007/s11207-008-9280-y
65. Lee CO, Arge CN, Odstrčil D, Millward G, Pizzo V, Quinn JM, Henney CJ (2013) Ensemble modeling of CME propagation. Sol Phys 285:349–368. https://doi.org/10.1007/s11207-012-9980-1
66. Li HC, Feng XS (2018) CESE-HLL magnetic field-driven modeling of the background solar wind during year 2008. J Geophys Res (Space Physics) 123:4488–4509. https://doi.org/10.1029/2017JA025125
67. Linker JA, Caplan RM, Downs C, Riley P, Mikic Z, Lionello R, Henney CJ, Arge CN, Liu Y, Derosa ML, Yeates A, Owens MJ (2017) The open flux problem. Astrophys J 848:70. https://doi.org/10.3847/1538-4357/aa8a70
68. Lugaz N, Temmer M, Wang Y, Farrugia CJ (2017) The interaction of successive coronal mass ejections: a review. Sol Phys 292(4):64. https://doi.org/10.1007/s11207-017-1091-6
69. Ma R, Angryk R, Riley P (2016) A data-driven analysis of interplanetary coronal mass ejecta and magnetic flux ropes. In: 2016 IEEE international conference on big data (Big Data), pp 3177–3186. https://doi.org/10.1109/BigData.2016.7840973
70. Ma R, Angryk RA, Riley P, Boubrahimi SF (2018) Coronal mass ejection data clustering and visualization of decision trees. The Astrophysical Journal Supplement Series 236(1):14. https://doi.org/10.3847/1538-4365/aab76f
71. Mackay DH, Yeates AR, Bocquet FX (2016) Impact of an L5 magnetograph on nonpotential solar global magnetic field modeling. Astrophys J 825(2):131. http://stacks.iop.org/0004-637X/825/i=2/a=131
72. MacNeice P (2009) Validation of community models: identifying events in space weather model timelines. Space Weather 7:S06004. https://doi.org/10.1029/2009SW000463
73. MacNeice P, Jian LK, Antiochos SK, Arge CN, Bussy-Virat CD, DeRosa ML, Jackson BV, Linker JA, Mikic Z, Owens MJ, Ridley AJ, Riley P, Savani N, Sokolov I (2018) Assessing the quality of models of the ambient solar wind. Space Weather 16(11):1644–1667. https://doi.org/10.1029/2018SW002040
74. Manchester W, Kilpua EKJ, Liu YD, Lugaz N, Riley P, Török T, Vršnak B (2017) The physical processes of CME/ICME evolution. Space Sci Rev 212(3):1159–1219. https://doi.org/10.1007/s11214-017-0394-0
75. Mann IR, Di Pippo S, Opgenoorth HJ, Kuznetsova M, Kendall DJ (2018) International collaboration within the united nations committee on the peaceful uses of outer space: framework for international space weather services (2018–2030). Space Weather 16(5):428–433. https://doi.org/10.1029/2018SW001815
76. Markidis S, Lapenta G, Rizwan-uddin (2010) Multi-scale simulations of plasma with iPIC3D. Math Comput Simul 80(7):1509–1519. https://doi.org/10.1016/j.matcom.2009.08.038
77. Mays ML, Taktakishvili A, Pulkkinen A, MacNeice PJ, Rastätter L, Odstrčil D, Jian LK, Richardson IG, LaSota JA, Zheng Y, Kuznetsova MM (2015) Ensemble modeling of CMEs using the WSA-ENLIL+Cone model. Sol Phys 290:1775–1814. https://doi.org/10.1007/s11207-015-0692-1
78. Mays ML, Thompson BJ, Jian LK, Colaninno RC, Odstrčil D, MÃűstl C, Temmer M, Savani NP, Collinson G, Taktakishvili A, MacNeice PJ, Zheng Y (2015) Propagation of the 7 January 2014 CME and resulting geomagnetic non-event. Astrophys J 812(2):145. http://stacks.iop.org/0004-637X/812/i=2/a=145
79. McGranaghan RM, Bhatt A, Matsuo T, Mannucci AJ, Semeter JL, Datta-Barua S (2017) Ushering in a new frontier in geospace through data science. J Geophys Res Space Phys 122(12):12,586–12,590. https://doi.org/10.1002/2017JA024835

80. McGranaghan RM, Mannucci AJ, Wilson B, Mattmann CA, Chadwick R (2018) New capabilities for prediction of high-latitude ionospheric scintillation: a novel approach with machine learning. Space Weather 16(11):1817–1846. https://doi.org/10.1029/2018SW002018
81. Mehta PM, Linares R (2018) A new transformative framework for data assimilation and calibration of physical ionosphere-thermosphere models. Space Weather 16(8):1086–1100. https://doi.org/10.1029/2018SW001875, https://agupubs.onlinelibrary.wiley.com/doi/abs/10.1029/2018SW001875
82. Meng X, van der Holst B, Tóth G, Gombosi TI (2015) Alfvén wave solar model (AWSoM): proton temperature anisotropy and solar wind acceleration. Mon Not R Astron Soc 454(4):3697–3709. http://dx.doi.org/10.1093/mnras/stv2249
83. Merka J, Merkova D, Odstrčil D (2007) A step toward data assimilation in solar wind research. J Atmos Sol Terr Phys 69:170–178. https://doi.org/10.1016/j.jastp.2006.07.012
84. Mok Y, Mikić Z, Lionello R, Downs C, Linker JA (2016) A three-dimensional model of active region 7986: comparison of simulations with observations. Astrophys J 817:15. https://doi.org/10.3847/0004-637X/817/1/15
85. Morgan H (2015) An atlas of coronal electron density at $5R_{\odot}$. I. Data processing and calibration. Astrophys J Suppl Ser 219:23. https://doi.org/10.1088/0067-0049/219/2/23
86. Morgan H, Druckmüller M (2014) Multi-scale gaussian normalization for solar image processing. Sol Phys 289:2945–2955. https://doi.org/10.1007/s11207-014-0523-9
87. Morgan H, Taroyan Y (2017) Global conditions in the solar corona from 2010 to 2017. Sci Adv 3(7):e1602056. https://doi.org/10.1126/sciadv.1602056
88. Morley SK, Welling DT, Woodroffe JR (2018) Perturbed input ensemble modeling with the Space Weather Modeling Framework. Space Weather 16(9):1330–1347. https://doi.org/10.1029/2018SW002000
89. Müller D, Marsden RG, St Cyr OC, Gilbert HR (2013) Solar Orbiter: exploring the sun-heliosphere connection. Sol Phys 285:25–70. https://doi.org/10.1007/s11207-012-0085-7
90. Murray SA (2018) The importance of ensemble techniques for operational space weather forecasting. Space Weather 16(7):777–783. https://doi.org/10.1029/2018SW001861
91. Murray SA, Henley EM, Jackson DR, Bruinsma SL (2015) Assessing the performance of thermospheric modeling with data assimilation throughout solar cycles 23 and 24. Space Weather 13(4):220–232. https://doi.org/10.1002/2015SW001163
92. Nerger L, Hiller W, Schroter J (2005) Pdaf - the parallel data assimilation framework: experiences with Kalman filtering. In: Zwieflhofer W, Mozdzynski G (eds) Use of high performance computing in meteorology, pp 63–83. 11th Workshop on the use of high performance computing in meteorology, european ctr medium range weather forecasts, reading, England, 25–29 Oct 2004. https://doi.org/10.1142/9789812701831_0006
93. Nguyen G, Aunai N, Fontaine D, Pennec EL, den Bossche JV, Jeandet A, Bakkali B, Vignoli L, Blancard BRS (2019) Automatic detection of interplanetary coronal mass ejections from in situ data: a deep learning approach. Astrophys J 874(2):145. https://doi.org/10.3847/1538-4357/ab0d24
94. Norquist DC (2013) Forecast performance assessment of a kinematic and a magnetohydrodynamic solar wind model. Space Weather 11(1):17–33. https://doi.org/10.1029/2012SW000853
95. Norquist DC, Meeks WC (2010) A comparative verification of forecasts from two operational solar wind models. Space Weather 8(12):S12005. https://doi.org/10.1029/2010SW000598
96. Olshevsky V, Bacchini F, Poedts S, Lapenta G (2019) Slurm: fluid particle-in-cell code for plasma modeling. Comput Phys Commun 235:16–24. https://doi.org/10.1016/j.cpc.2018.06.014
97. Owens MJ (2018) Time-window approaches to space-weather forecast metrics: a solar wind case study. Space Weather 16:1847–1861. https://doi.org/10.1029/2018SW002059
98. Owens MJ, Riley P (2017) Probabilistic solar wind forecasting using large ensembles of near-Sun conditions with a simple one-dimensional "upwind" scheme. Space Weather 15:1461–1474. https://doi.org/10.1002/2017SW001679

99. Owens MJ, Horbury TS, Wicks RT, McGregor SL, Savani NP, Xiong M (2014) Ensemble downscaling in coupled solar wind-magnetosphere modeling for space weather forecasting. Space Weather 12(6):395–405
100. Owens MJ, Lockwood M, Barnard LA (2017) Coronal mass ejections are not coherent magnetohydrodynamic structures. Sci Rep 7(1):4152. https://doi.org/10.1038/s41598-017-04546-3
101. Owens MJ, Lockwood M, Riley P (2017b) Global solar wind variations over the last four centuries. Sci Rep 7:41548. https://doi.org/10.1038/srep41548
102. Owens MJ, Riley P, Horbury TS (2017c) Probabilistic solar wind and geomagnetic forecasting using an analogue ensemble or "similar day" approach. Sol Phys 292:69. https://doi.org/10.1007/s11207-017-1090-7
103. Pesnell WD (2015) Predictions of solar cycle 24: how are we doing? Space Weather 14(1):10–21. https://doi.org/10.1002/2015SW001304
104. Petrie G, Pevtsov A, Schwarz A, DeRosa M (2018) Modeling the global coronal field with simulated synoptic magnetograms from Earth and the Lagrange points L_3, L_4, and L_5. Sol Phys 293(6):88. https://doi.org/10.1007/s11207-018-1306-5
105. Pevtsov AA, Bertello L, MacNeice P, Petrie G (2016) What if we had a magnetograph at Lagrangian L5? Space Weather 14(11):1026–1031. https://doi.org/10.1002/2016SW001471
106. Pizzo V, Millward G, Parsons A, Biesecker D, Hill S, Odstrčil D (2011) Wang-Sheeley-Arge-Enlil cone model transitions to operations. Space Weather 9:S03004. https://doi.org/10.1029/2011SW000663
107. Pizzo VJ, de Koning C, Cash M, Millward G, Biesecker DA, Puga L, Codrescu M, Odstrčil D (2015) Theoretical basis for operational ensemble forecasting of coronal mass ejections. Space Weather 13:676–697. https://doi.org/10.1002/2015SW001221
108. Pizzo VJ, Koning C, Cash M, Millward G, Biesecker DA, Puga L, Codrescu M, Odstrčil D (2015) Theoretical basis for operational ensemble forecasting of coronal mass ejections. Space Weather 13(10):676–697. https://doi.org/10.1002/2015SW001221
109. Raissi M, Karniadakis G (2018) Hidden physics models: machine learning of nonlinear partial differential equations. J Comput Phys 357:125–141. https://doi.org/10.1016/j.jcp.2017.11.039
110. Raissi M, Perdikaris P, Karniadakis GE (2017) Machine learning of linear differential equations using Gaussian processes. J Comput Phys 348:683–693. https://doi.org/10.1016/j.jcp.2017.07.050
111. Raissi M, Perdikaris P, Karniadakis G (2018) Numerical gaussian processes for time-dependent and nonlinear partial differential equations. SIAM J Sci Comput 40(1):A172–A198. https://doi.org/10.1137/17M1120762
112. Raissi M, Perdikaris P, Karniadakis G (2019) Physics-informed neural networks: a deep learning framework for solving forward and inverse problems involving nonlinear partial differential equations. J Comput Phys 378:686–707. https://doi.org/10.1016/j.jcp.2018.10.045
113. Reginald NL, Davila JM, St Cyr OC, Rastaetter L (2014) Evaluating the uncertainties in the electron temperature and radial speed measurements using white light corona eclipse observations. Sol Phys 289:2021–2039. https://doi.org/10.1007/s11207-013-0467-5
114. Reiss MA, Temmer M, Veronig AM, Nikolic L, Vennerstrom S, Schöngassner F, Hofmeister SJ (2016) Verification of high-speed solar wind stream forecasts using operational solar wind models. Space Weather 14:495–510. https://doi.org/10.1002/2016SW001390
115. Richardson IG (2018) Solar wind stream interaction regions throughout the heliosphere. Living Rev Sol Phys 15:1. https://doi.org/10.1007/s41116-017-0011-z
116. Riley P, Linker JA, Mikić Z (2013) On the application of ensemble modeling techniques to improve ambient solar wind models. J Geophys Res (Space Physics) 118:600–607. https://doi.org/10.1002/jgra.50156
117. Riley P, Mays ML, Andries J, Amerstorfer T, Biesecker D, Delouille V, Dumbović M, Feng X, Henley E, Linker JA, Möstl C, Nuñez M, Pizzo V, Temmer M, Tobiska WK, Verbeke

C, West MJ, Zhao X (2018) Forecasting the arrival time of coronal mass ejections: analysis of the CCMC CME scoreboard. Space Weather 16:1245–1260. https://doi.org/10.1029/2018SW001962
118. Ritter B, Meskers AJH, Miles O, Rußwurm M, Scully S, Roldán A, Hartkorn O, Jüstel P, Réville V, Lupu S, Ruffenach A (2015) A space weather information service based upon remote and in-situ measurements of coronal mass ejections heading for Earth. A concept mission consisting of six spacecraft in a heliocentric orbit at 0.72 AU. J Space Weather Space Clim 5(27):A3
119. Rudy SH, Brunton SL, Proctor JL, Kutz JN (2017) Data-driven discovery of partial differential equations. Science Advances 3(4):e1602614. https://doi.org/10.1126/sciadv.1602614
120. Schrijver CJ, Kauristie K, Aylward AD, Denardini CM, Gibson SE, Glover A, Gopalswamy N, Grande M, Hapgood M, Heynderickx D, Jakowski N, Kalegaev VV, Lapenta G, Linker JA, Liu S, Mandrini CH, Mann IR, Nagatsuma T, Nandy D, Obara T, Paul O'Brien T, Onsager T, Opgenoorth HJ, Terkildsen M, Valladares CE, Vilmer N (2015) Understanding space weather to shield society: a global road map for 2015–2025 commissioned by COSPAR and ILWS. Adv Space Res 55:2745–2807. https://doi.org/10.1016/j.asr.2015.03.023
121. Schultz C (2011) Space weather model moves into prime time. Space Weather 9:S03005. https://doi.org/10.1029/2011SW000669
122. Schunk RW, Scherliess L, Eccles V, Gardner LC, Sojka JJ, Zhu L, Pi X, Mannucci AJ, Wilson BD, Komjathy A, Wang C, Rosen G (2014) Ensemble modeling with data assimilation models: a new strategy for space weather specifications, forecasts, and science. Space Weather 12:123–126. https://doi.org/10.1002/2014SW001050
123. Shen F, Wang YM, Shen CL, Feng XS (2017) On the collision nature of two coronal mass ejections: a review. Sol Phys 292(8):104. https://doi.org/10.1007/s11207-017-1129-9
124. Shprits Y, Kondrashov D, Chen Y, Thorne R, Ghil M, Friedel R, Reeves G (2007) Reanalysis of relativistic radiation belt electron fluxes using CRRES satellite data, a radial diffusion model, and a Kalman filter. J Geophys Res Space Phys 112(A12):A12216. https://doi.org/10.1029/2007JA012579
125. Sirignano J, Spiliopoulos K (2018) DGM: a deep learning algorithm for solving partial differential equations. J Comput Phys 375:1339–1364. https://doi.org/10.1016/j.jcp.2018.08.029
126. Siscoe G (2006) A culture of improving forecasts: lessons from meteorology. Space Weather 4:S01003. https://doi.org/10.1029/2005SW000178
127. Siscoe G (2007) Space weather forecasting historically viewed through the lens of meteorology, p 5. Springer, Berlin. https://doi.org/10.1007/978-3-540-34578-7_2
128. Siscoe G, Baker D, Weigel R, Hughes J, Spence H (2004) Roles of empirical modeling within CISM. J Atmos Sol Terr Phys 66:1481–1489. https://doi.org/10.1016/j.jastp.2004.03.028
129. Sokolov IV, van der Holst B, Oran R, Downs C, Roussev II, Jin M, Manchester I WB, Evans RM, Gombosi TI (2013) Magnetohydrodynamic waves and coronal heating: unifying empirical and mhd turbulence models. Astrophys J 764(1):23. http://stacks.iop.org/0004-637X/764/i=1/a=23
130. Spence H, Baker D, Burns A, Guild T, Huang CL, Siscoe G, Weigel R (2004) Center for integrated space weather modeling metrics plan and initial model validation results. J Atmos Sol Terr Phys 66:1499–1507. https://doi.org/10.1016/j.jastp.2004.03.029
131. Stanier A, Chacón L, Chen G (2019) A fully implicit, conservative, non-linear, electromagnetic hybrid particle-ion/fluid-electron algorithm. J Comput Phys 376:597–616. https://doi.org/10.1016/j.jcp.2018.09.038
132. Strugarek A, Janitzek N, Lee A, Löschl P, Seifert B, Hoilijoki S, Kraaikamp E, Isha Mrigakshi A, Philippe T, Spina S, Bröse M, Massahi S, O'Halloran L, Pereira Blanco V, Stausland C, Escoubet P, Kargl G (2015) A space weather mission concept: observatories of the solar corona and active regions (OSCAR). J Space Weather Space Clim 5(27):A4
133. Sutton EK (2018) A new method of physics-based data assimilation for the quiet and disturbed thermosphere. Space Weather 16(6):736–753. https://doi.org/10.1002/2017SW001785

134. Talagrand O (1997) Assimilation of observations, an introduction (data assimilation in meteology and oceanography: theory and practice). J Meteorol Soc Jpn Ser II 75(1B):191–209
135. Thomas SR, Fazakerley A, Wicks RT, Green L (2018) Evaluating the skill of forecasts of the near-Earth solar wind using a space weather monitor at L5. Space Weather 16(7):814–828. https://doi.org/10.1029/2018SW001821
136. Tiwari D, Wadhvani R (2018) A review of time series modelling over coronal mass ejections (CMEs). In: 2018 2nd International Conference on Inventive Systems and Control (ICISC), pp 65–70. https://doi.org/10.1109/ICISC.2018.8398913
137. Tokumaru M (2013) Three-dimensional exploration of the solar wind using observations of interplanetary scintillation. Proc Jpn Acad Ser B 89(2):67–79. https://doi.org/10.2183/pjab.89.67
138. Tokumaru M, Fujiki K, Iju T (2014) North-south asymmetry in global distribution of the solar wind speed during 1985–2013. J Geophys Res Space Phys 120(5):3283–3296. https://doi.org/10.1002/2014JA020765
139. Tokumaru M, Satonaka D, Fujiki K, Hayashi K, Hakamada K (2017) Relation between coronal hole areas and solar wind speeds derived from interplanetary scintillation measurements. Sol Phys 292(3):41. https://doi.org/10.1007/s11207-017-1066-7
140. Tokumaru M, Shimoyama T, Fujiki K, Hakamada K (2018) Rarefaction of the very slow (<350 km/s) solar wind in cycle 24 compared with cycle 23. J Geophys Res Space Phys 123(4):2520–2534. https://doi.org/10.1002/2017JA025014
141. Trichas M, Gibbs M, Harrison R, Green L, Eastwood J, Bentley B, Bisi M, Bogdanova Y, Davies J, D'Arrigo P, Eyles C, Fazakerley A, Hapgood M, Jackson D, Kataria D, Monchieri E, Windred P (2015) Carrington-L5: the UK/US operational space weather monitoring mission. Hipparchos 2(12):25–31
142. Verbeke C, Mays ML, Temmer M, Bingham S, Steenburgh R, Dumbović M, Núñez M, Jian LK, Hess P, Wiegand C, Taktakishvili A, Andries J, (2019) Benchmarking CME arrival time and impact: progress on metadata, metrics, and events. Space Weather 17(1):6–26. https://doi.org/10.1029/2018SW002046
143. Vourlidas A (2015) Mission to the Sun-Earth L_5 Lagrangian point: an optimal platform for space weather research. Space Weather 13:197–201. https://doi.org/10.1002/2015SW001173
144. Vourlidas A, Howard RA, Plunkett SP, Korendyke CM, Thernisien AFR, Wang D, Rich N, Carter MT, Chua DH, Socker DG, Linton MG, Morrill JS, Lynch S, Thurn A, Van Duyne P, Hagood R, Clifford G, Grey PJ, Velli M, Liewer PC, Hall JR, DeJong EM, Mikic Z, Rochus P, Mazy E, Bothmer V, Rodmann J (2016) The wide-field imager for solar probe plus (WISPR). Space Sci Rev 204(1):83–130. https://doi.org/10.1007/s11214-014-0114-y
145. Wang C, Hajj G, Pi X, Rosen IG, Wilson B (2004) Development of the global assimilative ionospheric model. Radio Science 39(1):RS1S06. https://doi.org/10.1029/2002RS002854
146. Wang J, Chen J, Du J, Zhang Y, Xia Y, Deng G (2018) Sensitivity of ensemble forecast verification to model bias. Mon Weather Rev 146:781–796. https://doi.org/10.1175/MWR-D-17-0223.1
147. Weinzierl M, Mackay DH, Yeates AR, Pevtsov AA (2016) The possible impact of L5 magnetograms on non-potential solar coronal magnetic field simulations. Astrophys J 828(2):102. http://stacks.iop.org/0004-637X/828/i=2/a=102
148. Weinzierl M, Yeates AR, Mackay DH, Henney CJ, Arge CN (2016) A new technique for the photospheric driving of non-potential solar coronal magnetic field simulations. Astrophys J 823:55. https://doi.org/10.3847/0004-637X/823/1/55
149. Wiltberger M, Weigel RS, Gehmeyr M, Guild T (2005) Analysis and visualization of space science model output and data with CISM-DX. J Geophys Res Space Phys 110(A9):A09224. https://doi.org/10.1029/2004JA010956
150. Winske D, Yin L, Omidi N, Karimabadi H, Quest K (2003) Hybrid simulation codes: past, present and future-a tutorial. Springer, Berlin, pp 136–165
151. Wold AM, Mays ML, Taktakishvili A, Jian LK, Odstrčil D, MacNeice P (2018) Verification of real-time WSA-ENLIL+Cone simulations of CME arrival-time at the CCMC from 2010 to 2016. J Space Weather Space Clim 8:A17. https://doi.org/10.1051/swsc/2018005

152. Wold AM, Mays ML, Taktakishvili A, Jian LK, Odstrčil D, MacNeice P (2018b) Verification of real-time WSA-ENLIL+Cone simulations of CME arrival-time at the CCMC from 2010 to 2016. J Space Weather Space Clim 8(27):A17. https://doi.org/10.1051/swsc/2018005
153. Yang Y, Shen F, Yang Z, Feng XS (2018) Prediction of solar wind speed at 1 AU using an artificial neural network. Space Weather 16(9):1227–1244. https://doi.org/10.1029/2018SW001955
154. Yu HS, Jackson BV, Hick PP, Buffington A, Odstrčil D, Wu CC, Davies JA, Bisi MM, Tokumaru M (2015) 3D reconstruction of interplanetary scintillation (IPS) remote-sensing data: global solar wind boundaries for driving 3D-MHD models. Sol Phys 290(9):2519–2538. https://doi.org/10.1007/s11207-015-0685-0

Printed by Printforce, the Netherlands